ICANN '93

W0043862

ICANN '93

Proceedings of the International Conference
on Artificial Neural Networks
Amsterdam, The Netherlands
13–16 September 1993

Edited by
Stan Gielen and Bert Kappen

Springer-Verlag
London Berlin Heidelberg New York
Paris Tokyo Hong Kong
Barcelona Budapest

Stan Gielen and Bert Kappen
Dutch Foundation for Neural Networks
University of Nijmegen
Geert Grooteplein 21
6525 EZ Nijmegen
The Netherlands

ISBN-13: 978-3-540-19839-0   e-ISBN-13: 978-1-4471-2063-6

DOI: 10.1007/ 978-1-4471-2063-6

British Library Cataloguing in Publication Data
A catalogue record for this book is available from the British Library

Library of Congress Cataloging–in–Publication Data
International Conference on Artificial Neural Networks (3rd: 1993: Amsterdam, Netherlands)
    ICANN '93: International Conference on Artificial Neural Networks, Amsterdam, The Netherlands, 13–16 September 1993 / edited by Stan Gielen and Bert Kappen.
        p.   cm.
    Includes bibliographical references and index.

    1. Neural networks (Computer science) —Congresses.   I. Gielen, Stan, 1952–   . II. Kappen, Bert, 1958–   . III. Title.
    QA76.87.I57   1993   93–5522
    006.3—dc20   CIP

Apart from any fair dealing for the purposes of research or private study, or criticism or review, as permitted under the Copyright, Designs and Patents Act 1988, this publication may only be reproduced, stored or transmitted, in any form or by any means, with the prior permission in writing of the publishers, or in the case of reprographic reproduction in accordance with the terms of licences issued by the Copyright Licensing Agency. Enquiries concerning reproduction outside those terms should be sent to the publishers.

© in individual papers held by the authors or their employers

© Springer-Verlag London Limited 1993

The use of registered names, trademarks etc. in this publication does not imply, even in the absence of a specific statement, that such names are exempt from the relevant laws and regulations and therefore free for general use.

The publisher makes no representation, express or implied, with regard to the accuracy of the information contained in this book and cannot accept any legal responsibility or liability for any errors or omissions that may be made.

Typesetting: Camera ready by author

34/3830-543210   Printed on acid-free paper

Preface

This book contains the proceedings of the International Conference on Artificial Neural Networks which was held between September 13 and 16 in Amsterdam. It is the third in a series which started two years ago in Helsinki and which last year took place in Brighton. Thanks to the European Neural Network Society, ICANN has emerged as the leading conference on neural networks in Europe.

Neural networks is a field of research which has enjoyed a rapid expansion and great popularity in both the academic and industrial research communities. The field is motivated by the commonly held belief that applications in the fields of artificial intelligence and robotics will benefit from a good understanding of the neural information processing properties that underlie human intelligence. Essential aspects of neural information processing are highly parallel execution of computation, integration of memory and process, and robustness against fluctuations. It is believed that intelligent skills, such as perception, motion and cognition, can be easier realized in neuro-computers than in a conventional computing paradigm. This requires active research in neurobiology to extract computational principles from experimental neurobiological findings, in physics and mathematics to study the relation between architecture and function in neural networks, and in cognitive science to study higher brain functions, such as language and reasoning.

Neural networks technology has already lead to practical methods that solve real problems in a wide area of industrial applications. The clusters on robotics and applications contain sessions on various sub-topics in these fields.

These proceedings contain an up-to-date overview of research and applications, at universities as well as in industry. The large number of contributions have passed through a rigorous process of reviewing. Our program committee has been invaluable in this process.

We would like to thank all organizers and volunteers for their invaluable help that has made the conference possible. In particular, thanks are due to the Dutch Foundation for Neural Networks, the European Commission, Stichting Informatica Onderzoek Nederland, Shell Research and Siemens Nederland for financial support.

Nijmegen, 1993   Stan Gielen
                                                          Bert Kappen

Contents

Visuo-motor interaction—oral contributions

Visuo-motor interaction—poster contributions

Contents   ix

The visual system—oral contributions

The visual system—poster contributions

Dynamics of single neurons—oral contributions

Dynamics of single neurons—poster contributions

ROBOTICS

Robot vision—oral contributions

Robot vision—poster contributions

Robot control—oral contributions

Robot control—poster contributions

COGNITIVE CONNECTIONISM

Neural networks, natural language and artificial intelligence —oral contributions

Neural networks, natural language and artificial intelligence —poster contributions

Natural language and speech recognition
—oral contributions

Natural language and speech recognition
—poster contributions

PHYSICAL AND MATHEMATICAL THEORY

Novel architectures and learning rules—oral contributions

Novel architecture and learning rules—poster contributions

Stochastic dynamical systems—oral contributions

Stochastic dynamical systems—poster contributions

Contents   xvii

Selforganization—oral contributions

Selforganization—poster contributions

Dynamical systems—oral contributions

Dynamical systems—poster contributions

Attractor neural networks—oral contributions

Attractor neural networks—poster contributions

Learning and generalization—oral contributions

Learning and generalization—poster contributions

APPLICATIONS

Industrial applications—oral contributions

Industrial applications—poster contributions

Pattern recognition I—oral contributions

Pattern recognition I—poster contributions

Neural hardware and software—oral contributions

Neural hardware and software—poster contributions

PLENARY CONTRIBUTIONS

DYNAMIC COUPLING IN CORTICAL NEURAL NETWORKS

Ad AERTSEN

Inst. for Neurocomputing, Ruhr-University, PO Box 102184, W-4630 Bochum, Germany

Electrophysiological studies of cortical function on the basis of simultaneous, separable multiple-neuron recordings show that interactions between cortical neurons may strongly depend on stimulus- and behavioral context. Moreover, the interactions may show dynamics on several different time scales, with time constants of modulation as low as tens of milliseconds. These findings point at the need to distinguish between anatomical connectivity and functional coupling in cortical networks. Underlying mechanisms and functional implications of such dynamic coupling are discussed.

INTRODUCTION

Ever since the times of Sherrington [1] and Hebb [2], neurobiologists have pursued the idea that neurons do not act in isolation, but rather that they form into assemblies for various computational tasks (see also [3] for an interesting, early formulation of this concept). Over the years, a number of different, somewhat conflicting definitions of *'neuronal assembly'* have been proposed, phrased in terms of anatomy, of shared function (e.g. motor), of shared stimulus response, etc. (for a review, see [4]). Our operational definition for the cell assembly: near-simultaneity or some other specific timing relation in the firing of the participating neurons, is based on the fact that the synaptic influence of multiple neurons converging onto others is much stronger if they fire in (near-) coincidence. Thus, synchrony of firing is directly available to the brain as a potential code for information processing, as embodied in the concept of the 'synfire chain' [5,6].

The notion that the functional organization of the cortex is based upon interactions within and among groups of cells in large neural networks is supported by the anatomical structure and, in particular, by the massive connectivity of this part of the brain [7]. Until recently, however, there have been very few physiological data that directly address the cell assembly hypothesis. Neither the study of global activity in large populations of neurons, nor the recording of single neuron activity allow for a critical test of this concept. Rather, one seeks to observe the activities of many separate neurons simultaneously, preferably in awake, behaving animals, and to analyze these 'multi-neuron activities' for possible signs of (dynamic) interactions between the neurons. Results of such analyses may then be used to draw inference regarding the processes taking place within and between hypothetical cell assemblies. In this paper we will review some of the evidence produced in these experiments and discuss their functional interpretations.

DYNAMICS OF INTERACTIONS BETEEN CORTICAL NEURONS

Recent years have shown a growing interest in the simultaneous, separable recording of spike trains from groups of neurons (currently some 10-30 at a time) and the study of the interactions taking place within and between such groups [5,6,8-10]. Experiments of this kind have become increasingly feasible, thanks to various improvements in recording methodology (multiple micro-electrodes, optical recording, spike sorting devices). Above all, it has become necessary to develop new computational tools to analyze and interpret the enormous flow of information coming from such multi-neuron experiments.

The conventional method to analyse neural interactions is to cross-correlate spike trains, usually from pairs of neurons, under some appropriate stimulus conditions, and to inspect the results for departures of 'independence' [11,12]. Peaks and troughs in the correlograms, after comparison with appropriate control measurements, are characterized on the basis of parameters describing their shape (e.g. symmetry, width, sign) and location on the time axis. On the basis of such descriptors, the type and precision of temporal synchrony exhibited by the neurons are then interpreted in terms of their 'functional coupling'. Recent developments in analysis methodology have considerably expanded the scope of these studies. It is now possible to examine cooperativity in larger groups of neurons [13-15], to study the dynamic properties of the firing correlation between any two neurons in fine detail [16,17] and to reveal spatio-temporal firing patterns among single and multiple neurons [18,19]. Particularly, the *Joint-PSTH* [17] was designed to highlight the detailed time structure of firing correlation among two neurons, and its possible time-locking to a third event, such as a stimulus or behavioral event. Appropriate normalization of the Joint-PSTH enables us to distinguish between contributions due to stimulus- or behavior-induced modulations of the individual neuron firing rates, and those from interneuronal spike correlations.

Dependence on Stimulus Context

Using these new tools, we have investigated the interactions among neurons in a number of different cortical areas [20-23]. The salient and somewhat surprising result of such direct assembly observation is that the functional coupling between cortical neurons may strongly depend on the stimulus context and/or behavioral state. Moreover, the cooperativity among these neurons may exhibit dynamics on several different time scales, with time constants of modulation as low as tens of milliseconds. Two examples from studies in different cortical areas may illustrate these points. The first one is drawn from experiments on cat area 17 neurons done in the Krüger laboratory [24]. Figure 1 shows the results of Joint PSTH analysis of the activity of two such neurons, recorded simultaneously during presentation of a bar stimulus, moving back and forth over a series of trials, 4 seconds each. The movement was perpendicular to the bar orientation and proceeded in opposite directions from 0-1.8 and 2.2-4 seconds, respectively; inbetween, the stimulus was stationary (1.8-2.2 s). The two panels show the 'raw' (left) and the normalized Joint PSTH (right) and the associated marginal distributions for the selected two neurons. Within each panel, the individual PST histograms are shown along the horizontal and vertical axes. The density of coincident firing is indicated in the Joint PSTH matrix, according to the grey code above the matrix. At the right of each panel are respectively the PST coincidence histogram (along the diagonal rising to right) and the conventional cross-correlogram (perpendicular to the diagonal and descending to right). The PST coincidence histogram measures the counts near the diagonal of the Joint matrix and represents the time-locked average of near-coincident firings of the two neurons in the same sense as the ordinary PST histogram represents the stimulus time-locked average of the individual neuron's firings.

In the 'raw' JPSTH panel (left) we note that the individual PST histograms show a strong time-locked increase in firing for both neurons as the stimulus enters their receptive fields, but only during the second direction of motion. These are direction (as well as orientation) selective neurons. The Joint PSTH matrix shows a considerable hill at a location matching

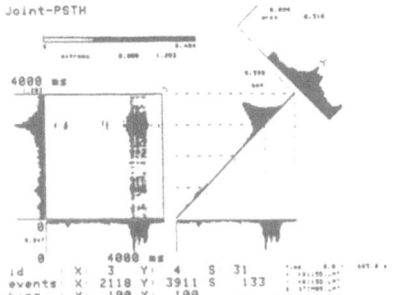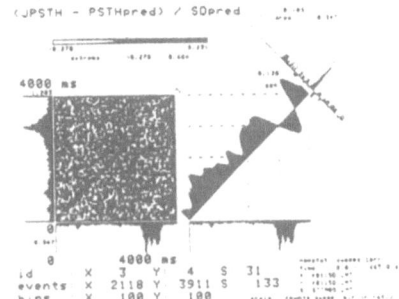

Figure 1. Stimulus-locked dynamic correlation of neuronal firing for two neurons recorded simultaneously from the cat visual cortex (area 17) during repeated presentation of a moving bar stimulus (modified from [20]). For further explanation see text.

the PST peaks, and a corresponding peak is visible in the PST coincidence histogram. In other words, individual, joint and near-coincidence firing rates all are increased during a portion of the second direction of movement. We refer to this phenomenon of correlated rates of firing as 'rate coherence' [25]. By contrast, none of these quantities is increased as the stimulus passes through the receptive field in the first (and opposite) direction of movement. The second panel (right) contains the normalized JPSTH, corrected for the effects of stimulus time-locked modulation of the individual firing rates. The normalized PST coincidence histogram along the diagonal reveals that during the first direction of movement, there is a positive level of residual near-coincidence firing, which is approximately constant, even when the stimulus passes through the receptive field region of the neurons. By contrast, during the second direction of movement, there is a lower than expected amount of near-coincidence firing just as the individual firing rates go up in response to the stimulus: the normalized correlation has gone negative, meaning that the neurons became de-correlated. Thus, Fig. 1 demonstrates that near-coincidence firing can be strongly modulated by the stimulus. These two neurons are repeatedly switching from a condition of ongoing co-firing to a stimulus-related period of strong anti-coincidence: when one fires, the other has a strong tendency to be more silent than expected from the individual firing rates. We refer to such detailed correlations of individual spikes, exceeding rate modulation effects of any one or both participating neurons, as 'event coherence' [25].

In the example shown here, the switching from a condition of co-firing to one of anti-co-firing seems to be associated, at least roughly, with the transition of low to high firing rate of the two observed neurons (compare the time course of the PST histograms with that of the normalized PST coincidence histogram in Fig. 1). Note, however, also the brief, but clear reduction of co-firing (almost down to zero-level) slightly before 2000 ms, i.e. immediately after the stimulus bar stopped moving. In this case, the modulation in co-firing was *not* associated with a corresponding change in either of the two neurons' individual firing rates. We conclude that correlation of firing among neurons may change dramatically and fast, even without any appreciable sign in the probability of firing of any of the individual neurons. We have also observed the converse phenomenon, i.e. strong modulation of single neuron firing rates without a significant modulation in the rate of residual coincidence firing. In general, therefore, rate coherence and event coherence present two instances of synchronization of firing at different levels of temporal acuity; in any specific case they may be, but must not be, correlated.

Dependence on Behavioral State

We have observed similar phenomena in other data. A second example comes from multi-neuron recordings in the prefrontal cortex in awake, behaving monkeys, made by Eilon Vaadia and Hagai Bergman in the Abeles laboratory [22]. Data was collected from different time sections of spike trains from two frontal cortex neurons during performance of a delayed localization task. The first section comprises 600 ms immediately following the GO-signal, the second one covers an interval of 600 ms, starting 900 ms after the monkey's hand left the target key (after hitting the correct location, and the monkey received its reward) and returned to its resting position.

Comparison of the normalized Joint-PST histograms in both panels of Fig. 2 reveals considerable differences in the interaction patterns among the neurons, both in the time-averaged correlograms and, even more so, in their detailed time course displayed along the diagonal. Following the GO-signal (left), the time-averaged correlation is characterized by a strong, 'bi-phasic' interaction pattern: a narrow peak for negative delays, accompanied by a wider trough for positive delays. Interestingly, the positive peak does not reflect an ongoing and stable interaction, but is the net result of two extremely short-lasting instances of discharge synchronization, occurring in coincidence with the onset and offset of firing of one of the two neurons (compare the PSTH along the y-axis), with no interaction whatsoever outside these two short intervals. Similarly, the negative interaction changes dramatically as time proceeds. The time course of modulation, however, is quite different: a sharply increasing and gradually decaying negativity, with the dominant contribution originating from the onset phase of the first neuron, shortly after the first peak of positive interaction, when also the second neuron (PSTH along the x-axis) elevates its firing rate. During the second time section, however, separated from the first one by a mere 1.5 s, these same two neurons exhibit a completely different type of correlation (right). In contrast to the earlier result, the overall interaction now is characterized by an asymmetric, damped oscillator type correlation, with the main peak straddling the origin. Again, inspection of the diagonal region reveals distinct signs of dynamic modulation of the interaction. Although in this case the increased noise level (presumably due to the considerable reduction in firing rates of both neurons) prohibits an unequivocal parsing into separate components, there are hints of two non-overlapping oscillatory subpatterns (extending roughly from 900-1100 ms and from 1200-1400 ms), possibly associated with concurrent features in the firing rate of one of the neurons (compare the PSTH along the x-axis).

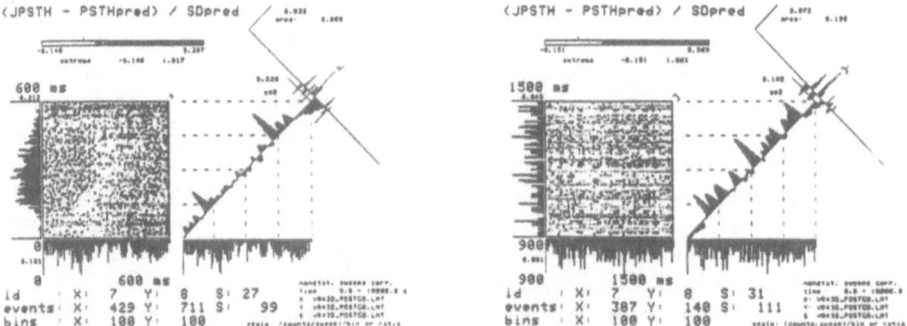

Figure 2. Behavior-related dynamic correlation of two simultaneously recorded neurons from the frontal cortex of an awake monkey during performance of a delayed localization task (modified from [21]). For further explanation see text.

These examples clearly demonstrate that cortical neurons' activities may exhibit rapid modulations of discharge synchronization that are related to stimulus context and behavioral state. These modulations -which could not be inferred from single neuron observations- may switch the neurons' firing behaviour from being mutually incoherent into a particular coherent state of joint synchrony, or, alternatively, from one particular pattern of mutual coherence into a different one. Each such pattern may last for only a few tens to hundreds of milliseconds. Finally, the observed modulations in synchronized firing may be, but are not necessarily associated with changes in either of the neurons' individual firing rates. These phenomena appear to be robust across different regions in the brains among a variety of animal species [26]. Thus, we conclude that dynamic cooperativity is an emergent property of neuronal assembly organization in the brain.

DYNAMIC COUPLING: MECHANISMS AND FUNCTIONAL IMPLICATIONS

These findings suggest that the usual concept of neurons with static interconnections of fixed or only slowly changing efficacy (during learning, for example) is no longer appropriate. Instead one should distinguish between *structural* (or anatomical) *connectivity* on the one hand and *functional coupling* (or effective connectivity) on the other [20,27]. Whereas the former can presumably be described as (quasi) stationary, the latter may be highly dynamic and context-sensitive. These findings raise a two-fold question: what is the nature of the underlying mechanisms, and what are the functional implications?

In a number of theoretical studies we have sought to elucidate these issues. Several different mechanisms may be invoked to mediate the transition from static, anatomic connectivity to dynamic, functional coupling. On the one hand, the underlying mechanism may be *local*, as in von der Malsburg's proposal of rapid modulation of synaptic efficacy [28,29]. On the other hand, more *global* network effects might be involved, as in Sejnowski's notion of the 'skeleton filter' [30]. In a series of theoretical studies, using computer simulations and analytical calculations on artificial neural networks with various types of architectures, we could demonstrate that considerable and rapid changes in functional coupling may, in fact, arise without any associated changes in anatomical connectivity [31-33]. Rather, such rapid modulations can be induced by 'dynamic convergence' of activity from the entire network onto the observed neurons, in particular by temporal variations of the rates [34] and the internal coherence [35,36] of background firing (see also [37,38]). Such global mechanisms do not have to invoke intrinsic rapid modulations of synaptic efficacy; obviously they do not exclude them as an additional mechanism either. Further experimental evidence is required to clear this issue.

Whatever the underlying mechanisms, the experimentally observed, stimulus-driven and behavior-related modulations of functional coupling form an interesting feature of cortical organization. They are the signature of an ongoing process of dynamical and activity-related 'linking' and 'unlinking' of neurons in the cortical network. This process may have interesting functional implications at different levels of observation. At the single-neuron level, it might explain how even little specificity in anatomical connections could be dynamically sorted out to yield the complex functional properties that have been observed for cortical neurons. Thus, it might provide a mechanism for the physiologically measured context-dependence and intrinsic dynamics of receptive fields in central sensory neurons [39-42]. At the multiple-neuron level, dynamic coupling might account for coherence variations in a spatially distributed neural code. Several recently observed phenomena in cortical activity point at possible candidates for such a distributed code. One example is the observation of stimulus-specific oscillatory events in the cat visual cortex, with coherence properties that may extend over wide ranges of cortex [43-45]. A second is the relative exuberance of detailed and behavior-related spatio-temporal spike patterns in cortical activity, pointing at the presence of 'synfire reverberations' [46,47]. Dynamic coupling is, almost by definition, a natural candidate to mediate the general process of 'temporal binding'; an example is the task of object recognition by the visual nervous system. Recent model work along these lines [e.g. 48-53] demonstrated that stimulus-related modulations of activity coherence, possibly differentiated into rate correlation and spike synchronization [25], may indeed subserve such binding functions (see also [54]). Finally, at the level of the organization of perception and action, modulation of functional coupling in interconnected neural networks might provide a mechanism for the selection and successive ignition of neural assemblies within and across such networks. Spatio-temporal variation of input activity, carried onto the target networks by divergent-convergent projections, could modulate the activity levels in these networks and, hence, provide the means to select and dynamically switch from activation of one cell assembly to the next. Such 'threshold control'-like [55,56] mechanisms for the generation of 'phase sequences' [2] of cell assemblies have been invoked in recent theories of learning on the basis of effects of actions, presumably mediated by cortico-striatal interactions [57-59].

Summarizing, the highly dynamic interplay of activity and connectivity in the cortex gives rise to an ongoing process of functional reorganization. Everchanging groups of neurons, each one recruited for brief periods of time, become co-activated and again de-activated, following each other in rapid succession. It is our conjecture that this dynamic reorganization provides the substrate to implement the neural computations involved in 'higher brain function', including the capacity to perceive, to behave and to learn.

ACKNOWLEDGEMENTS

This paper summarizes results from a number of ongoing collaborations. The contributions by Michael Erb, Hartmut Neven, Hubert Preissl and Stefan Rotter, as well as those by Moshe Abeles, George Gerstein, Jürgen Krüger, Günther Palm, Eilon Vaadia and their colleagues are gratefully acknowledged. Multi-neuron spike train data were kindly made available by Jürgen Krüger and Eilon Vaadia. This reseach was supported by the Max-Planck-Institute for Biological Cybernetics (Tübingen, the German Ministry for Science and Technology (BMFT), and the German-Israel Foundation for Research and Development (GIF).

REFERENCES

1. Sherrington, C.: Man on his nature. The Gifford Lectures, Edinburgh 1937-38. Cambridge: University Press, 1941
2. Hebb D (1949) *The organization of behavior. A neuropsychological theory.* Wiley, New York
3. James W (1890) Psychology (Briefer Course). In: Andersen JA, Rosenfeld E (eds)(1989) Neurocomputing. MIT Press, Cambridge, MA
4. Gerstein GL, Bedenbaugh P, Aertsen AMHJ (1989) Neuronal Assemblies. IEEE Trans Biomed Engineering 36: 4-14
5. Abeles M (1982) *Local cortical circuits. An electrophysiological study.* Springer, Berlin
6. Abeles M (1991) *Corticonics. Neural circuits in the cerebral cortex.* Cambridge University Press, Cambridge, UK
7. Braitenberg V, Schüz A (1991) *Anatomy of the cortex. Statistics and geometry.* Springer, Berlin
8. Eggermont JJ (1990) *The correlative brain. Theory and experiment in neural interaction.* Springer, Berlin
9. Gerstein GL, Bloom MJ, Espinosa IE, Evanczuk S, Turner MR (1983) Design of a laboratory for multi-neuron studies. IEEE Trans Systems, Man and Cybernetics SMC-13: 668-676
10. Krüger J (1983) Simultaneous individual recordings from many cerebral neurons: techniques and results. Rev Physiol Biochem Pharmacol 98: 177-233

11. Perkel DH, Gerstein GL, Moore GP (1967) Neuronal spike trains and stochastic point processes. II. Simultaneous spike trains. Biophys J 7: 419-440
12. Aertsen AMHJ, Gerstein GL (1985) Evaluation of neuronal connectivity: sensitivity of cross correlation. Brain Res 340: 341-354
13. Gerstein G, Perkel D, Dayhoff J (1985) Cooperative firing activity in simultaneously recorded populations of neurons: detection and measurement. J Neurosci 5, 881-889
14. Gerstein G, Aertsen A (1985) Representation of cooperative firing activity among simultaneously recorded neurons. J Neurophysiol 54, 1513-1527
15. Aertsen A, Bonhoeffer T, Krüger J (1987) Coherent activity in neuronal populations: analysis and interpretation. In: *Physics of Cognitive Processes*, pp 1-34. Caianiello ER (ed). World Scientific Publishing, Singapore
16. Palm, G, Aertsen AMHJ, Gerstein GL (1988) On the significance of correlations among neuronal spike trains. Biol Cybern 59: 1-11
17. Aertsen AMHJ, Gerstein GL, Habib MK, Palm G (1989) Dynamics of neuronal firing correlation: modulation of "effective connectivity". J Neurophysiol 61: 900-917
18. Dayhoff JE, Gerstein GL (1983) Favored patterns in spike trains. I. Detection, II. Application. J Neurophysiol 49:1334-1348, 1349-1363
19. Abeles M, Gerstein GL (1988) Detecting spatiotemporal firing patterns among simultaneously recorded single neurons. J Neurophysiol 60: 909-924
20. Aertsen AMHJ, Gerstein GL (1991) Dynamic aspects of neuronal cooperativity: fast stimulus-locked modulations of 'effective connectivity'. In: *Neuronal Cooperativity*, pp 52-67. Krüger J (ed). Springer, Berlin
21. Aertsen A, Vaadia E, Abeles M, Ahissar E, Bergman H, Karmon B, Lavner Y, Margalit E, Nelken I, Rotter S (1991) Neural interactions in the frontal cortex of a behaving monkey: Signs of dependence on stimulus context and behavioral state. J f Hirnforschung 32: 735-743
22. Vaadia E, Ahissar E, Bergman H, Lavner Y (1991) Correlated activity of neurons: a neural code for higher brain functions? In: *Neuronal Cooperativity*, pp 249-279. Krüger J (ed). Springer, Berlin
23. Vaadia E, Aertsen A (1992) Coding and computation in the cortex: single-neuron activity and cooperative phenomena. In: *Information Processing in the Cortex: Experiments and Theory*, pp 81-121. Aertsen A, Braitenberg V (eds). Berlin, Heidelberg, New York, Tokyo: Springer
24. Krüger J (1982) A 12-fold microelectrode for recording from vertically aligned cortical neurones. J. Neurosc. Meth. 6: 347-350
25. Neven H, Aertsen A (1992) Rate coherence and event coherence in the visual cortex: a neuronal model of object recognition. Biol Cybern 67: 309-322
26. A review on experimental results, based on recordings from a variety of regions in the CNS of different animal species, made in several different laboratories, is currently in progress (Gerstein G, Aertsen A, et al., in prep.)
27. Aertsen A, Preissl H (1991) Dynamics of activity and connectivity in physiological neuronal networks. In: *Nonlinear dynamics and neuronal networks*, pp 281-301. Schuster H (ed). VCH Verlag, Weinheim
28. von der Malsburg, C. (1981) The correlation theory of brain function. Internal report 81-2. Max-Planck-Institute for Biophysical Chemistry, Göttingen (FRG)
29. von der Malsburg, C. (1986) Am I thinking assemblies? In: Palm, G., Aertsen, A. (eds.): Brain Theory, pp. 161-176. Berlin, Heidelberg, New York: Springer
30. Sejnowski, T.J. (1981) Skeleton filters in the brain. In: Hinton, G.E., Anderson, J.A. (eds.): Parallel Models of Associative Memory, pp. 189-212. Hillsdale: Lawrence Erlbaum Assoc. Publishers
31. Erb M, Palm G, Aertsen A, Bonhoeffer T (1986) Functional versus structural connectivity in neuronal nets. In: *Strukturbildung und Musteranalyse*, p 23. Proc 9th Cybernetics Congress (DGK). Göttingen (FRG)
32. Erb M, Aertsen A, Palm G (1989) Functional connectivity in neuronal systems: context-dependence of effective network organization does not require synaptic plasticity. In: *Dynamics and plasticity in neuronal systems*, p 445. Elsner N, Singer W (eds). Thieme, Stuttgart, New York
33. Erb M, Aertsen A (1992) Dynamics of activity in biology-oriented neural network models: stability at low firing rates. In: *Information Processing in the Cortex: Experiments and Theory*, pp 201-223. Aertsen A, Braitenberg V (eds). Berlin, Heidelberg, New York, Tokyo: Springer
34. Boven K-H, Aertsen A (1989) Dynamics of activity in neuronal networks give rise to fast modulations of functional connectivity. In: *Parallel processing in neural systems and computers*, pp 53-56. Eckmiller R et al (eds). Elsevier Science Publishers
35. Bedenbaugh PH, Gerstein GL, Boven K-H, Aertsen AMHJ (1988) The meaning of stimulus dependent changes in cross correlation between neuronal spike trains. Soc Neurosci Abstr 14: 651
36. Bedenbaugh PH, Gerstein GL, Aertsen AMHJ (1990) Dynamic convergence in neural assemblies. Soc Neurosci Abstr 16: 1224
37. Bernander O, Douglas RJ, Martin KAC, Koch C(1992) Synaptic background activity determines spatiotemporal integration in single pyramidal cells. Proc Nat Acad Sci 88: 11569-11573

38. Rapp M, Yarom Y, Segev I (1992) The impact of parallel fiber background activity on the cable properties of cerebellar purkinje cells (preprint)
39. Aertsen, A.M.H.J., Johannesma, P.I.M. (1981) The spectro-temporal receptive field. A functional characteristic of auditory neurons. Biol. Cybern. 43: 133-143
40. Eggermont, J.J., Aertsen, A.M.H.J., Hermes, D.J., Johannesma, P.I.M. (1981) Spectro-temporal characterization of auditory neurons: redundant or necessary? Hearing Res. 5: 109-121
41. Aertsen in brain theory
42. Dinse, H.R., Krüger, K., Best, J. (1990) A temporal structure of cortical information processing. Concepts in Neuroscience 1: 199-238
43. Eckhorn, R., Bauer, R., Jordan, W., Brosch, M., Kruse, W., Munk, M., Reitboeck, H.J. (1988) Coherent oscillations: a mechanism of feature linking in the visual cortex? Multiple electrode and correlation analysis in the cat. Biol. Cybern. 60: 121-130
44. Gray, C.M., Singer, W. (1989) Stimulus-specific neuronal oscillations in orientation columns of cat visual cortex. Proc. Natl. Acad. Sci. USA 86: 1698-1702
45. Gray, C.M., König, P., Engel, A.K., Singer, W. (1989) Oscillatory responses in cat visual cortex exhibit inter-columnar synchronization which reflects global stimulus properties. Nature 338: 334-337
46. Abeles M, Berman H, Margalit E, Vaadia E (1993) Spatio-temporal firing patterns in the frontal cortex of behaving monkeys (submitted)
47. Abeles M, Prut Y, Bergman H, Vaadia E, Aertsen A (1993) Integration, synchronicity and periodicity. In: Brain Theory: Spatio-Temporal Aspects of Brain Function. Aertsen A (ed) Amsterdam, New York: Elsevier Science Publ. (in press)
48. Sompolinsky H, Golomb D, Kleinfeld D (1990) Global processing of visual stimuli in a network of coupled oscillators. Proc. Natl. Acad. Sci. USA 87: 7200-7204
49. Sompolinsky H, Golomb D, Kleinfeld D (1991) Cooperative dynamics in visual processing. Physical Rev A 43: 6990-7011
50. Sporns O, Gally JA, Reeke GN Jr, Edelman GM, (1989) Reentrant signaling among simulated neuronal groups leads to coherency in their oscillatory avtivity. Proc Natl Acad Sci USA 86: 7265-7269
51. Schillen TB, König P (1991) Stimulus-dependent assembly formation of oscillatory responses. 2: Desynchronisation. Neural Comp 3: 167-178
52. König P, Schillen TB (1991) Stimulus-dependent assembly formation of oscillatory responses. 1: Synchronisation. Neural Comp 3: 155-167
53. Eckhorn R, Reitboeck HJ, Arndt M, Dicke P (1990) Feature linking via a synchronization among distributed assemblies: simulations of results from cat visual cortex. Neural Comp 2, 293-307
54. Johannesma P, Aertsen A, van den Boogaard H, Eggermont J, Epping W (1986) From synchrony to harmony: Ideas on the function of neural assemblies and on the interpretation of neural synchrony. In: Palm G, Aertsen A (eds) Brain Theory, pp 25-47. Springer, Berlin Heidelberg New York
55. Braitenberg, V. (1978) Cell assemblies in the cerebral cortex. In: Heim, R., Palm, G. (eds.): Theoretical Approaches to Complex Systems. Lecture Notes in Biomathematics, Vol. 21, pp. 171-188. Berlin, Heidelberg, New York: Springer
56. Palm, G. (1982) Neural assemblies. An alternative approach to artificial intelligence. Studies in Brain Function, Vol. 7. Berlin, Heidelberg, New York: Springer
57. Miller R (1988) Cortico-striatal and cortico-limbic circuits: a two-tiered model of learning and memory function. In: Information Processing by the Brain: Views and hypotheses from a cognitive-physiological perspective, pp 179-198. Markowitsch H (ed). Bern: Hans Huber Press
58. Wickens J (1992) The contribution of the striatum to cortical function. In: Information Processing in the Cortex: Experiments and Theory, pp 271-284. Aertsen A, Braitenberg V (eds). Berlin, Heidelberg, New York, Tokyo: Springer
59. Plenz D, Aertsen A (1993) The basal ganglia: minimal coherence detection on cortical activity distributions, In: The Basal Ganglia IV. New Ideas and Data on Structure and Function. Percheron G, McKenzie JS, Féger J (eds), New York: Plenum Press (in press)

Keeping Neural Networks Simple

Geoffrey E. Hinton and Drew van Camp

Department of Computer Science
University of Toronto
10 King's College Road
Toronto M5S 1A4, Canada

Abstract

Supervised neural networks generalize well if there is much less information in the weights than there is in the output vectors of the training cases. So during learning, it is important to keep the weights simple by penalizing the amount of information they contain. The amount of information in a weight can be controlled by adding Gaussian noise and the noise level can be adapted during learning to optimize the trade-off between the expected squared error and the information in the weights. We describe a method of computing the derivatives of the expected squared error and of the amount of information in the noisy weights in a network that contains a layer of non-linear hidden units. Provided the output units are linear, the exact derivatives can be computed efficiently without time-consuming Monte Carlo simulations.

Introduction

For a supervised neural network to generalize well, there must be less information in the weights than there is in the output vectors of the training cases. Researchers have considered many possible ways of limiting the information in the weights:

- Limit the number of connections in the network (and hope that each weight does not have too much information in it).

- Divide the connections into subsets, and force the weights within a subset to be identical. If this "weight-sharing" is based on an analysis of the natural symmetries of the task it can be very effective (Lang, Waibel and Hinton (1990); LeCun 1989).

- Quantize all the weights in the network so that a probability mass, p, can be assigned to each quantized value. The number of bits in a weight is then $-\log p$, provided we ignore the cost of defining the quantization.

- Use some other method of limiting the number of effective bits in each weight.

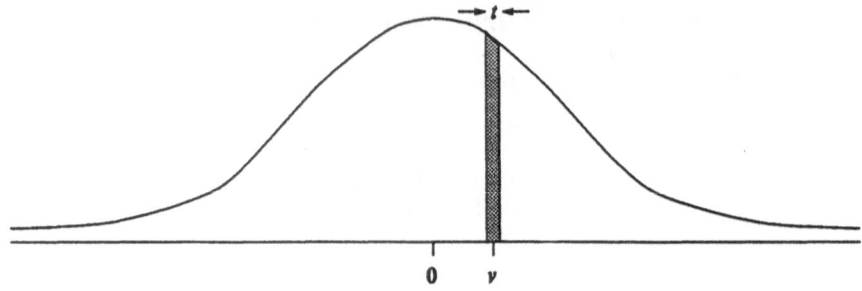

Figure 1: This shows the probability mass associated with a quantized value, v, using a quantization width t. If t is much narrower than the Gaussian distribution, the probability mass is well approximated by the product of the height and the width, so the log probability is a sum of two terms. The $\log t$ term is a constant, and if the distribution is a zero-mean Gaussian, the log of the height is proportional to v^2.

Applying the Minimum Description Length Principle

When fitting models to data, it is always possible to fit the training data better by using a more complex model, but this may make the model worse at fitting new data. So we need some way of deciding when extra complexity in the model is not worth the improvement in the data-fit. The Minimum Description Length Principle (Rissanen, 1986) asserts that the best model of some data is the one that minimizes the combined cost of describing the model and describing the misfit between the model and the data. For supervised neural networks with a predetermined architecture, the model cost is the number of bits it takes to describe the weights, and the data-misfit cost is the number of bits it takes to describe the discrepancy between the correct output and the output of the neural network on each training case. We can think in terms of a sender who can see both the input vector and the correct output and a receiver who can only see the input vector. The sender first fits a neural network, of pre-arranged architecture, to the complete set of training cases, then sends the weights to the receiver. For each training case the sender also sends the discrepancy between the net's output and the correct output. By adding this discrepancy to the output of the net, the receiver can generate *exactly* the correct output.

Coding the data misfits

To apply the MDL principle we need to decide on a coding scheme for the data misfits and for the weights. Clearly, if the data misfits are real numbers, an infinite amount of information is needed to convey them. So we shall assume that they are very finely quantized, using intervals of fixed width t. We shall also assume that the data misfits are encoded separately for each of the output units.

The coding theorem tells us that if a sender and a receiver have agreed on a probability distribution that assigns a probability mass, $p(\Delta y)$, to each possible quantized data misfit, Δy, then we can code the misfit using $-\log_2 p(\Delta y)$ bits. If we want to minimize the expected number of bits, the best probability distribution to use is the correct one, but any other agreed distribution can also be used. For convenience, we shall assume that for output unit j the data misfits are encoded by assuming that they are drawn from a zero-mean Gaussian distribution with standard deviation, σ_j. Provided that σ_j is large compared with the quantization width t, the assumed probability of a particular data misfit between the desired output, d_j^c on training case c and the actual output y_j^c is then well approximated by the probability mass shown in figure .

$$p(d_j^c - y_j^c) = t \frac{1}{\sqrt{2\pi}\sigma_j} \exp\left[\frac{-(d_j^c - y_j^c)^2}{2\sigma_j^2}\right] \tag{1}$$

Using an optimal code, the description length of a data misfit, $d_j^c - y_j^c$, in units of $\log_2(e)$ bits (called "nats") is:

$$-\log p(d_j^c - y_j^c) = -\log t + \log \sqrt{2\pi} + \log \sigma_j + \frac{(d_j^c - y_j^c)^2}{2\sigma_j^2} \tag{2}$$

To minimize this description length summed over all N training cases, the optimal value of σ_j is the root mean square deviation of the misfits from zero.[1] Using this value of σ_j and summing over all training cases the last term of equation 2 becomes a constant and the data misfit cost is:

$$C(\text{data-misfit}) = kN \;+\; \frac{N}{2}\log\left[\frac{1}{N}\sum_c (d_j^c - y_j^c)^2\right] \tag{3}$$

where k is a constant that depends only on t.

Independently of whether we use the optimal value or a predetermined value for σ_j, it is apparent that the description length is minimized by minimizing the usual squared error function, so the Gaussian assumptions we have made about coding can be viewed as the MDL justification of this error function.

A simple method of coding the weights

We could code the weights in just the same way as we coded the data misfits. We assume that the weights of the trained network are finely quantized and come from a zero-mean Gaussian distribution. If the standard deviation, σ_w, of this distribution is fixed in advance, the description length of the weights is simply proportional to the sum of their squares. So during learning we can minimize the total description length of the data misfits and the weights by minimizing the sum of two terms:

$$C = \sum_j \frac{1}{2\sigma_j^2}\sum_c (d_j^c - y_j^c)^2 \;+\; \frac{1}{2\sigma_w^2}\sum_{ij} w_{ij}^2 \tag{4}$$

where c is an index over training cases.

---

[1] If the optimal value of σ_j is to be used, it must be communicated before the data misfits are sent, so it too must be coded. However, since it is only one number we are probably safe in ignoring this aspect of the total description length.

This is just the standard "weight-decay" method. The fact that weight-decay improves generalization (Hinton, 1987) can therefore be viewed as a vindication of this crude MDL approach in which the standard deviations of the gaussians used for coding the data misfits and the weights are both fixed in advance.[2]

An elaboration of standard weight-decay is to assume that the distribution of weights in the trained network can be modelled more accurately by using a mixture of several Gaussians whose means, variances and mixing proportions are adapted as the network is trained (Nowlan and Hinton, 1992). For some tasks this more elaborate way of coding the weights gives considerably better generalization. This is especially true when only a small number of different weight values are required.

However, this more elaborate scheme still suffers from a serious weakness: It assumes that all the weights are quantized to the same tolerance, and that this tolerance is small compared with the standard deviations of the Gaussians used for modelling the weight distribution. Thus it takes into account the probability density of a weight (the height in figure) but it ignores the precision (the width). This is a terrible waste of bits. A network is clearly much more economical to describe if some of the weight values can be described very imprecisely without significantly affecting the predictions of the network.

Mackay (1992) has considered the effects of small changes in the weights on the outputs of the network *after* the network has been trained. The next section describes a method of taking the precision of the weights into account during training so that precision can be traded against both probability density and data misfit.

Noisy weights

A standard way of limiting the amount of information in a number is to add zero-mean Gaussian noise. At first sight, a noisy weight seems to be even more expensive to communicate than a precise one since it appears that we need to send a variance as well as a mean, and that we need to decide on a precision for both of these. As we shall see, however, the MDL framework can be adapted to allow very noisy weights to be communicated very cheaply.

When using backpropagation to train a feedforward neural network, it is standard practice to start at some particular point in weight space and to move this point in the direction that reduces the error function. An alternative approach is to start with a multivariate Gaussian distribution over weight vectors and to change both the mean and the variance of this cloud of weight vectors so as to reduce some cost function. We shall restrict ourselves to distributions in which the weights are independent, so the distribution can be represented by one mean and one variance per weight.

The cost function is the expected description length of the weights and of the data misfits. It turns out that high-variance weights are cheaper to communicate but they cause extra variance in the data misfits thus making these misfits more expensive to communicate.

---

[2]It is clear from equation 4 that it is only the ratio of the variances of the two Gaussians that matters. Rather than guessing this ratio, it is usually better to estimate it by seeing which ratio gives optimal performance on a validation set.

The expected description length of the weights

We assume that the sender and the receiver have an agreed Gaussian prior distribution, $P(.)$, for a given weight. After learning, the sender has a Gaussian posterior distribution, $Q(.)$, for the weight. We describe a method of communicating both the weights and the data misfits and show that using this method the number of bits required to communicate the posterior distribution of a weight is equal to the asymmetric divergence (the Kullback-Liebler distance) from $P(.)$ to $Q(.)$.

$$C(w) = \int Q(w) \log \frac{Q(w)}{P(w)} dw \qquad (5)$$

The "bits back" argument

To communicate a set of noisy weights, the sender first collapses the posterior probability distribution for each weight by using a source of random bits to pick a precise value for the weight (to within some very fine tolerance t). The probability of picking each possible value is determined by the posterior probability distribution for the weight. The sender then communicates these precise weights by coding them using some prior Gaussian distribution, $P(.)$, so that the communication cost of a precise weight, w, is:

$$C^*(w) = -\log t - \log P(w) \qquad (6)$$

t must be small compared with the variance of $P(.)$ so $C^*(w)$ is big. However, as we shall see, we are due for a big refund at the end.

Having sent the precise weights, the sender then communicates the data-misfits achieved using those weights. Having received the weights and the misfits, the receiver can then produce the correct outputs. But he can also do something else. Once he has the correct outputs he can run whatever learning algorithm was used by the sender and recover the exact same posterior probability distribution, $Q(.)$, that the sender collapsed in order to get the precise weights.[3] Now, since the receiver knows the sender's posterior distribution for each weight and he knows the precise value that was communicated, he can recover all the random bits that the sender used to collapse that distribution to that value. So these random bits have been successfully communicated and we must subtract them from the overall communication cost to get the true cost of communicating the model and the misfits. The number of random bits required to collapse the posterior distribution for a weight, $Q(.)$, to a particular finely quantized value, w, is:

$$R(w) = -\log t - \log Q(w) \qquad (7)$$

So the true expected description length for a noisy weight is determined by taking an expectation, under the distribution $Q(.)$:

$$C(w) = \langle C^*(w) - R(w) \rangle = \int Q(w) \log \frac{Q(w)}{P(w)} dw \qquad (8)$$

---

[3]If the sender used random initial weights these can be communicated at a net cost of 0 bits using the method that is being explained.

For Gaussians with different means and variances, the asymmetric divergence is

$$G = \log \frac{\sigma_p}{\sigma_q} + \frac{1}{2\sigma_p^2} \left[\sigma_q^2 - \sigma_p^2 + (\mu_p - \mu_q)^2 \right] \tag{9}$$

The expected description length of the data misfits

To compute the data-misfit cost given in equation 3 we need the expected value of $(d_j^c - y_j^c)^2$. This squared error is caused partly by the systematic errors of the network and partly by the noise in the weights. Unfortunately, for general feedforward networks with noisy weights, the expected squared errors are not easy to compute. Linear approximations are possible if the level of noise in the weights is sufficiently small compared with the smoothness of the non-linearities, but this defeats one of the main purposes of the idea which is to allow very noisy weights. Fortunately, if there is only one hidden layer and if the output units are linear, it is possible to compute the expected squared error exactly.

The weights are assumed to have independent Gaussian noise, so for any input vector we can compute the mean $\mu(x_h)$, and variance, $V(x_h)$, of the Gaussian-distributed total input, x, received by hidden unit h. Using a table, we can then compute the mean, $\mu(y_h)$ and variance, $V(y_h)$, of the output of the hidden unit, even though this output is not Gaussian distributed. A lot of computation is required to create this two-dimensional table since many different pairs of $\mu(x_h)$ and $V(x_h)$ must be used, and for each pair we must use Monte Carlo sampling or numerical integration to compute $\mu(y_h)$ and $V(y_h)$. Once the table is built, however, it is much more efficient than using Monte Carlo sampling at runtime.

Since the noise in the outputs of the hidden units is independent, they independently contribute variance to each linear output unit. The noisy weights, w_{hj}, also contribute variance to the output units. Since the output units are linear, their outputs, y_j, are equal to the total inputs they receive, x_j. On a particular training case, the output, y_j, of output unit j is a random variable with the following mean and variance:

$$\mu(y_j) = \sum_h \mu(y_h)\mu(w_{hj}) \tag{10}$$

$$V(y_j) = \sum_h \left[\mu^2(w_{hj})V(y_h) + \mu^2(y_h)V(w_{hj}) + V(y_h)V(w_{hj}) \right] \tag{11}$$

The mean and the variance of the activity of output unit j make *independent* contributions to the expected squared error $\langle E_j \rangle$. If the desired output of j on a particular training case is d_j, $\langle E_j \rangle$ is given by:

$$\langle E_j \rangle = \langle [d_j - y_j]^2 \rangle \tag{12}$$
$$= [d_j - \mu(y_j)]^2 + V(y_j) \tag{13}$$

So, for each input vector, we can use the table and the equations above to compute the exact value of $\langle E_j \rangle$. We can also backpropagate the exact derivatives of $E = \sum_j \langle E_j \rangle$ provided we first build another table to allow derivatives to be backpropagated through the hidden units. As before, the table is indexed by $\mu(x_h)$ and $V(x_h)$ but for the backward pass each cell of the table contains the four partial derivatives that are needed to to convert the output derivatives of h into its input derivatives using the equations:

$$\frac{\partial E}{\partial \mu(x_h)} = \frac{\partial E}{\partial \mu(y_h)} \frac{\partial \mu(y_h)}{\partial \mu(x_h)} + \frac{\partial E}{\partial V(y_h)} \frac{\partial V(y_h)}{\partial \mu(x_h)} \tag{14}$$

$$\frac{\partial E}{\partial V(x_h)} = \frac{\partial E}{\partial \mu(y_h)} \frac{\partial \mu(y_h)}{\partial V(x_h)} + \frac{\partial E}{\partial V(y_h)} \frac{\partial V(y_h)}{\partial V(x_h)} \tag{15}$$

Stochastic binary threshold units

When using the backpropagation algorithm, it is essential that the output of a hidden unit is a smooth function of its input. This is why the hidden units use a smooth sigmoid function instead of a linear threshold function. With noisy weights, however, it is possible to use a version of the backpropagation algorithm described above in networks that have one layer of linear threshold units. The noise in the weights ensures that the probability of a threshold unit being active is a smooth function of its inputs. As a result, it is easier to optimize a whole Gaussian distribution over weight vectors than it is to optimize a single weight vector.

Discussion

There is a correct, but intractable, Bayesian method of determining the weights in a feed-forward neural network. We start with a prior distribution over all possible points in weight space. We then construct the correct posterior distribution at each point in weight space by multiplying the prior by the probability of getting the outputs in the training set given those weights. Finally we normalize to get the full posterior distribution. Then we use this distribution of weight values to make predictions for new input vectors.

In practice, the closest we can get to the ideal Bayesian method is to use a Monte Carlo method to sample from the posterior distribution. This could be done by considering random moves in weight space and accepting a move with a probability that depends on how well the resulting network fits the desired outputs. Neal (1993) shows how the gradient information provided by backpropagation can be used to get a much more efficient method of obtaining samples from the posterior distribution. The major advantage of Monte Carlo methods is that they do not impose unrealistically simple assumptions about the shape of the posterior distribution in weight space.

Time-consuming Monte Carlo simulations can be avoided by finding a single locally optimal point in weight space and constructing a full covariance Gaussian approximation to the posterior distribution around that point. The alternative method proposed in this paper is to use a simpler Gaussian approximation (with no off-diagonal terms in the covariance matrix) but to take this distribution into account *during* the learning. With one layer of non-linear hidden units, the integration over the Gaussian distribution can be performed exactly and the exact weight derivatives can be computed efficiently.

It is not clear how much is lost by ignoring the off-diagonal terms in the covariance matrix. David Mackay (personal communication) has shown that if standard backpropagation is used to find a single, locally optimal point in weight space and a Gaussian approximation to the posterior weight distribution is then constructed around this point, the covariances between different weights are significant. However, this does not mean that the covariances are significant when the learning algorithm is explicitly manipulating the Gaussian distribution because in this case the learning will try to force the noise in the weights to be independent. The pressure for independence comes from the fact that the cost function will overestimate the information in the weights if they have correlated noise. We are currently performing simulations to see if this pressure does indeed suppress the covariances.

Acknowledgements

This research was funded by operating and strategic grants from NSERC. Geoffrey Hinton is the Noranda fellow of the Canadian Institute for Advanced Research. We thank David Mackay, Radford Neal, Chris Williams and Rich Zemel for helpful discussions.

References

Hinton, G. E. (1987) Learning translation invariant recognition in a massively parallel network. In Goos, G. and Hartmanis, J., editors, *PARLE: Parallel Architectures and Languages Europe*, pages 1–13, Lecture Notes in Computer Science, Springer-Verlag, Berlin.

Lang, K., Waibel, A. and Hinton, G. E. (1990) A Time-Delay Neural Network Architecture for Isolated Word Recognition. *Neural Networks*, **3**, 23-43.

Le Cun, Y., Boser, B., Denker, J. S., Henderson, D., Howard, R. E., Hubbard, W. and Jackel, L. D. (1989) Back-Propagation Applied to Handwritten Zipcode Recognition. *Neural Computation*, **1**, 541-551.

Mackay, D. J. C. (1992) A practical Bayesian framework for backpropagation networks. *Neural Computation*, **4**, 448-472.

Neal, R. M. (1993) Bayesian learning via stochastic dynamics. In Giles, C. L., Hanson, S. J. and Cowan, J. D. (Eds), *Advances in Neural Information Processing Systems 5*, Morgan Kaufmann, San Mateo CA.

Nowlan. S. J. and Hinton, G. E. (1992) Simplifying neural networks by soft weight sharing. *Neural Computation*, **4**, 173-193.

Rissanen, J. (1986) Stochastic Complexity and Modeling. *Annals of Statistics*, **14**, 1080-1100.

PRINCIPLES FROM
NEUROBIOLOGY

Memory and selforganization
—oral contributions

THE AUTOASSOCIATIVE HYPOTHESIS PLACES CONSTRAINTS ON HIPPOCAMPAL ORGANIZATION

Alessandro Treves[a,b] and Edmund T Rolls[a]

[a] Dept of Expl Psychology, University of Oxford, South Parks Rd, Oxford OX1 3UD, UK
[b] SISSA - Biophysics, via Beirut 2-4, 34013, Trieste, Italy

Abstract

We consider the theory that the hippocampus operates as an intermediate term buffer store during consolidation of long-term memory in neocortical areas, and that the crucial role in such operation is played by the CA3 region, which acts as an autoassociative memory network. We extend here previous work, which suggested ways in which the theory placed constraints on the organization of the CA3 region, by indicating how it could also imply constraints informing the organization of other hippocampal regions, as well as of hippocampal return projections to the neocortex.

1. THE AUTOASSOCIATIVE MEMORY HYPOTHESIS

1.1. Hippocampal function

Different lines of evidence, including behavioural impairments in human patients and in animals, *in vivo* recordings of the activity of hippocampal neurones and, recently[9], PET imaging of brain activity, all point to a role of the hippocampus and related structures in the formation of certain types of memories. The information contained in these memories has been characterised as predominantly *spatial* in rodents, for example[6], and as *declarative*, or more specifically *episodic*, in humans[8]. Remote memories tend to be resistent to hippocampal damage, indicating hippocampal involvement in the formation and consolidation of the memory traces, rather than in their permanent maintenance. The length of the period over which the hippocampus is required remains an open question, but tentative evidence suggests characteristic times ranging from a few weeks in rodents to years in humans.

1.2. The role proposed for CA3

The above data can be organised in a conceptual framework by combining the salient features of hippocampal anatomy with the notion[4] of an associative memory network. Different elements of a single percept, such as an episode or a spatial scene, that has to be stored in memory, are given separate neural representations as patterns of neuronal activity in the appropriate neocortical areas. Thanks to the converging connectivity from neocortex to hippocampus, those separate representations can generate in turn a single, highly compressed neural representation in the (relatively small) CA3 area, where all elements of information may be present in a single network. The CA3 pattern of activity may be stored on the modifiable synaptic efficacies of recurrent collateral connections by an associative type of plasticity, experimentally found to be prominent in the hippocampus. At a later time a partial cue presented on the inputs to CA3, such as a distribution of activity on the afferent fibers, corresponding to a small subsets of the elements originally present, may be sufficient, as illustrated by formal models[3], to elicit retrieval of the whole CA3 representation. The CA3 output can then be fed back via a few diverging backprojection stages to the neocortical areas, where it can help consolidate truly long-term memory storage, possibly concurrently with the reorganisation of the information in a different form (e.g., from episodic to semantic).

1.3. The operation of the CA3 network

As we have reported to ICANN'92, the theoretical view summarised here[7] leads to a quantitative analysis of the operation of the CA3 network, which has brought us to associate different elements of the circuitry with different tasks during storage and retrieval of the CA3 representation, and hence to formulate predictions directly testable with behavioural experiments[13]. In brief, we have extended previous formal models of autoassociative memory retrieval[2], to make them applicable to the network formed by the recurrent collateral synaptic connections between pyramidal cells of CA3[11, 12]. This has produced estimates of the number of different memories that can be held at the same time by the network, of the total amount of information retrievable, and therefore of how much information can be extracted, at the output of CA3, from each retrieved pattern of activity. With the assumption that the system is set up so as to efficiently utilise its capacity for retrieval, this results in an estimated lower bound on how much information has to be contained in each CA3 activity pattern during the storage phase:

$$i_p > a\ln(1/a) \tag{1}$$

where a is the sparseness of the stored representation, defined in terms of the probability distribution of the firing rate η of a cell:

$$a = \frac{<\eta>^2}{<\eta^2>}. \tag{2}$$

In order to endow the neural representations of new percepts to be stored, with an information content that satisfies this bound, the CA3 cells have to be driven, as we have shown previously, by afferent inputs with the same characteristics shown by the mossy fibers. We have concluded therefore that the *mossy fiber* system plays a crucial role during the *storage*, in CA3, of new information.

We have also shown that during retrieval of previously stored information, instead, an input system with very different characteristics is required. The argument in this case is based on the constraint that a partial cue carried by the afferent axons, realised as a distribution of activity weakly correlated with that present during storage of a particular memory, must be converted into a signal onto the dendrites of CA3 pyramidal cells, retaining a substantial amount of that correlation. That favors a synaptic system with the caracteristics of the perforant path to CA3, and we have thus posited that the *perforant path* afferents (which must also be coactivated during *storage* in order to enable associative modifications at their synapses) must be the ones that convey the information-rich input which initiates the *retrieval* process via the activation of recurrent collaterals.

2. THE ORGANIZATION OF THE OTHER HIPPOCAMPAL FIELDS

2.1. The granule cells of the dentate gyrus

The granule cells of the dentate gyrus, which give rise to the mossy fiber projection to CA3, are those, according to our hypotheses, that furnish CA3 cells with the input which mainly determines their firing activity during storage of new information. There are different reasons why the output of the granule cells may be particularly appropriate as the main input to CA3 pyramidal cells in the storage phase. In part these reasons have to do with the organization of the mossy fiber synaptic system, as mentioned above, and with the fact that the distribution of activity among granule cells appears to be extremely sparse[5]. In part, however, they have to do with the operation of the dentate network itself, i.e. with the way activity in the granule cells is elicited by the perforant path afferents to the dentate gyrus.

As elaborated elsewhere[7], the perforant path to granule cells synaptic system, with its prominent associative plasticity, would share some of the characteristics of a competitive network. In such a network the activity of the output cells is determined by feedforward inputs mediated by associatively modifiable synaptic efficacies, followed by nonlinear filtering produced in part by the single-cell input-output transduction characteristics, and in part by lateral inhibitory connections, via interneurons, between the output cells. In the dentate, a further source of non-linearity, acting in a different manner, is to be identified with the activation characteristics of NMDA receptors, which underlie synaptic modification. A competitive network is a simple type of unsupervised learning system, and the net result of the operation of such a network is a recoding of the inputs to the hippocampus into inputs with different statistical distributions. The recoding has been characterised as removing redundancy and producing sparser and better categorised patterns of activity. In particular, if synaptic modification is long lasting, it may underlie the formation of codes for regularly occurring *combinations* of active inputs that might need to participate in different episodic memories.

Moreover, the nonlinearity of the processing carried out by the dentate can be a way to assign for example a third, independent representation to inputs which, at the previous (perforant path) stage, were essentially a *linear combination* of two other inputs. To make a concrete example, a percept that combines a particular smell (e.g. of cigar) with a particular sight (e.g. of a black coat) can be given a separate representations at the dentate stage, linearly independent from the representations of the cigar and the black coat alone. This may be a prerequisite in order to associate, at the CA3 stage, the combined cigar-black coat percept with other elements of an episodic memory, without interference from memories that contain the individual cigar or black coat elements separately. The existence of a network capable of performing a degree of competitive categorisation, preceding the autoassociation network, might underlie in part the fact that the removal of the whole hippocampus appears to cause deficits in configural learning tasks, as proposed by some authors[10].

These notions might thus provide the basis for a quantitative analysis of a model of dentate gyrus processing, that would show the extent to which the proposed functions constrain the organization and the plasticity of dentate circuits.

2.2. The pyramidal cells of CA1

The approach we consider is based on the notion that, until each memory has been consolidated in its final form and integrated in the neocortical store, the intermediate term CA3 store is responsible for preserving all the information associated with the memory, or at least as much of it as the humam or animal will be able to remember at a later time. It is precisely this notion that motivates us to consider the information content, in quantitative terms, of patterns of activity retrievable from CA3, and how the circuitry can be optimised in order to maximise the total information in bits stored and retrievable from the system. It is important to note that alternative views, which only ascribe to the hippocampus a *pointer* role, in directing recall from neocortical areas where memories would already be stored even before consolidation, do not require that extensive amounts of information be retrievable from CA3 itself. In fact, the number of bits that such views would assume has to be stored in CA3 is very low, of the order of the logarithm in base 2 of the total number of memories served by the hippocampal system.

The amount of information associated with each percept, which can be kept in memory in CA3, is limited by the number of (recurrent collateral) synapses, by the parameters (mainly the sparseness) of the coding, and by the need to store concurrently as many other memories as possible. We propose, then, that the organisation of the stage following CA3, i.e. of CA1, is optimised so as to prevent further loss of information, after the massive but necessary reduction of information content along the way, up to the output of CA3. One way in which the need to avoid dispersion of information would constrain CA1 circuitry, is that it would require that each CA1 pyramidal cell perform a regular sampling of CA3 activity, by receiving approximately the same number of Schaffer collaterals as other CA1 cells. Morover, the synpses on the Schaffer collateral would need to be associatively modifiable, to optimise the signal-to-noise ratio during retrieval. In these ways, different activation patterns of CA1 would come to reflect accurately different activation patterns in CA3, minimising the effects both of statistical variations in the connectivity from CA3 to CA1, and of information loss due to interference during retrieval from CA3.

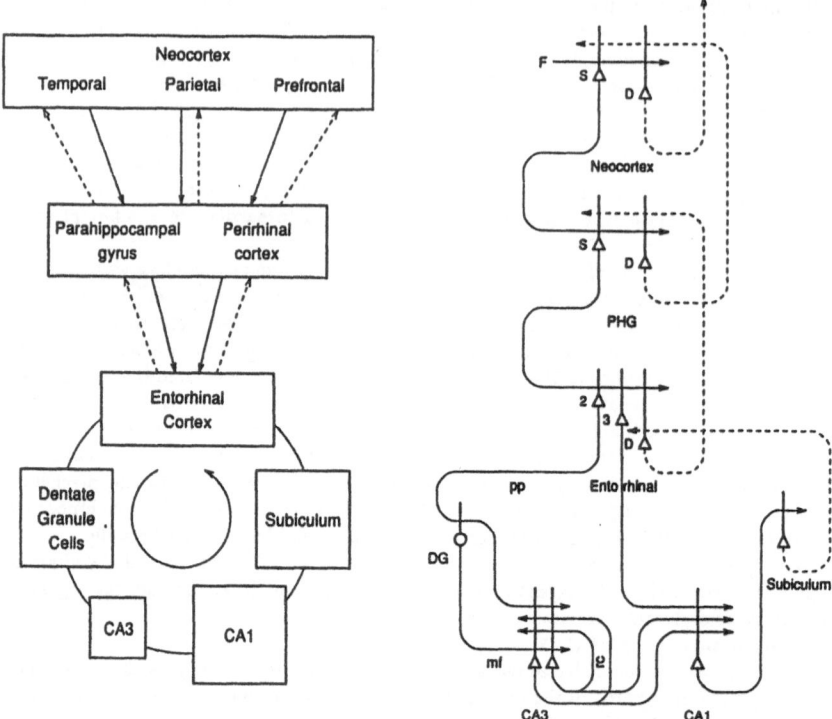

Figure 1: Block diagram of the system of connections between neocortex and hippocampus (left); and schematic representation of some of the main neuronal populations participating in the system, with their position in the layers (right). The thick lines above the cell bodies represent dendrites; solid lines feedforward projections, and dashed lines backprojections. Standard abbreviations.

Besides the Schaffer collaterals, CA1 pyramidal cells also receive a fraction (about $1/3$[1]) of their afferents directly through perforant path projections from entorhinal cortex. These fibers appear to originate mainly in layer 3 of entorhinal cortex, from a population of cells only partially overlapping with that (mainly in layer 2) which gives rise to the perforant path projections to the dentate granule cells and to CA3 pyramidal cells. This alternative set of projections might indicate a need to furnish the system that receives the CA3 output with information closely related to that available at the input. From a quantitative point of view, then, one may understand the role of the perforant path to CA1 in simple terms. During recall, when CA3 retrieves and transmits to CA1 an information-reduced representation of the whole memory, CA1 is able to integrate it with an information-richer representation limited however to the partial cue that was used to produce recall.

Finally, we note that the role proposed fo the CA1 stage is not limited to preserving existing information content, as a network of the CA1 type can also operate a recoding of that information in a different form. In particular, we have seen that the patterns of activity emerging in CA3 result from the coactivation of several representations, corresponding to all the elements of the percepts to be stored, e.g. elements A, B, C, D and E of percept $\mathcal{P}$. CA1, again operating along lines similar to those of a competitive net, would allocate to the memory a new, single firing pattern, where the *conjunction* of the different elements, rather the elements by themselves, is represented. This recoding would not imply that the CA1 representation is sparser than the CA3 one, as both the representations of A, B, C, D and E in CA3 and that of $\mathcal{P}$ in CA1 would be distributed rather than local. It might imply, however, a more efficient coding, useful in accessing the neocortical areas during consolidation, with a limited number of fibers.

3. THE BACKPROJECTIONS TO NEOCORTEX

3.1. Hippocampal capacity limits and the gradient of retrograde amnesia

The theory we are considering distuinguishes between two different phases of storage of episodic information. In the first phase, information is stored in real time, as it arrives, in the CA3 network, which therefore takes a sort of *snapshot* of sensory inputs[7]. In the second phase, that of consolidation, repeated recall of episodic memories from the hippocampal system gradually allows storage of the same information, possibly reorganised along different lines, in its permanent neocortical stores. Retrograde amnesia, i.e. amnesia for events that occurred *before* hippocampal traumas or lesions, is accounted for as the abrupt interruption, due to hippocampal knock-out, of the consolidation process. The *gradient* of retrograde amnesia corresponds to the gradual consolidation of progressively older memories, and a measure of the gradient thus yields a direct measure of the average time over which new memories are kept in the hippocampal buffer store. Once this measure has been obtained, it is possible to combine it with the estimated capacity of the CA3 system, i.e. the maximum number of memories that can be held simultaneously in store, to derive a bound on the average rate of storage of new memories in the hippocampus, and of their transfer to neocortex. For example, let us assume that new memories are found to be dependent on the integrity of the hippocampal formation, in the monkey, for a period of 5 weeks, $\approx 50,000$ minutes[8]. Assuming also that, once all effects of noise and correlation between memories are taken into account, a calculation yields, say, 50,000 memories as the capacity limit for the the monkey CA3 network, one immediately derives a bound of about a new memory per minute for the rate of storage in CA3. This is clearly an average measure, well below the maximal rate, which is constrained by the time scales for synaptic plasticity at the relevant sites, and which could be around one new memory per second. Arousal systems such as the cholinergic system would take care of selecting, through temporally selective but spatially diffuse facilitation, new episodic information to be stored, out of the vast amounts of incoming information.

3.2. Transfer of memories to neocortical stores

The consolidation process depends on the ability of the hippocampal system, upon recall of episodic memories, to selectively reactivate those neocortical pyramidal cells, in different neocortical areas, that were active during the original sensory experience. In the absence of the feedforward inputs, this task falls upon the backprojections that, through various diverging stages, trace back the pathways from hippocampus to neocortex. A key constraint that this hypothesised operation imposes is that the backprojecting synapses be associatively modifiable: only in this case would the hippocampal representation, which has been assigned to the memory in CA3 and CA1, be able to recall activity in the pyramidal cells participating in the original neocortical representations. Further, in order to exploit correctly the associative modifiability of these synapses, backprojecting fibers have to be activated also during the original experience of the event, so that the synapses may record the Hebbian conjunctions between activation of backprojecting axons and depolarization of postsynaptic cells by feedforward inputs. It is interesting to note that a high density of NMDA receptors has been found in the upper neocortical layers, where backprojecting fibers typically terminate. It is also interesting that, due to shunting by conductances opened by feedforward and inhibitory inputs, during the original event the backprojecting fibers may only be able to furnish just the presynaptic factor for Hebbian modifications, while later, during recall, a similar level of backprojecting activity may also be able, in the absence of other inputs, to drive on its own the pyramidal cells' firing activity.

Finally, a quantitative argument also suggests contraints on the minimum number of backprojecting fibers required to contact each neocortical cell. By using a model of multilayer heteroassociative recall[12], one can estimate the number of memories that may be recalled from a particular area as

$$p_{max} \simeq \frac{C^{BP}}{a\ln(1/a)} \times \text{slowly varying factor} \qquad (3)$$

where C^{BP} is the number of backprojecting synapses per cell, a is the sparsity of the neocortical representation, and the factor depends on the detailed structure of the rate distribution and on the connectivity pattern, but is broadly speaking in the order of $0.2 - 0.3$[12]. The scaling is precisely the same as that found between the maximum number of memories that can be retrieved from CA3 and the number of recurrent collateral connections per CA3 cell. Therefore, even allowing for the fact that not all neocortical areas might be interested in the information contained in each episodic memory, one can conclude that in order for the recall mechanism to work all the way up to neocortex, C^{BP} has to be of the same order, if smaller than, C^{RC}, i.e. probably a few thousands connections per cell. This is quite a stringent constraint as it imposes that a significant fraction of all inputs to neocortical cells must be of hippocampal origin; at the same time, it may account for the multiplicity of backprojecting stages observed: if backprojections were directly from CA3, or CA1, to neocortex, an enormous divergence would be required to allow each of the very many neocortical cells to receive enough contacts from the limited hippocampal cell populations.

Thus, an analysis of the recall hypothesis is found to constrain both the plasticity and the quantitative anatomy of the backprojection system.

Acknowledgements. The research described here was supported in part by the Medical Research council; by an EEC BRAIN grant; by a Human Frontier grant; and by the Interdisciplinary Laboratory of SISSA.

References

[1] D. G. Amaral et al, Progr. Brain Res. **83**, 1-11 (1990).

[2] D. J. Amit, Modelling Brain Function (Cambridge Univ Press, NY, 1989).

[3] J. J. Hopfield, Proc. Natl. Acad. Sci. USA **79**, 2554-2558 (1982).

[4] D. Marr, Phil. Trans. Roy. Soc. Lond. **B 262**, 23-81 (1971).

[5] S. J. Y. Mizumori, B. L. McNaughton, C. A. Barnes and K. B. Fox, J. Neurosci. **9**, 3915-3928, (1989).

[6] L. Nadel *et al*, Hippocampus **1**, 221-292, (1991).

[7] E. T. Rolls, in Neural models of plasticity, J. H. Byrne and W. O. Berry eds., 240-265 (Academic press, San Diego, 1989).

[8] L. R. Squire, Psychol. Rev. **99**, 195-231, (1992).

[9] L. R. Squire, J. G. Ojemann, F. M. Miezin, S. E. Petersen, T. O. Viden and M. E. Raichle, Proc. Natl. Acad. Sci. USA **89**, 1837-1841, (1992).

[10] R. J. Sutherland and J. W. Rudy, Hippocampus **1**, 250-252, (1991).

[11] A. Treves, Phys. Rev. A **42**, 2418-2430 (1990).

[12] A. Treves and E. T. Rolls, Network **2**, 371-397 (1991).

[13] A. Treves and E. T. Rolls, Hippocampus **2**, 189-199, (1992).

METASTABILITY OF NETWORK ATTRACTOR AND DREAM SLEEP

Mitsuyuki Nakao   Kazuhiko Watanabe
Yoshinari Mizutani   Mitsuaki Yamamoto

Neurophysiol. and Bioinform. Sci., Dept. of Inform. Eng., Fac. of Eng.
Tohoku Univ., Sendai 980, JAPAN

E-mail:nakao@ecei.tohoku.ac.jp FAX & TEL:+81-22-263-9437

Session:Principles from Neurobiology:Associative memories in biology

Abstract   We have found that single neuronal activities in various brain regions commonly exhibit the dynamics transition from the white to the 1/f spectral profiles during sleep cycle in cats. Simulations using the neural network model showed that the global inhibitory input could induce that transition. Especially, under the weak inhibition (the physiological situation of dream sleep), the metastability of the network attractor was found to be dominant. In this paper, the metastable behavior of the network is investigated for the symmetry and the asymmetry structures, which can be interpreted by the simple schemes based on the potential energy. The metastability could be a cue to understand the function of dream sleep.

INTRODUCTION

Concerning rapid eye movement sleep(REM), also named as "dream sleep", it is known that the brain is activated similar to the waking state though closed against the outside world, and dreaming periods seem to be concentrated in this peculiar sleep. This suggests the close relation between REM and the memory and learning. However, the physiological mechanism mediates between them has not yet been found.

In cat's central nervous system, we have found the following phenomena concerning the dynamics of single neuronal activities during sleep cycle. i) During slow-wave sleep(SWS), neurons showed almost flat power spectral density (PSD) profiles. ii) During REM, dynamics of neuronal activities showed 1/f-like PSD profiles. This phenomenon has been found in various regions of the brain. Pharmacological study suggested that the globally working serotonergic system is associated with the dynamics transition[2].

We successfully simulated the dynamics transition by the neural network models including a globally applied inhibitory input which is supposed to be mediated by serotonergic system[1]. Especially, concerning the 1/f dynamics during REM, the metastability of the network attractor was suggested. Under the simulation condition that the 1/f dynamics prevails, the network repeatedly wanders among metastable equilibrium states and stays them for a while. This behavior, you might say, resembles to dreaming that memorized patterns are sequentially recalled in a non-contextual way. Thus, the metastability of the network attractor is worth studying as a possible candidate underlying the function of dreaming and REM. In this paper, the metastable behavior of the network is further investigated in the state space for the symmetry and the asymmetry networks, respectively. The Markovian properties of the state transition process is analyzed to understand the statistics of the metastable behavior in the light of the geometry of the network attractor.

NEURAL NETWORK MODEL

Used neural network model consists of fully interconnected neuron-like elements (abbreviated as "neuron")[1]. For the i-th neuron, the evolution rule of its state is defined as follows.

$$u_i(t+1) = \sum_{j=1}^{N} w_{ij}x_j(t) - h + \varepsilon_i(t+1) \quad (1)$$

$$x_i(t+1) = g(u_i(t+1)) \quad (2)$$

$$g(x) = \begin{cases} 1 & , x \geq 0 \\ 0 & , x < 0 \end{cases} \quad (3)$$

$$i = 1, 2, \cdots, N$$

where $N(=100)$ denotes the number of neurons contained in the network, and t is a discrete time. $\{\varepsilon_i(t)\}$ denotes mutually independent, zero-mean white Gaussian noises with an identical variance σ^2, $h(>0)$ represents the inhibitory input. Besides, synaptic weights $\{w_{ij}\}$ are defined by the correlation between memorized patterns as is the case of an associative memory. Number of memorized patterns is denoted by M. This enables the parametric control of the network attractor.

As easily known in the symmetry network, each neuron has both of inhibitory and excitatory synapses. On the other hand, in the asymmetry neural network, permitted sign of a synapse depends on the attribute of a neuron, i.e, excitatory or inhibitory. Here, the synaptic weights of the asymmetry neural network are determined following Shinomoto's rule[3]. The attribute of a neuron is assigned at random on each neuron with certain proportions ρ(excitatory) and $1 - \rho$(inhibitory).

SIMULATION RESULTS

Here, time sequences of the network state are presented only for the weakly inhibited condition under which most of PSD of time series:$\{x_i(t)\}$ $(i = 1, \cdots, N)$ have 1/f-like profiles, corresponding to REM. The network state means the binary vector of length N:$\mathbf{x}(t) = (x_1(t), x_2(t), \cdots, x_N(t))$. The evolving operation is performed in the asynchronous (cyclic) manner[5]. The memorized patterns and the initial states are given as equiprobable binary random sequences.

In Figs.1 and 2, the brief sequences of network state are presented picked up from the total sequence of length 10000 in which initial transient sequences are excluded, where a dot indicates $x_i(t) = 1$. The corresponding behavior of the maximum direction cosines in each time are also shown in the middle. The direction cosine here represents the "similarity" between the current network state $\mathbf{x}(t)$ and one of certain equilibrium states $\mathbf{x}^*$ which are found in the network under no perturbation. Five equilibrium states and their reversed patterns are picked up for the references, which are mutually distant each other in terms of the direction cosine. Actually, even under this condition, only the "0" state is an global attractor. Therefore, equilibrium states are searched under slightly weaker inhibition. In the bottom trace in each figure, the closest reference state to the current state among the selected ones is shown, which is determined by the magnitude of the direction cosine.

In Fig.1, independent of network structure, sequences of network state explicitly show that ordered and disordered patterns alternatively appear with varied durations. The direction cosine shows that these ordered patterns almost correspond to one of the reference states. On the other hand, the disordered states correspond to the vicinity of the "0" state where the random pattern prevails directly driven by the noise. In other words, the network state hops from a metastable equilibrium state to the "0" state, and then hops back, intermittently. This behavior is repeated under the weakly inhibited condition.

In Fig.2, multiple metastable states are observed to be involved in the sequence of the network state, where number of memorized patterns: M=30. The structural properties of the network attractor is expected to be different from the above networks with M=20.

These behavior can be schematically understood in terms of the potential energy function of the network[5] as shown in Fig.3. A two well potential scheme is supposed to be roughly applicable in the case of Fig.1. That is, there are one metastable state at a higer energy level and the "0" state at a lower level. In contrast for the network in Fig.2, multiple metastable states including the "0" state are supposed to be surrounded by the high potential barrier as shown in Fig.3. Naturally, under the strong inhibition (results are omitted here), the "0" state becomes highly attractive: a single well potential scheme is available (see Fig.3C).

According to the mathematical theory concerning the metastability with a one-dimensional two well potential model, the distribution of staying time in each state obeys the exponential function whose parameter depends on the height of the potential barrier between them from the bottom potential of each state[4]. Concerning the networks in Fig.1, ten sets of simulations are performed to obtain the distribution of staying time in each state. Figure 4 shows semi-logarithmic plots of the obtained distributions. As clearly seen from these plots, straight lines can be fitted to them, which means each distribution can be approximated by the exponential. The slope of the fitted line corresponds to $\log p_{ii} (i = 1, 2)$, where p_{ii} denotes a one-step transition probability from the state i to i. Since these results well coincide with the mathematical theory, $\log p_{ii}$ corresponds to the height of the potential barrier from the bottom of the well. Both for the symmetry and the asymmetry, the slope of the distribution of the staying time in the metastable state is steeper than in the "0" state, which shows the "0" state is a global attractor. This simple scheme could be extended to understand the statistical features in the case that the multiple

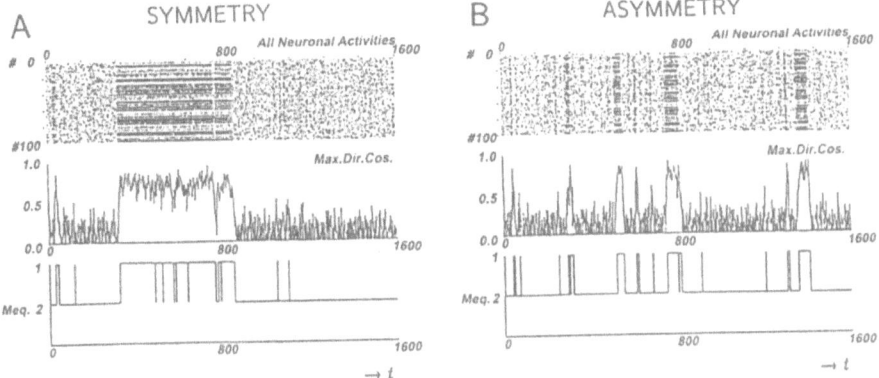

Fig.1 Behavior of network state in the state space.
A Symmetry network: $h = 0.500, \sigma = 0.260, M = 20$
B Asymmetry network: $h = 0.680, \sigma = 0.330, M = 20, \rho = 0.6$
Explanation for each trace is given in the text.

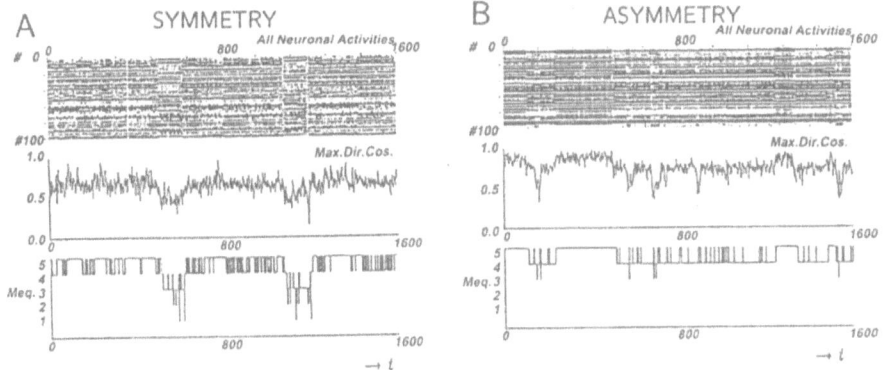

Fig.2 Behavior of network state in the state space.
A Symmetry network: $h = 0.430, \sigma = 0.230, M = 30$
B Asymmetry network: $h = 0.400, \sigma = 0.200, M = 30, \rho = 0.5$
Explanation for each trace is given in the text.

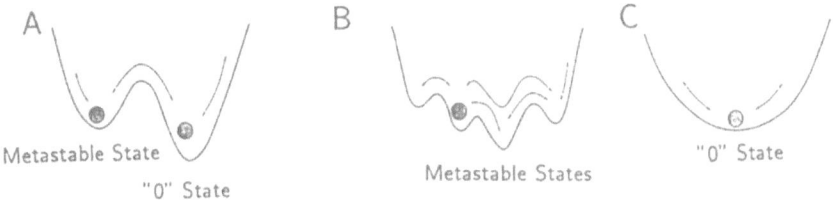

Fig.3 Schematic representations of network potential energy.
A weakly inhibited case (the "0" state and the metastable state in Fig.1)
B weakly inhibited case (the multiple metastable states in Fig.2)
C strongly inhibited case

metastable states are involved. Although the potential energy for the asymmetry network has been not yet known, the above scheme is supposed to be roughly applicable.

Actually, other metastable states in the vicinities of the "0" state and the metastable states under consideration are supposed to be also involved. These hidden metastable states behind the simplified scheme contribute to the 1/f neuronal activities, which contain variations in time with varied time scales.

CONCLUSION

In this study, the metastable properties of the network attractor have been investigated, which are associated with the globally applied inhibition, corresponding to REM. The rough sketch for the metastable behavior has been given in terms of the potential energy of the network.

The metastability of the network attractor during dream sleep is intriguing properties which could underly the function of dreaming and REM. This might be a new concept on the biological significance of dreaming.

Acknowledgements   This research is partly supported by a Grant-in-Aid for Scientific Research from the Ministry of Education, Science and Culture of Japan, No.04246203.

References

[1] M.Nakao, K.Watanabe, T.Takahashi, Y.Mizutani, M.Yamamoto, "Structural properties of network attractor associated with neuronal dynamics transition" *Proc. IJCNN '92*, vol.III, pp.529–534, 1992.

[2] H. Mushiake, T. Kodama, K. Shima, M. Yamamoto, H. Nakahama, "Fluctuations in spontaneous discharge of hippocampal theta cells during sleep-waking states and PCPA-induced insomnia" *J. Neurophysiol.*, vol.60, pp. 925–939, 1988.

[3] S. Shinomoto, "A cognitive and associative memory" *Biol. Cybernetics*, vol.56, pp. 1–10, 1987.

[4] C.Kipnis, C.M.Newman, "The metastable behavior of infrequently observed, weakly random, one-dimensional diffusion processes" *SIAM J.Appl.Math.*, vol.45, pp.972–982, 1985.

[5] J. J. Hopfield, "Neural networks and physical systems with emergent collective computational abilities" *Proc. Natl. Acad. Sci. USA*, vol.79, pp. 2254–2258, 1982.

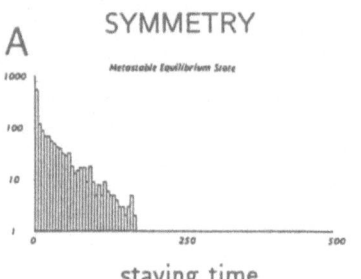
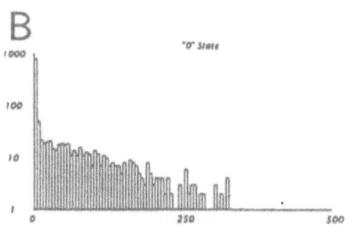
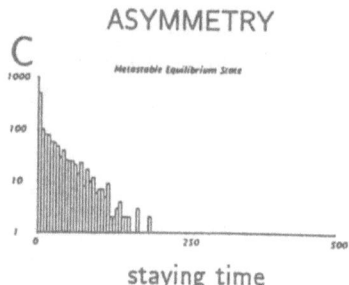
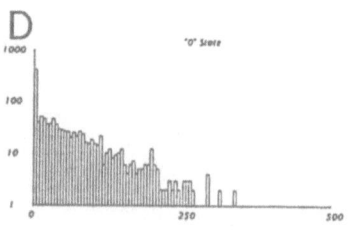

Fig.4 Distributions of staying time in metastable states.
Symmetry network:A the metastable state B the "0" state
Asymmetry network:C the metastable state D the "0" state

Somatosensory Cortical Maps:
Reorganization Following Post-Ontogenetic Plasticity -
Experiments and Theory

Sylvie Wacquant [1], Frank Joublin [1], Friederike Spengler [2]
Ben Godde [2], Hubert R. Dinse [2]

[1] La3i/LCIA, NeuroVision, INSA de Rouen, BP8, F-76131 Mt-St-Aignan.
[2] Institut für Neuroinformatik - Theoretische Biologie, Ruhr-Univ. Bochum (RUB), Germany

Abstract

The adult mammalian cortex maintains a substantial post- ontogenetic plasticity even after the critical developmental period. A few hours of Intra Cortical Micro Stimulation (ICMS) or of Paired Peripheral Tactile Stimulation (PPTS) leads to an overall reorganization of cortical representational maps within a region of up to 1 mm cortex. These changes are paralleled by alterations of geometric-topologic related parameters such as receptive field (RF) size, RF location, RF overlap and response latencies and of functional aspects such as tuning and transfer characteristics. This functional plasticity is interpreted as the modulation of fast plastic, probably Hebbian types of synapses in interactive and overlapping cortical and subcortical networks, serving as a part of dynamic lifelong adaptational mechanism of sensory processing. Based on a multi-level network, we propose a model which is composed of skin, thalamus and somatosensory area. The model is built upon the following hypothesis: (1) an inhibition of cortical layer VI by ICMS which improves latency responses of thalamic cells, (2) a lateral extension of stellate cells in layer IV in association with inhibitory basket cells to form a DOG operator modified by Hebbian rules during ICMS, and (3) a maxicolumnar organization of SI which mediates the modulation of tuning responses.

1.   Basic assumptions

Neural systems organize behavior in naturally occurring environments. However, environments and the constraints they impose change on a variety of time scales. Therefore, each system operating in such an environment must preserve considerable life-long adaptive capacities. During the last years, in many adult species long-term cortical reorganization following damages to the peripheral receptor arrays of the skin, cochlear or retina were well documented, proving that input-deprived cortical areas are occupied by the representation of neighboring sensory fields after weeks and month of recovery [1-7]. In addition, numerous manipulations such as classical conditioning, behavioral training, intracortical microstimulation and prolonged natural sensory stimulation have been shown to act at various time scales on remodelling cortical maps and receptive fields, suggesting that this type of learning- and experience-induced plasticity might serve as a dynamic lifelong adaptational mechanism of sensory processing [8-13].

In search of basic elements which can provide the operational basis for a description of informational processing, we choose single neuron receptive fields (RFs) within a columnar organization. To rely on RFs is mostly because they provide a link between the outside world and neuron activity that can be assessed by available recording techniques. In this view, the term RF is used in a very broad meaning: RFs contain rules for mapping physical events onto representations based on neural activity.

2.   Survey of short-term post-ontogenetic plasticity

Several hours of ICMS was shown to be highly effective in studies of short-term plasticity offering the advantage of investigating in acute experiments plastic capacities and constraints of adult sensory and motor cortex itself, regardless of effects from the sensory periphery and the ascending pathways [10]. In addition, PPTS was shown to be similar effective. As a rule, both manipulations induce short-term plastic reorganization by affecting geometric aspects such as receptive field (RF) size, location and overlap, timing (response latencies and durations) as well as functional properties such as frequency and intensity tuning and interneural interactions [11-13]. As to the time course of ICMS-induced plasticity, early effects can be detected after 15 to 20 min, but reach maximum after 2 to 3 hours. Experiments also demonstrated the full reversibility about 6 hours after terminating ICMS.

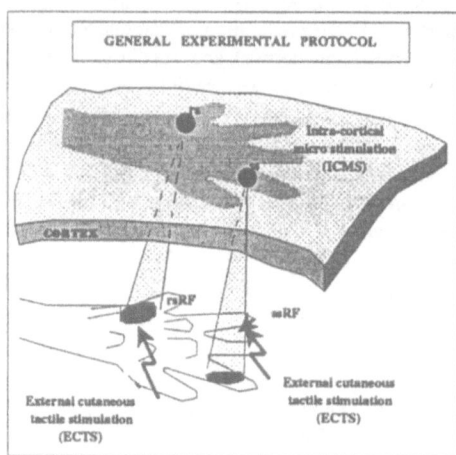

Fig. 1   Experimental ICMS protocol

The basic methodological approach is shown in Fig. 1: Two independent glass-microelectrodes were inserted at various distances of up to 1 mm into primary somatosensory cortex (SI), at so-called recording sites (rs) and stimulations sites (ss), where ICMS was delivered. The corresponding low-threshold cutaneous RFs were defined as the rsRF and ssRF and were stimulated with computer-controlled tactile stimulators. The main results following 4 selected protocols out of 8 tested are illustrated schematically in Fig. 2. The recording sites (rs- cells) up to 800 microns away from the ICMS stimulation site (ss- cell) were expanded by including selectively the ss-cell RF, dependent on the distance from the ss cell (Protocol 1). Similarly, the ss-cell RF is enlarged by integration of the surrounding receptive inputs.

Accordingly, the fine-grained topography of the hindpaw representation is replaced by a coarse representation of multiple skin sites, dominated by the receptive field of the stimulation site.

Response latencies at rsRFs to ss stimulation are normally clearly delayed. Under post-ICMS conditions, they are considerable shortened and now come to match those measured at the ssRF (Protocol 2). The functional transfer properties were also differentially affected: The rs-cell, which displays low excitability and selectivity under control to ssRF stimulation, adopts the control ss-cell's tuning after ICMS (Protocol 3), while the ss-cell looses selectivity due to broadening of tuning characteristics (Protocol 4).

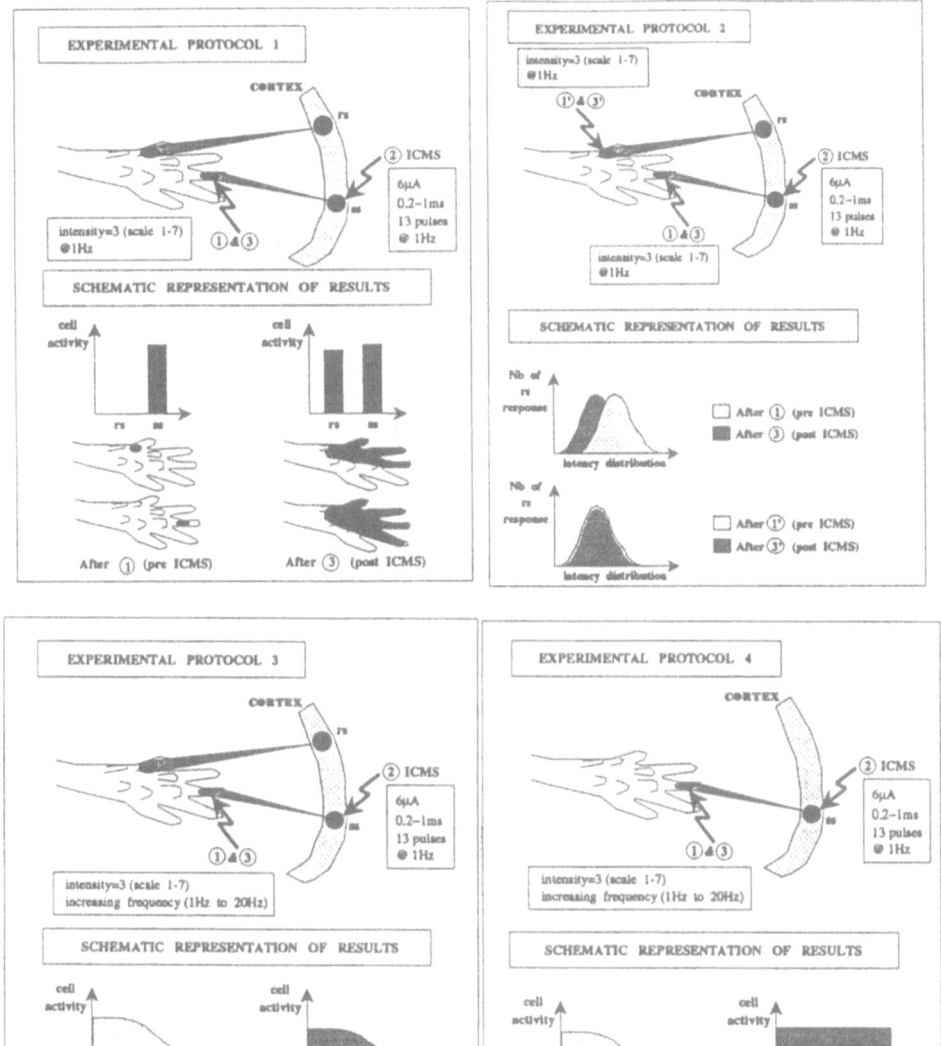

Fig. 2

Schematic illustration of the main results of four selected experimental protocols

Although long-term cortical plasticity, either lesion- or learning related, can be supposed to be accompanied by anatomical changes, short-term cortical reorganization may be exclusively due to modulations of synaptic efficacy in existing cortical and subcortical connections, resulting in the formation of new representational neuronal groups or assemblies.

3.   Theory and simulations.

An elementary three-level-network model of the main pathway of the rat somatosensory system was developed (Fig. 3) comprising skin (first level), thalamus (second level) and somatosensory area (third level). For reasons of simplification and computational costs, the pre-thalamic hierarchical organization of the RFs is reduced to a large dendritic arborization of the thalamic relay cells directly on the simulated skin. A bi-directional one-to-one connection is made between cells of thalamus and layers IV and VI of SI columns [14]. The basic unit of the cortex used in the model is the cortical column. In order to explain the observed plasticity, we rely mainly on neurons of layers IV and VI within this columnar organization.

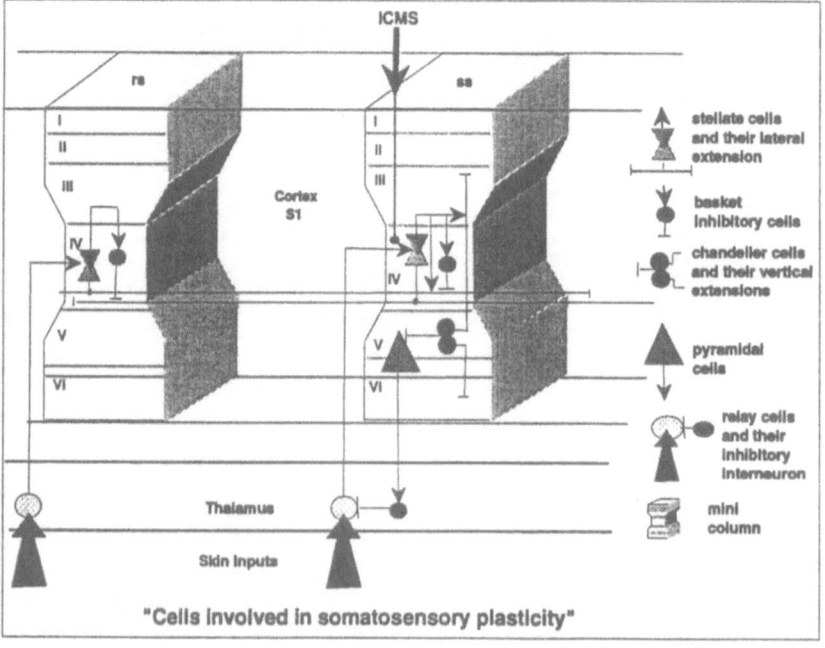

Fig. 3

The structure of the three-level network model

The first fundamental hypothesis concerns protocol 2 and assumes that ICMS or even PPTS performed on the ss-cells of layer IV act as an inhibition of pyramidal cells of layer VI, inducing a disconnection of the cortico-thalamic feedback. This disconnection is assumed to have the same effects on the thalamic transfer functions as a decortication procedure that causes an increase of the excitability of both early and late response components of the relay cells [15,16].

Protocol 1 reveals a modification in the geometric properties of the RFs of both ss- and rs-cells. We make the hypothesis that each stellate cell in layer IV has a lateral extension (LE) of its dendritic arborization large enough to cover the inter-electrode separation [17]. This LE associated with inhibitory and excitatory synapses along its axis will play the role of a DOG operator. A short distance excitatory influence will be exerted by direct glutamate synapses with other stellate cells. Over larger distances, inhibitory influences will be mediated by inhibitory basket (GABA-ergic) interneurons. The third main hypothesis postulates that a long-lasting excitation of a particular stellate cell (ss) will enlarge the excitatory influence of the DOG by means of Hebbian rules. This modification affects the entire extension of ss-cells and only the ss region of rs lateral extension. This could explain results of protocol 1.

Protocols 3 and 4 are concerned with modifications in the tuning responses of the ss and rs cells. The last hypothesis deals with this aspect and postulates a maxicolumnar organization incorporating the systems of minicolumns [18] that are specialized for different frequency or intensity tunings. As a consequence, a stimulation in ssRF leads post-ICMS to a tuned response of the ss-cell, which is then compensated by the neighboring cells that are assumed to be differentially tuned and which now exert an excitatory influence on the ss-cell.

4.   Outlook and conclusions

The computer simulations of this model are run on a specific simulator tool GALIEN [19]. First simulation results are fully compatible with the experimental data according to the simplifications made.

Simulated maps of the geometric aspects of RF reorganization show an enlargement of ssRF and the shift of rsRFs in the direction of the ss. In the simulations, the spontaneous neural activity accounts for the reversibility of the ICMS effects some time after terminating stimulation.

Simulations of long-term cortical plasticity were used to illustrate the capabilities of neural systems to reorganization or to adapt following lesions along the peripheral sensory pathways [20,21]. Our simulations and experiments show a similar degree of reorganizational capacities of short-term plasticity among sensory areas following repetitive stimulation.

The interpretation of such plastic behavior remains speculative to some extent and requires further studies, but is compatible with the experimental data: accordingly, this type of plasticity reflect a lifelong adaptational mechanism to changing constraints of an environment. The possible behavioral relevance of those changes were recently investigated in psychophysical experiments in humans using similar stimulation protocols which revealed a significant increase of spatial discrimination performance [13]. It appears tempting to attribute the observed cortical magnification to an increase of computational resources for the stimulated zone. However, spatial and temporal discrimination performance rely on different mechanism that are differentially affected, which might be reflected in the loss of neuronal identity in terms of tuning properties.

Supported by the DFG, DAAD and Conseil Regional de Haute-Normandie.

5. References

[1] Merzenich MM, Nelson RJ, Stryker MP, Cynader MS, Schoppmann A, Zook JM (1984) J Comp Neurol 224: 591-605;

[2] Merzenich MM, Recanzone G, Jenkins WM, Allard T Nudo RJ (1988) in: Neurobiology of neocortex. Dahlem Konferenzen 1988, Wiley, pp 41;

[3] Calford MB, Tweedale R (1988) Nature 332, 446-447;

[4] Pons, T.P., Garraghty, P.E. & Mishkin, M. (1988) Proc. Natl. Acad. Sci. 85, 5279-5281;

[5] Robertson D, Irvine DRF (1989) J Comp Neurol 282: 456-471;

[6] Kaas JH, Krubitzer LA, Chino YM, Langston AL, Polley EH, Blair N (1990) Science 248: 229;

[7] Gilbert CD, Wiesel TN (1992) Nature 356, 150-152;

[8] Weinberger NM, Ashe JH, Metherate R, McKenna TM, Diamond DM, Bakin J (1990) Concepts Neurosci 1, 91-132;

[9] Recanzone GH, Merzenich MM, Jenkins WM, Grajski K, Dinse HR (1992) J Neurophysiol 67: 1031-1056;

[10] Recanzone GH, Merzenich MM, Dinse HR (1992) Cerebral Cortex 2: 181-196;

[11] Dinse HR, Recanzone G, Merzenich MM (1990) in: Eckmiller R, Hartmann G, Hauske G (eds) Parallel Processing in Neural Systems and Computers. Elsevier, pp 65-70;

[12] Spengler F, Dinse HR (1992) Soc Neurosci Abstracts 18: 345;

[13] Godde B, Spengler F, Dinse HR (1993) in: Elsner N, Heisenberg M (eds) Gen - Gehirn - Verhalten, Thieme, in press;

[14] Saporta S, Kruger L (1977) J Comp Neurol 174: 187;

[15] Angel A, Clarke KA (1975) J Physiol 249: 399-421;

[16] McCormick DA, von Krosigk M (1992) PNAS 89: 2774;

[17] Jensen KF, Killackey AP (1987) J Neurosci 7: 3529;

[18] Mountcastle VB (1978) An Organizing Principle For Cerebral Function: the unit module and the distributed system. in: The Mindful Brain. (ed) Schmitt FO. MIT Press, Cambridge, Mass. pp 7-50;

[19] Joublin F, Wacquant S, Debrie R, (1991) Presentation d'un outil d'etude de reseaux bases sur la colonne corticale. in: Neuro-Nimes'91 pp 609-623;

[20] Pearson JC, Finkel LH, Edelman GM (1987) J Neurosci 7: 4209-4333;

[21] Grajski KA, Merzenich MM (1990) Neural Computation 2: 71-84

Adequate Input for Learning in Attractor Neural Networks

Daniel J. Amit[1] and Nicolas Brunel

INFN, Sezione di Roma, Istituto di Fisica

Università di Roma, La Sapienza, P.le Aldo Moro, Roma

Abstract

In the context of learning in attractor neural networks (ANN) we discuss the issue of the constraints imposed by the requirement that the afferents arriving at the neurons in the attractor network from the stimulus, compete successfully with the afferents generated by the recurrent activity inside the network. We simulate and analyze a two component network: one representing the stimulus, the other an ANN. We show that if stimuli are correlated with the receptive fields of neurons in the ANN, and are of sufficient contrast, the stimulus can provide the necessary information to the recurrent network to allow learning new stimuli, even in very disfavored situation of synaptic predominance in the recurrent part.

Attractors are playing an increasingly significant role in the interpretation of cortical activity, as well as in the description of cognitive phenomena[1, 2]. CA3 region in hippocampus is anatomically a candidate for attractor dynamics[3], since the recurrent (lateral) connectivity is very intensive. CA3 has the other noteworthy features: the typical number of recurrent synapses per CA3 neuron is about four (4) times higher than the number of synapses per neuron coming from the main input source – the perforant path (PP); the synapses from recurrent contacts are typically closer to the soma than those coming from the perforant path; there is a secondary input route which seems to be carrying the same information as the PP and communicates it indirectly, via mossy fibers, to the same area CA3; without the mossy fibers CA3 function, as a learning system, is significantly impaired.

Treves and Rolls[3] conclude that if the input through each synapse in the two sets of synapses (input and recurrent) is equal in amplitude, then the high ratio of recurrent to PP synapses precludes the possibility of learning new items, because of the interference of the recurrent activity.

We study a simple network which respects all the constraints listed above, namely the ratio of the synaptic numbers is the same and the average strength of **excitatory** synapses in the recurrent network and from the input to the recurrent network is equal. Our network is composed of two parts: an input layer and a recurrent part. The input layer (IN) is a two-dimensional grid of N_I neurons and presents binary patterns of activity to the recurrent network. In this layer there are no interconnections and no dynamics. Stimuli are presented on this network by activating a fraction f_I of the neurons with a fixed firing spike rate ν_I; the remaining neurons have zero firing rate. The recurrent network (RN), mimicking CA3, is composed of N_R neurons, and stores a set of p binary random patterns η^μ ($\mu = 1, \ldots, p$). f_R is the coding rate (the fraction of neurons active in an attractor) in

---

[1]On leave of absence from Racah Institute of Physics

this network. The active neurons in a recalled memory fire at a rate ν_R; the remaining neurons have zero firing rate. Memory of these patterns is expressed on a randomly chosen subset of $C_R N$ synapses out of the set of $(N-1)N$ possible synapses. The $(N-1-C_R').V$ remaining links are set to zero. The subset of links expressing the memory, $\{J_{ij}^R\}$, is set on top of a uniform excitatory background J^0, according to the Willshaw prescription[4]: if there exists at least one pattern μ such as $\eta_i^\mu = \eta_j^\mu = 1$ the synaptic value J_{ij}^R is set to $J^0 + J^R$. Otherwise it stays at J^0. For this prescription the probability for any connection being $J^0 + J^R$, in the subset of modifiable links, is $c_R = 1 - (1 - f_R^2)^p$ [4]. However, the average recurrent connectivity in RN is C_R. It is the average excitatory synaptic strength $c_R(J^0 + J^R) + (1 - c_R)J^0$ which will be set equal to the typical synaptic strength arriving from the input.

Each of the N neurons of RN has a receptive field on IN centered on one of the IN neurons, and receives, on average, C_I afferent connections from neurons in IN. This average connectivity will satisfy the constraint $C_I/C_R = K$, where $K \sim 0.25$ in ref [3]. The neurons of RN are placed on a grid identical to that of IN. The position, $\vec{x}_i$, of the neuron in IN at the center of the receptive field of neuron i in RN, is chosen to be the coordinate of neuron i in the identical grid of RN. If neuron j is at position $\vec{x}_j$ in IN the link J_{ij}^I, to neuron i in RN, will be present with probability

$$P(J_{ij}^I \neq 0) = \frac{C_I}{2\pi\rho^2} \exp\left(-\frac{(\vec{x}_i - \vec{x}_j)^2}{2\rho^2}\right) \tag{1}$$

where ρ measures the size of the receptive field. The existing links have uniform efficacies, i.e. $J_{ij}^I = J^I$.

Distribution of depolarization in the recurrent network. The contribution of the recurrent synapses to the current arriving in neuron i of RN can be expressed as $h_i^R = \sum_j J_{ij}^R \nu_j^R$ where ν_j^R is the activity of neuron j in RN. We take bimodal distributions for the neuronal activities in the attractors: the active neurons have $\nu_j^R = \nu_R$ and the others $\nu_j^R = 0$. When RN is in one of the memorized attractors, in the absence of a stimulus, we have for example, $\nu_i^R = \nu_R \eta_i^1$ for all neurons i in RN. The resulting distribution of the afferent currents is a bimodal distribution: neurons in the foreground of the retrieved memory (those for which $\nu_i = \nu_R \eta_i^1 = \nu_R$) have high afferent currents and subsequently will have elevated spike rates, while the neurons in the background ($\nu_i = \nu_R \eta_i^1 = 0$) receive low currents and would be essentially quiescent. This state of activity, in absence of a new stimulus, can be stabilized provided the neurons have an effective threshold lying in the gap between the two peaks.

The minimal requirement for a new stimulus would be that upon presentation of the new stimulus, a subset of neurons in RN, different from the retrieved memory η^1, that had low afferent currents from the recurrent activity, have combined afferents (recurrent plus input) significantly higher than those of the neurons in the foreground in the absence of the stimulus.

The incoming current, due to a stimulus presented on IN, to a neuron i of the recurrent network is given by $h_i^I = \sum_j J_{ij}^I \nu_j^I$ where ν_j^I is the activity of neuron j in IN.

We consider stimuli activating a circular spot of neurons on the grid, of radius r. Inside the circle the fraction of active neurons p_s is higher than the average fraction of active input neurons f_I: $p_s = f_I + \lambda(1 - f_I)$ and $0 \leq \lambda \leq 1$ is the contrast of the spot. Outside the circle the fraction of active neurons p_b is lower than f_I in order to keep the global activity fixed, and we have $p_b = (1 - \lambda)f_I$. Neurons which are activated by the stimulus have equal frequencies: $\nu_j^I = \nu_I$. Fig.1 shows the variation of the incoming current to a given RN neuron with the distance d (in grid units) from the center of the spot to the

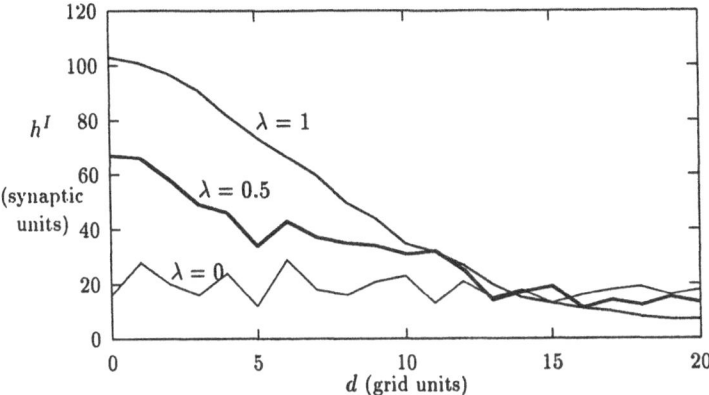

Figure 1: Tuning curves of RN neurons vs displacement of stimulus, for three stimuli with different λ. Parameters: $N = 1600$ (40 times 40 grid), $f_I = 0.1$, $\nu_I = 1$, $J^I = 1$, $C_I = 200$, $r \sim 7$.

center of the receptive field of the neuron, for different values of the contrast λ. This is a bona fide tuning curve.

We performed simulations in the two extreme cases $\lambda = 0$ (input without spatial structure) and $\lambda = 1$ (full spot). For $\lambda = 0$, the integrated input current in any neuron of RN is the sum of random uncorrelated variables. Thus when the number of incoming channels is large, the current from the input becomes a random gaussian variable. In this case the effect of the recurrent connections eliminates all the information contained in IN, since the distribution of the total currents is just the shifted recurrent one, dominated by the stored pattern. For $\lambda = 1$, we considered the two cases $\rho > r$ and $\rho < r$. In the case $\rho > r$, the distribution of both incoming and total currents is similar to the uniform case. The reason is that the size of the spot is small compared to the size of the receptive fields, thus the difference of incoming currents between a neuron close to the spot and a neuron far from it will be small, compared to the difference of currents between foreground and background neurons in the attractor. In the opposite case (Fig.2), $\rho < r$, the few neurons in RN that are close to the spot receive their input only from the active neurons of the spot and thus receive a much higher current than the neurons that are far from it. In this case the neurons in RN that receive the highest total currents are not anymore the neurons that were active before the presentation of the stimulus, but rather those that 'see' the most active region in IN. The information contained in the input is preserved, making the learning of the new pattern possible. This we have verified in simulations.

One can arrive at a quantitative estimate of the constraints. The difference of the mean input currents arriving at neurons of RN that 'see' the stimulus and those that do not is, for $\rho \ll r$, $\Delta\langle h^I \rangle = \lambda C_I J^I \nu_I$. When C_I is very large the fluctuations can be ignored, because the difference between the mean currents is of order C_I, while the fluctuations are of order $\sqrt{C_I}$. The difference between the average recurrent afferent currents of neurons in the background and in the foreground is for the Willshaw prescription $f_R C_R J^R \nu_R (1 - \Delta)$, where $\Delta = (c_R - f^2)/(1 - f^2)$. The condition for learnability is that for a sufficient number of background neurons the afferent current be driven above the average current arriving at a foreground neuron whose receptive field is away from the spot illuminated by the

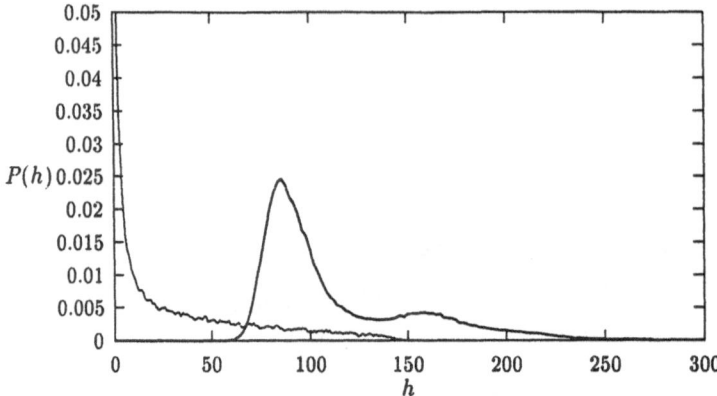

Figure 2: Distribution of incoming and total currents in an RN neuron, for $\lambda = 1, \rho = 4 < r$ and the same parameters as in Fig.1. Light curve: incoming current distribution. Bold curve: total current distribution.

stimulus to be learned. Substituting the constraint $C_I/C_R = K$, we find that the stimulus is learnable if:

$$\lambda > \lambda_0 = \frac{f_R(1 - \Delta)J^R \nu_R}{K J^I \nu_I}. \tag{2}$$

If this condition is satisfied, neurons that have the highest contribution in the distribution of the currents will be the ones that receive their inputs from the high activity spot of the IN layer, regardless of whether they were in the foreground or in the background before the presentation of the stimulus. Taking $K = 0.25$ as in [3], $\nu_R = \nu_I$, and $J^R = J^I$, the condition becomes: $\lambda > 4 f_R(1 - \Delta)$. In other words, the higher the advantage of recurrence over input connectivity (i.e the lower K), the higher must the contrast of the spot be for the stimulus to be learned. Not a very great surprise.

The effective threshold of the neurons in the recurrent network has to increase when a correlated pattern is presented. Otherwise, both the foreground neurons and the ones that see the high-activity regions in IN will be active. If the threshold increases, the network will separate these two types of neurons and may, in principle, be able to learn the 'new' pattern. We have investigated a simplified dynamics of the recurrent network. Given a an inhibitory reaction which controls the overall spike rates in the recurrent network, the distribution of depolarizations created by an effective stimulus leads from a pre-existing reverberation in an attractor, in the absence of a stimulus, to a new attractor representing the stimulus quite faithfully (Fig.3). This attractor is not a fixed point, since the synaptic matrix is not symmetric, but the states it wanders over cover a very small space with a large overlap with the new stimulus.

This study concludes that whatever the difficulties of learning in hippocampus may be, they do not represent a universal limitation on learning in ANN's. Usually, cortical communication is organized with receptive fields. Coding rates in attractors may be rather low, so the constraint on the level of contrast, λ in the stimulus is not very severe. This constraint becomes even less binding when one allows a higher coding rate on the input line than in the ANN, higher spike frequencies in neurons carrying the stimulus, or higher rates on the neurons of the recurrent network, due to the dynamics ensuing the new distribution

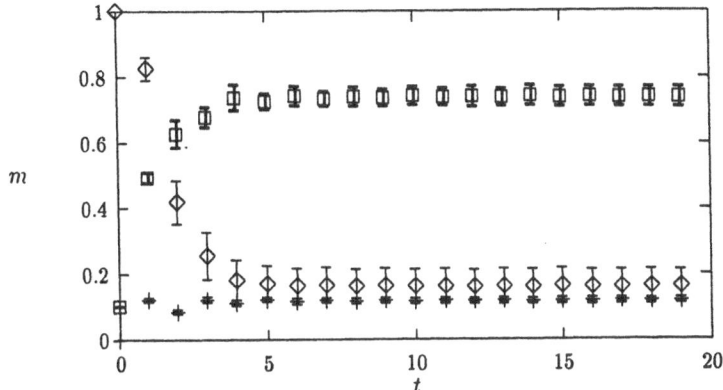

Figure 3: Temporal evolution of activity in RN: overlaps with memorized pattern, background and stimulus, respectively: m_+ ($\Diamond$), m_0 (+) and m_s ($\Box$), averaged over 100 stimuli. $N = 900$ neurons (30×30 grid), both for IN and for RN. The recurrent and input connectivities are, respectively: $C_R = 400$, $C_I = 100$. Coding rates: $f_R = f_I = f = 0.1$; synaptic efficacies: $J^0 = 0.9$, $J^I = J^R = 1$; activity rates in IN and RN: $\nu_I = \nu_R = 1$; receptive fields: $\rho = 3.5$ grid units. The stimulus is a circular spot with $r \sim 5$ (determined approximately by $N f_I$), $\lambda = 1$.

of afferent currents.

References

[1] Miyashita Y and Chang HS 1988 *Nature* **331** 68, *Nature*, and ibid **335** 817

[2] Amit DJ 1989 *Modeling Brain Function* (Cambridge University Press, NY)

[3] Treves A and Rolls E T 1992 *Hippocampus* **2** 625

[4] Willshaw D J 1969 *Nature*, **222** 960.

Memory and selforganization
—poster contributions

NEUROBIOLOGICAL MODELLING AND STRUCTURED NEURAL NETWORKS

Brückner, B.   Zander, W.

Institute of Neurobiology Magdeburg
Department acoustics, learning, speech
P.O.Box 1860, O-3010 Magdeburg, Germany
e-mail: brueckner@ifn-magdeburg.dbp.de

Abstract

We are presenting a coupled modelling concept which is suitable to reproduce sensory systems in a plausible way. The idea is to connect a biological relevant Neural Network as a filter with a structured formal Neural Network to classify the filtered data. Because the biological filter is very flexible and self organized it has the capacity to make segmentation and feature extraction of input data like a sensory pathway. So, temporal processes can be represented in a sequence of prefiltered segments. The produced time sequences are suited for the analysis by a multi-stage construction of the classification network, which has a modified Hypermap structure. The presented system is succesfully used for speech recognition of different subjects.

1. Introduction

Although every biological sensory system is very specialized to a determined kind of stimulus, there is a basic principle of function. First, the stimulus has been converted into an exitation, which is relayed and processed through a series of nodes before it arrives the corresponding areas of the sensory cortex [1]. Then the filtered information is processed on the cortical level. As a first abstraction step it is possible to seperate sensory processing in filtering (processing on the sensory pathway) and perception (cortical processing). There are existing model setups for the complex and more difficult neurobiological part of perception, but they can't produce good results without a suitable input. Our idea is to connect a biological relevant neural network, which has to work like a sensory filter, with a structured artifical neural network for classification. Because the classifying network has a multi-stage construction (Fig.1) it is suitable to analyse temporal data like speech.

filter　　　　　　　buffer　　　　　　　hypermap

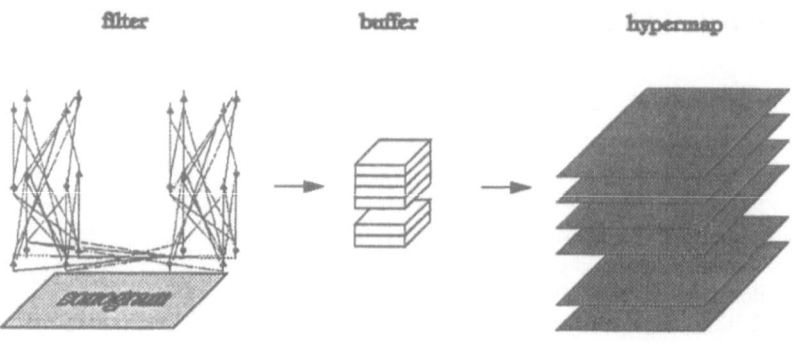

Figure 1: Overall structure of the model for speech recognition
The neurobiological filter sends segments of data to the input vector, where they are
buffered. If an initial segment is detected by the filter, then the input vector is processed
by the hypermap.

2. Neurobiological modelling of the filter

One task of the sensory pathway is feature extraction [2] and in this connection the
segmentation of temporal inputs. For example in the middle of the auditory pathway
the extraction of phonems is implemented [3]. Basis for such functions are modaly-
specific neurons and special sensory circuits with divergence and convergence between
and within sensory nodes and with feedback [1]. For modelling such conditions we use a
biological relevant model in accordance with [4] and [5]. It is based on a biological relevant
model of neurons and synapses, whose instances can be connected and parameterized
in a flexible data structure. The developed model of the neuron is operating on the
abstraction level of the membran potential. All essential properties of the neuron on
this level i.e. summation, cable properties, condenser properties, tiredness of synapses as
well as adaption are reproduced in the model. The network produces biological relevant
courses of potential (Fig.2) and is not linear and stable without exception.

3. Classification by Hypermap

In [6] we defined the modification of the Hypermap Architecture. There we described
the learning of context-dependent data, whereby the context is generated from the input
data. With our modification it is also possible to learn time-dependent data in the form
of time sequences [7]. Each part of a time sequence is related to the corresponding level
in the buffered input vector (Fig.1), which forms a time hierarchy. The learning process
has one step for each level in the hierarchy. The recognition of continuous speech signals
is possible by a sequentiation process. Each detected segment is related to a level of the

network hierarchy where recognition is done. The initial segment of a speech component is related to level 0.

The speech signal is transformed into segments by a segmentation process. The input vector consists of a concatenation of these segments for one speech component, like a word. The different levels of the input vector are trained and form a hierarchy of encapsulated subsets which define different generalized stages of classification.

4. Results

In our experiments with speech we got time sequences of segments derived from a segmentation process. These segments represent transients and the time sequence represents in this sense one complete speech component, i.e. a word or part of a sentence. The algorithm handles different numbers of segments in the input vector.

For our experiments, which are not yet finished, the described architecture for classification is used for experiments with speech recognition. The subjects were asked to repeat the same words. The network was trained to classify these words and within one word the different persons (Fig.2). First of all the digitized speech signal is prepared by FFT. The segmentation process followed is a filter algorithm indicating the segment boundary. This algorithm is independent of the Hypermap Architecture and has a closed relationship to neurobiological mechanisms in the audiotory cortex.

What you get from segmentation is a time-dependent sequence of segments distributed to the levels in the input vector. The trained network is able to classify speech components from continuous speech signal of various subjects.

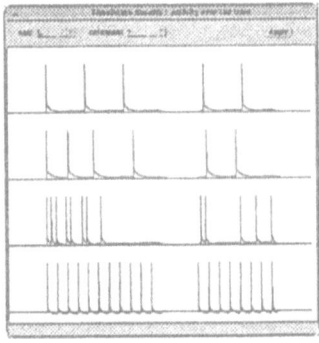 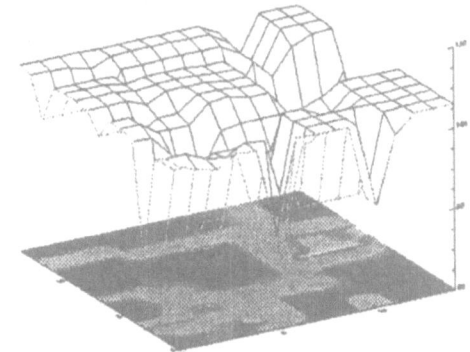

Figure 2: The left picture shows courses of potential, in which detected features for the filtering process are coded. On the right the global error surface (3. level) after learning of 2 words in form of time sequences of segments is shown. The network topology forms a single cluster for the initial segment of a word and within this cluster subclusters on the lower levels for the segments followed.

Taking account of previous studies with conventional self-organizing maps, our presently available data reveal an improved classification by about 20 percent.

Acknowledgements

This work is supported by the DFG grant Br 1289/1-1.

References

[1] Gordon M. Shepherd. *Neurobiology*. Qxford University Press, New York, Oxford, 1988.

[2] H. Handwerker. Allgemeine sinnesphysiologie. In R. F. Schmidt and G. Thews, editors, *Physiologie des Menschen*, pages 186–206. Springer-Verlag, Berlin, 1987.

[3] W. D. Keidel. *Biokybernetik des Menschen*. Wissenschaftliche Buchgesellschaft, Darmstadt, 1989.

[4] A. Richter, W. Zander, and E. Körner. Using parametric controlled structured networks to approach neural networks to neurosciences. In Kohonen et al. [13], pages 1451–1454.

[5] W. Zander, B. Brückner, and G. Brankatschk. Simulation of biological relevant neural networks. In *Contributions to the 21th Göttingen Neurobiology Conference*. Georg-Thieme-Verlag, 1993. in press.

[6] Bernd Brückner, Marcella Franz, and Andreas Richter. A modified hypermap architecture ... In I.Aleksander and J.Taylor, editors, *Artificial Neural Networks,2*, pages 1167–1170, Amsterdam, 1992. Elsevier Science Publishers.

[7] Bernd Brückner and Wiebke Zander. Classification of speech using a modified hypermap architecture. submitted to WCNN'93.

[8] B. Brückner, M. Franz, and A. Richter. Analyse und klassifizierung von eeg-daten mittels strukturierter neuronaler netze. In *Tagungsband 36.Internat.wiss. Koll. der TH Ilmenau*, pages 694–699, Ilmenau, 1991.

[9] H. Ritter, T. Martinetz, and K. Schulten. *Neuronale Netze*. Addison-Wesley, Bonn, 1990.

[10] Teuvo Kohonen. The hypermap architecture. In Kohonen et al. [13], pages 1357–1360.

[11] Jari Kangas. Time-dependent self-organizing maps for speech recognition. In Kohonen et al. [13], pages 1591–1594.

[12] T. Kohonen. The self-organizing map. *Proceedings of the IEEE*, 78(9):1464–1480, 1990.

[13] T. Kohonen, K. Mäkisara, O. Simula, and J. Kangas, editors. *Artificial Neural Networks*, Helsinki, 1991. Elsevier Science Publishers.

Model Analysis of Associative Learning in The Photoreceptor of Marine Mollusc, *Hermissenda Crassicornis*

Hidetoshi Ikeno and Shiro Usui

Department of Information and Computer Sciences,
Toyohashi University of Technology, Toyohashi 441, Japan

Abstract

A marine mollusc, *Hermissenda Crassicornis*, has an ability to learn the association of conditioned stimulus (CS) and unconditioned stimulus (UCS). In this learning, it was shown that some physiological changes, such as decrease of K^+ conductances and a long lasting depolarization (LLD) after light stimulus, were caused in the type B photoreceptor (BPR). However, the underlying mechanism of these changes has not been exhibited experimentally yet. In this paper, a BPR model based on the ionic current and cytosolic Ca^{2+} mechanism was presented in order to reveal the relation of these phenomena. As a result of model simulation, it was suggested that LLD would not be mediated by K^+ conductance changes, but by the change of cytosolic Ca^{2+} mechanism. Furthermore, it was shown that the dependency of inter-stimulus interval (ISI) between CS and UCS on the acquisition of memory could be yielded by an increase of cytosolic Ca^{2+} concentration in BPR.

1 Introduction

Biological nervous systems realize the abilities for information processing with their learning and memory mechanisms. The plasticities of neuron and synapse have been considered to be basic mechanisms for the memory facilities. Recently, principle and mechanism of the plasticity have received much attention not only from neuroscience but also from neuro-engineering.

A nervous system of marine mollusc, *Hermissenda Crassicornis*, has been used for the study of associative learning as a model of real neural network[1]. For this animal, light stimulus and rotation-induced turbulence were used for the association as conditioned and unconditioned stimulus, respectively. As a result of learning, reduction of K^+ currents and light response changes in BPR were observed through physiological experiments. From these experimental results, it was obscurely regarded that the amplification and prolongation of light response were generated by the reduction of K^+ currents[1].

We constructed a BPR model based on the ionic currents to analyze the background mechanism of associative learning[2]. In the model, time and voltage dependent ionic currents were described by Hodgkin-Huxley type equations and a model of light induced current was estimated from flash light responses. In this paper, a model of cytosolic Ca^{2+} mechanism, which affects the characteristics of I_A and I_C, have been introduced for the detail analysis. Simulation results pointed out that the amplitude of light response could be increased by reduction of K^+ currents. On the other hand, LLD could not be caused by a decrease of K^+ conductance, but by the change of dynamics of Ca^{2+} mechanism.

Then, we assumed that the acquisition of learning in BPR was proportional to the product of cytosolic Ca^{2+} concentration ($[Ca^{2+}]_i$), its time derivative ($d[Ca^{2+}]_i/dt$) and transmitter binding amount[3]. As a result of model simulation under this postulation, similar ISI characteristics with the experiments could be reproduced. Hence we concluded that the learning was progressed by the synergistic effect of these two mechanisms, and ISI characteristic was caused by the time course of cytosolic Ca^{2+} increase.

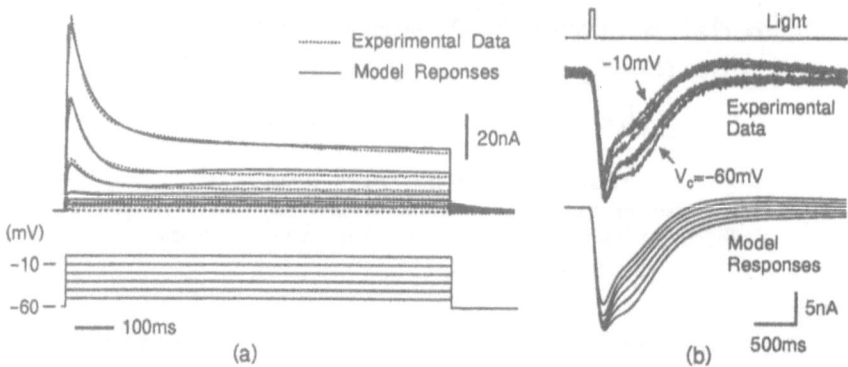

Figure 1: Comparisons of model responses and experimental data. (a) Voltage dependent ionic current was simulated under the dark condition (i.e. $I_{lgt}=0$). (b) I_{lgt} to flush light were represented (top) for the various membrane potentials together with the model responses (bottom).

2 Ionic Currents and Cytosolic Ca²⁺ Model

BPR model was constructed with two components, voltage-dependent and light-induced ionic currents and cytosolic Ca^{2+} mechanisms.

(1) Ionic Currents and Cellular Responses

BPR light response was prescribed by the ionic currents in soma. Three types of K^+ currents (I_A, I_C, I_K), an L type of Ca^{2+} current (I_{Ca}) and a light induced current (I_{lgt}) were observed. Time and voltage dependency on I_A and I_C are represented by Hodgkin-Huxley type equations. It was enough to represent only I–V characteristic for I_{Ca}, because it had little time dependency. As I_K could be neglected below 0 mV of membrane potential, it was omitted in the model.

Three components of light induced current were found in BPR[9]. Thus, I_{lgt} was reconstructed with light inducing fast and slow Na^+ currents, and the suppression of I_C by an increase of $[Ca^{2+}]_i$. Each current waveform was described by a composition of exponential functions, and the parameters were determined by fitting to the experimental data.

The nonlinear parameter estimation technique was practically used for fitting these model responses to the experimental data and estimated model parameters. Simulation results of membrane current and light induced current under the voltage clamp condition were found to agree with the experimental data as shown in Fig.1.

(2) Cytosolic Ca²⁺ Mechanism

Cytosolic Ca^{2+} does not affect the amplitude of I_A and I_C only, but also plays an important role as a second messenger to generate the membrane phosphorylation. In order to clarify the background mechanism of cellular responses, we introduced the cytosolic Ca^{2+} mechanism into the BPR model.

The mathematical model was constructed with following four major components influencing $[Ca^{2+}]_i$. Calcium current I_{Ca} and the release from endoplasmic reticulum are well-known as crucial mechanisms for uptake of $[Ca^{2+}]_i$. Calcium pump is working for stabilizing $[Ca^{2+}]_i$ [6]. Furthermore, in BPR, the light induced Ca^{2+} release was observed by Connor and Alkon[7]. Time constants and other parameter values were determined from the available references[6],[8], because the quantitative value of $[Ca^{2+}]_i$ has not been measured yet in *Hermissenda* BPR.

3 Characteristic Changes by Associative Learning

The conspicuous depressions of I_A and I_C in BPR of learned animal was revealed by voltage-clamp experiment[9]. On the other hand, increases of amplitude and duration were observed in the light response[10]. From these experimental results, it has been obscurely considered that the changes of light response were yielded by decreases of these K^+ currents.

We used BPR model to simulate the light response change caused by decreases of I_A and I_C. As the results of simulation, it was shown that the amplitude of light response was amplified by decreases of K^+ currents. However, the time course of the light response was not changed, and LLD was not observed. It became clear that the continuation of intracellular condition in the light was required for the change of time dependency. Thus, LLD-like response was reconstructed by introducing a change of down-take time course in Ca^{2+} mechanism (Fig.2). The result suggested that the dynamics of cytosolic Ca^{2+} mechanism might be changed by an acquision of associative learning.

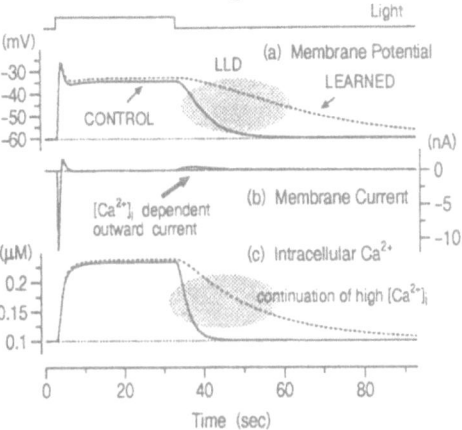

Figure 2: LLD like light response was caused by a change of time dependency in the cytosolic Ca^{2+} mechanism. LEARNED response was clearly prolonged in comparison with CONTROL (a) by suppression of outward current at the light stimulus off (b). High $[Ca^{2+}]_i$ continued after turning off the light stimulus (c).

4 Acquisition of Associative Memory

From the physiological experiments, Crow proposed a hypothesis that the phosphorylation of membrane, which yields characteristic changes of BPR, was activated by the synergistic of Ca^{2+} increase and transmitter binding[3]. On the other hand, it was shown by the behavioral experiments that the acquisition of memory should have strong temporal specificity on CS and UCS presentations[4]. In order to represent the temporal specificity of BPR associative learning, we constructed a model based on Crow's hypothesis.

Two principal components, an effect of transmitter from the hair cell and a learning rule, were added to the BPR model. Action potential in the hair cell caused by UCS was assumed to function as a trigger of the transmitter release. In the model, frequency of action potential was described by a linear function of stimulus amplitude based on our experimental measurements. The time course of transmitter release and binding was approximated by a composition of two first-order dynamical systems. It was shown that the activation of intracellular phosphorylation was proportional to the product of amount of bound transmitter, $[Ca^{2+}]_i$ level and its time derivative ($d[Ca^{2+}]_i/dt$). Learning ratio, i.e. acquisition of memory, was defined by a time integration of the activation of phosphorylation.

Simulations were carried out under the various CS conditions. Duration of UCS and inter-trial interval were set to 3 sec and 45 sec respectively. Duration of CS and turn-on timing were changed in each condition. Leaning ratio was measured at 30 sec after 50th trial. The result of simulation revealed that the learning ratio was maximum under the condition of turning on CS followed 1 second by UCS (Fig. 3). Similar characteristic for ISI was reported by Matzel et al. experimentally[4], however it was shown here for the first time that the ISI characteristics could be caused by a delay of response in $[Ca^{2+}]_i$ increase (Fig.3(c), arrow).

5 Conclusions

In this paper, an ionic current based model was constructed for investigating the associative learning in *Hermissenda* BPR. It was suggested that the decrease of K^+ conductance did not affect the

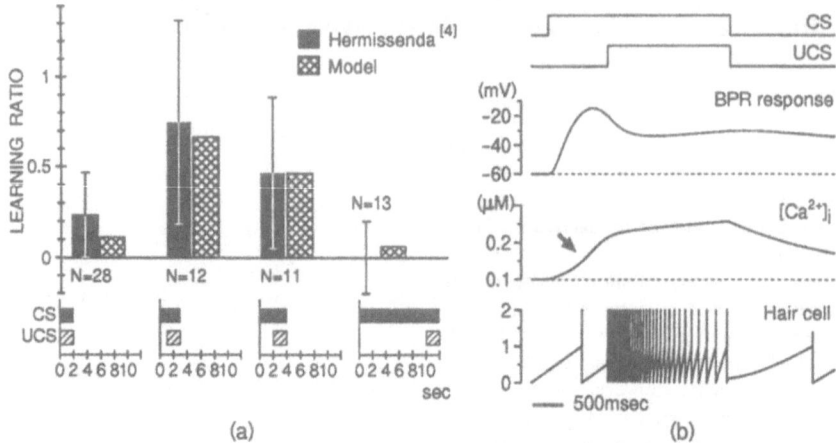

(a)   (b)

Figure 3: ISI characteristics was represented by simulation. (a) Showing similar ISI dependency in the learning ratio between Hermissenda[4] (Mean ± Standard Error, N: Number of Samples) and the model. Acquisition of learning is maximized under the ISI=1sec condition. (b) BPR, hair cell and $[Ca^{2+}]_i$ model responses were generated by CS and UCS in ISI=1sec condition. There is a delay in $[Ca^{2+}]_i$ increase (arrow).

duration of light response, and LLD was generated by a change of time dependency on cytosolic Ca^{2+} mechanism. The ISI characteristic for acqustion of memory was represented under the learning rule based on Crow's hypothesis, and it should be caused by a delay of $[Ca^{2+}]_i$ increase.

Werness et al. recently presented LLD and ISI on the abstract model[11]. However the basic mechanisms for these characteristics were not elucidated by their model. In this regard, the ionic current model, based on the experimental data, could be useful for the analysis of physiological mechanisms of associative learning.

References

[1] Alkon, D. L., "Memory traces in the brain", Cambridge University Press, 1987.

[2] Ikeno, H., Sakakibara, M., and Usui, S., "Reconstruction of molluscan type B photoreceptor responses based on the ionic currents", *IEICE Transactions*, **Vol. J72-D-II**, No. 12, 2094–2102, 1989 (in Japanease).

[3] Crow, T., "Cellular and molecular analysis of associative learning and memory in *Hermissenda*", *TINS*, **Vol. 11**, No. 4, 136–142, 1988.

[4] Matzel, L. D., Schreurs, B. G., Lederhendler, I. and Alkon, D. L., "Acquisition of conditioned associations in *Hermissenda*: Additive effects of contiguity and the forward interstimulus interval", *Behav. Neurosci.*, **Vol. 104**, No. 4, 597–606, 1990.

[5] Alkon, D. L. and Sakakibara, M., "Calcium activates and inactivates a photoreceptor soma potassium current", *Biophys. J.*, **Vol. 48**, 983–995, 1985.

[6] Yamada, W.M., Koch, C. and Adams P., "Multiple channels and calcium dynamics", in *Methods in Neural Modeling*, MIT press, 97–133, 1989.

[7] Connor, J. A. and Alkon, D. L., "Light- and voltage- dependent increases of calcium ion concentration in molluscan photoreceptor", *J. Neurophysiol.*, **Vol. 50**, 745–752, 1984.

[8] Connor, J. A. and Nikolalopoulou, "Calcium diffusion and buffering in nerve cytoplasm", *Lecture on Mathmatics in the Life Sciences*, **Vol. 15**, 79–101, 1982.

[9] Alkon, D. L. and Sakakibara, M., "Calcium activates and inactivates a photoreceptor soma potassium current", *Biophys. J.*, **Vol. 48**, 983–995, 1985.

[10] West, A., Barnes, E. and Alkon, D. L., "Primary changes of voltage responses during retention of associative learning", *J. Neurophysiol.*, **Vol. 48**, No. 5, 1243–1255, 1982.

[11] Werness, S. A., Fay, S. D., Blackwell, K. T., Vogl, T. P. and Alkon, D. L., "Associative learning in a network model of *Hermissenda Crassicornis* I. Theory", *Biol. Cybern.*, **Vol. 68**, 125–133, 1992.

A NEURAL NETWORK MODEL FOR MOTOR SHAPES LEARNING AND PROGRAMMING

VELAY J.L.[1], GILHODES J.C.[1], ANS B.[2], COITON Y.[1]

1) Laboratoire de Neurobiologie Humaine, Université de Provence
C.N.R.S. U.R.A. 372, 13397 MARSEILLE CEDEX 13, France.
2) Laboratoire de Psychologie Expérimentale, Université Pierre Mendès France
C.N.R.S. U.R.A. 665, BP 47, 38040 GRENOBLE CEDEX 09, France.

Abstract

A neural network consisting of a sensorimotor module associated with a dynamic memory (DM) is presented. The *sensorimotor module* is made of a Kohonen layer where exteroceptive and proprioceptive sensory information are combined on a functional map. The sensory layer controls a motor layer which drives the effectors of either a numerically simulated arm or an artificial jointed arm. After a learning phase, the arm is able to perform goal-directed movements. The *dynamic memory* is designed to be able to learn, memorize and execute temporal sequences. One of the main features of DM is that the repetition of learned sequences is triggered by the occurrence of an item relating to its identity, and the rhythm at which the model produces the response sequence can be controlled and freely modulated by a sub-system mimicking attentional processes. The association of both modules give rise to a neural network that can learn spatial shapes and translate them in motor terms by means of a simulated or robotic jointed arm, at any point in its working space (translation invariance) and at any suitable amplitude scale (size invariance).

INTRODUCTION

Most of our every day gestures involve the participation of numerous effectors which are co-ordinated in both time and space. Writing and drawing, for example, are probably based on centrally represented motor programs, the spatio-temporal characteristics of which are invariable. These motor outputs can be executed in various formats and sizes without thereby undergoing any change in their characteristic basic geometrical proportions. Moreover, they can be repeated using a different set of effectors from that with which they were initially learned. For example, a child who has learned to write on a horizontal sheet of paper, using mainly his fingers and wrist, will also be spontaneously able to write on a vertical blackboard, using his wrist, elbow and shoulder muscles. It is proposed here that a memory of a dynamic type provides a suitable representation for the structure responsible for planning a movement and is able to account for the above motor behaviors, assuming it is associated with a sensorimotor module capable of performing goal-directed arm movements. It is the reason why a neuromimetic model for a sensorimotor system, which was developed in a previous study [1,2] was connected to a neuromimetic model for the learning and memorizing of temporal sequences [3,4,5]. In order to complete this two-fold model assembly, we developed several modules in which the two networks in question were interfaced. Both modules will be first briefly described and the global model will then be presented.

The sensorimotor module

This model consists of a Kohonen layer [6] (K in figure 1) where sensory information of two types, exteroceptive and proprioceptive, is combined on a functional map. The cells are interconnected in such a way that each unit is linked via fixed excitatory connections to its nearest neighbours and via fixed inhibitory connections to the others. In this competitive situation, a single activity focus emerges, the site of which depends on the input afferent pattern. The structure of this layer served here mainly to provide both a non linear sensorimotor coupling and a plurimodal signal integration. The sensory layer controls a motor layer which drives the effectors of either a numerically simulated arm or an artificial jointed arm. The proprioceptive sensors code the angular values at the joints; the exteroceptive information specifies the position of any point in the arm working space.

The sensorimotor module is shaped during a learning period, when the arm randomly explores its own working space. It is assumed that during this phase, the exteroceptive signals will inform the network about the position of extremity D of the arm in the working space. At each new arm position, a pair of sensory signals (E,P) influences the adaptive connections with which sensory layer K is equipped. On the other hand, layer K participates in developing the adaptive links with the motor layer AM1, which is an associative memory, and the sensorimotor coupling is thus built up. Here a simple classical error correction rule is used, where the required output is induced by a specific activatory burst M delivered by the random generator. At some stage in the exploration of the arm's working space, the single activity focus in layer K begins to specifically encode the bi-modal information consisting of the arm posture P and position E of its extremity D. Any changes in the arm position are accompanied by changes in the site of the activity focus in the layer. This layer moreover undergoes some self-organization as the result of which it reflects the topological relationships implicitly contained in the inputs it receives. When the learning is complete, the co-ordinates of any target T in the working plane can be specified on the basis of the exteroceptive information. After a learning phase, the arm is able to perform movements which are aimed towards either a fixed target (pointing movements) or a moving target (tracking movements). In this case, the arm movement is such that the distal extremity D gradually approaches the target T. Because of the model's structure , movements of the arm towards the target consist of a series of successive steps. If the target begins to move, the arm then performs a tracking movement. Simulations were carried out on arms with two or three degrees of freedom moving in two- and three- dimensional space.

The dynamic memory module (DM)

This model was designed to be able to learn, memorize and execute temporal sequences. It consists essentially of five layers. The inputs and outputs of all the layers consist of pluridimensional patterns of neural activity with either graded or discrete values. The model receives inputs of two kinds: those of the first type, $\Delta E(t)$ represent a temporal sequence of neuronal events which require to be learned; and the inputs of the second type (ID) relate to the identity of the ongoing sequence. Each of these layers is a robust, self-sustained system which cannot be modified unless a resetting is induced by an inhibitory control process (A, I) which updates the whole architecture at regular pre-determined intervals.

During the *learning phase*, the input layers sample and recode the afferent signals $\Delta E(t)$ and ID, transforming them into binary patterns. The output S(t) thus conveys, in a recoded form, the time pattern of the current input $\Delta E(t)$. Due to the intervention of lag modules, all the successive states of the sequence S(t) accumulate in a temporary memory which at each instant contains the entire history, both recent and less recent, of the previous states of S(t). Another layer integrates this current time span (the length of which is variable) into the context of the recoded identity of the input sequence. Several sequences, each associated with its own separate identifier (ID), can be stored in this way in the permanent memory after being processed several times.

In the sequence *recall phase*, it is the occurrence of the identifier ID which triggers the recall of its associated sequence. This temporal pattern is reproduced at the output S(t) in a different format from that of the original sequence $\Delta E(t)$. The permanent memory reconstructs the items piecewise by means of the identifier and the temporary memory.

The global model

Figure 1 gives the general architecture of the model in which the sensorimotor module is combined with the dynamic-memory module DM. The two modules were connected by means of the interface layers AM2, AM3, and Σ. The associative memory AM3 learns to convert the localized format of the output S delivered by module DM into a graded activity conveyed by the exteroceptive signal ΔE. After the learning phase, layer AM3 is able to reconstruct the information ΔE even if it receives no specific sensory inputs to this effect. The second stage in the interface module is an additive device denoted Σ which is assumed to have pre-wired input connections and which additively combines both types of exteroceptive information: absolute E and relative ΔE. The associative memory AM2 implements a new relation between sensory layer K and the exteroceptive position information E. Learning in this layer is assumed to take place concomitantly with that in the sensorimotor module, and is set up in a very similar way to that of the symmetrical layer AM1 on the motor side. The specific forcing input here is the exteroceptive information E, whereas the distributed adaptive input is a collateral of the localized activity in layer K. Changes in the weighting are governed by the same error correction learning rule. Introducing this layer can prove to be useful in situations where the exteroceptive sensory flow is interrupted by the attentional switch V : when this occurs, the layer

continuously reconstructs the virtual exteroceptive information about the position E of the distal extremity D
of the artificial arm on the basis of the currently active focus in layer K.

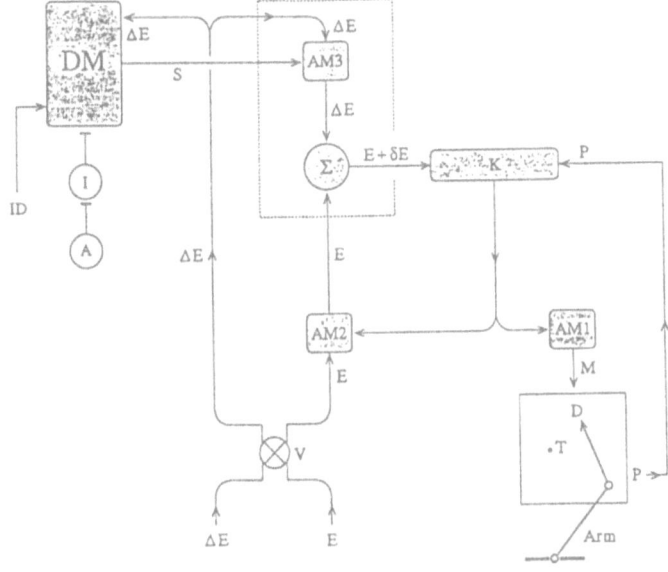

Figure 1 : The general architecture of the model. The left hand side mainly processes the exteroceptive
difference information ΔE in the dynamic memory module DM. The right hand side deals mainly with the
exteroceptive position information E within the sensorimotor module. The interface module, which is shown
here in a dotted frame, additively combines the two types of exteroceptive sensory information, the flow of
which can be interrupted by attentional valve V.

Learning of motor shapes
Once the pre-structuring of the various modules has been completed, the whole system is ready to learn
spatial forms, transformed into sequences of significant temporal cues, and to reproduce them in motor
terms. The study of the shapes is achieved by the exteroception which sequentially specifies the co-ordinates
of the most noteworthy points (for instance, the vertices of a geometrical figure). These points correspond to
position information encoded by vectors denoted by the symbol E. It is hypothesized that the exteroception is
in addition capable of providing the transient quantities δE, which are proportional to the differences
between successive positions E. Learning the contour of a shape will then consist of sequentially coding its
most outstanding points and storing in the dynamic memory the temporal sequence ΔE(t) thus generated.
Temporal sequence ΔE(t) is processed in the dynamic memory DM as described above in parallel with the
identifier (ID) of the pattern being learned, such as "triangle", "square", etc....

Reproduction of motor shapes
The motor production by the arm of a shape memorized in exteroceptive terms ΔE occurs as follows. The
occurrence of a given identifier ID triggers the production of the corresponding temporal sequence $S(t_n)$,
where t_n is the successive updating of the module DM; the rhythm of the updating is controlled by
attentional system A. The associative memory AM3 converts the sequence of localized activities $S(t_n)$ into a
graded sequence $\Delta E(t_n)$, which is then added to the position information E in the summator Σ. At the
beginning, the sensorimotor module is in an equilibrium configuration [P(0), E(0)] corresponding to any arm
position whatever. When the first element of sequence $\Delta E(t_1)$ begins to reach the summator Σ sensory layer
K receives the input activation $E(0) + \delta E(t_1)$, specifying the position of a virtual target, and an arm
movement directed towards this target is triggered. The sensorimotor module is consequently stimulated by a
series of virtual targets. The arm responds by performing a tracking movement, and could be said to have

54

repeated "from memory" the previously learned pattern. An identified shape is thus sequentially produced at the motor level in a size which depends on the frequency at which the dynamic memory is updated. This frequency must be maintained constant throughout the motor shape production period to ensure that the resulting product is not distorted. If the updating of module DM takes place at a low frequency, the final shape will tend to be large sized since the arm will continue to track the moving target in the same direction for long periods. Conversely, if the updating frequency is high, the final pattern will tend to be small-sized. Generally speaking, it is via the control exerted by attentional system A on the dynamic memory's output rhythm that the size of the figure produced in fine can be adjusted as required. The velocity of the arm movement can furthermore be controlled.

Several simulations have been carried out, both with the numerically simulated and the robotic arm, in which the whole model architecture has been implemented. Figure 2 illustrates the main behavioural properties of the model. After a period during which a given shape was presented to the model at a single point in space, the artificial arm was able to draw this shape anywhere in its working space (*translation invariance*) and in whatever size was required (*size invariance*).

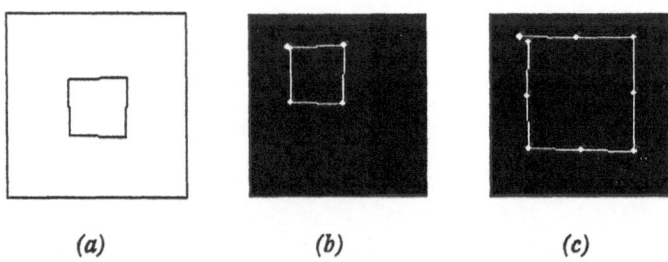

(a)   *(b)*   *(c)*

Figure 2 : An example showing the **dimensional transformation invariance of the motor pattern** produced by the arm after previous memorization. a) The pattern memorized. b) The path taken by the extremity of the arm while producing the memorized pattern. c) The path taken by the arm's extremity while producing the memorized pattern with a larger amplitude.

CONCLUSION

The most noteworthy feature of the neuromimetic model described here is its ability to quickly learn and reproduce sequences of movements at a freely adaptable rhythm. This ability is fairly commonplace among living creatures, but has rarely been dealt with in the connectionist literature. We have demonstrated that the ability to contract or expand the *time scale* of a sequence learned with the dynamic memory in exteroceptive terms resulted at the motor level in the production of forms which could be either reduced or enlarged on the *spatial scale*. Another consequence of the basic principle adopted here according to which the format in which a pattern is memorized does not necessarily determine that delivered by the effector system which translates the pattern into motor activity, is that the memorized pattern can be repeated by any of the body effector systems; the interface module connected to the dynamic memory could easily stimulate several sensorimotor units subserving separate effector systems.

References
1- COITON, Y., GILHODES, J.C., VELAY, J.L. & ROLL, J.P. (1991) A neural network model for the intersensory coordination involved in directed movements. *Biological Cybernetics*, 66, 167-176
2- GILHODES, J.C., COITON, Y. & VELAY, J.L. (1991). Sensorimotor space representation: a neuromimetic model. In J. Paillard (Ed.), *Brain and Space* (pp. 433-445). Oxford University Press.
3- ANS, B. (1990a). Associative learning in a neuromimetic network with local competitions. *Comptes rendus de l'Académie des Sciences, Paris, Série III*, 310, 127-132.
4- ANS, B. (1990b). Neuromimetic model for storage and recall of temporal sequences. *Comptes rendus de l'Académie des Sciences, Paris, Série III*, 311, 7-12.
5- ANS, B., COITON, Y., GILHODES, J.C. & VELAY, J.L. A neural network model for temporal sequence learning and motor programming., *Neural Network*, submitted.
6- KOHONEN, T. (1988). *Self-organization and associative memory.* Berlin, Heidelberg: Springer Verlag.

LEARNING THROUGH ADAPTIVE VALUE:
A MODEL WORKING IN A VARIABLE ENVIRONMENT

Murciano, A.; Zamora, J.

Dpt. Matemática Aplicada (Biomatemática). Facultad CC. Biológicas.
Universidad Complutense de Madrid.

ABSTRACT

Adaptive value (AV) is an environment-dependent variable. Therefore learning through AV shows different behaviors depending on the environment in which the model grows. In the presented model: 1) the model receives mobile visual stimuli and it should center them in the visual field, 2) the neurons responsible for the movement of eyes, execute a mapping of the visual field taking into account the AV for each movement, 3) AV is drawn from the interaction of biological like layers evolutionarily chosen, without any planning or supervision throughout the full learning process and 4) mapping changes as a function of the environment in which the model operates being able to revert when this changes.

INTRODUCTION

When modelling life organism behavior, it is necessary to provide any tentative model with adaptive parameters. These parameters endow the models with the necessary ability to answer to a wide variety of situations. To reach this purpose, any proposed model has to be conceived with those architectures able to represent the AV. AV, then, is the key to catch the meaning of any environmental condition, endowing the model to give an adequate answer in every case. This behavior enhances the chances of life organism survival. It is reflected in the artificial system as an increase in the efficiency in performing the required task. Therefore, once the model learns to carry out a given task in a definite environment, it reaches a maturity state. It should be desirable that the model be able to modify its working way from this maturity to another one, according to new environmental conditions to preserve its efficiency degree.

Models with this kind of behavior will have a high efficiency degree in working on-line at changing and/or unknown environments. Several models have reported that this kind of learning based upon AV is possible [1]. The refinement degree of the AV neuronal structures will reflect the complexity degree of the required tasks. We have previously experience on the high efficiency of an AV based model fed with fixed visual stimuli [2]. In the present work, the required task have been further modified according to different movements of the stimuli. The result is a new evaluation for AV which includes adaptive environment-dependent thresholds.

THE NETWORK

The model has three layers of neurons (Fig. 1) representing input processor (input layer, n x n square matrix), oculomotor activity (motor layer) and evaluation of adaptive value for each motor decision (value layer).

56

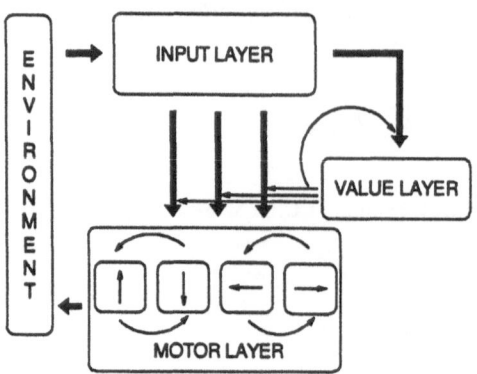

Figure 1. Architecture of the Model. Input layer receives stimuli from environment. The activation is diffused through the layer. There are two types of connections from this layer: 1) connections to Motor Layer (ML) 2) connections to Value Layer (VL). The first one leads to a motor response modifying the environment. The second one leads to the evaluation of the last movement adaptive value. This value modifies heterosynaptically the connection weight from IL to ML and the adaptive threshold for each connection between IL and VL.

Input layer receives stimuli from environment and connections intralayer produce an activity diffusion space-temporally limited to neighboring neurons. Finally, it produces an input pattern by means of this diffusion according to the next expression:

$$I_{ij}(t) = K_1 \ I_{ij}(t-1) + K_2 \sum_{x,y \in E} I_{xy}(t-1)$$

where K_1 and K_2 are adjustable parameters representing the stimulus permanence level and the activity diffusion coefficient respectively, I_{ij} the activity of the input i,j neuron, and E is the neighborhood of the neuron ij.

Input neurons activity is transmitted through connections to motor layer. These are modifiable, excitatory and with full connectivity. Motor layer is made of four sublayers innervating antagonistic ocular muscles in four directions (up, down, left and right). The motor output is obtained according to:

$$M_m = \sum_{1 \le i,j \le n} w_{ij,m} \ I_{ij}$$

where M_m is the activation value of the motor neuron m and $w_{ij,m}$ the connection weight from input neuron i,j to motor neuron m. The final movement is obtained from a competition cycle among motor neurons innervating opposite directions. The $w_{ij,m}$ term is modified according to:

$$w_{ij,m} = f \left(w_{ij,m} + \alpha \ w_{ij,m} \ I_{ij} \ M_m \ V + D \right)$$

where α is a learning coefficient, V the adaptive value of the last movement, D is a decay term and

$$f(x) = \begin{cases} 0 & if \ x < 0 \\ x & if \ 0 \le x \le 1 \\ 1 & otherwise \end{cases}$$

Value layer receives signal from the input layer through connections between both strata. These connections have fixed weights. These inversely increase to the Manhattan distance from each input cell to the center of the visual field. Such distribution of value is coherent with the connection pattern among the visual receptors and the ganglionar cells (bipolar cells mediated) vary depending on the proximity degree between any area and the central point of the retina [3][4]. Then,

$$W_{ij,v} = 1 - \frac{\left| i - \frac{n+1}{2} \right| + \left| j - \frac{n+1}{2} \right|}{n}$$

where $(n+1/2, n+1/2)$ are the central coordinates of the visual field, and $w_{ij,v}$ the connection weight from the input neuron i,j to the value cell v.

The value layer output represents the adaptive value increment of the movement decided for each time,

$$V(t) = \sum_{1 \leq i,j \leq n} w_{ij,v}(t) \, I_{ij}(t) - \sum_{1 \leq i,j \leq n} I_{ij}(t-1) \left(w_{ij,v}(t-1) + \theta_{i,j} \right)$$

where θ_{ij} is the input neuron i,j adaptive threshold obtained from

$$\Delta \theta_{ij} = K_3 \, V(t) \, I_{ij}(t-1)$$

Note that from this expression can be deduced that the overlearning is avoid when the implemented function is near from its optimum. When the environment change, adaptive thresholds collaborate to form a new distribution of the network weights. Then, the model can perform new global functions according to new environmental conditions.

SIMULATION AND RESULTS

Tests of the model were developed with mobile visual stimuli. The movement of the objects was designed to be specific for each environment. Several tests were performed training the network in a given environment. Suddenly, we change the environment (the movement) and look at the effect of this new situation upon the weight and adaptive threshold distributions. We also established as a measurement of the network efficiency the increase of stimulus centering rate average for each movement of eyes. We referred it as degree of performance.

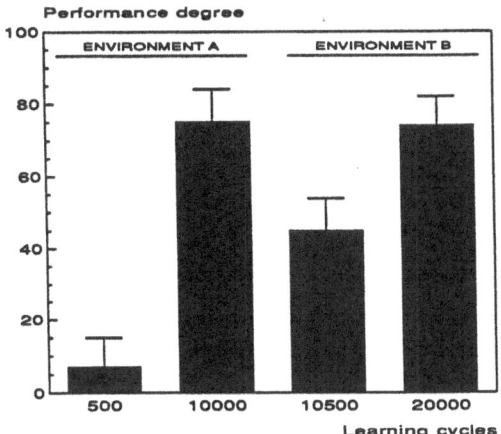

Figure 2. Degree of performance at four network states: 1) earliest learning phase, 2) end of learning in environment A (movement down, right), 3) environment change and 4) end of learning in environment B (movement up, left).

58

This measurement increase along the learning phase when the environment remains constant. When the environment change, initially there is a worse degree of performance, but during the learning in the new situation, this parameter increase until become similar to that in constant environment (Fig. 2). In this way, the network finishes the adaptation process to the new environment. In figure 3, we can observe the adaptive threshold distribution. It is modified in order to obtain an adequate weight distribution to perform a correct mapping of visual field. It also remains enough flexible to adapt to new environmental conditions.

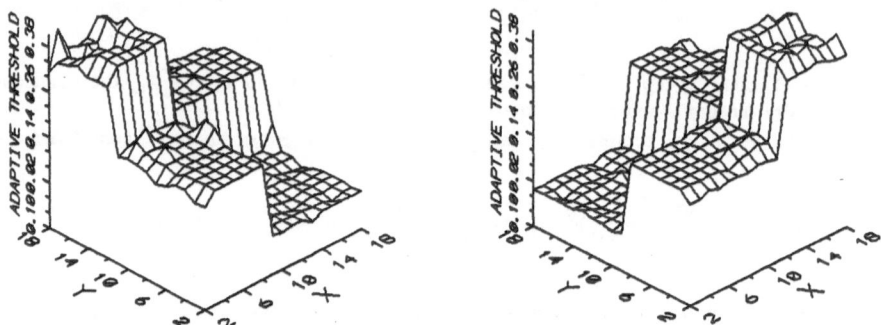

Figure 3. Adaptive threshold distribution. At the left, distribution of θ_{ij} at end of learning in environment A. At the right, thresholds have reverted from last distribution to current one in environment B.

CONCLUSIONS

The model shows the ability to learn in a environment-dependent way. It avoids overlearning and reaches a stable configuration of weights in constant environments, nevertheless, this state does not eliminate the flexibility to learn in the presence of new situations. Therefore, this stabilization gives the model a chance to answer to multivariate set of inputs, avoiding saturate states. We expect that similar models adapt themselves to, not only changing environments, but different stimuli, and they can specialize in definite task giving place a cooperative behavior.

REFERENCES

[1] Reeke, G.N.; Sporns, O.; Edelman, G.E., 1990. Synthetic Neural Modelling: The "Darwin" Series of recognition automata. *Proc. IEEE.* 78(9):1498-1530.

[2] Murciano, A.; Zamora, J.; Reviriego, M. 1993, A model for centering visual stimuli through adaptive value learning. To appear in Proceedings of IWANN93. Springer-Verlag.

[3] Schwartz, E.L., 1977. Spatial mapping in the primate sensory projection: Analytic structures and relevance to perception. *Biol. Cyber.* 25:181-194.

[4] Hubel, D.H.; Wiesel, T.N., 1977. Functional architecture of macaque monkey visual cortex. *Pro. Roy. Soc. Lon.* 198:1-59.

IMPROVING CATEGORIZATION WITH CALM MAPS

Ed Lebert & R. Hans Phaf

Psychonomics Department

University of Amsterdam, Roetersstraat 15

1018 WB Amsterdam, The Netherlands

Abstract

The Categorizing And Learning Module (CALM) represents different patterns on different nodes through a competitive learning procedure. We study an extension of CALM that enforces a topological structure on the representations. The main difference with Kohonen's self-organizing feature map is that no external regulating mechanisms are needed to learn a stable map. Simulations show that this CALM Map, in comparison to the standard CALM module, improves categorization because the stretching property of CALM Maps enables a continuous process of separation, whereas CALM will eventually commit itself to a once obtained categorization.

1 Introduction

Categorization problems are often related to interference effects in learning. Several factors play a role in interference, such as the intrinsic overlap and separation in the data set, the order of pattern presentation, and the type of learning procedure. In this paper we study categorization and interference in CALM networks. CALM (Categorizing And Learning Module [5]) has been proposed as a building block for modular network models. Competition between alternative representations in CALM already reduces interference because it enhances non-overlapping features of different patterns. A novelty dependent exploration (elaboration) process further increases the search for new representations. It turns out, however, that this latter mechanism is less useful when highly similar (i.e., less novel) patterns are presented. Separation in CALM is a complicated function of the distance between and overlap of different patterns as well as of the size of the module and the dimensionality of the patterns. Particularly in larger modules and with larger patterns, even a large Euclidean distance between patterns may not be sufficient for a stable distinct categorization.

To reduce interference, we modified CALM to stretch representations along the module in an ordered fashion, similar to Kohonen's self-organizing feature map [3]. When the order of patterns is preserved, this is generally referred to as topological self-organization. It is expected that the stretching process in the CALM Map will continue with every presentation of a pattern set, whereas the standard CALM is not able to improve its categorization after a certain point.

2 Topology preserving maps

A widely used network for learning topology conserving mappings is Kohonen's self-organizing feature map [3]. In this model nodes are arranged according to some type of 1, 2 or 3-dimensional neighborhood of connectivity (e.g., a line, grid, or cube). Input patterns are compared to the weight vector of every node resulting in the selection of a best-matching or "winning" node. Weights corresponding to nodes that are neighbors of the winning node are also modified according to a "winner-take-most" scheme. As a result, similarity between input patterns will be mapped into proximity of activated nodes.

To obtain maximal stabilisation in the Kohonen map, usually two heuristics are applied: a) the neighborhood size of activated nodes is reduced gradually during adaptation, and b) the learning parameter is adjusted externally. This requires prior knowledge about the adjustment schedules of the neighborhood function and the learning rate. Kohonen describes the choice of a decreasing sequence of learning rates as very "subtle". Such expertise is not necessary when processes capable of *automatically* self-adjusting global network parameters are implemented. CALM seems well suited for this purpose, because it already incorporates a novelty dependent learning rate, which automatically decreases during processing. Before we explain how the first heuristic can also be satisfied, we describe CALM in somewhat more detail.

3 CALM Map

A standard CALM is a "winner-take-all" learning module, in which the competition process is performed by intramodular interactions between excitatory Representation nodes (R-nodes) and inhibitory Veto nodes (V-nodes). The strong inhibitory weights from V-nodes to all non-matched R-nodes impose a veto-effect on the R-nodes. Input arrives along modifiable intermodular connections to the R-nodes. An Arousal node (A-node), which receives connections from both R and V-nodes weighs the amount of competition, which is a measure of the novelty of the presented pattern. The A-node activates an External node (E-node), which spreads random activations to prevent "deadlocks" in the competition, and controls the modular learning rate. Figure 1 shows a schematic drawing of CALM, in which the node types, and some intra- and intermodular weights are depicted.

CALM incorporates two modes of learning: elaboration learning and activation learning. Elaboration is dependent upon the amount of competition among the R-V node pairs. If a pattern is not yet represented in the module it will generally elicit much competition, because many nodes are simultaneously activated by the pattern. This gives rise to a high arousal level at the A-node, yielding an increased learning rate, and relatively large random pulses from the E-node, which facilitates the resolution of competition. A pattern, which has already been learned well, activates its corresponding node without much competition and only strengthens its representation. This constitutes activation learning, which is characterized by a relatively low learning rate. Learning in CALM, thus, has the effect of reducing the competition with repeated presentation of the pattern set, whereby elaboration learning is gradually replaced by activation learning.

One approach (CALSOM), reported by Murre [4], to implement a neighborhood of activated nodes, is to use a *fixed* type of graded inhibition. He applied a linear inhibition gradient, but did not obtain maximal separation of representations. In the approach we present here, a convex inhibition gradient (e.g., part of a Gaussian function) will be applied, for which it has been proven that, when the "full width at half height" of the Gaussian equals the number of neurons, convergence is optimal [1]. In the CALM Map discussed here only one-dimensional topologies will be considered (e.g., a line or a ring).

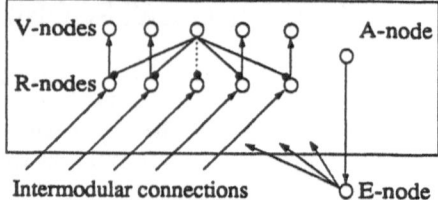

Figure 1: *CALM module with equal inhibitory weights (-10.0) from Veto nodes to non-matched Representation nodes, and a higher weight (-1.2) to the matched Representation node. Intermodular connections are modifiable. (Not all intramodular connections are shown.)*

The parameters had mainly the same values as in the standard CALM module [5]. Some changes were made in the learning rate u_t, which depends on the activation a_E of the E-node according to $u_t = d + w_{uE}a_E$, where d (the base rate of learning), and w_{uE} (a virtual weight from the E-node to the learning parameter) were reduced to 0.0001 and 0.0005 respectively. The weights to the A-node were also adjusted: the excitatory weights from R-nodes were set to 0.3, and the inhibitory weights from V-nodes to -0.7. New parameters were introduced for the inhibition gradient, which was determined by $Aexp[-\frac{(i-j)^2}{2\sigma^2}]+B$ with $A = 8.8$, $B = -10.0$, $\sigma = 3.0$, where $(i-j)$ denotes the distance between the ith R-node and jth V-node. To prevent boundary problems when nodes are organised in a line, we used a slightly smaller inhibition gradient for the first and the last node ($\sigma = 3.9$). Contrary to Kohonen's approach, the parameters of the Gaussian do not change during presentation of a pattern set.

4 Functioning of CALM Map

To illustrate the ordering process in the CALM Map, nine patterns were presented for 25 iterations each to a module of size 9 (the nodes were organised in a line). At every presentation, the order of the patterns was randomized. Between presentations activations were initialized (to zero). Due to the graded inhibition, the competition process first yields a broad, shallow activity pattern around the central R-V pair. As a consequence of the increasing intermodular weights, the R-nodes close to the winning node gain higher activations, resulting in higher inputs to V-nodes and even stronger inhibition of the other nodes. The "activity bubble" thus auto-

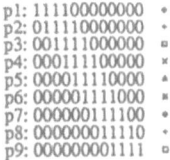

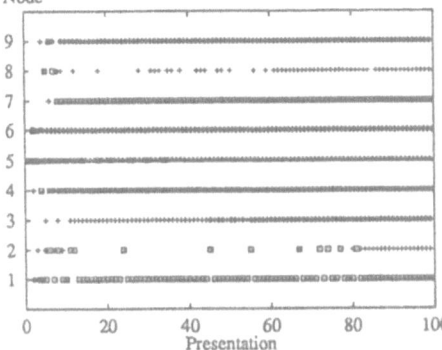

Figure 2: *Performance of CALM Map. Development of categorization for pattern set {p1...p9} containing a topological order.*

matically decreases during all phases of pattern presentations. In Figure 2 the pattern set and the categorization results are shown. After about 80 presentations the patterns are unidimensionally ordered.

The distinction between elaboration and activation learning can be seen to conform nicely to Kohonen's account of the computational requirements for topological self-organization. He describes two subsequent phases in the development of a map. In contrast to the standard CALM module all patterns are initially represented on one (central) node. The patterns are then roughly split up over the Representational nodes. This constitutes the *ordering phase*. After the initial order is established, the network gradually approaches a maximal spacing of representations. According to Kohonen this constitutes the *convergence phase*. The ordering phase may correspond to elaboration learning, and the convergence phase to activation learning.

5 Simulations

We performed a series of simulations to investigate the stretching process in CALM Maps and compared its behavior to the standard CALM module with respect to the size of the module, the overlap and the Euclidean distance between patterns. Each pattern set was presented for 400

times in both modules. All results were averaged over 5 replications. The CALM Map generally needed fewer iterations to converge upon a single node than CALM. In all simulations the maximum number of iterations was kept constant at 25 iterations for the CALM Map and at 50 iterations for the standard CALM module.

To study the influence of module size on categorization, a pattern set similar to the one in Figure 2, was presented to modules of size 5, 10, 15, 20, and 25. Categorization results are shown in Figure 3. The number of uncommitted nodes, which corresponds to the number of nodes with double representations (when there are as many patterns as nodes), slightly increases in larger modules, but on the whole the results for both CALM and CALM Map do not differ substantially. Interestingly, if the number of nodes equals twice the number of patterns, representations are separated maximally (without double representations) such that committed and uncommitted nodes alternate over the nodes of the module. Closer inspection of the weights revealed that uncommitted nodes actually interpolated the representations of the neighboring nodes. Due to the absence of topological stretching, CALM, of course, could not show such interpolation. Nevertheless, this simulation shows that the global categorization in CALM Maps does not differ from standard CALM modules with a fair amount of overlap between the patterns.

To investigate the role of overlap we varied the number of shared activations in the data set. Pat-

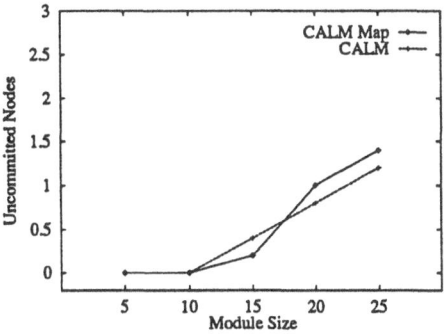

Figure 3: *Categorization as a function of size of the module. The number of patterns equals the size of the module.*

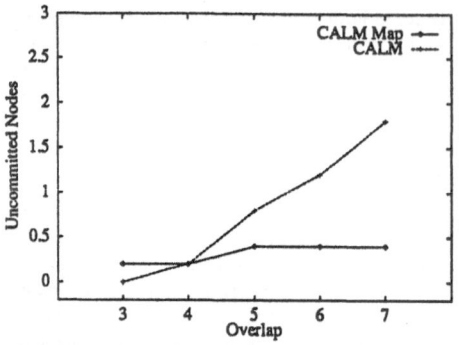

Figure 4: *Categorization as a function of overlap between patterns.*

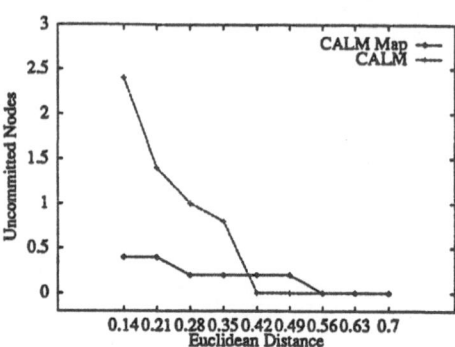

Figure 5: *Categorization as a function of distance between patterns.*

tern sets, each containing 11 patterns, were constructed with overlap 3, 4, 5, 6, and 7 (see Figure 2). Input activations were 0.5, and the size of the module was 11. Figure 4 illustrates the results of this simulation. In contrast to the standard CALM module, overlap hardly affected categorization performance in the CALM Map. As a function of overlap, categorization is clearly better in CALM Maps than in CALM.

The (Euclidean) distance between patterns may also play a role in categorization. In the third simulation a pattern set with a constant overlap of 3 activations was used, but the distance between patterns was varied. The activations of the nodes in the input patterns ranged between 0.10 and 0.50 in steps of 0.05. The number of patterns equaled the number of nodes in the module (11), but a ring topology without edges was used in the CALM Map. Figure 5 shows that particularly for small distances the CALM Map seperates representations better than the standard CALM module.

6   Conclusion

CALM Maps show some advantages above both Kohonen's self-organizing feature map and the standard CALM module. In contrast to Kohonen's map [3], it needs no external regulating mechanisms, but is capable of automatically adjusting its learning rate (to prevent twisted topologies), and shrinks its "activity-bubble" without adjusting the inhibitory weights. In comparison to CALM (and CALSOM [4]) categorization is improved in CALM Maps for highly similar patterns due to the stretching process which

continuously enhances the separation of representations. A nice spin-off of this process is the interpolation of intermediary representations on uncommitted nodes. The introduction of topological ordering in a competitive learning procedure seems to enhance the capabilities for unsupervised learning of arbitrary patterns. The frequent occurrence of topological mappings in the nervous system [2] may, thus, illustrate that such a stretching process presents an optimal and continuous differentiating power for unsupervised learning in real neural systems.

Acknowledgements

We wish to thank dr. Cees van Leeuwen and Henk Tijssen for their helpful advice.

References

[1] E. Erwin, K. Obermayer, and K. Schulten. Self-organizing maps: Stationary states, metastability and convergence rate. *Biological Cybernetics*, 67:47–55, 1992.

[2] C.D. Gilbert and T.N. Wiesel. Columnar specificity of intrinsic horizontal and cortico-cortical connections in cat visual cortex. *The Journal of Neuroscience*, 9:2432–2442, 1989.

[3] T. Kohonen. *Self-organization and associative memory.* Springer-Verlag, Berlin, 2nd edition, 1988.

[4] J.M.J. Murre. *Learning and categorization in modular neural networks.* Harvester Wheatsheaf, Hemel Hempstead, U.K., 1992.

[5] J.M.J. Murre, R.H. Phaf, and G. Wolters. CALM: Categorizing and learning module. *Neural Networks*, 5:55–82, 1992.

A Simple Selforganizing Neural Network Architecture for Selective Visual Attention *

D. Heinke, H.-M. Gross

Technical University of Ilmenau, Division of Neuroinformatics
O-6300 Ilmenau, POB 327, Germany
e-mail: dietmar@informatik.tu-ilmenau.de

Abstract

We present a simple neural network architecture which autonomously learns how to control a data driven selective attention process. In order to control the selective attention process a biologically plausible position coding is used which leads to fuzzy representations of position. An associative memory learns the connections between subsequent positions und local features. The result of presenting simple Real-World color images to the neural network architecture is shown.

1 Introduction

Because of the problem of combinatorical explosion a massively parallel data stream, e.g. a complex Real-World scene, can't be processed completely in parallel. Thus, it is necessary to transform a complex spatial data structure into a spatio-temporal data sequence [1] [2]. In order to achieve such type of transformation a data driven selective visual attention was introduced in [1].

The present paper is focused on an architecture for knowledge based control of a data driven selective attention process through a simple self-organizing neural network. After presenting Real-World scenes to the Neural Network architecture the selective attention process should scan the objects of a scene successivly. Hence, this behavior could be interpreted as autonomous knowledge acquisition about objects in a visual scene – without supervised training of single objects in a special learning phase [2].

2 The Neural Network Architecture

2.1 An Overview

Figure 1 shows the architecture of the whole system. In [1] a dynamical network for a data driven selective visual attention process was introduced. In the present work we use a modified version of that *selective attention network* (SAN). SAN generates an activity distribution which corresponds to the position of the actual "focus of attention". On the one hand this activity distribution is used to extract the area of attention out of the input scene and on the other hand it is used for a absolute position coding. Hence, we use activity distribution as a general position coding principle.

In order to be positional invariant the position coding of SAN is transformed into a relative position coding through the *position transformation* (PT). That means we get a coding of the distance and the direction between two successingly following foci of attention. In addition the PT maps relative

---

*supported by BMFT, Grant No. 413-5839-01 IN 101D - NAMOS-Project

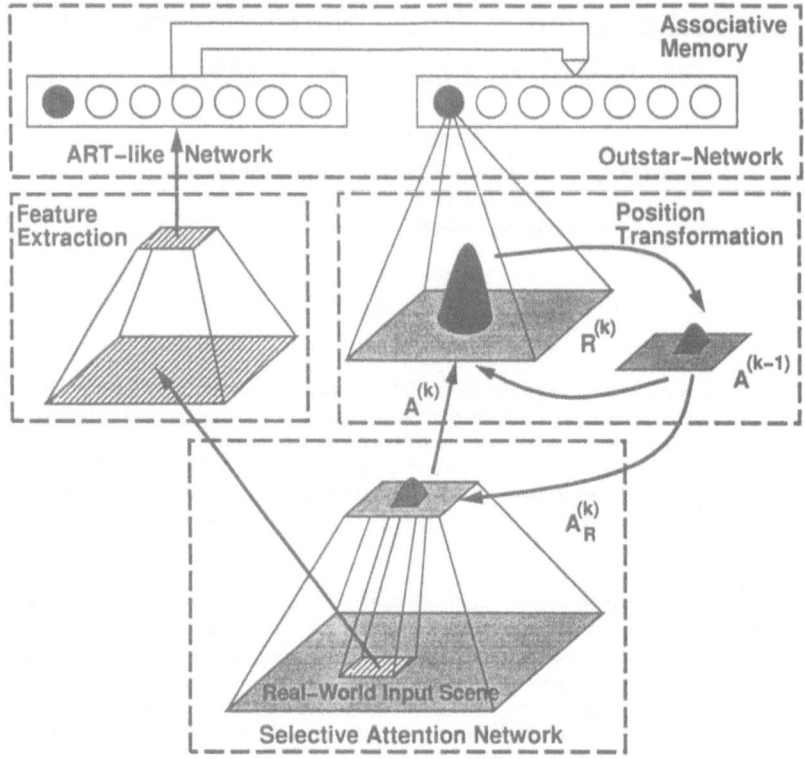

Figure1: The neural network architecture.

position codings into absolute codings. Through this mapping the SAN can be controlled in a top-down manner by the associative memory.

The *associative memory* (AM) is the central part of the whole system. It is able to learn the connection between the relative position coding and the output of the *feature extraction* of the focus area. It is continously learning using the statistics of the SAN behavior which are based on the scenes. Because in different scenes the objects are not always at the same position the frequency of inter-object moves is negligible. Thus, the AM can only learn and stabilize the moves within the objects of a scene. In addition the AM learns by successful control of the SAN. The strength of control increases with its success.

2.2 Feature Extraction

The feature extraction of the SAN consists of two multiresolution pyramids operating on a special color space, the WMM-space. The first pyramid is based on Laplace-filtering in the W-domain, the intensity image. The second pyramid uses the Euclidean distance to the average hue of the image in the $M_{rg}M_{yb}$ domain [1]. In order to have a strict local feature extraction in the focus area we simply use the WMM-space in an additional multiresolution pyramid.

For improving the feature extraction we currently discuss Eigenvectors to get a texture based enhancement [3].

2.3 Position Transformation

The position transformation uses two mapping rules (see Fig. 1). The first rule maps two successive SAN position codings $\mathbf{A}^{(k-1)}, \mathbf{A}^{(k)} \in IR^n \times IR^n$ into a relative position coding $\mathbf{R}^{(k)} \in IR^{2n} \times IR^{2n}$

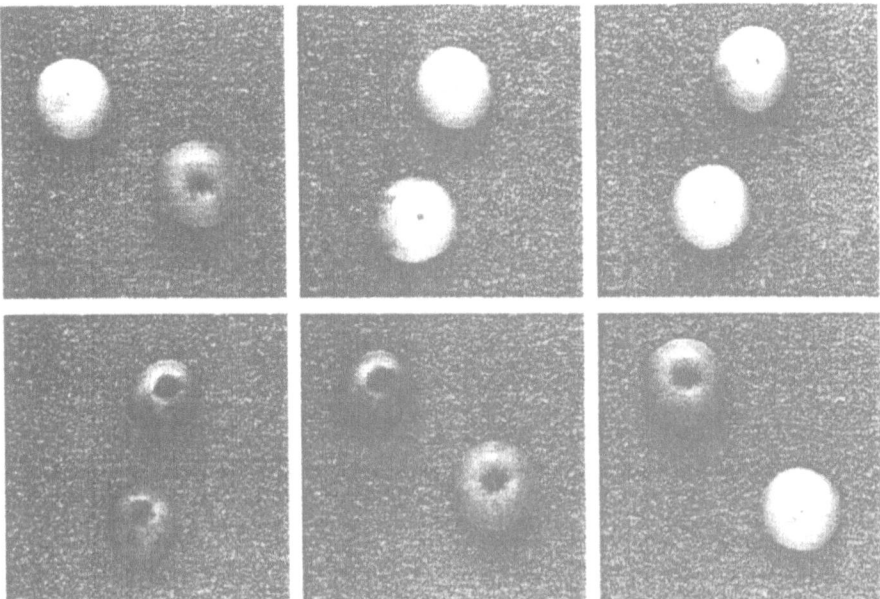

Figure2: The Figure shows examples of typical input images. The different objects are placed at different positions in a scene. This is the essential prerequisit for a selforganizing learning within the associative memory (see text).

through correlation:

$$r_{io}^{(k)} = \sum_{j=0}^{n-1} \sum_{l=0}^{n-1} a_{jl}^{(k-1)} a_{i+j,o+l}^{(k)}$$

The second mapping rule used in the opposite direction transforms a relative position coding $\mathbf{R}^{(k)} \in IR^{2n} \times IR^{2n}$ into a absolute position coding $\mathbf{A}_R^{(k)} \in IR^n \times IR^n$ through convolution:

$$a_{io}^{(k)} = \sum_{j=0}^{n-1} \sum_{l=0}^{n-1} a_{jl}^{(k-1)} r_{i-j,o-l}^{(k)}$$

Hence, we use a biological plausible way of a implicit representation of positions. In addition this coding principle is a fuzzy representation, which makes the whole system more robust and can be simply learned by the AM.

2.4 Associative Memory

The *associative memory* (AM) consists of two neural networks. The first network is similar to ART2 [4] and represents the features extracted in the focus area in a sparse coded way. This representation forms a suitable input for the second neural network, an outstar network [5].

The outstar-network learns relative position codings which are frequently met with a defined output of the ART2-like network. Its weight matrix $\mathbf{W} \in IR^p \times IR^{2n} \times IR^{2n}$ is accessed through: $r_{ij} = w_{Kij}$, where K is the actually active output node of ART2 and p the number of the output nodes. The result of this operation is mapped through the position transformation and controls the SAN through a simple additive superposition. The weights of the outstar network are adapted in the following way:

$$\tau \frac{dw_{Kij}}{dt} = -w_{Kij} + d_{ij}^{(k)},$$

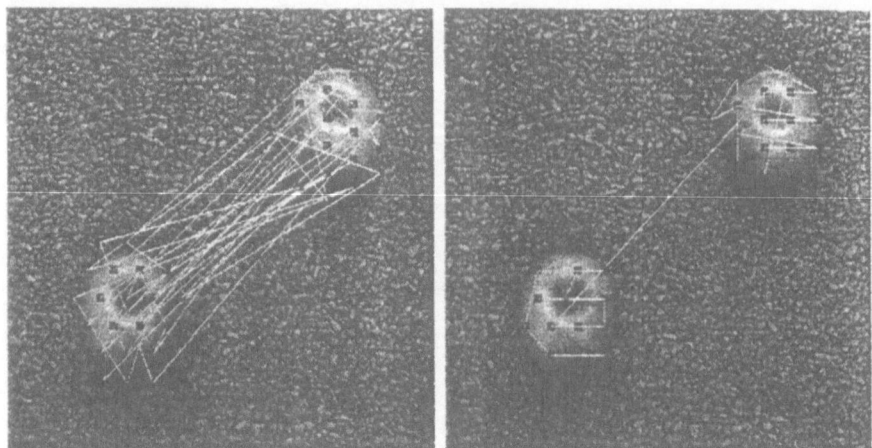

Figure3: The left part of the Figure shows the behavior of the neural network architecture with only data driven selective attention. The right part shows the behavior controlled by the associative memory after presenting 50 different images to the neural network architecture.

where $D^{(k)} \in IR^{2n} \times IR^{2n}$ is the normalized relative position coding: $D^{(k)} = R^{(k)}/\|R^{(k)}\|$. This learning rule leads to gradually increasing control of SAN by the AM.

3 Results

The whole neural network architecture was confronted with 50 different color images. Fig.2 depicts examples of them. All scences consist of different kind of fruits in different positions. Fig.3 compares the behavior of the neural network architecture before being confronted with the 50 images and afterwards. The "known" objects are scanned successivly because of the dominating control of the data driven selective process by the AM.

4 Conclusion

Currently we extent the present approach through the concept of efficiency [2]. This concept will be applied to two areas: At first the dynamic of the SAN can be used to achieve a more efficient use of processing time. At second the neural network architecture learns how to perform an efficient scan process within a given set of images. In this frame other top-down control mechanism than simple additive superposition are discussed.

References

[1] H.-M. Gross, E. Koerner, H.J. Boehme, T. Pomierski, "A Neural Network Hierachy for Data and Knowledge Controlled Selective Visual Attention", Proc. of ICANN92, 1, 825-828, 1992.

[2] H.-M. Gross, H.-J. Böhme, D. Heinke, R. Möller, T. Pomierski, "Steuerung parallel-sequentieller Verabeitungsprozesse und Strukturierung dynamischer Repräsentationen", will be published by Springer-Verlag.

[3] T. Pomierski, H.-M. Gross, D. Wendt, "A Distributed Multicolumnar System for Primary Cortical Analysis of Real-World Scenes", this volume.

[4] G.A. Carpenter, S. Grossberg, "ART2: self-organization of stable category recognition codes for analog input patterns", Applied Optics, 26, 23, 1987.

[5] S. Grossberg, "Some Networks That Can Learn, Remember, and Reproduce Any Number of Complicated Space-Time Pattern I", J. of Mathematics and Mechaniscs, 19, 1, 53-91, 1969.

Detection of Coincidences and Generation of Hypotheses – a Proposal for an Elementary Cortical Function

R. Moeller, H.-M. Gross

Technical University of Ilmenau, Division of Neuroinformatics
O-6300 Ilmenau, POB 327, Germany
e-mail: Ralf.Moeller@Informatik.TU-Ilmenau.DE

Abstract

A framework for a model of elementary cortical function is presented which is based on the detection of coincident events by intracortical processing. By extracting coincidences knowlegde about the rules of interaction with the environment can be acquired. Vice versa this knowledge can be used for the generation of anticipatory hypotheses and hypotheses about ambiguous data. Some basic functions of visual recognition could be explained by this model if recognition is considered to be in strong connection with motoric action.

1. Introduction

Our model is insprired by the former work of BARLOW who supposed that the detection of 'suspicious coincidences' in afferent data streams could be a basic cortical function [1, 6]. Coincident events occur more often combined than it can be expected from their single probabilities; they therefore should reflect causal connections of reality. Coincidence detectors operate on information from sensory sources and internal states, e.g. between visual impression and tactile stimuli or between generated motor commands and information from proprioceptive sensors.

2. Evidence from physiology

Detection of coincidences is supposed to be a function performed everywhere in the cerebral cortex in a similar manner. There is some evidence from physiology for a common cortical function, even if the developed cortex shows significant differences in the architectonic features of the areas. The structure of the developing protocortex is relatively uniform [5], only afferent and efferent fibers are arranged specifically. Afferent inputs cause structural and functional modifications of the areas according to the *spatiotemporal structure of the data*. If fibers of a certain modality are redirected to an area which normally processes another modality, the new function can be performed by this area. In terms of our hypotheses: cortical areas own all equipment necessary for the detection of a wide range of coincidences. What coincidences the area specializes in is decided by area-specific inputs; e.g. if especially temporal relations have to be considered, units providing a specific time behaviour will be integrated and others will be discarded. Because it is impossible to detect coincidences between all channels, the information streams to be combined are defined genetically; each area can only detect coincidences on a combination of channels that proved to be necessary during evolution [1]. Experiments with kittens [7] demonstrated the variability of coincidence detection within the genetic frame; some cells showed responses to a combination of signals that was never found in normal kittens.

3. Detection of coincidences and generation of hypotheses

Up to a certain stage of processing coincidence detecting units only *signal* the occurence of a known combination of events they are specialized in to higher levels of recognition (e.g. formation of orientation–sensitive cells [4]). In higher levels events which have been detected to be coincident seem to feed back mutually. An example: If you start moving, your sensors are regularly confronted with a connection of two impressions: all visual features move with growing velocity in certain directions and sensors of your body report the 'feeling' of acceleration. Because this combination is a main property of our physical world it occurs in (almost) all cases; it is then integrated into the internal model of the world. Sitting in a train you are now confronted with a well-known phenomenon: if a train next to yours drives up, you see the change of the visual features and *feel* the acceleration of your train although there is none; the occurence of *one* of the events sets up a hypothesis about the others. We don't move ourselves but we have a *conception* about movement derived from the coincidence between visual impression and the sensation of real movement. In our terminology this can be called a *completing hypothesis*.

Figure 1 shows the neural basis of the detection of coincidences and the generation of hypotheses. If two events often occur *at the same time*, a *completing hypothesis* can be created (figure 1, III) by symmetrical excitatory connections between 'hypotheses–cells', even if only *one* information is present. Two events in a *fixed temporal relation* that reflects a causal connection, e.g. a visual impression of an object together with a certain motor command and a tactile information when a limb touches the object, should result in asymmetrical weights between the 'hypotheses–cells' (figure 1, IV). Presentation of the 'cause–event' sets up an *anticipatory hypothesis* about the 'effect–event', whereas it is impossible to draw the opposite conclusion.

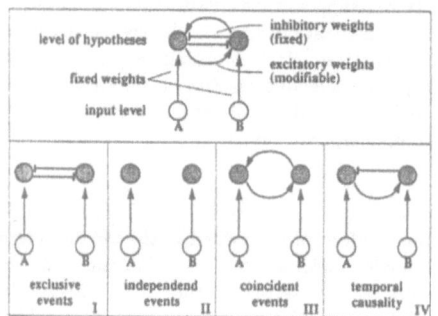

Figure1: **Upper part:** Two 'hypotheses–cells' are mutually connected by fixed inhibitory and modifiable excitatory weights. The efficiency of the excitatory connections is high, if the events A and B are coincident. **Lower part:** Possible interaction of 'hypotheses–cells'. **(I)** Exclusive events. Excitatory weights vanish. **(II)** Independend events. Excitatory and inhibitory effect are balanced. **(III)** Coincident events. Excitatory effect predominates. **(IV)** Event A often precedes B. Weights become asymmetric.

'Hypotheses–cells' transmit input signals as well as hypotheses. If HEBB–like learning is applied for the modifikation of the synaptic connections between them, excitation by hypotheses and by input information has to be distinguished in the learning rule. Excitation of one 'hypotheses–cell' by another, which is not excited by 'real' input, should not be interpreted as a coincidence of two events. A possible explanation could be the assumption of *separated ranges* for input activity (above a certain level) and activity induced by hypotheses (below this level) [2]. Strengthening of synapses is restricted to cases, when both the presynaptic and postsynaptic cell are excited above this level. Distinct pathways (e.g. axo–somatic synapses for input transmission and axo–dendritic for hypotheses) could be postulated to preserve the source of activation through all processing levels.

Interconnected 'hypotheses–cells' take part in two major generation processes of hypotheses:

1. *Generation of sequences of hypotheses:* We assume the events A and B to be 'coincident' (in this case they occur in a fixed temporal relation) as well as B and C (see figure 2, left part). If only A occurs, A's 'hypotheses–cell' is excited by the input and itself excites the 'hypotheses–cell' of B: a hypotheses about B is set up. B itself, which is *a hypothesis only* because the input cell of B is not active, is able to set up a hypotheses C and so on — a sequence of events can be predicted this way.

2. *Harmonization of hypotheses on ambiguous data:* Ambiguous data (such as the retinal image of one eye) lead to different sets of hypotheses about the real situation, some of them being *consistent*, the others *inconsistent*. If a set of hypotheses is inconsistent, there are active 'hypotheses–neurons' inhibiting each other. In the simplest case these cells represent complementary events (see **figure 2**, right part). In a relaxation process inconsistences could be solved by switching on or off some of the 'hypotheses–cells', a process comparable to relaxation in feedback associative memories. In **figure 2** event A and B favour C, whereas D is coincident with the complementary event of **C**. A hypotheses including an event and its complementary event is inconsistent and should be changed by deactivating one of the 'hypotheses–cells'. Inconsistences like the one shown in the right part of **figure 2** are solved by majority decision.

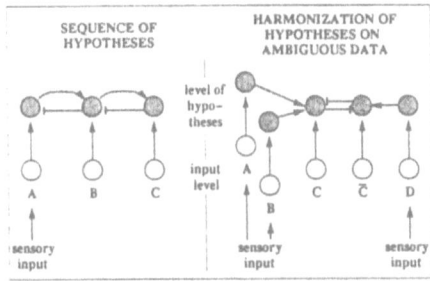

Figure2: The two major generation processes of hypotheses. **Left part:** Generation of sequences brings about the prediction of a chain of subsequent events. In this case 'hypotheses'–cells excite each other. **Right part:** The harmonization of hypotheses solves inconsistences in sets of hypotheses on ambiguous data. Inhibitory and excitatory effects between 'hypotheses–cells' cause a relaxation process towards a more consistent set of hypotheses.

In our model philosophy the 'recognition' of a sensory situation and the selection of appropriate behaviour is based on the generation of sequences of hypotheses (**figure 3**). Supposing the following coincidences have been detected formerly: if in a sensory situation S1 a motor command M1 is executed, a new situation S2 is set up, the same being valid for S2-M2 and S3 as well as S2-M3 and S4. S3 and S4 at the end of different sequences of hypotheses are assumed to be coincident with sensory situations of a genetically determined negative (pain) or positive (pleasure) meaning for the living being. Starting now from a given situation or event S1, different sequences are induced (in parallel or successively) by random activation of the motor command units below a level necessary for execution. In the hypothetical situation S2 the generation process can take two different ways in dependence of the random excitation of M2 or M3. If the chain of predicted events ends at a *negative impression* (S1-M1→S2-M2→S3), all motor command neurons which took part in the generation of this special sequence are suppressed by negative feedback, that is, actions supporting the real course predicted by the sequence of hypotheses cannot be executed. If the event is coincident with a *positive impression* (S1-M1→S2-M3→S4), the sequence is preferred for execution due to accumulation of excitatory feedback at the motor command neurons. The selection of appropriate behaviour in a given situation could be realized this way.

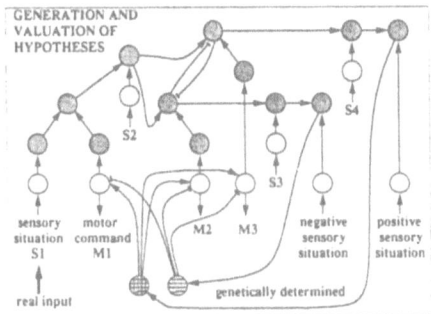

Figure3: The selection of appropriate behaviour in a given situation S1. If in this situation a motor command M1 is executed, an new situation S2 emerges. The execution of M2 in the situation S2 leads to a new situation S3, which itself was found to be coincident with a negative impression. In this case all motor commands involved in the sequence are suppressed. If M3 is executed in the situation S2, a 'positive' situation S4 arises — the corresponding motor commands become candidates for execution.

Random stimulation of motor command neurons appears in two forms. Stimulation *above* a level necessary for execution provokes 'sensory reactions' of the physical world. That is the way to detect coincidences between an action and its sensory consequences. If the corresponding coincidences have already been detected, stimulation *below* this level does not entail any action but may predict the sensory consequences of it in a certain situation.

4. Visual recognition

Recognition of *shape* is considered to be a basic function of the visual system: classification of an object requires the recognition of its functional properties [3], functional properties can only be recognized if the shape of the object is perceived. In our approach, visual perception of shape at higher processing levels can be explained by the detection of stable relations between visual information and information from other senses (multimodal processing) during activities in the environment. Visual information is connected to impressions like 'time-to-contact', when a certain movement is executed, or tactile information, when an object is grasped. Once having detected coincidences of this kind, visual information itself is sufficient to characterize the shape *by a sum of related actions and impressions which are typical for it*. The internal representation of perception of shape would be in our model a multitude of sequences of hypotheses about the consequences of different actions, induced in parallel or successively by the given visual information (a distributed representation of sensory and motor information relevant for the situation). These actions don't have to be executed, they only describe the visual scenery, but those actions, which are a starting point of one of the 'positive' sequences of hypotheses, are candidates for execution as described above.

Applying our approach to perception of shape could help to avoid problems arising from an artificial separation between recognition and generation of behaviour. If both parts are separated, a 'unit for recognition' has to analyze the visual information and to convert it into any descriptive code, whereas another 'unit for generation of behaviour' converts this code into appropriate behaviour. First, the descriptive code can be more compact than the visual information itself, but its interpretation is not necessarily simpler. Second, the conversion into a descriptive code could be a detour. If an object partially covered by another shall be grasped, knowledge about covering has to be applied, covering has to be expressed in a desriptive code, and by use of knowlegde about appropriate movements for grasping in the case of covering, motoric commands have to be derived from the code. It seems to be much simpler to characterize the visual scenery *immediately* by a sum of hypotheses about the consequences of actions possible in this situation; from this set of sequences of hypotheses appropriate behaviour is chosen.

Our model hypothesis should be understood as a first alternative approach to a general model of complex visual perception; perhaps it could be possible to explain some processes using more elaborate networks (e.g. those capable of generalizing detected coincidences) composed of the simple units described in this paper.

References

[1] Barlow,H.B. Cerebral cortex as model builder. In Rose,D. and Dobson,V.G., editors, *Models of the Visual Cortex*, chapter 4, pages 37–45. John Wiley & Sons Ltd, 1985.

[2] Koerner,E., Gross,H.-M., and Tsuda,I. Holonic Processing in a Model System of Cortical Processors. In *Biological Complexity and Information*. World Scientific, Singapore, 1990.

[3] Mallot,H.A., Kopecz,J., and Seelen,W.v. Neuroinformatik als empirische Wissenschaft. *Kognitionswissenschaft*, 3(1):12–23, 1992.

[4] Malsburg,C.v.d. Self-organization of orientation sensitive cells in the striate cortex. *Kybernetik*, 14:85–100, 1973.

[5] O'Leary,D.D.M. Do cortical areas emerge from a protocortex ? *TINS*, 12(10), 1989.

[6] Phillips,C.G., Zeki,S., and Barlow,H.B. Localisation of function in the cerebral cortex: past, present and future. *Brain*, 107:327–361, 1984.

[7] Spinelli,D.N. and Jensen,F.E. Plasticity: The mirror of experience. *Science*, 203:75–77, 1979.

DIVA: A Self-organizing Neural Network Model for Motor Equivalent Speech Production

Frank H. Guenther[1]
Department of Cognitive and Neural Systems, Boston University
111 Cummington Street, Boston, MA 02215

Abstract

This paper describes a model of speech production called DIVA that highlights issues of self-organization and motor equivalent production of phonological units. The model uses an action-perception cycle to learn two mappings between three levels of representation. Data on the plasticity of phonemic perceptual boundaries motivates a learned mapping between phoneme representations and vocal tract variables. A second mapping between vocal tract variables and articulator movements is also learned. To achieve the flexible control made possible by the redundancy of this mapping, desired directions in vocal tract configuration space are mapped into articulator velocity commands. Because each vocal tract direction cell learns to activate several articulator velocities during babbling, the model provides a natural account of the formation of coordinative structures. Model simulations show automatic compensation for unexpected constraints despite no previous experience or learning under these constraints.

Overview of the DIVA model

Production of an acoustic signal that invariantly conveys to listeners a particular phoneme is carried out with large variability in articulator movements from one instance to the next (e.g., [1],[3],[5]). The process of producing an invariant result in a motor system despite large variations in the contributions of individual components from trial to trial is called motor equivalence. This paper investigates the relationship between phonological units and the motor actions that realize them, focussing on how motor plans to produce phonemes can be learned, and how these plans can lead to motor equivalent phoneme production.

Figure 1a is a block diagram of the current model, which is named DIVA because a key component is a transformation from Directions (in vocal tract configuration space) Into Velocities of Articulators. This model assumes the existence of a neural representation of phonological units, or phoneme map, for production of speech. These units are each associated with a *plan* consisting of a set of synaptic weights that encode target values of vocal tract variables determining key acoustic properties of the vocal tract. These plans are then converted into appropriate articulator movements. The chosen articulator movements depend on context and external conditions, thus resulting in motor variability as seen in human speech. In addition, DIVA assumes the existence of a phoneme recognition system that transforms an appropriate incoming acoustic signal into a representation of the corresponding phonological unit. The model contains two learned mappings, indicated by filled semicircles in the figure: a mapping between phoneme representations and corresponding vocal tract configurations (or motor plans), and a mapping between vocal tract variables and articulator movements that realize desired vocal tract configurations.

The levels of representation in DIVA are very similar to those of the speech production model of [6]. However, the nature of the mappings between between levels differ. Furthermore, mappings in [6] were predefined by the modelers. In DIVA, emphasis is placed on how these mappings self-organize; that is, how can an infant's speech production system learn parameters governing this complex dynamical system through self-generated babbling *without* an external teacher?

Mapping from phoneme representations to vocal tract configurations

The variability in articulator positions seen during production of the same phoneme speaks against the explicit control of the spatial positions of speech articulators. Instead, direct, invariant

---

[1]Supported in part by grants NSF IRI-90-24877 and AFOSR F49620-92-J-0499.

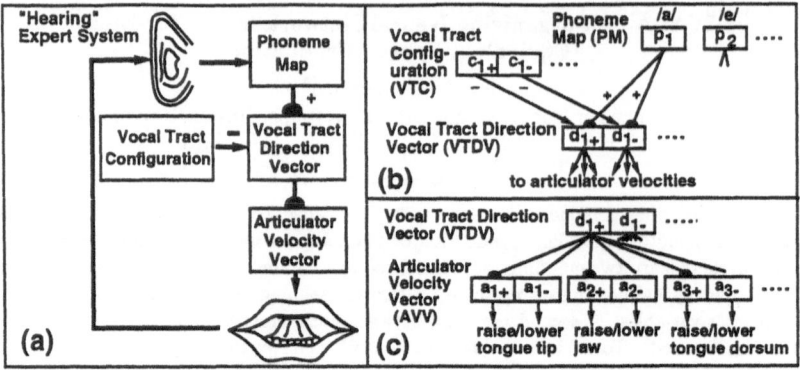

Figure 1: (a) Overview of the DIVA model. (b) Mapping from phoneme representations to vocal tract configurations. (c) Mapping from directions in vocal tract configuration space to articulator velocities. See text for details.

control of higher-level variables such as bilabial separation or tongue body constriction seems to be used. This result is not surprising; whereas these higher-level variables, or vocal tract variables, directly correspond to the acoustic properties of the vocal tract, the effect of an individual articulator is dependent upon the locations of other articulators. An efficient controller should utilize the flexibility afforded by the redundant set of articulators to invariantly produce acoustic information under a variety of circumstances.

Phoneme sets differ from language to language; that is, sounds that are perceptually distinct for speakers of one language may be indistinguishable for speakers of another language. This result implies that the mapping from acoustic information to phoneme representations is not fully defined at birth but instead organizes through learning. Therefore, the mapping from phoneme representations to plans for their production must also be tuned through learning.

Figure 1b schematizes a portion of the mapping from phoneme representations to vocal tract configurations in DIVA. This figure includes three stages of the system block diagram of Figure 1a: the phoneme map stage, the vocal tract configuration stage, and the vocal tract direction vector stage. These stages and their interactions will now be described in more detail.

The collection of cell activities p_i form the phoneme map (PM) stage in the model. These cells represent phonemes in the following way: if the model currently "hears" itself producing a given phoneme, the activity p_i of a cell corresponding to that phoneme is set to 1, with the activities of all other cells set to 0. The binary nature of these phoneme representations is consistent with data showing categorical perception of phonemes in humans. The model does not attempt to simulate the process of auditory recognition of phonemes. For the purpose of learning mappings for speech production, it suffices to activate appropriate PM cells whenever vocal tract configurations corresponding to the phonemes emerge. This is performed by an expert system that looks for vocal tract configurations to determine which, if any, phoneme is being produced, then activates the appropriate PM cell. The ranges of vocal tract configurations corresponding to phonemes roughly approximate the vocal tract profiles of speaking humans.

The ensemble of cell activities c_{i+} and c_{i-} represent the current configuration of the vocal tract. The variables c_{i+} and c_{i-} form an antagonistic pair defining the value of the ith vocal tract variable. In Figure 1b, only c_{1+} and c_{1-}, corresponding to *tongue tip constriction degree*, are shown.

The *plan* for a phoneme is stored as a set of synaptic weights (filled semicircles in Figure 1b) that specify a vocal tract configuration appropriate for that phoneme. In Figure 1b, the portion of the plan corresponding to a single vocal tract variable is shown for the phoneme /a/.

The vocal tract direction vector (VTDV) stage activities represent the difference between the current vocal tract configuration and the set of weights, or plan, projecting from the PM cell of the phoneme being produced. During babbling, this difference facilitates tuning of the phoneme plans as described below. During intentional phoneme production, the corresponding p_j cell is activated, presumably by a higher-level mechanism specifying a sequence of phonemes. The activities of the d_i cells specify the required direction and distance of movement in vocal tract configuration space required to achieve the planned configuration. The next section describes how this direction information is mapped into articulator movements.

Babbling plays a central role in learning the two mappings in DIVA. A cycle linking perception and production is completed by randomly activating articulator movements at the articulator velocity vector (AVV) stage to produce speech sounds; the perceived sounds are then mapped to the motor actions that produced them. During babbling, "accidental" production of a phoneme causes activation of the corresponding PM cell. Tuning of the plan for a phoneme is carried out by first formulating a difference between the set of weights projecting from the active PM cell and the current vocal tract configuration, then adjusting the weights to zero this difference. This is a variant of the vector associative map (VAM) mechanism [6] and utilizes a simple learning law that depends only on pre- and post-synaptic activity levels.

Mapping from vocal tract directions to articulator velocities

The second mapping learned by the model allows VTDV activity to command articulator movements coded by the AVV stage. Part of this mapping is illustrated in Figure 1c.

The AVV stage consists of cell activities a_i that code movement velocities of the following speech articulators: tongue dorsum, tongue tip, jaw, upper lip, and lower lip. Each cell in an antagonistic pair a_{i+}/a_{i-} is meant to correspond roughly to contraction/relaxation of a single muscle or small synergy of muscles controlled as a unit for *all* movements. This differs from coordinative structures as discussed below, wherein a group of muscles is controlled together only for particular movements.

When the VTDV-to-AVV mapping has been tuned, each VTDV cell projects through non-zero weights to all AVV cells that act to reduce the activity of the VTDV cell. Conversely, the pathways projecting to AVV cells that do not act to reduce the VTDV cell's activity have zero weight. For example, Figure 1c shows the projections from d_{1+}, corresponding to a desired increase in tongue tip constriction degree, to the AVV cells. Large semicircles at the ends of these projections indicate a large synaptic weighting in the pathways to a_{1+}, a_{2+}, and a_{3+}, corresponding to raising the tongue tip, jaw, and tongue dorsum, respectively. Each of these movements acts to increase the tongue tip constriction degree and thereby reduce the activity of d_{1+} during production of a given plan. Together, these movements form a task-specific *coordinative structure* (e.g., [2]; [5]), or group of muscle activations controlled as a unit. Each VTDV cell learns a different coordinative structure, and the set of AVV activities produced by a given VTDV activity pattern act to change the vocal tract configuration to zero VTDV activity. When VTDV activity is zeroed, the current VTC matches the plan specified by the currently active PM cell, thus producing the desired phoneme.

By simultaneously activating all of the AVV cells that can be used to achieve a desired vocal tract configuration, the system becomes robust to environmental or internal constraints that affect a subset of the articulator movements. For example, perturbation of the jaw while producing a phoneme that requires a decrease in tongue tip constriction degree will make raising the jaw ineffective, but raising the tongue body and tongue tip still work to carry out the phoneme's plan. This leads to an important insight: in order to achieve flexible control, DIVA must activate a coordinative structure for each VTDV cell; by learning coordinative structures through babbling (as described below), DIVA gives an account of how such structures naturally arise in a self-organizing system that efficiently uses redundant articulators.

The same babbling phase used to tune the map from the PM stage to the VTDV stage is used to tune the map from the VTDV stage to the AVV stage. Random movement of articulators results in changes in vocal tract configuration, and, consequently, changes in the corresponding VTC and VTDV cell activities. Changes in VTDV cell activity drive learning such that articulator movements that reduce VTDV activity gain synaptic weight. Again, the learning law employed depends only on pre- and post-synaptic activities.

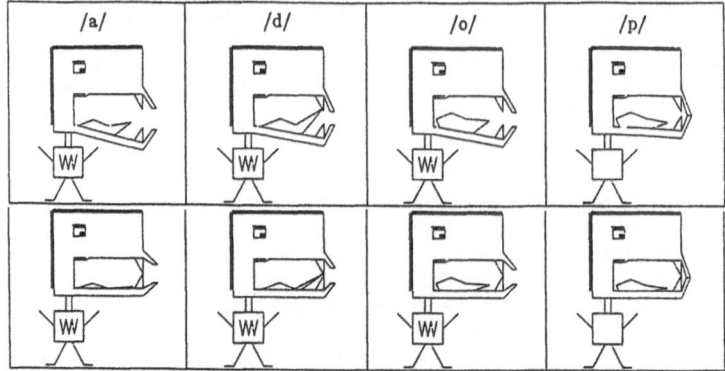

Figure 2: Simulation of the utterance /adop/ with jaw free (top row) and jaw clenched (bottom).

Model simulations

DIVA was implemented in a computer simulation that produces animation sequences of the articulators during production of user-specified phoneme strings. The model successfully learned 29 phonemes; limitations of the simplified articulators prevented production of additional phonemes.

Each training trial consisted of the following events: (1) each AVV cell was randomly activated to a level of 1.0 with probability 1/3, causing motion of several articulators. (2) the resulting "babble" was analyzed by the hearing expert system to determine if any phonemes were being produced, (3) the PM-to-VTDV and VTDV-to-AVV synaptic weights were adjusted. 20,000 trials were used to learn the 29 phonemes. Little attempt to optimize training speed was made.

After training, the model's competence was tested under several conditions, including bilabial stop consonant production under lip and jaw perturbations (c.f. experimental data from [1],[3],[5]) and speech with the jaw clenched. Despite the fact that no training was done under these constraining conditions, the model automatically used remaining articulatory degrees of freedom to compensate for the constrained articulators.

The top row of Figure 2 shows "snapshots" of the animation sequence during production of the utterance /adop/ with the jaw free to move. In this case, the jaw is opened to aid in achieving the low tongue dorsum position of the vowels /a/ and /o/, and it is raised back up to aid in achieving the bilabial closure of the consonant /p/. The bottom row of Figure 2 shows the model successfully performing the same phrase, but this time with the jaw "frozen" in the closed position. It is important to emphasize that no training was done with the jaw clenched. As the figure illustrates, the lack of jaw movement was compensated by increased movement of the tongue to achieve the appropriate vocal tract configurations.

References

[1] Abbs, J. H. and Gracco, V. L. (1984). Control of complex motor gestures: Orofacial muscle responses to load perturbations of lip during speech. *Journal of Neurophysiology*, 51(4), 705–723.

[2] Easton, T.A. (1972). On the normal use of reflexes. *American Scientist*, 60, 591–599.

[3] Folkins, J.W. and Abbs, J.H. (1975). Lip and jaw motor control during speech: Responses to resistive loading of the jaw. *Journal of Speech and Hearing Research*, 18, 207–220.

[4] Gaudiano, P. and Grossberg, S. (1991). Vector associative maps: Unsupervised real-time error-based learning and control of movement trajectories. *Neural Networks*, 4, 147–183.

[5] Kelso, J.A.S., Tuller, B., Vatikiotis-Bateson, E., and Fowler, C.A. (1984). Functionally specific articulatory cooperation following jaw perturbations during speech: Evidence for coordinative structures. *Journal of Experimental Psychology: Human Perception and Performance*, 10, 812–832.

[6] Saltzman, E.L., and Munhall, K.G. (1989). A dynamical approach to gestural patterning in speech production. *Ecological Psychology*, 1(4), 333–382.

Adaptive Non-Uniform A/D Conversion Achieved with an Unsupervised Learning Rule Maximizing Information-Theoretic Entropy

Marc M. VAN HULLE*

Department of Brain and Cognitive Sciences
Massachusetts Institute of Technology
Cambridge, MA 02139, USA

Abstract

Quantization is at the heart of analog-to-digital (A/D) conversion. Two criteria have been widely used for designing quantizers: the minimization of the average distortion due to quantization, and the maximization of information-theoretic entropy i.e. ensure that each of the quantization regions is used equally frequently in encoding the input signal [1]. In general, these two criteria are not equivalent and a particular quantizer is only optimal with respect to a given design criterion [3,4]. To help prevent performance degradation in case of non-stationary input signals, several adaptive quantization schemes have been developed. These schemes attempt to maintain near-optimal performance by matching quantization to the short-term characteristics of the input signal [2]. However, these schemes can only compensate for slowly varying input characteristics and e.g. are not suited for performing adaptive waveform quantization of speech signals.

In this contribution, a novel unsupervised competitive learning rule, called Boundary Adaptation Rule (BAR) [4], is presented for performing adaptive non-uniform A/D conversion. BAR is completely different from other unsupervised competitive learning rules since it maximizes information-theoretic entropy explicitly. In this way, it outperforms other unsupervised learning rules in generating an equiprobable quantization of the analog signal range. Two versions of BAR are introduced, with different computational requirements and speeds of convergence: a simple rule with time complexity $\mathcal{O}(k)$, with k the number of quantization intervals, and a fast rule, called fast BAR (FBAR), with time complexity $\mathcal{O}(1)$.

Using FBAR, an application to adaptive waveform coding of speech signals is considered. The signals originate from the TIMIT data base. For reasons discussed in the presentation, a fixed gain of 10 is used for normalizing the speech signals. The signal-to-quantization-noise ratio (SNR) is determined and compared with that obtained using a uniform and a μ-law quantizer ($\mu = 255$). The average SNR attained with FBAR clearly outperforms those of the other quantizers: 29.61 dB instead of 16.05 and 19.57 dB, respectively, for a 5 bit converter. To achieve a similar result with a uniform and a μ-law quantizer, 7.33 bits and 6.65 bits are required, respectively. Finally, in case (non-adaptive) Huffman coding is used before transmission, the average bit rate of our A/D converter decreases to 3.5 bits for an average SNR of 29.61 dB! This clearly demonstrates the advantage of performing A/D conversion adaptively with FBAR.

Acknowledgements

The author wishes to thank Prof. M. Jordan, Massachusetts Institute of Technology, Department of Brain and Cognitive Sciences, for helpful discussions. He also wishes to thank Dr. D. Martinez, LAAS-CNRS, Toulouse, France, for the unsupervised competitive learning simulations. The author is a senior research assistant of the National Fund for Scientific Research (Belgium). He is also supported by a Fulbright-Hays grant-in-aid and a NATO research grant.

References

[1] Ahalt, S.C., Krishnamurthy, A.K., Chen, P., & Melton, D.E. (1990). Competitive learning algorithms for vector quantization. *Neural Networks*, 3, 277-290.

[2] Gersho, A., & Gray, R.M. (1991). *Vector quantization and signal compression*. Boston: Kluwer.

[3] Ueda, N., & Nakano, R. (1993). A competitive & selective learning method for designing optimal vector quantizers. *Proc. 1993 IEEE Int'l Conf. on Neural Networks*, San Francisco, Vol. III, 1444-1450.

[4] Van Hulle, M.M., & Martinez, D. (1993). On an unsupervised learning rule for scalar quantization following the maximum entropy principle, *Neural Computation* (in press).

*E-mail: marc@psyche.mit.edu Present address: Laboratorium voor Neuro- en Psychofysiologie, K.U.Leuven, Campus Gasthuisberg, Herestraat, B-3000 Leuven, BELGIUM

Optimal Topology-Preservation Using Self-Organising Logical Neural Networks.

G. Tambouratzis & T.J. Stonham

Electrical Engineering Department, Howell Building, Brunel University,
Uxbridge, UB8 3PH, Middlesex, U.K.

Abstract.

The topology-preservation characteristics of a self-organising system are studied in this paper. The system consists of a logical neural network with a structure based on the discriminator network and a method of training that presents certain similarities to Kohonen's self-organising maps. In particular, the optimal neighbourhood size for the most accurate preservation of topological relationships is investigated. Experimental results presented in this paper indicate optimal ranges of values for some of the system's parameters.

1. Introduction.

A wide variety of different neural network models have been proposed as vehicles for pattern recognition tasks. One of the most interesting and most widely used ones is the Self-Organising Map which has been proposed by Kohonen [3]. This network consists of a set of neurons which are arranged so as to form a regular structure, usually in a one- or two-dimensional space. During training, the input patterns are assigned to the neurons in a way which replicates the topological relationships of pattern classes in the input space. This model has been successfully applied to a large number of practical applications of which speech recognition and Artificial Intelligence optimisation problems are only two examples.

A neural network which presents similarities to Kohonen's self-organising maps has been developed [4],[5]. Its structure is based on the discriminator logical neural network. The network is trained by an unsupervised-learning adaptive algorithm which has been designed specifically for logical neural networks and employs a neighbourhood concept similar to that of Kohonen's self-organising maps. Hence, the proposed model combines the speed in operation and ease of implementation of logical neural networks with the minimal supervision requirements of self-organising systems and the topology-preservation capabilities of Kohonen's maps.

The proposed model - as most other neural network models - has a number of parameters which may be used to adjust its behaviour to the given task. Consequently, obtaining a good performance from the neural network requires defining a suitable operating point by "tuning" these parameters. To achieve that, the user has to resort to a lengthy and computationally-intensive trial-and-error procedure. If, however, parameter values (or ranges of values) for which the system performance is optimised were known in advance, this "tuning" operation could be considerably reduced or even completely eliminated. As neural networks are most frequently applied to problems with a huge solution space, it suffices for the system to consistently generate solutions of a sufficiently-high quality instead of the optimal solution. The objective of this paper has been to define values of system parameters for which the performance is optimised. The parameter which has been studied in particular is the optimal width of the neighbourhood in the training phase.

2. Self-Organising Network Description.

The discriminator [1] consists of a number of functions, each of which samples a number of n pixels from the input image. Each logic function comprises 2^n memory locations which correspond to the possible combinations of the binary pixels sampled by the function. These memory locations have a capacity of a single bit, where a binary number reflecting whether the tuple combination has occurred during training is stored. The basic discriminator structure has been modified to adapt it to unsupervised-learning tasks [4],[5]. The main change involves extending the memory locations in the functions so as to enable the storage of a k-bit number (where k>1 and in most applications equal to 8). This number represents the relative frequency of occurrence of each tuple combination and is constantly updated during training by the adaptive learning algorithm.

The self-organising network comprises a number of discriminators, each of which may be used to represent a different pattern class. The discriminators are interconnected to form a one-dimensional (or higher-dimensional) structure in a manner similar to Kohonen's self-organising maps [3]. During learning, the interconnections define a neighbourhood of discriminators, the neighbourhood size being reduced as training progresses. All discriminators compete for each input, the winning discriminator together with its neighbours being adapted towards the input by applying the adaptation rule, which has been designed specifically for discriminator-type networks [4]. The adaptation rule consists of shifting units of information within each function so that, following the rule's application, the function better matches the current input. For each function, the content of address a_1 which is designated by the current input is increased by a given amount, while the content of address a_2 with maximum Hamming distance from a_1 and non-zero content is decreased by the same amount. According to this learning process, the sum of all address contents in each function remains constant throughout the training phase. Hence, the adaptation rule prevents the saturation of the discriminator (probably the most important shortcoming of the discriminator network) and allows the network operation to extend over training phases of unlimited length.

The winning discriminator is the one which generates the highest response to the current input among all discriminators, provided that it is well matched to the input. This matching is achieved by embedding in the selection process the **distribution constraint** [5] which prevents the clustering of fundamentally-different patterns, ensuring that as many discriminators as possible are used in the class formation.

A **forgetting mechanism** has also been incorporated so as to maximise the system efficiency [6]. This mechanism is in effect a reverse application of the learning rule, with the addresses that possess the highest contents in each function having their contents reduced, other functions' contents being similarly increased while the sum of all contents remains constant. This mechanism is used to progressively "clear" discriminators that have remained unused for a long period, erasing any out-of-date knowledge that may have been stored in them during training and making them available for storage of other pattern classes in the future.

The topology-preserving characteristics of the proposed system are due to the neighbourhood training. During training, adjacent discriminators are adapted towards the same patterns, this process resulting in the formation of topological maps in a way similar to that described in [3]. One would expect that the extent to which topology relationships are preserved depends on the width of the neighbourhood compared to the network diameter. If for example the width is constantly equal to 1 (the minimum possible value, with only a single neuron adapted to each input pattern), no ordering will take place. As reported in [5], using a relatively small neighbourhood size (approximately 15% of the network diameter), the system exhibits considerable topology-preservation characteristics. On the other hand, if the width is equal to the network diameter, all neurons will be similarly adapted towards every input and no meaningful learning will occur.

3. Experimental Results.

The proposed system has been applied to a specific task in order to investigate the range of values for which the system performance is optimised. This task should remain computationally tractable while being sufficiently large so as to generate results representative of the system's behaviour. The selected task is related to the character separation and recognition problem described in [6], the network being expected to separate different pattern classes while creating a map which preserves the most important topological relationships.

The system consisted of a number of discriminators arranged in a one-dimensional string. The one-dimensional structure was chosen as it generated the most amenable results for analytical evaluation. The data set consisted of 24 capital letters of a single font, stored as 16x24 arrays of binary pixels. The patterns were presented to the system in sequence for a number of training iterations. In each iteration, the input patterns were contaminated by an amount of noise (6.25% of the pixels) in order to make the classification task more realistic. The self-organising network was made up of 8-tuple functions, which sampled each pixel of the input retina exactly once. As noted in [5], the system performance had been found to be better when the ratio of discriminators-to-characters was 3-to-2. Consequently, the simulated system consisted of 36 discriminators, providing an adequate level of redundancy while also being fairly economical in terms of memory requirements.

For one-dimensional maps, the neighbourhood width is always an odd number as there are two equidistant (left and right) neighbours for each discriminator. The training phase was split up in two distinct sub-phases. In the first one, the neighbourhood was initially set to its largest value and was gradually reduced, in order to map the topological relations of pattern classes on the discriminator map. The reduction of the neighbourhood width in the first sub-phase was "linear". For example, for a maximum neighbourhood width equal to 5 discriminators, the neighbourhood width was 5 for the first half of the first sub-phase and 3 for the remaining half. In the second sub-phase, the neighbourhood was set to its minimum width (equal to 1, only a single discriminator being adapted to each input pattern) in order for the discriminators to be fine-tuned and any conflicts (cases where more than one classes had been assigned to the same discriminator) to be resolved. In the presented series of experiments, the first sub-phase had a duration of 200 iterations while the second one continued for 1800 iterations (giving a total training duration of 2000 iterations), provided that the system had settled at the end. If the system had not settled, the training was allowed to continue for as many iterations as required. The system was considered to have settled if all character-to-discriminator assignments remained unaltered for at least 100 iterations.

For each width size, the system was simulated with 20 different initial settings, these affecting the random initialisation of discriminators as well as the noise contamination of training patterns. The results obtained were averaged in an effort to eliminate any random variations and obtain results representative of the system's behaviour. In order to evaluate the topology-preservation characteristics of the system, the **similarity measure** S_m [6] was used. This has been defined as the sum of the number of identical pixels for each pair of adjacent characters in the discriminator string. The average similarity measures obtained with the different neighbourhood widths are illustrated in figure 1. Also included in the figure are the critical values derived from the statistical analysis [6] performed on the population of all possible character mappings. These are values of the similarity measure which exceed 50%, 90% and 99.5% of all possible mappings. As shown in figure 1, when the width was set to 1, the resulting mappings were virtually random. This is expected as there is effectively no neighbourhood training in that case. As the width was increased, the similarity measure rose, and for widths between 13 and 17 discriminators attained its highest value. Afterwards, it started falling, as demonstrated for a width of 21 discriminators. This drop is attributable to the fact that as the neighbourhood width is increased, an increasing amount of training is performed for each pattern, and this may cause the storage of useless information in certain discriminators. The ratio of neighbourhood width to network diameter for which the topology-preservation is maximised ranges from 1-to-3 to 1-to-2. This result is in accordance to the 1-to-2 ratio quoted in [2] as an empirical choice for Kohonen maps. It is worth noting that in all simulations of the system, all classes were successfully separated, illustrating the effectiveness of the self-organising algorithm. Also, for neighbourhood widths equal to or larger than 5, the generated mappings were considerably better than 99.5% of all possible mappings, indicating that the system preserved the most important topological relationships.

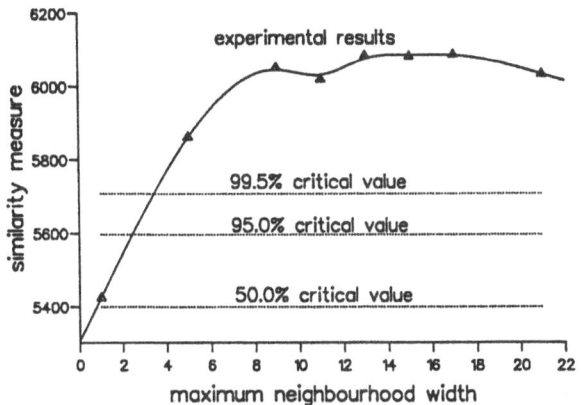

Figure 1. Topology-preservation for different neighbourhood sizes (averaged over 20 runs).

4. Conclusions.

In this paper, the effect of the neighbourhood size on the topology-preservation characteristics of a logical neural network-based self-organising system has been investigated. The system has been shown to retain the most important topological relationships, the extent to which this is achieved depending on the neighbourhood size used. A range of neighbourhood sizes for which the topology-preservation characteristics are optimised has been determined.

References.

[1]. Aleksander, I. & Morton, H. (1990) *An Introduction To Neural Computing.* Chapman and Hall, England.
[2]. Kohonen, T. (1989) *Self-Organisation and Associative Memory* (3rd edition). Springer-Verlag, Heidelberg.
[3]. Kangas, J.A., Kohonen, T., & Laaksonen, J.T. (1991) Variants of Self-Organising Maps. *IEEE Transactions on Neural Networks*, Vol. 1, No. 1, pp. 93-99.
[4]. Tambouratzis, G. & Stonham, T.J. (1992a) A Logical Neural Network that Adapts to Changes in the Pattern Environment. *Proceedings of the 11th IAPR International Conference on Pattern Recognition*, The Hague, Netherlands, August 1992, Vol. 2, pp. 46-49.
[5]. Tambouratzis, G. & Stonham, T.J. (1992b) Implementing Hard Self-Organisation Tasks Using Logical Neural Networks. In Aleksander, I., & Taylor, J. (eds.) *Artificial Neural Networks-2*, Vol. 1, pp. 643-646. North-Holland, Amsterdam.
[6]. Tambouratzis, G. & Stonham, T.J. (1992c) Evaluating the Topology-Preservation Capabilities of a Self-Organising Logical Neural Network. *Pattern Recognition Letters* (in print).

Incorporation of Neurobiological Aspects of *Aplysia*'s Associative Conditioning in Neural Networks for on-line Pattern Detection[*]

M.H. Spigt[1], D.S. Brée[1], M. Nielen[2]

[1]Univerisity of Manchester, Oxford Road, Manchester M13 9PL, UK, email: spigtt@cs.man.ac.uk
[2]Utrecht University, PO box 80151, 3508 TD Utrecht, The Netherlands

In this study anatomical characteristics of a neurobiological model of associative conditioning (i.e. *Aplysia*, a marine mollusc) were used for the development of a novel neural network architecture for the recognition (and on-line classification) of temporal patterns. Compared to the performance of a standard Back Propagation network, the results indicate that this novel architecture is a step forward towards the construction of a reliable mechanism for time-series analysis.

The startingpoint was a standard 3-layer fully connected feed-forward architecture, consisting of 24 input units, 3 hidden units and 1 output unit. In experiment 2 and 3, direct connections between the last 12 input units and respectively only the output unit and only the hidden layer were defined. These connections were disabled for training. The input layer was divided into two independent parts, both of 12 units. This division opened the opportunity of introducing spatial and temporal independence in the input. While the first 12 units cover characteristics of $t = -1$, the second 12 units cover characteristics of $t = 0$. Following this procedure the input to the first 12 units can be compared with the 'to be Conditioned Stimulus' in the *Aplysia* model, and the second 12 units can be compared with the 'Unconditioned Stimulus' in the *Aplysia* model, while the output is defined along the 'Unconditioned Stimulus'. Then, by testing the network by stimulating only the first 12 units, an enhanced output would be expected.

In order to test the proposed neural network approach, a dairy herd environment was chosen. Mastitis, an udder infection, causes a qualitative and quantitative loss of milk. An important variable that can be measured during milking is the electrical conductivity (EC) of the milk. When a teat of the udder, a quarter, is infected, the composition of the milk changes, and therefore the EC in that quarter.

Quarter EC-series, measured every 5 seconds during the morning- and evening-milkings, were used for the construction of datasets. It was decided to use only the first 12 elements of an EC-series, because a lot of information is located in the beginning of EC-series.

The trainingset was constructed as follows: 12 measurements of EC-data of the morning-milking of the day before clinical mastitis was observed (i.e. $t = -1$) were combined with the 12 measurements of EC-data of morning-milking of the day clinical mastitis was actually observed (i.e. $t = 0$). The trainingset consisted of 51 healthy and 11 mastitic EC-series.

The testsets consisted of both healthy and (sub)-clinical EC-series: I: data of sub-clinical EC-series from the morning milking of the day before clinical mastitis was observed ($t = -1$). II: data from the evening milking of that day ($t = -1$). III: data of clinical EC-series from the morning milking of the day on which clinical mastitis was actually observed ($t = 0$). IV: data from the evening milking of that day. ($t = 0$)

Experiment 1. Standard Back Propagation architecture with no extra connections.

Experiment 2. See experiment 1, except that for the second 12 units only to the output unit connections exist with a fixed value of 0.01.

Experiment 3. See experiment 1, except that for the second 12 units only to the hidden units connections exist with a fixed value of 0.01.

TABLE 1. The values indicate the number of correct or faulty classified EC-series.

testset	exp.1				exp.2				exp.3			
	I.	II.	III.	IV.	I.[a]	II.[a]	III.	IV.	I.[b]	II.	III.	IV.
healthy C.	6	6	36	*	7	11	37	*	8	10	32	*
f.	6	7	15	*	5	2	14	*	4	3	19	*
mastitic C.	6	8	8	10	10	8	10	11	11	8	10	12
f.	6	5	5	3	2	5	3	2	1	5	3	1

c.: correct classification, f.: faulty classification, *: not available.

[*] This study was sponsored by SKBS. SKBS (foundation for knowledge based systems) is a Dutch foundation that stimulates research in the field of KBS and technology transfer between universities and industry

[a] significant at 5 % level for testset I and II; I: $\chi^2 = 2.84$, p-value 0.04 (Fisher's exact test); II: $\chi^2 = 4.06$, p-value 0.04

[b] significant at 1 % level for testset I: $\chi^2 = 6.40$, p-value 0.005 (Fisher's exact test).

DESCRIPTION ON THE USE OF THE AUTOGENERATIVE NODAL MEMORY MODEL (ANM) AS CONTROLLING ELEMENT OF AN AUTONOMOUSLY RESPONSIVE SYSTEM

Frank A. MONACO

VITROCISET SPA - Via TIBURTINA 1020 - 00156 ROME ITALY
Tel. 0039 6 4169524 - Fax. 0039 6 4169480 - Telex 620482 VITSEL I

ABSTRACT

The AUTOGENERATIVE NODAL MEMORY Model (ANM) described, in detail in various preceeding articles has been used in various experiments where a single parameter at a time has been shown to the network as training sequence. In this, and various other articles the results of the synergistic effects of simultaneously stimulating a single ANM network by more than one input dominium parameter, and its effects on the autogeneration and autostructuring of the ANM will be illustrated.

When one analyzes the tight relationship between input data (CAUSE), senses (PROCESSING) and structure (EFFECT), so strong is their interaction and the effect on the resulting network that doubt arises on the actual correctness in the selection of; input stimuli, environment, and preprocessing algorithms, essentially we have found that:

1) Any selection paradigm will "color" the probability spectrum density of the ANM network structure.
2) There is no universally "better" way to select these paradigms.

Each system seems to develope its own form of interactivity with the external environment. In this view the ANM is now being used as a controlling element for an autonomous device which will hopefully develope spontaneous responses to simultaneous multiple stimuli from a non prestructured, non cooperating external environment.

THE EXPERIMENT

In order to test this, a system which mocks the human system is being developed. It is not a priori known what information is to be absorbed from the environment by an adaptive interacting system in order to respond coherently. The goal of this experiment is to determine the (or better a) structure that is capable of spontaneously performing this. It is NOT the intention of the experiment to reproduce an artificial life form, something probably still well beyond current technological capabilities.

The ANM based Autonomously Responsive System which is being developed, is shown in its basic constituent blocks in figure 1: and its interaction route with the EXTERNAL ENVIRONMENT is shown.

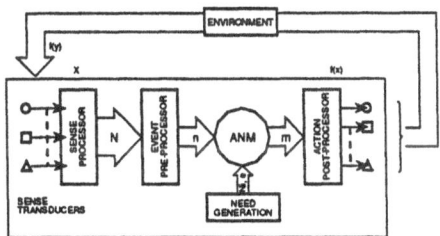

Figure 1. Block Diagram of Autonomously Responsive Unit its Interaction with External World.

The system is under construction since August 1992 and the main physical structure, with the ANM unit, and the SOUND sense input portion of the PREPROCESSOR have already been built and tested.

Details of the construction, analysis of the many problems solved and tests are extensively described in other papers currently being published. Figure number 2 shows the front view of the ANM based Autonomously Responsive System while Figure 3 shows its side view. The reference shown on the right is 1. 5 meters long. The top rack unit contains the complete Autogenerative Nodal Memory unit (ANM) while the bottom rack will contain the SENSE and EVENT Pre-Processors, NEED Generation block and ACTION Post-Processor. Currently only the AUDIO input portion is present.

Figure 2. Front View

Figure 3. Side View

Human Memory - Neurocomputer (MeNeCe Project): Structure For Reverbation Of The Information In N-Peaked Nets (In STMemory)

Simeon Jordanov Mrchev
ul. "Yordan Yovkov" 21A, 8600 Jambol, Bulgaria

Abstract

The author proposes the knowledge-base of the artificial intelligent systems to be simulated by neuro-network modelling of the human memory, i.e. theoretically to be build artificial MEMORY-NEUROCOMPUTER (MeNeCo) of three types: consecutive, hybrid and parallel, with future programme or optico-cholographic, or microelectronic or bioelectronic, or super-conductive realization. All versions of MeNeCo will: 1. keep and work in information in a statistically complex kind; 2. Automatically rate logical conclusions and summaries; 4. Find out, get to know, identify by passive and active reproduction-reminding; 5. Modelling and simulation of model visual and auditory memories; 6. Ensure, partial or full the other intellectual processes (thinking, speech, perception, attention, imagination and so on)., and a complex super-network of neurocomputers; 7. And other.

MeNeCo's versions will be created on the models of "NEURONGLII" system (analogical-discreet model of activity); dynamics of plasticity and adaptation, specialization of: neuro-column structures, the spatial and stochastic temporary processes and connections of the structure, self-organization in substratis and determinant, probable meaning) and SUB-CELLULAR NEURONIC SYSTEM (with model of ionic canals and the synapses the mediators, neuropeptides and nucleonic acids, and so on. MeNeCe will architectically present the whole sphere with: CENTRE (radialy-basic structure). ENVIRONMENT (columnic-nestlike networks). PERIPHERY (divergenting and convergentic staphyline-clasteric structures), with described below structure for reverbation of the information in N-peaked nets (by analogy with human Short-Term Memory - STM). Knowledge of MeNeCe will simulate both aspects (psycho-linguistic and anatomo-physiological) of the associativity [1, 2].

The informational mnemo-process (human STM) in the cortex-undercortex brain structures as models are commutational net of connections in type of connected oriented symmetric mnemograph (which peaks and arcs correspond to the "neuroknots" and input-output sheaves and canals) defined as follows:
Definition 1: Neuro-Knots for Commutation;
Definition 2: Neuro-Canals and Sheaves of Neuro-Canals;
Definition 3: Reliability of the Neuro-Canals and the Sheaves;
Definition 4: Intensity, Attractiven, Duration of the Conversation - Mutual Activity;
Definition 5: A Net Model of the Analysis of the Connections and the Distribution of the Flows of Information
Definition 6: Displaying the Structure and Loading on a Computer

References:
1. S.J. Mrchev, Human Memory Modelling: Memory-Neurocomputer (Ipart) X Vsesoyuznoi s mejdunarodnim uchastiem konferenttsii po neurocybernetike (posvyashchenoi pamyati A..B. Kogana), September 23-26, 1992, Rostov on Don, Russia (organizer: Prof. O. Chorayan)
2. S.J. Mrchev,... Memory-Neurocomputer (2part), Int. Conf. on Model. Probl. in Bionics - BIOMOD 92, June 92, St. Petersburg, Russia (organizer: Prof. Dr. Leonid P. Kraismer).

Visuo-motor interaction
—oral contributions

Neural Representation of Saccadic Eye Movements in Monkey Superior Colliculus

JOHN VAN OPSTAL AND BERT KAPPEN

Department of Medical Physics and Biophysics, University of Nijmegen,
Geert Grooteplein Noord 21, 6525 EZ NIJMEGEN, The Netherlands.

Tel.: 31 - 80 613834   Fax.: 31 -80 541435   E-mail: johnvo@mbfys.kun.nl

Abstract

Just prior to and during a saccadic eye movement a large population of neurons in the midbrain superior colliculus (SC) is recruited, representing a desired vectorial eye displacement in a topographically organized motor map. The SC transmits its output to horizontal and vertical burst cells in the brainstem, which generate the appropriate high-velocity commands for the oculomotor plant. Substantial experimental evidence suggests that saccade generation is under continuous feedback control. An important problem for models of the saccadic system is, how the spatial representation of the SC output is transformed into the temporal code of brainstem cell populations. This problem is even more intriguing, when the possibility is considered of a dynamic feedback loop which includes the SC. Recent experimental findings support such a possibility. In this paper we describe the properties of a two-dimensional neural network model for the monkey SC, which incorporates a number of realistic features known from neurophysiology and, in addition, is inside the dynamic feedback loop for saccade generation. We show that, in order to generate goal-directed saccades, both the feedforward and feedback synaptic connections of the SC neurons should be organized topographically. Several interesting emerging properties of the model are discussed.

1. Introduction

Saccades are extremely fast eye movements that allow for accurate foveation of a target of interest. Visually-evoked saccades have stereotyped dynamical characteristics, such as a nonlinear saturating relation between saccade amplitude and maximum eye velocity and a straight-line relationship between amplitude and saccade duration (the so-called main sequence). Several cortical and subcortical areas which are involved in saccade generation have been identified and it is now thought that the final common pathway for all saccade-related commands involves the participation of neurons in the midbrain superior colliculus (SC) and several distinct populations of cells in the brainstem.

The SC (see e.g. [7] for review) is a layered structure, with saccade-related neurons in its deeper layers. These cells are recruited whenever a saccade belongs to a restricted range of vectorial eye displacements (the cell's movement field) and there appears to be a topographical representation of saccade amplitude (R) and of saccade direction (Φ) (the collicular motor map [5]). Prior to a saccade, a large population of cells, centred at the appropriate location in the motor map, is recruited with a size and shape that are roughly translation invariant [9].

An important class of saccade-related cells in the brainstem, which gets direct input from the SC, is formed by the saccadic burst neurons. The high-frequency burst of activity of these cells is closely related to the temporal properties of the horizontal and vertical eye velocity components [8][11]. Quantitative analysis of the firing characteristics of these burst cells has yielded a functional relationship between instantaneous firing rate and dynamic motor error, which is the ongoing error that remains in order to foveate the target [8]. It has been argued that the operation of these cells is under continuous feedback control [6][8].

In a recent study, Waitzman and collegues [12] have analyzed presaccadic neurons in the intermediate and deep layers of monkey SC. Up till then, it had been customary to classify collicular

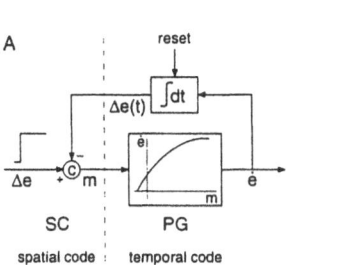
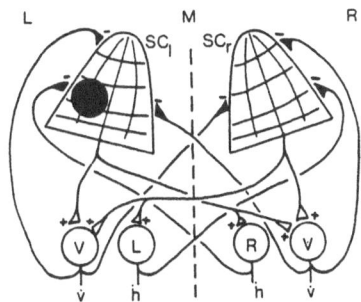

Fig. 1. (A): The Waitzman *et al.* model [12]. In this scheme, the SC clipped cells, c, generate a dynamic motor error signal, $m(t)$, by taking the difference between a static input signal, Δe, which represents desired eye displacement, and current eye displacement, $\Delta e(t)$, which is obtained by integrating the burst cell velocity output, $\dot{e}(t)$. (Adapted from [12]). (B): Scheme incorporating the connectivity of the colliculi to the left and right sides of the brainstem. Based on the movement field properties of burst cells, the following connections can be inferred: each SC is connected to *contralateral* horizontal burst cells, but projects *bilaterally* to the vertical bursters.

neurons as Visual, Visual-Movement, Movement, or Quasi-Visual, based on their *presaccadic* activation properties [7]. In the Waitzman study, however, these neurons are classified according to their relationship with the *offset* of saccades belonging to their movement field. In this new classification scheme, two main types of pre-saccadic collicular neurons are distinguished:

1. *Clipped Cells*: these cells exhibit, in the decaying part of the saccade-related burst, a tight relation with the dynamics of the saccade, whenever it is directed into the center of the movement field. The clipped cell stops firing just before the optimal saccade ends.

2. *Nonclipped Cells*: The activity of these cells has no obvious relation with saccade dynamics nor with its offset.

Waitzman *et al.* [12] suggested that the clipped cells are inside the local feed-back loop for saccade generation and that the code these cells carry, signals the *dynamic motor error* of the eye. They proposed a simple black-box scheme which puts the clipped cells inside the feedback loop (see Fig. 1A).

When the Waitzman *et al.* scheme of Fig. 1A is taken seriously, two main problems will have to be solved: (1) How is the *spatial code* of the SC translated into the horizontal/vertical velocity signals of the brainstem burst cells? This problem has become known by the name of *Spatial-to-Temporal Transformation* (STT). (2) How are the horizontal/vertical brainstem signals fed back into the spatial map of the SC? This is the reverse problem of a *Temporal-to-Spatial Transformation* (TST). The proposed black-box scheme of Fig. 1A does not provide an explicit answer to these questions. In a recent paper we have proposed a two-dimensional extension of this conceptual model, thereby incorporating a number of realistic properties of both the SC motor map and the brainstem saccade generator (see Fig. 1B) and have subsequently analysed the emerging properties of such a model [10].

2. Features and Properties of the Collicular Model

First, a single-cell version of our model is briefly described, in order to illustrate the model's properties. It consists of one active collicular clipped cell, connected to *linear* horizontal and vertical burst cells through feedforward weights $\vec{f} = (f_h, f_v)$, and receiving feedback connections $\vec{b} = (b_h, b_v)$, from the velocity generators. The cell is supposed to integrate the total input activity. Thus, the output of the clipped cell at time t is determined by:

$$c(t) = \int_0^t (E(t') - b_h \dot{h} - b_v \dot{v}) \cdot dt' \tag{1}$$

where $E(t)$ is the excitatory input to the cell at time t, and the output of the burst cells is given by $\dot{h} = f_h \cdot c(t)$ and $\dot{v} = f_v \cdot c(t)$, respectively. The excitatory input to the cell is described by a square wave with intensity ΔE and duration D, which is identical for each SC cell and may represent phasic input from unclipped cells or extracollicular sources. So-called pause neurons, which act as an inhibitory gate on the burst cells, are inhibited at time τ, thus enabling the operation of the feedback circuitry.

Since the burst neurons must encode horizontal and vertical eye velocity, the total horizontal and vertical eye displacement (h, v) is given by:

$$h = \int_\tau^\infty \dot{h}(t) \cdot dt \qquad \text{and} \qquad v = \int_\tau^\infty \dot{v}(t) \cdot dt \tag{2}$$

We have shown [10] that in order for the model to generate a goal directed eye movement, the synaptic weights from and to the clipped cell are determined by:

$$f_h = f \cdot \cos(\Phi) \quad \text{and} \quad b_h = \frac{\Delta E \cdot D}{R} \cdot \cos(\Phi)$$
$$f_v = f \cdot \sin(\Phi) \qquad b_v = \frac{\Delta E \cdot D}{R} \cdot \sin(\Phi) \tag{3}$$

where f is an (arbitrary) synaptic gain and (R, Φ) correspond to the polar coordinates of the desired eye displacement vector. The simplest solution is obtained when both f and the total input excitation to the cell, $\Delta E \cdot D$, are taken independent of its location in the motor map (translation invariance, see e.g. [9]). Note, that the feedforward gain f has no influence on saccade amplitude. In fact, this parameter only modulates saccade velocity [10]. Thus, amplitude and velocity of saccades can be tuned independently (see e.g. [3]).

It should be noted, that all neurons in the model are linear. However, given the model's architecture, as described by Eqn. (3), it can be shown that the nonlinear main sequence characteristics of saccades (see above) emerge automatically [10]. For example, peak eye velocity of the model is determined by the following relation,

$$\dot{R}_{pk} = \frac{R}{D}[1 - \exp(-\frac{f \cdot \Delta E \cdot D^2}{R})] \tag{4}$$

which saturates at $\dot{R}_{pk} = \Delta E \cdot D \cdot f$.

In the two-dimensional ensemble coding scheme, the saccadic burst neurons receive input from a large population of active clipped cells, instead of from a single cell (see Fig. 2). In what follows, the essential features of this scheme will be described, omitting the mathematical details which can be found elsewhere [10].

The input to the motor map of the SC is topographically organized. It is modeled by a translation-invariant Gaussian excitation of duration D (rectangular profile), height ΔE and spatial width σ_{pop}, which is superimposed on a complex-logarithmic collicular motor map. This map relates the anatomical coordinates (u, v) (in mm) of each clipped cell to its optimal saccade vector (R, Φ) (deg) by

$$u = B_u \cdot \ln(R/R_0) \qquad v = B_v \cdot (\frac{\pi\Phi}{180}) \tag{5}$$

with $R_0 = 1$ deg, $B_u = 1$ mm and $B_v = 1$ mm/rad.

The output of the clipped cells is projected in a weighted fashion to the horizontal and vertical burst cells. This stage provides the spatial-to-temporal transformation (STT). The burst cells sum the clipped cells' weighted spikes (weights $f_{h,i}$, $f_{v,i}$) and transform these into a horizontal (and vertical) eye velocity signal. The input/output characteristic of both populations of burst cells is assumed to be *linear*.

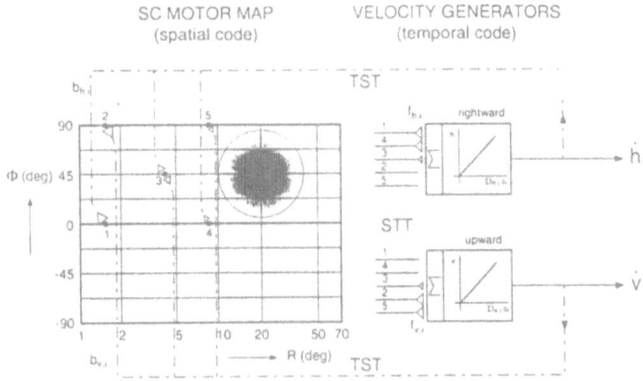

Fig. 2: Schematic drawing of the two-dimensional population model which incorporates feedback through the *SC*. Clipped cell i has feedforward connections, $(f_{h,i}, f_{v,i})$, with brainstem horizontal and vertical burst cells, respectively, and receives inhibitory feedback of strength $(b_{h,i}, b_{v,i})$. These strengths are specified by Eqn. (3). From [10].

The velocity outputs of the burst cells are projected back to the clipped cells in the motor map of the SC. Each clipped cell i receives inhibitory feedback from these velocity generators, weighted by an amount ($b_{h,i}$ and $b_{v,i}$) which is solely dictated by the cell's location in the motor map and by a constant gain factor, proportional to $\Delta E \cdot D$, according to Eqn. 3 (the temporal-to-spatial transformation, TST).

The burst cells are inhibited by pause neurons, which act as an external gate on the saccade generator. The model implicitly assumes a recurrent inhibitory connection between these neurons and the burst cells.

It should be noted that the shape of the saccadic velocity profile is influenced by the temporal profile of the input burst (in our simulations: rectangular shape), the input/output characteristic of the burst cells (here: linear) and the properties of the oculomotor plant. We have not attempted to mimic detailed saccade velocity profiles in our simulations. Therefore, we have not included a model of the oculomotor plant, and eye position is taken directly as the integated output of the burst cells.

Extending the single-cell model of Eqn. 1 to the ensemble coding scheme of Fig. 2, the activity of each clipped cell, i ($\in 1 \ldots N$), is then determined by the following equation:

$$c_i(t) = \text{Max}[\int_0^t dt' \cdot \{E_i(t') - \sum_{j=1}^N c_j(t')(f_{h,j}b_{h,i} + f_{v,j}b_{v,i})\}, \quad 0] \qquad (6)$$

In general, this equation cannot be solved analytically. However, we could show that the above-mentioned properties of the single-cell version of the model (e.g. its main sequence behaviour, see above) carry over to this more realistic population model [10].

3. Main Properties and Results
An important consequence of the proposed model architecture is, that saccade amplitude gain is determined by the total number of excitatory input spikes, $\Delta E_k \cdot D$, for the *longest-firing* clipped cell, k, only (c.f. Eqn. (3) for the single-cell model). As a result, large changes in the width of the excitatory input profile do *not* significantly alter saccade amplitude. This interesting property has two important consequences: First, it makes the model especially robust to lesions or noise in the population activities. For example, 70 deg amplitude saccades can still be generated by the model, despite the fact that half the Gaussian excitatory input no longer contributes to the saccade because of border truncation. As expected from Eqn. (4), these very large saccades tend to be slower than

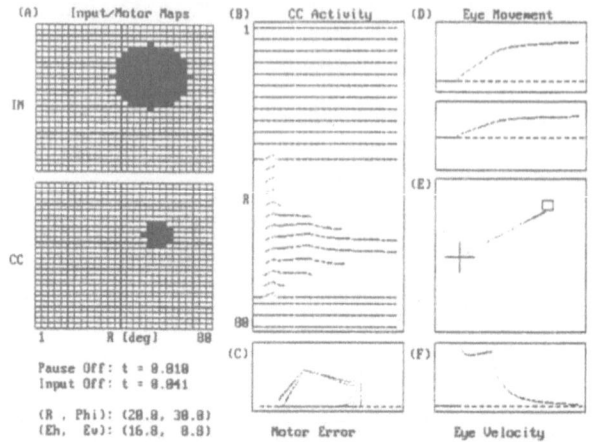

Fig. 3: Simulation of saccade to $[R, \Phi] = [20, 30]$) deg. (**A**): Input excitation and motor map. (**B**): Cross-section along the amplitude representation of the motor map at $\Phi = 30$ deg (Eqn. 5; see Fig. 2), showing activity of 27 clipped cells as a function of time. (**C**): Activity of 4 representative cells of (B) as a function of motor error. (**D**): Horizontal and vertical eye position as a function of time. (**E**): (h, v) plot of eye position. (**F**): Instantaneous eye velocity (i.e. burst cell activity).

e.g. 60 deg amplitude saccades. Similarly, significant lesions in the clipped cell population lead only to minor amplitude and directional changes of the saccades, but they will be slower. This result is in line with recent findings in the literature [2][4]. Second, this property lies at the basis of the model's behaviour to electrical double stimulation. In our paper we have shown that the model can respond with a saccade that is the *weighted average* response of the individual stimulation effects, where the relative stimulation intensities act as weighting factors [10]. This nonlinear property of the saccadic system has already been documented two decades ago [5].

Fig. 3 shows a simulation of the model's behaviour during a saccade to target coordinates $(R, \Phi) = (20, 30)$ deg. The right-hand side of the figure shows the actual saccade trajectory (Figs. 3D,E,F). The left-hand side provides a 'snap-shot' of the spatial extent of the input (Fig. 3A, top) and clipped-cell (Fig. 3A, bottom) activity at time D (set to 40 msec in this simulation), superimposed on the motor map of the left SC. The central part of this figure provides more detailed information on the temporal properties of the clipped cells' responses (Figs. 3B,C).

The activity of the clipped cells exhibits the same property as observed by Waitzman *et al.* [12], namely that the decaying part of the clipped cell's burst is linearly related to the dynamic motor error of the eye. Our simulations show that this is true for cells close to the *center* of the active population (see Fig. 3C). Clipped cells located near the population periphery have much shorter burst durations than the central cells and do not show this 'Waitzman' property (see Figs. 3B,C and also [7]).

It should be noted that, apart from rectification, all cells in the model have *linear* input-output characteristics. In our paper we show that the nonlinear main sequence relations as well as the weighted averaging to double stimulation result primarily from the model's architecture.

4. Conclusions

The main conclusions of our study are:

1. It is possible to extend the conceptual black-box scheme of Waitzman *et al.* [12] (Fig. 1A) to a relistic two-dimensional model of the motor colliculus (Fig. 1B) with a relatively simple synaptic connection scheme.

2. In order to generate goal-directed saccades, the inhibitory feedback from the horizontal and vertical velocity generators should be arranged topographically, where the strength of the synaptic weights decreases as $1/R$ in the motor map (Eqn. 3). The feedforward excitatory weights determine saccade direction by their relative strength, whereas the absolute value of these connections acts as a velocity gain.

3. In the population of active clipped cells, the longest-firing cell determines the final metrics of the saccade. For peaked input profiles (e.g. a Gaussian), the longest-firing cell is close to the population maximum. The recruitment of an ensemble of neurons in the SC renders it relatively insensitive to lesions and noise. In fact, larger recruited populations result in *faster*, not in larger saccades.

4. As a result of the topographical synaptic arrangement, the overall dynamics of the saccades qualitatively agree with saccade main-sequence properties, such as a straight-line amplitude-duration and saturating amplitude-peak velocity relationship. Our analysis shows that these properties do *not* depend on a nonlinearity in the saccadic burst neurons, as e.g. in the well-known Robinson model [6][8].

5. As a further emergent property, the model is able to generate weighted averaging saccades in response to electrical double stimulation under a wide range of input conditions.

6. These qualitative properties are not critically affected by the precise shape of the motor map, nor by the afferent excitation profile.

References

[1] Hepp K., Henn V., Vilis T. and Cohen B.: Brainstem regions related to saccade generation. In: *The neurobiology of saccadic eye movements.* (Eds. Wurtz R.H. and Goldberg M.E.), Elsevier Science Publishers, pp. 105-212 (1989).

[2] Hikosaka O. and Wurtz R.H.: Modification of saccadic eye movements by GABA-related substances. I. Effects of muscimol and bicuculline in monkey superior colliculus. *J. Neurophysiol.* **53:** 266-291 (1985).

[3] Jürgens R., Becker W. and Kornhuber H.H.: Natural and drug-induced variations of velocity and duration of human saccadic eye movements: Evidence for control of the neural pulse generator by local feedback. *Biol. Cybernet.* **39:** 87-96 (1981).

[4] Lee C., Rohrer W.H. and Sparks D.L.: Population coding of saccadic eye movements by neurons in the superior colliculus. *Nature* **332:** 357-360 (1988).

[5] Robinson D.A., Eye movements evoked by collicular electrical stimulation in the alert monkey. *Vision Res.* **12:** 1285-1302 (1972).

[6] Robinson D.A.: Oculomotor control signals. In: *Basic mechanisms of ocular motility and their clinical implications.* (Eds. Bach-y-Rita P. and Lennerstrand G.) Pergamon Press, Oxford, pp. 337-374 (1975).

[7] Sparks D.L.: Translation of sensory signals into commands for control of saccadic eye movements: role of primate superior colliculus. *Physiol. Rev.* **66:** 118-171 (1986).

[8] Van Gisbergen J.A.M., Robinson D.A. and Gielen S.: A quantitative analysis of generation of saccadic eye movements by burst neurons. *J. Neurophysiol.* **45:** 417-442 (1981).

[9] Van Opstal A.J., Van Gisbergen J.A.M. and Smit A.C., A comparison of saccades evoked by visual stimulation and collicular electrical stimulation.*Exp. Brain Res.* **35:** 800-812 (1990).

[10] Van Opstal A.J. and Kappen H., A two-dimensional ensemble-coding model for spatial-temporal transformation of saccades in monkey superior colliculus. *Network*, in press, (1993).

[11] Vilis T., Hepp K., Schwarz U. and Henn V.: On the generation of torsional and vertical rapid eye movements in the monkey. *Exp. Brain Res.* **77** 1-11 (1989).

[12] Waitzman D.M., Ma T.P., Optican L.M. and Wurtz R.H.: Superior colliculus neurons mediate the dynamic characteristics of saccades. *J. Neurophysiol.* **66:** 1716-1737 (1991).

A Self-organizing Neural Network for Learning A Body-centered Invariant Representation of 3-D Target Position

Daniel Bullock, Douglas Greve, Stephen Grossberg,
and Frank H. Guenther[1]

Department of Cognitive and Neural Systems, Boston University
111 Cummington Street, Boston, MA 02215

Abstract

This paper describes a self-organizing neural network that learns a body-centered representation of 3-D target positions. This representation remains invariant under head and eye movements, and is a key component of sensory-motor systems for producing motor equivalent reaches to targets [1]. Learning requires no teacher, instead utilizing information gained from an action-perception cycle in which head movements are made while a stationary target is foveated. Because the spatial representations used relate closely to neck anatomy, the network learns very rapidly, converging after foveating only 200 targets.

Forming a head-centered representation of target position

The bilaterally symmetric organization of the body provides a simple and direct source of information for computing absolute position of a fixated target with respect to the observer's head. When both eyes binocularly fixate a target, the point of intersection of the lines of gaze may be used to compute the absolute distance and direction of the fixation point with respect to the head. Such extraretinal information may also be used to complement visual processing to derive better estimates of the absolute distance and direction of visually detected but non-fixated objects.

Figure 1 describes the geometry of 3-D target localization in terms of spherical coordinates that are closely related to the 3-D representations proposed here. The origin of this coordinate system, called the cranial egocenter, lies at the midpoint between the two eyes. Thus the representation is "cyclopean". The head-centered horizontal angle or azimuth, θ_H, and the vertical angle or elevation, ϕ_H, measure deviations from straight-ahead gaze, and R_H measures distance from the head. Evidence for such a coordinate system and a full description of a neural network model which forms a neural representation of target position based on this coordinate system is given in [1].

Learning a body-centered representation of target position

We now address the formation of a body-centered representation of 3-D target positions using the above-mentioned head-centered representation coupled with information concerning the position of the head with respect to the torso. The adaptive computational strategy embodied by this network makes use of signals generated automatically during the typical behavioral sequence associated with changes of visual fixation. The following occurs in a typical learning episode: (1) The representation of a novel, initially non-foveal visual target

[1]This work was supported in part by grants NSF IRI-87-16960, NSF IRI-90-24877, and AFOSR F49620-92-J-0499.

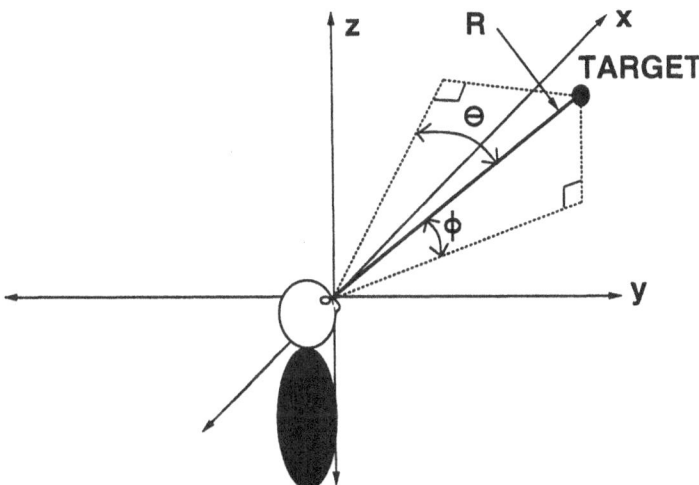

Figure 1: Spherical coordinate frame for specifying a target position with respect to the head. This coordinate frame is related to the head-centered and body-centered representations of space described herein.

wins an internal competition that determines the next target to be foveated, and a saccade is made to this target; (2) Information regarding head-centered target location is combined with information about neck muscle states to yield a stored estimate of target location relative to the body; (3) Neck muscles rotate the head (either randomly or such as to point the nose toward the target) while the eyes make a vestibular ocular reflex (VOR)-mediated counter rotation to ensure continued foveation during the head movement; (4) During the head movement and ocular counter-rotation, both internal representations of the target's location in head coordinates and internal representations of neck angles change while the stored representation of target position in body coordinates remains constant.

If the network that combines head-centered representation and neck angle information to yield an estimate of target location in body coordinates is well-tuned, then its estimate will remain invariant during the head rotation and ocular counter-rotation. If it is not well-tuned, then a discrepancy will develop during the head rotation between this network's current estimate and the estimate stored prior to the head rotation. This discrepancy may then serve as an error signal capable of directing a learning process that improves the network's knowledge, stored in its synaptic weights, about how to combine neck angle and head coordinate signals to estimate target positions relative to the body. The stage that registers the discrepancy in our model is called a difference vector (DV) stage, because errors are registered on a component by component basis. The Vector Associative Map (VAM) of [2] is a neural mechanism that carries out DV-based learning; the current network uses a variant of VAM learning. This learning process requires no teacher, and combines mechanisms known to be separately available *in vivo*.

The body-centered representation approximates a spherical coordinate frame similar to the one shown in Figure 1. The body-centered origin is the same as the origin of the head-centered system when the head is pointed straight ahead. The body-centered frame also uses the same three spherical coordinates as the head-centered system, denoted by (θ_B, ϕ_B, R_B) in the body-centered frame. When the head is pointed straight ahead, the head-centered

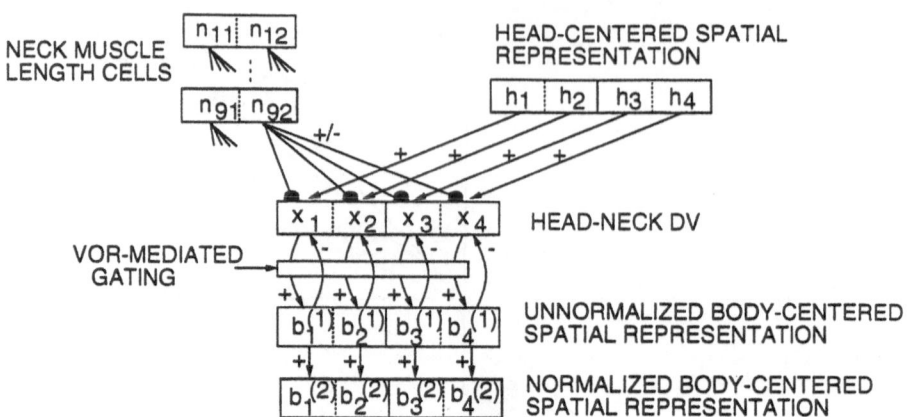

Figure 2: Network for learning transformation from a head-centered spherical coordinate representation to a body-centered spherical coordinate representation of target position.

representation (θ_H, ϕ_H, R_H) is identical to the body-centered representation (θ_B, ϕ_B, R_B). When the head is moved from straight ahead, however, the head-centered frame moves with the head while the body-centered frame remains stationary.

The choice of these coordinate frames results from an investigation of the physiology of head-neck systems in humans and other vertebrates. The biomechanics of the neck vertebrae favor rotations of the head around preferred axes [7]. Movements along one preferred axis correspond to changes in θ_N (i.e., side-to-side or horizontal movements), whereas movements along the other preferred axis correspond to changes in ϕ_N (i.e., vertical movements). Movements along other axes, for example tilting the head to one side, are much more constrained by the biomechanics of the neck. Further evidence for the biological importance of these preferred axes comes from [6], which showed that separate neural circuits were used to control horizontal and vertical head movements in the barn owl.

The importance of these results in the current context is as follows. Learning to discount head movements in the body-centered representation consists of compensating for changes in head position by negating the resulting changes in the head-centered representation of a fixed target position. In other words, $(\theta_B, \phi_B) = (\theta_H, \phi_H) + (\theta_{cor}, \phi_{cor})$, where $(\theta_{cor}, \phi_{cor})$ is a learned correction based on neck muscle information. (The third coordinate, distance from the head, changes relatively little with head movement; see [5] for further discussion.) When the transformation network is properly tuned, this correction is nearly linearly related to the head movement defined according to the preferred axes. This linear relation between head movements and the required correction to the head-centered representation allows very fast and accurate learning of the correction. The relationship between head movements and other head- and body-centered coordinate frames, such as Cartesian, is much more complex, making the transformation from a head-centered representation to a body-centered representation far more difficult to learn.

Although head position can be derived from neck muscle length information, an organism cannot without learning use this neck muscle information to accurately compensate for head movements when forming a body-centered representation. This is because the relationship between any one neck muscle length and head position is dependent upon details of the neck anatomy that vary from individual to individual and can change with time (e.g., due to growth). Therefore, the organism must *adaptively* find parameters that allow neck muscle length information to compensate for changes in head position. The network described here rapidly and successfully finds these parameters without the aid of an external teacher. Instead, network construction capitalizes on the fact that the positions of fixed objects with respect to the body do not change while the head moves, allowing the organism to internally generate teaching signals.

Figure 2 illustrates the network. There are five main neural population types in this network: (1) neck muscle length populations with activities n_{ji} ($1 \leq j \leq 9$, $1 \leq i \leq 2$), (2) head coordinate representation populations with activities h_i ($1 \leq i \leq 4$), (3) head-neck Difference Vector (DV) populations with activities x_i ($1 \leq i \leq 4$), (4) unnormalized body coordinate representation populations with activities $b_i^{(1)}$ ($1 \leq i \leq 4$), and (5) normalized body coordinate representation populations with activities $b_i^{(2)}$ ($1 \leq i \leq 4$). Each head-centered representation population projects with a fixed-weight connection to the corresponding DV population. Each neck muscle length population projects to every DV population through an adaptable-weight synaptic connection, indicated by filled semicircles in Figure 2. Furthermore, VOR-mediated gating modulates the interactions between the DV populations and the unnormalized body-centered representation populations. The learning law in the simulations is as follows:

$$\frac{d[z_{ijk}]}{dt} = -\epsilon x_k[-E z_{ijk} + n_{ij}] \tag{1}$$

where z_{ijk} is the strength of the synaptic connection between neck muscle length activity n_{ij} and DV activity x_k, ϵ is a learning rate parameter, and E is a decay rate parameter. Thus, learning is local, i.e. it depends only on the pre- and post-synaptic cell activities, not on activities at distant points in the network.

The following steps were used to train the network: (1) Initialize all weights to 0.0. (2) Choose a random initial head position (θ_N, ϕ_N). (3) Choose a random target position (θ_T, ϕ_T). (4) Foveate a new target (i.e., adjust h_i so that $\theta_H = \theta_T - \theta_N$ and $\phi_H = \phi_T - \phi_N$) and store this target into the body coordinate populations $b_i^{(1)}$ and $b_j^{(2)}$. This corresponds to breaking VOR fixation to foveate and store a new target. Target storage is carried out by the transient opening of the gated excitatory pathways from the populations x_i to the populations $b_j^{(1)}$. This gating action occurs each time VOR fixation is broken. (5) Choose a new head position while remaining foveated on the current target (i.e., change n_{ij} and adjust h_i accordingly to keep $\theta_H + \theta_N = \theta_T$ and $\phi_H + \phi_N = \phi_T$). This step corresponds to moving the head while using VOR to keep the target foveated. (6) Adjust the weights from the neck muscle length populations to the head-neck DV populations according to equation (1). (7) If more trials remain, repeat steps 3-7.

Despite the use of simple, local learning laws and no external teacher, the system globally self-organizes to perform the transformation from a head-centered to a body-centered representation of 3-D target positions. The left half of Figure 3 shows the internal representation (first box) and actual target position (second box) during a head movement after 20 learning trials. As the head moves, the internal representation of the target position also moves, even though actual target position remains fixed. After foveating only 200 targets, however, the network has learned to invariantly represent the body-centered target position despite large head movements, as shown in the right half of Figure 3.

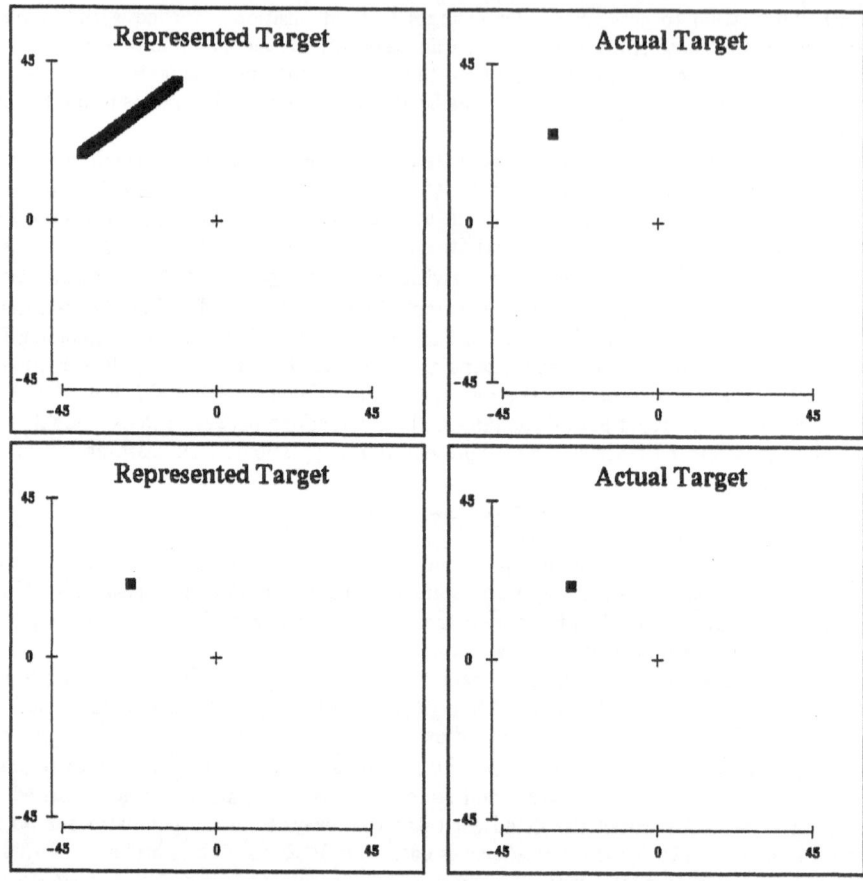

Figure 3: **Top row.** Results after 20 learning trials. The first box shows the internally represented body-centered target position as the head is moved through over 30° of both horizontal and vertical angle. The second box shows the actual target position. The change in represented target position as the head is moved indicates that the network has not yet learned to invariantly represent body-centered target position. **Bottom row.** Results after 200 learning trials. The internal representation is now invariant under head movements.

References

[1] Bullock, D., Grossberg, S., and Guenther, F. H. (1993). A self-organizing neural model of motor equivalent reaching and tool use by a multijoint arm. *Journal of Cognitive Neuroscience*. In press.

[2] Gaudiano, P. and Grossberg, S. (1991). Vector associative maps: Unsupervised real-time error-based learning and control of movement trajectories. *Neural Networks*, 4, pp. 147–183.

[3] Greve, D., Grossberg, S., Guenther, F. H., and Bullock, D. (1993). Neural representations for sensory-motor control, I: Head-centered 3-D target positions from opponent eye commands. *Acta Psychologica*, 82, pp. 115–138.

[4] Grossberg, S., Guenther, F., Bullock, D., and Greve, D. (1993). Neural representations for sensory-motor control, II: Learning a head-centered visuomotor representation of 3-D target positions. *Neural Networks*, 6(1), pp. 43–67.

[5] Guenther, F. H., Bullock, D., Greve, D., and Grossberg, S. (1993). Neural representations for sensory-motor control, III: Learning a body-centered representation of 3-D target position. Technical Report, Boston University Center for Adaptive Systems.

[6] Masino, T. and Knudsen, E. I. (1990). Horizontal and vertical components of head movement are controlled by distinct neural circuits in the barn owl. *Nature*, 345, pp. 434–437.

[7] Vidal, P. P., de Waele, C., Graf, W., and Berthoz, A. (1988). Skeletal geometry underlying head movements. In Cohen, B. and Henn, V., (eds.): *Representation of Three-Dimensional Space in the Vestibular, Oculomotor, and Visual Systems*, pp. 228–238. New York: New York Academy of Sciences.

Dynamic Field Approach to Target Selection in Gaze Control

Klaus Kopecz, Christoph Engels and Gregor Schöner
Institut für Neuroinformatik – Universität Bochum
D-4630 Bochum – Germany

Abstract:
A functional, three–level model for one–dimensional saccadic gaze control is proposed which includes the processing of information about multiple targets. These are represented within a continuous, topographic field with nonlinear internal dynamics. The dynamic target representation is coupled to a second dynamic field which spatially codes the current eye movement plan and, through its own nonlinear dynamics, makes decisions in cases of competing targets. This field resides in a feedback loop with the current gaze direction, the dynamics of which is modeled at a third level. In the limit case of only a single target the three–level model approximates the functional mass–spring–like dynamics of gaze control. Field dynamics contribute to memorization and the selection of a unique saccadic goal. We show how, in the presence of two simultaneously presented targets, averaging and bistable responses as well as response biases emerge as a consequence of topographically ordered projections between functional levels. The relation of the model to electrophysiological data is discussed.

Based on results from behavioral studies on saccade metrics a functional model of saccadic gaze control is proposed. Three levels are introduced that support: (1) memorization of multiple targets, (2) decisions about the actual saccade to be carried out and (3) conversion from spatial prescriptions into gaze trajectories. It is well–known, that the effective coordinates underlying the memorization of saccadic targets are nonretinocentric [1,2], and that interaction between two competing targets leads to average or bistable responses of first saccades, depending on inter–target spacing [3]. To model distance dependent interaction in combination with nonretinocentric memory, we introduce a topographic representation of targets in an allocentric frame of reference. Mathematically, this representation is described as a continuous, scalar field $\phi(y)$, where y is the spatial coordinate (only gaze shifts in one dimension and monocular control are considered). One target at position y_T will be mapped onto a localized distribution $\overline{\phi}(y - y_T)$ centered around y_T, so that its maximum yields an estimate of target position. Assuming that an extended region is activated is consistent with the idea of a population code. To afford memorization, persistence of $\overline{\phi}(y - y_T)$ is required even when the target is presented for a short time only. These conditions can be fulfilled by defining dynamics of the field $\phi(y)$ which yields localized, stationary solutions even in the absence of external inputs, independent of geometrical boundary conditions. Dynamics of this type have been studied in physics and chemistry [4], or as "neural fields" [5]. In all formulations characteristically a fast, long–range inhibitory process is combined with a slower, more localized activation process. In the following, we use a system inspired by the analytical work of Amari [5]:

$$\tau \frac{d}{dt}\phi(y,t) = -\phi(y,t) + \int_R K(y - y') \, S[\phi(y',t)] \, dy' + F_T(y,t) \tag{1}$$

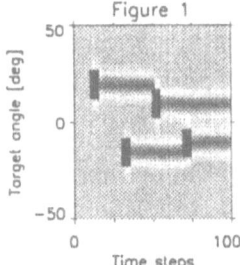

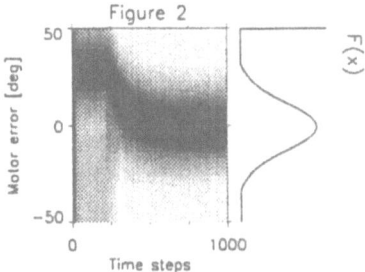

$K(y)$ is a "Mexican-Hat" shaped Kernel leading to short–range excitation and long–range inhibition and the function $S(\phi)$ is a sigmoid shaped nonlinearity. The range of integration, R, extends over the allowed range of y. The Kernel is parametrized as $K(y) = K_0 [G(y, \sigma_1) - r G(y, s \sigma_1)]$, with $0 < r < 1$ and $s > 1$. $G(y, \sigma_1)$ is an area–normalized Gaussian of width σ_1. The nonlinearity is chosen as $S(\phi) = (1 + \exp[-\beta(\phi - \phi_0)])^{-1}$. The autonomous part of eqn. (1) exhibits over a wide range of parameters localized, stationary solutions coexistent with a stable, homogeneous solution of approximately vanishing activity. This property does not depend critically on the concrete choice of K or S, as long as "Mexican-Hat"- and sigmoid shape, respectively, are maintained. The additive force $F_T(y, t)$ is assumed to be centered around external target specifications and is transient in nature (e.g., derived from visual signals by a temporal filter with a differential component). It leads to local switching from the homogeneous state to a localized, stationary peak. This peak represents the memorized information about a desired target. In simulations shown here, we choose $\tau = 1$, $K_0 = 12$, $\sigma_1 = 2°$, $r = 0.6$, $s = 2$, $\beta = 2$, $\phi_0 = 3$, $R = [-50°, 50°]$.

Fig. 1 shows the activation, coexistence and competition of target entries. The gray levels are scaled to maximal black and minimal white values of ϕ. Targets are specified through a force profile extending over 14° width which is present for a duration of 50 time units (these numbers are chosen as examples only and have no functional consequence for those properties of the model studied here). In the figure, a first target appears at $t = 100$ and 20° gaze angle, evoking a corresponding stationary peak. A second target appears at $t = 300$ and $-15°$, leading to two coexisting peaks. At $t = 500$ and 10°, a third peak is evoked close to the first. Mutual interaction leads to extinction of the earlier peak. Similarly, a fourth stimulus at $t = 700$ and $-10°$ extinguishes the second entry. In general, the extinction of an existing peak by a new target specification depends on its distance on the representation and the amplitude and width of the new input.

Activity at the target level couples into a second functional level, dealing with control and *decisions*. Motivated by electrophysiological findings [6], it is chosen as a field, $\psi(x)$, with x given in retinocentric coordinates. The dynamics of ψ is chosen similar to eqn. (1) with the additional constraint of exhibiting only a single–peaked stationary, autonomous solution, whose location should code for the dynamic motor error [7]. This can be implemented by replacing the negative part of the "Mexican--Hat" by a global, uniform inhibition:

$$\tau \frac{d}{dt}\psi(x,t) = -\psi(x,t) + G_0 \int_R G(x-x',\sigma_2)\,S[\psi(x',t)]\,dx' - H_0 \int_R S[\psi(x',t)]\,dx' + F(x,t)$$

(2)

Again, the autonomous part exhibits the desired type of stationary solutions over a wide range of parameters. G_0, σ_2 and H_0 are set to 12, 10° and 250, respectively. The motor error itself is assumed to be coded by the center of gravity, $\bar{x}$, of $S[\psi(x)]$ ($\bar{x}$ is well–defined, due to $S(\psi) > 0$). The dynamics of the current gaze angle z including plant and control is modeled by a second order dynamical system into which $\bar{x}$ couples as a driving force:

$$\ddot{z} + \rho\dot{z} = \omega^2 \bar{x}$$

(3)

The functional implications of this mass–spring dynamics will be discussed below.

The additive force $F(x,t)$ in eqn. (2) represents the activity projected from the target level, shifted by the coordinate transformation, $x = y - z$, from allocentric to retinocentric coordinates. The projection rule from the target level to the decision level is

$$F(x,t) = F_0 \int_R G(x+z-y',\sigma_3)\,S[\phi(y',t)]\,dy' \quad ,$$

(4)

with $F_0 = 5$ and $\sigma_3 = 12°$. Eqn. (4) combines a diffuse projection with the desired transformation of coordinates. Note, that this projection rule closes a feedback loop between the motor error level and the effector variable z. The functional purpose of the force $F(x,t)$ is the specification of current motor error. This is illustrated in **fig. 2** where the temporal evolution of the field $\psi(x)$ under *open loop conditions* (z fixed) is shown in the presence of a single target that appears at $t = 200$. Its projection, $F(x)$, is indicated to the right. The localized peak initially located at $x = 30°$ moves to the local maximum of $F(x)$ (the time scale is 10 times smaller than in fig. 1). This local maximum search is a generic property of the mathematical setting used here [5]. The main restriction is, that the specifying force $F(x,t)$ is small compared to intrinsic contributions of the autonomous dynamics, so that localized peaks preserve their identity. In this limit, the time scale of relaxation to the local maximum is determined by the coupling constant F_0.

The entire closed loop system has stable control properties if the peak on the motor error level relaxes to a local maximum of $F(x,t)$ fast compared to the time scale of gaze movement. More precisely, three time scales must be distinguished: (1) The fast time scale of the intrinsic field dynamics, governs relaxation to localized peaks. Here, this scale is set to unity through the choice of τ. (2) The intermediate time scale determined by the input strength F_0 governs the relaxation of localized peaks to local maxima of motor error specifications. (3) The slow time scale of gaze shifts is determined by the dynamics of eqn. (3) of the effector variable z. For stable closed loop behavior we set minimally $\rho = 1$ and $\omega = 0.1$, so that the gaze dynamics are approximately a factor of 10 slower than the fast time scale (in the limit of strong damping ρ the relevant time scale is approximately given by $1/|\omega|$). If the time scales of the three functional components (representation of targets, decisions, effector dynamics) are separated the dynamical analysis of individual components carries over to the entire closed–loop system. Under this condition and in the case of a single target the three dynamical levels represent a simple mass–spring

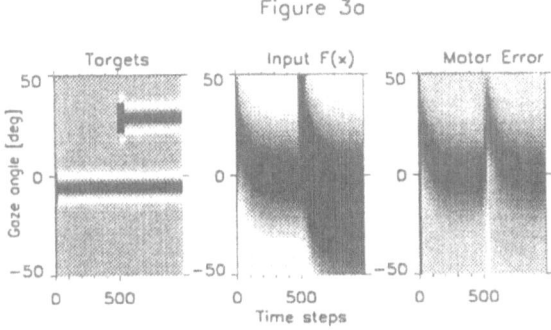

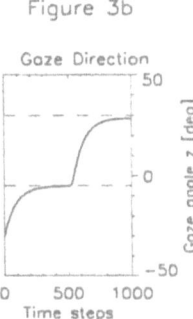

Figure 3a

Figure 3b

system. This can be seen as follows: Because the coded motor error follows the input $F(x)$ adiabatically, we have $\bar{x} = y_T - z$ and therefore $\ddot{z} + \rho\dot{z} + \omega^2(z - y_T) = 0$, so that y_T specifies an equilibrium point. Therefore, the dynamics of z converts the spatial information y_T into the actual movement trajectory [8].

Fig. 3a shows the time evolution of $\psi(y)$, $F(x)$ and $\phi(x)$ under *closed loop conditions* for two targets presented at times $t_1 = 0$ and $t_2 = 500$ at angles $y_1 = -5°$ and $y_2 = 30°$ respectively. The initial gaze direction is $z(t=0) = -30°$, with vanishing initial velocity and the time scale is identical to fig. 1. After the first target entry, the motor error relaxes from initially 25° to 0°. It is shifted to 30° during the transient phase of the second target specification, followed again by relaxation to zero error. Note the persistence of the first target entry and the final, bimodal shape of $F(x)$. **Fig. 3b** shows the resulting time course of the gaze angle z.

The case of two targets presented simultaneously at $t=0$ with $y_1 = -5°$ and $y_2 = -20°$ is shown in **fig. 4**. The initial gaze angle is 10° and motor error has relaxed to 0°. With this inter-target spacing, the specifying force $F(x,t)$ is unimodal (only the final $F(x)$ is indicated to the right), initially centered around $\bar{x} = -12.5°$. leading to an averaging response.

In **fig. 5** the position of the first target is unchanged, but the second target is now presented at $y_2 = -40°$ (the target level is not shown). Now, $F(x,t)$ is bimodal, specifying two potential targets, but the peak on the motor error representation is trapped by the initial $-15°$ specification (initial gaze angle is still 10°) which is closer to the initial 0° motor error than the necessary gaze shift of $-50°$ to y_2. Hence, the response is almost completely (in the absence of stochastic influences) biased towards the closest specification. The final gaze position is shifted a little bit towards the internal representation of the second target at y_2 due to the superposition of the two specifying projections (see plot of gaze direction).

The situation changes if the width of the specification $F_T(y)$ around the second target is nearly doubled (**fig. 6**). In that case the motor error specification $F(x,t)$ becomes unimodal during the period of transient input into the target level, with the maximum located near $-40°$, leading to a corresponding shift of the motor error peak. After the transient period, $F(x,t)$ again becomes bimodal. but now the motor error follows the maximum initially located at $-40°$. For a some critical width of the second target and in

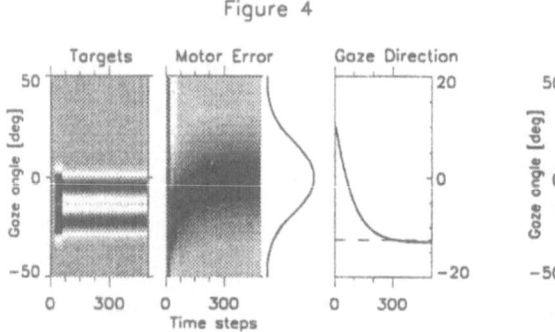

Figure 4

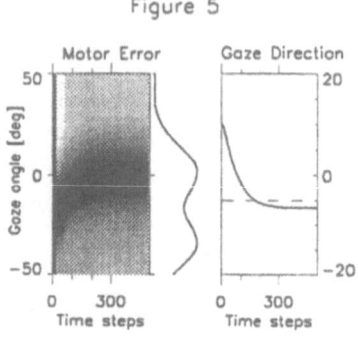

Figure 5

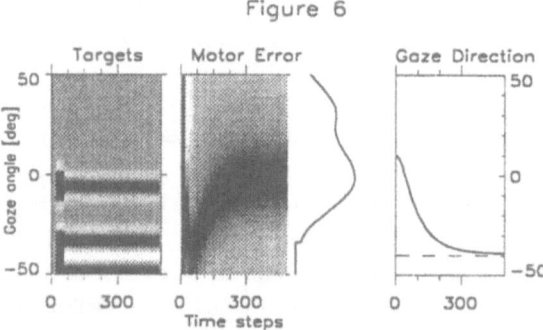

Figure 6

the presence of stochastic influences, an ensemble of identical trials will yield a bimodal distribution of final gaze positions located around y_1 and y_2 respectively, which is the basic feature of bistable responses. Note, that the input to the target level exceeds a critical width, so that the intrinsic field dynamics selects two representatives for the second target. These are close to each other, so that their projection to the motor error level is unimodal and represent one error specification.

The simulations shown in figs. 4,5 and 6 capture the basic phenomenology of averaging, response bias and bistable responses [3]. Note, that information about targets shown in parallel or serially persists at the target level. This is a prerequisite for scanning entire configurations of targets with multiple gaze shifts.

Although the model presented here is based on behavioral studies and refers to functional aspects of saccadic motor behavior some of the ideas might be of relevance also for understanding the underlying neurophysiology. An operational relation between the functional dynamic fields and neuronal response can be established through the concept of population code. Points in the dynamic field can be mapped onto points in cortical and collicular neuronal layers by using neurophysiologically plausible coordinates. In this study, a retinocentric system was chosen at the decision level in order to enable this mapping. From a purely functional point of view allocentric coordinates may lead to a simpler description in which an activated region directly specifies the equilibrium point of the mass–spring level and the coordinate transformation requiring internal feedback becomes unnecessary. Retinocentric coordinates implies a feedback architecture with a zone of activity moving during a gaze shift as observed by Munoz et. al. [7] through the activity of tectoreticular and tectoreticulospinal neurons in the Superior Colliculus of cats. On the other hand, the underlying coordinates used here for the representation of targets are chosen to be allocentric for simplicity. Direct electrophysiological evidence for a representation of this kind is still lacking. It was shown recently by Droulez and Berthoz [9] that targets might be represented in retinocentric coordinates, if updating by feedback of velocity information is employed to generate the observed effectively allocentric representation.

The interplay of "neuronal" coordinate systems and internal dynamics in the three-level model may throw light on the question "what is coded by the activity of neurons at specific locations". In the interpretation of their data, Munoz et al. [7] conclude that the maximum of a moving peak of activity codes instantaneous gaze error. In our model this would be an adequate description only in the limit case of very fast relaxation of the peak on the decision level to prescriptions from the target level. In general, the position of the moving peak lags behind the position of the prescription (which is identical to instantaneous gaze position here). Therefore, "neurons" respond maximally at gaze amplitudes *smaller* than expected by determining optimal response locations with targets presented under the static condition of a stationary visual axis. Hence, within our framework, the concept of "coding dynamic motor error" must be viewed as an approximation and a detailed experimental analysis of the timing relation between peak activity and gaze amplitude would be required to establish the functional role of neuronal activity at different stages of eye movement planning and control.

References:
[1] L.E. Mays and D.L. Sparks. *Science*, 208:1163–1165, 1980.
[2] D. Pélisson, D. Guitton and D.P. Munoz. *Exp. Brain Res.*, 78:654-658, 1989
[3] F.P. Ottes, van Gisbergen J.A.M., and J.J. Eggermont. *Vision Res.*, 24:1169–1179, 1984.
[4] A.S. Mikhailov. *Foundations of Synergetics I.*. Springer, Berlin, 1990.
[5] S. Amari. *Biol. Cybern.*. 27:77–87, 1977.
[6] D.L. Sparks, R. Holland and B.L. Guthrie. *Brain Res.*, 113:21-34, 1976.
[7] D.P. Munoz, D. Guitton and D. Pélisson. *J. Neurophysiol.*. 66:1642-1666, 1991.
[8] G. Schöner. In: S.P. Swinnen, H. Heuer, J. Massion and P. Casaer (eds.). *Interlimb Coordination: Neural, Dynamical, and Cognitive Constraints*. Academic Press (in press).
[9] J. Droulez and A. Berthoz. *Proc. Natl. Acad. Sci.*, 88:9653-9657, 1991

This work was supported by the Bundesministerium für Forschung und Technologie, Germany.

DIFFERENCES IN SYNAPTIC INPUT AND EXCITABILITY BETWEEN SUPERFICIAL AND DEEP PYRAMIDAL CELLS IN THE CAT SENSORIMOTOR CORTEX

J.F.M. van Brederode, and W.J. Spain. Depts. of Biological Structure and Physiology and Biophysics, University of Washington and Seattle Veterans Administration Medical Center, Seattle, U.S.A.

Abstract

In this study we recorded stimulus-evoked intracellular synaptic potentials in superficial and deep pyramidal neurons of the cat sensorimotor cortex. All cells studied received non-NMDA receptor-mediated fast depolarizing excitation, while some deep pyramidal cells also showed a slower NMDA-mediated input. Superficial pyramidal cells showed a triphasic response characterized by both fast and slow hyperpolarizing inhibition. The latter response was absent in deep pyramidal cells. These results suggest that pyramidal cells from different cortical layers receive different types of synaptic input. These differences are likely to have consequences for the way excitatory and inhibitory inputs interact and the kind of operations that can be performed by cortical neurons.

The likelihood of action potential generation in neurons at the axon hillock is determined by the temporal and spatial interaction between the inhibitory and excitatory inputs, dendritic geometry, and the intrinsic membrane properties of the postsynaptic membrane. It is well known that intrinsic membrane properties differ among neocortical neurons (Spain et al. '91; Connors and Gutnick '90), but less is known about differences in synaptic input and excitability. From detailed anatomical reconstructions of the distribution of asymmetric (presumed excitatory) and symmetric (presumed inhibitory) synapses onto the cell bodies and dendrites it is clear that different varieties of cortical pyramidal cells differ with regard to the number, proportion, and distribution of asymmetric and symmetric synapses (for review see Fairen and Farinas, '92). The aim of these studies, therefore, was to investigate *in vitro* the synaptic inputs onto different types of pyramidal cells in cat sensorimotor cortex evoked by extracellular stimulation. Our studies indicate that deep and superficial pyramidal cells differ with regard to the relative proportion and types of synaptic input they receive.

Full details of the surgical pocedures, slice preparation and intracellular recordings have been described previously (Spain et al. '87). Briefly, slices were obtained from the lateral cruciate cortex (area 4) of pentobarbitone-anesthetized cats. After opening the skull and removing the dura, a small block of sensorimotor cortex was dissected and sliced into 350-400 μm coronal sections. Slices were maintained at 36°C in a holding chamber. Recordings were performed after the slices were transfered to an submerged-type recording chamber. Identification of layer V and the large Betz cells located in this layer was performed as described earlier (Spain et al. '87, '91). For recordings from superficial pyramidal cells the recording electrode was positioned between 200 and 400 μm below the pial surface which corresponds to the location of layers 2/3 in the cat sensorimotor cortex based on Nissl-stained stained sections (unpublished observation). The firing characteristics in response to intracellular current injection of the impaled neurons in layers 2/3 suggest that they were all pyramidal cells (Connors et al. '88; Connors and Gutnick '91). Intracellular recording microelectrodes contained 2M $KMeSO_4$ in 30 mM MOPS buffer (pH=7.2) and recordings were performed in active bridge mode or in discontinuous mode in current clamp according to procedures described elsewhere (Spain et al. '87). Synaptic potentials were evoked by placing a sharpened, insulated monopolar tungsten electrode into the slice in the white matter (WM) or in one of the cortical layers. Membrane current and potential were recorded on tape and after digitizing they were further analyzed by computer. All drugs were bath applied.

We recorded from a total of 41 neurons of which 28 were located in layer 5 and 13 in layer 2/3. All recorded cells fired trains of action potentials in response to long-duration depolarizing intracellular current injections and exhibited other electrophysiological characteristics typical for in vitro recordings of Betz cells (Spain et al. '91) or superficial pyramidal cells (Connors et al. '88) in the cat neocortex. Nearly all layer 5 neurons responded to extracellular stimulation with a graded, transient depolarization (fast excitatory postsynaptic potential or fEPSP) from resting membrane potential (Fig.1A). At high stimulus intensities this fEPSP gave rise to an orthodromic action potential. Layer 2/3 cells, in contrast, responded to extracellular stimulation with a tri-phasic synaptic potential (Fig. 1B), consisting of a fEPSP, a fast inhibitory postsynaptic potential (fIPSP) and slow inhibitory postsynaptic potential (sIPSP).

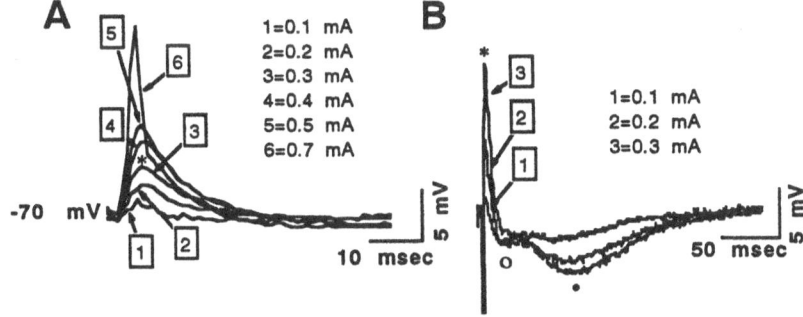

Figure 1. Examples of synaptic potentials recorded intracellularly from layer 5 (A) and layer 2/3 pyramidal cells (B) in response to increasing stimulus strength. Note the graded increase in fEPSP (*) amplitude with increasing stimulus strength in A. At the highest stimulus strength (0.7 mA) an orthodromic spike (truncated) was evoked in this cell. The cell in B shows a graded tri-phasic response consisting of a fEPSP (*), fIPSP (0) and sIPSP (•).

In both layer 5 and layer 2/3 cells fEPSP amplitude and duration increased with hyperpolarization (Fig. 2A, B, and D). At holding potentials just below spike threshold the fEPSP in layer 5 cells was followed by either a small (1-3 mV) fIPSP (Fig. 2B), or a slow excitatory postsynaptic potential (sEPSP; Fig. 2A). The amplitude of the sEPSP increased with depolarization. The fEPSP in both superficial and deep cells was abolished by non-NMDA receptor blockade with CNQX (Fig. 2C). After complete blockade of excitatory neurotransmission with bath application of both CNQX and the NMDA receptor blocking agent AP-5 focal stimulation near the soma unmasked isolated fIPSP's in all layer 5 cells tested (Fig. 2C). Comparisons of the traces recorded with CNQX alone and CNQX + AP-5 in combination revealed a small NMDA-component to the depolarizing synaptic potential in layer 5 cells even at resting membrane potential (Fig. 2C). The fIPSP of the tri-phasic potential in layer 2/3 cells reversed at potentials near rest and was abolished by application of the GABA$_A$ antagonist bicuculline (data not shown), indicating that this was a GABA$_A$ receptor- mediated potential, while the sIPSP usually did not reverse even at very negative holding potentials (<-90 mV), suggesting that this was a K-dependent postsynaptic potential (Fig. 2D). Complete pharmacological blockade of excitatory neurotransmission in layer 2/3 cells resulted in biphasic fIPSP-sIPSP sequences (data not shown).

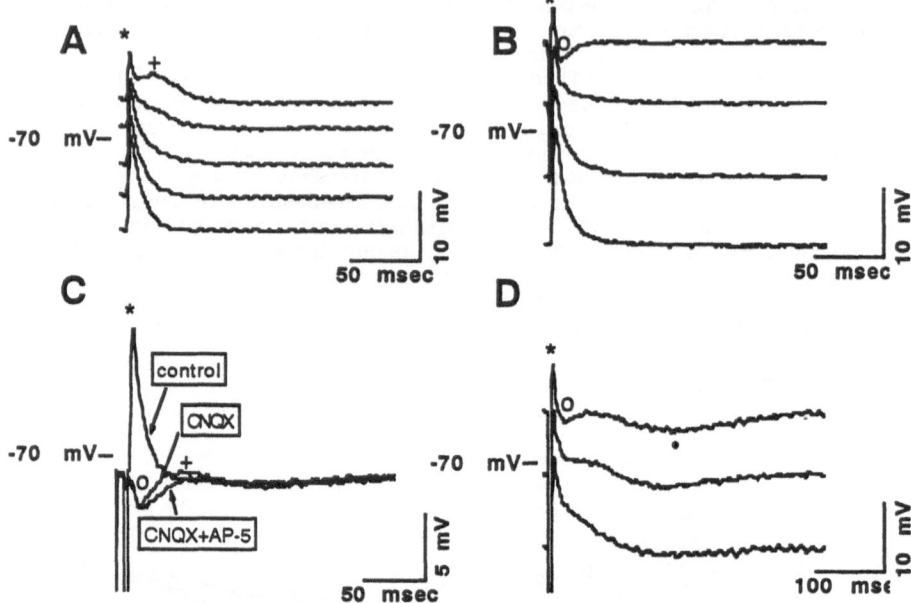

Figure 2. Examples of synaptic potentials recorded in layer 5 (A-C) and layer 2/3 pyramidal cells (D). Resting membrane potential in A, B, and D was altered with steady injection of hyperpolarizing or depolarizing current through the recording electrode. Extracellular stimulation evoked fEPSP's (*) in all layer 5 cells. These were followed either by a sEPSP (+) as shown in (A) or a fHSP (o) as in (B). Blockade of non-NMDA receptors by 10 μM CNQX (C) resulted in a response characterized by a fEPSP followed by a sEPSP. Additional blockade of NMDA-mediated synaptic potentials with 50 μM AP-5 (C) results in a fEPSP only. Synaptic responses elicited in layer 2/3 cells (D) were characterized by a fEPSP (*), fIPSP (o) and sIPSP (•).

We found that perfusion of the slices with an GABA$_A$ antagonist resulted in long-lasting stimulus-evoked bursts in both superficial and deep pyramidal cells (Fig. 3). These bursts tended to be shorter in superficial pyramidal cells in which they were terminated early by a large hyperpolarization following the burst. These results indicate that in both superficial and deep pyramidal cells in the cat sensorimotor cortex GABA$_A$ mediated inhibition functions to dampen excitability in response to synaptic input.

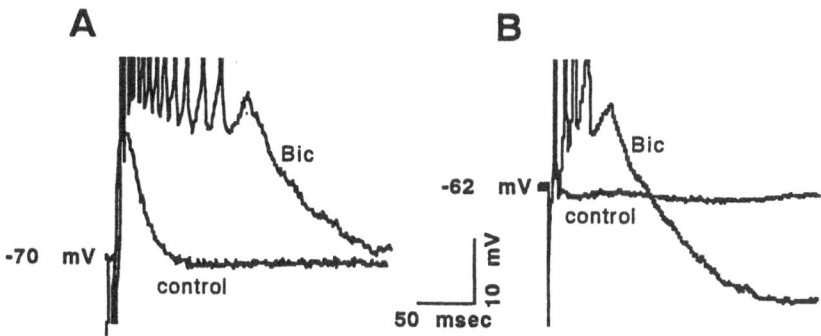

Figure 3. Examples of the effects of 10 μM bicuculline (bic) on synaptic potentials evoked in layer 5 cells (A) and layer 2/3 pyramidal cells (B). Spikes are truncated at the top.

In summary, we found that superficial and deep cortical pyramidal cells differ with regard to the types of excitatory and inhibitory input they receive, In layer 5 cells the fast inhibitory synaptic input is usually masked at resting membrane potential by a large non-NMDA and NMDA mediated excitatory potential. In addition, layer 5 pyramidal cells lack the long-lasting hyperpolarizing inhibition which is typical for superficial pyramidal cells (Connors et al. '88; this study). However, preliminary results suggest that layer 5 cells do posses GABAB receptors, as evidenced by the large, long-lasting hyperpolarizations that we recorded in response to droplet application of the GABAB agonist baclofen near the soma. Superficial pyramidal cells show little or no NMDA component to the excitatory postsynaptic potential under these recording conditions. These differences in synaptic input taken together with different intrinsic membrane properties of the postsynaptic membrane between deep and superficial pyramidal cells (Connors and Gutnick '90; Spain et al. '91) suggest that they play different roles in cortical microcircuits.

References

1   Connors, B.W., Malenka, R.C. and Silva, L.R. (1988). Two inhibitory postsynaptic potentials , and GABAA and GABAB receptor-mediated responses in neocortex of rat and cat. J. Physiol. (Lond.) 406:443-468.
2.   Connors, B.W. and Gutnick, M.J. (1990). Intrinsic firing patterns of diverse neocortical neurons. TINS 13:99-104.
3.   DeFelipe, J. and Farinas, I. (1992). The pyramidal neuron of the cerebral cortex: morphological characteristics of the synaptic input. Progr. Neurobiol. 39:563-607.
4.   Koch, C., Douglas, R. and Wehmeier, U. (1990). Visibility of synaptically induced conductance changes: theory and simulatons of anatomically characterized cortical pyramidal cells. J. Neurosci. 10:1727-1744.
5   Spain, W.J., Schwindt, P.C., and Crill, W.E. (1987). Anomalous rectification in neurons from cat sensorimotor cortex in vitro. J. Neurophysiol. 57:1555-1576
6.   Spain, W.J., Schwindt, P.C., and Crill, W.E. (1991). Post-inhibitory excitation and inhibition in layer V pyramidal neurones from cat sensorimotor cortex. J. Physiol. (Lond.) 434:609-626.

Funded by the Veterans Administration Merit Review, Seattle VA Medical Centre, WA 98108 and the National Eye Institute Grant EY01208-18.

Visuo-motor interaction
—poster contributions

An Adaptive Sensory Fusion Approach for the Superior Colliculus

Hans-Martin R. Arnoldi

Institut für Medizinische Informatik und Systemforschung

GSF - Forschungszentrum für Umwelt und Gesundheit, GmbH

Neuherberg, Postfach 1129

85758 Oberschleißheim, F.R.Germany

arnoldi@gsf.de

Abstract: Fundamental questions arise when trying to combine different sensory information into a coherent representation. The answers relate to every domain within cognitive science. A well-established brain structure to approach these questions is the Superior Colliculus. Visual and auditory stimuli merge into a common coordinate system to initiate saccadic eye movements. The coordinate transformation of head-centered auditory stimuli into a retinotopic or motor error coordinate system was modeled in a neural network. Auditory and visual space are short-term connected by every physical stimulus in the environment. This association was used in order to develop the transformational network through experience. Simulations with local Hebbian learning also revealed characteristics of the fusion process concerning eye position signals and topographic maps.

1. Introduction

How does a cognitive system use information from different sources and modalities yet still yield the most appropriate behavior? It must ensure the coherence of different inputs. A particular brain structure to attack a solution to this sensory fusion - based on neurophysiological and psychophysical findings - is the Superior Colliculus (SC). Visual, auditory, and somatosensory sensory inputs merge in the deeper layers of the SC into one motor error coordinate system and initiate saccadic eye movements.

The present work [Arnoldi, 1993] concentrates on the connection of auditory and visual information in the SC. Several different models have been proposed to transform the head-centered auditory stimulus into a different coordinate system [Groh and Sparks 1992; Anderson and van Essen 1987]. However, these models approach the problem in an engineered kind of network and fail to account for the plasticity with which the auditory map develops in the SC. The learning issue for a coordinate transformation was previously addressed in parietal cortex simulations [Zipser and Andersen 1988]. A neural network learned to transfer a retinotopic stimulus into a head-centered coordinate system. This back propagation network utilizes a signal about current eye position. It produces receptive fields in the hidden layers significantly similar to those in the parietal cortex. Mazzoni, Andersen, and Jordan [1991] recently applied a biologically less controversial learning rule to the same transformation with identical results.

Very strong feedback projections are not confirmed for the auditory pathway to the SC. Such an information flow is essential for reinforcement learning [Barto, 1985]. This processing and learning scheme would also give rise to an enormous collicular network with a neurophysiologically questionable number of synapses. Therefore another approach was chosen to simulate convergence of different modalities in the SC. Auditory space is transformed separately in horizontal and vertical dimension. Unique representations develop for each combination of auditory input and eye position signal in a layer of internal representations. Each internal representation can be associated with an activity pattern in the output layer. A teaching input provides this pattern. It models the visual signal ascending from the retina.

Tonically discharging neurons were analyzed in detail. In the parietal cortex these neurons were reported to reflect the influence exclusively related to eye position. The model neurons discharge linearly with the orbital angle of the eye. The incorporation of this frequency code substantially complicates the transformation process. In its combination with the place code for the auditory stimulus it severely shapes the neuron responses. It seems a valid hypothesis that this tonic influence produces those very characteristic receptive fields, previously addressed as spatial gain fields [Zipser and Andersen 1988].

2. Eye Position Signals

The brain requires information about eye position for a rapid transformation of the head-centered auditory stimulus. Initial saccades to auditory targets have a latency of 300 msec [Lueck et al., 1990] and they have to overcome large misalignments between auditory and visual space. Therefore a relaxation process would be too slow to match visual and auditory inputs within this very brief period of time.

Studies of the posterior parietal cortex repeatedly reported about eye position signals. In this visual area some of the cells solely respond to the orbital angle of the eye and change their activity "linearly with eye position" [Andersen et al., 1990]. The parietal cortex is supposed to accomplish the reverse transformation of the SC, namely mapping a retinotopic signal into a head-centered coordinate system [Zipser and Andersen, 1988]. In their back propagation model Zipser and Anderson represent eye position through neurons which increase their firing rate monotonically according to horizontal and vertical orbital angle of the eye. As a result of the simulations very characteristic response properties develop in the hidden layer units. Eye position signals and input from the retina converge in these units in a way analogous to neurons in the parietal cortex. Both have "a planar spatial gain field" [Andersen et al., 1990]. This receptive field property is a significant, experimentally determined characteristic of cells in the parietal cortex. It is a similarity between experiment and simulation which is often interpreted as a proof that back propagation learning finds the same solutions as natural neural networks [Hinton, 1992].

The fusion in the SC might exploit an eye position signal which is equivalent to the signal recorded in the parietal cortex. Therefore the same eye position signal was analyzed in the work presented. Within an unsupervised paradigm a one-dimensional output layer was used to develop a topographic map of horizontal eye position. Output layer neurons learned strictly locally at each synapse in accordance with pre- and postsynaptic activity. Note that information about eye position may naturally form a normalized pattern. For every neuron activated by one linear function of certain slope and intercept, an antagonistic neuron might exist. Added together, both neuron activations always result in the same sum of total activity. A similar representation of information is possible in the brain where the eye is moved by antagonistic muscle pairs.

The output layer neurons interacted in soft competition through a Mexican hat distribution. A strict winner-takes-all was not implemented in order to force output layer neurons in a topological organization.

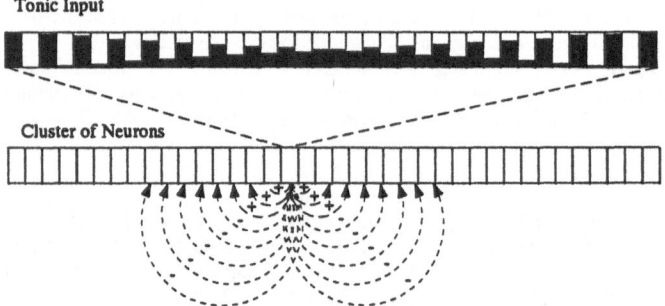

Different eye position were successively presented to the network and learned after contrast enhancement in the output layer. However, the network did not produce feature detectors for intermediate eye positions. Different eye position patterns were trained, but none of the neurons in the output layer reacted selectively to intermediate positions. In contrast output layer neurons were exclusively selective to the extreme eye positions at the medial or lateral possible orbital angle.

Of course those input patterns are not normalized within the Euclidean norm. Patterns are normalized in the sum of their activity but not in their vector length. To investigate the influence of input patterns with a normalized vector length a squared function was used to map eye position into neuron activities. This function can be explained as "a rationale to approximate the cumulative response of a group of eye position neurons" [Andersen and Zipser, 1988]. Still all output layer neurons responded optimally to the two eye positions at left or right extreme. None of the neurons developed a selectivity for intermediate positions. A principal component analysis of the eye position data revealed only two positive eigenvalues of the correlation matrix. This indicates that the signal is difficult to classify.

Although these are surprising results they can be interpreted in a consistent theory. Compared to the responses of hidden layer units in the back propagation simulations, those neurons in the output layer show a compatible behavior. Consider the spatial gain field of a feature detector for an intermediate eye position. It cannot be planar but has to have a peak of activity at the very eye position the neuron codes. The neurophysiological findings of planar spatial gain fields [Andersen et al., 1990] support the simulation results in the unsupervised paradigm. Eye position signals shape the spatial gain fields to be planar. Every learning rule which has the freedom to form an internal representation will develop planar spatial gain fields for those eye position signals. Therefore back propagation simulations do not gain much neurophysiological support through the similarity of its results with experimentally determined data.

Eye position signals are one form of a tonic input activity. But this frequency code is not restricted to the ocular-motor system. It might be used all over the brain either for command signals to the muscles or for feedback signals in return. Therefore the fusion of a frequency code and a place code might be a general problem to be solved for a comprehensive integration of information. An approach to perception will have to integrate the active component of behavior sooner or later. Then perceptual codes have to interact with those regulating the muscles.

3. An adaptive approach:

Several neurophysiological and psychophysical data constrain possible explanation of the sensory fusion process in the SC:

1. Auditory receptive fields of neurons in the deeper layers of the SC shift with eye position [Jay and Sparks, 1987]. Eye position modulates auditory responses in the SC of monkeys. Later these effects were also reported for cat SC [Hartline et al., 1989].

2. Auditory receptive fields in the SC develop through experience. This plasticity was confirmed through several reports for barn owls [Knudsen et al., 1984], ferrets [King et al., 1988], and guinea pigs [Withington-Wray et al., 1990a].

3. Auditory responses in the SC interact with visual input in a coherent manner. An auditory stimulus which is spatially as well as temporally coincident with a visual stimulus will be enhanced, otherwise suppressed [Meredith and Stein, 1986]. Psychophysical experiments also report this functional link [Lueck et al., 1990].

4. Saccadic eye movements are coded in a distributed representation in the SC. Either vector summation model [McIlwain, 1976] or population averaging model [Lee et al., 1988] stress this widespread activity in a population of cells.

5. Distinct cues of horizontal and vertical auditory localization suggest an independent processing in the brain [Konishi, 1986].

The proposed neural network takes advantage of these constrains. Horizontal and vertical components of the auditory stimulus are merged in separate internal representation with the respective component of the eye position signal. The orbital angle of the eye is encoded in neurons with different linear activation functions. These activation functions have the same slope, but different offsets. The neurons form a one-dimensional topographic map of eye position. Every eye position corresponds to an edge of activation distributed over several neurons in this map. Neurons in a higher layer are selective to this edge in the eye position map. A simple on-center-off-surround distribution of the receptive field produces this selectivity. This pattern of connectivity is well-know from retinal ganglion cells. It provides a plausible mechanism to transform a frequency code into a place code.

Auditory input to each layer of internal representation is represented by a gaussian activity on a one-dimensional map. This pattern of activity results from interaural time differences and intensity differences in the nucleus laminaris [Konishi, 1986]. Responses to auditory stimuli in the layer of internal representations can either be self-organized or designed to reflect the topology of its inputs.

In the deeper layers of the SC auditory and visual stimuli are integrated into a common coordinate system. These layers are simulated by the output layer of the neural network. In general an auditory stimulus originates from a cause which can also be perceived as a visual stimulus. This visual stimulus creates the teaching input in the SC. It can be associated with the activity initiated by the internal representation of an auditory stimulus. It is this coincidence which instructs the auditory map in the SC.

The learning paradigm in the output layer [Bienenstock et al., 1982] is not purely Hebbian as it has to explain the very short, crucial period during which a topographic auditory map develops in the SC. Experimentally determined auditory receptive fields in the SC of very young animals are "typically large, occupying most of the contralateral hemifield" [Withington-Wray et al., 1990b]. An auditory stimulus excites a significant area of SC cells before its influence is shaped by experience. Interactions between visual and auditory stimuli were established in several studies [Meredith and Stein, 1986]. In the developmental phase the visual stimulus will enhances some responses to the auditory stimulus, but it will depress others. Those neurons with enhanced activity increase their synaptic weights. If a neuron's activity is inhibited below its average response, the synaptic weights will decrease.

4. Results: In the simulations it proved impossible to associate internal representations with the visual teaching signal through simple Hebbian learning. This learning resulted in an error rate of 50%. A forced association is possible with the "Widrow-Hoff" error correction procedure [Widrow and Hoff, 1960]. Error correction is a simplification of those learning processes in the SC which are influenced by inhibitory and excitatory interactions. The network generated the correct peak of activity in the output layer for 93.25 % of the tests. All of the errors were due to edge effects.

The output layer always responds with a distributed neural code. Several efforts failed to learn a localized activity in the output layer. However, this is in accordance with experimental findings in the SC [McIlwain, 1976]. In the simulations it is due to the great overlap of internal representations that a grandmother cell activity could not represent the target for a correct association.

5. Discussion: A possible mechanism is presented in order to map head-centered auditory stimuli in a retinotopic coordinate frame. The number of combinations between both modalities determines the number of neurons in the layer of internal representation. With more than few modalities this might lead to a combinatorial explosion of required neurons. Therefore different extensions to the proposed network are possible. A feedback signal from the output layer might help to organize the internal representations more economically. Input interactions on the dendrite of output layer neurons could also be exploited. Especially an eye position signal which connects close to the soma can shunt auditory excitation. Therefore it easily produces "multiplicative" effects between different inputs as reported experimentally [Meredith and Stein, 1986].

Synchronization of neural responses [Singer, 1990] releases from the requirement to code different combinations with explicit detectors. Oscillatory response patterns were reported for latencies of pursuit eye movements [Pöppel and Logothetis, 1986] and for neurons in the avian optic tectum [Neuenschwander and Varela, 1990]. But the synchronizing sensory fusion also has to account for the neurophysiological findings that auditory receptive fields shift with eye position [Jay and Sparks, 1987]. It will be difficult but rewarding to simulate synchronization between frequency and place coding in a neural network. More theoretical research on the influence of tonic input is needed in this context to prove the proposal that tonic input shapes the neuron responses characteristically into spatial planar gain fields.

Bibliography:
Andersen, R.A. and Zipser, D. [1988]. *Can. J. Physiol. Pharmacol*, 66:488-501.
Andersen, R.A., Bracewell, R.M., Barash, S., Gnadt, J.W., Fogassi, L. [1990] *J Neurosci*, 10:1176-1196.
Anderson, C.H. and Van Essen, D.C. [1987]. *Proc. Natl. Acad. Sci. USA*, 84:6297-6301.
Arnoldi, H.-M. R. [1993]. Master's Thesis, Dep. of Cognitive and Linguistic Sciences, Brown University.
Barto, A.G. [1985]. *Hum. Neurobiol.* 4, 229-256
Bienenstock, E.L., Cooper, L.N., and Munro, P.W. [1982]. *J Neuroscience*, 2:32-48.
Groh, J.M. and Sparks, D.L. [1992]. *Biological Cybernetics*, 67(4):291-302
Hartline, P.H., King, A.J., Kurylo, D.D., Northmore, D.P.M., and Vimal, R.L.P. [1989]. *Invest. Ophthalmol. Visual Sci. (Suppl.)*, 30:181.
Hinton, G.E. [1992]. *Scientific American*, 267(3):145-151.
Jay, M.F. and Sparks, D.L. [1987]. *J. Neurophysiology*, 57(1):35-55.
King, A.J., Hutchings, M.E., Moore, D.R., and Blakemore, C. [1988]. *Nature*, 332:73-76.
Knudsen, E.I., Esterly, S.D., and Knudsen, P.F. [1984]. *J Neuroscience*, 4(4):1001-1011.
Konishi, M. [1986]. *Trends Neurosci*, 4:163-168.
Lee, C., Rohrer, W.H., and Sparks, D.L. [1988]. *Nature*, 332:357-360.
Lueck, C.J., Crawford, T.J., and Kennard, C. [1990]. *Exp Brain Res.*, 82:149-157.
Mazzoni, P., Andersen, R.A., and Jordan, M.I. [1991]. *Proc. Natl Acad. Sci. USA*, 88:4433-4437.
McIlwain, J.T. [1976]. *Int. Review of Physiology, Neurophysiology II*, 10:223-248.
Meredith, M.A. and Stein, B.E. [1986]. *Brain Res*, 365:857-873.
Neuenschwander, S. and Varela, F.J. [1990]. *Soc. Neurosci. Abstr.* 16:47.6
Pöppel, E. and Logothetis, N. [1986]. *Naturwissenschaften* 73:276-277.
Singer, W. [1990]. *Concepts in Neuroscience*, 1(1):1-26.
Widrow, B. and Hoff, M.E. [1960]. 1960 *IRE WESCON Convention Record, New York: IRE,pp. 96-104*
Withington-Wray, D.J., Binns, K.E., and Keating, M.J. [1990a]. *Dev. Brain Res.*, 51:225-236.
Withington-Wray, D.J., Binns, K.E., and Keating, M.J. [1990b]. *Eur. J. Neurosci.*, 2:682-692.
Zipser, D. and Andersen, R.A. [1988]. *Nature*, 331:679-684.

A Neural Network Model for Spatial Information Representation

Ryuta HOSAKA   Takashi NAGANO

Department of Industrial and System Engineering

College of Engineering, Hosei University

3-7-2, Kajino-cho, Koganei, Tokyo, 184, JAPAN

e-mail ryu@keiei.hosei.ac.jp

ABSTRACT

In this paper, we propose a neural network model that forms a two-dimensional spatial relation map self-organizingly. Cues for spatial relations between objects are given by efference copy signals of saccadic eye movements. The model is able to code the relative positions of objects existing simultaneously in the visual field in spite of its simple structure. The model was simulated on a computer to be shown to have the desired behavior.

1. INTRODUCTION

Neural networks for spatial vision[1] have been much less studied than those for pattern vision though spatial vision is one of the main function of the visual system. One of the authors proposed a model for the self-organization of a spatial map[2]. It may not, however, be plausible in the actual visual nervous system because it needs too many neurons to represent the map of many objects. In this paper, we propose a new model that codes relative spatial relations of many objects into a two-dimensional array of neurons. The two-dimensional model of a spatial map is supported by the psychological findings[3] which show that human memorizes spatial information as a two-dimensional 'cognitive map'.

2. MODEL

2.1 Structure of the model

The model is composed of five functionally different layers EM, EC, EP, SR and OR as shown in Fig. 1.

Neurons in EM give signals for saccadic eye movements. It is known that the signal for eye movements is mainly composed of two one-dimensional components, i.e., the vertical component and the horizontal component[4]. EM is composed of the four neurons: $EMx+$, $EMx-$, $EMy+$ and $EMy-$. Each of these neurons gives output impulses the number of which is proportional to the magnitude of each component signal. $EMx+$ and $EMx-$ ($EMy+$ and $EMy-$) give the positive and the negative directional signals of the horizontal (vertical) component.

EC is composed of two parts: ECx and ECy. Each part is composed of a pair of one-dimensional arrays of neurons. ECx and ECy represent the horizontal and vertical components of a fixation point respectively. ECx and ECy receive inputs from the corresponding neurons in EM that code the vertical and the horizontal components of eye movements respectively. To explain the behavior of the model simply and clearly, consider a simplified model of EC that represents the shift of a fixation point caused by an one-dimensional and one-directional eye movement. The structure of the one-dimensional and one-directional model is shown in Fig. 2(A). Each neuron has two excitatory inputs from the left neighboring neuron and from the neuron *EMx+*, and has inhibitory inputs from all of the neurons on its right side, and has an excitatory self-recurrent connection. All the synaptic weights are assumed to be 1 and its threshold is 2. So each neuron in *EC* fires only when it simultaneously receives the two excitatory inputs and no inhibitory signals. Once a neuron fires, it keeps its activation by the self-recurrent connection until it receives inhibitory inputs from the next neuron on the right side. In this array of neurons, only one neuron can keep firing. A fixation point is represented by the position of a firing neuron. In this array of neurons, neuron firing shifts to the right neighboring neuron by one impulse from *EMx+*. Suppose a neuron keeps firing, which means the eye is being fixated at a point. Then, the neuron *EMx+* fires to cause an eye movement. At this time, all neurons in *EC* receive impulses. In this case only the neuron which has been receiving excitatory inputs from its left neighboring neuron, that is, the neuron which exists on the right side of the firing neuron, starts firing. The neuron that has been firing stops firing by the inhibitory input from the neuron that has started firing. One shift of firing is caused by one impulse from *EMx+*. So, the magnitude of shift is proportional to the number of impulses from *EMx+*. As a result, this one-dimensional and one-directional model can represent the one-dimensional and one-directional shift of a fixation point. The structure of *ECx* is shown in Fig. 2(B). It is composed of a pair of the models in Fig. 2(A). One of the paired model receives excitatory inputs from *EMx+* which gives output impulses for the movement to the positive direction and the other from *EMx-* which gives output impulses for the movement to the negative direction. There are excitatory mutual connections between corresponding neurons in the two arrays. As a neuron in one of the two arrays fires, the corresponding neuron in the other array also fires because of the mutual connections. Inhibitory connections from a neuron to all the neurons on its left side in one of the arrays are also given to the corresponding neurons in the other array. So the movement of the firing neuron in one array is always copied into the other array. Consequently, *ECx* can represent one-dimensional eye movements to either the positive direction or the negative direction. *ECy* has the same structure and function as *ECx*.

EP restores the two-dimensional position of a fixation point from two one-dimensional signals given by *ECx* and *ECy*. In this layer, neurons are placed in a two-dimensional array. There are no

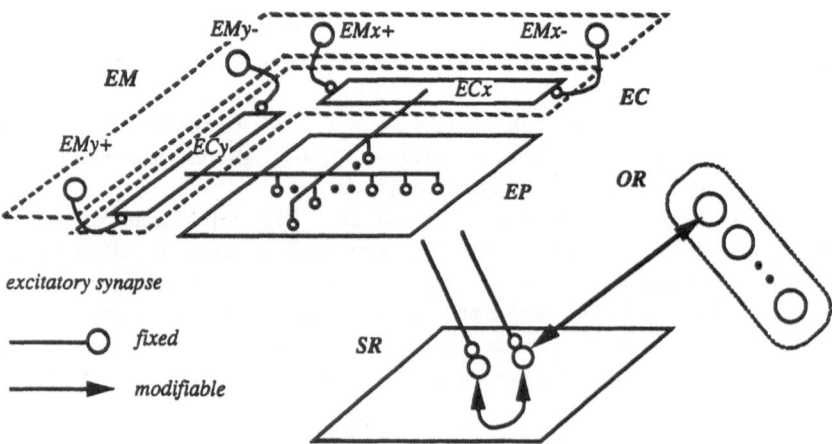

Fig.1 Schematic diagram of the model

mutual connections between neurons in this layer. Each neuron receives two excitatory inputs from EC, one from ECx and the other from ECy, in such a way as shown in Fig 1. Only the neuron fires that receives inputs from both ECx and ECy simultaneously.

SR represents spatial relations of objects existing simultaneously in the visual field. This layer is also an two-dimensional array of neurons which have one-to-one correspondence to the neurons in EP. A neuron in SR receives an excitatory input from the corresponding neuron in EP. Each neuron in SR has mutual connections with modifiable weights to all the other neurons in SR. The weights of these mutual connections are modified by the hebbian learning rule. Neurons in SR are assumed to have the three activation levels[5] : E0(null activation state), E1(moderate activation state) and E 2(high activation state). When a neuron in SR receives an input from EP, its activation level rises to E 2, but falls down to E1 before the next input from EP comes into SR. The neuron keeps its activation level E1 for a comparatively long time enough for several inputs from EP to come into SR. The connection between the neurons in SR that are firing at activation level E1 or E2, are reinforced by the hebbian learning rule. In this way, spatial relations of objects that exist simultaneously in the visual field, are coded in SR by scanning the objects one by one.

OR is a set of "grandmother cells". An object is represented by the corresponding neuron in OR which fires specifically to the object. While an object is being gazed, the corresponding neuron in OR keeps firing and mutual connections between the neuron in OR and the neuron in SR with activation level E2 are reinforced by the hebbian learning rule. These connections represent the relations between objects and their positions.

2.2 Acquisition of spatial relations

How can spatial relations between objects be acquired in the model by sequential saccades? This will be explained more concretely by using the example shown in Fig. 3. Suppose that the first fixation point is 1, the second is 2 and the third is 3. First, the network learns the position of object 1. That is, while object 1 is being gazed, its position is represented by the firing of a neuron in EP, and then the corresponding neuron in SR fires at level E2. The neuron in OR representing the object 1 also fires at the same time. So, the connection between the firing neuron in OR and the firing neuron in SR with activation level E2 is reinforced. Then the fixation point is shifted by a saccade and the next target (object 2) is gazed. The network learns the position of object 2 in the same way. The connection between the two neurons in SR representing the corresponding positions of objects 1 and 2 respectively is reinforced. The same things occur when the fixation point is shifted from object 2 to object 3. The mutual connections between the three neurons in SR representing the positions of objects 1, 2 and 3 respectively are all reinforced because the neuron keep activation level E1 for a while. Consequently the network acquires spatial relations between three objects that existing simultaneously in the visual field.

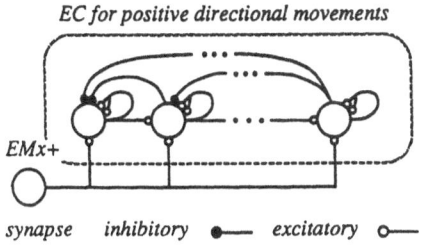

Fig.2(A) One-directional and one-dimensional
model of EC layer

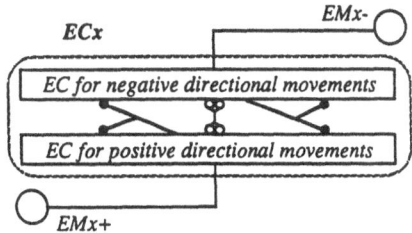

Fig.2(B) Model of ECx

2.3 Recall of spatial relations

Recall of spatial relations acquired in the model is done as follows. When one of the objects whose spatial relations have been already learned is gazed again, the spatial relations between the objects are recalled as follows. The activation of the "grandmother cell" of the gazed object causes activation of the corresponding neuron in SR. Then the neuron in SR activates all the other neurons in SR which represent the positions of the rest objects because those neurons are strongly connected mutually.

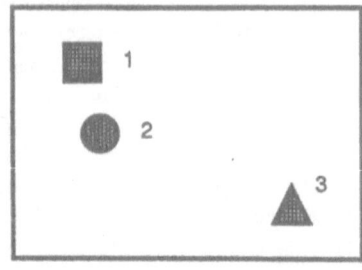

Fig.3 Positions of three objects used in a computer simulation of the model

3. SIMULATION

To confirm that the behavior of the model is the expected one, we tested the model on a computer with the simple example shown in Fig 3. In this simulation, the three objects were gazed one by one according to the number in Fig 3. As a result, the model learned the spatial relations between the objects. That is, the connections between the neurons in SR representing the positions of the three objects and the corresponding neurons in OR were reinforced. The connections between the neurons in SR representing the positions of these three objects were also reinforced. After the acquisition of the spatial relations, one of the objects was presented to the model again. The corresponding neuron in SR fired and caused firing of the rest two neurons in SR. The neurons in OR corresponding to the rest two objects were also activated. So it was shown that the model could recall the memorized spatial relations.

4. DISCUSSION

It was shown that the proposed model can successfully memorize spatial relations between objects existing simultaneously. The model can acquire the spatial relations of all the objects that exist simultaneously irrespective of the order of scanning because neurons in SR hold activation level at E1 for a long time once they are activated and all the connections between neurons with activation level E1 are reinforced in this model.

This research was supported by Grant-in-Aid #04246107 for Scientific Research on Priority Areas on "Higher-Order Brain Functions", the Ministry of Education, Science and Culture of Japan.

REFERENCES

[1] For example, Hartman G. (1992) *Motion Induced Transformations of Spatial Representations: Mapping 3D onto 2D*. Neural Networks, Vol. 5, pp.823-834

[2] Hirahara M., Nagano T. (1992) *A Neural Network for Fixation Point Selection Based on Spatial Knowledge*. Artificial Neural Networks 2, pp. 903-906

[3] Spoehr K. T., Lehmkule S. W. (1982) *Visual Information Processing*. pp.263-267, W. H. Freeman.

[4] Pitts W. H., McCulloch W. S. (1947) *How we know universals: The perception of auditory and visual forms*. Bull. Math. Biophys. 9, pp.127-147

[5] Isabelle Otto, et. al. (1992) *Direct and Indirect Cooperation between Temporal and Parietal Networks for Invariant Visual Recognition*. J. Cog. Neurosci., Vol.4, No.1, pp.35-57

A Dynamical Model for the Generation of Curved Trajectories

P. Morasso†, V. Sanguineti†, and T. Tsuji‡

† Department of Informatics, Systems, and Telecommunications, Genoa University, Italy

‡ Faculty of Engineering, Hiroshima University, Japan

Abstract - In the framework of a central hypothesis of kinematic invariance, we propose a model (ξ-model) which is a non-linear dynamical system capable of generating, as motor primitives, a family of curved trajectories. The model links shape and speed by means of a suitable time base generator that drives two equations: a linear-speed equation and a turning-speed equation.

1. Introduction

The kinematics of human movements has remarkably invariant features in a wide range of timing, loading, and "postural" conditions. The question, then, is what we can infer about the underlying neural processes. The first dichotomy is between models (like the λ-model [3]) that privilege the properties of the musculo-skeletal apparatus and models that attribute the main morphogenetic role to more central planning processes. Although we think that the viscous-elastic properties of muscles are essential for skilled movements, particularly as regards fine compliance control, the *peripheral hypothesis* of kinematic invariances is not powerful enough, in our opinion, to capture the complexity of the topic and, in any case, it leaves open the question of where and how complex muscle patterns are generated.

For a *central hypothesis* of kinematic invariance, on the other hand, there is the opposite danger of reducing it to an abstract curve fitting exercise, quite uncoupled from the musculo-skeletal reality. From this point of view, we think that the central generator of kinematic patterns should be viewed as a *dynamical system* and not as some kind of *static* mechanism, like in the minimum-jerk model [4] or the power-law model [6]. Moreover, there is a *compositionality* problem: since complex trajectories are obviously composed as ordered sequences of discrete motor commands, which are the basic primitives? *Global* models, like the power law model, are good (perhaps, too good) for complex, endless trajectories but are not plausible for simple reaching movements, whereas *local* models, like the minimum-jerk model, (over-)privilege straight trajectories.

The model that we propose (ξ-model) is somehow in the middle: it is a non-linear dynamical system that, in the framework of a central hypothesis of kinematic invariance, is capable of generating, as motor primitives, a family of curved trajectories that include straight lines as special cases. The dynamics of the model quite constraints the range of possible dynamic behaviours: differently from the VITE model [2], that can generate any kind of speed profile by an appropriate choice of the $\gamma(t)$ function ($\dot{x}(t) = \gamma(t)(x_f - x)$), the ξ-model is intrinsically based on a symmetric mechanism and it only has 3 free parameters of immediate cognitive significance.

2. The mathematical model

The shape of a trajectory depends on the way in which curvature varies with the curvilinear coordinate. The model links shape and speed by means of a suitable time base generator that drives two equations: a linear-speed equation and a turning-speed equation.

116

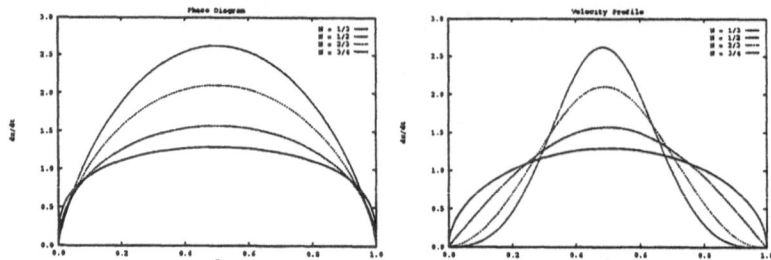

Figure 1: System dynamics corresponding to different values of e

2.1. The time base generator

The time base generator is a scalar dynamical system $\dot{\xi} = f(\xi)$ in the normalized variable ξ which is supposed to generate smooth sigmoidal signals $\xi = \xi(t)$ from $\xi(0) = 0$ to $\xi(T_f) = 1$ with a bell-shaped velocity profile and a desired (finite) movement duration T_f [1].

A class of models $f(\xi)$ that satisfy these conditions is given by $f(\xi) = \gamma\left[\xi(1 - \xi)\right]^e$ with $e \in (0,1)$ [2]. Figure 1 plots the phase space diagram and the velocity profile corresponding to different values of e. As regards the exponent e, it can be shown that the condition $e \geq 2/3$ is necessary for the 3rd time derivative of $\xi(t)$ (jerk) to be defined at $t = 0$ and $t = T_f$. In these conditions, the equilibrium configurations of the dynamical system do not satisfy the Lipschitz condition because $|df/d\xi| \to \infty$: $\xi = 1$ behaves as a *terminal attractor* and $\xi = 0$, which is unstable, is a *terminal repeller* [1]. In summary, the time base generator used in the simulation is described by the equation:

$$\dot{\xi} = \gamma[\xi(1 - \xi)]^{2/3} \tag{1}$$

Remarkably, this model can be approximated with a simple neural network, presented in Figure 2, which consists of 4 neurons: $N1, N2, N3, N4$. $N1$ and $N2$ are additive neurons, $N3$ and $N4$ are multiplicative neurons. All of them have a sigmoidal activation function with high gain ($y = g(x) = (1 - e^{-kx})/(1 + e^{-kx})$, with $k \gg 1$). $N3$ has a sufficiently long time constant for approximating it with an integrator, while the others have sufficiently short time constants to consider only the steady-state components. The network is a dynamical system in the ξ state variable and it is easy to derive the following state equation: $\dot{\xi}(t) = \gamma g(g(\xi)g(1 - \xi))$. Since the sigmoidal function is monotonic, the state function $g(g(\xi)g(1 - \xi))$ can be approximated by a function $f(\xi(1 - \xi))$ that, similarly to a power function with an exponent smaller than unity, is very steep in the origin and thus guarantees a finite T_f. The γ parameter has a double function: it controls the speed and it can be used to reset the time base generator and make it excitable for subsequent activation cycles.

2.2. Generation of curved trajectories

The linear speed v and the turning speed ω can be linked to a common goal (reaching a target with a desired orientation) by using the same time base generator and a pair of error measures: the linear error Δr and the angular error $\Delta\theta$. The former one is simply the norm of the difference between the final and the current position: $\|x_f - x(t)\|$. With this notation, it is possible to demonstrate that the following linear speed equation

$$v = \Delta r \, \dot{\xi}/(1 - \xi) \tag{2}$$

---

[1] $\dot{\xi}$ should tend to zero for $t \to 0$ and $t \to T_f$. T_f is given by $\int_0^1 d\xi/f(\xi)$ and a sufficient condition for attaining a finite value is that $f(\xi)$ is infinitesimal of order n, $n < 1$, for both $\xi \to 0$ and $\xi \to 1$.

[2] For this class of functions, it can be shown that movement duration T_f is inversely proportional to the gain factor γ: $\gamma(e) = 1/T_f\Gamma^2(1 - e)/\Gamma(2 - 2e)$

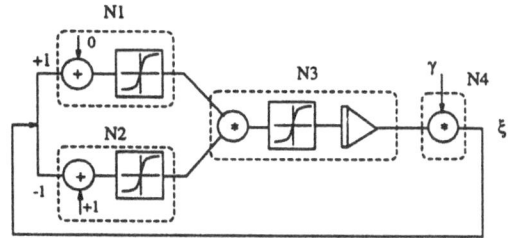

Figure 2: Neural time base generator

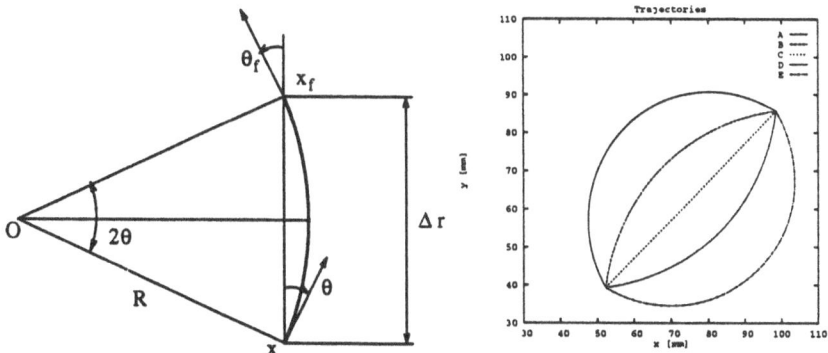

Figure 3: Circular trajectory (left) and simulated trajectories (right)

allows $x(t)$ to reach x_f at the same time in which $\xi(t)$ reaches 1, provided that the turning speed equation allows Δr to decrease in a monotonic way. In the limit case of a straight trajectory, this is equivalent to a simple interpolation with no rotation: $x(t) = x_0 + (x_f - x_0)\xi(t)$ [3]. For general movements, in which the initial and/or the final directions are not aligned with $x_f - x_0$ vector, a sufficient condition for the trajectory to hit the target with the right direction is that at some point in the trajectory (i.e. for $\xi < 1$), the running point $x = x(t)$ reaches the configuration which characterizes a circular motion: $\theta(t) = -\theta_f$ (see Fig. 3, left). In this case, the following relation between ω and v holds: $\omega = v/R$ (R is the radius of the circle) and since $\Delta r = 2R \sin \theta$, we get $\omega = 2 \sin \theta \, \dot{\xi}/(1 - \xi)$. In general, we can write a turning speed equation

$$\omega = \Delta\theta\dot{\xi}/(1 - \xi) \tag{3}$$

where the angular error term is given by $\Delta\theta(t) = 2 \sin \theta(t)$ in the limit case of a circular motion. If the current conditions do not support a circular motion, then a possible strategy is to smoothly drive $\theta(t)$ towards such symmetric condition. The two criteria (approaching the condition of symmetry and following a circular path) can be combined in the following hierarchical way ($\Delta\theta_{sym} = -(\theta + \theta_T)$):

$$\Delta\theta = \Delta\theta_{sym} + 2 \sin \theta e^{-\Delta\theta_{sym}^2/2\sigma^2} \tag{4}$$

It can be seen that while the error term for symmetry is large, the equation reduces to $\Delta\theta = \Delta\theta_{sym}$. This equation works just like a feedback control law to attain a symmetric configuration where the target and present directions of the trajectory are symmetric with respect to the line between the initial and target points. Once the symmetric configuration is attained, the symmetry error vanishes and the motion equation of the angular velocity ω reduces to $\Delta\theta = 2 \sin \theta$. Consequently,

---

[3]It is interesting to note that, in this form, if we put $\gamma(t) = \dot{\xi}/(1 - \xi)$, the ξ-model is a special case of VITE.

118

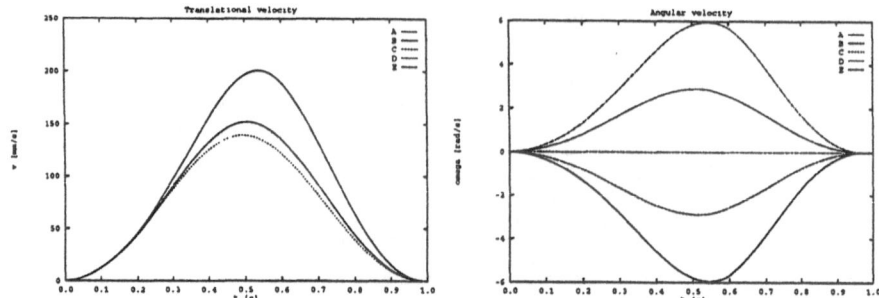

Figure 4: Simulated translational velocity (left) and angular velocity (right)

once the symmetric configuration is attained, the ξ-model can generate the circular trajectory. The trajectory terminates at the target point where the translational velocity becomes zero. This guarantees the convergency of the ξ-model for curved movements. The value of σ determines the relative weights of the two components of the error function. To verify the convergency of the ξ-model, computer simulations have been performed. Fig. 3, right shows examples of the generated trajectories by the ξ-model where the 2/3 power law is used. The initial and target points are fixed, while several initial and target directions satisfying the symmetric configuration condition are used. The corresponding translational and angular velocities of the generated trajectories are indicated in Fig. 4. The spatial profiles of the generated trajectories are almost circular and the velocity profiles are almost bell-shaped. The ξ-model can generate the curved trajectory as well as the straight trajectory as a single stroke movement.

3. Conclusions

In conclusion, the ξ-model consists of three equations: the time base generator (1) and the two speed equations (2,3). It can generate a large class of curved movements, parametrized by Δr_0, θ_0 and θ_f. In a preliminary experiments, in which we recorded hand movements with a digitizing tablet, we estimated the range of values of the angular differences that produce trajectories with unimodal speed profiles. We are also in the course of integrating the ξ-model, which dictates the behaviour of the *end effector*, with a self-organized neural model [5] of a kinematic chain which is actually carrying out the multi-joint coordination task.

References

[1] J. Barhen, S. Gulati, and M. Zak. Neural learning of constrained nonlinear transformations. *IEEE Computer*, 6:67–76, 1989.

[2] D. Bullock and S. Grossberg. VITE and FLETE: Neural modules for trajectory formation and postural control. In W.A. Hershberger, editor, *Volitional Action*, pages 253–297. North-Holland/Elsevier, Amsterdam, 1989.

[3] A.G. Feldman. Functional tuning of the nervous system with control of movement or maintenance of a steady posture. ii. controllable parameters of the muscle. *Biophysics*, 11:565–578, 1966.

[4] T. Flash and N. Hogan. The coordination of arm movements: an experimentally confirmed mathematical model. *Neuroscience*, 7:1688–1703, 1985.

[5] P. Morasso and V. Sanguineti. SOBoS – A Self-Organizing Body Schema. In I. Aleksander and J. Taylor, editors, *Artificial Neural Networks*, Amsterdam, 1992. North-Holland.

[6] P. Viviani and C. Terzuolo. Trajectory determines movement dynamics. *Neuroscience*, 7:431–437, 1982.

Functional Organisation in the Cerebellum

L.N.Kalia*

Neural Systems Engineering, Dept. of Electrical Engineering,
Imperial College, Exhibition Road, London SW7 2BT, U.K.

The structure of the cerebellum is examined and the emergent functional organisation deduced. This yields the observation that the cerebellum exhibits two forms of modularity of importance in artificial neural network models. Emphasis is placed on the definition of the individual cerebellar module, on the inhibitory mechanisms which lead to efficient modular action, and on how parallel fibres serve to link together, or coordinate, the activity of different modules.

It is proposed here that the cerebellar module should be defined as a *composite* of the two organisational schemes (vertical and parasagittal) which *coexist* in the cerebellum. In this way the cerebellar module is seen as a complex adaptive control system in its own right displaying feedback and feedforward characteristics.

Roles are proposed for the inhibitory interneurons: the Golgi cell feedback loop is seen as performing thresholding, for the purpose of detecting time-locked stimuli, and also, as a way of imbuing temporal sensitivity to the representation surface provided by the granule cells. Laterality in basket cell connectivity is interpreted as a means of performing competition between parallel fibre beams. Three possible functional roles for the basket cell circuit are that it serves (i) as part of an attention mechanism acting on the cerebellar cortical surface; (ii) as a target-setting mechanism and (iii) as a means of making the Purkinje cells sensitive to the normalised form of the spatial pattern of the parallel fibre wavefront. From the localised nature of stellate cell connectivity it is proposed that they are responsible for the functional parcellation of the Purkinje cell dendritic tree into competing blocks.

The hypothesis is put forward that cerebellar control of coordination arises from the coordination of activity in its constituent modules which, in turn, is brought about by the parallel fibres which supply information to many modules simultaneously *and* provide a given module with contextual information regarding the activity in other modules. In this way the individual modules are able to act in a *learned* synergy. From an evolutionary perspective, the parallel fibre projection can lead to the augmentation of existing modules as new modules are incorporated into the cerebellar structure.

The interplay between feedforward and feedback schemes in defining the cerebellar module and the concept of coordinated multi-modular action may be a neurobiological concept relevant to problems in control engineering. Artificial neural networks are an ideal paradigm for investigating the interplay between structure and function, as described in this paper, embodying as they do a structural approach to computation.

References

[1] S.Grossberg *Adaptive Resonance in development, perception and cognition.* In S.Grossberg (ed.) *Mathematical psychology and psychophysiology.* Providence, RI: American Mathematical Society 1981.

[2] M.Ito *The Cerebellum and Neural Control* New York: Raven Press 1984.

---

*I should like to thank the UK Science and Engineering Research Council for their financial assistance

Activation and Contraction of a Muscle

D.G. Rüegg, L. Studer, J.-P. Gabriel
Institute of Physiology and Mathematics
Université of Fribourg
Rue du Musée 5, CH-1700 Fribourg, Switzerland

Abstract

A parametric model of a pool of motor units (MU) of a skelettal muscle is developed. The model is based on the following assumptions: (1) The size principle is respected, (2) only steady state conditions and no time dependency are considered, (3) the motoneuronal membrane is homogeneous. An equation could then be developed for the relation between the input and the EPSP current in a motoneuron. In line with experimental data, the discharge frequency of model MUs is linearly related to the EPSP current, an equation with an exponential term describes how the MU contraction force depends on the discharge frequency, and the sum of the MU contraction forces gives the whole muscle force. Incorporating into the model the experimental finding that, over part of the working range, there is a linear input-output relation of the pool, all the parameters of the model are determined. The model predicts motoneuronal pool properties which can be tested experimentally.

Introduction

A muscle is composed of motor units (MU) consisting of a motoneuron (MN) and all the muscle fibers which the MN is activating. During most voluntary and reflex contractions, MUs are recruited during force increase according to their contraction force (size principle, [1,2]) and the recruited MUs modulate their firing frequency and thus their contraction force with the activation level. Experimental data provided evidence that the input-output relation of the MU pool of a muscle is linear till about 60% of maximum force [3]. The aim of this study was to develop a parametric model of the MU pool taking into account the above experimental data and intrinsic properties of MUs.

The Model

Only muscles and thus MUs which are activated at a constant level are considered in the present model. The contraction force F of a steadily active muscle equals the sum of the MU contraction forces f(s,In) which depends on the input In to the motoneuronal pool and on the parametric tetanic contraction force s of the MUs.

(1) $F(In) = \int h(s)\, f(s,In)\, ds$

where h(s), the distribution of the MUs within the motoneuronal pool, is approximated by

(2) $h(s) = \exp[a - b\, s]$

with a and b are constants to be estimated [4].

If long-lasting currents of different intensity are injected into MNs the force - frequency relation is of exponential shape [5] starting at recruitment with frequency $freq_T$. It can be described by:

(3) $\quad f(s,freq) = \begin{cases} s\,\{1 - c\,exp[\,-\gamma(s)\;(freq - freq_T(s))\,]\,\} & \text{if } freq>=freq_T \\ 0 & \text{if } freq<freq_T \end{cases}$

with $\gamma(s)$ a parameter related to the motoneurons and c a constant. The firing frequency increases linearly with current strength in the physiologically important range [5]. The frequency - current relation was thus described by:

(4) $\quad freq(s,I) = \begin{cases} \theta(s)\,(I - I_T(s)) + freq_T(s) & \text{if } I>=I_T \\ 0 & \text{if } I<I_T \end{cases}$

were $\theta\psi\sigma(s)$ is the slope and I_T the threshold current at which the MU starts firing.

To facilitate the further development of the model, we introduce a virtual EPSP which is the size of an EPSP without interfering action potentials and afterhyperpolarisations. Supposing an ohmic cell membrane with conductance G_m and an input current I, the virtual EPSP V_{EPSP} is

(5) $\quad V_{EPSP} = \dfrac{I}{G_m}$

Using the finding that the voltage threshold T is independent of the MN [6] and that $T=I_T/G_m$, we get from (3), (4) and(5) the force - virtual EPSP relation

(6) $\quad f(s,V_{EPSP}) = \begin{cases} s\,\{1 - c\,exp[-\alpha\,(V_{EPSP} - T)]\} & \text{if } V_{EPSP}>=T \\ 0 & \text{if } V_{EPSP}<T \end{cases}$

where we found $\alpha = \gamma(s)\,\theta(s)\,\beta(s)\,G_m(s)$ to be independent of the MNs.

The motoneuronal membrane was taken to be homogeneous. The membrane conductance is divided in components which are independent of EPSPs and components which are the source of EPSPs. The component i of the ionic currents, which are independent of the EPSP, depends on the conductance G_i and the reversal potential E_i, and the current dependant on the EPSP on the conductance change G_{EPSP} due to the synaptic input and the reversal potential E_{EPSP} of the EPSP. Expressing all potentials with respect to the resting membrane potential, and applying Ohm's law, we get

(7) $\quad G_{EPSP}\,(V_{EPSP} - E_{EPSP}) + \sum_i G_i\,(V_{EPSP} - E_i) = 0.$

Since there is no EPSP induced current at rest ($V_{EPSP} = 0$ and $G_{EPSP} = 0$)

(8) $\quad \sum_i G_i\,E_i = 0.$

Taking into consideration that the resting membrane conductance G_m is a property of the MN, and that the virtual EPSP size, V_{EPSP}, and conductance change due to an EPSP, G_{EPSP}, are dependent on both the MN and the input In to the MN, we get

(9) $\quad V_{EPSP}(s,In) = \dfrac{G_{EPSP}(s,In)\,E_{EPSP}}{G_m(s) + G_{EPSP}(s,In)}$

with $G_m = \sum_i G_i$ the resting membrane conductance as used in (5).

A global input which consists of motor commands impinging on all MUs of the pool induces a conductance change which is expressed by $G_{EPSP}(s,In) = g(s)\,In$ where $g(s)$ is a gain factor which is only dependent on the MNs.

If the MUs are recruited following the size principle [1,2] the EPSP of the last MN which is recruited by an input $In(s)$ just reaches threshold ($T = V_{EPSP}(s,In(s))$). Using (6) and (9) and rplacing G_{EPSP}, we find for the MU force

$$(10) \quad f(s,In) = \begin{cases} s \{1 - c \exp[-\alpha \dfrac{T\,(E_{EPSP} - T)\,(In - In(s))}{T\,In + (E_{EPSP} - T)\,In(s)}]\} & \text{if } In \geq In(s) \\ 0 & \text{if } In < In(s) \end{cases}$$

which can be used in equation (1) to compute the muscle force. In(s) remains to be determined.

The experimental finding that the input-output relation of the MU pool of a muscle is linear up to about 60% of maximum muscle activation [3] was incorporated into the model by imposing a linear input-output relation over the whole recruitment range of the MUs:

$$(11) \quad In(s) = \frac{F(s)}{k} + I_0.$$

where F(s) is the muscle force at which all MUs till Mu s are activated, k an unknown proportionality factor and I_0 the input which is necessary to bring the first MU to firing threshold. Using that $\overline{F}(t) = F(In(t))$, introducing (11) in (10) and then in (1) we get for the muscle force:

$$(12) \quad F(t) = \int_0^t h(s)\,s\,\{1 - c\,\exp[-\alpha \frac{T\,(E_{EPSP} - T)\,(F(t) - F(s))}{T\,F(t) + (E_{EPSP} - T)\,F(s) + E_{EPSP}\,I_0\,k}]\}\,ds$$

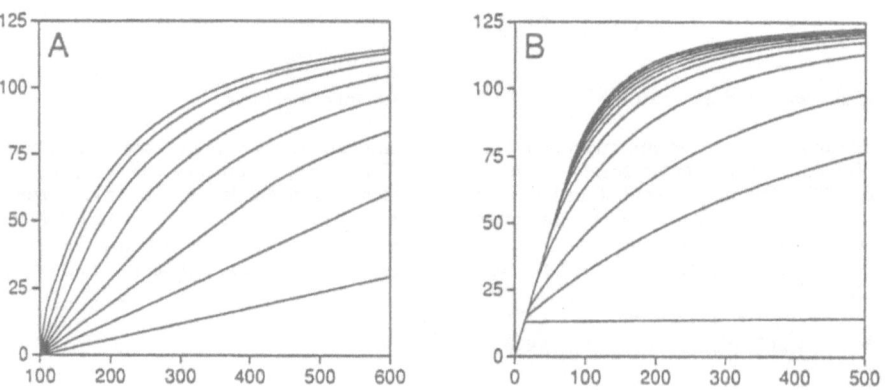

Figure 1: Muscle force as a function of input for different k I_0. Y-axis: Muscle fore in N, x-axis: Input (A) and input - I_0 (B) in arbitrary units.

Numerical Results

In equation (12) which can be integrated numerically, the parameters were initialized as follows: (a) a and b are set to 8.4 and -5.8 [4]. (b) It follows from equation (11) that c is 0.1 since the mean contraction force of a MU is 10% of its maximum force [7]. (c) It was found during the modelling process that α is independent of the MU type and equals 4.17. (d) The equilibrium potential of the EPSP E_{EPSP} was set to 60 mV in relation to the resting potential. (e) The voltage threshold T was set to 12 mV [6]. (f) It can be seen from equation (12) that the integrated muscle force does not vary independently with k and I_0 but only with the product of k and I_0. Integration was performed for different products. In Fig. 1A the results are presented in which I_0 was kept constant at 100 and the slope k was varied from 10^0 to 80^0. The relation between input to the motoneuronal pool and muscle force is linear in all curves as long as new MUs can be recruited with increasing input. As soon as all MUs are active, only their discharge rate is modulated which is governed by the intrinsic properties of the MUs. Muscle force is then approximating asymptotically its maximum.

The slope k determines MU discharge rate modulation during recruitment. With a large slope, MUs are activated without concomitant significant discharge increases. At a large angle of 80^0, the linear range is small and reaches only about 15 % of the maximum contraction force, whereas at an angle of 10^0 it reaches about 60 %. This relationship can be recognized better if I_0 varies and the slope is kept constant (Fig. 1B). In the extreme situation where I_0 is infinity, all MUs remain during recruitment at their threshold firing frequency $freq_T$ and muscle force reaches 10 % of its maximum value.

Discussion

The present results indicate that the relative contribution of recruitment and rate coding to muscle force are determined by one parameter in the model. This enables us to test the model since n peripheral small muscles, mainly recruitment occurs first and then rate coding thereas, in proximal large muscles, recruitment and rate coding occur in parallel [8]. In these different muscles, the properties of MUs during activation can be compared with those in the model with the appropriate size of the parameter I_0 k. Furthermore the linear range of the input - output relation of real can be compared with that of model muscles.

For future work, we will assume that peripheral and central input to MUs is organized similarly. This assumption will then allow us to simulate with the model experimental data about the interaction of peripheral input evoking monosynaptic H reflexes and of central input leading to voluntary muscle contractions.

References

[1]   Henneman, E., Principles governing distribution of sensory input to motor neurons. In F.O. Schmitt and F.G. Worden (Eds.) The Neurosciences: 3 rd. Study Program, MIT Press, Cambridge, 1974, pp. 281-291.

[2]   Desmedt, J.E., Size principle of motoneuron recruitment and the calibration of muscle force and speed in man. In J.E. Desmedt (Ed.) Motor Control Mechanisms in Health and Disease, Raven Press, New York, 1983, pp. 227-251.

[3]   Ruegg, D.G., Krauer, R. and Drews, H., Superposition of H reflexes on steady contractions in man, J. Physiol., 427 (1990) 1-18.

[4]   Harrison, P.J. and Taylor, A., Individual excitatory post-synaptic potentials due to muscle spindle Ia afferents in cat triceps surae motoneurones, J. Physiol., 312 (1981) 455-470.

[5]   Kernell, D., Functional properties of spinal motoneurons and gradation of muscle force. In J.E. Desmedt (Ed.) Motor Control Mechanisms in Health and Disease, Raven Press, New York, 1983, pp. 213-226.5

[6]   Pinter, M.J., Curtis, R.L. and Hosko, M.J., Voltage Threshold and Excitability Among Variously Sized Cat hindlimb Motoneurons, Journal of Neurophysiology, 50,No.3 (1983) 644-658.

[7]   Kernell, D., Rhythmic properties of motoneurones innervating muscle fibres of different speed in m. gastrocnemius medialis of the cat, Brain Research, 160 (1979) 159-162.

[8]   Kukulka, C.G. and Clamann, P., Comparison of the recruitment and discharge properties of motor units in human brachial biceps and adductor pollicis during isometric contractions, Brain Research, 219 (1981) 45-55.

The visual system
—oral contributions

CORRELATED NEURONAL ACTIVITY AND BEHAVIOR

Olaf Sporns, Giulio Tononi, and Gerald M. Edelman

The Neurosciences Institute, 1230 York Avenue, New York NY 10021, USA

Short-term correlations among neurons have been observed in different animals and brain regions. In a series of anatomically and physiologically based simulations of neuronal networks we have explored the emergence and the temporal dynamics of such correlations, and we have suggested potential functional roles for correlated neural activity in visual perception. A main focus of this paper is the role of short-term correlations in the control of behavior. Several examples of behavior that depend on the presence of a pattern of correlations among modeled neurons are discussed.

Several lines of experimental research have revealed that neurons in various parts of the brain can show temporally correlated activity at a time scale of tens to hundreds of milliseconds (**short-term correlations**). Recent studies have explored the temporal characteristics of these correlations, as well as their spatial patterns and dependence on various modes of sensory stimulation and behavior [1]. In some cases, temporally correlated neural activity can occur in the form of synchronized oscillations and can be observed over long distances in the brain [2,3]. Several models have suggested mechanisms of how neurons synchronize their activities, as well as potential roles for temporal correlations in the brain [e.g. 4,5]. In this short paper we review some results obtained with a series of models [6-10] closely based on the anatomy and physiology of the cortex. We focus on possible functional roles for short-term correlations in visual perception and in particular in the control of behavior.

Experiments [11] and modeling studies have shown that short-term correlations depend on the presence of reciprocal anatomical pathways. Such pathways are ubiquitous in the cerebral cortex [12] and it has been proposed [13,14] that they give rise to a process of **reentry**, the dynamic bidirectional exchange of neural signals along reciprocal anatomical projections. Correlations brought about by reentry can mediate **integration** of neural activity both within (**linking**) and between (**binding**) distributed cortical areas. These processes of linking and binding form important neural bases for perceptual phenomena such as grouping and figure-ground segregation. Our modeling studies demonstrate that correlated neural activity can also be used to control behavioral responses to complex sensory stimuli.

Local Correlations and Neuronal Groups

Orientation-selective neurons in the cat primary visual cortex show oscillatory discharge patterns when presented with an optimally oriented stimulus [2]. Neighboring neurons discharge in a synchronous fashion. Oscillatory neural activity has been observed in the motor cortex [15] and other cortical areas. These and other observations provide direct experimental evidence for the notion that the functional units of the cortex are **neuronal groups** [13], local collectives of several hundred to thousands of neurons that are more strongly interconnected, have similar response properties, and tend to synchronize their firing patterns. We have modeled neuronal groups as collections of excitatory and inhibitory cells in several computer simulations [6,7]. Within each modeled group, cells are sparsely and randomly connected; for example, a given excitatory cell connects to 10 percent of all other excitatory cells within the group. Generally, intra-group connections are stronger relative to those between groups. In the simulations locally correlated oscillatory activity is generated by the intra-group interactions between excitatory cells and inhibitory cells. The oscillation frequency depends critically on the temporal delay introduced by the recurrent inhibitory connections. In accordance with experimental results [16], the instantaneous frequency of the oscillations varies significantly. This is due to the fact that each neuronal group acts as a population oscillator, composed of many sparsely connected and partly independent neurons. The coherent activity of such neuronal groups is statistically more reliable than the activity of each of its constituent neurons and thus form the basis for the establishment of functionally significant correlations between distant groups.

Linking, Gestalt Laws and Short-Term Synaptic Modification

Further experiments in cat visual cortex revealed widespread patterns of correlations and provided support for the notion that these patterns depend on reentrant connectivity. When a single long light bar is moved across the receptive fields of spatially separated neurons with similar orientation specificity, cross-correlations reveal that their oscillatory responses are synchronized [3,17]. The synchronization becomes weaker if a gap is inserted into the stimulus contour and it disappears completely if two parts of the contour are moved separately and in opposite directions. Synchrony is established rapidly, often within 100 msec, and single episodes of coherency last for 50 to 500 msec. Frequency and phase of the oscillations vary within the range of 40-60 Hz and +/- 3 msec, respectively [16]. Synchronized neural activity was observed between cortical areas 17 and 18 (V1 and V2) [17,18] and between striate and extrastriate cortical areas [19]. Engel et al. [11] have shown stimulus-dependent correlations between neurons located in the two hemispheres of cat visual cortex in response to two simultaneously presented stimuli, one on each side of the visual midline. These correlations disappear when the corpus callosum is transected, a strong indication that their generation depends on the intactness of this reciprocal projection.

Temporal correlations as established by reentrant signaling may provide a key to the solution of a classical problem in visual perception, that of perceptual grouping and figure-ground segregation. These two processes, both of fundamental importance in perceptual organization, refer to the ability to group together elementary features into discrete objects and to segregate these objects from each other and from the background. At the beginning of this century, Gestalt psychologists thoroughly investigated the factors influencing grouping and the distinction between figure and ground. They described a number of perceptual laws (Gestalt laws), such as those of similarity, continuity, proximity, and common motion.

We have investigated the potential role of temporal correlations in perceptual grouping and figure-ground segregation in a model [8] which consisted of an array of orientation- and direction-selective neuronal groups forming a map of visual space. These groups are linked by a pattern of reentrant connections similar to that existing in visual cortex. An important feature of the model is that synaptic efficacies of reentrant connections can change on a short time scale (within tens of milliseconds). In general, the efficacy among correlated groups increases (with the effect of rapidly amplifying and stabilizing the correlations) and that among uncorrelated groups decreases. These changes are transient; in the absence of activity, the efficacy returns to its resting value within 100-200 milliseconds. In cat visual cortex, evidence for voltage-dependent short-term increases in the efficacy of horizontal connections has been found [20]. Fast modulation of synaptic efficacy has also been observed in other parts of the nervous system.

To test the responses of the model we exposed it to several stimuli composed of collections of light bars that resulted in perceptual grouping and figure-ground segregation in humans. Neuronal groups responding to light bars which composed an object are rapidly linked by coherent oscillations (within 60-100 msec after stimulus onset) and are segregated from other groups responding to elements of the background. We found that the ability to establish specific linking (grouping) is directly related to the ability of achieving segmentation. Accordingly, there is no coherency among groups responding to elements of the figure and others responding to elements of the background. In the model, synchronization after stimulus onset is rapid and - in accordance with perceptual data - occurs usually within 100-200 msec. Multiple coherent episodes of varying length may occur at different times in different trials. Furthermore, synchrony is transient (between 100-500 msec), and its offset is fast, as would be clearly required by the fact that the visual scene continuously changes due to eye movements. In the model, episodes of correlated activity coincide with transient enhancement of reentrant connectivity due to short-term changes in synaptic efficacy. This would suggest that the perceptual time scale of several hundred milliseconds is in part determined by voltage-dependent effects at cortical synapses.

Based on this computer model we suggest that the neural basis for the integration and segregation of elementary features into objects and background might be represented by the pattern of temporal correlations among neuronal groups mediated by reentry. In addition, since the resulting grouping and segregation are consistent with the Gestalt laws of continuity, proximity, similarity, common orientation and common motion, the model suggests that the specific pattern of reentrant connectivity in the visual cortex forms a possible neural basis for these laws.

Binding and Cortical Integration

In a more recent computer model we investigated the role of correlated neural activity within and between the multiple functionally segregated areas of the visual cortex [9]. We introduced a computational scheme that deals explicitly and efficiently with short-term temporal correlations among large numbers of units. The model receives visual input from a color camera and contains nine functionally segregated areas divided into three parallel anatomical streams for form, color, and motion; the areas are connected by reciprocal pathways. Altogether, 10,000 units are linked by about 1,000,000 connections between areas at different levels (forward and backward), areas at the same level (lateral), and within an area (intrinsic). Many of these connections are voltage-dependent, i.e. they incorporate a mechanism of short-term plasticity.

We studied two psychophysical phenomena (the perception of form from motion and motion capture) which, respectively, illustrate the constructive and correlative functions of reentry. We showed that the reentrant interactions between the motion and form streams can be used to construct responses to oriented lines from moving random dot fields (form from motion). In the model, the activity of a "higher" area of the motion stream (V5) is reentered into a "lower" area of the form stream (V1) and modifies its responses to an incoming stimulus. Motion capture involves the illusory attribution of motion signals from a moving object to another object (defined by chromatic or color boundaries) that is actually stationary. Based on the model, we propose that the basis for the perceptual effect of motion capture is the emergence of short-term correlations between units in the motion and color streams as a result of reentrant connections linking these streams.

The correlative properties of the model are explored further in simulations involving all three streams. When presented with a single object, the model solves the so-called "binding problem" and displays coherent unit activity both within and between different areas, including a non-topographic one. Two or more objects can be simultaneously differentiated. We show that coherent unit activity depends on the presence of reentrant pathways giving rise to widespread cooperative interactions among areas. Successful integration depends critically on the presence of short-term temporal correlations brought about by reentry. It does not require integration by a hierarchically superordinate area.

Correlations and the Control of Behavior

While it is generally accepted that changes in the firing frequency of neurons can influence and control behavior, it is less obvious that changing patterns of temporal correlations (even in the absence of any changes in firing frequency) can do the same. Since neurons are sensitive to temporal correlations in their inputs, the timing of activity in one set of neurons can have a well-defined effect on the activity of other neurons and can ultimately influence behavior. In order to demonstrate this point [10], we modified our grouping and segmentation model [8] by adding a single group, called the effector, which is connected uniformly to all groups within the direction-selective network. The effector is sensitive to temporal correlations in its inputs and is able to trigger a behavioral response (e.g. pressing of a lever). After training with a number of stimuli the model responds selectively to a coherent square moving in a particular direction. It does not respond to squares moving in other directions, or to moving disconnected edges. The correct behavior depends on the presence of the individual attributes of the square (relected in the pattern of active units) as well as their "connectedness" or "Gestalt" (reflected in the pattern of temporal correlations). Thus, the model shows in an elementary way how cooperative interactions leading to short-term correlations among neuronal groups may be used to integrate the relevant characteristics of the stimulus and to elicit an appropriate behavioral response. Supporting this notion are experimental results obtained from awake monkeys [21] which indicate that prefrontal neurons show correlated activity for short periods of time (250 msec) when a behavior occurs.

The cortical integration model [9] also demonstrated that integrated neural activity (i.e. patterns of correlations) can affect behavior. Successful integration is linked to an observable output, in this case a simulated foveation response. This response is used as a basis for conditioning. Foveation on the "correct" object (i.e. a red cross) will elicit a reward, mediated by activation of a saliency system that resembles diffuse projection systems in the brain, such as the monoaminergic and cholinergic systems. The simulated diffuse release of a modulatory substance from the saliency system regulates long-term synaptic changes in multiple cortical pathways. After conditioning, the model performs a behavioral discrimination of objects that requires the integration through reentry of distributed information regarding form, color, and location. This simulation links cortical integration to the immediate control of behavior as well as learning (conditioning). Short-term correlations may not only play a role in determining instantaneous behavioral responses to complex stimuli, but also in contributing to the stabilization of adaptive long-term synaptic changes.

Conclusion

Short-term correlations between neurons can be observed in many areas of the brain. Anatomically and physiologically detailed computer simulations show that intra- and inter-areal reentrant interactions can give rise to such temporal correlations. Short-term correlations can serve as a possible neural basis for several visual perceptual phenomena. They can also play a role in controlling behavioral responses, e.g. in the discrimination of complex (multi-attribute) stimuli.

Acknowledgement. This work was carried out as part of the Institute Fellows in Theoretical Neurobiology program at The Neurosciences Institute, which is supported through the Neurosciences Research Foundation. The Foundation received major support from the J.D. & C.T. MacArthur Foundation, the Lucille P. Markey Charitable Trust, and the van Ameringen Foundation. O.S. is a W.M. Keck Foundation Fellow.

[1] Basar E, Bullock TH (1992) *Induced Rhythms in the Brain.* Boston: Birkhäuser.

[2] Gray CM, Singer W (1989) Stimulus-specific neuronal oscillations in orientation columns of cat visual cortex. *Proc Natl Acad Sci USA* 86:1698-1702.

[3] Gray CM, König P, Engel AK, Singer W (1989) Oscillatory responses in cat visual cortex exhibit inter-columnar synchronization which reflects global stimulus properties. *Nature* 338:334-337.

[4] Sompolinsky H, Golomb D, Kleinfeld D (1990) Global processing of visual stimuli in a network of coupled oscillators. *Proc Natl Acad Sci USA* 87:7200-7204.

[5] von der Malsburg C, Buhmann J (1992) Sensory segmentation with coupled neural oscillators. *Biol Cybern* 67:233-242.

[6] Sporns O, Gally JA, Reeke GN Jr, Edelman GM (1989) Reentrant signaling among simulated neuronal groups leads to coherency in their oscillatory activity. *Proc Natl Acad Sci USA* 86:7265-7269.

[7] Sporns O, Tononi G, Edelman GM (1991) Dynamic interactions of neuronal groups and cortical integration. In: *Nonlinear Dynamics of Neural Networks* (Schuster HG, ed), pp 205-240. Weinheim, FRG: VCH.

[8] Sporns O, Tononi G, Edelman GM (1991) Modeling perceptual grouping and figure-ground segregation by means of active reentrant connections. *Proc Natl Acad Sci USA* 88:129-133.

[9] Tononi G, Sporns O, Edelman GM (1992) Reentry and the problem of integrating multiple cortical areas: Simulation of dynamic integration in the visual system. *Cerebral Cortex* 2:310-335.

[10] Tononi G, Sporns O, Edelman GM (1992) The problem of neural integration: induced rhythms and short-term correlations. In: *Induced rhythms in the brain* (Basar E, Bullock T, eds), pp 365-393. Boston, MA: Birkhäuser.

[11] Engel AK, König P, Kreiter AK, Singer W (1991) Interhemispheric synchronization of oscillatory neuronal responses in cat visual cortex. *Science* 252:1177-1179.

[12] Felleman DJ, Van Essen DC (1991) Distributed hierarchical processing in the primate cerebral cortex. *Cerebral Cortex* 1:1-47.

[13] Edelman GM (1978) Group selection and phasic re-entrant signalling: A theory of higher brain function. In: *The Mindful Brain* (Edelman GM, Mountcastle VB, eds), pp 51-100. Cambridge, MA: MIT Press.

[14] Edelman GM (1987) *Neural Darwinism.* New York: Basic Books.

[15] Murthy VN, Fetz EE (1992) Coherent 25- to 35-Hz oscillations in the sensorimotor cortex of awake behaving monkeys. *Proc Natl Acad Sci USA* 89:5670-5674.

[16] Gray CM, Engel AK, König P, Singer W (1992) Synchronization of oscillatory neuronal responses in cat striate cortex: Temporal properties. *Visual Neurosci* 8:337-347.

[17] Eckhorn R, Bauer R, Jordan W, Brosch M, Kruse W, Munk M, Reitboeck HJ (1988) Coherent oscillations: A mechanism of feature linking in the visual cortex? Multiple electrode and correlation analyses in the cat. *Biol Cybern* 60:121-130.

[18] Nelson JI, Salin PA, Munk MH-J, Arzi M, Bullier J (1992) Spatial and temporal coherence in cortico-cortical connections: A cross-correlation study in areas 17 and 18 in the cat. *Vis Neurosci* 9:21-37.

[19] Engel AK, Kreiter AK, König P, Singer W (1991) Synchronization of oscillatory neuronal responses between striate and extrastriate visual cortical areas of the cat. *Proc Natl Acad Sci USA* 88:6048-6052.

[20] Hirsch JA, Gilbert CD (1991) Synaptic physiology of horizontal connections in the cat's visual cortex. *J Neurosci* 11:1800-1809.

[21] Vaadia E, Ahissar E, Bergman H, Lavner Y (1991) Correlated activity of neurons: a neural code for higher brain functions? In: *Neuronal cooperativity* (Krüger J, ed), pp 249-279. Berlin: Springer.

Map Structure from Pinwheel Position

F. Wolf, K.Pawelzik, T. Geisel

Institut für Theoretische Physik and SFB 185 Nichtlineare Dynamik
Universität Frankfurt, Robert-Mayer-Str. 8, D-6000 Frankfurt/Main 11, Germany

D.-S.Kim, T.Bonhoeffer

Max Planck Institut für Hirnforschung, Deutschordenstraße 46,
D-6000 Frankfurt/Main, Fed. Rep. of Germany.

Correspondence to F. Wolf, e-mail: fred@chaos.uni-frankfurt.dbp.de

Abstract

We propose a mathematical description for the spatial organization of orientation preference in the visual cortex. In this approach the spatial pattern of orientation preference
is predicted from position and chirality of its singularities (i.e. "pinwheels"). The theory
is derived from a few phenomenological principles characterizing the qualitative structure
of orientation maps under the requirement of mathematical simplicity. A comparison
with optically recorded images of cortical maps suggests that orientation preference can
be predicted over a much larger spatial range than previously estimated on the basis of
correlation measurements.

Introduction

The precise layout of iso-orientation domains (IOD's) in the visual cortex recently became
accessible by optical imaging techniques[1, 2]. It turned out that responses to a given
stimulus orientation occur in patches which are organized in a circular manner around
centers where all orientations meet (Fig.1), the so called pinwheels[3]. The complete
sequence of orientations appears once around a single singularity, such that there are only
two kinds of pinwheels, distinguished by the two posible chiralities, and any IOD connects
two pinwheels of opposite chirality. We refer to the pinwheel centers as singularities
because in these points the preferred orientation (PO) changes discontinuously. Model
predictions have been compared to these experimental observations qualitatively or by
statistical means (e.g. correlation functions). In most cases these measures capture only
linear dependencies and tend to suggest an organization that is characterized by local
order and global disorder of the columnar systems in the visual cortex[4, 5, 6, 7, 8] by a
rapid decay of linear correlation functions.

The decay of correlation functions however does not demonstrate necessarily the absence
of global or long range order in columnar structures. More generally long range order

means the existence of rules that enable the prediction of a pattern over a spatial range considerably larger than its elementary length scale (here the extension of a cortical hypercolumn). 2-point correlation functions, as is clear from their relation to fourier analysis, are sensitive only to a periodic repetition of an elementary pattern, a form of order obviously absent in maps of PO. From a methodological point of view the prevalence of statistical measures in most studies at first sight seems appropriate because of the variability of cortical maps and the high degree of multistability exhibited by most models of map formation (for a critical discussion of this problem see e.g. [9]).

The approach employed in this contribution tackles both, the problem of variability and the question of long range order. We propose a mathematical description for the spatial organization of orientation preference on a phenomenological level. The theory focuses on the internal regularities of the pattern of preferred orientations. We exploit the formal analogy of the IOD's with electrostatic force lines, which connect point charges of different sign and equal absolute charge. Our field-analogy model (FAM) is derived from elementary principles and predicts the spatial pattern of POs given only the position and chirality of its singularities and the PO of an arbitrary point. By this the FAM reduces the variability between individual maps to their difference in pinwheel positions. Individual maps can be compared directly with the model by extracting the pinwheel data from the measured map and calculating the theoretical prediction resulting from that particular pinwheel configuration. Futhermore we show that given the pinwheel positions and chiralities, there is a way of predicting PO over a spatial range considerably larger than the correlation length of the pattern.

The Structure of Iso-Orientation Domains

The spatial pattern of orientation preference is given by a scalar function $\phi(\mathbf{x})$ assigning the preferred orientation ϕ to any cortical location $\mathbf{x}$. To separate the geometry of IODs from the assignment of POs to individual IODs, we first describe the pattern of preferred orientations by the direction field $\mathbf{f}(\mathbf{x})$ of the iso-orientation lines and in a second step assign preferred orientations to this set of predefined iso-orientation contours. $\mathbf{f}(\mathbf{x})$ is defined as the field of unit vectors that are oriented tangentially to the contours defined by $\phi(\mathbf{x})$ =const. and which by convention direct from the singularities of positive chirality to those of negative chirality. Inversely the iso-orientation contours are considered to be solutions to the differential equation

$$\frac{\partial \mathbf{x}}{\partial t} = \mathbf{f}(\mathbf{x}). \tag{1}$$

To predict $\mathbf{f}(\mathbf{x})$ from the distribution of singularities, we consider $\mathbf{f}(\mathbf{x})$ to be derived from a vector field $\mathbf{F}(\mathbf{x})$ by

$$\mathbf{f}(\mathbf{x}) = \mathbf{F}(\mathbf{x})/|\mathbf{F}(\mathbf{x})|$$

and look for the simplest choice of $\mathbf{F}(\mathbf{x})$ that is consistent with a given distribution of pinwheel positions $\mathbf{x}_i$ and chiralities q_i. Generally a vector field is determined by its sources $\nabla \mathbf{F}(\mathbf{x})$ and its density of circulation $\nabla \times \mathbf{F}(\mathbf{x})$ which are both scalar functions for a two-dimensional field. Considering the measured structures (compare Fig.1) we

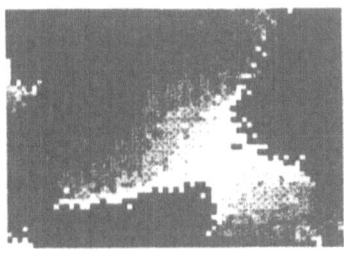 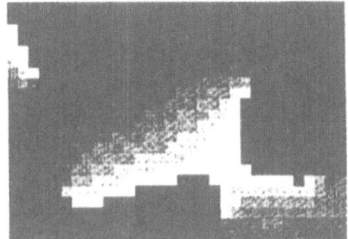

Figure 1: Section of an OPM obtained by optical imaging of intrinsic signals from A17 of cat(left) and the corresponding map cal culated according to the FAM. Prefered orientations are coded in 16 discrete greyvalues.

determine the field from four elementary principles:

(1) IODs do not form spirals, i.e.

$$\nabla \times \mathbf{F}(\mathbf{x}) = 0.$$

(2) Any IOD starts and ends at a singularity, i.e.

$$\nabla \mathbf{F}(\mathbf{x}) = \sum_i q_i \, \delta(\mathbf{x} - \mathbf{x}_i) + Q'(\mathbf{x}).$$

(3) There are no other sources besides the singularities, i.e.

$$Q'(\mathbf{x}) = 0.$$

(4) The complete sequence of orientations is represented only once in the vicinity of a single singularity, i.e.

$$q_i = \pm 1.$$

In conclusion $\mathbf{F}(\mathbf{x})$ is the rotation free vector field determined by

$$\nabla \mathbf{F}(\mathbf{x}) = \sum_i q_i \, \delta(\mathbf{x} - \mathbf{x}_i). \tag{2}$$

Equation (2) is solved by

$$\mathbf{F}(\mathbf{x}) = \sum_i q_i \, \frac{\mathbf{x} - \mathbf{x}_i}{|\mathbf{x} - \mathbf{x}_i|^2}$$

which is the mathematically simplest structure of IODs for a given distribution of singularities at $\mathbf{x}_i$ with chirality q_i which is consistent with the above principles.

Propagation of Orientation Preference

Formally there are many possibilities to assign PO's to the spatial arrangement of iso-orientation domains. The different choices of IOD differ in the way in which preferred

134

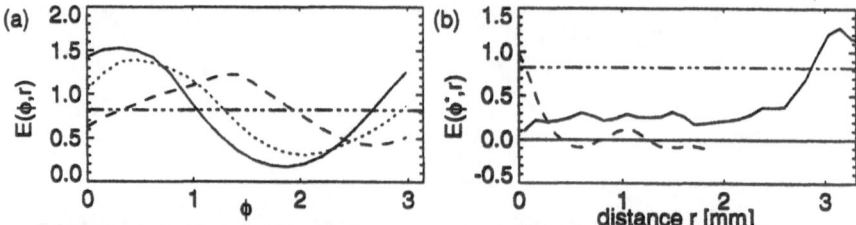

Figure 2: Quantifying the structure in a map measured in area 17 (cat) using the FAM.
a) Structure measure $E(\phi_t, r)$ depending on the test orientation ϕ_t at a single pixel for distances $r = 270\mu m$ (line), $r = 1100\mu m$ (dotte d), $r = 2700\mu m$ (broken line). Significant oscillations indicate order. b) Range of predictability for fixed test orientation ϕ_t^* and the usual c orrelation function $C(r) = <\tilde{\phi}(x)\tilde{\phi}(x+r)>_{|r|=r}$ (broken line). The error $E(\phi_t^*, r)$ (line) remains well below the the expectation for disorder $E_e = \pi^2/12$ until r b ecomes comparable to the size of the image.

orientations are mapped circularly around the singularities. For simplicity we assume isotropy of this mapping i.e. a linear dependency between the polar angle θ and the sequence of orientations ϕ

$$\phi = \phi_i + \frac{q_i}{2}\theta$$

around the center of the pinwheel which is in agreement with the experimental findings.

Determining ϕ on a single iso-orientation contour defines ϕ_i at the singularities connected by this contour. By the same mechanism the assignment of preferred orientations spreads from this singularitiy to the others in a snowball like fashion. For the actual computation of the complete OPM according to the FAM we first determine the fixed points h_j of $F(x)$, i.e. $F(h_j) = 0$. Then we connect the centers of negative chirality by the network of separatrices, which are given by the unstable manifolds of the hyperbolic fixed points h_j. Given the orientation preference at one point this procedure fixes the OP around each singularity because the OP is constant along the separatrices. $\phi(y)$ for an arbitrary point y finally can be determined by integrating $\frac{\partial x}{\partial t} = F(x)$ with $x(0) = y$ until meeting the next singularity of negative chirality. The OP under which this trajectory enters the singularity then equals $\phi(y)$. At this point it is important to note, that only the above choice of $F(x)$ guarantees the consistent assignment of preferred orientations to the force lines, i.e. the uniqueness of 'clocks' at the pinwheels.

Range of Predictability

The FAM implies infinite long range order. This becomes particularly clear from the fact that for a given pinwheel distribution a fit of the OP ϕ_t at a single and arbitrary point x_t fixes the whole map. In order to quantitatively test whether such long range order is indeed present in OPM we estimate the structure measure $E(\phi, r)$, which is the squared difference between predicted and measured OP's averaged over pixels at distance r from the test point x_t. In a first application to an image from area 17 of the cat we find excellent agreement (Fig. 2) except for deviations which occur at the borders of the image, an effect most probably due to the inadequacy of the periodic boundary conditions used.

Summary

We presented a mathematical characterization of cortical orientation maps. Given the observed chiralities and positions of the pinwheels, the predicted OPM is determined by the orientation preference at a single cortical location, which implies strong spatial order. The comparison with experimental observations from area 17 of the cat indicates that the pattern of orientation preference can indeed be predicted over a larger spatial range than previously estimated on the basis of correlation functions.

Bibliography

[1] Grinvald, A., Lieke, E., Frostig, R.D., Gilbert, C.D. & Wiesel, T.N., Functional architecture of cortex revealed by optical imaging of intrinsic signals, Nature 324,361-364 (1986).

[2] Blasdel, G.G., Salama, G., Voltage-sensitive dyes reveal a modular organization in monkey striate cortex, Nature 321,579-585 (1986).

[3] Bonhoeffer, T., Grinvald, A., Iso-orientation domains in cat visual cortex are arranged in pinwheel-like patterns, Nature 343,429-431 (1991).

[4] Obermayer, K., Schulten, K., & Blasdel, G.G., A comparison between a neural network model for the formation of brain maps and experimental data, in *Neural Information Processing Systems 4*, eds. Moody, E.M., Hanson, S.J. & Lippmann, R.P., (Morgan Kaufmann, 1992).

[5] Obermayer, K., Blasdel, G.G., & Schulten, K., Statistical-mechanical analysis of self-organization and pattern formation during the development of visual maps, Phys. Rev. A 45,7568-7589 (1992).

[6] Niebur, E., Wörgötter, F., Orientation columns from first principles, in *Computation and Neural Systems*, eds. Bower, J., Eeckmann, F., (Kluwer, 1993).

[7] Swindale, N.V., Matsubara, J. & Cynader, M.S., Surface organization of orientation and direction selectivity in cat area 18, J. Neurosci.7,1414-1427 (1987).

[8] Swindale, N.V., Cynader, M.S. & Matsubara, J., Cortical cartography: A two-dimensional view, in *Computational Neuroscience*, ed. Schwartz, E.L., (MIT-Press, Cambridge, 1990).

[9] Jones, D.G., Van Sluyters, R.C. & Murphy, K.M., A computational model for the overall pattern of ocular dominance, J. Neurosci.12,3794-3808 (1991).

Emergence of Transient Oscillations in an Ensemble of Neurons

H.-U. Bauer, K. Pawelzik, T. Geisel

Institut für Theoretische Physik and SFB Nichtlineare Dynamik
Universität Frankfurt, W-6000 Frankfurt/Main 11, Germany

Abstract

An interesting property of oscillatory responses in various areas of cat and monkey visual cortex as well as monkey sensorimotor cortex is that the local oscillations and spatial correlations do not persist during the whole stimulation. Instead they appear to switch between an oscillating/synchronized/regular state and a stochastic/uncorrelated/irregular state. We present a simple model for a local ensemble of neurons, which exhibits bistability of two such states, and switching between them. The bistability in the model is linked to a nonlinear lateral interaction function within the ensemble. In a spatially extended version of the model we also observe transient spatial correlations, which arise from a concurrent switching of several ensembles from the stochastic to the oscillatory state. Our model not only reproduces the high variability of the observed neuronal dynamics, but also has the potential for a very fast spatial synchronization due to simultanous switching. This latter aspect is important for an assessment of the possible role of oscillations and synchronization in vision algorithms.

1. Introduction

The experimental investigation of cortical 40-Hz oscillations and neuronal synchronization [1-4], their modeling and mathematical analysis [5-9], and the discussion of their possible role for feature binding [10-12] have been a major research topic in the last years. Numerous models linked the occurence of the oscillations to particular local neuronal circuitry, often involving inhibitory feedback loops. Other investigations focused more on the spatial synchronization aspect, discussing the dependence of synchronization on delays, for example [13]. A third avenue of research was more concerned with further analysis of the experimental data [14]. As a result of this latter work, the non-stationarity of the oscillations is becoming more and more prominent. In particular, an analysis of the data in terms of the time-resolved predictability [15] and in terms of a Hidden-State model [16-18], involving stochastic and oscillatory states, suggests, that the underlying local dynamics in the data is best characterized by a switching between an oscillatory state and a stochastic state. This alternating nature of the dynamics introduces a new challenge for models.

In this contribution, we present a mathematical model for the dynamics of a local neuronal ensemble, which can exhibit oscillatory and stochastic states upon stimulation. Since both of them can be simultanously attractive, the system is bistable. Under the influence of external noise, the ensemble can switch from one state to the other. This yields an overall dynamics which resembles closely the observed local dynamics.

We would like to add a few remarks on the functional aspects of our model. If neuronal populations exhibit a bistability, spatial synchronization must not be a process involving the decrease of phase differences between oscillating elements, but instead a simultaneous switching of several elements from the stochastic to the oscillatory state. Since in this way all the elements start out with the correct phase relation, the latter process can be very fast, whereas the former may take very long. This aspect of the alternation-picture is important for the discussion of the functional role of 40 Hz oscillations in vision, since the slow synchronization processes discussed so far seem to be incompatible with the fast figure-ground segmentation or Gestalt-perception in animals and humans. First simulations of a spatially extended version of our model support these arguments.

2. Model and Results

Let us now turn to the description of our bistable neuronal model. The description is reduced to the essentials, a longer version is given in a forthcoming paper [19]. We start by considering the threshold function $\Theta(\phi)$ of a single neuron after a spike,

$$\Theta(\phi) = \begin{cases} \infty & \text{if } \phi < t_{refr} \\ \Theta_0 \exp\left((\phi - t_{refr})/\tau_{refr}\right) & \text{if } \phi > t_{refr} \end{cases} \tag{1}$$

The threshold function is parametrized with the time ϕ after the last spike, which we will call phase of a neuron. Using $\Theta(\phi)$, an external input $I_{ext}(t)$ and a lateral input $I_{int}(t)$ due to corticocortical excitation within the subpopulation, we can then define a firing probability $p_f(\phi, I_{ext}(t), I_{int}(t))$ for the neuron to fire, provided its last spike was a time ϕ ago,

$$p_f(\phi, t) = \text{sigm}\left(I_{ext}(t) + I_{int}(t) - \Theta(\phi)\right). \tag{2}$$

In the next step we consider an ensemble of such individual neurons. The ensemble is characterized by the distribution $\tilde{\rho}(\phi, t)$ of internal phases ϕ (Fig. 1).

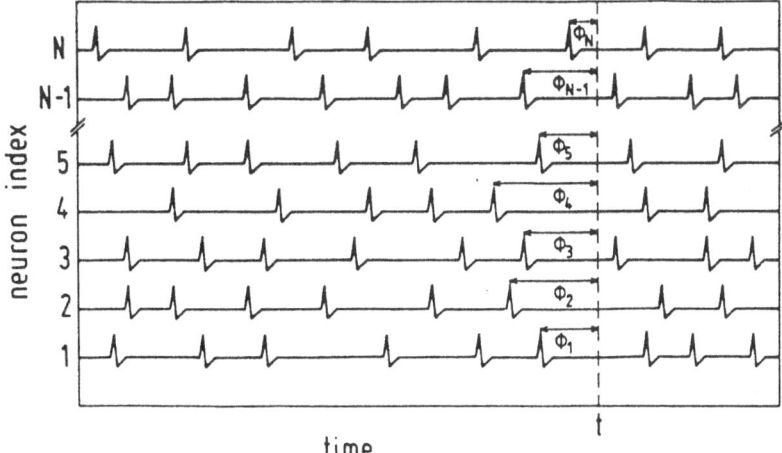

Fig. 1 Spike trains of neurons in an ensemble. Each neuron i is characterized by the internal phase $\phi_i(t)$ (time since last spike), the ensemble is characterized by the distribution $\rho(\phi, t)$ of the internal phases.

Discretizing the internal phases ϕ and the "external" time t, we transform $\tilde{p}(\phi, t)$ to $\vec{p}(j)$, a vector whose components i give the density of phases $\phi = i\Delta t$ at time $t = j\Delta t$. The vector has $T + 1$ components, with T sufficiently large such that the threshold function $\Theta(T)$, and consequently $p_f(T)$, do not change anymore. This vector evolves in time according to

$$
\vec{p}(j+1) = \begin{pmatrix} 0 & p_f(1,j) & p_f(2,j) & \cdots & p_f(T-1,j) & p_f(T,j) \\ 1 & 0 & 0 & & & \\ 0 & 1-p_f(1,j) & 0 & \cdots & & 0 \\ & & 1-p_f(2,j) & \ddots & & \vdots \\ & \vdots & & & 0 & 0 \\ & 0 & & \cdots & 1-p_f(T-1,j) & 1-p_f(T,j) \end{pmatrix} \vec{p}(j), \qquad (3)
$$

with the matrix incorporating the effects of firing (reset) via the firing probability $p_f(i,j)$.

It remains to define the lateral interaction in the subpopulation. Clearly only the firing fraction of the neurons interact, therefore we have

$$
I_{int} = w_{int} g(\rho_0), \qquad (4)
$$

with a cortico-cortical interaction function g. g could be linear, corresponding to simple "linear summation" neurons, but it could also contain a nonlinearity, reflecting either nonlinear channel properties like an NMDA-receptor synapse or other nonlinearities in the dendritic tree.

Iterating the dynamics (3) we find, that the distribution can evolve in two distinct ways, depending on the lateral interaction function g and the initialization. First $\vec{p}$ can relax to a fixed point, with a constant fraction of the neurons firing at a particular time. On the level of an individual neuron, this corresponds to a stochastic firing characteristic, in the experimental data this state corresponds to the irregular periods. Second the distribution can evolve according to a limit cycle, with a rather large fraction of the neurons in a narrow band of phases (Fig. 2).

We find parameter combination w_{ext}, w_{int} where both states coexist, i.e. we find bistability, if the interaction function is nonlinear, like $g(x) \propto x^2$ or $g(x) \propto x^4$. As was mentioned above, this does not seem to be an physiologically unreasonable assumption. The relation between nonlinearity and bistability can be analyzed in terms of a one dimensional map [19].

In the bistable regime some initializations of $\vec{p}$ lead to the oscillatory state, others to the stochastic state. Adding some noise to the external input the system can also switch between the two (Fig. 3). The transition probability $P(stoch \rightarrow osc)$ and $P(osc \rightarrow stoch)$ for these transitions can be determined from the time series by thresholding ρ_0. $P(stoch \rightarrow osc)$ is found to increase with external stimulation I_{ext}, the reverse probability $P(osc \rightarrow stoch)$ decreases (Fig. 4). These dependencies of the transition probabilities on stimulus parameters constitute a prediction of our model, which should be testable by analyzing the transitions probabilities in oscillation time series measured for varying stimulus parameters like bar length or difference between the actual and the optimal orientation of the bar. However, at this point we can already use an indirect measure for changes in the data due to changes in the stimulus parameters, the autocorrelation function. The increase of $p(stoch \rightarrow osc)$ and decrease of

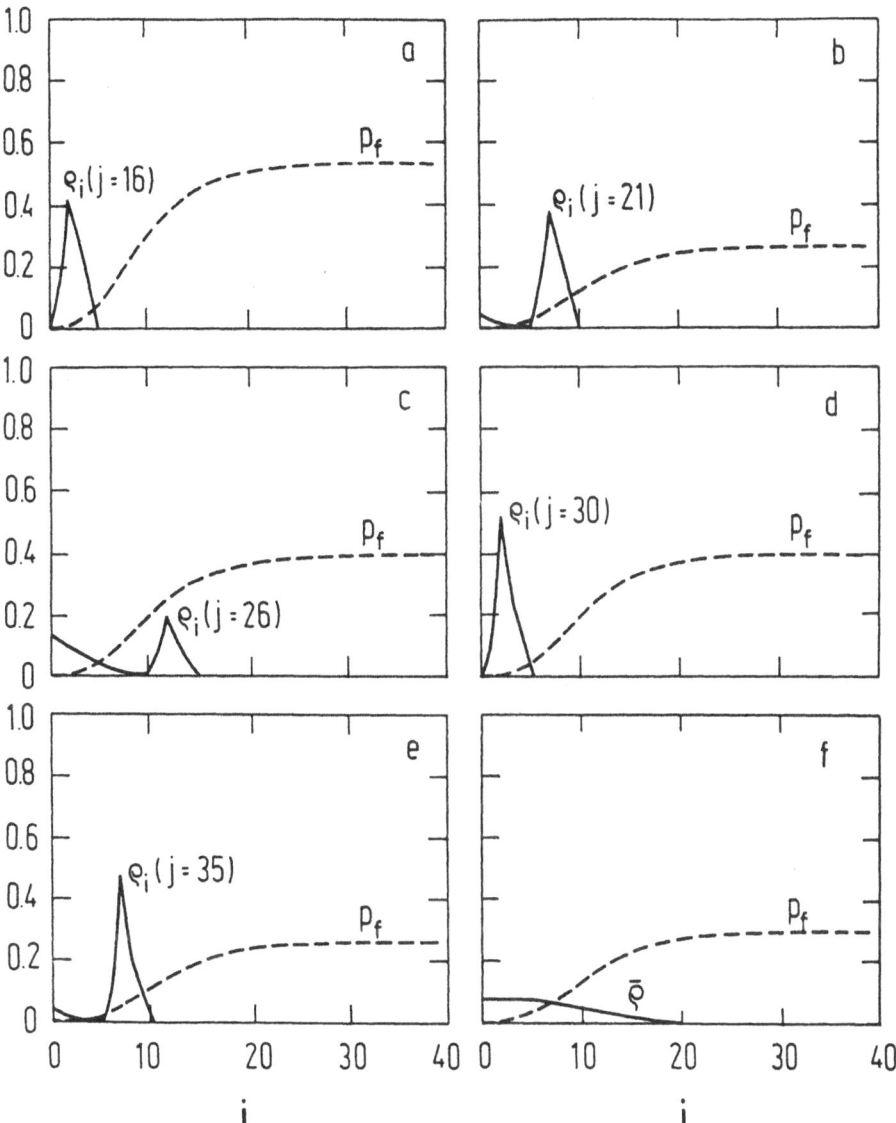

Fig. 2 In each Fig. a-f the distribution $\rho_i(j)$ is depicted as a function of the internal phase i at a constant j, together with the corresponding firing probability $p_f(i,j)$. Figs. 2a-e show a sequence of distributions which constitute the oscillatory state. Fig. 2f shows the fixed point $\bar{\rho}$ of the distribution. Parameters of the simulation were: $g(x) = x^4$, $w_{int} = 4200$, $I_{ext} = -1$, $\Theta_0 = 0$, $\Theta_1 = 5e^{2/5}$, $\tau_{refr} = 5$, $t_{refr} = 2$. The dynamics in a-e and in f differ only due to the different initilization.

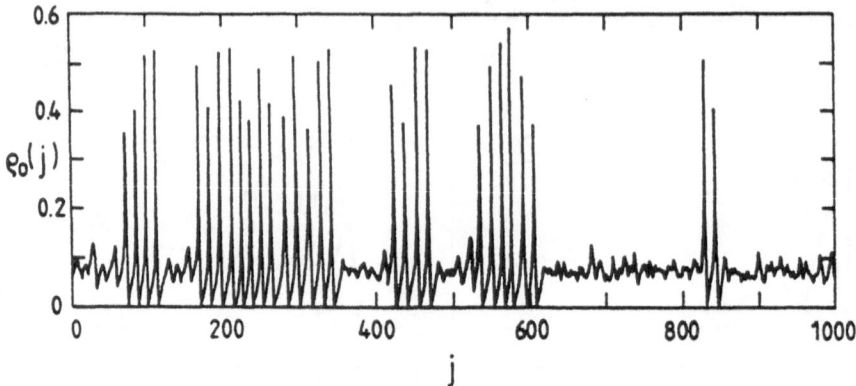

Fig. 3 Time series of the ensemble activity $\rho_0(t)$ for a system in the bistable regime, under the additional influence of external noise. The switching between the oscillatory and stochastic state is clearly visible.

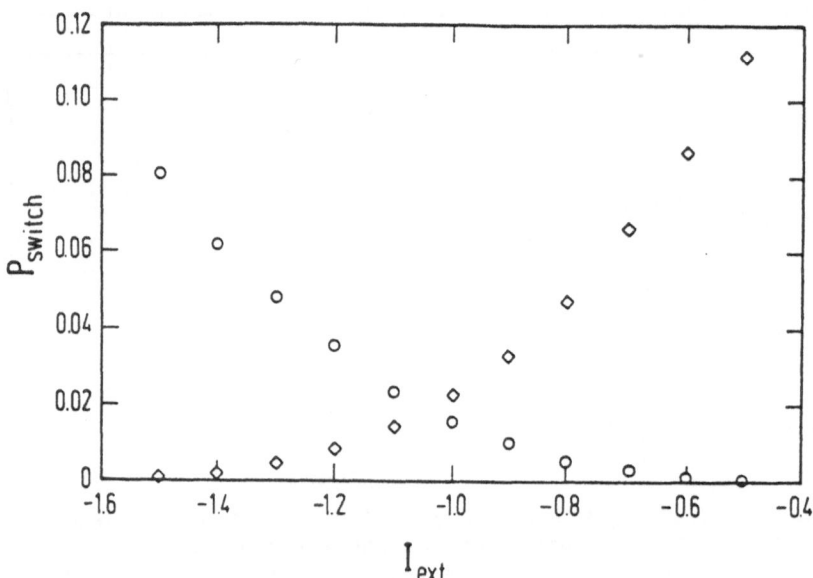

Fig. 4 Transition probabilities $p(o \rightarrow s)$ (circles) and $p(s \rightarrow o)$ (squares), as a function of the external input I_{ext}.

$p(stoch \rightarrow osc)$ with I_{ext} would result in more and longer oscillatory periods, which in turn would yield more pronounced modulations of the $\rho_0(t)$ autocorrelation function. On the other hand, such increases of the autocorrelation modulation amplitude have been observed, e.g. for binocular vs. monocular stimulation, or for variations of the stimulus orienation [20].

In this way our simple model captures not only the switching phenomenon but also other details inherent to the experimental data.

3. References

1. Gray C.M., König P., Engel A.K., and Singer W., Nature 338, pp. 334-337 (1989).

2. Eckhorn R., Bauer R., Jordan W., Brosch M., Kruse W., Munk M., and Reitboeck H.J., Biol. Cyb. 60, pp121-130 (1988).

3. Kreiter A.K., Singer W., Eur. J. Neurosci. 4, 369 (1992).

4. Murthy V.N., Fetz E.E., Proc. Nat. Acad. Sci. USA 89, 5670 (1990).

5. Schuster H.G., Wagner P., Biol. Cyb. 64, 77 and 83 (1990).

6. Niebur E., Kammen D., Koch C., in: Nonlinear Dynamics and Neuronal Networks, ed. H.-G. Schuster, VCH Weinheim, p. 173

7. Schillen T., König P., Neural Comp. 3, 2 155 and 167 (1991).

8. Bush P.C., Douglas R.J., Neural Comp. 3, 1 pp. 19-30 (1991)

9. Koch C., Schuster H.G., Neural Comp. 4, 211 (1992).

10. v.d.Malsburg C., *The correlation theory of brain function* Internal Report 81-2, Max-Planck-Institute for Biophysical Chemistry, Göttingen, F.R.G. (1981).

11. Shastri L., Ajjanagadde V., Tech. Rep. Dept. of Computer and Information Science, Univ. of Penn., Philadelphia, PA 19104 (1990)

12. Horn D., Sagi D., Usher M., Neural Comp. 3, 510 (1991).

13. Nischwitz, A., Glünder H., Oertzen, A., Klausner, P., in Aleksander, I., and Taylor, J. (eds.), Artificial Neural Networks II (Proc. ICANN 92), Elsevier, 851 (1992).

14. Gray C.M., Engel A.K., König P., Singer W., Vis. Neurosci. 8,337 (1992).

15. Pawelzik K., Bauer H.-U., Geisel T. *Switching between predictable and unpredictable states in data from cat visual cortex*, Proc. CNS 1992, Kluwer Academics, in print. Morgan Kauffman, in print.

16. Pawelzik K., Deppisch J., Bauer H.-U., Geisel T., Proc. NIPS 5 (1993).

17. J. Deppisch, K. Pawelzik, T. Geisel, this volume (1993).

18. J. Deppisch, K. Pawelzik, T. Geisel, submitted to Biol. Cyb. (1993).

19. Bauer H.-U., Pawelzik K., submitted to Physica D (1993).

20. Gray C.M., König P., Engel A.K., Singer W., Eur. J. Neurosci. 2, 607 (1990).

A Distributed Multicolumnar System for Primary Cortical Analysis of Real-World Scenes*

T. Pomierski, H. M. Gross, D. Wendt

Technical University of Ilmenau
Div. of Neuroinformatics
D-6300 Ilmenau, P.O.Box 327, Germany
e-mail: pomi@informatik.tu-ilmenau.de

Abstract

In this paper we present a distributed multicolumnar system using an intracolumnar principal component analysis (PCA) for a topology preserving mapping of real-world grey level distributions within a two-dimensional intercolumnar Kohonen Feature Map. A two-stage principal component analysis within each processing column allows a similarity preserving description with only a few highly effective fitting parameters suited for a local translation invariant processing.

1. Introduction

Cognitive abilities required for analysis or interpretation of real-world scenes can not be explained by pattern matching completely in parallel. Instead of this a controlled decomposition of a highly parallel and complex visual scene into a sequence of lower dimensional components (meaningful pieces) is more probable. Hereby both preattentive or data-driven and attentive vision mechanisms based on internal system knowledge are of decisive importance for controlling this decomposition process [1]. The regions of a visual scene that are of high interest for an active vision system because of their syntactical meaning (preattentive scene analysis) or because of data-driven activated internal hypothesis about the scene or single components (attentive or knowledge-based vision) will be referred to this paper as internal *regions of attention*. Central point of this paper is to propose a neural network architecture for distributed parallel analysis of internal *regions of attention* selected within a visual scene by the special attention mechanisms mentioned above. In the presented model concept the distributed analysis is realized by an array of cortical processing columns having a structural and functional similarity with the minicolumns localized at the primary visual cortex [7]. This columnar array analyzes parallel the *regions of attention* of the real-world input scene selected by preattentive or attentive mechanisms which are not subject of this paper. Each processing column is able to detect essential input features within the corresponding *analyzing field* inside the *region of attention* on the basis of complex receptive fields. These receptive fields can be structured unsupervised by an adaptive self-organizing process based on a neural motivated principal component analysis (PCA). Oja [3] demonstrated that a particular version of the Hebb rule leads to a synaptic weight vector that has a strong similarity with the principal component of the set of input vectors. As extension according to Sanger [5] the self-organization of principal components corresponding to the largest eigenvalues for input data sets selected randomly from various real-world scenes leads to sets of principal component arrays or receptive fields nearly identical in quality (see Fig. 1). These results lead us to the supposition, that the same sets of complex receptive fields for local input analysis are available to each cortical column independent of its localization within the visual cortex as well as within our simple model architecture. The focus of this paper is to present a new extended neural network approach that enables a local topology preserving principal component analysis within each processing column that guarantees a locally translation invariant description of the local *field of analysis* with the lowest rate of describing parameters. Based on all fitting parameters in all processing columns a mapping into a two-dimensional intercolumnar organized Kohonen Feature Map becomes possible.

*Supported by Ministry of Research and Technology (BMFT), Grant No. 413-5839-01 IN 101D - NAMOS-Project

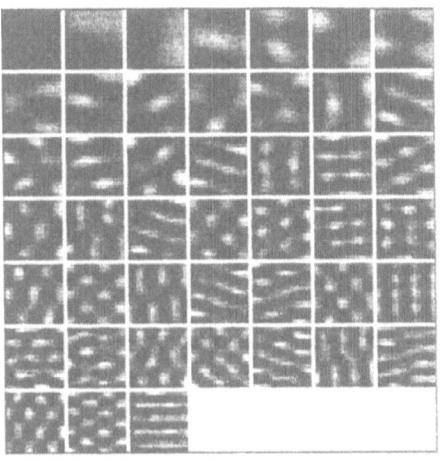

Figure 1: *(left) Self-organized receptive fields (principal component arrays)* $\mathbf{w}^{(1)}, ..., \mathbf{w}^{(i)}, ..., \mathbf{w}^{(n)}$ *structured in an unsupervised learning process by randomly selected input samples out of the shown real-world scene (right). This set of receptive fields is implemented within each processing column of the columnar array (for further explanation see text).*

2. Principal Component Analysis for Self-Organization of Receptive Fields

In the training period numerous images were obtained by recording real-world scenes and discretizing them to 512 pixel square images with 256 grey levels. No attempt was made to correct any optical irregularities. Local training patterns of 16 x 16 pixels were obtained by choosing an area within the image at random. Each training vector was normalized only to interval $<0, 1>$. This was possible because of the conformity of mean point vector $\mathbf{w}^{(0)}$ and first eigenvector $\mathbf{w}^{(1)}$ of real-world scene data distributions (see Fig. 3). So it is justifiable to indicate real-world scenes as ones without mean value. This is the prerequisite for adaptive unsupervised learning without a-priori knowledge about the data distribution of real-world scenes. For our simulations we used a single-unit rule like that proposed by Oja [3] in 1982.

$$w_j^{(1)}(t+1) = w_j^{(1)}(t) + \gamma(t)\, y(t)\, [x_j(t) - y(t)\, w_j^{(1)}(t)] \tag{1}$$

Here $\mathbf{x} = (x_1, ..., x_j, ..., x_m)^T$ denotes the real valued input normalized to the interval $<0, 1>$, y is the output signal, $w_j^{(1)}$ is the weight from input unit x_j to the single output unit y, and γ is the learning rate. This rule can be shown to produce a weight vector $\mathbf{w}^{(1)}$ corresponding to that eigenvector of the correlation matrix of all the inputs $\mathbf{x}$ which has a maximal eigenvalue. It extracts the principal component of the input data. The weight vector also tends to unit length. This rule was extended to i output units extracting the principal components in ascending sequence [5]. The learning process consists of two steps for each neuron i. First all parts of the input vector $\mathbf{x}^{(i-1)}$ already representable by weight vector $\mathbf{w}^{(i)}$ are subtracted.

$$\mathbf{x}^{(i)}(t) = \mathbf{x}^{(i-1)}(t) - y_i(t)\mathbf{w}^{(i)}(t),\ \mathbf{x}^{(0)} = \mathbf{x},\ i = 1, 2, ..., n \tag{2}$$

Afterwards the adaptation of all components $w_j^{(i)}$ of weight vector $\mathbf{w}^{(i)}$ takes place.

$$w_j^{(i)}(t+1) = w_j^{(i)}(t) + \gamma_i(t)\, y_i(t)\, [x_j^{(i)}(t) - y_i(t)\, w_j^{(i)}(t)],\ i = 1, 2, ..., n \tag{3}$$

The number of output units was determined by experiments showing that in context of our desired application 45 principal components are sufficient for extracting the essential information out of

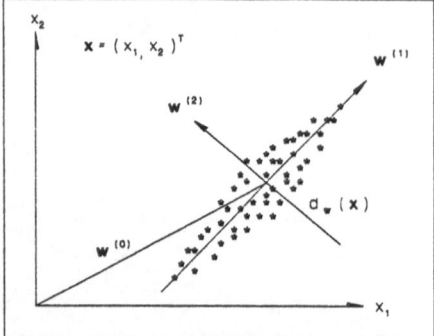

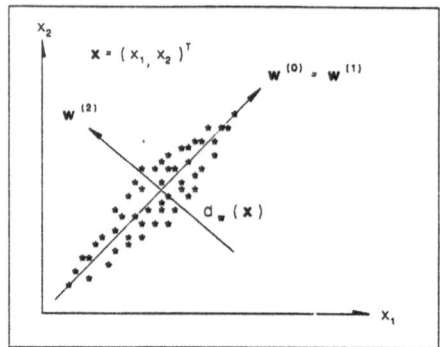

Figure 2: *Principal depiction of the possible correlation of two variables x_1 and x_2. The mean of the data distribution is not equal to zero. Accordingly, first it is necessary to extract the mean before determining a more favorable system of perpendicular coordinates.*

Figure 3: *Principal depiction of randomly selected pairs of grey-valued pixels x_1 and x_2 situated side by side within the visual scene in Fig. 1. The mean point vector $\mathbf{w}^{(0)}$ of the data distribution and the eigenvector corresponding to the largest eigenvalue $\mathbf{w}^{(1)}$ are identical here (for more explanations see text).*

16×16 square samples taken from real-world scenes (17% of all possible eigenvectors). The net was initialized by setting all the weights $w_j^{(i)}$ to small random values so that the square sum of the components of each weight vector $\mathbf{w}^{(i)}$ was approximately unity. Training consisted of applying randomly selected inputs and updating the weights. We used a total of 50.000 presentations. Because of the output feed-back to the input neurons:

$$\mathbf{x}^{(i)} = \mathbf{x}^{(i-1)} - y_i\mathbf{w}^{(i)}, \; i = 1, 2, ..., n \tag{4}$$

any weight modification of neuron i has consequences for neuron $i+1$. So it is necessary to give the weight vector $\mathbf{w}^{(i)}$ of neuron i the chance to stabilize before weight vector $\mathbf{w}^{(i+1)}$ of neuron $i+1$ gets able to learn effectively (see Fig. 4). That is the reason why convergence of weights is assisted by gradually reducing the learning rate in following way:

$$\gamma_0(t) = \frac{a}{t+1} \tag{5}$$

$$\gamma_i(t) = \gamma_{i-1}(t) \, v, \; i = 1, 2, ..., n \tag{6}$$

Here $\gamma = (\gamma_1, ..., \gamma_i, ..., \gamma_n)^T$ is the learning rate for the neurons $\mathbf{n} = (n_1, ..., n_i, ..., n_n)^T$, v the threshold parameter, t the learning step and a the starting value. This leads to a weight stabilization of all 45 neurons after 50.000 learning steps (see Fig. 1). This principal component analysis evoked by visual stimulation of real world and resulting in forming complex receptive fields is comparable in the broadest sense with the development of specialized receptive fields in the kitten visual cortex [6].

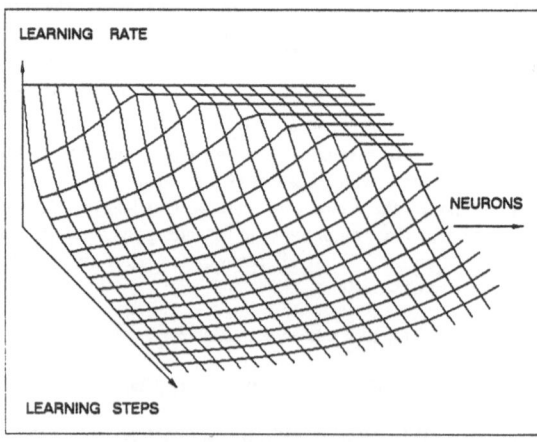

Figure 4: *Gradually reducing learning rate for neurons i. The mathematical description of the time varying and neuron dependent learning rate is presented in text.*

The ability of the self-organized receptive field set to extract nearly complete the information of any real-world scene is shown in Fig. 5. To code the images, each scene was segmented in $m \times m$ parts of $n \times n$ pixels without overlap. Each segment in the scene was analyzed by the complete receptive field set learned on the basis of a completely different visual scene (Fig. 1, right). In this way for each segment 45 fitting values for coding are determined. Fig. 5 shows the two scenes after reconstruction from the $m \times m \times 45$ fitting values.

Figure 5: *Two views of a printer are shown after reconstruction from fitting parameters derived from the set of receptive fields shown in Fig. 1. These principal components have been self-organized on the basis of the real-world scene shown at the right hand side of Fig. 1. Both reconstruction test scenes shown here have been first segmented and analyzed by the receptive fields of Fig. 1. and then reconstructed from 45 fitting values for each segment. It is not possible to find essential differences between the original and the reconstructed scenes, so it is justified to present only the reconstruction results.*

3. Intracolumnar Processing of Local Receptive Fields

In our model architecture (see Fig. 6) each processing column analyzes a definite local domain of the focus region, the so called *region of analysis*, by complex receptive fields self-organized by the proposed neural learning mechanism. These *regions of analysis* enclose in our model 32×32 square pixels for each column. Because of this field dimension 16×16 processing columns analyze parallel a *region of attention* of 152×152 pixels, as the overlapping between regions of analysis has been chosen to 75%. Presupposing grey-level distributions of the real-world scenes as additive superposition of periodic process parts, a convolution of the self-organized receptive fields (16×16 pixels) with the *regions of analysis* (32×32 pixels) is in accordance with an analysis of the period determined by the receptive field of first order $\mathbf{w}^{(1)}$. The motivation for this convolution is the following: There were no internal relations detectable between fitting vectors for *analyzing regions* shifted inside the scene by only one pixel. The convolution presents a way to escape from this dilemma. By equalizing local translations similar fitting vectors for similar grey-level distributions are generated at the same time. Naturally, this is connected with a loss of precision, and in addition, this represents an irreversible process of analysis. Each of the 16×16 columns determines the average of activation for all neurons with the same receptive fields situated within its *region of analysis*.

$$f_i = \frac{1}{N^2} \sum_{k=0}^{N-1} \sum_{l=0}^{N-1} \| \mathbf{w}^{(i)} \mathbf{x}_{kl} \|, \ i = 1, 2, ..., 45 \tag{7}$$

In this way for each column an averaged 45-dimensional vector **f** of fitting values can be determined, which describes the mean conformity of the complex receptive fields with their *regions of analysis*. Here $\mathbf{f} = (f_1, ..., f_i, ..., f_{45})^T$ denotes the averaged fitting vector, $\mathbf{w}^{(1)}, ..., \mathbf{w}^{(i)}, ..., \mathbf{w}^{(45)}$ denote the two-dimensional arrays of principal components (complex receptive fields), and **x** is the real-valued input of the *analyzing region*. The 1024 grey values of one *region of analysis* are reduced to 45 real fitting values, this is a compression rate of 95%. The first component of the fitting vector f_1 describing the steady component was not used further for analyzing the features independent of the mean grey-level. A simple clustering of the 16 × 16 fitting vectors determined in this way for different *regions of attention* was showing a conformity with the human visual feeling. Several tests with different *regions of analysis* showed a generally exponential decay within the fitting vectors. This was a good prerequisite for an additional principal component transformation of the (45 − 1) fitting values to 2 (x and y) per processing column describing the *region of analysis* sufficiently. A simple cluster analysis of this two-component fitting vectors showed a still sufficient description of the *analyzing regions*. Based on these results each column is transmitting its fitting values x and y into a two-dimensional intercolumnar Kohonen Feature Map [2]. All columnar activations within a *focus of attention* (as mentioned above we implemented 16 × 16 processing columns per focus) contribute to activation distributions within the two-dimensional feature map. This map could be understood as an alphabet of all possible local grey-level distributions within real-world scenes. In

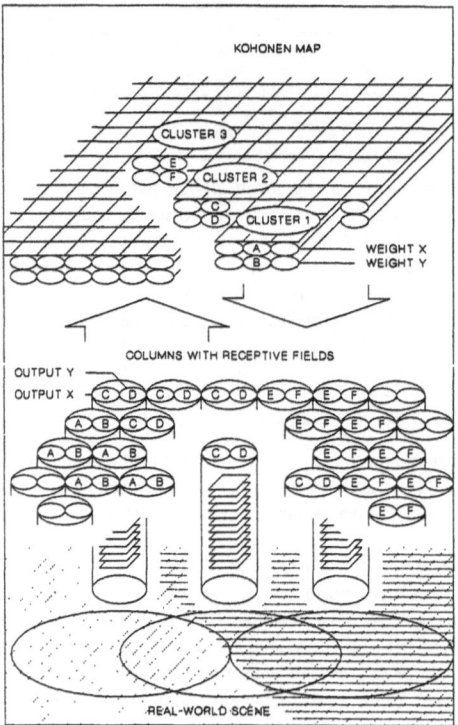

result of internal competition and selection processes within the Kohonen Map only one cluster can be activated in a particular region of the map for a certain time whereas the others are suppressed until this cluster is deactivated. Using a topographical correct reciprocal projection from the map to the multicolumnar system that columns contributing to the just winning cluster are activated sequentially (see Fig. 7). The neural implementation of the analyzing and the Kohonen-based selection subsystems is the subject of our present work.

Figure 6: *The distributed multicolumnar system for primary cortical analysis, consisting of a subsystem for the two-phase principal component analysis (see text) and a subsystem for sequential selection of topological adjacent columnar analyzing results. AB and EF are columnar activations (analyzing results) describing completely different grey-level distributions (for instance textures) in a complex real-world scene. The activation CD is indicating a border-region of overlapping visual structures AB and EF. This cluster is localized within the Kohonen Feature Map between neuron populations sensitive for the features AB and EF.*

4. Results and Conclusions

By using such a two-stage principal component analysis (PCA) it is possible to describe local grey level segments of real-world scenes by only two fitting parameters (x, y) sufficiently. It became possible to realize a topology preserving mapping within an intercolumnar feature map. An analyzing region overlap of 75% per column leads to very compact cluster-formed activations within the feature

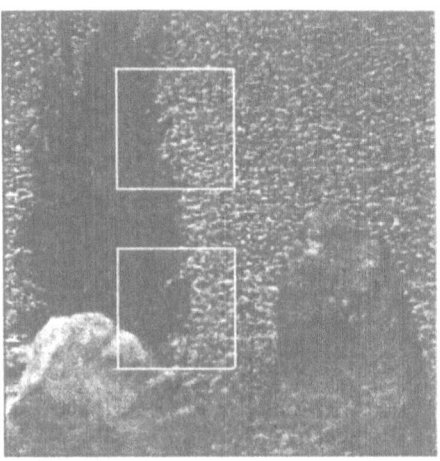 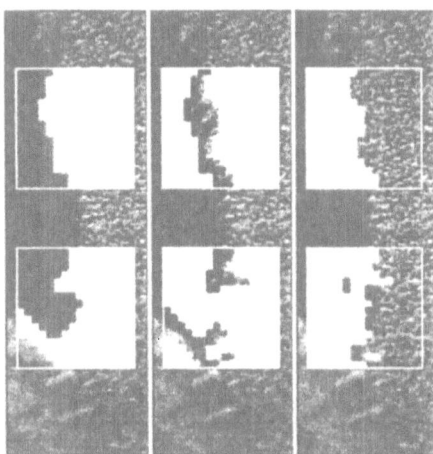

Figure 7: *(left) Real-world input scene. White frames show different regions of attention, that are processed by the distributed multicolumnar system. (right) Temporal developing of segmentation decisions within the regions of attention, based on the two-phase principal component analysis and the sequential selection in the Kohonen Feature Map. Similar regions of the attentional focus are selected and activated coherently time after time.*

map evoked by grey level distributions equal in quality inside the *region of attention*. The columnar architecture proposed in this paper has been tested at different real-world scenes. A typical result is illustrated in Fig. 7. It is interesting to establish that texture crossings independent of their direction are assigned here to one common cluster of mixed textures. In our further work this model architecture will be implemented for the analysis of coloured real-world scenes, organized in special colour difference spaces like that proposed in [1].

References

[1] **Gross, H. M., Koerner, E., Boehme, H. J., Pomierski, T.** A Neural Network Hierarchy for Data and Knowledge Controlled Selective Visual Attention. In Proceedings of ICANN92, vol. 1, pp. 825 - 828, Brighton, 1992.

[2] **Kohonen, T.** Self-Organization and Associative Memory, 3rd Edition, Springer Verlag, Berlin, 1989.

[3] **Oja, E.** A Simplified Neuron Model as Principal Component Analyzer. In J. Math. Biol., vol. 15, pp. 267 - 273, 1982.

[4] **Ritter, H., Schulten, K., Martinetz, T.** Neuronale Netze, 2. erweiterte Auflage, Addison-Wesley, Bonn, 1991.

[5] **Sanger, T. D.** Optimal unsupervised learning in a single-layer linear feedforward neural network. In Neural Networks, vol. 2, pp. 459 - 473, 1981.

[6] **Singer, W., Rauschecker, J. P.** The Effects of Early Visual Experience on the Cat's Visual Cortex and their Possible Explanation by Hebb Synapses. In J. Physiol., vol. 310, pp. 215 - 239, 1981.

[7] **Szenthagothai, J.** Theorien zur Organisation und Funktion des Gehirns. In Naturwissenschaften, pp. 303 - 309, 1985.

The visual system
—poster contributions

Singularities in Cortical Orientation and Direction Maps: Vortices, Strings, and Bubbles

André J. Noest

Utrecht Biophysics Institute, Dept. MFF, Utrecht University

Princetonplein 5, NL-3584-CC Utrecht, The Netherlands

Abstract:

The maps of orientation tuning in areas V1 and V2 contain many singular points ('vortices') around which orientation rotates by $\pm\pi$. By contrast, direction vortices ($\pm 2\pi$ rotation) appear in the motion processing area V5. These useful mappings of orientations/directions plus positions into two dimensions carry no evolutionary or developmental costs, since vortices are unavoidable and very robust structures in almost any 2D map of a rotationally periodic quantity in which smoothing competes with local noise. Vortices of opposite sign may annihilate or dissociate in pairs, depending on the noise level.

In many cases, the cell tuning has coupled orientation (π periodic) and direction (2π periodic) components. Interactions that favour alignment of both components can then produce additional singularity structures: Orientation-vortices must carry an odd number of 'strings', defined as the locus where the direction component inverts its sign. The strings can have tension, enabling them to bind opposite-sign vortices. In other parameter regimes, meandering strings may produce freely floating loops that enclose opposite direction "bubbles" in areas of aligned orientation.

1. Introduction

Neural responses in many areas of the visual cortex are to some degree selective for the orientation of visual contours, or the direction of visual motion. The cortical layout of the cell tuning parameters in the early stages of the visual system often shows a largely continuous mapping of visual locations, but numerous point singularities are found in the mapping of orientations or directions. In the areas V1 and V2, the singularities are known [1,2] to be vortices with winding number $n = \pm\frac{1}{2}$, i.e. points around which orientation rotates by π, in either the same or the opposite sense in which one encircles the singularity. The situation in the motion area V5 is less certain [3], but recent evidence [4] suggests the occurrence of vortices with $n = \pm 1$ (i.e. rotations of $\pm 2\pi$).

The functional advantages of such a layout (economy of wiring) tend to suggest that this may be a well-evolved adaptive design [1]. Yet, the analysis given here shows that this type of structure arises unavoidably from virtually any mechanism that favours smoothness of the map, but suffers from local random noise.

Models for the formation of these maps have been studied numerically [5,6], but it is not entirely clear which specific elements of the model are essential. At this time it does not seem possible to settle the matter by appealing to neurobiological data about the formation of the maps. Accordingly, this paper concentrates on the most basic aspects of the problem, namely the topology of the maps, and it relies on only very weak assumptions about the underlying biological processes.

2. Vortex singularities in orientation or direction fields

On the purely topological level, the robustness of an isolated vortex in an otherwise smooth orientation or direction map is easy to see. Consider first the case of directions.

When encircling any point, the field f of directions maps the closed path C to (part of) the set S of all possible directions. Both C and S have the topology of the circle. The image $f(C)$ must wind an integer number of times (perhaps zero) around S. This winding number assumes some final value n as C is contracted on a point of interest. Vortex singularities are those points for which $n \neq 0$. Local perturbations can split $|n| > 1$ vortices into lower-n ones, with conservation of the sum of all n. Thus an isolated unit vortex can not be removed by any local process whatsoever. It can only be annihilated through collision with a vortex of opposite n. These conclusions follow easily by considering the image $f(C)$ of a path encircling vortex pairs. Conversely, vortices can be created in opposite-sign pairs by local distortions of an initially smooth region of the map f.

For (planar) orientations, the reasoning above remains valid, but it is useful to count the winding-number by multiples of $\pm\frac{1}{2}$ to reflect the fact that orientation turns by π instead of 2π on paths around an elementary vortex.

The same general conclusions emerge from a less idealized analysis in which the local tuning properties are represented by a complex value $\psi = \rho \exp[i\phi]$. This choice allows one to add a measure ρ of the 'strength' of tuning for a direction ϕ or an orientation ($\phi/2$). To illustrate what happens during smoothing of a field of such quantities, it is useful to study the lowest-order differential equation that captures the essentials of the situation.

$$\partial_t \psi = \psi(1 - |\psi|^2) + \nabla^2 \psi.$$

The equivalent in terms of ρ and ϕ is the pair of coupled equations

$$\partial_t \rho = \rho\{1 - \rho^2 - (\nabla\phi)^2\} + \nabla^2 \rho$$

$$\partial_t \phi = \frac{2}{\rho}\nabla\rho \cdot \nabla\phi + \nabla^2\phi$$

One may note that this dynamics minimizes an energy $H(\psi) = \int -|\psi|^2 + \frac{1}{2}|\psi|^4 + |\nabla\psi|^2 dx$. The trivial stable solution ($\rho = 1, \phi = \phi_0 \in (0, 2\pi]; \forall x$) is obviously not of any interest here; it would imply blindness to all but the ϕ_0 orientation or direction. Appropriately, this solution is not reached under realistic conditions where a finite amount of noise is present.

The interesting solutions are the ones containing vortex-singularities. In the simplest case of an isolated vortex, the solution has circular symmetry about –say– the origin, so it is best written down using polar coordinates (r, θ) in the plane. The general form becomes

$$\psi(r, \theta) = \rho(r) \exp[i(\phi_0 + n\theta)]$$

with integer n and $\phi_0 \in (0, 2\pi]$. The amplitude $\rho(r)$ satisfies

$$\rho_{rr} + \frac{1}{r}\rho_r + \rho\{1 - \frac{n^2}{r^2} - \rho^2\} = 0$$

with boundary conditions $\rho(0) = 0$ and $\rho(\infty) = 1$. No closed form for $\rho(r)$ is known, but asymptotically one finds $\rho(r) = 1 - \frac{n^2}{2}r^{-2} + O(r^{-4})$ for $r \to \infty$, respectively $A_n r^{|n|} + O(r^{|n|+2})$ for $r \to 0$. Topologically (and dynamically), the only stable solutions are the unit-vortices ($|n| = 1$, with $A_1 = O(1)$). All other vortices dissociate spontaneously into n unit-vortices.

2.1 Vortex energetics

The presence of noise causes pairs of oppositely signed vortices to appear from initially smooth regions of the field. An equilibrium density of vortices will be established through the opposing process of pairwise vortex annihilation. At low 'temperature' T (noise level), vortex pairs will remain essentially bound together, whereas they will fill the space essentially freely at higher T. This qualitative picture can be supported by simple estimates of the energetics of vortices.

Inserting a single vortex into a domain of typical radius R, increases the energy by $\Delta H = 2\pi \log R + C_H$, where $C_H = O(1)$ is the vortex-core energy which depends on details of the discrete microstructure of field. The vortex can be positioned at a roughly equal cost in energy within a constant fraction of the area of the the domain, hence its entropy is $\Delta S = 2 \log R + C_S$, with $C_S = O(1)$ again a core-constant. The critical temperature T^* above which the vortex can move essentially freely is then estimated by requiring the Gibbs free energy $\Delta G = \Delta H - T \Delta S$ to be of order one as $R \to \infty$. Thus, $T^* \approx \pi$.

The same ΔG estimate is obtained by considering a vortex-antivortex pair separated by a distance R. Thus, vortex pairs dissociate when $T > T^* \approx \pi$.

3. String singularities in combined orientation/direction fields

In most cases of interest, each cell will have some combination of orientation-tuning and direction-tuning. Examples in the static domain are receptive fields (RF's) that combine odd- and even-symmetric terms (e.g. 'edge'-type and 'line'-type RF's), or purely odd-symmetric RF's that feed into a nonlinearity containing odd and even order terms. In addition, it is usual to find motion sensitive cells that combine orientation and direction selectivity.

To capture the essential behaviour of fields of quantities ψ which combine π- and 2π-periodic components, one can take the ϕ dependent factor in the interaction energy for pairs ψ_i, ψ_j at points i and j to have the form

$$\cos[2(\phi_j - \phi_i)] + A \cos[\phi_j - \phi_i] \; ; \; A \geq 0$$

At $A = 0$, one simply has the case of pure orientation, as treated above. For $A > 4$, the existence of a single energy minimum reduces the situation to that of a (slightly perturbed) direction-field, again as discussed above.

Interesting new phenomena occur for $0 < A < 4$. Then there are *two* energy minima, the main one at $\Delta \phi = 0$, and one at $\Delta \phi = \pi$ which lies $2A$ above the first. As before, the fundamental aspects of the resulting behaviour of the map are topological in nature.

Consider first a case with small A. The point-singularities in the field will still be half-vortices (as for $A = 0$), but one must now assign to each point around the vortex a value of the (2π-periodic) direction component, which is fixed to the (π-periodic) orientation modulo sign-inversion. Thus, an odd number of sign-inversions of the direction must occur on any small circle around a half-vortex. By shrinking the circle, one can conclude that these sign-inversions must occur on an odd number of lines, to be called "strings", that emanate from the vortex. The number k counts these strings.

The topology of the strings as discontinuities in direction allows:
- strings connecting half-vortex pairs,
- strings that loop back to the same ($k \geq 3$) vortex,
- strings forming freely floating loops, which enclose "bubbles".

3.1 Energetics of string-bound vortices

Neglecting the string contributions, the estimated critical temperatures T_h^* for half-vortex and T^* for integer-vortex dissociation (as above) are $T_h^* = \frac{1}{4} T^* = \frac{\pi}{4}(1 + \frac{A}{4})$. However, the presence of the strings introduces another critical temperature T_s^* into the problem. Above T_s^*, the effective 'tension' in a string vanishes, whereas below T_s^* the string tension quickly dominates the behaviour of the system by binding the half-vortices very strongly.

This prediction follows again from simple estimates. The energy cost of a string of length r is dominated by the contributions from the sign-inversion of the direction component across it. Thus, one has $\Delta H \approx 2Ar$. The dominant term in its entropy must also be proportional to r: $\Delta S \approx \alpha r$, with $\alpha = O(1)$. Thus, the string has a 'tension' (ΔG per unit of length), but this vanishes at a finite critical temperature $T_s^* = O(A)$

The value of A, which controls the relative position of T_s^* with respect to the other critical temperatures, then determines which of three scenarios applies as one raises the temperature. In all scenarios ($0 < A < 4$) the state at $T \to 0$ is characterised by (close) pairs of half-vortices, connected by (taut) strings, so that the correlation scale for directions is similar to that for of orientations. The scenarios are as follows.

•$T_h^* < T^* < T_s^*$: The half-vortex pairs are prevented from dissociation at their normal critical point T_h^*, because the string tension still binds them, until T^* is reached. Then integer-vortex pairs, which do not carry strings connecting them, will dissociate.

•$T_h^* < T_s^* < T^*$: Again, half-vortex pairs remain bound together by their strings, but now only up to T_s^* where the strings loose their tension.

•$T_s^* < T_h^* < T^*$: Now the string tension vanishes while the half-vortices remain bound by their normal pair interaction. This causes the strings to meander strongly; loops will then pinch off, forming freely floating bubbles of sign-inverted direction. Thus, the direction structure becomes randomized while the orientations remain correlated over rather large distances. The half-vortex pairs dissociate only as T is raised to T_h^*.

4. Conclusions

By concentrating on the topology of the singularities and the qualitative structure of the field in the different regimes that may be relevant for various specific models and parametrizations, a general overview has been obtained of the remarkable richness of phenomena that may occur in maps of combined orientation and direction selective units. This analysis shows that the functional advantages of a map rich in vortex singularities need not burden philogeny or ontogeny in any way with the costs of establishing specific mechanisms for its formation, since precisely these structures emerge unavoidably when orientation/direction maps evolve by any local smoothing process disturbed by noise.

Indications are that various examples of the phenomena found here do indeed occur in the visual cortex. Specifically, the 'meandering strings' regime seems to apply to area 18 in cat [2], and might perhaps be looked for in monkey V1 or V2, which is known to contain many half-vortices. Recent experiments in the motion processing area V5 suggest [4] that its structure corresponds to the regime of integer-vortices, but it seems worthwile to explore this in more detail.

References
1. G.G.Blasdel, J. Neurosci. 12 (1992), 3139-3161.
2. N.V.Swindale, J.A. Matsubara and M.S.Cynader, J.Neurosci. 7 (1987), 1414-1427.
3. T.D.Albright, R.Desimone and C.G.Gross, J. Neurophysiol. 51 (1984), 16-31.
4. R.B.H. Tootell and R.T. Born, Soc.Neurosci. Abstr. 17 (1991), 524
5. R.Linsker, Proc. Nat. Acad. Sci. USA 83 (1986), 8779-8783.
6. N.V.Swindale, Biol.Cybern. 66 (1992) 217-230.

A New Model for Spatial Frequency and Orientation Tuning in the Visual Cortex based on Delayed Inputs from the Retina

Philippe Grandguillaume

Institut des Neurosciences du C.N.R.S., Université Pierre-et-Marie Curie,
Département des Neurosciences de la Vision Active,
9, quai Saint-Bernard, 75005 Paris

Abstract. We propose a new model of the spatial frequency and orientation selectivity in the visual cortical neurons. It uses lateral inhibition and temporally segregated activations between cortical neurons, resulting in predictive coding. It is in agreement with some caracteristics of the architecture of the visual cortex of the monkey. The model is based on the segregation of the times of arrival in the visual cortex of signals originating from the retina and relayed in the lateral geniculate nucleus. This distribution of delays depends both on the type of ganglion cell, and of its position on the retina. Indeed the parallel channels into which the retinal image is decomposed are represented as delay lines; and the resulting spatially uniform and discrete distribution of delays is modulated by spatial factors determined by the geometry of the retina.

1. Introduction

We propose a model of spatio-temporal coupling in the visual system which results in tuning in spatial frequency and orientation in the visual cortex. Anatomical and electrophysiological studies have shown that the visual cortex is composed of several areas [1]. Each contains a representation more or less distorted of the visual field. Altogether, they form a network with a complex pattern of connectivity. Other major results concern the tuning of cortical neurons. The majority of neurons in the striate cortex and extra striate areas are narrowly tuned to orientation and spatial frequency. This is not the case in the retina and the lateral geniculate nucleus (LGN) [2,3]. This selectivity agrees with results of psychophysics according to which, in first stages of the visual system, information is processed in separate parallel channels [4]. The efficiency of this kind of decomposition is yet born out by its use in mathematical and algorithmical models for computer vision, where it leads to a multi-resolution analysis of the image [5]. However, the interactions between the different channels, and their relations to the architecture of the visual cortex remain unclear. Besides, to our knowledge, timing relations between the channels have not been fully taken into account. On the other hand, many models have been proposed to explain the selectivity to orientation in the cortex [6]. But, these models are also based on a description essentially spatial of the visual system.

We think that the efficiency of the visual system in term of image processing partly results from the optimum integration of the different levels of processing [7]. We intend to show the relevance of the delays originating in the retina to some caracteristics of the architecture of the visual cortex and of the neuronal processing.

2. Delayed lines in the visual system

This part concerns the spatio-temporal organization of the streams of signals running from the first stages of the visual pathway up to the cortex. We focus on two mecanisms.

2.1 Decomposition of the image signal along several channels

The first refers to the different parallel channels which have been identified in psychophysics, physiology and anatomy [4]. Two of these channels originate in the retina, named X and Y in the cat, and colour-opponent and broad-band in the monkey. Their properties derive of those of corresponding ganglion cells. The ganglion cells constitute the output stage of the retina, they transform amplitude-modulated signals into frequency modulated ones (sequence of spikes). In the retina, several types of cells perform parallel analog processings locally. Each class is characterised by a particular size of receptive field. Cells with different receptive field sizes sample the photoreceptors array along different scales. In accordance with the sampling principle of Shannon, their density varies in relation to the sampling rate. These different scales can be put in relation with different spatial frequencies in the signal projected onto the retina through the optics of the eye. In the fourier plan, the sampling rate ΔX, and the spatial frequency bandwidth Δf_S are linked by the relation: $\Delta X * \Delta f_S$ = Cte. These retinal caracteristics are present in X and Y ganglion cells: receptive fields of X cells are smaller than for Y cells which are more scattered than X cells. Besides, the conduction velocity of axons of Y cells is much higher than that of X cells: 30-40 m/sec versus 15-23 m/sec in the cat [8]. The mean number of spikes in output signals of X cells is also much higher than for Y ones (so the "messages" sent by X cells are longer than ones of Y cells). From the next processing stage, the corresponding channels can thus be represented as two sets of parallel delay lines, one with a short delay, the other with a long one. Short delay lines transmit short messages and have a low spatial frequency bandwidth; long delay lines transmit long messages, their bandwidth is high.

The axons of ganglion cells reach relay neurons in the lateral geniculate nucleus (LGN). In this structure, inputs of both channels are segregated within different layers. Measure of response time have shown that the temporal segregation between the two channels is kept at the output of the LGN. In the cat, rise time of Y neurons of the LGN (getting inputs from Y ganglion cells of the retina) are shorter than those of X neurons [9], so that the processing and transmission of low spatial frequencies is yet accelerated. Besides, inputs from both fast and slow channels in the LGN participate to create additional intermediate channels [10].

Each channel is associated a delay (from the retina to the cortex). The channel number n is represented by a set Ln, n≥2, of delay lines Ln(z) so that L1 corresponds to the shortest delay and the delay of Ln (z) is shorter than that of Ln+1(z). The sequence of sets (Ln) with increasing delays results from sampling of the retina at decreasing rates, so that the mean density (rapported to the retinal surface) of delay lines for each set Ln increases with n. The resulting distribution of delays within each patch of the retina is discrete.

2.2 Spatial modulation from the geometry of the retina

For each set Ln, the value of the delay of Ln (z) (supposed above implicitly constant for each set Ln) is modulated by spatial factors determined by the geometry of the retina.

Because the retinal portion of the axons of ganglion cells is not myelinated, the conduction velocity in the retina is much lower than it is in the optic nerve (about 3m/sec for Y cells and 1,6 m/sec for X cells in the monkey [11]). Thus, the time of transmission in the retina varies continuously in function of the distance run by the axon from the position of the ganglion cell on the retina up to the optic disk (OD), the spot of the retina deprived of photoreceptors to which all the fibers converge to form the optic nerve. Given the dimensions of the retina (it covers $1100mm^2$), and the minimal duration between two consecutive spikes (1ms), this range of velocities results in a desynchronization of output signals sensible from distances of about 3 mm for Y cells and 1,6 mm for X cells. Besides, the distance run by the axon is not the same in the nasal and temporal hemi-retinas. First, the OD is in the nasal hemi-retina, about 4mm from the fovea on the horizontal meridian, making the path longer for temporal cells than for nasal cells at the same excentricity. Besides, in the nasal hemi-retina, axons converge radially up to the OD, whereas in the temporal hemi-retina they follow an arcuate path to bypass the fovea [12]. These two different geometries result in different delays for signals originating at the same time in the two parts at the same excentricity; they also result in different orientations of the gradient of delays.

2.3 Differential propagation within the visual cortex

We call Vn a visual cortical area such that n=1 for the striate cortex, and the size of receptive fields of neurons of Vn+1 is superior to that of those of Vn in the same cortical layer and at the same excentricity. Vn sends feedforward connections to Vn+1 and possibly to other areas Vm with m>n+1, and recurrent connections to Vn-1 for n≥2 and possibly to other areas Vk with 2<k<m. A unit $U_{n, p, z}$ is specified by the area Vn to which it belongs, the channel Lp from which its excitatory inputs derive, and the position z of the center of its receptive field in a reference frame centered on the fovea.

In the hierarchical model of the visual cortex [1], feed-forward information is completely processed in the area Vn before being sent to the area Vn+1. Besides, temporally segregated input signals are not supposed to interact within an area in which they are relayed.

We propose another interpretation of the multi-area architecture of the visual cortex. In our model, the delay lines propagate across the successive areas, so that a unit $U_{n, p, z}$ is activated before a unit $U_{n, q, z}$ if p<q for any area Vn, and there is a temporal overlapping between the processings of input signals from different delay lines within the successive areas. Thus, within an area, local interactions between the different delay lines are possible. These interactions can be expressed by the following relation: Un, p, z ⇒ Un, q, z with p<q .

Besides, recurrent connections can be used during the processing between the different areas, corresponding to the relation: Un, l, z ⇒ Um, k, z with m<n and k<l

We propose that it is used to link vertically the analysis in a specific range of spatial frequency realized on each area. Indeed, analysis of low frequency components transmitted by the short delay line is completed on an area Vn with n≥2, in units having large receptive fields and so high positional incertitude (given we suppose the sampling principle is respected), while the analysis of high frequencies is performed in Vp with p<n (for instance V1) by units having smaller receptive fields (at the same excentricity) and low positional incertitude.

The vertical propagation of signals from successive sets of delay lines is horizontally modulated within each area (according to 2.2). Dynamic processings can result from this horizontal wave form in the intra-area network, thanks to horizontal connections.

Indeed, a caracteristic of the visual cortex compared to the LGN is the great number of horizontal connections, and specifically long range ones. They play a dynamic role, according to the relation:

$$U_{n, q, z} \Rightarrow U_{n, q, z+\delta z_{n,q}}$$

3. Cortical interaction between delayed inputs

In our model, the segregation of times of arrival of input signals on the visual cortex, coupled with lateral inhibition, results in tuning of neuronal response for different attributes of the retinal image.

3.1 Predictive coding and lateral inhibition

We consider the neurons belonging to a same visual area Vn. According to part 2, some of these neurons ($U_{n, p, z}$ with p and z allowed to vary separately) are activated before others. We propose that during the delay which separates times of arrival of the input signals, lateral inhibition is generated on the non-yet activated neuron from the earlier activated one. The effect of lateral inhibition is to increase the level of input activity necessary to make the neuron fire action potentials. We take the level of lateral inhibition equal to a linear function of the level of input activity of the earlier activated neuron. So, the output activity of the later activated neuron can be represented by the difference between the effective input and the predicted activity (given by lateral inhibition):

$$A_{out} = A_{in} - A_{pre} \qquad (1)$$

The theory of predictive coding has been applied to represent lateral inhibition in the retina [13]. Predictive coding is used in signal processing to compress efficiently a quantified signal, using linear prediction (the binary value of the next sample is predicted from the one of the past sample or a linear combination of past values) [14]. The transmitted signal reflects the level of error between the effective and predictive values. One major interest of predictive coding in neural networks is that it greatly increases the dynamic range of response of the neuron, while keeping the rate of activated outputs in the network at a low level.

In our model, we suppose that convergent input signal on a unit have to be synchronous in order to produce a possible output.

3.2 Tuning of spatial frequency

It is based on the interaction of neurons of a same area, getting excitatory inputs deriving from different channels, and with receptive fields analysing the same part of the visual field (the receptive field of an earlier activated neuron include those of later activated ones for different channels are associated to different sampling rates on the retina). In that case, the relation (1) becomes:

$$A_{out} (U_{n, q, z}) = A_{in} (U_{n, q, z}) - A_{out} (U_{n, p, z}) \quad \text{with } p < q \qquad (2)$$

In agreement with experimental results on the spatial frequency selectivity [3], we can represent a LGN neuron, or a ganglion cell of the retina as a low pass filter. fc being the cut-off frequency of this filter, its transfer function can be written $H_{fc} (p) = 1/ 1+\tau*p$ with $fc*\tau = Cte$. In 2.1, we have seen that the ganglion cells of the retina process the signal in parallel, so that lower harmonics are present in the response of all types of cell. We suppose that the amplitude coding is similar between different channels, and these differ only by the value of fc. The shorter is the delay associated to a channel, the smaller is the corresponding cut-off frequency. The relation (2) thus can be written:

$$H = H_{fc}^p - H_{fc}^q \quad \text{with} \quad fc^q < fc^p$$

This is the difference between two low-pass filters with different cut-off frequencies. Thus the resulting form of the tuning in spatial frequencies is a band-pass filter with maximum at the frequency $(fc^p- fc^q) / 2$. This shape of tuning curve is in agreement with experimental data on cortical neurons [3]. Inhibition could be generated directly by the cortical neurons activated first or via inhibitory interneurons. In this model, the tuning curve of the early activated neurons is not modified. Neurons in the layer IV of V1 which get inputs from the LGN have such tuning curves.

A possible neuronal circuit would then start from inhibitory interneurons of layer IVcα acting on neurons situated outside the layer IV and receiving excitatory inputs from neurons in IVa on which partly project the intermediate and slow channels from the LGN.

3.3 Tuning of orientation

Pharmacological experiments have shown that reversible suppression of lateral inhibition locally in a small region of the striate cortex strongly decreases the selectivity to orientation of neurons narrowly tuned in normal conditions [15].

In our model, it derives from the coupling of lateral inhibition with the horizontal modulation of delays defined in 2.2. Indeed, this supposed mecanism generates a non isotropic distribution of delays between neurons of a same channel (so with separate or only partially overlapping receptive fields). The relation (1) becomes in this case:

$$A_{out}(U_{n, p, z}) = A_{in}(U_{n, p, z}) - A_{out}(U_{n, p, z+dz})$$

Thus, if a spatially extended visual stimulus (like a bar shaped surface of constant color and constrast) projects over the two receptive fields, both neurons will get the same excitatory input, and the early activated will inhibit the delayed one which will remain silent. Now, rotating the bar so that it projects only on one of the two receptive-fields, excitatory inputs will differ, and the delayed neuron will not be inhibited. So, the response of the neuron is spatially oriented. Whereas delays separating inputs coming from the same hemi-retina could act only over relatively long distances, the delays separating inputs from each hemi-retina could act on very short distances.

Lateral inhibition could be produced on long range by a specific class of inhibitory interneurons. Indeed, these cells (called basket cells) have long myelinated laterally directed axons, their axonal arbors are distributed in patches far from the soma. Number of these neurons have been found in the area 18 of the cat [16].

4. Conclusion

We have proposed a formal model of processing in the visual cortex based on interactions between differentially delayed input signals. This model depends on activity, and on the coupling between space and time which originates in the retina, and on its propagation across the cortex. Lateral inhibition and delay input signals have led to an interpretation in term of predictive coding. Furthemore, the model leads to a new approach of coding of visual forms, which we are currently developping.

References

1. D.C. Van Essen, & J.H.R. Maunsell: Hierachical organization and functional streams in the visual cortex , TINS, 6,9 (1983)
2. R.L. De Valois, E.W. Yund, N. Hepler: The orientation and direction selectivity of cells in macaque visual cortex, Vis.Res. Vol. 22, 531-544 (1982)
3. R.L. De Valois, D.G. Albrecht, L.G. Thorell: Spatial frequency selectivity of cells in macaque visual cortex, Vis.Res. Vol. 22, 545-559 (1982)
4. P.H. Schiller & N.K. Logothetis: The color-opponent and broad-band channels of the primate visual system. TINS, Vol13, 10, 392-398 (1990)
5. S.G. Mallat: Multifrequency channel decompositions of images and wavelet models. IEEE transactions on acoustics, speech and signal processing, 37,2091-2109 (1989)
6. R. Linsker: Self-organization in a perceptual network. Computer, march, 105-117 (1988)
7. R. Linsker: Perceptual neural organization: some approaches based on network models and information theory, Annu.Rev.Neurosci. 13:257-81 (1990)
8. J. Stone, B. Dreher, A. Leventhal: hierrarchical and parallel mechanisms in the organization of visual cortex. Brain Res., 1, 354-394 (1979)
9. A.K. Sestokas, S. Lehmkuhle, K.E. Kratz: Visual latency of ganglion X and Y cells: a comparison with geniculate X and Y cells. Vision Res. Vol.27, 1399-1408 (1987)
10. D.N. Mastronarde: Nonlagged relay cells and interneurons in the cat lateral geniculate nucleus: Receptive-field properties and retinal inputs. Vis.neurosci., 8, 407-441 (1992)
11. P. Gouras: Antidromic responses of orthodromically identified ganglion cells in monkey retina. J. Physiol. , 204, 407-419 (1969)
12. Polyak: The Retina. University of Chicago Press (1941)
13. M.V. Srinavasan, S.B. Laughlin and A. Dubs: Predictive coding: a fresh view on inhibition in the retina. Proc.R.Soc.Lon.B 216,427-459 (1982)
14. C.W. Harrison: Experiments with linear prediction in television, Bell Syst.Tech.J. 765-783 (1952)
15. A.M. Sillito: Orientation selectivity and the spatial organization of the afferent inputs to the striate cortex. Exp. Brain Res., 41,9 (1980)
16. J.A. Matsubara, M.S. Cynader, and N.V. Swindale: Anatomical properties and physiologogical correlates of the intrinsic connections in cat area 18. J.Neurosci. 7,5,1428-1446 (1987)

Cascaded Intracortical Inhibition: Modeling Connection Schemes on a Large Scale Simulator

Eckart Nelle
and
Florentin Wörgötter

Institute of Physiology, Dept. of Neurophysiology
Ruhr-Universität Bochum
D-4630 Bochum, FRG

Abstract

Receptive field properties of simple cells in the primary visual cortex arise partly from the connection structure of the input paths from the LGN and within the visual cortex. We briefly present our modeling setup used to study various connection schemes, such as spatially opponent inhibition. In spite of its ability to explain certain receptive field properties, severe conceptual difficulties are associated with the biological implementation of the model. We introduce a new cascade-like connection scheme, which largely overcomes these problems and produces more realistic simulation results.

1 Introduction

The richness and variability of neural structures and mechanisms within vertebrate brains presents the modeler with a formidable task: On one hand the properties of single neurons form the basis of visual information processing; on the other hand the collective or emergent properties of possibly very many neurons combined within a neural assembly seem vital for the processing task as a whole. Large scale simulations on fast parallel computers can be used to model such big networks. Already the implementation of different connection structures, however, raises a number of conceptional problems that have to be solved prior to implementation.

In this study, we will first give a description of the neural network simulator we use to model and study aspects of visual information processing. We will then extensively discuss a certain type of intracortical connection scheme presenting the conclusions from our simulations. It will become clear, that this scheme has several drawbacks. Finally, we will extend the previous scheme, showing that a less specific connection pattern outperforms the original version, in particular with respect to biological realism.

2 The simulator

At the single cell level the substrate of our model is formed by Hodgin&Huxley-like neurons. The membrane potential evolves passively as determined by cell parameters, some of which are membrane capacitance, leakage conductance and leakage voltage. The membrane potential is driven by the synaptic input, which is represented by a sum of α-functions varying with time for both, the excitatory and the inhibitory input separately. If the membrane potential exceeds a threshold value, which is randomly drawn from a distribution, an action potential is generated and passed on with a specific delay for each path to all postsynaptic cells. In addition, a relative and absolute refractive period after the occurrence of a spike is taken into consideration.

At the network level, we concentrate on a 5° × 5° area of the visual field at about 4.5° retinal eccentricity. This part of the visual field is followed from the retina through the LGN to layer IV of area 17, in which we model a field of only 2.5° × 2.5° to avoid artificial boundary effects. Both, the ON and OFF pathways are included in retina and LGN; in the cortex about 1/4 of the inhibitory cells are modeled as simple cells. Altogether the simulator comprises 16384 cells of a monocular pathway in LGN and V1. A detailed description of the parameter ranges and equations used is given in [2]. The different processing stages in the visual system are reflected in the modular structure of the simulator, which runs mainly on workstations (IMAGE, RET, CONN, Fig. 1) and partly on a parallel supercomputer, a CM-2 (SIM). Figure 1 shows a schematic diagram of the complete simulator package. It consists of IMAGE, which computes the convolution of the stimulus with a hexagonal grid of DOG[1]-filters. In the next stage RET performs the conversion of the convolved image into spikes of retinal ganglion cells. As an independent step CONN determines the connection pattern and connection strength of cells in the LGN and of V1 (synaptic weight matrix), and, therefore, determines the local and global topography within LGN and cortex. The output files of these programs can be used repeatedly for a multitude of simulations performed with SIM. SIM is the part of the simulator which generates the activity pattern in the LGN and V1, using a parallel implementation on a Connection Machine [1]. In order to produce biologically plausible results, physiological parameters drawn from the literature are generally drawn from finite distributions rather than using single, fixed values. In addition realistic noise and spatial jitter are introduced at various stages. For a single orientation a simulated time of 10 seconds corresponds to about 1 hour of calculation time on the Connection Machine equivalent to an efficiency of about $0.22 \ s/cell = 3600 \ s/16384 \ cells$.

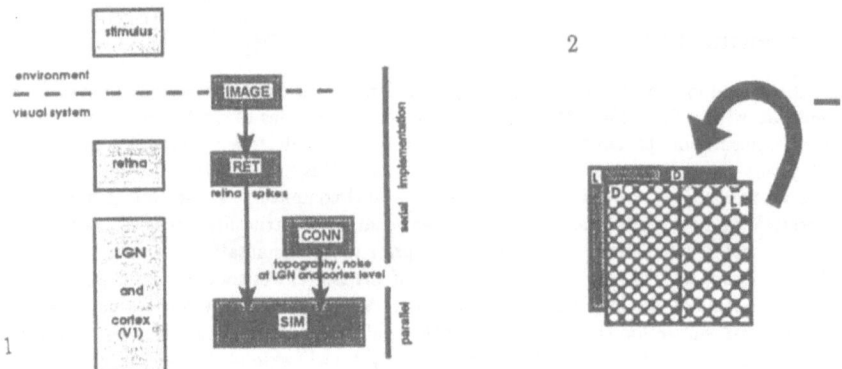

Figure 1: Simulator of the primary visual system. Right: the parts of the simulator; Left: their correspondence in the visual system.

Figure 2: Spatially opponent inhibition. L = light excited, i.e. ON subfield, D = dark excited, i.e. OFF subfield.

3 Thalamo-cortical connection schemes

We restrict our discussion to the effects of different coupling scenarios implemented through CONN. Already in the LGN effects are clearly visible: increasing the strengths of short range inhibitory connections within the LGN leads to an overall reduction of spike activity in the LGN, therefore also reducing the input to the cortex and suppressing the transmission of LGN-noise into V1. In addition, through competition between LGN cells an, in the extreme form unphysiological, winner-take-all process takes place, leaving many LGN cells without any activity. With more realistic

---

[1]DOG=differences of Gaussian

parameter settings we receive a sharpening of the centers of activity within the LGN which might correspond to the physiologically observed contrast enhancement found in LGN cells as compared to retinal cells.

4   Spatially Opponent Intracortical Inhibition

Passing from the LGN to V1, many more refined connection schemes have been proposed since Hubel and Wiesel's [3] first model for single cells in V1.

A recent example is the model of spatially opponent inhibition discussed by Ferster [4]. The receptive field of a simple cell in V1 is in this case shaped using the inhibitory input of another cortical cell with highly overlapping but antagonistic arrangement of subfields (see Fig. 2). The result is an ON excitation and OFF inhibition in ON regions of the receptive field, and an OFF excitation and ON inhibition in the OFF regions. If we combine several of such subfields in a rectangular array in a Hubel&Wiesel-like fashion, we obtain a fairly realistic description of the receptive field of a simple cell, including orientation specificity. This model was tested with the simulator and did not produce any unexpected results.

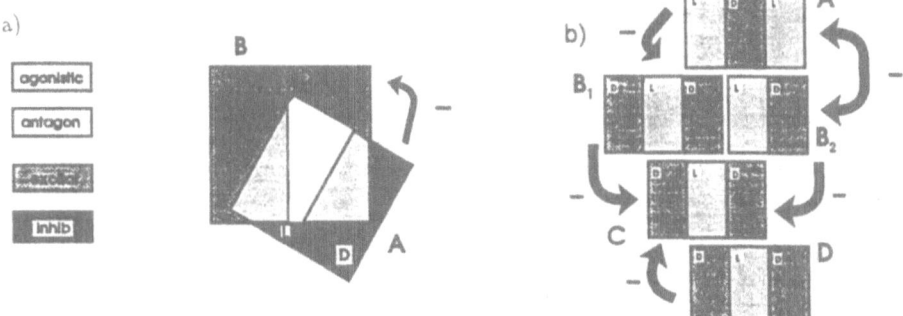

Figure 3: a. Spatially opponent inhibition with shifted, rotated and scaled subfields. b. Cascade-like connection.

There are, however, several problems already associated with the topography required for this model (for functional considerations see Tolhurst and Dean [5]). Even though the receptive field properties are mimicked nicely, the topographical arrangement of the overlapping subfields has to be very accurate. This would probably require a highly specialized synaptic reinforcement (or elimination) mechanism during development. Such a mechanism is hard to imagine because inhibitory connections are concerned, for which straightforward Hebbian mechanisms fail. Calculating the coverage of the visual field by cortical cells with realistic assumptions concerning cell numbers, receptive field scatter and jitter, etc. we estimated conservatively that only 0-5 inhibitory cells out of $\approx$ 3200 inhibitory cells per $degree^2$ would share the same receptive field with an accuracy in the overlap of only 75%. This problem can be to circumvented to some degree by tolerating larger translations and rotations, and possibly differences in size of the corresponding subfields (Fig. 3a), which seems to be physiologically more plausible than the strict overlap proposed in the original model. Even with this simplification, it remains a tremendous task during ontogenesis or subsequent learning to determine cortical cells with an appropriate receptive field overlap and to couple them correspondingly. Fig. 3a exemplifies this. With overlapping receptive fields of two cells in an arrangement where only cell A would inhibit cell B, we have to take four effects into account: 1) Only parts of the resulting receptive field of cell B would show the desired effect of "antagonistic" inhibition. (2) In other parts the inhibition would be "agonistic" which could lead to an unwanted elimination of spiking activity in these regions. (3) A third part of the receptive field would be purely excitatory, and (4) yet another region would be purely inhibitory. Obviously, in the most desired but unrealistic situation only a single antagonistic region would exist as in Fig. 2. For a more realistic case, the question has to be answered how a possible arrangement of this kind could be probed in order to have a measure of the effectiveness of the desired mechanism. Computationally this can be done by computing the sizes of the different regions and defining an arbitrary "measure of overlap" combining the four area values. This is a tedious task and requires significant computing time. We have defined several different measures of overlap which cover the transition from a topographically strictly defined connection scheme to a cascade-like connection as described below. The biologically relevant problem in this context is to find a set of computational criteria that would reflect a realistic developmental reinforcement or elimination of intracortical connections and yield realistic behavior of the cells. Due to temporal interactions between the subfields it is *a priori* not at all clear how such a combination of the regions would finally shape the spatio-temporal structure of the receptive field. As an example we could clearly observe comparatively small inhibitory side bands in the simulated cells arising from purely inhibitory regions, while the elimination of spiking activity in "agonistic" regions critically depends on the inhibitory synaptic weight.

5 Cascaded Spatially Opponent Intracortical Inhibition

It seems to be a rather unphysiological assumption to require a one-to-one spatially opponent structure. Instead, here we suggest a connection scheme with less restrictive alignment requirements that still produces realistic receptive fields. In this case the receptive subfields are arranged in a cascade-like manner. Fig. 3b shows parts of such an arrangement. For illustrational reasons the topographically overlapping receptive fields are drawn apart, and only optimal antagonistic overlaps are shown. From this diagram two effects can be seen that have also been observed in our simulations: (1) Significant and comparatively wide inhibitory sidebands arise as in cell C from the inhibitory action of cells B_1 and B_2 produced by nonbalanced OFF subfields at the borders of the receptive field. (2) As a secondary effect disinhibition is observed, as also can be seen in cell C, due to the inhibition from cell A which in turn inhibits the cells B_1 and B_2. Those secondary effects are rather weak and occur only at high spiking rates of the cortical cells.

We have combined this connection scheme in our simulations with the different measures of overlap as described above to obtain a realistic connection structure. The receptive fields observed in this way display the high degree of complexity also found in real cortical cells. Although we cannot be sure that such an arrangement actually exists, this makes us confident to further pursue this line of investigation. After having shown the gross spatial effects of mutual inhibition in this study, the next question concerns the linearity of the cell behavior as in the different models studied experimentally by Tolhurst and Dean which needs to be tested next.

References

[1] D. Brettle, E. Niebur, A Detailed Simulator of a Biological Neural Network Implemented on a Massively Parallel Supercomputer, submitted

[2] F. Wörgötter, C. Koch, A Detailed Model of the Primary Visual Pathway in the Cat: Comparison of Afferent Excitatory and Intracortical Inhibitory Connection Schemes for Orientation Selectivity. J. of Neuroscience 11(7), 1959-1979 (1991)

[3] D.H. Hubel, T.N. Wiesel, Receptive fields, binocular interaction and functional architecture in the cat's visual cortex. J. Physiol. (Lond) 160:106-154 (1962)

[4] D. Ferster, Spatially Opponent Excitation and Inhibition in Simple Cells of the Cat Visual Cortex. J. of Neuroscience 8(4), 1172-80 (1988)

[5] D. J. Tolhurst, A. F. Dean, The effects of contrast on the linearity of spatial summation of simple cells in the cat's striate cortex. Exp. Brain Research 79, 582-588. 1990

Hidden Assembly Dynamics and Correlated Neuronal Responses

K. Pawelzik, J.Deppisch, and T. Geisel

Institut für Theoretische Physik, Universität Frankfurt
Robert-Mayer-Str. 8, D-6000 Frankfurt/Main 11, Germany

Correspondence to K. Pawelzik, e-mail: klaus@chaos.uni-frankfurt.dbp.de

Oscillatory neuronal responses have been suspected to reveal assembly formation by synchronization. We analyse these data in terms of a hidden-state model, the parameters of which are estimated directly from experimental spike trains. The model characterizes the excitation dynamics of the network which is only stochastically reflected by the activity of the observed neurons. Our approach reproduces the observed spike dynamics and predicts the cross-correlations of units which belong to the same assembly. The application of our method to multi-unit activities from cat visual cortex reveals that for certain stimuli the recorded neurons indeed participate in a common dynamics, i.e. belong to the same assembly.

Introduction

Experiments in the visual cortex of the cat[1] and the monkey[2] seem to support the idea, that synchronization of oscillatory responses of different feature detectors is used for the representation of a common visual object[3]. This means that objects are represented not by an increased firing rate of certain neurons but instead by a collective excitation of a neuronal subpopulation which we call burst for convenience. The inclusion of dynamics into the assembly notion is advantagous when concidering segmentation and binding of features and much theoretical work currently attempts to demonstrate the usefullness of dynamics in perception[4]

The experimental evidence for the existence of such assemblies, however, is mostly based only on the correlation functions calculated from neurophysiological observables (multi-unit-activity and local field potential) which have fast decaying modulations[5]. Although these results clearly demonstrate that there is coherent neuronal activity which depends on various aspects of the stimuli, they do not unambigously support the assembly hypothesis. On the one hand the conclusion that a modulated auto-correlation function indicates the existence of an oscillator is not compelling because for this already refractoriness can be sufficient. On the other hand the presence of cross-correlations per se indicates coupling and does not proove assembly formation.

We here focus on the question to what extent the measured responses are consistent with the hypotheses that there is an underlying neuronal population which is defined by a unique dynamics. As a first approach to this problem we model the burst dynamics of the network by a renewal process to which the spiking single neuron is coupled stochastically. This phenomenological hidden state model (HSM) has already been shown to reproduce the dynamics of responses which can be measured at one electrode and served as a tool for quantifying the switching behaviour between oscillatory and stochastic states in models as well as in experimental data [6]. In this contribution we show, that the assumumption of an underlying dynamics can explain the correlated responses measured at *different* cortical locations. In particular our Ansatz predicts the cross-correlograms from multi-unit activities and thereby quantifies assembly formation which may depend on the stimulus.

The Hidden State Model

The stochastic dynamics of an isolated spiking neuron may be modeled by a renewal process[7]. For discretized neuron states such a description is equivalent to a stochastic automata with a simple structure (Fig. 1a). In this model the neuron's state is given by the time since the last spike ϕ and the parameters P_f are the probabilities for excitation i.e. for generating the next spike. This description neglects any memory in the neuron going beyond the last spike and is equivalent to the interspike-interval histogram P_h with the relation $P_h(t) = P_f(t) \cdot (1 - \int_o^t P_h(t')dt')$. From $P_h(t)$, the autocorrelation function then can be calculated via $\hat{C}(\tau) = P_h(\tau) + \int_0^\tau P_h(\tau)\hat{C}(\tau - t)dt$. Usually one has a sigmoidal shape of $P_f(\phi)$ which reflects refractoriness and easily can be derived from a threshold dynamics (comp. [8]).

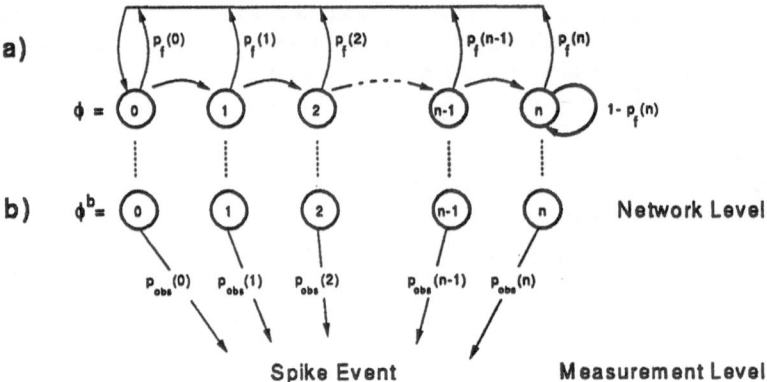

Fig. 1: The hidden-state model. a) A renewal dynamics. States ϕ are time steps since the last point event ($\phi = 0$). Parameters $P_f(\phi)$ denote transition probabilities per time step for returns to $\phi = 0$. b) Network dynamics and measurement. The occurence of a spike in a measurement is represented by the probabilities (per time step) $P_{obs}(\phi^b)$.

The Ansatz of a renewal process can also be applied for the description of the burst dynamics in networks of spiking neurons. In that case ϕ^b denotes the time since the last collective excitation and $P_f^b(\phi^b)$ gives the probability for generating the next excitation (burst). Despite of simple elements the network dynamics can be more complex which is reflected by a qualitative difference in the shape of the excitability $P_f^b(\phi^b)$ for the network as compared to $P_f(\phi)$ of the single neuron. E.g. a pronounced maximum in

$P_f^b(\phi^b)$ indicates a switching between an oscillatory and an irregular states[6], a property which hardly can be identified from the modulations in correlation functions [9].

A spike activity of an elementary neuron, however, only indirectly reflects the network excitation. In general, the neuron may remain silent or may contribute with one or more spikes to a burst. Furthermore there may be some uncorrelated background activity independent of the network state. All this can easily be included into a model by introducing the spike observability function $P_{obs}(\phi^b)$ which denotes the probability for observing a spike when the network is in state ϕ^b (Fig. 1b). Different neuron properties may then be reflected by different P_{obs}, e.g. the average firing rate is proportional to P_{obs} leaving the time structure unchanged. While P_f^b and P_{obs} are useful by themselves for interpreting the network dynamics[6] we here only note that a given HSM predicts the statistical properties of spike trains $\{s_t^i\}, t = 1, ..., T$ from neurons i which belong to the same network characterized by P_f^b. In particular the correlograms $C_\tau^{ij} = \langle s_t^i s_{t+\tau}^j \rangle$ which are used widely in neurophysiology[10] can be calculated from the corresponding HSM using

$$\hat{C}_\tau^{ij} = \sum_\phi \sum_{\phi'} P_{obs}^i(\phi') M_{\phi',\phi}^\tau P_{obs}^j(\phi) \rho(\phi), \tag{1}$$

where M denotes the Markov matrix of the model, $\langle ... \rangle$ denote the time average and $\rho = M\rho$ is the stationary state density. Note that cross-correlograms of neuronal responses at different locations $(i \neq j)$ which reflect the same assembly (identical P_f^b) are predicted from the model, even if the corresponding neurons differ in their individual properties $(P_{obs}^i \neq P_{obs}^j)$. A more detailed analysis of this model can be found in [11].

Predicting Neuronal Correlations

We use the HSM presented above for the analysis of experimental multi-unit activities (MUA) from the visual cortex of the cat measured with two electrodes located 6mm apart[1]. Because the HSM is a hidden-Markov model the parameters P_f^b and P_{obs}^i can be estimated directly from the multi-unit activities using the Baum-Welch algorithm[12](Fig. 2). The comparison of the autocorrelograms with the prediction (Eq. 1) indicates, that for coherent stimuli (long bar in [1]) the observed dynamics is captured by the HSM (Fig. 3a). Furthermore we find that for these stimulus conditions the estimated excitability functions P_f^b agree within error while the deviation of the P_{obs}^i essentially reflect the different firing rates. The similarity of the P_f^b's indicates comparability of the underlying dynamics but is not sufficient for concluding identity, i.e. the presence of one assembly. This, however, can be tested by comparing the cross-correlogram with the prediction of the HSM (Fig. 3b). For the mentioned stimulus conditions we find excellent ageement. For other stimuli cross-correlograms become flatter, indicating that different assemblies are involved. This is quantitatively reflected in the HSM either by different P_f^b for the two MUA's, or, in case of similar burst excitabilities by a significant deviation of the cross-correlogram from the prediction by the model (not shown).

Concluding we expect, that the Ansatz of a hidden excitability dynamics applies to many cases of correlated neuronal activity which emerges as a network phenomenon. In particular the extraction of HSM-parameters should help to operationalize the notion of assembly formation by synchronous neuronal acticity.

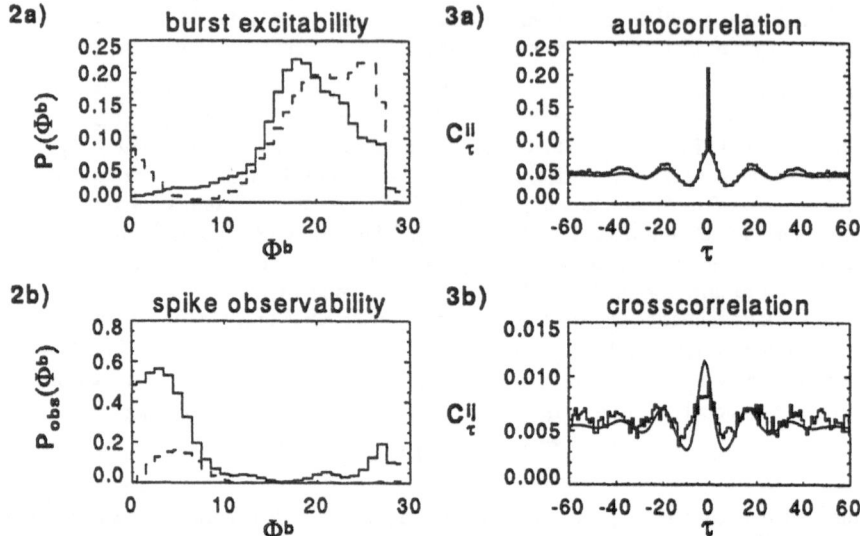

Fig. 2: The burst excitability $P_j^b(\phi^b)$ (a) and the spike observabilities $P_{obs}^i(\phi^b)$ (b) for electrode $i = 1$ (full line) and electrode $j = 2$ (broken line) extracted from the respective MUA's using the Baum-Welch algorithm. The individual couplings of the observed neurons to the network are reflected by different P_{obs}.

Fig. 3: Comparison of the HSM with the data. a) Autocorrellograms of the spike trains (histogram) compared to the prediction (line) of the HSM (full line in Fig. 2) via Eq. 1. The underestimation of the offset is due to the nonstationarity of the firing rate in the measurements. b) Cross-correlogram (histogram) and prediction from the HSM (line). We used the more significant P_j^b in Fig. 2 and rescaled the P_{obs}^i to the average firing rates. Shift, offset and modulation amplitudes agree well indicating that the neurons observed in the experiment participate in the same underlying burst dynamics.

[1] Gray C.M., König P., Engel A.K., and Singer W., Nature **338**, pp. 334-337 (1989).
[2] Kreiter A.K., Singer W., Europ. J. Neurosci. **4**, 369 (1992).
[3] Malsburg C.v.d., Int. Rep. 81-2, MPI Biophys. Chem. Göttingen FRG (1981).
[4] Schuster H.-G., *Nonlinear Dynamics and Neuronal Networks*, Weinheim (1991).
[5] Engel A.K., König P., Gray Ch.M, Singer W., Europ. J. Neurosc. **2**, 607-619 (1990).
[6] Pawelzik K., Bauer H.-U., Deppisch J., Geisel T., in Moody J.E., Hanson S.J., Lippmann R.P., Neural Information Processing 5, Morgan Kaufmann (1993).
[7] Perkel D.H., Gerstein G.L., Moore G.P., Biophys.J. **7**, 391 (1967).
[8] Bauer H.-U.,Pawelzik K.,Geisel T., this volume
[9] Schuster H.-G., Koch C., in Moody J.E., Hanson S.J., Lippmann R.P., Neural Information Processing 4, Morgan Kaufmann (1992).
[10] Eggermont J., *The Correlative Brain*, Springer (1990).
[11] Deppisch J.,Pawelzik K.,Geisel T., submitted to Biol. Cyb.
[12] Rabiner, L.R., Proc. IEEE **77**, 2 pp. 257-286 (1989).

A Model for Latencies in the Visual System

Guido Bugmann, John G. Taylor

Centre for Neural Networks, King's College London, London WC2R 2LS, United Kingdom
Fax (+44) 71 873 2017, Phone: (+44) 71 873 2234, email: gbugmann@oak.cc.kcl.ac.uk

Abstract

The visual system is modeled as a multilayer network of coincidence detecting neurons switching to a state of sustained firing after the production of their first spike. Inhibition from the next layer stops the sustained firing only when the information has been used (coincidence detected). Simulations of such a self-timed network show the propagation of a firing probability wave from layer to layer. The latency of the activity onset in each layer is determined by the level of the sustained activity and by the initial jitter of onset times in the first layer. Theoretical analyses confirm the observed relations. Our results indicate that the layer-to-layer latencies in the visual system may be mainly caused by the jitter in onset time in retinal ganglion cells.

1   Introduction

There are conflicting data on latencies in the visual system. Measurements based on electrical stimulations consistently indicate latencies of less than 2 ms per neuronal relay [1]. However, when visual stimulation is used, an average latency of 10 ms per neuronal relay is a generaly accepted value [2]. It is observed, for instance, between layer 4 and layers 2-3 in the striate cortex [3]. What is the cause for these longer latencies with visual input ?

It is assumed that electrical stimulations produce highly synchronized input spikes [4] which cause unnaturally short latencies. With visual stimulation, input spikes arrive asynchronously and it has been proposed that they must be integrated for appropriate intervals [3]. However, temporal integration is incompatible with the observed irregularity of the spike trains [5]. On the contrary, coincidence detection based on very short integration times, or possibly a synaptic saturation mechanism [6], seems required. Coincidence detection corresponds to a multiplicative, AND-type, function of the neuron which has been postulated for neurons in MST [7]. If we assume neurons operating as coincidence detectors, and thereby discard the temporal integration hypothesis, how can the long latencies be explained ?

In the present paper we describe a simple model of information processing in the visual system, based on coincidence detection, which can reconcile the above mentioned observations, producing long delays with visual stimulation and short ones with electrical stimulations. We make the following assumptions:

1) Neurons in the visual cortex are part of a pyramidal multilayer network where each neuron receives spikes from m distinct neurons in the preceding layer.

2) Neurons act as coincidence detectors, firing only if all m inputs provide a spike within a given time window.

3) Neurons receive excitatory feedback from local-circuit neurons and stay in a state of persistent activity ("ON" state) after producing their first spike, as sugested by physiological observations [4].

4) Neurons in the self-sustained ON-state are silenced by inhibitory feedback from their target neurons in the next layer. This feature ensures that neurons fire only during the minimum necessary time and prevents the loss of information from one layer to the next. Similar feedback inhibition schemes are used in models of speech production [8], olfactory recognition [9] and visual search [10].

Inhibitory feedback is usually assumed to be a local process [4], with a delay determined by the local circuitry. In our model, this would cause frequent failures in propagating the information, in contradiction to the systematic response of neurons to optimal visual stimuli [11]. However, it is not excluded for inhibitory feedback to have a dual role: i) controlling the local level of activity, in which case inhibition would be most apparent in pathological cases like electrical stimulation, ii) disabling a selection of neurons for computational reasons, in which case it would be determined by long-range projections.

2 Simulations

The simulations are performed with a pyramid of pRAM neurons [12]. These neurons can act as coincidence detectors, like leaky integrate-and-fire neurons [13], and are easier to analyse mathematically. The pRAM operates in discrete time-steps Δt with spike trains defined as sequences of 1's (a spike) and 0's. Each pRAM has $m = 4$ inputs and is set for the coincidence detection mode, firing only when all 4 input spikes are present during a time-step Δt. As soon as the first spike is produced, the pRAM is set to fire at each subsequent time step with a probability P_1. This produces a random spike train of frequency $f_1 = P_1/\Delta t$ characteristic of the ON-state and simulates the effect of local excitatory feedback. Such a random firing is consistent with findings that, at all frequencies, biological neurons fire with near Poisson distributions of interspike intervals [5].

The pyramidal network has 64 neurons in its input layer (Layer 0), divided in 16 groups of 4 neurons connected to 16 neurons in layer 1. These neurons are divided into 4 groups of 4 neurons connected to 4 neurons in layer 2. These 4 neurons are connected to a single neuron in layer 3.

The input neurons are initially set to fire randomly with a frequency $f_0 = P_0/\Delta t$. However, as soon as their first spike is produced, their frequency is set to f_1. So, the only role of f_0 is to produce a jitter in the starting time of the neurons in the first layer.

For a neuron in a layer $n + 1$ to fire there are 2 conditions: 1) all m neurons in layer n must be in the "ON" state and 2) a coincidence has to occur (coincidence probability $P_c = P_1^m$).

In biological neurons, coincidences are defined as spikes arriving in a given time window τ and occur with a probability $P_c' = f_s^m \tau^m$ [13] where f_s is the biological sustained rate. If we assume the coincidence probabilities to be equal in both systems, this gives us the relation between the time-steps and frequencies used in simulations and those in biological systems to be $f_1 \Delta t = f_s \tau$.

When a neuron in layer $n + 1$ has fired, the firing probability of all its m input neurons is set to zero. The figures 1 and 2 show the observed firing probabilities at each time-step in two cases, respectively without and with feedback inhibition. In the layers 0, 1 and 2, the feedback seems to reduce the firing probability. Actually, neurons are still firing at f_1 when they are in the "ON" state but they do so at different times in different runs. This is exemplified in figure 3 showing a spike raster of 30 runs in the case with feedback. There are no changes in the last layer which receives no feedback. After the onset time, the firing probability saturates at P_1.

We define the total latencies L_n as the time elapsed from the start of the simulation until a neuron in layer n as reached an average firing probability $P_1/2$. We define the relative latencies as $\Delta L_n = L_n - L_{n-1}$.

Figure 4 shows the latencies L_0 and ΔL_n for n=1, 2, 3 in the case of a fixed $P_0 = 0.05$ for various P_1. The relative latencies converge all to the same minimum value ΔL_{min} for large values of P_1. The theoretical analysis of the system (to be described in details in a later paper) predicts a $1/P_1^m$-dependence for ΔL_n. This is confirmed by simulations (the curve in figure 4 is: $\Delta L_1 = \Delta L_{min} - 0.8 + 0.8/P_1^4$). Figure 5 shows the minimum latencies ΔL_{min} (determined at $P_1 = 1$) for various values of P_0. The curve in the figure is the theoretical prediction $\Delta L_{min} = \log(1/m)/\log(1 - P_0)$.

3 Discussion

In our model we have not included transmission delays, synaptic delays or background potential effects [14]. Therefore the model provides a lower bound to latencies and is purely related to the performed computation. It shows two components to the computational latencies: 1) The initial jitter or time necessary for all m input neurons to be in the ON state. This depends on P_0 in the model. 2) The time for a coincidence to occur which depends on P_1.

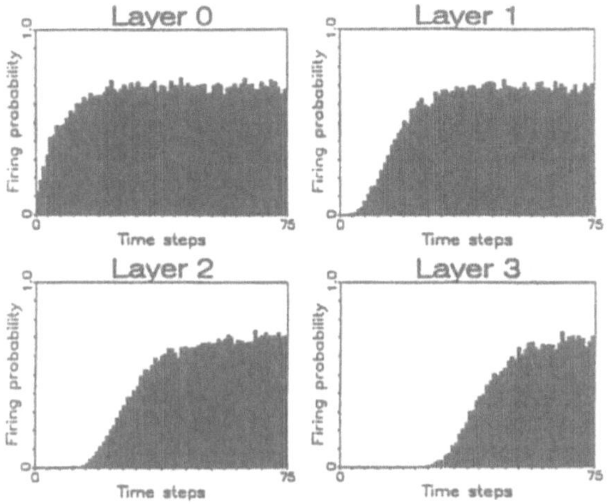

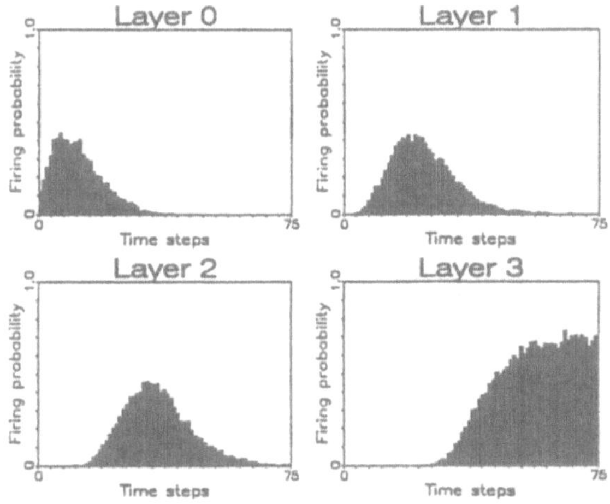

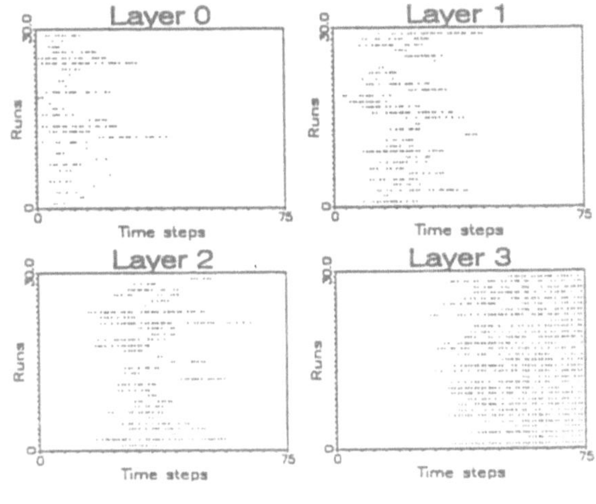

Layer 0
Firing probability / Time steps

Layer 1
Firing probability / Time steps

Layer 2
Firing probability / Time steps

Layer 3
Firing probability / Time steps

Layer 0
Firing probability / Time steps

Layer 1
Firing probability / Time steps

Layer 2
Firing probability / Time steps

Layer 3
Firing probability / Time steps

Layer 0
Runs / Time steps

Layer 1
Runs / Time steps

Layer 2
Runs / Time steps

Layer 3
Runs / Time steps

Figures 1,2,3. Top Left (1): temporal evolution of the firing probability in the case without reset by the inhibitory feedback. Middle left (2): same with reset. Bottom left (3): spike raster of 30 runs in the case with reset. Each dot indicates the time of occurence of a spike. Simulation parameters: $P_0 = 0.15$ $P_1 = 0.7$.

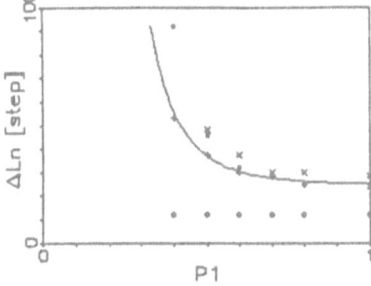

Figure 4. Relative latencies ΔL_n in dependence of the sustained firing probability P_1. Symbols: circle: L_0, cross: ΔL_1, square : ΔL_2, x: ΔL_3.

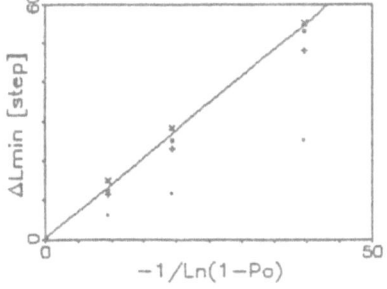

Figure 5. Minimum relative latencies in dependence of P_0. (The jiter in onset time in layer 0 decreases as P_0 increases). Symbols as in Fig. 4 except: dots: L_0.

The initial jitter, corresponding physiologically to the fluctuations in onset times in retinal ganglion cells [15], determines the absolute minimum in per-layer-computation time ΔL_{min}.

The gain in per-layer computation time due to the level of sustained activity P_1 becomes smaller and smaller as P_1 is increased. With a value $P_1 = 0.6 - 0.8$ most of the gain is already made (Fig. 4). Assuming a realistic biological time-window for coincidences of the order of 2ms, sustained frequencies of 300 - 400 Hz (as observed in biological neurons) would be sufficient for the computation time to approach its smallest possible value. As ΔL_{min} increases with the number of inputs, the spread of latencies of neurons in a same layer [3] could reflect differences in fan-in.

Preliminary investigations show that this model may also explain the psychometric curves obtained in visual masking experiments.

4   Conclusion

If the hypothesis of this model are valid, it indicates that the visual system uses maximum firing rates probably near to the optimum with regards to computation speed. The main limiting factor would therefore be the noise at the level of the sensory receptors which determines the latencies in all subsequent processing layers. Noise refers here to jitter in onset times, not to fluctuations in firing levels. A prediction of this model is that latencies in the visual system can be manipulated by modifying the onset-time jitter in retinal ganglion cells.

Aknowledgements
This work has benefitted from discussions with M.D. Plumbley and from the SERC grant GR/H22495.

References

1. **Ferster D. and Lindstrom S.** (1983) "An intracellular analysis of geniculo-cortical connectivity in area 17 of the cat", em J. Physiol., 342, 181-215.

2. **Thorpe S.J. and Imbert M.** (1989) "Biological constraints on connectionist modelling", in Pfeifer R. et al. (eds.) "Connectionnism in perspective", Elsevier, 63-92.

3. **Maunsell J.H.R. and Gibson J.** (1992) "Visual response latencies in striate cortex of the macaque monkey", J. Neurophysiol., 68, 1332-1344.

4. **Douglas R.J. and Martin K.A.C.** (1991) "A functional microcircuit for cat visual cortex", J. Physiol., 440, 735-769.

5. **Softky W.R. and Koch C.** (1993) "The highly irregular firing of cortical-cells is inconsistent with temporal integration of random EPSP's", J. Neurosci., 13, 334-350.

6. **Bugmann G.** (1992) "Multiplying with neurons: Compensation of irregular input spike trains by using time-dependent synaptic efficiencies", Biol. Cybern., 68, 87-92.

7. **Verri A., Straforini M. and Torre V.** (1992) "Computational aspects motion perception in natural and artificial neural networks", in Orban G.A. and Nagel H.H. (eds.) "Artficial and biological vision systems", Springer, 71-92.

8. **Houghton G.** (1990) "The problem of serial order: A neural network model for sequence learning and recall", in Dale R., Mellish C. and Zock M. (eds.) "Current research in natural language generation", Academic Press, London, 287-319.

9. **Granger R., Ambros-Ingerson J., Staubli U. and Lynch G.** (1990) "Memorial operation of multiple, interacting simulated brain structures", in Gluck M.A. and Rumelhart D. (eds) Neuroscience and Connectionist Theory, Lawrence Erlenbaum Associates, London, 95-129.

10. **Humphrey G.W. and Muller H.J.** (1993) "SEarch via Recursive Rejection (SERR): A connectionist model of visual search", Cognitive Psychology, 25, 43-110.

11. **Newsome W.T., Britten K.H. and Movshon J.A.** (1989) "Neuronal correlates of a perceptual decision", Nature, 341, 52-54.

12. **Gorse D. and Taylor J.G.** (1990) "A general model of stochastic neural processing", Biol. Cybern., 63, 299-306.

13. **Bugmann G.** (1991) "Summation and multiplication: Two distinct operation domains of leaky integrate-and-fire neurons", Network, 2, 489-509.

14. **Bugmann G.** (1992) "The neuronal computation time" in Aleksander I. and Taylor J.G. (eds) "Artificial neural networks II", Elsevier, 861-864.

15. **Levick W.R.** (1973) "Variation in the response latency of cat retinal ganglion cells", Vision Res., 13, 873-853.

A Neural Architecture for Textured Color Image Segmentation and Recognition

F. J. Diaz Pernas and Juan López Coronado

Department of Automatic Control and Systems
School of Industrial Engineering, University of Valladolid
Paseo del Cauce S/N, 47011, Valladolid, Spain

Abstract

We propose a neural architecture to segment textured color images using perceptual group-ings and to recognize the scene according to the obtained segmentation.

This architecture offers an approximation to the visual system structure related to the processing of color and texture informations,

Retinal ganglion cells → LGN → V1 → V2 → V4 → Inferior Temporal Cortex (IT).

This architecture is made of three main modules. The first module, called Color Opponent System (COS), transforms a RGB color signal into three signal, two chromatic ones as a result of red-green and blue-yellow color opponent processes, and an achromatic one, black-white. This system models the antagonist mechanisms, with *ON-center/OFF-surround* structures, that take place in the retinal glanglion cells. Each color opponent process is modelled as a bidimensional field with *shunting ON-center/OFF-surround* interactions.

The processing continues in a Chromatic Segmentation System (CSS) module. This mod-ule proposes an extension of the BCS/FCS visual model to achieve a preattentive color seg-mentation. The CSS extracts the real and illusory contours from the three channels and makes a perceptual grouping of the emergent textural characteristics. It is necessary to extract the contours in the chromatic channels because there can exist chromatic contours invisible to the achromatic channel.

In each channel a simple cells or oriented contrast detectors stage is processed. Our network uses receptive fields modelled by Gabor functions for these cells. The selection of this kind of functions has been favored by an important number of investigators that have pointed out that these functions approximate perfectly the shape of the receptive fields that striate cortex simple cells have. Four kind of receptive fields are used in our architecture , two of them come from even components and the other two from odd components of the Gabor function. The receptive fields used can be seen as two pairs of opposed polarity Gabor filters. Such opposed pairs exist in the visual system and would be required to preserve the information that would otherwise be lost by truncation.

Meanwhile, two FCS stages, regulated by the contour structure, activate both *filling-in* processes that allow a diffusion of the chromatic qualities in all directions except in those where there are strong contours. Each of these two FCS stages processes a chromatic channel (Red-Green or Blue-Yellow).

Finally, we propose a superior module to carry out the recognition process. This module, called MultiSignal Recognition Module (MSRM), is made up of a bidimensional architecture, 2D Multi-ART, based on the adaptive resonance theory (ART).

Dynamics of single neurons
—oral contributions

Activity-Dependent Modification of Intrinsic Neuronal Properties

L.F. Abbott*, G. LeMasson**, M. Siegel* and E. Marder**
Departments of Physics* and Biology**
and
Center for Complex Systems
Brandeis University, Waltham MA 02254 USA

Abstract

The processes that develop and maintain the intrinsic electrical properties of neurons are modeled by allowing the maximal strength of membrane conductances to be slowly varying functions of the intracellular calcium concentration. The resulting dynamic regulation of conductances allows model neurons to self-assemble their membrane currents and to react to and recover from external perturbations. This dramatically increases the stability of the model neuron and makes its intrinsic characteristics activity-dependent. For example, model neurons in a two-cell network spontaneously differentiate in response to each other's activity. In a spatially extended model neuron, the dynamic regulatory mechanism causes a non-uniform distribution of currents to develop in response to both the morphology and the activity of the neuron.

Introduction

Work on connectionist neural networks has focused largely on the effect of activity-dependent synaptic plasticity on network function [1]. At the same time, enormous experimental effort has gone into studying long-term potentiation and depression of synapses [2]. While the role of synaptic changes is of great importance to learning and memory, it should not be forgotten that the intrinsic characteristics of individual neurons can also be modified by activity [3]. Connectionist networks tend to treat individual neurons as relatively trivial dynamic systems. Real neurons are, of course, far from simple and they can exhibit a wide range of endogenous behaviors including tonic spiking, periodic firing in bursts and bistability between silent and firing states, even in the absence of synaptic input [4].

The intrinsic electrical characteristics of a neuron depend on the strength and distribution of its membrane currents. Modeling studies indicate that even small changes in current strengths or distributions can have a dramatic effect on neuronal behavior. In mathematical modeling terms, realistic neuron models have poor structural stability properties. In light of this fact, the robust stability of biological neurons is surprising and even more remarkable when we realize that the ion channels that conduct membrane currents are continually degrading and being replaced. Furthermore, neurons must adjust to their own growth and to possible changes in the extracellular environment.

The Model

In an attempt to understand how neuronal characteristics develop and are maintained in stable dynamic equilibrium, we have constructed models [5, 6] in which the maximal conductances of membrane currents are not fixed parameters as they are in most neuronal models, but are dynamic variables as they are in real cells. Knowledge of how ion channels are synthesized, transported, inserted and later removed from the membrane is far from complete. As a result, our models are highly simplified and somewhat conjectural. However, they are useful because they allow us to study what impact such processes have on the function of neurons and networks.

The basic idea for dynamic regulation of conductances is that the electrical activity of a neuron must play a feedback role in regulating its currents. It is difficult to imagine how a neuron could ever develop and maintain the membrane currents it needs to function properly without such a feedback loop. As a putative feedback element, we use intracellular calcium. Calcium enters a neuron through voltage-dependent channels and is removed or buffered by largely voltage-independent processes. As a result, the concentration of calcium inside the neuron is a sensitive indicator of its electrical activity [5, 7]. Calcium works much better than other ions for this purpose because intracellular calcium concentrations are very small, giving it the required sensitivity. Calcium is also a ubiquitous regulator of biochemical pathways and is implicated in modulation of gene expression [8], channel characteristics [9] and other activity-related processes. It is thus an ideal candidate for an activity-dependent feedback element.

The membrane potential V of the model neurons we study satisfies the partial differential equation [10],

$$C \frac{\partial V}{\partial t} = \frac{a}{2r} \frac{\partial^2 V}{\partial x^2} - I(g_i) \tag{1}$$

where C is the membrane capacitance per unit area, a is the radius of the neuronal cable at the point x and r is the resisitivity of the intracellular fluid. I is the total membrane current which depends on the parameters g_i that are the current "strengths", or more properly, maximal conductances we have been discussing. These are fixed parameters in conventional neuron models, but we choose to make them dynamic variables for the reasons discussed above. Specifically, we assume that the steady-state values of the g_i are calcium dependent and that the approach to the steady state is very slow. We take the steady-state values to be sigmoidal functions of the intracellular calcium concentration [Ca]. Thus,

$$\tau \frac{dg_i}{dt} = \frac{G_i}{1 + \exp[\pm([Ca] - C_T)/\Delta]} - g_i \tag{2}$$

where τ, G_i, C_T and Δ are parameters. A key element in the model is that changes of the g_i are slow, that is, τ is larger than any of the time constants characterizing other dynamic processes in the model. In equation (2), we use the plus sign for inward currents and the minus sign for outward currents. This choice of signs makes the sign of the feedback loop from activity to currents strengths negative and is therefore made on the basis of stability arguments [5, 6].

Since the intracellular calcium concentration plays such an important role, we must model it as well. Calcium enters the cell through voltage-dependent membrane currents that we will call collectively I_{Ca}, diffuses through the cell and is removed by a buffering mechanism that we model as exponential decay. Thus,

$$\frac{\partial [Ca]}{\partial t} = D \frac{\partial^2 [Ca]}{\partial x^2} - A I_{Ca} - B[Ca] \tag{3}$$

Here $D = 6 \times 10^{-6}$ cm^2/s is the calcium diffusion constant and $B = 1/600$ms is the calcium uptake rate. The factor A converts from Coulombs of current to moles/liter of calcium concentration. Because calcium enters through the surface of the cell while concentration refers to the volume inside the cell, A is proportional to the surface to volume ratio in the area around the point labeled by the coordinate x.

Single Compartment Model - Stability and Activity-Dependence

We have simulated the model outlined above in two different ways. First, we make a one compartment approximation. This amounts to ignoring the spatial derivatives in equations (1) and (3), making them ordinary differential equations. The results we show in the following three figures are based on a model of a neuron in the stomatogastric ganglion of the crab [11]. All of the maximal conductances g_i in this model are allowed to vary according to equation (2) above.

The figure below shows a plot of voltage versus time for the dynamically regulated model neuron. Initially, the model is adjusted to make the neuron fire in periodic bursts as shown by the activity at the start of the top trace. At the time indicated, we simulate an increase in the amount of extracellular potassium causing the neuron to go into a fast, tonic firing mode. Due to the increased activity, the intracellular calcium concentration increases and, through equation (2), this readjusts the maximal conductances g_i. The result of the readjustment is a return to a bursting mode of activity as seen in the bottom trace. Thus, the model neuron has recovered from an environmental change by regulating its conductances to preserve its pattern of electrical activity. Note that the model neuron producing the activity in the bottom trace has a completely different set of parameters than it had initially at the beginning of the top trace and that the new parameter values were found automatically by the regulatory scheme on the basis of feedback from activity through the intracellular calcium concentration. Note that, to speed up these simulations, dynamical regulation in the figures is faster than we expect it to be in real neurons.

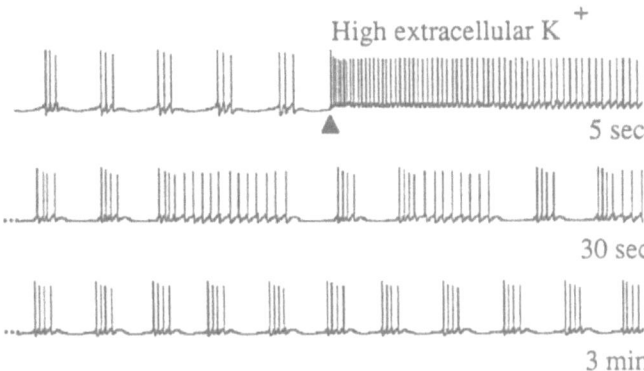

The next figure shows that the same regulatory scheme that produces stability to perturbations also induces activity-dependent intrinsic properties. Once again, the model neuron is initially a burster as seen at the beginning of the top trace. At the times indicated by bars, a pulse of external current is injected, increasing the activity of the neuron. This increased activity raises the intracellular calcium concentration thereby changing the maximal conductances g_i. We show this by stopping the pulses in the bottom trace. After the stimulation, the model neuron has different intrinsic properties, it no longer fires spontaneously in bursts, but is instead silent.

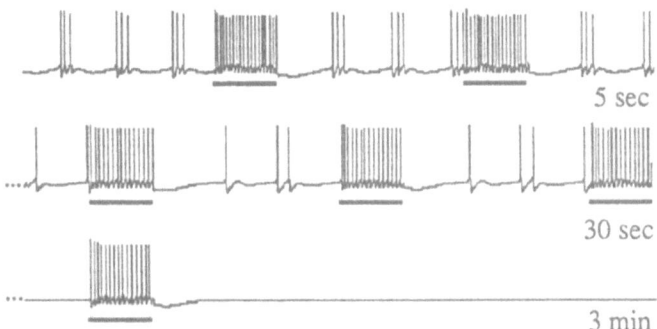

The fact that activity can shift the intrinsic electrical characteristics of these model neurons has important implications for networks. We show this in a simple two-cell network in the figure below. We begin, in panel A, with two bursting model neurons that are identical and uncoupled. In panel B, we show what happens after they are coupled electrically and allowed to interact for an extended period of time. Even though the two neurons are described by identical underlying models and the coupling is completely symmetric, their activity in the network is not identical. To see what is going on, we remove the electrical coupling and show in panel C the behavior of the two neurons immediately after the coupling is removed. Clearly neuron 1 has retained a bursting pattern of activity, but neuron 2 has changed from a bursting to a tonically firing cell. The coupling between the two neurons has shifted their intrinsic properties so that the two initially bursting neurons have spontaneously differentiated to form a circuit in which one acts as a pacemaker and the other as a follower.

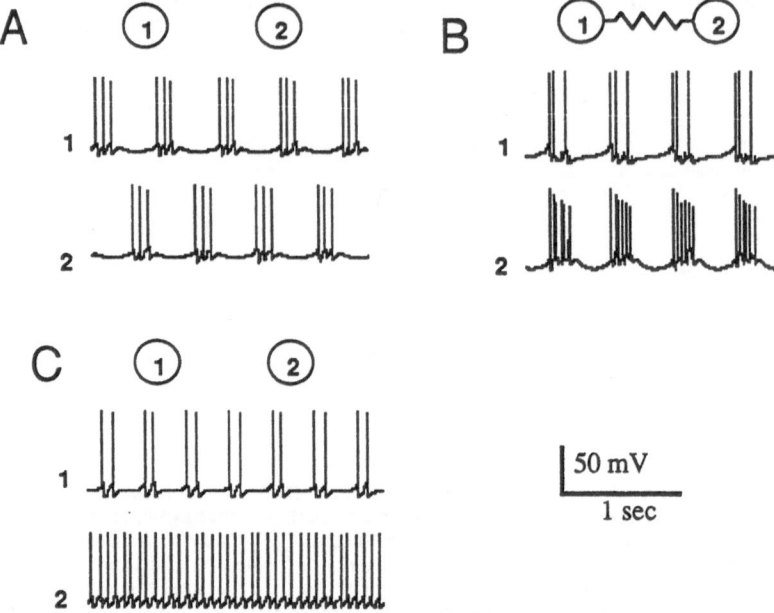

Multi-Compartment Model - Current Distributions Depend on Activity and Morphology

We have also studied the model in a multi-compartmental version with realistic spatial structure [12]. The top of the figure below shows a simple model neuron with a soma, axon and dendritic tree. We use only three active membrane currents for this spatially extended model, a fast sodium, a delayed-rectifier potassium and a calcium current. This allows the model to fire action potentials but it cannot fire spontaneously in bursts like the model considered in the last section. The calcium-dependent regulation rule (2) acts locally in this case. In other words, the calcium concentration within each compartment of the model determines the values of the g_i in that particular compartment. As a result of differences in the calcium concentration in different parts of the neuron, the g_i values will vary over the surface of the cell. Thus, the model produces a spatially structured set of membrane currents on the basis of the morphology of the cell as reflected by the local intracellular calcium concentration.

We begin the simulation by setting all three maximal conductance parameters g_i to zero. We then let the model evolve until a quasi-static set of maximal conductances is produced. During the simulation we impose a random pattern of excitatory synaptic inputs over the dendritic tree. At first, the model neuron is completely passive and no action potentials are produced. Eventually, however, a more or less realistic pattern of membrane conductances arises as shown in the following figure where we plot the maximal conductance parameter for the sodium current as a function of the position on the neuron. The sodium current is strongest in the area of the cell body and on the initial 20 microns of the axon. The axon has a uniform, intermediate value of g_{Na} and is capable of propagating action potentials. The sodium current is only weakly expressed on the dendritic tree. This produces the classic type of extended neuron where the dendritic tree integrates inputs passively and action potentials are initiated near the cell body and propagate uniformly out along the axon. This pattern of currents has arisen spontaneously from the model purely as a result of the shape and size of various elements of the neuronal structure.

The spatial distribution of membrane currents is affected by the electrical activity and synaptic input received by the model neuron. We have studied a more complex model neuron with both apical and basal dendrites [12]. If the synaptic input to one dendritic branch is stronger than to the other, we find that the more highly stimulated branch develops larger outward currents. The added leakage attenuates the input coming to the more highly stimulated branch thereby tending to equalize the synaptic input coming from both dendritic trees. This example shows that the distribution of membrane currents is affected by the pattern of synaptic input as well as by the morphology of the model neuron.

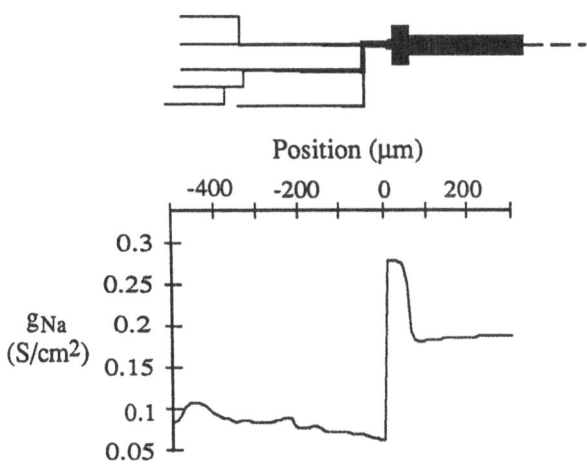

Conclusions

Conductance-based models can simulate the behavior of neurons quite realistically over time scales ranging from milliseconds to seconds. However, over longer time periods ranging from hours to days additional processes involving the synthesis, transport and modulation of ion channels become relevant. These can produce activity-dependent shifts in the intrinsic electronic properties of neurons that, along with synaptic plasticity, are an important element for network development, adaptation and learning. We have made a start at modeling these slower processes and have found that a number of interesting properties arise [5, 6, 12]:

1) Dynamically regulated model neurons can self-assemble the conductances needed to achieve a desired target electrical activity pattern.

2) When extracellular conditions change, the regulated model neurons can adjust to find a new set of conductances that restore the original pattern of activity.

3) The calcium-dependent feedback loop in our model causes the intrinsic properties of model neurons to shift in response to stimulation and to the presence of other neurons in a network.

4) Coupled neurons can differentiate spontaneously so that neurons described by an identical underlying model can develop different sets of conductances and can play different roles in the functioning of a network.

5) A non-uniform distribution of membrane currents can arise in a spatially extended model neuron in response to spatial variations of the intracellular calcium concentration caused by both the morphology and the synaptic input to the cell.

6) Synaptically driven shifts in the distribution of membrane currents tend to equalize synaptic inputs that are non-uniform over the dendritic tree.

These observations have important implications for networks. A role for the plasticity of the intrinsic properties of neurons has been suggested [13], but this form of plasticity is often overlooked by network modelers. The existence of activity-dependent intracellular processes [3, 8, 9] and their impact on model neurons [5, 6, 12] suggests that plasticity at nodes may be an important element in network learning.

References

1. Gluck, M.A. and Rumelhart, D.E. 1990 **Neuroscience and Connectionist Theory.** Lawrence Erlbaum, Hillsdale NJ.

2. Byrne, J.H. and Berry W.O. 1989 **Neural Models of Plasticity.** Academic Press, San Diego.

3. Franklin, J.L., Fickbohm, D.J. and Willard, A.L. 1992 Long-term regulation of neuronal calcium currents by prolonged changes of membrane potential. *J. Neurosci.* 12:1726-1735.

4. Marder, E. 1993 Modulating membrane properties of neurons: role in information processing. In Poggio, T.A. and Glaser, D.A. eds. **Exploring Brain Functions: Models in Neuroscience.** Wiley, Chichester. pp. 27-42.

5. LeMasson, G., Marder, E. and Abbott, L.F. 1993 Activity-Dependent Regulation of Conductances in Model Neurons. *Science* 259:1915-1917.

6. Abbott, L.F. and LeMasson, G. 1993 Analysis of Neuron Models with Dynamically Regulated Conductances. *Neural Comp.* (in press).

7. Ross, W.M. 1989 Changes in intracellular calcium during neuron activity. *Annu. Rev. Physiol.* 51:491-506.

8. Armstrong, R.C. and Montminy, M.R. (1993) Transynaptic control of gene expression. *Annu. Rev. Neurosci.* 16:17-30.

9. Kaczmarek, L.K. and Levitan, I.B., eds. 1987 **Neuromodulation. The Biochemical Control of Neuronal Excitability.** Oxford Univ. Press, NY.

10. For an introduction to conductance-based neuronal modeling see: Abbott, L.F. 1993 Single Neuron Dynamics: An Introduction. In **Proceedings of Capri School for Neuroscience** (in press).

11. Buchholtz, F., Golowasch, J., Epstein, I. and Marder, E. 1992 Mathematical model of an identified stomatogastric neuron. *J. Neurophysiol.* 67:332-340.

12. Siegel, M., Marder, E. and Abbott, L.F. 1993 Activity-dependent conductances produce nonuniform current distributions in spatially extended model neurons. (In preparation).

13. Levy, W.B., Colbert, C.M. and Desmond, N.L. 1990 Elemental adaptive processes of neurons and synapses: a statistical/computational perspective. In Gluck, M.A. and Rumelhart, D.E. eds., **Neuroscience and Connectionist Theory.** Lawrence Erlbaum, Hillsboro, N.Y. pp. 187-236.

The present address for G. LeMasson is: Laboratoire de Neurobiologie et Physiologie Comparees, Place du Dr. Peyneau, 33120 Arcachon, France.

Implications of Activity-Dependent Neurite Outgrowth for Developing Neural Networks

A. van Ooyen and J. van Pelt

Netherlands Institute for Brain Research, Meibergdreef 33, 1105 AZ Amsterdam, The Netherlands

Abstract

The presence of only two basic neuronal properties - a firing threshold and activity-dependent neurite outgrowth - is sufficient to cause a transient overproduction of connections in developing neural networks. This overproduction is enhanced by inhibition. Solely as the result of activity-dependent outgrowth and local cell interactions, the neuritic field of an inhibitory cell becomes smaller than that of an excitatory cell. Furthermore, a specific distribution of cell sizes is generated in the area surrounding an inhibitory cell.

Introduction and Summary

Among the many factors influencing the ultimate structure of the nervous system, electrical activity plays a pivotal role. Many mechanisms that determine neuronal connectivity, such as neurite outgrowth, synaptogenesis and cell death, have been found to be modulated by electrical activity. For example, electrical activity of the neuron reversibly arrests neurite outgrowth (or produces retraction). High levels of activity, resulting in high intracellular calcium concentrations, cause neurites to retract, whereas low levels of activity, and consequently low calcium concentrations, allow further outgrowth [1]. As a consequence of such activity-dependent processes, a mutual influence exists between the formation of synaptic connectivity and neuronal activity, i.e., a feedback loop exists between changes in network structure and changes in network activity. This feedback loop must be expected to have major implications not only for the structure of the mature network, but also for the stages a network goes through during its development. Neurons develop their mature form and connectivity pattern under influence of both intrinsic factors and environmental factors (e.g., local cell interactions). During this development, all kinds of structural elements such as neuritic extensions and synapses are initially overproduced (so-called overshoot phenomena) [2].

In this article, we have studied the implications of activity-dependent neurite outgrowth for network formation, using a model in which initially disconnected cells organize themselves into a network under the influence of their intrinsic activity. A neuron is modelled as a neuritic field, the growth of which depends upon its own level of activity, and neurons become connected when their fields overlap. In a purely excitatory network, we have demonstrated that activity-dependent outgrowth in combination with a neuronal response function with some form of firing threshold - which gives rise to a hysteresis effect - is sufficient to cause an overshoot with respect to connectivity or synapse numbers [3]. Here we show that in the presence of inhibition, overshoot can still take place and is in fact enhanced if inhibitory cells (inh-cells) grow at the same rate as excitatory cells (exc-cells). If, on the other hand, the development of inhibition is delayed (i.e., if the outgrowth rate of inh-cells is lower than that of exc-cells) overshoot is not enhanced, while now the growth curve of the number of inhibitory connections does not exhibit overshoot. An interesting emergent property of the model is that - solely as the result of simple outgrowth rules and local cell interactions - the size of the (dendritic) fields of inh-cells become always smaller than that of exc-cells, even if both types of cells have exactly the same growth properties. Furthermore, a specific distribution of cell sizes is generated in the area surrounding an inh-cell.

The results are robust and show certain similarities with findings in developing cultures of dissociated nerve cells.

The Model

We use a distributed network, with neuron dynamics governed by (shunting model):

$$\frac{dX_i}{dt} = -\frac{X_i}{\tau} + (A - X_i)\sum_k^N W_{ik}F(X_k) - (B + X_i)\sum_l^M W_{il}F(Y_l) \tag{1}$$

$$\frac{dY_j}{dt} = -\frac{Y_j}{\tau} + (A - Y_j)\sum_k^N W_{jk}F(X_k) - (B + Y_j)\sum_l^M W_{jl}F(Y_l) \tag{2}$$

where X_i and Y_j are the membrane potentials of the N exc-cells and M inh-cells, respectively, A and $-B$ are the saturation potentials, $1/\tau$ determines the rate of decay, and W_{ik}, W_{il}, W_{jk}, W_{jl} are the connection strengths (all $W \geq 0$). All potentials are relative to the resting potential, which is set to 0. The firing-rate function, F, is taken to be the same for exc- and inh-cells, and is given by:

$$F(x) = \frac{1}{1 + e^{(\theta - x)/\alpha}}. \tag{3}$$

where x is the membrane potential, α determines the non-linearity, and θ is the firing threshold. The small firing rate when the membrane potential is sub-threshold can be thought of as arising from spontaneous activity. In most of the simulations we took $A=B=1$, $\tau=8$, $\alpha=0.10$ and $\theta=0.5$. The results, however, do not depend on the exact choices of the parameters. Growing cells are modelled as growing circular fields, which might be conveived of as neuritic fields. When two such fields overlap, the corresponding neurons become connected with a strength proportional to the area of overlap:

$$W_{ij} = A_{ij}c \tag{4}$$

where $A_{ij} = A_{ji}$ is the amount of overlap ('number of synapses'; $A_{ii} = 0$) and c is a constant of proportionality ('synaptic strength'). Although in this way a symmetric network is built, the result do not depend on symmetry. The outgrowth of each individual cell depends upon its own level of electrical activity (in the same way for exc-and inh-cells):

$$\frac{dR_i}{dt} = \rho\left[1 - \frac{2}{1 + e^{(\epsilon - F(x_i))/\beta}}\right] \tag{5}$$

where R_i is the radius of the field of cell i, ϵ is the firing-rate at which $\frac{dR_i}{dt} = 0$, and β determines the non-linearity. Eqn(5) is just a phenomenological description of Kater's hypothesis [1] that the depolarization level of the neuron influences (via Ca^{2+} influx) its rate of outgrowth. Any other function for which $\frac{dR_i}{dt} > 0$ at low values of $F(x_i)$ and < 0 at high values will yield similar results. ρ is taken so small that the connectivity can be regarded as quasi-stationary on the time scale of membrane potential dynamics.

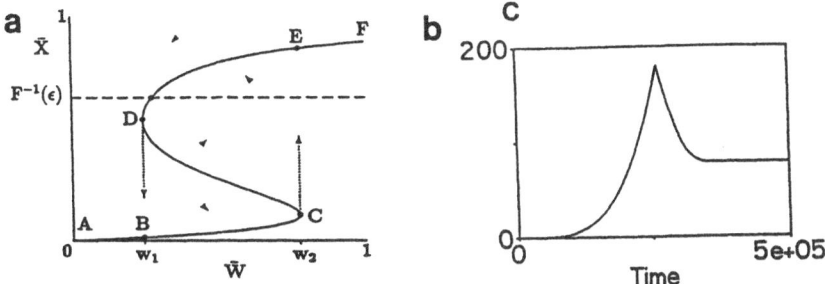

Fig.1. a Hysteresis. Steady state dependence ($\frac{d\bar{X}}{dt} = 0$) on $\bar{W}$. See text. **b** Overshoot in connectivity (C = total area of overlap = $\frac{1}{2}\sum_{i,j}^{N} A_{ij}$). Simulation with $c = 0.1$, $\epsilon = 0.60$, $N = 64$.

Results

Excitatory Network

Overshoot. To understand the occurrence of overshoot, first consider a purely excitatory network (i.e., M=0). For a given connectivity (**W** need not be symmetric) this network has convergent activation dynamics [4]; the equilibrium points are solutions of:

$$0 = -X_i/\tau + (1 - X_i)\sum_{j}^{N} W_{ij}F(X_j) \qquad \forall i \tag{6}$$

If the variations in X_i are small (relative to $\bar{X}$, the average membrane potential of the network), we find [3]:

$$0 \simeq -\bar{X}/\tau + (1 - \bar{X})\bar{W}F(\bar{X}) \tag{7}$$

Based on this approximation, the average connection strength $\bar{W}$ can be written as a function of $\bar{X}$:

$$\bar{W} = \frac{\bar{X}/\tau}{(1 - \bar{X})F(\bar{X})} \qquad 0 \le \bar{X} < 1 \tag{8}$$

which gives the steady state ($\frac{d\bar{X}}{dt} = 0$) dependence on $\bar{W}$ (Fig. 1a).

The steady states are stable on the branches ABC and DEF, and unstable on CD. This hysteresis loop (whose presence hinges upon a sigmoidal firing-rate function) underlies the emergence of overshoot. The $R_i's$, and therefore $\bar{W}$, are governed by the system itself, being under control of neuronal activity. The size of a neuritic field remains constant if $X_i = F^{-1}(\epsilon)$, where F^{-1} is the inverse of F. Thus, the equilibrium point of the system is the intersection point of the line $\bar{X} = F^{-1}(\epsilon)$ with the curve of Fig. 1a. Because the activity in the network is initially low ($\bar{W} = 0$), $\frac{dR_i}{dt}$ is positive and $\bar{W}$ increases, whereby $\bar{X}$ follows the branch ABC until it reaches w_2, where it jumps to the upper branch, thus exhibiting a phase transition from quiescent to activated state. The activity in the network is then however so high, that the neuritic fields begin to retract ($\frac{dR_i}{dt} < 0$) and $\bar{W}$ to decrease, whereby $\bar{X}$ moves along the upper branch from E to the intersection point. Thus, in order to arrive at an equilibrium point on DE, a developing network has to go through a phase in which $\bar{W}$ is higher than in the final situation, thus exhibiting a transient overshoot in $\bar{W}$ (Fig. 1b).

Local behaviour. In a purely excitatory network all the cells are exactly identical except for their position. Local variations in cell density, however, suffice to generate a great variability among individual cells, with respect both to their neuritic field size at equilibrium (Fig. 2)

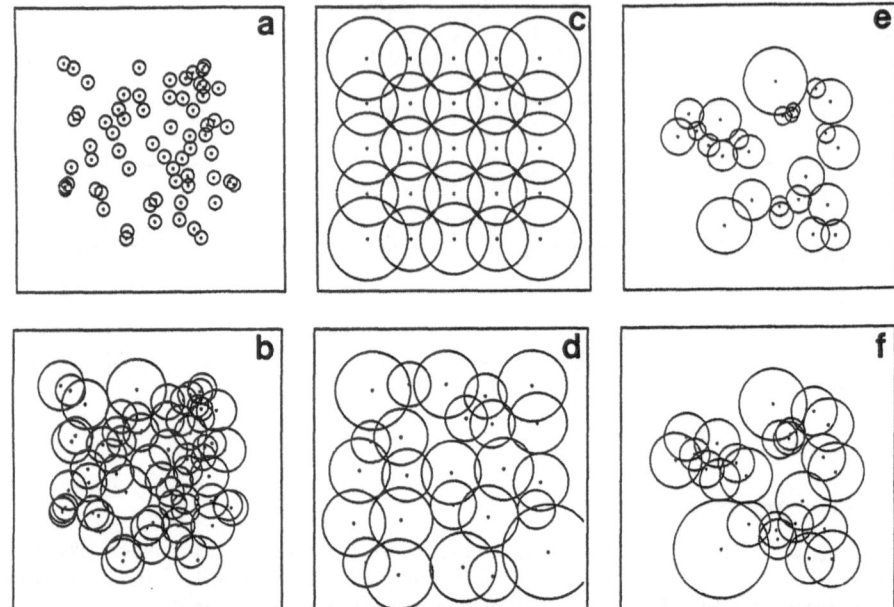

Fig.2. Excitatory networks. **a** Initial network and **b** at equilibrium. **c** At equilibrium, cells on grid. **d** Noise on grid. **e** High c results in sub-networks. **f** Same network but with lower c.

and to their developmental course of field size and firing behaviour. Cells surrounded by a high number of neighbouring cells tend to become small since a small neuritic field will already give sufficient overlap with other cells. In contrast, relatively isolated cells must grow large neuritic fields in order to contact a sufficient number of cells. One might say that the neuritic fields adapt to the available space so as to cover it optimally.

Mixed Network

Overshoot. In the presence of inh-cells, overshoot is enhanced (Fig. 3a). To counteract the effect of inhibition, a larger (excitatory) connectivity is necessary for the network to become activated. Also the equilibrium connectivity level must be higher. If inhibition is strong (many inh-cells or a high synaptic strength from inh-to-exc cells), the electrical activity in the network remains so low that the cells keep growing out (increasing the exc-exc and inh-exc overlap). With moderate inhibition, complicated interactions are possible between oscillatory activity and outgrowth (not shown).

Even without differences in local cell density, the presence of inhibition generates variability among individual cells, e.g., with respect to their firing behaviour. Cells that receive inhibition become activated later and retract later than cells that do not receive inhibition. For the overshoot curve this means that after the onset of network activity the average connectivity can still increase considerably (in contrast to what is found in purely excitatory networks), because there may be cells that are still growing out, while others are already retracting.

Some studies have indicated that inhibition develops later in time than excitation [5]. If the development of inhibition is delayed by giving inh-cells a lower outgrowth rate, overshoot is

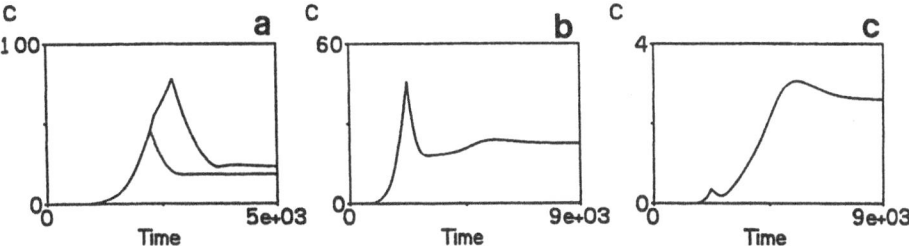

Fig.3. **a** Overshoot in exc-exc connectivity is larger with ($N = 14$, $M = 2$) than without ($N = 14$, $M = 0$) inhibition. **b** Exc-exc and **c** exc-inh connectivity when ρ of inh-cells is lower than that of exc-cells.

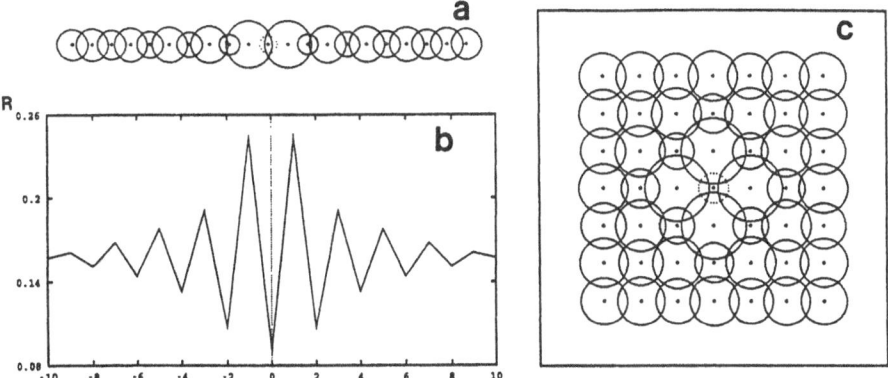

Fig.4. Mixed networks at equilibrium (torus boundary conditions). **a,c** Central cell (dashed line) is inhibitory. **b** Size of neuritic field (R) against position in network of **a**.

not enhanced (Fig. 3b), while now the growth curve of the number of *inhibitory* connections does not exhibit overshoot (Fig. 3c), because the inh-cells develop into a network that is already electrically active.

Local behaviour. Solely as the result of activity-dependent outgrowth and local cell inter-actions, the neuritic field of an inh-cell becomes smaller than that of an exc-cell. To illustrate this, consider a one-dimensional string of cells with only one inh-cell in the middle (Fig. 4a). Initially, all cells have the same size. At equilibrium, however, the inh-cell becomes the small-est cell, surrounded by two large exc-cells, which, in turn, are surrounded by smaller cells, and so on. Thus, even without local variations in cell density, a great and specific variability among individual cells is generated with respect to neuritic field size. Though more complex, essentially the same situation is obtained in the two-dimensional case (Fig. 4c). Clearly, the effect of an inh-cell is not restricted to its direct neighbours, but percolates through the network, so that a particular distribution of cell sizes is imposed. The mechanism causing cell sizes to differ is as follows. Since every cell tries to adjust its neuritic field so that $F(x_i) = \epsilon$ [eqn(5)], cells connected to an inhibitory cell must grow large neuritic fields to receive suf-ficient excitation. As a result, large exc-cells will surround an inh-cell, whereas the inh-cell itself can remain small because a small neuritic field will already give sufficient overlap with its large surrounding cells.

Conclusions and Discussion

The presence of activity-dependent neurite outgrowth was found to have implications both for the stages a network goes through during its development and for the structure of the stable network. During development, activity-dependent neurite outgrowth in combination with a nonlinear firing-rate is sufficient to cause an overshoot with respect to connectivity or synapse numbers. This overproduction is enhanced by the presence of inhibition (if it develops simultaneously with excitation). The results are robust under: different firing-rate functions (provided they have a type of firing threshold and low but none-zero values for sub-threshold membrane potentials, i.e., spontaneous activity); variance among neurons in all parameters; different neurite outgrowth functions; the way in which connections are defined; network size and different neuron models. Our model results are similar to those found in developing tissue cultures with respect to a transient overproduction in the numerical density of synapses [6] and the existence of a transition period wherein increasing electrical activity is associated with retraction of neurites [7]. Moreover, the model responds in a similar way to suppression or intensification of electrical activity [6]. In the model, the number of inhibitory connections exhibits no overshoot if inhibition develops later in time than excitation. The observation that in tissue cultures the putative inhibitory synapses (synapses on shafts) show hardly any overshoot, while the synapses on spines - which are probably mostly excitatory - show a clear overshoot [6], would thus be in agreement with a progressive increase in the ratio of effective inhibitory to excitatory synaptic activity during development, as suggested in [5].

The distribution of neuritic field sizes emerging in the stable network show that the fields of inhibitory cells become smaller than those of excitatory cells, even if both types of cells have exactly the same growth properties. Indeed, the (dendritic) fields of most of the inhibitory cells in the cortex are smaller than those of excitatory cells [e.g., 8]. The neuritic field size of the model cells adapts to the local cell density, much in the same way as the dendritic fields of ganglion cells in the retina [9].

References

[1] Kater SB, Guthrie PB, Mills LR (1990) Integration by the neuronal growth cone: a continuum from neuroplasticity to neuropathology. Progress in Brain Research 86: 117-128.

[2] Purves D, Lichtman JW (1980) Elimination of synapses in the developing nervous system. Science 210: 153-257.

[3] Van Ooyen A, Van Pelt J. Activity-dependent outgrowth of neurons and overshoot phenomena in developing neural networks. J. Theor. Biol. (in press).

[4] Hirsch MW (1989) Convergent activation dynamics in continuous time networks. Neural Networks 2: 331-349.

[5] Corner MA, Ramakers GJA (1992) Spontaneous firing as an epigenetic factor in brain development - physiological consequences of chronic tetrodoxin and picrotoxin exposure on cultured rat neocortex neurons. Dev. Brain Res. 65: 57-64.

[6] Van Huizen, F., Romijn, H. J., Habets, A. M. M. C. (1985) Synaptogenesis in rat cerebral cortex is affected during chronic blockade of spontaneous bioelectric activity by tetrodoxin. Dev. Brain Res. 19: 67-80.

[7] Schilling K, Dickinson MH, Connor JA, Morgan, JI (1991) Electrical activity in cerebellar cultures determines purkinje cell dendritic growth patterns. Neuron 7: 891-902.

[8] Parnavelas, JG (1984) Physiological properties of identified neurons. In: Cerebral Cortex, EG Jones and A Peters (eds.), Plenum Press, pp. 205-239.

[9] Wässle L, Peichl L, Boycott BB (1981) Dendritic territories of cat retinal ganglion cells. Nature 292: 344-345.

PCA Properties of Interneurons

Colin Fyfe
colin@cs.strath.ac.uk
Department of Computer Science
University of Strathclyde,
Glasgow,
Scotland.

Abstract

A review is given of a novel, very computationally simple form of artificial neural network: the network contains no weight decay or clipping of weights, yet is proved to be equivalent to Oja's Subspace Algorithm. An algorithm for using the network to calculate the actual Principal Components is given. Experimental results are given and compared to Oja's Weighted Subspace Algorithm. A set of biologically more feasible initial conditions are proposed and investigated both analytically and experimentally.

1 Introduction

Principal Component Analysis is a technique for finding the code which will encode as much information in a data set as possible in a given number of values. It is well known that the optimal solution to this encoding is to be found in terms of the eigenvectors corresponding to the largest eigenvalues of the input data covariance matrix. Neural networks provide a novel way to calculate principal components from on-line data.

Currently the most popular method of computing Principal Components using neural nets is unsupervised Hebbian learning with weight decay (e.g. [1, 4, 5, 6, 7, 8]).

It has been shown [3] that the crucial form of the Hebbian learning rule which makes the weights converge to the principal eigenvectors is

$$\Delta w_{ij} = \alpha x_j y_i - \gamma(w_{ij}) w_{ij} \tag{1}$$

where α is the learning rate, x_j is the jth input, y_i is the ith output and $\gamma(w_{ij})$ is some function of the weight, w_{ij}.

The critical point is that, to ensure that the weights converge to the principal subspace[1] of the input data, there must be, built into the learning rule, a weight decay term which is the product of a function of the weights and the weights themselves.

We can see a specific example of this rule in Oja's(1989) Subspace Algorithm:

$$\Delta w_{ij} = \alpha(x_i y_j - y_j \sum_k w_{ik} y_k) \tag{2}$$

which has been shown to converge to the subspace of the principal eigenvectors.

2 Interneuron Equivalence to Oja's Subspace Algorithm

It has been shown [2] that a layer of interneurons whose weights are modified using simple Hebbian learning can also be used to calculate principal components.

---

[1]The principal subspace is the space of points whose basis is those eigenvectors corresponding to the largest eigenvalues of the covariance matrix of the input data

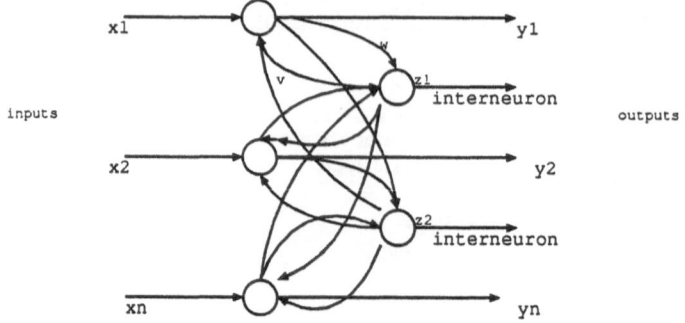

Figure 1: The interneurons sum the weighted y values which are themselves calculated by subtracting from the corresponding x values the weighted sum of the z values

The rules governing the dynamics of this network which is shown in Figure 1 are

$$\mathbf{y} = \mathbf{x} - \mathbf{W}\mathbf{z}$$
$$\mathbf{z} = \mathbf{W}^{\mathbf{T}}\mathbf{y}$$
$$\Delta\mathbf{W} = \eta\mathbf{y}\mathbf{z}^{\mathbf{T}}$$

There is no explicit weight decay, normalisation or clipping of weights in the model. If we consider the network as a transformation from inputs, $\mathbf{x}$, to interneuron outputs, $\mathbf{z}$, and consider the effects of these rules on individual neurons, we can show that the resultant network is equivalent to Oja's Subspace Algorithm. We have

$$y_i = x_i - \sum_k w_{ki} z_k \tag{3}$$

$$z_i = \sum_j w_{ij} y_j \tag{4}$$

Therefore, substituting equation (3) in the learning rule,

$$
\begin{aligned}
\Delta w_{ij} &= \eta y_i z_j \\
&= \eta (x_i - \sum_k w_{ki} z_k) z_j \\
&= \eta (x_i z_j - z_j \sum_k w_{ki} z_k)
\end{aligned}
$$

This last formulation of the learning rule is exactly the learning rule for Oja's(1989) Subspace Algorithm, given earlier. Experimental results have previously been given to substantiate this analysis[2].

3 Algorithm for PCA

The Subspace Algorithm has been improved with the following modification [6] to the learning rule

$$\Delta w_{ij} = \eta y_j (x_i - \theta_j \sum_{k=1}^{N} y_k w_{kj}) \tag{5}$$

Ensuring that $\theta_1 < \theta_2 < \theta_3 < ...$ allows the neuron whose weight decays proportional to θ_1 (i.e. whose weight decays least quickly) to capture the principal component of the variance. The second captures the next largest component, and so on.

W			W		
1.000	-0.036	-0.008	**1.054**	-0.002	-0.002
0.036	**0.999**	-0.018	0.002	**1.000**	0.001
0.010	0.018	**1.000**	0.003	-0.002	**0.954**
-0.002	-0.002	0.016	-0.001	0.001	-0.002
0.010	0.003	0.010	0.001	-0.001	0.000

Table 1: Results from the interneuron network (left) and from Oja (right).
Both methods find the principal eigenvectors of the input data covariance matrix. The interneuron algorithm has the advantage that the each vector is equally weighted.

This algorithm is local and homogeneous in that each neuron knows only its own value of θ_i. Analysis of the interneuron learning rule shows that, to simply insert a parameter, θ_i, would require computation at the level of the synapse. Whilst this may be biologically feasible and algorithmically simple to implement, a different algorithm is developed here which uses the fact that the proposed network already incorporates subtraction of values.

The algorithm is: the interneuron network evolves initially with only 1 interneuron: this interneuron finds the first principal component using the above learning rule. We then create a second interneuron. Since the process of training the first interneuron has found and *subtracted* the first principal component, the second interneuron will find the largest remaining principal component. Then the third interneuron is created etc..

To compare the results with Oja's (1992) Weighted Subspace Algorithm, the experiment reported in [6] was repeated with this algorithm. The results shown in Table 1 are from a network with 5 inputs each of zero mean random Gaussians, where x_1's variance is largest, x_2's variance is next largest, and so on. The advantage of using such data is that it is easy to identify the principal eigenvectors as the largest eigenvalue of the input data's covariance matrix comes from the first input, x_1, the second largest comes from x_2 and so on.

When the random Gaussians were generated, the y-values were initially taken to be equal to the x-values. Then the z-values are calculated and subtracted out of the x-values as the equations indicate. The simulation in [6] was carried out for 40000 iterations. The interneuron simulation allowed each interneuron to learn in 13000 iterations. The first interneuron learnt during the first 13000 iterations, the second learnt during the next 13000 and the third learnt during the last 13000 iterations. The results are shown in Table 1; the left set is from the interneuron network, the right ones from [6].

A mathematical analysis of the algorithm has been developed where convergence of the algorithm is proved in 3 stages.

1. The weights of a single interneuron are shown to converge to an eigenvector of the input data's covariance matrix

2. This eigenvector is shown to be that corresponding to the largest eigenvalue

3. The weights of the kth interneuron are shown to converge to the kth eigenvector

4 The VW Model

The results of the last section have one major drawback when considered as a model of biological systems: the weights of the connections from the interneuron,z, to the summing neuron,y, are assumed to be identical to those from the summing neuron,y, to the interneuron,z. This is biologically implausible. We now propose a model where these weights are initially different.

$$y = x - Vz \tag{6}$$
$$z = Wy = Wx \tag{7}$$

$$\Delta \mathbf{W} = \alpha_w \mathbf{y}\mathbf{z}^T \tag{8}$$

$$\Delta \mathbf{V}^T = \alpha_v \mathbf{y}\mathbf{z}^T \tag{9}$$

where the initial values of both $\mathbf{V}^T$ and $\mathbf{W}$ are small random numbers not correlated in any way with each other.

Note that both learning rules for $\mathbf{W}$ and $\mathbf{V}$ are identical up to the learning rate and use only simple Hebbian learning. Since we are interested in the weights to and from the interneurons, we will use the convention that $\mathbf{w_j}$ is the vector of weights into the ith interneuron and $\mathbf{v_j}$ is the vector of weights from the ith interneuron. Then $\mathbf{w_j}$ is a row of $\mathbf{W}$ while $\mathbf{v_j}$ is a column of $\mathbf{V}$. We consider the effect of these rules on a network with a single interneuron,z.

Let w_i be the weight of the connections from y_i to z and and v_i be that from z to y_i.

If the weights of a single interneuron converges to a limit, the expected weight change over a sufficiently long time will tend to zero. Given the usual approximations, particularly regarding the learning rate η, and using $< x >$ to indicate the average value of x over the time period,

$$
\begin{aligned}
< \Delta w_i >= 0 &\iff < \eta y_i z >= 0 \\
&\iff < y_i z >= 0 \\
&\iff < (x_i - v_i z)z >= 0 \\
&\iff < (x_i - v_i \sum_k w_k x_k) \sum_l w_l x_l >= 0 \\
&\iff < \sum_l w_l x_l x_i - v_i \sum_{k,l} w_k x_k x_l w_l >= 0 \\
&\iff \sum_l w_l C_{li} - v_i \sum_{k,l} w_k C_{kl} w_l = 0
\end{aligned}
$$

where C_{ij} is that element of the co-variance matrix of the input data $\mathbf{x}$ showing the co-variance between the ith and j^{th} elements. We note that the same criterion may be deduced from $< \Delta v_i >= 0$. If the weights of the interneuron are to converge, then the above must be true for all values of w_i. Therefore it may be written in matrix notation as

$$
\begin{aligned}
< \Delta \mathbf{w} >= 0 &\iff \mathbf{Cw} - (\mathbf{w}^T \mathbf{Cw})\mathbf{v} = 0 \\
&\iff \mathbf{Cw} = (\mathbf{w}^T \mathbf{Cw})\mathbf{v}
\end{aligned}
$$

Now it is a standard result that the co-variance matrix $\mathbf{C}$ is positive-semidefinite; and hence

$$\mathbf{w}^T \mathbf{Cw} = \gamma \geq 0 \tag{10}$$

where γ is a non-negative real number. Hence ,

$$\mathbf{Cw} = \gamma \mathbf{v} \tag{11}$$

Therefore, if $\mathbf{w}$ converges to an eigenvector of $\mathbf{C}$ (see below), then $\mathbf{Cw} = \lambda \mathbf{w}$ for some real number,λ), and so $\mathbf{v} = \alpha \mathbf{w}$, where α is a scalar; that is, $\mathbf{v}$ and $\mathbf{w}$ converge to the same eigenvector. Therefore, it is possible to apply the further analysis developed for the WW network and hence show that the ith interneuron converges to the ith eigenvector of the covariance matrix.

Experimental results, shown in Table 2 confirm this. It can be seen that both $\mathbf{v}$ and $\mathbf{w}$ converge to the same eigenvector, although the results are slightly less clear cut that in the previous algorithm. However, given the simplicity of this biologically inspired model, the results are extremely clear: any entity which used such a method would be able to extract the greatest amount of information from its environment with a minimal amount of interneurons using a very simple learning rule.

5 Convergence to an Eigenvector

It remains to show that the weights, if they converge, do so to an eigenvector. Assume that there is a solution of

$$\mathbf{Cw} = \gamma \mathbf{v} \tag{12}$$

Inter	neuron	model	VW	Model				
W			W			V		
1.000	-0.036	-0.008	**0.985**	-0.041	-0.003	**1.013**	-0.017	-0.024
0.036	**0.999**	-0.018	-0.019	**1.033**	0.031	-0.027	**0.965**	0.032
0.010	0.018	**1.000**	0.022	-0.032	**1.028**	0.020	-0.017	**0.969**
-0.002	-0.002	0.016	-0.024	-0.041	0.038	-0.007	-0.034	0.037
0.010	0.003	0.010	0.098	-0.007	-0.011	0.010	0.000	0.002

Table 2: Results from the interneuron network (left) with symmetric weights,W. and for the V and W vectors from the VW Model(see text)

where $\mathbf{w}$ is not an eigenvector nor the degenerate solution, $\mathbf{w} = \mathbf{0}$.

Let the eigenvectors of $\mathbf{C}$ be $\mathbf{c_1}, \mathbf{c_2}, ..., \mathbf{c_n}$. Then

$$\mathbf{w} = \sum_{i=1}^{n} w_i \mathbf{c_i}$$

Since $\mathbf{w} \neq \mathbf{0}$, there exists a direction $\mathbf{c_b}$, such that $w_b \neq 0$. Since $\mathbf{w}$ is not an eigenvector, there exists 1 other direction $\mathbf{c_a}$ with a non-zero component w_a.

Then

$$\mathbf{w} = w_a \mathbf{c_a} + w_b \mathbf{c_b} + \sum_{i \neq a,b} w_i \mathbf{c_i}$$

where $1 \leq a, b \leq n, a \neq b$, and

$$\mathbf{v} = v_a \mathbf{c_a} + v_b \mathbf{c_b} + \sum_{i \neq a,b} v_i \mathbf{c_i}$$

and from Equation 12,

$$\lambda_b w_b = \gamma v_b$$

$$\lambda_a w_a = \gamma v_a$$

Consider a disturbance of magnitude $\epsilon > 0$ in the direction of $\mathbf{c_a}$ i.e. a disturbance of ϵ_a. Then if $\mathbf{w}$ is a stable point of convergence of the weights, the expected change in the weights over time is zero. Therefore,

$$< \Delta \mathbf{w} > = 0 \iff \mathbf{Cw} - (\mathbf{w}^T \mathbf{Cw})\mathbf{v} = 0$$

$$\iff \mathbf{C}(w_a \mathbf{c_a} + w_b \mathbf{c_b} + \epsilon_a + \sum_{i \neq a,b} w_i \mathbf{c_i}) - \gamma'(v_a \mathbf{c_a} + v_b \mathbf{c_b} + \sum_{i \neq a,b} v_i \mathbf{c_i}) = 0$$

$$\iff \lambda_a w_a \mathbf{c_a} + \lambda_b w_b \mathbf{c_b} + \lambda_a \epsilon_a + \sum_{i \neq a,b} \lambda_i w_i \mathbf{c_i} - \gamma' v_a \mathbf{c_a} - \gamma' v_b \mathbf{c_b} - \gamma' \sum_{i \neq a,b} v_i \mathbf{c_i} = 0$$

where $\gamma' = (\mathbf{w} + \epsilon_a)^T \mathbf{C}(\mathbf{w} + \epsilon_a) \geq 0$ since $\mathbf{C}$ is a positive semi-definite matrix.

Now, considering the components of the transformation in the direction of $\mathbf{c_b}$,

$$\lambda_b w_b - \gamma' v_b = 0$$

Then, $\quad \gamma v_b - \gamma' v_b = 0$

Therefore, $\gamma = \gamma'$ since $v_b \neq 0$. Now, considering the components of the transformation in the direction of $\mathbf{c_a}$,

$$\lambda_a w_a + \lambda_a \epsilon - \gamma' v_a = 0$$

$$\gamma v_a - \gamma v_a + \lambda_a \epsilon = 0$$

$$\lambda_a \epsilon = 0$$

which is a contradiction. Hence there does not exist a non-zero, non-eigenvector solution to equation (12).

6 Conclusion

The interneuron network finds the principal components of the input data; it is characterised by

simplicity - there are no logistic or hyperbolic functions to be calculated; there is no additional computation within the learning rule; there is no sequential passing back of errors or decay terms.

homogeneity - every interneuron is performing exactly the same calculation as its neighbours; every summing neuron is performing exactly the same calculation as its neighbours.

locality of information - each interneuron uses only the information which it receives from its own connections; similarly with the summing neurons which calculate the y values

The VW model retains these properties while making a step towards a more biologically realistic model.

References

[1] Földiák, P (1992) Models of Sensory Coding *PhD Thesis, University of Cambridge*

[2] Fyfe, C (1993) Interneurons which Identify the Principal Component Subspace in *Recent Advances in Neural Networks, Conference Proceedings of BNNS'93*

[3] Miller,K. and MacKay,D., (1992) The Role of Constraints in Hebbian Learning submitted to *Neural Computation*

[4] Oja,E (1982) A Simplified Neuron Model as a Principal Component Analyser in *Journal of Mathematical Biology*

[5] Oja,E (1989) Neural Networks, Principal Components and Subspaces in *International Journal of Neural Systems*

[6] Oja,E., Ogawa,H., Wangviwattana,J.,(1992) PCA in fully Parallel Neural Networks in *Artificial Neural Networks,2, Conference Proceedings of ICANN'92, ed. Aleksander & Taylor*

[7] Rubner, J. and Schulten, K. (1990) Development of Feature Detectors and Self-organisation:A Network Model in *Biological Cybernetics*

[8] Sanger,T.D. (1990) Analysis of the Two-Dimensional Receptive Fields learned by the Generalized Hebbian Algorithm in Response to Random Input in *Biological Cybernetics*

TEMPORAL DISTRIBUTED PROCESSING - TDP :
A TIME-BASED PROCESSING SCHEME ACCOUNTS FOR TIME DEPENDENT RECEPTIVE FIELDS AND REPRESENTATIONAL MAPS

Hubert R. Dinse [1], Christoph E. Schreiner [2], Friederike Spengler [1],
Ben Godde [1], and Bernt Hartfiel [1]

[1] Institut für Neuroinformatik - Theoretische Biologie, Ruhr-Univ. Bochum (RUB), Germany
[2] Coleman Laboratory, W.M. Keck Center for Integrative Neuroscience, University of California, San Francisco, USA

Abstract

Based on the assumptions that the entire temporal structure of neuron responses carries significant information and that single cell receptive fields (RFs) and representational maps (RMs) typify representative stages of cortical processing, their dynamic properties were investigated in three modalities. The resulting time dependent receptive fields and representational maps are interpreted as specific adaptations to processing of inherently time-variant signals. Additionally, a time-based concept of temporal distributed processing (TDP) across subcortical and cortical substrates is presented that accounts for time-dependent RFs and RMs. Post-ontogenetic plastic adaptive processes are assumed to act within this scenario to provide sufficient flexibility to slow changing conditions in the environment and individual performance requirements. In our view, the stimulus-triggered emergence of complex spatio-temporal activity patterns are a key feature of sensory information processing. A hypothesis is outlined that attempts to map such spatio-temporal activity onto behavioral states.

1.   Observables - Time scales - Modalities

In search of basic elements which can provide the operational basis for a description of informational processing, we choose single neurons and as a next higher, macroscopic level, cortical areas. The reliance on RFs in this context is a consequence of the close link between the outside world and neuron activity that is reflected in RFs and can reliably be assessed by available recording techniques. In this view, the term RF is used in a very broad meaning: RFs reflect basic rules for mapping physical events onto representations based on neural activity. In addition, the relationship between RFs and maps is in most cases unambiguously defined. It is the neighborhood relationship of RFs, their topology, that establishes a systematic spatial representation of environmental aspects: a map.

To deal appropriately with the dynamics of the environment, there must be a match between processing dynamics with that of the dynamics in the stimulus space. Additionally, there must be plastic-adaptive processes [1-12]. Accordingly, a ten to several hundred millisecond scale, during which intrinsic RF dynamics become apparent will be studied as well as a minute to hour scale that deals with plastic-adaptive processes that overlay and interact with the faster scale.

Experiments were performed in anesthetized rats and cats with recordings made in striate and extrastriate visual areas, and in primary somatosensory and auditory cortices. Comparison of differences and similarities found in different modalities offer a unique way of differentiating modality specific modes of processing from those that are biased by constraints of cortical architecture and processing principles.

2.   Survey of time dependence of receptive fields

Time dependence of receptive field organization was investigated by applying the so-called time slice technique. The main idea behind this protocol is to avoid average or maximal measures of neuron responses, but to utilize and to rely on the entire response episodes. This procedure yields temporal sequences of RF descriptors (response planes (visual and somatosensory system), orientation tuning and length tuning (visual system), frequency-intensity planes and frequency tuning (auditory system)) which provide information about their instantaneous and actual content.

All experiments were performed using a *temporal stimulus isolation* protocol. Under these conditions, cells are studied by stimulating them from a *zero state* condition which prohibits interaction between successive stimuli. Therefore, the observed dynamics are interpreted as the manifestation of the intrinsic dynamics of the system undisturbed by timing constraints of the stimulation.

2.1.   Visual, auditory and somatosensory RF dynamics

Details of visual RF dynamics have been described elsewhere [13,14]. They consist of substantial changes of RF organization with time indicating non-separability of space and time as well as time dependence of tuning characteristics. Most significant are observations that cortical response selectivities, such as orientation selectivity or length preference, increase with time after stimulus onset following an initial unspecific state in which cells appear fairly invariant against changes of the stimulus. This gives rise to the notion of dynamic orientation selectivity and dynamic length preference. Most notable is the match between the emergence of anisotropic RF organization over time and the development of orientation selectivity which both evolve after a state of isotropic organization during which no or little selectivity can be found.

Similar analyses of time dependence of auditory RFs in the primary auditory cortex of cats revealed complex frequency and intensity related temporal changes in response specificities [15]. After an initial preference for high intensities and relatively wide frequency ranges, more restricted obliquely oriented bands of activity across the frequency-intensity planes were often observed. The oblique bands were found near the upper (high frequency) or lower (low frequency) edge of the conventional, time- averaged frequency tuning curves. When the data were re-plotted as frequency-time plots, the time-dependent frequency characteristics revealed frequency transitions and time- dependent multi-peaked tuning properties. The time slice technique was recently applied to study RF dynamics of neurons in the hindpaw representation of rat primary somatosensory cortex (SI). Up to 28 stimulations sites on the hindpaw were selected for computer controlled tactile stimulation, revealing a time dependent spread of activity resulting in complex spatio-temporal patterns [16]. Most notable was the gradual shift of activity starting at the distal part of the digit representation which then emerged further proximal or the emergence of focal zones of activity that then spread towards the peripheral portions of the time-averaged RFs, resembling very much the overall spatio-temporal behavior described for the visual and auditory system.

2.2.   Representational map dynamics of the somatosensory system

To unravel comparable dynamics acting at the macroscopic level of cortical maps, we used optical monitoring based on voltage sensitive dyes (RH 795, RH 414) in the somatosensory cortex of the hind-paw representation. This method allows real-time monitoring of two-dimensional, spatio-temporal activity distributions [17]. Data acquisition windows were set to obtain a coarse temporal resolution [18,19]. The spatial extent of the representational zones following point-like stimulation was different when recorded at different times after stimulation. Also, the zone of cortex being activated was considerably larger when longer integration times were used. The available data suggest wave-like patterns of activity, implying that cortical maps change on a time scale of tens of milliseconds. Temporal averaging cannot resolve these pattern but will fuse smaller zones of activity. As these spatio-temporal patterns are reproducible and stimulus specific, they must be regarded as part of a macroscopically detectable processing scheme.

2.3.   Common features of time-dependence and absolute timing

As a common finding in these three studied modalities, two response components reflecting two different states of responsiveness and selectivity were observed. The early components exhibit a fairly unspecific state paralleled by a simple, isotropic RF organization. In contrast, the succeeding later states display a sharper tuning indicating a higher level of selectivity paralleled by the emergence of an anisotropic RF organization. However, concerning the absolute timing, there were also clear differences. While the time constants of RF changes of the visual system are in the range of 80 to 200 ms, those of the auditory were much shorter (10 to 20 ms), while in the somatosensory system an intermediate timing between 10 and 50 ms was observed.

3.   No wait-states in the sensory pathway chains?

It might be tenable that within a pure feedforward system, RF dynamics can evolve due to specific distributions of latencies and thresholds. However, as cortical and subcortical systems are paragons for feedback [20-23], it appeared appropriate to incorporate feedback mechanisms from the very beginning. We therefore propose a framework that incorporates the contributions of feedforward, lateral interactions and feedback. It is based on experimental data on absolute timing conditions (response latencies and response durations) observed in different processing stages of the visual system and was called *Temporal Distributed Processing* - TDP.

To find the absolute timing conditions of processing along different stages of the visual system, we inspected visual response latencies and response durations following on-off stimulation in the lateral geniculate nucleus (LGN), area 17, area 18, 19, area PMLS (visual area in the posteromedial portion of the lateral suprasylvian sulcus) and area 7. According to the data there is a broad scatter of latencies over a range of 40 to more than 100 ms. There is also a broad scatter and wide range of response durations bet-ween 50 to up to more than 200 ms. Accordingly, there is a considerable overlap of responses suggesting that most of the neurons within different stages of a hierarchically organized sensory pathway are simul-taneously active when stimulated. The available data indicate that this holds also for the other modalities.

In the light of these findings, the chain of processing across a multi-staged sensory pathway seems to be activated in a temporally continuous fashion. The high percentage of simultaneously activated neurons, makes it rather unlikely that proceeding stages wait until the end of the processing of a previous one. What seems more plausible is a continuous interaction that links together all stages of the processing path including higher-order areas [13].

3.1.   Temporal Distributed Processing - TDP

The basic idea of the TDP concept is grounded on the lack of wait-states. In this scheme, illustrated in Fig. 1, the first available information about external stimulation is fed without delay through the entire pathway which represents the feedforward aspects of the response characteristics. During the continuation of the response, this behavior is repeated. However, after these initial response components, different types of lateral interactions and feedback mechanisms come into play. Among these are the thalamo-cortical feedback, massive intracortical feedback and the feedback from other and higher areas. The time required for this action to have an impact on the response characteristics at a given site might be variable

192

for each of the described mechanisms and might depend on the type of stimulation and on the modality. Taken together, these temporally delayed influences can be assumed to modify to a considerable but variable degree the informational content being processed or being represented. It seems therefore conceivable that neuronal responses are changed systematically over time making late components completely or at least significantly different from early ones. This is exactly what could be found experimentally: TDP represents a scheme that explains late and time-dependent selectivity. Inspection of neuron responses following stimulation with complex pattern or faces in IT (inferior temporal cortex which is regarded as an high-end substrate for visual processing), latencies between 80 and 100 ms are a common finding which is highly compatible with the TDP scheme [24-26]. Moreover, TDP provides a substrate that can be used for routing back memory-related information from higher areas. In view of the recent discussion of "memory-based" processing, late response components are most likely to carry such information [27].

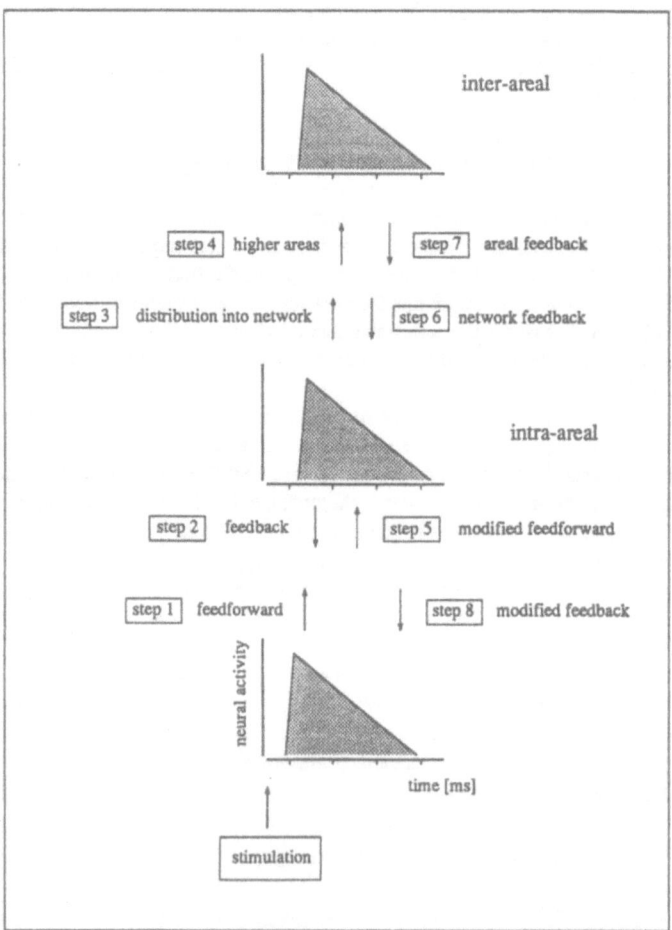

Fig. 1   Schematic illustration of the temporal distributed processing scheme (TDP). Schematic neuron responses are shown as post stimulus time histograms (PSTHs) for different stages of an assumed sensory pathway. Due to the time dependent action of feedforward components, lateral interactions, different types of feedback and temporally delayed and modified feedforward, the late response episodes are assumed to be significantly altered in their informational content compared to the early parts, independent of their lower firing rates.

4.   Functional impacts of time dependent RFs:
        Adaptation to inherently time-variant signals

At a first attempt to clarify the possible functional implications of time dependent RFs, it seems conceivable to assume that they represent specific adaptations for processing of inherently time-variant signals. In the case of the visual system, motion of either objects or of the system itself are typical examples of signal dynamics. In the case of the auditory system, frequency transitions of second-order formants in human speech resemble the dynamics observed during the changes of RF organization [15,28]. Video analysis of the rat walking pattern revealed time scales of sensorimotor behavior that are fully compatible with that observed during somatosensory dynamics [12].

From a general point of view, there are a number of arguments in favor of time-variant information processing, because it is essential for many processing tasks to keep information available over a longer period of time. This includes estimation and integration tasks, predictions, generation of continuity and short-term memory functions, which fill the gap between low-level processing dynamics and higher cognitive functions [29,30].

4.1.   Alterations of time dependent RFs by plastic reorganization

Using intracortical microstimulation (ICMS) and paired peripheral tactile stimulation (PPTS) as a tool for inducing short-term plastic reorganization, geometric and functional aspects of RF organization are altered such as RF size, location, overlap, latencies and tuning and transfer properties. Also, patterns of time-dependent RF organization are changed as well as interneural interactions [10-12,16,31].

4.2.   Mapping spatio-temporal activity onto behavioral states

In contrast to the other modalities, the somatosensory system is unique in that it is inherently related to motor action and behavior. This intimate relationship can be utilized in search of interpretations of spatio-temporal activity pattern in terms of behavioral significance and relevance. The main problem is to identify such patterns. The constraints imposed by the somatosensory system on the patterns available are extremely valuable in providing clues to identify them.

The video-based analysis of walking patterns of rats and the time course of representational map dynamics observed for rat extremities match to a surprisingly high degree. Under normal conditions, walking consists of a certain sequence of steps, causally related to a fixed sequence of neural events. It is hypothesized that the late response components anticipate in their spatial and temporal spread of activity behaviorally relevant patterns, making the expected and anticipated pattern easier to evolve. It appears conceivable that this requires some act of learning to establish the sequence. Accordingly, complex spatio-temporal patterns are a manifestation of highly relevant behavior that are created either during the critical developmental period or during reorganization of the system following adult posy-ontogenetic plasticity, or both. In either way, RF dynamics link learning and memory to normal, continuously on-going processing. They represent highly adaptive states based on acquired knowledge that are crucial for facilitating sequences of neural events that can anticipate fixed sequences of behavior-related action. However, these dynamics are highly variable and subject to plastic alterations. From the point of view of this hypothesis, re-learning would be required to establish new spatio-temporal patterns whose temporal order and sequence would represent again behaviorally relevant adaptations to changes in the environment or changes in the significance of the environmental events.

5.   Outlook and conclusions

So far, time dependence of RFs and representational maps have been found in a variety of modalities and species, indicating that the stimulus triggered emergence of complex spatio-temporal activity patterns are a key feature of sensory information processing. However, their functional roles are mostly speculative. The extensive changes of receptive fields dynamics during reorganizational plasticity link temporal aspects of informational processing strategies to plastic-adaptive processes. The ideas about a temporal distributed processing scheme (TDP) add a further time-based aspect. In this scenario, spatio-temporal patterns are generated as a synergetic process through the entire sensory pathways including very high processing stages, and are subject to changes due to plastic-adaptive processes along slower time scales.

6.   References and Acknowledgements

We are thankful for extensive discussions to our colleagues and friends, especially to A. Aertsen, K. Kopecz, M. Merzenich, G. Schöner and W. von Seelen. We acknowledge the support of M. Schäfer in the optical recording experiments and we are grateful to D. Plenz for valuable advice. The research was supported by the Deutsche Forschungsgemeinschaft DFG (Di 334/4-1, 334/5-1, Se 251/31-1, the Forschergruppe at the RUB Neurovision Ey 8/17-1,2), the Office of Naval Research ONR and a NATO collaborative research grant CRG 910626. Exchange visits between the UCSF and RUB were supported by the DFG and NATO.

[1] Merzenich MM, Nelson RJ, Stryker MP, Cynader MS, Schoppmann A, Zook JM (1984) J Comp Neurol 224: 591-605;

[2] Merzenich MM, Recanzone G, Jenkins WM, Allard T Nudo RJ (1988) in: Neurobiology of neocortex. Dahlem Konferenzen 1988, Wiley, pp 41;

[3] Calford MB, Tweedale R (1988) Nature 332, 446-447;

[4] Weinberger NM, Ashe JH, Metherate R, McKenna TM, Diamond DM, Bakin J (1990) Concepts Neurosci 1, 91-132;

[5] Kaas JH, Krubitzer LA, Chino YM, Langston AL, Polley EH, Blair N (1990) Science 248: 229;

[6] Gilbert CD, Wiesel TN (1992) Nature 356, 150-152;

[7] Recanzone GH, Merzenich MM, Jenkins WM, Grajski K, Dinse HR (1992) J Neurophysiol 67: 1031-1056;

[8] Recanzone GH, Merzenich MM, Dinse HR (1992) Cerebral Cortex 2: 181-196;

[9] Dinse HR, Recanzone G, Merzenich MM (1990) in: Eckmiller R, Hartmann G, Hauske G (eds) Parallel Processing in Neural Systems and Computers. Elsevier, pp 65-70;

[10] Spengler F, Dinse HR (1992) in: Elsner N, Richter DD (eds) Rhythmogenesis in neurons and networks. Thieme, pp 129;

[11] Spengler F, Dinse HR (1992) Soc Neurosci Abstracts 18: 345;

[12] Godde B, Spengler F, Dinse HR (1993) in: Elsner N, Heisenberg M (eds) Genes, Brains and Behavior, Thieme, in press;

[13] Dinse HR, Krüger K, Best J, (1990) Concepts in Neuroscience (CINS) 1: 199-238;

[14] Dinse HR, Krüger K, Mallot HP, Best J (1991) in: Krüger J (ed) Neuronal Cooperativity. Springer, pp 67-104;

[15] Dinse HR, Schreiner CE (1992) Soc Neurosci Abstracts 18: 150;

[16] Dinse HR, Godde B, Spengler F (1993) in: Elsner N, Heisenberg M (eds) Genes, Brains and Behavior, Thieme, in press;

[17] Grinvald A, Frostig RD, Lieke E, Hildesheim R (1988) Physiol Rev 68: 1285-1366;

[18] Dinse HR, Hartfiel B, Schäfer M, Krüger K, von Seelen W (1992) in: Elsner N, Richter DD (eds) Rhythmogenesis in neurons and networks. Thieme, pp 109;

[19] Hartfiel B (1993) Optische Registrierung neuronaler Aktivität an Cortex-Oberflächen mit spannungsabhängigen Fluoreszenzfarbstoffen. Fortschrittberichte VDI. VDI Verlag;

[20] Braitenberg V, Schüz A (1990) Anatomy of the Cortex, Statistics and Geometry. Studies of Brain Function. Springer;

[21] Freeman WJ, Skarda CA (1985) Brain Res Rev 10: 147-175;

[22] Eggermont JJ (1990) The Correlative Brain. Springer;

[23] von Seelen W, Mallot HA, Giannakopoulos F, (1987) Biol Cybern 56: 37-49;

[24] Perret DI, Rolls ET, Caan W (1982) Exp Brain Res 47: 329-342;

[25] Richmond BJ, Optican LM, Podell M, Spitzer H (1987) J Neurophysiol 57: 132-146;

[26] Tanaka K, Saito H, Fukada Y, Moriya M (1991) J Neurophysiol 66:170-189;

[27] Poggio T, Fahle N, Edelman F (1992) Science 256: 1018-1021;

[28] Liberman AM, Studdert-Kennedy M (1978) in: Handbook of Sensory Physiol. Springer, pp 143-178;

[29] Nakamura K, Mikami A, Kubota K (1992) NeuroReport 3: 117-120;

[30] Schöner G, Kopecz K, Spengler F, Dinse HR (1992) NeuroReport 3: 579-582;

[31] Wacquant S, Joublet F, Spengler F, Godde B, Dinse HR (in press) ICANN 1993.

Dynamics of single neurons
—poster contributions

Stochastic Specificity in Neural Interaction.

Vicente López, Juan A. Sigüenza, José R. Dorronsoro
Instituto de Ingeniería del Conocimiento (IIC),
Universidad Autónoma de Madrid, Cantoblanco, 28049 Madrid, SPAIN.

Santiago Carrillo-Menendez
Departamento de Matemáticas,
Universidad Autónoma de Madrid, Cantoblanco, 28049 Madrid, SPAIN.

Abstract

We present in this communication results from a dynamical model of two interacting stochastic neurons, simple enough to provide the understanding of basic mechanisms by which the stochastic nature of units enhances information processing capabilities. The active role played by noise or randomness in stochastic integrating and firing units is analyzed using a Markovian Chain approach. We conclude that randomness allows for robust specifity in unit response.

1 Introduction

Neuronal firing (generation of a train of action potentials) is an expression of the activity of a single neuron after integration of several postsynaptic potentials induced by presynaptic terminals. This neural activity is believed to be the building block from which neuronal processing of information has to be understood. Recordings from single sensory neurons have demonstrated that the intensity of a static stimulus can be coded in the firing rate of a sensory neuron. However, the possibility of phase encoding of information in neural activity is unclear since the prevalent stochastic character of neuron responses. Statistical analysis of experimentally obtained spikes always shows a significant random component in the interspike intervals. At present is not clear how the sensory information is encoded nor how its processing is affected by noise (whether the noise simply obscures it by introducing random interval errors, or whether it plays a deeper role). Since biological systems are based on noisy elements, it is worth exploring the possibility that the extremely efficient processing of information in the nervous system is possible not in spite of, but because the stochastic character of biological units. Several authors have already advanced in that direction [1] and also there exists an increasing number of theoretical studies where single neuron activity is approached as a stochastic process [2]. Less often, those studies are concerned with groups of interacting neurons. We present in this communication results from a model of two interacting stochastic neurons simple enough to provide the understanding of basic mechanisms by which the stochastic nature of units enhances information processing capabilities.

2 The model

Reductionism is a common practice when facing the understanding of basic paradigms. Information processing in the nervous system fails in this category. We consider in this work a reductionist two neuron model in which basic elements of neural activity are incorporated in a manner simple enough to allow for theoretical analysis. The dynamics (frequency and phase) and stochasticity of spike generation are considered. Every neuron in our model is assumed to be an stochastic integrating and firing unit. The activity of a unit i at time t is represented by its state $a_i(t)$. Possible states of a unit are discrete, and in the range from 1 to N_i. In absence of interaction with other units, the transition between states is simply governed by the following rule:

$$a_i(t+1) = a_i(t)+1 \quad \text{with probability } p_i$$
$$a_i(t+1) = a_i(t) \quad \text{otherwise}$$

for $a_i(t) \in \{1, N_i-1\}$. The transition from state N_i to state 1 is considered to be the neuron firing in which the neuron spike is produced and it occurs with certainty. The interaction between units is due to immediate spike transmission through the neural axon and produces a postsynaptic change in the receiving unit j, $a_j(t) = a_j(t) + \epsilon_{ij}$, i being the index of the unit firing at time t. If the receiving unit is already in state N_j (ready to fire) the postsynaptic change is irrelevant, otherwise with positive ϵ_{ij} (excitatory connection) the waiting time for occurrence of discharge in unit j decreases. The above stochastic model is Markovian since unit states at a given time only depend on the previous time step states. Quantization of time and unit states could be easily relaxed but theoretical analysis is simpler in the discrete version of the model.

The isolated unit i behaves as a stochastic oscillator firing at period T_i with probability $P_{N_i}(T_i)$ given by the negative binomial distribution. In the deterministic limit ($p_i = 1$) the period is always N_i. Otherwise, the mean firing time τ_i and its standard deviation σ_i are $1 + (N_i - 1)/p_i$ and $\sqrt{(N_i - 1)(1 - p_i)}/p_i$. The probability of finding unit i in state x at time Δt after the last firing ($x > \Delta t$), is also given by a well known discrete distribution, the binomial distribution.

In the general model of interacting units the situation is more complex since the firing period of every unit depends on the state of every other connected unit. In this communication we present the results for the simplest two neuron model. In Figure 1 an illustration of a portion of the two unit system state space is presented. Possible states are in a grid having N_1 and N_2 points per axis and oscillator period. Unit 2 is chosen to progress parallel to the abcisa and unit 1 perpendicular. Each unit fires every time it crosses the lines marking the deterministic period and producing a jump in the other unit of at least the magnitude of the synaptic strength. Filled dots are marking the situations where both units fire in synchrony. This situation happens every time a trajectory arrives to the positions marked with dashed boxes in the figure. Every time a spike is produced by unit i, unit j will fire if its state is less than ϵ_{ij} units far away from the firing transition ($a_j(t) < (N_j - \epsilon_{ij})$).

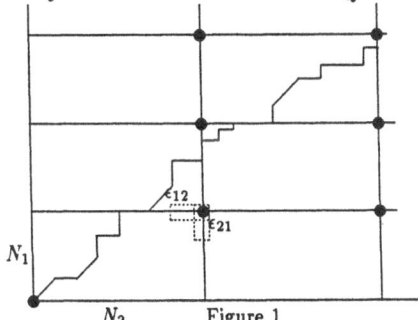

N_1

N_2 Figure 1

The trajectory displayed in Figure 1 starts in a common firing situation at the origin and repeats such synchrony after one intermediate firing of unit 2 and two of unit 1.

The variety of possible different units is generated in our model by two parameters, N_i and p_i, that determine the distribution of firing periods. The variety of possible different networks is generated, together with the component units, by the non-symmetric synaptic efficiency parameters, or synaptic weights, ϵ_{ij}.

3   Results

The presented two neuron model spans a large variety of situations. Among them we have found most interesting the analysis of synchronic firing of units for those cases where one unit (unit 1) fires more regularly and faster than the other ($p_1 > p_2$ and $N_1 < N_2$). This situation recalls the case where a neuron (unit 1) is actively firing as a response to an external stimulus and the other (unit 2) is, in absence of interaction, having an spontaneous and highly random low frequency firing.

Within the Markovian chain formalism, the average time between synchronic firing of units τ_0, and its standard deviation σ_0, can be easily calculated once the matrix of transition probabilities is built for the chosen model parameters. With respect to Figure 1, τ_0, represents the average time of trajectories that start at the origin and end up in the first encountered synchrony, some of them appearing in the Figure 1 as filled dots. It is also interesting the mean time needed for a first synchronic firing τ, and its standard deviation σ, when initial states of every unit are uniform and randomly distributed. We have investigated how the occurrence of low dispersion syncronic firing depends on the model parameters. The trivial case of very large synaptic weights ($\epsilon_{ij} \sim N_j$) is left aside and we have concentrated on the analysis of those situations where small weights are capable of producing low dispersion, synchronic firing of units. In Figure 2 we present a typical result in our scan of model parameters. For $N_1 = 5$, $N_2 = 70$, $p_1 = 0.95$, $p_2 = 0.5$, and $\epsilon_{21} = 2$, in Figure 2 τ_0 is plotted versus ϵ_{12}, and error bars correspond to the dispersion σ_0. It can be seen in Figure 2 the presence of small sets of ϵ_{12} for which unit 2 fires induced by unit 1 and with low dispersion.

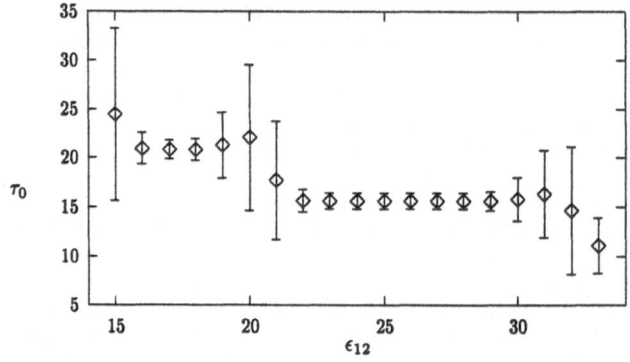

Figure 2

For the units used to calculate the plotted results, τ_1 and τ_2 are 5.2105 and 139, respectively, and the dispersions σ_1 and σ_2, are 0.4707 and 11.747. For the two unit system with $\epsilon_{12} = 0$ and $\epsilon_{21} = 2$, the values obtained for τ_0 and σ_0 are 233.237 and 149.034 , what indicates that the common firing of both units is not a regular pattern. On the other hand, with $\epsilon_{12} = 25$ we get that once unit 2 and unit 1 fired once together, unit 2 will keep firing whenever it receives the third spike from unit 1. This can be deduced from the $\tau_0 = 15.6316$ and $\sigma_0 = 0.81536$ values obtained for $\epsilon_{12} = 25$, since they exactly correspond to the average time and dispersion of unit 1 producing three spikes. On the other hand, a first coincidence of a common firing in this situation may appear quite soon since the values calculated for τ and σ are 7.29507 and 1.17062. That is, most likely the low dispersion activity of unit 2 will be observed within the first two fires of unit 1.

The common result among many of the calculations performed is the presence of small sets of synaptic ϵ_{12} weights for which low dispersion firing of the driven unit occurs, surrounded by synaptic weights for which unit 2 remains firing irregularly. This specificity in the synaptic weights is also obtained when varying the deterministic period of the driving unit for fixed weights and driven unit. In general, the specifity is not restricted to an isolated value of the synaptic weight or the driving frequency, but to a small set of adjacent values.

We could resume the results obtained in the large set of calculations performed as follows: an irregularly firing unit is induced to fire regularly by an interacting unit with more regular activity provided the adequate (and specific) synaptic efficiency and activity pattern of the inducing unit. The regular interspike period of the driven unit will depend on the synaptic weight and the driving interspike period.

These results can be better understood in the simpler limit case for which $p_1 = 1$, and therefore unit 1 fires with deterministic period N_1. In this limit case and for $\epsilon_{21} = 0$ we can provide a simple explanation for the selectivity in the driving of unit 2 to a regular firing. A schematic drawing of unit 2 state before producing the spike is displayed in Figure 3a. The gaussian shaped curve represents the density of probability of finding unit 2 in state x (close to N_2) at time t_m, when unit 1 fires. Unit

2 has arrived to the x states by spontaneous promotion and also as a result of $m-1$ promotions of size ϵ_{12} resulting from unit 1 spikes.

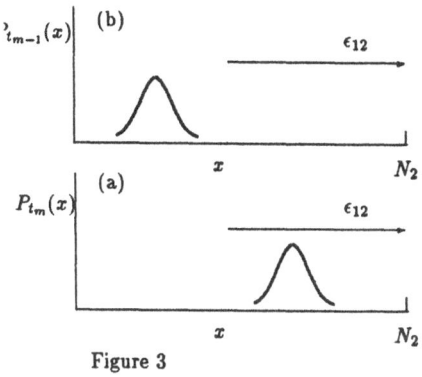

Figure 3

A complete synchronization, and therefore a regular firing of unit 2, would result of an ϵ_{12} large enough to promote unit 2 to N_2 from whatever x state would be at time t_m. In practice, we consider enough an ϵ_{12} covering a high percentage of the density curve. This is the origin for a lower bound in the size of ϵ_{12}. However, there is also an upper bound, since ϵ_{12} has to be small enough to avoid an irregular promotion to N_2 of unit 2 in the previous firing of unit 1. In Figure 3b the density of probability of unit 2 at time t_{m-1}, when unit 1 produces the previous spike, is displayed. In order to observe a regular firing of unit 2 the size of ϵ_{12} has to fall in between the two curves. A non trivial relation between N_1, N_2, p_2 and ϵ_{12} determines the values of either N_1 or ϵ_{12} capable of making unit 2 to produce a regular, high frequency, activity.

4 Discussion

Simple model networks with two deterministic interacting, integrating and firing units have been previously studied [3]. Results for such models talk about commensurability relations within frequencies and unstable pathways to phase locking. The same complex mechanisms for information processing may be expected from larger networks of deterministic periodically firing units. However, the results we found with the study of stochastic units suggests that noise or randomness may act as an stabilizing ingredient to allow for robust relation between the activity of connected neurons.

A trivial conclusion from the results would be that stochasticity precludes the possibility of phase encoding of information in the studied model. However, this is not the case. The unit with regular activity considered in our model could be either a neuron activated by external stimuli or by internal neurons. If so, its spike pattern response may be complex, although regular, and may in turn produce different complex, low dispersion, firing patterns in connected units. The preliminary results we have for larger networks support this idea. Therefore, the noise or randomness of every unit may act as an active background allowing for robust and stable processing of the information encoded in the time sequence of spikes.

5 Acknowledgements

This work has been supported by CICyT grants TAP92-117 and PB91-45 .

References

[1] Buhmann J. & Schulten K. Biol. Cybern. **56** (1987) 313. Horikawa Y. Biol. Cybern. **66** (1991) 19. Longtin A., Bulsara A. & Moss F. Phys. Rev. Lett. **67** (1991) 656.

[2] H. Tuckwell "Stochastic Processes in the Neurosciences", SIAM (1989).

[3] Budelli R., Torres J., Castigeras E. & Enrich H., Biol. Cybern. **66** (1991) 95.

A Computer Simulation Model of Backwards Feedback Across Synapse Via Arachidonic Acid

R.Lahoz-Beltra (1), A.Murciano (1), J.Zamora (1), F.Vico (2), J.M.Jerez (2),
S.R.Hameroff (3), J.E.Dayhoff (4)

(1)   Departamento de Matematica Aplicada, Facultad de Biologia,
       Universidad Complutense, Madrid 28040, Spain.

(2)   Departamento de Tecnologia Electronica. E.T.S.I. Telecomunicacion.
       Universidad de Malaga, Malaga 29013, Spain.

(3)   Advanced Biotechnology Laboratory, Department of Anesthesiology,
       University of Arizona, Health Sciences Center, Tucson, AZ 85724, USA.

(4)   Systems Research Center, University of Maryland, College Park, MD 20742, USA.

Abstract

Algorithms for artificial neural networks are usually developed assuming in the most of the models that information propagates fordward and backward across the neural network. Fordwards propagation is modeled easily in ANN and biologically plausible in biological neurons, however for backwards propagation the plausibility of the algorithms developed for ANN seems remote in biological neurons. Based on the presynaptic changes induced by the arachidonic acid released by postsynaptic neurons during long-term potentiation (LTP) in the dentate gyrus, we show a computer simulation model where a backward feedback is performed across local synapses by arachidonic acid. Our simulation model shows how arachidonic acid could be playing the role of retrograde messenger during LTP.

1. Introduction

Artificial neural networks (ANN) models were originally inspired by biological neurons assuming in the most of the models that information propagates fordward and backward across the neural network. Forwards propagation is modeled easily in ANN and is biologically plausible in biological neurons with action potentials that go from axons to dendrites and with a modulation of synaptic strengths [5]. However, for backwards propagation the plausibility of the algorithms developed for ANN seems remote in biological neurons. Back propagation is not the only ANN learning paradigm to incorporate *backward feedback* signals from the axon to the dendrite of the neuron. Other learning paradigms such as the sigma-pi [7] and RCE [6] architectures, and ART [3] networks also require backward signals. If we were interested to look for some plausibility of any of these paradigms into biological neurons [4] then we were assuming that a neural circuit is able to send backward feedback signals along each fordward connection.

At present experimental data support the hypothesis that a backward feedback could be performed in biological neurons across *local synapses* via nitric oxide, arachidonic acid, or another *retrograde messenger*. Bliss and col. [2] found in the dentate gyrus experimental results suggesting LTP is expressed, at least in part, by presynaptic mechanisms: arachidonic acid would be playing the role of retrograde messenger from the postsynaptic to the presynaptic side of the synapse. In this article we describe a computer simulation model of backwards feedback across synapse (local feedback) via arachidonic acid (Figure 1).

2. Model description

A computer simulation model is presented in which a backwards feedback across synapses is given by the interplay of the following biological elements: neurotransmitter, arachidonic acid, phospholipases, protein kinases, inositol triphosphate, Ca fluxes and buffering, K/Q and NMDA (postsynaptic receptors) open/closed channel transitions and their modulation, postsynaptic action potentials etc. Algorithms developed to simulate the molecular activity inside presynaptic and postsynaptic neurons have been implemented as a subroutine library written in Turbo C 2.0. Graphics showing intermediate results during simulation were developed using a real-time graphics subroutine library.

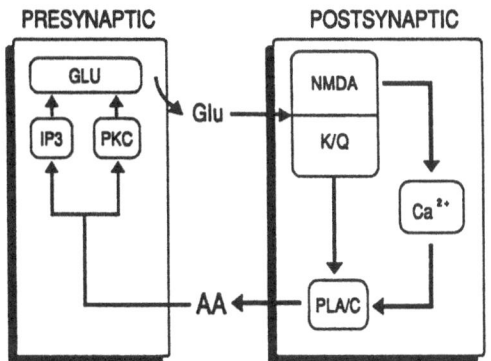

Figure 1. Schematic drawing illustrating the possible role of arachidonic acid (AA) as a retrograde messenger during LTP (Modified from Bliss and col. [2]).

We start considering the model is ruled by the following expression:

$$[Glu]_{sc} = K_1 [AA]_{sc} P + [Glu]_0$$

where $[Glu]_{sc}$ and $[AA]_{sc}$ represent the glutamic acid (Glu) and arachidonic acid (AA) concentrations respectively in the synaptic cleft, P is a state variable modeling the presence/absence (0/1) of stimuli and $[Glu]_0$ is the glutamate concentration in the sinaptic cleft when P=0 (resting state).

The $[AA]_{sc}$ dynamic was simulated based in the following evaluation function:

$$[AA]_{sc}(t) = K_2 [AA]_{sc}(t-1) + K_3 [AA]_{PLA}(t-1,t) + K_4 [AA]_{PLC}(t-1,t)$$

where $[AA]_{PLA}$ and $[AA]_{PLC}$ are the AA concentrations produced by phospholipase A and C activities respectively and released into the synaptic cleft. Phospholipase A and C activities have been modeled as follows:

$$[AA]_{PLA}(t-1,t) = K_5 A_{PLA}(t-1)$$

$$A_{PLA}(t) = K_6 A_{PLA}(t-1) + K_7 [Ca^{2+}]_{POST}$$

$$[AA]_{PLC}(t-1,t) = K_8 A_{PLC}(t-1)$$

$$A_{PLC}(t) = K_9 A_{PLC}(t-1) + K_{10} [Glu]_{sc}$$

where A_{PLA} and A_{PLC} are the phospholipase A and C activities respectively in the interval time (t-1,t). Phospholipase activities were modeled as follows: A_{PLA} considering that postsynaptic calcium is involved in its activation via calmoduline protein kinases, and A_{PLC} considering it is ruled by K/Q receptor which open/closed channel state is governed by $[Glu]_{sc}$. In the present model we do not consider the effect of substrate and product concentrations in phospholipase activities. Related to the Ca^{2+} concentration in the postsynaptic neuron, Ca^{2+} values were simulated using the following evaluation functions:

$$[Ca^{2+}]_{POST}(t) = [Ca^{2+}]_{POST}(t-1) - [Ca^{2+}]_{BUFFER}(t-1,t) + I_{Ca}$$

$$[Ca^{2+}]_{BUFFER} = K_{11} ([Ca_{2+}]_{POST}(t-1) - [Ca_{2+}]_{EQ})$$

where Ca_{BUFFER} term [5] has been introduced in the model to summarize the calcium activity inside the neuron reviewed by Blaustein [1]: the calcium buffered by high-affinity binding substances, organelles and neuronal plasmalemma. Ca_{EQ} is the term modeling the calcium concentrion in the resting state. Flux of calcium into the postsynaptic neuron (I_{Ca}) is carried out by NMDA receptors and it has been simulated considering the fact that the activation of NMDA receptors is triggered by depolarization (mediated by K/Q receptors) and glutamate binding ($[Glu]_{sc}$) to receptor:

$$I_{Ca} = NMDA [Glu]_{sc}$$

In the above function NMDA is a rate modeling the "availability" of NMDA-calcium channels, in other words, the rate of NMDA-calcium channels unblocked by Mg^{2+}. NMDA values were simulated considering a nonlinear relationship (logistic eq.) between SF (stimuli frequency) and NMDA rate:

$$NMDA = 1/1+exp(-(K (SF-c))$$

where K,c are constants. Note in this function if K is sufficiently large then the limiting form of the function is the discrete step function.

Since the flux of Na^+ is mediated by K/Q receptors whose single-channel state (open/closed) is only triggered by glutamate, then I_{Na} values were simulated using the following expression:

$$I_{Na} = K_{12} [Glu]_{sc}$$

Finally, the post-synaptic action potential (PSP) value generate by ion fluxes (I_{Ca} and I_{Na}) has been simulated by:

$$PSP = K_{13} ([Ca^{2+}]_{POST} + I_{Na})$$

3. Results

Our simulation model show how the interplay of molecular mechanisms involved in the induction of LTP results in a *positive feedback*, where the level of neurotransmitter (Glu) released by the presynaptic neuron in the synaptic cleft is increased as a consequence of the arachidonic acid released by the postsynaptic neuron. This mechanism bear a resemblance to the Hebb learning rule, since the presence of a retrograde messenger has the effect of increasing the neurotransmitter released, which may contribute to enhance the quality of the synapse (Figure 2).

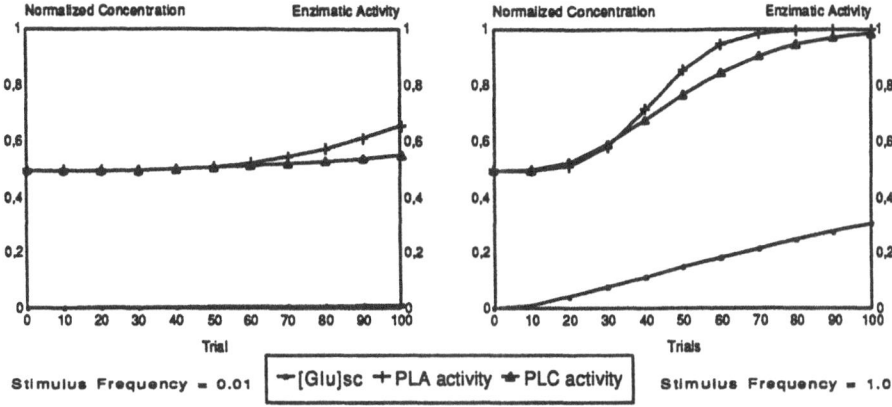

Figure 2. [Glu]sc and phopholipase activities for different stimulation frequencies.

An interesting question come up with the type of information that is hold by the retrograde messenger. For instance, we guess that the arachidonic acid released by the postsynaptic neuron could have the meaning to inform to the presynaptic neuron that the postsynaptic neuron is *aware* of a tetanic stimulation took place. Nitric oxide or other retrograde messengers could be playing the role of neural *molecular words* with different meaning. This and other questions will be treated in a future work.

4. References

[1] M.P.Blaustein. (1988). Calcium transport and buffering in neurons. TINS 11, No. 10:438-443.

[2] T.V.P.Bliss, M.P.Clements, M.L.Errington, M.A.Lynch, J.H.Williams. (1990). Presynaptic changes associated with long-term potentiation in the dentate gyrus. The Neurosciences 2:345-354.

[3] G.Carpenter and S.Grossberg. (1987). ART2: Self-organization of stable category recognition codes for analog input patterns. Applied Optics: 4919-4930.

[4] J.Dayhoff, S.R.Hameroff, C.Swenberg, R.Lahoz-Beltra. (1992). Biological plausibility of back-error propagation through microtubules. University of Maryland, SRC Technical Report TR 92-17: 1-79.

[5] R.Lahoz-Beltra, S.R.Hameroff, J.E.Dayhoff. (1992). Connection weights based on molecular mechanisms in Aplysia neuron synapses, in: Artificial Neural Networks, 2 (eds. I.Aleksander, J.Taylor) Elsevier Science Publishers: 869-872.

[6] D.L.Reilly, L.N.Cooper, C.Elbaum. (1982). A neural model for category learning. Biol. Cyber. 45: 35-41.

[7] D.E.Rumelhart, G.E.Hinton, R.J.Williams. (1986). Learning internal representations by error propagation, in: Parallel Distributed Processing (eds. D.E.Rumelhart, J.L.McClelland) Cambridge, Massachusetts: MIT Press.

Study of a Self-Learning Artifical Neuron Model

Gilles Vaucher

Supélec, BP 28, 35511 Cesson-Sévigné, France

E-mail : vaucher@supelec-rennes.fr

Giving a large autonomy to each cell and using a statistical average of the entries as a self-learning mechanism are not new ideas [1,2]. But linking statistical and linear dependencies through algebric properties is an approach which leads to biological interpretations.

During the training phase, the incoming of the input vector $\vec{X}$ produce basis' vectors distortions in such a way that the length of $\vec{X}$ increase or decrease (depending on the sign of the neuron's potential : Hebbian effect) while the direction of $\vec{X}$ stay constant (fig. 1). So, the basis vectors of the input space are not orthogonal and normalized ; the metric $M = ((m_{ij}))$, where m_{ij} is the scalar product of the ith by the jth basis vectors, gives an algebric form of the anisotropic neuron "perception" of its input space. The learning rule (for more details see [3]) is written :

$$\Delta M = \mu \frac{\nu}{\|\vec{X}\|^2}\left(P - \frac{T_P}{T_M}M\right) \text{ with } \quad \|\vec{X}\|^2 = {}^t XMX \,, \quad P = MX\,{}^tXM \qquad \begin{array}{l}\mu \text{ is a positive constant,}\\ \nu \text{ the neuron's potential and}\\ X \text{ the column matrix of the}\\ \text{input vector}\end{array}$$
$$T_P = trace\,P, \; T_M = trace\,M$$

where

For the output calculation, the algebric addings involve cell potential calculation reconsiderations. The scalar product $\nu = \vec{X}.\vec{W}$ is given by ${}^t XMW$ and $\vec{W}$ is extracted from M. We need to underline that the main inertia axis of M point to a direction which is a sort of statistical average of the input's vectors. Instead of defining $\vec{W}$ as a eigenvector associated to the greatest eigenvalue, we preferred to choose it in the eigenvectors set at random by assigning each of them a probability proportional to its associated eigenvalue.

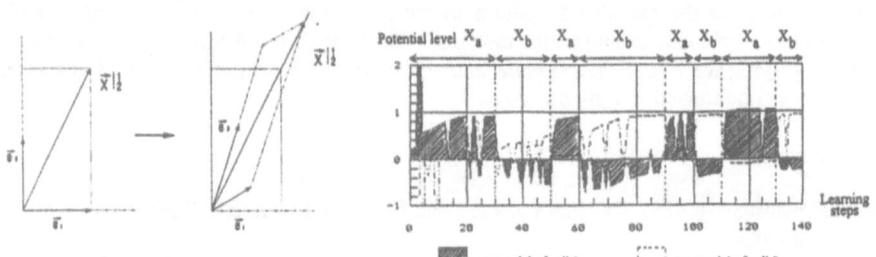

Fig.1 : Modification of M due the incoming of $\vec{X}$ (with $\nu>0$)

Fig.2 : Potential evolution of the two cells

Randomizing the potential calculation allows to have several neurons, connected to the same afferents, getting specialized in different directions of the input space due to lateral inhibitions (the specific treatment of the inhibition inputs is describe in [3]). As an example, let us consider two cells with the same two external inputs and inhibiting themselves mutually. The two following vectors are presented during the training phase : ${}^t X_a = (1,0)$ and ${}^t X_b = (0,1)$. For this type of algorithm, the good teacher always proposes the same vector until one of the cell has learned it and has forced the other cell to forget it (by inhibiting effect) ; then the other vector can be offered and be learned by the second cell (fig.2). At the end of the learning phase the system is leaded towards a stationary state ; statistically and with a great probability the cell 1 answers to X_a while the cell 2 reacts upon X_b.

As a discussion we would like to propose a biological interpretation. Introducing the metric M allows to take into account synaptic sensitivities (diagonal elements) but also inter-synaptic correlations (non-diagonal elements). So, thanks to M, it is possible to take into account some of the dentritic tree properties.

[1] B.P. Zeigler, *Theory of Modelling and Simulation*, Wiley-Interscience, New-York, USA, 1976
[2] E. Oja, *A simplified neuron model as a Principal Components Analyser*, Journal of Mathematical Biology, 15, 267-273, 1982
[3] G. Vaucher, *Contribution à l'enrichissement du modèle formel d'un neurone*, Rapport d'étude, Supélec, R-SI-GV-TH-A1.1, France, June 1992.

Simulation Study on Calcium-Activated Dynamics of Compartment Dendrite Model

Norihiro Katayama, Mitsuyuki Nakao, Yoshinari Mizutani, Mitsuaki Yamamoto

Graduate School of Information Science, Tohoku University, Sendai 980, JAPAN,
Telephone & Facsimile: +81-22-263-9437, E-mail: kata@yamamoto.ecei.tohoku.ac.jp

Recent physiological experiments have proved that a neuronal dendritic tree in the central nervous system (CNS) possesses the complex spatio-temporal dynamics. The dynamical properties of dendrite are expected to play an important role for the information processing in CNS. From this point of view, we construct a compartment model of an active dendrite in a discrete form in order to investigate its functional significance. The dynamics of the single compartment model is controlled mainly by the following two kinetic variables. One is the membrane potential which involves the several types of currents: T-type Ca^{2+}, L-type Ca^{2+}, and Ca^{2+}-activated K^+. The other is the intracellular Ca^{2+} concentration that is increased by Ca^{2+} influx and reduced by the cellular homeostatic system. The single compartment model responds in an oscillatory manner when a maintained current stimulation is applied. Figure 1 shows the relationship between the response firing frequency and the current intensity. There is the threshold and the optimal intensity of stimulus for the oscillatory response.

The compartment models of proximal and distal dendrites are made up by connecting longitudinally 10 thick and thin compartments, respectively. Each of the compartments are electrically coupled with the neighboring one. Figure 2 shows the firing frequency of the compartment dendrite models in response to the localized steady current stimulus. We observe the responses of the 1st compartment and 10th compartment, which are indicated by S and T, respectively. The stimulus is applied to S. For the proximal dendrite model, the responses are homogeneous among the compartments and their response firing frequency is lower than that of the single. For the distal, only S and its neighboring compartments fire with high frequency. It is worth noting that their firing frequency tends to be higher than that of the single in the range from 6.6 to $8\mu A/cm^2$, probably due to their spatial cooperation. It has been found that synapses from different regions tend to be located separately on the dendritic system. Therefore the neural information coming from the different regions is suggested to be processed in the location-dependent manner on the dendritic system.

Acknowledgements

We wish to thank *the Sumitomo Foundation* for their partly supporting our study. The author (*N.K.*) thanks *JSPS Fellowships for Japanese Junior Scientists*.

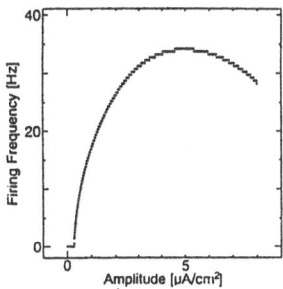
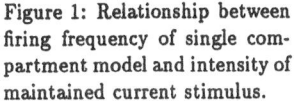
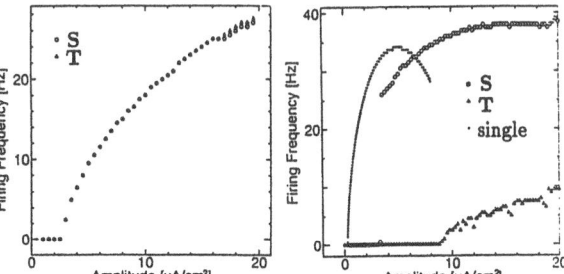

Figure 1: Relationship between firing frequency of single compartment model and intensity of maintained current stimulus.

Figure 2: Firing frequency of compartment dendrite model in response to localized steady current stimulus. *left*: proximal dendrite model. *right*: distal dendrite model.

On the Adapative Capabilities of Pulse-Coded Cable Neurons

Arno J. Klaassen[*] & Jaap Hoekstra[†]

[*]LIMSI-CNRS, B.P. 133, F-91403 Orsay-Cedex, France; email: arno@limsi.fr
[†]TU Delft, Dep. Electr. Eng., P.O. Box 5031, NL-2600 GA, Delft, the Netherlands

Abstract

We present various possibilities to implement adaptive behaviour of cable neurons and show their qualitative effect. We argue that incorporating these possibilities in local learning rules (or schemes) can account for adaptation that combines spatial and temporal properties. Experiments with a phasic XOR explain why a local learning rule based on detection of coincidence of a high local potential in the cable and the arrival of an input pulse, renders a network capable of performing the XOR function.

Neural networks that involve cable neurons, 1 bit delayed interconnections and group forming local learning rules could form the basis of neural systems with performant spatio-temporal information processing capabilities[1]. Nevertheless, the Achilles' heel of this approach seems to be the local learning rule. Work on local learning rules is just starting, but it is already shown that local learning rules, apart from group formation[2], can account for conditioning[3] and certain biological phenomena[4, 5]. And although no direct supervisor or gradient descent learning is possible, e.g. for quasi-supervised backprop at least error descent can be guaranteed[6].

In earlier work we demonstrated that, in a fixed network architecture, a simple local learning rule, which basically increments a weight upon detection of coincidence of a high post-synaptic potential and the arrival of an action potential at the very same synapse, could account for learning a tonic version of the classical XOR-problem with a low frequency input standing for a 0 and a high one for a 1 (3). In this paper we question why it does so and whether changing a weight is the sole or even the best way to implement adaptive behaviour. In fact, cable neurons provide for various a surplus of possibilities for adaptive behaviour as compared to add-multiply neurons. The latter can only be adaptive by changing their weights. Apart from this, the behaviour of cable neurons moreover is influenced by:

- The shape of the input pulse.

- The distance of an input from the soma, or more precisely from the axon hillock.

- In principle by all membrane variations. Here we will only consider change of R_i, the axial internal resistance, or R_m, the membrane resistance, as well as changing a neuron's geometry.

- By splitting an input in a number of equally or not equally delayed fractions that end in different synapses at different places.

Each kind of parameter has its own characteristic influence on a neuron's behaviour and might interfere for different purposes (e.g. synchronisation, control, etc.) and on different time scales (e.g. long for R_i, synapse placement or neuron layout; short for R_m, weight).

References

(1) A. J. Klaassen, *Computing with cables: towards massively parallel neuro computers.* PhD thesis, Delft University of Technology, Delft, the Netherlands, 1992.

(2) L. H. Finkel and G. M. Edelman, "Population rules for synapses in networks," in *Synaptic function* (G. M. . Edelman, W. E. Gall, and W. M. Cowan, eds.), pp. 711–757, New York: Wiley, 1987.

(3) A. J. Klaassen and A. Dev, "Learning pulse coded spatio-temporal neurons with a local learning rule," in *IJCNN*, vol. 1, (Seattle, WA), pp. 829–837, 1991.

(4) D. L. Alkon, "Calcium-mediated reduction of ionic currents: A biophysical memory trace," *Science*, vol. 226, pp. 1037–1045, Nov. 1984.

(5) D. L. Alkon, "Memory storage and neural systems," *Scientific American*, pp. 27–34, Jul 1989.

(6) Y. Qiao, "Learning in large optical networks," in *SPIE Aerospace Sensing. Science of Artificial Neural Networks.* (Orlando, FL), 1992.

[*]Support from CEE grant ERBCHBICT920116 is gratefully acknowledged.

A Local Approximation of the Cable Equation for Implementing a Local Interaction Model

Jaap Hoekstra
Delft University of Technology,
Dept. Electrical Engineering,
P.O. Box 5031, 2600GA Delft, The Netherlands
e-mail: jaap@neuron.et.tudelft.nl

This paper discusses a reverse engineering approach to artificial neural networks. The starting-point chosen for the reverse engineering process is the local interaction model. The engineering step models this local interaction model by an electronic ladder network, in which an input signal causes a potential pulse that rapidly decreases in amplitude during transport toward the end of the network. The mathematical model of this step consists of a system of coupled differential equations that solve the "cable equation". The computational model of this step consists of a local approximation of the cable equation, such that local learning can be computed by local computations. As a consequence of the model learning is local and dependent on the place where input signals enter the dendrite and on the temporal relation of these inputs.

The starting-point chosen for the reverse engineering process is the local interaction model described by Alkon [1]. local interaction between post-synaptic sites is assumed. I have taken some of Alkon's design principles as a basis for an electrical model, describing local interactions, that is simulated by a local approximation of the dendritic cable equation.

Models incorporating local learning rules allow increase and decrease of weights based on activities of connected nodes if the synapses are placed together on the same dendrite. These models model the biological dendrite by solving an equivalent electrical circuit, consisting of several compartments. Current models solve the electrical circuit model by the numerical calculation of a recurrence relation in which the current of a compartment is expressed in the value of the current in a next compartment, see for example [2]. In this way it is always necessary to solve the whole dendrite, even if we are only interested in local effects. In this paper an alternative mathematical description is used on bases of which local activities can be calculated by only simulating a part of the dendrite model.

The engineering step models this local interaction model by an electronic ladder network (for the dendritic cable) in which an input signal causes a potential pulse, $V_k(t)$, that rapidly decreases in amplitude during transport toward the end of the network (representing the cell body). The learning rule can now be formulated by:

$$\Delta\omega_{ij}(t) \sim \eta V_k(t)$$

where ω_{ij} is the weight between the input from node (neuron) j at the place k in the network, η is a learning rate.

Basic to a local learning scheme is that a voltage pulse induced at a specific compartment only affects other compartments within a limited range. In relation to the above described electrical model it means that the influence of a voltage pulse strongly decreases after a limited number of compartments. And on the other hand that the effect of neighboring compartments on a voltage pulse is only limited to a small number of them.

The compartmental voltage can be approached by

$$For : h \leq h_{max} :$$

$$V_k = V_{k-h} - \sum_{f=k-h+1}^{k} r_{f-1,f} \sum_{l=f}^{k_{max}} \left(c_{m_l} \frac{dV_l}{dt} + \frac{V_l}{r_{m_l}} - I_{j_l} \right)$$

Simulations show that for values of the capacitance and resistors not far besides biological plausible values a value of 5 for both k_{max} and h_{max} can be obtained.

The description above has an interesting feature. Recall that the model input was a one dimensional dendrite in which the transport direction of the charge was only toward the cell body, thus in one direction. There is, however, an influence of a limited number of capacitances from the potential in compartments between the compartment and the cell body. The height of the potential not only depends on previous (further away from the cell body) compartments, but also on how many following capacitances have to be charged. This last number also depends on stimuli present on following compartments. Thus, even in this model, the potential is determined by stimuli from two directions.

References

[1] D.L. Alkon, 'Memory Storage and Neural Systems', In: *Scientific American*, July 1989.

[2] **Methods in Neuronal Modeling**, C. Koch and I. Segev (Eds.), Cambridge MA: MIT Press, pp 63-97, 1989.

Effects of Glutamate Uptake on the Response Dynamics of the Retinal Horizontal Cell

Yoshimi KAMIYAMA[†], Tatsuya SUZUKI[‡], Hiroyuki ISHII[‡] and Shiro USUI[‡]

[†]*Department of Knowledge-based Information Engineering*
[‡]*Department of Information and Computer Sciences*
Toyohashi University of Technology
Toyohashi 441, Japan

Abstract

Glutamate is a principal neurotransmitter in lower vertebrate retina, which is employed by photoreceptors and bipolar cells[1]. Photoreceptors encode a light signal as a concentration change of glutamate. In the dark, glutamate is tonically released from the photoreceptor terminal into the synaptic cleft, where it acts on channels in postsynaptic horizontal cells. Light-induced hyperpolarization of photoreceptor causes a decrease of glutamate release, and as a result, membrane potential change is generated in the horizontal cells. Since there are no extracellular enzymes to inactivate the glutamate action in the synaptic cleft, glutamate is thought to be removed from the cleft by an uptake mechanism in photoreceptor, or by diffusion away which is uptaken by glial cells. However, the real-time process of the glutamate release has not been measured[2] because of a lack of experimental techniques for an intact retina.

Recently, we developed a computational method for estimating the input signal to a neuron by the ionic current model[3]. In the present study, we combined the method with the electrophysiological experiment to reveal the synaptic transmission between photoreceptors and horizontal cells. Voltage responses of the horizontal cell were measured from the isolated carp retina, and the glutamate uptake was controlled by the uptake inhibitor, βHA. From these recordings we estimated the time course of concentration change of glutamate in the synapse by reversing the dynamics with the ionic current model of the horizontal cell.

The present results suggest that the concentration change of glutamate at the synapse between the photoreceptor and horizontal cell is controlled by the independent release and removal mechanisms. When the glutamate uptake was inhibited by the blocker, the light onset of the concentration change was dramatically slowed down, while the light offset represented a little change. These results imply that the fast removal of glutamate is mainly realized by the uptake carrier in the photoreceptor and the slow removal reflects the diffusion of glutamate from the synaptic cleft. The unchanged dynamics in the light offset suggests that the release mechanism of glutamate is fast enough to be controlled by the voltage response of the photoreceptor which may not be changed in the presence of the blocker.

References

[1] Attwell, D., "The photoreceptor output synapse", Progress in retinal research, 9, 337–362 (1989)

[2] Clements, J.D., Lester, R.A.J., Tong, G., Jahr, C.E. and Westbrook, G.L., "The time course of glutamate in the synaptic cleft", Science, 258, 1498–1501 (1992)

[3] Usui, S., Ishii, H. and Kamiyama, Y., "An analysis of retinal L-type horizontal cell responses by the ionic current model", Neuroscience Res., Suppl.15, S91–S105 (1991)

ROBOTICS

Robot vision—oral contributions

Neural Networks for Robot Eye-Hand Coordination

F. C. A. Groen, B. J. A. Kröse
P. P. van der Smagt, M. G. P. Bartholomeus
Faculty of Mathematics and Computer Science
University of Amsterdam

A. J. Noest
Biophysics Research Institute
University of Utrecht

Abstract

In this paper learning the eye-hand coordination with Artificial Neural Networks is discussed as well the requirements and relevance to its practical application. A system to position an end-effector in 3D to grasp objects with an eye-in-hand camera system is presented both in simulation and in practice. The accuracy is discussed in relation to the number of learning samples and the adaptation period. In particular when objects have to be tracked high demands are encountered for the vision system. A vision system is presented which is able to detect moving targets in a cluttered dynamically changing environment, based on a multi-resolution scale. In is shown that also dynamic visual information can be used to control the robot acceleration.

1 Introduction

Artificial Neural Networks (ANNs) have a number of potential applications in the robotics field. Neural computational techniques have been applied in vision-based adaptive controllers [1, 2, 3, 4]. Wei and Hirzinger [5] described the off-line learning of a 6 degrees of freedom robot arm, and Pomerleau [6] the use of vision to control the steering of an AGV.

Neuro-computational methods, when applied in practice, have to compete with conventional adaptive control methods based on explicit physical models. Parameters of the model are adapted based upon the difference between the real output and the desired output. In this sense parameter estimation and learning in ANNs are closely related.

The a priori knowledge we have about the problem at hand forms an essential issue. This knowledge can be incorporated in the system by building a physical model of a robotic system, or can be learned. The physical model of a (sensor controlled) robot can be very complex and has to be simplified for practical applicability. As a result a practical model will always be an approximation of the reality, which may deviate considerably from the real system behaviour.

On the other hand, ANNs are quite capable of approximating arbitrary functions [7]. So in practice the difference between an ANN model and an adaptive physical control model can become very small. Both describe an approximation of the relation, while the parameters of the model are estimated from measured differences.

In sensor based systems the goal is expressed in the sensor domain. The only thing we know of the real world is what we perceive through the sensor. There are two essentially different sensor configurations:

1. sensors which are at fixed positions in the world. They perceive both the robot and the world. We will call these *world based sensors*;

2. sensors which are mounted on the robot or a moving cart. They perceive the world from a varying position and we will therefore call these *ego-centered*.

The perception from both sensors differs essentially. In the first case we have a static background, in which objects and the robot may move. To detect the differences in action between the robot and the goal (like gripping an object) it is essential that both the object and the robot are perceived by the static system. In the second case, because of the moving camera we have to cope with a dynamic background and the motion of the object and background differ in the images. However, all motion is expressed relative to the robot position as the sensor is ego-centered, and the sensor accuracy increases, when we approach a target.

Also in controlling the actuators two possibilities are present:

1. position oriented control: the actuator moves from point to point. Also when position set-points are incorporated in a feedback loop it still has a look-and-move strategy;

2. dynamic feedback control: not only the position but also time derivatives of the position are essentially used. In particular for the sensor data processing this method puts high requirement to the real-time aspects. Also, accurate estimation of time-derivatives from complex sensor measurements is often hard.

A third aspect is whether the learning is on-line or off-line. In an off-line learning situation, the learning procedure is performed once to estimate the model parameters, and there is no hard time limitation for the learning stage. This is completely different when the learning is on-line to create an adaptive system. Then special learning methods and high performance computer systems are needed to be able to process the learning samples in real-time.

2 What is needed for realistic systems?

Processing in real-time. This requires fast sensor data processing and fast learning algorithms to be able to process the samples in real-time. It proves that for instance the conjugate gradient method offers good possibilities to that end [8].

Insight in the accuracy of the system. A very important aspect is the accuracy that can be obtained with the system. The accuracy depends on the number of learning samples, the capacity of the ANN (related to its number of hidden units) and the learning method. Although these relations cannot be accessed it general, for a specific problem they can be found and optimum tradeoffs can be made [9].

The system should be tested in real practice. Systems should not only be evaluated in simulation. To test the system in simulation is very valuable, and has the same advantages as off-line robot programming. It requires rather good simulators of the sensing and the dynamics of the actuators. A simulation environment is only an approximation of reality, so the developed system in a simulation environment has to prove its robustness under real working conditions.

We will illustrate these aspects with the eye-hand coordination of a robot arm.

3 Problem definition of the robot arm

To position a robot gripper above an object so that it can be grasped, requires positioning in three dimensions. Because of the advantages of ego-centered sensors we use a single camera, which is mounted in the end-effector of the robot arm.

We restricted ourselves to movements where the camera is always looking down. From the camera image we can calculate: position, orientation and size of an object.

The robot configuration is sketched in figure 1. Joint angle θ_5 is not independent but expressed as a function of θ_2 and θ_3. Joint angles θ_4 and θ_6 are kept at a constant value. This results in joint values θ_1, θ_2 and θ_3 as the three degrees of freedom of the arm.

4 Learning the eye-hand coordination in the static case

Data coming in from the visual front-end has to be mapped onto joint angle displacements. For this mapping a multi-layer feed-forward network is used. The input consists of the current position $\vec{x}$

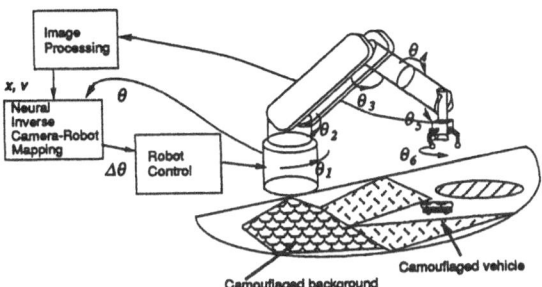

Figure 1: Sketch of the robot used in the experiments.

of the target in the image domain, the desired position $\vec{x}_s$ of the target, and the joint angles of the robot. The output of the feed-forward network consists of the joint angle displacements $\Delta\vec{\theta}$.

The object can be grasped when it is in the centre of the image with a certain size. Using an object with known size, the observed area e of the object is a measure for the distance to the object. We introduce $z = 1/e$ to make it compatible with the other vision measurements. The set-points in the sensory domain for the object to be grasped are now $\vec{x}_s = (x_s, y_s, z_s)$, where (x_s, y_s) indicates the centre of the camera image and z_s gives the desired area value. Although we can detect from the visual information that the set-point is reached, this information is not sufficient to define how the robot should move, as the object can be present at arbitrary positions in the workspace and the object position does not define the joint values (state) of the robot. So we have to add these joint values as inputs to be able to learn the eye-hand coordination. In our configuration it is sufficient to use only θ_2 and θ_3, as the desired mapping is independent of θ_1.

We could make now arbitrary moves which gives us samples of the vision vector $\vec{x}[n]$ before and the vision vector $\vec{x}[n+1]$ after the move $\Delta\vec{\theta} = \vec{\theta}[n+1] - \vec{\theta}[n]$. Together with the joint values $\vec{\theta}[n]$ we obtain successive learning samples of the mapping between the vision domain and the joint angles. This requires 8 inputs $(\vec{x}[n], \vec{x}[n+1], \theta_2, \theta_3)$ for the network and 3 outputs $\Delta\vec{\theta} = (\Delta\theta_1, \Delta\theta_2, \Delta\theta_3)$.

We can, however, use a priori information to reduce the number of inputs, which is important to increase the accuracy. For a given state of the robot $\vec{\theta}$ we want to know the joint angle displacement needed to bring the target to the centre of the camera image. Instead of learning the coordinate transformation of the image vector $\vec{x}[n]$ to the coordinates of the (due to the robot motion) rotated image vector $\vec{x}[n+1]$, we can directly calculate from $\vec{x}[n]$, $\vec{x}[n+1]$ and the rotation $R(\theta_1)$ about the robot base) the target position $\Delta\vec{x}$ from which the robot movement $\Delta\vec{\theta}$ will bring the target to the centre of the camera image

$$\Delta\vec{x} = \vec{x}[n] - R(\Delta\theta_1)\vec{x}[n+1]. \tag{1}$$

In this case we have only 5 input values and 3 output values, and the input $(\Delta\vec{x}, \theta_2, \theta_3)$ together with $\Delta\vec{\theta} = (\Delta\theta_1, \Delta\theta_2, \Delta\theta_3)$ forms a correct learning sample.

With these values a feed-forward network is trained. The network has as inputs the visual information $\Delta\vec{x}$ and the robot state. The inputs are connected via 27 hidden units to 3 output units, which give the displacement vector. The network is continuously trained, using the conjugate gradient method and adapts to changing situations.

In setting up the configuration of the neural network the optimal network configuration (number of hidden units and learning samples) has to be found. Vyšniauskas [9] showed that for this case the approximation error e depends on the number of learning samples N and the number of hidden units h by

$$e(N, h) = \frac{\gamma_{20}}{N^2} + \frac{\gamma_{11}}{Nh} + \frac{\gamma_{02}}{h^2}. \tag{2}$$

In the problem described above the coefficients were found from simulation experiments with values $\gamma_{20} = 14.08$, $\gamma_{11} = 19.9$ and $\gamma_{02} = 2.47$. In figure 2 the approximation error is given as a function of N and h.

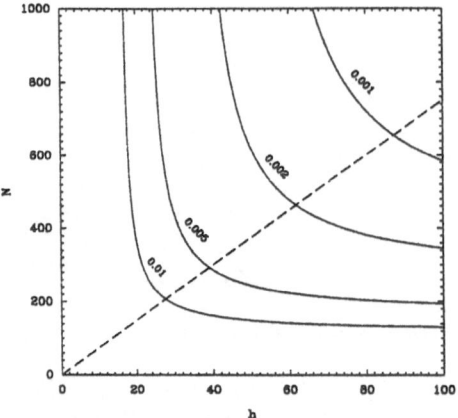

Figure 2: Contours of the approximation error function ε_a at different levels 0.001, 0.002, 0.005, 0.01. The minimal resources line for $k = 1$ is represented in dashed style.

Given the computational complexity of the used learning procedure optimal values of N and h can be obtained. For an error of 1 degree in the robot joint values, this results in optimal values $N = 200$ and $h = 27$.

To make a tradeoff between the number of learning samples and the accuracy required, we have to take into account the adaptation time t_a.

For real-time learning, the network learning and the feed-forward step through the network to generate a new robot move must be seen as two concurrent processes. Ideally, the learning time t_l does not exceed the time t_v needed for vision processing:

$$t_v \geq t_l = c_l h \tag{3}$$

where c_l is a constant. In one t_l period, for a single learning sample in a bin of samples the network gradient is computed; N of such updates are needed to give a new search direction for the conjugate gradient method. Thus, at least $N t_v$ seconds are needed to update the bin and, hence, the neural network.

When more accurate learning is needed, parallel to the neural network controlling the robot a network can be taught with more samples. After the learning time $N t_l$ the weights of the network are adjusted. Although, we are not able in that case to take each sample into account in the learning procedure, we are able to make an arbitrary tradeoff between accuracy and learning time. By monitoring the accuracy of the network, we can adapt the number of learning samples to accommodate a low accuracy in a changing situation and a high accuracy in a static situation. This approach is closely related to that of nested networks [10]. The total accuracy is limited by the inaccuracy of the vision system and the robot arm. This limits the maximum number of useful learning samples.

In figure 3 results are given from simulation experiments.

5   Vision needed for target tracking

The task of the visual subsystem is to locate the position of a possible target in the camera image. Simple methods such as blob-analysis or matched filters to locate objects can be applied. For target tracking, an important demand is that the system can still detect a moving target in a visually cluttered environment. In general there is only a weak definition of what constitutes a target: it should have a visible area and a certain speed. Because of the moving camera, the target can only be found by motion-induced segmentation. Thus sufficient velocity differences must exist between the target and the background, and one needs to measure quantitative local motion signals across the whole image.

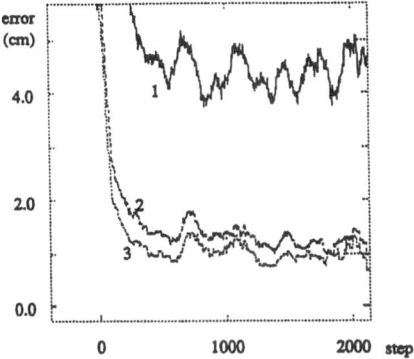

Figure 3: Grasping results with the simulated robot after one, two, and three steps (i.e., 0, one, or two feedback steps); the desired precision was 0.7 cm.

A scale-space framework is used for motion detection as one can not sample and process the images fast enough for the upper range of velocities required. Given a fixed sampling interval (multiple of 40 ms in our case) one can only detect faster motion reliably by using target structures at coarser scales.

In principle one could compute the (optical) velocity of a component from the spatial gradient $\partial_u I = \partial I / \partial u$ and the time derivative $\partial_t I = \partial I / \partial t$ as

$$v = -\frac{\partial I / \partial t}{\partial I / \partial u}. \tag{4}$$

However two problems are encountered in practice with this method: first, the spatial gradient should be sufficiently smooth for the scale on which we calculate the velocity, and secondly, there is a robustness problem: in regions with a low signal-to-noise ratio the problem of small divisors will produce essentially random velocities.

The first problem is solved by calculating the derivative of an Gaussian filtered image. The filtering suppresses the fine-structure in a well-controlled manner. The convolution kernels that perform this filtering are given by:

$$R_{n,m}(x, y, \sigma) = (\partial_x)^n (\partial_y)^m \frac{1}{2\pi\sigma^2} \exp(-\frac{x^2 + y^2}{2\sigma^2}). \tag{5}$$

The solution of the second problem is not to compute the velocity directly, but to derive from eq. (4) velocity detectors. These velocity detectors essentially check whether the time derivative is compatible for a certain velocity with the spatial gradient in a certain direction u. The response of the detector is given by the strength of the time and spatial derivatives (space-time contrast).

5.1 Velocity detectors

Across the whole image, we compute at each pixel the response of 12 motion detectors representing 24 different nominal velocities. Opposite nominal velocities are derived from one detector and encoded by one signed strength. The 12 detectors can be split in 3 spatial scale groups of 4 directions as given in figure 5. These scales are respectively 1, 2, and 4 times the basic spatial sampling rate. The corresponding nominal velocities are proportional to the spatial scales. The strength of a detector is given by

$$M = -\sqrt{(\partial_t I)^2 + (\partial_u I)^2} \, \text{sgn}(\partial_t I) \, \text{sgn}(\partial_u I) W(\partial_t I, \partial_u I), \tag{6}$$

in which a window function W tests the compatibility of the gradients

$$W(a, b) = \begin{cases} 1 & \text{if } \frac{1}{2}\sqrt{2} < |a/b| < \sqrt{2} \\ 0 & \text{otherwise.} \end{cases} \tag{7}$$

Each velocity detector results in a motion field. To locate the target, we search for the motion field in which any blob of acceptable size has the largest contrast (ratio of mean vote within the blob to global mean vote). The blob position is determined by computing the centre of gravity of the spatial support. Also a rough estimate of the target velocity can be obtained by a weighted sum of the nominal velocities within a blob, with the responses of the velocity detectors as weights.

In figure 4 an example of a cluttered image is given and in figure 5 the 12 motion field are given, with the responses of the motion detectors.

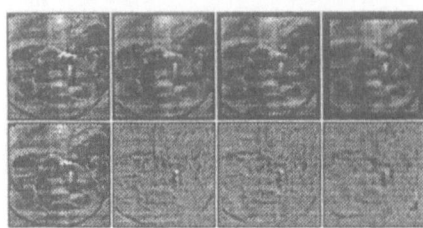

Figure 4: Left column: two images taken with a time interval of 40 ms. Upper row, columns 2–4: the multi-resolution representation. Lower row, columns 2–4: temporal derivatives at the three spatial scales, represented in a grey-level plot.

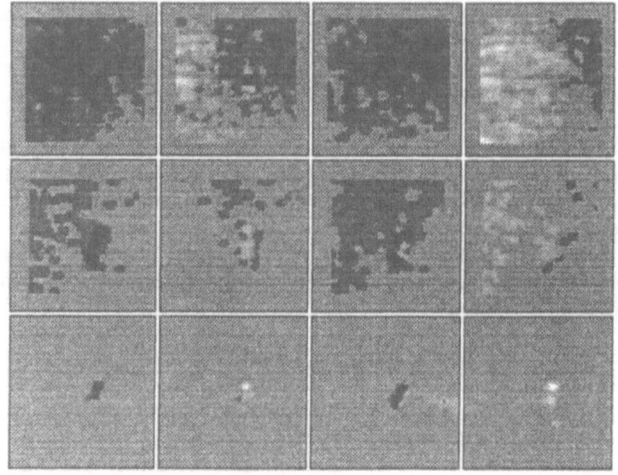

Figure 5: The 12 motion vote fields. The weights of the votes are shown in grey-level images, ranging from black to white.

Images are processed at a rate of 12.5 images/s with the Datacube Max-Video 20 image processor hardware. The neural network has been trained on a simple scene with static distinguishable object target (a white blob on a dark background). With the trained network the system is able to track a moving camouflaged vehicle when it is moving at velocities of about 0.2–0.4 m/s.

6 Learning in the dynamic case

So far the network was trained with position information pseudo-static) and no dynamic information was used. We may ask ourselves, what kind of dynamic visual information can be used to train a neural network on and instead of controlling the position controlling the acceleration to stop at the correct location?

Assume that the camera moves towards an object with constant acceleration $a(\tau) = a_0$ in the Cartesian domain. All quantities are expressed in the time τ where $\tau = 0$ is defined as the moment of contact between the end-effector and the object. Therefore, increasing t means decreasing τ as long as the camera is approaching the object. The vertical distance $d(\tau)$ and the vertical velocity $v(\tau)$ between the end-effector and the object is then

$$d(0) = d(\tau) + v(\tau)\tau + \tfrac{1}{2}a(\tau)\tau^2, \tag{8a}$$

$$v(0) = v(\tau) + a(\tau)\tau. \tag{8b}$$

When we measure the observed area $e(\tau)$ the area is inversely proportional to $d(\tau)^2$, with an unknown proportionality constant. However, we can derive that:

$$\frac{d(\tau)}{v(\tau)} = -2\frac{e(\tau)}{\dot{e}(\tau)} \tag{9a}$$

$$\frac{a(\tau)}{v(\tau)} = \frac{\ddot{e}(\tau)}{\dot{e}(\tau)} - \frac{3}{2}\frac{\dot{e}(\tau)}{e(\tau)}. \tag{9b}$$

Perfect deceleration means that the robot moves to the state where the distance between target object and the gripper is zero, and the robot is in rest. We can derive [11] that the optimal deceleration is given by

$$a_{\mathrm{opt}}(\tau) = \frac{v(\tau)^2}{2d(\tau)} \tag{10}$$

which means that for a given deceleration $a(\tau)$:

$$\frac{a(\tau)}{a_{\mathrm{opt}}(\tau)} = 4\frac{e(\tau)\ddot{e}(\tau)}{\dot{e}(\tau)^2} - 6. \tag{11}$$

Thus the error in acceleration in the Cartesian domain can be completely measured in the visual domain. However, since we do not want to use a model of the robot manipulator, but would rather learn the inverse kinematics from examples, we can neither measure nor control the acceleration of the manipulator in Cartesian domain, but only in joint space, and hence cannot use equation (11). When the system is in its optimal state however, $a(\tau) = a_{\mathrm{opt}}(\tau)$ and

$$s(\tau) \equiv \frac{e(\tau)\ddot{e}(\tau)}{\dot{e}(\tau)^2} - \frac{5}{4} = 0. \tag{12}$$

So $s(\tau)$ can be used as a set point for a visual controller. This method has been tested in simulation. A feed-forward neural network was trained with a simulated robot and vision system. Results are given in figure 6 and show that after 40 trials the network is capable of learning the deceleration. This opens up the possibility to train networks with visual time sequences of features of $\ddot{x}[n]$ as inputs and acceleration patterns that are successful.

7 Discussion

Applying Artificial Neural Networks for robot eye-hand coordination requires a tradeoff between accuracy and adaptation speed. In a two-step feedback a precision of 0.7 cm in a pyramidical visual reach space of base 100×65 cm and height 80 cm is reached. The adaptation period to become completely adapted to a new situation is in this case 40 s. Higher accuracies can be obtained at the expense of a longer adaptation period and not taken all samples into account as learning samples.

The requirements for the vision system are high to be able to perform on-line learning. Minimally 10 images have to be processed per second. Besides basic image processing operations (blob-analysis and matched filters) also more advanced methods can be performed at that speed. Using motion detectors based on a spatial multi resolution moving targets can be detected in cluttered environments. Although delays in these vision systems are small (in the order of 80 ms), they have to be taken into account. Current research focuses on neural computational techniques for predictive control in which also dynamic visual information is taken into account.

218

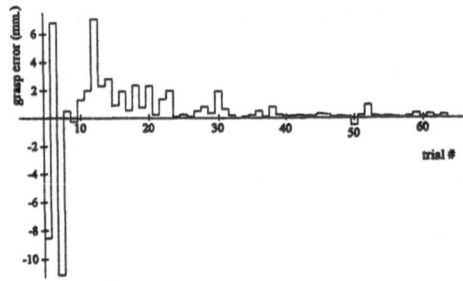

Figure 6: The error in mm in approaching the target in the z direction after the manipulator has been decelerated. On the vertical axis the distance between the end-effector and the object is plotted after deceleration; on the horizontal axis the trial number. On the average, seven feedback iterations are performed for each deceleration. Iterations 1 through 4 are not shown.

Acknowledgement

This work has been partly sponsored by the Dutch Foundation for Neural Networks.

References

[1] T. M. Martinetz, H. J. Ritter, and K. J. Schulten. Three-dimensional neural net for learning visuomotor coordination of a robot arm. *IEEE Transactions on Neural Networks*, 1(1):131–136, March 1990.

[2] W. T. Miller III. Real-time application of neural networks for sensor-based control of robots with vision. *IEEE Transactions on Systems, Man, and Cybernetics*, 19(4):825–831, August 1989.

[3] P. van der Smagt and B. J. A. Kröse. A real-time learning neural robot controller. In *Proceedings of the 1991 International Conference on Artificial Neural Networks*, pages 351–356, Espoo, Finland, June 1991.

[4] J. R. Cooperstock and E. E. Milios. An efficiently trainable neural network based vision-guided robot arm. In *IEEE International Conference on Robotics and Automation*, 1993. Submitted.

[5] G.-Q. Wei and G. Hirzinger. Learning motion from images. In *11th IAPR Conference on Pattern Recognition*, pages 189–192, The Hague, Netherlands, 1992.

[6] D. A. Pomerleau and D. S. Touretzky. Analysis of feature detectors learned by a neural network autonomous driving system. In F. C. A. Groen, S. Hirose, and C. E. Thorpe, editors, *Intelligent Autonomous Systems*, pages 572–581. IOS Press, 1993.

[7] K. Hornik, M. Stinchcombe, and H. White. Multilayer feedforward networks are universal approximators. *Neural Networks*, 2(5):359–366, 1989.

[8] P. van der Smagt. Minimisation methods for training feed-forward networks. *Neural Networks*, 1993. In press.

[9] V. Vyšniauskas, F. C. A. Groen, and B. J. A. Kröse. A method for finding the optimal number of learning samples and hidden units for function approximation with a feedforward network. In S. Gielen and B. Kappen, editors, *Proceedings of the International Conference on Artificial Nerual Networks*. Springer Verlag, 1993.

[10] A. Jansen, P. van der Smagt, and F. C. A. Groen. Nested networks for robot control. In A. F. Murray, editor, *Neural Network Applications*. Kluwer Academic Publishers, 1993. In press.

[11] P. van der Smagt, B. J. A. Kröse, and F. C. A. Groen. A self-learning controller for monocular grasping. In *Proceedings of the 1992 IEEE/RSJ International Conference on Intelligent Robots and Systems*, pages 177–182, Raleigh, N. C., June 1992.

Unsupervised Formation of Feature Detectors Using Residual Inputs

Petri Koistinen
Rolf Nevanlinna Institute
P.O. Box 26 (Teollisuuskatu 23)
FIN-00014 University of Helsinki, FINLAND
Email: pek@rolf.helsinki.fi

Abstract: The problem of designing feature detectors for low-level computer vision using unsupervised learning is considered. A new approach to the problem based on residual vector modeling is proposed. The residual vectors are obtained by positing a parametric model for the gray value function near an image point when the feature of interest does not appear in its neighborhood. A long residual vector indicates a poor fit with the model and thus suggests the presence of the feature of interest. The new approach is based on finding systematic structure in the distribution of the residual vector using unsupervised learning. Both principal component analysis and a form of the self-organizing map algorithm are used as learning methods. The approach is applied to the design of edge and corner detectors.

1 Introduction

Though the overall structure of the human visual system is genetically determined, it is apparent that many aspects of it must emerge by self-organization. Some researches have tried to model self-organization in biological low level (or early) vision—see Becker [2] for a survey of such research. E.g., in Linsker's simulations [7] a multi-layered network of adaptive units obeying simple, Hebbian-type rules developed orientation sensitive cells as a result of being driven by random input. In this paper we propose mechanisms whereby an artificial vision system could develop useful feature detectors; consequently we do not make any claims as to the biological plausibility of the mechanisms. Our feature detectors do not have to develop from scratch as in Linsker's simulations but we allow judiciously designed preprocessing of the image prior to adaptation in order to make learning easier. On the other hand we do not want to design the feature detectors completely from start to finish, because a certain degree of adaptability can be of advantage, e.g., when the image material or the preprocessing steps do not conform to our idealized models. Although only the self-organizing map (SOM) algorithm and principal component analysis (PCA) are used in this paper, our residual analysis considerations could be relevant also in connection with other unsupervised learning methods.

This paper is based on a much more detailed report [6].

2 A Framework for Designing Feature Detectors by Unsupervised Learning

One straightforward way to form feature detectors would be as follows. At each location r_i in the image f form a vector z_i comprising the image intensity values in a $(2N + 1) \times (2N + 1)$ window about the point r_i. Then one could use, e.g., a SOM to cluster the window vectors z_i, and after training, the feature values at position r_i could be the distances of the vector z_i from the reference vectors m_j. However, we do not consider such an approach to be practical. E.g., to describe an edge going through a point r_0 one needs at least three variables: the constant intensities at both sides of the edge and the orientation angle. To detect all possible edges we would have to have a code book big enough to contain a reference vector for each significantly different combination of

This work was done in the ESPRIT basic research project SUBSYM; the work was funded by TEKES.

these three variables. In addition, naive vector quantization of the raw image window vectors z_i produces many reference vectors that do not represent edges at all.

Instead we suggest the following scheme for testing whether the image window around point r_i has the property F. Our null hypothesis H_0, which we try to disprove, is that the image window does not have the property F. Moreover, we assume that we can guess a reasonable parametric model

$$Z = m(\theta) + \text{noise}, \qquad \text{if } H_0 \text{ true}$$

for the distribution of the data Z when H_0 is true. However, we do not assume prior information about the distribution of Z when H_0 is false. Instead we form an estimate $\hat{\theta}$ for the unknown parameter vector assuming H_0 is true and then look at the values of the residual $X = Z - m(\hat{\theta})$. If H_0 indeed is true, the residual vector is likely to be short. On the other hand, when the parametric model can not adequately describe the data, the residual vector is long and, hopefully, exhibits systematic structure. We model this residual structure (the outliers of the distribution of X) using an unsupervised algorithm, and then our test statistic indicates how well the estimated residual x_i matches the prototypical residuals of the outlier distribution. Similarities with the prototypes can be used as feature values.

So, the learning algorithm should only pay attention to the distribution of long input vectors. In addition, it should use a meaningful similarity measure. We use correlation-type similarity measures, i.e., we judge the residual x_i similar to the reference vector m_j if the inner product $x_i^T m_j$ or some related expression is large. The reference vectors are constrained to be unit vectors so that the correlation outputs can be directly compared with one another. One learning algorithm which meets our requirements is PCA, if only a few dominant eigenvectors of the correlation matrix of the residual vectors are retained. We have also use a form of the SOM algorithm which corresponds to the vector quantization criterion

$$\text{maximize } E[\max_{i=1,\dots,k} X^T m_i] \qquad \text{subject to } \|m_i\| = 1, \qquad i = 1, \dots, k.$$

Such an algorithm is presented in [5, Ch. II.F]. When the input vectors x_i are not normalized, this criterion has the property that the optimal reference vectors reflect the distribution of long input vectors while the distribution of short ones is ignored.

3 Edge Detectors

First we illustrate the approach with the problem of automatic formation of edge detector units. Our model for the image intensity variation inside a window centered at r_0 when there is no edge present is as follows

$$f(r + r_0) = c + n(r), \qquad \text{if } r_0 \text{ is not an edge point,}$$

where $n(r)$ is zero mean noise. Accordingly, we may estimate the parameter c as the mean value of the image within the window and then form the mean-removed residual vector $x = z - \mu(z)\mathbf{1}$, where $\mathbf{1}$ is a vector of all ones, $\mu(\cdot)$ denotes the mean value operator, and z denotes the image intensities inside the window represented as a vector.

It is easy to find structure corresponding to edges in this kind of residual data. E.g., typically the two dominating eigenvectors v_1 and v_2 of the correlation matrix of the residuals correspond to masks which perform differencing in two, spatially perpendicular directions—see Figure 1 for an example. Each little square represents a weight of a mask. An empty square corresponds to a positive weight and a filled square to a negative one and the size of a square is proportional to the absolute value of the corresponding weight.

A sensible way to assess the similarity of a given window vector z with these two eigenvectors would be to calculate the length of the projection of the mean-removed vector $z - \mu(z)\mathbf{1}$ on the subspace spanned by v_1 and v_2. This length works out to be $[(z^T v_1)^2 + (z^T v_2)^2]^{1/2}$ because the eigenvectors are also mean-removed. Notice that calculating the correlation $z_i^T v_j$ at each pixel simply means convolving the image with the mask v_j. Because of the properties of the eigenvectors,

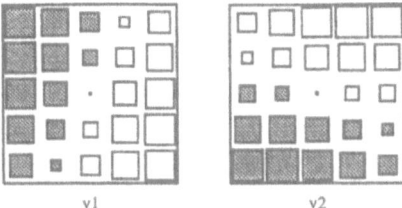

v1   v2

Figure 1: The two dominant eigenvectors v_1 and v_2 of the correlation matrix of the mean-removed residual vectors corresponding to 5×5 windows in the Lenna test image.

the projection length is (proportional to) an estimate of the gradient magnitude of the image function at the point of interest.

We have also experimented with the SOM algorithm corresponding to the vector quantization criterion

$$\text{maximize } E[\max_{i=1,\ldots,k}(X - \mu(X)1)^T m_i] \quad \text{subject to } \|m_i\| = 1, 1^T m_i = 0, \quad i = 1,\ldots,k.$$

The constraints can be enforced by first making a correction in the direction of the gradient, then by removing the mean from the modified weight vectors and last by normalizing the vectors to length one. As expected, the SOM converges very quickly to a configuration where all the weight vectors are clear edge detectors and where the orientation preference of the units changes in a smooth manner as a function of the position of the unit in the map, see [6].

Ideas similar to our use of the SOM for forming edge detectors have been proposed in vector quantization literature: in particular the scheme called mean/shape-gain VQ [3, Ch. 12.10] is closely related. Also, the use of a SOM algorithm with a correlation-type similarity function has been demonstrated previously in the task of autonomous formation of edge masks [1].

4   A Self-Organizing Corner Detector

The property that distinguishes corner points (or L-junctions) from edges is that the intensity distribution is 1D at an edge point but 2D at a corner point, i.e., at an edge point there is one dominant direction (the gradient direction) along which all the essential variation takes place whereas the intensity is approximately constant along any perpendicular section. So, if we fix the position of interest as r_0, our null hypothesis is that the intensity variation described in a local coordinate frame whose x axis is aligned with the gradient is a function of the first coordinate only, or symbolically

$$g(x,y) \ = \ S(x) + n(x,y), \quad \text{if } r_0 \text{ is not a corner point}$$
$$g(r) \ := \ f(R_\phi r + r_0),$$

where ϕ denotes the angle $\nabla f(r_0)$ makes with respect to the x axis and R_ϕ is the rotation matrix

$$R_\phi = \frac{1}{\sqrt{f_x^2 + f_y^2}} \begin{bmatrix} f_x & -f_y \\ f_y & f_x \end{bmatrix}.$$

It turned out that this model is easier to apply in a differentiated form, i.e., as

$$\nabla g(x,y) = [s(x),\ 0\,]^T + [n_x(x,y), n_y(x,y)]^T, \quad \text{if } r_0 \text{ is not a corner point.}$$

In practice, we form an estimate for ∇g by resampling ∇f on a small grid aligned with the local coordinate frame (see Figure 2 for an example) and then by transforming the result. The needed transformation $\nabla g(r) = R_\phi^T \nabla f(r')$ can be obtained by matching the Taylor expansions of f and g.

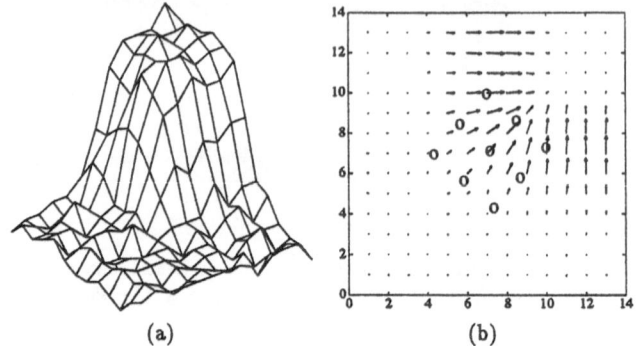

(a)   (b)

Figure 2: Example of a gray-level corner: (a) 3D plot of a corner (b) gradient field (arrows) and the resampling grid for one pixel (circles).

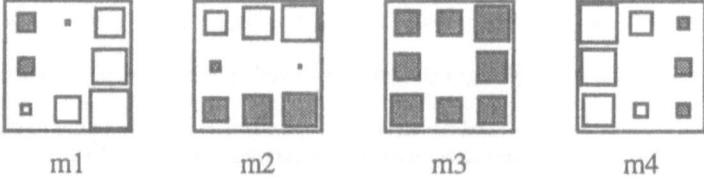

m1   m2   m3   m4

Figure 3: Four masks derived from residual data g_y.

We got interesting results by using as residuals the estimates g_y, which according to the model, should be small when there is no corner inside the window. The input was derived from a synthetic image containing four squares. Gradient direction was estimated with a 5×5 mask. At each pixel we had a 3×3 resampling grid whose grid spacing was one pixel. The values ∇f at the grid points were estimated using bilinear interpolation and then the value of ∇g was obtained from the transformation formula. The dominant eigenvector of the correlation matrix of the residuals turns out to be a differencing mask in the y direction (it looks practically the same as mask number two in Figure 3). So, that feature detector tries to form an estimate for the partial derivative g_{yy}. Expressed in terms of the partial derivatives of the unrotated image function f,

$$g_{yy} = \frac{f_{xx} f_y^2 + f_{yy} f_x^2 - 2 f_x f_y f_{xy}}{f_x^2 + f_y^2}.$$

This is exactly the "cornerness" measure k Kitchen and Rosenfeld proposed in one of the first papers on gray-level corner detection [4].

We also trained a SOM with the vector quantization criterion

$$\text{maximize } E[\max_{i=1,\ldots,k} X^T m_i] \qquad \text{subject to } \|m_i\| = 1, \qquad i = 1, \ldots, k$$

where X now denotes g_y. We had four weight vectors and the neighborhood interaction was turned off. The resulting weight vectors are shown in Figure 3. Mask number two responds at corner points and the remaining ones respond near edges. Figure 4 shows how this mask collection performs with a natural image of a toy block; we show the output of mask two at those locations where it responds more strongly than any other mask. There remain non-corner points where our corner detector gives non-zero output. However, it is easy to remove these spurious responses by thresholding.

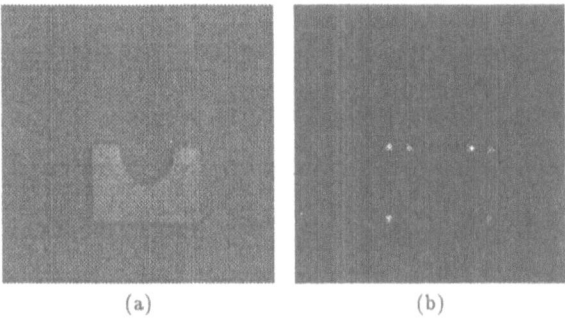

(a)   (b)

Figure 4: Corner detection for the toy block image: (a) a toy block, (b) corner detector output.

5 Conclusion

We have demonstrated an approach to forming feature detectors based on residual modeling using neural-type unsupervised learning algorithms. Our examples are detectors for edges and corners. The feature detector design framework does not have a direct model in the literature, and our approach to forming corner detectors seems to be completely new.

References

[1] N.M. Allinson, M.J. Johnson, and K.J. Moon. Digital realisation of self-organizing maps. In D.S. Touretzky, editor, *Advances in Neural Information Processing Systems I*, pages 728–738. Morgan Kaufmann Publishers, 1989.

[2] S. Becker. Unsupervised learning procedures for neural networks. *International Journal of Neural Systems*, 2(1 & 2):17–33, 1991.

[3] A. Gersho and R.M. Gray. *Vector Quantization and Signal Compression*. Kluwer Academic Publishers, 1992.

[4] L. Kitchen and A. Rosenfeld. Gray-level corner detection. *Pattern Recognition Letters*, 1:95–102, 1982.

[5] T. Kohonen. The self-organizing map. *Proc. of the IEEE*, 78(9):1464–1480, September 1990.

[6] P. Koistinen and L. Holmström. A framework for the design of feature detectors by self-organization. Research Reports A10, Rolf Nevanlinna Institute, 1993.

[7] R. Linsker. Self-organization in a perceptual network. *Computer*, pages 105–117, March 1988.

Geometry-Driven Diffusion: Coupled Diffusion Maps as a Model for Excitatory and Inhibitory Behaviour in Vision

M. Proesmans, E.J. Pauwels, L.J. Van Gool, T. Moons and A. Oosterlinck

ESAT-MI2, K.U.Leuven, K. Mercierlaan 94, 3001 Leuven, Belgium

Abstract

We investigate the potential of coupled non-linear diffusion processes to provide an efficient and implementable model for the processing that takes place in the first stages along the visual pathway, and study their role in edge preserving smoothing and noise suppression. The basic idea is that several maps undergo coupled development towards an equilibrium state. These maps could e.g. contain intensity, local edge strength, range, or another quantity. All these maps, including the edge map, contain continuous rather than all-or-nothing information, following a strategy of least commitment. Each of the approaches has been developed and tested on a parallel transputer network.

1   Introduction

The problem facing a sensory device in general and the visual system in particular is that given an arbitrary input signal, it needs to convert this signal into a new but less noisy signal, in which all the salient features are still present and, if at all possible, enhanced. Furthermore, to be of practical use, processing-speed is crucial. Biological systems tackle this problem through parallelism, one of the basic paradigms underlying Artificial Neural Networks (ANNs). However a major obstacle is the massive connectivity required in classical ANNs, which makes it very hard indeed to design an efficient hardware implementation of the computation schemes. The implementation of the basic algorithms in hardware is crucial to attain the required speed. At this stage it is fair to say that ANNs have only rarely been able to produce convincing processing results on images of realistic dimensions and complexity.

It is for these reasons that we have turned to the study of "computational paradigms" for which *restricted connectivity* suffices: processing cells need only communicate with their immediate neighbours and global connectivity is no longer required. From a biological viewpoint there is much to be said for such a configuration. In fact, neurological research has clearly shown that the brain has highly specialized (and probably largely "prewired") cortical areas where the spatial organisation is optimized (e.g. retinotopically) in order to keep the connectivity low and simple. In contrast to traditional neural networks, connections between layers turn out to be spatially restricted and reciprocal, and in particular within-layer connections seem to be essential.

In an attempt to implement such computational characteristics, we have focussed our attention on *systems of coupled non-linear diffusion equations, governing the evolution of important image features, such as the grey-level value, edge indicators, range, . . .* These systems of coupled diffusion maps (CODIMs) are interpreted as a phenomenological model for the processes that take place along the visual pathway. The rationale underlying this approach can be traced back to various models in (developmental) biology and neurophysiology: the idea is that a pattern evolves from an initial configuration due to the simultaneous action of excitatory and inhibitory influences. A combination of competition and coorperation allows the system to converge to a final and consistent configuration in different coupled feature maps. Desirable and salient features are enhanced while at the same time the random noise is filtered out or at least reduced. The CODIMs are natural candidates to model biological behaviour since their diffusion component allows for a (weighted) averaging of the input signal over a region of a certain extent (the receptive field), thus realising a reduction in the complexity of the signal, whereas the non-linear reaction components provide a flexible medium through which all sorts of excitatory and inhibitory behaviour can be modelled.

2 First order diffusion equations

2.1 Introduction and motivation

Recently several authors (cfr. e.g. [2, 5] and the references given therein) have rekindled the interest in regularization of images using energy functionals and the associated evolution equations. From a conceptual point of view, the simplest way to achieve regularization is to change the original signal g into a new signal f which minimizes some cost- or energy-functional. Such a functional will try and find an optimal position between two conflicting requirements: smoothness on the one hand and faithfulness to the input-signal on the other. A protypical example of this sort of functional was introduced and studied extensively by Mumford and Shah [1]. Unfortunately, their functional is non-convex and usually has a large number of local minima which makes the search for a global minimum extremely time-consuming.

Prompted by these limitations, Shah [5] proposed the following system of coupled diffusion equations:

$$\frac{\partial f}{\partial t} = v^2 \nabla^2 f - \frac{1}{\sigma^2}(f - g) \qquad \frac{\partial v}{\partial t} = \rho \nabla^2 v - \frac{v}{\rho} + 2\alpha(1 - v)\|\nabla f\|, \qquad (1)$$

where v is an edge-map or -indicator and therefore proportional to the probability of there being an edge at that position in the image. This results in two CODIMs: one for the intensity (f) and one for the edgemap (v). Evolution proceeds rather efficiently to a stable solution, albeit one that may look noticeably blurred. This dynamics cannot enhance edges as such, but it does give a way to describe the presence of a discontinuity not just by its gradient. In fact v can be considered as a region of interest along the discontinuity lines. The width of this region is determined by the smoothness coefficient ρ. Moreover, v behaves non-linearly with respect to the contrast (gradient) in the image, and turns out to be more convincing than the normal gradient approach. The smoothing however, results in a loss of localization of the boundary.

Instead of smoothing, Perona & Malik [3] proposed a method to sharpen the edges locally by making the diffusion coefficient c in the heat-equation dependent on the local image geometry:

$$\text{P\&M:} \qquad \frac{\partial f}{\partial t} = div(c(f)\nabla f) \qquad (2)$$

It is clear that choosing $c = 0$ at edges in the image, and $c = 1$ elsewhere, encourages smoothing within the uniform region in preference to smoothing across the boundaries of such regions. Of course, these boundaries are not known in advance. The most obvious solution is to choose c locally as a function of the gradient: $c(f) \equiv c(\|\nabla f\|)$. Possible functions are $c(f) = \exp(-(\|\nabla f\|/K)^2)$ and $c(f) = (1 + (\|\nabla f\|/K)^2)^{-1}$. This algorithm does succeed in sharpening edges (for more details we refer to [3, 4]).

This however is not the whole story: an essential aspect of edge sharpening is the simultaneous preservation of edge contrast. If one starts off with a simple picture consisting of two intensity plateaus connected by a gently sloping ramp, the above-mentioned evolution equations will increase the slope of the ramp, but it may well happen that simultaneously contrast difference over the edges diminishes gradually, until the difference is no longer discernable. As a consequence, the process should be halted at a stage where an acceptable compromise between edge sharpness and contrast is established, otherwise the image structure might be lost altogether.

2.2   Systems of coupled, non-linear diffusion equations

Comparing eqs.(1) and (2) one sees that in each case we are dealing with diffusion-like, non-linear evolution equations. This is the starting point of the approach we intend to take: it is our aim to investigate the potentials of *systems of coupled non-linear diffusion equations governing the evolution of important image features (such as grey-level, edge-indicator, range, etc....)* Such systems have several characteristics which make them especially attractive:

- The evolution equations are local and proceed to their stationary limit without the need for a difficult and time-consuming global search;

- Moreover, these evolution equations are easily parallelizable which speeds up the processing.

- The non-linearity of the equations allows for non-trivial and non-proportional responses to the input;

- Coupling the different equations makes information transfer between the different features possible and ensures internal consistency of the final result.

Since proving general results for such systems is difficult, we will look at a number of concrete examples which illustrate the aforementioned points.

It would for instance be interesting if we could include the discontinuity function v into the edge sharpening process. This could be done by changing the first equation of (1) by the one of Perona&Malik (2), adding a term proportional to $f - g$ to facilitate convergence (as was done by Nordström [2]):

$$\frac{\partial f}{\partial t} = div(c(f)\nabla f) - \frac{1}{\sigma^2}(f - g), \qquad \frac{\partial v}{\partial t} = \rho\nabla^2 v - \frac{v}{\rho} + 2\alpha(1 - v)\|\nabla f\| \qquad (3)$$

This approach does indeed converge for a wide range of σ's. At this stage this characteristic has only been observed experimentally: after a reasonably small number of iterations the process practically always seems to have stabilized and no change is observed if the process is continued.

If we introduce the discontinuity function v into the diffusion coefficient for the grey-level f, we arrive at the following system of equations:

$$\frac{\partial f}{\partial t} = v \cdot div(c(f)\nabla f) - \frac{1}{\sigma^2}(f - g), \qquad \frac{\partial v}{\partial t} = \rho\nabla^2 v - \frac{v}{\rho} + 2\alpha(1 - v)\|\nabla f\| \qquad (4)$$

Notice that now the information transfer between f and v is bi-directional (reciprocal). This process will encourage edge sharpening in regions nearby discontinuities (see Fig. 1). These regions are obviously determined by v and the smoothing parameter ρ. Within the interior pixel values are forced to stick more closely to their original values. Compared to (3) the result turns out to be more realistic in the sense that the intensity profile is not abruptly subdivided into intensity planes. On the other hand, due to the edge sharpening, the discontinuity function clearly shows some refinement across the boundaries, which is in favour of its localization.

As another illustration of the edge-enhancement capacities of the proposed systems we reproduce the image of an arterial network (see Fig. 2) that has been processed using the system

$$\frac{\partial f}{\partial t} = \nabla(c_1\nabla f) - \frac{(1 - v)}{\sigma^2}(f - g), \qquad \frac{\partial v}{\partial t} = \rho\nabla(c_2\nabla v) - \frac{1}{\rho}v + \frac{1}{2}\alpha v(1 - v)\|\nabla f\| \qquad (5)$$

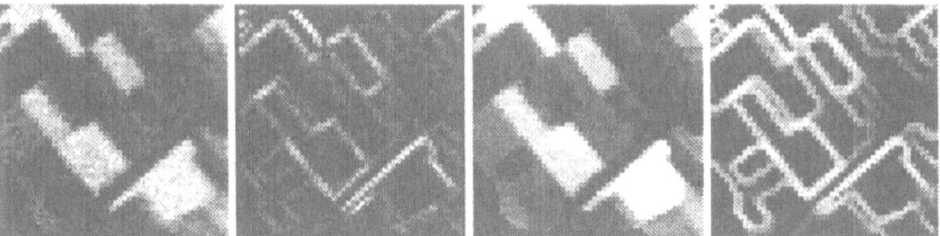

Figure 1: Original (2 left) and processed (2 right) SPOT-image and their corresponding edge maps, using CODIM (4)

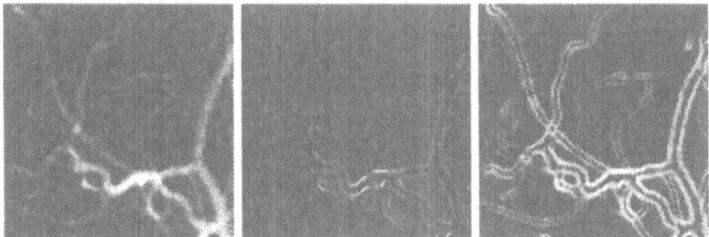

Figure 2: Original image (left) togehter with unprocessed (middle) and enhanced (using CODIM (5) edge-map for arterial network (right).

3 Second order diffusion equations in one dimension

3.1 Introduction

The above smoothing algorithms have the disadvantage that

- edges can be sharpened, but contrast (measured as the difference in intensity across the edge) is not enhanced;

- the result shows rather poor response in the neighbourhood of ramp edges (oversegmentation).

In an attempt to alleviate these drawbacks we turn to so-called "second order smoothing" where one aims to minimize a functional involving the second derivative of the image-function. More precisely the objective is to minimize the integral

$$F(f) = \int_\Omega \left[\left(\frac{\partial^2 f}{\partial x^2} \right)^2 + \sigma^2 (f - g)^2 \right] dx$$

(To keep the analysis that follows as transparent as possible we will restrict ourselves for the rest of this paper to the one-dimensional case $\Omega = I\!R$.) Since the gradient no longer appears in the functional, jumps in the first derivative are no longer penalized, so that continuous piecewise linear solutions (e.g. ramps) become possible. Moreover, this functional has another attractive feature in that it is able to produce contrast enhancement. Unfortunately, its Euler-equation involves a fourth-order differential, which causes the corresponding evolution equation based on this ODE, to be too noise sensitive. For this reason we introduce a new function β which is meant to be an approximation of $\partial f / \partial x$. Of course, this entails changing the cost-functional to one in which the first derivative of β plays the role of the second derivative of f and which involves an extra term penalizing large discrepancies between β and $\partial f / \partial x$:

$$F_\lambda(f,\beta) = \int_R \left[\left(\frac{\partial \beta}{\partial x} \right)^2 + \sigma^2 (f-g)^2 + \frac{1}{\lambda^2} \left(\beta - \frac{\partial f}{\partial x} \right)^2 \right] dx \qquad (6)$$

The Euler-equations for the approximating functional are given by the system of ODEs

$$\begin{cases} f'' - \beta' - \sigma^2 \lambda^2 (f-g) = 0 \\ \lambda^2 \beta'' + f' - \beta = 0 \end{cases} \qquad (7)$$

The behaviour of this coupled set of equations is that it will generate under- and overshoots in the neighbourhood of a discontinuity, as a cnsequence of which this discontinuity will be smoothed much less than it would using an ordinary smoothing approach.

3.2 Smoothing and edge-enhancing using second-order diffusion

Incorporating the above-mentioned second-order terms in the diffusion equations yields the following functional

$$\int_\Omega \left((\frac{\partial \beta}{\partial x})^2 + \frac{\xi^2}{2} (\beta - \frac{\partial f}{\partial x})^2 + \frac{\sigma^2}{2} (f-g)^2 \right) dx$$

and corresponding system of PDEs:

$$\frac{\partial f}{\partial t} = d_1 \nabla^2 f - \sigma^2 (f-g) - s_h \frac{\partial \beta}{\partial x}$$

$$\frac{\partial \beta}{\partial t} = d_2 \nabla^2 \beta - \xi^2 (\beta - \frac{\partial f}{\partial x})$$

(To be correct, the PDEs only correspond to the evolution equations of the functional if $s_h = d_1 = \xi^2 > 0$ and $d_2 = 1$. However we have allowed these constants to take different values in order to have more parameters that we can fine-tune to the problem at hand.) Notice how once again we arrive at a system of coupled non-linear diffusion equations for features (in this case β and f). As pointed out earlier, the behaviour of this coupled set of equations is that it will generate under- and overshoots in the neighbourhood of a discontinuity.

This model can be improved when edges in the first order derivative (i.e. β) are sharpened. In that case a piecewise linear profile can be recovered from an noisy original one. The equations are

$$\frac{\partial f}{\partial t} = d_1 \nabla^2 f - \sigma^2 (f-g) - s_h \frac{\partial \beta}{\partial x}$$

$$\frac{\partial \beta}{\partial t} = d_2 \, div(c(\beta)\nabla\beta) - \xi^2 (\beta - \frac{\partial f}{\partial x}) \qquad (8)$$

The diffusion process behaves rather well, e.g. on a ramp edge (cfr. Fig. 3), even if the noise amplitude in β is larger than the discontinuity steps, this is due to the coupling with f.

Contrast enhancement can be obtained by sharpening f, while retaining the shooting parameter.

$$\frac{\partial f}{\partial t} = d_1 div(c(f)\nabla f) - \sigma^2 (f-g) - s_h \frac{\partial \beta}{\partial x}$$

$$\frac{\partial \beta}{\partial t} = d_2 \nabla^2 \beta - \xi^2 (\beta - \frac{\partial f}{\partial x}) \qquad (9)$$

The diffusion process on a noisy step edge clearly results into a Mach-band profile (see Fig. 4). The contrast enhancement can easily be adapted depending on the applications. However, like the Perona algorithm, the response on ramp edges is poor.

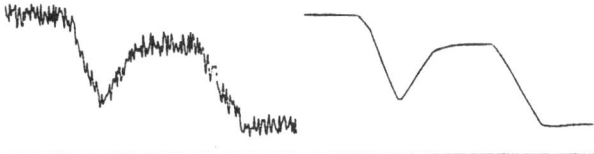

Figure 3: A system driven by second-order information can handle noisy ramp edges.

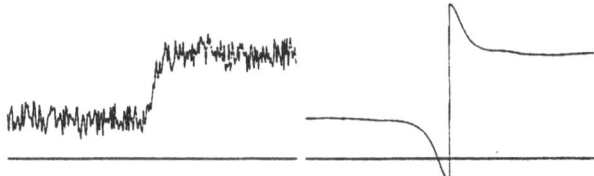

Figure 4: Systems driven by second-order derivatives can generate over- and undershoots as with the Mach band effect.

4   Conclusion

In this paper we have shown how systems of coupled non-linear diffusion equations provide a powerful and flexible conceptual framework for many aspects of image enhancement. Such systems exhibit many attractive characteristics: they are highly parallelizable and are therefore natural candidates for VLSI-implementation. Time-consuming global searches through high-dimensional spaces are no longer necessary. Their coupling ensures internal consistency of the final result and the non-linearity allows for a rich and interesting dynamics. In short, they open up an interesting new avenue of research in low-level image processing.

References

[1] David Mumford and Jayant Shah: Optimal Approximation by Piecewise Smooth Functions and Associated Variational Problems. *Comm. on Pure and Applied Math.* Vol.42, 1989, pp. 577-685.

[2] N. Nordström: Biased Anisotropic Diffusion: A Unified Regularization and Diffusion Approach to Edge Detection. Image and Vision Computing, Vol.8, No.4. 1990.

[3] P. Perona and J. Malik: Scale-Space and Edge Detection Using Anisotropic Diffusion. *IEEE Transactions on Pattern Analysis and Machine Intelligence,* Vol.12, No.7, July 1990.

[4] M. Proesmans, E.J. Pauwels, L.J. Van Gool, T. Moons and A. Oosterlinck: *Image Enhancement using Geometry-Driven Diffusion.* ESAT-MI2 Internal Report, 1992.

[5] J. Shah: Segmentation by non-linear diffusion. Proc. IEEE CVPR 91, Hawai, 1991.

SPIN
Learning and Forgetting Surface Classifications with Dynamic Neural Networks

Herman Keuchel, Ewald von Puttkamer & Uwe R. Zimmer

University of Kaiserslautern - Computer Science Department - Research Group Prof. E. v. Puttkamer
P.O. Box 3049 - W6750 Kaiserslautern - Germany
Phone: ...49 631 205 2624 - Fax: ...49 631 205 2803 - Telex: 4 5627 unikl d
e-mail: uzimmer@informatik.uni-kl.de

This paper refers to the problem of adaptability over an infinite period of time, regarding dynamic networks. A never ending flow of examples have to be clustered, based on a distance-measure. The developed model is based on the self-organizing feature maps of Kohonen [6], [7] and some adaptations by Fritzke [3]. The problem of dynamic surface classification is embedded in the SPIN project, where sub-symbolic abstractions, based on a 3-d scanned environment is being done.

1. Survey

First the framing project and the concrete problem context is discussed in short (chapter 2). Then in chapter 3 the network structure and associated aspects and problems are shown in detail, supported by simulation results (chapter 4).

2. SPIN-Project

At the actual state of research the project SPIN (from Spatial Perception to Identification with Neural networks) is based on the data of a 3-d scanning device and designed to reach a stage of abstraction where convex clusters of surfaces are generalized, completed and classified.

2-1. Main strategies

The system-design is based on some main principles, which have in common that none of them is in contradiction to a biological system. It is not intended to find the best fitting model for the lower levels of the mammal object-recognition-system, but obviously implausible features should be avoided.

a. Learning instead of preprogrammed models

The internal world model should be build up from a flow of examples scanned from the outer world. The "pre-programmed" knowledge is reduced to elementary features that should be searched for (here: edges).

b. Hierarchy

The main structure is pipeline-oriented instead of being controlled by a central instance (see the Neocognitron by Fukushima [5] for a good example of this strategy).

c. Symmetry

The general purpose processes (like classification or completion) should be quite similar at the different hierarchical levels.

d. Extensive use of feedback

A strict hierarchy is not as useful as it could be, if the different layers are not connected in both directions.

d-1 Error feedback

The back-propagated error messages correct decisions on lower processing stages, so it is not necessary to find always the best answer immediately. Lower stages may take the most likely way, knowing that there is another instance, which will give a negative feedback, if this decision was wrong.

d-2 Focus-of-interest feedback

At the start-up time of the whole system, lower components are triggered by local "instincts", whereas later on the activity of the lower components is more and more initiated by a focus-of-interest, generated by higher stages.

e. Parallelism

The above described strategies lead straight forward to forms of parallelism of rough granularity and by the use of neural networks at several stages to parallelism at a finer granularity.

2-2. Focus on surface-classification

The concrete problem area that will be discussed in this article, is defined now.

Surface representation

Given a laser-range scanner and several preprocessing steps, both beyond the focus of this paper, which produce a surface representation, built upon the curvatures at the borders of the surfaces. The curvatures are calculated by projecting the border (at each border point) on two orthogonal two-dimensional planes and then regarding the 2-d-curvatures on these planes (called the surface- and the border-curvature). The orientations of these planes are defined at each border point by the orientation of the local border tangent and the local surface normal: Both planes have to include the border tangent and one of them (the plane to determine the surface-curvature) has to include the surface normal (this has to be approximated or to be solved analytically). Each surface is then described by the concatenation of two vectors; one consisting of the surface-curvatures at m_1 equidistant points along a whole cycle along the border and another consisting of the border-curvatures at m_2 equidistant points along the same distance.

This may be regarded as just another surface vector-representation, but with the main aspect that the form of the surface is described by the curvatures at the border only, i.e. the characteristics e.g. at the middle of the surface are not detected at all. The idea beyond this representation is the assumption that surfaces in indoor environments may be sufficiently captured by the characteristics at their borders.

The task

Based on the representation described above, the continuous flow of scanned and pre-processed surfaces is to be clustered (or classified) based on the euclidian distance of the surface-vectors.

3. Network Model

The part of SPIN discussed in this paper requests a network model, with features listed below:

- Unsupervised clustering
- Dynamical number of clusters
- Forgetting by time or frequency of access
- Flexibility over an infinite period of time

There is only a small number of well known networks that might be used for such purposes (e.g. ART-models by Carpenter & Grossberg [2], Self-organizing feature-maps by Kohonen [6], [7], GAL by Alpaydin [1]). But all show limited abilities in at least one of the mentioned points.

3-1. Dynamic Network Model

The base of the following network-model is the self-organizing feature-map model by Kohonen [6]. This well known structure is extended by the possibility of adding and removing new cells. This work was being done by Fritzke in 1991 [3], [4]. Although this is already published, we will show the main aspects of this extension in short form before we discuss our adaptations.

Generalization & Learning

The representation of the surfaces as shown above is a vector in a m-dimensional real-valued vector-space. These vectors are mapped to an array S of cells c, each attached to an m-dimensional position vector $pos(c)$. The cells in S are connected in a triangular structure (the original Kohonen-model uses a rectangular structure). An input vector x is then mapped onto the cell with the smallest distance to it (in the Euclidian norm). This cell is called *bmu* (best matching unit) and this part is the classification.

$$\forall c \in S :$$
$$\| pos(bmu) - x \| \leq \| pos(c) - x \| \qquad (1)$$

For learning purposes the cell bmu and it's topological neighbours in the triangular structure are moved towards the input vector (by a fraction e_{bmu} and $e_{neighbour}$ of the distance). This training step is repeated n_d-times.

The described procedure results (if well defined) in a *topology-preserving* (i.e. adjacent input vectors are mapped on adjacent cells) and *distribution-preserving* map (i.e. the relative density of the cells approximates the probability density $P(X)$ of the input vectors).

Growing cell-structures

Up to here, the number of cells in S is constant, i.e. the choice of the number of cells that should represent adequately the probability-distribution of the input vectors has to be known in advance (Additionally the initial position of the cells is critical regarding the ability and speed of learning). To overcome this restriction a mechanism of expanding the structure is introduced. The initial structure is a single triangle, with arbitrary values of the three cells at the corners. In each learning step the distance between the input vector and the *bmu* is added to an error-variable associated with the found *bmu*. After n_d learning steps the cell with the maximal error-variable is detected. This cell is called *bs* (black sheep). Then the farthest direct neighbour in the structure is detected and called f. The new cell is inserted in the middle between them:

$$\text{pos}(c_{new}) = \frac{1}{2} \cdot (\text{pos}(bs) + \text{pos}(f)) \quad (2)$$

This new cell must be connected with surrounding cells in order to keep the triangular structure. The error-variable is initialized as a mean value of the error-variables of all d direct connected neighbours:

$$\text{err}(c_{new}) = \frac{1}{d+1} \sum_{i=1}^{d} \text{err}(\text{neighbor}_i) \quad (3)$$

The error-variables of the neighbours are reduced according to that amount:

$$\forall i = 1, ..., d:$$

$$\text{err}(\text{neighbor}_i) = \frac{d}{d+1} \text{err}(\text{neighbor}_i) \quad (4)$$

Shrinking cell structures

Starting from a single triangle and expanding the structure as described above will produce a *connected* structure trying to approximate the probability density $P(X)$. This might be the inadequate model because $P(X)$ can consist of several distinct regions with $P(X)=0$ between them. So there is a need for a procedure to disconnect the structure if necessary.

The basic idea is to find cells which are positioned in areas with $P(X)=0$ and to remove them. As the indicator of this constellation, the number of classifications without being *bmu* is recorded for each cell. If this number k for a certain cell exceeds the value in (5) this cell is removed and so are those of it's neighbours necessary to return to a structure of triangles (p_s means the probability of keeping needed cells and n is the actual number of cells)

$$k > n \cdot n_d \cdot \left(1 - (1 - p_s^{1/n})^{1/(n_d+1)}\right) = k_r \quad (5)$$

For a more detailed description of the algorithms so far see e.g. [3], [4].

3-2. Extensions & Adaptations

A number of problems with the above network-structure have being found regarding the concrete restrictions of our surface-clustering and classification. The central aspect here is the fact that our learning set is not fixed but consists of a continual flow of examples. So the task is not to model the best approximation of a limited set of input vectors, but to find a representation of the probability distribution and a good clustering of the *most recently* presented surfaces. Additionally the classification-function has to be accessible *all the times* (after a certain amount of learned surfaces), i.e. the structure should change smoothly from one state to the next while learning.

Buffering the flow of examples

The number of produced examples per time-interval is smaller than the number of surfaces that might be learned during this interval. So the received examples are buffered in a FIFO-list and each example is being learned several times.

Limited growing and adaptation

In the original model the learning phase simply stops when the required accuracy is reached. The SPIN-project does not know of an end of learning, so there must be another definition of stability.

Assuming a tolerated error of $d_{accuracy}$ and an input vector x. If the bmu resulting from the classification of the vector x fulfils the equation (6)

$$d_{bmu} = \| x - pos(bmu) \| \leq d_{accuracy} \quad (6)$$

then the learning for this cell is switched to a modified learning scheme: The fraction of bmu correction e_{bmu} is reduced by the factor Δe_{bmu} and the bmu is now moved towards x by this reduced fraction of $\Delta e_{bmu} \cdot e_{bmu}$. The direct neighbours are not corrected at all. Additionally this classification does not increment n_d, i.e. the generation of new cells is delayed.

If these classifications continue over some period of time, the moving of cells will become slower and no new cells are created. The modified learning scheme is reset to normal learning when one classification does not fulfil (6), i.e. e_{bmu} is set to the original value, etc.

Summarizing this modification, two main effects are important:

a. The growing of the network is stopped (or at least slowed down, depending on the classification error), although the flow of examples may still continue.

b. The network structure (i.e. the positions of the cells) is stabilized, when the examples fit into the clustering built up.

Cautious removal

The removal of cells in the original algorithms causes the removal of neighbouring cells, in order to return to a structure of triangles. This procedure does not care about the importance of these neighbouring cells, so often used cells might disappear. If the relative density of cells is large, then the remaining cells might fulfil most of the further oncoming classifications. But in a sparse clustering, i.e. each cell represents a whole isolated class (e.g. with $P(X)=0$ at it's borders), the functionality of these cells can not be adequately approximated by the remaining ones. And even worse, these cells will be generated again in the further process, because the classification-errors in this area will rise significantly. On the way to this rebuilt structure, the remaining cells are corrected in large steps because of large classification-errors.

In some constellations we found a cyclic behaviour, i.e. even in the moment when the structure was build up again, one of the cells in this area is being removed, and the procedure starts again. This unstable and discontinuous behaviour can not be tolerated in our system.

The solution we have chosen is based on the idea of accepting "over-classification" under some circumstances, but not deleting cells, which are used. This means concrete that each time a cell is detected as not being used for a certain period of time, it is only removed when all the cells which would be removed in order to keep the structure of triangles intact are also detected as unused. As a result of this manipulation, the network will consist of more cells than necessary, and so one might think of effects like swapping between two cells in situations where one cell would be sufficient or other instability problems. But in our simulations we have never observed such situations.

Speed of forgetting

In the original model the value of k_r might be approximated as (see equation (5)):

$$k_r \leq \lceil n \cdot n_d \rceil \, ; (0 \leq p_s \leq 1) \quad (7)$$

A value of p_s near 1 implies a good approximation of the vectors in the current learning set. But as our learning set is dynamic, another aspect arises: What happen to a learned cell, when the generating examples are deleted from the learning set. In our simulations this cell is removed or even massively corrected in a couple of minutes, i.e. there is absolutely no long-term memory.

In order to create a possibility to determine slower forgetting then implied by $p_s=1$, we extended the definition of k_r for values of $p_s>1$ (Notice that p_s is not a probability in this case):

$$k_r = \begin{cases} \left\lceil n \cdot n_d \cdot \left(1 - (1 - p_s^{\frac{1}{n}})^{\frac{1}{n_d+1}} \right) \right\rceil & ; p_s \leq 1 \\ \lceil n \cdot n_d \cdot p_s \rceil & ; p_s > 1 \end{cases} \quad (8)$$

So arbitrary long storage-times of presented surfaces can be implemented. One might think of additional information stored for every cell in S, like frequency of access or time since last access, etc. pp., in order to find a better choice of cells to delete, but this ideas are not tested here.

3-3. Parameters

This section is intended to give an idea of the meaning of some parameters as well as showing up the ranges which appears to be useful to us.

Network size

The size of the network structure is not determined directly by an upper or lower bound, but is implied by the required accuracy of the classification. Therefore this parameter needs not to be tuned at all (because it is not there).

Accuracy of classification

The accuracy of classification is the euclidian distance between the example-vector and the found bmu (see equation (6)). So the first stage of learning is reached when all the vectors of the learning set are in between of at least one of the hyper-spheres with radius $d_{accuracy}$ around the cells of the network. Then the modified learning-phase is entered, i.e. no new cells are created and the speed of moving and the number of cells moved in each step are reduced (until the classification errors rises again). The network is called "stable", when the value of e_{bmu} remains below a certain limit e_{stable}.

Choosing $d_{accuracy}$ too small results in a one-to-one mapping of the actual learning set and the cells in the structure, i.e. one cell is created for each vector in the learning set. On the other hand a large value of $d_{accuracy}$ means a small number of created cells and a large tolerated classification-error. Finding the "optimal" value depends widely on the purpose of the system and the used vector-representation of the examples.

Moving fractions e_{bmu}, $e_{neighbour}$ & Δe_{bmu}

Simulations have shown good results with values in range of (9) for e_{bmu}.

$$\frac{1}{20} \leq e_{bmu} \leq \frac{1}{5} \tag{9}$$

A "working range" for $e_{neighbour}$ is shown in (10), but there are some conditions, which one has to keep in mind.

$$\frac{1}{50} e_{bmu} \leq e_{neighbor} \leq \frac{1}{20} e_{bmu} \tag{10}$$

A small value for $e_{neighbour}$ might result in moving of cells not preserving topology in the net-work-structure, because one cell, which is marked as the actual bmu could be moved over a long distance, without moving the surrounding cells in an adequate manner. Large values for $e_{neighbour}$ are dangerous too, because if a neighbouring cell of a critical-cell is the bmu, the correction distance of this critical-cell could be larger than in the case that the critical-cell is the bmu itself.

Δe_{bmu} is limited by the following idea. When the value of e_{bmu} remains below the limit e_{stable} the network is said to be stable. But this should imply that every vector of the learning set is classified at least once with a classification error smaller than $d_{accuracy}$ (in (11) $|Learning\ set|$ is the number of example vectors in the learning set).

$$e_{bmu} \cdot \Delta e_{bmu}^{|Learning\ set|} > e_{stable} \tag{11}$$

or

$$\left(\frac{e_{stable}}{e_{bmu}}\right)^{1/|Learning\ set|} < \Delta e_{bmu} < 1 \tag{12}$$

But these are only the trivial limits. Both bounds would not result in a reasonable learning behaviour, because a value of Δe_{bmu} too near to 1 produces a very slow convergence of the network without reaching a better accuracy and a value too near to the lower bound makes the system rest too early i.e. in sub-optimal positions of the cells. The simulations have shown good results for values in the range of:

$$\Delta e_{bmu} = \left(\frac{e_{stable}}{e_{bmu}}\right)^{\frac{1}{c|Learning\ set|}} ; c \in [3,6] \tag{13}$$

Removal of cells

Our simulations have shown an uncritical behaviour (with the new learning scheme) with values of p_s larger than 0.5. Each increase of this value depends only on the amount of time a learned example vector (or class of vectors) should remain in the network, without the need to "refresh" the corresponding cell with this example vector(s).

Creation of new cells

The factor n_d, which determines the number of learning steps before a new cell is inserted, should be chosen in a way that there can be "just

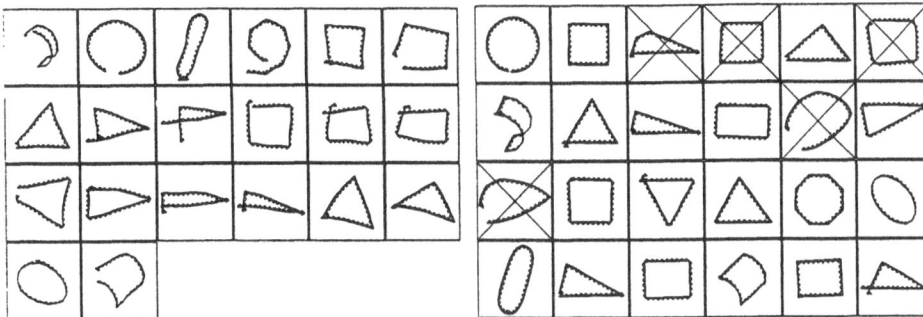

figure 1 : Some of the input surfaces

figure 2 : Generated classes

enough" probability-density-information accumulated before a position for a new cell is determined. Twice the number of expected classes works as a rough approximation, but obviously this "just enough" depends widely on the distribution of the example vectors.

4. Simulations

A flow of examples consisting of 20 different types of surfaces, which are all quantisized in 32 values for the segment curvature and in 16 entries for the surface curvature was simulated. The number of generated different examples is 2000. Noise is added in the form that every orientation of the bordering surface pieces is manipulated in the range of $\pm 6°$ for the segment orientation and $\pm 9°$ for the surface orientation. A part of the generated examples is shown in figure 1. As a result of adding noise, some surfaces are not closed, but this does not influence the generality of the test-set. After 6500 learning steps (1′ 50″ on a MC68030 processor at 40 MHz clock) 19 different classes are generated (n_d=40, e_{bmu}=0.1, $e_{neighbour}$=0.005, Δe_{bmu}=0.998). In figure 2 the surface-classes are shown, with unused cells crossed out. The shown example is representative, i.e. all of the well defined test-sets have caused such a network behaviour.

5. Conclusion

An open problem in our structure is that most of the surfaces are learned relatively fast, and only a small number (one or two) of not still learned surfaces keeps the learning-phase running for quite a long time. A learning focused on these

"problem-vectors" might be a solution, but this has not yet been tested in detail.

As a quite general technique for dynamic clustering problems (and based on the encouraging simulations), we are about to use this network on problems like topologic environment mapping, based on sparsely preprocessed sensor-informations, i.e we are trying to get additional refinements through further real applications.

References

[1] Alpaydin Ethem - 1.5.1991
GAL: Networks that grow when they learn and shrink when they forget
Technical Report 91-032

[2] Carpenter Gail A., Grossberg Stephen - 1987
A Massively Parallel Architecture for a Self-Organizing Neural Pattern Recognition Machine
Computer Vision, Graphics, and Image Processing 37, 54-115 (1987)

[3] Fritzke Bernd - 2.7.1991
Unsupervised clustering with Growing Cell Structures
Proc. of the IJCNN-91 Seattle (IEEE)

[4] Fritzke Bernd - 1991
Let It Grow - Self-Organizing Feature Maps with Problem Dependent Cell Structure
Proc. of the ICANN-91 Helsinki

[5] Fukushima Kunihiko - 1.3.1988
A Neural Network for Visual Pattern Recognition
Computer, March '88, pp. 65-75

[6] Kohonen Teuvo - 1.6.1990
Statistical Pattern Recognition Revisited
Advanced Neural Computers / R. Eckmiller (ed.)

[7] Kohonen Teuvo - 1984
Self-Organization and Associative Memory
Springer - Berlin, Heidelberg, New York, Tokyo

Robot vision—poster contributions

Motion Parallax from Catastrophies in Scale-space

T.M.H. Dijkstra, E. Argante and C.C.A.M. Gielen

Lab. of Medical Physics and Biophysics, University of Nijmegen
P.O. Box 9101, NL-6500 HB, Nijmegen, The Netherlands

Abstract

We present a way to extract affine motion parallax structure from time-varying images. In our method we construct scale-space by convolving the image with gaussians of all scales and then track the scale-space structure around annihilation-points, where spatial derivatives of the image disappear catastrophically because of the gaussian blurring. We show that the position change of these points corresponds to the image velocity and the scale change is related to the expansion of the image. The change in structure around these points is related to the rotation and deformation of the image. Simulations show our method to be robust against noise.

Introduction

An important capability of the visual system is to convey 3D information about the environment. When the observer is moving relative to the environment the irradiance pattern on the retina is changing in a way that contains this information. The temporal change of features in the irradiance pattern is called optical flow. The 3D structure of the environment can be deduced from spatial variations of the optical flow called motion parallax. The optical flow field is continous for smooth objects but is generally discontinuous at the rim of objects. We restrict ourselves to continous flow fields, but will later point out enhancements whereby this restriction can be lifted. In an infinitesimal neighbourhood the motion parallax is an affine transformation i.e. a translation plus an arbitrary linear transformation. Affine transformations can be decomposed [1] into a translation and three transformations with simple geometrical interpretations: expansion, rotation and deformation. These first order differential invariants characterise local motion parallax in a coordinate-free manner.

Many ways of extracting the optical flow field from time-varying imagery have been proposed, but they are generally sensitive to noise. To remedy this they generally propose averaging of the noisy flow field over a fixed spatial scale. In a recent paper [2], where scale is added a priori, it is shown that motion parallax can be estimated from the outputs of filters, that take a gaussian derivative of the image. But the relation between filter-outputs on different scales is not developed and algorithm is quite sensitive to noise.

Our method will make use of scale-space, which has been proposed as a structure to analyse static images on multiple scales [3]. The idea is to add a scale-parameter to the description of the image in such a way that there will be less features at a coarser scale. With the requirements of linearity and translation-invariance one can show that the unique structure that fullfils these requirements is the structure that results from convolving the image with gaussians of all scales. Equivalently one can say that scale-space results from submitting the image to a diffusion-proces, because the gaussian is the solution of the diffusion-equation. The most important property of scale-space is the disappearing of image features (e.g. a pair of edges meet and disappear) as we go to coarser scales. The points where this takes

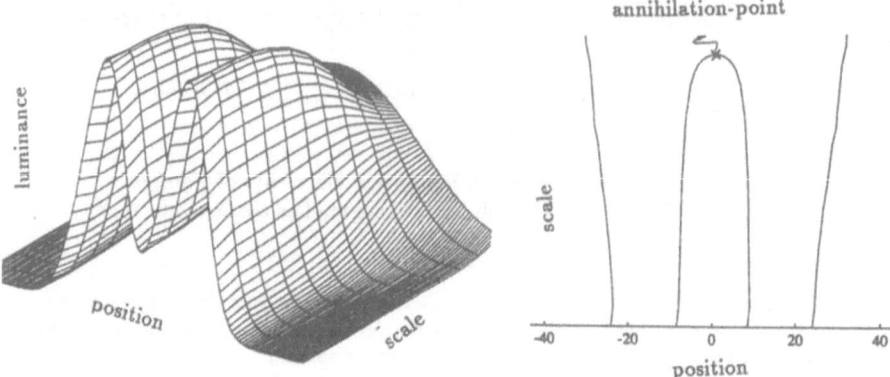

Figure 1: Example of one-dimensional scale-space, the image being a sum of two gaussians with different centres but otherwise identical. Left: the luminance as a function of position and scale. Right: the derived fingerprint i.e. the position where the second spatial derivative of luminance is 0

place are called annihilation-points and their scale is the coarsest scale on which the image features are present. One can associate the part of the image between the two features with an annihilation-point. This is illustrated in Fig. 1b, where we have plotted the positions of the edges in Fig. 1a in scale-space. These plots are called fingerprints. Because they are averages over large parts of the image, annihilation-points are unsensitive to noise. The structure of the fingerprint around an annihilation-point can be described with the tools of catastrophe-theory. For a two-dimensional image around the point of annihilation of two edges this structure is a cylinder. When we have time-varying imagery it is obvious that both the position and the scale-space structure around the annihilation-point will change dependent on the motion parallax.

Theory

Our method naturally decomposes in two parts: translation and divergence are estimated from the position-change of the annihilation-point of two edges in scale-space whereas rotation and deformation are estimated from the structure around it.

Translation and Divergence

Let the luminance-pattern of the part of the image belonging to a certain annihilation-point undergo a small translation and divergence:

$$\vec{r}(t + dt) = \vec{r}(t) + \vec{v}\, dt + D\, I\, \vec{r}(t)\, dt$$

with $\vec{v}$ the translation, D the divergence and I the identity-matrix. Denote the position of annihilation-point by $(\vec{r}_A, \sigma_A)$. Then the velocity and divergence are given by (Fig. 2):

$$\vec{v} \;=\; \frac{\partial \vec{r}_A}{\partial t}$$

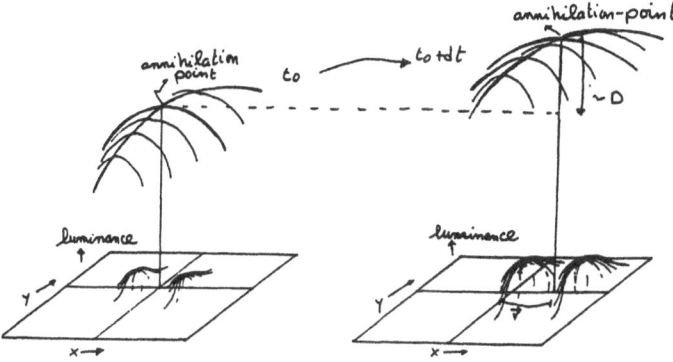

Figure 2: Estimation of translation and divergence. Left: time t_0. Right: time $t_0 + dt$. Lower parts show the image, upper parts show the fingerprint near the annihilation-point. Image velocity is indicated by a shift of the center of the two edges, image divergence by their enlargement

$$D = \frac{\partial \log(\sigma_A)}{\partial t}$$

Rotation and Deformation

Let the luminance-pattern of the part of the image belonging to a certain annihilation-point undergo a small rotation and deformation:

$$x(t + dt) = x(t) + (-Ry(t) + S_x x(t) + S_y y(t))dt$$
$$y(t + dt) = y(t) + (Rx(t) + S_y x(t) - S_x y(t))dt$$

with rotation R and deformation $(S_x, S_y) = (S \cos \alpha, S \sin \alpha)$. The fingerprint, which is a cylinder in the neighbourhood of the annihilation-point, can be characterised by two parameters, the angle of the principal directions relative to the x-axis (ϕ) and the non-zero curvature. If we denote the change in direction of the principal curvatures of the fingerprint as $\dot{A}$ and the change of curvature as $\dot{K}$, we find:

$$\dot{A} = R + S\sin 2(\phi - \alpha)$$
$$\dot{K} = S\cos 2(\phi - \alpha)$$

These relations do not completely determine the rotation and deformation. This means that rotation and deformation cannot be estimated fully from just two edges: more structure (e.g. from another edge) is needed.

Simulations

We have performed simulations to test the sensitivity to noise of our procedure to estimate translation and divergence. In Fig. 3 we show the results of a one-dimensinal scale-space

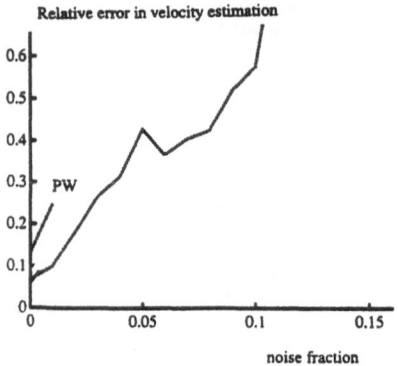
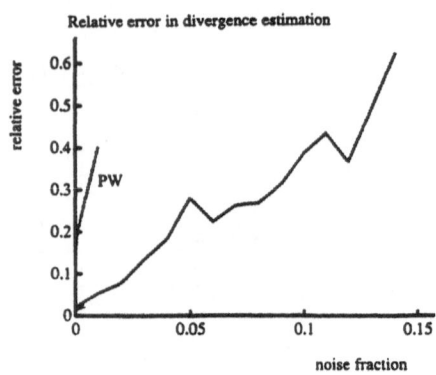

Figure 3: Results of simulation of our algorithm to estimate translation and divergence from a noisy luminance structure, PW = results from Werkhoven & Koenderink [2]

which was both translated and expanded. The luminance-pattern was a sum of two gaussians and the power of the noise was a fraction of the maximal luminance. We compared our results with [2], where translation and divergence were applied to an image separately. The results show our algorithm to perform better.

Discussion

We have shown a way to extract motion-parallax from a spatio-temporal luminance distribution by means of scale-space. The method could extract translation and divergence completely from the annihilation-point of a pair of edges, but could only give restrictions on rotation and deformation. The method only estimates motion-parallax at places of significant luminance-change. For the one-dimensional case we have shown estimation of translation and divergence to be robust against noise.

A problem with this way of extraction is that it compares scale-space from two static images and that large image-features disappear at a very coarse scale. However, large image-features need not be coalescent with objects. For the human visual system motion is a very strong cue for segmentation of an image. A possible enhancement of our method would be to add temporal aspects to the diffusion-process. This way it would also be possible to lift the restriction of smooth optical flow fields. Annihilation-points in this hypothetical spatio-temporal diffusion-process would then be the points where coherently moving patches disappear due to the blurring.

This work was partly supported by the Foundation for Biophysics of the Netherlands Organisation for Scientific Research (NWO).

References

[1] Koenderink, J J & van Doorn, A J (1975). Invariant properties of the motion parallax field due to the movement of rigid bodies. *Optica Acta*, 22, 773–791.
[2] Werkhoven, P & Koenderink, J J (1990). Extraction of motion parallax structure in the visual system I & II. *Biol. Cybern.*, 63, 185–199.
[3] Witkin, A (1983). Scale space filtering. *Proc. Int. Joint Conf. on Artificial Intelligence*, Karlsruhe, Germany, 1019–1023.

Stability and Convergence Control
in Cooperative Integration Networks

Scott T. Toborg
Hughes Research Laboratories
3011 Malibu Canyon Rd., Malibu, CA 90265 USA
E-mail: toborg@maxwell.hrl.hac.com

Abstract

This paper addresses the problem of how to control stability and convergence of multiple strongly coupled cooperating neural networks. The problem arises because a normally stable individual network can become unstable due to feedback from other interacting networks. The paper briefly summarizes the conditions for stability and more importantly, details procedures for controlling the rate of system convergence. Convergence rate control is important so that no single network unfairly dominates computation. These types of problems are conveniently formulated in terms of regularization with weak continuity constraints. Cooperative Integration networks have been applied to problems such as: integrating multiple early vision modules, multispectral processing, and color segmentation.

1 Introduction

An important problem in vision and image processing is determining how to combine multiple data sources (modules) to enhance overall system performance. One prototypical problem is the combination of multiple early vision modules. The integration of multiple early vision processes (e.g., stereo, motion, color, texture, etc.) can be used to reconstruct robust 3-D surface properties in spite of partially invalid *a prior* constraints. Data integration can be used for many similar problems such as multispectral processing and color segmentation.

Techniques have been developed for combining (integrating) multiple data sources using Markov Random Fields (MRFs) or regularization with weak continuity constraints [1, 3, 9]. MRFs or weak continuity constraint formulations result in complex, nonconvex minimizing energy functionals. Neural networks have been defined to perform the energy minimization [7]. Resulting iterative systems have different network dynamics depending on how the energy functionals are processed.

The dynamics of cooperative integration approaches are best addressed by first defining some key concepts. Clark and Yuille introduced a useful taxonomy for describing many different types of data fusion/vision integration techniques [2]. In general, approaches can be broken down into two main classes: *weak* and *strong coupling*. Weak coupling is characterized by the combination of data from multiple independent sources in a way that does not affect the operation of the contributing modules. In this case there is no feedback between data sources and interactions occur only through feedforward operations.

On the other hand, *strong* coupling approaches allow information from one data source to influence or change the other data source's model or constraints. A special case of strong coupling is defined when there exists a feedback loop such that one module modifies the prior constraints of the others and vice versa. This type of interaction is called *recurrent strong coupling*. The algorithms described in this research use a form of recurrent strong coupling that could be termed *cooperative coupling*. In this case, interactions occur cooperatively within modules (to compute continuous values and discontinuities) and between modules (using the discontinuity fields of different modules).

For example, a typical energy functional for early vision integration can be divided into four terms: data, smoothing, line process penalty, and intermodule coupling. For more concrete illustration, consider a simple 1-D system of k modules all computing intensity edges but using different input images I^k (e.g., different colors or sensors):

$$E^k = \int \{\beta^k(f^k - I^k)^2 + \lambda^k(\nabla f^k)^2\}dA + \alpha^k \int dl^k + C_{jk}(1 - l^j)l^k + \int_0^l g^{-1}(l)dl^k \qquad (1)$$

with the first integral evaluated over piecewise continuous regions bounded by discontinuities, and the second along the length of all discontinuities. The first term encourages the smoothed intensity estimate f to be close to the sparse, noisy input image I (if possible). The confidence in the data is modulated by β. The second term imposes smoothness on the solution, with $\sqrt{\lambda}$ representing a characteristic length or scale constant. The third term penalizes the formation of discontinuities with a cost of α per unit length. The forth term is the *coupling* term that modulates the influence one module has on another. Interactions between modules are made through their discontinuities l^j, l^k. Finally, the gain term is used in the minimization process to avoid local minima by "deterministically annealing" discontinuity variables. These formulations naturally extend to 2-D systems.

2 Stability in Cooperative Integration Networks

Stability is typically not an issue in multimodule feedforward systems since each stage equilibrates before interacting with the next stage (See for example [3]). In the case of multiple *cooperating* networks, stability is a greater concern. While individual modules might be stable, feedback from other modules can destructively perturb equilibria. Nevertheless, the stability of cooperative integration networks can be established by extending Lyapunov arguments used for single network systems. A dynamical system can be defined for solving equation 1 similar to Hopfield [6]. In this case, system state changes are equal to the negative gradient of the total system energy defined by equation 1:

$$\frac{df^k}{dt} = -\frac{\partial E^k(f^k, l^k)}{\partial f^k} \qquad (2)$$

$$\frac{dm^k}{dt} = -\frac{\partial E^k(f^k, l^k)}{\partial l^k}$$

with $l^k = g(m^k)$ and $g(\cdot)$ is a monotonic, nonlinear transfer function with gain parameter μ:

$$g(x) = \frac{1}{1 + e^{-2\mu x}} \qquad (3)$$

Stability can be established using Lyapunov arguments. (For simplicity the superscript k is dropped but arguments clearly extend to multimodule systems.) First, by examining the time evolution of E we see:

$$\begin{aligned}
\frac{dE}{dt} &= \frac{dE}{df}\frac{df}{dt} + \frac{dE}{dl}\frac{dl}{dt} \\
&= -\left(\frac{dE}{df}\right)^2 + \left(\frac{dE}{dl}\right)g'(m)\dot{m} \\
&= -\left(\frac{dE}{df}\right)^2 - \left(\frac{dE}{dl}\right)^2 g'(m)
\end{aligned}$$

which shows that E always decreases since

$$g'(m) = \frac{2\mu e^{-2\mu m}}{(1 + e^{-2\mu m})^2}$$

is always positive. Equation 1 is bounded from below and is continuously differentiable near an equilibrium point. Therefore, by Lyapunov's theorem [5], it is globally asymptotically stable. It is the gradient dynamics that assure system stability. Other types of dynamics may not have the same properties.

Simplifications to these stability arguments can be made by decoupling the dynamics of interacting variables. For example, if variables in a dynamic system have very different relaxation times, an *adiabatic approximation* can be made by assuming that fast relaxing variables are at equilibrium with respect to the slow relaxing variables. This idea can be used to simplify stability conditions in multimodule systems. Namely, if l^j and l^k in equation 1 are adiabatic with respect to both f^j and f^k and if l^j and l^k are in equilibrium with respect to each other, then the system 2 equations are clearly stable. This is because if l^j and l^k are in equilibrium, then they can be treated as constants wrt f^j and f^k whose dynamics are independent of each other.

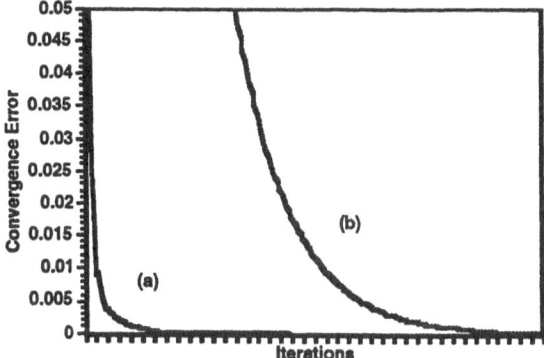

Figure 1: Experimental comparison of convergence rates for line process (a) and continuous process variables (b) shows that line processes converge in much fewer iterations. This empirically supports the adiabatic approximation used to establish stability in cooperating systems.

Currently, arguments for the adiabatic approximation are backed mainly by empirical evidence [8]. Figure 1 experimentally supports the adiabatic approximation by comparing the relaxation rates for line (l) and continuous processes (f) for a single module. More research is needed to establish rigorous mathematical justification.

3 Convergence Control for Unbiased Network Interactions

Another more immediate problem is the need to control the convergence rates of different component modules so that all relax at roughly the same rate. Why is this important? From experiments coupling edge detection and optical flow, we found that the edge detection module often converged faster than the optical flow module. Consequently, intensity edges unfairly biased the computation of optical flow discontinuities. This is undesirable in the cases involving "the leopard in the jungle" problem where it is impossible to segment the object correctly without motion information. In these cases, we want the strong optical flow information to influence the formation of intensity edges to better segment the moving object.

To encourage unbiased interactions between early vision modules, the rates of convergence need to be the same. The primary variable for controlling convergence is the gain parameter μ of the sigmoid function defined in Eq. 3. When this parameter reaches a predefined maximum, all line processes have been forced to on or off states. If the gain in one module rises faster than another, the module with the highest gain tends to have a greater bias over other modules. Simply making the gain equivalent in all modules does not solve the problem. This is because the value of the line process is also determined by other parameters and these parameters may be on completely different scales for different vision modules (e.g., values for the smoothing weight (λ) are much different between intensity edges and optical flow modules).

To alleviate this problem, it is necessary to take a closer look at the convergence behavior for each module. A lower bound on the rate of convergence for a given module is determined from the largest eigenvalues of the iteration matrix derived from the discrete update equations [4]. Consequently, the convergence of the whole system is determined by the module containing the largest eigenvalues. To normalize system convergence, we use an estimate of the eigenvalues for each module (proportional to λ) and scale the faster modules so their rate of convergence approximates the slowest module. Empirical experiments show that this scaling procedure does effectively equalize the convergence rates of different modules [9]. While this procedure implements *unbiased* intermodule coupling, there may be applications requiring more general dynamics. For instance, the convergence of each module could be adapted to implement a desired system behavior. For example, allow intensity edges to dominate **unless** there is significant motion information.

4 Conclusions

Stability and convergence issues are important in cooperating networks because feedback from interacting modules complicates system dynamics. Stability of cooperating systems with gradient dynamics can be shown using Lyapunov arguments. Also, by decoupling network dynamics using an adiabatic approximation, stability can be assured. In cooperating systems the convergence rates of different modules must be controlled to allow equitable interactions. Procedures based on eigenvalue analysis provide some reasonable mechanisms for control. These ideas can be expanded to allow more flexible multinetwork dynamics.

References

[1] A. Blake and A. Zisserman. *Visual Reconstruction.* MIT Press, 1987.

[2] J. Clark and A. L. Yuille. *Data Fusion for Sensory Information Processing Systems.* Kluwer Academic Publishers, 1990.

[3] E. Gamble, D. Geiger, T. Poggio, and D. Weinshall. Integration of vision modules and labeling of surface discontinuities. *IEEE Trans.*, SMC-19(6):1575–1581, 1989.

[4] L. A. Hageman and D. M. Young. *Applied Iterative Methods.* Academic Press, 1981.

[5] M. W. Hirsch and S. Smale. *Differential Equations, Dynamical Systems, and Linear Algebra.* Academic Press, 1974.

[6] J. Hopfield. Neural networks and physical systems with emergent collective computational abilities. *Proc. National Academy of Science, Biophysics, USA*, 79:2554–2558, 1982.

[7] C. Koch, J. Marroquin, and A. Yuille. Analog "neuronal" networks in early vision. AI Memo 751, MIT Artificial Intelligence Laboratory, June 1985.

[8] A. Rangarajan and R. Chellappa. Generalized graduated non-convexity algorithm for maximum a posteriori image estimation. TR 149, USC Signal & Image Processing Institute, Dec. 1989.

[9] S. T. Toborg and K. Hwang. Cooperative vision integration through data-parallel, neural computations. *IEEE Transactions on Computers*, 40(12):1368–1379, 1991.

Towards a Neural Architecture for Unified Visual Contrast and Brightness Perception [1]

Heiko Neumann, H. Siegfried Stiehl

Universität Hamburg, FB Informatik, AB Kognitive Systeme
Bodenstedtstr.16, D-2000 Hamburg 50, Germany

Abstract. In this contribution a brief overview is given on a computational architecture that serves as a framework for a unified theory for contrast, contour and luminance perception. The computational mechanisms utilize a center-surround antagonism based on shunting interactions which allow to multiplex local contrast as well as luminance data. This data is demultiplexed at a later stage to provide separate input to parallel but interacting processing streams for contrast and brightness information, respectively. The approach unifies the models for BCS and FCS architecture as developed by Grossberg and colleagues. Within the architecture proposed here a process for brightness reconstruction is suggested which consists of three functional stages, the activities of which are controlled by the accompanied contrast channel which in turn is build by segregation and pooling of data from the input data stream.

1. Introduction

Recent neurophysiological and neuroanatomical findings in vision research suggest the existence of two parallel afferent processing streams. These so-called ON- and OFF- (or, synonymously, brightness- (B-) and darkness- (D-)) channels transmit visual information on projection fibres that are segregated up to at least the striate cortex (V1, area 17). The retinal processing can be adequately explained on a functional level at the ganglion cell layer, whose cells comprise a center-surround antagonistic lateral coupling. The segregated ON- and OFF-channels receive separate input from activations generated by lateral interaction described in terms of on-center/off-surround (on/off) and off-center/on-surround (off/on) receptive field (RF) profiles. The major axonal projections make up a processing path that enter LGN and subsequently connect to area 17 in the visual cortex. It is now commonly accepted that mechanisms responsible for form and brightness perception can be mainly found in various cortical areas (e.g. [4, 1]). In this paper a processing architecture for coherent contrast and brightness perception is defined that unifies the architecture of the BCS and FCS, which has been developed by Grossberg and colleagues as part of the FACADE theory (e.g. [7]). It provides a new hypothesis concerning the functional role of coexisting parallel ON- and OFF-channels (see [4] for a recent discussion), suggests a brightness reconstruction process which utilizes three functional stages, and provides a novel simplified computational solution to cortical processing hitherto explained by orientation sensitive single cells in area 17.

2. Center-surround Antagonism and ON/OFF- (B/D-) Channels

Based on an investigation of response properties of retinal ganglion cells ([3]) a computational model based on a center-surround interaction has been derived that accounts for various phenomena. An outcome of utmost importance for our modeling approach is that ganglion cells — in addition to their facilitation of contrast detection — do also (minimally) respond to homogeneous illumination. In modeling the cell response with shunting membrane equations of the type $\dot{y}_i = -Ay_i + (B - $

---

[1] Work has been partially supp. by the German Fed. Dept. of Res. and Technol. (BMFT), grant 413-5839-01 IN 101 C/1

y_i)net$_i^+$ $-$ $(C + y_i)$net$_i^-$ (where net$^{+/-}$ denote some input activation, and constants A, B, C define the passive decay and upper and lower saturation points, respectively) we can account for both this observed functional property and various additional functional phenomena, such as activity saturation, featural noise suppression and reflectance processing ([6]). Physiological data suggests that the spatial extents of on/off and off/on RF profiles are symmetric such that weighted inputs that form center and surround enter into both the B- and the D-channel. Using these weightings to describe ON- and OFF-activities (J_i^c, J_i^s), and furthermore utilizing the above mentioned membrane equation and solving for steady state, we arrive at

$$
\begin{aligned}
^{ON}y_i &= \gamma(J) \cdot \max[B \cdot (J_i^c - J_i^s) + (B - C) \cdot J_i^s, 0] \qquad \text{and} \\
^{OFF}y_i &= \gamma(J) \cdot \max[B \cdot (J_i^s - J_i^c) + (B - C) \cdot J_i^c, 0], \qquad \text{with} \quad \gamma(J) = (A + J_i^c + J_i^s)^{-1}. \quad (1)
\end{aligned}
$$

Net activations J^c and J^s are generated by convolution of the stimulus luminance distribution I with weighting functions λ^c and λ^s, respectively, each of scalable width. As one can easily recognize, both channels encode a scaled version of a difference-of-low-pass (DoLP) filtered version of the input stimulus and, in addition to that, both channels also contain a scaled version of the low-pass filtered stimulus. Therefore, the shunting properties of a center-surround antagonism not only provide the additional properties of saturation levels but at the same time also allow for an encoding of a scaled low-pass filtered version of the luminance distribution.

3. Segregation of Contrast- and Brightness&Darkness- (B&D-) Paths

As briefly sketched in the previous section, such a pair of independent paths serves as a transmission and processing channel capable of multiplexing three data streams to allow for subsequent segregation into separate representations for parallel channels for local a) ON- and b) OFF-contrast activities and c) a representation of (low spatial frequency) brightness&darkness (B&D) information. Starting from the steady state equations for the ON- and the OFF-channel a simple set of mechanisms realizes the above motivated *segregation* into selective processing streams for different specialized purposes such as contrast detection and brightness reconstruction. ON-OFF competition eliminates the DC-components from both channels to generate representations of DoLP filtered luminance. These suggested data streams represent local ON and OFF contrast information which feeds forward to generate activities $^{ON}z_i^s$ and $^{OFF}z_i^s$, respectively. Ground level activity s_i, as a low-pass filtered version of the input luminance distribution, is generated in the B&D channel by simply adding the ON- and OFF-channel responses. Since the activity distribution s_i is only a low-pass filtered version of the input I, the activities are suggested to be *contrast enhanced* for generation of sharply bounded region outlines in the visual perception of object surfaces. Such a compensatory mechanism for modulation of B&D channel activity is postulated to be realized by excitatory and inhibitory interactions from both the ON- and OFF-contrast channels with diffused activity distributions $^{ON}z_i^s$ and $^{OFF}z_i^s$, respectively (see fig. 1 for an overview of the overall architecture). Based on empirical findings from stabilized image experiments a permanent diffusive activity within the B&D system has been postulated which local diffusion coefficients are controlled by the activities of the contrast (BC) system ([2, 5]). This *diffusion* of local activity within this final brightness representation stage is modeled by a simple local averaging process. The generation of local activation in the B&D channel is described by the following iterative form:

$$
u_i = \frac{s_i + E\,^{ON}z_i^s + v_i}{D + F\,^{OFF}z_i^s} \qquad \text{and} \qquad v_i = \sum_{j \in N_i} (u_j - u_i) P_{ij}, \qquad (P_{ij} = \rho/(1 + \varepsilon(z_i + z_j))) \quad (2)
$$

It is postulated here that blurring of activities in these separate channels occurs which can be formalized as a convolution with a spatial low-pass filter (e.g. a Gaussian). Due to the separation of the ON- and OFF channels, selected activities in the response distribution show sensitivity to contrast polarity ascribed to cortical *simple* cells. The properties of cortical *complex* cells can now be easily synthesized by both again smearing and additively combining (spatial pooling) the ON- and OFF-channel simple cells responses. These activations — which in coincidence with the empirical evidence show neither specificity to contrast polarity nor selectivity to local phase (i.e. responsiveness to odd/even symmetry) — form the basis for contrast detection and localization (e.g. [9]). The responses of these simple and complex cells can be described by the following equations:

$$^{ON/OFF}z_{i\varepsilon}^s = \sum_j {}^{ON/OFF}z_j^s \cdot \lambda_{ij\varepsilon} \quad \text{and} \quad z_i^c = \sum_j ({}^{ON}z_{j\varepsilon}^s + {}^{OFF}z_{j\varepsilon}^s) \cdot \lambda_{ij\varepsilon} \qquad (3)$$

where $\lambda_{ij\varepsilon}$ is a blurring function with elongation in direction ε to account for orientation selectivity of simple cells[2]. The z^c-activities feed into a nonlinear feedback network which basic architecture has been developed by Grossberg&Mingolla (e.g. [8]) and extended in functionality by Neumann&Stiehl (e.g. [10]). This subsystem in the overall processing architecture enables emergent generation of closed contour compartments, spontaneous groupings, subjective contour generation as well as corners and junctions (being salient features in the stimulus arrangement). The generated z-activities enter as inhibitory components into the local mechanisms for controlling the conductivities of the diffusion layer in the B&D system.

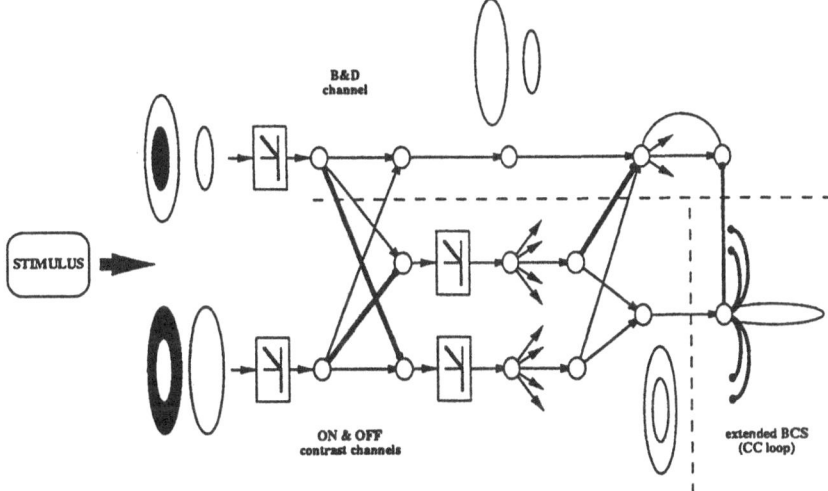

Figure 1: Overview of the computational architecture (thin lines with arrows define exitatory, bold lines ending with bold dots define inhibitory connections); for a functional description, see text

4. Computational Experiments

A set of selected computational experiments have been performed in order to demonstrate the functional behavior of the proposed architecture. The stimuli used for computational experiments include simple contrast variations (e.g. bars) as well as intensity patterns used in psychophysical experiments that support the generation of e.g. spontaneous dot groupings, Mach bands and simultaneous contrast phenomenon. Figure 2, as one selected example, shows the result of resulting brightness pattern generated for an input that obeys the simultaneous contrast phenomenon.

5. Further Work

Within the present paper a framework for a computational architecture has been proposed that accounts for various constraints derived from psychophysical data and electrophysiological as well as

---

[2]Please note that for the specification of the 2D case, the z-activations which enter in eq. (2) are defined by the sum of all activities at one spatial location taken over the range of discrete orientations.

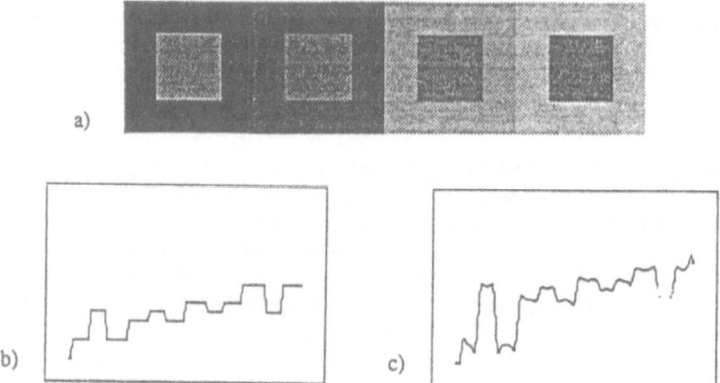

Figure 2: Brightness reconstruction that accounts for the simultaneous contrast phenomenon, (a) original intensity distribution and (b) selected luminance profile, (c) activity distribution at the second processing stage in the B&D channel

neuroanatomical findings. Ongoing research work has now been concentrated on the implementation of a 2D version of the preprocessing stage described in this paper and its subsequent integration with the competitive-cooperative feedback architecture as described in [10].

References

[1] H.B. Barlow. Understanding natural vision. In O.J. Braddick and A.C. Sleigh, editors, *Physical and Biological Processing of Images*, pages 2 – 14. Springer, Berlin, 1983.

[2] H.D. Crane and T.P. Piantanida. On seeing reddish green and yellowish blue. *Science*, 221:1078 – 1080, 1983.

[3] C. Enroth-Cugell and J.G. Robson. Functional characteristics and diversity of cat retinal ganglion cells. *Investigative Ophtalmology and Visual Science*, 25:250 – 267, 1984.

[4] A. Fiorentini, G. Baumgartner, S. Magnussen, P.H. Schiller, and J.P. Thomas. The perception of brightness and darkness — relations to neuronal receptive fields. In L. Spillmann and J.S. Werner, editors, *Visual Perception — The Neurophysiological Foundations*, chapter 7, pages 129 – 161. Academic Press, San Diego (FL/USA), 1990.

[5] H.J.M. Gerrits and G.J.M.E.N. Timmerman. The filling-in process in patients with retinal scotoma. *Vision Research*, 9:439 – 442, 1969.

[6] S. Grossberg. How does a brain build a cognitive code? *Psychological Review*, 87:1 – 51, 1980. (reprinted in: S. Grossberg. "Studies of Mind and Brain". R.S. Cohen and M.W. Wartofsky (eds.). Boston Studies in the Philosophy of Science, Vol. 70. D. Reidel Publishing Co., Boston (USA), 1982.).

[7] S. Grossberg. Cortical dynamics of three-dimensional form, color, and brightness perception: I. Monocular theory. *Perception and Psychophysics*, 41(2):87 – 116, 1987.

[8] S. Grossberg and E. Mingolla. Neural dynamics of perceptual grouping: Textures, boundaries, and emergent segmentation. *Perception and Psychophysics*, 38(2):141 – 171, 1985.

[9] D. Hubel and T.N. Wiesel. Functional architecture of macaque monkey visual cortex. *Proceedings Royal Society of London*, 198:1 – 59, 1977.

[10] H. Neumann and H.S. Stiehl. Emergent segmentation of monocular visual invariants for space perception. In *Proc. Int. Joint Conf. on Neural Networks (IJCNN-92), Vol. I-IV*, pages (III) 266 – 271, Baltimore (Md/USA), June 7 - 11 1992. IEEE.

Fuzzy Kohonen Clustering Networks for Reducing Search Space in 3-D Object Recognition

Eric Chen-Kuo Tsao

Inst. of Computer Sci. & Information Eng., National Chung Cheng Univ., Chiayi, Taiwan, R.O.C.

Hong-Yuan Liao

Inst. of Information Science, Academia Sinica, Taipei, Taiwan, R.O.C.

3-D object recognition is a process of matching an object to a scene description to determine the object's identity and/or its pose (position and orientation) in space. The problem of object recognition can be separated into two closely related subproblems - that of model building and that of recognition. Object recognition using multiple-view approach is achieved by finding a best match between the description of object projection and one of the model descriptions in the database. An object is modeled by its 2-D projections as seen from a set of predetermined viewpoints uniformly distributed on a view sphere enclosing the object. In our experiments, an icosahedron (twenty-sided) is used to approximate the view sphere. The method to search the model database is crucial to the performance of the recognition process. To recognize an object, a coarse-to-fine search strategy is adopted to reduce the processing time. At the coarse search stage, a small set of views most similar to the object in the scene is located. In most cases, this step vastly decreases the search space. The fine (or exhaustive) search process starts with these located views and compares the topology of the object in the scene with the projections rendered by the neighboring viewpoints of the located viewpoints until a best match is found. In this paper, a FKCN model which is an integration of Fuzzy c-Means (FCM) and Kohonen clustering networks (KCNs) is proposed to exploit the parallelism at the coarse search stage in object recognition, where the learning rate for Kohonen updating is proposed to be :

$$\alpha_{ik,t} = \left(u_{ik,t}\right)^{m_t}$$

where $u_{ik,t} = \left(\sum \left(\|x_k - v_i\|_A / \|x_k - v_j\|_A\right)^{2/(m-1)}\right)^{-1}$ by minimizing $\sum_i \sum_k u_{ik}^m (\|x_k - v_i\|_A)^2$, but weighting exponent m is a function of t, the iterate number, i.e. $m = m_t = (m_0 + t\Delta m)$ (an increasing strategy). Feature vectors are first extracted from each view. Then a FKCN clusters these feature vectors to get cluster centers, each of which is regarded as a prototype of some similar views. Finally, a feature vector from an input view is compared to the cluster centers instead of all views and finds the potential candidates which are clustered into the closest cluster center in order to achieve the goal of reducing search space.

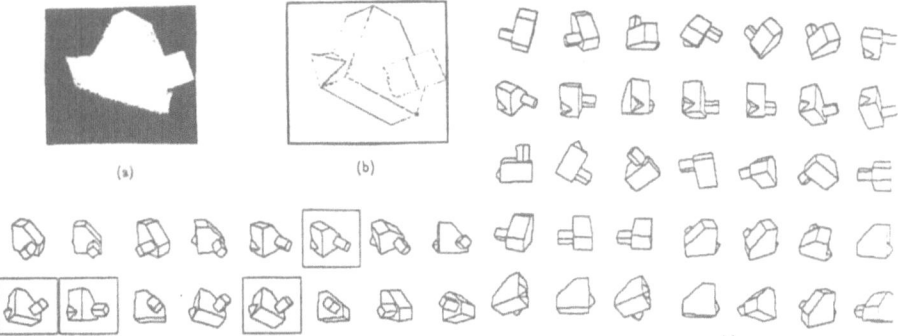

Figure (a) A range image; (b) result after image segmentation; (c) four of the most similar viewpoints located by the FKCN.

An Active Resistor Mesh Embedding Cortical Visual Processing

Luigi Raffo, Silvio P. Sabatini, Giacomo Indiveri, Daniele D. Caviglia, Giacomo M. Bisio

DIBE-University of Genova-Via Opera Pia 11/A-16145 Genova-ITALY

In preattentive vision a vast arrays of intercommunicating, identical processes are carried out for perceiving edges by texture differences, depth mapping, computing optical flow, recovering local surface structure, segmenting images, etc. To perform these tasks in real time, analog hardware devices are required. Information is distributed, being mapped directly in the electrical variables, and computation is carried out massively in parallel with high efficiency and speed. Two steps can be distinguished when formalising these computations: 1) mapping the image onto an intermediate abstract representation by means of appropriate receptive fields and 2) arranging information in functional maps, thus generating useful image descriptors for subsequent processing. Relating these computations to the signal processing capabilities of active resistor meshes, may ensure the most efficient use of hardware. In this paper we extend the approach of [1] to not *homogeneous anisotropic meshes*. The mesh has different orientation-selective neurons organized on the same layer, to emulate the orientation domains in the mammalian visual cortex. In a discrete model, the orientation map is defined as a 2D array in which every node is associated with a preferred orientation ranging from 0 to π radians [2]. Such orientation maps allow the fusion of information from different channels.

The starting point is an isotropic square grid network. The input image is a 2D distribution of current sources at the nodes of the grid. The shape of the node voltage response to a current impulse is similar to a Gaussian function centered around the excited node the convolution width depends on the conductance ratio (node-to-node)/(node-to-groud). To generate elongated Gaussian functions oriented along 0, $\pi/4$, $\pi/2$, $3\pi/4$ directions we consider four schemata characterized by the presence of an extra positive conductance. This resistor makes the network anisotropic and the shape of the voltage response loses its circular symmetry assuming an elongated form along the specific corresponding directions. Other orientations can be considered. By example, arranging two different schemata in a chess-board configuration, we obtain operators elongated along the intermediate directions. Spatial frequency selectivity of the whole network can be improved by resorting to Gabor-like convolution kernels. To this purpose, we add clustered inhibitory (negative) conductances along the direction orthogonal to the one to which the node is selective.

In order to design resistive networks that resemble the structure of real continuous varying orientation maps, we varied the pattern of interconnections in a continuous fashion. In this way, the voltage response of the network to a current impulse at a node presents anisotropic Gabor-like shapes with orientations varying from node to node in accordance to the orientation map. In this way, the resistor mesh, through local interconnections, can combine a limited set of operators enough to cover a wide range of features, and achieve functional interactions among the operators themselves.

The network can detect texture differences in an image using the responses of cells selective to the various orientations. If the test image is composed of repeated oriented elements, the cells in the regions of the map selective to that orientation have the strongest output. We have considered both synthetic and netural textured images. The network reveals itself particularly robust. The choice of the value of the resistances is not critical since the final orientation of the convolution kernels depends more on the topology of connections than on single resistance values.

References

1. H. Kobayashi, et al. *IEEE J. Solid State Circuits*, 26:738–748, May 1991.
2. W.T. Baxter and B.M. Dow. *Biol. Cybern.*, 61:171–182, 1989.

A Fast BCS/FCS Algorithm for Image Segmentation

José L. Contreras-Vidal and Mario Aguilar
Center for Adaptive Systems and
Department of Cognitive and Neural Systems
Boston University

Abstract

A fast and efficient segmentation algorithm based on the Boundary Contour System/Feature Contour System (BCS/FCS) of Grossberg and Mingolla [3] is presented. This implementation is based on the FFT algorithm and the parallelism of the system.

Introduction.

Grossberg and Mingolla [3] proposed a neural network model for image segmentation which has proven very useful in explaining various psychophysical data (e.g. neon color spreading). However, its application to computer vision or image processing has not been wide spread. Recently, this model has been applied to segmentation problems which had been impossible to treat with standard techniques [2].

We attribute this lack of extensive application of the model to the apparent complexity of the mechanism and notation published. We hope to dimystify this model by providing a description in terms of vector notation for each stage of the system. In addition, fast algorithms to operate this system are not readily available and its operation has not been explained in applied terms. In this paper, we present an optimized algorithm based on the BCS/FCS. This implementation takes advantage of the inherent parallelism in the BCS/FCS system, and of the fast fourier transform (FFT) algorithm to solve the equations in terms of convolutions in the frequency domain. This provides the means for efficient preprocessing of images for computer vision and image processing.

The following section introduces the model. A later section will describe a simplified and efficient algorithm of the boundary extraction system.

BCS/FCS system

This system consists of two parallel pathways which extract high frequency (HF) information (e.g. boundaries, edges, or lines) and low frequency (LF) information (e.g. luminance or color) respectively. Both channels have noise suppression properties. In particular, the HF channel reconstructs, emphasizes, and extracts boundaries from the image. The LF channel performs filtering of "salt-and-pepper" and wide-band noise. This system contrast-enhances the image and suppresses the noise without blurring the image. This is accomplished by fusing the HF channel with the LF channel.

The processing stages of the HF channel consists of 1) contrast-enhancement (via isotropic filters or on-center/off-surround competition), 2) line/edge orientation detection (via anisotropic filters), 3) local competition among orientations and across space to define the locally coherent orientation to be supported, and 4) boundary or line completion (via spatial cooperation among filters of similar orientations).

The LF channel, on the other hand, consists of a single stage in which both the image data and its boundaries (from the HF channel) are combined. This allow for the filling in of textural values within segmented grouping in the image. Filling in is implemented by the spread of activation values across units; the spread is constrained by the boundaries established by signals from the HF channel. Therefore, this stage counteracts the effects of pure high-pass filtering approaches by improving the signal-to-noise ratio of the image, while preserving and enhancing its boundaries.

The HF channel will be emphasized because its edge enhancement, boundary completion, and segmentation stages can by themselves provide sufficient information for segmentation or as silhouettes inputs to existing algorithms.

Fast BCS/FCS Algorithm

Notation

A:	Decay rate.
B:	Small bias constant.
C:	Cooperative filter (e.g. bipolar kernel defined as a thresholded product of left and right shifted gaussian kernels.
D:	$A_1 e^{-\frac{(x-y)^2}{\sigma_1}} - A_2 e^{-\frac{(x-y)^2}{\sigma_2}}$: Concentric filter. $\sigma_2 > \sigma_1, A_1 > A_2$
E:	Difference of gaussians.
F:	Sum of gaussians.
$f()$, $h()$:	Non-linear signal transfer functions (eg. threshold-linear).
G^k:	2-D Gabor wavelets (or any other oriented filter).
I:	Identity matrix.
J^k:	Activity matrix of second stage (line/edge detection).
k:	Orientation.
M:	Maximal activity level.
N:	2 × (image dimension + filter dimension).
P:	Image of size $m \times m$
S:	Sum of gaussians.
V^k:	Feedback activity matrix.
W^k:	Activity matrix of lower third stage (competition across space).
X:	Activity matrix of first stage (contrast-enhancement).
Y^k:	Activity matrix of upper third stage (competition among orientations).
Z^k:	Activity matrix of fourth stage (boundary completion).
ω:	Winner-take-all.
$\otimes$:	Convolution in frequency domain.

Table 1 shows the equations which define the HF system and the degree of complexity of the system based on the FFT algorithm. On the other hand, the flow diagram in figure 1 illustrates the interactions among the various stages. In the first stage of processing (Noise Suppression), the image is contrast-enhanced by a feedforward shunting network. A shunting network can be implemented as the convolution of the image with an isotropic filter, followed by normalization of energy.

253

Second, the image is filtered through a set of oriented or anisotropic filters (Gabor filtering). For this, we have selected Gabor wavelets to filter the image because they have shown to provide complementary information useful when segmenting ambiguous areas[1].

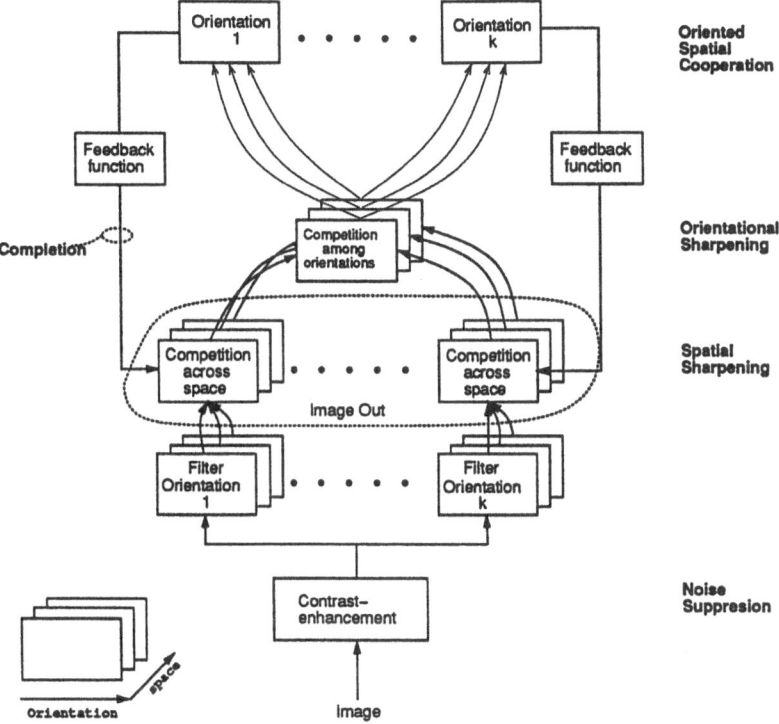

Figure 1: Flow diagram of HF system.

This process leads to a stage of competition across space for a given orientation (spatial sharpening). Its output is fed to a layer of competition among orientations at a given position (orientational sharpening). It is in the first competitive stage that bottom-up and top-down information converge. Activity in this layer is biased by feedback from higher stages to reinforce spatially coherent orientations (boundary completion). Also, the input is contrast-enhanced and boundary edges extracted. The output of the first competitive stage, when stable, is the actual output of the system.

The fourth stage (cooperation or boundary completion) consists of cooperation among coherent orientations across space. The neighborhood of cooperation is defined by a kernel around a given cell expanding in the orientation of that cell. This cooperation leads to boundary reinforcement and completion through its feedback to third stage (see table 1 for the definition of the feedback term (V^k) in the last equation). In simulations, this signal is fedback until the difference between activations at two succesive times are less than a preset constant ($\epsilon \ll 1$). A thresholded product of left and right shifted gaussian kernels is used to insure inward completion of boundaries [3].

	Equations	Complexity based on FFT
Contrast-enhancement	$X = \frac{D \otimes P}{A I + S \otimes P}$	$2N \log N + 2 N^2$
Gabor filtering	$J^k = G^k \otimes X$	$k (2N \log N + N^2)$
Spatial sharpening	$W^k = \frac{B I + M J^k + V^k}{I + M J^k \otimes D}$	$k (N \log N + N^2)$
Orientational sharpening	$Y^k = \frac{W^k \otimes E}{I + \sum_m W^m \otimes F}$	$k (3 N \log N + 2N^2)$
Cooperation	$Z^k = \omega(f(Y^k) \otimes C^k)$	$k (2N \log N + N^2)$
Feedback	$V^k = \frac{h(Z^k)}{I + \sum_m Z^m \otimes D}$	$k (N \log N + N^2)$
Output	$\sum_k W^k$	

Table 1: Equations and approximate complexity analysis in terms of convolutions in the HF channel algorithm. Convolution is defined in the frequency domain. Image and kernels are assumed to be available in the frequency domain. Therefore, their forward FFT are not included in the complexity analysis.

Acknowledgments

The authors wish to thank the following agencies for their partial support of their research. J. Contreras-Vidal is on leave from Monterrey Institute of Technology (ITESM, México) and received a fellowship from CONACYT (63462). M. Aguilar receives support from DARPA (contract AFOSR 90-0083) and Office of Naval Research (contract ONR N00014-91-J-4100).

References

[1] Aguilar, J.M. and Contreras-Vidal, J.L. (1992). Image Segmentation through Gabor-based Neural Networks. In *Proceedings of the SPIE's Aerospace and Remote Sensing Conference: Applications of Neural Networks*, Orlando, Fl.

[2] Cruthird, D., Gove, A., Grossberg, S., Mingolla, E., Nowak, N., and Williamson, J. (1992). Processing of synthetic aperture radar images by the boundary contour system and feature contour system. In *Proceedings of the Wang Conference on Neural Networks for Learning, Recognition, and Control*, Boston, MA.

[3] Grossberg, S. and Mingolla, E. (1985). Neural dynamics of form perception: Boundary completion, illusory figures, and neon color spreading. *Psychological Review*, **92**, 173-211.

Robot control—oral contributions

Neural Architecture for Robot Planning

P. Morasso†, V. Sanguineti†, and T. Tsuji‡

† *Department of Informatics, Systems, and Telecommunications, Genoa University, Italy*

‡ *Faculty of Engineering, Hiroshima University, Japan*

1. Introduction

Human motor skills depend, to a large extent, on two factors: (i) the intrinsic compliance of the musculo-skeletal system and (ii) its kinematic redundancy. The former factor allows a smooth modulation of contact forces during manipulation as well as a partial compensation of disturbances during trajectory formation. The latter is necessary for incorporating different task constraints in the same motor plan. In particular, experimental investigations of human arm movements [3] showed that the arm (even of a deafferented monkey, deprived of any kinesthetic/somesthetic feedback) returned toward an intermediate position (between the initial and final one) when the arm was temporarily displaced before the onset of the target. This suggests the hypothesis that the central nervous system plans a movement in terms of a sequence of equilibrium points, or *virtual trajectory* [5] but does not tell us anything about the central process which is responsible for producing it, particularly in the case of a redundant system. The main idea of this paper is that in complex robotic manipulators which, similarly to the human arms, are characterized by mechanical compliance and kinematic redundancy, it is quite useful for the planner to establish an analogy beween the real-time gradient-descent process determined by the mechanical potential field and a similar process associated with the dynamics of a neural computational engine which produces the virtual trajectories. In fact, the reaction of the arm to a contact/disturbance force is a gradient-descent in the elastic potential field of the musculo-skeletal system and this does not imply just a single motion pattern but a family of responses, indexed by the stiffness level of the different muscle groups, thus allowing a run-time adaptation of the *reactive-part* of the plan. Analogously, the run-time adaptation of the *trajectory-formation part* of a plan can be achieved by a gradient-descent process in a computational potential field modulated according to different task constraints. In this paper, we show how artificial potential fields can be expressed by means of a self-organized map [8], which also represents a *forward model* of the manipulator, and how gradient-descent can be performed in real-time on this map.

2. Self-organized forward model of redundant manipulators

Redundant manipulators imply an ill-posed inverse kinematic problem, because the mapping from sensory stimuli to motor variables is one-to-many. Although a *direct inverse modelling* approach can solve the problem on the base of *self-supervised learning* with a suitable training set[1], a better solution breaks down the learning process into two phases [6]: (i) in the first phase, the self-supervised strategy is used in order to learn the motor-to-sensory transformation (a *forward model* of the robot), which is always well defined, irrespectively of the degree of redundancy; (ii) in the second phase, the inverse sensory-to-motor transformation (an *inverse model* of the robot) is trained

---

[1]Self-supervised learning is a strategy for learning the model of a system or its inverse and it consists of generating pseudo-random input patterns, applying them to the system, measuring the response patterns and using the input-output pairs as training set of the model: either a direct/forward model (capable to predict the response given the stimulus) or an inverse/backward model (capable to estimate the stimulus that would produce the observed response).

by combining two criteria, one which aims at getting a global identity mapping[2] and another one which attempts to minimize some additional cost index. In this way, it is possible to have inverse models which are tuned according to specific task needs and are able to integrate different sensory channels. In the implementation of this concept proposed by [6], the forward model as well as the controller are multi-layer networks trained with the well-known technique of back-propagation. This is simple and efficient but can only work in an off-line manner and thus does not allow run-time task adaptation. The approach proposed here exploits the same forward-modelling concept and two-phase adaptation method but attempts to introduce an on-line element: (i) the forward model of the body is learned off-line as a self-organized map, while (ii) the synthesis of the inverse sensory-motor transformation is performed at run-time exploiting computational features of the self-organized forward model that allow a local computation of the gradient field. In this way, it is possible to take into account task-dependent constraints that are not fixed but are indeed specified on-line.

3. Off-line learning of the forward model

The forward model is defined by an input *motor* vector μ, which is the set of joint angles, and an output *sensory* vector β, which stores all the observables dependent upon μ, as the coordinates of the end-effector $\mathbf{x}_{ef}$ in the workspace or visual features of the robot images from a stereo-pair of cameras. We represent such a model $\beta = \beta(\mu)$ by means of a self-organized cortical map [1] which consists of a single layer or *neural field* F of M processing elements (PE's): They operate in parallel on a common input vector $\mu \in \mathbf{M} \subset \mathbf{R}^N$ and their activation function is the normalized Gaussian or *softmax* function

$$U_i(\mu) = \frac{G(\|\mu - \bar{\mu}_i\|)}{\sum_j G(\|\mu - \bar{\mu}_j\|)} \qquad (1)$$

(the $G(\cdot)$'s are Gaussian functions of equal variance and the norm is L_2) which has been used in the field of regression and classification [7, 4] and is a type of hyper radial basis function [10, 9]. PE_i's have limited receptive fields, centered around *preferred vector prototypes* $\bar{\mu}_i$'s, where the activation function peaks. The distribution of activities on the field for a given input pattern is also known as coarse or *population code* of that pattern.

Learning is performed by means of self-supervised soft competitive learning:

$$\begin{aligned} \Delta\bar{\mu}_i &= \eta_1 \left(\mu_k - \bar{\mu}_i\right) U_i(\mu_k) \\ \Delta\bar{\beta}_i &= \eta_2 \left(\beta_k - \bar{\beta}_i\right) U_i(\mu_k) \end{aligned} \qquad (2)$$

which is based on a training set of self-generated pseudo-random patterns $(\mu_k, \beta_k : k = 1, 2, ...)$ and carries out a smooth distribution of prototype vectors on the neural field with optimal statistical properties[3]. Moreover, the forward model is approximated by the following formula:

$$\beta = \beta(\mu) \approx \sum_i \bar{\beta}_i U_i(\mu) \qquad (3)$$

which was demonstrated [13] to be a minimum-variance estimator.

As an example, let us consider a simple planar arm with 2 degrees of freedom and let us restrict the sensory part β only to the 2D position vector of the end-effector. While performing a simulation of this model with a cortical map of $M = 200$ processing elements, training was carried out by generating 4000 pseudo-random postures, uniformly distributed in a restricted area of the configuration space. Such a training set was presented 10 times to the network, reducing the learning rates (η_1 and η_2 of Equations 2) from 0.3 to 0.05 and the variance $\sigma_\mu{}^2$ of the softmax

---

[2]The cascade of the inverse model + forward model should be able to generate a motor command, in response to a given sensory stimulus, in such a way that the predicted sensory consequences of the command can reproduce, as closely as possible, the same stimulus.

[3]The learning rule can be derived by minimizing the cross-entropy between the probability density function of μ and its approximation by means of a Gaussian mixture, with the Gaussian centers in $\bar{\mu}_i$'s [2].

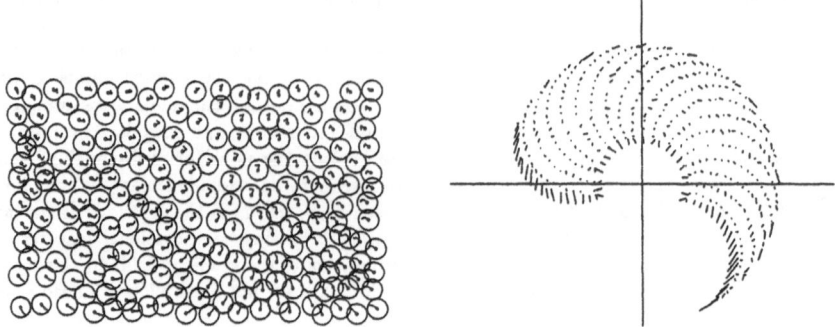

Figure 1: Left: Trained cortical map of a 2 dof arm. Right: Distribution of error vectors on the workspace.

function from 0.2 to 0.004. Figure 1(left) shows the trained cortical map, where each PE is represented by means of a circle containing a small picture of the arm in the learned configuration $(\tilde{\mu}_i, \tilde{\beta}_i)$. For the trained map, we also tested the accuracy of reconstructing β by means of (3) as follows: (i) an arm configuration μ with the corresponding end-effector position β is chosen, (ii) the population code is computed, (iii) the reconstruction function (3) is applied, yielding an estimate $\hat{\beta}$, and (iv) the error vector $e = \beta - \hat{\beta}$ is computed, that is the displacement between the original and the estimated β. Figure 1(right) shows the distribution of these error vectors over the whole workspace for a map of 200 PE's.

4. Run-time gradient descent

A potential field $\varepsilon = \varepsilon(\mu)$, as any smooth function of μ, can be represented by means of a distributed representation which uses the same population code and the same interpolation mechanism of the forward model:

$$\varepsilon(\mu) \approx \sum_i \tilde{\varepsilon}_i U_i(\mu) \tag{4}$$

where the $\tilde{\varepsilon}_i's$ are samples of the potential field which are assigned to each processing element in relation to its preferred sensory-motor pattern $(\tilde{\mu}_i, \tilde{\beta}_i)$.

The gradient-descent strategy requires to integrate the equation

$$\dot{\mu} = -\gamma \nabla \varepsilon(\mu) \tag{5}$$

and this can be implemented in the cortical map by using the following result:

$$\dot{\mu} = \gamma \sum_i (\mu - \tilde{\mu}_i) \tilde{\varepsilon}_i U_i(\mu) \tag{6}$$

which can be derived by writing the equation for the gradient vector $\nabla \varepsilon(\mu) \approx \sum_i \tilde{\varepsilon}_i \partial U_i(\mu) / \partial \mu$ and then taking into account the structure of the softmax function (with variance σ^2):

$$\frac{\partial U_i(\mu)}{\partial \mu} = -\frac{1}{\sigma^2} (\sum_j \tilde{\mu}_j U_j - \tilde{\mu}_i) U_i(\mu) \approx -\frac{1}{\sigma^2} (\mu - \tilde{\mu}_i) U_i(\mu) \tag{7}$$

The block diagram of figure 2 summarizes the simple circuitry which carries out the local computations outlined above, particularly as regards equations (3) and (6) and allows the cortical map to carry out gradient-descent. It can work in parallel and can support real-time provided that gradient-descent is allowed to operate near equilibrium. This is a key computational constraint which can be satisfied by a strategy of *local-incremental gradient-descent*, i.e. by a time-varying

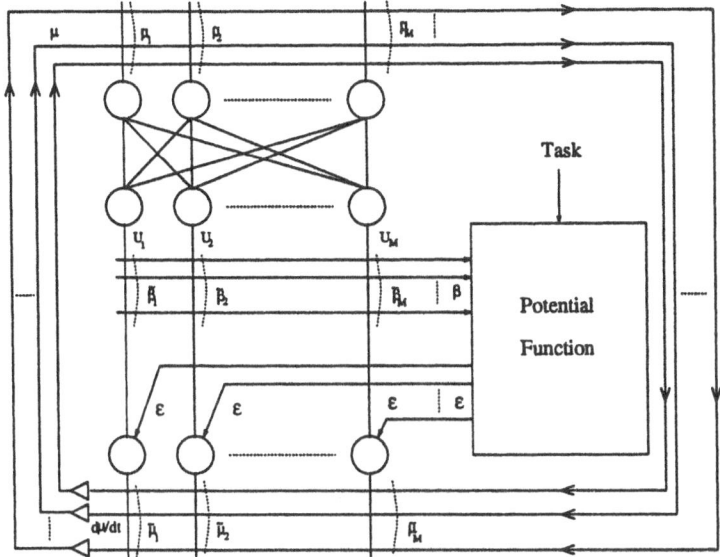

Figure 2: Block diagram of the cortical map gradient-descent network. The neurons in the top bank have a Gaussian activation, in the middle bank perform lateral gating inhibition, and in the bottom bank are simply multiplicative. The triangles are a bank of integrators.

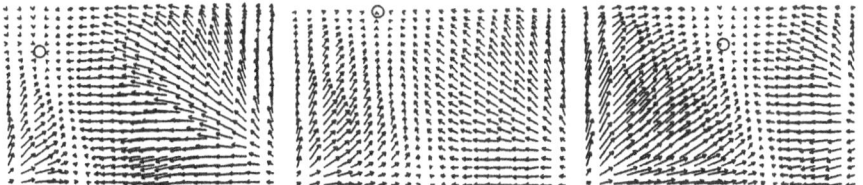

Figure 3: Three time frames of the target-distance gradient field snapped when the virtual target is at the initial, intermediate, and final position.

potential field which varies in such a way that its minimum stays close to the current position. It is important not only for implementing the real-time constraint but also for avoiding a typical pathology of gradient-descent mechanisms, i.e. getting trapped into local minima. In particular, this can be obtained by a target generation mechanism that smoothly shifts a virtual target $x_v = x_v(t)$ from the initial hand position x_0 to the real target position x_T and a target potential function which simply measures the Euclidean distance between the end-effector and virtual-target positions: $\varepsilon^{tgt} = \varepsilon^{tgt}(\mu, t) = \frac{1}{2} \| x_v(t) - x_{ef}(\mu) \|^2$. A model of the target generation mechanism is discussed elsewhere in the book [12]. Figure 3 shows the evolution over time of the field for a simple targeting movement.

5. Integration of different task constraints

The use of potential fields is a powerful technique for representing task constraints of different nature and defined in different coordinate frames. We previously considered a field for the representation of targets which operates as an attractor on the cortical representation of the end-effector. This is effectively a kinematic inversion mechanism and it operates equally well with redundant and non-redundant systems, implicitly computing an inverse matrix: the inverse Jacobian of the direct kinematic function, for non-redundant systems, or the Moore-Penrose pseudo-inverse matrix, for redundant systems. One of the nice features of performing the inversion via gradient-descent in a cortical map is its computational robustness, even in the vicinity of kinematic singularities, such as the boundary of the workspace, which tend to make unstable conventional inversion methods. The robustness comes from two elements: (i) the nature of the cortical map organization and training limits the domain of the generable motor patterns, whichever is the synergy formation mechanism, to a smooth area covering the set of training patterns; (ii) the gradient-descent mechanism implies that the flow of motor patterns is, in any case, a smooth path in the cortical map. For example, if the target x_T is outside the workspace, then the generated virtual trajectory *hits* the boundary of the workspace and then smoothly *slides* on it until it reaches the point closest to the target.

In fact, potential fields are general purpose tools for the specification of tasks, namely for representing both the *attraction* to a desirable state and the *repulsion* from undesirable or dangerous states. For example, a task component which aims at staying away from a dangerous joint configuration $\hat{\mu}$ can be represented by means of a repulsive field of the type $\varepsilon^{rep} = \varepsilon^{rep}(\mu) = g(\| \mu - \hat{\mu} \|^2)$, where $g(.)$ is a suitable monotonically decreasing function.

Attractive or repulsive fields can be computed from measures performed in different coordinate frames and the crucial point is that the fields can be superimposed on the same cortical map provided that the forward kinematic model embedded in the map allows to perform such measurements. The additivity of the task-related fields is the fundamental concept in our model for the integration of different task constraints as well as for the composition of complex tasks that attempt to reach several objectives at the same time, exploiting kinematic redundancy. An example is the task of keeping the end-effector as parallel as possible to its initial orientation while it is following an assigned path. This can be solved by defining, in addition to the *target-potential* above, a *parallelism-potential*[4] $\varepsilon^{par} = \varepsilon^{par}(\mu) = \frac{1}{2} \| f(\mu) - f(\mu(t_0)) \|^2$, and by carrying out a gradient-descent in the combined field: [5]

$$\varepsilon = \varepsilon(\mu) = k_1 \varepsilon^{tgt}(\mu) + k_2 \varepsilon^{par}(\mu) \tag{8}$$

where k_1 and k_2 weight the relative contributions of the two fields. This problem was simulated in relation with the cortical map of a 3 degree-of-freedom planar arm (map size: 900 $PE's$) and then we performed two experiments of reaching: in the first one, only the target reaching task was specified and the potential function ε contained the pure target component, computed in the end-effector space; in the second experiment, the initial configuration as well as the final target were the same, but we added the parallelism-potential with a small relative weight (0.1). Figure 4 (left) shows a simulation run of the first experiment: the end-effector follows the planned straight trajectory, smoothly changing its absolute orientation. Figure 4 (right) shows the simulation of the second experiment. As is apparent, the trajectory of the end-effector is maintained and its orientation is significantly more stable than in the previous case. A further example is given by obstacle avoidance. Let us suppose that while tracking a virtual target in a cluttered environment a robot might hit an obstacle unless it exploits its redundant degrees of freedom. In this case there should be a repulsive *obstacle-potential* $\varepsilon^{obs} = \varepsilon^{obs}(\mu)$ in addition to the attractive *target-potential*.

In the presentation above we assumed that the samples of the potential functions $\bar{\varepsilon}_i$'s are somehow "downloaded" on the PE's of the cortical map. The discussion of how this can be done is outside the scope of this paper. A possible model, which is based on a further layer of maps, is discussed in [11].

---

[4]The function $f(.)$ is the sum of the components of the argument vector, which corresponds, in the case of a planar arm, to the absolute orientation of the end-effector. $\mu(t_0)$ is the initial configuration vector.

[5]We can note that the combined field is "sharper" than the original target field which allows infinite equilibrium configurations if the system is redundant (the null-space of the transformation). The second field reduces the dimensionality of this space, increasing the cost of configurations which do not match that constraint.

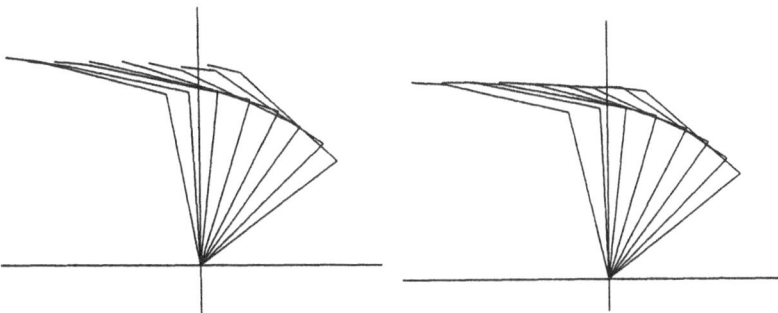

Figure 4: Reaching movements with a planar 3 degree-of-freedom arm. Left: only the target constraint is present. Right: for the same target, a constraint of parallelism is added.

References

[1] S. Amari. Dynamical stability of formation of cortical maps. In M.A. Arbib and S. Amari, editors, *Dynamic interactions in neural networks: models and data*, pages 15–34. Springer-Verlag, Berlin, 1989.

[2] M. Benaim and L. Tomasini. Competitive and Self-Organizing algorithms based on the minimization of an information criterion. In T. Kohonen, K. Makisara, O. Simula, and J. Kangas, editors, *Artificial Neural Networks*, Amsterdam, 1991. North-Holland.

[3] E. Bizzi, N. Accornero, W. Chapple, and N. Hogan. Posture control and trajectory formation during arm movements. *J. Neuroscience*, 4:2738–2744, 1984.

[4] J.S. Bridle. Probabilistic interpretation of feedforward classification network outputs, with relationships to statistical pattern recognition. In F. Fogelman Soulie and J. Herault, editors, *Neurocomputing*, volume NATO ASI F-68, pages 227–236. Springer Verlag, Berlin, 1989.

[5] N. Hogan. An organizing principle for a class of voluntary movements. *J. Neuroscience*, 4:2745–2754, 1984.

[6] M. I. Jordan and D. E. Rumelhart. Internal world models and supervised learning. In L. Birnbaum and G. Collins, editors, *Machine Learning: Proceedings of the Eighth International Workshop*, San Mateo, CA, 1991. Morgan Kaufmann.

[7] J. Moody and C. Darken. Fast learning in networks of locally-tuned processing units. *Neural Computation*, 1:281–294, 1989.

[8] P. Morasso and V. Sanguineti. SOBoS – A Self-Organizing Body Schema. In I. Aleksander and J. Taylor, editors, *Artificial Neural Networks*, Amsterdam, 1992. North-Holland.

[9] T. Poggio and F. Girosi. Networks for approximation and learning. *Proc. of the IEEE*, 78:1481–1495, 1990.

[10] M. J. D. Powell. Radial basis functions for multivariable interpolation: a review. In J. C. Mason and M. G. Cox, editors, *Algorithms for Approximation*. Clarendon Press, Oxford, 1987.

[11] V. Sanguineti and P. Morasso. Models of cortical maps. In *5nd Italian Workshop on Parallel Architectures and Neural Networks*, Vietri sul Mare, Italy, 1993. in press.

[12] V. Sanguineti, T. Tsuji, and P. Morasso. A dynamical model for the generation of curved trajectories. In *International Conference on Artficial Neural Networks*, Amsterdam, 1993. in press.

[13] D. F. Specht. Probabilistic neural networks. *Neural Networks*, 3:109–118, 1990.

From Situations to Actions: Motion Behavior Learning by Self-Organization[1]

J. Heikkonen, P. Koikkalainen and E. Oja
Lappeenranta University of Technology
Department of Information Technology
P.O. Box 20, 53851 Lappeenranta, Finland

Abstract

We show that the self–organization principle, implementable in artificial neural networks, is highly useful in connection with autonomous robots. Equipped with a self–organizing controller, a mobile robot can learn sensor–action type of behaviors that are difficult to realize otherwise. It is shown that the sensory information from several sources can be combined such that the obtained representation is directly applicable for higher level operations like navigation and obstacle avoidance. The performance of the approach is demonstrated in two examples: one, where the sensor information guides the robot to move around a corner; and the other, where the robot must navigate between two points avoiding obstacles.

1. Introduction

Industrial robots have shown their strength in stationary environments, where the given task can be programmed by using a deterministic algorithm. However, there are a great number of potential applications where such a control strategy is hard to realize because of the unforeseen combinations of sensory inputs from the surrounding environment. This problem of combining sensory data to higher level actions can be seen in the larger framework of adaptive systems, and from the methodological point of view it can be asserted that these type of tasks are best handled by using self–organization principles. The drawback is that generally self–organization, due its non–deterministic characteristics, is difficult to apply as part of a classical control system. As a solution we propose the use of behavior–based robots [2,3] instead of traditional control.

In contrast with the mainstream of robotics, where the information processing system is put together as a sequence of modules, the processing of the behavioral robot is organized as a parallel set of task–achieving modules. This gives several advantages. First, it removes the necessity of intermediate processing, which traditionally is required to fill up the gap between the lower level processing dealing with physical data, and higher levels doing symbolic processing. Second, it is easy to program, since the whole system can be made of parallel, reactive network–level units. Third, it copes very well with neural networks and related techniques that are finding their way to practical applications also in many other branches of applied information science, supported by an increasing number of hardware and software products.

In addition to sensor–actuator mappings that can be learned from a representative set of trainings, the autonomous robot must be able to train itself. Thus, it must be able to examine its surroundings and to overcome problems that result from weak and incomplete training examples. This introduces two requirements that are hard for many learning techniques. First, it sets strong demands for the generalization properties of the algorithms, as also discussed in [4] in connection with the function–learning problem. And second, the robot must be able to learn on–line and update its experiences without a separate training phase.

Self–organizing and competitive methods match best our requirements of error tolerance, and Kohonen's Self–Organizing Map (SOM) [5,7] is probably the best known neural network for the task. It has also been applied successfully to other similar high dimensional problems, such as speech recognition [6], pattern recognition [9], and robot control [11].

For demonstration purposes we have chosen simulated examples to exhibit sensory based motion control on corridor intersections where typical low–level behaviors are "go–straight", "turn–left"

1. *This work is supported by TEKES under SUBSYM Esprit Basic Research Action*

and "turn–right". As a more complex case, a self–organizing controller that learns to navigate and avoid obstacles is presented. It implements a trial and error learning strategy to improve its performance when new type of collision situations are observed.

We are interested in situations where the mobile robot perceives only the portion of the environment in front of it. It has eight distance sensors, covering a front view of 180 degrees up to a certain maximum distance R. These give the primary sensory information. In addition, the robot can measure its turning angle between consequent time instants in cases when it is trained by moving it along a trajectory. For the obstacle avoidance task, the robot has also tactile sensors for detecting the collisions.

2. Learning Motion Behaviors from Examples

There are several motion behaviors needed when navigating in typical corridor intersections. In order to learn "turn–left", "turn–right" and "go–straight" behaviors, we must figure out what circumstances in the environment will cause the behavior, what environment changes may happen during their execution, and how the robot should react to them. In our simulated robot, the only source of invariant description for motion is the sensory information collected from the surrounding world. Thus, the aim is to automatically select the most convenient set of descriptive primitives for motion behavior by showing several examples from the motion of interest. As an example, if a "turn–left" behavior is learned at a particular corridor intersection, the invariant learned primitives should make this "turn–left" behavior applicable to another intersection with a different shape. This generalization ability is vital because we cannot expect to obtain examples from every possible situations i.e. different corridor intersections.

In the following, we show how "go–straight", "turn–left" and "turn–right" behaviors can be learned from examples by using the learning and representation power of a Self–Organizing Map. In order to support the architecture of behavior–based robots, we are using a separate SOM for each behavior.

The learning of a motion behavior can be divided into two stages. First, a teacher shows several example movements of a particular behavior. During these training lessons, while the robot is moving, its learning system records the current sensory information and the corresponding change of motion. After showing these examples, the system has acquired a sequence of vectors $x_t = (d_{t1}, \ldots, d_{t8}, a_t)$ over instants of time $t = 0, \ldots, T - 1$, where $d_{tj} = r_{tj}/R$, $j = 1, \ldots, 8$, give the current distance sensor measurements r_{tj} at time t, normalized by the maximum value R, and a_t is a scalar which gives the corresponding change of the heading direction. The value of a_t is determined as follows: if the robot is turned left A_L degrees with respect to its current heading direction $\vec{V}_t$, then $a_t = 0.5 - A_L/360$; and if the robot turns right A_R degrees, then $a_t = 0.5 + A_R/360$. It should be noted that x_t is composed of quantities which may be perceived using only the robot's own local distance and direction sensors; no other knowledge is needed.

Next, the measured quantities $x_t = (d_{t1}, \ldots, d_{t8}, a_t)$ describing one behavior are used as an input to a SOM. The idea is that the SOM is presented with a random sequence of such observation–action pairs and it learns a mapping which predicts the action (turning angle) from a given sensory input (range measurements). We will use a SOM, denoted V_M, whose units have a nine–dimensional weight vector $W_i = (w_{i1}, \ldots, w_{i9})$. During the training process the best matching unit b_t for each x_t is found using equation (1)

$$\sum_{j=1}^{8} \left(w_{b_tj} - d_{tj} \right)^2 = \min_{i \in V_M} \sum_{j=1}^{8} \left(w_{ij} - d_{tj} \right)^2. \tag{1}$$

Note that only the d_{tj} components are used in finding the best–matching unit, not the a_t. Then the best matching unit and units in its topological neighborhood $Ne(b_t)$ are updated according to the Kohonen rule [7]:

$$W_r^{new} = W_r^{old} + \Psi_{rb_t}(x_t - W_r^{old}) , \ r \in Ne(b_t) \cup b_t. \tag{2}$$

In updating also the 9th component corresponding to a_t is changed. There $0 \leq \Psi_{rb_t} \leq 1$ is a prespecified adjustable gain, a function of distance $s(b_t, r)$ between the units r and b_t. This function Ψ_{rb_t} has its maximum at b_t and its value decreases as distance s increases. During training $Ne(i)$ and Ψ_{rb_t} are varied so that the neighborhoods of each neuron and gain function slowly decrease with time.

In practice, the basic SOM training is computationally costly in high dimensional input spaces, making our experimenting with on–line learning unnecessarily time–consuming. Therefore we have used here a Tree Structured SOM (TS–SOM) [8], which is a fast variant of the original SOM.

After learning is complete, the SOM corresponding to a particular behavior can be employed to control the motion of the robot. For instance, if a robot has decided that it should perform the "turn–right" behavior then the SOM which has learned this task controls the motion of the robot. The motion control strategy is straighforward: at a particular time instance t, the range measurements $(d_{t1},...,d_{t8})$ perceived from the surrounding world are first fed to the activated SOM, and the weight vector W_{b_t} of the best–matching unit b_t is found using Eq. (1). Now its 9th component $w_{b,9}$ specifies the next heading direction $\vec{V}_{t+1}$, and a new movement can be performed.

The upper left picture in Figure 1 shows a simulated corridor intersection where ten example trajectories were given for each of the "turn–left", "turn–right", and "go–straight" behaviors. Here only two sample trajectories of the "turn–left" event are drawn for illustration purposes. Three SOMs of 64 neurons in two–dimensional discrete lattice were taught according to the given principles. The number of training samples acquired from example paths were 730, 715, and 652 for "turn–left", "turn–right", and "go–straight" behaviors, respectively. The robot moved three pixels ahead at each time step in the environment of 256×256 pixels.

The upper right picture in Figure 1 presents resulting paths obtained using a "turn–left" SOM as a controller for the simulated mobile robot. For every trajectory, a starting point (white ball) and a starting direction (white tick) were given, and each trajectory was plotted according to the movement of the robot. The bottom left and right illustrations in Figure 1 show resulting trajectories of "turn–right" and "go–straight" behaviors. These trajectories clearly show that after teaching, the SOMs can control successfully the motion of a mobile robot in this task.

Next, we will demonstrate the generalization ability of our motion learning approach. In Figure 2 there are several different corridor intersections. Each trajectory gives the motion of the simulated robot controlled by exactly the same "turn–left" map that was used in the previous demonstration. According to these results, our approach supports generalization from a given set of example paths, and the learned behavior can successfully guide the robot in diverse intersections not seen before.

3. Learning to Avoid Obstacles During Navigation

Obstacle avoidance means the ability to avoid collisions with obstacles in the task space of a mobile robot. Even if the map of the environment is modeled a priori, i.e. there are only static obstacles with known coordinates, obstacle avoidance behavior is needed, because it is often difficult to get accurate metric information in a space and thus a collision might occur. In dynamically changing domains with both moving and static obstacles the role of obstacle avoidance is, of course, more important than in static domains, because a robot should respond immediately to its sensory inputs from the unknown and changing environment.

The collision avoidance learning proposed here is based on trial and error and self–organizing principles. The SOM is utilized to learn an effective collision avoidance behavior on the basis of the robot's own experiences. This minimizes the need of human guidance during learning and operation. The basic idea is that if the robot comes to a state which leads to a collision, the SOM is trained to give a better action to this situation, preventing this collision in the future. Thus this scheme is close to reinforcement learning (see for instance [1] and [10]). The proposed approach allows continuos learning during the robot's lifetime and thus it makes the robot adaptable into new situations and unknown environments, because no a priori knowledge about the locations of the obstacles is needed. Only local sensor information and a direction to the goal are required.

During navigation at a particular time instance t, the learning system can record the current range sensor state $d_{tj} = r_{tj}/R$, $j = 1,...,8$. In addition, the relative direction to the goal is measured. A normalized scalar β_t is used in practice to indicate this, and its value is determined as follows: the

 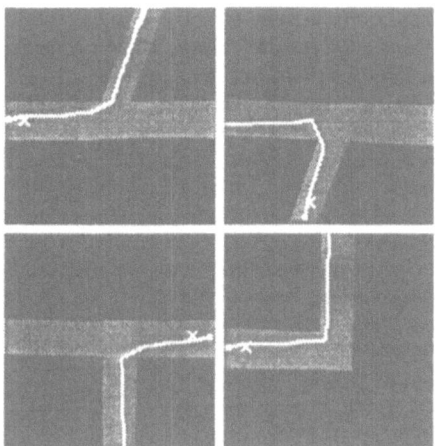

Figure 1. A corridor intersection where ten example trajectories for each "turn–left", "turn–right" and "go–straight" behaviors have been shown. See text for details.

Figure 2. Several intersections and "turn–left" trajectories obtained using "turn–left" net as a controller for a mobile robot. See text for details.

angle B_t between the current heading direction $\vec{V}_t$ and the current direction to the goal $\vec{D}_t$ is calculated in degrees, and if the heading direction $\vec{V}_t$ is on the left–hand side with respect to $\vec{D}_t$ then $\beta_t = 0.5 - B_t/360$ else $\beta_t = 0.5 + B_t/360$.

The control variable is the change in direction a_t from the previous instant. Now the change in direction is given relative to the direction to the goal, which is known by the robot. Knowing the direction to the goal $\vec{D}_t$ and parameter a_t, the next heading direction $\vec{V}_{t+1}$ is obtained by determining the angle A_t between $\vec{V}_{t+1}$ and $\vec{D}_t$: $A_t = 360(0.5 - a_t)$ (in degrees). The sign of A_t determines the side of the next heading direction $\vec{V}_{t+1}$ with respect to vector $\vec{D}_t$. If $A_t < 0$ then $\vec{V}_{t+1}$ is on the right–hand side of the vector $\vec{D}_t$, else it is on the left–hand side.

In the obstacle avoidance behavior learning system a SOM denoted by V_A is used. Each unit of the map V_A has a 10–element weight vector $W_i = (w_{i1},...,w_{i10})$. The weights $(w_{i1},...,w_{i8})$, w_{i9} and w_{i10} correspond to to distances $(d_{t1},...,d_{t8})$, the scalar β_t and corresponding change of direction a_t, respectively. Initially (at $t = 0$) weights $(w_{i1},...,w_{i8})$ are initialized to random values between 0 and 1, and w_{i9} and w_{i10} are set to 0.5. This means that at the beginning the robot moves straight towards the goal.

In order to decide the next movement from the map under operation, the range measurements $x_t = (d_{t1},...,d_{t8})$ and the relative direction to the goal β_t are taken at time t. They are presented as an input to the map V_A. The best–matching unit $b_t \in V_A$ for x_t is found according to Eq. (3) which weights equally the distance measurements and the direction to the goal.

$$\left[\sum_{j=1}^{8}\left(w_{b,j} - d_{tj}\right)^2\right] + 8\left(w_{b,9} - \beta_t\right)^2 = \min_{i\in V_A}\left\{\left[\sum_{j=1}^{8}\left(w_{ij} - d_{tj}\right)^2\right] + 8\left(w_{i9} - \beta_t\right)^2\right\} \quad (3)$$

The tenth component $w_{b,10}$ of the weight vector W_{b_t} of the best–matching unit specifies now the next heading direction $\vec{V}_{t+1}$ as outlined above, and a movement can be performed.

Training the map is trial–and–error type, totally based on collisions. These are detected by two tactile sensors situated on the left and right front corners of the simulated robot. If a collision occurs with an obstacle, the robot is first backed–up M steps according to the M latest movements which have

266

been stored in short–time memory. The SOM V_A is then trained to improve the performance using the M latest states as follows: if the left tactile sensor touched an obstacle, then the SOM is trained using the training vectors $x_k = (d_{k1},...,d_{k8},\beta_k,a_k + A(k))$ with $k=t,...,t-M-1$. If the right tactile sensor touched the obstacle, the training vectors are $x_k = (d_{k1},...,d_{k8},\beta_k,a_k - A(k))$. The idea is to associate new movement directions to the sensory values so that the collision would be prevented. The function $A(k)$ is a prespecified positive function of time k. The $A(k)$ may be constant for all k, but it is preferable that it has its maximum at $k = t$ and its value decreases as k decreases. A suitable choice for $A(k)$ is $A(k) = A_{max}/(t - k + 1)$, where A_{max} gives the maximum value for $A(k)$ at $k = t$.

It turns out that best results are obtained if we also use training vectors obtained by mirroring the collision situation. There are two reasons for including these mirrored situations into the training set. First, they speed up learning by increasing the size of the learning set, and second, they assure that the robot learns either turning left or right in front of an obstacle, instead of learning one type of obstacle avoidance behavior only. Thus, in case the left sensor touched the obstacle, vectors $\underline{x}_k = (d_{k8},...,d_{k1},1.0 - \beta_k,1.0 - a_k - A(k))$, $k=t,...,t-M-1$ are added to the training set, and if the right sensor touched, the new training vectors are $\underline{x}_k = (d_{k8},...,d_{k1},1.0 - \beta_k,1.0 - a_k + A(k))$.

The performance of this obstacle avoidance learning strategy was examined in three test environments (see Fig. 3) by randomly selecting 500 start and goal locations of navigation missions. Three SOMs (one SOM for each environment) of 100 neurons in a two–dimensional discrete lattice were used and trained according to the trial–and–error principles given above. The parameters M (number of back–up steps) and A_{max} (increment/decrement maximum in the movement direction) were set to 15 and 0.15, respectively, and the robot moved two pixels ahead at each time step. The number of unsuccessful navigations (one or more collisions) were calculated and tests with the same 500 missions were repeated with on–line learning, until the number of unsuccessfull navigations did not decrease any more from the previous test round.

In the first environment with three obstacles (Fig. 3), 2 test rounds of 500 missions each were performed. In the first round the robot collided with obstacles in 9 missions, but after the first round, it performed all the next 500 navigations successfully. Similar results were also obtained in the second environment with six obstacles (Fig. 3) : during the first round there were 10 missions with collisions, but the second test round was performed without any collisions. In the last test environment with 14 obstacles (Fig. 3), the numbers of unsuccessful navigations during test rounds 1 to 4 were 72, 32, 20, and 0, respectively. The above results clearly show how the performance of the collision avoidance behavior improved during the robot's lifetime. The robot can learn quite fast the obstacle avoidance behavior even in rather complex domains.

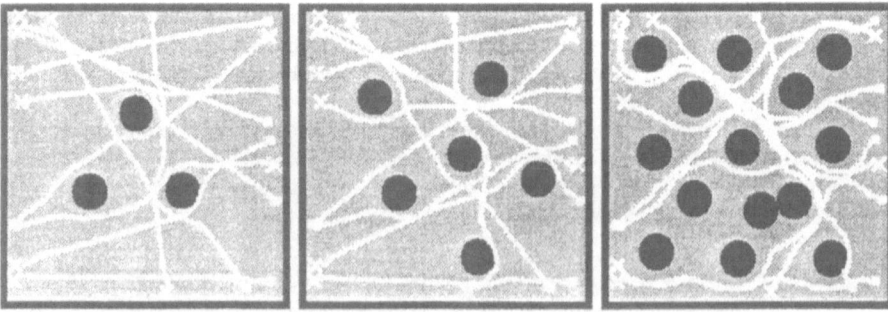

Figure 3. Test environment 1 (left picture), 2 (middle picture) and 3 (right picture) for obstacle avoidance learning and ten example movements.

The generalization ability of the obstacle avoidance behavior learning was tested in two environments with diverse obstacles (see Fig. 4). The SOM that was trained for obstacle avoidance in environment 3 (Fig. 3) in the previous test was used to control the motion of the robot. The results

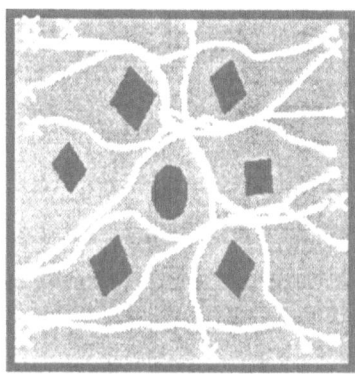 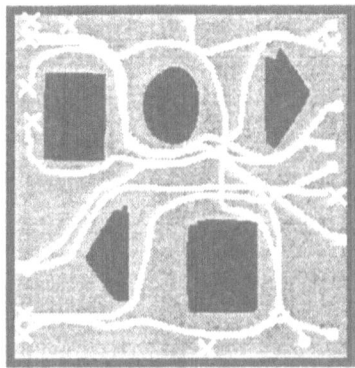

Figure 4. Test environment 1 (left picture) and 2 (right picture) for obstacle avoidance behavior testing and ten example paths.

were excellent: during 500 randomly generated navigation missions no collisions occurred. This experiment demonstrates that even if the shapes and the sizes of the obstacles are changing, the SOM based system manages to control successfully the motion of the mobile robot.

4. Conclusion

We have proposed a self–organizing strategy to control a mobile robot which must learn sensor–action behaviors by using its own observations in addition to the given examples. The Self–Organizing Map was used to build these mappings for different combinations of sensory inputs. There are two important characteristics that provide the basic reason for using SOM in these tasks. First, the SOM proved to be efficient in simplifying problems by introducing mappings from high dimensional input spaces to lower dimensional space while preserving the topological relations between inputs. Second, the SOM is capable of making good generalizations in situations that are not covered explicitly by the training set. In sensor–action tasks this kind of generalization is vital because it is usually impossible to provide examples for every situation that may occur when the mobile robot, for example, is freely moving around in real–world environment. The experiments reported here demonstrate the power of self–organization in learning the desired behaviors.

REFERENCES

[1] Barto, A. G, Sutton, R. S., Anderson, C. W., 1983, Neuronlike Adaptive Elements that Can Solve Difficult Control Problems, IEEE Trans. on Systems, Man and Cybernetics, Volume 13, 834–846.

[2] Brooks, R. A., 1986, A Robust Layered Control System for a Mobile Robot, IEEE Journal of Robotics and Automation, Volume 2, Number 1, 14–23.

[3] Brooks, R. A., 1991, New Approaches to Robotics, Science, Volume 253, Number 13, 1227–1232.

[4] Kaelbling, L. P., 1991, Foundations of Learning in Autonomous Agents, Robotics and Autonomous Systems, Vol. 8, 131–144.

[5] Kohonen, T., 1982, Self–organized Formation of Topologically Correct Feature Maps, Biological Cybernetics, 43, 59–69.

[6] Kohonen, T., 1988, The "Neural" Phonetic Typewriter, IEEE Computer, March, 11–22.

[7] Kohonen, T.,1989, Self–Organization and Associative Memory, Springer–Verlag, Berlin, Heidelberg.

[8] Koikkalainen, P., Oja, E., 1990, Self–Organizing Hierarchical Feature Maps, In proc. IJCNN–90, International Joint Conference on Neural Networks, San Diego, CA, 279–284.

[9] Lampinen, J., Oja, E., 1990, Distortion Tolerant Feature Extraction with Gabor Functions and Topological Coding, In proc. INNC 90, Paris, 301–304.

[10] Mahadevan, S., Connell, J., 1992, Automatic Programming of Behavior–based Robots Using Reinforcement Learning, Artifial Intelligence, 55, 311–365.

[11] Ritter, H., Martinez, T., Schulten, K., 1989, Topology Conserving Maps for Learning Visuomotor–Coordination, Neural Networks, Volume 2, Number 3, 159–168.

Application of Q-Learning in Robot Grasping Tasks [1]

Thomas Wengerek and Helge Ritter

Department of Information Science
Bielefeld University
D-4800 Bielefeld, FRG

email: wengerek@techfak.uni-bielefeld.de and helge@techfak.uni-bielefeld.de

Abstract: Reinforcement learning plays a major part in the adaptive behaviour of autonomous robots. But in real-world environments reinforcement learning techniques, like Q-learning, meet great difficulties because of rapidly growing search spaces. We explored the characteristics of discrete Q-learning in a high-dimensional continuous setup consisting of a simulated robot grasping task. Very simple sensors in this setup allow only a rather coarse identification of the actual "physical" state, thus leading to consequences known as "perceptual aliasing". We identified parameters - especially the sensory sampling-rate - directly controlling the grade of generality of the policies to be learned. Actually in case of the more general policies performing the raw positioning effects of ambiguity can be suppressed. So the system can find feasible grasping positions after few explorative actions and we suggest, that a neural net supporting Q-learning could improve the overall performance significantly.

1   Introduction

Using reinforcement learning with delayed rewards to solve complex tasks in continuous state and action spaces, for instance grasping an object by a robot, bears great problems caused by tremendously large search spaces and by ambiguous states. In this paper we illustrate first results about features of conventional Q-learning in a real-world-like workspace. This is the first step to develop a reliable and efficient learning scheme based on neural network tools to overcome these known problems of reinforcement learning.

Q-learning is an incremental, Monte-Carlo form of asynchronous Dynamic Programming [5, 3, 1], that enables *adaptive optimal control*, which faces the central antagonism between system identification (i.e. exploration) and control. In adaptive control prior knowledge of the system (i.e. the transition probabilities) underlying the Markovian decision process is not given. Actions may not be optimal on the basis of prior assumptions, but in the limit as experience accumulates - expressed by *action values* $Q(x,a)$ for action a in state x - the policy should approach an optimal policy. The learning process consists in propagating credit, carried by the action values, to every visited state. This credit is an estimate about the expected future reinforcement, but discounted by the number of steps to reach the reinforcement, if the actual policy is used.

A central issue in this paper is learning under restricted sensory information such that a complete and unambiguous system state observation is not possible. This is a characteristic feature of many real-world situations and leads to effects that have been termed "perceptual aliasing" and investigated in the context of Q-learning, e.g. by [7].

We also emphasize that there is a strong dependency between "sensational" states, in which the learning and control process takes place, and the control objective grounded in the "physical" world, for which suitable sensations have to be chosen to fulfill its demands. So a subsumption architecture [2] to coordinate control objectives, supplementary to each other, seems to be plausible. Other important questions in the field of reinforcement learning are how to cut down the search space into tractable pieces and how to combine the sub-policies [6].

---

[1]This work was supported by the German Ministry for Research and Technology (BMFT) under grant number ITN9104AO. Any responsibility for the content of this publication is with the authors.

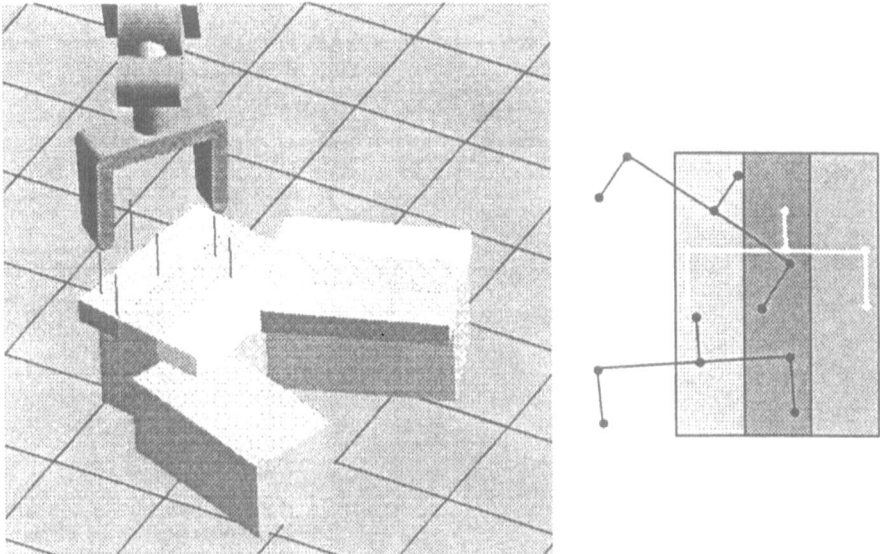

Figure 1: (*left*) The workspace with the robot-arm and the sensors (vertical strokes) mounted on the gripper and an object consisting of three bricks. For two of them the regions are added. (*right*) Schematic topview of a scene with only one brick with the three regions and the sensors (dots) linked by vertices. Shown are several positions of the gripper above the brick. In black: two different positions but related to the same sensational state and therefore causing ambiguity. In white: a possible target position.

2   The Robot Grasping Simulation

We consider the following task: a robot-arm with a two-jaw gripper has to find suitable positions above a brick or an object consisting of several bricks connected with each other (Fig.1). The gripper carries six "sensors" which provide information about the relative position of the gripper with respect to the object. Each sensor probes the scene along a vertical ray (indicated by a stroke in Fig.1(*left*)) and is assumed to be able to discriminate between four types of regions: two types of edge regions (indicated by a light grey and a middle grey stripe in Fig.1(*right*)) and one type of central region (the dark grey stripe) and none of them (i.e. the complement of the regions) as the fourth type. The geometrical arrangement of the sensors in our simulation is not optimized with respect to the defined regions which belongs to the more general question what sensors or combination of sensors should be chosen for what tasks opening another wide field of investigation. Note that the stripes in Fig.1 are only meant to exist for the eye of the viewer. The actual object is given by the union of the three "bricks" beneath them and it would be the task of a suitable sensory system to somehow extract the described region information from the scene - e.g. by using tactile or visual information or a combination of both.

Now the information about which sensor recognizes which region is a possible coding scheme to determine coarsely the real geometrical position of the gripper relatively to the object. Therefore the more sensors and the more detailed regions there are the better is the determination of the geometrical position. Because each of the six available sensors recognizes one out of four different states, we end up with a space of $4^6 = 4096$ discriminable sensational states. These states cannot completely describe the real "physical" states causing therefore "perceptual aliasing" effects. That means the occurrence of situations in which identical actions performed in different "physical" states but identical sensational states (Fig.2(*right*)) can lead to different sensational successor states and therefore corrupt Q-learning. Reward is delivered only for those very few special relations between sensors and regions that specify a feasible grasping position (examples for these target positions are shown in Fig.1(*right*) and in Fig.2). "Actions" consist of movements of the robot-arm relative to the actual position of the gripper with respect to the so called "tool coordinate system", that is stationary with respect to the gripper. Actually, we allow six possible actions (+/ − x− or +/ − y−translation or clockwise or counterclockwise rotation), that can be alternatively executed with respect to the tool frame until at least one sensor changes its state. Therefore the compound space of sensational states and actions consists of $6 * 4^6 = 24576$ states, in which the creation of policies by Q-learning takes place.

Figure 2: Shown is a typical grasp-situation, but now with another object, different from that in Fig.1(*left*). After a policy to find grasps was generated by a learning process, which consists of a sequence of 50 randomly chosen starting positions, this solution for a new starting position was found already in the first trial.

3 Learning to Grasp

In our experiments we consider the rather complex task to find suitable grasping positions for the large class of objects consisting of several bricks. Now if in a randomly chosen starting position no sensor touches the object, control is taken over by another module, that moves the arm in the direction of the center of mass of the object until at least one sensor broadcasts contact.

The conditions for trial-based Q-learning in these experiments were the following: an object was constructed and a first starting position was chosen randomly, then maximally a given number of trials was performed beginning each time at this actual starting position; each trial allowed maximally a given number of actions to be executed (in the following called "steps"); a new trial began if the given number of steps was consumed or if the target was found or if the gripper lost contact with the object. Further the number of trials was given, that had to be consecutively successful (in the following called "break-condition") before learning in a new starting position began. We introduced this condition to force the system to generate in a controlled way more or less stabilized solutions. After that a new starting position was randomly chosen and learning was continued while exploiting and adapting the already gained experience for that new situation in the same manner as before.

We observed that the rate, at which the sensors are read out, (in the following called "sampling-rate") has to be controlled suitably to prevent that sensors are located too closely at region-boundaries, when a state is going to be established. With different sampling rates one can switch from fine-grained to coarse-grained state detection and vice versa, so it seems to be the most important parameter mediating between the discrete sensational space and the continuous workspace. We implemented the sampling-rate in the simulation by choosing for each degree of freedom, i.e. the actions introduced above, a stepsize Δ. Whenever the current movement along the respective degree

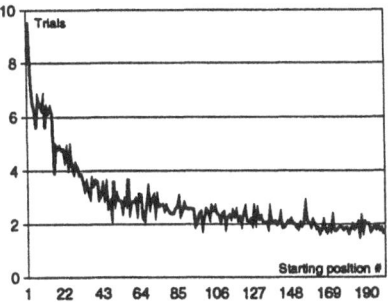 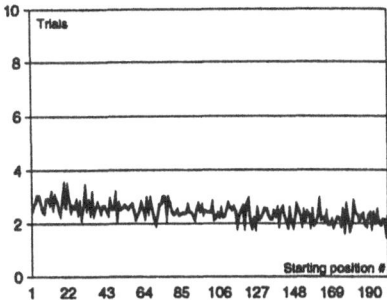

Figure 3: Mean values of required trials to grasp the object over 100 independent runs against sequences of 200 random starting positions with sampling-rate= 5 and break-condition= 1. (*left*)Learning from scratch. At the end of the learning process quasi-reactive (less than two trials are required in the mean) behavior can be reached. (*right*)Using experience learned with another object.

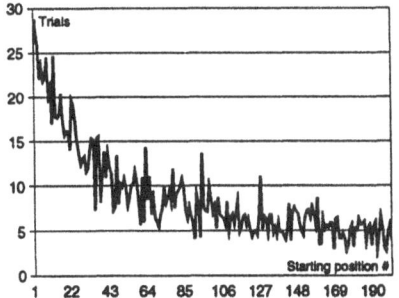 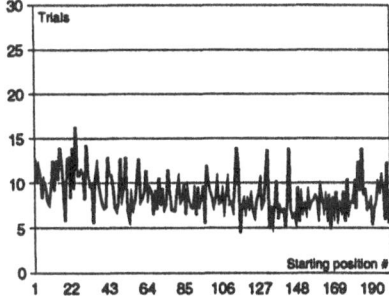

Figure 4: As in Fig.3 but with sampling-rate= 30 and break-condition= 1.(*left*)Learning from scratch. (*right*)Using experience learned with another object.

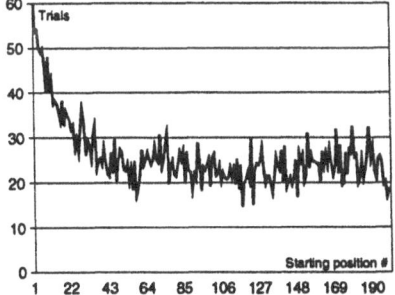 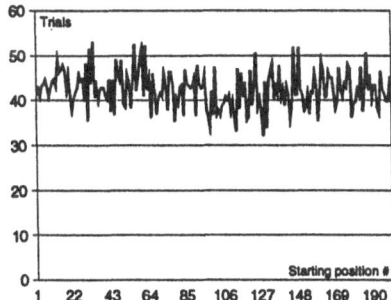

Figure 5: As in Fig.3 but with sampling-rate= 30 and break-condition= 5. Note the large gap in the performance at the end of the learning processes of the two related experiments, that is caused by learning policies specialized to actual situations.(*left*)Learning from scratch. (*right*)Using experience learned with another object.

272

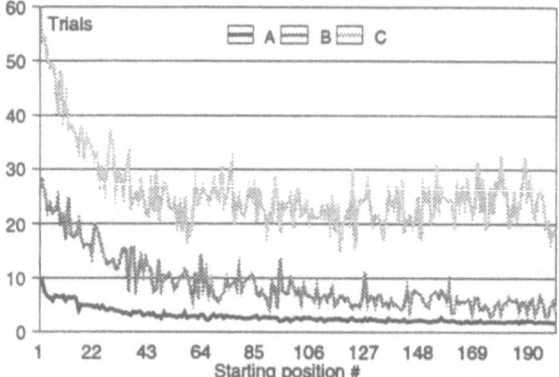

Figure 6:
To elucidate the scaling behavior of the learning processes controlled by sampling-rate and break-condition the curves of Figs.3(*left*)-5(*left*) are shown together here in one plot.   (A) sampling-rate= 5, break-condition = 1; (B) sampling-rate= 30, break-condition= 1; (C) sampling-rate= 30, break-condition= 5.

of freedom has progressed by Δ, the sensors are read out and the possibility for a change of the actual sensational state is given. In case of a change a new (possibly different) action can be chosen. Otherwise successive repititions of the former movements are executed with stepsize Δ until a predefined limit has been exceeded or a change of the sensational state has been detected while repeating the movements. Also we observed, that the performance depends critically on another parameter (called "temperature" [1]), which controls how explorative the system behaves while learning. In our setup too much exploration (in adaptive control termed "system identification") could destroy previous experience because of "perceptual aliasing". This is the reason for splitting the learning phase into parts each time related to a distinct starting position. The third important parameter we identified was the break-condition introduced above.

The effects of sampling-rate and break-condition are reported in the following experimental results (Figs.3-5), where the mean values of the number of required trials to find a grasping position are plotted against sequences of 200 randomly chosen starting positions. The other parameters of trial-based Q-learning were chosen once and left unchanged afterwards: maximally 100 trials per starting position, maximally 50 steps per trial and 100 runs with different random number seeds were performed to get mean values of required trials. Figs.3(*left*)-5(*left*) show the results for learning from scratch. To investigate the generalization abilities of policies further experiments were done (Figs.3(*right*)-5(*right*)). These consisted in using policies as a-priori experiences, that were learned before in setups with other objects, but in each case with respectively identical combinations of the learning parameters.

The asymptotic mean number of required trials to reach a grasping position at the end of each learning process (shortly termed "required trials"), increases with increasing sampling-rate (Figs.3(*left*)-4(*left*)) and increases too with increasing break-condition (Figs.4(*left*)-5(*left*)). With sampling-rate= 5 and break-condition= 1, see Fig.3(*left*), *quasi-reactive behaviour* can be reached after learning, because just about two trials are needed to find a grasp. In this case the robot finds permanently new solutions for a given starting position and because of the coarse-grained state detection the positioning of the gripper is performed only coarsely. With higher sampling-rate on the other hand, for instance sampling-rate= 30 and break-condition= 1, see Fig.4(*left*), the required trials to find a grasp are increased up to about five. But now the robot performs fine positioning and "perceptual aliasing" begins to play a mayor part. This tendency to need more trials is strengthened, if the break-condition is increased to five consecutive successful trials(Fig.5(*left*)), which forces now the learning-system to generate something like a *special purpose policy* for the actual starting position with restricted applicability to other situations. Therefore the required trials are further increased up to about 25 trials. But in all these cases the main part of learning is done after about the 50*th* starting position. To provide a better impression of the scaling of the performance governed by the two control parameters, sampling-rate and break-condition, the curves of Figs.3(*left*)-5(*left*) are shown together in one plot in Fig.6.

4 Discussion

For testing the generalization abilities of the policies learned in the situations of Figs.3(*left*)-5(*left*) the objects to be grasped were exchanged by other objects but the gained experiences were used from the beginning. The results are shown in Figs.3(*right*)-5(*right*). The first observation is, that the gap of required trials between the related cases increases with increasing sampling-rate and break-condition. This fits well into our former interpretation, because the policies in the cases with higher sampling-rate and break-condition are not only more adapted to actual starting positions but also to some characteristics of the objects, which were previously used to learn them. So one can say that in this case *policies more specialized to distinct situations* are created for doing the fine-positioning. The second observation is, that learning with small sampling rate (Figs.3(*right,left*)) seems to *extract a basic policy* applicable with high performance (few required trials) to a large proportion of starting positions. That means a policy, which is not highly adapted to special situations and to special objects, directing coarse movements of the gripper and that is nearly independent from corrupting effects caused by "perceptual aliasing".

So we can summarize. Restricted sensory information is a central problem for many tasks in the field of robotics. Adaptive neural networks have been proven successful in partially overcoming these problems by their ability to create transformations and world models through learning. However, as we move to more and more demanding tasks (such as, e.g. extraction of sensory information for controlling grasping movements, see [4]), the chance for ambiguities and inaccuracies inevitably increases and additional strategies become necessary to guide exploration and to allow learning and control even in the presence of such difficulties. Focusing on the task of controlling grasping movements, we have investigated in the present paper Q-learning for this purpose and found that already conventional, discrete Q-learning is able to create control-policies applied to continuous spaces, in spite of ambiguous sensory information. We identified the sensory sampling rate as the most important control parameter, that on the one hand can suppress the corrupting effects of "perceptual aliasing" by generating general policies performing the raw positioning and on the other hand by generating more specialized policies for doing the fine positioning, but then being fully aware of effects of ambiguity. By choosing a compromise between these extremes feasible grasping positions can be found after few explorative trials that finally adapt the current policy in the actual situation. And we suggest, that totally reactive behaviour can be reached, if Q-learning itself would be supported by the generalization abilities of a neural network.

References

[1] Barto A.G.,Bradtke S.J.,Singh S.P.(1991) Real-Time Learning and Control using Asynchronous Dynamic Programming. Dept. of Computer Science, University of MA, Amherst, Technical Report 91-57.

[2] Mahadevan S.,Connell J.(1991) Scaling Reinforcement Learning to Robotics by Exploiting the Subsumption Architecture. In: Proc. of the 8th Int. Conf. on Machine Learning, Morgan Kauffman.

[3] Sutton R.S.(1990) Integrated Architectures for Learning, Planning, and Reacting Based on Approximating Dynamic Programming. In: Proc. of the 7th Int. Conf. on Machine Learning, San Mateo, Ca: Morgan Kauffman.

[4] Meyering A.,Ritter H.(1992) Learning to recognize 3D-Hand Postures from Perspective Pixel Images. In:Artificial Neural Networks 2, I. Aleksander, J. Taylor (eds.), Elsevier Science Publ.

[5] Watkins C.J.C.H.(1989) Learning from delayed Rewards. PhD Thesis. University of Cambridge, England.

[6] Wengerek Th.,Ritter H.(1992) Reinforcement Learning and Subtasks. In:Artificial Neural Networks 2, I.Aleksander,J.Taylor (eds.), Elsevier Science Publ.

[7] Whitehead S.D.,Ballard D.H.(1990) Active Perception and Reinforcement Learning. In: Proc. of the 7th Int. Conf. on Machine Learning, San Mateo, Ca: Morgan Kauffman.

I/O–Stability for Robot Control with a Global Neural Net Inverse Model in the Feedback Loop *

M. Jansen, J.R. Beerhold, and R. Eckmiller
Department of Computer Science VI, Neuroinformatics
University of Bonn, FR Germany
Tel.: ++49-228-550-349   FAX: ++49-228-550-425   e–mail: jan@nero.uni-bonn.de

Abstract A set of Neural Networks learns matrix components representing mass–coupling, coriolis, and friction forces in an inverse robot model. While obtaining training data from specific random trajectories, the inverse model becomes highly precise over the entire robot working range. Global model precision and subsequently L_∞–stability for a controller using the networks in the feedback loop following the learning phase are established by a new method. The approach is demonstrated for a simulated planar 4–joint–machine.

Introduction

The optimal way to achieve dynamic control of invertible nonlinear MIMO–Systems is to use an inverse model as feedforward controller. This method is referred to as "nonlinear decoupling" [3], "feedback linearization" or "computed torque control" [1].

For a simulated rigid planar 4–joint–robot (4JM), we trained a given set of neural networks to learn such an inverse model in an identification phase.

As compared to classical approaches of dynamic systems identification, the neural network needs neither a difficult and computation–time consuming analytical model structure (e.g. Newton–Euler dynamics), nor laborious parameter measurements (often done following partial robot decomposition). Other approaches using neural networks (e.g. [5], [6]) are restricted to the formation of just a "local" inverse model in the neighbourhood of a single training trajectory, or do not even address this problem of the range of validity of the neural inverse model.

In contrast, we present a novel approach consisting of a robust neural controller structure, specific random training trajectories, and a method for analysis that provides highly accurate tracking control and model precision over the robot's whole working range except for a small border zone. This is the basis for the subsequent global input/output stability guarantee. Part of the results have recently been summarized in [2].

The approach shares some drawbacks with all model–based robot control architectures: i.e. the difficulty in treating controller delay, elasticity, and gear play. For our current application to the real physical 4JM, different amendments to the controller such as a neural

*) Supported by Ministry for Research and Technology (BMFT–Project SENROB), under grant 01 IN 105 A/O

state–predictor and polynominal extrapolation on position measurements to avoid delays, or fast active (possibly adaptive) damping, are currently investigated. These techniques are, however, not the topic of this paper. Finally it should be mentioned that we neglect discrete time effects, as is usual for a relatively smoothly moving robot with a smooth control law and a high sampling rate.

Neural Networks: Type and Structure

In the presented example Multi–Layer–Perceptrons (MLP) are used. To overcome the problem of the high controller input dimension (angles $\vec{\theta}$, angular velocities $\dot{\vec{\theta}}$, accelerations $\ddot{\vec{\theta}}$), the controller was structured into three separate networks, being responsible for torque–components from different physical sources. According to the robot dynamics (eqn.1), the networks learn the matrix components representing inertia and coupling forces M, coriolis and centrifugal forces C, and friction forces $\vec{f}$, respectively.

$$\vec{\tau} = M(\vec{\theta})\ddot{\vec{\theta}} + C(\vec{\theta})\dot{\Theta}\dot{\vec{\theta}} + \vec{f}. \tag{1}$$

$\vec{\tau}$ is the torque vector, M is an $[n \times n]$ matrix, C is a formal $[n \times n^2]$ matrix, and $\dot{\Theta}$ is an $[n^2 \times n]$ matrix with the vectors $\dot{\vec{\theta}}$ in the diagonal and zeros elsewhere; n is the number of joints [*].

A common approach [6] uses neural networks outside the feedback loop. In this case, given a neural controller after the training phase, stability is easy to guarantee.

On the other hand, such a control system is not very robust, because distortions are still acting on a nonlinear, coupled closed–loop transfer function. Furthermore, this method requires desired angular acceleration values at the input, which in many practical digital systems are difficult to obtain with sufficient resolution.

In contrast, the presented control system (Fig.1) employs the inverse model inside the feedback loop and thereby overcomes the above restrictions. Stability guarantee, however, becomes more difficult. We, for the first time, demonstrate stability for this efficient control structure.

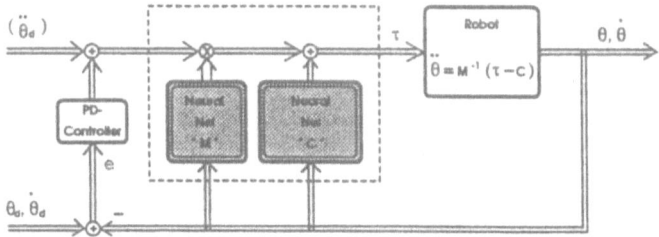

Figure 1: The neural dynamic control architecture: A structured control net inside the closed feedback loop. (The nets for compensation of static and viscose friction are not depicted)

---

[*]) For the sake of simplicity, gravity forces are not considered. They are not present in the planar 4JM, but could easily be introduced into all following considerations.

Training, Performance, and Stability

Using a small MLP with less than 700 weight parameters for the whole controller structure, the network was trained as direct inverse while the robot was PD–controlled according to Fig.2. The backpropagation (BP) learning rule was used, based on single–step errors.

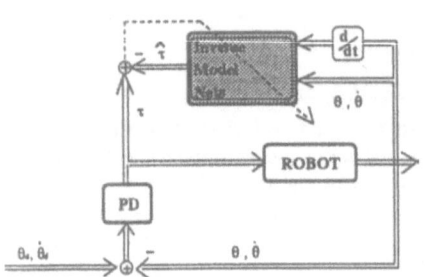

To improve interpolation properties, to minimize required memory, and to facilitate the subsequent stability test, a special "pruning" method for the network connections was implemented: each synapse carries a "life energy" corresponding to the low–pass filtered absolute value of the product of it's weight and it's input. A synapse "dies", if this value falls below a given threshold. A slight weight decay is also implemented. This simple local mechanism leads to the pruning of a synapse if it has constantly low weight or input values.

Figure 2: Training of a "direct inverse"

High tracking accuracy within a certain region was already achieved for repeated training with data from only one single trajectory. However, tracking accurary and stability could not be guaranteed so far over the entire defined working range.

To solve this problem, suitable random training trajectories were generated: the robot had to move to equally distributed random angular locations in its working range. Minimal–torque–change profiles were generated as desired trajectories between two random points. This training scheme meets two requirements for the training of an MLP as global inverse model: firstly, the training patterns are changing quickly enough in the network 's input space (i.e. $\vec{\theta}$) and secondly, they are well distributed in this space.

With such a training the model error turned out to become very small over the complete working range inspite of a small margin. This was numerically checked by systematic measurements of $M^{-1}(\vec{\theta})$ and $\hat{M}(\vec{\theta})$ on a regular 3-dimensional grid in the angular working range of the planar 4JM.

Input/Output Stability Guarantee after the Learning Phase

Using these measurements, L_∞–stability was analysed following [7]. Based on functional analytic methods the boundedness of the temporal maximum (∞–norm) of the output error vector norm, given the maximum input values, was established. The derivation is based on conditions for robustness against mismatch between the network estimation and the real matrix values. Only a brief outline of the mathematical derivation can be presented here: According to Fig.1 and eqn.1 the closed loop system dynamics are:

$$\ddot{\vec{\theta}} = M^{-1}(\vec{\theta}) \left(\hat{M}(\vec{\theta}) \left(\ddot{\vec{\theta}}_d + \vec{v} \right) + \Delta C(\vec{\theta})\Theta\dot{\vec{\theta}} + \Delta F\dot{\vec{\theta}} - \vec{f}_s \right), \tag{2}$$

M is the network estimation for M and ΔC and ΔF are the network approximation errors (estimated minus real value) for the coriolis matrix and the viscose friction (diagonal) matrix, respectively. $\vec{f}_s$ is the static friction. $\vec{v}$ is the output of a PID or PD feedback controller.

Eqn.2 now is expressed in the state error variables $\vec{e}_1 := \vec{\theta} - \vec{\theta}_d$, $\vec{e}_2 := \dot{\vec{e}}_1$ and the controller output $\vec{e}_3 := \vec{v} = R_1\vec{e}_1 + R_2\vec{e}_2$ (R_1 and R_2 are feedback controller gains) and an additive perturbation ψ:

$$\dot{\vec{e}}_2 = \vec{e}_3 + \vec{\psi}, \quad \vec{\psi} = \left(M^{-1}(\vec{\theta})\hat{M}(\vec{\theta}) - I\right)\left(\vec{v} + \ddot{\vec{\theta}}_d\right) + M^{-1}(\vec{\theta})\left(\Delta C(\vec{\theta})\dot{\Theta}\dot{\theta} + \Delta F\dot{\theta} - \vec{f}_s\right) \quad (3)$$

From the first part of this equation the closed loop disturbance transfer matrices $\mathsf{G}_i(p)$, defined by def.4 are easily calculated.

$$\vec{e}_i(p) = \mathsf{G}_i(p) \cdot \vec{\Psi}(p) \qquad i = 1, 2, 3 \qquad \vec{e}(p), \vec{\Psi}(p): \text{ Laplace transforms of } \vec{e}(t) \text{ and } \vec{\psi}(t). \quad (4)$$

The error ∞–norms can be established using $\qquad |\vec{e}_i(t)|_\infty \leq \beta_i \cdot |\vec{\psi}(t)|_\infty \qquad (5)$

where β is the 1–norm (i.e. the integral over time) of the spectral matrix norm $\|\mathsf{G}_i(t)\|$, and $\mathsf{G}_i(t)$ contains the impulse responses corresponding to the disturbance transfer functions $\mathsf{G}_i(p)$. (Possible dirac impulses at the beginning of these responses are considered in the integral). For matrix norm arithmetics see [8].

Now norm bounds on the disturbance $\vec{\psi}$ must be found. Applying the triangular inequality and Schwarz's inequality to the norm of the disturbance (eqn.3) leads to

$$|\vec{\psi}| \leq \|M^{-1}\hat{M} - I\|\left(|\vec{e}_3| + |\ddot{\vec{\theta}}_d|\right) + \|M^{-1}\|\left(\|\Delta C\|\left(|\vec{e}_2|^2 + |\dot{\vec{\theta}}_d|^2\right) + \|\Delta F\|\left(|\vec{e}_2| + |\dot{\vec{\theta}}_d|\right) + |\vec{f}_s|\right)$$
$$(6)$$

Substituting $\vec{\psi}$ in eqn.5 with these upper bounds yields an inequality system in the variables $|\vec{e}_1|_\infty$, $|\vec{e}_2|_\infty$, $|\vec{e}_3|_\infty$ and $|\vec{e}_2|_\infty^2$ that is solved for a quadratic inequality in $|\vec{e}_2|_\infty$. It determines upper bounds on $|\vec{e}_2|_\infty$, if the corresponding quadratic equation has zeros and $|\vec{e}_2|_\infty$ is initially smaller than the smaller one of the two zeros. The equation system further shows that if $|\vec{e}_2|_\infty$ is bounded, then $|\vec{e}_1|_\infty$ and $|\vec{e}_3|_\infty$ are bounded. Further considering the properties of β_i for suitable PD or PID controllers (Using design charts e.g. [4] or considering limit cases) it turns out that the condition for zeros can be met if upper bounds on $\|M^{-1}\|, \|\Delta C\|, \|\Delta F\|$ and $|\vec{f}_s|$ are found and if:

$$\boxed{\|M^{-1}(\vec{\theta}) \cdot \hat{M}(\vec{\theta}) - I\| < 1 \quad \text{for all } \vec{\theta}.} \quad (7)$$

Results

Fig.5 depicts the time course of the maximum and the mean value of this criterion on a grid in the 4JM working range following the above described training procedure. The critical value of 1 was first passed after 3500 positionings from one random point to another (ca. $40min.$ of simulated manipulator movement time). Values down to 0.4 have been reached in longer simulations. The density of the measurement grid was determined using estimates on the network derivatives $\partial \hat{M}/\partial \theta_k$, $k = 2, 3, 4$ and determining bounds on their effect on the deviation of the stability criterion values (eqn.7) between the measurement points. For example, few synapses with low values lead to few required measurements.

Fig.3 depicts the position accuracy with a simple PD–controller on a test trajectory, and Fig.4 shows the performance after learning of 3500 positioning movements. The test task was to "draw a house" (total time: 4s) with 30cm baseline, consistent of straight lines with trapezoidal velocity profiles.

Starting with the base joint, the 4 joint lengths and weights were: (15cm, 1.7kg),(15cm, 1.3kg),(30cm, 1kg) and (17cm, 0.65kg). The simulated robot had no gears. Sampling time was $2ms$.

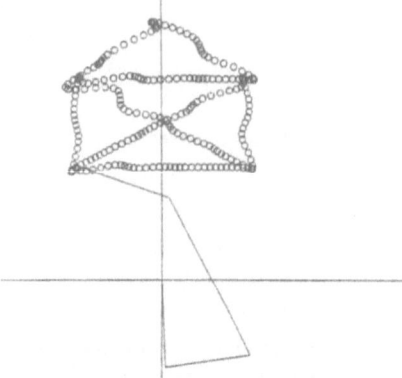

Figure 3: Tracking accuracy before training: mean square angle error $= 4.7 \cdot 10^{-5}$

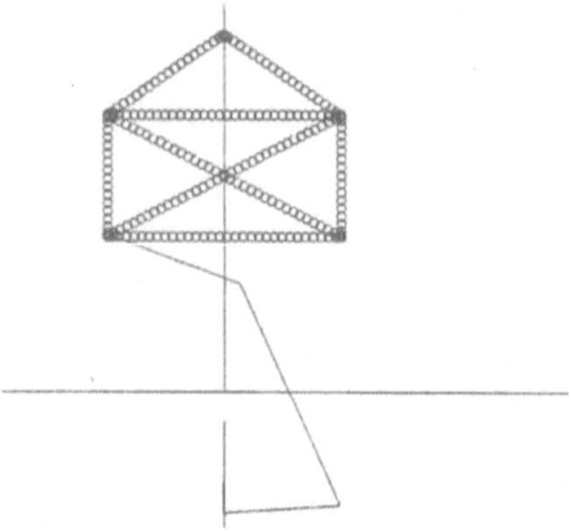

Figure 4: After training: mean square angle error $= 8.5 \cdot 10^{-9}$

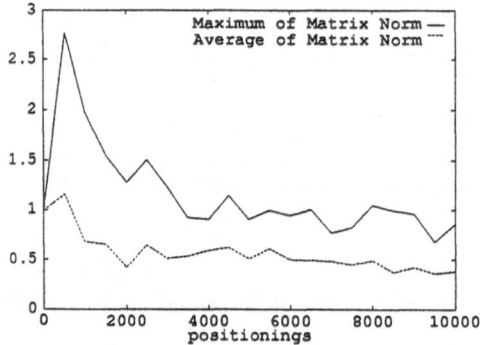

Figure 5: Maximum and average value of the stability criterion for all $\vec{\theta}$ in the working range. Values < 1 mean stability

Discussion

A set of Neural Networks was shown to achieve high model precision in the robot working range and thus to guarantee stable and precise control, when used as controller inside a feedback loop. Further work has to be done to reduce the number of required measurements, such that the analysis method is applicable to a real physical robot. New techniques to generate training trajectories for faster and monotonous convergence are investigated.

Conclusion

Taking additional mechanisms to treat unmodeled dynamics into account (like for all model-based control architectures), but profiting from the ease of achievement of a accurate global inverse model by training a Neural Network, the presented approach appears very promising for the application to control of high speed robots.

References

[1] J. J. Craig. *Introduction to Robotics: Mechanics and Control.* Addison Wesley, Reading, 1986.

[2] R. Eckmiller. J. Beerhold, G. Fahner. N. Goerke, J. Hakala, M. Jansen, B. Kreimeier, and H.W. Werntges. Neural network applications for robot control. In *Proc. of BMFT Status-Seminar Neuro-Informatics, Maurach, Oct 1992.* in Press 1993.

[3] E. Freund. Fast nonlinear control with arbitrary pole–placement for industrial robots and manipulators. *Int. J. Robotics Res.*, 1(1):65 – 78, 1982.

[4] W. M. Grimm. *Robustness Analysis and Synthesis of Model–based Robot Control*. Number 202 in VDI Fortschrittsberichte. VDI–Verlag, Düsseldorf, 1990.

[5] M.I. Jordan and D.E. Rumelhart. Forward models: Supervised learning with a distal teacher. Occasional Paper 40, MIT, 1989.

[6] M. Kawato. Feedback–error–learning neural networks for supervised motor learning. In R. Eckmiller, ed., *Advanced Neural Computers*, pages 365 – 372. Elsevier Science Publisher, Amsterdam 1990.

[7] M.W. Spong and M. Vidyasagar. Robust nonlinear contol of robot manipulators. In *Proc. 24th IEEE Conf. on Decision and Control, Ft. Lauderdale*, pages 1767–1772, 1985.

[8] M. Vidyasagar. *Nonlinear Systems Analysis*. Prentice Hall, Engelwood Cliffs, 1978.

Robot control—poster contributions

A Self-Organizing Neural Network for Robot Motion Planning*

Jules M. Vleugels   Joost N. Kok   Mark H. Overmars

Department of Computer Science, Utrecht University
Padualaan 14, P.O. Box 80.089, 3508 TB Utrecht, the Netherlands.

Abstract: The robot motion planning problem asks for determining a collision-free path for a robot moving amidst a fixed set of obstacles. In this paper we present an approach that combines an extended Kohonen network and deterministic techniques to solve this problem. The network constructs a graph of possible motions of the robot, which is then searched for a shortest path connecting given source and goal configurations. The method has been implemented and, compared to existing methods, achieved better results in typical planar motion planning problems. It can also be easily generalized to higher-dimensional spaces.

1 Introduction

Kohonen networks have recently been applied to various problems in the field of robotics, like the inverse kinematics problem [7] and the grasping problem [8]. A different problem that arises in the design of autonomous robots is the *motion planning problem*: given a robot R and an environment containing a fixed set of obstacles, find a path from some configuration *source* to some configuration *goal*, such that R can travel freely from *source* to *goal* without colliding with any of the obstacles. The motion planning problem has received considerable attention over the past years (see [5] for an overview of various techniques). However, there has so far been little research in the application of neural networks in this problem.

A solution to the motion planning problem can be constructed in either the *work space* of the robot (usually $\mathbb{R}^2$ or $\mathbb{R}^3$) or the *configuration space* consisting of all possible configurations of the robot. The configuration space is usually of higher dimension than the corresponding work space (e.g., a planar robot moving amidst pla-

nar obstacles using both translation and rotation has a three-dimensional configuration space of $\mathbb{R}^2 \times [0, 2\pi[$). Constructing a path in the configuration space however is often easier because this reduces the robot to a single point. Many existing approaches involve the construction of a representation of the *free space*, i.e., the subset of the space where the robot does not collide with any of the obstacles (this can be done in either the work space or the configuration space). The main drawback is that, due to the complexity of the free space, these approaches tend to be slow.

A growing network as described by Fritzke [4] could be used to learn a mapping of the free space by presenting only configurations that do not cause the robot to collide with any of the obstacles to the network. However, in experimenting with this approach we found that the number of nodes needed to cover the free space with a satisfactory precision becomes very large for more difficult problems (especially when dealing with three- or higher dimensional configuration spaces). This number of nodes can be reduced by having the network learn only parts of the free space that are useful for finding paths among the obstacles. E.g., the classifier network described in [6] could be used to approximate the boundaries of the obstacles. Our approach goes one step further by learning both a representation of the obstacle boundaries and a collection of possible paths between those boundaries. The way in which this is done takes advantage of the fact that in the motion planning problem, unlike most applications of Kohonen networks, it is usually possible to obtain information other than that given by the input vectors. This paper describes the basic ideas behind the method and gives some experimental results. For a more detailed description and more experimental evidence justifying our performance claims, we refer to [9]. For easier understanding, the algorithm is presented as if working in a two-dimensional work space; however, it works (practically) without modification for higher dimensional configu-

*This research was partially supported by ESPRIT Basic Research Action No. 6546 (project PROMotion) and by the Netherlands Organization for Scientific Research (NWO).

ration spaces. In fact, the results in section 3 have been obtained using a three-dimensional configuration space.

2 The network structure

Similar to [4] our network graph consists of triangles with one node in each corner, and is able to extend itself by adding nodes to its graph, allowing it to gradually increase its precision where necessary. With every node N_i a position vector w_i (indicating a configuration of the robot) and a classification label c_i (*safe* if N_i's position vector is located in the free space, *unsafe* otherwise) is associated. Nodes that are labeled *safe* represent configurations that are safe to place the robot in (*safe configurations* for short) and are intended to build a graph of paths along which the robot can travel. Similarly, nodes labeled *unsafe* represents configurations that cause the robot to collide with at least one of the obstacles; they will be used to approximate the obstacle boundaries. The network is initialized to a regular grid in the configuration space, where every node is initially given its correct classification. This can be determined with a general intersection test for planar polygons (i.e., the robot and the obstacles).

In the learning phase the network receives random input configurations of the robot; with every input its correct classification is given (again this can be determined with an intersection test). The different behavior of the two kinds of nodes is achieved by allowing only safe nodes to compete for unsafe inputs and vice versa. The best-matching node N_k (which we will call the *bmu*) then adjusts its position vector w_k to the input v according to Kohonen's learning rule:

$$w_k^{new} = w_k^{old} + \varepsilon(v - w_k^{old}) \qquad (1)$$

Also the nodes in a small neighborhood of N_k are allowed to adjust their position vectors by a small amount. In both cases, adjusting a position vector w_i is allowed only if this does not violate N_i's classification c_i; w_i is not adjusted otherwise. The learning parameter ε depends on the label of N_i as follows:

- if $c_i = safe$ we take $-1 < \varepsilon < 0$ causing the node to be 'pushed away' from the input which is located in an obstacle. The safe nodes will therefore move to positions away from the obstacles. The connections

between safe nodes can then be seen as possible paths along which the robot can travel.

- if $c_i = unsafe$ we take $0 < \varepsilon < 1$, thus attracting the node towards the input which is located in the free space. Since the unsafe nodes are not allowed to violate their classification, they will settle to positions on the obstacle boundary.

However, two problems arise when doing this:
1. If an unsafe node N_i is located in the middle of an obstacle and is 'surrounded' by other unsafe nodes that are located in the same obstacle, it will have a small chance of winning a competition because for almost every input v there is a different node N_j nearer to v than N_i is (since v is located outside the obstacle). Therefore its weight vector w_i will never be adjusted. This problem does not occur with the safe nodes since the probability of a safe node winning a competition is decreased by adjusting its position vector. The sketched phenomenon is well-known with Kohonen networks. To prevent this, we equipped the unsafe nodes with a conscience mechanism [3], causing nodes that win the competition relatively often to feel 'guilty' about this.

2. The safe nodes are pushed away from the input vectors. Simply applying this however would (in the end) move all safe nodes to a single position: the configuration that is furthest from all obstacles. To prevent this, the neighbors of a safe node produce an additional repulsive force on it.

Since locally a high precision can be required to move the robot through small passages, nodes are added to the network during the learning phase where necessary. This is done using two heuristic arguments:

1. As in [4], each node N_i has an *error* variable e_i attached to it. Every time w_i is adjusted, the distance $|w_i^{new} - w_i^{old}|$ is added to e_i. After n learning steps, the node with the highest error value has a new neighbor added to it (*error-based adding*). Unfortunately, adding nodes this way turned out not to be sufficient for this problem. As already mentioned, the network can obtain additional information from the scene by classifying (arbitrary) configurations. This information can then be used in the second way of adding nodes:

2. When a node N_k is *bmu*, a small number of configurations $s_1, s_2, \ldots, s_m$ ($m = 3$ in our imple-

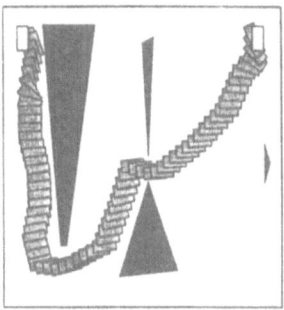

Figure 1: Three different test scenes.

mentation) between N_k and each direct neighbor N_i with the same classification as N_k are sampled. A new node is added if the classification of some s_j is different from c_k (*scene-based adding*); this means that a connection between two safe obstacles cuts through an obstacle, or that a connection between unsafe nodes is partly located in the free space.

When the learning phase is finished, the safe nodes (along with their connections) can be regarded as a road map of possible paths along which the robot can travel. The unsafe nodes are used to extend this map by moving them across the obstacle boundaries and into the free space. This is done by adding to each w_i the average of w_j for all of its direct safe neighbors N_j. The effect is that the unsafe nodes will become located somewhere between its direct safe neighbors (which are located in the free space), thus representing additional configurations at which the robot can safely be placed.

The resulting road map can then be searched for a path $source = N_{i_1}, N_{i_2}, \ldots, N_{i_n} = goal$ connecting a given *source* and *goal* configuration. In addition, the shortest path (according to the network) can be found using a slightly modified A* algorithm [2]. However, since there is no guarantee that a connection between two safe nodes will not cut through an obstacle, a slightly blown-up copy of the robot is placed at a large number of successive configurations on every edge of the path thus found. If any such placement between two safe nodes N_i and N_j causes a collision, the graph is searched again for a different path not containing that particular connection. An example of part of the network after 1000 learning steps is shown in figure 2. *Unsafe* nodes are drawn in white, *safe* ones in black.

3 Experiments and results

The algorithm has been implemented on a Silicon Graphics Indigo workstation (based on a R3000 processor with a performance of 25.4 SPECMARKS). Although the basic idea is very simple, a lot of effort has been put in improving the actual implementation. The program was optimized with regards to the data structure representing the network (the nodes are stored in an uniform hashed K-d tree based on [1], which allows for determining the best-matching node in expected logarithmic time) and collision-checking between the robot and the obstacles (this is done with an optimized library of geometric routines developed at our department; it is ftp-able from archive.cs.ruu.nl). However, little work has so far been spent on optimization of the program code; we expect that doing so will result in a considerable increase in speed.

The program was tested extensively on a large number of motion planning scenes which tend to be difficult for many approaches. Figure 1 shows three such scenes, along with a solution found by the program. The first scene is relatively simple. A rectangular robot has to move from the top left to the top right around a few obstacles; it can only move through the passages in the middle at a certain angle. The path took 3.4 seconds to compute, using an initial network of 400 nodes. The second scene is more complicated. An L-shaped robot has to move from the bottom left to the top right, avoiding the closely spaced obstacles. The scene requires a lot of rotation and many passages are narrow. The path was computed in 11.3 seconds, using an initial network of 600 nodes to account for the higher precision needed. The third scene consists of a triangular robot moving amidst a large number of small obstacles. The path was computed in 5.7 seconds, again using an initial network of 600 nodes because the robot has little freedom to move.

To test the learning behavior of the network, 100 source and goal configurations were chosen at random. The *success rate* for a scene is defined as the relative number of configurations between which the program is able to construct a path. Figure 3 shows the success rate for the three scenes shown in figure 1, set against the number of learning steps. The results for the leftmost scene are indicated with a dotted line. The networks performance very quickly converges towards a success rate of 1.0, indicating it perfectly 'knows' the whole scene. The performance for the second scene is shown with a dashed line. Again the network converges quickly, but towards a success rate of only 0.95. This can be explained by the fact that the scene contains configurations from which the robot cannot escape (e.g., the small part of free space in the upper right corner). The same applies to the third scene where the performance converges towards approximately 0.95. In this case, the high complexity somewhat lessens the convergence speed.

An advantage of combining the network with a graph search algorithm to find a shortest path is that the resulting motions are in general very short and smooth. This is already illustrated by the motions shown in figure 1.

4 Conclusions

We described a new approach to solving the motion planning problem. An extended Kohonen network is combined with algorithmic techniques to approximate a road map between a given set of obstacles, which is then used to determine a shortest path connecting given configurations. The method has been implemented and results show that, compared to conventional techniques [5], it is faster and yet renders short and smooth paths. In addition, it can be easily generalized to higher-dimensional spaces.

References

[1] J.L. Bentley. Multidimensional binary search trees used for associative learning. *Comm. ACM*, 18:509–517, September 1975.

[2] T.H. Cormen, C.E. Leiserson, and R.L. Rivest. *Introduction to Algorithms*. MIT Press. 1989.

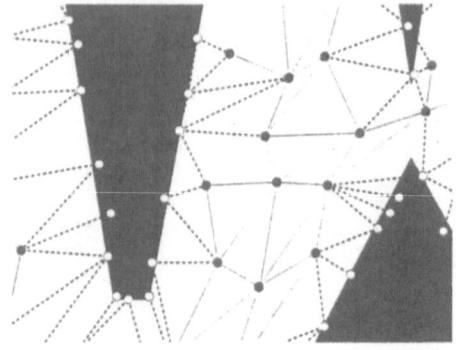

Figure 2: An intermediate state of the network.

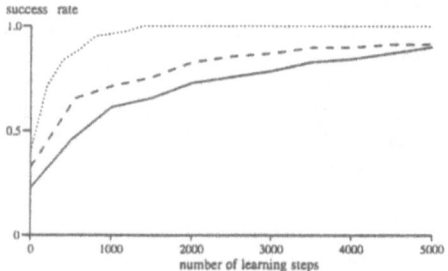

Figure 3: Some experimental results.

[3] D. Desieno. Adding a conscience to competitive learning. In *Proc. Int. Conf. Neural Networks*, volume 1, pages 117–124. IEEE Press, July 1988.

[4] B. Fritzke. Let it grow — self-organizing feature maps with problem dependent cell structure. In *Artificial Neural Networks*, volume 1, pages 403–408. North-Holland, Amsterdam, 1991.

[5] J.-C. Latombe. *Robot Motion Planning*. Kluwer Academic Publishers, Boston, 1991.

[6] P. Morasso, A. Pareto, and V. Sanguineti. SOC: A self-organizing classifier. In *Artificial Neural Networks*, volume 2, pages 1223–1226. Elsevier Science Publishers, 1992.

[7] H.J. Ritter, T.M. Martinetz, and K.J. Schulten. *Neural Computation and Self-Organizing Maps*. Addison-Wesley, 1992.

[8] P. van der Smagt, B.J.A. Kröse, and F.C.A. Groen. A self-learning controller for monocular grasping. In *Proc. 1992 IEEE/RSJ Int. Conf. Intelligent Robots and Systems*, pages 177–182. Raleigh, N.C., June 1992.

[9] J.M. Vleugels, J.N. Kok, and M.H. Overmars. Robot motion planning using a self-organizing neural network. Technical report, Dept. Computer Science, Utrecht University, to appear.

EVOLVED RECURRENT DYNAMICAL NETWORKS USE NOISE

Dave Cliff [1,2] and Inman Harvey[1] and Philip Husbands[1]
[1]School of Cognitive and Computing Sciences
[2]Neuroscience IRC, School of Biological Sciences
University of Sussex, BRIGHTON BN1 9QH, U.K.
davec or philh or inmanh, all @cogs.susx.ac.uk

Abstract

We report on studies of recurrent dynamical artificial neural networks developed using artificial evolution (i.e. an extended genetic algorithm). The networks control a simulated visually guided robot, which is an accurate model of a real robot. The genetic algorithm generates useful network architectures from an initial set of randomly-connected networks. During evolution, uniform noise was added to the activation of each neuron. After evolution, we studied two evolved networks, to see how their performance varied when the noise range was altered. Significantly, we discovered that when the noise was eliminated, the performance of the networks degraded significantly: the networks *use* noise to operate efficiently.

1 Introduction

This paper discusses studies of recurrent dynamical artificial neural networks, developed using artificial evolution (i.e. an extended genetic algorithm). The architecture of the networks is under control of a genetic algorithm: both the connectivity of the network, and the number of "hidden" units[1] are under genetic control. The genetic algorithm starts with an initial set of randomly-generated network architectures, and produces useful networks after a number of "generations". The methods we employ allow for arbitrary connectivities, and the evolved networks are often highly recurrent. Our neuron model uses real-valued activations in the range $[0, 1]$, and includes delays on connections between units – so the networks can be considered as continuous dynamical systems.

We simulate these recurrent dynamical networks using fine-time-slice approximation techniques. Also, because both biological neurons and hardware-implemented artificial neurons are limited-precision devices, we add uniform noise from a distribution $[-n, +n] : n \in \mathbf{R}$ to the activation of each unit in the network on every time-step. We refer to this injected noise as *internal noise*, to distinguish it from other sources of noise in the system, discussed further below. During evolution of the networks, the internal noise distribution was constant (i.e. n was fixed at a set level). Once the evolutionary process had produced useful networks, we examined how the performance of the networks altered as n was varied. We found that some networks were fairly resilient to increases in n (i.e. to greater ranges of internal noise distribution), while other networks exhibited a rapid loss of performance as n increased. Most significantly, we found that in the special case $n = 0$, i.e. in the absence of internal noise, the performance of the networks dropped significantly. From this we infer that the evolved networks *used* noise in order to operate at maximum efficiency. This may be surprising to some readers, given that noise is often seen as a problem, rather than a feature.

The networks were evolved to control an accurate simulation of a wheeled robot built at Sussex. The real robot has a number of tactile sensors (bumpers and whiskers), and has two independent drive wheels, with a third free-wheel castor giving tripod stability. The wheels can rotate either forwards or reverse, and at full or half speed. The wheels are configured so that the robot can rotate on the spot, if required. The simulation model also includes two simple oriented photo-receptors, which could readily be constructed from discrete components such as phototransistors or light-dependent resistors. The photoreceptors each give a signal proportional to the amount of light in their "field of view", and are referred to as the left and right "eyes". In addition to the internal noise, the simulation model introduces noise in the sensors, motors, and kinematics (e.g.

---

[1]A *hidden unit* is one not directly involved in input/output to/from the network.

in modelling the actions occurring when the robot bumps into an object). In this paper, only the effects of varying the internal noise are examined.

The genetic algorithm is used to create networks which control the robot, and make it perform visually-guided tasks. Because it only has two simple eyes, only simple tasks can be performed. The networks discussed in this paper guide the robot to the centre of a circular arena which has white floor and ceiling, with black walls. Current research is directed at extending our work so that more complex visual sensors can be used, allowing for more sophisticated visually-guided tasks to be performed.

The rest of this paper first introduces the neuron model used in our networks, before discussing the analysis of varying the noise level. It is beyond the scope of this paper to give full details of the genetic algorithm employed, or of the simulation system. However, these issues are fairly unimportant in the context of studying the final evolved networks. For an introduction to genetic algorithms, see [4], and for details of the genetic algorithm employed in our work see [5]. For details of the robot simulation see [3], and for details of the background to this work, see [7, 6]. For further analysis of the two networks studied below, see [2]. This paper is abridged from [1].

2 The Neuron Model

The neuron model we have employed in our work to date has separate channels for excitation and inhibition. Values propagate along links between units, and are all real numbers in the range $[0, 1]$. All links are subject to a delay Δt. Unusually, the inhibition channels operate as a 'veto' or 'grounding' mechanism: if a unit receives *any* inhibitory input, its excitatory output is reduced to zero (but it can still inhibit other units). Excitatory input from sensors or other units is summed: if this sum exceeds a specified veto threshold t_v, the unit produces an inhibitory output. Independently, the sum of excitatory inputs has uniform noise (distribution: $[-n, +n] \in \mathbf{R}$) added, and is then passed through an excitation transfer function, the result of which forms the excitatory output for that unit, so long as the unit has not been inhibited.

More formally, the excitation transfer function T takes the form:

$$T(x) = \begin{cases} 1 & \text{if } x \geq t_u \\ 0 & \text{if } x \leq t_l \\ (x/(t_u - t_l)) - (t_l/(t_u - tl)) & \text{otherwise} \end{cases}$$

where t_l and t_u are lower and upper threshold levels. The veto output function $\mathcal{U}$ takes the form:

$$\mathcal{U}(x) = \begin{cases} 1 & \text{if } x \geq t_v \\ 0 & \text{otherwise} \end{cases}$$

where t_v denotes the veto threshold, and the veto input function $\mathcal{V}$ is:

$$\mathcal{V}(x) = \begin{cases} 0 & \text{if } x > 0 \\ 1 & \text{otherwise} \end{cases}$$

Because there are separate excitation and inhibition channels, two connectivity matrices are required: a veto matrix V (where each element $v_{i,j}$ indicates whether unit i vetoes units j), and an excitatory matrix W, with elements w_{ij}. Then, if $o_{ej}(t)$ and $o_{vj}(t)$ respectively denote the excitatory output and veto output from unit j at time t, and $N_j(t)$ denotes the internal noise injected to unit j at time t, the output channels from unit j can be expressed as:

$$o_{vj}(t) = \mathcal{U}(\sum_{\forall i} w_{ij} e_{vi}(t - \Delta t))$$

$$o_{ej}(t) = \mathcal{V}(\sum_{\forall i} v_{ij} o_{vi}(t - \Delta t)) \cdot T(N_j(t) + \sum_{\forall i} w_{ij} o_{ei}(t - \Delta t))$$

In most of our work, we have used: $t_l = 0.0$; $t_u = 2.0$; $t_{io} = 0.75$; and noise $n = 0.1$; for all units. The dynamic properties of these units are simulated using asynchrous fine time-slice approximation techniques, with random variations in time-cycling to counteract periodic effects. Significantly, we have found that this neuron model is sufficiently sophisticated that there has been no need to introduce variable weights on the links between units, or variable delays: in all our work, weights and delays are all set at one, for all links in the network. Nevertheless, we are actively investigating the use of placing such parameters within evolutionary control.

3 Varying the Noise

The typical behaviour of a robot controlled by an evolved network is that it finds its way to the centre of the circular arena, and then stays there by spinning on the spot. This behaviour was evolved by assigning each network a "fitness value"; a scalar value $\in \mathbf{R}$ indicating how long the robot spent at or near the middle of the arena. As is demonstrated in [3], this is sufficient to evolve controllers for visually-guided behaviours: no explicit specification of visual processing is required. The robots are evaluated by positioning them at a random orientation and location in the world. The random process governing the starting location is biased so that the robots always start close to the walls of the arena. Once at the initial position, each robot is monitored for the same fixed amount of time, during which its fitness value is calculated. For further details of this process, see [2]. For the purposes of this discussion, it is sufficient to note that, if the robot spent all its time at the centre, it would receive a score of 100. But, because each robot's randomly chosen initial position is always some distance from the centre, this maximum score can never be reached: an optimum controller would score about 85 points. A robot which never moved would score less than one.

We examined the performance of two of the highest scoring control networks,[2] referred to as C1 and C2: after 100 generations of evolution, both networks managed an *average* score of around 65 (peak scores were nearer 80). These are the scores obtained with $n = 0.1$. Both networks were then tested with different values of n, varying from $n = 0$ (i.e. no noise) to $n = 1.0$ (i.e. noise uniformly distributed in the range $[-1.0, 1.0]$). For each value of n, the network was evaluated 80 times, and the average score taken. Results from these tests are shown in Figure 1.

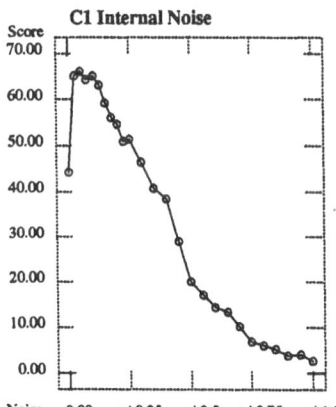
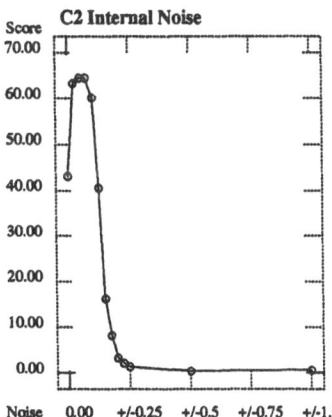

Figure 1: Results from testing with varying noise distributions. Left graph is for the C1 controller network; right graph is for the C2 network. Ordinate is average performance score over 80 trials. See text for discussion.

As can be seen from the graphs, both controllers show their peak performance close to $n = 0.1$, which was the value used in evolution. The performance of C1 degrades fairly gracefully as n increases, whereas C2 rapidly fails. However, a much more significant observation is the fact that both controllers exhibit a significant loss of performance when $n = 0$; this is discussed further below.

---

[2]Space restrictions prevent illustration of the networks here: see [1] for further details.

4 Discussion and Conclusion

Further examination of the results indicates that the drop in performance when noise is eliminated is due to the recurrent dynamical nature of the networks: the recurrency implies that the network architectures contain feedback loops at a number of levels. That is, it is common to see a unit with connection(s) to itself, or two mutually excitatory units, or cycles of excitatory links incorporating several units. In such cases, low levels of internal noise may build up over time by a process of accumulation through feedback loops. However, because the noise distribution is centred on zero, it is also possible that these high levels of activity could then be driven downwards by injections of negative noise: therefore, when noise is injected into networks of the type we use, so-called "drunkard's walk" phenomena emerge; where activation values 'wander' between upper and lower bounds; and in certain feedback configurations, circuits with quasi-periodic oscillatory activity can be seen to evolve. For theoretical analysis supporting these arguments, see [8].

So, when the noise is eliminated, any parts of the neural circuitry which act as accumulators or oscillators in the presence of noise will be rendered ineffective, and the performance of the controllers is consequently impaired. In both the C1 and C2 networks, the drop in *average* performance was due to an increase in the number of near-zero scores: peak scores were still high, but under certain conditions the absence of noise allowed the activity in the network to fall to such an extent that the robot was rendered immobile. Put more formally, the noise helps the state trajectory of the controller system from becoming trapped on attractors which correspond to inactivity or unproductive behaviours. In this sense, it is realistic to describe the networks as *using* noise to produce useful behaviours.

References

[1] D. T. Cliff, I. Harvey, and P. Husbands. Incremental evolution of neural network architectures for adaptive behaviour. Technical Report CSRP 256, University of Sussex School of Cognitive and Computing Sciences, 1992.

[2] D. T. Cliff, P. Husbands, and I. Harvey. Analysis of evolved sensory motor controllers. Technical Report CSRP 264, University of Sussex School of Cognitive and Computing Sciences, 1992.

[3] D. T. Cliff, P. Husbands, and I. Harvey. Evolving visually guided robots. In J.-A. Meyer, H. Roitblat, and S. Wilson, editors, *Proceedings of the Second International Conference on Simulation of Adaptive Behaviour (SAB92)*, pages 374–383. MIT Press Bradford Books, Cambridge, MA, 1993.

[4] D. E. Goldberg. *Genetic Algorithms*. Addison Wesley, 1989.

[5] I. Harvey. Species adaptation genetic algorithms: A basis for a continuing SAGA. In F.J. Varela and P. Bourgine, editors, *Towards a Practice of Autonomous Systems: Proceedings of the First European Conference on Artificial Life (ECAL91)*, pages 346–354. M.I.T. Press — Bradford Books, Cambridge MA, 1992.

[6] I. Harvey, P. Husbands, and D. Cliff. Issues in evolutionary robotics. In J.-A. Meyer, H. Roitblat, and S. Wilson, editors, *Proceedings of the Second International Conference on Simulation of Adaptive Behaviour (SAB92)*, pages 364–373. M.I.T. Press — Bradford Books, Cambridge MA, 1993.

[7] P. Husbands and I. Harvey. Evolution versus design: Controlling autonomous robots. In *Integrating Perception, Planning and Action, Proceedings of 3rd Annual Conference on Artificial Intelligence, Simulation and Planning*, pages 139–146. IEEE Press, 1992.

[8] P. Husbands, I. Harvey, and D. T. Cliff. Analysing recurrent dynamical networks evolved for robot control, 1993. To Appear in: *Proceedings IEE conference on artificial neural networks (ANN93)*, Brighton, 1993. Also available as University of Sussex School of Cognitive and Computing Sciences Technical Report CSRP265.

The Bellmann Mapping Machine for Nonlinear Approximation in Control Policy Space

G. Fahner, R. Eckmiller
University of Bonn
Dept. of Computer Science VI (Neuroinformatics)
FR Germany
Tel.: ++49-228-550-364   FAX: ++49-228-550-425   e-mail: gerald@nero.uni-bonn.de

Abstract We propose a novel scheme, named 'Bellman Mapping Machine' (BMM), that aims to extend the scope of reactive robot controllers towards more complex tasks. BMMs are implemented by shallow feed forward networks, that receive as input the compound information about desired action task, present robot state, and short-term predicted cluttered constraints.

The street-crossing problem serves as a test-bed for our performance studies: the task there is to generate optimal goal-directed robot motor trajectories, that avoid collisions with moving obstacles, at the same time respecting the robot's dynamics limitations. We supervise some novel higher order neurocontroller with optimal control examples as computed by Dynamic Programming. We find very efficient representations of the underlying optimal control policy space, as well as sensible generalization to new control situations.

1. Introduction Flexible and robust control of robots acting in rapidly changing environments must give special emphasis to real-time sensorimotor integration. Computationally shallow, inexpensive reactive system designs are preferable over iterative global search methods, particularly when robot position constraints become time-variant in the presence of moving obstacles.

A branch of reactive system designs suited for obstacle avoidance emanates from the idea of furnishing Euklidian space with spatial potential fields [5]. Along with the representational simplicity of these schemes there go however limitations, such as emergence of local minima for entangled obstacle constellations. In addition, spatial field representations fail to take into account dynamics limitations of the robot, and lack temporal reasoning qualities. Every-day experience in crossing busy streets, as an example, makes obvious, that overcoming these representational inadequacies would distinctly enhance the survivability of inert robots exposed to speedy obstacles.

A profound demand on robot control is formulated as the 'kinodynamic path planning problem' [2]: given a robot system, and some desired goal, find a cost-optimal trajectory, that avoids moving obstacles, while respecting bounds on robot velocities and accelerations.

Can this formidable task be handled by shallow circuitry at all? And are feed forward nets powerful enough to represent the underlying control policy spaces? Or does the temporal dimension of the task demand on relaxation-like controller designs, equipped with internal feedback loops and sequential processing capabilities [6, 7]?

In this paper we try to give partial answers to these questions, largely based on simulation results. For the task discussed above, we implement the BMM as an adaptively structured, nonlinear feed forward neural net that serves as an approximator of the optimal control policy mapping. We investigate the capacity of these distributed, and sparse representations, to *generalize from control rule examples*. In section 2 of this paper we introduce robot and environment models. The proposed sensorimotor system design and functionality is described in section 3. Section 4 discusses our method to obtain optimal control examples by means of Dynamic Programming (DP). The neurocontroller model is then introduced. In section 5, results of extensive simulations are reported. The paper concludes with a discussion of the capabilities and limitations of the proposed approach.

2. Environment and Robot Models The world around the robot is a two-dimensional scene, occupied by square obstacles moving all in parallel to the y-axis, with randomly choosen discretized x-positions. and with a continuous velocity spectrum. The environment's state is given by a list reporting position, and velocity of each obstacle i. The environment dynamics is given by

$$y_i(t+1) \;=\; y_i(t) + v_i \,. \tag{1}$$

A point-like robot of unit mass confined to some interval along the x-axis, and obeys a discretized position-/velocity spectrum: $\mathcal{X} \in \{0, ..., 8\}$; $\dot{\mathcal{X}} = \{-1, 0, 1\}$. At each time step, a motor command $u \in \ddot{\mathcal{X}} = \{-1, 0, 1\}$ is applied to the robot. Dynamic equations are given by

$$\begin{aligned}
\dot{x}_r(t+1) &= \dot{x}_r(t) + u(t) \\
x_r(t+1) &= x_r(t) + \dot{x}_r(t+1) \,.
\end{aligned} \tag{2}$$

Notice that the set of admissible motor commands depends on the present robot state.
The above settings for our test-bed sketch a robot with limited dynamics, which has to plan ahead in time, in order to avoid fluctuating numbers of obstacles, that cross its baseline in ever new constellations. The situation is similar to that of a pedestrian crossing a busy street (Figure 1).

3. Sensorimotor System Design and Functionality BMM receives as input the compound information about desired action task, present robot state, and cluttered spatiotemporal constraints as extending over some limited future time interval (Figure 2). The latter information is not immediately availabe from the environment. In [4], a pre-processing device, denoted there as Perception Module, that implements the necessary environment model, was discussed. From sensory information it computes short-term forecasts of future obstacle positions. In effect, it assembles some robo-centric constraints vector, whose the components label the occupancy state of those spacetime cells, that are within reach for the robot within a finite planning horizon.
BMM acts as an inverse model within a closed-loop control cycle. It generates state-dependent robot motor accelerations, that aim to move the robot towards the desired goal position as specified by the action task input, while respecting the constraints vector.

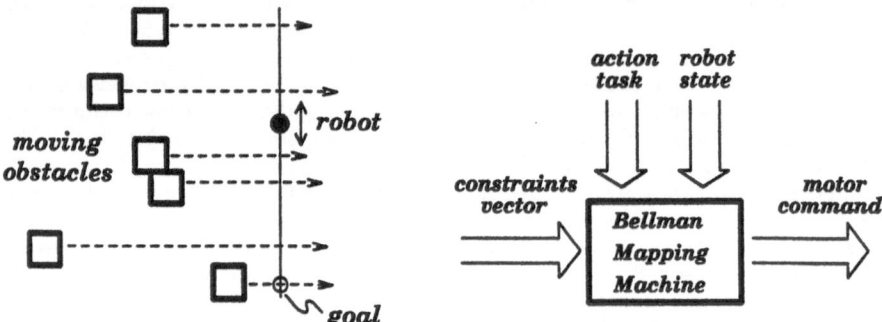

Figure 1: Street-Crossing Problem Figure 2: Bellman Mapping Machine

4. Computation and Approximation of Bellman Machines Firstly, we realize computation of the optimal control policy by DP [1]. Secondly, we use supervised learning to distribute examples of this 'Bellman mapping' over the neural BMM.
At every timestep t, DP determines a sequence of motor commands minimizing some cost functional. Here we use the quadratic finite-horizon version:

$$cost_{\{u(t),...,u(t+HORIZON)\}} = \sum_{k=0}^{HORIZON} (x_r(t+k) - x^o)^2 + c\, u(t+k)^2 \,, \tag{3}$$

with $x_r(t+k)$ given by the dynamics eqns.(2). By x^o, we denote the desired robot position. Deviations from this, as well as costly accelerations, are punished by higher costs. Collision-free progress of the near-future solution trajectory is guaranteed by restricting search to cells labelled 'free' as in the constraints vector. Training targets are constituted by that optimal present motor actions $u^{opt}(t)$, for which the minimum is attained in eqn.(3).

BMM is implemented as a single layer of parsimonious Higher Order Neurons (parsiHONs) [3], computing outputs $y_i \in (0,1)$; $i = 1,2,3$. Recently, parsiHON classifiers with high degree of sparsity were shown to emerge most efficient representation, and robust generalization, for difficult nonlinear benchmark problems.

Target values for each single neuron are given by $y_i^{des} = 1$, if motor-action i is the optimal policy, otherwise, $y_i^{des} = 0$. At recall time, the motor command of choice is labelled by the index of the most active neuron. As input, each neuron receives a bit-vector $\mathbf{x} = x_1, ..., x_N \in \{-1,1\}^N$, the components of which encode the compound information as enumerated above, in a suitably discretized version. The single motor neuron's estimate is given by some nonlinear expression

$$y^{est}(\mathbf{x}) = sigmoid \left[\sum_{\beta} w^{\beta} \eta_{\beta}(\mathbf{x}) \right] ; \ w^{\beta} \in I\!R , \ \text{with} \ \eta_{\beta} \equiv \prod_{j=1}^{N} x_j^{\beta_j} \ ; \ \beta = \beta_1 ... \beta_N \in \mathcal{R} \equiv \{0,1\}^N ,$$

where the sum computes an expansion containing some modest number of terms that model higher order interactions among inputs. By means of structural adaptation learning, as discussed in the above reference, each neuron is independently adapted to examples of the Bellman mapping, thereby identifying some (hopefully) sparse set of *relevant* higher order dependencies as present in the mapping.

5. Simulation Results The neurocontroller was trained with respect to two alternative long-term goals ($x^o \in \{0,8\}$), such that one bit was sufficient to encode the external task switch. Examples of optimal DP control actions were assembled from a run over several thousend time steps of the simulated environment (fairly crowded with moving obstacles). At each time step t, optimal controls $u^{opt}(t)$ were computed for any possible robot state (9 positions × 3 velocities, encoded by 6 bits), and for both desired goal positions. Spatio-temporal constraints on near-future robot positions were represented by spacetime cells, labelled -1, or 1, according to their predicted future occupation states. This representation was defined exactly on that cone-shaped cell region that was in reach for the robot within the short-term horizon chosen: $HORIZON = 3$. Taking into account that the maximum speed of the robot was one cell per time step, spatio-temporal constraints imposed during the time-interval $[t+1, t+3]$ were represented by a constraints vector of dimension $3+5+7$. Taking all the compound information together, the input dimension for the neurocontroller was 22 bits.

We generated a data base containing about $20,000$ unique optimal control examples, as computed by DP for the robot in general states. Several training runs were performed with parsiHONs of sizes between 83 and 110 terms (or rather parameters), for up to $14,000$ training examples. In all cases, we limited training to a maximum of $1,000$ epochs. In most cases, this was sufficient for successful training set classification for any of the three neurons ($y_i < .2$ for $y_i^{des} = 0$, and $y_i > .8$ for $y_i^{des} = 1$; $i = 1,2,3$). In the less successful cases, just a few errors occured, and additional robustness stemming from the winner-takes-all decision rescued fault-freeness of the majority vote. Our findings show, that the optimal policy mapping can be represented with extreme sparsity by the chosen network model.

To test generalization of the neurocontroller, we employed the unused part of the data base, the patterns of which describe obstacle constellations not present in the training set. Extensive simulations with varying training set sizes reveal a distinct decrease of test error with increasing training set size. Very impressively, for a load of $10,000$ training patterns, the networks policy estimates on unseen patterns differed from the true optimal policy for just about 1 times out of 100, at an average. These results corroborate, that sensible generalization emerges from sparse modelling of the Bellman mapping.

To get some insight into the amount of nonlinearity present in the problem, we counted the number of terms which carry a given order. The resulting distribution has its maximum at order 3, exhibits many terms of orders 4, 5, 6, and higher, and finally decreases to zero for orders exceeding 10. These findings hint on the high-order character of the Bellman mapping for the kinodynamic path planning problem.

5. Conclusions Sparse representation of control laws is desirable when table look-up becomes impracticable (Bellman's 'curse of dimensionality'), and when iterative computation of optimal policies becomes too expensive, or conflicting with real-time requirements. Some mechanism of generalization, which turns already acquired control skills over to new task instances, can distinctly improve the survivability of sensory driven robots. For these reasons it is urgent to investigate the competence of neurocontrol for efficient distributed representation, and for robust generalization of optimal control policies.

Here, we focused on a new type of shallow feed forward neurocontroller for local kinodynamic trajectory planning. An advantage with feed forward nets is their low-latency recall, and their relatively quick learning, as compared to recurrent networks. However, from theoretical considerations concerning the non-local nature of the related connectedness predicate [6], the problem under focus is expected to be hard for feed forward nets, when scaled up. Even for limited time-horizons, complex, nonlinear, and jumplike optimal control policies must be faced, due to constraint-induced bifurcations of optimal phase-space trajectory bundles. We met the required mapping complexity with a powerful novel classifier model supporting effective computation, and automatic identification, of the relevant nonlinearities inherent in the mapping. We found extremely parsimonious distributed representations of optimal control policies, indicating that some compact set of important high-order features determines the optimal control. The neural BMMs emerged excellent generalization to new control task encounters.

We encourage use of feed forward neurocontrol for approximation of Bellman mappings obeying cluttered constraints, but care must be taken that the models support efficient representation of high-order nonlinearities. For growing time-horizons, it is expected that feed-forward neurocontrol will run into limitations [7]. Some deficiency of our approach is its burden with increasing constraints vector dimension for growing planning-horizons. This objection is weakened, however, when considering partially unmodelled natural environments, where long-term planning is not feasible due the absence of globally disposable constraints.

References

[1] R. E. Bellman (1957). *Dynamic Programming*. Princeton University Press.

[2] B. Donald (1989). *Near-Optimal Kinodynamic Planning for Robots With Coupled Dynamic Bounds*, Proc. IEEE Int. Conf. on Robotics and Automation, pp. 958-969.

[3] G. Fahner, N. Goerke, R. Eckmiller (1992). *Structural Adaptation of Boolean Higher Order Neurons: Superior Classification with Parsimonious Topologies*, Proc. ICANN, Brighton, UK, Vol. 1, pp. 285-288.

[4] G. Fahner, and R. Eckmiller (1992). *Learning Spatio-Temporal Planning from a Dynamic Programming Teacher: Feed-Forward Neurocontrol for Moving Obstacle Avoidance*. In Giles, C. L., Hanson, S. J., and Cowan, J. D. (eds.), Advances of Neural Information Processing Systems 5. San Mateo, CA: Morgan Kaufmann Publishers.

[5] Khatib (1985), *Real-time Obstacle Avoidance for Manipulators and Mobile Robots*, Proc. IEEE Int. Conf. on Robotics and Automation, pp. 500-505.

[6] M. Minsky, S. A. Papert (1969). *Perceptrons*. Cambridge: The MIT Press.

[7] P. Werbos (1992). *Approximate Dynamic Programming for Real-Time Control and Neural Modeling*. In D. White, D. Sofge (eds.) Handbook of Intelligent Control, pp. 493-525. New York: Van Nostrand.

A real-time Robot demonstration
controlled by the BSP400 Neurocomputer

Jan N.H. Heemskerk & Patrick T.W. Hudson
Leiden University, Unit of Experimental and Theoretical Psychology
P.O. Box 9555, 2300 RB Leiden, The Netherlands
E-mail: HMSKERK@rulfsw.LeidenUniv.nl, Tel: (31) 71/273631, Fax: (31) 71/273619

ABSTRACT

An actual implementation of a real-time, neural network controlled, robot car is presented in this paper. The simple car consists of two motors and 4 light sensors. Supervised learning behaviour of the car is achieved by using a neural network with adaptive connections. The car can be taught to avoid obstacles. The controlling neural network is implemented on the BSP400 neurocomputer, a Brain Style Processor with 400 nodes. A subset of the digital nodes in the BSP400 are connected by fixed weights to form logical circuits in order to re-train the car. In this way cooperative computation of both 'logical' and 'neural' processes are integrated into one system.

1. Three basic parts of the demo

In order to create a real-time system that exploits behaviour, a minimum of three basic parts are required. A car supplied only with eyes cannot be said to see if it does not respond to light; a control mechanism and a physical environment are necessary to provide the car with the ability to behave in response to light signals. The three parts will now be described.

1. The robot car, depicted in Figure 1.a, is sized 15*15*6 cm, and consists of two simple motors, each controlled by a 5V DC signal. These motors enable the car to drive straight on, to the left, or to the right. Four light sensors attached to the top of the car function as 'eyes' that look Forwards, Backwards, Left, and Right. This represents the absolute minimum in a system which would be said to see.

2. The environment is a restricted area in the real-world. The car is limited to drive only on a flat surface. As it can only sense light, light sources are the only activators in this world. These can be concealed by placing obstacles between the car and the light sources itself.

3. The control mechanism enables the system to transfer and process information that is being received by the car. If light hits a sensor, a 5V DC signal is fired into a neural network. The architecture of this neural network will be described in section 2. The network is implemented on the BSP400, a dedicated neurocomputer with 400 simple microprocessors [2]. These processors are distributed over 25 Module boards. The BSP400 is also equipped with a number of Input/Output boards that serve as direct interaction connections to the real-world. The light sensors and motors of the robot car are connected to one of these I/O-boards of the BSP400. The robot car was constructed to demonstrate the BSP400 in real-time operation. It provides a system that is adaptive in a non-simulated world in which unpredictable changes might occur that cannot be captured by simulated environments. These properties are referred to as situatedness and embodiment, and are necessary for the development and understanding of autonomous robots, see for instance [4, 6].

A neural network topology can be implemented one-to-one on the processors of the BSP400. This means that every processor forms a node. The processors calculate in parallel and the interconnections are time-multiplexed over a communication bus. The activation and learning rule implemented in the processors are based on the rules of the CALM (Categorizing And Learning Module) paradigm [4,5]. One of the main characteristics of these rules is locality. The learning rule is Hebbian, which means that only pre- and post-synaptic activations are considered for adjusting the weights. The resolution of the weights and the internal activations is 8 bits. The output activations of the processors are binary.

A neural iteration on the BSP400 takes about 5ms. In every iteration all nodes calculate a new activation, adjust the specified learning weights, and send their activation to the other nodes. This update speed of 200 Hz, with all inputs and outputs processed simultaneously, is suitable for controlling a real-time application like the car.

2. The neural network

The control mechanism is implemented as a neural network. A simplified version of the network is drawn in Figure 1.a. The inputs to the network are the connections of the four light sensors and two supervisory inputs to the motor nodes. These enable a human teacher to train the car to perform a special behaviour in specific situations. The motor nodes of the network (*M1* and *M2*) are the only two outputs. They are directly connected to the motors of the car.

The four (binary) light sensors (Left, Right, Front, and Back) form the input layer of the network. In total 15 input situations can occur: 4 sensors on/off means $2^4 = 16$ situations. The situation where none of the sensors receives light will not be considered. Each of the 15 input possibilities can be explicitly encoded in one specific node by connecting more sensors to a node and by adding lateral inhibitory weights between the nodes in the second layer. This second layer can be viewed as a lateral inhibition vision layer wherein the pre-wired (anti-competition) mechanism results in only one active (winner) node. In this way each input situation is encoded in the firing of a pre-determined node.

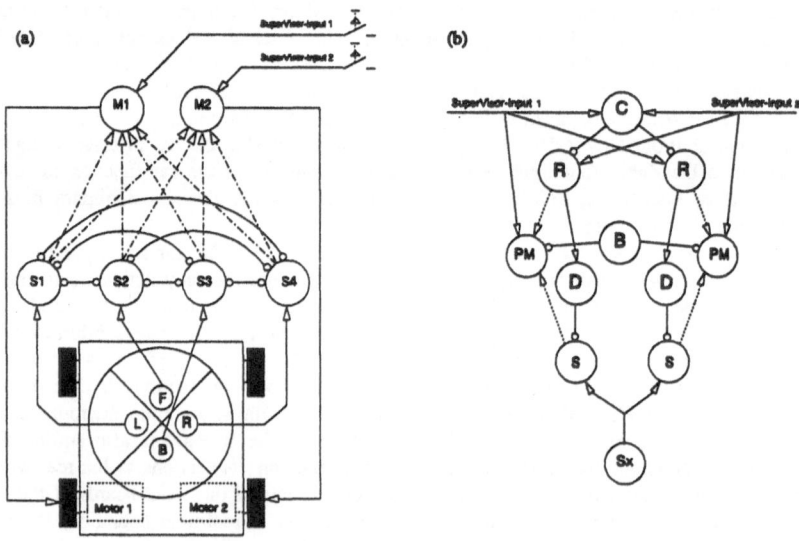

Fig.1.a. The robot car connected to a simplified version of the neural network.
Fig.1.b. One of the intermediate networks for the learning and re-learning building block.

The arrows in Figure 1.a denote excitatory (positive) connections, while the circles denote inhibitory connections. Here, for reasons of exposition, only four instead of fifteen 'sensor-nodes' (*S1*, *S2*, *S3*, and *S4*) are shown as connected to the four light sensors. The normal connections in the figure are fixed, their weights are non-adjustable. All dashed connections are learning (adjustable). The two supervisory inputs are implemented by simple push buttons which are also connected directly to the output (motor) nodes.

In the initial state, all learning weights in the network are set to zero. In other words, the network starts without any knowledge, except that which is restrained by the basic network architecture. The car can be manually controlled in any direction by pushing the supervisory buttons. As soon as light hits a sensor, the weight of the connection between the active sensor node and any active motor node(s) increases according to the Hebbian learning rule. In this way, several behaviours can be trained by manually controlling (training) the car. It is, for instance, possible to teach the car a 'find-light' procedure. In this case, it will always move towards a light from any position. After the first training no further user-interaction is necessary.

In the case where two or more light sources are used, the number of sensor nodes must be increased up to 15, then every possible input situation can be encoded. With this network it is possible to train the car to avoid obstacles. If the car is driving from one light source to another and an obstacle is placed in front of the car, it will stop. Now it can be trained to turn either left or right until it sees the target light source again and continues its (pre-taught) way. In the next trial the car will automatically avoid colliding with an obstacle without the help of a teacher.

This network architecture is sufficient to learn every input/output relation. However, with the implemented learning rule it is impossible to *re-learn* a different behaviour. The amount of weight changing Δw_{ij} is determined by the CALM learning rule:

$$\Delta w_{ij}(t+1) = \mu.a_i.[(K - w_{ij}(t)).a_j - L.w_{ij}(t).\sum_{f,f\neq j} w_{if}(t).a_f]$$

where a_i denotes the activation of node i. The value of μ determines the learning rate; L and K are constants that limit the maximal weight values; w_{ij} is the learning weight of the connection from node j to node i, and $\sum w_{if}(t).a_f$ is the weighted background activation (Grossberg term). The original CALM learning rule operates with continuously valued activations and weights. In the discrete BSP400 implementation the activations are binary and the (positive) learning weights range from 0 to 127 (two's complement, one byte representation). An adaptive weight can only decrease if one, or a number of, its 'neighbour' connections is more important. When the car is behaving in its environment the correlated sensor and motor nodes will be active simultaneously. In order to decrease the weight the (pre-synaptic) sensor node must be inhibited and another node must be activated. This can only be achieved by adding 'logical' circuits between the supervisory (motor) nodes and the sensor nodes. These intermediating circuits consist of 'neural' and 'logical' nodes.

Each sensor node (Sx) is connected to such an intermediate network as depicted in Figure 1.b. The sensor node (Sx) is split up into two separate sensor nodes, one for each direction. The (dashed) learning connection between such a sensor node (S) and a pseudo-motor node (PM) has a 'competing' weight that will be more important when a supervisor input activates the relearning node (R). The sensor node will be switched off during supervision by the inhibition caused by the delay node (D). This adds to the value of the right hand term of the learning rule and so causes Δw_{ij} to be negative. The bias node (B) provides the pseudo-motor node from becoming active as a result of the small positive remaining weight of the re-learned connection. The control node (C) operates like a binary AND gate. It ensures that the network can also strengthen the connections between both sensor and pseudo-motor nodes. The pseudo-motor nodes of all intermediating circuits are connected to two logical OR nodes. These drive the left and right motors.

The intermediate networks function as building blocks. We have implemented a complete network consisting of eight of these intermediating networks. The total number of logical and neural nodes is 96. Four networks encode the situation: 'light on a specific sensor', the others encode 'light on a specific combination of sensors'. The network enables the car to learn any reaction to any input situation. The supervisor can always change this behaviour by simply processing the two supervisory buttons. The BSP400 allows this processing to take place at the full operating speed.

3. Discussion and future perspectives

It can be said that after training, the distal activators of the world, such as moving lights, (unexpected) moving obstacles, and the moving of the robot itself, are proximally represented by the sensor circuits. The network receives an input stimulus and generates a sequence of outputs that result in a 'shifting' of the cars sensors in the direction of the light source. This results in a 'find-light' or 'avoid-obstacle' behaviour. A first attempt to describe this behaviour in schema-theoretical terms is made in [3].

In the continuation of this project, we aim to add a 'reverse gear' and other sensors to the car. Sonar and touch enable us to implement reinforcement learning. If the car hits an object it can be 'punished' by decreasing the weights responsible to the car driving in this direction. During the time the car avoids obstacles it will be 'rewarded' and a correlation will be made between its sonar signals and the motor control.

More complex behaviour can be taught without requiring a more complex environment. The control mechanism will probably grow in number of nodes. The number of available nodes on the BSP400 will still be sufficient for the time being, there is no time penalty attached to increasing number of nodes in use. One advantage of the use of a neurocomputer is that there is no principle difference between I/O and other processing nodes. This allows large numbers of I/O elements (e.g., retinal cells) to be processed in parallel. Another advantage is that no distinction has to be made between the learning and the performance phase. The robot always learns.

The apparent complexity of the demonstration behaviour lies in the speed of operation and learning, together with the perception of the behaviour by observers who see more than there is.

Acknowledgements

The investigations were supported in part by the Foundation for Computer Science in the Netherlands (SION) with financial support from the Dutch Organization for Scientific Research (NWO). We would like to thank Marcel Rieck and Danny Kunst for implementing and testing the network.

References

[1] Brooks, R.A. New Approaches to Robotics. *Science*, 253, 1991, 1227-1232.
[2] Heemskerk, J.N.H., J.M.J. Murre, J. Hoekstra, L.H.J.G. Kemna and P.T.W. Hudson, The BSP400: a modular neurocomputer assembled from 400 low-cost microprocessors. In: *Artificial Neural Networks: Proceedings of the International Conference on Artificial Neural Networks (ICANN-91) Espoo Finland*, T. Kohonen, K. Mäkisara, O. Simula and J. Kangas (Eds), Elsevier Science Publishers B.V. (North-Holland), 1991, 709-714.
[3] Heemskerk, J.N.H. & F.A. Keijzer. A real-time neural implementation of a schema driven toy-car. (To be presented at the workshop on 'Schemas and Neural Networks: Integrating Symbolic and Subsymbolic Approaches to Cooperative Computation', 19-20 October 1993, LA.)
[4] Murre, J.M.J. *Learning and Categorization in Modular Neural Networks*. Hemel Hempstad: Harvester Weatsheaf, and Hillsdale, NJ: Lawrence Erlbaum, 1992.
[5] Murre, J.M.J, R.H. Phaf, and G. Wolters, CALM: Categorizing And Learning Module, *Neural Networks*, 5, 55-82, 1992.
[6] Verschure, P.F.M.J., B.J.A. Kröse & R. Pfeifer. Distributed adaptive control: The self-organization of structured behavior. *Robotics and Autonomous Systems*, 9, 1992, 181-196.

FIRST RESULTS ON STABLE ADAPTIVE ROBOT CONTROL WITH RBF NETWORKS *

J.R. Beerhold, M. Jansen, R. Eckmiller
University of Bonn
Dept. of Computer Science VI (Neuroinformatics) FR Germany
Tel.: xx49-228-550-349   FAX: xx49-228-550-425   e-mail: beerhold@nero.uni-bonn.de

Abstract

We present a method to use RBF networks for stable adaptive robot control, with the task to track prescribed joint trajectories. The controller architecture is based on the concept of feedback linearization. The ·RBF networks serve as the model-based controller part. The learning rule is derived from a Lyapunov-based stability criterion to assure global stability at all times. Bounds on network approximation errors, which are essential to validate the stability proof, are established with the help of multidimensional sampling theory. Training the networks with point-to-point movements between randomly generated locations leads to a trajectory independent controller over the complete working range. Simulation results for a planar 4-joint manipulator (4JM) are given to demonstrate the performance of this control method.

1 Introduction

Feedback linearization is a very attractive approach to the control of highly nonlinear systems[1, 6]. The basic idea is to algebraically transform a nonlinear system dynamics into a linear one, so that linear control techniques can be applied. Such a transform exists, if the system meets certain assumptions, as all common robot manipulators do.

In the neural control literature this approach is known as 'inverse modeling', i.e. the controller employs a neural network to approximate the inverse of the unknown system dynamics [2, 3]. Combined with a linear feedback-controller, good control performance could be achieved for selected MIMO systems, provided that the approximation was sufficiently accurate, without, however, proving stability. Only for SISO systems, adaptive control methods, which provide global stability during on-line adaptation were recently developed [4, 5].

In this paper we present a neural net approach for MIMO systems, which assures global stability even during adaptation of the network within the feedback loop.

Consider a planar robot manipulator with n rotational joints with the following inverse dynamics:

$$\vec{\tau}(t) = \mathbf{M}(\vec{\theta}) \cdot \ddot{\vec{\theta}} + \mathbf{C}(\vec{\theta}, \dot{\vec{\theta}}) \cdot \dot{\vec{\theta}}(t) + \mathbf{F}(\vec{\theta}) \cdot \dot{\vec{\theta}}(t) \tag{1}$$

where $\tau(t), \vec{\theta}(t), \dot{\vec{\theta}}(t), \ddot{\vec{\theta}}(t)$ are the vectors of input torques, joint positions, velocities, and accelerations, respectively. $\mathbf{M}(\vec{\theta})$ denotes the robot's inertia matrix, which is nonsingular and thus invertible, $\mathbf{C}(\vec{\theta}, \dot{\vec{\theta}})$ is the matrix of coriolis- and centrifugal terms, and $\mathbf{F}(\vec{\theta})$ is the diagonal matrix of friction terms.

*) Supported by Federal Ministry for Research and Technology, under grant 01 IN 105 A/0

The controller's task is to force the robot joint-positions $\vec{\theta}(t)$ and velocities $\dot{\vec{\theta}}(t)$ to track a desired trajectory $\vec{\theta}_d(t)$ and its derivative $\dot{\vec{\theta}}_d(t)$. The desired trajectory and its derivative are assumed to be smooth and bounded.

2 Controller Design with RBF Networks

The control structure described in this paper employs neural networks to approximate the elements of the inertia, coriolis and friction matrices, $M_{ij}(\vec{\theta})$, $C_{ij}(\vec{\theta}, \dot{\vec{\theta}})$ and $F_{ij}(\vec{\theta})$, respectively. We use gaussian radial basis function networks with the basis functions spread on a regular grid and the same variance for each neuron. Approximation errors are defined as:

$$\Delta f(x) = \hat{f}(x) - f(x) = \tilde{w}^T \cdot \Phi(x) - \delta f(x) \tag{2}$$

where $\tilde{w}^T$ is a vector of weight errors to the optimal weights and $\delta f(x)$ is an inherent approximation error. The control law is based on a particularly suited representation of the manipulator's dynamics [1], as it is highly desirable to incorporate prior knowledge about the system under consideration into the control design. Consider the robot's equation of motion (1), where the matrix $C(\vec{\theta}, \dot{\vec{\theta}})$ is not uniquely defined. It exists a particular definition, so that the relation $\dot{M}(\vec{\theta}) - 2C(\vec{\theta}, \dot{\vec{\theta}})$ is skew-symmetric holds. This representation is used and exploited in the derivation of the stability proof. To reduce the input space of the networks, $C(\vec{\theta}, \dot{\vec{\theta}})$ is furthermore expressed as $C(\vec{\theta}, \dot{\vec{\theta}}) = C^*(\vec{\theta}) \cdot \dot{\Theta}$, which is possible due to inherent system properties. The following error measure is defined:

$$\vec{s}(t) = \dot{\vec{\theta}}(t) - \dot{\vec{\theta}}_d(t) + \Lambda(\vec{\theta}(t) - \vec{\theta}_d(t)) \tag{3}$$

with extension: $\vec{s}_\Delta(t) = \vec{s}(t) - \Gamma \, \mathrm{Sat}(s_i(t)/\gamma_i)$, where sat$(x)$ is the saturation function. An appropriate control law to achieve the desired task is:

$$\vec{\tau}(t) = \hat{M} \cdot \ddot{\vec{\theta}}_r + \hat{C} \cdot (\dot{\vec{\theta}}_r + \Gamma \, \mathrm{Sat}(\frac{s_i}{\gamma_i})) + \hat{F} \cdot \dot{\vec{\theta}} - K_{pd} \cdot \vec{s} - K_{sl}(\vec{s}) \cdot \mathrm{Sat}(\frac{s_i}{\gamma_i}) \tag{4}$$

with $\dot{\vec{\theta}}_r(t) = \dot{\vec{\theta}}_d(t) - \Lambda(\vec{\theta}(t) - \vec{\theta}_d(t))$. The first two terms represent the feedforward part of the controller, driven by the derivatives of the desired output $\vec{\theta}_d$. The negative feedback term $K_{pd} \cdot \vec{s}(t)$ is added to stabilize the system and $K_{sl}(\vec{s}) \cdot \mathrm{Sat}(s_i/\gamma_i)$ is a sliding feedback term, the purpose of which is to compensate inherent approximation errors [5, 4].

The resulting controller structure is depicted in figure 1.

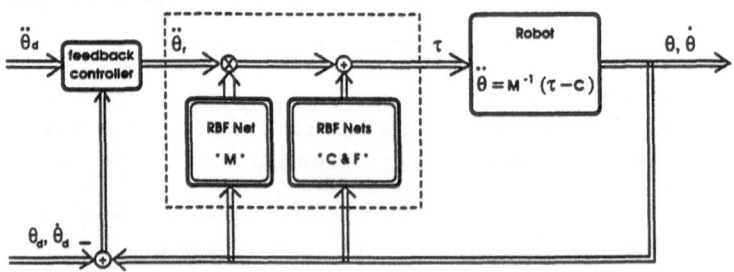

Figure 1: closed loop controller architecture (auxiliarly control components not shown).

3 Stability Analysis

Using the Lyapunov function candidate:

$$V(t) = \frac{1}{2}[\vec{s}_\Delta^T \cdot M \cdot \vec{s}_\Delta + \sum_{i=1}^{n} \frac{1}{\eta_f} \sum_{l=1}^{N} \tilde{w}_{i l_f}^2 + \sum_{i=1}^{n} \sum_{j=1}^{n} \frac{1}{\eta_h} \sum_{l=1}^{N} \tilde{w}_{i j l_h}^2 + \sum_{i=1}^{n} \sum_{j=1}^{n} \sum_{l=1}^{n} \frac{1}{\eta_c} \sum_{m=1}^{N} \tilde{w}_{i j l m_c}^2] \quad (5)$$

and applying the following adaptation rules and controller gains:

$$\dot{\tilde{w}}_{i l_f} = -\frac{1}{\eta_f} s_{\Delta_i} \dot{\theta}_i \Phi_{l_f} \quad (6)$$

$$\dot{\tilde{w}}_{i j l_h} = -\frac{1}{\eta_h} s_{\Delta_i} \ddot{\theta}_{r_i} \Phi_{l_h} \quad (7)$$

$$\dot{\tilde{w}}_{i j l m_h} = -\frac{1}{\eta_c} s_{\Delta_i} \dot{\theta}_j (\dot{\theta}_l - s_{\Delta_l}) \Phi_{m_c} \quad (8)$$

$$\bar{k}_i . \geq \mu_i + |\sum_{j=1}^{n} \delta H_{ij} \ddot{\theta}_{r_j} + \sum_{j=1}^{n} \delta C_{ij} (\dot{\theta}_j - s_{\Delta_j})| \quad , \quad \mu_i \geq 0 \quad (9)$$

it could be shown here that $\dot{V}(t) \leq -\vec{s}_\Delta^T \cdot K_{pd} \cdot \vec{s}_\Delta$ for $|s(t)_i| \geq \gamma_i$ at all times $t \geq 0$. Furthermore is $\dot{V}(t) \equiv 0$ for $|\vec{s}(t)| \leq \gamma_i$. From the above considerations it can be concluded, that if $\vec{s}_\Delta(t)$ and all $\tilde{w}$ are bounded at time $t = 0$, they remain bounded for all $t \geq 0$.

4 Error Bounds and Simulation results

A simulation of a four-joint planar manipulator was used to demonstrate the performance of this control method. We assume a robot whose masses are located at the distal end of each link and whose joint ranges are limited to $\pm 60°$ relative to a predefined initial position.

To specify the inherent errors δH_{ij} and δC_{ij}, the network approximation properties were investigated with the help of multidimensional sampling theory. The network operation, described by $f_N(x) = \sum_{i=1}^{\infty} w_i \cdot \Phi(x, \xi_i)$ can be regarded as reconstruction of spatially sampled data[4]. The learning task is thus, roughly spoken, to find optimal weights w_i, which match the values of the underlying function $f(x)$ on a discrete subspace of the input space. It can be shown analytically that the inherent approximation error can be made arbitrarily small, if the network parameters are properly chosen.

For the robot simulated in this work, calculations yielded that networks with 11, respectively 13 neurons per dimension result in inherent approximation errors less than 10^{-2}, respectively 10^{-3}. These errors take also into account, that only a small number of neurons contribute to the network output. More detailed, for a network with a total of $11^3 = 1331$ neurons, only 25 neurons have to be evaluated during each recall. This makes our approach particularily appealing for real-time tasks.

The networks were trained with point-to-point movements between randomly generated locations, to achieve a trajectory independent controller over the complete working range. The tracking error on a test trajectory was reduced to $\approx 4 \cdot 10^{-4}$ rad after only 500 movements. Furthermore, the controller could cope with changes in systems parameters (e.g. payload), since the adaptation was continously operational.

Tracking accuracy at the beginning of a training session and after 5000 movements is illustrated in figure 2.

300

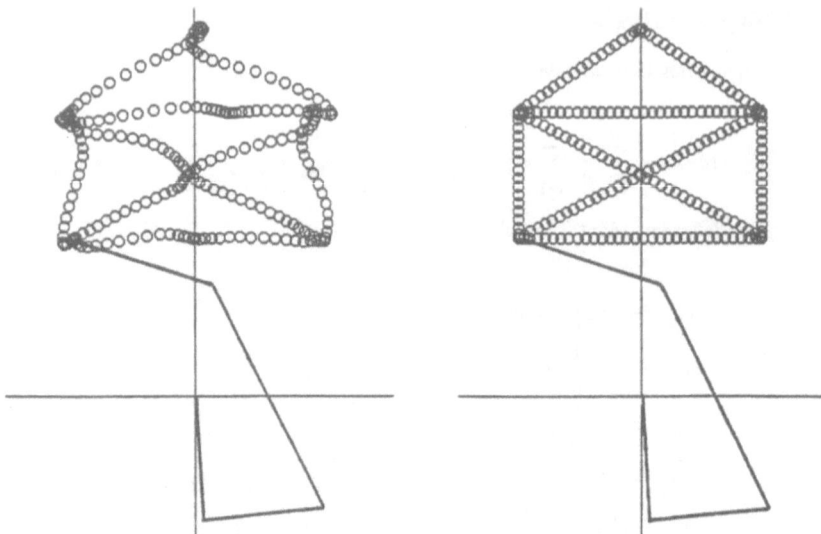

Figure 2: Tracking performance on a test trajectory before training and after 5000 movements. The tracking error has decreased from $\approx 10^{-2}$ rad to $\approx 10^{-4}$ rad.

5 Summary

It has been shown that RBF networks can be efficiently used for stable adaptive robot control. In particular, global stability could be assured from the very beginning of the on-line training session. Network size and resulting approximation uncertainty have been derived analytically, which is essential for the validation of the stability proof. During recall, only a small number of neurons, typically 25, contribute to the network output, making our approach feasible for real-time tasks. Training was performed on-line with a 'random-movement' method that led to a trajectory independent controller. This training method is also feasible for real robots.
Further work will concentrate on the application of the proposed control structure to real manipulators.

References

[1] J.J.Slotine, W.Li. *Applied Nonlinear Control.* Prentice-Hall, Englewood Cliffs, 1991.

[2] M.Kawato. *Optimization and learning in neural networks for formation and control of coordinated movement.* ATR Technical Report, 1990.

[3] M.Jordan, D.Rumelhart. *Forward models: supervised learning with a distal teacher.* Center for Cognitive Science, Technical Report (Occ. paper) 40, 1990.

[4] R.M.Sanner, J.J.Slotine. *Direct adaptive control using gaussian networks.* Technical Report NSL-910303, 1991.

[5] E.Tzirkel-Hancock, F.Fallside. *Stable control of nonlinear systems using neural networks.* Technical Report 81, Cambridge University, Cambridge, 1991.

[6] E.Freund, H.Hoyer. *Das Prinzip nichtlinearer Systementkopplung mit der Anwendung auf Industrieroboter.* Regelungstechnik. 28: 80-87, 1980.

Fuzzy Inference, Radial Basis Functions, and Control of Flexible Robotic Manipulators

Jorge I. Arciniegas   Krzysztof J. Cios   Adel H. Eltimsahy

University of Toledo
Toledo, OH 43606, U.S.A.

Abstract

Fuzzy inference systems can be regarded as linear models, i.e. they can be represented by a linear combination of a set of basis functions. These systems are structuraly similar to those in the class of neural networks consisting of a collection of locally-tuned units. Therefore, coming up with the appropriate set of fuzzy rules to generate the corresponding fuzzy basis function expansion is analogous to the problem of efficiently placing the centers of the neural network receptive fields. This is accomplished by using an efficient orthogonal least squares algorithm. The technique is applied to control a two link manipulator consisting of one rigid link and one flexible link. The results are shown and analyzed.

Introduction

A fuzzy-logic controller characterizes a control process in linguistic terms using of a collection of "IF ... THEN ..." statements whose meaning is modelled by fuzzy sets. The control action taken by the controller can be characterized by a linear model belonging to a class of fuzzy inference systems. This type of system is structurally similar to the class of neural networks composed of a collection of locally-tuned units. In other words, both systems can be realized by a linear aggregation of basis functions corresponding to radial basis functions in a neural network, and fuzzy basis functions in a fuzzy inference system. It is known that a crucial factor in determining the accuracy and complexity of radial basis function neural networks is to find the location of the receptive field centers of the processing units. In this paper we point out the similarity between both kinds of systems and show that as in the case of its neural network counterpart, the accuracy of a fuzzy inference system depends on the correct distribution of basis functions in the domain of the underlying function. An efficient orthogonal least squares algorithm is used to accomplish this. The technique is used to develop a controller for a flexible robotic manipulator. The resulting controller is evaluated in terms of its ability to control the position and vibration of the links.

Fuzzy-logic Controllers

A fuzzy-logic controller characterizes a control process by using a collection of N conditional "IF... THEN ..." statements. The goal of the controller is to achieve or maintain a desired state of the controlled process. In this context, the antecedents of the conditional statements refer to the set of state variables, while the consequents describe the control action. The conditional statements have the following form:

$$R_i = \text{IF } e_1 \text{ is } A_{1,i} \text{ AND } e_2 \text{ is } A_{2,i} \text{ AND ... AND } e_m \text{ is } A_{m,i} \text{ THEN } u \text{ is } B_i \text{ ; } i = 1, 2, \ldots, N \quad (1)$$

$e_j, j = 1,2,\ldots,m$, refer to the inputs, and u to the output of the control process. They represent the state and control variables, respectively. $A_{j,i}$ and B_i are linguistic terms belonging to the term set $T(x_j)$ and $T(x_u)$, where x_j and x_u are linguistic variables representing the magnitudes of the jth state variable and control-signal, respectively, with the corresponding universes of discourse E_j and U. The conditional statements R_i, are fuzzy relations $(A_{1,i} \text{ AND} A_{2,i}\ldots \text{AND} A_{m,i}) \rightarrow B_i$ defined on $E_1 \times E_2 \times \ldots \times E_m \times U$, and express the relation among the m state variables e_j and the control variable u. These statements are combined using the connective ALSO as in

$$R = R_1 \text{ ALSO } R_2 \text{ ALSO } \ldots \text{ ALSO } R_N \quad (2)$$

where R is a fuzzy relation which is the combination of the individual fuzzy relations in the system. The R_i relations can be evaluated in different ways [6]. Using correlation-product encoding [4] on the ith rule and interpreting the ALSO in equation (2) as union results in the following membership function:

$$\mu_R(e_1', e_2', \ldots, e_m', u') = \bigcup_{i=1,\ldots,N} (\prod_{j=1}^{m} \mu_{A_{j,i}}(e_j')) \cdot \mu_{B_i}(u) \tag{3}$$

where $\cup$ denotes the pairwise maximum operator.

As shown in Figure 1, a fuzzy inference engine performs a mapping from fuzzy sets in E to fuzzy sets in U. The fuzzy sets in E are obtained by means of a fuzzification operator such as the singleton fuzzifier which maps a crisp point x_o into a fuzzy set A with membership function $\mu_A(x)$ equal to one at point x_o and zero elsewhere.

Given a set of input fuzzy sets A_j to the inference system, the consequent fuzzy set B is inferred using the sup-t compositional rule of inference, i.e. $B = (A_1 \text{ AND } A_2 \text{ AND... AND } A_m) \text{ o } R$, where R is the compact fuzzy relation of equation (2), o is the sup-t composition, and t denotes an operator in the class of triangular norms. Using the product operator and singleton fuzzifier, the membership function of B is written as follows:

$$\mu_B(u') = \bigcup_{i=1,\ldots,N} (\prod_{j=1}^{m} \mu_{A_{j,i}}(e_j)) \cdot \mu_{B_i}(u') \tag{4}$$

B is a fuzzy set which represents the possibility distribution of the required control action. In terms of on-line control, however, what is needed is a crisp command. This is achieved by the block marked "defuzzifier" in Figure 1. A defuzzifier generates a crisp control action that best characterizes the inferred fuzzy set B. The most commonly used defuzzification strategy is the centroid method which generates the center of gravity of the possibility distribution of the inferred control action. Taking the output fuzzy sets as distinct singletons β_i yields the following:

$$u = \sum_{i=1}^{N} \beta_i \cdot \prod_{j=1}^{m} \mu_{A_{j,i}}(e_j) \Big/ \left(\sum_{i=1}^{N} \prod_{j=1}^{m} \mu_{A_{j,i}}(e_j) \right) \tag{5}$$

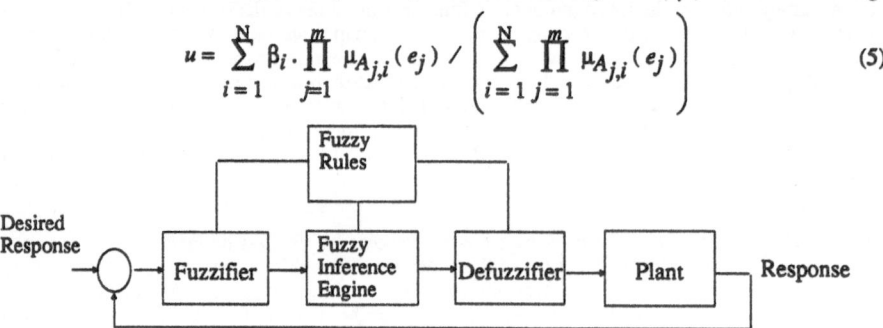

Figure 1. Structure of a fuzzy-logic controller.

Fuzzy Inference Systems and Radial Basis Functions Neural Networks

Equation (5) defines a mapping from the set of m state variables to a control action. This equation can be written in the form of a (fuzzy) linear model as shown below:

$$u = \sum_{i=1}^{N} \beta_i \cdot \Psi_i(e) ; \quad \Psi_i(e) = \prod_{j=1}^{m} \mu_{A_{j,i}}(e_j) \Big/ \left(\sum_{i=1}^{N} \prod_{j=1}^{m} \mu_{A_{j,i}}(e_j) \right) \tag{6}$$

The set of functions $\Psi(.)$ in equation (6) is called a fuzzy basis function, [7], or just basis functions, [3]. It has been proven that fuzzy inference systems, i.e. systems characterized by equation (6), can uniformly approximate any real continuous function on a compact set to arbitrary accuracy [3],[5],[7].

It is interesting to note the analogy between fuzzy inference systems and a class of neural networks known as radial basis functions (RBF). The structure of this type of neural networks is modelled as a linear combination of radial basis functions:

$$f(\xi) = \sum_{I=1}^{K} \lambda_I \, \Phi_I \, (\| \, \xi_I - \upsilon_I \| \,) \tag{7}$$

where f is a function being approximated by the neural network, $\Phi_I(.)$ are radial basis functions, $\xi \subset \Re^N$, $\| \, \|$ is a norm, λ_I are weighing parameters, and υ_I are the centers of the receptive fields of the κ units. There are two components in this network which need to be determined: the weighing parameters λ_I and the receptive field centers υ_I. A crucial factor in determining the accuracy as well as the complexity of radial basis function neural networks is finding the location of the receptive field centers. Once this problem is resolved, training of the network reduces to finding the values of the parameters of the linear expansion of equation (7).

Similarly, it is clear that the accuracy of the fuzzy-inference systems of (6) depends on the correct distribution of the basis functions inside the domain of the underlying control function. In other words, the basis functions should span the space defined by the cross product of the state variables of the controlled process. The problem that we are facing now is one of finding a sufficiently large number of fuzzy basis functions as well as their correct location in the domain of the desired control function. As in the case of radial basis functions neural networks, a number of approaches exist. The simplest solution corresponds to using one basis function for each point in the training set of data. Even though such approach is simple, it is not practical in real-life nonlinear applications where the amount of available data points can be very large. Also, such approach would correspond to a fuzzy inference system having a rule for each possible point in the input space. Another approach recognizes the weakness of the first and chooses to arbitrarily select a subset of the training data points to determine the set of basis functions. The disadvantage here is that the resulting fuzzy inference system might not be able to suitably span the domain of the intended control function. The approach which we have taken in this paper is to determine an optimal distribution of basis functions by using the orthogonal least squares learning algorithm [1],[2]. Consider the following mathematical functional model:

$$\Upsilon(t) = \psi_1(t) \, \theta_1 + \psi_2(t) \, \theta_2 + \ldots + \psi_K(t) \, \theta_K = \sum_{I=1}^{K} \psi_I(t) \, \theta_I \tag{8}$$

where Υ is an observed variable, $\theta_1, \theta_2, \ldots, \theta_K$ are unknown parameters, and $\psi_1, \psi_2, \ldots, \psi_K$ are known functions which may depend on other variables. In statistics the variables $\psi_1, \psi_2, \ldots, \psi_K$ are called the regression variables and the model itself is referred to as the regression model. The similarity between equations (6), (7) and (8) can be immediately seen. In the strict regression sense, an error term should be included in the regression model of equation (8).

$$\Upsilon(t) = \psi_1(t)\theta_1 + \psi_2(t)\theta_2 + \ldots + \psi_K(t)\theta_K + \varepsilon(t) = \sum_{I=1}^{K} \psi_I(t) \, \theta_I + \varepsilon(t) \tag{9}$$

where the error $\varepsilon(t)$ and the regressors $\psi_I(t)$ are uncorrelated. Clearly, the last equation can be rewritten in a matrix form as follows:

$$Y = \Psi \, \Theta + E \tag{10}$$

From a practical point of view, we need to obtain pairs of observations and regressors, and then determine the parameters $\theta_1, \theta_2, \ldots, \theta_K$ to satisfy a certain performance. Naturally, the linear least squares technique would solve this problem. An initial set of regressors is obtained by using the entire available set of data. Choosing an optimal set of basis functions reduces the problem to one of selecting a subset of significant regressors from the original set. This can be accomplished by using the orthogonal least squares method of Chen et.al., [1], [2]. This method transforms the original set of regressors into an orthogonal set of basis vectors which spans the same space spanned by the original set of regressors. It then selects a subset of these regressors by observing the amount of the output variance which is explained by each regressor.

Flexible Manipulator Control

The orthogonal least squares algorithm has been used to synthesize a controller for a two link robotic manipulator consisting of one rigid and one flexible links. For simplicity, we used independent fixed-gain PID controllers to control the motion of the manipulator along a set of reference trajectories obtained from a 4-3-4 trajectory planner. Figure 2 shows a set of reference

304

trajectories for the flexible link of the manipulator. Figure 3 shows the actual trajectories followed by the link. It is seen that there is good trajectory following performance; however, the velocity of the link experiences some deterioration and there is definite degradation of the acceleration. Note that such acceleration would result in an increase of the vibration of the link. The information contained in Figures 2 and 3 was collected and used to synthesize a fuzzy controller. The orthogonal least squares technique was used to obtain a reduced but effective system. Figure 4 shows the performance of the manipulator when the size of the original expansion is reduced to 20% of its original size. A position error can be noticed as a result of utilizing the fuzzy inference system to control the manipulator. This error is of the order of 1.5%. Nevertheless, it is also observed that the velocity profile of the link does not deteriorate as much as it does in the conventional case. Even more important in our case is the significant improvement in the acceleration of the flexible link. It is observed that the acceleration did not jerk as much as it did before. This significantly reduces the vibrations of the link.

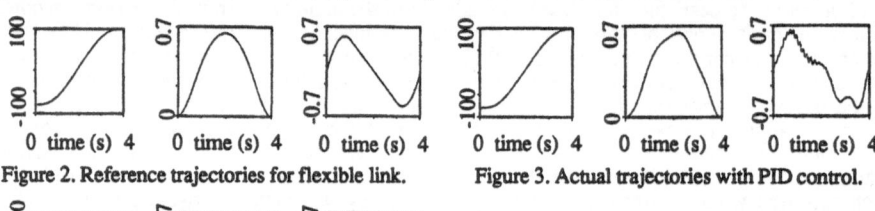

Figure 2. Reference trajectories for flexible link.   Figure 3. Actual trajectories with PID control.

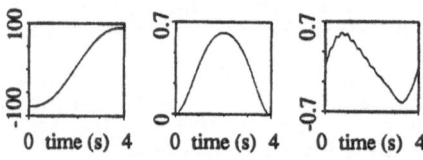

Note: Figures 2, 3, and 4 depict position in deg, velocity in deg/sec, and acceleration in deg/sec^2 (from left to right).

Figure 4. Actual trajectories with fuzzy control.

Conclusions

On top of achieving an acceptable manipulator performance, it is necessary to point out the advantages of this compromising control strategy. First, a fuzzy inference system can be obtained in a very short time. All that is needed is a statistically representative set of input output pairs in the space defined by the state variables and output of the desired control system. Second, a computationally expensive controller can always be simulated even though it cannot always be implemented in real time. The technique shown in the paper is a compromising alternative in the sense that it can be used to approximate any desired control function. Furthermore, the resulting fuzzy inference system very easily performs real-time control at a very low computational cost. There is no need for a special purpose hardware or expensive computer resources to implement this type of systems. Finally, fuzzy inference systems can be easily altered by adding linguitic rules.

References

[1] Chen S., Billings S., Luo W., "Orthogonal Least Squares Methods and their Application to Non-Linear System Identification", Int. J. Control, vol. 50, no. 5, 1873-1896, 1989.

[2] Chen S., Cowan C., Grant P., "Orthogonal Least Squares Learning Algorithm for Radial Basis Function Networks", IEEE Trans. on Neural Networks, vol.2, no. 2, 302-309, 1991.

[3] Jou Chi-Cheng, "On the Mapping Capabilities of Fuzzy Inference Systems", Proc. Joint Conference in Neural Networks, 708-713, 1992.

[4] Kosko B., Neural Networks and Fuzzy Systems: A Dynamic Systems Approach to Machine Intelligence, Prentice-Hall, 1992.

[5] Kosko B., "Fuzzy Function Approximation", Proc. Joint Conference in Neural Networks, 209-213, 1992.

[6] Lee C, "Fuzzy Logic in Control Systems: Fuzzy Logic Controller", IEEE Trans. on Systems, Man, and Cybernetics, 404-435, 1990.

[7] Wang L., "Fuzzy Basis Functions, Universal Approximation, and Orthogonal Least Squares Learning", IEEE Trans. on Neural Networks, vol. 3, no. 5, 807-814, 1992.

A Recurrent Trajectory Storage Network with Parceling of the Workspace

N. H.-R. Goerke, C. M. Müllender, R. Eckmiller

Department of Computer Science VI - Neuroinformatics
University of Bonn,
Römerstr. 164, W-5300 Bonn 1, F. R. Germany
e-mail: goerke@nero.uni-bonn.de

Abstract

A prestructured neural network, capable of learning and generating smooth multi-dimensional desired trajectories was designed for the control of robotic manipulators. The recurrent network structure combines adaptive parceling of the workspace with a second order differential equation for the relation between force vector and fingertip position.
Network performance for storage (using a modified δ-rule) and generation of a given 2-dimensional trajectory was successfully tested. Simulation results showed that additional implementation of adaptive parceling significantly improved the storage accuracy relative to learning of accelerations with fixed parceling.

Introduction

Adaptive neural oscillators as adjustable central pattern generators [1], embedding fields [6], and biology inspired pulse processing neural networks with spatial information coding [2, 3, 5], have been suggested in the literature.
In contrast, the presented approach is specially designed to meet several requirements for trajectory storage in robotics: smoothness in position, continuity in velocity, and infinitely low speed at begin and end of the trajectory. The internal representation should allow for many arbitrary trajectories an efficient storage with respect to their individual characteristics. Part of the results has recently been published [4].
Furthermore this approach is open for subsequent trajectory modification in response to obstacle signals. In this paper, however, obstacles will not be considered.

Network Design

To reproduce a desired trajectory, a virtual fingertip was driven through the workspace by means of locally constant force vectors. Learning a trajectory was realized by learning this force sequence stored as a synaptic weight vector $\mathbf{W}$.
For each trajectory to be stored, the spatio-temporal workspace was individually partitioned into N spatio-temporal parcels, defined by the weights of N-neurons that make up the *Parcel-layer* (P-layer). Depending on the complexity of the trajectory and the resources of neurons and synapses, parcel topology was adapted in a special parceling cycle.
The network was initially designed for storing and generating a single 2-dimensional trajectory. Expanding the task to higher dimensions simply requires a linear increase of the amount of synapses.

A desired ideal trajectory $\hat{S}$ is a 2-dimensional vector with the components $\hat{x}(t), \hat{y}(t)$, (first and second temporal derivative: $\hat{S}', \hat{S}''$). The example trajectory of the handwritten letter **b** (fig. 1), consists of 800 sampled floating point values between 0.0 and 1.0 for each coordinate.

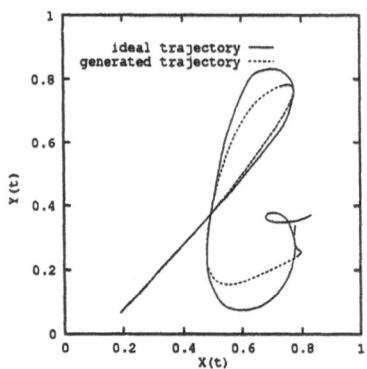

Fig. 1: generated trajectory vs. ideal trajectory "b".
Generated trajectory represents intermediate result following 60 acceleration learning steps with 15 fixed parcels.

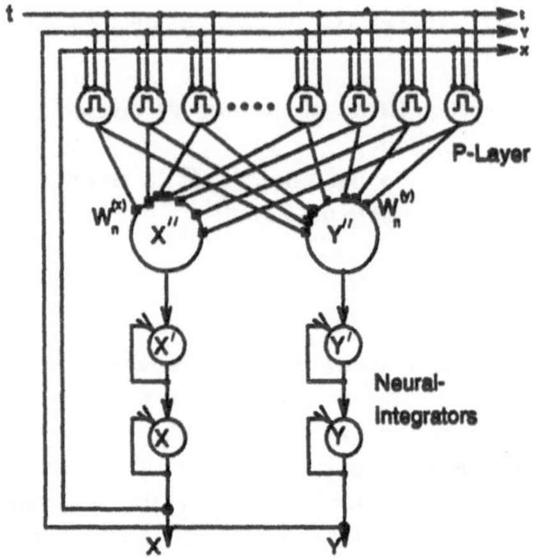

Fig. 2: Recurrent neural network (2-dimensional case) with P-layer, weights $\mathbf{W}_n$ for the force vector, and neural integrators.

The approach to generate a continuous trajectory within the network is to integrate a sequence of accelerations $S'' = (x''(t), y''(t))$ twice, yielding a sequence of velocities $S' = (x'(t), y'(t))$, and a sequence of spatial positions $S = (x(t), y(t))$. The vector of accelerations can be interpreted as a force vector driving a virtual fingertip through the workspace. Each integration is performed by two recurrent linear neurons for each component. The integration of S'' via S' to S for one component can be described as:

$$x'_t = x'_{t-1} + x''_t \cdot \Delta t \qquad x_t = x_{t-1} + x'_t \cdot \Delta t \qquad x_t \text{ denotes the value of x at time t.} \qquad (1)$$

The task of learning a desired trajectory is equivalent to the task of finding a set of adequate accelerations. Unlike the trajectory itself, the sequence of accelerations $S''(t)$ doesn't have to be continuous, but can be a piecewise constant function stored as N-dimensional synaptic weight vectors $\mathbf{W}^{(x)}, \mathbf{W}^{(y)}$. Piecewise constant accelerations S'' are close enough to the ideal second derivative $\hat{S}''$, to result in a fairly good approximation of the ideal trajectory $\hat{S}$, because the constant accelerations will be integrated to parabolas of order two.

Depending on the actual position S and the course of time t the spatio-temporal position $\mathbf{R} := (S, t)$ of the virtual fingertip, is governing the activation state of the neurons in the P-layer (eqn. 2).

$$x''_t = \sum_{n=1}^{N} w_n^{(x)} \cdot out_n(t), \qquad \text{with} \quad out_n = \left\{ \begin{array}{ll} 1.0 & \text{if } \mathbf{R} \text{ is inside parcel } n \\ 0.0 & \text{otherwise} \end{array} \right. \qquad (2)$$

Neuron n of the P-layer is fully active ($out_n = 1$) if the actual position of the virtual fingertip $\mathbf{R}$ lies in the parcel represented by neuron n; in all other situations neuron n remains silent ($out_n = 0$). A parcel is defined by a set of lower and upper boundaries for each dimension of the expanded workspace, which are stored as synaptic weights of the P-layer neurons (fig. 2).

The differences between desired ($\hat{S}$) and generated trajectory (S) following sufficient learning steps are caused by the piecewise constant function of accelerations and the non-perfect neural integrators. For evaluating the capabilities of this approach we concentrated our efforts to solely temporal instead of spatio-temporal parcel bounderies.

Learning of Accelerations

Learning the accelerations is an iterative procedure that updates the weights $\mathbf{W}^{(x)}$ after each epoch (duration of the trajectory), using the cumulative δ-rule (eqn. 3):

$$_{new}w_n^{(x)} =_{old} w_n^{(x)} + \eta \cdot \delta_n^{(x)} \quad \text{with} \quad \delta_n^{(x)} = \sum_{t=0}^{t=end} \Delta x_t'' \cdot out_n(t), \quad \text{and} \quad \Delta x_t'' := \hat{x}_t'' - x_t'' \quad (3)$$

$\delta_n^{(x)}$ is the cumulated difference $\Delta x_t''$ between the ideal and the estimated acceleration for the x-component with respect to parcel n. Cumulation takes place for each dimension separately during the time, the trajectory $\mathbf{R}$ is in a given parcel ($out_n = 1$, see eqn. 2). The time-averaged Euclidian distance between ideal and estimated position is referred to as average error $E_a :=< \hat{S} - S >$.

In addition the differences for velocity ($\Delta x_t'$) and position (Δx_t) were also incorporated into the cumulation to further reduce E_a:

$$\delta_n^{(x)} = \sum_{t=0}^{t=end} \left(\gamma \Delta x_t'' + \beta \Delta x_t' + \alpha \Delta x_t \right) \cdot out_n(t) \quad \text{with} \quad \Delta x_t' := \hat{x}_t' - x_t', \quad \Delta x_t := \hat{x}_t - x_t \quad (4)$$

Optimal parameters (α, β, γ) depend on number, position and size of the parcels as well as on the complexity of the trajectory. All simulations presented in this paper were performed with $\alpha \equiv \beta \equiv \gamma \equiv 1.0$.

Adaptive Parceling

An adaptive scheme to optimize parcel position and size with a given number (N) of P-neurons was employed for minimizing E_a with respect to a given trajectory $\hat{S}$.

Specifically, every 105 steps of learning of accelerations those two P-neurons for adjacent parcels with the most similar force vectors were combined to one P-neuron for a single enlarged parcel. The freed P-neuron was assigned to represent part of that parcel with the largest position error, in order to reduce the average error E_a (fig. 3). Then the parceling cycle was restarted.

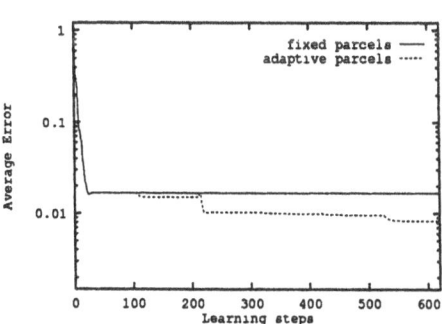

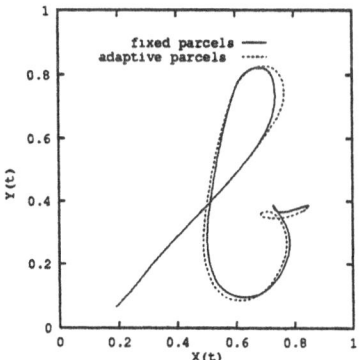

Fig. 3: Learning curves for a network with 25 P-layer neurons. One parceling cycle every 105 learning steps.

Fig. 4: The generated trajectory: 25 fixed parcels vs. 25 adaptive pacels after 6 parceling cycles (630 acceleration learning steps).

The results for the sample trajectory with and without adaptive parceling are visualized as learning curves (fig. 3) and generated trajectories (fig. 4).

The average error E_a was significantly reduced by adaptive parceling (fig. 3), which is recognizable by a higher resemblance to the ideal trajectory (fig. 4 vs. fig. 1).

Discussion and Conclusion

The results clearly demonstrate the capability of the proposed novel network topology for trajectory storage and generation. The network was specifically designed for future processing of static and dynamic obstacle signals, in order to modify the trajectory during re-generation.

The proposed mechanism of adaptive parceling, which currently refers only to time, will be expanded to spatial parameters.

Furthermore, the recurrent network structure can be easily expanded to higher dimensions by adding two integration neurons and $2N$ synapses per additional dimension.

In conclusion, the proposed network is a powerful module for path planning and obstacle avoidance capable of real-time processing.

This work was supported by Federal Ministry for Research and Technology (BMFT) under grant 01 IN 105 A/O (SENROB).

References

[1] K. Doya, Y. Shuji, A neural network model of temporal pattern memory: adaptive neural oscillator using continuous-time back-propagation learning. Neural Networks, 2: 375-385, 1989.

[2] R. Eckmiller, Neural nets for sensorimotor trajectories. IEEE Control Systems Magazine, 9: 53-59, 1989.

[3] R. Eckmiller, N. Goerke, J. Hakala, Neural networks for internal representation of movements. In: Neural Networks, Theory and Applications, R. J. Mammone, Y. Zeevi, (eds.), Academic Press, San Diego, pp 97 - 112, 1991.

[4] R. Eckmiller, J. Beerhold, G. Fahner, N. Goerke, J. Hakala, M. Jansen, B. Kreimeier and H.W. Werntges, Neural network applications for robot control. In: Proc. of BMFT Status-Seminar Neuroinformatics, Oct. 1992, In Press 1993.

[5] N. Goerke, M. Schöne, B. Kreimeier, R. Eckmiller, A network with pulse processing neurons for generation of arbitrary temporal sequences. In: Proc. IEEE Int. Joint Conf. Neural Networks, June 1990, Vol. III, pp 315 - 320, San Diego, 1990.

[6] S. Grossberg, Some networks that can learn, remember, and reproduce any number of complicated space-time patterns. J. Math. and Mechan., 19: 53 - 91, 1969.

Node Allocation and Topographical Encoding NATEnet for Inverse Kinematics of a 6–DOF Robot Arm *

J. Hakala   R. Eckmiller

Department of Computer Science VI (Neuroinformatics)

University of Bonn, F.R. Germany

Tel.: ++49-228-550-349   FAX: ++49-228-550-425

e-mail: hakala@nero.uni-bonn.de

Abstract

In order to continuously map desired tool center point (TCP) movements onto the corresponding set of joint angles for a 6–DOF robot arm without the need for repeated calculation of the inverse Jacobian we present an adaptive neural network with the following features:

- topographical encoding of the input for smooth interpolation and fast adaptation

- error–driven node allocation for efficent representation of input patterns with respect to the required mapping operation

- a prestructured network taking advantage of linearities within inverse kinematics

Combining these properties, a small, but efficent network architecture evolves, which solves the differential inverse kinematics problem.

Introduction

Inverse kinematics requires the mapping of a path specified for the TCP onto the corresponding joint angles of the robot arm. The nonlinear mapping of the actual position in joint space and the desired movement in workspace onto the movement of joint angles is generally solved using linear algebra involving the Jacobian which represents the local derivatives of the TCP position with respect to joint angles [4]. Our approach tackles the problem directly using approximation of the inverse Jacobian without the need to invert matrices. In fact, the network learns the inverse Jacobian.

Radial basis functions using topographical encoding of inputs are well known for their rapid adaptation. A general drawback for their use is the limitation to low dimensions of the input. Several authors tried to overcome this problem using node allocation [5, 3]. Some alternative allocation scheme [2] was developed earlier, and successfully applied to the inverse kinematics of a redundant four–joint robot arm within a planar working area. The aim of this paper is to demonstrate the applicability of the proposed scheme to more degrees of freedom.

*Supported by Ministry for Research and Technology (BMFT–Project SENROB), under grant 01 IN 105 A/O

Error–driven node allocation

Topographical encoding of input patterns $\underline{x}$ generates activity a_i for neuron i using Gaussian basis functions $a_i = exp(-\frac{r_i^2}{\sigma_i^2})$ with width σ_i and distance $r_i = ||\underline{x} - \underline{x}_i||$ of the receptive field center $\underline{x}_i$ to the actual input pattern $\underline{x}$.

Topographical encoding is a distributed representation of the input pattern that ensures smooth interpolation [1].

The overall result y^k of the required mapping is generated using normalized linear summation of the output–pattern attached to each neuron.

$$y^k = \frac{\sum_{i=1}^{N} y_i^k a_i(\underline{x})}{\sum_{i=1}^{N} a_i(\underline{x})} \tag{1}$$

The sum runs over all neurons N. Adaptation of the output y^k towards a desired output $\hat{y}^k$ due to an observed error $e^k = \hat{y}^k - y^k$ involves a simple delta–rule: $\Delta y_i^k = \gamma \, a_i(\underline{x}) \, e^k$ with learning rate γ.

A severe problem is the efficent placement of receptive field centers $\underline{x}_i$. Without knowledge about the required mapping it is not possible to find an overall satisfying solution to that problem beforehand. One issue is the representation of input–pattern due to their statistics. There are several methods e.g. vector quantization, data clustering and feature maps to approximate the probability distribution of occuring pattern with the respective density of neurons placed. These methods are not very well suited for mappings with uniform distribution of input–pattern but varying output.

An alternate approach is to use the error as criterion whether the already found solution is satisfying or not.

In our approach a node is allocated everytime the real–valued error–measure $\mathcal{M}_e(\hat{\underline{y}}, \underline{y}, \underline{x})$ is above a threshold T_e i.e. $\mathcal{M}_e(\hat{\underline{y}}, \underline{y}, \underline{x}) > T_e$ and the most active neuron is less active than a threshold T_a i.e. $\max_{i=1}^{N} \{a_i(\underline{x})\} < T_a$.

If both conditions are fulfilled simultaneously a node is allocated with center $\underline{x}_{N+1} = \underline{x}$, output $y_{N+1}^k = \hat{y}^k$, and width $\sigma_{N+1} = \lambda \min_{i=1}^{N} \{||\underline{x}_{N+1} - \underline{x}_i||\}$, and every width of the already allocated neurons is reset to $\sigma_i = \lambda \min \{||\underline{x}_{N+1} - \underline{x}_i||, \frac{\sigma_i}{\lambda}\}$. λ allows for adjustable interpolation properties. Chosing λ close to zero approximates nearest neighbor table lookup with fixed outputs within each Voronoi–cell. Choosing λ much larger than one takes global properties of the mapping for each node into account but gradient learning becomes more difficult since there are more parameters to adjust simultaneously [6].

Earlier approaches [5, 3] never altered the width σ_i after allocation of the nodes resulting in a hierarchical order of decreasing width for the allocated nodes. In our approach the local density of neurons and the receptive field width are strongly coupled resulting in a democracy of importance for all nodes.

If the allocation conditions are not fulfilled a usual adaptation step using the delta–rule is done. The threshold for the error T_e depends on the number of learning steps s i.e. $T_e(s) = e_0 \, exp(-\frac{s}{\tau}) + e_f$ decaying exponentiallly to a final desired error e_f. The threshold for the activity T_a is kept fixed during learning.

Using this scheme, areas with large errors are covered first, and it is possible to suspend allocation of new neurons after a satisfying behaviour is reached or all the available resources came into use.

Prestructured network for inverse kinematics

Using robot arms requires a mapping of a desired path for the TCP in work–space $\underline{x}(t)$ to a path given in joint–angles $\underline{\Theta}(t)$. The variables describing the path are the position and orientation of the TCP. Differential inverse kinematics uses the desired velocity $\dot{\underline{x}}_{pos}$ and angular velocity $\dot{\underline{x}}_\omega$ of the TCP $\dot{\underline{x}}(t) = (\dot{\underline{x}}_{pos}(t), \dot{\underline{x}}_\omega(t))$ and the initial configuration of the robot $\underline{\Theta}_0$ to generate the desired path.

The general solution for the differential inverse kinematics problem for the non–redundant case is:

$$\dot{\underline{\Theta}} = \mathcal{J}^+(\underline{\Theta})\,\dot{\underline{x}} \tag{2}$$

where $\mathcal{J}^+$ is the inverse of the Jacobian $\mathcal{J}$.

$$\underline{\Theta}(t) = \int_{t_0}^{t} \dot{\underline{\Theta}}(t')dt' = \int_{t_0}^{t} \mathcal{J}^+(\underline{\Theta}(t'))\,\dot{\underline{x}}(t')dt' \tag{3}$$

with $\underline{\Theta}(t_0) = \underline{\Theta}_0$

Earlier approaches to train a neural network to find $\underline{\Theta}(t)$ ($\dot{\underline{\Theta}}(t)$ respectively) given a path $\dot{\underline{x}}(t)$ used the current joint angles and the desired TCP velocity as input. But in order to simplify the task for the network we use the linear relationship between $\dot{\underline{x}}$ and $\dot{\underline{\Theta}}$ (eq. 2) to prestructure the network as a linear perceptron with weights representing the entries of the inverse Jacobian $\mathcal{J}^+$ [2, 7]. The configuration dependance for the inverse Jacobian is incorporated by a network generating these weights. This approach decreases the number of inputs to the number of joints (Fig. 1) and increases the outputs to the number of entries of the inverse Jacobian $\mathcal{J}^+$.

NATEnet as described above is used to approximate $\mathcal{J}^+(\underline{\Theta})$. The output y^k for each node is J_{ij}^+.

Training NATEnet as described above is not possible since the values of the inverse Jacobian are usually not known. To overcome this problem the error for the TCP is used to produce an estimate for the error in joint coordinates which is used to alter the weights of NATEnet. The error–measure used for the decision about node allocation is

$$\mathcal{M}(\dot{\underline{\hat{x}}}, \underline{\Theta}, \dot{\underline{\Theta}}) = \frac{1}{||\dot{\underline{\hat{x}}}_{pos}||}\left(||\dot{\underline{\hat{x}}}_{pos} - \dot{\underline{x}}_{pos}^{net}(\underline{\Theta}, \dot{\underline{\Theta}})||\right) + \frac{1}{||\dot{\underline{\hat{x}}}_\omega||}\left(||\dot{\underline{\hat{x}}}_\omega - \dot{\underline{x}}_\omega^{net}(\underline{\Theta}, \dot{\underline{\Theta}})||\right) \tag{4}$$

where $\dot{\underline{x}}^{net}$ is the TCP movement generated by the network output $\dot{\Theta}$ for configuration Θ.

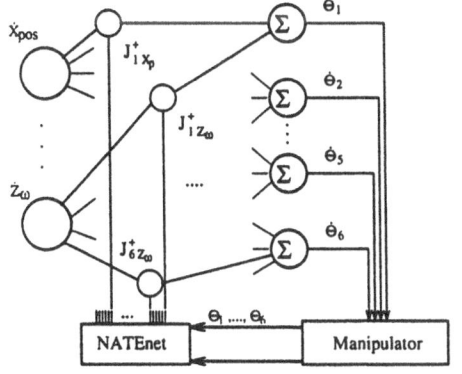

Fig 1. Prestructured Network for inverse kinematics of a 6–DOF robot arm. NATEnet has 6×6 outputs for approximation of the inverse Jacobian.

Results

The robot arm used for our simulations is a Siemens manutec r2 with 6 rotational joints. The r2 is a robot with a spherical wrist i.e. the last three axis intersect at one point (fig. 2).

312

We trained NATEnet using random positions within the workspace of the 6–DOF robot arm. Positions near singular points are excluded from training. For comparison a network with a fixed number of nodes was trained. The average error measure $M = \overline{\mathcal{M}}$ is shown in fig. 3.

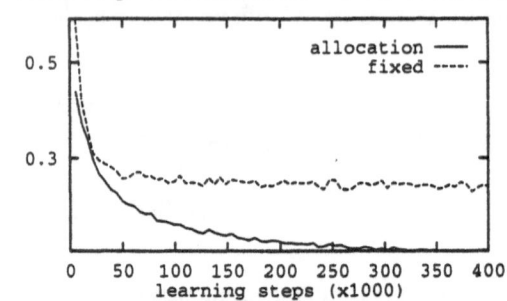

Fig 2. 6–DOF robot arm

Fig 3. average of the error measure $M = \overline{\mathcal{M}}$ vs. learning steps for a network using node allocation and a fixed number of nodes

Conclusions

Combining the use of a–priori knowledge to prestructure the network with error–driven node allocation and topographical encoding facilitates training of a network to approximate the inverse kinematics solution of a 6–DOF robot arm. The solution found using node–allocation compares favourably to a solution with the same number of nodes on a rectangular lattice. The proposed approach can deal with redundant robot arms[2]. Acceleration of simulated networks of NATEnet type on sequential and coarse–grained parallel computers for real–time applications shows promising first results.

References

[1] J. Hakala. HENAMnet homogeneous encoding for approximation of mappings. In Kohonen, Mäkisara, Simula, and Kangas, editors, *Artificial Neural Networks*, pages 1319–1322, Elsevier Science Publ., Amsterdam, 1991.

[2] J. Hakala, H. W. Werntges, J. R. Beerhold, and R. Eckmiller. Node allocation and topographical encoding (NATE) for inverse kinematics of a redundant robot arm - learning without a teacher. In I. Aleksander and J. Taylor, editors, *Artifical Neural Networks*, pages 615 – 618, Elsevier Science Publ., Amsterdam, 1992.

[3] V. Kadirkamanathan, M. Niranja, and F. Fallside. Models of dynamic complexity for time–series prediction. In *ICASS 92*, 1992.

[4] Y. Nakamura. *Redundancy and Optimization*. Advanced Robotics. Addison Wesley, Menlo Park CA, 1991.

[5] J.C. Platt. Learning by combining memorization and gradient descent. In Lippman, Moody, and Touretzky, editors, *Proc. Advances in Neural Information Processing Systems 3*, pages 714–720, Morgan Kaufman, San Mateo, 1991.

[6] D. F. Specht. A general regression neural network. *IEEE Transactions on Neural Networks*, 2:568–576, 1991.

[7] D.Y. Yeung and G.A. Bekey. On reducing learning time in context–dependent mappings. *IEEE Transactions on Neural Networks*, 1992.

Learning Optimal Control Using Neural Networks
N. Alberto Borghese
Center Neural Engineering University Southern California Los Angeles, CA 90089-0782
I.N.B. - C.N.R. , Via Mario Bianco 9, 20131 Milano, I
Borghese@imisiam.siam.mi.cnr.it

Introduction

Optimal Control is one of the most valuable techniques to determine the desired control action. In this framework, a movement can be evaluated by two parameters: its effectiveness (how well the goal of the movement is achieved) and its cost (how much energy, acceleration etc. is required to perform that movement); the control sequence generated by the system will be the one that minimizes these two parameters that constitute a Global Cost. The most important feature of these algorithms is that there is not an a-priori idea on which Control and which Path the system will choose, but these are the outcome of the process of optimisation itself [1]. Hereafter, an improved version, named Local Dynamic Programming, of the Heuristic Dynamic Programming (HDP) approach proposed by Werbos, is described along with its application to a reaching task.

Theoretical background

In the framework of optimal control, the challenge faced is to relate a global measure taken throughout the movement (the *Global Cost*) to a local error that can be used to tune the Controller at every time step; in connectionist literature, the *"Temporal Credit Assignment"* problem. In Dynamic Programming this is solved introducing a new function, the*Cost-to-go* , that memorizes the optimal Global Cost through time and states. It can be demonstrated that the sequence of Controls that minimizes the Cost-to-go at each time step, also minimizes the Global Cost. The system that can solve this problem is constituted of a two-level hierarchical adaptive structure (figure 1). The *Plant* is the model of the system that is controlled, in our case the kinematics equations that relate position, velocity, acceleration and jerk of the hand considered as a point moving along a direction in the three-dimensional space. The *Controller* at each time step, k, outputs the control, u_k, as a function of the actual state and time, and of the set of synaptic weights, wb. The Plant and the Controller constitute a recurrent network that generates the current trajectory. The *Cost* module outputs a local cost, U_k, and it is a function of only local variables, in our case the square of the jerk at time k: it may be regarded as a tactic function involved in a short term minimization process. The *Critic* is a function of state and time, and of a set of synaptic weights, wa. It outputs the evaluation of the current state in terms of the Cost-to-go from that state to the end of the movement, given a certain sequence of controls and states. It may therefore be regarded as a strategic function, related to the achievement of the global task. The Critic is needed only in the learning phase: its role is to observe the trajectory, to provide an evaluation of it in terms of the Cost-to-go. Classical dynamic programming is computationally very expensive; the cost-to-go (J_k) and the control (u_k) can be very complex functions of the state and they should be represented for the whole state space. For these reasons approximated methods have been introduced. In HDP [6,7], J and u are approximated by two-layered Neural Networks (NN) and gradient descent procedure, that in sigmoidal NN is efficiently implemented by the *back-propagation* algorithm, is adopted to tune the synaptic weights. The learning is obtained running the system only forward in time and it is carried out in two steps iteratively; every iteration named n, is the execution of a movement from the initial state $x_o/k=0$ to the final state $x_N/k=N$.

1) *Forward pass.* The system is run forward in time, the activity of the elements of control and critic networks are computed as well as the next-state function but not memorized like in BTT approaches. At each time step, the updating of the weights is computed as follows. (a) For the Critic network, dwa are computed such that J(xk,k/wa(n)) equals J(xk+1,k+1/wa(n-1)) + Uk where xk+1 is obtained from the Plant using uk=g(xk,k/wb(n)). (b) For the Control, dwb are computed such that uk=g(xk,k/wb(n)) minimises J(xk+1,k+1,wa(n)). With the first rule (a), the system teaches the critic to learn the cost-to-go associated to the different states xk/k touched during the movement. The second rule (b) allows the Controller to increase its performances by decreasing the overall cost from xk/k to the end of the movement. The gradient of the Critic respect to the Controller is computed and used as an "error" to tune the weights wb of the Controller itself. It is obtained by first computing the gradient of the Cost-to-go relative to its input, the state at k+1, (back-propagation-to-activation [4]), by back-propagating the obtained gradient through the Plant and, finally, through the Controller network. The weights are adjusted consequently.

2) *Adaptation.* The updating of the weights collected throughout the entire trajectory is added to the actual value of the weight to generate a new weights value.

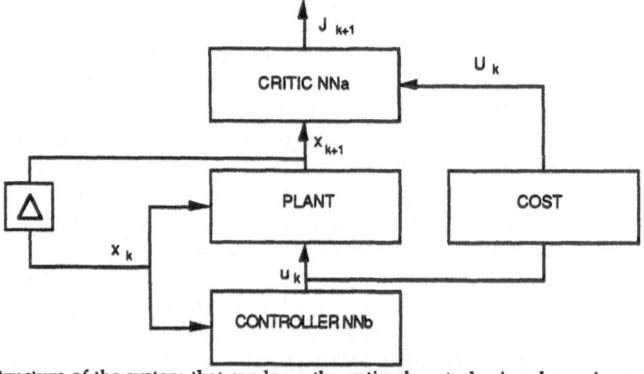

Figure 1. The structure of the system that can learn the optimal control using dynamic programming: the Plant and the Cost function are in analytical form while the Critic and the Controller are two neural networks. The second level constituted of the Critic and the Cost is used only in the learning phase.

Local dynamic programming (LDP)

In tuning the controller as prescribed by the HDP algorithm, not only the absolute value of the cost-to-go, but also its value around the actual trajectory is required to reliably compute the gradient of the Cost-to-go relative to its input state. Using HDP in its original formulation, this cannot be achieved unless an exploration of almost the entire input space is carried out. When this gradient has not been correctly established, the system may converge to a sub-optimal value and even diverge. We propose here an approximation of Dynamic Programming, LDP, that allows to use the HDP approach and obtain a more robust convergence thanks to the better approximation of the gradient of the Critic. In LDP the state space close to the actual trajectory is explored sampling the Cost-to-go in a volume around the actual trajectory. To achieve this, we start the trajectories in parallel from x_0 and from a set of close positions. Since the system kinematics is given by continuous functions, the perturbed trajectories will remain close to the original trajectory. The exploration of the cost-to-go on the set of the resulting trajectories allows to have a good approximation of the Cost-to-go in an entire region of space. The number of trajectories to be explored in parallel for every forward pass is 1 + 2 * S, where S is the dimension of the state space. With this procedure, we can determine with a good approximation, not only the absolute value of the Critic, but also its local gradient. Given the choice of running the system forward, we question the importance of the weights update obtained from the early stages of the trajectories in the first movements. Therefore a weighing of the updating of the weights that depends on the number of executed movements has been introduced. At the beginning of the learning the updates collected in the latter stages of the trajectory are heavily weighted and very little the ones collected in the early stages. As learning progresses the weight on the updating computed in the first stages is incremented and becomes equal to that of the last stages.

Results

The system has been implemented within the NSL simulation environment [5]. The Critic and the Controller are implemented as two-layered Neural Networks with sigmoidal units, fully interconnected. In order to learn the optimal control, we follow the strategy proposed in [4]: the system first learns to maximize the smoothness of the movement and after that it learns to reach the desired final state. This is achieved by using different weights for the cost associated to smoothness and to deviations from the final state. In the first phase, the sum of the squares of the jerk output during the movement is heavily weighted. The system forces the Controller to decrease the output jerk to zero, which causes the hand to move very little from its initial position. At the end of this first phase, position, velocity, acceleration and jerk maintain a value very close to zero. In the other phases, the weight on the Cost associated to the deviation from the final state, is increased in discrete steps keeping the Cost-to-go far from the lower saturation level of the output of the Critic network. In these phases the Critic monitors a big Global Cost that is due mainly to the deviation from the final desired state, and the system forces the Controller to modify its output to get the state reached at the end of the movement closer to the desired final position.

In figure 2 the trajectories of the state (position, velocity and acceleration) and of the control (jerk) at the end of the learning are plotted versus time. They are indistinguishable from the analytical solution and show the classical minimum jerk profile [2]. In the bottom row, we show the time course of the activity in the hidden units of the Controller. Their activity is highly modulated throughout the entire movement. We also explicitly observe that none of them is directly correlated either with position, or with velocity or with acceleration alone throughout the entire movement but only their ensemble can output the correct controller as a function of the input variables and can be seen as "primitives" of the movement.

This highly modulated activity in the hidden units is absent in the first phase of learning. The system is initialised generating very small random weights which correspond to a small positive or negative value of the jerk throughout the movement. As this is integrated three times in the plant, it is sufficient to drive the hand about five times away from the desired final position. In the first phase of the learning, the system minimises the squared of the jerk and the critic decreases the output of the controller to zero. This results in position, velocity and acceleration trajectories that slightly deviate from zero. This great change in the final position, velocity and acceleration is accomplished by a small change in the activity of the hidden units of the controller. The big change is to learn an effective time modulation of the activity in the hidden units like that shown in figure 2.

Conclusion

Optimal control is a natural way to describe the formation of an elementary motor act in humans where the goal would be dictated by the internal needs and the smoothness by the evolution. DP is an effective way to relate actual consequences to past actions translating global minimisation problems into local ones which are mathematically much more tractable. The use of LDP with a local exploration of the

316

space allows a convergence to an optimal solution in a more robust way. The system is constituted of a two-level hierarchical adaptive structure that allows a separation of the learning phase (Critic) from the execution phase (Controller-Plant loop). The Critic may have also the role of re calibrating the Controller when the performances degrade. This property can make the system robust with respect to changes in the environment and to ageing. It has also implications for system intelligence, where intelligence is defined as the ability to output the best long-term behaviour depending on the circumstances.

10/20/92,model:FF,frame:Trial_A

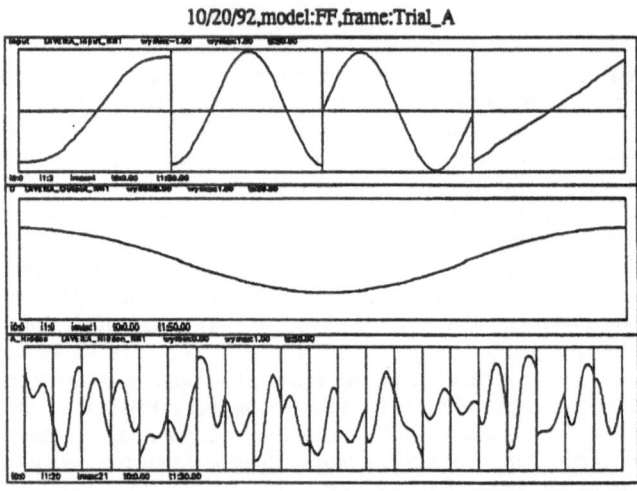

Figure 2. Trajectories of the Plant and the Controller at the end of the learning phase. Upper row: the trajectories of the state (position, velocity and acceleration and time) plotted versus time. Middle row: the trajectory of the output of the Controller (jerk). Bottom row: the activity in the 21 hidden units of the controller network plotted over time.

Acknowledgements

This work was mainly carried out at CNE of USC where N.A. Borghese was supported by CNR grant n. 203.15.3/1990.

References

[1] Bryson, A.E. Jr. and Ho, Y.C. (1969). Adapted optimal control. Waltham, MA: Ginn and Co.

[2] Hoff BR (1992) A Computational Description of the Organization of Human Reaching and Prehension, Technical Report USC-CS-92-523, USC-CNE-92-02, University of Southern California -- Center for Neural Engineering, Los Angeles, CA.

[3] Jordan MI, Rumelhart DE (1992) Forward models: Supervised learning with a distal teacher, Cognitive Science.

[4] Kawato M, Maeda Y, Uno Y, Suzuki R (1991) Trajectory Formation of Arm Movement by Cascade Neural Network Model Based on Minimum Torque-change Criterion, technical report, TR-A-0056.

[5] Weitzenfeld A (1991) NSL: Neural Simulation Language Version 2.1, CNE-TR 91-05, University of Southern California, Center for Neural Engineering.

[6] Werbos PJ (1990a) A menu of designs for reinforcement learning over time, Neural Networks for Control WT Miller, RS Sutton, PJ Werbos eds. Cambridge MIT Press, 67-95.

[7] Werbos PJ (1990b) The consistency of HDP applied to a simple reinforcement problem. Neural Networks, 3: 179-189 .

A boolean net as an adaptive and universal robot control

F. Bini Verona, F. E. Lauria, M. Sette and *S. Visco*
Università di Napoli, Dipartimento di Scienze Fisiche,
Mostra d'Oltremare, Pad. 19, 80125 Napoli (Italy)

We consider a McCulloch and Pitts, also called Boolean, net in the Caianiello formalism, as a set of "node assemblies", to implement an Hebbian rule [1]. So we obtain a system learning, by examples, how to control a robot in the execution of task sequences, without any limitation on their length.

A node assembly is a subset of the network nodes with the following properties: first, it contains just one node, the assembly Hebbian node, receiving at least both one Hebbian connection (an "Hebbian connection" is a connection with associated a coupling coefficient computed by applying the Hebbian rule) and one connection with a coupling coefficient greater than its threshold, ie, a firing connection. Second, it contains just two output nodes, the assembly first and second output, generating either Hebbian connections, if they impinge onto another assembly Hebbian node, or the network output connections, ie, those controlling the primitives. By construction, the assembly output nodes are active in the given order. Third, an assembly remaining node activity is a temporary pointer to the, different, assembly containing the Hebbian node activated by the signal last generated in the former output nodes. Actually, as the assembly address space is knowledge based, the remaining nodes execute what could be called "address matching". By construction, once the training is terminated and the coupling coefficients are frozen, an assembly output node channel impinges, at the most, on the Hebbian node of just one assembly. The activity in the network output nodes is the command to execute an elementary task, thus, the output assembly activity controls the robot actuators in the execution of a primitive.

Let us consider the net controlling a robot arm with the six primitives: **left**, or **l**, **right**, or **r**, **up**, or **u**, **down**, or **d**, **move**, or **m**, and **write**, or **w**. The first four are the command to move the robot hand one unity in the said directions. The last two control the hand movements in a direction perpendicular to the paper sheet, ie, away or toward the latter. That is, when it is near the paper sheet and executes an **l**, **r**, **u**, or **d** command, its pencil leaves an, horizontal or vertical, unitary sign. Instead, when it is away from the paper it simply moves, relative to its environment. The net has two independent inputs: respectively, the visual and the auditory channel. This net can be trained to compute any general recursive function, provided the system I/O, ie, the robot actuators have been chosen accordingly. We have simulated a thousand node net, containing sixty assemblies and up to thirteen primitives on the Vax cluster of the *Dipartimento di Scienze Fisiche della Università di Napoli*. As an example, we discuss its training as the control of an arm learning the drawing of the alphanumeric characters.

The net we have simulated has been successfully trained to control the robot arm in drawing some alphanumeric characters as sequences of horizontal and vertical unitary bars. To output the digit 1, 0, we use the input sequence 'one', 'zero', and we call 1-assembly, 0-assembly, the assembly so activated. Now, we are training the robot to concatenate sequences, e. g. to write 10 or 110 when asked as 'ten' or 'eleven hundred', respectively.

References

[1] F. E. Lauria and M. Sette. "A general approach to learning of task sequences". In I. Aleksander and J. Taylor eds. "Artificial neural network, 2". Elsevier Sc. Pub. (1992) 475-478.

Transforming Occupancy Grids Under Robot Motion

Joris W.M. van Dam[1] and Ben J.A. Kröse and F.C.A. Groen

Faculty of Mathematics and Computer Science, University of Amsterdam
Kruislaan 403, NL-1098 SJ Amsterdam, The Netherlands
dam@fwi.uva.nl

If an autonomous robot uses a local, robot-centered model of its environment, it is essential that this model is transformed every time the robot moves. In our application, an *occupancy grid* is used to model the robot's environment (see [1]). The transform of occupancy grids involves the computation of numerous intersecting volumes of original and transformed grid cells. If no knowledge is available on the sizes and shapes of the cells, the fastest method to compute these volumes is Monte Carlo estimation. We define a neural network that implements this estimation procedure for all volumes in parallel (see [2]). Each cell of the occupancy grid is represented by a neuron and the transform of the grid is represented by a single layer feed forward network. This paper describes how such a network learns to transform an occupancy grid for a single robot move.

In our network model, neurons x_j^A in the input layer represent the cells in the original occupancy grid and neurons x_i^B in the output layer represent the cells in the transformed grid. The activation of each neuron gives the *probability* of the corresponding cell being occupied. Taking this into account, it can be derived that the activation of the output neurons is given by:

$$x_i^B = 1 - \prod_j (1 - w_{ij} x_j^A) \tag{0.1}$$

The weights of the network determine the transform and are learned for a given robot move. Starting with a randomly initialised network with each output unit connected to all input units, examples of grids and their correct transforms are generated. With these examples, the weights can be learned using the δ-rule. With the activation function from (0.1), the δ-rule rule gives:

$$\Delta w_{ij} = \gamma (\hat{x}_i^B - x_i^B) \cdot x_j^A \cdot \prod_{k \neq j} (1 - w_{ik} x_k^A) \tag{0.2}$$

with $\hat{x}_i^B$ the desired output value taken from the correct transformed grid.

The learning speed of the network is determined by the number of occupied cells (active input neurons) in the learning samples. If we take only few cells j with $x_j^A = 1$, equation (0.2) gives that only few weights are updated. However, if we take too many $x_j^A = 1$, the *magnitude* of the update will be suppressed by the product term in equation (0.2). On the other hand, we may expect that as learning proceeds, we will get many $w_{ik} = 0$ so that the effect of suppression will decrease in time. We defined an optimisation procedure which calculates the optimal number of occupied cells per learning sample as a function of the size of the grid and of the average number k for which $w_{ik} \neq 0$.

Results show that the network is indeed capable of learning the correct weight values. Furthermore, if the number of occupied cells per learning sample is optimised using the procedure mentioned above, the network converges within 4000 iterations for a 256x256 grid (!!), whereas if no optimisation is performed, the network fails to converge within 20000 learning steps.

References

[1] Alberto Elfes. *Using Occupancy Grids for Mobile Robot Perception and Navigation.* Proceedings of the 1989 IEEE Int. Conference on Robotics and Automation.
[2] J.W.M. van Dam, B.J.A. Kröse, F.C.A. Groen, *A Neural Network that Transforms Occupancy Grids by Parallel Monte Carlo Estimation.* submitted to CSN'93.

---

[1]The investigations were supported by the Foundation for Computer Science in the Netherlands (SION) with financial support from the Netherlands Organization for Scientific Research (NWO)

Complex Tasks and Robots

A. Martinengo [1], M. Campani [2], and V. Torre [1]

[1] Dipartimento di Fisica and [2] Consorzio I.N.F.M.

Via Dodecaneso 33, 16146 Genova, Italy

Robots have to move and survive in a variety of environments. In order to do so they have to avoid obstacles, find a collision free trajectory and cope with erroneous information; therefore, they must have rather efficient feedback loops, very similar to animal reflexes (Brooks, 1986). Real robots, however, have also to accomplish useful tasks, which are likely to have a reasonable degree of complexity. The purpose of this research is to perform experiments, in an unfriendly environment, with a robot having to accomplish tasks, which from a computational point of view may be rather complex, such as the exploration of a maze, the recovery of its structure and the planning of some nearly optimal trajectories through the maze. The robot used in the experimentation moves in a laboratory where a maze was constructed by sticking black stripes on the floor. The vision system of the robot is composed of a single uncalibrated T.V. camera pointing to the floor. In order to obtain almost real time performances, the perceptual world of the robot has been simplified, and is composed of black stripes and silhouettes over a flat floor with reflexes and shadows. A network of three computers is used to control the system behaviour. The processing of images is quite simple: the gray levels of the image are binarized, edges are extracted, and edge chains are approximated with a collection of straight segments. The robot has to recognize the direction of black stripes and the presence of junctions and silhouettes. The recognition of the silhouettes is based on the analysis of edge segments and takes into account the deformations due to perspective projection. In order to process the visual information correctly the system requires the setting of several parameters. By using an automatic learning procedure, given the lighting conditions, the material of the floor and of the black stripes, it was possible after a period of supervised learning to obtain a set of parameters, which could be used rather reliably. When the visual information has been processed, the robot has to move according to its task and its perception of the visual world. Two different strategies have been used: often, it is not necessary to follow the black stripes much carefully; in these cases the velocity of the wheels is continuously updated without halting the robot. In the presence of a junction, a better control is necessary: the robot has to reach a complete stop and adjust its position and heading direction according to its perception. During the navigation along a single stripe, the robot uses the first strategy; when a junction has been detected the robot switchs to the second. The main task concerns the exploration of the maze and the recovering of its structure, and has been implemented using a slightly modified version of the standard depth first search algorithm; by using this new strategy the number of edges that have to be followed to complete the exploration of the maze has been reduced. The experiments performed with standard low cost hardware show that the robot follows the lines of the maze in a crude way, but at a remarkable speed (0.2 meters/sec); moreover, since the vision system is almost uncalibrated, when the viewing camera is tilted from the nearly optimal position, the robot is still able to explore the maze, although with a more irregular trajectory. The robot is also able to accomplish its tasks, even if its wheels are maliciously slid or an erroneous stripe is shown for a while. Two factors contribute to this robust behaviour: the short time required for closing the optomotor feedback loop (200–500 msec) and a control strategy largely based on qualitative measurements. A way to evaluate the relative difficulty is to measure the time spent by the system solving different problems. Given the available computing power the robot spends a large amount of time in the control of its motion, and a significant fraction of the total time in the communication between the different computers. This time can be drastically reduced by a proper hardware mounted on board of the robot. The system will spend a significant fraction of the total time reasoning and planning only when the combinatorial complexity of the problem to be solved becomes relevant, for instance in the case of a very complex maze, composed of hundreds of junctions. In the presence of mazes with a limited number of junctions, the robot spends most of its time seeing and moving. These experiments suggest that the hard problem in the control of real robots, performing standard inspection procedures or monitoring, is perception and motor control.

COGNITIVE
CONNECTIONISM

Neural networks, natural language and artificial intelligence
—oral contributions

NN Approaches to Natural Language: Context and Trends

Anton Nijholt
University of Twente, P.O. Box 217,
7500 AE Enschede, The Netherlands

Abstract. This is a short introduction to topics and problems in natural language processing that are of interest for connectionist research. We recall the natural language processing (NLP) problem and conclude that until now no promising approaches from a neural network (NN) point of view are available. It does not mean that there is no progress in this field. Rather than attacking the complete problem it has become customary to search for subtasks for which a NN approach is useful. In addition, the sometimes rather ad hoc and exotic approaches to language analysis have been replaced by fundamental research into sequence recognition, finite state automata (FSA) simulation and general research on grammatical inference with NNs. We do not confine ourselves to NN approaches, but survey a range of approaches to NLP and provide the framework in which connectionist NLP can be embedded and evaluated.

1. Introduction

This is a short introduction to topics, problems and approaches in NLP that are of interest for connectionist research in this area. The emphasis is on typed natural language. Speech recognition with NNs has become an important research area and many comparisons with conventional methods have been made. Conventional and successful methods for speech processing are not knowledge-based, that is, rather than being based on theories of linguistic processing and/or psychological processing they are statistically based. For NLP the situation is different from that of speech processing. In speech recognition continuous patterns are transformed into symbols. In NLP symbols are transformed into symbols. In speech recognition NN approaches can be evaluated against powerful and mathematically based approaches (e.g., HMMs). In NLP there are no successful existing approaches, only starts to them. Successful research in speech recognition or, e.g., the conversion from English words into phonemes can only be matched by NN approaches to restricted NLP subtasks.

NNs have been used for a number of subtasks. Not all of them require 'high-level' processing of information and not all of them require the same amount of linguistic knowledge. So, we can mention speech recognition, grapheme to phoneme translation (NNs that learn to read aloud), hand-written and printed character recognition (OCR) and the use of linguistic knowledge to improve results (lexical, syntactic or co-occurrence information), kana-kanji conversion, modelling of the mental lexicon, word and word class prediction (the categorisation of words into syntactic categories), word sense disambiguation, PP-attachment resolution, syntactic analysis (parsing), robust parsing (complete sentences with missing words), interpreting imprecise expressions, resolve 'garden path sentences', resolving of lexical and syntactic ambiguity for sentence interpretation, unification and parsing of unification grammars, using NNs built from definition texts in machine readable dictionaries for word sense disambiguation, case role assignment, the determination of referents of pronouns in sentences, learning the past tense of verbs, learning and assignment of linguistic stress, resolving the ambiguity between idiomatic and non-idiomatic meanings, etc. Sometimes the results are intended to be part of larger, often hybrid (combining symbolic and subsymbolic approaches), systems, e.g., an information retrieval system, a machine translation system or a natural language interface. For such applications only toy-domain results have been obtained. Hybrid architectures have also been introduced for word sense disambiguation and for parsing.

2. Natural Language Understanding

In a well known textbook (Allen[1987]), the different traditional forms of knowledge that are needed to understand sentences are distinguished: phonetic and phonological, morphological, syntactic, semantic, pragmatic and world knowledge. Finer distinctions have been made. In a dialogue participants have intentions behind utterances, they are planning their contribution to the dialogue. Hence, knowledge about participants and their plans is another source that has to be called on when processing sentences. In addition, texts and conversations have structure. Theory has been developed to model possible structures and to relate plans and communication goals to utterances in these structures. Another source of knowledge that has to be used.

Generally, in order to understand a sentence three phases of processing are distinguished. In the parsing phase a sentence is processed into a structural description using syntactic and morphological knowledge, in the semantic interpretation phase the structural description is mapped, using semantic knowledge, into a logical form which represents the (literal) meaning of the sentence (independent of the context), and in the contextual interpretation phase this representation is mapped into a final representation of the effects of understanding the sentence. These phases suggest a sequential and/or incremental processing of sentences and a modular and procedural approach. However, grammars for (subsets of) natural languages are - whatever formalism is used - very large and highly ambiguous. This ambiguity causes computational problems. Too many syntactic parses are constructed before semantic analysis can take place. Several approaches to 'interleaved' syntactic and semantic analysis can be found in the literature. However, none of these approaches has become a standard. Semantic grammar and case frame grammar parsing have been introduced. From Artificial Intelligence (AI) knowledge based ideas have been imported. Schank's conceptual dependency parsing almost bypasses syntax. The use of frames and scripts and related knowledge representation formalisms allow the interpretation of sentences in a context that gives priority to world knowledge rather than syntactic or semantic knowledge. The obvious problem, hardly addressed in AI, is the choice of contexts of interpretation during sentence processing.

There exist a lot of formal approaches to aspects of language use and language description. There does not exist an integrated approach and the approaches that can be distinguished do not cover all facets and are often developed independently of each other. Morphology and syntax are reasonably understood phenomena. This does not mean that syntax is a computationally manageable problem. Large grammars describing a considerable part of a natural language are rare. Because of the large number of rules they are difficult to extend and change without loosing consistency. Although there are many competing grammatical formalisms, syntactic formalisms are often based on context-free-like formalisms with added attributes or features which enforce agreement (e.g., auxiliary-verb agreement) and complements (subcategorization of the verb, e.g., whether it allows no NPs, one object NP or an object and an indirect object NP to follow).

2.1 Architectures for Language Analysis

Literature on the architecture of natural language understanding systems is scarce. A distinction has to be made between architectures that are introduced from a cognitive or psycho-linguistic point of view, that follow from software design considerations or that are based on new or existing computational models and that hope to profit from new or yet to develop computer architectures. It is interesting to note that, at least in the case of syntactic analysis, the introduction of new paradigms in programming (parallel, logic, functional and object-oriented) also has led to new approaches to - and architectures for - the parsing problem.

From a psycholinguistical point of view computer models of human sentence processing should be consistent with theories developed in these fields. That is, with a model it should be possible to simulate phenomena of human sentence processing. Psycho-linguistic experiments (e.g. on the correspondence between the sequence of psychological processes and the sequence of operations whereby a grammar generates a structural description) and theoretical results (e.g., the time and space complexity of recognition and parsing) can play a role in determining the psychological plausibility of parsing models. Moreover, the design of these models can be inspired by these results. Theoretical topics that play a role are, e.g., weak and strong generative capacity, learnability, succinctness of description and parsing efficiency (Berwick & Weinberg[1982], Abney & Johnson [1991]). Experimental data can tell us how lexical decisions are made, whether humans parse deterministically or whether on hearing a word all its senses are activated in the human brain. Much can be learned from deviations of ordinary language use. Being able to model this deviation, e.g. of an aphasia patient, can be highly informative.

Human sentence processing has been explained with serial models (e.g., Miller[1962], Fodor et al [1974], Frazier et al[1978]), where there is a 'syntax-first' approach and the syntactic processing task must be successful before semantic processing can begin, which in turn precedes pragmatic processing. If in this model of linear interaction between levels of knowledge higher level information cannot be used to influence decisions at lower levels, this approach leads to an explosion of legal syntactic possibilities. For that reason models of interleaved syntactic-semantic processing have been developed (e.g., cascaded or multi-level augmented transition networks) and models where different knowledge sources interact with each other to obtain the meaning of a word or sentence. In a heterarchical system natural language processing tasks are assigned to different 'processes' and every knowledge source interacts with every other. Examples include systems where a task is the construction of a noun phrase (NP) or where a task is the full syntactic analysis of a sentence. In a blackboard model the multiple sources can process in parallel and co-operatively by means of a globally accessible 'blackboard' on which they can write and read intermediate results. In these models parallelism is task-oriented.

In an influential paper Marslen-Wilson[1975] reported about experiments that gave psychological evidence to the idea that processing at each level of natural language description can constrain and guide simultaneous processing at other levels. That is, " . . . *evidence that sentence perception is most plausible modelled as a fully interactive parallel process: that each word, as it is heard in the context of normal discourse, is immediately entered into the processing system at all levels of description, and is simultaneously analysed at all these levels in the light of whatever information is available at each level at that point in the processing of the sentence.*" Parallelism in agreement with this view can be obtained by considering words as active agents that interact with each other and possibly with other knowledge sources in order to obtain a meaning representation of a sentence. In 'word expert parsing', when reading a sentence, each word triggers a 'word expert', a program attached to a word. This word expert starts processing and by interacting with others and some higher-order processes it tries to determine the intended meaning of the word in its context. Hence, it will ask questions to other word experts and it is ready to answer questions posed by others.

Illustrative is the difference between the Fodor book of 1974 and the well known Clark & Clark book of 1977. In the latter we see a much more creative approach to comprehension. Well known is their strategy in which subtasks are assigned to little demons who co-operate to fulfil the comprehension task.

2.2 No Comprehensive Computational Models
A returning discussion, which we only mention now, is the difference between a holistic and a reductionist view of language. In the holistic view language processes should be described as cognitive strategies employed by the brain. Knowledge used by these strategies is shared with other strategies, e.g. for vision. A model of a phenomenon of the brain, such as language, is achieved from a model of the brain as a system as a whole. In the reductionist view the individual phenomena are modelled and there is not necessarily an account of the relation between phenomena in a unified whole. In the connectionist literature 'holistic' is often used to express the interaction between all possible types of knowledge in solving problems.

Whatever viewpoint, lack of convincing results suggest that computational models of real natural language and its use will not be available in the near future. For that reason researchers in NLP started looking for applications that allow restricted language use. A natural language interface to a database containing information about flight times has no need to be able to conduct a dialogue with a user about sports events. The domain provides the pre-specified context in which questions, remarks and commands can be interpreted. Examples of commercial NLP systems exist which indeed owe their success to a well-chosen domain of application.

The lack of success led to criticism on the rule based approach towards natural language. It is assumed that the huge amount of linguistic data and the necessary interaction between knowledge sources during analysis do not allow a manageable exhaustive description. Conclusions that have been drawn differ. It has led to discussions about a possible evolutionary account of language universals, about the necessity of being able to decide between grammatical and non-grammatical strings, about the necessity of complete syntactic trees, about the use of statistical methods, and, starting in the 1980s, the use of NNs and genetic algorithms in language analysis. The question has also been raised whether not more attention should be paid to grammars that describe performance rather than competence. Hence, there have been proposals and arguments for describing natural languages with FSA since recursion in language use is always highly restricted.

2.3 Regular versus Context-Free/Context-Sensitive: Performance versus Competence?
There have been hot discussions about the relationships between the generating power of the grammars from the Chomsky hierarchy and natural languages. Examples of language constructs from sometimes well known, but more often less well known or exotic languages have been used to make statements about the non-context-freeness of natural languages. Proofs have been criticised because the arguments involved not only syntactic, but also semantic constraints. Approaching the question from a 'performance' viewpoint, it has been noticed that ordinary language users (non-linguists) never need unbounded recursion (embedding) or sentences of unbounded length. See e.g. Miller[1962], Suppes[1972], Schütze & Reich[1990] and Hasida[1990]. For example, Suppes remarks that *"From the standpoint of empirical application, one of the more dissatisfying aspects of the purely formal theory of grammars is that no distinction is made between utterances of ordinary length and utterances that are arbitrary long, for example, of more than 10^{50} words. One of the most obvious and fundamental features of actual spoken speech or written text is the distribution of length of utterances, and the relative sharp bounds on the complexity of utterances, because of the highly restricted use of embedding or other recursive devices."*, while Schütze and Reich notice that *"While it is still a point of contention, recent carefully controlled experiments suggest that the human syntactic mechanism, without*

semantic or pragmatic cues, and without the aid of pencil and paper to construct sentences, does have a sharp level of one or two levels of embedding." And, as has been remarked by Rumelhart & McClelland in their PDP books: *"As we have already seen, one can make an arbitrary computational machine out of linear treshold units, including, for example, a machine that can carry out all the operations necessary for implementing a Turing machine; the one limitation is that real biological systems cannot be Turing machines because they have finite hardware. ..... We have not dwelt on PDP implementations of Turing machines and recursive processing engines because we do not agree with those who would argue that such capabilities are of the essence of human computation ..."* The emerge of NNs in the area of NLP has given a new impetus to this discussion. One reason is that a NN has a fixed size, and that in several approaches to NN processing of natural language this size determines the maximum of sentence length that can be handled. A second reason is that current NNs can only model limited embedding in sentences.

2.4 Probabilistic Parsing Approaches

In 1949 science advisor Warren Weaver, impressed by Shannon's *Mathematical Theory of Communication*, suggested to use a statistical (information-theory-like) approach to the description of language (*"entropy speaks the language of language"*) and even to language translation. However, Chomsky's cognitive approach in the early fifties was taken more seriously. The last five years have seen a growing interest in statistical methods and corpus-based linguistic research. A corpus can be used to provide examples and test material, but also to provide a basis of a statistical model of language. The decrease in interest in these models after the late 1950s has been explained by pointing at the 'anti-empiricist, anti-numerical, pro-symbolic trend in the Zeitgeist' during these years (Liberman[1991]). However, in the 1990s we see statistical approaches not only to speech processing, but also to tasks as message parsing, machine translation or even the translation of English specifications (prescribing the behaviour of a telephone switching system) into an automated testing language. Liberman argues that enormous corpora will be needed to obtain statistical models, since rather than modelling the regularities of speech and language, in this approach we are modelling regularities of the world.

Simple surface properties, such as which word will probably follow another word or will follow a sequence of *n* words, can be described using statistical information (nth order Markov chains). Although useful for some tasks, this information does not help in finding the most probable constituent structure of a sentence. Structure is described with the rules of the grammar. Hence, if we confine ourselves to syntax, then rather than designing formal grammars that are based on our intuitions about correct sentences, grammars can be based on a corpus of sentences that once have been uttered or written down. The corpus is finite and it allows the distribution of probabilities to grammar rules or parse actions (e.g., shift and reduce actions in a shift-reduce parser). This leads to the introduction of stochastic grammars and parsers. The statistical information can guide the parsing process and reduces the number of syntactic trees that have to be considered. Clearly, the parsing process can be guided to a wrong solution. The hope is that the resulting grammar or parser behaves well on sentences that are not present in the corpus. Assigning and inferring probabilities for rules of context-free grammars according to this idea can be found in, e.g., Fu & Booth[1975] and Ophoff [1992]. In criticism on the stochastic grammar approach the localness and context-independentness of assigning probabilities to rules is emphasised. A general problem is also the question whether a corpus is representative enough to have a high enough level of statistical significance. Clearly, the same problem arises with corpus-based NN learning of NLP tasks.

3. AMBIGUITY, NEURAL NETS, NEW APPROACHES

It is useful to mention a characteristic and influential paper from the mid-1980s. In Waltz & Pollack[1985] sentence interpretation is done by using concurrently a conventional chart parser and NN disambiguation. It has been very influential because it was concerned with the main problem in NLP (ambiguity), despite that its contents only centred around two example sentences: *John shot some bucks* and *The astronomer married a star*. These sentences illustrated the intended disambiguation process using semantic and contextual nodes in the network. For example, if the context node 'hunt' was activated, the animate 'bucks' node received more activation than the node that represented the 'money' interpretation of 'bucks'. Similarly, 'married' activates a 'spouse' node which in turn can activate a 'movie-star' node, but not a 'celestial body' node in the network. Determining the correct sense of a word is a fundamental problem in NLP and is a problem that cannot be isolated. It requires the interaction of many sources of yet not completely known knowledge. Therefore, it is also this interaction and integration that can be said to be a fundamental problem in NLP. Mutual constraints and competing hypotheses ask for parallel, simultaneous, access to all information. Models for this type of processing were not available in (computational) linguistics (see section 2.1) and therefore NNs that can excel at integrating different constraints, were expected to provide a break-through and a change in the undue attention for syntax to the detriment of semantics and pragmatics.

The enthusiasm for using neural networks for NLP tasks has led to a great research activity that has produced a large quantity of papers on small subtasks using sometimes exotic and ad hoc defined versions of existing neural networks. Mostly considerations how to integrate the results in embracing tasks are not available and learning is not part of the research. Generalisation, let alone large scale generalisation (scaling-up of networks) is hardly considered. The blind application of NNs or NN-like models to example problems is an other characterisation of this period. Primitive intuitive and extremely reductionist models of cognitive tasks appeared and disappeared. Most well known NNs (back propagation, Hopfield, Kohonen, Boltzmann) have been used to attack NLP problems. Despite this negative characterisation, fundamental issues have emerged. To mention a few: local vs. distributed representation, coding of semantic micro-features, the representation of concepts, compositionality, modularity, systematicity, sequence recognition and grammatical inference. In this paper the latter two, where we do not want to restrict ourselves to syntax but also assume the inclusion of semantic aspects, are considered to be the most important issues at this moment in connectionist NLP. The extent to which a theory succeeds in constructing an adequate grammar on the basis from finite samples from the language could be critical for the choice between alternative linguistic theories.

As we mentioned in section 2.2 no comprehensive computational models of language processing are available. Criticism on the rule-based approach and in particular generative grammar has been amplified by the arrival of NN approaches. Unfortunately, until now no alternative is available. What has become available are ideas in which the effects of language use achieve emphasis (rather than the structure of utterances), attempts to design a 'Cognitive Grammar' that asks for connectionist modelling (Langacker[1987]), Sentence Gestalt and Story Gestalt theories that are based on ideas about cue-based constraint satisfaction, evolutionary and biological accounts of the language faculty in which self-organising plays an important role and attempts to embed ideas of Humberto Maturana in a cognitive linguistic approach to language processing. Nevertheless, some of the ideas can possibly give us more firm grip to important issues than has become possible from all connectionist approaches to word sense disambiguation and the interaction between syntax, semantics and world knowledge during language comprehension. Learning, learning from examples, learning systematics (structure) in language from temporal patterns and language structure as the product from self-organising are the main issues. The ability to analyse (understand) and produce language follows from the 'devices' that are learned. Expressed in an other way, there is no need to learn to parse if you are a parser.

In the next section we survey results on learning grammar or recogniser. In addition we have a few words on genetic and NN parsing without reference to a learning algorithm for the parser.

4. LANGUAGE PROCESSING AND LEARNING

Formal models of language acquisition and discussions about them can be found in the literature since the early 1960s. Language learning means learning a grammar or a recogniser. Gold[1967] proved that if a learner can have access to both grammatical and ungrammatical strings a grammar can be induced for the decidable languages ('language identification in the limit'). This class includes the regular, context-free and context-sensitive languages. If only access to grammatical strings is possible then they can not be learned exactly. However, approximations to correct grammars are possible. It has also been shown that from a sample of sentences a most probable stochastic grammar can be induced. Learning from a sample of sentences according to these algorithmic methods is intractable in practice because of learning time. They are also psychologically implausible. Heuristic procedures have been introduced to deal with both aspects (Fu & Booth[1975] or Solomonoff[1964], who was concerned with recursive context-free rules).

In this research the 'syntax-first' approach translated into poor attention to semantics. An other way of dealing with plausibility or efficiency can be the inclusion of meaning. In the 'Cognitive Theory' of developmental psycholinguistics it is assumed that in learning by children, meaning inferred from non-linguistic context plays an important role in learning syntax. It is easier, as has been confirmed by experiments, to learn syntax if semantic information is available. Anderson[1976, 1977] has given semantic versions of Gold's techniques. He also introduced semantic heuristics, e.g. for relating a meaning structure to a constituent structure. Well known is the learning of an augmented transition network from pairs of sentences and semantic structures. In Wexler et al[1975]. the assumption is made that humans entertain innate hypotheses about grammars. This allows them to design models for learning Chomsky's transformational grammar.

4.1 Genetic Algorithms and Simulated Annealing

Parsing can be viewed as searching the space of all possible parse trees for a tree that fits the current input sentence. Uncontrolled 'generate and test' methods were already proposed in the late 1950s, but were abandoned for efficiency reasons. An example of an annealing parser is in Sampson[1988]. The approach is part of statistics-based natural language research in which no clear-cut grammatical/ungrammatical distinction of sentences is assumed. In Sampson's approach possible tree structures are generated at random with the aim to choose a tree with a high plausibility. For that reason it is necessary to have an evaluation function, to define a class of local changes that can be made to a tree and to define an annealing schedule. Tree evaluation is based on statistics. The local changes to trees consist of selecting a node at random, disconnect its subtree and locate it somewhere else. For the latter operation some constraints are defined. Similar ideas can be found in De Weger[1990]. Instead of simulated annealing, a genetic algorithm is adapted for context-free language analysis. Usually, in genetic algorithms, codings representing solutions of problems consist of bit strings. For parsing, other schemes can be more natural. Here the search space consists of inorder representations of syntactic trees. The evaluation function which defines the fitness of a tree is based on the occurrences and the positions of words. Mutation (replacing a random subtree by a randomly created parse tree with the same root) and (tree) cross-over (from two random trees new trees are constructed by swapping subtrees) are defined.

Generating tree structures from elementary parts and evaluating them with respect to a given sentence is also the approach to parsing presented by Kempen & Vosse[1988]. Another similarity is the use of simulated annealing. Syntactic trees are constructed out segments. A segment consists of two nodes connected by an arc. The arc is labelled with a syntactic function (Head, Subject, etc.) and feature matrices are associated with the nodes of an segment. By unification of feature matrices a foot and a root node of a segment can be merged and a concatenation of segments is obtained. Similarly, roots of segments can be unified. Parsing is syntactic tree formation out of segments. Metaphorically, this can be compared with the biosynthesis of proteins. In a 'test tube' - called the Unification Space - nodes will continuously hit upon each other, each time feasibility of unification is checked, and the probability of unification depends on the activation levels of the nodes and a fitness function. It matches word order in the parse tree with the input. It can take into account non-syntactic factors. Due to activation level decay merged nodes may separate and stronger unifications become possible.

The previous paragraphs were concerned with the mapping from sentences to parse trees. Genetic algorithms, as probabilistic optimisation algorithms, have also been used to learn (induce) machines that accept languages corresponding to certain grammars as well as learning grammars generating certain languages (grammatical inference). These problems are NP-complete, that is, we can not expect to find an efficient algorithm that learns the correct machine or grammar (Gold[1978]). However, heuristic search techniques allow efficient processes to approximate the correct solution. Genetic algorithms have been used to learn descriptions for FSA for accepting regular languages (Zhou & Grefenstette[1986]) and for learning descriptions of context-free grammars (Wyard[1991]) or pushdown automata (PDA) that accept context-free languages (Sen & Janakiraman[1992]). Learning requires a given set of legal and illegal sentences of the language. In the latter paper the search space consists of codings of (deterministic) PDA into bit strings that can be manipulated by the recombinant operations of the genetic algorithm. The evaluation function that measures the quality of the bit string first decodes the bit string to a set of PDA rules and then the PDA is simulated on the input string. Penalties are given if simulation does not lead to acceptance. For each PDA (bit string) the fitness follows from the accumulation of penalties over all training instances.

4.2 Learning Connectionist Finite State and PushDown Parsers

NNs as an implementation tool for parsing (and unification) algorithms have been studied. Traditional CYK and Earley algorithms have been given connectionist implementations. None of these methods seem to fit in 'genuine' NN theory. Often they are based on localist representations and often there is an attempt to obtain structures that resemble parse trees. Rather than sequential processing of the input the methods require that complete sentences are presented to the nodes of the network. Besides the traditional methods many other examples of mappings of context-free grammars or parsers on connectionist architectures appear in the literature. They are not always 'learning' architectures. Lucas and Damper[1990] is an exception, but their grammars do not allow recursion and therefore the languages are of finite cardinality. This is accepted by pointing out advantages of being able, unlike HMMs and (stochastic) regular languages, to represent hierarchical structure directly. Other attempts to stay close to a connectionist context-free grammar or recogniser (PDA) formalism can be found in Giles et al[1990] (a hybrid system that learns to behave as a PDA for an inferred context-free grammar). These methods will hardly be considered here. Instead we turn our attention to attempts to learn NN implementations of grammars and associated parsers.

Grammar inference is of interest for NLP, but also for other research areas. In syntactic pattern recognition, grammar inference is learning the patterns and parsing is classification. As such, neural networks can be implementations of syntactic pattern recognition algorithms, where patterns are regular or context-free. FSA have found many applications. For example, they can be used to model the interaction of a robot with its environment or, more generally, to model behavioural patterns. Hence, inferring FSA is important for various disciplines. Time or sequential information to be processed by a FSA requires temporal representations in connectionist networks. Fixed length sliding windows on the input, delayed transitions, copying parts of the network and recurrent nets are among the presented solutions to this problem. For the serial processing of strings recurrent BP networks, recurrent self-organising maps and versions of Hopfield networks have been introduced. The idea to use 'memory units' in BP networks to represent a preceding state of the network was put forward by Jordan. In sequence recognition or production the network's history has to be recorded. Recurrent connections can take care of that. Jordan copies the output nodes back to 'state nodes' that together with standard input nodes are connected to the hidden units. The hidden units can work on their previous output. In Elman's more powerful simple recurrent network (SRN) the hidden units are recurrently connected to themselves. The prediction of the next word in a string becomes possible by SRNs and the grammatical correctness of simple sentences can be determined. In Cleeremans et al[1990] the relation between FSA and SRNs is discussed. See also Moisl[1992]. Unfortunately, the behaviour of SRNs (and other BP networks) is a problem because of their instable characteristics. Moreover, SRNs ignore structure. Pollack's Recursive Auto-Associative Memory (RAAM) is an approach to distributedly representing recursive data structures. SRNs have been taught to build RAAM representations of parse trees with an on-line parsing technique.

From the start, research on Hopfield models has been concerned with storing and retrieving sequences of states (see Drossaers[1992a] for a review). Early models were not powerful enough to model FSA. The number of FSA states could only be very limited, a superset of the intended set of strings was recognised or only strings of identical inputs were recognised. In Drossaers[1992b] it is argued that in modelling human NLP with a PDA there can be a limit on the number of items on the stack. This makes it possible to convert the PDA into a (deterministic) FSA. Since Drossaers is also able to design a Hopfield model that simulates a FSA it has become possible to model limited embedding in context-free languages with a Hopfield model (see also Drossaers' paper in this volume). A self-organising learning algorithm is part of future research.

Sequence recognition with Kohonen feature maps is less developed than with BP or Hopfield nets. Combinations of feature maps, buffering mechanisms and symbolic techniques have been proposed for handling (simple) sequences. In Scholtes[1993] (where more references can be found) Recurrent Kohonen Feature Maps are introduced that use a multi-level organisation of feature maps. They have been tested for tasks such as sentence structure derivation, word prediction and FSA simulation. Convergence is problematic because of the unstable character of the process. Hoekstra & Drossaers (this volume) have introduced an integrated Kohonen network for handling sequences (and long-distance dependencies). Their aim is to let the network learn to simulate an FSA that is obtained from a PDA with a finite-length stack.

4.3 BEYOND LEARNING SYNTAX

Various research attempts deal with language learning that includes semantics. That is, in sequence recognition not only syntactic, but also semantic predictions can be made and revised. The networks learn (mostly in BP nets) to map sentences into a syntactic/semantic representation (case frame parsing). See e.g. Jain[1991] for robust spoken language parsing with a modular recurrent BP net. An extension of the parser has been used in a speech-to-speech translation system (a conference registration dialog task). Also the impressive work of Miikkulainen[1991,1992] should be mentioned here. He has used a variety of simple and modular NNs (BP and feature maps) for learning to parse into case role representations (with different types of limited embeddings) and for processing stories (building paraphrases of script-based stories).

5. CONCLUSIONS

Hopefully the period of useless and ugly results from research based only on modest knowledge of NNs and easy examples from the application domain can be closed. A new period should be entered in which the newness of the paradigm, which invited to start from scratch in a domain that despite a lot fundamental research has not yet yield satisfying models of language description and use, should be subordinate to all results and approaches that are already available. For that reason we presented in the previous sections a selection of topics of/and views on NLP that in our view are important for framing research approaches in connectionist NLP. Obviously, as also may have become clear from previous sections, we do not mean that with the emerge of NNs approaches to the NLP problem should be as they were. NNs allow new forms of NLP and smooth the way for other approaches (biologically-inspired language learning and processing and Darwinistic models). Rather we mean that NNs should be model-based and that in the application domain itself we have to find the fundamental ways to treat the problems. There we have to give high-level accounts of NLP and associated architectures that allow NNs as tools of implementation. That these high-level accounts may be non-rule-based or surmount the traditional borders between scientific disciplines can only be hailed.

6. LITERATURE

Abney, S.P. et al [1991]. Memory requirements and local ambiguities of parsing strategies. *J. of Psycho-Ling. Review.*

Anderson, J. [1976]. *Language, Memory, and Thought.* Erlbaum, Hillsdale, NJ.

Anderson, J. [1977]. Induction of augmented transition networks. *Cog. Sci.*, 1, 125-157.

Berwick, R.C. & A.S. Weinberg [1982]. Parsing efficiency, computational complexity, and the evaluation of grammatical theories. *Linguistic Inquiry* 13, 165-191.

Clark, H.H. & E.V. Clark [1977]. *Psychology and Language.* Harcourt Brace Jovanovich, Inc.

Cleeremans et al [1989]. Finite state automata and simple recurrent networks. *Neural Computation* 1, 372-381.

Drossaers, M. [1992a]. Hopfield models as non deterministic finite-state machines. *COLING '92*, Nantes, 113-119.

Drossaers, M. [1992b]. Neural-network acceptors. Memoranda Informatica 92-36, University of Twente.

Elman, J.L. [1990]. Finding structure in time. *Cognitive Science* 14, 179-211.

Fu, K.S. & T.L. Booth [1975]. Grammatical inference: Introduction and survey. I: *IEEE Trans. on Systems, Man and Cybernetics*, SMC-5 (1), 95-111. II: SMC-5 (4), 409-423.

Fodor, J.A., T.G. Bever & M.F. Garrett [1974]. *The Psychology of Language.* McGraw-Hill, New York.

Frazier, L. and J. Fodor [1978]. The sausage machine: a new two-stage-parsing model. *Cognition* 6, 291-325.

Giles, C.L. et al [1990]. Higher order recurrent neural networks and grammatical inference. In: *Advances in Neural Information Processing Systems 2.* D.S. Touretzky (ed.), Morgan Kaufmann, Cal.

Gold, E.M. [1967]. Language identification in the limit. *Infor. Control* 16, 447-474.

Gold, E.M. [1978]. Complexity of automaton identification from given data. *Infor. Control* 37, 302-320.

Hasida, K. [1990]. A constraint-based approach to linguistic performance. *COLING '90*, Helsinki, 149-154.

Jain, A.N. [1991]. Parsec: A connectionist learning architecture for parsing spoken language. Ph.D. Thesis, CMU.

Kempen, G. & T. Vosse [1989]. Incremental syntactic tree formation in human sentence processing. Manuscript.

Langacker, R.W. [1987]. *Foundations of Cognitive Grammar. Vol.I. Theoretical Prerequisites.* Stanford U. Press.

Liberman, M.Y. [1991]. The trend towards statistical models in natural language processing. In: *Natural Language and Speech*, E. Klein & F. Veltman (eds.), Springer-Verlag, 1-7.

Lucas, S.M. & R.I. Damper [1990]. Syntactic neural networks. *Connection Science* 2, 195-220.

Marslen-Wilson, W.D. [1975]. Sentence perception as an interactive parallel process. *Science* 189, 226-228.

McEnery, T. [1992]. *Computational Linguistics.* Sigma Press, U.K.

Miikkulainen, R. [1990]. Script recognition with hierarchical feature maps. *Connection Science* 2, 83-101.

Miikkulainen, R. [1991]. NLP with modular PDP networks and distributed lexicon. *Cognitive Science* 15, 343-399.

Miller, G.A. [1962]. Some psychological studies of grammar. *American Psychologist* 17, 748-762.

Moisl, H. [1992]. Connectionist finite state natural language processing. *Connection Science* 4, 67-91.

Nijholt, A. [1993]. Parallel approaches to context-free language parsing. Chapter 2 in: *Parallel Natural Language Processing*, U. Hahn and G. Adriaens (eds.), Ablex Publishing Corporation, Norwood, New Jersey.

Ophoff, H.R. [1992]. The quest for probabilistic parsing. M.Sc. Thesis, University of Twente, IPO Report #852.

Pinker, S. [1979]. Formal models of language learning. *Cognition* 7, 217-283.

Sampson, G. [1988]. A stochastic approach to parsing. *COLING '88*, Budapest, 151-155.

Scholtes, J.C. [1993]. *Neural Networks in NLP and Information Retrieval.* Ph.D. Thesis, University of Amsterdam.

Schütze, C.T. & P.A. Reich [1990]. Language without a central pushdown stack. *COLING '90*, 64-69.

Sen, S. & J. Janakiraman [1992]. Learning to construct pushdown automata for accepting deterministic context-free languages. SPIE Proceedings Series, Vol. 1707 *Applications of AI X: Knowledge-Based Systems*, 207-213.

Solomonoff, R. [1964]. A formal theory of inductive inference. *Infor. Control* 7, 1-22 & 224-254.

Suppes, P. [1972]. Probabilistic grammars for natural languages. In: Davidson & Harman (eds.), *Semantics of Natural Language*, 741-762; D. Reidel Publishing Company, Dordrecht-Holland.

De Weger, M. [1991]. Generalised adaptive search: analysis of codings and extension to parsing. Manuscript.

Wexler, K. et al [1975]. Learning-theoretic foundations of linguistic universals. *Theoret. Ling.*, 2, 215-253.

Wyard, P. [1991]. Context free grammar induction using genetic algorithms. 4th *Intl. Conf. on Genetic Algorithms.*

Zhou, H. & J.J. Grefenstette [1986]. Induction of finite automata by genetic algorithms. In: Proc. of the 1986 *IEEE International Conference on Systems. Man and Cybernetics*, Atlanta, GA, 170-174.

Integration of Artificial Neural Networks and Dynamic Concepts to an Adaptive and Self-Organizing Agent

Csaba Szepesvári[1] and András Lőrincz
Department of Photophysics, Institute of Isotopes of the Hungarian Academy of Sciences
Budapest, P.O. Box 77, Hungary H-1525
[1] János Bolyai Institute of Mathematics, Attila József University of Szeged,
Szeged, Hungary, H-6720
E-mails: szepes@obelix.iki.kfki.hu, lorincz@obelix.iki.kfki.hu

A brain model based alternative to reinforcement learning is presented. It integrates artificial neural networks (ANN) and knowledge based (KB) systems into one unit or agent for goal oriented problem solving. The agent may possess inherited and learnt ANN and KB subsystems. The agent has and develops ANN cues to the environment for dimensionality reduction in order to ease the problem of combinatorial explosion. Here, a dynamic concept model is forwarded that builds cue-models of the phenomena in the world, designs dynamic action sets (concepts) and make them compete in a neural stage to come to a decision. The agent works under closed-loop control. Here we examine a simple robotic-like object in a two dimensional non-probabilistic world.

Introduction

Our starting point, or recipe, in the design of knowledge based system (KBS) and artificial neural network (ANN) integration is to take self-organizing ANN's for sizing down the tremendous amount of information from the external world to a set of 'high' or 'low' bits. It is this reduced set 'internal representation' that KBS could try to work by searching for rules, sets of rules and sequences of rules.

The working mechanism we present here is an alternative to reinforcement learning (RL) [1]. Our main concern regarding RL has two components: (i) in a real world of competitive and collaborative autonomous agents only conditional probabilities and not probabilities may be defined, (ii) it is difficult or impossible to measure the goals or the cost of actions. There is, however, a natural building principle of autonomous agents that we intend to use, and this principle relies on the existence of dangers in the real world: The more spare time the agent has to explore the external world, the less chance it has for falling into unexpected, multiple and thus unsolvable traps and the more chance it has of survival. That is, we start by assuming that spare time should be optimized to allow exploration. The consequence is that problems should be solved as soon as possible. Trying to solve problems simultaneously and in a time saving fashion determines the problem (or goal) priority through the explored structure of problem hierarchy in a natural fashion. The goal priority is always a function of the explored world and the actual and detected state of the agent. Nevertheless, we do not establish goal priority, but it is self-defined in a dynamic fashion. The agent tries to build action sets (set of concepts) by activation spreading based on experienced situations that is realized with the help of ANN methods [2, 3]. In parallel, there is competition between concepts [4] by selecting the winning concept at time-out leading to an action and thus further experience. The actual decision reflects the actual goal priority.

Self-organizing cues to the environment: Dimensionality reduction [4]

If there are stable features in the external world then, for example, a categorizing system may serve well. The categories could be inherited or learnt. If inherited, then adaptivity is severely decreased. For category learning self-organizing artificial neural networks are needed since there are cases with no supervisor who could pair inputs to outputs [5, 6]. The self-organizing model of Szepesvári, Balázs and Lőrincz [7] provides spatial filters for position estimation of extended objects in a self-organizing fashion. Spatial filters are useful as a means of reducing the high resolution image to a few positions. Special filters allow quasi-positioning of objects. Recognition of quasi-positioned objects may be performed by several methods (see e.g. [8, 9]).

We consider self-organizing learning as the slow learning process of an infant [10]. The building up of the KB system that works on the outputs of the dimension reducing self-organizing ANN's may be taken as maturity learning.

Building concepts of the KB system

From the system's point of view, the world is made of two parts: the internal world, i.e. the system itself, and the external world – the environment. However, from the point of view of operation there is no need to make this distinction, we need three categories for the internal representation of the world:

(i) Internal representation of objects, whether they are internal or external. Bits represents objects or features of objects. We shall call the set of bits of the internal representation at a given instant an I-state.

(ii) Operators of the system. We assume that the system can influence the environment, can move, may eat, grasp, etc. Every operator is associated with a state: if the state is set 'high' then the operator is activated, if it is set 'low' then the operator is inactive. We shall call the set of operators at a given instant an O-state.

(iii) Goals of the system: these are special states that can be either 'high' or 'low' that initiate search procedures in the I-state and the O-state to find routes to fulfil them. We might say that: the system is in a 'neutral state', 'it is hungry', 'it is interested', 'it is suffering', etc.

The KB system is built under the following principles: The I-states are the ones the KBS should work on. The task is then to experience I-O-I time sequences. The I-O-I triplets that were encountered contain all the information the system can collect, and the task is to process this information in order to fulfil the goals. It is the environmental feedback that validates the set of collected 'rules'. The rules may be used to plan future actions.

It is worth mentioning that operators are not mappings on the I-state space but relations and if one is collecting the I-O-I triplets for a given operator O, one cannot expect the influence of operator O to be predicted with certainty nor with a given probability. Even more, the task is to make decisions having one or just a few examples and thus the use of a probabilistic approach is severely restricted.

Concepts, and the competition of concepts

This system was designed for problem solving. We formulated problems as goals. Goals may be achieved by I-O-I series. If an I-O-I series is found that promises the achievement of a goal, then we say we have a first concept to match the goal. In practice the actual concept is the organizing of I-O-I triplets. If we have a single concept then we proceed accordingly. However, the system might develop a set of concepts. In this case one has to decide what concept ought to be followed. An attractive way of solving this problem is the suggestion of Csányi [4] on the competition of concepts. Decision is made with the help of a competitive ANN in a similar way to Thagard [2] and Maes [3].

Thagard used hypothesis as ANN units and designed an ANN that made consistent ANN subsets to compete. Maes made a decision between preprogrammed behaviours with a competitive network and has shown that the resulting behaviour selection is data driven (opportunistic), goal-driven (the more a behaviour contributes to goals the higher its activation level), sensitive to goal conflicts and it has a certain inertia or bias (toward previous behaviour sequences). The decision making ANN model presented here differs in the ANN units since it is dynamic: experience develops a network of *I-O-I* triplets and determines behaviours.

Implementation of the model

The model has three major parts: the dimension reducing interface system, the cognitive system, and the decision system. The implementation is quite general; it is the interface system and the set of inherited rules of the cognitive system that are the only problem dependent parts. The example we present here corresponds to a simplified robotic like object. The adaptive part of the cognitive system, and the decision system are independent of this special example. For this robotic like object (for expedience we shall call it *Robject*) we simulate the environment, consisting of objects causing pain and objects allowing energy refilling. It is easy to define hunger in this case: Robject is hungry if its energy is low. Robject will be programmed to avoid painful situations.

Robject's world

We assume that Robject has well developed self-organizing eyes with spatial filters [7]. In the example we are treating, Robject has 25 spatial filters. It is assumed that Robject has a categorizing system that can work on filtered images [9]. The simulated world fits Robject by assuming that all objects of the world are of spatial filter size. Now, we may model Robject's world as a grid. To every grid position there belongs one of three categories: the empty position, food at that position, or a certain blockage at that position.

Robject's world is shown in Figure 1. Robject is represented by the sign '>'. It means that Robject is at grid point $X = 2, Y = 2$ and is facing right (Orient 1). Robject's mouth is at the leading edge of the face (not shown). The 'Suck' operator should be 'high' to acquire food. Another condition for eating is that the food should be in front of Robject . Food is denoted by '@'. If eating is successful then energy is refilled to 1.00. In other instances energy is constantly decreasing. Robject can move forwards or backwards on the grid, and can turn left or right. Robject cannot enter a grid point if it is occupied by food or by blockage ('#'). Any trial results in pain. Robject's eye is a 'chameleon' eye, Robject can see its 5×5 grid environment. Robject is in the center of the environment. A vector of 24 components represents Robject's environment.

Working mechanisms of Robject's mental architecture

Robject has three main parallel components, shown in Fig. 2: the cognitive space (CS) that contains the short term memory (STM), the decision system (DS) and the goal system (GS).

The GS is the very core of the whole system. It is preprogrammed not adaptive and determines the behaviour of Robject . The working mechanism of Robject may be expressed as learning from successful and unsuccessful shorter or longer time sequences. The inherited properties of GS are: (i) it is able to designate goals, (ii) it is able to determine if the planned goal is achieved, (iii) it is able to detect pain.

The basis of CS is a dynamic directed graph of I-O-I triplets with weighted connections. The set is built and erased by the STM of CS. The STM chains the time sequence of consecutive *I-O-I* triplets in a FIFO (first in first out) finite length buffer. The STM is controlled by GS: (i) if the actual

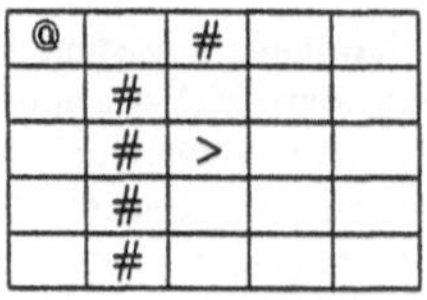

334

@		#		
	#			
	#	>		
	#			
	#			

Time Step: 7, Energy: 0.9, Mouth: 0 Pain: 0
Eye: @-#— -#— -#— -#— -#—
Open:0 Close:0 Suck:0 Fore:1 Back:0 Left:0 Right:0
Goals; FillEnergy:0 RemovePain:0 (Explore:1)
RuleNum:4 Orient:1 (X=2,Y=2) FromDatabase 0

Figure 1: Robject's world

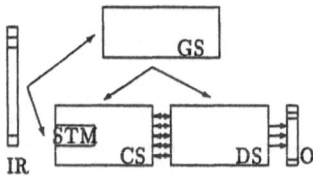

Figure 2: Robject's architecture

goal is fulfilled, then STM tries to lengthen the corresponding *I-O-I* triplet path of CS, strengthens the connections along the path and associates the actual goal to the triplets, (ii) if the goal was not achieved after the planned chain of actions then STM weakens the connections of the chain, (iii) if the chain of actions leads to pain at any step, then the last *I-O-I* triplet is communicated to CS and STM is erased. Strengthening and weakening mean adding or subtracting 1 from the actual connection strength. Connection strengths are limited to positive values. On reaching zero, the connection disappears. Useless *I-O-I* triplets (with no connections) were dropped time to time. The role of STM is to build the directed graph from the goals backwards. The procedure described here saves memory and keeps only the important information for achieving goals.

It is a KB system that works on the dynamic set of CS. In order to describe this KB processor, let us suppose that the dynamic set of CS is not empty and let us call an actual state with an actual goal a problem. The role of the KB processor is to build *I-O-I* chains that correspond to the present problem. Such chains should start from *I-O-I* triplets having initial states that contain relevant information for the problem. The KB processor presented here is weak at this point; we applied a simple Hamming distance on the set of *I*-states to determine the relevant initial *I-O-I* set, called *kernel*. Decision shall be made between elements of the kernel.

The next task of the KB system is to find possible outcomes starting from this kernel. Two distinct ways seem to be appropriate: (i) search for memorized, connected, successful *I-O-I* triplets, (ii) search for unconnected *I-O-I* triplets whose initial state matches one of the final states of the *I-O-I* triplets in the kernel (not implemented here). This procedure could be continued up to time-out. The selected part of CS will be called *extended kernel*. DS keeps working on the growing extended kernel. We simulate that DS and CS work parallely.

The decision system (DS) is based on the idea of the competition of concepts [4] generated in cognitive space. We have constructed the competition in a way similar to Thagard's ECHO system. The ECHO system has two parts: (i) builds up coherence relations between hypotheses, here these are *I-O-I* triplets; (ii) utilizes a competitive algorithm [11] for decision making. In our case coherence relations correspond to connection strengths that are inversely proportional to the degree of diversity, i.e. to the number of possible choices leading to the goal. This plays a normalizing role.

Decision is made at time-out, when the highest activity node of the kernel is chosen. If the activity of that node is higher than a certain threshold, then the triplet is accepted, otherwise an inherited weighted random choice is made between situation-allowed operators. More sophisticated inherited strategies, in fact, full inherited KB systems may be incorporated at this point. In the special case of Robject the operator 'move forward' has a higher probability than the others. This 'inherited' property speeds up Robject's learning as it turns less frequently and explores larger

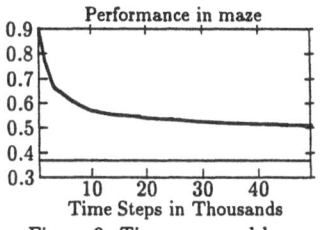

Figure 3: Time averaged hunger ratio

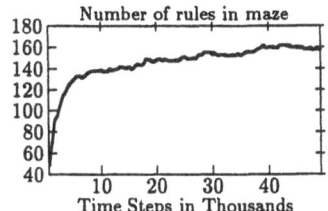

Figure 4: Average number of rules vs. time

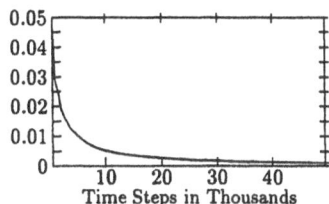

Figure 5: Probability of steps resulting in pain

spaces. The selected operator is then tried and the *I-O-I* list is updated.

Experimental results

Robject's world has 'periodic boundary conditions'; this means that if one leaves the world downwards one comes back from the top and similarly on the two sides. Robject could see the whole world with itself in the center. Robject's energy was decreased by 0.1 at every step. In case of eating it was set to 1.0, the maximum value. If it decreased to 0.3, Robject became hungry and started to search for operations to cease its hunger. Robject felt pain if it hit the food. To consume the food Robject had to face it, stand in front of it, and had to suck.

In a set of experiments Robject was placed in a 'maze' (see Fig. 1). Figures 3-5 show Robject's results in the 'maze'. Having eaten, new food was generated at a given position. The theoretical minimal value for performance, the ratio of cases when Robject was hungry to the total number of cases, was 0.37 with the number of rules being 144. Robject collected 157 rules and reached a 0.51 value. Steps resulting in pain disappear very quickly.

Conclusions

We have suggested goal oriented behavior as an alternative to reinforcement learning. In spite of the fact that the theoretical background of our model is far from complete the combination of self-organizing cues to the environment and competing dynamic concepts seems to be a useful and adaptive scheme for goal oriented problem solving.

We are grateful to Dr. Klára Konrád and Dr. Iván Futó for helpful discussions.

References

[1] A.G. Barto, S.J. Bradtke, and S.P. Singh. Real-time learning and control using asynchronous

dynamic programming. Technical report 91–57, Computer Science Department, University of Massachusetts, 1991.

[2] P. Thagard. Explanatory coherence. *Behavioral and Brain Sciences*, 12:435–467, 1989.

[3] P. Maes. Learning behavior networks from experience. In *Toward a Practice of Autonomous Systems, Proceedings of the First European Conference on Artificial Life*, pages 48–57. MIT Press, Cambridge, Mass., 1992.

[4] V. Csányi. *General Theory of Evolution*. Akadémiai Kiadó, Budapest, 1982.

[5] G.A. Carpenter and S.A. Grossberg. Massively parallel architecture for self-organizing neural pattern recognition machine. *Computer Vision, Graphics, and Image Processing*, 37:54–115, 1987.

[6] P. Földiák. Learning invariance from transformation sequences. *Neural Computation*, 3(2):194–200, 1991.

[7] Cs. Szepesvári, L. Balázs, and A. Lőrincz. Topology learning solved by extended objects: a neural network model. submitted to Neural Computation, 1993.

[8] K. Fukushima. Character recognition with neural networks. *Neural Computing*, 4:221–233, 1992.

[9] T. Fomin and A. Lőrincz. On the potential of hebbian and anti-hebbian learning. *submitted to Neural Networks*, 1993.

[10] D.O. Hebb. *The Organization of Behavior*. John Wiley and Sons, New York, 1949.

[11] S. Grossberg. Competitive learning: From interactive activation to adaptive resonance. *Cognitive Science*, 11:23–63, 1987.

Learning Fuzzy Production Rules For Approximate Reasoning In Connectionist Production Systems

Nikola K. Kasabov
Department of Information Science
University of Otago, P.O.Box 56 Dunedin, New Zealand
email:nkasabov@otago.ac.nz, Fax:+(3)4798311, Tel:+(3)4798319

Abstract

The paper presents a connectionist method for learning fuzzy production rules from data. The fuzzy concepts (fuzzy linguistic values) in which every attribute is quantized, are assumed as known by their membership functions. The method extracts, from a set of data, fuzzy rules of the form of IF <conditions> THEN <actions>, where a relative coefficient of importance is assigned to every condition element. Two more coefficients of uncertainty that characterise partial match of the condition elements, i.e. a noise tolerance coefficient, and a data sensitivity factor, are learned as well. A certainty degree coefficient for every rule is also learned. The learned fuzzy rules suit very much, and were indeed designed for, the inference mechanism built in a Connectionist Production System NPS developed for partial match and approximate reasoning.

1.Introduction

Connectionist realization of production systems is very important for the future development of knowledge engineering because it could bring all the benefits of the connectionist approach to AI systems and symbolic computation [10,9,4,5]. The present paper is focused on learning production rules and fuzzy rules in particular which can be interpreted in connectionist production systems. Connectionist methods for learning simple fuzzy rules are published in [1,2,3,6,8]. This paper introduces a method of learning more sophisticated fuzzy productions of the form of:

$$\text{Ri:IF Ci1(DIi1) and Ci2(DIi2) and...and Cim (DIim) THEN Ai1,Ai2,..,Ail(NTCi,DSFi,CDi)} \quad (1)$$

where: C_{ij} (i=1,2,...,n;j=1,2,...,m) are fuzzy propositions of the form of "X is B" B" (X is a fuzzy variable, B - its fuzzy value); A_{ij} are either fuzzy or non-fuzzy propositions; DI_{ij} are relative degrees of importance; NTC_i and DSF_i are respectively a noise tolerance coefficient and a data sensitivity factor which characterise the partial match of the condition part of the rule R_i; CD_i is a certainty factor of the rule. The context of those coefficients of uncertainty is the same as introduced in NPS3 [5] and explained in the next section.

2. The NPS3 architecture and its type of production rules with uncertainties

The NPS architecture, first described in [4] uses a local representation. It consists of three neural subnets: PM - production memory; WM - working memory, and VBS - variable binding space. PM consists of n neurons, where n is the number of productions. WM is a matrix of k.s neurons, where k is the number of the attributes and s the number of their possible values in the condition elements of the productions. All the condition elements are of the form of <attribute-value> pair. VBS is a matrix of $s.n_1$ neurons, where n_1 is the number of productions with variables. $n_2 = n - n_1$ is the number of productions without variables. WM provides the external inputs to PM and VBS. The functional phases of the NPS are similar to those of a classical symbolic production systems (match, select, act), though realized in a connectionist way. Here, the match and select phases are mixed in one, which is different when comparing to the symbolic production systems. The act phase consists of updating the WM after opening two gates - one from the PM, and another - from the VBS. Productions in NPS3 are represented in the following form:

$$R_i: \text{IF } C_{i1}(DI_{i1}) \ \& \ C_{i2}(DI_{i2}) \ \& ... C_{im}(DI_{im}) \ \text{THEN } A_{i1}, A_{i2}, .., A_{il}(\Theta_i, P_i, \beta_i, CD_i) \qquad (2)$$

where C_{ij} is an <attribute-value> pair and A_{ij} is one of the actions: assert, or delete(-) a WM fact. The number of conditions and actions in a production is arbitrary and each C_{ij} may have or may not have a variable. DI_{ij} is a degree of importance of the jth condition element in the ith rule R_i, which is a relative one within the rule. The attached degrees of importance can be used for partial match e.g. if one important fact is not present in the WM, then the rule can not fire, but a lack of unimportant fact for a particular rule should not stop firing the rule. Every fact is represented in the WM by a certainty degree - a number between 0 and Ie (0 and 1 by default) representing how certain is the presence of the fact at the moment. In another mode of NPS3, negative and unknown facts can be represented as well. Three inference control coefficients Θ_i, P_i, β_i can be attached to every rule R_i. Θ_i is a noise tolerance coefficient. The greater the value of Θ_i, the more resistant to the noise among the facts is the rule R_i. P_i is a data sensitivity factor which represents how sensitive is the rule towards the presence of the facts. Θ_i and P_i control the partial matching. β_i is a reactiveness factor for the rule R_i. It is realised by the power coefficient in the sigmoidal activation function in the neurons of PM. It shows how high should be the activation level of a sufficiently matched by the present facts rule R_i. The certainty degree coefficient CD_i defines the range for the certainty degrees of the inferred by this rule facts.

PM is a network of an additive type. Every successful match of the facts and the productions correspond to a stable state of the network. The following formula for calculating the external input I_i to the ith neuron in the production memory PM is used:

$$I_i = \begin{cases} Ie \cdot ((S_i - \Theta_i)/(Ie - \Theta_i))^{P_i}, & \text{if } S_i \geq \Theta_i \\ 0, & \text{otherwise.} \end{cases} \qquad (3)$$

where: S_i is the total support to the ith neuron in PM from the existing facts in the WM.

3. Reasoning over Fuzzy Production Rules in NPS3

The main idea of reasoning over fuzzy production systems in NPS3 is that certainty degrees attached to the facts in NPS3 could contextually represent membership coefficients when dealing with fuzzy values. A successful match of the current data causes firing a rule, which, according to its consequent, contributes to the certainty degrees of the inferred facts which may be fuzzy or non-fuzzy [5]. The following fuzzy rules for a bank loan approval, introduced and hardware implemented by Lim and Takefuji [7], have been used here as an experimental example:

Rule 1:IF Score is High & Ratio is Good & Credit is Good THEN Decision is Approve;

Rule 2:IF Score is Bad & Ratio is Bad OR Credit is Bad THEN Decision is Disapprove,

where the fuzzy values in the rules are represented graphically in Figure 1 by their membership functions as given in [7].

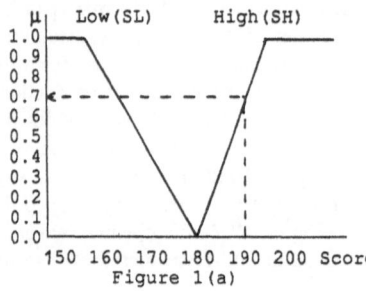

Figure 1(a)

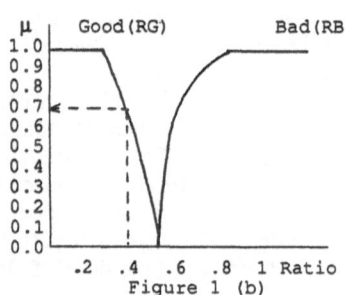

Figure 1 (b)

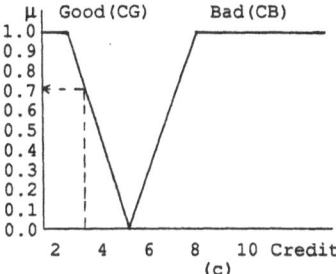
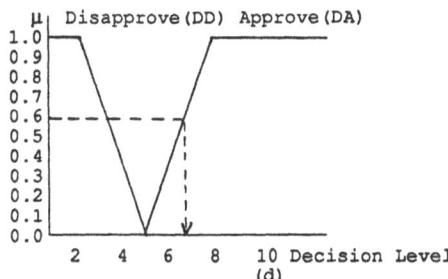

Figure 1. A graphical representation of the fuzzy membership functions in the FXLoan (Lim & Takefuji, 1990) expert system: (a) Score-High/Low; (b) Ratio-Good/Bad; (c) Credit-Good/Bad; (d) Decision-Approve/Disapprove. Fuzzyfication of one client's data and defuzzyfication of an inferred result DA=0.6 are also shown.

When values for a new client (for example Score = 190, Ratio = 0.4, Credit = 3) are supplied, their corresponding membership coefficients can be extracted from the graphics in Figire 1 (fuzzyfication), which provide the certainty degrees of the input facts for the NPS3 program. After a run of it, the certainty degrees for the two output facts Decision Approve and Decision Disapprove are inferred. A centroidal defuzzyfication [11] can be applied if it is necessary to obtain an exact solution for the Decision Level (a number between 0 and 10 as shown in Figure 1 (d)).

4. A Connectionist Method of Learning Fuzzy Production Rules

A procedure is developed for learning fuzzy rules in the form (1) from a set of data. The membership functions of the fuzzy concepts (labels) into which the data will be quantized are supposed to be known. As a pre-processing, the data are fuzzyfied into the fuzzy concepts. The procedure consists of 6 steps explained and illustrated below:

Step 1. A neural network is constructed according to the present data . The first layer consists of nodes representing all fuzzy values (concepts) B_{ij} (i=1,2,...,k ; j=1,2.,..., s) into which k independent variables (attributes) of the data set are adequately represented. Every fuzzy variable can be represented by any number s of fuzzy values (labels). The second layer represents all, or only some, of the possible combinations between the input fuzzy concepts that are expected to appear in the fuzzy rules to be learned. The third, output layer, represents all the possible l consequent actions (propositions) in the rules. In Figure 2(a) a general architecture of a network for learning fuzzy rules in the form of (1) is given.

In order to be able to compare learned fuzzy rules with a-priori existing fuzzy rules for the bank loan approval problem given above, a set of 6 data items was generated with the use of these rules. The generated set of raw data (Score, Ratio, Credit) and their corresponding membership values after fuzzyfication (SH, SL, RG, RB, CG, CB) are shown in Table 1. The architecture of the network to be trained with this data set is shown in Figure 2 (b).

Table 1

	Score	SH	SL	Ratio	RG	RB	Credit	CG	CB	Decision	DA	DD
1	200	1	0	0.1	1	0	0	1	0	10	1	0
2	150	0	1	1	0	1	10	0	1	0	0	1
3	155	0	1	0.7	0	1	2	1	0	1	0	1
4	150	0	1	0.3	1	0	8	0	1	2	0	1
5	195	1	0	1	0	1	9	0	1	1	0	1
6	200	1	0	0.1	1	0	10	0	1	2	0	1

Figure 2. The network architecture for learning fuzzy rules of type (1): (a) a general architecture; (b) the architecture used for learning fuzzy bank loan rules according to the training data given in Tabl.1

Step 2. Training the network. A supervised training is applied to the constructed in Step1 network with an use of the Back-propagation algorithm over the set of training data. The output node DA which represents the Decision Approve output fuzzy value and the connections (after training) from the input nodes to the only two intermediate nodes which support the activation of this output are shown in Figure 3.

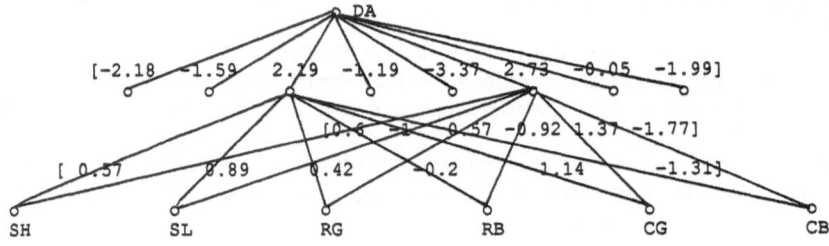

Figure 3. A part of the trained network for learning fuzzy rules from Table 1.

Step 3. Extracting an initial set of rules $\{r_j\}$ from the trained network. All the connections to a neuron Ni, which contribute significantly to its possible activation (their values are over a defined threshold), are picked up and their corresponding hidden nodes, which represent a combination of fuzzy input values, are analyzed further more. Only the input fuzzy values which support activating the chosen hidden nodes, are going to be used in the antecedent part of a constructing rule. As an initial relative degree of importance of the antecedent fuzzy propositions, the corresponding weights of the connections between the input neurons and the intermediate ones are taken. The weight from the hidden neuron to the output neuron is attached as an initial degree of importance of the whole antecedent part of a rule. The following initial fuzzy rules are extracted for Decision Approve from the trained network shown in Figure 3: r1: [SH(0.6) & RG(0.57) & CG(1.37)] (2.73) -> DA; r2: [SH(0.57)& RG(0.42) & CG(1.14)] (2.19) -> DA. Other 5 initial rules are also extracted for Decision Disapprove: r3: [SL(1.47) RB(1.54) CB(2.60)] (3.28) -> DD; r4: [SL(1.39) RB(1.05) CB(1.71)] (2.25) -> DD; r5: [SL(1.18) RB(0.9) CB(1.9)] (1.89) -> DD; r6: [SL(0.95) RB(0.85) CB(1.45)] (1.54) -> DD; r7: [SL(0.65) RB(0.63) CB(1.17)] (1.29) -> DD.

Step 4. Aggregating the initial rules into rule-clusters. All initial rules $\{r_{i1}, r_{i2}..., r_{ip}\}$, which have same condition elements and same action elements, are aggregated into one rule - Ri. A relative degree of importance DIij is calculated for every condition element Cij of the rule Ri as a normalized sum of the initial degree of importance of the corresponding antecedent elements in the initial rules r_{ij} multiplied by the corresponding whole antecedent degree of importance. The following two rules R1 and R2 were obtained by aggregating the 7 initial rules from Step 3:

R1: SH(1.2) & RG(1) & CG(2.5) -> DA; R2: SL(1.1) & RB(1) & CB(1.8) -> DD;

In the learned rules R1 and R2 the antecedent elements are connected only by "AND" connectives. In general "OR" connectives can be learned by this method if there are more groups of similar initial rules which have same consequent. Every group then is represented by one aggregated rule.

Step 5. Defining the certainty degrees CDi of the extracted in Step 4 rules Ri (i=1,2,...,n). Here the case, when the action part of the rule Ri consists of a single element, will be discussed. CDi is defined as the magnitude of the reaction of the output neuron Ni (which corresponds to a rule Ri), when the maximum membership coefficients for the antecedent fuzzy values are fed as inputs to the network. For the case of the two rules R1 and R2 above, we define CD1=0.99 and CD2=1.0.

Step 6. Defining the noise tolerance coefficients NTCi and the data sensitivity factors DSFi for the extracted rules Ri. Following the semantic meaning of the NTCi described in section 3 and implemented in NPS3 as shown in formula (3), a NTCi here is considered to be a parameter which estimates the threshold, up from which every small increase in the degree of the fuzzy input facts causes non-zero value of the net input signal Ui to the neuron Ni. As NTCi in NPS3 is a real number of the [0,1] interval, the whole k-dimensional input space X ={(X,X2,...,Xk)} is to be quantized into one-dimensional space {(S)} of the facts' support to Ni in the interval of [0,1]. For this purpose either a formula S= F(X1,X2,...,Xk, W1, W2) , where W1 and W2 are the matrices of connection weights, or an one dimensional Kohonen quantization map can be used. By changing the values of the input vectors (X1,X2,...,Xk) in such a way that their corresponding quantized values Sj monotonously increase, and for every value Sj calculating the total input signal Uij to the neuron Ni in the trained network, we can obtain the NTCi as the value Si for which: Ui = 0 if Sj = Si, Uij > 0, for Sj > Si. The value of the data sensitivity factor DSFi can be calculated analytically when the formula F which calculates the net input signal Ui to the neuron Ni in the trained network and formula (3) are made equivalent. The process of defining the noise tolerance coefficient NTCi and the data sensitivity factor DSFi for rule Ri are base on partial mapping of the trained network to the network in NPS3 which realises the production memory PM. For the bank loan example, the following values were obtained experimentally: NTC1=0.4; DSF1=1.5; NTC2=0.2; DSF2=1, thus the following rules of the form of (1) were finally learned for the experimental case:

R1: SH(1.2) & RG(1) & CG(2.5) -> DA (0.4, 1.5, 0.99);
R2: SL(1.1) & RB(1) & CB(1.8) -> DD (0.2, 1.0, 1.0).

5. Connectionist production systems that learn and interpret fuzzy rules

The two learned rules R1 and R2 were run on the NPS3 simulator [5] for four clients' input data. The results are shown in Table 2 where the results obtained by the a-priori given fuzzy rules [7], run on NPS3 over the same test input data are given for comparison. The inferred membership values for the two consequent fuzzy concepts DA and DD are almost the same.

Table 2

Client	Score	SH	SL	Ratio	RG	RB	Credit	CG	CB	Learned Rules DA	DD	A-priori Rules DA	DD
1	190	.7	.0	.4	.7	.0	3	.7	.0	.7	0	.6	0
2	160	.0	.8	.5	.0	.7	7	.0	.7	0	1	0	1
3	170	.0	.5	.5	.0	.7	8	.0	.1	0	1	0	1
4	175	.0	.2	.41	.3	.0	6	.0	.3	0	0	0	0

When comparing the rules obtained by the described method and the original rules we can notice that in the second original rule an "OR" connective is used. The obtained rules represent contextually this connective by the relative degrees of importance attached to every condition element. The fuzzy variable Credit obviously plays in this case a dominant role for the decision process. Using coefficients of importance seems to be an universal method for representing any connectives between condition elements in a set of rules.

An architecture of a connectionist production system that learns and interprets fuzzy production systems is shown in Figure 4.

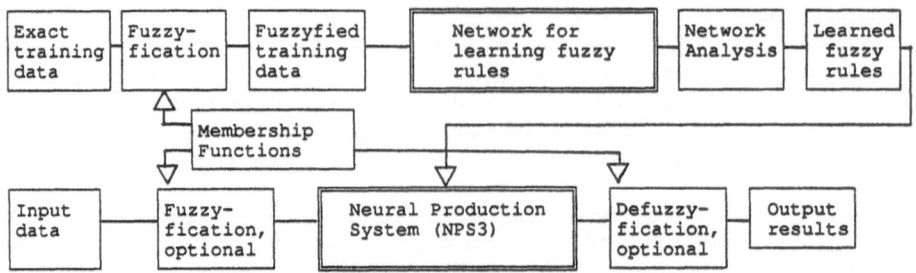

Figure 4. An architecture of a connectionist system that learns and interprets fuzzy rules

6. Conclusions

The paper introduces a procedure for learning fuzzy rules with four coefficients of uncertainty for the purpose of approximate reasoning in a connectionist production system. It has been empirically proved that the procedure learns successfully correct fuzzy rules and captures much of the present in the data uncertainty. An architecture of a production system that learns and interprets fuzzy productions is presented. It is based on the suggested learning method and the neural production system NPS3 [5]. The paper appeals that the production systems, enriched by the connectionism and fuzzy logic, are more a powerful paradigm for knowledge engineering that they have been before.

7. References

[1] d'Alche-Buc,F., Andres,V., Nadal, J.-P. (1992) Learning Fuzzy Control Rules with a Fuzzy Neural Network, in: I.Aleksander and J.Taylor (Eds). Artificial Neural Networks,2. Elsevier Science Publ.B.V., pp.715-719.

[2] Hyun-Jung Yi, Kyung-Whan Oh (1992) Neural Network-based Fuzzy Production Rule Generation and its Application to an Approximate reasoning Approach. Proceedings of the 2nd International Conference on Fuzzy Logic & Neural Networks (Iizuka, Japan), pp.333-336.

[3] Isshiki,H., Endo, H. (1992) Learning Expert's Knowledge by Neural Network and Deduction of Fuzzy Rules by Powell's Method, in: Proceedings of the 2nd International Conference on Fuzzy Logic and Neural Networks (Iizuka, Japan), pp.95-98.

[4] Kasabov, N., S.Shishkov (1992) On the problem of connectionist production systems - models and their implementation, in: I. Alexander and J. Taylor (eds.) Artificial Neural Networks II. Elsevier Science Publishers, 699-702.

[5] Kasabov, N., S.Shishkov (1993) A Connectionist Production System with a Partial Match and Its Use for Approximate Reasoning, to appear in a Special Issue of "Connection Science".

[6] Kosko B. (1992) Neural Networks and Fuzzy Systems: A Dynamical Approach to Machine Intelligence. Prentice Hall.

[7] Lim M. and Y.Takefuji (1990) Implementing fuzzy rule-based systems on silicon chips, IEEE Expert, February, 31-45.

[8] Mukaidono, M., Yamaoka,M (1992) A Learning Method of Fuzzy Inference Rules with Neural Networks and its Application, in :Proc. 2nd Intern. Conf. on Fuzzy Logic & Neural Networks (Iizuka, Japan), pp.185- 187.

[9] Sun, R. (1992) On Variable Binding in Connectionist Networks. "Connection Science", vol.4, No2, pp.93-124.

[10] Touretzky D. & G.Hinton (1988) A Distributed Connectionist Production System. Cognitive Science, vol.12,pp.423-466.

11. Terano, T., Asai, K., Sugeno, M. (1992) Fuzzy Systems Theory and Its Applications, Prentice Hall.

A Representational Architecture for Nonmonotonic Inheritance Structures

Mikael Bodén
Department of Computer Science,
University of Skövde
Box 408, S-541 28 Skövde, Sweden
mikael@his.se

Ajit Narayanan
Department of Computer Science,
University of Exeter
EX4 4PT Exeter, UK
ajit@dcs.exeter.ac.uk

Abstract

This paper describes a connectionist system for representing and reasoning with multiple inheritance structures with exceptions. The representational architecture has three characteristics. First, it merges relational with taxonomic representations. Secondly, it handles conflicts generated by exceptions and the use of multiple superclasses. Thirdly, it uses fully distributed representations. One novel feature is that, since the distributed representation of an entity is influenced by its position in the inheritance structure, representations of assertions are influenced by the context of the entities. An extension to the model which implements and makes use of *confluent* inference is described.

1   The problem of structured representation

Ideally, object-centred representations (e.g. frames, semantic networks) should be hierarchic, with subclasses and instances having no more than one class to which they belong. But commonsense reasoning is littered with examples of *multiple inheritance*, where a class takes some or all properties of two or more classes (e.g. electric guitars have properties of electrical devices and of guitars). Also, commonsense reasoning makes widespread use of *exceptions*: whilst birds fly and penguins are birds, penguins don't fly. The issue here for logicians is that of *nonmonotonicity*. Formally, a system is *monotonic* if a formula p is derivable from a set of premises P and if p is also derivable from each superset of P. If p is *Tweety flies* and $P1$ is {*Tweety is a bird, birds fly*}, then since p is no longer derivable from a superset of $P1$, namely, $P2 = \{$ *Tweety is a bird, birds fly, Tweety is a penguin, penguins are birds, penguins don't fly*}, we have nonmonotonicity. Attempts have been made by logicians to extend FOL so that defaults and exceptions can be handled (e.g. [11]).

Another common way of representing type/token information symbolically is by forming class structures. Such structures can be represented as semantic networks, and inheritance can then be viewed as the process of collecting properties along appropriately labelled paths between nodes. Such path-based systems allow exceptions to be immediately represented, since there is a difference between an *inheritable* property and *inherited* property. An inheritable property is one for which an inheritance path exists (i.e. not preempted or removed by some blocking mechanism). For a property to be inherited, an *interpretation* has to be made of conclusions supported by paths so that inconsistent properties can be identified. The distinction between derivable and derived (inheritable and inherited) is not so clearly identifiable in logic-based approaches because rules of inference in logic are themselves tautologous (necessary): conclusions must be true if their premises are true. Path-based approaches have semantic problems with structures involving multiple superclasses, however (just as logic-based approaches do). There is currently no standard or intuitively appealing way of dealing with nonmonotonic inheritance structures [14].

2   Structured representations

The term 'categorization' is used in the psychology literature to refer to the process of grouping objects together which are alike. *Categories* are usually defined or described using feature similarities and/or geometrical distances between categories in conceptual space [15, 13]. Both types of approach can be used in connectionist systems. For instance, the *domains* account [12], where concept centrality is computed on the basis of covariation among microfeatures, is a feature-based approach. Another feature-based approach for representing type/token information relies on the use of *subvectors* [5]: inheritance can be implemented via token representations if the distributed representation of a token contains a subvector which refers to its type.

On the other hand, if we assume that closely related entities have closely related representations (rather than share subsets of features), the associational linkage between those entities emerges from the geometric position in the encoding space. For instance, one approach here is for representations of entities in a *single inheritance hierarchy* to be computed with respect to their position in the hierarchy according to some formula. Attributes can be attached to isa-related entities (inheritance) through an attractor neural network [16]. To obtain *meaningful distributed representations*, one can either code semantic features of an entity as microfeatures or let the processing task (e.g. associating semantically similar entities) decide what representation is appropriate for a specific entity [9]. In the latter case, the process consists of minimizing the error at the output by modifying the input according to an appropriate learning rule. This context-sensitive feedback is sometimes referred to as confluent inference [3]. The problem with this, as with many other systems, is that not all entities might be known at processing time. This makes it difficult to evaluate just how sufficiently different the input has to be. Yet, human nonmonotonic reasoning is characterized by the apparent effortlessness with which new information, including information about new entities, is handled.

The only connectionist system that explicitly attempts to model multiple inheritance is Cottrell's localist representation of default logic [4, 11]. Unfortunately, two problems with default logic (DL) concern non-cumulativity and difficulty in handling arbitrary new information. Cumulativity is the idea that adding a theorem of a set of premises to those premises should not change the derivable formula. Default logic has been shown not to exhibit this logically desirable property [8]. Default logic also cannot handle arbitrary new information in a way that guarantees that a belief set remains consistent [2]: we must know in advance where the exceptions are going to arise and code these in the logic, i.e. we cannot, as in the path-based approach, add new exceptions as they come along. A third problem, and perhaps the most damning in the eyes of AI researchers, is that DL cannot by itself signify preferences between conflicting default inferences, whereas we humans usually do.

3   Representing and handling class-based information

We chose to use the RAAM compressor [10] as an entity representation encoder as well as an assertion encoder. It has been shown that the resulting representation will be a *reduced description* of the original expression [7, 1, 6]. RAAMs produce superpositional distributed representations by making use of regularities and correlations in the input data. This results in compressed data structures. Since RAAM supports a decoding phase the compressed representation is always unique and reflects the regularities signifying the input set. If the resulting representational space is analyzed, *clusterings* of points can be identified, where closely clustered points evolve from similar input. The representations are produced as follows. A number of entities positioned in a multiple inheritance structure of medium complexity get their representations by first arbitrarily[1] choosing an initial activity pattern for each entity and then transforming this into a *position-dependent* representation by compressing them in the context of their neighbours. For example, if *Tweety* is located immediately next to *penguin* which in turn is located next to *bird*, the representation of *Tweety* is produced by compressing the initial representations of *bird*, *penguin* and *Tweety*. The representation of *penguin*

---

[1] We are not concerned here with semantically evaluable units and therefore there is no need to appeal to microfeatures.

will have similarities with that of *Tweety* since the representation of *penguin* is produced by compressing the representations of *bird* and *penguin* and so is computed from a similar representational context. Using a binary RAAM network we can feed the *positional* context as a binary tree, e.g. *((R(bird) R(penguin)) R(Tweety))*, where *R* is a function returning the initial arbitrary patterns of individual entities. We encode the entities with respect to their taxonomical position, and thereby eliminate the need for representing *bi-directional* relations (taxonomical relations, e.g. *isa, has as part*) explicitly.

After the entity representations are produced, the relational (*directional*) information concerning the entity is generated. (Typical directional relations are *loves, eats, can* and *cannot*.) This is also achieved by a binary RAAM encoder. Hence, the representation of *Tweety eats fish* will consist of compressing *Tweety* and *fish* separately, taking into account their position in the taxonomic structure, then combining the two resulting representations together, and finally associating the resulting representation with *eats*, resulting in the tree *((Tweety fish) eats)*. The main purpose of this representational architecture is for an associative network to establish *class sensitivity*, enabling entities to inherit properties belonging to their superclasses by associating an assertion, such as *(Tweety fish)*, with an assertion with direction, such as *((Tweety fish) eats)*. Since input representations are encoded with respect to their constituents and they in their turn are encoded with respect to their position in the inheritance structure, the associative net can generalise with respect to a certain class typicality — a form of role-binding.

Given the relation in the assertion as the architectural goal, the associative network will implement a 'normal' mapping for defaults and an 'exceptional' mapping for exceptions. Say, for example, *bird can fly*. This would be coded as an assertion *bird fly* and then be associated with *can* to produce an influenced representation of *bird can fly*. Further, we know *penguin cannot fly*. First, the representation of *penguin* will be similar to that of *bird* since they are located next to each other. Then we associate the assertion *penguin fly* with *cannot* to derive the influenced representation *penguin cannot fly*. Now, since *Tweety* is a penguin and therefore is close to *penguin* representationally, the associative network will most probably map to a geometric point in the representational space closer to relational assertions *X cannot fly* than to relational assertions *X can fly*. On the other hand, *eagles are birds* in that they fly. The distributed representation for *eagle* will be influenced by *bird* and not *penguin*. Without retraining, the associative network produces *eagle can fly* since *eagle* is close to *bird* representationally. We let the representations of the last step, the associative network, be fed back into the input representations, by letting the network develop its normal learning parameters (weights) as well as its input according to an extended learning scheme such as FGREP [9]. This will influence the input representation to reflect regularities and correlations concerning property inheritance within the associative network.

Using two RAAM nets for the two encoding phases, and one feedforward net for associating an assertion with an assertion with direction, the overall influencing process can be achieved in the following way (see Figure 1).

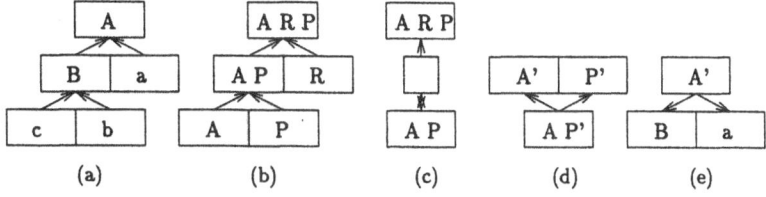

Figure 1: The five encoding phases when learning takes place in the confluent model

(a) Identify the list of superclasses for a given type or token *a*; *b* and *c*. Let this ordered list (*a, b,* and *c*) be compressed into *A* with a binary, sequential RAAM. This is repeated for each type and token in the hierarchy.

The next step (b) is performed by a second RAAM net and involves encoding assertions using the above encoded entities. This is always done in two steps. First we encode a simple assertion, $A\ P$, and then we use this to encode an assertion with direction, $A\ R\ P$, i.e. we attach a relation to the assertion.

The associative network (c) is used to implement a mapping between an assertion and an assertion with direction. This is a feedforward net with an extended learning algorithm which will adjust its weights and its input according to how well it learns this mapping.

(d) The new input, $A\ P'$, is then decoded using the second RAAM net, into its confluently influenced constituents, A' and P'.

(e) To achieve both confluent influence and class-based representation grounding, we adjust weights in the class-encoder (the first RAAM net) to produce the constituents from (d). That is, we do not decode a class-representation, such as A, into its constituents, because that would lead to a moving input problem (atomic representations must be static).

4 Exceptions with multiple inheritance

The main problem with multiple inheritance structures is that an entity can belong to several superclasses, e.g. *Nixon* is (at the same time) a *republican*, a *quaker* and, let's say, a *colonel*. The question now is whether the binary tree form of assertion described above is adequate. For instance, does it matter whether we encode *Nixon*'s representation as *(((R(republican) R(colonel)) R(quaker)) R(Nixon))* or *(((R(quaker) R(republican)) R(colonel)) R(Nixon))*? A cluster analysis performed on the resulting representation showed that the later we present the pattern to the network the more influence it will have on the resulting representation. This is acceptable if we introduce preferences among the superclasses, e.g. *Nixon* is mainly thought of as a *republican* and the representation of *Nixon* can therefore be computed by compressing the tree *(((R(quaker) R(colonel)) R(republican)) R(Nixon))*. Another solution would be to extend the RAAM encoder to take an arbitrary number of tokens as input, a variable-width RAAM, which will give us the freedom of placing superclasses, at the same level in the structure, on the same level in the tree, so that each part of the input has equal chance of equal influence. We have taken the former approach since the possibility of expressing preference will make it easier to analyse the results.

The efficacy of the model was tested by means of issuing typical inheritance-based queries to the system and noting the results. The same experiment was done with a model with no feedback from the associative network, and with the above described confluent model. In the simple model we got correct answers when we had syntactically and semantically clear situations. In unclear situations we got mixed results: it seems that the *property* inheritance realized in the system prefers some kind of shortest-path method. This cannot be compared with shortest-path in symbolic systems since the associative network is sensitive to all kinds of representational features, e.g. vector angles, euclidean distances, vector lengths and individual unit activation. In the confluent model we noticed a tendency towards semantically correct output by means of studying the frequencies of a class to have a particular property. If subtypes or tokens of a class A have a stronger tendency of having a property P than subtypes or tokens of B having a property *not* P, it is more likely to inherit P in a multiple inheritance situation where a subtype or token C inherits properties of both A and B. In addition to this, we have analysed three different situations where we think the property inheritance mapping contributes with information using the confluent model[2]. This enables us to account for the "whole picture", including correlations and regularities in the example set of the associative network, when creating representations. These situations are shown as classical multiple inheritance graphs in Figure 2.

The first graph represents an extension of the problematic Nixon diamond. *Nixon* (N) is a *quaker* (Q), a *republican* (R) and a *colonel* (C). Cs and Rs are not P, but Qs are. What property

---

[2]All tokens and assertions were represented by 4 units, set to values between 0 and 1. We used 2 hidden units in the associative network which was trained using FGREP [9] extended with the above described feedback mechanisms.

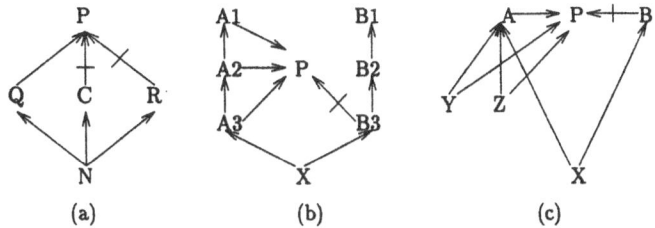

Figure 2: Situations where the "whole picture" contributes with semantics

should N prefer? A shortest-path method would fail since it is the same distance between N and its classes. If we look more closely at the graph we notice that not being P is more frequent (or "normal") and perhaps should this be reflected when investigating what properties N has. When performing this experiment we noticed exactly this[3]. The associative network "senses" this correlation and confluently modifies its input accordingly. In this multiple inheritance situation we encoded N by $((((R(Republican)\ R(Colonel))\ R(Quaker))\ R(Nixon))$. Cluster analysis showed that the order in this case did not matter with respect to the final output.

In the second graph, X inherits from two classes, $A3$ and $B3$, which disagree on property P. Also, $A3$ is a subclass of $A2$ which is a subclass of $A1$ and $B3$ is a subclass of $B2$ which is a subclass of $B1$. However, it is more frequent that classes in the A-branch have property P. The simulation bore out this prediction: it is more likely that P applies to X as well [4].

In the third graph X inherits from two classes, A and B, which disagree on P. In other subclasses of A it is more frequent to have property P, and it is therefore (perhaps) more likely that we attach this property to X. Again, the results of the experiment pointed in this direction[5].

5 Conclusions

As stated earlier, we use euclidean distance for checking results. This is by no means the only way of measuring the "confidence" of the model, given a certain output. We have noticed similarity correlations in activity vector angles and lengths. Further research will concentrate on identifying the correct analytical tools for handling nonmonotonic connectionist representations. Our connectionist approach as it stands avoids some of the problems with classical representations when representing ambiguous knowledge. Whereas the symbolic camp represents objects in such a way that knowledge of, rather than about, an object's superclasses is used for linking objects together, our approach is to bypass the problem of inferential distance ordering along paths and instead rely on similarities in the representations of entities. Such similarities are generated through a systematic coding of entities according to their taxonomic position as well as of the relations such entities enter into. The resulting representations are then confluently re-entered into the system so that inheritance regularities and correlations are strengthened. Our claims is that our model is thereby more likely than a classical model to manage what in classic symbolic terms would be called syntactically ambiguous situations.

---

[3] In euclidean distance the output differed by 0.145 to N being not P, by 1.431 to N being P.

[4] The distance between the output and 'A has property P' was 0.789, compared to 0.916 between the output and 'A has not P'.

[5] The distances between the output and 'X has P' and 'X has not P were 0.407 and 0.979 respectively.

References

[1] M. B. Bodén and A. Narayanan. A connectionist model of nonmonotonic reasoning: Handling exceptions in inheritance hierarchies. In *Connectionism in a Broad Perspective: Selected Papers from the Swedish Conference on Connectionism - 1992*, University of Skövde, 1993. Ellis Horwood.

[2] G. Brewka. *Nonmonotonic Reasoning: Logical Foundations of Commonsense.* Cambridge University Press, 1991.

[3] L. Chrisman. Learning recursive distributed representations for holistic computation. *Connection Science*, 3(4), 1991.

[4] G. W. Cottrell. Parallelism in inheritance hierarchies with exceptions. In *Proceedings of IJCAI 85*, 1985.

[5] G. E. Hinton. Implementing semantic networks in parallel hardware. In Hinton and Anderson, editors, *Parallel Models of Associative Memory*. Lawrence Erlbaum, 1981.

[6] G. E. Hinton. Mapping part-whole hierarchies into connectionist networks. *Artificial Intelligence*, (46):47–75, 1990.

[7] G. Lee, M. Flowers, and M. G. Dyer. Learning distributed representations of conceptual knowledge and their application to script-based story processing. *Connection Science*, 2(4), 1990.

[8] D. Makinson. General theory of cumulative inference. In *Proceedings of the Second International Workshop on Nonmonotonic Reasoning*, 1989.

[9] R. Miikkulainen and M. G. Dyer. Forming global representations with extended backpropagation. In *Proceedings of the Second Annual International Conference on Neural Networks IEEE*, 1988.

[10] J. B. Pollack. Recursive distributed representations. *Artificial Intelligence*, (46), 1990.

[11] R. Reiter. A logic for default reasoning. *Artificial Intelligence*, 13, 1980.

[12] A. Robins. The distributed representation of type and category. *Connection Science*, 4(1), 1989.

[13] E. Rosch. Principles of categorization. In Collins and Smith, editors, *Readings in Cognitive Science*, chapter 3, pages 312–322. Morgan Kaufmann, 1988. also in *Cognition and Categorization*, 1978.

[14] D. S. Touretzky. *The Mathematics of Inheritance.* Research Notes in Artificial Intelligence. Pitman, 1986.

[15] A. Tversky. Features of similarity. In Collins and Smith, editors, *Readings in Cognitive Science*, chapter 3, pages 290–302. Morgan Kaufmann, 1988. also in *Psychological Review*, 1977, 84.

[16] M. Usher and E. Ruppin. An attractor neural network model of semantic fact retrieval. In *Proceedings of IJCNN*, 1990.

Neural networks, natural language and artificial intelligence
—poster contributions

SPECTRAL TIMING and INTEGRATION of MULTIMODAL SYSTEMIC PROCESSES

Jean P. Banquet[1] and José L. Contreras-Vidal[2]

Center for Adaptive Systems and Cognitive and Neural Systems Department

Boston University

111 Cummington Street, Boston, MA 02215

Abstract

A spectral timing system based on cell populations responding to inputs at different rates learns any biologically plausible interval between events sequentially delivered. This system interacts with three other systems, a pattern recognizer, a sequencer, and a probability system, in a complex architecture devoted to context evaluation. The spectral timer seems to be essential to the coordination and integration of multimodal brain processes.

1 INTRODUCTION

Life situations where timing and temporal ordering of events cooperate are common, not only in perception and cognition, but also in motor control. This is also true in the case of artificial systems devoted to sequential pattern recognition as in speech or writing, or motion pattern generation, as in robotics.

The phenomenological evidence of coordination and integration between perceptual, cognitive and motor processes rises the question of the unicity or multiplicity of the timers and sequencers in the brain. Up to now, the most conspicuous evidence for an automatic timer comes from recordings of granule-CA3 neurons in the hippocampus during conditioning (Berger et al., 1980). Conversely, there seems to be evidence for several sequencing systems. In particular, beyond the cortical sequential ordering system in the inferior convexity and sulcus principalis of the frontal lobe, a specific motor sequencer seems also to exist in the cerebellum. The multiplicity of these systems requires an ultimate integrating system, for coordinating and synchronizing the supposed perceptual, motor and cognitive subsystems.

There are several scales of biological and mental time. The scale evaluated here is far away from the longer scales involved in the circadian or ultradian biological rhythms; it is closer to, still different from the fast local oscillations occurring in small cell populations. These processes imply different mechanisms, and implement different functions. One of the most typical examples of small time scale processes is the synchronization of cortical cells in olfactory or visual cortex, as a possible way for feature linking processes at primary cortex level (Gray and Singer, 1991). Timing and sequencing studied here take place at an intermediate level, relevant for the coordination of systemic processes. As such they suppose specific and independent neural nets to perform them.

1.1 Timing and classical conditioning

Grossberg and Merrill (1992) have integrated such a spectral timer, (based on an array of cell populations responding at different rates) in a classical conditioning circuit. They were thus able to explain a large corpus of data from the associative learning literature, concerning the ability of the animal to accurately time the delay of a goal object, based on its previous experience in a given situation. This timing ability controls the right balance between exploratory and consumatory behaviors, and seems to be essential for the animal survival.

[1]Supported by INSERM, and DGA/DRET Grant # 911470/A000/DRET/DS/DR

[2]On leave from Monterrey Institute of Technology, México, and by CONACYT # 63462.

1.2 Timing, Temporal conditioning, and cognitive-motor processes

The paradigm we used in humans to explore cognitive-motor processes is a formal analog of *temporal conditioning*. There is no distinct CS. The equivalent of the US is presented in Bernoulli series at regular intervals. After some practice, a cognitive response related to priming and preparation starts prior to the onset of the stimulus. Thus, the factors manipulated concern not only the learning of a time interval, but also cognitive and motor processes. It becomes thus possible to address the problem of timing in the context of a human complex cognitive task, involving perceptual, cognitive and motor processes, embedded in a learning paradigm. And we emphasize the role of timing in the *coordination* and *integration* of these multimodal processes, during learning. A central timing system could also contribute to enhance not only the coordination, but also the *synchronisation* between perceptual, motor, and cognitive processes. Indeed, perfect synchronisation is not possible at the level of performance, since motor and cognitive processes depend partly on perceptual evaluation of the events. Conversely, there is great flexibility in priming, whose specificity is modulated by the amount of preliminary anticipated knowledge about events and task requirements. Therefore, preparation processes seem to be entirely parallel and synchronous, at perceptual, motor and cognitive levels. Obviously, one of the functions of this synchronisation between the different processing modes during preparation, is to further improve parallel at the detriment of serial processing, during performance. This claim is somewhat congruent with the shift from controlled serial to automatic parallel processes , during task learning. Anticipation-preparation is not the unique consequence of learning, but is still one of the most important ones, when it comes to take into account the context for improving performance.

1.3 Associative learning and priming-preparation modes

Different types of temporal contexts at various levels of complexity and/or integration are learned during the practice of a cognitive task with a choice RT, when the stimuli are sequentially delivered in Bernoulli series. This contextual learning can be considered as a particular case of *associative learning*.

The simplest context is formed by a pair of events. And in sequences of repetitive stimuli, the simplest relation presenting some character of constancy is the time interval that separates two stimuli. If the ISI is constant in the series, or presents only two or a few modes, this information seems to be very rapidly learned, after a few stimuli.

There is a correspondence between a type of temporal context and a specific mode of priming-preparation. The most elementary type of priming, related to time evaluation, is a non-specific increase in the level of activation or alertness. This information indicates when the next event will appear in future time. But it does not give any indication on the nature of the event. Therefore, the corresponding priming must affect any type of possible events.

The third major point derived from experimental results concerns the strong interactions between the different types of contexts. In particular, the ability by the brain to learn sequences and probability is strictly dependent on the duration of the interval between events.

2 SPECTRAL TIMING NETWORK: DESIGN PRINCIPLES

Extra constraints must be brought to bear on the model system designed to implement complex cognitive-motor behaviors. One of the most conspicuous constraints derives from the capability of the brain to receive and process an enduring flow of information, sequentially delivered at variable

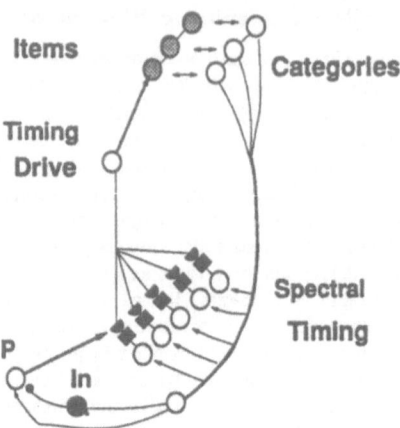

Figure 1: Neural design for the timing circuit. Key: In, inhibitory interneuron; P, Now Print signal; Timing Drive, output response of expected timing.

speeds and loads. So doing, the brain is still able to extract and learn the latent regularities, in the midst of noise and irrelevant variations. Then it can perform a specific adaptive response, and/or initiate an integrated global behavior.

Several design problems had to be solved in the network that emulates these processes: -The equivalence of each stimulus, and in fact its bivalence as both a CS and a US suppose parallel pathways to the spectral cell population; -The continuous flow of inputs prevents any specific coupling by pairs, since any stimulus belongs simultaneously to two pairs, one with the preceding and one with the following stimulus; -Also, the continuous flow of inputs did not allow a resetting of the system after each trial, as it could be done after pair presentations. The reset had to be built into the system. In spite of these new constraints, the design principle of spectral timing is robust enough to support a system that functions correctly even under severe conditions.

3   SYSTEM'S ARCHITECTURE AND SIMULATIONS

The spectral timer is made of a battery of 30 nodes (empty cycles). Each node is endowed with its own dynamical range. Thus, the entire population cooperates to cover the entire plausible range. This spectral decomposition and learning of a time interval by the node populations is made possible by a double gating of these node activations: - First, by a *depletable neurotransmitter* gate (filled squares). - Second, by LTM weights that sample and learn the activation spectrum corresponding to a particular time interval. Learning is modulated by a transiently active Print (P) signal, turned on by sufficiently large and rapid increments of the input activation. The transformation of the input to a sharp phasic activity results from the interaction with a slow inhibitory interneuron, In.

The global response of the system comes from the integration of the activities of all the doubly gated activation signals. This timing drive serves to prime the other structures of the architecture, as the categorizer and the sequencing system.

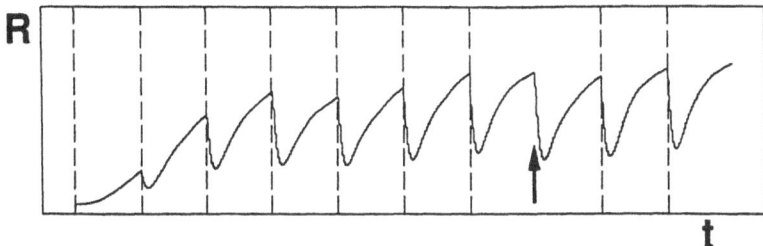

Figure 2: The module for evaluation of the temporal interval (ISI) learns the time interval between stimuli in a sequence of events. After learning, the system is capable of anticipate a response even in the case of an omission in the input sequence. The peak response correspond to the time of presentation of the stimulus being attended to.

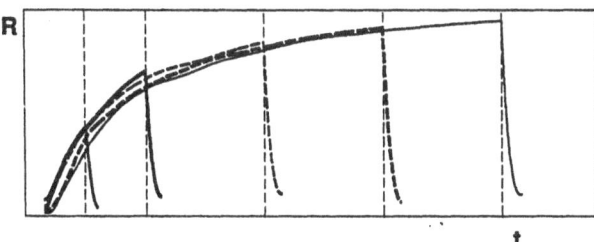

Figure 3: Output of the timing module as a function of ISI's ranging from 250 to 2000 msec. After learning these different ISI's, the output of the module anticipates the expected occurrence of the input, and presents a maximal activity at the time of this occurrence even in the case of input omission.

4 References

[1] Banquet, J.P. and Grossberg, S. (1987). *Applied Optics*, 26, 4931-4946.

[2] Banquet, J.P., and Contreras-Vidal, J.L. (1992). In: *Proc. of the ICANN'92 International Conference on Artificial Neural Networks*. Brighton, UK, 4-7 september 1992.

[3] Banquet, J.P., and Contreras-Vidal, J.L. (1992). *In Proc. of the IJCNN International Joint Conference on Neural Networks*, Baltimore, Maryland, June 7-11, 1992. Vol.I, pp. 541-546.

[4] Berger, T.W., Berry, S.D., and Thompson, R.F. (1980). In *Brain Research*, 193:229-248.

[5] Grossberg, S., and Schmajuk (1989). *Neural Networks*, 2, 79-102.

[6] Grossberg, S. and Merrill, J.,(1992). *Cognitive Brain Research*,1, 3-38. *Neural Networks*, 2, 79-102.

[7] Acknowledgments: Thanks to Gail Carpenter and Steve Grossberg for their helpful comments at different stages of this research.

NET-TO-RULE TRANSFORMATION USING PENALTY FUNCTIONS

Reinhard Blasig, Kaiserslautern University, Germany
Supported by Siemens AG, München
Present address: Siemens AG, Corporate Research and Development
  Otto-Hahn-Ring 6, 8000 München 83, Germany.

Abstract
This paper deals with backpropagation networks trained to perform a classification task on Boolean or continuous input. Our objective is to transform the net into a set of comprehensible rules. While most transformation methods perform a one-step approach being carried out on the network when training is finished, we utilize a penalty function to impose restrictions on the weight parameters. These restrictions adapt the network to the expressive power of the type of rules we choose to describe the network behavior. The resulting net will thus be transformable into an **equivalent** set of rules.

1   Introduction

There have been several attempts to transform networks into an understandable set of rules (see [6] and references therein). This is motivated by a large number of applications, where one is not only interested in an adaptive system to provide good learning and generalization performance, but where one also wants to understand, why and how the system arrives at its output. The objective is to get an insight into the structure of the network and thereby of the underlying classification task. The notorious difficulties in interpreting backpropagation networks arise for mainly two reasons:

- Neural nets are well known for their distributed representation of information; so in order to transform a net into a concise and comprehensible rule set one has to find a way of condensing this information without substantially changing it.
- In the case of backpropagation networks a continuous activation function determines the nodes' output depending on their activation. However, the dynamics of the node activation functions have no counterpart in the context of rule-based descriptions.

To overcome these difficulties we modify the backpropagation error function E_C (i.e. network's classification error) by introducing a penalty term E_P to yield the total backpropagation error

$$E_T = E_C + \lambda * E_P. \tag{1}$$

2   The Penalty Term

The term E_P shall have two effects on the network weights. First a weight decay component aims at reducing network complexity by pushing a (hopefully large) fraction of the weights to 0. The smaller the net, the more concise the rules describing its behavior will be. As a positive side effect, this component will incorporate the notion of "Occam's razor", which says that simple networks are more likely to exhibit good generalization than complex ones.

Secondly, the penalty term shall minimize the transformation error caused by transcribing the network into a set of rules.[1] Adopting the common approach that each non-input neuron represents

---

[1]Throughout this paper we will use rules of the type IF $<$ *premise* $>$ THEN $<$ *conclusion* $>$, where *premise* is a Boolean expression and *conclusion* is Boolean valued.

one rule, there would be no transformation error if the neurons' activation function was a threshold function; the Boolean node output would then indicate, whether the conclusion is drawn or not. But since backpropagation neurons use continuous activation functions like $o = \tanh(a)$ to transform their activation value a into the output value o, we are left with the difficulty of interpreting the continuous output of a neuron. Thus our penalty term is designed to return a high value for those neurons of the backpropagation net, whose behavior cannot be well approximated by threshold neurons, i.e. whose activation value is likely to fall into the nonsaturated region of the tanh-function.[2] For a better understanding of our penalty term, one has to be aware of the fact that the IF-THEN rule type with Boolean premise and conclusion restricts a rule set to essentially calculate a Boolean function. It can easily be shown that any such function can be calculated by a net of threshold functions containing one (sufficiently large) hidden layer. This is still true if we restrict the connection weights to the values $\{-1, 0, 1\}$ and the node thresholds to be integers (see e.g. [3]). In order to transfer this scenario to nets with sigmoidal activation functions and having in mind that the activation values of the sigmoidal neurons should always exceed ± 3, we demand the nodes' biases to be an odd multiple of ± 3 and we restrict the weights w_{ji} to

$$w_{ji} \in \{-6, 0, 6\}. \tag{2}$$

We shortly comment on the practical problem that sometimes bias values about as large as $\pm 6 m_i$ (m_i being the fan-in of node i) may be necessary to implement a certain Boolean function. This may slow down or even block the learning process. A simple solution to this problem is to connect each non-input node to another m_i so called "bias units" with a constant output of 1. These connections are also subject to the penalty function E_P. Due to these additional weights it is sufficient to demand for the bias values

$$b_i \in \{-3, 3\}. \tag{3}$$

Now we can define penalty functions that push the biases and weights to the desired values. Obviously E_b (the bias penalty) and E_w (the weight penalty) have to be different:

$$E_b(b_i) = \begin{cases} |-3 - b_i| & \text{for } b_i < 0 \\ |3 - b_i| & \text{for } b_i \geq 0 \end{cases} \tag{4}$$

$$E_w(w_{ji}) = \begin{cases} |-6 - w_{ji}| & \text{for } w_{ji} \leq -\Theta \\ |w_{ji}| & \text{for } |w_{ji}| < \Theta \\ |6 - w_{ji}| & \text{for } w_{ji} \geq \Theta \end{cases} \tag{5}$$

The parameter Θ determines whether a weight should be subject to decay or pushed to attain the value 6 (or -6 respectively). We defined the penalty to be the absolute difference between actual and desired values. Figure 1 displays the graphs of the penalty functions.

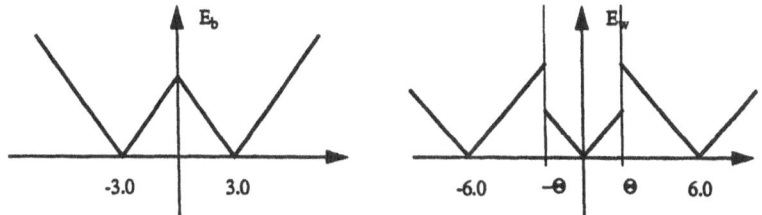

Figure 1: The penalty functions E_b and E_w.

The value of Θ is chosen according to the objective that only those weights should exceed this value, which almost certainly have to be nonzero to solve the given classification task. Since we initialize the network with weights uniformly distributed between -0.5 and 0.5, $\Theta = 1.5$ works well at the beginning of the training process. The penalty term then has the effect of a pure weight decay.

[2]We take $|a| > 3$ with $|o| = |\tanh(a)| > 0.9$ as the saturated regions of the tanh-function.

When learning proceeds and the weights converge, we can slowly reduce the value of Θ, because superfluous weights will already have decayed. So after each sequence of 100 training patterns Θ was multiplied by a factor of 0.995. Observation showed that weights which once exceeded the value of Θ quickly reached 6 or -6. On the other hand, there were relatively few cases where a high weight was reduced again to a value smaller than Θ. Accordingly, the number of weights in $\{-6, 6\}$ successively grows in the course of learning, and the criterion to stop training thus influences the number of nonzero weights, which in turn determines the complexity of the rule set.

We decide on the end of training by means of cross validation. However, we do not examine the cross validation performance of the trained net, but that of the corresponding rule set. This is accomplished by calculating the performance of the original net with all weights and biases replaced by their optimal values according to the formulas 2 and 3.

The weighting factor λ of the penalty term (see formula 1) is critical for good learning performance. We pursued the strategy to start learning with $\lambda = 0$, so that the network parameters first move into a region where the classification error is small. If this error falls below a prespecified tolerance level L, λ is incremented by 0.001. The factor λ goes down by the same amount when the error grows larger than L.[3] By adjusting the weighting factor every 100 training patterns, we keep the classification error close to the tolerance level. The choice of L of course depends on the learning task. As a simple heuristic, L should be slightly larger than the classification error attainable by a non-penalized network.

3 Application to the Prediction of interest rates

We demonstrate the applicability of our rule extraction method on the task of predicting German interest rates. Since we have to deal with continuous valued input, we supplement the network by a discretization layer to yield Boolean input of thermometer code style [2]. We point out that in contrast to pure Boolean learning algorithms (see e.g. [1, 4]), which can also be endowed with discretization facilities, here the discretization process is fully integrated into the learning system, as it will be adapted by the backpropagation algorithm.

The data comprises a total of 226 patterns, which we distribute randomly on three sets: training set (60%), cross-validation set (20%) and test set (20%). The input represents the monthly development of 14 economic time series during the last 19 years, the Boolean target indicates, whether the interest rates have gone up or down during the six months succeeding the measurement of the input data. The time series include e.g. month of the year, the income of private households or the amount of German investments on banks of foreign countries. It has been proven useful for some time series not to take the raw feature measurements as input, but the difference between two succeeding measurements; this is advantageous if the underlying time series show only small changes relative to their absolute values. All series were normalized to range from -1 to $+1$.

We used a network containing a discretization layer of two neurons per input dimension. So there are 28 discretization neurons, which are fully connected to the 10 hidden nodes. The output layer consists of a single neuron. Since our data set is relatively small, the intention to obtain simple rules is not only motivated by the objective of comprehensibility, but also by the notion that we cannot expect a large rule set to be justified by a small amount of training data. In fact, our penalty function managed to push about 90% of the weights to zero; only three of the hidden neurons were used. Nevertheless the prediction error on the test set could be reduced to 25%. This compares to an error rate of about 20% attainable by a standard backpropagation network with one hidden layer of ten neurons. So we sacrificed 5% of prediction performance to yield a very compact net, that can be easily transformed into a set of equivalent rules. The following table shows a part of the generated rules[4] to give a flavor of what the rules look like. If the rules resulting from the transformation process produce contradicting predictions for a given input, the final decision will be made according to a majority vote.

---

[3]Negative λ-values are not allowed.

[4]We shall not publish the complete rule set here.

```
IF (at least 2 of { change in private income      <  0.36,
                    change in foreign investments > -0.32  })
THEN (interest rates will rise)
ELSE (interest rates will fall).

IF (at least 3 of { change in business climate    <  0.34,
                    treasury bonds yields         > -0.35,
                    treasury bonds yields         >  0.30,
                    change in foreign investments <  0.31  }
THEN (interest rates will fall)
ELSE (interest rates will rise).
```

Remember that the network input had been normalized to [−1, +1]. Thus we can roughly interpret an expression like "$x < -0.32$" as "x is small" or (if x represents a difference series) as "x is decreasing". In effect, the first rule tells that interest rates will rise, if the disposable income of German private households is constant or going down AND the amount of German foreign investments does not decrease. To make a more exact use of the thresholds learned by the discretization layer, one has to reverse the transformations and normalizations applied to the original time series.

4   Conclusion and Future Work

This paper advocates the utilization of penalty terms in order to prepare a network for rule extraction. We propose a specific penalty function which not only reduces the network size to yield a concise rule set. More importantly, it adapts the network to the expressive power of the type of rules which we wish to obtain. Consequently, net-to-rule transformation will generate a rule set equivalent to the underlying net.

The applicability of our approach is demonstrated on the task of predicting German interest rates. A number of rules have been generated that exhibit a good performance on this difficult task. However, an unpenalized backpropagation net will produce slightly better results. We regard this as the cost of making the network comprehensible, because the penalty term imposes several severe restrictions on the net: beyond minimizing the number of nonzero weights, the weights are restricted to a small set of distinct values. Last but not least, the simplification of sigmoidal to threshold units also affects the net's computational power [5].

Our main objective is to alleviate these strong restrictions and thus improve on the learning and generalization quality of the network and consequently of the generated rules. This will of course result in more complex rules, demonstrating the tradeoff between comprehensibility and performance.

References

[1] R.M. Goodman, C.M. Higgins, J.W. Miller: "Rule-Based Neural Networks for Classification and Probability Estimation", *Neural Computation*, Vol. 4, pp. 781-804, 1992.

[2] P.J.B. Hancock: "Data Representation in Neural Nets: an Empirical Study", *Proc. Connectionist Summer School*, 1988.

[3] J. Hertz, A. Krogh, R.G. Palmer: "Introduction to the Theory of Neural Computation", Addison-Wesley 1991.

[4] M. Mezard, J.-P. Nadal: "Learning in Feedforward Layered Networks: The Tiling Algorithm", *J. Phys. A: Math. Gen. 22* 1989, pp. 2191-2203.

[5] W.Maass, G.Schnitger, E.D.Sontag: "On the Computational Power of Sigmoids versus Boolean Threshold Circuits", *Proc. 32nd Annual IEEE Symp. on Found. of Comp. Science.*, 1991.

[6] G. G. Towell, J. W. Shavlik: "Interpretation of Artificial Neural Networks: Mapping Knowledge-Based Neural Networks into Rules", In: *NIPS 4*, Morgan Kaufman, 1992.

Teaching Homing Behaviour to a Neural State Machine

Richard G. Evans
Neural Systems Engineering, Electrical Engineering Dept.,
Imperial College, London SW7 2BT
rgevans@ee.ic.ac.uk

ABSTRACT

A logical sparsely connected auto-associative network (GNU) is used to model homing behaviour of an organism in an environmental simulation. The evolution of a solution can be seen as a form of auto-associative recall in which a subset of feedback pathways pass through an environmental function. Three phases of homing behaviour are identified in the simulation results.

Introduction

The use of auto-associative networks to solve combinatorial optimisation problems is well established. Hopfield networks have been used to solve the Travelling Salesman problem [4]. These techniques have been found to be effective in obtaining good solutions to problems for which exact solution time is NP complete. These problems are static in that the function to be optimised does not vary with time and the network is structured accordingly.

The current work is an attempt to modify the technique to derive solutions to dynamic problems where a solution evolves under the influence of a changing external input. This results in a Neural State Machine (NSM) architecture. A network of this kind can control the behaviour of an organism in an environment, in particular to produce homing towards a goal location, see fig 2.

GNU architecture

The NSM is implemented as a Generalising Neural Unit (GNU) as described in [1]. GNU architectures are characterised by a number of parameters.

W is the number of inputs to the network
K is the number of neurons in the network
N is the number of external inputs to each node
F is the number of feedback connections to each node

Where F<K the network can be viewed as sparsely connected. In Fig 1, a network is illustrated with W=4, K=4, N=3, F=3.

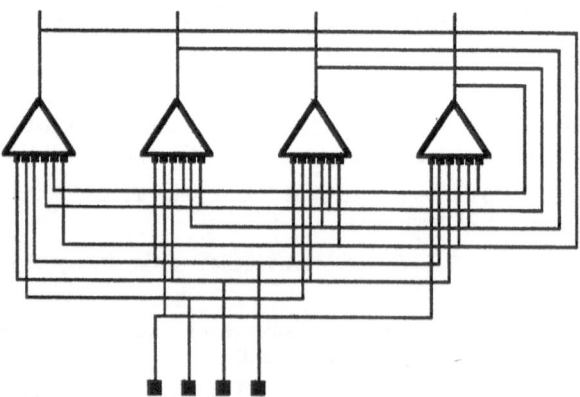

Fig 1. GNU structure

Neurons are logical

The current network is logical in that each neuron implements an arbitrary logical function of its inputs. This functionality can be viewed as that of a Random Access Memory where the truth table of the function is defined by the contents of individual memory locations.

Neurons are probabilistic

In addition, neurons are stochastic in that a given location in the memory can contain a 0, 1 or U. A U value when addressed by inputs results in either a 0 or 1 being output with equal probability. Neurons with this property are known as PLN or Probabilistic Logic Neurons.

Hamming distance generalisation

Generalisation is produced in PLNs by the use of spreading. This technique spreads a trained value to other uncommitted minterms within a specified hamming radius R. An uncommitted minterm is one currently containing a value of U. A PLN with this property is referred to as a GRAM or generalising RAM. This form of generalisation is necessary for the creation of attractors in state space and so to ensure that a GNU can function as an auto-associator.

Problem

The network controls the movement of an organism through a 32x32 binary environment. The problem is to make the organism home to a specific location within the environment on the basis of input from an 8x8 region in the vicinity of the organism. The organism moves up to one pixel at a time in any of 8 directions.

Network

The network used is a GNU with W=64, K=68, N=64, F=32, R=64. The network outputs form two state variables, a move variable and a location hypothesis variable, see fig 2.

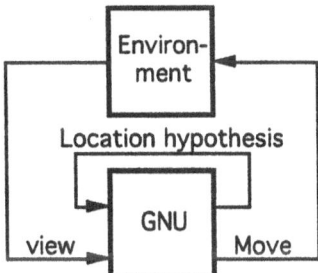

Fig 2. Network organisation

Move variable

The move variable is fed to the environment simulator and causes the organism to move through the environment. It is encoded by the output of neurons 0..3, the nodes represent the vectors for North, East, West and South. The selected vectors are summed to produce the resultant movement.

$$M = \Sigma_{i=0,3} V_i$$

$$V_0=(0,-1) \quad V_1=(1,0) \quad V_2=(-1,0) \quad V_3=(0,1)$$

Location variable

The GNU is trained using supervised learning so that its neural state represents a hypothesis as to its location within the environment after moving. This variable is made up of an X and Y co-ordinate, each of which is thermometer encoded in the range 0..32.

Training

A supervised learning algorithm is used to generate an association for each location in the environment relating the current view and location, with the desired move and the expected new location. An example is illustrated in fig 3. The algorithm trains in an inward spiral so that the locations around the goal are trained last.

 + 15,14 -> SouthEast + 14,13

Fig 3. Training example

Results

The simulation was initialised with the organism at an arbitrarily chosen location and the internal state was initialised with noise. The resulting squares visited by the organism are shown in fig 4. The evolution of the error in location hypothesis E is visible in fig 5. E represents the hamming distance between the values of the location variable L and the values required to represent the actual location A.

$$E = \sum_{i=0,63} (L_iA_i + (1-L_i)(1-A_i))$$

The organism was observed to wander randomly until its location hypothesis converged to reflect its actual position, see t=26, fig 4. Random wandering was characterised by a continuously varying E. The organism then homed towards its goal by the shortest possible route during which the value of E remained at or very near to zero.

It was also possible to observe a behaviour we shall label as 'misguided homing' where due to the small size of the viewing window, the organism mistook a view for an identical view in a different part of the environment and moved accordingly. This can be observed in Fig 4 for time 5≤t≤8 where the system hypothesised it was at square A when it was actually at square B, it could be seen to move towards the East at this point which is correct given this hypothesis.

This 'misguided homing' produced a temporary plateau in the value of E equal to the distance in coordinate values between the hypothesised and actual locations. This short term stability was maintained until the evolving view failed to match expectations whereupon random wandering re-occurred.

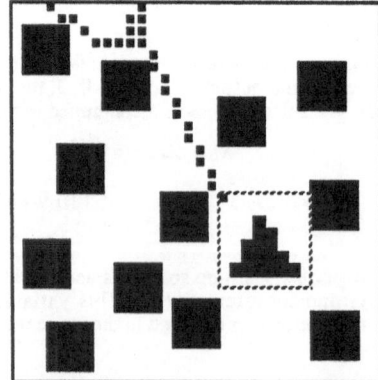

Fig 4. Observed trajectory of organism

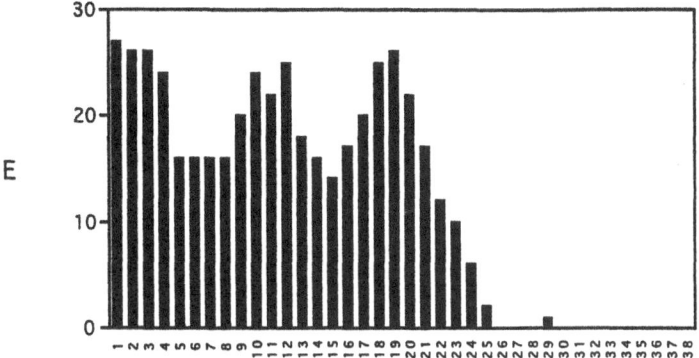

Fig 5. Observed convergence of location hypothesis

Conclusions

A logical sparsely connected autoassociative network (GNU) was used to model homing behaviour of an organism in an environmental simulation. Three phases of homing behaviour were identified in the simulation results. A location error measure was proposed that was seen to characterise the different phases of homing behaviour, namely homing, misguided homing and random wandering.

References

[1] Aleksander, I. (1990). "Weightless Neural Tools: Towards Cognitive Macrostructures" the CAIP Neural Network Workshop, Rutgers University New Jersey.

[2] Wong K.Y.M., Sherrington D. (1988) "Storage properties of Boolean Neural Nets", Proc. NEuro88, Paris.

[3] Hopfield, J.J. (1982) "Neural networks and physical systems with emergent collective properties", Proc. Nat. Acad. Sci. USA **79**, 2554-2558.

[4] Hopfield J.J., Tank D.W. (1985) "Neural computation of decisions in optimisation problems", Biological Cybernetics **52**, 141-152.

[5] Nelson, R.C. (1991) "Visual homing using an associative memory" Biological Cybernetics **65**, 281.

[6] Gauthreaux S.A. Jr, (ed.) "Animal migration, orientation and navigation" , Academic Press.

NEURAL/ICONIC UNDERSTANDING OF THE VISUAL WORLD

Igor Aleksander[1]
Helen Morton[2]

[1]Neural Systems Engineering Group
Department of Electrical and Electronic Engineering Department
Imperial College of Science Technology and Medicine
London SW7 2BT UK

[2]Department of Human Sciences
Brunel University
Uxbridge UB8 3PH UK

Abstract

The key element presented in this paper is an adaptive state machine in which neurons form the state variables and these are sufficient in number for the state to be "iconic". That is, the state reflects events in the world in a many-to-some manner. Early processing forms part of this model. The model is a variant of the previously defined Neural State Machine Model (NSMM). The variant presented in this paper has the ability of functioning in the vision domain both in a rule-based way and an adaptive, neural way. Examples are given of integrated models of concept binding, learning concepts from widely different objects and the memory of scenarios. This is a brief summary of a wide-ranging group of research topics in visual understanding that are enabled by the NSMM technique.

1. Introduction

Much of the work discussed in this paper is a "principle of predicted experience". As an example, this relates to the fact that a person seeing a cup and saucer has an internal presentation that recreates the entire experience of interacting with the seen objects - picking up the cup, sipping from it, hearing the clink as it's put back on the saucer, and so on. So the principle of predicted experience implies the existence of a mechanism that stands aside from the physics of the processing of signals from sensor cells. This mechanism must be capable of retrieving experience of considerable richness and detail. This experience should also be capable of being evoked from other sensory modes: spoken language or (in this case) perhaps the smell of coffee or the sound of clinking crockery in the distance. We now wish to show that the NSMM can, operating according to the principle of predicted experience, acquire the necessary expertise for 'understanding' the visual world.

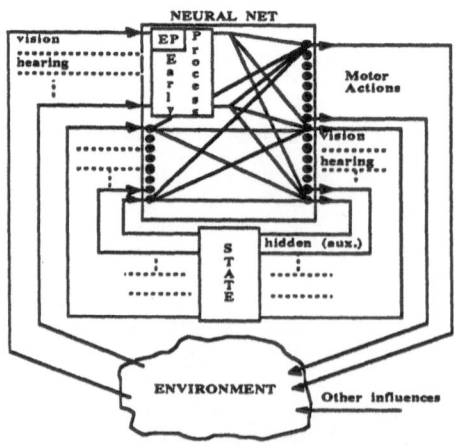

Fig. 1. A neural state machine model (NSMM) for cognitive studies.

2. The Neural State Machine Model

Fig. 1 shows a new development of the original neural state machine first described in by Aleksander [1]. Each of the right-hand dots in the "neural net" represent a neuron. They all receive inputs from the input interface and any architectural bias is determined by a connection matrix from these inputs to the neurons. One group is devoted to providing motor actions on the environment, while another group forms the internal state in which the iconic representations are made to occur. This is partitioned into a number of sensory modalities (all five if needs be) and makes provision for auxiliary state variables on which concepts such as ordinality and duration can be represented. One of the features of this model is the presence of an early processing block (EP). This recognizes the discoveries of Hubel and Wiesel [2] and points to the fact that iconic states which may occur in the brain need not be at the raw sensory level. It can be hypothesized that the EP controls the creation of iconic state patterns. This has the implication that the inner world of the NSMM takes place in the codes generated by the EP.

3. Binding.

One of the properties of learning models that worries cognitive psychologists is how does a system bind the right labels to the right concepts. Connectionists have had some success in demonstrating very specific architectures that bind concepts in a particular, technological way (see Shastri and Feldman [3], for example). Here we can demonstrate that the NSMM provides some general principles that carry out this binding as a natural property. Returning to the example of the cup, how could an NSMM get to represent this as a cylindrical, smooth and hollow object? Clearly this must come from some form of simultaneous experience of sensations from different sensory modalities followed later by the learning of verbal descriptions of these attributes. Babies grasp things and develop a knowledge of the way they feel not only to the hand, but also to the mouth and tongue. This is a feedback operation in the NSMM where the action loop is involved in setting up the tactile experience at the same time as the visual one.

What in this procedure is learned and what is instinctive can be put aside for the moment. What is very clear is that the earliest experiences will lead to the simultaneous exposure to objects from several sensory modalities. In NSMM terms vision, touch and smell, occurring together will generate multi sensory images in the state area of the model. As an example say that the available objects are a billiard ball (b) a cubic steel paperweight (p) and a tennis ball (t). Also say that the first experience is that of a tennis ball. Let the visual experience result in data V(t) being trained into the visual part of the state simultaneously with the words *tennis-ball* and the tactile experience T(t) and the shape experience S(t). As usual, these form a stable pattern in the state field so that presentation of one or more of the attributes causes a recall of the others through pattern completion. Now say that the paperweight is 'learned' leading to a stable pattern V(p), *paperweight*, T(p) and S(p). At a later stage the organism is 'told' that the *tennis ball* is *rough* and *round* and that the *paperweight* is *smooth* and *cubic* the stable patterns are:

V(t),*tennis ball,* T(t), S(t), *rough, round*

V(p), *paperweight,* T(p), S(p), *smooth, cubic*

As the two objects have no common attributes, given an unknown object such as a billiard ball which has ostensibly T(b)=T(p) and S(b)=S(t) when questioned, it would not be able to retrieve either of the two patterns. However, if the first two training objects had been

V(t),*tennis ball,* T(t), S(t), *rough, round*

V(b), *billiard-ball,* T(b), S(b), *smooth, round*

any object x with S(x)=S(t)=S(b), would reliably find the verbal descriptor "*round*" bound to the correct shape experience. In other words, given an attribute, either spoken or sensed, the NSMM will retrieve a series of learned patterns all of which have that attribute in common and discover the sensation (in the case of a given verbal description) or the verbal description (in the case of a given sensation). Therefore, appropriate binding relies on encounters with appropriate, unambiguous training sets, a requirement that connectionism shares with symbolic learning algorithms. Two points are worth noting. First, the effect of ambiguity in learning has been observed in children who (for example) on having been shown three blue flowers and four red ones, when shown three red flowers might report that they were blue. Second, the NSMM combines set-theoretic concept learning with pattern reconstruction, which is a contrast to the prevailing view that object naming is the progression through Marr's 3-process model. The NSMM approach is seen to integrate the neural and the symbolic - the binding process above being found in symbolic work, with the neural enhancing the process as a result of generalization.

4. Classifying similar objects that are different.

A question often asked in conjunction with visual object identification is how objects that look very different and possibly have no obvious features in common, can be identified as belonging to the same class, sometimes even if they have never been seen before. One simple fact is that certain object classes just do have a wide variety of unrelated shapes associated with them, and some form of cataloguing will always be necessary. In trying to approach this through the NSMM, the factor that stands out is that there may be many mechanisms for generalization that come into play in creating models for object classes with many differing elements. It is the 'principle of predicted experience' that is helpful in bringing not only the shape of an object into the procedure for recognising it, but also its function and a prediction of what would happen were it to be handled or used in some way. In NSMMs this becomes learned and objects become classed according to the prediction of experience principle. So if a cup is turned upside down, the predicted experience is that it still is a cup, and that action does not alter the meaning of the object. But if a ball is squashed into a disk the predicted experience is that the class of the object will change to that of a disc. The point we are making is that invariance must be learned and retrieved as a predicted experience - a natural task for the NSMM.

5. Scene understanding from a dynamic standpoint.

A known feature of the output of the early visual system of living things is that nothing ever stands still. Eyes are darting about heads are moving, things in the living world are rushing about. And yet the sensation we have is a calm one that both easily distinguishes motion and is unaffected by it when this is necessary. Fig. 2 is used to support an explanation based on the NSMM.

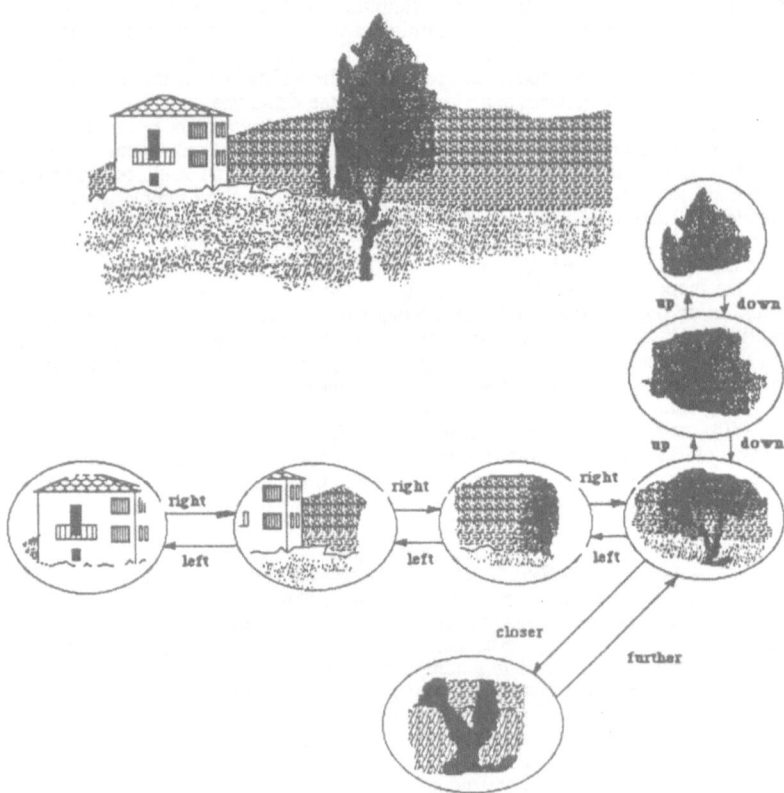

Fig. 2. Making sense of images from a roving eye.

The first factor that we can explain relates to the ocular-motor reflex - the eyes can fix on an object and retain that fixation despite head movements. In NSMM terms this is saying that movement detected by the EP feeds directly to action outputs that activate eye muscles so as to keep the gaze fixed. This is a classical control system that has been modelled as if it were an engineering feedback process. What we need to discuss is the process of changing the gaze from one fixation point to another. It is known that when gazing at an image the eyes focus on centers of interest in the image for a short time and then jump or "saccade" to another. The choice of where to jump is determined by some event in the visual field. It should be remembered that the eye, unlike a television camera, favours in resolution the centre of its field of view (the foveal area). This is capable of a great degree of detail which we have suggested in fig. 2 as the content of iconic states. The output of the EP not only contains this foveal data which creates stable iconic states, but also it contains information as to where other areas of attention might be. So, in any state in fig. 2 any iconic state is accompanied by data which (probabilistically) determines where the next point of the gaze should be. When an extensive scene is memorized (or, indeed, a newly met face is memorized, where saccadic scanning is targeted on salient features such as the mouth, the eyes, etc..) all the information is available to retain the relative positions of the foveal images as a state structure. What we have described is an embryonic approach to a theory of visual memory which takes account of the highly dynamic nature of vision. Several comments come to mind. First, the state structures that retain visual experience need not be limited to transitions that involve saccadic eye movements. Any movement can be involved. So "getting closer" is a state link that could be remembered by an intentional move towards an object. Indeed visions of car journeys or walks over the hills are fully covered by the mechanisms we have discussed so far. The only difference is the fact that the "next point of interest" may be generated in ways other than a peripheral vision signal in the EP. Areas for further investigation may occur in the field of view, requiring getting closer (or, indeed, further, if a greater perspective is required). In some way this borders closely on the evocation of a homunculus that is making decisions about where to look next. We stress as firmly as possible that this is not the case. Every perceptual state carries with it options for the action output of the NSMM which could either be executed through random choice, or, if the state carries some intention with it (e.g. part of the state is a code that requires the discovery of an exit from a space) the investigation will be constrained by this intention.

6. Conclusions.

Iconic representations in the NSMM solve problems related to the understanding of the visual world. The model presented in this paper benefits from the inclusion of an early processor of raw data, which generates the iconic representations that form the state structure of the system. This relates to a process of planning previously demonstrated for the NSMM, which is here applied to visual understanding through a learned "principle of predicted experience". It simply draws attention to the fact that high-level genralization needs to be learned and can be learned by the automaton. A few examples are given in the areas of linguistic binding to iconic states, the classification of images that are topologically different and the creation of inner scenarios from a roving visual sensors. This is part of ongoing research which is going to be reported in full in Aleksander and Morton [4]. This includes application of the NSMM to studies of short and long term visual memory, the memory capacity of weightless implementations of the model and a proper consideration of the cognitive role of mental imagery. An important final conclusion is that this work is in the domain of computation which is not neural/symbolic hybrid, but neural/symbolic integrated.

References

[1] Aleksander, I. An automata-theoretic assessment of the cognitive debate. In: Aleksander and Taylor(Eds), Artificial Neural Networks II, 625-630. Amsterdam: Elsevier.(1992)

[4] Aleksander, I. and Morton, H.B. Neurons and symbols: the stuff that mind is made of. London: Chapman and Hall (in press), (1993).

[2] Hubel, D.H. and Wiesel, T.N. Receptive fields of single neurons in the cat's striate cortex. Jour. of Physiology, 148, 574-591, (1959).

[3] Shastri, L. and Feldman J.A. Neural nets, routines and semantic networks. In Sharkey N.E. (ed) Advances in cognitive science. Chichester: Ellis Horwood, pp 158-203. (1986)

CONNECTIONIST 'SYMBOL' SYSTEMS: COGNITION AS THE SUM OF ANALOGY, EXEMPLAR MANIPULATION AND LANGUAGE

Ronald Lemmen, Department of Philosophy, Utrecht University,
Heidelberglaan 8, 3584 CS UTRECHT. Email: lemmen@phil.ruu.nl

Abstract

According to the classical symbolist view, cognition is symbol manipulation. In order to be a serious rival to symbolism, connectionism has to come up with a viable alternative to symbols and symbol manipulation. My main claim is that the notion of symbol manipulation should be replaced with that of exemplar manipulation, making analogy the central subject for cognitive science. I further claim that language is needed to make the step from concrete to abstract cognition. As a result of the important role of language, our cognitive architecture can best be described as a 'mixed system', combining symbolic and connectionist processes.

Introduction

According to Newell & Simon [15], being a Physical Symbol System (PSS) is necessary and sufficient for general intelligent action. It is therefore claimed that the explanation of intelligent behavior should be at the symbol level and that explanations at lower levels are merely directed at implementation issues [6]. In connectionism the PSS Hypothesis is contested. It is generally argued that cognition can only be understood if subsymbolic processes are also taken into account. In this paper, I assume that for general intelligent action it is necessary, but not sufficient to be a PSS.

The problem now is to give a connectionist account of the nature of symbols. In symbolism, the account is given in terms of conventional programming languages. "List-processing produced a model of designation, thus defining symbol manipulation in the sense in which we use this concept in computer science today" [15]. This model of designation really is the 'pointing' to memory addresses where a full description of the data structure resides. This pointing is not an option in connectionist models. The *content* addressability of memory is often put forward as one of the main strengths of connectionism, but what about symbolic power?

Reduced descriptions

The unavailability of pointers is particularly felt when one tries to implement variables in connectionist models. Hinton [8] has tried to remedy this by making use of roles and role fillers: variables are identified with roles and the values of these variable with fillers. The distributed representations of some set of units can stand for different role-specific representations. This way, a variable can take on different values at different times. Still, the situation is somewhat removed from that in symbolic AI, in which symbols can be seen as small representations of objects providing remote access to a fuller representation of the same object. For this and other reasons [9], Hinton calls for 'reduced descriptions': small representations that can be expanded to full representations of the same objects. Such reduced descriptions are developed by Pollack in his RAAM architecture:

They combine apparently immiscible aspects of well-understood representations: They act both like feature vectors ... and like pointers ... Even further, they act like compositional symbol structures [but] unlike symbol structures, they can be easily compared, and do not have to be taken apart in order to be worked on. Recursive distributed representations may thus lead to a reintegration of the syntax and semantics at a very low level. [16]

Exemplars and analogy

Should one expect connectionism to offer a solution for variable binding? In a sense, this is only necessary if one takes neural networks to be mere implementational models of classical symbol systems. This is the case with Hinton's work which honours the classical semantic distinctions, like those between AGENT, RELATIONSHIP and PATIENT [17].

To see what an alternative could be, it should be realized that the power of classical symbols is really not that they can *designate*, but that they are easily *manipulated*. The point of this paper is to argue that a viable connectionist alternative to the traditional symbol manipulation is exemplar manipulation. To get an idea of what is meant by this, consider the way in which Feynman used to determine the truth value of statements about topology:

> I had a scheme, which I still use today when somebody is explaining something that I'm trying to understand: I keep making up examples. For instance, the mathematicians would come in with a terrific theorem, and they're all excited. As they're telling me the conditions of the theorem, I construct something which fits all the conditions. You know, you have a set (one ball)—disjoint (two balls). Then the balls turn colors, grow hairs, or whatever, in my head as they put more conditions on. Finally they state the theorem, which is some dumb thing about the ball which isn't true for my hairy green ball thing, so I say, "False!" [5].

As becomes readily clear from this example, exemplar manipulation can best be seen as an internalization of our abilities to physically manipulate objects and to perceive the effects of such manipulations [4, 20]. (It is quite interesting to note that our language skills also seem to be derived from object manipulation skills [1, 7]). Exemplar manipulation is more commonly referred to as mental simulation or reasoning with mental models [12].

How much of cognition can be understood as exemplar manipulation? There are several reasons to suppose that exemplars play a very central role in human cognitive processes. To give some:

(1) Memory is distributed and superpositional. Storage of specific memory traces leads automatically to the construction of schemata and prototypes [14, 20]. As a consequence, when something is recognized, what happens is that a coherent set of (micro-)features is activated (a feature vector in activation space): an exemplar. For the dynamics of memory it doesn't matter whether the exemplar represents something that the system really has encountered before or whether it is some sort of 'blended' memory. To a first approximation it can be assumed that with the activated exemplar a relevant precedent for the current situation has been selected. "Precedents" can be actions, options, remindings, etc. [22]. Operations like these are related to case-based reasoning and explanation based learning.

(2) It can be further argued that recognition is always a matter of the recognition of analogy, since it usually is not the case that some stimulus pattern is identically repeated. The point of recognition is that something is recognized even if it doesn't look exactly the same as some previously encountered object or event. The central role of analogy has been forcefully put forward by several researchers (e.g., [2, 3, 19]). When processes are based on analogy, they are based on exemplars.

The significance of pointing to analogy lies in the fact that the system in general cannot simply do what it previously did in similar cases but always has to adapt its behavior to the peculiarities of the present situation. If you understand analogy, you understand exemplar manipulation.

(3) In her work on prototypes, Rosch [18] has shown that much of cognition is based on prototypes. Even knowledge of abstract things like numbers seems to be tainted with prototype effects: when people think of an odd number they tend to think of 7.

That is, when there are no contextual pressures to take another number instead. Such pressures, that can lead to a slippage from one concept to another are investigated by Hofstadter [10, 11].

(4) People are not ideally rational, as is evidenced by the findings of Tversky and Kahneman [21] about judgment under uncertainty. They have uncovered several biases in judgment caused by so-called 'anchoring' or by heuristics like judgment by representativeness or availability. These are typical examples of the omnipresence of exemplars. An example of the heuristic of representativeness is the fact that when a coin is tossed, "people regard the sequence H-T-H-T-T-H to be more likely than the sequence H-H-H-T-T-T, which does not appear random, and also more likely than the sequence H-H-H-H-T-H, which does not represent the fairness of the coin". Judgment insensitivity to, e.g., Bayes rate, sample size, and predictability seem all to be rather natural consequences of memory as exemplar generator.

Language

Like object manipulation, exemplar manipulation is bounded by the concreteness of the manipulative operations. Language offers the additional mechanism to make the step from 'concrete cognition' to 'higher cognition'. Abstraction is an attainment that comes with language.

People often use language to instruct themselves what to do. To understand the function of this, language should be seen as an instrument to influence other people's behavior, instead of a way of communicating meanings. By way of linguistic self-instructions people are able to influence their *own* behavior and escape from the rather concrete operations of exemplar manipulation. Exceptions to the rules coded in the exemplars can be brought to bear. This idea is related to that of Smolensky's 'conscious rule interpreter' [4].

Language brings you yonder. What exemplars are selected as relevant to the situation is determined by the complete context the system finds itself in. But language can also be used to expand the concrete context with a linguistic one. This is another way in which it is possible to take distance from the concrete context and do more abstract thinking.

Eventually, processes that originally needed the guidance of linguistic instructions, and were therefore executed in several steps, can sometimes be compressed into a single step. In Hinton's words [9]: rational inference, requiring several settlings to stable network states, can be turned into an intuitive inference, requiring just a single settling of the network. This is a kind of chunking process, in which linguistic instructions are used for supervised learning.

Conclusion

Much of symbol manipulation (including language) should be understood as an internalization of object manipulation. It has been called exemplar manipulation in this paper. Exemplar manipulation is closely tied with the notion of analogy. Leaving language aside, exemplar manipulation and analogy are the most central notions for cognitive science. A cognitive science program that completely concentrates on other aspects of cognition, assuming that matters of analogical thinking can be solved some time in the future is bound to fail.

Language is an important tool in abstract thought; it triggers certain subsymbolic processes, it expands the context and thus makes the processing less dependent on the concrete situation, and, of course, it is the vehicle of reasoning. Furthermore it is possible to use language as a teacher in supervised learning.

Symbolism has rightly been accused of being guided too much by our linguistic abilities. Yet, symbolism might be right about great parts of cognition in that there is 'real symbol manipulation' and there are explicit rules that state how the symbols (words and exemplars) are to be manipulated. This does not describe all of our mental architecture, however, but only a superficial layer, that is laid on top of an 'animal-brain'.

Following Clark [4], we may conclude that some psychological explanation is properly given at the level of conventional symbolic programs, but other phenomena (Clark mentions creative laps, flashes of insight, jokes coming to us, analogical understanding, perception, fast expert problem solving) seem to require connectionist psychological models. In other words: the mind is best seen as a mixed system.

References

[1] Calvin, W.H., 1991. *The Ascent of Mind: Ice Age Climates and the Evolution of Intelligence*, New York: Bantam Books.

[2] Chalmers, D.J., French, R.M., and Hofstadter, D.R. 1992. High-level Perception, Representation, and Analogy: A critique of artificial intelligence methodology, *Journal of Experimental AI* 4: 185-211.

[3] Churchland, P.M. 1989. *A Neurocomputational Perspective*, Cambridge, Mass.: MIT Press.

[4] Clark, A. 1989. *Microcognition: Philosophy, Cognitive Science, and Parallel Distributed Processing*, Cambridge, Mass.: A Bradford Book, The MIT Press.

[5] Feynman, R. 1986. *Surely You're Joking Mr. Feynman!* New York: Bantam Books.

[6] Fodor, J.A. & Pylyshyn, Z.W. 1988. Connectionism and Cognitive Architecture: A Critical Analysis. *Cognition* 28: 3-71.

[7] Greenfield, P.M. 1991. Language, Tools and Brain: the ontogeny and phylogeny of hierarchically organized sequential behavior, *Behavioral and Brain Sciences* 14: 531-595.

[8] Hinton, G.E. 1984. *Distributed Representations* (Tech. Rep. CMU-CS-84-115). Pittsburgh: Carnegie-Mellon University, Dep. of Computer Science.

[9] Hinton, G.E. 1990. Mapping Part-Whole Hierarchies into Connectionist Networks, *Artificial Intelligence* 46: 47 - 76.

[10] Hofstadter, D.R. 1984. The Copycat Project: An Experiment in Nondeterminism and Creative Analogies. *MIT AI laboratory memo* 755.

[11] Hofstadter, D.R. and Mitchell, M. 1992. An Overview of the Copycat Project. In K.J. Holyoak and J. Barnden (eds.) *Connectionist Approaches to Ananlogy, Metaphor, and Case-Based Reasoning*, Norwood, NJ: Ablex.

[12] Johnson-Laird, P. 1983. *Mental Models: Towards a Cognitive Science of Language, Inference, and Consciousness*, Cambridge: Cambridge University Press.

[13] McClelland, J.L. and Rumelhart, D.E. and the PDP Research Group 1986a. *Parallel Distributed Processing Vol. 2*, Cambridge, Mass.: MIT Press.

[14] McClelland, J.L. and Rumelhart, D.E. 1986b. A Distributed Model of Human Learning and Memory. In: McClelland and Rumelhart 1986a.

[15] Newell, A. and Simon, H. 1981. Computer Science as an Empirical Inquiry: Symbols and Search, In: J. Haugeland (ed.), *Mind Design*, Cambridge, Mass.: MIT Press.

[16] Pollack, J.B. 1990. Recursive Distributed Representations, *Artificial Intelligence* 46: 77-105.

[17] Quinlan, P. 1991. *Connectionism and Psychology*, New York: Harvester Wheatsheaf.

[18] Rosch, E. 1978. Principles of Categorization. In: E. Rosch & B.B. Lloyd (eds.) *Cognition and Categorization*. Hillsdale: Lawrence Erlbaum (pp. 27-48).

[19] Rumelhart, D.E. 1979. Some Problems with the Notion of Literal Meanings. In A. Ortony (ed.) *Metaphor and Thought*, Cambridge: Cambridge University Press (pp. 78-90).

[20] Rumelhart. D.E., Smolensky, P., McClelland, J.L., and Hinton, G.E. 1986. Schemata and Sequential Thought Processes in PDP Models. In: McClelland and Rumelhart 1986a.

[21] Tversky, A. and Kahneman, D. 1977. Judgment under Uncertainty: heuristics and biases. In: P.N. Johnson-Laird and P.C. Wason (eds.), *Thinking*, Cambridge: Cambridge University Press.

[22] Waltz, D.L. 1991. Eight Principles for Building an Intelligent Robot. In: J.-A. Meyer & S.W. Wilson, *From Animals to Animats*, Cambridge, Mass.: MIT Press (pp. 462-4).

Symbol-Manipulation with Attractor Neural Networks

Frank van der Velde

Unit of Experimental and Theoretical Psychology, Leiden University
Wassenaarseweg 52, 2333 AK Leiden, The Netherlands, Bitnet: VDVELDE@HLERUL55

A fundamental issue in the implementation of symbol-manipulation with neural networks is the unlimited productivity of symbol-production systems. Because each realistic neural network is a finite-state system [1], unlimited productivity cannot be implemented with neural networks. Therefore, the only symbol-production which seems possible with neural networks is the production of a regular (finite-state) language. This has serious consequences for the generation of cognitive behaviour with neural networks, in particular natural language processing, which requires the productivity of non-regular production systems [2]. If it is not possible to implement such production systems in neural networks, neural network theory will remain on the level of behaviourism.

However, it should be realized that a digital computer is a finite-state machine as well. Yet, non-regular languages are successfully implemented on a digital computer. Therefore, actual symbol-production with neural networks should be compared with actual production on a digital computer. A good example is the production of a context-free language, which is the first non-regular language in the Chomsky hierarchy (many computer languages are context-free). Any actual production of a context-free language on a digital computer results in a regular subset of the context-free language because, after all, the computer is a finite-state machine. However, the subset can be increased by enlarging the working memory of the computer, without changing the production-program, that is, the context-free grammar itself.

Similar considerations should be made for symbol-production with neural networks [3]. Thus, implementation of, for instance, a context-free language with neural networks is possible if the actual set of sentences produced can be increased by changing the working memory without changing the production-scheme itself. This means that the working memory is an identifiable element of the system and can be changed without changing the system as a whole. Attempts have been made to implement a variable storage with neural networks by a feedback from the output neurons to the input neurons. Thus, information stored in the connection weights is combined with new information and again encoded in the connection weights. However, this procedure results in limited production even in practice, because the connection weights will soon become saturated.

I will show neural network memories which can be increased arbitrarily without changing the system as a whole. With these memories non-regular languages, such as context-free languages, can be produced with neural networks as well. The symbols are distributed patterns. Rules of the language (e.g., $S \rightarrow AB$, $S \rightarrow AC$, $C \rightarrow SB$, $A \rightarrow a$, $B \rightarrow b$) are stored as concatenated patterns in an attractor neural network (ANN). After presentation of the symbol (pattern) S to the left segment of the ANN, the network will settle in a pattern representing a whole rule. The other symbols of the rule (e.g., A or B) can either again be presented to the network, producing the 'words' a or b, or they can be stored in a working memory (WM), to be used later for symbol production. The WM is another ANN consisting of three segments. One segment contains the stored symbols, the other segments contain additional patterns needed to retrieve a sequence of symbols in the correct order. For instance, the stack $AABB$ is stored as r_1Ar_2, r_2Ar_3, r_3Br_4, r_4Br_5. The patterns r_i are created randomly in the WM. After presentation of r_5 to the right segment of the WM, the pattern r_4Br_5 will emerge. Shifting r_4 from the left to the right and presenting it to the WM will produce r_3Br_4. Repeating in this fashion will retrieve the stack $BBAA$ in that order. The WM can be increased without changing the system as a whole. Thus, the competence of symbol-manipulation with neural networks equals that of classical production systems. In actual behaviour, however, there are differences between the systems, which shows the importance of the implementation in the generation of rule-like behaviour.

[1] Van der Velde, F. (1993). Is the brain an effective Turing machine or a finite-state machine? *Psychological Research/ Psychologische Forschung*, **55**, 71-79.

[2] Prince, A. & Pinker, S. (1988). Rules and connections in human language. *TINS*. **11**, 195-202.

[3] Van der Velde, F. (1993). Computing with patterns. (in preparation).

A Consideration on Visual Strategy of Fovea and Saccadic Movement from Experimental Results

Yoshiaki Ajioka

Department of Information and Control Engineering, Toyota Technological Institute
2-12-1 Hisakata, Tempaku-ku, Nagoya 468 Japan

Presently, my most interesting aspect on human visual perception was a visual strategy of fovea and eye movement, particularly saccadic movement, because Adaptive Junction, which is a continuous-time asymmetric neural network, has suggested that some kinds of optical illusion can be easily explained in terms of the visual strategy, by simulating duck-rabbit illusion and interblocking. In this work, then, an experimentation that more than one hundred subjects see either a duck-rabbit figure or some similar figures was executed for verifying the suggestion.

I briefly explain here about this experimental method using a questionnaire survey.

1. Four figures such as a duck-rabbit figure, a figure except an eye, a figure except rabbit's nose, a figure except duck's mouth were prepared.

2. These figures are magnified within the size of 8 cm by 8 cm.

3. The questionnaire consists of three papers representing directions, two questions and one of the above figures.

4. First question requires to answer what the figure shows at a glance (within one second).

5. Second question requires to answer what the figure shows after watching it (for more than twenty seconds).

6. Experimental subjects can take and move these questionnaire papers freely.

Now, from some experimental results, three figures except a feature seem not to make many experimental subjects occur duck-rabbit illusion. Let's compare these results with one another.

First, although the figure except an eye roughly represents the contour of an object, almost all subjects answer a bird at a glance and more than fifty percents of subjects also answer a bird after watching. Under consideration of the effect of contour, the duck-rabbit figure trends to be recognized as a bird. It is not difficult for us to guess that this trend is very remarkable in the case of a glance. Next, an eye of the duck-rabbit figure seems to excite rabbit's nose because many subjects can not find out the nose despite that the figure except the eye includes it, and furthermore all subjects can not recognize the figure except rabbit's nose as a rabbit. In short, the subjects can not recognize a rabbit without rabbit's nose.

By the way, I am afraid many subjects occurred duck-rabbit illusion while watching the figure except duck's mouth. If the removed area was so large as it hid either one of duck's bills, many experimental subjects would answer only a rabbit. However, only the result of the figure except duck's mouth indicates that more than fifty percents of experimental subjects can recognize the figure as a rabbit at a glance. This result suggests not only that duck's mouth is important for recognition of a duck but also that the eye in the center of the figure excites two neighbor features, duck's mouth and rabbit's nose. In brief, the nearer two features are with each other, the interaction of them is stronger, but a direction from one feature to another seems not concerned with recognition. If a human visual perception system had all possible neurons for each combination of neighbor features, it would need the astronomical figures of neurons and synapses. Thus, it seems that saccadic movement avoids this problem and furthermore the next feature from the eye decides a first recognition result of duck-rabbit illusion.

Concluding this work, these experimental results support the idea that duck-rabbit illusion is caused by the visual strategy of fovea and saccadic movement. This idea certainly shows a body plays an important role for cognitive behavior.

ICONIC LANGUAGE REPRESENTATION IN A RECURSIVE NEURAL SYSTEM

Igor Aleksander[1]
Helen Morton[2]

[1]Neural Systems Engineering Group
Department of Electrical and Electronic Engineering
Imperial College of Science Technology and Medicine
London SW7 2BT UK

[2]Department of Human Sciences
Brunel University
Uxbridge UB8 3PH UK

Abstract

In a totally reduced view of the interaction of a learning machine with a "world", such a machine can receive world-state information, it can act on the world and it can receive data from the world that controls learning. We see the learning machine as a classical state machine whose states are its constructs of the experienced world. Learning is a process of developing state structure that is related to taking appropriate actions in the world. There are only two ways that states can acquire an assignment - unsupervised and supervised. The difficulty with unsupevised learning is that should the machine ever be required to describe some of its own constructs there will need to be internal decodings of the arbitrarily assigned states. We believe that this does not remove the need for some states having arbitrary assignments but the only way that the machine can have an "understanding" of the world is through the following alternative way of assigning values to states. It is suggested that the states of the world are best represented through a many-to-some relationship between the state variables of the world and the state variables of the state machine. As the perception of the states of the world can only occur through some senses our definition of an *iconic state* is 'a state assigned to the state variables of the state machine in such a way that it represents the sensory pattern that is generated by a world state'.

Three key points are made in this paper. The first is that "iconic" states arise naturally in all recursive neural nets which are trained on world states in one of a limited possible ways. The second point is that this leads to a new paradigm of cognitive representation. Language is seen as a vehicle for retrieving iconic representations in a recursive system - this is dubbed the "iconic hypothesis". The third point is that while a similar idea has been presented as a "symbol grounding problem" [1], the "iconic hypothesis" goes further in suggesting that a recursive neural system can operate in both a symbolic fashion and use grounded internal states. To illustrate these points we introduce an architectural concept (the Neural State Machine Model - NSMM) which allows a clear formalisation of the concept of iconic representations. Examples of the application of this concept to representation of visuo-linguistic data are given.

Details of this approach may be found in [2].

References

[1] Harnad, S. The Symbol Grounding Problem. *Physica D*. **42,** 335 - 346 , 1990.

[2] Aleksander, I. and Morton, H.B. *Neurons and symbols: the stuff that mind is made of.* London: Chapman and Hall   (in press), (1993).

Miniature Language Acquisition Tasks Using Dynamic Weightless Systems

Nicholas J. Sales

Neural Systems Engineering Research Group, Department of Electrical and Electronic Engineering, Imperial College of Science, Technology and Medicine, London SW7 2BT.
email: cyber@ee.ic.ac.uk

ABSTRACT

The GNU weightless Artificial Neural Network (ANN) [2] has properties which are important for the modelling of language acquisition in humans. One property, described in this paper, is the ability to successfully learn associations between symbols that have been embedded in large amounts (up to 85%) of random noise.

The "Symbol Grounding Problem" has figured large in language acquisition research for many years, and it was noted by Harnad [3], that connectionist methods might be employed to make feasible a solution to this problem.

Work is under way to model acquisition of perceptually grounded symbols, and folk physics, (common sense interaction with the physical world), using dynamic recurrent weightless systems [6]. This paper examines a simple example of acquisition tasks using weightless systems.

The system whose behaviour is investigated here is a GNU (General Neural Unit), constructed using multiple GRAM's (Generalising RAM's) [1]. There are many results for this system, e.g., Ntourntoufis' work, [4], but the behaviour of the system in language acquisition tasks has not been investigated before. The properties investigated here are directly relevant to our approach to the study of language acquisition (outlined in [5]).

Experiments were carried out using simulated recurrent GNU's, composed of between 4 and 8 GRAM's. Simple 4 and 8 step sequences were easily learned by the networks. The sequences were then interspersed, first with noisy versions of themselves, then with randomly generated noise. The networks still learned the sequences with 100% accuracy, even when correct sequences composed only 15% of the total training inputs. Individual steps in the sequences were then dispersed through a long sequence of random noise, but the networks still succeeded in learning the sequences with 100% accuracy, with only about 10% of the training inputs being relevant.

Such properties are important to our model of language acquisition, because they are exactly the kind of learning processes one could expect children to go through, when first learning the simple connections between nouns and referents. Many parts of linguistic input are essentially ignored or incompletely understood, but the child understands that whenever "horse" is used indicatively, there will be a horse present.

REFERENCES

1. Aleksander, I., "Ideal Neurons for Neural Computers", in Parallel Processing in Neural Systems and Computers 1990, pp. 225-228, Elsevier Science Publishers, Amsterdam, 1990.
2. Aleksander, I., "Neural Systems Engineering: Towards a Unified Design Discipline", Computing & Control Engineering Journal, pp. 259-265, November 1990.
3. Harnad, S., "The Symbol Grounding Problem", Physica D, vol. 42, pp. 335-346, 1990.
4. Ntourntoufis, P., " Capacity and Retrieval Properties of an Auto-associative G.N.U", (Abstract only) in Proceedings of IJCNN '91 Seattle, IEEE, New York., 1991.
5. Sales, N. J., "The Essentials Of Language Acquisition: What Do We Need To Study", NSEIR/NJS#2/92, Neural Systems Engineering Research Group Internal Report, London, December 1992.
6. Sales, N. J., "Motivating The Use Of Neural Systems For Language Acquisition: Can Grounding Of Symbols And Development Of Folk Physics Provide A Framework For Language", NSEIR/NJS#3/92, Neural Systems Engineering Research Group Internal Report, London, December 1992.

Activity Curvature: A new Approach to Perception

Konrad Weigl

INRIA B.P.93 06902 Sophia-Antipolis Cedex FRANCE

weigl@sophia.inria.fr

Abstract

We present a new approach to perception: The perception of objects by organisms is the deformation the objects cause to differentiable manifolds of sensory filters in function space; this deformation is characteristic for the manifold *and* the object perceived, and thus does depend on the organism and the outside stimulus. This warping by the stimulus can be considered as analogous to the warping caused by a large gravitational body to the four-dimensional spacetime manifold in its neighborhood.

We model a layer of the sensory cortex as a manifold in function space, e.g. a layer of simple cells of the primate visual cortex as a manifold of Gabor filters $g(x, \phi^j)$ [2], with a parameter vector ϕ^j. The derivative vectors $\frac{\delta}{\delta \phi^j}$ form then a natural base for local tangent spaces to the manifold [3].

A stimulus A(x) entering through the pathways x generates an activity potential

$f(x, \phi^j) = \; < A(x)|g(x, \phi^j) > \; = \sum_{x=1}^{m} A(x)g(x, \phi^j)$ expressible in the local base of $\frac{\delta}{\delta \phi^j}$. We can thus define an *activity metric* for that manifold, via a metric tensor $g_{\mu\nu}$, analogous to the well-known Fisher Information Metric of the manifold of probabilities [1], but derived from $f(x, \phi^j)$:

$$g_{\mu\nu} \;=\; \sum_{x=1}^{m} \frac{\delta(A(x)g(x,\phi^j))}{\delta\phi^\mu} \frac{\delta(A(x)g(x,\phi^j))}{\delta\phi^\nu} \;\Rightarrow\; g_{\mu\nu} \;=\; \sum_{x=1}^{m} A(x)^2 \frac{\delta f(x,\phi^j)}{\delta\phi^\mu} \frac{\delta f(x,\phi^j)}{\delta\phi^\nu} \tag{1}$$

$A(x)$, through changing $f(x, \phi^j)$ from its resting state, changes the metric, and thus causes the *activity curvature* of the manifold to change: The local bases warp differently as we move along the intrinsic coordinate axes defined by the parameter vector. This warping/curving can be expressed by the Riemannian Curvature Tensor, computable as follows: For a given input $A(x)$, we obtain $g_{\mu\nu}$ through (1). The connection coefficients $\Gamma^\mu_{\alpha\beta}$ and the Riemannian Curvature Tensor $R^\alpha_{\beta\gamma\tau}$ will then be:

$$\Gamma^\mu_{\alpha\beta} \;=\; \sum_\nu \tfrac{1}{2} g^{\mu\nu} \left(\frac{\delta g_{\nu\alpha}}{\delta\phi^\beta} + \frac{\delta g_{\nu\beta}}{\delta\phi^\alpha} - \frac{\delta g_{\alpha\beta}}{\delta\phi^\nu} \right) \quad R^\alpha_{\beta\gamma\tau} \;=\; \frac{\delta\Gamma^\alpha_{\beta\tau}}{\delta\phi^\gamma} - \frac{\delta\Gamma^\alpha_{\beta\gamma}}{\delta\phi^\tau} + \sum_\mu \Gamma^\alpha_{\mu\gamma}\Gamma^\mu_{\beta\tau} - \sum_\mu \Gamma^\alpha_{\mu\tau}\Gamma^\mu_{\beta\gamma} \tag{2}$$

As a simple example [4], we consider a manifold of gaussians, two-dimensional because we have two parameters, mean and standard deviation: We can now simply compute the coordinate independent scalar curvature Sc of such a manifold via the equations above as function of different inputs $A(x)$ and location on the manifold:

$$Sc = \sum_\nu \frac{g_{1\nu} R^\nu_{212}}{g_{11}g_{22} - g_{12}g_{21}} \tag{3}$$

Besides potential applications to model results obtained by brain recordings, we conjecture that the brain itself uses that curvature to perceive the environment, disregarding the reaction of individual neurons or clusters, instead integrating the reaction of the whole manifold/layer. One could conceive the brain measuring the curvature by measuring the relative acceleration of nearby geodesics, for example by waves travelling over the cortex, or measuring relative distances by comparing locally the neuronal activity.

We could then interpret habituation, learning, etc., as prewarping of the manifold.

Keywords: Biological Vision, Perception, Differential Geometry, Riemannian Spaces, Gabor functions.

[1] Caianiello, E.R., Systems and Uncertainty: A Geometric Approach, in: Caianiello, E.R., Aizerman, M.A., Topics in the General Theory of Structures, D.Reidel, Publ., 1987

[2] Daugman, J., Six Formal Properties of Two-dimensional Anisotropic Visual Filters: Structural Principles and Frequency/Orientation Selectivity, IEEE trans. SMC, vol. SMC-13, Sept/Oct 1983

[3] Misner, C.W., Thorne, K.S., and Wheeler, J.A., Gravitation, Freeman & Co., Pub., N.Y. 1970

[4] Weigl, Konrad, A new Approach to Perception: Activity Curvature, Tech. Rep. INRIA, to be published.

An Outline for a Theory of the Emotions

L. Michalis

King's College, London, U.K.

I.   The Phenomenolgy of Emotion

Emotion is "a way of acting" [3]. Acting that is directed toward the self when action in the external world is of no avail, but which in any case is "intended as a transformation of the world, not of the self" [3]. In that lies the deception of emotion, its magical quality as well.

Emotion, in the context of Sartre's theory of the emotions arises in a world charged with negative affect and seeks to avert, mask or elude the lack of instrumental means for reaching to an end. It does so by switching actión on to another plane, that which in the theory is referred to as the "non-instrumental (magical) field of unreality". [3]

The contrast between our perception of the world as it has been experienced in the past and its present form apart from bringing about, or forcing a new awareness of the self in a changed environment, it becomes a source of tension. This tension comes to signify the organism's need, need more than desire, for a way to adapt to, and master the conditions of a new environment. It may be positive or negative depending on the nature of the conditions, or changes the organism is trying to adapt, its strength varying according to, and being a measure of the organism's perceived ability or the perceived difficulty of adapting to the new conditions.

Emotional behaviour arises in reaction to a rising tension as all action to an end is inhibited by internal or external constraints. It (emotion) acts to resolve the tension by resorting on to another form of behaviour, and in a way that masks, or eludes any sense of inferiority. By so transforming (our consciousness of) the world, each of the emotions constitutes a "distinct mechanism of self deception". [3]

II.   The Neurobiological Correlates of Emotional Behaviour.

There is evidence - anatomic (electro) physiological, and pharmacological - for the localization of tension in the cingulate gyrus [2]. Afferents from the cingulate gyrus project both to the amygdala and the hippocampus. Feedback signals from the hippocampus act to control the activity in the cingulate cortex in relation to the outcome of the comparison (in the subiculum) between actual and expected events.

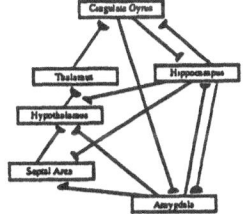

Increased activity in the cingulate gyrus, in response to the breakdown of adaptation - coded in the hippocampus by the mismatch between the actual and expected outcome of an action, facilitates activation of the amygdala by the cortex. Cortical inputs to the amygdala come from the sensory representation of external stimuli associated with the breakdown of behaviour.   Amygdalo-hippocampal interactions, superimposed over intrinsic connections in the amygdala then act not only to select an appropriate emotional reaction, but also to calibrate weighted signals from the cortex to the amygdala so that these signals could, at a later stage elicit the same (emotional) response without mediation from the hippocampus.

References

1.   J.A. Gray, (1982) The Neuro-psychology of Anxiety, Oxford, Clarendon Press.

2.   B. Kissin, (1992), Conscious and Unconscious Programs in the Brain, Plenum Press.

3.   J.P. Sartre (1962), Sketch for a Theory of the Emotions, Methuen & Co Ltd.

4.   J.G. Taylor and M. Reiss (1992), Docs the Hippocampus Store Temporal Patterns Neural Network World 3-4/92, pp 365-384.

Natural language and speech recognition—oral contributions

αβ-TDNN Implement "Fuzzy" Connectionist Time Alignment in Speech Recognition

Patrick Haffner

FRANCE TELECOM, CNET LAA/TSS/RCP

BP 40, 22301 LANNION, FRANCE

Abstract

This paper presents the αβ-TDNN, an extension to Multi-State Time Delay Neural Networks (MS-TDNNs) [4] with a "fuzzy" alignment procedure. We describe several ways to incorporate a Forward-Backward (αβ) alignment procedure in a αβ-TDNN architecture. This appears to be theoretically more consistent with the Back-Propagation training procedure and more robust than the standard MS-TDNN. With the αβ-TDNN, simple modelling assumptions about time alignment suggest choices in the architecture and the objective function which leads to improvements in performance. In particular, we show the possibility to train αβ-TDNN with a global one-pass algorithm based on Maximum Mutual Information (MMI): this system was trained on a speaker independent isolated word recognition task, without any bootstrapping, within a reasonable learning time and with good performances.

1. Proper Modelling vs. discrimination

The most common systems used for speech recognition are currently Hidden Markov Models (HMM). The objective function one wants to maximise at the word level is Maximum Likelihood Estimation (MLE). MLE does not make any explicit discrimination between words, as it assumes that the model is correct. To take into account the possibility of incorrect modelling assumptions, several systems based on discriminative training have been proposed.

In the framework of Markov Modelling, MMI [2] can cope, to some extend, with incorrect modelling assumptions, however it is mostly applied as a fine tuning procedure after MLE.

Neural Networks (NNs) [3]are known as highly discriminant classifiers, which do not make strong data modelling assumptions. To extend their classification capabilities to the word level, the alignment procedure has to be integrated into the connectionist architecture. In the standard MS-TDNN [4], each word output unit accumulates the state scores over the optimal path. No modelling assumption is made about the alignment process. In theory, this allows for a variety of extensions, which, by adding new parameters to the alignment layer, can improve the discrimination power of the system. In practice, we found that Gradient Back-Propagation through the optimal alignment path, as it is done with MS-TDNNs, is not satisfactory in terms of temporal credit assignment. For instance, one state in the sequence tends to take all the credit, and becomes more and more prominent as learning proceeds. To cope with this problem, one must bootstrap the system on pre-segmented data, or impose minimum durations on states.

To train a system that includes an alignment procedure, we face the following dilemma. On the one hand, if we make modelling assumptions, they will constrain the learning procedure to produce a solution which satisfies those assumptions, but which may be far from the optimal one, if those assumptions are not correct. On the other hand, when no modelling assumptions are made, the system may have the possibility to reach a more optimal solution. But the loosely constrained Back-Propagation procedure is likely to find a local minimum, as the objective function to minimise is much more complex than in static NNs.

In HMM/NN hybrid systems [5], a NN produces state scores which are subsequently aligned with an HMM. What they attempt to do is to make modelling assumptions on the alignment part of the system only. Unlike most hybrid systems where connectionist procedures replace some parts of a pre-existing system, we would like to adopt the "reverse" approach, to start from a purely connectionist system, where no modelling assumption is made, and to see what constraints on the architecture and the parameters can be added to make learning more robust, and more likely to converge towards a desired minimum.

In section 2, we show how to make MS-TDNN alignment procedure continuous, to obtain $\alpha\beta$-TDNN. In sections 3, we add some simple modelling assumptions. Section 4 shows how $\alpha\beta$-TDNNs are a robust and continuous generalization of MS-TDNNs. Section 5 interprets our training procedure in terms of MMI.

2. Adding a "fuzzy" time alignment to MS-TDNNs

The first step is to enrich the MS-TDNN architecture, to make its Back-Propagation procedure fully consistent on a theoretical standpoint.

Standard MS-TDNNs back-propagate the error gradient to the units in the optimal path (We call *alignment path* a sequence of one state per time frame). This is logical, as only the outputs of these units are taken into account to calculate the word score. However, the Back-Propagation procedure does not consider alignment paths whose scores may be very close to the optimal. Moreover, it does not account for the fact that, after modification of the weights, the optimal path may be different.

The solution is to take into account the *output scores O(S)* of all the alignment paths S (corresponding to the state trajectory $(s_1,...,s_T)$) allowed by the word W, weighted by their *likelihood factors F(S)*, so that the score of the word W is

$$< O >_W = 1/Z_W \sum_{S \in W} F(S) \cdot O(S) \quad \text{with } Z_W \text{ a normalization factor}$$

2.1. HMM approximation

As no modelling assumptions are made here, we cannot connect $F(S)$ to the probability of some event. In the restricted case of Hidden Markov Modelling, we would have:

$F(S) = \prod_t a_{s_{t-1} \to s_t} P(X_t / s_t)$ where $a_{s_{t-1} \to s_t}$ is a state transition probability and $P(X_t / s_t)$ is the probability to observe the speech signal X_t given the current state s_t

So to take $F(S)$ as the product of state likelihood factors $f_i(t)$ is reasonable:

$$F(S) = \prod_t a_{s_{t-1} \to s_t} f_{s_t}(t) \text{ and the normalization factor } Z_W = \sum_{S \in W} F(S)$$

2.2. MS-TDNN generalization

Imagine we start from a standard MS-TDNN which sums frame level scores: $o_i(t)$ is the output of state i at time t and $O(S) = \sum_t o_{s_t}(t)$ is the output of the MS-TDNN aligned with path S.

The "fuzzy" MS-TDNN output appears as a weighted average of different MS-TDNN outputs for different optimal paths. If the MS-TDNN classifier is applied an infinite number of times to the same sequence, each time picking an alignment path according to the distribution $P(S)/Z_W$, the resulting average output would be $<O>_W$.

2.3. Connectionist implementation and optimization of the alignment layer

The simple assumptions made in section 2.1 and 2.2 lead to efficient and very general connectionist implementation of the alignment procedure.

$$<O>_W = 1/Z_W \sum_{s \in W} F(S) \left(\sum_t o_{s_t}(t) \right)$$

$$<O>_W = 1/Z_W \sum_t \left(\sum_{s \in W} F(S) . o_{s_t}(t) \right)$$

$$<O>_W = 1/Z_W \sum_t \sum_i \left(\sum_{s \in W} F(S) . \delta_{s_t, i} \right) o_i(t) \text{ where } \delta_{i,j} = 1 \text{ if } (i = j) \text{ and } 0 \text{ otherwise}$$

$\sum_{s \in W} F(S) . \delta_{s_t, i}$ is the sum of the likelihoods of all the paths going through state i at time t. It can be computed iteratively with an efficient Forward-Backward algorithm:

$$\sum_{s \in W} F(S) . \delta_{s_t, i} = \alpha_i(t) . \beta_i(t) \text{ with } \alpha_i(t) = f_i(t) \sum_j \alpha_j(t-1) . a_{j \to i}$$

$$\text{and } \beta_i(t) = \sum_k \beta_k(t+1) . a_{i \to k} . f_k(t+1)$$

Bridle [1] showed that α and β can be computed in purely connectionist manner, with a recurrent NN. We call $\alpha\beta$-TDNN the global connectionist system where:

(i) The $o_i(t)$ and $f_i(t)$ are computed from the outputs, at each time frame t, of a frame-level TDNN, which takes as input the speech signal between *t-delay* and *t+delay*.

(ii) A recurrent $\alpha\beta$ layer implements the Forward-Backward algorithm described here.

(iii) The outputs of the system are the word outputs $<O>_W$

As its output is a continuous function of all its parameters, the system can be trained as a classifier which maximises the correct word output with a global gradient back-propagation algorithm. In particular, another Forward-Backward pass computes the partial derivatives $\partial\alpha_i(t)/\partial f_j(t')$ and $\partial\beta_i(t)/\partial f_j(t')$. Those derivatives allow to measure how a change in the weights (and consequently in the TDNN outputs) affects the alignment paths. This is a major improvement compared to MS-TDNNs (and many other HMM/NN hybrid systems) where the consequences of weight changes over the alignment a of sequence are unpredictable.

3. Modelling Assumptions

In the previous section, we described the $\alpha\beta$-TDNN, an extension to MS-TDNNs whose alignment procedure appears to be theoretically more consistent with the Back-Prop training procedure. Now we make some simple modelling assumptions about time alignment, which suggest choices in the architecture and the objective function to improve performance.

In our first modelling hypotheses, the TDNN frame level outputs are interpreted as the posterior probabilities of the states: $y_i(t) = P(s_t = i / X_t)$[3]. According to Bayes rule:

$$P(s_t = i / X_t) / P(i) = P(X_t/s_t = i) / P(X_t)$$

So to take $f_i(t) = y_i(t) / P(i)$ should approximate the HMM observation probability $P(X_t/s_t = i)$ (divided by $P(X_t)$, a factor which is independent of the state, and which is eliminated by the Z_W normalization).

We describe in the next sections choices in the state outputs as functions of the TDNN output activations $y_i(t)$ which are based on similarities with HMMs.

4. The *weighted average* model

In the section 3 formalism, the outputs would approximate log(probabilities):

$$o_i(t) = log(f_i(t)) \text{ and } O(S) = log(F(S))$$

In this case, $<O_W> - log(Z_W)$ is the entropy of the path distribution $P(S)/Z_W$.

$$\sum_{S\in W} \frac{F(S)}{Z_W} \cdot log\left(\frac{F(S)}{Z_W}\right) = <O>_W - \sum_{S\in W} \frac{F(S)}{Z_W} \cdot log(Z_W) = <O>_W - log(Z_W)$$

This function is very similar to the auxiliary function the Baum-Welch algorithm to maximise in HMMs training.

In our first experiments, the $\alpha\beta$-TDNN was trained like the standard MS-TDNN. The objective function requires the output for the correct word to be the largest one. The most efficient way to implement that is with the Classification as a Figure of Merit (CFM) error[6]. This approach is very straightforward, and allows to define a training criterion which matches very closely the testing conditions.

Experiments on a 750 speakers isolated digit database (spoken over telephone and pre-segmented) show that the CFM trained $\alpha\beta$-TDNN yields a 1.15% error rate which matches the best that could be obtained with a standard MS-TDNN. In the case of $\alpha\beta$-TDNN, no bootstrapping is needed. The learning and the alignment procedures cooperate in a very consistent way, in the sense that the system converges toward a solution where states are given equal importance by the alignment procedure. The alignment procedure is more complex in the case of the $\alpha\beta$-TDNN, but its computational load of the Forward/Backward procedure implemented in the $\alpha\beta$ layer is still much smaller than the one required by the forward pass in the frame-level TDNN.

5. The *probability sum* model

Here, the TDNN output activations are only used to compute the likelihood factors F(S), whose role is no longer to average the outputs O(S), but to give an estimate of the word probability. With O(S) constant equal to 1, we simply have:

$$< O >_W = \sum_{S \in W} F(S) \Big/ \sum_{All\ S} F(S)$$

In this simpler model, $<O>_W$ can be interpreted as the conditional probability of the word W given the speech input X: P(W/X). The interpretation proposed in section 2.2 is no longer valid and to interpret this system as a MS-TDNN generalization is less straightforward, but still possible. If we assume, as in Viterbi trained HMMs, that for each word W, it is sufficient to compute the probability of the optimal path $S_{opt(W)}$, then this probability can be connected to the output of the MS-TDNN.

$$< O >_W = O(S_{opt(W)}) \Big/ \sum_{W'} O(S_{opt(W')}) \quad \text{with } O(S_{opt(W)}) = \sum_t \log(y_{s_t}(t)) \text{ the MS-TDNN output}$$

We try here to maximise the information given by the model about the word W:

$$MMI = \log(P(W_C / X)) = \log(<O>_W) = \log(\sum_{S \in W} F(S)) - \log(\sum_{ALL\ S} F(S))$$

We find as in [2] the Mutual Information to be a difference of two terms, one accounting for the correct model and the other for all the models, including the incorrect ones. Experiments implement those ideas in the $\alpha\beta$-TDNN. On the same digit recognition task as quoted in Section 4, the error rate is 1.05%. The difference with Section 4 is not really significant, and we found both models to have very similar learning behaviours. Our results are only partial, as the $\alpha\beta$-TDNN have not been extended to connected speech yet, but, compared to the standard MS-TDNN, it appears to be more robust and gives access to a variety of architectural designs that is even richer. The possibility to train a speech recognition system (here $\alpha\beta$-TDNN) with a global one-pass algorithm based on Maximum Mutual Information (MMI) is very interesting.

6. Concluding remarks about our modelling assumptions.

6.1. What can we expect from them ?

Unlike HMMs, hybrid systems do not attempt to build a complete production model of the speech signal. They only build a model of how a sequence of states is produced by a word. Then, at a given state of the sequence, the NN tells whether the corresponding speech input can, or cannot, correspond to this state. The NN output is shown to converge, at the end of the training phase, towards the probability of this state given the input only when is explicitly trained, with a proper error criterion[6], to perform this association. This means that the NN should be trained to classify states at each time frames, and not words, as we do with $\alpha\beta$-TDNNs. In our case, there is not even a bootstrapping phase to train the state-level TDNN to output probabilities. We can only say that, among the possible solutions, there is one which corresponds to our model, but it is not necessarily the only minimum of our error criterion.

6.2. A weakly constrained NN architecture

A numerical analysis of the outputs of the TDNN shows that, for instance, activations which are supposed to approximate probabilities do not sum up to 1. Additional constraints in the architecture, which are supposed to improve the fitting of the NN outputs to the values expected from the model, do not help significantly.

- It is recommended to observe the constraint $\Sigma_i \, p(s_t = i / X_t) = 1$. This can be implemented by replacing, at the output of the TDNN, the standard Sigmoid function, with the *SoftMax* function: $f(x_j) = exp(x_j) / \Sigma_i \, exp(x_i)$. However, Gradient Back-Propagation with the SoftMax function still has some instability problems, so we obtain a better performance by applying the normalisation at the word level.

- The NN output should be divided by the prior $p(s_t = i)$. However, with word models where states tend to be given equal importance, and whose prior probability is not known, this division does not seem to be necessary.

6.3. Improved experimental results

Section 3 and 4 show choices in the architecture which are motivated by modelling assumptions. We also experimented with other architectures whithout inderlying models, with lower performance:

- The *"probability sum"* model gave us the idea to use MMI as a error criterion, which is the error which gives the best performance.

- The *"weighted average"* model suggests that we should have $O(S) = log(F(S))$. We tried many other relations between O and F, with lower performance.

[1] Bridle, J.S. Alpha-nets: a recurrent neural network architecture with a hidden Markov Model interpretation". Speech Communication, 9, 83-92.

[2] P.F Brown, "The Acoustic Modelling Problem in Automatic Speech Recognition", *PhD* Thesis, Carnegie Mellon University, May 1987

[3] Gish, H, "A probabilistic approach to the understanding and training of neural network classifiers". In *Proc IEEE ICASSP* 1990, pp. 1361-1364, Albuquerque.

[4] Haffner, P.,Waibel, A. "Multi-State Time Delay Neural Networks for Continuous Speech Recognition" in *Advances in Neural Information Processing Systems 4*, Morgan Kaufmann, San Mateo, 1992.

[5] Renals, S. and Morgan, N. "Connectionist Probability Estimation in HMM Speech Recognition ". TR-92-081, ICSI, Berkeley, December 1992.

[6] Hampshire, J. and Pearlmutter B. "Equivalence Proofs for Multi-Layer Perceptron Classifiers and the Bayesian Discriminant Function", in *Proc of the 1990 Connectionist Models Summer School*, Morgan Kaufmann, San Mateo, 1992.

CONTINUOUS SPEECH RECOGNITION PREDICTIVE SYSTEMS

A. Mellouk*, P. Gallinari**

* LRI, UA 410 CNRS,	** LAFORIA UA CNRS 1095
Bat. 490	Tour 46-00 Boite 169
Université Paris Sud	Université Paris 6
91405 Orsay cedex	4 place Jussieu
France	75252 Paris cedex 05 France
mellouk@lri.lri.fr	gallinari@laforia.ibp.fr

ABSTRACT

This paper presents a discriminative neural prediction system for continuous speaker independent speech recognition. We first compare different neural predictors for modeling speech production. We then propose new criteria for discriminative training. These networks are incorporated into a complete speech recognition system where they cooperate with other modules (grammar model, correction rules and dynamic time warping). Our best systems allow to reach 74,9% accuracy on TIMIT which compares well with other state of the art systems, while being less complex and easier to implement.

1. INTRODUCTION

Neural networks (NN) have shown promising results when embedded into hybrid speech recognition (SR) systems. Recently several hybrid models have been proposed for continuous speech recognition in order to improve current recognizers performances. They differ by the organization of the overall recognizer and the techniques they use. For now, there is no agreement about which approach should be used and these different directions are developed in parallel.

The neural predictive approach to speech modelization allows to built systems which are easy to implement and offer good performances. Different models have been proposed which share the same basic ideas [1, 2, 3, 4, 5]. In these systems predictive NNs are used for low level modeling of words or phonemes. Sequences of matching scores between models computed outputs and reference templates are then processed through dynamic programming (DP) techniques. The best path allows to perform segmentation and the corresponding error is back propagated through the prediction models which allows to train the system parameters. Continuous speech recognition can be performed by linking elementary phoneme models. These hybrid systems perform non-linear prediction, do not make stationarity assumptions about data and take easily into account context information. Production systems are usually trained according to non discriminant criteria. We have proposed in [5] a discriminant predictive system for continuous speech recognition. It is based on simple approaches for both prediction and discrimination. We extend this work here by studying different neural predictors for speech modelization and new criteria for discriminative training. We then propose new systems with increased performances. The basic system is presented in section 2, sections 3, 4 and 5 discuss neural prediction and discrimination. In 6 we present tests on TIMIT and a comparison to reference SR systems.

2. BASIC SYSTEM

In our system [5], several modules cooperate in order to achieve continuous speech recognition (fig. 1). Each phonemic entity is modelled by a Phonemic Model (PM). Each PM is composed of three NNs for modeling the successive parts of the phoneme. This phonemic modelization is a trade-off between model accuracy and correct statistical estimation of the models parameters. These NNs will be called state models (SM), transitions are allowed from any SM to itself or to its right neighbors, PMs are thus LR2 models. After processing a sentence each PM prediction is time aligned with reference patterns through DP. The DP finds the best match between target sequences of frame labels and the actual output of the PMs by summing the frame-by-frame euclidean distance. In the best path computation, we use an inter-model cost C_{mn} between models m and n which takes into account information from a left to right bigrams grammar. The bigrams probability distribution, P_{mn}, was computed from the cooccurence matrix of the phonemes. To incorporate the available grammar knowledge, we use $C_{mn} = \beta + \log P_{mn}$ where β is a bias term used to imbalance deletion and insertion errors [6]. The correction module implements simple rules which incorporate duration constraints into the phonemic models. The winning model is the valid model with the highest score.

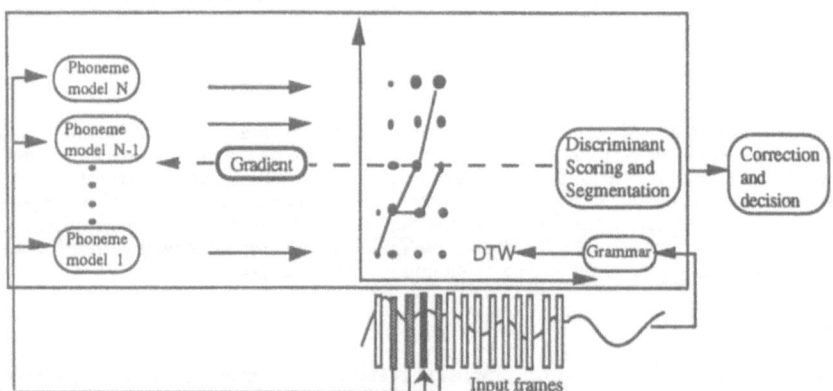

Figure 1 : the hybrid system. Each phoneme model is a set of three NNs; "DTW" is a Dynamic Programming module; "Correction" implements correction rules on the code sequence delivered by DTW. "Grammar " allows to compute an inter-model cost matrix used in DTW.

3. PREDICTION MODELS

In the system, predictive NN models for SR are trained to predict frame at time t from context frames (fig.2) . They allow to represent dynamic speech features. Although most predictors in speech systems are left to right (LR), other predictors may be used as well and give better performances [7]. We have tested here LR, right to left (RL), and left & right to centre (LRC) predictive systems (fig. 2). In all our experiments, we have used three context frames (fig. 2) which appear to be a good compromise.

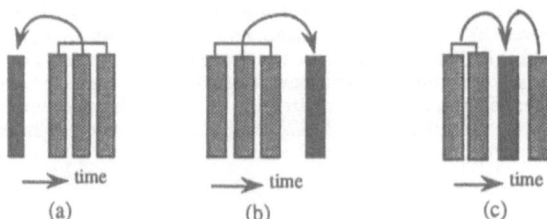

Figure 2 : predictive systems (a) right to left (3), (b) left to right (3), (c) left & right to centre (2-1). Context frames appear in grey and current frame in black.

Table 1 compares the performances of the three predictors. LR and RL systems have the same global behavior but left context seems to be more important for accurate modelization. A good compromise is to use a combined model (LRC) which uses left and right contexts, and gives more weight to the left. This model performs an interpolation rather than a prediction. The main improvement is in the reduction of substitutions. Training criterion for these three models is criterion DC2 and tests have been performed on DR1 (see sections 5 and 6 below). Similar behavior has been fond for other criteria as well.

	LR (3)	RL (3)	LRC (2-1)
Substitutions (%)	13,5	17,6	9,4
Insertions (%)	3,5	3,9	2,9
Deletions (%)	5,7	7,9	4,6
Correct (%)	80,8	74,5	86,0
Accuracy (%)	77,3	70,6	83,1

Table 1: comparison of three NN predictive models (LR, RL and LRC). Each model used three context frames. The "Correct" rate is defined as 100% - (Substitutions + Deletions). The "Accuracy" rate is defined as Correct - insertions. Tests have been performed on DR1.

4. INTERPRETATION OF THE MODEL

The predictive model can be given a simple statistical interpretation. Speech is modelized as being produced at each time by one state from a finite set. In each state emissions are supposed to come from a stationary process which is modelized by a non linear auto-regressive process of order three. The global model is therefore nonstationary. It can be described alternatively as a non linear autoregressive model whose parameters are time varying.

Recognition is performed by inverse filtering of the speech signal by the models and scoring of the local errors along an optimal path. Segmentation is performed by dynamic programming. Training the model consists in optimizing the SMs parameter. It is performed by a two steps iterative algorithm : in the first step, SMs parameters are optimized for a given path, in the second, segmentation is performed for fixed parameters values.

Compared to HMMs, in this model SMs transitions are not trained and are therefore imposed by the SM model topology. The phoneme models we have used are simple L-R three states models. Inter PMs transitions are estimated from the corpus using bigrams scores. Observations are not considered independent which allows for a better modelization of the signal dynamics.

Similar ideas have been used for modeling more accurately the speech dynamic in the recent development of linear predictive HMMs [8].

The emission model can be described by a non linear auto-regressive process driven by a noise source whose parameters are allowed to vary according to the current state:

$$x_t = F_i(x_T) + \varepsilon_t$$

where F_i is the function computed by the ith SM, ε_t is a noise whose realizations at different times are supposed independent, and x_T is the prediction context for x_t (e.g. frames x_{t-1} to x_{t-3} for the above LR model). Let us suppose for simplification that the noise is gaussian with zero mean.

In this case $P(\varepsilon_t) = P(x_t / x_T, s_t) = \dfrac{1}{2\pi^{n/2} |\Sigma_{s_t}|^{1/2}} e^{-\frac{1}{2}(x_t - F_i(x_t))^t \Sigma_{s_t}^{-1}(x_t - F_i(x_t))}$

where s_t is the state at time t and n the dimension of the input vector.

The conditional likelihood of the sequence $(x_1, ..., x_L)$ is :

$$P(x_1{}^L / s_1{}^L) = \prod_t P(x_t / x_T, s_t)$$

where $s_1{}^L$ is the sequence of state models corresponding to the path.

Optimization is performed jointly on the predictors parameters and on the segmentation i.e. on $P(x_1{}^L, s_1{}^L)$. E. Levin [9] has shown that for similar models this corresponds to the optimization of an approximate maximum likelihood corresponding to the best DP path.

We have used in our tests the euclidean distance between desired and predicted output, which corresponds to the assumption of a white noise. This assumption is valid only if the data have been properly scaled and if the components of the observation vector are uncorrelated. Although this is not fully justified in the present case this greatly simplifies the computations and allows for more robust estimates of the parameters.

5. DISCRIMINATIVE CRITERIA

Predictive models for classification are usually trained independently one from the other on the different classes. NNs allow to perform discriminative training at the expense of a reasonable extra computation cost. The basic idea for discriminative training is the following. Let D_i denote the euclidean distance (prediction error) between the desired frame and actual outputs of SM number i and let us define the score of model k by :

$$P_k = \frac{e^{-D_k}}{\sum_j e^{-D_j}} \tag{1}$$

where the summation is over all the SMs. Viterbi algorithm may be used to compute the best path using the P_ks. P_k incorporates discriminant information between the SMs at the frame level. Combination of the P_ks corresponding to the successive frames and to the different models allow to introduce discrimination at the phoneme or word levels. Our algorithms use adaptive gradient rules so as to move, for each predicted frame, the score of the correct model towards 1 and others towards 0. However, different criteria may be defined using the P_ks, resulting in very different behavior. For simplicity, we describe below informally the effects of 4 criteria which have been tested on TIMIT (DCi holds for Discriminant Criterion n° i).

• DC1 draws the SM with highest score from the desired PM towards 1 (non discriminant).

• DC2 draws the highest score SM from the desired PM towards 1 and all other SM towards 0.

• DC3 draws the highest score SM from the desired PM towards 1 and SMs from all other PM towards 0.

• DC4 draws the SM with highest score from the desired PM towards 1 and the highest score SM among all other PM towards 0.

Criterion 1 performs non discriminative training and modifies only the parameters of the winning SM from the correct class, criterion 2 discriminates between all SMs and their parameters are changed accordingly, criterion 3 performs an inter phoneme model discrimination, but no intra model discrimination, i.e. non winning SMs from the correct PM are not changed. Criterion 4 which takes into consideration only two of the SMs at a given time, allows to approximate more closely than other DCs the misclassification rate and concentrates on models whose prediction is close to the decision frontier [10]. Other criteria may be used, as well as other mean risk approximations. Models trained with discriminative criteria do not optimize the maximum likelihood of the sequence but approximations of the misclassification risk. As discrimination has been implemented at the frame level without taking into account temporal information, this is a rather raw approximation but as will be seen below it works pretty well in pratice and the extra computational cost is reasonable.

6. TESTS

We have performed tests on the december 88 DARPA TIMIT continuous speech database. The best conventional systems for phoneme recognition are based on sophisticated multiple codebooks and context dependent HMMs, e.g. CMU SPHINX system [11]. Good results with NN systems on TIMIT have been obtained by Robinson using the Cambridge Recurrent Error Propagation Network (REPN) [6]. In order to compare our results to these baseline systems we have used the same experimental conditions as those reported in [5, 6, 11]. The 64 TIMIT phonetic classes were thus mapped onto a set of 39 phonetic classes [11]. A first evaluation on the sx sentences of dialect DR1 (185 sx sentences, 37 speakers) has been performed. We used 27 speakers for training and 10 for testing. Input signals corresponding to 16 LPCC coefficients were computed on 25,6 ms Hamming windows. All our PM were initialized using the rough labelling provided by TIMIT. NN parameters inside each PM were also initialized using a simple segmentation of the data. The grammar information and the bias (β) have been used during training and DTW scoring.

Table 2 compares the different discriminative criteria on DR1. It is clear that they have very different effects on the performance of the system. DC4 is the best one both in performances and in computation time.

We have also used the entire TIMIT to validate the system on a large database. There are 420 speakers in total, 400 for training, 20 for testing. Eight sentences were used per speaker (the sx and si sentences). Comparison with reference systems appears in table 3.

DNPS-LRC (2-1)	Correct(%)	Insertion(%)	Substitution(%)	Deletion(%)	Accuracy(%)
DC1	76,5	5,1	16,4	7,1	71,4
DC2	86	2,9	9,4	4,6	83,1
DC3	89,2	2,2	7,1	3,8	87
DC4	90,6	1,5	5,9	3,3	89,1

Table 2 : comparison of discriminant criteria on DR1. Notations for these criteria are similar to those in section 4. DNPS means Discriminant Neural Predictive Systems, the "LRC" indicates that the predicted frame is between the context frames.

Systems	Correct(%)	Insertion(%)	Substitution(%)	Deletion(%)	Accuracy(%)
SPHINX	73,8	7,7	19,6	6,6	66,1
REPN	78,6	3,6	15,0	6,4	75,0
DNPS-LR (3) (DC2)	74,5	5,9	18,7	6,8	68,6
DNPS-LRC (2-1) (DC2)	77,1	5,5	16,9	6,0	71,6
DNPS-LRC (2-1) (DC3)	79,0	4,9	15,6	5,4	74,1
DNPS-LRC (2-1) (DC4)	79,7	4,8	15,0	5,3	74,9

Table 3 : Comparison of Discriminant Neural Predictive Systems (DNPS) with other phone recognizers, LR and LRC hold for Left to Right and Left & Right to Centre predictors. DCi is the discriminant criterion which has been used in each case. SPHINX holds for context-dependent SPHINX [9]. REPN is the recursive network proposed by Robinson [6].

Our systems performances are similar to REPN and above SPHINX. By using combined predictive and discriminative systems, we have considerably improved upon the performances of our previous DNPS-LR technique based on DC2 optimization [5].

7. CONCLUSION

This paper has described the use of predictive neural networks with discriminative training algorithms for the task of speaker-independent phoneme recognition from continuous speech. It explores different types of predictors and discriminant criteria. The resulting systems are competitive with the best HMM technology. Further research will include different contextual informations, the evaluation of more sophisticated phonemic models, and new discriminant criteria for sequence discrimination.

Acknowledgements : all the tests have been performed with SN2 neural networks simulator, many thanks X. Driancourt for fruitful discussions.

8. REFERENCES

[1] Tebelskis J., Waibel A., Petek B., Schmidbauer O.: " Continuous speech recognition using linked predictive neural networks", ICASSP 91, pp 61-64, 1991.
[2] Iso K., Watanabe T.: " Large vocabulary speech recognition using neural prediction model", ICASSP 91, pp 57-60, 1991.
[3] McDermott E., Katagiri S.: "Prototype-based discriminative training for various speech units", ICASSP 92, pp 417-420, 1992.
[4] Mellouk A.:"Quelques approches pour la reconnaissance automatique de la parole continue", Tech. Rep. 715 Univ. Paris-Sud/ LRI. Dec 1990.
[5] Mellouk A., Gallinari P.:"A discriminative neural prediction system for speech recognition", to appear at ICASSP 93.
[6] Robinson T.:" Several improvements to a recurrent error propagation network phone recognition system", Tech. Rep. CUED/F-INFENG/TR.82, Cambridge Univ. Eng. Dept, Sept. 1991.
[7] Kawabata T.: "Predictor codebook for speaker-independent speech recognition" , ICASSP 92, pp I.353-I.356, 1992.
[8] Kenny P. , Lennig M., Mermelstein P. : "A linear predictive HMM for vector valued observations with applications to speech recognition", IEEE Trans. on ASSP, vol. 38, N° 2, Feb. 90.
[9] Levin E. :"Hidden control neural architecture modelling of non linear time varying systems and its applications", IEEE trans. on NN, Vol. 4, N° 1, Jan 93.
[10] Driancourt X., Gallinari P.: "A speech recognizer optimaly combining learning vector quantization, dynamic programming and multi-layer perceptron", ICASSP 92, pp 609-612, 1992.
[11] Lee K. F., Hon H-W.: "Speaker-independent phone recognition using hidden markov models" , IEEE Trans. ASSP, Vol 37, no 11. pp 1641-1648, 1989.

Handling Context-Dependencies in Speech by LVQ

Jyri Mäntysalo, Kari Torkkola*, and Teuvo Kohonen

Helsinki University of Technology, Laboratory of Information and Computer Science
Rakentajanaukio 2 C, SF-02150 Espoo, FINLAND   (jyri@nucleus.hut.fi)
* Institut Dalle Molle D'Intelligence Artificielle Perceptive (IDIAP),
C.P. 609, CH-1920 Martigny, SWITZERLAND   (karit@idiap.ch)

Abstract

In the framework of phonemic speech recognition using codebooks trained by Learning Vector Quantization (LVQ) together with Hidden Markov Models (HMMs), a novel way to model context-dependencies in speech is presented. We use LVQ to map acoustic contextual data into context-independent phonemic form. The contextual data is in the form of concatenated averages of successive short-time feature vectors. This mapping eliminates the need to employ context dependent phonemic HMMs and the difficulties associated therein. Instead, simpler context-independent discrete observation HMMs suffice. We report excellent results for a speaker dependent task for Finnish.

1   Introduction

To tackle the variability induced by context, the most successful speech recognition systems reported in the literature employ context-dependent subunit models, phoneme models among others, [8] [1]. Due to large variations in acoustic realizations of phonemes caused by context, usually a separate model is constructed for a sub-unit in each context that causes its realization to be different. Construction of such models is a tedious task and requires considerable linguistic knowledge and/or sophisticated methods. Also, because of the large number of emerging sub-unit models, sparse training of the models may be a problem with a limited amount of training data. Tying of parameters may, however, alleviate some of these problems [12].

In this paper, we present experiments whose purpose is to demonstrate that context effects can be handled at a level lower than Hidden Markov Models (HMMs). To this end, we use the Learning Vector Quantization (LVQ) methods, including the latest developments [4, 6, 5, 7]. The idea is to employ LVQ to map acoustic data, including context in time, into context-independent phonemic form, and to interpret the symbolic data further by using context-independent HMMs. The mapping can also be looked from the pattern recognition point of view. Hereby LVQ is performing a task of phoneme classification on the basis of contextual information.

The resulting discrete data is decoded into phoneme sequences by context-independent HMMs. In addition to mapping results provided by LVQ, we try to exploit vector quantization errors as an aid to decoding. We also applied LVQ to the segmentation task, whereby the context vectors are classified into two classes only, representing the stationary regions and transitions, respectively. The outputs from LVQ are interpreted as output symbols of discrete-observation HMMs with multiple observation streams [3]. The models also include quantization errors as a separate observation stream.

The rest of this paper is organized as follows. Section 2 presents the experiments in finding the best-performing form of context data for the mapping. In Section 3 we describe the decoding by HMMs and additional information that can be derived using LVQ to aid the decoding. Finally, in Section 4 we draw some conclusions.

2 Finding the best representation of context data

To construct a mapping from acoustic data into phonemic form, we first made experiments to find the best performing *representation* of context data. By the best performance we mean here the highest accuracy as a phonemic classifier. In this work, the pattern vectors to be classified are formed by both averaging and concatenating short-time feature vectors within a time domain window. In this way we obtain a very large number of different patterns that represent the same "phoneme". Their classification needs high computing capacities, which are provided by neural-network methods, such as LVQ here. Since there exist no known theoretical way of deciding the combination that will give the best performance with the given data, we have tried to find the form experimentally.

2.1 Speech data and preprocessing

In the following experiments the same small Finnish speech database has been used as in [11]. It contains four repetitions of a set of 311 words (1737 phonemes) spoken by three male Finns. Three sets were used for training, and the remaining one for testing. Four independent runs were made by leaving one set at a time for testing. Thus all recognition results presented are averages of 12 runs. As features we use 20-component mel-scale cepstra computed every 10 milliseconds in the way described in [2]. The cepstral vectors are thereafter liftered with raised sine [11].

2.2 LVQ training procedure

We have first run an excessive amount of preliminary experiments to find good parameters to LVQ for this specific task. We used the public domain software package *Lvq_Pak* [7] throughout these experiments. The following learning scheme proved to give the best results on the average: 1) Codebook initialization so that each class contains an identical number of codebook vectors. 2) Balancing the number of codebook vectors so that medians of the shortest distances between codebook vectors are approximately equal. 3) OLVQ1-algorithm using 10000 iterations and learning-rate initial value $\alpha_i(0) = 0.3$. 4) LVQ1-algorithm using learning-rate initial value $\alpha(0) = 0.2$, which is decreased linearly during the learning process. The total number of iterations is 100 times the number of codebook vectors.

2.3 Experiments with concatenation and averaging

Our preliminary experiments consisted of finding the effect of pure averaging or pure concatenation of successive short-time feature vectors into one context vector. Next we tested combined concatenation and averaging of successive short-time feature vectors into one context vector. Since it is obvious that farther from phoneme centers speech signal contains more irrelevant information to classification, it seems reasonable to use wider averages outside the centers. Table 1 lists some of the best experiments. The details can be found in [9]. Context vector forms in the table are as follows. I is a single short-time feature vector, [] means average of two, [.] average of three, [..] average of four, and so on. For example, [..][] means an average of four feature vectors concatenated with an average of two. Compared to a single short-time cepstral vector, the best context data representation increases phoneme classification accuracy as much as from 87.0% to 99.0%. The representation of the last row was chosen for further experiments.

representation of context	dim.	cb:500 correct%	cb:1000 correct%	cb:2000 correct%
I	20	87.0	86.9	86.6
[....]	20	93.0	93.4	93.1
IIIIII	120	95.2	96.0	96.0
[] [] []	60	95.5	96.0	95.9
[][][][][][][][][][]	220	95.7	97.9	98.8
[.][.][.][.][.][.][.][.]	140	95.9	97.9	98.8
[..][][][][..]	100	97.0	97.9	98.2
[..][..][][][][..][..]	140	97.1	98.5	99.0

Table 1: *Phoneme recognition results with pre-segmented locations.*

3 Application to HMM-based phonemic decoding

3.1 What information can be derived using LVQ?

By training with phoneme centers only we can gain additional information for segmentation, because the chosen context vector covers a relatively long period of time domain. We can thus assume that transitions, since they are represented as long context vectors, will appear substantially different from center parts of phonemes. This will in turn result in high quantization errors at phoneme borders, since the codebook vectors represent phoneme centers. Our solution is thus to slide the fixed time domain window, that covers a period of 220 milliseconds, (Fig. 1) over all of the speech signal and to compute phoneme classification by LVQ for its center position. This results in a so-called quasiphoneme sequence with a phoneme symbol and the corresponding quantization error every 10 ms. This error can then be used as an aid to segmentation.

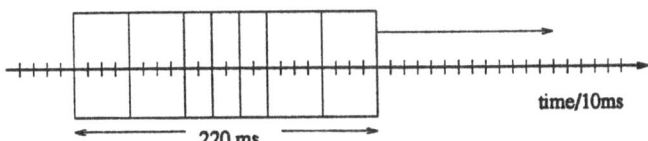

Figure 1: *The selected 140-dimensional context vector*

To obtain even more explicit information about phoneme locations, we have made experiments with training another codebook for a two-class problem: to discriminate phoneme centers from transitions between phonemes. For this codebook, the context vector form and the number of codevectors were kept the same as in the most successful phoneme recognition experiment, i.e., the context vector of Fig. 1, and 2000 codevectors. The codebook is trained using exactly the same procedure, but with examples from phoneme centers and phoneme borders. The best accuracy we obtained in this phoneme/border classification problem was 94.6%. This segmentation task really emphasizes the advantage of using context windows. Using only short-time cepstral vectors, the accuracy was as low as 75.4%, as expected.

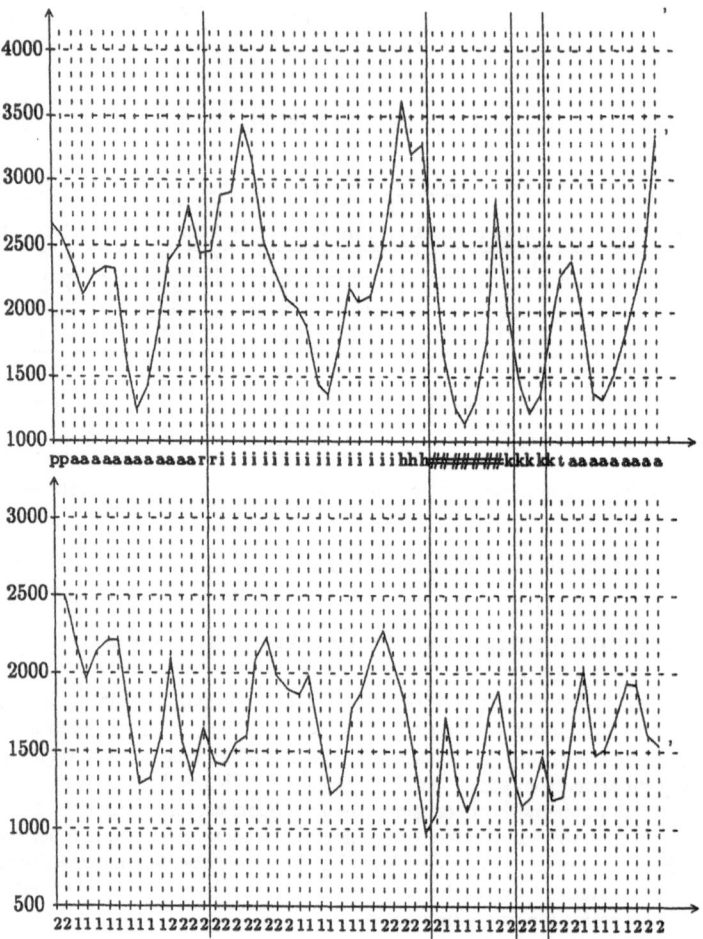

Figure 2: *Segmentation information derived by LVQ. See text.*

Figure 2 illustrates all this information available for segmentation for the Finnish word 'aika'. The top graph displays the recognized quasiphonemes together with their quantization errors. Manual segmentations (not used in the experiments) are shown as vertical lines (the phonemes /a/, /i/, /k/, /a/). Character '#' denotes silence. Respective segmentation classifications are shown in the lower graph. '1' and '2' stand for stationary regions and transitions, respectively. Note that the diphthong /ai/ is here considered as a concatenation of two vowels.

3.2 Using HMMs to exploit information produced by LVQ

In our approach, the above mentioned classification results are interpreted as output symbols of discrete observation HMMs. The details of the basic configuration can be found in [11]. Each phoneme is modeled as a separate context independent Markov source. A simple four-state left-to-right topology has been used. Output symbol distributions have been defined for state transitions within the phoneme models and they are trained using the standard maximum-likelihood approach [10].

The main differences between a standard discrete observation (vector quantization) HMM–approach and our approach here are the following: 1) LVQ-codebook is used to map contextual data into context independent form. Another interpretation is that LVQ is used to compute frame-synchronous phoneme *classification*, not to *represent* the speech signal with minimum distortion. 2) Codebook quantization errors are included in the HMMs as a separate input stream. 3) An additional codebook is used for segmentation as a separate input stream. We are thus using three separate parallel streams in the HMMs. For the quantization errors, both 8-level quantized values and continuous density probability distributions with large constant variance were tried.

Table 2 shows the final results of the experiments divided in four main groups: 1) Using traditional short-time cepstral vectors as the basis of short-time phoneme classification (STF). The codebook size was 500. 2) Using a large LVQ-codebook of 2000 codevectors to map high-dimensional context vectors into phoneme classes (FON). Quantization error of this codebook is modeled as a separate stream, either as a discrete (dQE) or a continuous density distribution (cQE). 3) In addition to the above, using a large segmentation LVQ-codebook also with high-dimensional context vectors to produce a third parallel stream of information that consists of only two possible symbols (SEG). 4) Combine the two-stream information from the two above mentioned large codebooks into a single stream in the following way (FON/SEG). For each phoneme, there will be two alternative symbols, a symbol for the center part, which is outputted whenever the segmentation codebook produces the center symbol, and a symbol for the transitional part, outputted whenever the segmentation codebook produces the transition symbol.

Method	correct%	errors%
STF	93.9	9.5
FON	97.1	4.6
FON, dQE	97.0	4.1
FON, cQE	97.6	3.8
FON, SEG	98.4	3.2
FON, SEG, dQE	98.1	3.2
FON, SEG, cQE	98.4	2.9
FON/SEG	98.5	3.0
FON/SEG, dQE	98.0	3.2
FON/SEG, cQE	98.5	2.8

Table 2: *Experimental results with different HMM-codebook variations. See text.*

4   Conclusions

In this research we have applied the LVQ methods, including the latest developments [4, 6, 5, 7], to the task of Finnish speaker-dependent speech recognition. Our aim was to produce phonemic transcriptions of spoken utterances. LVQ was used to map contextual acoustic information into context-independent phonemic form, further decoded by context-independent discrete observation HMMs. We experimented with various ways to form a high-dimensional context vector by both averaging and concatenating short-time feature vectors within a time domain window. This method of taking local phonemic context into account was shown to give excellent results when applied to classification of Finnish phonemes.

Compared to using short-time feature vectors, classification accuracies increased from 87 % up to 99 % with the best form of context vector. We can also observe that the sizes of

LVQ-codebooks must be increased in accordance with the dimensionality and the amount of context included in the vectors. It seems to be obvious that in this way we are able to capture some context-dependent phoneme variations in a large codebook automatically without the need to impose any further linguistic knowledge about phonemes in different contexts.

We also applied the LVQ methods with context vectors to phoneme border/center detection task and exploited this additional two-class time-series information successfully by using it as a separate HMM codebook, or as a combined codebook with the main LVQ-codebook.

The lowest error percentage that we obtained was approximately 3%, when the starting and ending points of words were supposed to be known. This figure shows a significant improvement when compared to results obtained in real-time operation, based on classification of short-time feature vectors [11]. Naturally, the computing load is now multiplied, calling for massively parallel hardware.

References

[1] Paul Bamberg and Laurence Gillick. Phoneme-in-context modeling for DRAGON's continuous speech recognizer. In *Proceedings of the DARPA Speech and Natural Language workshop*, pages 163–169, Hidden valley, Pennsylvania, USA, 1990.

[2] Stephen B. Davis and Paul Mermelstein. Comparison of parametric representations for monosyllabic word recognition in continuously spoken sentences. *IEEE Transactions on Acoustics, Speech, and Signal Processing*, 28(4):357–366, August 1980.

[3] V. N. Gupta, M. Lennig, and P. Mermelstein. Integration of acoustic information in a large vocabulary word recognizer. In *Proceedings of the IEEE International Conference on Acoustics, Speech and Signal Processing (ICASSP87)*, volume 2, pages 697–700, Dallas, Texas, April 1987.

[4] Teuvo Kohonen. *Self-Organization and Associative Memory*. Springer-Verlag, Berlin-Heidelberg-New York-Tokio, 3 edition, 1989.

[5] Teuvo Kohonen. Improved versions of learning vector quantization. In *Proceedings of the International Joint Conference on Neural Networks*, volume I, pages 545–550, San Diego, June 1990.

[6] Teuvo Kohonen. The self-organizing map. *Proceedings of the IEEE*, 78(9):1464–1480, 1990.

[7] Teuvo Kohonen, Jari Kangas, Jorma Laaksonen, and Kari Torkkola. LVQ_PAK: A program package for the correct application of Learning Vector Quantization algorithms. In *Proceedings of the International Joint Conference on Neural Networks*, volume I, pages 725–730, Baltimore, June 1992. IEEE.

[8] Kai-Fu Lee. Context-dependent phonetic hidden Markov models for speaker-independent continuous speech recognition. *IEEE Transactions on Acoustics, Speech, and Signal Processing*, 38(4):599–609, April 1990.

[9] Jyri Mäntysalo, Kari Torkkola, and Teuvo Kohonen. LVQ-based speech recognition with high-dimensional context vectors. In *Proceedings of the International Conference on Spoken Language Processing (ICSLP92)*, volume I, pages 539–542, Banff, Alberta, Canada, Oct. 12-16 1992.

[10] Lawrence R. Rabiner. A tutorial on hidden Markov models and selected applications in speech recognition. *Proceedings of the IEEE*, 77(2):257–286, 1989.

[11] Kari Torkkola, Jari Kangas, Pekka Utela, Sami Kaski, Mikko Kokkonen, Mikko Kurimo, and Teuvo Kohonen. Status report of the Finnish phonetic typewriter project. In *Proceedings of the International Conference on Artificial Neural Networks (ICANN-91)*, pages 771–776, Espoo, Finland, June 24-28 1991.

[12] S.J. Young. The general use of tying in phoneme-based HMM speech recognisers. In *Proceedings of the IEEE International Conference on Acoustics, Speech and Signal Processing (ICASSP92)*, volume I, pages 569–572, San Francisco, CA, USA, March 23-36 1992.

Natural language and speech recognition—poster contributions

An Analytically Transparent Network for Sequence Recognition

Marc F.J. Drossaers

Computer Science Dept., University of Twente, P.O Box 217, 7500 AE Enschede, The Netherlands.

abstract

In this article technical details of a neural network for sequence recognition are presented. The network is powerful enough to simulate finite-state acceptors, while its analysis is much simplified, compared to the standard Hopfield model. Also its processing speed is optimal, and the presence of mixture states is externally controllable. The network is robust under synaptic noise, it has error correcting properties and it can recover from gross input errors.

1 Introduction

An important aspect of natural language processing with neural networks is judging the legality of a sequence of input words against a grammar. This way syntactical constraints can be used in other linguistic analyses, like semantic analysis, and thus contribute to the machine-understanding of an utterance.

Hopfield model-like neural networks [6][2] have the nice property of analytical transparency: they can be analytically analyzed in great depth. However, models for sequence recognition[1] that were proposed by several authors in this area were computationally not very strong. The network proposed by Kleinfeld [7], for instance, recognizes much more sequences than it should, and a similar network proposed by Amit [1] recognizes only sequences of the same 'stimulus'.

On the other hand, proposals by e.g. Servan-Schreiber et al. [9] and Moisl [8] show that neural networks can be very well used for recognition tasks. However, they use networks that learn by a multi-layer back-propagation algorithm. Multi-layer back-propagation learning algorithms are analytically opaque in that there can be no convergence theorems for these algorithms. This fact implies in particular that there

---

[1]Sequence recognition is here understood to be equivalent to what string acceptance is in the context of formal languages.

can be no theory that maps each finite-state machine onto a neural network that is guaranteed to learn to simulate it, within an a priori specified error-rate, for networks that exploit that kind of learning algorithm.

In this article a neural network is presented that is both computationally powerful enough to simulate finite-state acceptors and analytically transparent. Linguistic aspects, among which a correctness result, were discussed in [5]. Here the results of an analysis of the dynamics of a slightly revised version of the network are presented. Some verification of the theoretically founded predictions was provided by computer simulation of which a sample is given also.

In the following, the network's definition, and an informal description of its operation is given in section 2. In section 3 an energy function, and the corresponding free-energy function are presented, after which the effect of synaptic noise on the size of the overlap parameter is discussed in section 4. Simulation results then follow in section 5.

2 The Network

The network consists of N formal neurons $S_i \in \{0,1\}$, $i = 1, \ldots, N$.

$$S_i(t) = \begin{cases} 1 & \text{if} \quad h_i(t) > U \\ 0 & \text{if} \quad h_i(t) < U, \end{cases}$$

where U is a threshold, and $h_i(t)$ is the total neuronal input at time t. Under noisy dynamics a neuron fires with probability:

$$P(S_i = 1) = \frac{1}{1 + \exp[-(h_i/T - U)]},$$

where T models synaptic noise. $P(S_i = 0) = 1 - P(S_i = 1)$. The threshold is noise free, only the input is subject to synaptic noise. The total neuronal input at time t consists of an external and a temporal contribution:

$$h_i(t) \equiv h_i^e(t) + h_i^t(t),$$

$$h_i^e(t) = \sum_{j=1}^{N} J_{ij}^e S_j^e(t),$$

$$h_i^t(t) = \sum_{j=1}^{N} J_{ij}^t \bar{S}_j(t),$$

$$\bar{S}_j(t) \equiv \frac{\tau - \theta}{\theta} \int_0^\infty S_j(t - t') w(t') dt',$$

$$w(t) = \begin{cases} \frac{1}{\tau - \theta} & \text{if } \theta < t < \tau \\ 0 & \text{otherwise} . \end{cases}$$

S_i' denotes a neuron in another but identical network. The time averaged prior neural activity $\bar{S}_j$ is determined as in [10], accept that here the averaging does not cover current activity: $\tau \equiv (2t_a + t_d + 3t_r)/N$ and $\theta \equiv (t_a + t_r)/N$, where t_a is the period the network is active, t_d is a delay period in between active periods, and t_r is the transition time for the network. $t_a \leq t_d$.

The synaptic coefficients $J_{i,j}^e$ and $J_{i,j}^t$ are obtained by an Adaline learning process, and satisfy:

$$\lambda^t \xi_i^\nu = \sum_{j=1}^{N} J_{ij}^t \xi_j^\mu, \quad \forall \langle \{\xi^\nu\}, \{\xi^\mu\} \rangle \text{ and} \quad (1)$$

$$\lambda^e \xi_i^\mu = \sum_{j=1}^{N} J_{ij}^e \zeta_j^\mu, \quad \forall \langle \{\xi^\mu\}, \{\zeta^\mu\} \rangle, \quad (2)$$

where $\langle , \rangle$ denotes an ordered pair, $\{\xi^\mu\} = (\xi_1^\mu, \ldots, \xi_N^\mu)^T$ and $\{\zeta^\mu\}$ is an activity pattern of the network that provides external input. $\{\lambda^t \xi^\nu\} = \{h^{t,\nu}\}$ is called a temporal image of prior activity, and $\{\lambda^e \xi^\mu\} = \{h^{e,\mu}\}$ is called an input image of external activity. λ^e, λ^t are the relative strengths of the neuronal interactions. In eqs. (1), (2) it is assumed that the temporal input patterns $\{\xi^\mu\}$ and the external input patterns $\{\zeta^\mu\}$ are linear independent, and that the learning algorithm terminated with zero error. The storage capacity of the network is given by $\alpha_c = 1$, see [4] for reflections on this point, so the pattern index is restricted to $1 \leq \mu \leq N$, $J_{ii}^x = 0$ for $x = e, t$. The components of the N-bit vectors to be stored are chosen from $\{0, 1\}$ with probability:

$$P(\xi_i^\mu) = a\delta(\xi_i^\mu - 1) + (1 - a)\delta(\xi_i^\mu),$$

see [3]. Here a is the bias of the activity patterns. It is related to the average number of active neurons as $\lim_{N \to \infty} \frac{1}{N} \sum_{i=1}^{N} \xi_i^\mu = a$. If $a \neq \frac{1}{2}$ the pattern is biased. Usually a is chosen small. The Adaline learning algorithm was preferred over the Kohonen pseudo-inverse storage prescription, because it requires only input patterns to be linear independent.

The threshold is either $\lambda^e + 1/2\lambda^t$ or $1/2\lambda^e$, depending on whether there is temporal input or not. A quantitative measure for the turning point will be given below. Usually $\lambda^e > \lambda^t$, see

[5]. In the analyses below it is assumed that $N \to \infty$, whereas p, the number of stored transitions, is kept fixed. In this network neurons are updated asynchronously.

The operation of the network can be described as follows. Suppose the neurons receive input over the temporal connections. The neurons that also receive external input become active. This lasts for at most a period θ, after which the external input is removed and the activity dies out within a period t_r. The network is then silent for a period t_d during which the previous temporal image decays, and the network builds up a new temporal image.

In a noise free network temporal or external input images cannot generate activity by themselves. Due to the minimal correlation between the activity patterns (small a), an erroneous external input image can not lead to significant amounts of activity in the network, given the presence of some temporal image. The simultaneous presence of a temporal image and an external input image of the same pattern is a necessary condition for network activity. External input thus provides a way to control the occurrence of mixture states.

3 Energy Functions

For the network a (free) energy function can be defined that describes its behavior during a period that covers either an active period or a delay period, with its preceding transition time.

3.1 The Energy Function

In the network a new equilibrium state is only dependent on input that is constant over the relevant time span, its energy is:

$$E = -\sum_{i=1}^{N} h_i S_i + U \sum_{i=1}^{N} S_i.$$

The change of the network's energy, ΔE, as a consequence of some neuron S_i changing its state is $(\Delta S_i = S_i(t + 1) - S_i(t))$:

$$\Delta E = -\Delta S_i (h_i - U),$$

so the energy function decreases with every change in the network's state during relaxation. To determine the value of an energy minimum in case of high threshold: assume that the network is in a state $\{S\} = \{\xi^1\}$. Then there are $h_i^e = \lambda^e \xi_i^1$ and $h_i^t = \lambda^t \xi_i^1$, where the implicit assumption $\{\bar{S}\} \in \{0, 1\}$ is sanctioned by the law of large numbers. So:

$$E = -\sum_{i=1}^{N} S_i((\lambda^e + \lambda^t)\xi_i^1 - U) = -\frac{1}{2}\lambda^t a N.$$

The fact that activity is not immediately fed back to the neurons minimizes the relaxation time: one update per neuron suffices.

3.2 The Free-Energy

The construction of a free-energy function is simplified by the fact that the network always immediately samples its states according to an equilibrium distribution. The energy can be written as:

$$E = -\sum_{\mu=1}^{p}\sum_{i=1}^{N} h_i^{\mu}\xi_i^{\mu}S_i + U\sum_{i=1}^{N} S_i,$$

where S_i was actually quadratic. The partition function Z then becomes:

$$Z = TR_S \exp\left[\sum_{i=1}^{N} S_i\left(\beta h_i \xi_i - U\right)\right], \qquad (3)$$

where $\beta = 1/T$, $h_i = \{h_i^{\mu}\}$, and $\xi_i = \{\xi_i^{\mu}\}$. The trace is of the form: $e^{cx} + e^{c(1-x)} = e^c + 1$ with $x \in \{0,1\}$. So:

$$Z = \exp\left[-\beta aN f(\beta, h)\right], \text{ where}$$

$$f(\beta, h) = -\left\langle\left\langle\frac{1}{a\beta}\ln\left(1 + \exp\left[\beta h\xi - U\right]\right)\right\rangle\right\rangle \quad (4)$$

is the free energy. The double angular brackets express the fact that the free energy is self-averaging.

The overlap $\langle m\rangle$ (the single angular brackets denote a thermal average) is:

$$\langle m^{\mu}\rangle = -\frac{\partial f(\beta, h)}{\partial h^{\mu}}$$

$$= \left\langle\left\langle\frac{1}{2a}\xi^{\mu}\left(1 + \tanh\left[\frac{1}{2}\left(\beta h\xi - U\right)\right]\right)\right\rangle\right\rangle.$$

From eqs. (3) and (4) it follows that we have at the saddle point:

$$\langle m^{\mu}\rangle = \frac{1}{aN}\sum_i \xi_i^{\mu}\langle S_i\rangle.$$

4 Size of the Overlap

In the noiseless case the overlap parameter $m^{\mu} \in [0,1]$ is $m^{\mu} = \frac{1}{aN}\sum_{i=1}^{N}\xi_i^{\mu}S_i$. The size of the overlap is independent of the number of patterns the network state has a nonzero overlap with. This fact is the motive behind the choice for $S_i \in \{0,1\}$.

If the network receives noisy input, take:

$$\frac{1}{aN}\sum_i \xi_i^{\mu}\langle S_i\rangle = \frac{1}{aN}\sum_i \xi_i^{\mu}S_i m = mm^{\mu} = \langle m^{\mu}\rangle,$$

where m is the size of a thermal correction. Assuming that $\overline{S}_j = \langle S_j\rangle \approx \xi_j^{\nu}$, and $\langle\{\xi^{\mu}\},\{\xi^{\nu}\}\rangle$ was stored in the temporal synapses, then:

$$\left\langle h_i^{t,\mu}\right\rangle = \sum_j^N J_{i,j}^t\langle S_j\rangle = \sum_j^N J_{i,j}^t S_j m = \lambda^t \xi_i^{\mu} m,$$

and similarly for external input images. Now assume that the external input consists of k input images, that the temporal input consists of l images, all corrected by m, and that the external input and the temporal input have n images in common. Then the vectors of neuronal inputs

can be written as (with the nonzero contributions first):

$$h_i^e = m(\underbrace{\lambda^e\xi_i^1,\ldots,\lambda^e\xi_i^k}_{k},\underbrace{0,\ldots,0}_{p-k})^T, \text{ and}$$

$$h_i^t = m(\underbrace{\lambda^t\xi_i^{k+1},\ldots,\lambda^t\xi_i^{k+1+l}}_{l},\underbrace{0,\ldots,0}_{p-l})^T.$$

The size of the overlap is then:

$$m_{k+l} = \frac{1}{2a(k+l)}\times$$

$$\left\langle\left\langle z_{k+l}\left(1 + \tanh\left[\frac{1}{2}\left(\beta m\left(\lambda^e z_k + \lambda^t z_l\right) - U\right)\right]\right)\right\rangle\right\rangle,$$

where $z_k^i = \sum_{\mu=1}^{k}\xi_i^{\mu}$, $z_l^i = \sum_{\nu=k+1}^{k+l}\xi_i^{\nu}$. Next the average overlap is split into three parts, a part concerning the patterns that occur twice in the input, m_n, a part for the rest of the external input, m_{k-n}, and a part for the rest of the temporal input, m_{l-n}. Contributions to the average exist only if $\xi_i^{\mu} = 1$. This happens, on average, e.g. in the first case, anN times. So we obtain:

$$m_n =$$
$$\frac{1}{2}\left(1 + \tanh\left[\frac{1}{2}\left(\lambda^e\left(\beta m - 1\right) + \lambda^t\left(\beta m - \frac{1}{2}\right)\right)\right]\right)$$

$$m_{k-n} =$$
$$\frac{1}{2}\left(1 + \tanh\left[\frac{1}{2}\left(\lambda^e\left(\beta m - 1\right) - \frac{1}{2}\lambda^t\right)\right]\right)$$

$$m_{l-n} =$$
$$\frac{1}{2}\left(1 + \tanh\left[\frac{1}{2}\left(-\lambda^e + \lambda^t\left(\beta m - \frac{1}{2}\right)\right)\right]\right).$$

Under noisy dynamics and above a critical noise level T_c subsequent overlaps form a monotonic decreasing sequence due to the recursive relation in the network between those subsequent overlaps. For increasing values of λ^e, λ^t the overlap m reduces to insignificant values ($m \leq 0.5$, see below) increasingly faster, and for higher values of T_c. Computation of these critical noise levels is complicated by the form of the expression for the neuronal firing probability. However, for $\lambda^e, \lambda^t \to \infty$ the critical noise levels T_c (absolute upper limits) for the various overlap parameters can be determined as follows. Consider the expression for m_n. If we let $\lambda^e, \lambda^t \to \infty$ either $m_n = 1$ or $m_n = 0$, depending on the sign of the argument of the tanh. If m_n is computed iteratively starting with $m = 1$, and $T \to \infty$ starting at a small value, the critical noise level is reached when:

$$2\beta m - 1\frac{1}{2} = 0 \Leftrightarrow \beta = \frac{3}{4} \Rightarrow T_c = 1\frac{1}{3}.$$

The other two cases are equal and yield:

$$\beta m - 1\frac{1}{2} = 0 \Leftrightarrow \beta = 1\frac{1}{2} \Rightarrow T_c = \frac{2}{3}.$$

Computations showed that for large λ^e, λ^t above T_c $m = 0$ within two iterations. In case of a pattern that only one source of input has an overlap with, this means that the pattern has lost its causal power the next active period of the network. So if $\frac{2}{3} < T < 1\frac{1}{3}$ the noise works beneficial on the operation of the network and network activity depends only on the cooperation of external and temporal input.

The threshold level is:

$$U(t_a^i) = \begin{cases} \frac{1}{2}\lambda^e & \text{if} \quad \forall\mu : m^\mu(t_a^{i-1}) < 0.5 \\ \lambda^e + \frac{1}{2}\lambda^t & \text{if} \quad \exists\mu : m^\mu(t_a^{i-1}) \geq 0.5, \end{cases}$$

where t_a^i is the i-th active period in a sequence recognition run. The two-valued threshold allows the network to recover from serious input errors. If external input does not generate activity with sufficient overlap with known patterns, no temporal image is generated, and the threshold is lowered. The network can then accept any input, which restarts the recognition run. Assuming that all active neurons contribute equally to a temporal image, the network can correct up to $\frac{1}{2}aN$ missing neurons in a pattern in a temporal image, assuming that the external input image is perfect.

5 Simulation

An instantiation of the presented network was simulated for recogntion of the very simple sequence abcd. In the network $N = 400$ and $a = 0.1$. The graphs below show the size of the overlap parameter for subsequent inputs at $T = 0.4$ and $T = 0.8$ respectively. In both simulation runs the input was abcd. The program

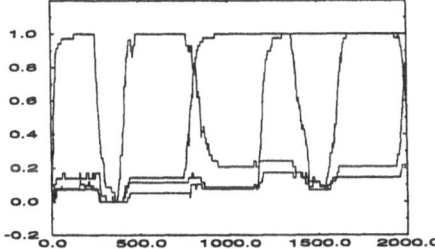

Figure 1: *The network with $T = 0.4$, $\lambda^e = \lambda^t = 10$. The noise is too low to eliminate the causal consequences of input images that occur only once in the total input.*

simulates noisy input from a completely stable external network. This means that the highest critical noise level changes to $T_c = 2$ in the case $\lambda^e, \lambda^t \to \infty$.

6 Conclusions

An analysis of a computationally powerful neural network for sequence recognition was presented. Analyzing this network is much simpler than analyzing a standard Hopfield model. The computation of critical noise levels is however complicated by the particular form of the

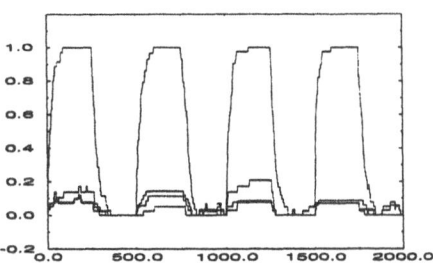

Figure 2: *The network with $T = 0.8$. $\lambda^e = \lambda^t = 10$.*

expression for the neuronal firing probabilities. Here the theory needs further development. The network is fast, and the presence of mixture states is externally controllable. The network is robust under synaptic noise, can cope with degraded input, and recovers from gross input errors. In the near future the network is to be extended with a self-organizing learning algorithm. Neural networks that recognize sequences without having learned to do so in an unsupervised fashion, do not add anything to existing conventional methods.

References

[1] D.J. Amit. *Proc. Natl. Acad. Sci. USA,* 85:2141–2145, 1988.

[2] D.J. Amit, H. Gutfreund, and H. Sompolinsky. *Physical Review A,* 32(2):1007–1018, 1985.

[3] J. Buhmann, R. Divko, and K. Schulten. *Physical Review A,* 39(5):2689–2692, 1989.

[4] E. Domany, J.L. van Hemmen, and K.Schulten. *Models of Neural Networks.* Springer, Berlin, 1991.

[5] M.F.J. Drossaers. In *Proceedings of COLING '92,* pages 113–119, 1992.

[6] J.J. Hopfield. *Proc. Natl. Acad. Sci. USA,* 79:2554–2558, 1982.

[7] D. Kleinfeld. *Proc. Natl. Acad. Sci. USA,* 83:9469–9473, 1986.

[8] H. Moisl. *Connection Science,* 4(2):67–91, 1992.

[9] D. Servan-Schreiber, A. Cleeremans, and J.L. McClelland. In D.S. Touretzky, editor, *Advances in Neural Information Processing systems I,* Los Altos (Cal.), 1989. Morgan Kaufmann.

[10] H. Sompolinsky and I. Kanter. *Physical Review Letters,* 57(22):2861–2864, 1986.

CONCEPTUAL CLUSTERING USING A CONNECTIONIST APPROACH

Adélaïde Stévenin, Patrick Gallinari

LAFORIA IBP- UA CNRS1095
Tour 46-00, Boite 169
Université Pierre et Marie Curie
4 Place Jussieu - 75252 Paris cedex 05 - France
stevenin@laforia.ibp.fr, gallinari@laforia.ibp.fr

Abstract

We describe a connectionist system for natural language processing in database query applications. It is a production system for conceptual decoding of task specific information and concept prediction. It has been designed for extracting semantic knowledge from text input. The system has been validated on an Air Travel Information System decoding task. It offers good performances with only a small number of parameters. Predicted concepts may be used as an intermediate step in a speech understanding system.

1. Introduction

Connectionist techniques have aroused widespread interest as a tool for low-level tasks such as image or speech processing. They have, however, been less successful for symbolic and high-level processing. Current work in the domain takes its inspiration from symbolic AI, reproducing simple functions of expert systems [1] or building simple hybrid numeric-symbolic systems. Important progress has been made recently and new research directions have emerged. Most current systems, however, still face severe limitations.

Our model does not mimic symbolic systems, but tackles the problem of numeric-symbolic processing via a pure Neural Network (NN) approach. In developing this system, we have targeted a class of applications and shown the possibility to extract symbolic information by training NN systems on low level data.

This approach may be used for a broad class of problems. It has been used here for Natural Language (NL) processing in the limited domain of airline reservation with text input. Our system associates concept to words and produces an intermediate conceptual representation of the sentence. Similar systems may be built for other database retrieval applications.

Real-life applications raise many problems in NL processing. Grammars must have very high complexities. Grammatically incorrect sentences or speech recognizer errors are difficult to handle. NNs directly extract information from raw data through automatic training. They may offer advantages over pure linguistic approaches since NNs are trained on real situations. Further, they do not rely on grammars, they are less subject to the grammatically incorrect sentences problem and no expertise is required. They easily take into account additional knowledge sources or contextual information. NNs implement a unified formalism for different levels in language processing, allowing for an easy integration of the corresponding processing steps.

The paper is organized as follows : in 2 we describe the task and discuss current methods, in 3 we present our approach and its evaluation on a text database and in 4 conclude on further research.

2 . The decoding task

Our airline reservation task uses a domain-specific sub-language. Its difficulty lies in dealing with spontaneous speech and with a great number of potential users. The task corresponds to a part of the standard evaluation domain ATIS (Air Travel Information

System) chosen by DARPA for spoken language systems [2]. It deals with general problems such as generic database query and interactive problem solving.

Our system extracts semantic knowledge from text sentences producing concept classes associated to the input words. It decomposes text sentences from the domain task into conceptual constituents, like those shown in Table 1.

QUERY	reserver	**ARRIVAL**	*aero2	**DEP-TIME**	enpartancede
	svp	**DEP-DATE**	le		*aero1
ID-PASS	pour		*day		a
	*first-name		*nday		*h
	*name		*month		h
DEPARTURE	*aero1				

*Table 1: conceptual segmentation of the pre-processed sentence "réservez svp pour Sylvie Fujol un Paris Toulouse le mardi dix-huit janvier en partance de roissy a seize heures". symbol * has been used for task variables.*

The conceptual approach presents the following advantages :
- many ambiguities can be avoided since some interpretations do not make sense for the task (each parameter can only be filled-in once)
- syntactic issues which do not affect the task can be ignored.

We make the same hypothesis as [3] : "The meaning of a sentence can be expressed by a sequence of basic meaning units and there is a sequential correspondence between each of these units and a sub-sequence of the words pronounced".

We have conceived our system both for the association of concepts to words and for prediction of concepts. Predictions can be used to reduce the effective search space during acoustic decoding. Concept prediction is much easier than word prediction because in a limited domain task, the number of concepts is often finite even though the number of instances may be very large.

2.1. Current approaches

We quote below two approaches representative of important research directions in this area.

[3] proposes a speech understanding system based on statistical representation of task specific semantic knowledge which uses Hidden Markov Models. This system presents many similarities with ours. Both rely on statistical feature extraction and allow an easy integration of successive processing levels for speech understanding. Our approach, however, takes the context into account more easily, and performs accurate prediction with a much smaller number of parameters. It takes advantage of the NN capacity to predict which has already been used for speech recognition [4].

The European SUNDIAL ESPRIT project also aims at task oriented dialogues. The system developed in [5] is more linguistically than statistically oriented. The dialogue level provides contextual information to the linguistic component in the form of predictions. This approach requires more linguistic expertise than ours.

2.2. System architecture

For conceptual decoding and prediction, a system must both be able to take word and concept contexts into account, and to deal with two major problems: the simultaneous use of information from different knowledge levels and time dependent data.

• Merging knowledge levels. At the first level of analysis, linear sequences of words exhibit structures that show how words relate to each other. At the second level, subsequences of the first structure are mapped to objects in the task domain. These two structural levels are linked, hence the advantage of a connectionist structure that integrates both. Our system will resist grammatically incorrect sentences and uses concept structure. It will however

402

be able to use word context to reduce the arborescent structures of analysis. Processing these levels simultaneously allows it to use all the available knowledge.
• Time problem. Basic NN models deal poorly with temporal AI tasks like natural language parsing. Networks with recurrent connections attempt to remedy this situation. Training algorithms are, however, not efficient enough for real world problems. Partially recurrent networks are good compromises: they can be trained efficiently to recognize and sometimes reproduce sequences.

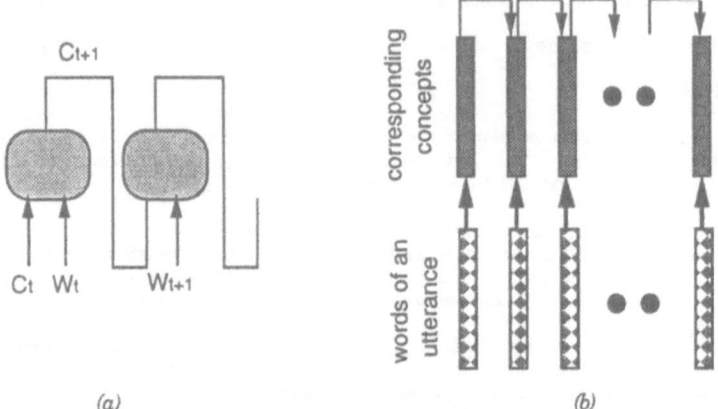

(a)   (b)

figure 1: (a) The structure of the recurrent network, for prediction concept C(t+1) is computed from C(t) and word w(t) . (b) The Network unfolded in time grows until the end of of the utterance.

After testing several modular and recurrent architectures [6,7] for solving respectively the knowledge integration and the time problems, we have designed a system which uses past output (concepts) as well as current input (words) to compute the current output (fig.1). Recurrent connections in this system implement a linear recursive filter upon the quasi-linear output of the feedforward network. In standard filters, a sample of the output signal $g(k)$ at time k is given in terms of past as well as future input $f(t)$. To include the recursive property, a weighted sum of the past outputs can be added. The general formula (1) with $f(k)$ an input signal at time k, and $g(k)$ the corresponding output is :

$$g_m = \sum_{n=-N}^{N} b_n f_{m-n} - \sum_{n=1}^{N} a_n g_{m-n} \quad (1)$$

The recurrency here does not appreciably complicate the training. We used an adapted version of back-propagation.

3 . Evaluation

Our system has been tested on a conceptual segmentation and prediction task. The data comes from a text re-transcription of the french SUNDIAL database [3]. The final corpus consisted of some 300 transcribed dialogues, involving utterances from 26 users, and 16 concepts for the representation of the task-knowledge. We decided to consider only the utterance that corresponds to the system open question "Formulate your demand" which corresponds to about 200 utterances. Any task parameter can interfere here, and the question does not determine the structure of the answer.
To preprocess the sentences we have adapted to our problem classical choices in the ATIS domain [3]. It mostly consists of grouping some words (verbs are taken as infinitives), associating others (articles and the following word), converting compound phrases into hyphenated compound expressions (en date du-> endatedu). Also, to limit vocabulary

during testing, we grouped under labeled variables the parameters whose values are subject to unpredictable changes (the passenger identity is not in a dictionary).
Two thirds of the database was used for learning after manual segmentation, and one third for testing.

k-first candidates	Association		Prediction	
	correct %	C. I. 95%	correct %	C.I. 95%
1	88.24%	[86.99,89.38]	84.20%	[82.76,85.55]
2	94.15%	[93.22,94.96]	93.26%	[92.24,94.15]
3	96.22%	[95.45,96.87]	95.42%	[94.55,96.15]
4	98.00%	[97.42,98.46]	96.86%	[96.12,97.46]
5	98.57%	[98.06,98.95]	97.80%	[97.17,98.30]

Table 2 : Results of cross validation for association and prediction. Different performances have been computed by considering the answer correct if among the k highest output for k = 1..5.

Results (table 2) show very good performances for both tasks. This shows that our connectionist system can extract symbolic information from low level data without using domain expertise. The architecture reflects the imbrication of the words-context and the concept-context structure and uses both type of information through recurrent connections. To demonstrate the need for this, we have tested the system without recurrent connections and have obtained significantly lower results.

4 . Discussion

This paper has described the use of predictive NNs whose recurrent part has a linear recursive filter structure, for a conceptual clustering task. It validates a new approach to concept extraction which may be an alternative or a complement to more traditional developments.
Further research should include integration of a NN speech recognizer [4] and the use of the conceptual information to enhance speech recognition.

Acknowledgments : we would like to thank Cap Gemini Innovation for allowing us to use the text version of the SUNDIAL french database for test purposes and S.V.B. Aiyer for helpful comments on the manuscript.

5 . References

1   Towell G.G. , Shavlik J.W., Noordewier M.O.(1990)-*"Refinement of Approximate Domain Theories by Knowledge-based Neural Networks"* Proc. eight nat. conf. on A.I. 861-866, Boston.
2   Price P.J.(1990)-*"Evaluation of Spoken Language Systems : The ATIS domain"* Proc. 3rd DARPA Workshop on Speech and Natural Language.
3   Pieraccini R., Tzoukermann E., Gorelov Z., Gauvain J.L, Levin E., Lee C.H, Wilpon J.G(92)-*"A Speech Understanding System Based on Statistical Representation of Semantics"*, ICASSP 92, San Francisco CA.
4   Mellouk A., Gallinari P.(93)-*"A discriminative neural prediction system for speech recognition"*, to appear in ICASSP 93.
5   Charpentier F., Andry F., Choukri K., Gavignet F., Magadur J-Y.(1992)-*"A Speech Understanding and Dialogue System for Flight Reservation in French"*, ICSLP, Alberta.
6   Jordan M.(1986)- *"Serial Order : a parallel distributed processing approach"*, ICS report 8604, UCSD.
7   Elman J.(1988)- *"Finding Structure in Time"*, UCSD report.

An Extended Kohonen Feature Map for Sentence Recognition

Aarnoud Hoekstra and Marc F.J. Drossaers
University of Twente, Dept. of Computer Science,
P.O. Box 217, 7500 AE Enschede, The Netherlands.

abstract

In this paper an extended Kohonen feature map is proposed that is able to store sequences of input patterns. The extended map is used in a natural language application: the recognition of sentences produced by a regular grammar. This network is able to learn to simulate a finite-state machine for the grammar, given examples of legal sentences from the grammar.

1 Introduction

In natural language processing (NLP) the notion of sentence recognition is important, it is a judgement on the grammaticality of a sentence. A sentence is recognized if it takes a recognizer from a starting state to an accepting state. The Kohonen map has already been used in NLP applications such as speech recognition [4, 5, 8] and categorization of words and their semantics [7]. The maps used for speech recognition applications were capable of handling sequences (utterances of words), but consisted of a Kohonen map on top of a memory. These networks cannot handle long-distance dependencies, syntactical relations between distant sequence elements are ignored. This paper proposes an integrated network: a Kohonen map that is extended in such a way that it can handle sequences and long-distance dependencies. The model was tested with an NLP task, consisting of storing and recognizing sentences generated by a regular grammar.

The body of this paper starts with the introduction of the extended Kohonen feature map (EKFM), followed by a description of the parameter tuning process for the map. The ensuing section gives a viewpoint on the recognition of sentences generated by natural language grammars, followed by a section on the mapping of a deterministic finite-state automaton (DFA) onto the EKFM. The paper ends with an example and some conclusions.

2 The Extended Kohonen Feature Map

A Kohonen feature map [6] consists of N neurons arranged as a k-dimensional lattice. Here it is assumed that $k = 2$, i.e. the neurons are arranged in a 2-dimensional grid, where the indices of neurons are denoted by 2-D vectors. Each neuron receives its external input via n-input lines. The output of a neuron s is determined by calculating a similarity measure between the weights of that neuron and the external input. Usually the Euclidean distance is chosen as similarity measure:

$$A_s^{ex} = \|\mathbf{w}_s^{ex} - \mathbf{v}\| = \sqrt{\sum_{i=1}^n (w_{s_i}^{ex} - v_i)^2}, \quad (1)$$

where $\mathbf{w}_s$ is the weight vector of neuron s and $\mathbf{v}$ is the input pattern. A winner is chosen according to:

$$\mathbf{r} = \arg\min \|\mathbf{w}_s^{ex} - \mathbf{v}\|,$$

where $\mathbf{r}$ is the index of the winning neuron. The weights of the neurons are updated in the direction of the presented input pattern. A weight update takes the relative position of a neuron s into account with respect to the winning neuron $\mathbf{r}$. This is done by the neighbourhood function $h(\mathbf{r}, \mathbf{s}, t)$:

$$\Delta \mathbf{w}_s^{ex}(t) = \varepsilon(t) h(\mathbf{r}, \mathbf{s}, t)(\mathbf{v}(t) - \mathbf{w}_s^{ex}(t)),$$
$$\mathbf{w}_s^{ex}(t+1) = \mathbf{w}_s^{ex}(t) + \Delta \mathbf{w}_s^{ex}(t),$$

where $\varepsilon(t)$ is the learning rate, a value that linearly decays in time [3]. The Kohonen map is unable to store sequences of input patterns, therefore the map is extended with additional connections. These connections are referred to as internal connections, they connect each neuron with all other neurons. With each internal connection an internal weight w_{ij}^{in} is associated, where i and j are indices of neurons. A neuron i now has two associated activations: A_i^{ex} for the external activation (see equation (1)) and A_i^{in} given by:

$$A_i^{in} = \underline{1}([\sum_{j=1}^N w_{ij}^{in} \cdot s_j] - \theta),$$

with $\underline{1}(x)$ the Heaviside function, w_{ij}^{in} the internal weight from neuron i to neuron j and θ a threshold (0.5). The state of a neuron may either be active ($s_j = 1$) or inactive ($s_j = 0$). The network state S is the vector of states of the individual neurons. Internal weights and external weights are used to invoke transitions from one active neuron to the next active neuron.

Learning sequences of input patterns is done by computing A_j^{in} and A_j^{ex}, thus having computed the possible successors of the prior active neuron (those that have $A_j^{in} = 1$) and their distance to the input pattern. The internal weights are updated as follows:

$$w_{ij}^{in}(t+1) = f(w_{ij}^{in}(t) + \Delta w_{ij}^{in}(t)),$$

where $\Delta w_{ij}^{in}(t)$ is the weight update, $f(x)$ a function that keeps the weights between 0 and 1 given by:

$$f(x) = \underline{1}(x)x + \underline{1}(x-1)(1-x),$$

and $w_{ij}^{in}(0)$ some random value, $0 \leq w_{ij}^{in}(0) \leq 1$. The weight update algorithm uses *associative reward-penalty* [3] and the learning rule is given by:

$$\Delta w_{ij}^{in}(t) = \\ s_i(t-1)s_j(t)\alpha + s_i(t-1)(1-s_j(t))A_j^{in}(t)\gamma + \delta,$$

where α is the reward, γ the penalty with $|\gamma| < \alpha$ and δ a constant that is used to weaken the connections that are hardly ever used. The value of $s_j(t)$ is determined by the current mode of the network which is either *learning* or *recognition*. In the learning mode $s_j(t)$ is given by:

$$s_j(t) = \begin{cases} 1 & if \ A_j^{ex} \leq A_k^{ex} \ \ \forall k \neq j \\ 0 & otherwise. \end{cases} \quad (2)$$

This means that a neuron j becomes active if it has the least distance to the input pattern (a possible successor). During recognition $s_j(t)$ is determined as follows:

$$s_j(t) = \begin{cases} A_j^{in} & if \ A_j^{ex} \leq A_k^{ex} \ \ \forall k \neq j \\ 0 & otherwise, \end{cases} \quad (3)$$

Neuron j becomes active if it has a link to the prior active neuron $w_{ij}^{in} \geq \theta$ and it has the least distance to the input pattern. The state evolving after applying equation (3) may be correct or incorrect. A correct state S contains a single active neuron, the network could 'choose' a successor from the set of possible successors. Otherwise the network ends up in an incorrect state that was caused by either misclassification or trying to recognize a non-existing sequence.

3 Tuning the Network Parameters

The network model uses several parameters. First there are the 'standard' Kohonen parameters such as:

$$\begin{aligned} h(\mathbf{r}, \mathbf{s}, t) &= \exp(-t/m_t * d(\mathbf{r}, \mathbf{s})/m_d * \delta_1), \\ \varepsilon(t) &= \alpha_0(1 - t/m_t), \\ \sigma(t) &= m_d \exp(-t^2/(\delta_2 * m_t^2)), \\ d(\mathbf{r}, \mathbf{s}) &= \max(|r_1 - s_1|, |r_2 - s_2|), \end{aligned}$$

where m_t is the maximum number of steps the network runs, m_d the maximum distance in the map, $d(\mathbf{r}, \mathbf{s})$ is the distance between neuron $\mathbf{s}$ and neuron $\mathbf{r}$. The function $\sigma(t)$ is used to determine the maximum distance at a certain time t. If the distance between two neurons is less than $\sigma(t)$ then the neighbourhood is calculated, otherwise $h(\mathbf{r}, \mathbf{s}, t) = 0$. The learning rate $\varepsilon(t)$ linearly decreases in time from a starting point $\alpha_0 = 1$. Usually the network runs for 10,000 steps, $m_t = 10,000$. The constant δ_2 can be calculated by: $\delta_2 = (m_t^2 * \log(1/m_d))/(m_t - 0.1m_t)^2$ it is used to ensure that at time $t = 0, \sigma(t) = m_d$ and at $t = m_t, \sigma(t) = 0$. Constant δ_1 is used for similar reasons, it ensures that at $t = 0, h(\mathbf{r}, \mathbf{s}, t) = 1$ and at $t = m_t, h(\mathbf{r}, \mathbf{s}, t)$ is small, $\delta_1 = -1/2 * m_d * \log(0.01)$ when $m_t = 10,000$. The extra parameters used by the EKFM are harder to estimate, possible values are:

$$\begin{aligned} \alpha &= 0.5, \\ \gamma &= -0.1, \\ \delta &= -0.001. \end{aligned} \quad (4)$$

If the reward α has a large value the network will update the internal weights for some successor strongly, but if that neuron never wins again and receives no penalty it may be the case that the strong update will not be forgotten. On the other hand if α is too small connections of successors may never receive enough reward to keep the weight $\geq \theta$. The penalty γ is closely related to α, if $|\gamma| = \alpha$ and the connections receive equally often a reward as a penalty, the net update will be zero. However if $|\gamma|$ is too large compared to α a connection may end up in receiving only penalties. Experiments show that the above values (4) are satisfactory. The decay factor δ is preferably equal to $1/m_t$ because then all weights that are never updated or hardly ever updated will go to 0 even if they are 1 initially, the value 0.001 was chosen because during experiments this was the smallest computational unit.

Initially the internal weights fluctuate since the external weights are not yet organized. As the external weights stabilize the internal weights also become more stable because patterns of sequences are more often classified on the same neurons.

4 Natural Language Recognition

It is here assumed that a natural language grammar has a regular backbone, and that therefore a large part of a natural language can be recognized using a push-down automaton (PDA). The human performance in natural language use is characterized by a very limited degree of center-embedding, therefore a PDA that has a bound on the number of items on the stack can be used to recognize a natural language. Consequently a non-deterministic finite-state acceptor (NFA) M' can be defined that has the PDA's set of stack configurations as its set of states [2], from which a deterministic finite-state acceptor (DFA) can be derived. A restriction that the grammars must fulfill is that they are in 2-standard form with a minimal number of quadratic productions: productions of the form $A \rightarrow bCD$ where b is a terminal and C and D are variables. Such a grammar can be seen as a minimal extension of a right-linear grammar. Within such grammars quadratic productions provide for the center-embedding. Since such grammars have a minimal number of quadratic productions, acceptance by a PDA defined for such grammars requires minimal use of (stack) memory and thus generates a minimal number of stack states. To maintain this minimal use of memory a restriction to one-state PDAs that accept by empty stack is also required: when a PDA is mapped onto an NFA, the information concerning its states is lost, unless it was stored on the stack. For every such PDA an NFA (and consequently a DFA [1]) can be defined for which it can be proven [2] that it accepts the same language as the bounded PDA does.

5 Simulation of a DFA by the EKFM

A DFA M is defined as: $M = (Q, \Sigma, \delta, q_0, F)$ where Q is a finite set of states of M, Σ is the input alphabet, $\delta : Q \times \Sigma \rightarrow Q$ is a transition function that gives a new state given a prior state and an input symbol $x \in \Sigma$, q_0 the start state for M and $F \subseteq Q$ is the set of accepting states. M can be mapped onto an extended Kohonen feature map FM in the following way:

1. The input alphabet Σ is mapped onto the neurons of FM, i.e. clusters of neurons (with at least one neuron per cluster) represent an element of Σ: $x \in \Sigma$ is repre-

sented in the network by a normal distribution f_x, where f_x can be used to generate occurrences, v^x, of x. An occurrence of x is mapped onto neuron i if $||v^x - w_i^{ex}|| = \min ||v^x - w_k^{ex}||$ $\forall k \neq i$ where v^x is a 2-dimensional vector.

2. State q_0 is mapped onto neuron 1, a sequence in the network always starts at neuron 1.

3. F is mapped onto the set of neurons that are active when a legal sequence has ended. The sequence can be made cyclic by point extension, i.e. a sequence is then preceded and followed by a point (for an example see [2]).

4. The function δ is mapped onto the weights w^{in}. A weight $w_{ij}^{in} = 1$ if there exists a transition $\delta(q, x)$ with $q \in Q$ and $x \in \Sigma$, in terms of the network FM: $w_{ij}^{in} = 1$ if $s_i(t) = 1$ and $j = \arg\min ||v^x - w_k^{ex}||$

5. The states $q \in Q$ are mapped onto the different correct states $S(t)$ of the network FM.

A network FM that simulates a DFA can be constructed using the rules above, however the network has not learned to simulate the DFA by self-organization. Extracting a DFA from examples is done by generating legal sequences and offering them to the network. The network will extract from those sequences a DFA by self-organization. The external weights will cover the input space Σ by classifying the different $x \in \Sigma$ on different clusters of neurons, assuming that the representation of each x is drawn randomly from independent normal distributions. The internal weights will be modified in such a way that the sequences that were offered to the network will be memorized, and thus memorizing $\delta(q, x)$ for different q and x.

Two notoriously hard cases are a DFA that recognizes sequences of identical symbols of a specified length, and a DFA that accepts unbounded sequences. If there is a grammar G that generates sequences of the form a^n with $1 \leq n \leq 4$. The EKFM can recognize these sequences if for every a in the sequence another neuron is used. This can be achieved by defining a distribution f_a, that has a large variance. This way it can be guaranteed that at least 4 neurons will be used to cover f_a, assuming the network consists of enough neurons. Hence input symbols a that differ temporally will be classified on different neurons covering f_a. The network will then accept a sequence if it is of the form a^n with $1 \leq n \leq 4$. On the other hand the network must be able to learn to accept sequences of the form a^n with unbounded n. This causes no problems since the network will automatically do that.

The network can also learn to recognize sequences of the form $a^n b^m$ with for example $1 \leq n \leq 4$ and no restrictions on m. This can be done since a and b are drawn from different distributions say f_a and f_b and it can be demanded that a has at least 4 neurons covering f_a. The occurrences for b do not need to be different for all the bs in the sequence because the network automatically generalizes to b^m, the EKFM can recognize sequences of the form $a^n b^m$. From the forgoing it can be seen that the EKFM can handle long-distance dependencies since the network can handle sequences of any length.

6   An Example

A regular grammar was constructed that produces sequences of the form $.ac^n d.$ and $.abd.$ with no restrictions on n. The representations of the input symbols were drawn from 5 different distributions (see figure 2). The distributions represent, clockwise starting at the lower left corner, $., c, d, b$ and a. Figure 2 also displays the way the neurons are distributed over the different distributions. The network was trained with 2,000 sequences (19,546 input patterns) for 10,000 epochs. In figure 1 the transitions the network found between different clusters of neurons are displayed, where neuron 16 is the accepting neuron. The maximum score obtained in the correct recognition of the learn set was 97.15% and 97.283% for a test set of 6000 sequences, with in both cases $\theta = 0$ during recognition.

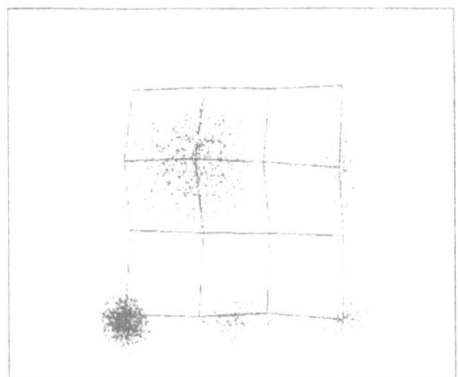

Figure 2: distributions and the external weights.

7   Conclusions

If a natural language grammar is looked at as regular, a DFA accepting such a grammar can be defined. The DFA can be learned by the extended Kohonen feature map in a self-organizing manner. This makes it possible to do syntactical sentence analysis using such a neural network. The EKFM is able to store sequences without loss of the long-distance dependencies in the sequence. Also the EKFM provides an integrated learning scheme, both the internal weights and external weights are trained simultaneously.

References

1. Aho, A.V., Sethi, R., Ullman, J.D. *Compilers Principles Techniques and Tools.* Addison-Wesley, Reading: Mass., 1986.

2. Drossaers, M.F.J. Memoranda Informatica 92-36, University of Twente.

3. Hertz, J.A., Palmer, R.G., Krogh, A.S. *Introduction to the theory of neural computation.* Addison-Wesley, Reading: Mass. 1991.

4. Kangas, J. In: *IJCNN, International Joint Conference on Neural Networks*, pp. 331-336. IEEE, 1990. vol 2.

5. Kangas, J. In T. Kohonen (ed.), *Proceedings of the 1991 International Conference ICANN-'91.* North-Holland: Amsterdam. 1991.

6. Kohonen, T. *Self-Organization and Associative Memory.* Springer-Verlag: Heidelberg. 3rd edition, 1989.

7. Miikkulainen, R. PhD thesis, University of California, Los Angeles. 1990.

8. Zandhuis, J.A. Technical report MPI-NL-TG-4/92, Max Planck Institute for Psycholinguistics, Nijmegen. 1992.

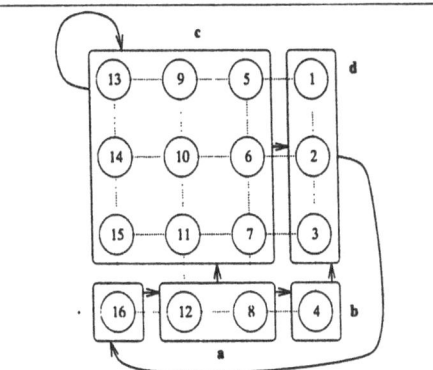

Figure 1: the transitions the network found.

Neural Nets that Discuss:

A General Model of Communication Based on Self-Organizing Maps

Timo HONKELA

Technical Research Centre of Finland

Laboratory for Information Processing

Lehtisaarentie 2 A, FIN-00340 Helsinki, Finland

E-mail: tho@tik.vtt.fi

There have been attempts to develop systems in which a collection of artificial agents co-operate communicating with each other to solve problems in a particular domain. In designing such systems one has to decide how autonomous the agents are, what is the level of communication between them, do all the agents have a common model of the environment, and what is the nature of that model. In this paper, the requirements for the most advanced levels of autonomy and communication are studied. The advantages of unsupervised learning models, particularly Kohonen's self-organizing maps, as a basis for intelligent communication between multiple agents, are presented.

1. COMMUNICATION BETWEEN AGENTS

A central issue in distributed artificial intelligence is the study of multi-agent systems. A collection of entities, called agents, interact by co-operation, by coexistence, or by competition. The use of multiple agents may be motivated by the need to treat distributed knowledge or activity. The distribution may be caused e.g. by geographical reasons, or by natural functional distribution of the problem. The most primitive communication between agents is restricted to some finite set of fixed signals with fixed interpretations [1]. In the primitive case, all the agents have a common model of their environment. Thus, there are no signal interpretation problems. An advanced level of multi-agent co-operation makes high-level communication necessary between the agents. Each agent has its own model or view of the environment, too. The differences in the models motivate development of learning abilities of the agents. Each agent could learn to interpret the messages from other agents. There are at least two basic approaches to machine learning: symbolic inductive inference and learning in artificial neural networks.

Brazdil and Muggleton [2] show how to use symbolic inductive inference as a means to learn to relate terms in a multiple agent communication. They have shown how to overcome language differences between agents automatically in a situation where the agents do not have the same predicate vocabulary. The system consists of a number of separate agents that can communicate. Each agent has certain perceptive, communicative and reasoning capabilities, being able to (1) perceive a portion of the given, possibly simulated world, (2) accept facts and rules from another agent, (3) formulate queries and supply them to another agent, (4) respond to queries formulated by another agent, (5) interpret answers provided by another agent, (6) induce rules on the basis of facts, and (7) integrate knowledge. These capabilities are limited, though. The use of a symbolic model of the environment is closely related to the model-theoretic approaches in defining the semantics of formal languages. The problem lies in the fact that the meaning of an expression (queries, responses) in a natural domain is fuzzy and changing, biased by the particular context.

2. UNSUPERVISED LEARNING AS A BASIS FOR AUTONOMY OF AGENTS

A number of models exists where a neural network is taught to analyse natural language expressions (an overview is presented e.g. in [3]). Most of the models concentrate on syntactical analysis. Many experiments are based on the assumption that a symbolic syntactic theory is suitable as a background theory for the learning. Therefore a network is taught giving a suitable amount of ordered pairs. In this kind of supervised learning algorithms, a correct output (description of the syntactical structure) is needed for the example inputs. The requirement that the correct output would be available is clearly unrealistic especially when semantics is concerned (see e.g. [4]). Unsupervised learning algorithms are capable of identifying salient information in the input vectors and finding similarities among the inputs so that they are taxonomized into potentially useful groupings. It is possible to use Kohonen's self-organizing map (SOM) algorithm to create a semantic map giving the system a set of words in their sentential context [5]. Extensions to the algorithm exist which make it possible to process input sentences with varying length [6]. Self-organizing feature maps have also been used to model the interpretation of imprecise expressions [7].

3. A MODEL OF COMMUNICATION BETWEEN AUTONOMOUS AGENTS

In multi-agent communication, agents pass expressions to each other. The topic of the communication may be perceived by both of the agents (Figure 1b), or only by the one passing the expression (c). It is also possible that the agent receives input concerning the phenomenon without any descriptions from other agents (a).

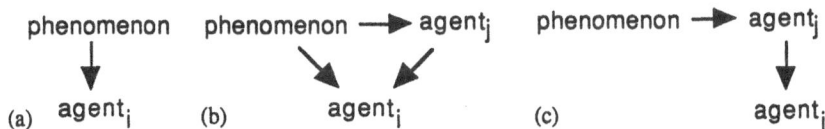

Figure 1. Basic sources of information for agents.

A semantic map is formed by presenting symbols in context during the learning process [8]. Similarity between the items of the expressions is then reflected through the similarity of contexts. The input vector for an agent maximally consists of (1) the perceived phenomenon, (2) information of the utterer, and (3) the actual expression (Figure 2).

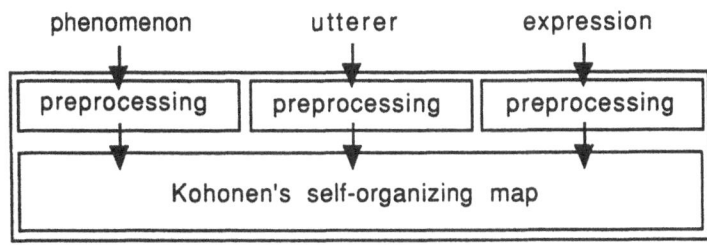

Figure 2. A model of an agent relating expressions and phenomena taking into account the utterer. The input data vector is a concatenation of the components.

The level of naturalness of the model may vary. In one extreme, an agent has straightforward access to some visual images and acoustic signals.

The unsupervised learning algorithm makes it possible for the agents to be autonomous: each of them has a model of its own for its environment, and is able to relate the expressions to the phenomena they refer to. It is also possible that an agent receives any combination of the components of the input vector. Receiving input concerning only the phenomenon (e.g. pictorial images), an agent forms internal representation of these "experiences" without any reference to symbolic descriptions. This kind of internal model may even be a basis for thinking without language.

If an agent receives input which contains both the image and the expression describing it, the self-organizing map associates these components of the input appropriately. In [5] the norm of the context part of the input vector (phenomenon component) predominated over the symbol part (expression component). If both components are handled equally, also the descriptions influence on the topology of the map. This may be considered to be a process in which the agent's view of the external world is influenced by the categorisations it is given by the other agents. The influence is strong especially in the early stages of the self-organization.

In a negotiation, the mutual understanding is based on the intersubjective agreement on the meanings of the expressions. If all three components — phenomenon, expression describing it, and the utterer of the expression — are included during the learning process, an agent forms a model which takes into account the personal differences in the interpretation of the expressions. An example of the results of such process is shown in Figure 4.

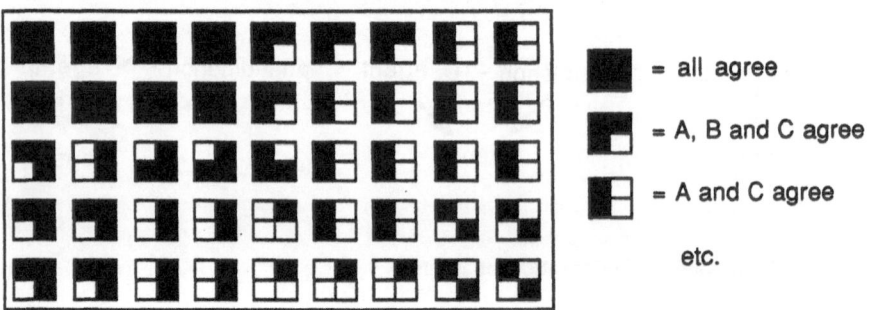

Figure 4. Topological map obtained with a learning set which contained five features and a response by four agents concerning a single expression (10000 presentations). For simplicity, only binary truth values were used. As an example, the upper left corner corresponds to the cases in which all the four agents agreed on the status of the expression.

The same map contains information on the topology of the features of the external reality and the mapping between the expressions and the reality. A semantic map for a set of expressions is shown e.g. in [5], [6] and [7]. The phenomenon component of the input vector is 5-dimensional and therefore only the average of each reference vector is visualized in Figure 5. Some simulations were made using even 900-dimensional input vectors showing same kind of properties in the self-organization.

Figure 5. The average of each 5-dimensional phenomenon component of the reference vectors.

4. CONCLUSION

It has been shown how autonomous agents can learn to communicate taking into account the individual differences in the interpretation of the expressions. The interpretation is based on the results of an unsupervised learning process devised by Kohonen's self-organizing map. The answer to a pair of important questions — how symbols receive their meanings and how communication is possible in spite of the inherently subjective nature of understanding — seems to be somewhat closer.

REFERENCES

[1] Werner, E.: Cooperating Agents: A Unified Theory of Communication and Social Structure. Distributed Artificial Intelligence, Volume II, Gasser, L. and Huhns, M.N. (eds.), Pitman, London, 1989, pp. 3-36.

[2] Brazdil, P. and Muggleton, S.: Learning to Relate Terms in a Multiple Agent Environment. Proceedings of Machine Learning - EWSL-91, Springer, Berlin, 1991, pp. 424-439.

[3] Honkela, T.: Connectionism and Linguistics. The 1992 Yearbook of the Linguistic Society of Finland, M.Vilkuna (ed.), Linguistic Society of Finland, Helsinki, 1992, pp. 9-33.

[4] Churchland, P.M.: A Neurocomputational Perspective: The Nature of Mind and the Structure of Science. MIT Press, Cambridge, Massachusetts, 1989.

[5] Ritter, H. and Kohonen, T. : Self-Organizing Semantic Maps. Biological Cybernetics, 61, 1989, pp. 241-254.

[6] Scholtes, J.C.: Kohonen Feature Maps in Natural Language Processing. University of Amsterdam, Department of Computational Linguistics, Amsterdam, 1991.

[7] Honkela, T. and Vepsäläinen, A.M.: Interpreting Imprecise Expressions: Experiments with Kohonen's Self-Organizing Maps and Associative Memory. Proceedings of the International Conference on Artificial Neural Networks (ICANN-91), Elsevier Science Publishers, 1991, pp. 897-902.

[8] Kohonen, T.: The Self-Organizing Map. Proceedings of IEEE, vol. 78, no. 9, 1990, pp. 1464-1480.

Neural Network and Nearest Neighbor Comparison of Speaker Normalization Methods for Vowel Recognition

Gail A. Carpenter * and Krishna K. Govindarajan †

Center for Adaptive Systems and Department of Cognitive and Neural Systems
Boston University, 111 Cummington Street, Boston, MA, 02215-2411, USA

Abstract

Fuzzy ARTMAP and K-Nearest Neighbor (K-NN) categorizers were used to evaluate intrinsic and extrinsic speaker normalization methods by training and testing on disjoint sets of speakers of the Peterson-Barney database. Intrinsic methods included one nonscaled, four psychophysical scales (bark, bark with end-correction, mel, ERB), and three log scales, each tested on four different combinations of the frequencies F_0, F_1, F_2, F_3. Four extrinsic schemes were tested in conjunction with the intrinsic methods: centroid subtraction across all frequencies (CS), centroid subtraction for each frequency (CSi), linear scale (LS), and linear transformation (LT). Categorizers showed similar trends, with K-NN performing better but requiring more storage. The optimal intrinsic method was bark scale, or bark with end-correction, using differences between all frequencies (BDA). The order of performance for extrinsic methods was LT, CSi, LS, and CS, with ARTMAP performing best using BDA; and K-NN choosing psychophysical measures for all except CSi.

Speaker Normalization

Human listeners are able to identify as a single phoneme a wide variety of speech signals produced by different speakers in different contexts. For example, the vowel /æ/ is recognized despite the fact that the average F_1 formant frequency is 660 Hz for males and 1010 Hz for children [10]. *Speaker normalization* is a general term used to describe the process whereby a listener compensates for individual characteristics of a speech signal in order to extract invariant features needed to identify the sound.

This paper describes a procedure that can be used to make systematic comparisons of the many speaker normalization schemes that have been proposed in recent decades. To evaluate a given normalization method, the 1520 vowel token vectors of the Peterson and Barney (1952) database, each consisting of the fundamental frequency (F_0) and the first three formants (F_1, F_2, F_3), are preprocessed using that method. Normalized inputs from about 30% of the speakers (10 males, 9 females, and 5 children), corresponding to 480 vectors, are used to train three different classifiers, a neural network (fuzzy ARTMAP [3]) and two K-nearest neighbor systems[4]. The remaining test data set is then presented to each classifier, which tries to identify each as one of ten vowel sounds. The normalization scheme in question is evaluated in terms of the number of correct test set identifications made by each of the classifiers. Speaker independence is required since the test set inputs and the training set inputs are generated by disjoint sets of speakers. Comparative evaluations of 160 different normalization schemes were carried out using this method.

The two main classes of normalization methods are *intrinsic* and *extrinsic* [9]. Intrinsic normalization uses only the information present in each vowel token. Extrinsic normalization uses information from several vowel tokens of a given speaker. Intrinsic normalization methods include psychophysical measures, such as bark differences [11], logarithm measures [1, 7], and logarithms of formant ratios [7]. Extrinsic normalization methods include centroid subtraction across all frequencies (CS) [1, 9], centroid subtraction for each frequency (CSi) [1, 9], linear scale (LS) [6], and linear transformation (LT) [13].

We would like to thank R. Watrous for providing the Peterson-Barney database and for useful discussions on the linear transformation extrinsic method; and Ah-Hwee Tan for help with the fuzzy ARTMAP code.

*Supported in part by British Petroleum (BP-89-A-1204), DARPA (AFOSR-90-0083 and ONR-N00014-92-J-4015), the National Science Foundation (NSF- IRI-90-00530), and the Office of Naval Research (ONR-N00014-91-J-4100).

†Supported in part by the Air Force Office of Scientific Research (AFOSR-F49620-92-J-0225), DARPA (AFOSR-90-0083), and the National Science Foundation (NSF- IRI-90-00530).

Fuzzy ARTMAP and K-Nearest Neighbor Algorithms

Fuzzy ARTMAP [3] is a supervised neural network algorithm that learns to map (transformed) frequency vectors to vowel categories. ARTMAP clusters frequency vectors on-line in one module (ART_a) and vowel categories in a second module (ART_b). An intervening map field (F^{ab}) adaptively associates frequency categories to vowel categories. Performance was compared with that of K-nearest neighbor (K-NN) algorithms [4], using both city block (L_1) and Euclidean (L_2) metrics. The K-NN algorithm chooses a vowel category based on the K training points that lie nearest to a test point. Preliminary simulations on different normalization methods were used to choose parameters for the two different recognition methods. Fuzzy ARTMAP parameters for all the simulations were: $\bar{\rho}_a = 0.0$, $\alpha = 0.1$, and $\beta = 1.0$. For the K-NN systems, the number of neighbors (K) was fixed at 10 throughout.

Intrinsic Normalization Methods

For the intrinsic normalization schemes, eight normalization scales were compared: one nontransformed (N) scale; four psychophysical scales: bark scale (B) [15], bark scale with end-correction (Be) [12], mel scale (Mel) [5], and equivalent rectangular bandwidth scale (ERB) [8]; and three log measures: a semitone scale ($log_{1.06}$), natural log scale (log_e), and log base 10 scale (log_{10}).

The bark scale (B) transforms $F_0 \ldots F_3$ to $F_0' \ldots F_3'$ according to the equation: $F_i' = 13.0 * \arctan(0.76 * F_i/1000) + 3.5 * \arctan(F_i/7500)^2$, where F_i is the ith frequency, in Hz. Bark scale with end-correction (Be) adjusts the low frequencies before converting to them to bark scale: frequencies below 150 Hz are increased to 150 Hz; frequencies between 150 and 200 Hz are reduced to $0.8F_i + 30$; and frequencies between 200 and 250 Hz are increased to $1.2F_i - 50$. The mel scale (Mel) corresponds to the transformation: $F_i' = 2595 \log_{10}(1 + F_i/700)$. Finally, the equivalent rectangular bandwidth (ERB) scale is calculated by: $F_i' = 11.17 * \log_e((F_i + 312)/(F_i + 14675)) + 43$. The three logarithmic measures consist of the semitone scale, $F_i' = \log_{1.06}(F_i)$, the natural logarithm scale, $F_i' = \log_e(F_i)$, and the log base 10 scale, $F_i' = \log_{10}(F_i)$.

Each of the eight normalization scales was tested with four different combinations: only the first two formants $[F_1', F_2']$; the fundamental and all three formants $[F_0', F_1', F_2', F_3']$; the three differences $F_1' - F_0', F_2' - F_1', F_3' - F_2'$ (Diff Subset); and all six difference combinations $F_1' - F_0', F_2' - F_0', F_3' - F_0', F_2' - F_1', F_3' - F_1', F_3' - F_2'$ (Diff All). Syrdal and Gopal (1986) proposed the Diff Subset method with bark scale with end-correction. Nearey and colleagues [1, 9] and Miller and colleagues [7] proposed using log scaled frequency ratios, which correspond to the differences between the log coverted frequencies. Combining the 8 vowel space scales and the 4 frequency combinations, 32 intrinsic methods were tested.

Extrinsic Normalization Methods

For the extrinsic methods, adaptation to a speaker was superimposed on each of the 32 intrinsic normalization methods. Four types of extrinsic normalization were tested: centroid subtraction across frequencies (CS), centroid subtraction for each frequency (CSi), linear scale (LS), and linear transformation (LT). The CS method finds the mean frequency value ($\bar{F}$) across all transformed frequencies of all the vowels of a given speaker and subtracts this value from F_i' [1, 7, 9]. The CSi method extends the CS method by computing the centroid ($\bar{F_i}$) for each transformed frequency and subtracting this value from F_i'. The CLIH2 method [9], and CLIH3 method [1] are functionally equivalent to the CSi method in a log vowel space. The linear scale (LS) approach [6] finds the minimum and maximum frequency values for each F_i' across all vowels of a given speaker, then rescales each frequency to the range [0,999]. In the LT method [13], a linear transformation matrix $\mathcal{A}$ is obtained which transforms each speaker's frequencies into some prototypical frequency values. New frequencies are linear combinations of the original transformed frequencies: $F_i'' = \sum_{k=0}^{3} \alpha_{ik} F_k' + \beta_i$. The matrix $\mathcal{A}$ is derived using the LMS algorithm [14] to minimize the mean squared error between a given speaker's fundamental and formant frequencies and the mean fundamental and formant frequencies across all speakers for each vowel. In all, 128 extrinsic normalization schemes were tested: 4 speaker adaptations x 4 frequency combinations x 8 scales.

414

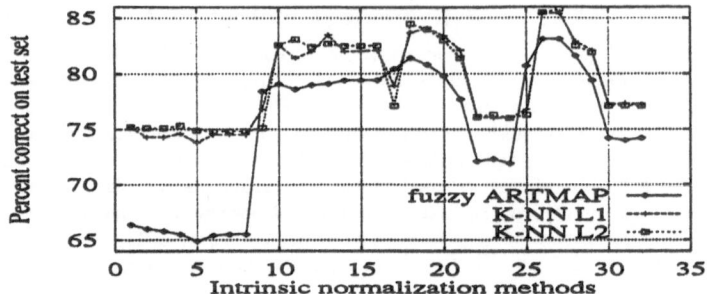

Figure 1: Test set performance of fuzzy ARTMAP and K-NN for intrinsic normalization methods # 1-32, which are identified in Table 1.

Comparative Evaluation of Normalization Methods

The three pattern recognition systems (fuzzy ARTMAP, L_1 K-NN, and L_2 K-NN) generally agreed on which normalization methods gave better predictive performance on test set data. K-NN tended to outperform fuzzy ARTMAP by a few percent (Figure 1). However, improved performance achieved by K-NN comes at a cost of storing all 480 training vectors. Fuzzy ARTMAP coded between 22 and 135 F_2^a nodes, providing a compression of 3.5 to 21.8 compared to the storage requirements of K-NN. Table 1 and Figure 1 show fuzzy ARTMAP and K-NN performance on the 32 intrinsic normalization methods. Similar analysis of the four extrinsic schemes has also been carried out [2].

Vowel	$[F_1, F_2]$			$[F_0, F_1, F_2, F_3]$			Diff Subset			Diff All		
Space	\multicolumn{12}{c}{Fuzzy ARTMAP}
	Id	%	F_2^a	Id	%	F_2^a	Id	%	F_2^a	Id	%	F_2^a
N	1	66.4	123.1	9	78.4	63.5	17	80.4	55.8	25	80.7	57.5
B	2	66.0	123.7	10	79.1	61.6	18	81.4	56.3	26	83.1	43.9
Be	3	65.8	123.1	11	78.6	63.9	19	80.8	54.8	27	83.1	43.4
Mel	4	65.5	124.3	12	79.0	62.2	20	79.8	57.1	28	81.6	46.3
ERB	5	64.9	124.8	13	79.1	62.3	21	77.7	66.1	29	79.4	49.4
$log_{1.06}$	6	65.4	122.0	14	79.4	60.7	22	72.1	73.2	30	74.2	58.9
log_e	7	65.5	121.9	15	79.4	60.6	23	72.3	72.5	31	74.0	58.8
log_{10}	8	65.5	122.1	16	79.4	60.8	24	71.9	73.9	32	74.2	58.9
Vowel	\multicolumn{12}{c}{K-NN}
Space	Id	L_1 %	L_2 %	Id	L_1 %	L_2 %	Id	L_1 %	L_2 %	Id	L_1 %	L_2 %
N	1	75.2	75.2	9	76.8	75.1	17	78.9	77.1	25	76.8	76.3
B	2	74.3	75.1	10	82.6	82.6	18	83.7	84.5	26	85.5	85.5
Be	3	74.3	75.1	11	81.4	83.1	19	84.1	84.0	27	85.4	85.8
Mel	4	74.6	75.3	12	82.0	82.4	20	83.4	83.0	28	82.9	82.5
ERB	5	73.8	74.9	13	83.5	82.7	21	82.1	81.4	29	82.1	81.9
$log_{1.06}$	6	74.5	74.8	14	82.0	82.5	22	76.1	76.1	30	77.2	77.1
log_e	7	74.5	74.8	15	82.0	82.5	23	76.0	76.3	31	77.3	77.1
log_{10}	8	74.5	74.8	16	82.1	82.5	24	76.0	76.0	32	77.2	77.1

Table 1: Fuzzy ARTMAP and K-NN test set performance with intrinsic normalization.

The psychophysical measures (B, Be, Mel, ERB) outperformed the log measures in most cases. For all the intrinsic and extrinsic methods, fuzzy ARTMAP performed best using bark, or bark with end correction, Diff All (Table 1). Although K-NN optimal performance varied more, these classifiers also chose the psychophysical measures in all cases except for the extrinsic scheme CSi. For the intrinsic and LS extrinsic method, K-NN chose the bark Diff All method. For the CS extrinsic method, K-NN chose ERB $[F_0, F_1, F_2, F_3]$. For the LT method, L_1/L_2 K-NN performed best with Mel/ERB $[F_0, F_1, F_2, F_3]$. Finally, for the CSi method, K-NN chose the log scales $[F_0, F_1, F_2, F_3]$.

While the LT method has the best performance, it requires vowels that are labeled *a priori* to obtain the transformation matrix $\mathcal{A}$. Thus, for speaker-independent machine vowel recognition, LT requires the user to say an initial specified utterance containing the requisite vowels. The other three extrinsic methods do not require these vowels to be labeled. Thus, the second best method (CSi) may be the best candidate for prototype human and machine perpeption systems, since CSi does not require as much *a priori* knowledge as LT, its computational demands are less, and its performance is almost as good.

References

[1] Assmann, P. F., Nearey, T. M., and Hogan, J. T. (1982). "Vowel identification: Orthographic, perceptual and acoustic aspects," J. Acoust. Soc. Am. **71**, 975–989.

[2] Carpenter, G. A. and Govindarajan, K. K. (1993), "Speaker normalization methods for vowel recognition: Comparitive analysis using neural network and nearest neighbor classifiers," CAS/CNS Techical Report, Boston University, Boston, MA.

[3] Carpenter, G. A., Grossberg, S., Markuzon, N., Reynolds, J. R., and Rosen, D. B. (1992). "Fuzzy ARTMAP: A neural network architecture for incremental supervised learning of analog multidimensional maps," IEEE Trans. on Neural Networks **3**, 698–713.

[4] Dasarathy, B. V. (1991). *Nearest Neighbor (NN) Norms: NN Pattern Classification Techniques* (IEEE Computer Society Press, Los Alamitos, CA).

[5] Fant, G. (1973). *Speech Sounds and Features* (MIT Press, Cambridge, MA).

[6] Gerstman, L. J. (1968). "Classification of self-normalized vowels," IEEE Trans. on Audio and Electroacoustics **AU-16**, 78–80.

[7] Miller, J. D. (1989). "Auditory-perceptual interpretation of the vowel," J. Acoust. Soc. Am. **85**, 2114–2134.

[8] Moore, B. C. J. and Glasberg, B. R. (1983). "Suggested formulae for calculating auditory-filter bandwidths and excitation patterns," J. Acoust. Soc. Am. **74**, 750–753.

[9] Nearey, T. M. (1989). "Static, dynamic, and relational properties in vowel perception," J. Acoust. Soc. Am. **85**, 2088–2113.

[10] Peterson, G. and Barney, H. L. (1952). "Control methods used in a study of the vowels," J. Acoust. Soc. Am. **24**, 175–184.

[11] Syrdal, A. K. and Gopal, H. S. (1986). "A perceptual model of vowel recognition based on the auditory representation of American English vowels," J. Acoust. Soc. Am. **79**, 1086–1100.

[12] Traunmüller, H. (1981). "Perceptual dimension of openness in vowels," J. Acoust. Soc. Am. **69**, 1465–1475.

[13] Watrous, R. L. (1993). "Speaker normalization and adaptation using second-order connectionist networks," IEEE Trans. on Neural Networks, 4, pp. 21–30.

[14] Widrow, B. and Stearns, S. D. (1985). *Adaptive Signal Processing* (Prentice Hall, Englewood Cliffs, NJ).

[15] Zwicker, E. and Terhardt, E. (1980). "Analytical expression for critical-band rate and critical bandwidth as a function of frequency," J. Acoust. Soc. Am. **68**, 1523–1525.

Speech Recognition by Hierarchical Segment Classification

Holger Behme, Wolf Dieter Brandt, Hans Werner Strube
Drittes Physikalisches Institut der Universität Göttingen
Bürgerstr. 42–44, D-37073 Göttingen, Germany
Phone: (+49) 551 397731, FAX: (+49) 551 397720
email: behme@up3spr2.gwdg.de or brandt@up3spr2.gwdg.de

Abstract

A neural network for speech processing is presented. Its complex architecture, incorporating self-organizing feature maps [1], allows the construction of a hierarchy of layers, where each layer operates on a larger time scale and deals with higher units of speech, like phonemes, syllables, word parts and so on. Tasks the network has to deal with include representation of speech, segmentation of the speech signal and classifying segments.

1. Introduction

Speech consists of hierarchic levels of objects: phonemes, syllables, words, phrases and so on. In the field of computer speech recognition, the basic level is the time frame, determined by the sampling rate and the preprocessing. We present an artificial neural network constructed to do speech processing using the hierarchy mentioned above. Each "layer" of the network (in a far more complex sense than in perceptrons) performs the integrative step from one hierarchy level to the one above it. The basic structure of all layers is the same, just the meaning of the input and the time scale are different. One important constituent of such a layer is a self-organizing feature map, which is connected to a segmentation unit and to segment classifiers. The network performs a kind of "high level preprocessing", not the recognition itself.

2. Outline of the Network

As mentioned above, each layer of the network has the same architecture. It consists of a self-organizing feature map, wherein the input vectors are encoded, a segmentation unit, which cuts the incoming sequence of (encoded) input vectors into segments, and several segment classifiers, which react to segments of the input sequence and produce one output vector for each segment. These output vectors form the input for the next level, which thus operates on a larger time scale. Starting point of this hierarchy is a sequence of speech spectra using the Bark (mel) scale. As all levels of the network are essentially identical, the first layer may serve as an example (cf. Fig. 1).

The Bark spectra are fed into a two-dimensional self-organizing feature map. This kind of "phonotopic map" has been used already for several kinds of speech processing [1,5,6,7]. Each Bark spectrum has a representing neuron within the map. The sequence of spectra is transformed into a path through the map. This path is cut into segments by the segmentation unit, as described below, and each segment is then presented to the segment classifiers. Each classifier has been trained to recognize a special class of segments and reacts to a given segment with an activity representing the similarity between the class centroid and the given segment. These activities of all the classifiers are fed as a vector into the next layer of the network, being the input for the next feature map.

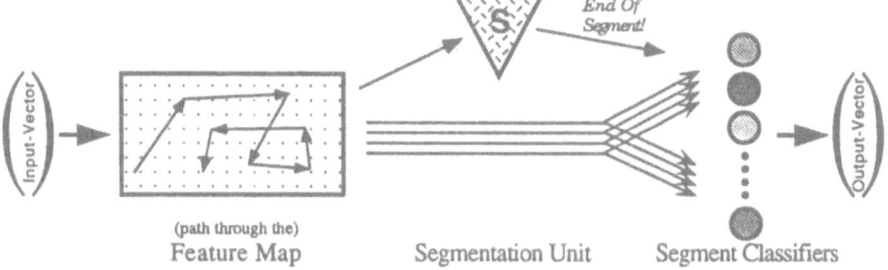

End Of
Segment!

Input-Vector

Output-Vector

(path through the)
Feature Map   Segmentation Unit   Segment Classifiers

Figure 1: First level of the network. Bark spectra are transformed into a sequence of representing neurons within the phonotopic map (path through the map). The Segmentation Unit detects the segment boundaries and sends a signal to the Segment Classifiers which then produce an output vector. The output is fed into the map of the next level. This is the basic structure of all network levels.

In this way the described hierarchy of levels is constructed. The first level detects units of roughly the length of phonemes (but not necessarily real phonemes), the next one syllables, and so on. At the top of the hierarchy, there has to be a recognizer of a different kind with some "knowledge" about *what* has to be recognized. For simple single-word-recognition, a usual linear perceptron with 1-of-N coding on top of two or even one network level has proven to be sufficient. For the processing of continuous speech, some kind of expert system (or "expert network") with knowledge about syntax and semantics on top of more levels will be more suitable.

If such a high-level "intelligent" recognizer is part of the system, feedback information in the form of expectations for coming segments can be incorporated into the network in an easy way.

3. Segmentation

The task is: split a given path through the map into segments in a consistent manner. That is, the paths for different versions of the same word should be split up "in the same way". For example, the word 'segment' should always be split up like 'seg-me-nt', and not sometimes like 'se-gm-ent' or something else.

The basic idea can best be visualized for the first layer, the phonotopic map. Paths within the phonotopic map usually consist of stationary parts representing vowels, during which the path does not leave the neighbourhood of a few cells, and transition paths, when the path jumps to different parts of the map. Segmentation is done by distinguishing 'short jumps' from 'long jumps' and taking the latter as segment boundaries.

With $S = (s_1, s_2, \ldots)$ being the sequence of winner cells in the path through the map, we have the sequence of "jump distances" $J = (j_{12}, j_{23}, \ldots)$, where $j_{t,t+1}$ is the distance (within the map) between neuron s_t and neuron s_{t+1}. We define a *critical jump distance* j_{crit} separating short and long jumps, and a *minimum segment length* l_{min}. The basic segmentation algorithm then is: "if $j_{t,t+1} \geq j_{crit}$ and segment length $l(t) \geq l_{min}$, then the end of the segment is reached." For a more sophisticated approach, "higher order jumps" can be taken into the calculation, e.g. "if $j_{t,t+n} \geq j_{crit}$ and segment length $l(t) \geq l_{min}$, then the end of the segment is reached" would be a segmentation rule of *order n*. In this case, the segment border is placed at an appropriate position between t and $t + n$.

For higher levels, a more general approach is needed. From a more universal point of view, it is not 'short' and 'long', but 'occurring frequently' and 'occurring infrequently' or 'probable' and 'improbable', which gives the segmentation cues.

In a purely neural picture, there would be N^2 additional units, reacting selectively to the N^2 possible jumps. The output of these units is weighted with their "improbability" (e.g. high for long jumps, low for short jumps in the first layer) and added up in a segmentation unit, which gives the "end-of-segment" signal when a critical activity is reached.

It is important to remember that the segments do not have to have a meaning such as half-syllables or syllables, but just have to be consistent.

4. Segment Classification

At each level, there exists a fixed number of segment classifiers. Each segment classifier has synaptic connections to and receives input from all feature map units of that layer. (Additionally, there may be some feedback connections from the level above.) Each incoming segment produces an activity pattern in the map, as shall be explained below. When the segmentation unit gives the "End of Segment!"-signal, each segment classifier produces an output activity according to the input pattern in the map and its synaptic connections. Then, the activities of all classifiers are fed up as the input to the next layer.

Let c_i denote the activity of map neuron i. Then the activity a_m of classifier m is computed as $a_m = \sum_i c_i w_{im}, i = 1 \ldots N, m = 1 \ldots M$, where w_{im} is the synaptic strength from map neuron i to classifier m. These connections may be inhibitory ($w_{im} < 0$), but always fulfill $\sum_i w_{im}^2 = 1$.

To form the input pattern $\{c_i\}$, in each time frame not only the best matching feature map cell but the four best matching cells (denoted by $i_1 \ldots i_4$) are taken into account (see also [4]). Their activity is increased inversely proportional to their match: if $\bar{x}$ denotes the input vector (spectrum) and $\bar{y}_{i_n}$ the weight vector of cell i_n, its activity c_{i_n} is computed as

$$c_{i_n} = \frac{\|\bar{x} - \bar{y}_{i_1}\|}{\|\bar{x} - \bar{y}_{i_n}\|} \text{ for } n = 1 \ldots 4 \quad , \quad c_{i_n} = 0 \text{ else.}$$

These activities are integrated over the segment, and at the end of the segment, the $\{c_i\}$ are scaled to fulfill $\sum c_i = 1$.

This procedure has several advantages: by "widening" the "narrow" path formed by the best matching cells only, we take advantage of the topology preserving property of the feature map. Neighbouring neurons (i.e. "second winners") represent similar features. Training and test patterns may differ just by the order of first and second (, third,...) winner, which is only a small difference in the map but would make a big difference if only the winner neuron would be used. The scaling of the $\{c_i\}$ at the end on the segment has the effect of time normalization.

We tested different algorithms for training the w_{im} in an unsupervised way. The most promising method seems to be a mixture of a Hebb/Anti-Hebb rule inspired by [3] and the 'Neural Gas' algorithm of [2]. For a given segment, the a_m are computed and ordered such that $a_1 \geq a_2 \geq \ldots \geq a_M$. Then the weights are updated according to

$$\Delta w_{im} = \alpha e^{-m/\lambda} c_i a_m, \qquad \alpha, \lambda \to 0 \text{ as } t \to \infty.$$

The optimum number of segment classifiers can be determined by means of an add-and-delete rule: segment classifiers which are never winners and always have about the same output activity are obviously superfluous. On the other hand, if for some segments no classifier shows high output activity, additional classifiers are needed.

5. Simulation Results

Although the network is designed to deal with word-spotting, our first test results were obtained from a single word recognition experiment. This method has the advantage of being comparable to classical recognition methods. We used a two-level network with either a simple two-layer perceptron or a DTW algorithm for recognition and a database with 40 different German words designed for robot control in 10 repetitions.

In the speaker-dependent case using the perceptron we were not able to reach rates better than 95%, but we can use these results to prove the functionality of our system in general. Using a (non-neural) DTW-algorithm instead of the perceptron produced recognition rates of up to 99%.

The results for speaker-independent recognition are presented in the table. We compared recognition of the Bark spectra using DTW with recognition using the output vectors of the first network layer. (There were not enough speech data for training a HMM recognizer.)

Although the results for the network are still somewhat below DTW, we see the potential of improving these rates further. Additionally, the network can be directly applied to fluent speech, as cannot be done with DTW. This is the field of our present work.

method	recognition rates
for single-speaker:	
network+perceptron	68.3%
network+DTW	96.0%
DTW	99.8%
for multiple-speaker:	
network+DTW	70.4% (51%–80%)
DTW	85.5%

We used analog differentiated speech signals, sampled at 10 kHz, Fourier transformed (Hamming window, length 51.2 ms, shift 20 ms) and mapped to the Bark scale (19 channels). As network parameters, we used a 20×30 feature map and 30 segment classifiers. The training parameters for 50 segment classifiers were $\alpha_{start} = 0.7$ and $\lambda_{start} = 30$; the segmentation unit parameters were: critical jump distance $j_{crit} = \sqrt{81}$, minimum segment length $l_{min} = 2$ and order of jumps $n = 5$.

6. Conclusion

We presented a complex, multi-level network for speech processing. It constructs a hierarchy of units and time scales similar to the hierarchy found in human speech. The number of hierarchy levels can be chosen according to the recognition task and the desired final unit size. The network solves the problems of data representation, segmentation and classification of the segments for further processing.

It should be noticed that the network is self-organizing and unsupervised to a high degree. As self organization is a main feature of natural neural systems, we think that this feature should be encorporated into such models as the one presented here as far as possible.

Acknowledgement: this work has been supported by the German Federal Ministry of Science and Technology (grant no. 01 IN 108 A/2).

References

[1] Teuvo Kohonen (1988): "Self-Organization and Associative Memory", 2nd ed., Springer Series in Information Sciences.

[2] Thomas Martinez, Klaus Schulten (1991): "A 'Neural Gas' Network Learns Topologies", in: T. Kohonen et al (Eds.) *Artificial Neural Networks — Proceedings of ICANN-91*, Elsevier, 397–400

[3] Jeanne Rubner, Klaus Schulten, Paul Tavan (1990): "A Self-Organizing Network for Complete Feature Extraction", in: Eckmiller, R., Hartmann, G. und Hauske, G. (Eds.), *Parallel Processing in Neural Systems and Computers*, Elsevier, 365–369

[4] P. Tavan, H. Grubmüller, H. Kühnel: "Self-organization of associative memory and pattern classification: recurrent signal processing on topological feature maps", *Biological Cybernetics* 64 1990, 95–105

[5] H. Behme, H. W. Strube (1990): "A neural net for recognition and storing of spoken words.", in: Eckmiller, R., Hartmann, G. und Hauske, G. (Eds.), *Parallel Processing in Neural Systems and Computers*, Elsevier, 379-382.

[6] W. D. Brandt, H. Behme, H. W. Strube (1993): "Segmentierung fließender Sprache durch topologieerhaltende Vektorquantisierer", to be published in *Fortschritte der Akustik – DAGA '93* (DPG-GmbH, Bad Honnef 1993)

[7] H. Behme, W. D. Brandt, H. W. Strube (1993): "Spracherkennung durch Segmenterkennung über Merkmalskarten", to be published in *Fortschritte der Akustik – DAGA '93* (DPG-GmbH, Bad Honnef 1993)

Visualization and Classification of Voice Quality with the Self–Organizing Map

Tapio Hiltunen, Lea Leinonen and Jari Kangas
Laboratory of Information and Computer Science,
Helsinki University of Technology, 02150 Espoo, Finland

The self–organizing map was applied to classify and visualize normal and pathological voice qualities. A two–dimensional feature map was computed from speech samples (the long vowel [a:] cut from normally spoken finnish words) of 40 men with or without voice disorders of laryngeal origin. The speech was digitized with 22 kHz sampling rate. Short–time power spectra were calculated every 10 ms from 23–ms Hamming–windows by 512–point FFT. The 16–dimensional feature vectors were calculated by windowing the power spectra in frequency dimension. The windows were sinc–functions positioned unevenly across the spectrum. They performed effectively an uneven decimation of 256–dimensional power spectra to 16–dimensional feature vectors.

The speech samples were judged in a listening experiment with respect to 11 semantic features. The window positioning in the feature extraction were adjusted with the aid of the semantic feature vectors. The similarity measure between the sets of semantic and spectral feature vectors was obtained by comparing the matrices of pairwise euclidean distances.

The map was organized according to the degree of voice pathology and according to the hoarseness and roughness of the samples. Thus, the map grouped the samples according to the spectral features significant for the perception of voice disorder. The shape of the trajectories conveyed information of the temporal evolution of the spectrum. The mean length of shifts in a trajectory was correlated with the roughness of the voice.

The method is speaker independent and can be performed in real time. It is being developed for the clinical measurement of voice. Its ability to provide immediate visual feedback can be useful in speech therapy.

Weighted Distance Measure for Speaker-Independent Digit Recognition with Hidden-Control Neural Network

Dou-Seok Kim and Soo-Young Lee*
Department of Electrical Engineering
Korea Advanced Institute of Science and Technology
373-1 Kusong-dong, Yusong-gu, Taejon 305-701, Korea (South)

Hidden control neural network (HCNN) incorporates dynamic state transition of the hidden Markov model (HMM) with Viterbi segmentation. [1]   In this paper we report a different hidden-state representation and weighted distance measure for better recognition rates.

The HCNN is a recurrent network, of which output signals at time t are fed back to input for time t+1.   To control the state transitions, another vector $c(t)$ is added in the input layer and consists of "hidden-control" neurons.   Provided the hidden-control vector $c(t) = C_i$ at time t, $c(t+1)$ is restricted to be either C_i or C_{i+1}. Adaptive learning consists of "re-estimation" of the synaptic weights and "segmentation" of the control vector at each iteration epoch.   The "re-estimation" process is the same as error back-propagation learning of standard multi-layer Perceptron, while the "segmentation" process involves Viterbi algorithm.

The HCNN is applied to Korean digit recognition.   Since some of the Korean numbers sound quite similar with only one syllable, e.x. "i", "il" and "chil", and "sam" and "sa", its speaker-independent recognition is regarded as a hard problem.   We had collected 64 sets of speech data from 14 males. Fourteen of them spoke 5 times, and the other 2 spoke only 2 times.   Six speakers are randomly selected from the 14, and 18 sets of speech data, i.e. 3 sets for each of the selected speaker, are used for training.   All other data sets are used for testing.   Cepstrum coefficients up to 14th order and difference cepstrum coefficent, power, and difference power are calculated as speech features.   Also 8 states are assumed for each word.

In the original HCNN only one component of the hidden-control vector was non-zero. [1]   To provide more natural transition we made two adjacent elements are non-zero and states C_i and C_{i+1} share one common non-zero element.   It increases recognition rate by 2.3% from 90.6%.   Further improvement is made by introducing weighted distance measure.   Histograms of ceptsrum coefficients show much smaller variance for higher order coefficients than lower order coefficients.   To utilize full domain on feature space the coeficients need to be normalized.   Since diagonal terms usually dominate in covariance matrix, a weighted distance with only the diagonal terms is used. [2]   For the isolated Korean digit recognition applications the HCNN with this weighted distance measure shows 97.4% accuracy, which is 2.3% better than HCNN with popular Euclidean distance and 3.5% better than HMM.

Acknowledgement: This research was supported by Korea Advanced Institute of Science and Technology.

References

[1] E. Levin, "Modeling time varying systems using hidden control neural architecture," Proc. NIPS-3, pp. 147-154, 1991.
[2] Y. Tohkura, "A weighted cepstral distance measure for speech recognition," IEEE Trans. ASSP, pp. 1414-1422, 1987.

---

* Currently on sabbatical leave from the KAIST, and stays at Angewandte Optik, Universität Erlangen, Stadudtstr. 7/B2, D-8520 Erlangen, Germany.

Modulation-Frequency Encoding of Speech with Applications to Neural Speech Recognizers

M. Paping, H. W. Strube, T. Gramss[†]

Drittes Physikalisches Institut, University of Göttingen
Bürgerstrasse 42–44; D-37073 Göttingen; Germany
Tel: +49(551)397731; Fax: +49(551)397720
e-mail: paping@up3spr2.gwdg.de

Most of the present speech recognition systems make use of a signal representation based on parameter frames equidistant in time. We propose a new method of signal representation which is motivated by linguistic, psychoacoustic and neurobiological facts.

The basic idea of our approach is a nonlinear segmentation of the time signal. Starting with a frame-based Bark spectrogram (stepsize 10 ms, 40 ms Hamming window) the segment boundaries are defined by the local minima of the smoothed modified loudness. We get the modified loudness by summing up the weighted frequency components of each time frame of the Bark spectrogram. The length of the resulting segments range from demisyllables to syllables. Seen from a linguistic point of view there are several reasons for choosing such large segments. One advantage may be that coarticulation does not affect syllables so much as it affects phonemes.

For each of these segments a feature vector is calculated. Motivated by recent results in modern brain research we perform another Fourier transformation on each frequency channel of the Bark spectrogram. In doing so we get a signal representation in the three-dimensional modulation frequency space. This kind of sound encoding is found in the auditory midbrain of several mammals and birds. Since the temporal structure of a signal is encoded in a redundant way now, it is possible to integrate in time over the segments defined above. As shown in previous papers this can be done without causing too much loss of information.

Now a speech signal is represented by a sequence of feature vectors, whose number of components is constant and thus independent of the length of the respective speech segment. This implies a local time normalization. The average length of the segments corresponds well to the time window of about 200 ms up to which an acoustical event can be integrated in the human auditory system.

Before starting recognition experiments we focused on the segmentation problem. The aim was to get a phonetically consistent segmentation: different versions of an utterance should be separated at phonetically equivalent points. For this task a special optimization procedure has been developed. The free parameters were the weighting window, used to calculate the modified loudness, and the degree of smoothing. We found that best results were reached with a trapezoidal window.

An artificial neural network is then used to perform isolated word recognition. Because of the feature vectors' high dimension a two-layer perceptron is sufficient for that purpose, rendering computation very efficient. In the training mode the simple δ-rule is used to change the synaptic plasticities in such a way that each feature vector of a given word is mapped onto the output cell representing this word. In the test mode each feature vector of a given utterance is applied to the neural network, resulting in a sequence of probability distributions from the output layer. From this sequence the recognized word can be estimated by multiplying all probabilities of each word. The winner cell is then determined by the product with maximal activity.

The capability of the recognition system was tested on the SPINA data base which consists of 62 isolated German words (five male and five female speakers). The speaker dependent recognition rates ranged from 90.7% to 99.4% (average 94.7%).

This new approach is suitable for being extended to processing continuous speech.

[†]current address: Physics of Computation Lab., Californian Institute of Technology, Pasadena, USA

Functional Compositionality: a *G.N.U.* Approach *

Shan H. Parfitt

Neural Systems Engineering, Department of Electrical and Electronic Engineering,
Imperial College, Exhibition Rd, London, SW7 2BT, UK.
email: s.parfitt@ic.ac.uk[†]

Fodor and Pylyshyn's famous 1988 polemic [Fodor and Pylyshyn 1988] on the use of Artificial Neural Networks (ANNs) for natural language modelling prompted a number of researchers, including van Gelder [1989 and 1990], Pollack [1990], Chalmers [1990], Davies [1991], Goschke [1991], Touretzky [1991] and Niklasson [1992] to undertake work which has already gone quite some way to answering the formers' important claims that ANNs are not capable of concatenative compositionality or systematicity. Smolensky [1987] argued successfully, that (in Van Gelder's words), "Connectionist representations can be compositional without being strictly combinatorial in the Classical sense". Van Gelder went on to show that Classical concatenative compositionality is just one form of compositionality, and that one strength of ANNs lies in another, which he calls functional compositionality. Its existence can be demonstrated through an analysis of the hidden representations which form during the training of both feedforward and recurrent multi-layer perceptrons. These representations are a point of particular interest because of their potential to enable the development of effective models of natural language which do not rely upon the combinatorial syntax and traditional, explicitly algorithmic methods of Computational Linguistics. In his well-known 1990 paper, "Finding structure in time", Jeffrey Elman [1990] investigated the ability of a Simple Recurrent Network (SRN) to make predictions about language-like sequences. The present work examines the performance of a weightless equivalent of the SRN (a General Neural Unit, or GNU) [Aleksander 1990 and 1991, Lucy 1991, Ntourntoufis 1991] on the same task. The GNU used consisted of 16 external and 16 internal GRAMS, each external GRAM outputting to one output unit, and each internal GRAM feeding its output back to 1 internal input node. There were, correspondingly, 32 input units (16 external and 16 internal). Each GRAM received input from a random subset, n, of the input nodes, such that the proportion of inputs received by a GRAM from external and internal input nodes was equal [Aleksander 1991]. An experimental corpus was generated to be as similar as possible to that used by Elman. The size of n was varied between $n = 4$ and $n = 22$. The following range of results was obtained.

n = 4: correct wordclasses = 15.72%; correct words = 4.4%.
n = 8: correct wordclasses = 21.46%; correct words = 7.0%.
n = 16: correct wordclasses = 46.63%; correct words = 33.63%.
n = 20: correct wordclasses = 71.66%; correct words = 65.67%.
n = 22: correct wordclasses = 82.0%; correct words = 77.0%.

If the corpus or language upon which an SRN is trained contains few strings, both an SRN and GNU should be capable of perfect or near-perfect next-word prediction. Elman showed that, for a large corpus, an SRN loses the ability to correctly predict the next word, but attains sufficient generalisation to be able to correctly predict likely next wordclass. It is rather difficult to make a direct comparison between the % correct predictions of the two types of model: in Elman's, since each word is represented by one positive bit, the response can be interpreted as a set of approximate probabilities, whereas in the other, it is not possible to distinguish more than one pattern in the response, because the nature of a weightless system makes distributed patterns preferable. Nevertheless, for larger n, the GNU learns to predict correct next word in the majority of cases, its performance greatly exceeding that of the SRN. For smaller n, it learns to predict next wordclass at well above chance, but does less well than the SRN. The auxiliary representations were passed through a hierarchical clustering program[1]. For a similar training set, the nature of the auxiliary representations obtained using a GNU are rather different from those obtained using an SRN, the former appearing to be largely random. However, it is important to note that their random nature did *not* prevent the network from being able to make predictions from the training data. Whilst there was no apparent similarity in the representations for similar tokens, functional compositionality is nevertheless present, as demonstrated by the figures for correct next wordclass prediction. In fact, the results suggest that, in order for functional compositionality to exist, internal representations which are ordered in state-space may not be a fundamental requirement, in weightless systems at least. It is not at present clear, however, whether this would remain the case for increasingly large languages.

---

*I would like to thank: Igor Aleksander, Jeffrey Elman, Andreas Stolke and members of the Neural Systems Group.
[†]A technical report describing the work may be obtained at this address.
[1]Cluster V. 2.5 by: Yoshiro Miyata (miyata@boulder.colorado.edu), Andreas Stolcke (stolcke@icsi.berkeley.edu), Steve Omohundro (om@icsi.berkeley.edu) and Kim Daugherty (kimd@gizmo.usc.edu).

PHYSICAL AND
MATHEMATICAL THEORY

Novel architectures and learning rules—oral contributions

Competitive Hebbian Learning Rule Forms
Perfectly Topology Preserving Maps

Thomas Martinetz

Siemens AG

Corporate Research and Development

Neural Networks Group

8000 München 83, Germany

Abstract: The problem of forming perfectly topology preserving maps of feature manifolds is studied. First, through introducing "masked Voronoi polyhedra" as a geometrical construct for determining neighborhood on manifolds, a rigorous definition of the term "topology preserving feature map" is given. Starting from this definition, it is shown that a network G of neural units i, $i = 1, ..., N$ has to have a lateral connectivity structure $\mathbf{A}$, $A_{ij} \in \{0, 1\}$, $i, j = 1, ..., N$ which corresponds to the "induced Delaunay triangulation" of the synaptic weight vectors $\mathbf{w}_i \in \Re^D$ in order to form a perfectly topology preserving map of a given manifold $M \subseteq \Re^D$ of features $\mathbf{v} \in M$. The lateral connections determine the neighborhood relations between the units in the network, which have to match the neighborhood relations of the features on the manifold. If all the weight vectors $\mathbf{w}_i$ are distributed over the given feature manifold M, and if this distribution resolves the shape of M, it can be shown that Hebbian learning with competition leads to lateral connections $i - j$ ($A_{ij} = 1$) that correspond to the edges of the "induced Delaunay triangulation" and, hence, leads to a network structure that forms a perfectly topology preserving map of M, independent of M's topology. This yields a means for constructing perfectly topology preserving maps of arbitrarily structured feature manifolds.

1. Introduction

Topology preserving feature maps play an important role in a variety of natural as well as artificial neural information processing systems [1-3]. By projecting input patterns onto a network of neural units such that similar patterns are projected onto adjacent units and, vice versa, such that adjacent units code similar patterns, a representation of the input patterns is achieved which in postprocessing stages allows one to exploit the similarity relations of the input patterns. Examples of topology preserving feature maps in the nervous system are the retinotopic map in the visual cortex [4], the mapping from the body surface onto the somatosensory cortex [5], or the tonotopic maps in the auditory cortex [6]. As components of artificial neural information processing systems topology preserving feature maps have been applied successfully in speech processing [7, 8], image processing [9], and robotics [10].

A number of neural network models for adaptively forming topology preserving feature maps have been proposed [11-14]. A model which provides a very compact procedure and, therefore, has found widespread application in artificial neural information processing systems is Kohonen's self-organizing feature map [2, 13]. This algorithm requires that one first chooses a graph (network) G, usually a one-, two-, or three-dimensional lattice; in a subsequent adaptation step, pointers (synaptic weight vectors) $\mathbf{w}_i$ which are assigned to the vertices (neural units) i of G are distributed over a given feature manifold $M \subseteq \Re^D$ in such a way, that (i) pointers lie on M, and (ii) pointers of vertices which are adjacent in G are assigned to locations which are close on M. To obtain a topology preserving map, it is necessary to choose a graph G, the topological structure of which matches the topological structure of the given feature manifold M.

In many applications, however, the feature manifold M is a submanifold of a high-dimensional space and may neither be known *a priori* nor topologically simple enough for prespecifying a correspondingly structured graph G. For these applications it would be highly desirable to have a procedure which adapts the topology of the graph G to the topology of the given manifold M. An approach to this problem has been introduced by Kohonen and coworkers [15]. They take the minimum spanning tree between the pointers $\mathbf{w}_i$ as the graph G. Another approach has been proposed by Fritzke [16]. His approach distributes two-dimensional, triangular cell structures over the manifold M for forming an appropriate graph G. In this paper we will describe an approach which is based on the so-called Delaunay triangulation [17] of the pointers $\mathbf{w}_i$ and employs a competitive version of the Hebb rule for forming the graph G. A preliminary version was presented in [18]. Starting from a rigorous definition of topology preservation, which is given in the next section, we show that the approach presented forms network structures G which preserve the topology of given feature manifolds completely.

2. A rigorous definition of topology preservation

Given a feature manifold M. Which graphical structure forms a perfectly topology preserving map of M? To answer this question, we first have to define exactly when topology preservation is given. The problem is that adjacency of vertices i in a graph G is clearly defined; however, a definition for adjacency of pointers $\mathbf{w}_i$ on M which is in agreement with our intuitive understanding of topology preservation is not obvious. This is why in previous contributions on topology preserving feature maps the interpretation of "topology preservation" has usually been left to the reader's intuition.

An exception is the trivial one-dimensional case. Obviously, two points $\mathbf{w}_i, \mathbf{w}_j \in \Re$ are neighboring if there is no point $\mathbf{w}_k$ in between. Expressed in terms of Voronoi polyhedra an equivalent definition is: two points $\mathbf{w}_i, \mathbf{w}_j \in \Re$ are neighboring if their Voronoi polyhedra V_i, V_j are adjacent, i.e., if $V_i \cap V_j \neq \emptyset$ with

$$V_i = \{\mathbf{v} \in \Re^D \mid \|\mathbf{v} - \mathbf{w}_i\| \leq \|\mathbf{v} - \mathbf{w}_j\| \; j = 1, \ldots, N\} \qquad i = 1, \ldots, N. \tag{1}$$

In these terms a generalization to higher dimensional embedding spaces $\Re^D$ is straightforward. However, since we need a definition of neighborhood of points *on a manifold* M, we first introduce the *masked Voronoi polyhedron*. The masked Voronoi polyhedron $V_i^{(M)}$ is the part of V_i which is also part of M, i.e., $V_i^{(M)} = V_i \cap M$. The superscript indicates the dependence of the masked Voronoi polyhedron on the given manifold M. By using the neighborhood of the masked Voronoi polyhedra $V_i^{(M)}$, $V_j^{(M)}$ instead of the neighborhood of the Voronoi polyhedra V_i, V_j for determining the neighborhood of the points $\mathbf{w}_i$, $\mathbf{w}_j$ on M, we ensure that two points $\mathbf{w}_i$, $\mathbf{w}_j$ are called *adjacent on M* only if they do not belong to disconnected regions of M. This leads to the following definition:

Definition 1 *Let $M \subseteq \Re^D$ be a given manifold and $S = \{\mathbf{w}_1, \ldots, \mathbf{w}_N\}$ be a set of points $\mathbf{w}_i \in M$. The Voronoi polyhedra of S are denoted by V_i, $i = 1, \ldots, N$. Two points $\mathbf{w}_i, \mathbf{w}_j \in M \subseteq \Re^D$ are adjacent on M if their masked Voronoi polyhedra $V_i^{(M)} = V_i \cap M$, $V_j^{(M)} = V_j \cap M$ are adjacent, i.e., if $V_i^{(M)}$ and $V_j^{(M)}$ share an element $\mathbf{v} \in M$ or, equivalently, if $V_i^{(M)} \cap V_j^{(M)} \neq \emptyset$ is valid.*

Each masked Voronoi polyhedron is part of the manifold M, and the set of all masked Voronoi polyhedra forms a complete partitioning of the manifold M; i.e., $M = \bigcup_{i=1}^{N} V_i^{(M)}$ is valid.

With this definition the term "topology preserving feature map" can be formulated rigorously:

Definition 2 *Let G be a graph (network) with vertices (neural units) i, $i = 1, \ldots, N$ and edges (lateral connections) defined by the adjacency matrix $\mathbf{A}$, $A_{ij} \in \{0, 1\}$, $i, j = 1, \ldots, N$. Let*

$M \subseteq \Re^D$ *be a given manifold of features* $\mathbf{v} \in M$ *and* $S = \{\mathbf{w}_1, \ldots, \mathbf{w}_N\}$ *be a set of pointers (synaptic weight vectors)* $\mathbf{w}_i \in M$, *each of which is attached to a vertex* i *of the graph* G. *Let each feature* $\mathbf{v}$ *of the manifold* M *be mapped onto that vertex* i *whose pointer* $\mathbf{w}_i$ *is closest to* $\mathbf{v}$. *The graph* G *with its vertices* i *assigned to the locations* $\mathbf{w}_i \in M$ *forms a topology preserving map of* M, *if pointers* $\mathbf{w}_i, \mathbf{w}_j$ *which are adjacent on* M *belong to vertices* i, j *which are adjacent in* G, *and, vice versa, if vertices* i, j *which are adjacent in* G *are assigned to locations* $\mathbf{w}_i, \mathbf{w}_j$ *which are neighboring on* M.

3. Induced Delaunay triangulations as perfectly topology preserving maps

Assuming the pointers $\mathbf{w}_1, \ldots, \mathbf{w}_N$ which are attached to the vertices i, $i = 1, \ldots, N$ of a graph G are distributed over the given manifold M. The graph G forms a topology preserving map of M, if vertices i, j and only vertices i, j whose corresponding masked Voronoi polyhedra $V_i^{(M)}$ and $V_j^{(M)}$ are adjacent are connected by an edge $i - j$ ($A_{ij} = 1$). The graphical structure which connects those and only those vertices i, j whose corresponding Voronoi polyhedra V_i and V_j are adjacent is the so-called *Delaunay triangulation*[1] [17]. Analog to the definition of the Delaunay triangulation $\mathcal{D}_S$ of a set of points $S = \{\mathbf{w}_1, \ldots, \mathbf{w}_N\}$, which is based on the Voronoi polyhedra $V_1, \ldots, V_N$ of S, we define the *induced Delaunay triangulation* $\mathcal{D}_S^{(M)}$ based on the *masked* Voronoi polyhedra $V_1^{(M)}, \ldots, V_N^{(M)}$:

Definition 3 *Let* $M \subseteq \Re^D$ *be a given manifold and* $S = \{\mathbf{w}_1, \ldots, \mathbf{w}_N\}$ *be a set of points* $\mathbf{w}_i \in M$. *The induced Delaunay triangulation* $\mathcal{D}_S^{(M)}$ *of* S, *given* M, *is defined by the graph which connects two points* $\mathbf{w}_i, \mathbf{w}_j$ *if and only if their masked Voronoi polyhedra* $V_i^{(M)}, V_j^{(M)}$ *are adjacent, i.e., by the graph whose adjacency matrix* $\mathbf{A}$, $A_{ij} \in \{0, 1\}$, $i, j = 1, \ldots, N$ *has the properties*

$$A_{ij} = 1 \quad \Leftrightarrow \quad V_i^{(M)} \cap V_j^{(M)} \neq \emptyset. \tag{2}$$

We obtain the result that a graph G forms a perfectly topology preserving map of a feature manifold M, if and only if it is the induced Delaunay triangulation of the set S of pointers $\mathbf{w}_i \in M$. This is illustrated in Fig. 1. In (a), (b), (c), and (d) the given manifold M, which is disconnected and is the same in all four examples, is depicted by the two shaded areas. Only in (d), where the graph G is the induced Delaunay triangulation of the points $\mathbf{w}_i$, two vertices are connected by an edge if and only if their masked Voronoi polygons are adjacent. Only in (d) the graph G forms a perfectly topology preserving map of the given manifold M.

4. Competitive Hebbian rule

In the following we assume a set of neural units i, $i = 1, \ldots, N$ which develop *lateral connections* between each other, starting from being unconnected initially. A neural unit connects itself with another unit by developing a *synaptic link* to this unit. The lateral connections are described by a connection strength matrix $\mathbf{C}$ with elements $C_{ij} \in \Re_0^+$, $i, j = 1, \ldots, N$. The larger a matrix element C_{ij}, the stronger is the synaptic link from unit i to unit j. Only if $C_{ij} > 0$, we regard neural unit i as being connected with unit j. If $C_{ij} = 0$, neural unit i is *not* connected with unit j. Negative values for C_{ij} do not arise.

The basic principle which governs the change of interneural connection strength has first been formulated by Hebb [19]. According to Hebb's postulate a presynaptic unit i increases the strength of its synaptic link to a postsynaptic unit j if both units are concurrently active, i.e., if both activities do correlate. In its simplest mathematical formulation Hebb's rule is described by the equation

$$\Delta C_{ij} \propto y_i \cdot y_j, \tag{3}$$

---

[1] The Delaunay triangulation is an important structure in computational geometry, particularly for solving proximity problems.

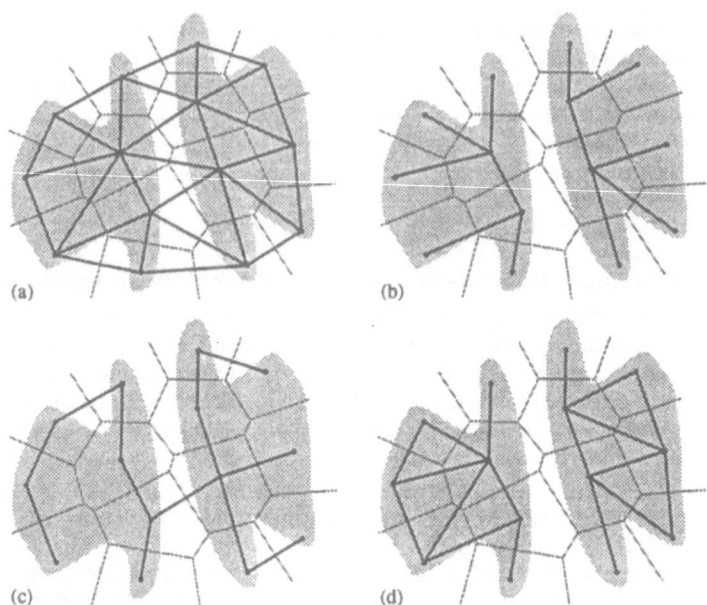

(a)

(b)

(c)

(d)

Figure 1: Illustration of our definition of topology preserving maps. In the four examples the given manifold M is disconnected and depicted by the two shaded areas. In (a) the graph G is the Delaunay triangulation of the pointers $\mathbf{w}_i$ (the location of each pointer is marked by a dot). The resulting map of M we do not consider as being topology preserving since some vertices which are connected in G belong to masked Voronoi polygons which are not adjacent on M. In (b) the opposite case is shown. The graph G does not define a topology preserving map of M since some vertices which belong to adjacent masked Voronoi polygons are not connected in G. In (c) the graph G is the minimum spanning tree of the points $\mathbf{w}_i$, as it would be suggested by the approach of Kangas et al. [15]. In (d) the graph G is the induced Delaunay triangulation. Pointers and only pointers the Voronoi polygons of which are adjacent *on* M, i.e., pointers the *masked* Voronoi polygons of which are adjacent, are connected. Only in this case the graph G forms a perfectly topology preserving map of the given manifold M.

in which the change of the strength C_{ij} of the synaptic link from unit i to unit j is linearly proportional to the presynaptic activity y_i and to the postsynaptic activity y_j. The quantities y_i, $i = 1, \ldots, N$ denote the output activities of the neural units i.

We will employ the Hebb rule in a form which incorporates the novel aspect of *competition* among the synaptic links. Again to each neural unit i a weight vector $\mathbf{w}_i \in \Re^D$ is assigned. Further, we assume that each neural unit i, $i = 1, \ldots, N$ receives the same afferent input patterns $\mathbf{v} \in \Re^D$. The weight vector $\mathbf{w}_i$ determines the center of the receptive field of unit i in the sense that with the reception of an input pattern $\mathbf{v}$ the output activity y_i of unit i is the larger the closer its $\mathbf{w}_i$ is to $\mathbf{v}$. In mathematical terms, we assume that $y_i = R(\|\mathbf{v} - \mathbf{w}_i\|)$ is valid, with $R(.)$ being a positive and continuously monotonically decreasing function, e.g., a Gaussian.

Applying the Hebb rule in the simple form as given in eq. (3) yields the rather trivial result that each neural unit i develops connections to all the other units $j \neq i$ with lateral connection strengths C_{ij} which are simply proportional to the overlaps of the receptive fields $R(\|\mathbf{v} - \mathbf{w}_i\|)$ and $R(\|\mathbf{v} - \mathbf{w}_j\|)$. The strength of the synaptic link between two units i and j is simply monotonically and continuously decreasing with the distance between $\mathbf{w}_i$ and $\mathbf{w}_j$. However, as in many systems governed by self-organizing processes, also the connectivity pattern which evolves on the set of neural units becomes significantly more structured if we introduce competition. In a winner-take-all network, for example, the units compete with each other based on their output activities, which finally leads to an adaptation only of the weights of the unit with the highest output activity. Without competition, all the units would behave alike and no specialization of the units, as it is characteristic for winner-take-all networks, would evolve.

Analog to the competition among the units in a winner-take-all network we introduce competition among the synaptic links. Instead of being based on the output activities of the neural units itself as in a winner-take-all network, in our model the competition among the synaptic links is determined by the *correlated* output activities Y_{ij}, the correlations of the output activities of all pairs of pre- and postsynaptic units. In the quantitative formulation given below, the correlated output activities are determined by $Y_{ij} = y_i \cdot y_j$, according to the Hebb rule (3). Keeping the analogy to winner-take-all networks, with the presentation of an input pattern $\mathbf{v}$ only the synaptic link $i - j$ whose "activity" $Y_{ij} = y_i \cdot y_j$ is highest is modified. Instead of changing the connection strengths C_{ij} according to the Hebb rule (3), in the following we will employ a winner-take-all or competitive version of (3), determined by

$$\Delta C_{ij} \propto \begin{cases} y_i \cdot y_j & \text{if } y_i \cdot y_j \geq y_k \cdot y_l \quad \forall \, k, l = 1, \ldots N \\ 0 & \text{otherwise.} \end{cases} \tag{4}$$

Instead of connecting each unit with all the other units, we will show that the competitive Hebb rule (4) forms a connectivity structure among the neural units i, $i = 1, \ldots, N$ which corresponds to the Delaunay triangulation of the weight vectors $\mathbf{w}_1, \ldots \mathbf{w}_N$. More precisely, we will show that if we present sequentially input patterns $\mathbf{v}$ with a distribution $P(\mathbf{v})$ which has support (is nonzero) everywhere on $\Re^D$, then the elements C_{ij} of the connection strength matrix $\mathbf{C}$ obey asymptotically

$$\theta(C_{ij}(t \to \infty)) = A_{ij} \qquad i, j = 1, \ldots, N \tag{5}$$

with $\theta(.)$ as the Heaviside step function and A_{ij} as the elements of the adjacency matrix $\mathbf{A}$ of the Delaunay triangulation of the points $\mathbf{w}_1, \ldots \mathbf{w}_N$, for which

$$A_{ij} = \begin{cases} 1 & \text{if } V_i \cap V_j \neq \emptyset \quad (V_i, V_j \text{ are adjacent}) \\ 0 & \text{if } V_i \cap V_j = \emptyset \quad (V_i, V_j \text{ are not adjacent}) \end{cases} \tag{6}$$

is valid. V_i, V_j again denote the Voronoi polyhedra of $\mathbf{w}_i$, $\mathbf{w}_j$.

For the proof we introduce the *second order Voronoi polyhedra* V_{ij}, $i, j = 1, \ldots, N$. The second order Voronoi polyhedron V_{ij} is given by all the $\mathbf{v} \in \Re^D$ for which $\mathbf{w}_i$ and $\mathbf{w}_j$ are the two closest points of S, i.e., V_{ij} is defined by

$$V_{ij} = \{\mathbf{v} \in \Re^D \mid \|\mathbf{v} - \mathbf{w}_i\| \leq \|\mathbf{v} - \mathbf{w}_k\| \ \wedge \ \|\mathbf{v} - \mathbf{w}_j\| \leq \|\mathbf{v} - \mathbf{w}_k\| \ \forall k \neq i, j\}. \tag{7}$$

As V_i, also V_{ij} forms a convex polyhedron. We see from (7) that the competitive Hebb rule connects two units i, j only if $V_{ij} \neq \emptyset$ is valid. Only if $\mathbf{w}_i$, $\mathbf{w}_j$ are the two points which are closest to the presented input pattern $\mathbf{v}$, $Y_{ij} = y_i \cdot y_j$ is the highest correlated output activity. We will prove that $V_{ij} \neq \emptyset$ is valid if and only if the corresponding first order Voronoi polyhedra V_i, V_j are adjacent, i.e., if and only if $V_i \cap V_j \neq \emptyset$. Then, in case $\int_{V_{ij}} P(\mathbf{v}) d\mathbf{v} \neq 0$ holds for each $V_{ij} \neq \emptyset$, the connections generated by the competitive Hebb rule form the Delaunay triangulation of the set $\mathbf{w}_1, \ldots, \mathbf{w}_N$ [2].

Theorem 1 *For a set $S = \{\mathbf{w}_1, \ldots, \mathbf{w}_N\}$ of points $\mathbf{w}_i \in \Re^D$ the relation*

$$V_i \cap V_j \neq \emptyset \ \Leftrightarrow \ V_{ij} \neq \emptyset \tag{8}$$

is valid. V_i denotes the first order Voronoi polyhedron of point $\mathbf{w}_i$, and V_{ij} denotes the second order Voronoi polyhedron of the points $\mathbf{w}_i$, $\mathbf{w}_j$.

---

[2]The following theorem and its proof has been formulated together with Philippe Dalger and Benoit Noël [20].

Proof: If $V_i \cap V_j \neq \emptyset$ is valid, there is a $\mathbf{v} \in \Re^D$ with $\mathbf{v} \in V_i$ and $\mathbf{v} \in V_j$. Then we obtain $\|\mathbf{v} - \mathbf{w}_i\| = \|\mathbf{v} - \mathbf{w}_j\| \leq \|\mathbf{v} - \mathbf{w}_k\|$ for all $\mathbf{w}_k \in S$ and, therefore, $\mathbf{v} \in V_{ij}$, i.e., $V_{ij} \neq \emptyset$, is valid. The reverse implication follows by contradiction if we assume $V_{ij} \neq \emptyset$ and $V_i \cap V_j = \emptyset$ being valid. For each $\mathbf{v} \in V_{ij}$ the points $\mathbf{w}_i$ and $\mathbf{w}_j$ are the two nearest neighbors. Without loss of generality we can assume that for all $\mathbf{v} \in V_{ij}$ the point $\mathbf{w}_i$ is the nearest neighbor. Otherwise, against our assumption, there would be a $\mathbf{v} \in V_{ij}$ for which $\|\mathbf{v} - \mathbf{w}_i\| = \|\mathbf{v} - \mathbf{w}_j\|$ and, therefore, also $\mathbf{v} \in V_i \cap V_j$ were valid. Then, $V_{ij} \subseteq V_i$ follows. V_{ij} is given by all the $\mathbf{v} \in V_i$ which are closer to $\mathbf{w}_j$ than to all the other $\mathbf{w}_k \in S/\{\mathbf{w}_i, \mathbf{w}_j\}$. Hence, V_{ij} is bounded by hyperplanes, each of which is perpendicular to the connecting line between $\mathbf{w}_j$ and the respective $\mathbf{w}_k \in S/\{\mathbf{w}_i, \mathbf{w}_j\}$. For each hyperplane, $\mathbf{w}_j$ belongs to the half space which contains V_{ij}. Hence, $\mathbf{w}_j \in V_{ij}$ and, therefore, $\mathbf{w}_j \in V_i$ is valid. However, since also $\mathbf{w}_j \in V_j$, we obtain $\mathbf{w}_j \in V_i \cap V_j$ which is a contradiction to our assumption.

In the following we will consider pattern distributions $P(\mathbf{v})$ which have support not on the entire embedding space $\Re^D$, but only on a submanifold M. In these cases for some $V_{ij} \neq \emptyset$ the integral $\int_{V_{ij}} P(\mathbf{v}) d\mathbf{v}$ might vanish, with the result that the edge $i-j$ will not be established by the competitive Hebb rule. In these cases the competitive Hebb rule does not form the entire Delaunay triangulation, but only a subgraph of it.

5. Competitive Hebbian rule forms induced Delaunay triangulations

The competitive Hebb rule (4) constructs the full Delaunay triangulation of a set of points $\mathbf{w}_1, \ldots, \mathbf{w}_N$ only if each Voronoi polyhedron of second order V_{ij} is, at least partially, covered by the density distribution $P(\mathbf{v})$. If we define a given feature manifold M as being the manifold of $\Re^D$ on which $P(\mathbf{v})$ is non-zero, two units i, j become connected if and only if $V_{ij} \cap M \neq \emptyset$. Hence, if the manifold M forms a submanifold which does not cover each Voronoi polyhedron of second order, the Delaunay triangulation will evolve only partly. If the distribution of the points $\mathbf{w}_i$ is dense on M in a sense we will define below, the subgraph of the Delaunay triangulation which is formed by the competitive Hebb rule will be the *induced* Delaunay triangulation which was introduced in Section 3.

Definition 4 *Let $S = \{\mathbf{w}_1, \ldots, \mathbf{w}_N\}$ be a set of points $\mathbf{w}_i$ which are distributed over a given manifold $M \subseteq \Re^D$. The distribution of the points $\mathbf{w}_i \in M$, $i = 1, \ldots, N$, is dense on M, if for each $\mathbf{v} \in M$ the triangle $\triangle(\mathbf{v}, \mathbf{w}_{i_0}, \mathbf{w}_{i_1})$ formed by the point $\mathbf{w}_{i_0}$ which is closest to $\mathbf{v}$, the point $\mathbf{w}_{i_1}$ which is second closest to $\mathbf{v}$, and $\mathbf{v}$ itself lies completely on M, i.e., if $\triangle(\mathbf{v}, \mathbf{w}_{i_0}, \mathbf{w}_{i_1}) \subseteq M$ is valid.*

A distribution of points $\mathbf{w}_i$ is dense on M according to the above definition, if the distribution is dense, in the common sense, compared to the topological structure of M. The distribution of the points $\mathbf{w}_i$ has to have a density which resolves the details of the shape of the submanifold M. If for each sample point $\mathbf{v} \in M$ there is a closest point $\mathbf{w}_{i_0}$ and a second closest point $\mathbf{w}_{i_1}$ such, that the triangle $\triangle(\mathbf{v}, \mathbf{w}_{i_0}, \mathbf{w}_{i_1})$ lies completely on M, the distribution of the $\mathbf{w}_i$ is dense on M. If the distribution is homogeneous, the distribution becomes dense simply by increasing the number N of points $\mathbf{w}_i$.

With Definition 4 we obtain the main theorem:

Theorem 2 *Let i, $i = 1, \ldots, N$ be a set of vertices (neural units). Let $M \subseteq \Re^D$ be a given manifold of features $\mathbf{v} \in \Re^D$ and $S = \{\mathbf{w}_1, \ldots, \mathbf{w}_N\}$ be a set of pointers (synaptic weight vectors) $\mathbf{w}_i \in M$, each of which is attached to the corresponding vertex (neural unit) i and defines the center of the receptive field $R(\|\mathbf{v} - \mathbf{w}_i\|)$ of i. If the distribution of the pointers $\mathbf{w}_i \in M$ is dense on M, then the edges (lateral connections) $i - j$ which are formed by the competitive Hebb rule define a graph (network) G which corresponds to the induced Delaunay triangulation $\mathcal{D}_S^{(M)}$ of S and, hence, forms a perfectly topology preserving map of M.*

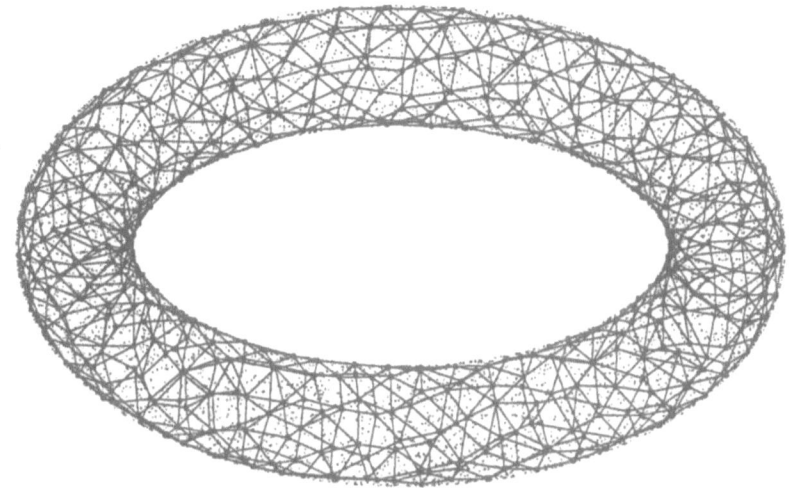

Figure 2: A topology preserving map of a torus, formed by the competitive Hebb rule. The pointers, the locations of which are marked by the large dots, were distributed over the given manifold M, i.e., the torus, by the "neural gas" algorithm [18, 21] in a preprocessing stage. Then the pointers stay fix and the edges are formed by the competitive Hebb rule. The small dots depict already presented patterns $\mathbf{v} \in M$. The few edges of the induced Delaunay triangulation which are still missing have small masked Voronoi polyhedra of second order and would emerge if further input patterns were presented.

Proof: Analog to Theorem 1 we prove the above theorem by showing that

$$V_i^{(M)} \cap V_j^{(M)} \neq \emptyset \;\Leftrightarrow\; V_{ij}^{(M)} \neq \emptyset \tag{9}$$

is valid, with $V_{ij}^{(M)} = V_{ij} \cap M$ as the masked Voronoi polyhedron of second order.

If $V_i^{(M)} \cap V_j^{(M)} \neq \emptyset$ is valid, there is a $\mathbf{v} \in M$ with $\mathbf{v} \in V_i$ and $\mathbf{v} \in V_j$. Then we obtain $\|\mathbf{v} - \mathbf{w}_i\| = \|\mathbf{v} - \mathbf{w}_j\| \leq \|\mathbf{v} - \mathbf{w}_k\|$ for all $\mathbf{w}_k \in S$ and, therefore, $\mathbf{v} \in V_{ij}^{(M)}$.

If $V_{ij}^{(M)} \neq \emptyset$ is valid, there is a $\mathbf{v}^* \in V_{ij}^{(M)}$ with $\triangle(\mathbf{v}^*, \mathbf{w}_i, \mathbf{w}_j) \subseteq M$, since the distribution of the pointers $\mathbf{w}_1, ..., \mathbf{w}_N$ is dense on M. For each $\mathbf{v} \in V_{ij}^{(M)}$ the points $\mathbf{w}_i$ and $\mathbf{w}_j$ are the two nearest neighbors. Without loss of generality we assume that for $\mathbf{v}^*$ the point $\mathbf{w}_i$ is the nearest neighbor. Since for each $\mathbf{u} \in \overline{\mathbf{v}^*\mathbf{w}_j}$ the point $\mathbf{w}_j$ is either the nearest or the second nearest neighbor of $\mathbf{u}$, and since for $\mathbf{u} = \mathbf{v}^*$ the point $\mathbf{w}_i$ is closest and for $\mathbf{u} = \mathbf{w}_j$ the point $\mathbf{w}_j$ is closest to $\mathbf{u}$, there is a $\mathbf{u}^* \in \overline{\mathbf{v}^*\mathbf{w}_j}$ for which $\|\mathbf{u}^* - \mathbf{w}_i\| = \|\mathbf{u}^* - \mathbf{w}_j\|$ is valid. Hence, we obtain $\mathbf{u}^* \in V_i$, $\mathbf{u}^* \in V_j$, and $\mathbf{u}^* \in \triangle(\mathbf{v}^*, \mathbf{w}_i, \mathbf{w}_j) \subseteq M$, which yields $M \cap (V_i \cap V_j) \neq \emptyset$ or, equivalently, $V_i^{(M)} \cap V_j^{(M)} \neq \emptyset$.

With the theorem above we have shown that the competitive Hebb rule forms perfectly topology preserving maps, supposed the distribution of the points $\mathbf{w}_i$ is dense on the given feature manifold M. In Fig. 2 we show a simulation example. The manifold M is a torus. To obtain $\mathbf{w}_i \in M$ for each i, $i = 1, ..., N$, the pointers $\mathbf{w}_i$ are distributed over M in a preprocessing stage, e.g., by a pattern driven vector quantization procedure like the "neural gas" algorithm [18, 21], which leads to a homogeneous distribution of the pointers $\mathbf{w}_i$ on M. After having distributed the pointers, the connectivity structure is formed by the competitive Hebb rule. Simply by sequentially presenting patterns $\mathbf{v} \in M$ and each time connecting those two units i, j which have the highest correlated output activity $Y_{ij} = y_i \cdot y_j$, a connectivity structure evolves which defines a perfectly topology preserving map and reflects the dimensionality and topological structure of the manifold M, i.e., of the torus.

5. Discussion

We showed how the term "topology preserving map" can be defined rigorously based on *masked Voronoi polyhedra* and *induced Delaunay triangulations*. Both the masked Voronoi polyhedra as well as the induced Delaunay triangulation of a set of points depend on the shape of the given feature manifold M. We showed that the induced Delaunay triangulation $\mathcal{D}_S^{(M)}$ as a particular subgraph of the full Delaunay triangulation $\mathcal{D}_S$ forms a perfectly topology preserving map of the manifold M. We proved rigorously and demonstrated through a computer simulation that a competitive version of the Hebb rule forms induced Delaunay triangulations and, hence, yields perfectly topology preserving maps of feature manifolds. Necessary is a distribution of the receptive field centers $\mathbf{w}_i$ of the neural units i which is dense enough to resolve the shape of the manifold M. If the manifold $M \subseteq \Re^D$ fills the embedding space $\Re^D$ completely, then the competitive Hebb rule forms the full Delaunay triangulation $\mathcal{D}_S$ as a perfectly topology preserving map of M. If M is only a submanifold of $\Re^D$, then the competitive Hebb rule forms a subgraph of $\mathcal{D}_S$, i.e., the induced Delaunay triangulation $\mathcal{D}_S^{(M)}$, as a perfectly topology preserving map of M.

Acknowledgement

I am indebted to Klaus Schulten for many fruitful discussions and for supporting this work. I thank Benoit Noël and Philippe Dalger for pointing out the connection between the competitive Hebb rule and second order Voronoi polyhedra.

References

[1] Knudsen EI, du Lac S, Esterly SD (1987), Ann. Rev. Neurosci., 10:41-65.

[2] Kohonen T (1990), Proceedings of the IEEE, 78:1464-1480.

[3] Ritter H, Martinetz T, Schulten K (1992), Neural Computation and Self-Organizing Maps. Addison-Wesley, Reading, Mass., 1992.

[4] Hubel DH, Wiesel TN (1974), J. Comp. Neurol., 158:267-294.

[5] Kaas JH, Nelson RJ, Sur M, Lin CS, Merzenich MM (1979), Science, 204:521-523.

[6] Suga N, O'Neill WE (1979), Science, 206:351-353.

[7] Kohonen T, Mäkisara K, Saramäki T (1984), Proc. 7th Int. Conf. on Pat. Recog., Montreal, pp 182-185.

[8] Brandt WD, Behme H, Strube HW (1991), Fortschritte zur Akustik– DAGA 91, Bad Honnef, DPG-GMBH, Germany, pp 1057-1060.

[9] Nasrabadi NM, Feng Y (1988), IEEE Int. Conf. on Neural Networks, San Diego, 1988, pp 1101-1108.

[10] Ritter H, Martinetz T, Schulten K (1989), Neural Networks 2(3):159-168.

[11] Willshaw DJ, von der Malsburg C (1976), Proc R Soc London B 194:431-445.

[12] Takeuchi A, Amari S (1979), Biological Cybernetics, 35:63-72.

[13] Kohonen T (1982), Biological Cybernetics, 43:59-69.

[14] Durbin R, Mitchison G (1990), Nature, 343:644-647.

[15] Kangas JA, Kohonen TK, Laaksonen JT (1990), IEEE Transactions on Neural Networks 1(1):93-99.

[16] Fritzke B (1991), Artificial Neural Networks, vol I, Kohonen T. et al. (eds.), North Holland, Amsterdam, pp. 403-408.

[17] Delaunay B (1934), Bull. Acad. Sci. USSR (VII), Classe Sci. Mat. Nat., pp 793-800.

[18] Martinetz T, Schulten K (1991), Artificial Neural Networks, vol I, Kohonen T. et al. (eds.), North Holland, Amsterdam, pp. 397-402.

[19] Hebb D (1949), Organization of Behavior. Wiley, New York.

[20] Dalger P, Noël B, Martinetz T, Schulten K (1992), Beckman Institute Technical Report TB 98-01, University of Illinois at Urbana-Champaign.

[21] Martinetz T, Berkovich S, Schulten K (1992), IEEE Transactions on Neural Networks (in press).

Approximating Optimal Information Transmission using Local Hebbian Algorithms in a Double Feedback Loop

Mark D. Plumbley

Centre for Neural Networks, King's College London
Department of Mathematics, Strand, London WC2R 2LS, UK
Tel: +44 (0)71-873 2214; Email: M.Plumbley@oak.cc.kcl.ac.uk

Abstract

Maximising mutual information (MI) under various constraints has been suggested as a goal for neural networks in a perceptual system. Networks using Hebbian algorithms have been found to be suitable for optimising MI with either input or output noise. In this paper we show that a double feedback loop network, using local Hebbian algorithms, can approximate the characteristics required for optimizing MI with *both* input and output noise. This represents a better approximation than simply orthonormalising the principal subspace.

1   Introduction

Recently, interest has been expressed in using information theory to investigate unsupervised learning in perceptual systems. Linsker [3] suggested that a perceptual system should adapt itself to maximize the mutual information (MI) between its input and its output.

The type of network which optimizes information transmission depends considerably on the noise sources assumed to be present in the system. If the only noise source is one of independent equal-variance (*spherical*) Gaussian noise on the input components, then a linear N-input M-output network maximizes MI when the M outputs span the same subspace as the M principal eigenvectors of the input [5]. On the other hand, if the only noise source is one of spherical Gaussian noise on the *output* components, with a power cost preventing the output from being amplified without bound, MI is optimized when the output components are decorrelated and of equal variance [6].

The case of noise on both input and output has been analyzed in the frequency domain [1, 5]. This predicts a *whitening filter* which equalizes output power spectral density (equivalent to equalizing output variance) when the input signal-to-noise ratio (SNR) is large, but which cuts off completely below a certain critical input SNR, with intermediate behaviour between these two extremes. The close match that Atick and Redlich [1] found between this theoretical approach and the measured sensitivity curves for the human visual system is very suggestive that optimization of information transmission, with built-in power-like costs, is important for the efficient operation of real perceptual systems.

Linear unsupervised neural networks using local Hebbian algorithms have already been discovered which find optimum MI condition for single noise sources, i.e. with noise on *either* the input or the output. In this paper we attempt to construct a network with a local Hebbian algorithm which approximates an optimization of MI with both input *and* output noise.

2   Previous Network Algorithms

Consider a linear network with N-dimensional input vector $\mathbf{x}$, weight matrix W, and N-dimensional output vector $\mathbf{y}$ such that $\mathbf{y} = W\mathbf{x}$ (Fig. 1(a)). Several neural network algorithms have been proposed over the last decade or so to allow such a network to extract one or more principal components of the input, or at least the subspace spanned by those principal components. Most of these are based on a principal component analysis (PCA) neuron due to Oja (see e.g. [4]). Any

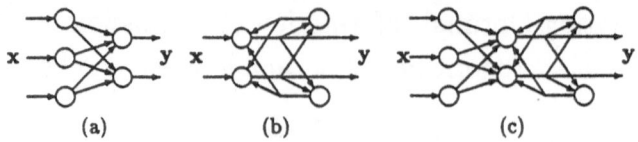

Figure 1: Previous information optimising networks.

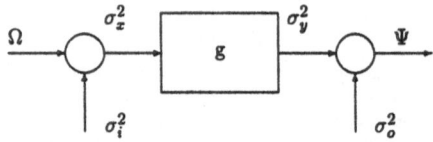

Figure 2: Block diagram showing information source Ω, destination Ψ, and noise sources.

one of these algorithms is sufficient to maximize MI in a network subjected to (spherical Gaussian) input noise only.

With spherical Gaussian noise on the *output*, and an output power cost proportional to $\sum_i \sigma_{y_i}^2 = \mathrm{Tr}(\Sigma_y)$, where $\Sigma_y = E(\mathbf{y}\mathbf{y}^T)$ is the ouptut covariance matrix, MI is optimized if the output signal is also spherical Gaussian [6], i.e. $\Sigma_y = \beta I$. This can be achieved using e.g. a network with inhibitory interneurons (Fig. 1(b)), where we have $\mathbf{z} = V^T\mathbf{y}$ and $\mathbf{y} = \mathbf{x} - V\mathbf{z}$ i.e. $\mathbf{y} = (I + VV^T)^{-1}\mathbf{x}$ after an initial transient. In this case, the algorithm

$$\Delta V = \eta_V(\mathbf{y}\mathbf{z}^T - \beta V) = \eta_V(\mathbf{y}\mathbf{y}^T - \beta I)V \qquad (1)$$

is suitable, since it converges when $\Sigma_y = \beta I$ as required, provided the variances of all the input components (the eigenvalues of Σ_x) are greater than β. The variance of any input component which is less than β will be left unchanged at the output: in effect the network 'decouples' itself from these components [6].

In an earlier paper [7], the author approached the problem of combining these solutions for single noise sources for the case of both input and output noise. A combination of a modified PCA stage and the inhibitory interneuron stage generalized from the Barrow and Budd [2] Automatic Gain Control (AGC) network was suggested (Fig. 1(c)) with the local Hebbian algorithms

$$\Delta W = \eta_W(\mathbf{y}\mathbf{x}^T - \alpha W) \qquad \Delta V = \eta_V(\mathbf{y}\mathbf{z}^T - \beta V). \qquad (2)$$

This network does cut off small variance components, and fixes large variance input components to $\sigma_y^2 = \beta$, as required for an optimal filter, but the mid-range characteristics need improving.

3 Mid-range Optimal Filtering

Fig. 2 shows a 1-input 1-output linear system with input and output noise and gain g. We use the Lagrange multiplier technique to maximize the mutual information (MI) $I(\Psi, \Omega)$ across this system with power cost σ_y^2. Thus we should maximize the function

$$J = I(\Psi, \Omega) - (\lambda/2)\sigma_y^2 = (1/2)\left(\log(\sigma_y^2 + \sigma_o^2) - \log(g^2\sigma_i^2 + \sigma_o^2) - \lambda\sigma_y^2\right) \qquad (3)$$

where λ is a Lagrange multiplier. Differentiating and equating to zero, and scaling so that $\sigma_i^2 = \sigma_o^2 = 1$, the optimal value for σ_y^2 satisfies

$$\sigma_y^2\left[(\sigma_x^2)^2 - (\sigma_x^2 + \sigma_y^2)(1 + \lambda(\sigma_y^2 + 1)\sigma_x^2)\right] = 0 \qquad (4)$$

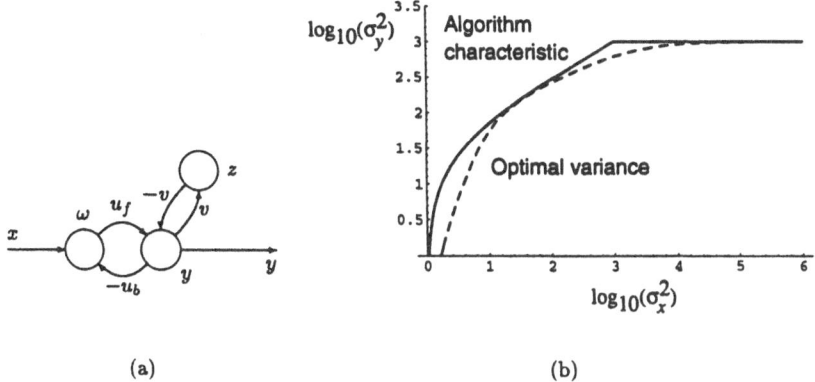

Figure 3: Double feedback loop network: (a) layout; (b) characteristic

If $\sigma_x^2 \gg 1/\lambda$ and $\sigma_x^2 \gg 1$ then the optimal value for the output noise is approximately constant for a given value of λ, i.e.

$$\sigma_y^2 \approx 1/\lambda - 1 \approx 1/\lambda \quad \text{if } \lambda \ll 1 \tag{5}$$

Alternatively, if $\sigma_x^2 \leq 1 + \lambda$, the output variance σ_y^2 should be zero, so the system should shut off completely. Both of these ranges are already handled by the PCA/inhibitory interneuron combination (Fig. 1(c)) considered in the previous section.

For $\lambda \ll 1$ a mid range appears between these two extremes. If we consider the case $1/\lambda \gg \sigma_x^2 \gg 1$, we find that the optimal value for output variance approximates

$$\sigma_y^2 \approx \left(\sigma_x^2/\lambda\right)^{1/2}. \tag{6}$$

In a previous paper [6] the author observed that the variance of the inhibitory interneuron in a decorrelating network approximately varies with the square root of the input variance, which is precisely the behaviour we require for this mid range. This leads us to suggest a *double* feedback loop (Fig. 3(a)): the first to generate the low input variance cutoff and square-root-variance behaviour in the mid range, with the second to limit the upper bound to the output variance in the high input variance regime.

4 Double Feedback Loop Operation

When the activations in the network of Fig. 3(a) have settled, we have that

$$z = vy \tag{7a}$$
$$y = u_f\omega - vz \tag{7b}$$
$$\omega = x - u_b y \tag{7c}$$

from which we can deduce, for example, that

$$(1 + v^2)y = u_f\omega. \tag{8}$$

We propose to use the following local Hebbian algorithms with weight decay

$$\Delta u_f = \eta_u(\omega y - \alpha_f u_f) \tag{9a}$$
$$\Delta u_b = \eta_u(\omega y - \alpha_b u_b) \tag{9b}$$
$$\Delta v = \eta_v(yz - \beta v) \tag{9c}$$

where η_u and η_v are update rates for the weights in the first and second loops respectively. To simplify the analysis for the purposes of this paper, we shall assume that $\eta_v \gg \eta_u$, so v adapts much faster than u_f and u_b (this makes stability conditions simpler, but does not affect stationarity conditions).

Consider now the conditions for (9a)–(9c) to be stationary on the average: we approximate this by setting their expectations to zero. (See e.g. [4] for an example of a more rigorous approach to this type of problem using stochastic approximation.) Firstly, equating the expected value of (9c) to zero and substituting in (7a) gives us

$$E(y^2)v - \beta v = 0$$

i.e.

$$\sigma_y^2 = \beta \quad \text{or} \quad v = 0. \tag{10}$$

Now, equating the expected value of (9a) to zero we get

$$E(\omega y) = \alpha_f u_f$$

which, multiplying both sides by $(1 + v^2)$ and substituting in (8) gives us

$$E(\omega^2)u_f = \alpha_f u_f(1 + v^2)$$

so

$$\sigma_\omega^2 = \alpha_f(1 + v^2) \quad \text{or} \quad u_f = 0. \tag{11}$$

Finally, equating the expected value of both (9a) and (9b) to zero, we get $\alpha_f u_f = \alpha_b u_b$, i.e.

$$u_b = (\alpha_f/\alpha_b)u_f. \tag{12}$$

Let us now consider the implications of these convergence conditions (10)–(12). One thing to notice immediately is that there is a *choice* of conditions, due to the 'or's in (10) and (11). We shall see in a moment that these lead to three (rather than four) convergence regimes, depending on the input variance σ_x^2.

Firstly, choose the condition $u_f = 0$ from (11). We can immediately see from (7c) that $\sigma_\omega^2 = \sigma_x^2$ in this case. In addition, we also have $u_b = 0$ from (12), and also that

$$\sigma_y^2 = 0 \tag{13}$$

from the stationarity of (9a) (or (9b)). Thus, provided $\beta \neq 0$, in (10) we must have $v = 0$ since $\sigma_y^2 = 0 \neq \beta$: this has removed one of our choices. Perturbation analysis reveals that this regime is only stable for $\sigma_x^2 \leq \alpha_f$.

This time, choose $\sigma_\omega^2 = \alpha_f(1 + v^2)$ from (11) but $v = 0$ from (10): consequently $\sigma_\omega^2 = \alpha_f$. Substituting (7b) into (7c) and rearranging, we get $\omega(1 + u_b u_f) = x$ so

$$\sigma_x^2 = \sigma_\omega^2(1 + u_b u_f)^2 = \alpha_f(1 + u_b u_f)^2. \tag{14}$$

Substituting in (12) gives us

$$(\sigma_x^2/\alpha_f)^2 = 1 + (\alpha_f/\alpha_b)u_f^2 \approx (\alpha_f/\alpha_b)u_f^2 \tag{15}$$

for $\sigma_x \gg \alpha_f$, which with (7b) gives us

$$\sigma_y^2 \approx \alpha_b(\sigma_x^2/\alpha_f)^{1/2}. \tag{16}$$

Our final choice gives us

$$\sigma_y^2 = \beta \tag{17}$$

which implies (after a little manipulation) that the input variance must satisfy

$$\sigma_x^2 = \alpha_f(1 + \beta/\alpha_b^2)(1 + v^2) \tag{18}$$

i.e. (17) is valid for $\sigma_x^2 \geq \alpha_f(1 + \beta/\alpha_b^2)$.

To summarize, this algorithm (9a)–(9c) produces three regimes of behaviour for the output variance σ_y^2, depending on the range of the input variance σ_x^2. These are:

1. Small input variance $\sigma_x \leq \alpha_f$, for which $u_f = u_b = 0$ so $\sigma_y^2 = 0$;

2. Intermediate input variance $\alpha_f \leq \sigma_x^2 \leq \alpha_f(1+\beta/\alpha_b^2)$, for which $v = 0$ and $\sigma_y^2 \approx \alpha_b(\sigma_x^2/\alpha_f)^{1/2}$; and

3. Large input variance $\sigma_x^2 \geq \alpha_f(1 + \beta/\alpha_b^2)$, for which $v \neq 0$, leading to $\sigma_y^2 = \beta$.

These are similar to the three regimes of behaviour required for the optimal filter described in section 3, especially for small λ where the approximations are valid.

For a particular value of λ in (4), a reasonable approximation can be obtained using $\alpha_f = 1+\lambda$ to fix the same cutoff condition, $\beta = 1/\lambda$ to fix the same asymptote, and $\alpha_b = (\alpha_f/\lambda)^{1/2} = (1+1/\lambda)^{1/2}$ to get similar mid-range behaviour. Fig. 3(b) compares the output variance produced by this algorithm with the optimal value of σ_y^2 for an example case of $\lambda = 10^3$.

5 Discussion

In sections 3 and 4 we have concentrated on the single-input, single-output case to investigate in detail. However, the same approach can be extended in a relatively straightforward manner to a network with N inputs and M outputs (with $M \leq N$).

With N inputs, we need to consider an orthogonal basis for the input components, perform a principal component analysis on the inputs, and optimize the orthogonal components independently according to (4), using the same value of λ for each component. If the number of outputs M is less than the number of inputs N, we simply ignore the $N - M$ input components with smallest input variance (i.e. the *minor* components). This will give us the optimal transform for the N input to M output transform.

The network of Fig. 3(a) can also be modified into an N-input M-output network in a similar manner. The two convergence regimes where the output loop of weights v units z is decoupled (the equivalent of $v = 0$) reduces to the single feedback loop that has already been considered by the author in a previous paper [6]. It was observed that such a network will reject input components below the cut-off limit, and produce an approximately square root variance output for others, but without explicitly performing a principal component analysis at any stage. It is possible to show that the second feedback loop does indeed limit larger components of the input to have variance no larger than β, but again without necessarily performing a principal component analysis to do so. We shall consider this in more detail in a later paper.

6 Conclusions

We have seen that a double feedback loop network, using only local Hebbian algorithms with weight decay, can learn to reasonably approximate the characteristics required for optimal transmission of information with both input and output noise, and an output power cost. This represents an improvement for mid-range input variance values over simply orthonormalising the principal subspace of the input.

Acknowledgements

I would like to thank John Taylor and Guido Bugmann for many useful comments and discussions about this paper and related work. The author is supported by a Temporary Lectureship from the Academic Initiative of the University of London.

References

[1] J. J. Atick and A. N. Redlich. What does the retina know about natural scenes? *Neural Computation*, 4:196–210. 1992.

[2] H. G. Barrow and J. M. L. Budd. Automatic gain control by a basic neural circuit. In I. Aleksander and J. Taylor, editors, *Artifical Neural Networks, 2.* Elsevier, 1992.

[3] R. Linsker. Self-organization in a perceptual network. *IEEE Computer*, 21(3):105–117, Mar. 1988.

[4] E. Oja. Principal components, minor components, and linear neural networks. *Neural Networks*, 5:927–935, 1992.

[5] M. D. Plumbley. On information theory and unsupervised neural networks. Technical Report CUED/F-INFENG/TR.78, Cambridge University Engineering Department, UK, 1991.

[6] M. D. Plumbley. Efficient information transfer and anti-Hebbian neural networks. *Neural Networks*, 1993. (in press).

[7] M. D. Plumbley. A Hebbian/anti-Hebbian network which optimizes information capacity by orthonormalizing the principal subspace. In *Proceedings of the IEE Conference on Artificial Neural Networks, ANN'93*, Brighton, UK, May 1993. (To appear).

Time-Varying Neural Networks for Large Tasks

Bert de Vries
David Sarnoff Research Center and
the National Information Display Laboratory
CN 5300, Princeton, NJ 08543-5300
bdevries@sarnoff.com / ph. +1-609-734-2456

Abstract

Large neural nets are required to solve large dimensional problems. In this paper we propose a concept where the required size of the neural net is decreased at the cost of additional computation during the recall phase. We use time-varying weights in order to increase the capacity of the neural net without increasing its size. The idea is illustrated in a word recognition framework, but the concept extends to broader contexts as well.

1· Introduction

For illustrative purpose we will use a speech recognition task in this paper, but any other large dimensional task would serve as well. We will try to develop a neural network for a word recognition problem. Let us first recapitulate how to design such a classifier from a pattern recognition perspective. In the simplest case, we create a number of word prototypes for each word class. Next we need a rule for comparing a given word pattern (A) to all word prototypes, which leads to a (scalar) distance measure between A and the prototypes. A simple (and good) mechanism assigns the class that contains the prototype with minimal distance to A. The number of word prototypes that we need depends on the complexity of separating the various classes. In particular for speech recognition tasks, good decision boundaries may be highly convoluted or equivalently, the number of required word prototypes is large. This is due to the large spectral as well as temporal variability within word classes, as a result of a number of factors, including speaking rate, gender, intentional context and many others.

Now let us develop a neural net classifier for this task. Roughly speaking, the number of prototypes is proportional to the number of hidden units in a neural net. In order to store all prototypes corresponding to various speaking rates, gender types, intonations, etcetera, requires a unrealistically large network.

As a general strategy, the number of prototypes (hidden nodes) can be drastically reduced if we do not store prototypes corresponding to various speaking rates, but compensate for speaking rate during recall[1]. The current mechanisms of choice for speaking rate compensation are dynamic time warping (DTW, a dynamic programming technique) and Hidden Markov Modelling (HMM, where speaking rate compensation involves the Viterbi algorithm, a stochastic dynamic programming technique). Several mixed DTW/NN ([3]) and HMM/NN models ([1]) have been proposed. In effect, these architectures reduce capacity

---

1. Recall refers to the recognition (retrieval) phase.

requirements for the neural net at the cost of additional computation during recall.

In this paper we propose a similar strategy, but the additional computation during recall is an integrated part of the neural network model. The idea is to create a (relatively small) network where speaking rate (or other sources of variability) compensation occurs by on-line adaptation of the network parameters. Next, we present a linear memory structure that allows for temporal re-alignment by on-line parameter adaptation. We also present an example where this memory structure is utilized in a neural net for a word recognition task.

2· The Time-alignment Filter

The time-alignment (TA) filter is a linear adaptive memory structure that generalizes the tapped delay line for storage of temporal patterns. Refer to Figure 1 for the following exposition. The filter outputs $y_k(t)$, k=1,...,K, hold memory traces of the input signal $u(t)$. Let us first consider the case where the filter parameters μ_k=0 for k=1,...,K. In this case, the outputs $y_k(t)$ equal the (internal) tap variables $x_k(t)$. The tap-to-tap transfer for $x_k(t)$ is given by

$$x_k(t) = \mu_0 x_k(t-1) + (1-\mu_0) x_{k-1}(t-1), \tag{1}$$

or $X_k(z) = ((1-\mu_0)/(z-\mu_0)) X_{k-1}(z)$ in the z-domain, where we defined $x_0(t) \equiv u(t)$. For $\mu_0 = 0$, this is just a tapped delay line. For $0 < \mu_0 < 1$ however, the tap-to-tap transfer becomes a leaky integrator with gain μ_0. The cascade of leaky integrators with equal gain μ_0 is also called a gamma delay line. While the memory depth of the regular tapped delay line is fixed (and equals K, the number of taps), the memory depth for the gamma delay line can be adapted by parameter μ_0. It is shown in de Vries ([2]), that the memory depth for a K-th order gamma delay line can be estimated by K/(1-μ_0). Thus, for increasing μ_0, the effective delay increases.

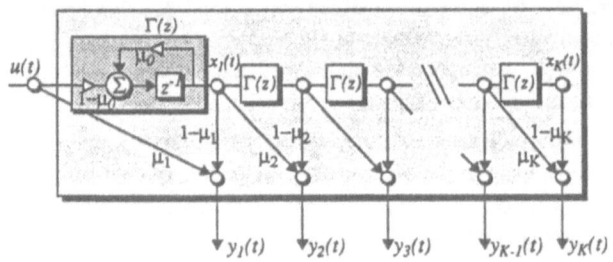

Figure 1 . The time-alignment memory filter.

The gamma delay line can be compared with a homogeneous rubber band that spans the delay domain. By stretching the band (increase μ_0), the depth of the memory increases, but as the band remains homogeneous, the distances between the taps increase proportionally and uniformly for all taps. For speech signals however, variations in speech rate create a necessity for adaptive *tap-dependent* resolution. Tap-dependent modulations of depth and resolution are implemented by variation of the parameters μ_i, i = 1,...,K. The connection pattern between $x_k(t)$ and $y_k(t)$ defines the boundaries of the tap-dependent modulations. A particular example which demonstrates the tap-dependent resolution is displayed in Figure 2. Note that the parametrization $\{\mu_0, \mu_1, \mu_2, \mu_3, \mu_4\} = \{0, 0, 1, 1, 0\}$ leads to a tapped delay line with unequal tap delays. This feature is very interesting when dealing with speech where sometimes signal values change slowly (vowels) and sometimes rapidly (consonants).

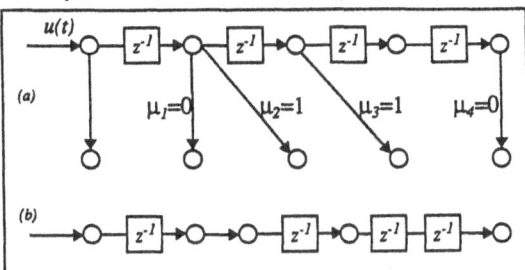

Figure 2 . Example of TA filter. Structure (a) is equal to structure (b).

3• Dealing with Temporal Variation

The problem with speech is that apart from spectral variation, a spatial pattern recognition problem, there is also a large amount of temporal variation for one word (class). Temporal variation occurs naturally and is due to changes in speaking rate and prosodics at various levels. In a word recognition problem, we distinguish between two kinds of temporal variation: the overall *duration* of a word may vary as well as the duration of various phonemes (vowels, consonants, etc., the *phonetic duration problem*) that constitute a word will vary significantly within a word class. To compensate for or nullify temporal variation in a speech recognition task is sometimes referred to as the *time-warp* problem or *time-alignment* problem.

Let us analyze why neural nets do not perform well in an environment with unpredictable temporal variation. In a neural net, we store in the weight vector w some kind of average of the words that are presented during training. Thus, the weights are tuned to the average speaking rate in the training set. It is important to realize that we can never store *unpredictable* aspects of speech (such as speaking rate). We conclude that independent of whether we use a recurrent or feedforward net, for constant weights, an additive net such as $x_i(t) = \sigma(\sum_{j,k} w_{ijk} x_j(t-k))$ cannot be invariant to the speaking rate. By invariance we mean that $x(t)$ would not depend on the speaking rate.

In order to achieve a speaking rate invariant network we need to augment the net with a speaking rate dependent (set of) variable(s) $m(t)$ that modulate(s) the states $x(t)$. $m(t)$ may modulate the states $x(t)$ directly or indirectly by modulating the weights w (which makes the weights time-varying, since $w=w(m(t))$). Note that $m(t)$ (and therefore $w(t)$) changes at least as fast as the rate of change of the speaking rate. An example of state modulation by a variable that measures the speaking rate was proposed by Sun et al. ([4]). This network is entirely invariant to the speaking rate.

Here we propose an alternative model. The idea is to let the parameters μ of the time-alignment filter be *fast adaptive* so as to compensate for time-varying speaking rate. There are *three* time-scales in this processing scheme. The slowest time scale relates to the adaptation of the weights w, that store the average word prototypes. We call w the *slow weights*. The input signal $u(t)$ and neural states $x(t)$ change faster. The time-alignment parameters $\mu(t)$ are adapted

444

at an even faster time scale than the states: At every time step t, we adapt $\mu(t)$ so as to maximize a performance measure (for the word recognition, maximize the net output $y(t)$). The parameters $\mu(t)$ are called *fast weights*.

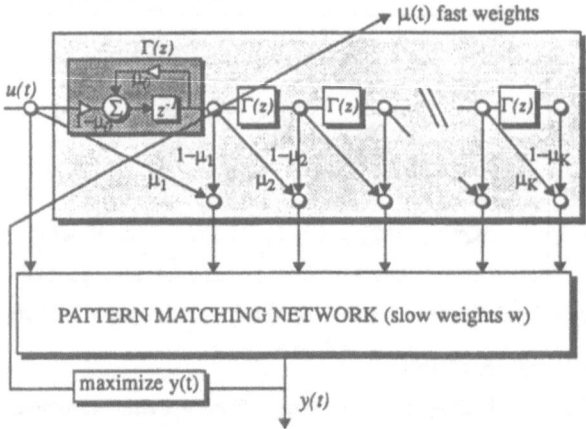

Figure 3 . A fast adaptive dynamic neural net architecture during recall for word recognition.

How this all fits together in a practical example is illustrated in Figure 3. The word recognizer consists of a cascade of a time-alignment memory filter and a (static) pattern matching (PM) network. The pattern matching network is non-linear and parametrized by slow weights w. The slow weights w are adapted only during the training phase. A layered feedforward net or a one-layer higher order net are good choices.

The speech spectral input $u(t)$ cannot be just windowed and fed to the PM net however. The unpredictable temporal variation will sharply degrade recognition performance. Thus as a pre-processing stage we insert a *fast adaptive* time-alignment network. At every time step t (or every few time steps), the network adapts $\mu(t)$ so as to maximize $y(t)$, in other words, to obtain the best possible temporal alignment of $u(t)$ parametrized by $\mu(t)$. It is assumed here that $y(t)$ holds a measure of fit for a particular word. This scheme is very similar to a dynamic time-warping (DTW) pre-processing stage. Such an approach, DTW followed by a pattern matching network, has been proposed and performs very promising ([3]). The scheme that we propose is different in that the time-warp stage is performed by an adaptive neural net itself. As a result, we do not need an dynamic programming pre- or post-processing algorithm. Instead, the time-warping is implemented by a gradient descent procedure such as backpropagation. There are several interesting features regarding this architecture. Note that the amount of temporal warp allowed is easily regulated by limiting the excursions of $\mu(t)$; for instance we can require $|\mu_i(t)| \leq \alpha < 1$ where α holds a warping bound. Also, the *structure* of the allowed time warps can be changed by altering the connection pattern between the gamma delay line and the filter outputs in the time-alignment memory filter.

4• Conclusions

In this paper we propose to use time-varying neural nets as a means of reducing network size. We illustrate the ideas in a speech recognition context, but they apply to a broader context.

In general, it is possible to trade (spatial) capacity requirements of the neural net for an additional computational load during recall. Time-varying weights provide a means of increasing the network capacity without increasing its size. The fundamental problem of course is *how* to vary the weights during recall. Here we optimize a recall performance index (maximize output), but other schemes are conceivable as well. For example, for predictable variability (speaking rate is unpredictable), it is possible to encode the time-dependency of the weights in the training phase.

Acknowledgments

This research was supported by the Advanced Projects Agency of the Department of Defense and was monitored by the Air Force Office of Scientific Research under Contract No. F49620-92-C-0072. The United States Government is authorized to reproduce and distribute reprints for governmental purposes notwithstanding any copyright notation hereon.

References

[1] Boulard H. and Wellekens C.J., Links between Markov models and multilayer perceptrons, *IEEE transactions on pattern analysis and machine intelligence 12-12*, 1167-1178, 1990.

[2] De Vries B., Temporal processing with neural networks--the development of the gamma model, Ph.D. Dissertation, University of Florida, 1991.

[3] Sakoe H., Isotani R., Yoshida K., Iso K., and Watanabe T., Speaker independent word recognition using dynamic programming neural networks, *Proc. ICASSP-89*, 29-33, 1989.

[4] Sun G.Z., Chen H.H., Lee Y.C. and Liu Y.D., Time warping recurrent neural networks and trajectory classification, *IJCNN proceedings*, Baltimore, MD, I-431-436, 1992.

A 'SELF-REFERENTIAL' WEIGHT MATRIX

J. Schmidhuber
Institut für Informatik
Technische Universität München
Arcisstr. 21, 8000 München 40, Germany

ABSTRACT. *Weight modifications in traditional neural nets are computed by hard-wired algorithms. Without exception, all previous weight change algorithms have many specific limitations. Is it (in principle) possible to overcome limitations of hard-wired algorithms by allowing neural nets to run and improve their own weight change algorithms? This paper constructively demonstrates that the answer (in principle) is 'yes'. I derive an initial gradient-based sequence learning algorithm for a 'self-referential' recurrent network that can 'speak' about its own weight matrix in terms of activations. It uses some of its input and output units for observing its own errors and for explicitly analyzing and modifying its own weight matrix, including those parts of the weight matrix responsible for analyzing and modifying the weight matrix. The result is the first 'introspective' neural net with explicit potential control over all of its own adaptive parameters. A disadvantage of the algorithm is its high computational complexity per time step which is independent of the sequence length and equals $O(n_{conn} log n_{conn})$, where n_{conn} is the number of connections. Another disadvantage is the high number of local minima of the unusually complex error surface. The purpose of this paper, however, is not to come up with the most efficient 'introspective' or 'self-referential' weight change algorithm, but to show that such algorithms are possible at all.*

1. INTRODUCTION

In contrast to traditional machine learning systems, humans do not appear to rely on hard-wired learning algorithms only. Instead, they tend to reflect about their own learning behavior and *modify* it and tailor it to the needs of various types of learning problems. To a degree, humans are able to learn how to learn. The thought experiment in this paper is intended to make a step towards 'self-referential' machine learning by showing the theoretical possibility of 'self-referential' neural networks whose weight matrices can learn to implement and improve their own weight change algorithm, without any significant theoretical limits.

Structure of the paper. Section 2 starts with a general finite, 'self-referential' architecture involving a sequence-processing recurrent neural-net (see e.g. Robinson and Fallside [2], Williams and Zipser [8], and Schmidhuber [3]) that can potentially implement any computable function that maps input sequences to output sequences — the only limitations being unavoidable time and storage constraints imposed by the architecture's finiteness. These constraints can be extended by simply adding storage and/or allowing for more processing time. The major novel aspect of the system is its 'self-referential' capability. The network is provided with special input units for explicitly observing performance evaluations (external error signals are visible through these special input units). In addition, it is provided with the basic tools for explicitly reading and quickly changing *all* of its own adaptive components (weights). This is achieved by (1) introducing an address for each connection of the network, (2) providing the network with output units for (sequentially) addressing *all* of its own connections (including those connections responsible for addressing connections) by means of time-varying activation patterns, (3) providing special input units whose activations become the weights of connections currently addressed by the network, and (4) providing special output units whose time-varying activations serve to quickly change the weights of connections addressed by the network. It is possible to show that these unconventional features allow the network (in principle) to compute any computable function mapping *algorithm components* (weights) and *performance evaluations* (e.g., error signals) to *algorithm modifications* (weight changes) – the only limitations again being unavoidable time and storage constraints. This implies that algorithms running on that architecture (in principle) *can change not only themselves but also the way they change themselves, and the way they change the way they change themselves, etc.*, essentially without theoretical limits.

Connections are addressed, analyzed, and manipulated with the help of differentiable functions of activation patterns across special output units. This allows the derivation of an *exact* gradient-based *initial* weight change algorithm for 'introspective' supervised sequence learning. The system starts out as *tabula rasa*. The initial weight change procedure serves to find improved weight change procedures – it favors algorithms (weight matrices) that make sensible use of the 'introspective' potential of the hard-wired architecture, where 'usefulness' is solely defined by conventional performance evaluations (the performance measure we use is the sum of all error signals over all time steps of all training sequences).

A disadvantage of the algorithm is its high computational complexity per time step which is independent of the sequence length and equals $O(n_{conn} log n_{conn})$, where n_{conn} is the number of connections. Another disadvantage is the high number of local minima of the unusually complex error surface. The purpose of this paper, however, is not to come up with the most efficient 'introspective' or 'self-referential' weight change algorithm, but to show that such algorithms are possible at all.

2. THE 'INTROSPECTIVE' NETWORK

Throughout the remainder of this paper, to save indices, I consider a single limited pre-specified time-interval of discrete time-steps during which our network interacts with its environment. An interaction sequence actually may be the concatenation of many 'conventional' training sequences for conventional recurrent networks. This will (in theory) help our 'self-referential' weight matrix to find regularities among solutions for *different* tasks. The network's output vector at time t, $o(t)$, is computed from previous input vectors $x(\tau), \tau < t$, by a discrete time recurrent network with n_I input units and n_y non-input units. A subset of the non-input units, the 'normal' output units, has a cardinality of $n_o < n_y$.

z_k is the k-th unit in the network. y_k is the k-th non-input unit in the network. x_k is the k-th 'normal' input unit in the network. o_k is the k-th 'normal' output unit. If u stands for a unit, then f_u is its differentiable activation function and u's activation at time t is denoted by $u(t)$. If $v(t)$ stands for a vector, then $v_k(t)$ is the k-th component of $v(t)$.

Each input unit has a directed connection to each non-input unit. Each non-input unit has a directed connection to each non-input unit. There are $(n_I + n_y)n_y = n_{conn}$ connections in the network. The connection from unit j to unit i is denoted by w_{ij}. For instance, one of the names of the connection from the j-th 'normal' input unit to the the k-th 'normal' output unit is $w_{o_k x_j}$. w_{ij}'s real-valued weight at time t is denoted by $w_{ij}(t)$. Before training, all weights $w_{ij}(1)$ are randomly initialized.

The following features are needed to obtain 'self-reference'. *Details of the network dynamics follow in the next section.*

1. The network receives performance information through the *eval units*, which are special input units. $eval_k$ is the k-th eval unit (of n_{eval} such units) in the network.

2. Each connection of the net gets an address. One way of doing this is to introduce a *binary* address, $adr(w_{ij})$, for each connection w_{ij}. This will help the network to do computations concerning its own *weights* in terms of *activations*, as will be seen later.

3. ana_k is the k-th *analyzing unit* (of $n_{ana} = ceil(log_2 n_{conn})$ such units, where $ceil(x)$ returns the first integer $\geq x$). The analyzing units are special non-input units. They serve

to indicate which connections the current algorithm of the network (defined by the current weight matrix plus the current activations) will access next (see next section). A special input unit for reading current weight values that is used in conjunction with the analyzing units is called *val*.

4. The network may modify any of its weights. Some non-input units that are not 'normal' output units or analyzing units are called the *modifying units*. mod_k is the k-th modifying unit (of $n_{mod} = ceil(log_2 n_{conn})$ such units). The modifying units serve to address connections to be modified. A special output unit for modifying weights (used in conjunction with the modifying units, see next section) is called $\triangle$. $f_\triangle$ should allow both positive and negative activations $\triangle(t)$.

2.1. 'SELF-REFERENTIAL' DYNAMICS AND OBJECTIVE FUNCTION

I assume that the input sequence observed by the network has length $n_{time} = n_s n_r$ (where $n_s, n_r \in \mathbf{N}$) and can be divided into n_s equal-sized blocks of length n_r during which the input pattern $x(t)$ does not change. This does not imply a loss of generality — it just means speeding up the network's hardware such that each input pattern is presented for n_r time-steps before the next pattern can be observed. This gives the architecture n_r time-steps to do some sequential processing (including immediate weight changes) before seeing a new pattern of the input sequence.

In what follows, *unquantized variables are assumed to take on their maximal range*. The network dynamics are specified as follows:

$$net_{y_k}(1) = 0, \quad \forall t \geq 1: \quad x_k(t) \leftarrow environment, \quad y_k(t) = f_{y_k}(net_{y_k}(t)),$$

$$\forall t > 1: \quad net_{y_k}(t) = \sum_l w_{y_k l}(t-1) l(t-1), \tag{1}$$

The network can quickly read information about its current weights into the special *val* input unit according to

$$val(1) = 0, \quad \forall t \geq 1: \quad val(t+1) = \sum_{i,j} g[\|ana(t) - adr(w_{ij})\|^2] w_{ij}(t), \tag{2}$$

where $\|\ldots\|$ denotes Euclidean length, and g is a differentiable function emitting values between 0 and 1 that determines how close a connection address has to be to the activations of the analyzing units in order for its weight to contribute to *val* at that time. Such a function g might have a narrow peak at 1 around the origin and be zero (or nearly zero) everywhere else. This essentially allows the network to pick out a single connection at a time and obtain its current weight value without receiving 'cross-talk' from other weights.

The network can quickly modify its current weights using $mod(t)$ and $\triangle(t)$ according to

$$\forall t \geq 1: \quad w_{ij}(t+1) = w_{ij}(t) + \triangle(t) \, g[\, \|adr(w_{ij}) - mod(t)\|^2 \,]. \tag{3}$$

Again, if g has a narrow peak at 1 around the origin and is zero (or nearly zero) everywhere else, the network will be able to pick out a single connection at a time and change its weight without affecting other weights.

Objective function and dynamics of the eval units. As with typical supervised sequence-learning tasks, we want to minimize

$$E^{total}(n_r n_s), \quad where \quad E^{total}(t) = \sum_{\tau=1}^{t} E(\tau), \quad where \quad E(t) = \frac{1}{2} \sum_k (eval_k(t+1))^2,$$

where

$$eval_k(1) = 0, \quad \forall t \geq 1 : eval_k(t+1) = d_k(t) - o_k(t) \text{ if } d_k(t) \text{ exists, and } 0 \text{ else.} \quad (4)$$

Here $d_k(t)$ may be a desired target value for the k-th output unit at time step t.

3. INITIAL LEARNING ALGORITHM

The following algorithm[1] for minimizing E^{total} is partly inspired by (but more complex than) conventional recurrent network algorithms (e.g. Robinson and Fallside [2]).

Derivation of the algorithm. We use the chain rule to compute weight increments (to be performed *after* each training sequence) for all *initial* weights $w_{ab}(1)$ according to

$$w_{ab}(1) \leftarrow w_{ab}(1) - \eta \frac{\partial E^{total}(n_\tau n_s)}{\partial w_{ab}(1)}, \quad (5)$$

where η is a constant positive 'learning rate'. Thus we obtain an *exact* gradient-based algorithm for minimizing E^{total} under the 'self-referential' dynamics given by (1)-(4). To reduce writing effort, I introduce some short-hand notation partly inspired by Williams [7]. For all units u and all weights w_{ab}, w_{ij} we write

$$p_{ab}^u(t) = \frac{\partial u(t)}{\partial w_{ab}(1)}, \quad q_{ab}^{ij}(t) = \frac{\partial w_{ij}(t)}{\partial w_{ab}(1)}. \quad (6)$$

To begin with, note that

$$\frac{\partial E^{total}(1)}{\partial w_{ab}(1)} = 0, \quad \forall t > 1 : \frac{\partial E^{total}(t)}{\partial w_{ab}(1)} = \frac{\partial E^{total}(t-1)}{\partial w_{ab}(1)} - \sum_k eval_k(t+1)p_{ab}^{o_k}(t). \quad (7)$$

Therefore, the remaining problem is to compute the $p_{ab}^{o_k}(t)$, which can be done by incrementally computing all $p_{ab}^{z_k}(t)$ and $q_{ab}^{ij}(t)$, as we will see. We have

$$p_{ab}^{z_k}(1) = 0; \quad p_{ab}^{z_k}(t+1) = 0; \quad p_{ab}^{o_k}(t+1) = -p_{ab}^{o_k}(t), \text{ if } d_k(t) \text{ exists, and } 0 \text{ otherwise, } (8)$$

$$p_{ab}^{val}(t+1) = \sum_{i,j} \{ \ q_{ab}^{ij}(t)g[\|ana(t) - adr(w_{ij})\|^2)] \ + w_{ij}(t) [\ g'(\|ana(t) - adr(w_{ij})\|^2) \times$$

$$\times 2 \sum_m (ana_m(t) - adr_m(w_{ij}))p_{ab}^{ana_m}(t) \ ] \ \} \quad (9)$$

(where $adr_m(w_{ij})$ is the m-th bit of w_{ij}'s address),

$$p_{ab}^{y_k}(t+1) = f'_{y_k}(net_{y_k}(t+1)) \sum_l w_{y_k l}(t)p_{ab}^l(t) + l(t)q_{ab}^{y_k l}(t), \quad (10)$$

where

$$q_{ab}^{ij}(1) = 1 \text{ if } w_{ab} = w_{ij}, \text{ and } 0 \text{ otherwise,} \quad (11)$$

$$\forall t > 1 : \ q_{ab}^{ij}(t) = q_{ab}^{ij}(t-1) + p_{ab}^{\triangle}(t-1)g(\|mod(t-1) - adr(w_{ij})\|^2) +$$

$$+2 \triangle (t-1) g'(\|mod(t-1) - adr(w_{ij})\|^2) \times \sum_m [mod_m(t-1) - adr_m(w_{ij})]p_{ab}^{mod_m}(t-1). \quad (12)$$

According to (8)-(12), the $p_{ab}^j(t)$ and $q_{ab}^{ij}(t)$ can be updated incrementally at each time step. This implies that (5) can be updated incrementally at each time step, too. The storage complexity is independent of the sequence length and equals $O(n_{conn}^2)$. The computational complexity per time step (of sequences with arbitrary length) is $O(n_{conn}^2 log n_{conn})$.

---

[1]It should be noted that in quite different contexts, previous papers have shown how one net may learn to perform appropriate lasting weight changes for a second net [4] [1]. However, these previous approaches could not be called *'self-referential'* — they all involve at least some weights that can *not* be manipulated other than by conventional gradient descent.

4. CONCLUSION

The thought experiment presented in this paper is intended to show the theoretical possibility of certain kinds of 'self-referential' weight matrices. The network I have described can, besides learning to solve problems posed by the environment, also use its own weights as input data and can (in principle) learn new algorithms for modifying its weights in response to the environmental input and evaluations. This effectively embeds a chain of *'meta-networks'* and *'meta-meta-...-networks'* into the network itself.

5. ACKNOWLEDGEMENTS

Thanks to Mark Ring, Mike Mozer, Daniel Prelinger, Don Mathis, and Bruce Tesar, for helpful comments. Parts of this paper are based on previous publications [6][5]. This research was supported in part by a DFG fellowship to the author, as well as by NSF award IRI–9058450, grant 90–21 from the James S. McDonnell Foundation, and DEC external research grant 1250.

References

[1] K. Möller and S. Thrun. Task modularization by network modulation. In J. Rault, editor, *Proceedings of Neuro-Nimes '90*, pages 419–432, November 1990.

[2] A. J. Robinson and F. Fallside. The utility driven dynamic error propagation network. Technical Report CUED/F-INFENG/TR.1, Cambridge University Engineering Department, 1987.

[3] J. H. Schmidhuber. A fixed size storage $O(n^3)$ time complexity learning algorithm for fully recurrent continually running networks. *Neural Computation*, 4(2):243–248, 1992.

[4] J. H. Schmidhuber. Learning to control fast-weight memories: An alternative to recurrent nets. *Neural Computation*, 4(1):131–139, 1992.

[5] J. H. Schmidhuber. An introspective network that can learn to run its own weight change algorithm. In *Proc. of the Third International Conference on Artificial Neural Networks, Brighton*. IEE, 1993. Accepted for publication.

[6] J. H. Schmidhuber. A neural network that embeds its own meta-levels. In *Proc. of the International Conference on Neural Networks '93, San Francisco*. IEEE, 1993. Accepted for publication.

[7] R. J. Williams. Complexity of exact gradient computation algorithms for recurrent neural networks. Technical Report Technical Report NU-CCS-89-27, Boston: Northeastern University, College of Computer Science, 1989.

[8] R. J. Williams and D. Zipser. A learning algorithm for continually running fully recurrent networks. *Neural Computation*, 1(2):270–280, 1989.

Novel architectures and learning rules—poster contributions

Neural Network Complexity Reduction
Using Adaptive Polynomial Activation Functions

F. Piazza, A. Uncini, M. Zenobi

Dip. di Elettronica ed Automatica, Università di Ancona
Via Brecce Bianche - 60131 Ancona, Italy
e-mail: upfm@eealab.cineca.it

Abstract

In this paper, a new neural network structure is proposed that presents a low complexity and is particularly suitable for digital hardware realization. The proposed structure is based on the "polynomial neuron", a classical additive neuron but with a polynomial activation function. Such a neuron is then used to build multilayer networks trainable with an algorithm very similar to the Back Propagation. The adaptive polynomial neural networks (APNN) allow a good reduction in terms of dimensions and computational complexity both in learning and in forward phase compared with traditional MLPs using sigmoidal activation functions. Many experiments have been extensively carried out both on pattern recognition and data processing problems. The relationship of the APNNs with the polynomial Adaline and the Volterra expansion is also discussed.

Introduction

In the last years, the artificial neural networks have been successfully applied to the solution of many real problems of pattern recognition and data processing. In both kinds of applications, the structural and computation complexity of the involved networks is one of the major hindrance to a broad spreading of the neural techniques.

A first way to reduce the complexity of the traditional approaches is to design particular implementations of common neural networks which greatly decrease the computational burden. It is well known that the Multilayer Perceptron (MLP) [1] is one of the most popular neural network (NN) models, together with the Back Propagation (BP) learning algorithm that is almost universally used for its training. A good example of reducing the complexity of a digital MLP, is reported in [4], where the complexity is reduced by constraining the synaptic weights to be simple powers of two.

A second way to reduce the complexity of the traditional approaches, is to introduce different and powerful neural paradigms, able to get performance comparable with those obtained by the traditional models, but with smaller and/or simpler networks. For the MLP case, it is well known that digital architectures have received increasingly attention in these years, due to the immediate availability of the VLSI technology for their design and implementation. In such architectures, as in many other cases of digital realization of neural networks, the activation function of the neurons is implemented by the use of a look-up table. Thus it could be feasible to choose suitable (in general more complex) activation functions which allow to solve a given problem with a smaller network, hence reducing the complexity and the implementation costs. Extensive studies have been carried out on the capabilities and behaviours of the neuron activation functions in a MLP environment [see for ex. 5].

In this paper, a new neural network structure is presented that has a low complexity and is particularly suitable for digital hardware realization, especially in VLSI devices. The proposed structure is based on the "polynomial neuron", a classical additive neuron but with a polynomial activation function. Such a neuron is then used to build multilayer networks trainable with an algorithm very similar to the Back Propagation. These modified MLPs (with polynomial neurons) do not exhibit more complexity in digital implementations than the classical MLPs, since the activation functions are realized with look-up tables, nevertheless they are usually smaller in terms of layers and neurons per layer than the classical MLPs on a large class of problems. It can be shown that the proposed networks are equivalent to polynomial Adaline [2,3] in which some relationships exist among the polynomial coefficients; therefore they implement a truncated and constrained Volterra series expansion.

The Proposed Architecture

The scheme of the architecture proposed in this paper is presented in Fig. 1.

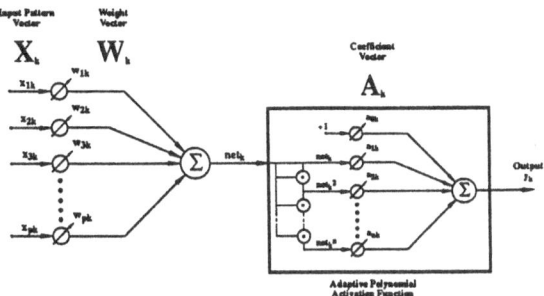

Fig. 1. A processing unit of APNN.

In order to analyze the behavior of the proposed structure, let's examinate a single neuron with two inputs and a 2nd degree polynomial nonlinearity. The two inputs, x_1 and x_2, multiplied by the respective weights, w_1 and w_2, and then summed together, become the argument of the nonlinearity, a quadratic polynomial with coefficients a_0, a_1 and a_2:

$$f\left[(w_1x_1 + w_2x_2)\right] = a_0 + a_1(w_1x_1 + w_2x_2) + a_2(w_1x_1 + w_2x_2)^2 =$$

$$= a_0 + a_1w_1x_1 + a_1w_2x_2 + 2a_2w_1w_2x_1x_2 + a_2w_1^2x_1^2 + a_2w_2^2x_2^2$$

after doing the following positions:

$$\alpha_0 = a_0; \quad \alpha_1 = a_1w_1; \quad \alpha_2 = a_1w_2; \quad \alpha_3 = 2a_2w_1w_2; \quad \alpha_4 = a_2w_1^2; \quad \alpha_5 = a_2w_2^2;$$

we obtain the expression:

$$\alpha_0 + \alpha_1x_1 + \alpha_2x_2 + \alpha_3x_1x_2 + \alpha_4x_1^2 + \alpha_5x_2^2$$

which is similar to the input-output relationship established by a second-order polynomial network [2], [3]. As it is well known [3], the polynomial Adaline or Padaline [2] can be trained to realize an approximate Taylor's series expansion of the Bayesian discriminant function, thereby implementing the optimum Bayes classifier. The particular structure of the polynomial Adaline has also a strict relationship with the Volterra series expansion of a non-linear functional. For data processing purpose, in fact, it can be successfully employed to implement approximate realization of non-linear systems with memory. In this case, the Padaline is characterised by a system order, i.e. the polynomial degree, and a memory, both tied to the number of the Volterra kernels associated with the truncated series. The structure proposed in this paper, however, does not fully realize the polynomial Adaline input-output relationship, since several constraints exist among the α_k coefficients of the expansion. In fact the polinomial free parameters are the a_k coefficients, which are much less in number than the α_k. This allows to greatly reduce the complexity of the network with respect to the original polynomial Adaline when the order and the memory of the truncated Volterra series are high. On the other hand, the presence of such constraints could reduce the computational capabilities of the resulting structure. In fact the nonlinear link between the coefficients of the Volterra expansion and the parameters of the network does not allow to obtain analytically the optimal $\mathbf{A}$ and $\mathbf{W}$ vectors. Anyway, experimental results point out that the BP algorithm, even if it cannot guarantee the optimal solution, converges to an adequate solution with an extremely reduced variance with respect to MLPs. Moreover, since the network performance depend often on the number of overall free parameters, APNNs exibit a remarkable save in terms of complexity, when digital implementations are involved, requiring a considerably smaller number of connections, as it can be seen from Fig. 2.

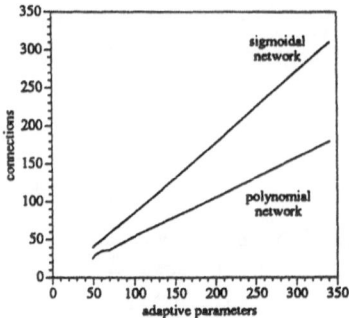

Fig. 2. No. of synapses (connections) vs no. of adaptive parameters for the network used in the experiments.

Learning Algorithm

The proposed algorithm is an extension of the traditional Back-Propagation procedure. At each step, the updating of both the synaptic weights and the coefficients of the polynomial activation function is performed, according to the rule:

$$W_{k+1} = W_k + \Delta W_k; \qquad A_{k+1} = A_k + \Delta A_k;$$

where W_k and A_k are respectively the weight vector and the coefficient vector. The objective of the procedure is to minimize the MSE (mean-square error), which in this case is a function not only of the weights, but also of the coefficients of the nonlinearity. At time index k we have:

$$\varepsilon = (d - y);$$

with d as the desired output and y as the output of the network:

$$y = f(a, net) = \sum_{i=0}^{n} a_i \, net^i; \qquad net = X^T W$$

The minimization of MSE with the steepest-descent method, performed altering the weights and coefficients vectors in the direction corresponding to the negative of the measured gradient:

$$\nabla(\varepsilon^2) = [\frac{\partial \varepsilon^2}{\partial W}, \frac{\partial \varepsilon^2}{\partial A}]^T = [\frac{\partial \varepsilon^2}{\partial w_1}, \frac{\partial \varepsilon^2}{\partial w_2}, \cdots \frac{\partial \varepsilon^2}{\partial w_p}, \frac{\partial \varepsilon^2}{\partial a_0}, \frac{\partial \varepsilon^2}{\partial a_1}, \cdots \frac{\partial \varepsilon^2}{\partial a_n}]^T;$$

where

$$\frac{\partial \varepsilon^2}{\partial w} = -2\varepsilon [\frac{\partial f}{\partial net}] x; \qquad \frac{\partial \varepsilon^2}{\partial a_j} = -2\varepsilon [\frac{\partial f}{\partial a_j}];$$

with

$$\frac{\partial f}{\partial net} = \sum_{i=1}^{n} i \, a_i \, net^{(i-1)}; \qquad \frac{\partial f}{\partial a_j} = net^j.$$

Finally we obtain:

$$\Delta W = 2\eta\varepsilon [\frac{\partial f}{\partial net}] X; \qquad \Delta A = 2\eta\varepsilon [\frac{\partial f}{\partial a_0}, \frac{\partial f}{\partial a_1}, \frac{\partial f}{\partial a_2}, \cdots \frac{\partial f}{\partial a_n}]^T;$$

where η is the learning rate. In some cases, to optimize the convergence time, two different learning rates have been used, one for weights and another one, smaller, for the polynomial coefficients.

Experimental Results

In order to estimate the performances of neural networks with an adaptive polynomial activation function with respect to the traditional MLPs, several problems available in literature have been tested.

One is the example number 4 proposed by Narendra and Parthasarathy [6] and even used by Specht [7], and concerns the identification of a nonlinear system in the form:

$$y_p(k+1) = f\ [\ y_p(k), y_p(k-1), y_p(k-2), u(k), u(k-1)\ ].$$

Using first the network proposed in [6] made of 20 neurons in the first hidden layer and 10 neurons in the second hidden layer, the system identification has been carried out for 100 times with an input pattern of 10,000 samples of noise uniformly distributed within the interval [-1.2, 1.2]. The procedure has been repeated for a network with adaptive polynomial activation function, made of 30 neurons with a 4th degree polynomial and 1 output neuron with a 2nd degree polynomial. This configuration has been chosen so that the two networks approximatively have the same amount of adaptive parameters. The typical plot of the Mean Square Error (MSE) for the two networks is shown in Fig. 3.

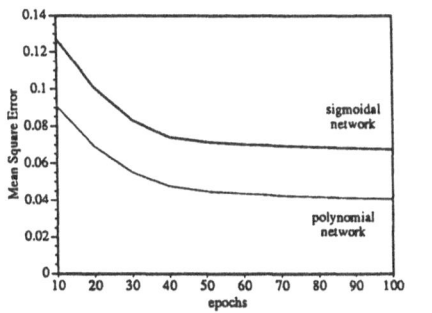
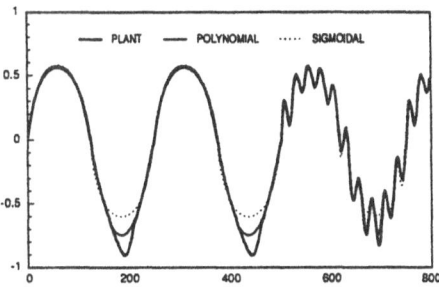

Fig. 3.   (a) Plot of MSE during the identification procedure;
(b) after training, outputs of the plant, the polynomial and the sigmoidal networks, when the input is the test signal of [6].

The procedure has been applied to other networks with several configurations and consequently with a different number of adaptive parameters. The result is that, at a parity of performances, neural networks with adaptive polynomial nonlinearity require a considerably smaller number of connections, as pointed out before.

APNNs have shown an excellent behavior on problems of pattern recognition and data processing. Their results are comparable with those obtained by classical MLPs, but a significantly smaller amount of adaptive parameters is needed. With respect to the known polynomial structures, for which a certain knowledge of the characteristics of the discriminant function is required, APNNs don't need any particular a priori information.

References

[1] D. E. Rumelhart, J. L. McClelland, and the PDP Research Group, Parallel Distributed Processing (PDP): Exploration in the Microstructure of Cognition (Vol. 1), MIT Press, Cambridge (MA), 1986.
[2] M. Caudill, "The polynomial Adaline algorithm," Comput. Lang., Dec 1988.
[3] D. F. Specht, "Probabilistic neural networks and the polynomial Adaline as complementary techniques for classification", IEEE Trans. on Neural Networks, Vol. 1 No. 2, March 1990.
[4] M. Marchesi, G. Orlandi, F. Piazza, A. Uncini: "Fast neural networks without multipliers", IEEE Trans. on Neural Networks, Vol. 4 No. 1, January 1993.
[5] K. Hornik, M. Stinchcombe, H. White, "Multilayer feedforward networks are universal approximators", Neural Networks, Vol. 2, pp. 359-366, 1989.
[6] K. S. Narendra, K. Parthasarathy, "Identification and control of dynamical systems using neural networks," IEEE Trans. on Neural Networks, Vol. 1, No. 1, March 1990.
[7] D. F. Specht, "A general regression neural network", IEEE Trans. on Neural Networks, Vol. 2, No. 6, November 1991.

FIELDNET, A Dynamic Network For Pattern Classification

A.J.M. Russel   Th.E. Schouten

University of Nijmegen, Faculty of Mathematics and Computer Science, Department of Real-Time Systems, Toernooiveld 1, 6525 ED Nijmegen. e-mail: albertr@cs.kun.nl, ths@cs.kun.nl.

1 Abstract

A new kind of neural network (FIELDNET) is introduced. It is trained with supervised data and grows its single hidden layer (also called its codebook) during learning. Each hidden neuron belongs to an output class and has a kind of force field around it, which determines how well an input pattern belongs to its class. Only one learning parameter is present. The performance of FIELDNET is compared with three other neural networks on three real-world data sets. It will be shown that FIELDNET is capable of very fast learning while achieving high classification rates.

2 Introduction

The advantage of using neural networks for classification tasks is that no explicit knowledge about the underlying classification problem is necessary. It is sufficient to present the network a learn set of examples from which it will extract a mapping that classifies that learn set as good as possible. Depending on the information content of the learn set and the neural network model used, the trained network will be able to classify with a certain correctness new patterns that were not present in the learn set. FIELDNET is able to find a mapping that classifies any learn set 100% correctly. It does this by building a codebook, which is a subset of the learn set, during the training of the network. Only one learning parameter needs to be adjusted to obtain the best performance on the test data. The training method is very fast compared to backprop networks, faster than Kohonen and has about the same speed as ARTMAP networks. First the network architecture is described followed by the learning algorithm. Then the used real-world data sets are given followed by the neural network implementations used for comparison. Obtained results are given and discussed.

3 Network architecture

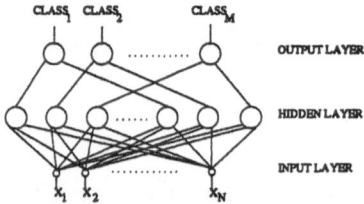

Figure 1: FIELDNET architecture

The network consists of three layers (figure 1). The input layer only distributes the input pattern $(x_1, \ldots, x_N)$. A neuron in the hidden layer (the codebook) represents a selected training pattern of a certain class, therefore that class is explicitly associated with that neuron. Each hidden neuron is fully connected with the input layer and has a weightvector $(w_1, \ldots, w_N)$. The output Y of a hidden neuron i is calculated as: $Y_i = F(\sum_{j=1}^{N} |w_{ij} - x_j| + \delta)$ with $F(x) = x^{-decayrate}$. The infinitesimal constant δ is necessary in order to prevent having to calculate the nonexistent value of $F(0)$ if the input pattern is identical to the weight vector. The hidden neuron calculates a kind of force that depends on the norm between its weight vector and the input vector. The decay rate controls the decrease of the force field and is the only variable in the model. The output layer consists of neurons that act as summation devices. For each class there is one classification neuron which is only connected to the neurons in the hidden layer that represent the same class. If the hidden layer consists of K neurons, the output O of a classification neuron i is calculated by:

$$O_i = \sum_{j=1}^{K} a_{ij} Y_j \text{ with } a_{ij} = \begin{cases} 1 & \text{if class}_{\text{output neuron}} = \text{class}_{\text{hidden neuron}} \\ 0 & \text{otherwise} \end{cases}$$

The classification neuron with the highest output designates the class which the network associates with the input pattern.

4 Learning algorithm

Before the network has learned anything, it only consists of the input layer and the output layer. The codebook (hidden layer) will be formed in the process of training with the learn set. This training occurs in so called epochs, in which every pattern from the learn set is presented exactly once to the network. The stimulation with such a pattern results in the calculation of the output values of the classification neurons and thereby in the classification of the pattern by the network. If the classification is right nothing is changed. If the classification is wrong, a neuron is added to the codebook with its weight vector identical to the input pattern and its class attribute identical to the input pattern class. As soon as a pattern is added to the codebook it is excluded from the learn set. Training is completed if the codebook hasn't increased in size during an epoch. The learn set will then be classified 100% correctly by the network, with exception of ambiguous patterns that are identical but belong to different classes.

5 Data sets

Three publically available real-world data sets for classification problems have been selected and divided into learn and test sets. The first data set [1, 2] (indicated by LET) consists of 26 capital letter classes taken from 20 different fonts, each letter was randomly distorted to produce 20000 patterns in total. For each pattern 16 primitive numerical attributes (integer range 0 to 15) were extracted from a black and white scan. The first 16000 patterns are used for training, the remaining 4000 for testing. In [2] maximal 82.7% of the test patterns was classified correctly using a Holland-style adaptive classifier trained on the training set. The second data set [1, 3] (indicated by CAN) is obtained from a clinical study of breast cancer. There are 2 classes (benign and malignant) and there are 9 numerical attributes (integer range 1-10) per pattern. The reported studies [1] obtain a performance of 92.2% to 95.9% on the selected test set, ranging from 33% to 50% of the patterns. For this study 16 patterns with a missing attribute are removed from the total set of 699 patterns. Every third pattern of each class is placed in the test set, the remaining patterns form the training set. The third data set [4] (indicated by DIG) consists of handwritten digits from 49 writers, spacially normalized to 32 by 32 binary pixels. In [5] a correct classification of 92.1% was given using 27 selected writers for training and 22 writers for testing. For writer independent recognition the first 34 writers are used for training and the remaining 15 writers for testing.

6 Neural networks used for comparison

Three different kinds of neural networks are used to compare FIELDNET with: backpropagation [6] (indicated by NBP), Kohonen [7] (indicated by NKOH) and ARTMAP [9, 10] (indicated by NART). NBP [11] contains many additions to the standard backprop method and has been implemented on a parallel transputer system. For all the above data sets various architectures were tried with many parameter settings (which were changed by hand during the learning) to obtain the best results. NKOH [12, 8] uses the Ritter and Schulten method for learning and NART [13] uses Fuzzy-ART/ART2a as basis for the ARTMAP implementation. For NKOH and NART less tests were done to obtain the reported results.

7 Results

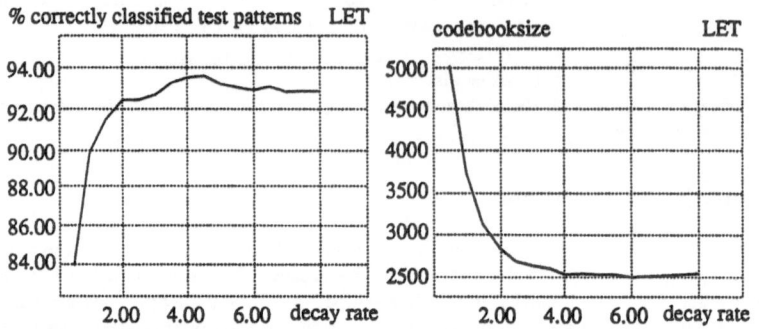

Figure 2: LET data, test set classification and codebook size as function of the decay rate.

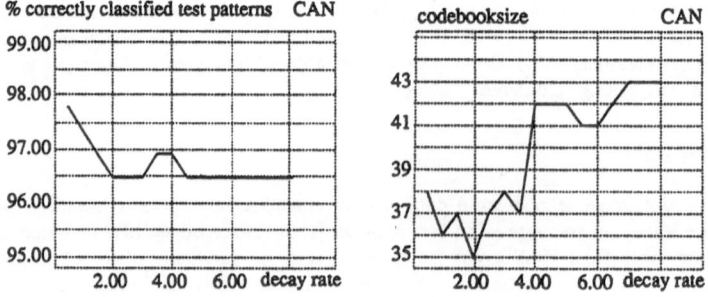

Figure 3: CAN data, test set classification and codebook size as function of the decay rate.

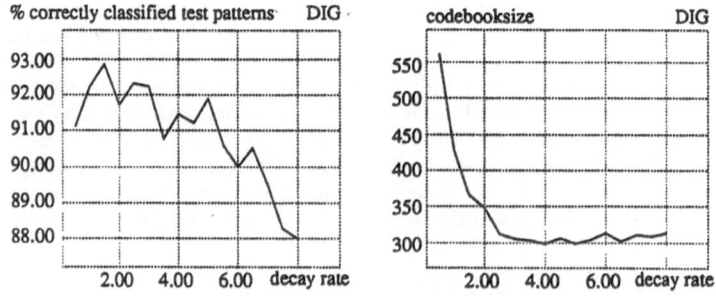

Figure 4: DIG data, test set classification and codebook size as function of the decay rate.

The percentage correctly classified patterns of the test set and the size of the codebook (the number of hidden neurons) are given as function of the decay rate in figures 2 to 4 for the three data sets.

The decay rates giving maximum performance on the test set are 0.5, 4.5 and 1.5 respectively. The codebook sizes are 8%, 16% resp. 16% of the learn set sizes. In table 1 the percentage of correctly classified test patterns for the three data sets is given for the four neural networks. The learning times needed by FIELDNET,NART,NKOH and NBP roughly relate as 1 : 1 : 100 : 10000.

	LET	CAN	DIG
FIELDNET	93.4%	97.5%	92.7%
NBP (Backpropagation)	94.8%	96.9%	95.3%
NKOH (Kohonen)	71.9%	97.8%	84.9%
NART (ARTMAP)	95.8%	97.8%	85.7%
reported in literature	82.7%	95.9%	92.1%

Table 1: test set classification

8 Conclusions

FIELDNET is a simple dynamic network with a fast learning method and a good performance on the used data sets. An advantage compared with other network models is that only one parameter needs to be adjusted to obtain the best results on the test sets.

References

[1] Murphy, P.M., Aha D.W. "UCI repository of machine learning databases", Irvine, University of California, Department of Information and Computer Science (anonymous ftp to ics.uci.edu in pub/machine-learning-database)

[2] Frey P.W., Slate D.J. "Letter Recognition Using Holland-style Adaptive Classifiers" Machine Learning Vol 6, 1991

[3] Mangasarian O.L., Wolberg W.H. "Cancer diagnosis via linear programming" SIAM News, vol 23, pp 1-18, 1990

[4] National Institute of Standards and Technology, "fl3" subset of the NIST Special Database 1, obtainable from sequoyah.ncsl.nist.gov in pub/data

[5] Garris M.D., Wilkinson R.A., Wilson C.L. "Methods for Enhancing Neural Network Handwritten Character Recognition", International Joint Conference on Neural Networks, Volume 1, IEEE Seattle, July 1991

[6] Rumelhart et al. Parallel Distributed Programming (MIT Press 1986)

[7] Kohonen Self-Organization and Associative Memory (Springer-Verlag, 1988)

[8] Ritter H.J, Martinetz Th.M., Schulten K.J. "Topology-Conserving Maps for Learning Visuo-Motor-Coordination", Neural Networks 2 (1989)

[9] Carpenter G.A., Grossberg S., Reynolds J.H. "ARTMAP Supervised real-time learning and classification of nonstationary data by a self-organizing neural network", Neural Networks 4 (1991)

[10] Carpenter G.A., Grossberg S., Rosen D.B., "FuzzyART: Fast Stable Learning and Categorization of Analog Patterns of an Adaptive Resonance System", Neural Networks 4 (1991)

[11] Th. E. Schouten, Internal Communication, University of Nijmegen

[12] L. G. Vuurpijl, Internal Communication, University of Nijmegen

[13] P. H. Willems, "Parallel implementation and evaluation of the ART Neural Networks", Master thesis in preparation, University of Nijmegen

REDUCING THE RATIO BETWEEN LEARNING COMPLEXITY AND NUMBER OF TIME VARYING VARIABLES IN FULLY RECURRENT NETS

J. Schmidhuber
Institut für Informatik
Technische Universität München
Arcisstr. 21, 8000 München 40, Germany

ABSTRACT. *Let m be the number of time-varying variables for storing temporal events in a fully recurrent sequence processing network. Let R_{time} be the ratio between the number of operations per time step (for an exact gradient based supervised sequence learning algorithm), and m. Let R_{space} be the ratio between the maximum number of storage cells necessary for learning arbitrary sequences, and m. With conventional recurrent nets, m equals the number of units. With the popular 'real time recurrent learning algorithm' (RTRL), $R_{time} = O(m^3)$ and $R_{space} = O(m^2)$. With 'back-propagation through time' (BPTT), $R_{time} = O(m)$ (much better than with RTRL) and R_{space} is infinite (much worse than with RTRL). The contribution of this paper is a novel fully recurrent network and a corresponding* exact *gradient based learning algorithm with $R_{time} = O(m)$ (as good as with BPTT) and $R_{space} = O(m^2)$ (as good as with RTRL).*

1 INTRODUCTION

Architecture. The basic architecture considered in this paper is the one of a traditional fully recurrent sequence processing network. The network has n non-input units and $n_x = O(n)$ input units. Each input unit has a directed connection to each non-input unit. Each non-input unit has a directed connection to each non-input unit. Obviously there are $(n_x + n)n = n_{conn} = O(n^2)$ connections in the network. The k-th input unit is denoted by x_k. The k-th non-input unit is denoted by y_k. The k-th output unit is denoted by o_k (the n_o output units are a subset of the non-input units). The connection from unit j to unit i is denoted by w_{ij}. For instance, one of the names of the the connection from the j-th input unit to the the k-th output unit is $w_{o_k x_j}$.

Dynamics of conventional recurrent nets. To save indices, I consider a *single* discrete sequence of real-valued input vectors $x(t)$, $t = 1, \ldots, n_{time}$, each with dimension n_x. In what follows, if $v(t)$ denotes a vector then $v_k(t)$ denotes the k-th component of $v(t)$. If u denotes a unit, then $u(t)$ denotes the activation of the unit at time t. w_{ij}'s real-valued weight at time t is denoted by $w_{ij}(t)$. *Throughout the remainder of this paper, unquantized variables are assumed to take on their maximal range.* The dynamic evolution of a traditional discrete time recurrent net (e.g. [2], [7]) is given by

$$x_k(t) \leftarrow environment, \quad net_{y_k}(1) = 0, \quad y_k(t) = f_{y_k}(net_{y_k}(t)), \quad net_{y_k}(t+1) = \sum_{units\ l} w_{y_k l}(t) l(t), \quad (1)$$

where f_i denotes the semi-linear activation function of unit i, and where $w_{y_k l}(t)$ is *constant* for all t.

Typical objective function. There may exist specified target values $d_k(t)$ for certain outputs $o_k(t)$. We define

$$e_k(t) = d_k(t) - o_k(t) \quad if\ d_k(t)\ exists,\ and\ 0\ otherwise.$$

The supervised sequence learning task is to minimize (via gradient descent)

$$E^{total}(n_{time}), \quad where\ E^{total}(t) = \sum_{\tau=1}^{t} E(\tau), \quad where\ E(t) = \frac{1}{2}\sum_k (e_k(t))^2. \quad (2)$$

Complexity of traditional recurrent net algorithms. Let m be the number of time-varying variables for storing temporal events. Let R_{time} be the ratio between the number of operations per time step (for an *exact* gradient based supervised sequence learning algorithm), and m. Let R_{space} be the ratio between the maximum number of storage cells necessary for learning arbitrary sequences, and m.

The fastest known exact gradient based learning algorithm for minimizing (2) with fully recurrent networks is 'back-propagation through time' (BPTT, e.g. [3]). With BPTT, $m = n$, and $R_{time} =$

$O(m)$. BPTT's disadvantage is that it requires $O(n_{time})$ storage – which means that R_{space} is infinite: BPTT is not a *fixed-size storage* algorithm. The most well-known *fixed-size storage* learning algorithm for minimizing $E^{total}(n_{time})$ with fully recurrent nets is RTRL [2][8]. With RTRL, $m = n$, $R_{time} = O(m^3)$ (much worse than with BPTT) and $R_{space} = O(m^2)$ (much better than with BPTT).[1]

Contribution of this paper. The contribution of this paper is an extension of the conventional dynamics (as in equation (1)) plus a corresponding exact gradient based learning algorithm with $R_{time} = O(m)$ (as good as with BPTT) and $R_{space} = O(m^2)$ (as good as with RTRL). The basic idea is: The $O(n^2)$ weights themselves are included in the set of time-varying variables that can store temporal information ($m = n + O(n^2) = O(n^2)$). Section 2 describes the novel network dynamics that allow the network to actively and quickly manipulate its own weights (via so-called *intra-sequence* weight changes) by creating certain appropriate internal activation patterns [2] during observation of an input sequence[3]. Section 3 derives an *exact* supervised sequence learning algorithm (for creating *inter-sequence* weight changes which affect only the *initial* weights at the beginning of each training sequence) forcing the network to use its association building capabilities to minimize $E^{total}(n_{time})$. The gradient-based learning algorithm for inter-sequence weight changes takes into account the fact that intra-sequence weight changes at some point in time may influence both activations and intra-sequence weight changes at later points in time.

2 NOVEL DYNAMICS

I will keep the architecture and the objective function from section 1 but I will modify the system dynamics. Recall that unquantized variables are assumed to take on their maximal range. For our single training sequence with n_{time} discrete time steps, the system dynamics (explanation follows below) are defined by

$$x_k(t) \leftarrow environment, \quad net_{y_k}(1) = 0, \quad y_k(t) = f_{y_k}(net_{y_k}(t)), \quad net_{y_k}(t+1) = \sum_l w_{y_k l}(t)l(t), \quad (3)$$

$$w_{ij}(1) \leftarrow initialization, \quad w_{ij}(t+1) = \sigma_{ij}\left[w_{ij}(t) + g(j(t))h(i(t+1))\right], \quad (4)$$

where σ_{ij} is a differentiable function (e.g. for limiting the weight on w_{ij} to a given interval), and g and h are differentiable monotonic functions (the 'threshold approximators', to be explained below).

Equation (3) is just the conventional recurrent net update rule (1). Unlike with conventional recurrent nets, however, the weights do *not* remain constant during sequence processing : Equation (4) says that connections between units active at successive time steps are immediately strengthened or weakened essentially in proportion to pre-synaptic and post-synaptic activity. These *intra-sequence* weight changes are modulated by the non-linear functions g and h and may be negative (anti-Hebb-like) or zero as well as positive. Let us assume that all input vectors and all f_i are such that all units can take on only activations between 0 and 1. g and h are meant to specify the upper and lower thresholds that determine how strongly units have to be excited or inhibited to contribute to intra-sequence weight changes. A reasonable choice for g and h is one where g and h are strongly negative only if their argument is close to 0 and are strongly positive only if their argument is close to 1. Both g and h should return values close to 0 for arguments from the largest part of the interval between 0 and 1. This implies hardly any intra-sequence weight changes for connections between units that have non-extreme activations during successive time steps.

---

[1]It should be noted that there is a *fixed-size storage* algorithm (a hybrid between BPTT and RTRL) with $R_{time} = O(m^2)$ (better than with RTRL, worse than with BPTT) and $R_{space} = O(m^2)$ (much better than with BPTT, same as with RTRL) ([7][4]).

[2]Certain biological evidence is consistent with the idea of fast weight changes ('dynamic links', see [6]). In [1], for instance, it is shown that the effective connectivity between certain neurons may change drastically within a few 10 msec.

[3]The active weight changing capabilities represent a similarity to the system described in [5], which is based on two separate modules – one for learning to control fast weight changes of the other one. Unlike with this previous approach, however, the system described herein does not require two separate modules. It can learn to manipulate *its own* weights.

The overall effect is that only connections between units that are exceptionally active or exceptionally inactive during successive time steps can be significantly modified. *Intra-sequence* weight changes essentially occur only if the network 'pays a lot of attention' to certain units by strongly exciting them or strongly inhibiting them. Weights to units that are not 'illuminated by adaptive internal spotlights of attention' essentially remain invariant and participate only in 'automatic processing' as opposed to 'active intra-sequence learning'. The remainder of this paper derives an exact gradient-based algorithm designed to adjust the system (via *inter-sequence* weight changes) such that it creates appropriate *intra-sequence* weight changes at appropriate time steps.

3   SUPERVISED LEARNING ALGORITHM

The following algorithm for minimizing E^{total} is partly inspired by conventional recurrent network algorithms (e.g. [2], [7]). The notation is partly inspired by [8].

Derivation. Before training, all *initial* weights $w_{ab}(1)$ are randomly initialized. The chain rule serves to compute weight increments (to be performed *after* each training sequence) for all initial weights according to

$$w_{ab}(1) \leftarrow w_{ab}(1) - \eta \frac{\partial E^{total}(n_{time})}{\partial w_{ab}(1)}, \tag{5}$$

where η is a constant positive 'learning rate'. Thus we obtain an *exact* gradient-based algorithm for minimizing E^{total} under the dynamics given by (3) and (4).

We write

$$q_{ab}^{ij}(t) = \frac{\partial w_{ij}(t)}{\partial w_{ab}(1)}, \quad \forall \text{ units } u : p_{ab}^{u}(t) = \frac{\partial u(t)}{\partial w_{ab}(1)}. \tag{6}$$

(Recall that unquantized variables are assumed to take on their maximal range.)

First note that

$$\frac{\partial E^{total}(1)}{\partial w_{ab}(1)} = 0, \quad \forall t > 1 : \frac{\partial E^{total}(t)}{\partial w_{ab}(1)} = \frac{\partial E^{total}(t-1)}{\partial w_{ab}(1)} - \sum_k e_k(t) p_{ab}^{o_k}(t). \tag{7}$$

Therefore, the remaining problem is to compute the $p_{ab}^{o_k}(t)$, which can be done by incrementally computing all $p_{ab}^{z_k}(t)$ and $q_{ab}^{ij}(t)$:

$$p_{ab}^{z_k}(1) = 0, \quad p_{ab}^{x_k}(t+1) = 0, \tag{8}$$

$$p_{ab}^{y_k}(t+1) = f_{y_k}'(net_{y_k}(t+1)) \sum_l \frac{\partial}{\partial w_{ab}(1)} [l(t) w_{y_k l}(t)] =$$

$$f_{y_k}'(net_{y_k}(t+1)) \sum_l \left[w_{y_k l}(t) p_{ab}^l(t) + l(t) q_{ab}^{y_k l}(t) \right], \tag{9}$$

where

$$q_{ab}^{ij}(1) = 1 \text{ if } w_{ab} = w_{ij}, \text{ and } 0 \text{ otherwise}, \tag{10}$$

$$q_{ab}^{ij}(t+1) = \sigma_{ij}'(w_{ij}(t+1)) \left[q_{ab}^{ij}(t) + h'(i(t+1)) p_{ab}^i(t+1) g(j(t)) + h(i(t+1)) p_{ab}^j(t) g'(j(t)) \right]. \tag{11}$$

According to equations (8)-(11), variables holding the $p_{ab}^j(t)$ and $q_{ab}^{ij}(t)$ values can be updated incrementally at each time step. This implies that (5) can be updated incrementally, too. With non-degenerate networks, the algorithm's storage complexity is dominated by the number of variables for storing the $q_{ab}^{ij}(t)$ values. This number is independent of the sequence length and equals $O(n_{conn}^2)$. Since $m = O(n_{conn})$, $R_{space} = O(m^2)$ (like with RTRL). The computational complexity per time step also is $O(n_{conn}^2)$ – essentially the same as the one of RTRL. Since $m = O(n_{conn})$, however, $R_{time} = O(m)$ (like with time-efficient BPTT and unlike with RTRL's much worse $R_{time} = O(m^3)$).

4 CONCLUDING REMARKS

I have described a novel fully recurrent network that may choose to behave like a conventional fully recurrent net. In addition, however, the novel net may choose to use its own weights for storing temporal events. The network can do so by creating and directing 'internal spotlights of attention' to cause intra-sequence weight changes that may help the system to achieve its goal (defined by a conventional objective function for supervised sequence learning). The corresponding exact gradient based learning algorithm turns out to have the same ratio between number of learning operations per time step and number of time-varying variables as the time-efficient BPTT algorithm. In addition, it turns out to have the same ratio between maximum storage and number of time-varying variables as the space-efficient RTRL algorithm.

Of course, since fast weights and unit activations are different kinds of variables, I am subsuming different things under the expression 'time-varying variable'. I expect that some problems may be more naturally solved using information processing based on time-varying unit activations, other problems may be more naturally solved using information processing based on fast weights (e.g. certain kinds of temporal variable binding problems, see [5]). Careful experimental investigations of the mutual advantages and disadvantages of both kinds of time-varying variables are needed (experiments are also needed for analyzing different reasonable choices for functions like g, h, σ) but are beyond the scope of this short paper and will be left for the future.

5 ACKNOWLEDGEMENTS

This work was supported in part by a DFG fellowship to the author, as well as by NSF award IRI-9058450, grant 90-21 from the James S. McDonnell Foundation, and DEC external research grant 1250.

References

[1] A.M.H.J. Aertsen, G.L. Gerstein, M.K. Habib, and G. Palm. Dynamics of neuronal firing correlation: Modulation of "effective connectivity". *Journal of Neurophysiology*, 61:900–917, 1989.

[2] A. J. Robinson and F. Fallside. The utility driven dynamic error propagation network. Technical Report CUED/F-INFENG/TR.1, Cambridge University Engineering Department, 1987.

[3] D. E. Rumelhart, G. E. Hinton, and R. J. Williams. Learning internal representations by error propagation. In *Parallel Distributed Processing*, volume 1, pages 318–362. MIT Press, 1986.

[4] J. H. Schmidhuber. A fixed size storage $O(n^3)$ time complexity learning algorithm for fully recurrent continually running networks. *Neural Computation*, 4(2):243–248, 1992.

[5] J. H. Schmidhuber. Learning to control fast-weight memories: An alternative to recurrent nets. *Neural Computation*, 4(1):131–139, 1992.

[6] C. v.d. Malsburg. Technical Report 81-2, Abteilung für Neurobiologie, Max-Planck Institut für Biophysik und Chemie, Göttingen, 1981.

[7] R. J. Williams. Complexity of exact gradient computation algorithms for recurrent neural networks. Technical Report Technical Report NU-CCS-89-27, Boston: Northeastern University, College of Computer Science, 1989.

[8] R. J. Williams and D. Zipser. A learning algorithm for continually running fully recurrent networks. *Neural Computation*, 1(2):270–280, 1989.

464

Deletion of Trained Patterns by Incremental Learning in Artificial Neural Network using Fahlman-Lebiere Learning Algorithm

Masanori HAMAMOTO[†]   Joarder KAMRUZZAMAN[††]   Yukio KUMAGAI[†††]

[†]Kashima Syst. Dept.	[††]Dept. of Elect. & Electr.	[†††]Dept. of Comp. Scie. & Syst. Engg.
Sumitomo Metal Industries Ltd.	Bangladesh Univ. of Engg. & Tech.	Muroran Inst. of Tech.
3 Ooazahikari Kashima-cho	Dhaka 1000	27-1 Mizumoto-cho
Ibaragi, 314 Japan	Bangladesh	Muroran, 050 Japan

Abstract --- In this paper, we describe a method for the deletion of some of the trained patterns which are previously learned in the already-trained network in order to be adapted in a changeable environment. The deletion is performed by inhibiting all the actual outputs for the trained patterns to be deleted. Our approach is to realize the deletion by incremental learning without destroying the previously trained network but by incorporating some new hidden units so as to inhibit all the actual outputs for the trained patterns to be deleted. Fahlman-Lebiere (FL) learning algorithm is particularly suitable for this purpose since this algorithm can gradually add the required number of new hidden units. Previous studies show that the addition of categories and patterns can be performed by incremental learning [9], [10]. In this paper, we apply FL algorithm to the deletion of some of the trained patterns by incremental learning. Investigation shows that FL network realizing this deletion task has better generalization ability than Backpropagation (BP) network due to the attainment of well-saturated hidden outputs in FL network. By performing the deletion, the generalization ability of the resultant network does not degrade in comparison with the network before deletion.

1. INTRODUCTION

The usual practice in multilayer neural networks is to retrain a new network when the environment has to be changed. In the previous study [9], [10] we showed that one alternate way to do this is to retain the information acquired by the already-trained network which performs a specific pattern classification task and then add new hidden units to realize the newly defined task. Previous study dealt with a kind of incremental learning needing more categories to be learned with the newly added training set. In this paper, we deal with the deletion of some of the trained patterns by incremental learning in which all the actual outputs for the trained patterns to be deleted are trained to be inhibited. Here, the incremental learning can be performed as the weights between the input and hidden layer of the previously trained network are not modified. Thus the previously trained network will retrain the feature extraction capability already learned and the newly added hidden units will gain the necessary additional capability to realize the task.

To perform above mentioned task, standard Backpropagation learning algorithm [1] is not suitable since in this case the required number of newly added hidden units has to be pre-specified and a priori specification of the number of hidden units sufficient to realize the task is rather difficult, and trial and error is the only solution. Thus we used a learning algorithm that can gradually add new hidden units one by one and stop creating new hidden units when the desired task is realized.

Several algorithms that gradually build network topology have been reported, e.g., perceptron based pocket algorithm by Gallant [2], tiling algorithm by Mezard and Nadal [3], upstart algorithm by Frean [4], cascade-correlation learning algorithm by Fahlman and Lebiere [5], CASQEF, CASER and CASLLM algorithms by Littmann and Ritter [6], [7]. However, the algorithms proposed in [2]-[4] have been mainly analyzed for a single output unit and use of these algorithms for multiple output units might need an excessively large network. The approach in [6] and [7] needs more sweeps, more cascaded hidden units and additional powers of their activity values than cascade-correlation algorithm to achieve similar results for XOR problem. The basic idea of cascade-correlation learning algorithm proposed by Fahlman and Lebiere [5] is that, to realize any input-output relationship, a network is constructed by adding hidden units one by one based on maximizing the correlation between the residual error at the outputs and the output of candidate hidden unit. This algorithm learns very quickly [5], and is capable of constructing near-optimum network automatically [5], builds a network with high generalization ability [8] and can be applied to an incremental learning with increased category [9], [10].

In this paper, we trained the network which perform a specific pattern classification by FL algorithm and BP algorithm, both having similar network architecture. We then applied FL algorithm to realize the deletion of some of the trained patterns by incremental learning. And we made a comparison between the generalization properties of the networks trained by FL and BP algorithm.

2. DESCRIPTION OF THE ALGORITHM

Learning by Fahlman-Lebiere algorithm [5] is divided in two stages, namely, learning of output units and learning of hidden units. This algorithm begins with a two layer network of input and output units and then continue adding hidden units until the designer is satisfied with the network. In this paper, our approach is to generate a three-layer network similar in structure to a BP network by this algorithm which is constructed in the following way [8]-[10]. To construct a layered network, all the weights between input and output units are first set to zero and from then on, all the weights except the biases of the output units are kept frozen. Fig. 1 shows network construction as a hidden unit is added during learning. In Fig. 1, the unit whose output is not connected to the active network is called candidate hidden unit. At this stage, the candidate hidden unit's input weights are adjusted to maximize S, i.e., the sum of the absolute values of covariances C_j between its output value and the observed errors at the output unit j by gradient ascent.

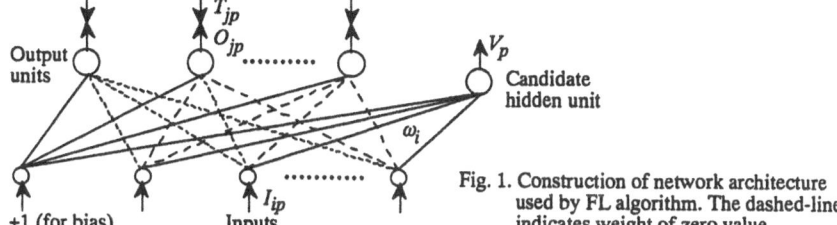

Fig. 1. Construction of network architecture used by FL algorithm. The dashed-line indicates weight of zero value.

Equations on learning of hidden unit :

$$S = \sum_j |C_j| \quad , \quad C_j = \sum_p (V_p - \overline{V})(E_{jp} - \overline{E_j}) \quad , \quad E_{jp} = T_{jp} - O_{jp} \quad , \quad V_p = f(N_p) \quad ,$$

$$\overline{V} = \sum_p V_p / M \quad , \quad \overline{E_j} = \sum_p E_{jp} / M \quad , \quad N_p = \sum_i \omega_i I_{ip} \text{ (including bias)} \quad ,$$

where T_{jp}, O_{jp} and E_{jp} are the target output, actual output and error respectively at output unit j for training pattern p. N_p, f and V_p are the net input to the candidate hidden unit, its activation function and output respectively, M being the total number of training patterns. I_{ip} is the input that the candidate hidden unit receives from input unit i corresponding to pattern p and ω_i is the weight between them. In order to maximize S, computing $\partial S / \partial \omega_i$ in a manner similar to the derivation of Backpropagation rule [1], weight change for ω_i is given by

$$\Delta \omega_i = \eta \sum_j \sum_p \sigma_j (E_{jp} - \overline{E_j}) f'(N_p) I_{ip} \quad ,$$

where $\sigma_j = 1$ $(C_j > 0)$, $\sigma_j = -1$ $(C_j < 0)$, and η is the learning rate.

When S stops improving, the candidate hidden unit is connected to the output units and its incoming weights are frozen. Next the algorithm begins output stage learning to reduce the network error and modifies the weight between hidden unit (the newly added hidden unit and pre-existing hidden units, if any) and the output unit along with the biases of the output units by delta rule. After the network error, i.e., sum-squared error at the output layer for the whole training set, reaches an asymptote and if the designer is not satisfied with the network behavior, the algorithm generates a new hidden unit. This process, which has two stages as stated above, is repeated until the designer is satisfied with the network. A three-layer feedforward network is constructed by this process.

3. SIMULATION RESULTS AND DISCUSSIONS

One of the essential properties of neural network for real world application is its ability to generalize any pattern not included in the training set. Previous work in [8] shows that generalization ability of a network trained by FL algorithm is significantly better than that of a similar network trained by BP algorithm. And also for performing incremental learning with increased category, FL network generalizes far better than BP network [9], [10]. In the present work, we deal with the deletion of some of the already-trained patterns and explore the possibility of achieving this through a kind of incremental learning in which all the patterns to be deleted are trained to produce completely inhibited outputs (e.g., zero output if the output activation is bounded between 0 and 1). This was investigated using both FL and BP algorithm, and a performance comparison between these networks are made.

The experiment we performed was a capital letter (A~Z) and small letter (a~z) recognition task. Each character was 8×8 black and white pixels, and the component of each input vector consisted of +1 or -1, +1 representing the black pixel. Target vectors were binary vectors [0,1] and were locally represented ("A" and "a" belong to same category and so on). Activation function used for the output and hidden units was sigmoidal function within the range (0,1). Initially a network was trained with the standard training set (A~Z,a~z). To realize this task (26 categories, 52 patterns), FL algorithm began with a minimal network of 64 inputs, 26 output units and finally generated 6 hidden units, i.e., a 64-6-26 network was constructed (we call it "standard FL network"). Similarly, three-layer networks with 6 hidden units were trained by BP algorithm (we call it "standard BP network").

466

Given that a network is already trained with standard training set and now if a task of realizing deletion of small letter (a~z) is to perform in which all the target outputs for those patterns are to set at '0' (inhibition), one possible way is to retrain a new network as described above. In this case, starting with a new network destroys the previous information gained by already-trained network. In this paper, our approach to do this is to keep the weights between input and hidden layers in the previously trained network fixed and then extend this network by adding new hidden units. Fig. 2 illustrates the network construction for realization of the deletion of small letters (a~z). For this task, FL algorithm is particularly suitable since this algorithm gradually adds on new hidden units without being pre-specified by the designer. To perform this task, FL algorithm began with a 64-6-26 standard FL network and then added 2 new hidden unit. Thus the resultant 64-8-26 FL network (we call it "deleted FL network") is constructed without destroying the weights between input and hidden layers in standard FL network.

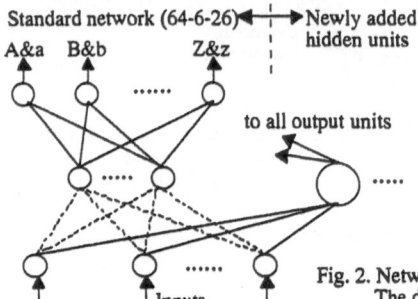

Fig. 2. Network construction for realization of the deletion. The dashed-line indicates fixed weight.

The same incremental learning task can also be performed by using BP algorithm through the newly added hidden units. However, in this case the number of newly added hidden units must be pre-specified. To make a performance comparison with FL algorithm, we constructed a deleted BP network similar in architecture to the deleted FL network described above. To perform this task, a 64-8-26 deleted BP network was constructed by adding 2 new hidden unit to the 64-6-26 standard BP network. The generalization abilities of the resultant networks trained by FL and BP algorithms were investigated as follows.

In order to form test patterns, training patterns of capital letter was corrupted by noise. This noise changes white pixel (-1) into black (+1), or black pixel into white. All the training patterns of capital letter were corrupted by changing 1, 2, 3, and 4 pixels. All of the possible test patterns were formed by injecting noise upto 4 bits (in total 17551636 test patterns). Recognition ability was investigated under two different recognition criteria as stated below.

Criterion 1 : Upon presenting a test pattern to the network, maximum value and the second maximum value at the output layer were detected. If the maximum value was greater than 0.9 and the second maximum value lower than 0.1, the test pattern was recognized to belong to the category represented by the output unit of maximum value. This was a rather severe criterion.

Criterion 2 : Upon presenting a test pattern to the network, maximum value at the output layer was detected and the test pattern was recognized to belong to the category represented by the output unit of maximum value. This was a rather mild criterion.

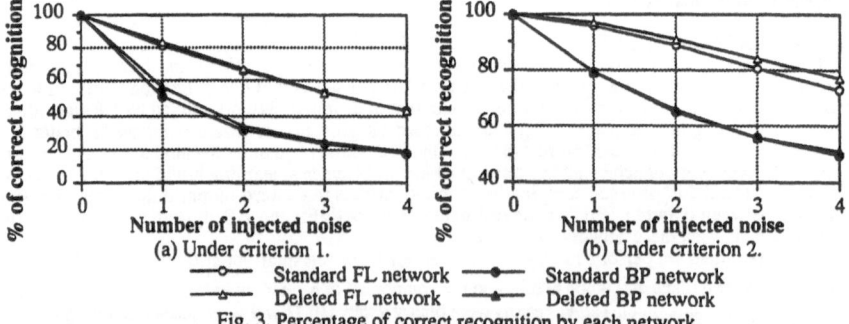

Fig. 3. Percentage of correct recognition by each network.

Fig. 3 shows the percentage of correct recognition by each network under each recognition criterion. From the simulation results, it is obvious that generalization ability of FL network is always far better than that of BP network. And there is little difference of generalization ability between the standard and deleted networks. This means that, as revealed in our experiment, the performance of the network remains unchanged by realizing deletion of some of trained patterns through the incremental learning described earlier.

The reason of better performance in FL network than that of BP network can be explained by making a comparative study of the behavior of hidden units in both the networks. It can be reasonably assumed that if the hidden units attain saturation for almost all of the training patterns, noise will be well absorbed by the hidden units. Then hidden layer can effectively filter out the noise and the network is expected to have good generalization ability [8]-[10]. Fig. 4 shows frequency distribution of the net input to the hidden units for the training patterns of capital letter after the completion of learning in the deleted FL and BP networks. In FL network, hidden units have the tendency to converge to the extreme value and the hidden units attain saturation for almost all the training patterns. Here, we considered a hidden unit to be saturated if its net input was greater than +2.5 or less than -2.5. In contrast, in BP network almost all the training patterns except a few use intermediate values of hidden units for internal representation and this greatly reduces the generalization ability. Similar trends in the saturation of hidden outputs in response to the training patterns are observed in standard networks trained by FL and BP algorithms. In both standard and deleted networks, it is the saturation of hidden outputs in FL network that makes this network perform far better than BP network.

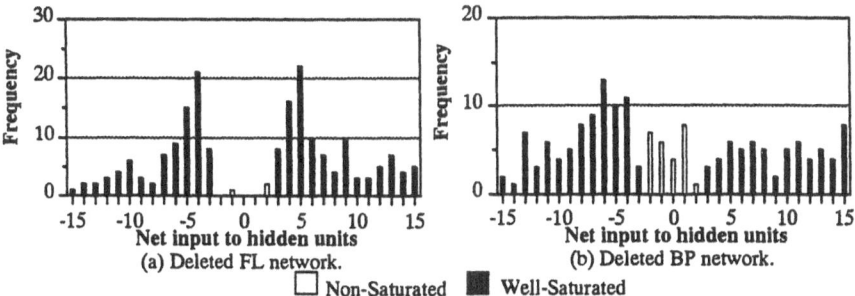

(a) Deleted FL network. (b) Deleted BP network.

☐ Non-Saturated ■ Well-Saturated

Fig. 4. Frequency distribution of the net input to hidden units for the training patterns of catipal letter.

4. CONCLUSIONS

In this paper, we show that deletion of some of the trained patterns in a already-trained network is possible through a kind of incremental learning in which the to-be-deleted patterns are trained to inhibit completely all the output units while retaining the already-rained network. This was done by both FL and BP algorithms. Simulation results show that the resultant FL network has generalization ability better than BP network, and the performance of the network is unchanged by deletion. The underlying fact is that in FL network hidden outputs in response to the training set attain well-saturated values, and thus the hidden layer which acts as a filter against corrupted disturbances makes FL network more robust than conventional BP network.

REFERENCES

[1] D. E. Rumelhart, G. E. Hinton and R. J. Williams, "Learning Internal Representations by Error Propagation", in *Parallel Distributed Processing*. MIT Press, vol. 1, pp. 318-362, 1986.

[2] S. I. Gallant, "Perceptron-Based Learning Algorithms", *IEEE Trans. Neural Networks*, vol. 1, no. 2, pp. 179-191, 1990.

[3] M. Mezard and J. -P. Nadal, "Learning in Feedforward Neural Networks : the Tiling Algorithm", *J. Phys.*, A: Math. Gen. 22, pp. 2191-2203, 1989.

[4] M. Frean, "The Upstart Algorithm : A Method for Constructing and Training Feedforward Neural Networks", *Neural Computation*, vol. 2, pp. 198-209, 1990.

[5] S. E. Fahlman and C. Lebiere, "The Cascade-Correlation Learning Architecture", in *Advances in Neural Information Processing Systems*, D. S. Touretzky, Ed. Los Altos, CA: Morgan Kaufmann, vol. 2, pp. 524-532, 1990.

[6] E. Littmann and H. Ritter, "Cascade Network Architectures", in *Proc. of IEEE/INNS Int. Joint Conf. Neural Networks*, Baltimore, vol. II, pp. 398-404, 1992.

[7] E. Littmann and H. Ritter, "Cascade LLM Networks", in *Proc. of Int. Conf. Artificial Neural Networks*, Brighton, vol. 1, pp. 253-257, 1992.

[8] M. Hamamoto, J. Kamruzzaman and Y. Kumagai, "A Study on Generalization Properties of Artificial Neural Network Using Fahlman and Lebiere's Learning Algorithm", in *Artificial Neural Networks, 2*, I. Aleksander and J. Taylor, Eds. Amsterdam: North-Holland, vol. 2, pp. 1067-1070, 1992.

[9] M. Hamamoto, J. Kamruzzaman, Y. Kumagai, "Network Synthesis and Generalization Properties of Artificial Neural Network Using Fahlman and Lebiere's Learning Algorithm", in *Proc. of the 35th Midwest Symposium on Circuits and Systems* , Washington, D. C., vol. 1, pp. 695-698, Aug. 1992.

[10] M. Hamamoto, J. Kamruzzaman, Y. Kumagai and H. Hikita, "Incremental Learning and Generalization Ability of Artificial Neural Network Trained by Fahlman and Lebiere's Learning Algorithm", *IEICE Trans. Fundamentals*, vol. E76-A, no. 2, pp. 242-247, Feb. 1993.

CASCADE NEURAL NETWORK DEVELOPED FOR TIME SERIES PREDICTION

Igor Grabec,
Faculty of Mechanical Engineering , University of Ljubljana
61000 Ljubljana, Slovenia, p. o. box 394, fax-38-61-218567

ABSTRACT
This article presents a theoretical basis for the development of a cascade neural network. A definition of the cascade stems from the hierarchical expansion of a general dynamical law used for the time series description and corresponds to a creation of a multilayer neural network. A particular layer predicts changes in the time series from a certain number of past data. The network complexity, determined by the dimension span and the population size of the neurons, is increasing until a proper prediction performance is achieved on a given time series. The adapted network is then applicable for a forecasting. The operation of the proposed architecture is demonstrated by the examples including a regular and a chaotic time series.

FORMULATION OF THE CASCADE NETWORK
Let us consider a dynamical phenomenon which can be described by a variable $x(t)$, observable at discrete times $t=1,2,....$ The fundamental problem of a forecasting is to formulate a method, or a mathematical model, by which forthcoming changes $\Delta x(t)$ of this variable can be predicted from a series of past observations $\{x(t),x(t-1),x(t-2),...\}$. For this purpose a vast class of linear as well as nonlinear models have already been studied.[1,2] Modeling by neural networks (NN) developed recently [3,4] appears to be very advantageous, but as in the other fields of NN applications there appears a fundamental problem of a proper selection of the network architecture. The objective of this article is to show how this problem can be solved by creating and adapting the NN architecture to a particular phenomenon by following an analysis stemming from a general description of dynamical phenomena. For this purpose we assume that the changes of the variable $x(t)$ can be formally described by the expression :

$$\Delta x(t+1) = x(t+1) - x(t) = G(x(t), x(t-1),...; v(t)) \tag{1}$$

in which the generating function G generally denotes some nonlinear mapping. Beside a series of past values of variable x we have included into the dynamical law the vector $v(t)=(v_1(t),.., v_s(t))$ to provide for the description of non-autonomous phenomena. Our task is to formulate a procedure by which the generating function G can be modeled from the past series $\{x(t'); t' \leq t\}$ without a priori information about the dimension of the state vector $s(t)=(x(t), x(t-1),..)$ needed in the description of the dynamical law (1). This dimension can be either estimated by any of the existing methods of non-linear dynamics and deterministic chaos [5] or it can be adaptively determined by the following consideration. Generally, the most simple for modeling are one-dimensional phenomena determined by a function G that depends only on the first component of the state vector $s(t)$. Numerical examinations show that various higher order mappings can often be approximately described by using only one-dimensional generating function.[4,5] We therefore conjecture that it could be quite generally of advantage to expand the generating function into hierarchy of terms including an increasing number of components of the state vector $s(t)$[5]

$$G(x(t),x(t-1),...;v) = G_1(x(t);v) + G_2(x(t),x(t-1);v) + G_3(x(t),x(t-1),x(t-3);v) + ... \tag{2}$$

We split also the changes of the variable x into contributions:

$$\Delta x(t+1) = \Delta x_1(t+1) + \Delta x_2(t+1) + \Delta x_3(t+1) + \ldots \tag{3}$$

corresponding to ever more complex dynamical laws :

$$\Delta x_1(t+1)=G_1(x(t);v) \quad ; \quad \Delta x_2(t+1)=G_2(x(t),x(t\text{-}1);v) \quad ; \quad \ldots \tag{4}$$

However, such a splitting is arbitrary as long as we do not specify the procedures for hierarchical modeling of functions G_i. For this purpose we invoke the results of previous articles on the modeling of chaotic time series[2-4] and treat a dynamical law

$$\Delta x_k(t+1)=G_2(x(t),x(t\text{-}1),.., x(t\text{-}k+1);v) \tag{5}$$

as a generator of points

$$z(t)=(\Delta x_k(t+1), x(t), x(t\text{-}1),.., x(t\text{-}k+1); v(t)) \tag{6}$$

in a state space of dimension $d_z=k+1+s$. The corresponding probability density function can be estimated from a given time series of N sample values by using the Parzen's window approach [3,4,7] as

$$f(z) = \frac{1}{N} \sum_{t=1}^{N} w(z - z(t), \sigma) \tag{7}$$

Here the vector $z(t)$ describes a multidimensional parameter of a formal neuron, and w denotes a radial basis, or window function. The parameter σ in it characterizes the width of the receptive field of a neuron. For the sake of simplicity we assign here the same σ to all neurons. The information about the phenomenon is thus represented by the set of given sample vectors $\{z(t), t=1...t_r\}$ determined by the recorded past time series while the model is indirectly specified by the probability density function (7). It is applicable for the estimation of the forthcoming changes of time series. Let us for this purpose assume that a partial information about the phenomenon is given by a truncated vector

$$z'(t)=(?, x(t), x(t\text{-}1), \ldots , x(t\text{-}k+1) ; v(t)) \tag{8}$$

in which the question mark "?" indicates the missing component $\Delta x_k(t+1)$. Its value can be optimally estimated from the vector $z'(t)$ by the conditional average estimator describing a non-parametric regression[4,7]

$$\Delta \hat{x}_k(t+1) = \frac{\sum\limits_{s=1}^{N} \Delta x_k(s+1)\ w(z'(t)\text{-}z'(s),\sigma)}{\sum\limits_{i=1}^{N} w(z'(t)\text{-}z'(i),\sigma)} \tag{9}$$

The estimator is a basis of an associative recall performed by a neural network composed of N d_z-dimensional neurons.

In accordance with this estimator we propose the following cascade modeling :
I. From a given time series $\{x(t), t=1...t_r\}$ we form the prototype vectors $z_1(t)=(\Delta x(t+1), x(t); v(t))$ representing the parameters of the neurons in the first layer. By using the conditional average estimator (9) we predict for each $z'_1(t)$ the corresponding value $\Delta \hat{x}_1(t+1)$ which is treated as the output of the first layer of a neural network. However, due to incomplete modeling of the time series by the probability distribution function in a space of vectors z_1, that corresponds to the dynamical law $\Delta x_1(t+1)=G_1(x(t);v)$, the estimated forthcoming value $\Delta \hat{x}_1(t+1)$ does not exactly coincide with the true value $\Delta x_1(t+1)$.
II. A crucial step in in the development of the cascade NN is then done by assuming that the difference $\Delta x_1(t+1) - \Delta \hat{x}_1(t+1) = \Delta x_2(t+1)$ can be predicted by the next layer of the cascade according to the dynamical law $\Delta x_2(t+1)=G_2(x(t),x(t\text{-}1);v)$ which can be modeled by using the prototype vectors $z_2(t)=(\Delta x_2(t+1),x(t),x(t\text{-}1);v(t))$. They represent the parameters of neurons in the second layer. By these prototypes we can estimate the value $\Delta \hat{x}_2(t+1)$ and define the difference $\Delta x_2(t+1) - \Delta \hat{x}_2(t+1) = \Delta x_3(t+1)$ which is the basis for the formation of the III layer, and so on.

By using the outputs from the successive layers the time series predictor can be represented as:

$$y(t) = x(t) + \Delta\hat{x}_1(t+1) + \Delta\hat{x}_2(t+1) + ... \tag{10}$$

The prediction error can be estimated by the square of $\Delta\hat{x}_r(t+1)$ in the final layer summed over the given time series. We generally expect that the prediction error decreases with the number of layers similar as in examples demonstrated later. The growing of the NN can be stopped when a desired forecasting accuracy is achieved.

EXPERIMENTS

A multilayer NN with the proposed architecture was simulated on a PC. From a given time series the layers of the network were formed by the described batch-type adaptation. At each level the prediction performance was estimated and the network growing was stopped when the total error of a layer got below 10^{-3} of the span of x .

Fig. 1 shows the results obtained by a regular time series $x(t) = c\, t$. The upper diagram of the figure represents the given time series, the middle one shows the time dependence of the differences for each layer r, and the last one represents the time series reproduced as well as predicted by the network. In this case the time series is well reproduced by only one layer. It is characteristic, that the trend is correctly described for $t > t_r$, which is not the case when only Parzen's estimator with spherical basis functions and without calculating the differences Δx is applied.

Fig. 2 shows the results obtained by a chaotic time series generated by the logistic map $x(t+1) = 3.9\, x(t)\, [1-x(t)]$. In this case the network needs more layers in order to reproduce the given time series with the desired accuracy. The map is one-dimensional, therefore the first layer quite accurately performs it. The convergence of the performance accuracy to 0 as well as a good quantitative agreement of reproduced and given time series is evident. However, a long-term prediction leads to divergence between given and predicted series with increasing time due to inherent instability of a chaotic process.[4,6]

CONCLUSIONS

The method proposed in this article is applicable for the prediction when the time series is generated by a stationary process. In this case one can expect that the model built on a preceding part of a time series is not essentially influenced by the time shift. One problem that has not been mentioned here is how to deal with an increasing number of sample values in the time series. In order to avoid saturation of computer memory a fixed number of prototypes can be adapted by the self-organization procedure developed in the previous studies of time series prediction.[4,8] The number of neurons included in the modeling can be limited by employing the experimental error δx of measurement of variable x. The number of prototypes is then allowed to grow until their mutual separation in the corresponding phase space is greater than the error δx.

REFERENCES

1. M. West, J. Harrison, Bayesian Forecasting and Dynamic Models, Springer-Verlag, Berlin,1989
2. H. D. I. Abarbanel, Phys. Rev. A, 41, 1782-1807 (1990)
3. K. Stokbro, D. K. Umberger, J. A. Hertz, Complex Systems, 4, 603-622 (1990)
4. I. Grabec, Artificial Neural Networks, Proc. ICANN, I: Espoo, Finland, 1991, Ed. T. Kohonen et àl, N. Holland, Vol.1, 151-156, II: Brighton, UK, 1992, Ed. I. Aleksander, J. Taylor, Elsevier Sci. Pub., 379-382, & Neural Network World, 2, 607-614 (1992)
5. J. Deppisch, Vorhersagen chaotischer Zeitreihen mit neuronalen Netzen, Diplomarbeit, Physikalisches Institut, Theoretische Physik, Universität Würzburg, 1990
6. I. Grabec in : B. Souček and IRIS Group, Dynamic, Genetic and Chaotic Programming, The sixth-generation, J. Wiley & Sons, INC, New York, 1992, p-p 145-164 & 471-500
7. I. Grabec, W. Sachse, J. Appl. Phys., 69, 6233-6244, (1991)
8. I. Grabec, Biological Cybernetics, 63, 403 (1990)

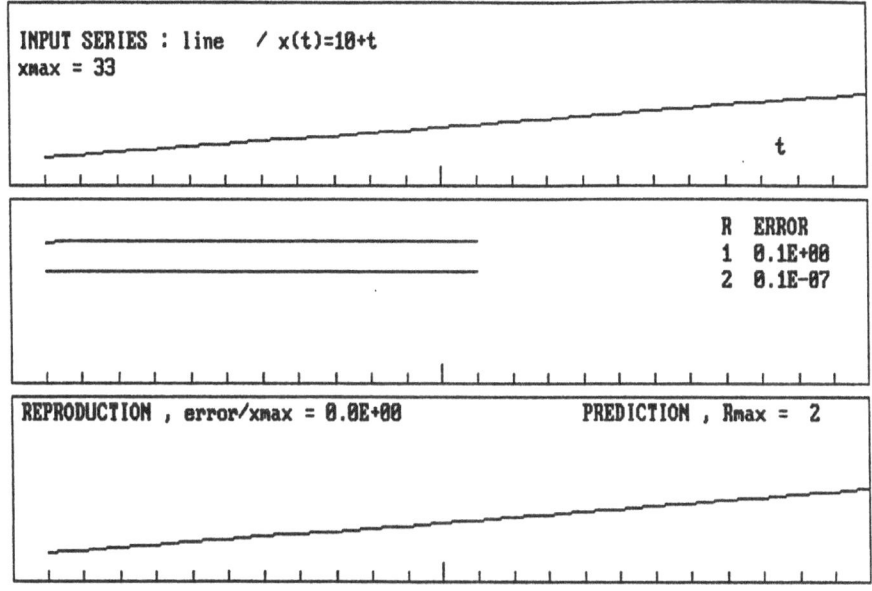

Fig. 1 Prediction of the regular time series

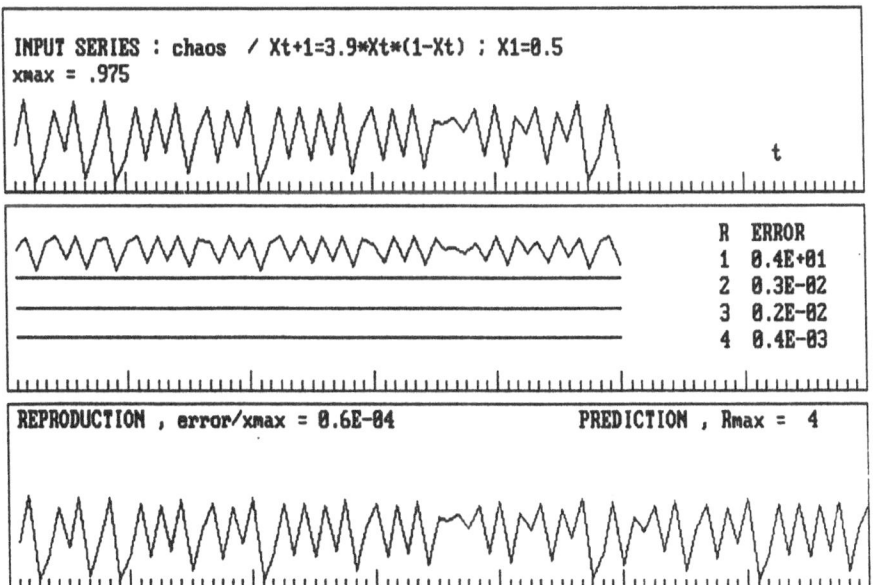

Fig. 2 Prediction of the chaotic time series generated by the logistic map.

On the Information Capacity of Auto-associative RAM-based Neural Networks

Antônio de Pádua Braga[*]

Imperial College, Dept. of Electrical and Electronic Engineering, London SW7 2BT, England

E-mail : APBRAGA@EE.IC.AC.UK

Abstract

The network studied in this work is based on the idea of a General Neural Unit (GNU), which can be configured as a feed-forward network for pattern classification or as a feedback system with associative memory properties [1]. Storing patterns in a GNU in the auto-associative mode consists on creating re-entrant states in which the network stabilises in the retrieval phase as a function of the excitation pattern. When the system is not fully interconnected, a collision may occur between patterns ξ_r and ξ_s if, after storing ξ_r, storing ξ_s makes ξ_r no longer stable. Collisions cause the formation of dynamic attractors instead of fixed and stable attractors. The cycle size of the dynamic attractor formed increases exponentially with the number of nodes where collisions occurred [2]. Therefore, the prediction of the expected number of collisions is an important issue when estimating the storage capacity [3]. A statistical approach to estimate the number of collisions of such networks is presented, allowing the prediction of the information capacity as a function of the main parameters of the system.

The results show that, for a training set with fixed size, the probability of happening collisions decreases when the connectivity of the network increases. This means that, in a network with k nodes that should store t patterns, it is possible to change the expectation of collisions by simply changing the number of inputs each node receives from the other nodes or from the input field. It can also be observed from the results that, for fixed sizes of training sets and for fixed connectivity, the probability of happening collisions tends to zero when the number of nodes increases. In other words, in a large network with low connectivity it is possible to store patterns with great confidence. It was also defined the *Degree of Confidence* of the network by which it is possible to define previously how reliable is the process of storing patterns as a function of the main parameters of the system.

The work presented a method to estimate the expected number of collisions in the storage phase of GNUs as a function of the number of nodes, the connectivity and the number of inputs each node receives from the input field. Fully connected systems in which collisions do not appear have higher costs as the amount of memory needed increases exponentially with the number of nodes. Lower cost systems with low connectivity are needed when the number of nodes increases very much. With the expressions developed it is possible to define the values of the main parameters to predict the expected number of collisions and the storage capacity for a given application. The results show that low connectivity systems are feasible, allowing the hardware design of large networks with current state-of-the-art commercial memory technology.

References

[1] Aleksander I. : 'Neural Systems Engineering : towards a unified design discipline?', *IEE Computing and Control Engineering Journal*, (1):259-265,1990.

[2] Braga A.P. and Aleksander I. : 'Sparse Connections and Retrieval Performance of Recurrent RAM-based Neural Networks', *Submitted for publication*, April 1993.

[3] Braga A.P. : 'Predicting collisions and storing patterns in GNUs', Internal Report, Department of Electrical and Electronics Engineering, Neural Systems Group, Imperial College London, December 1992.

---

[*] On leave from Universidade Federal de Minas Gerais, Brazil. Supported by CNPq grant 202286/91.6.

Modified CMAC Neural Network Architectures for Nonlinear Dynamic System Modelling

R. Dunay, G. Horváth Department of Measurement and Instrument Engineering
Technical University of Budapest, Budapest, Hungary H-1521.

Recently many neural network structures have been suggested to model nonlinear dynamic systems or to apply them in time series analysis. In these applications the output to be determined depends not only on the current input but on the previous input and/or output values too. The suggested various neural structures are mainly based on error-backpropagation multilayer networks. The main drawback of these networks is the extremely slow learning capability. This slow trainability is the most serious obstacle against the use of these networks in real time adaptive applications.

CMAC neural network [1],[2] is a real alternative to backpropagated multilayer networks in static nonlinear function approximation. In this paper some extensions of the CMAC network are presented which form from the inherently static network a dynamic neuron net. While the advantageous properties of the original CMAC network are preserved, the modified networks will be more useful to model nonlinear dynamic systems. CMAC network can be considered as an associative memory which performs two subsequent mappings. The first one projects an input space point into a binary association vector with C active bits and zeros elsewhere. These active bits select C weights, stored in a memory. The second mapping calculates the output of the network by summing up the selected C weights. In the association vector notation the output is calculated by a linear combiner with binary inputs. The first mapping is responsible for the nonlinear property of the whole system, the second or output mapping is a linear function of the association vector. The two mappings are implemented by a two-layer network. The first layer is responsible for mapping the input points to the association vectors; this mapping is fixed, the second layer implements the linear combination where the weights can be modified during training using LMS rule.

To form a dynamic net from the CMAC network some time dependence have to be built into it. FIR or IIR filters can be used in the output layer of the modified CMAC network instead of single weights; the input layer is not changed. (1) describes the operation of a single input single output FIR-CMAC network, (2) is the weight updating equation:

$$y(k)=\sum_i z_i(k)=\sum_i W_i^T(k)\ X_i(k) \quad (1) \qquad W_i(k+1)=W_i(k)-\mu\frac{\partial E(k)}{\partial W_i(k)}=W_i(k)+2\mu\epsilon(k)X_i(k) \quad (2)$$

where $X_i(k)=[x_i(k),x_i(k-1),\dots x_i(k-N_i)]^T$ is a vector of the delayed binary outputs of the first layer of the CMAC network, $W_i(k)=[w_i(0),w_i(1),\dots w_i(N_i)]^T$ is the i-th filter coefficient vector and $z_i(k)$ is the output of the i-th FIR filter; $y(k)$ is the output of the network and k is the discrete time index. Because of the nonlinear mapping of the first layer, in spite of using linear FIR filters in the output layer, the whole network operates as a piecewise linear filter. Every time instant different bits will be active in the association vector which means that the filters that contribute to the output may be different from step to step. The learning rule of (2) is similar to the rule of the FIR backpropagation network [3], however, in that case such simple equation can be obtained only for the last layer, the learning rules for the hidden layers are much more complex. IIR extension of the CMAC network is also possible, however, in this case many advantageous properties of the CMAC/FIR-CMAC networks will be lost. Applying IIR structures instead of FIR elements the whole network cannot be built without multipliers. What is more, because of the feedback paths at every branch of the output layer, after a few steps of operation, most of the IIR filters will be working, highly increasing the computational burden. One of the best features of the CMAC network, that - because of the sparse coding of the association vector - only a few weight values have to be summed up to get the output, will also be lost. Instead of using IIR filters in every branch of the output layer there are two other possibilities to get IIR behaviour: (*) a single IIR filter can be added to the output of the network, where a FIR-CMAC network is followed by a filter with trainable feedback weights q(m), (**) or a feedback FIR-CMAC network containing only a single feedback path can be used. The derivation of the learning rule is similar to that of the FIR case, but here recursive equations are obtained.

In the paper we present the detailed derivation of the weight updating equations for both the FIR and IIR extensions and we show some simulation results. The results show that the solutions are practically identical with or very close to the backpropagation based solutions [4], but the simulation times necessary to get the results were very different. Not only the time of one learning step but the number of learning cycles were significantly reduced.

References

[1] J. S. Albus, "A New Approach to Manipulator Control: The Cerebellar Model Articulation Controller (CMAC)," *J. of Dynamic Systems, Measurement and Control* pp. 220-227. Sept. 1975.

[2] W. T. Miller, "CMAC: An Associative Neural Network Alternative to Backpropagation," *Proceedings of the IEEE* vol. 78. No. 10. pp. 1561-1567. Oct. 1990

[3] Eric A. Wan, "Temporal Backpropagation for FIR Neural Networks," *Proc. of the 1990 IEEE International Joint Conference on Neural Networks*, pp. I-575-I-580.

[4] S. K. Narendra, K. Pathasarathy, "Identification and Control of Dynamical Systems Using Neural Networks," *IEEE Trans. Neural Networks*, vol.1. pp. 4-27. 1990.

Preliminary Results on Adaptively Trained Neural Networks

C. E. Pedreira and N. M. Roehl
Catholic University , PUC-Rio , DEE
C.P. 38063 , 22453-900 Rio de Janeiro Brazil

Real world applications often involve time varying models. There is an intrinsic difficulty in dealing with this sort of models, specially when one is concerned with nonlinear syster·. Layered neural networks have been successfully used in a variety of relevant problems when invariance assumptions can be properly made. On the other hand, very little can be found in the literature concerning time varying systems. We propose a new procedure to adjust weights in Neural Networks suitable for this type of models with no necessity of retraining. One of the main features of our approach concerns the designer flexibility to control a trade off problem between fitting new incoming data and causing minimum damage to the information related to the original data set. We desire to keep the error related to the late incoming data within a pre established tolerance, while maximizing the information incorporated by original training data. In this way, the designer is able to judge the relevance he or she wants to attribute to the latest data.

Given a trained network and a new data pair, our goal is to determine a new weight set such that the following associated energy function is minimized, subject to the new data constraint.

$$\text{Minimize} \quad E = \frac{1}{2} \sum_{i=1}^{N} \| d_i - y_i \|^2 \tag{1}$$

(P_ε)

$$\text{subject to} \quad -\varepsilon^l \le (d_{N+1}^l - y_{N+1}^l) \le \varepsilon^l, l=1\ldots o \tag{2}$$

where $\varepsilon^l \in (0 , \text{Min} (d_{N+1}^l, (1 - d_{N+1}^l))$, $l = 1\ldots o$ is the associated tolerance, y and d are the output and the target, respectively. Problem P_ε represents a trade off between the old system information and the new data, that is presumably reflecting a variance in the original plant. In order to improve analytic and implementation tractability, we linearize constraint (2), and define the feasible region as the intersection between a linearity constraint and the hyperspace defined by (2). If parameter ε is set null this constraint becomes identical to the one imposed in [1]. In this way, our problem is in fact a generalization of the one proposed in [1], where a rigid constraint is used in order to oblige the network output to exactly match the latest income data. By using a flexible constraint, the designer should choose the larger ε the less slow the process. This seems to be quite reasonable since for rapid varying processes one does not expect an exact forecast for the last income data. Nonlinear programming techniques are used in order to properly calculate the new weight set. We can prove that the proposed algorithm does not drift. Finally, we point out, that our approach allows the designer to choose how much he or she is prepared to pay in order to match the new data. The smaller the chosen parameter ε, the better the last data will be matched, but more damage will be imposed to the previously trained network.

REFERENCES:
[1] Park D.C., El-Sharkawi M.A. and Marks II R.J., An Adaptively Trained Neural Network, IEEE Trans. N.Nets, Vol 2, N.3, May 1991.

Networks for Learning and Differentiating
an Input-Output Mapping

Artur KORGUL

Military Academy of Technology, Warsaw, Poland

01-489 Warszawa ul.S.Kaliskiego 2

A scheme is described which allows learning and differentiating an input-output mapping $y=f(x)$ from a set of sparse data points $\{x_i,y_i;\ i=0,1,2,\ldots,n\}$. The key of the method consists in solving the inverted problem of approximation in two steps, as follows: use of parabolic splines $P_i(x)$ to interpolate the unknown nodal gradients $\phi^{\bullet}(x_i)$ of continuons function $y^{\bullet}=f^{\bullet}(x)$ being considered as a smooth approximation of input-output relation $y=f(x)$ in the bounded region $x \in [x_0,x_n]$; design of the symmetrical operator for integrating the combined local basis gradients $\phi^{\bullet}(x) = \sum_{j=0}^{n} \phi_j^{\bullet}\ P_j(x)$ and thus reconstructing the real-valued function of examined mapping. For this purpose the approximating formula (i.e. similar to this developed in [1]) was used in following form:

$$|y_i^{\bullet}|=\frac{1}{2}\ |(y_0+\ y_n)+ \int_{x_0}^{x_1}\phi^{\bullet}(x)dx-\int_{x_1}^{x_n}\phi^{\bullet}(x)dx| \equiv \sqrt{\left(\ y_0\ +\ \int_{x_0}^{x_1}\phi^{\bullet}(x)dx\right)\cdot\left(\ y_n\ -\ \int_{x_1}^{x_n}\phi^{\bullet}(x)dx\right)}$$

This equation was transformed to the linear and vectorizable procedure for differentiating the unknown real-valued function, taking the final form:

$$\begin{cases} \phi_0^{\bullet} = \dfrac{y_{-1} - y_0}{h} \quad or \quad \phi_0^{\bullet} = \dfrac{y_{-1} - y_1^{\bullet}}{2h} \\ [\underline{\phi}^{\bullet}]_{(n\cdot 1)} = [\mathbb{D}]_{(n\cdot n)}\ \cdot\ [2\underline{y}^{\bullet} - \underline{y}_0 - \underline{y}_n - \phi_0^{\bullet}\cdot\underline{a}_0]_{(n\cdot 1)} \end{cases}$$

where: $h=(x_n-x_0)/n$; n-is an even natural number; y_{-1} - additional "bias" value of output variable measured beyond the range of practical interest. This equation suggests a novel procedure for fitting the function in the sense of first-order proximity and it may be easily outlined as the best approximation feed-forward network with one hidden layer. Due to uniqueness of the best approximation, proposed configuration has constant structure and weights of connections. This novel architecture although being discrete in its own nature, preserves the "internal compactness" of shared information. Such kind of property was admited as yet for physical continuum (e.g. analog signal) only and it was not considered for discrete information processing. This procedure was used in modelling and simulation of target tracking process [2].

1. Korgul A.: The new computing method for countercurrent exchange phenomena simulation. Proceed. Int. AMSE Conf."Modelling & Simulation", Sorrento (Italy) Sept.29 - Oct.1; 1986, Vol.3.4, 41-54.

2. Korgul A.: Developing of the detailed algorithms for fighter homing at the air targets. Military Academy of Technology, Warsaw 1992; in polish.

Design vs. Training of Neural Machines.

Harriet Somers
Alan Wood
The Department of Computer Science, University of York, Heslington, York.

Abstract.

This paper presents an investigation of the balance between designing structure into a machine composed of neural components, and training to produce an architecture. The training of recurrent neural networks scales badly, largely because of the high number of connections introduced by increasing the number of nodes.

As the size of the problem increases, so the number of hidden nodes needed increases. However as the network gets larger, an increasing proportion of the connections will be redundant in the final solution. The example investigated here is using neural networks to emulate basic digital circuits. If the conventional electronic solutions were built out of neural components, most of the connections would be absent. Therefore pruning neural networks before conventional training will reduce the training complexity, and also the training time.

First the problem of training recurrent networks to behave like flip-flops is considered, and then extended to include the larger problem of binary counters. Training times and reliability are compared for each problem, using fully recurrent networks, and those in which specific connections are severed before training begins. The training characteristics are compared for those networks differing in the amount of structure imposed prior to training.

COUNTEREXAMPLE OF WITSENHAUSEN UNDER SET-BOUNDED MODEL
OF UNCERTAINTY AND ITS NEURAL NET SOLVER

K.W. Wojciechowski[*], M.A. Brdys, Z.R. Swider[**]

School of Electronic and Electrical Engineering, University of Birmingham
Birmingham B 15 2TT. Tel. 021 414 4354, Fax 021 414 4291

[*] *On leave from Silesian Technical University, Gliwice,*
[**] *On leave from Technical University of Rzeszow.*

Key words. Control law synthesis, bounded model of uncertainty, neural networks, numerical optimization.

Abstract. The paper considers the Witsenhausen's counterexample [5] under not probabilistic but set-bounded model of uncertainty [4]. Neural networks are employed to determine structure [2],[3] of the nonlinear optimal control law [2],[6] in order to overcome a key problem of functional minimization. Efficient numerical algorithm is designed to solve the problem by minimization with respect to the neural network weights. The paper delivers comprehensive numerical results which are quite new in field of nonlinear control law synthesis. Based on these results a discussion of different aspects of control under set-bounded model of uncertainty is presented.

References
[1] Brdys, M.A. & Ulanicki, B. (1990). Separation principle in optimizing control of state-constrained dynamical systems under bounded uncertainty. Lecture Notes in Control and Information Science, INRIA 144.
[2] Cybenko, G. (1989). Approximation by superposition of a sigmoidal function. Mathematics of Control, Signals and Systems, 2, 303-314.
[3] Hornik, K. (1991). Approximation capabilities of multilayer feedforward networks. Neural Networks, 4, 251-257.
[4] Schweppe, F.C. (1968). Recursive state estimation. Unknown but bounded errors and system inputs. IEEE Transactions on Automatic Control, 13, 408-414.
[5] Witsenhausen, H.S. (1968). A counterexample in stochastic optimum control. SIAM Journal on Control, 6, 131-147.
[6] Wojciechowski, K.W. (1992). Control law synthesis under bounded uncertainty. Case of noncentered set T. Archives of Control Sciences, 1-2, 83-99.

A Modified Learning Algorithm for Backpropagation Network

Yao Zhang, Grant E. Hearn and Pratyush Sen
Dept. of Marine Technology, The University of Newcastle upon Tyne
Newcastle upon Tyne NE1 7RU, U.K.

ABSTRACT

A modified backpropagation (MBP) algorithm is proposed in this paper. The MBP algorithm is based on the idea that both the error E_p and its change ΔE_p can be applied as information to update weights. It can be seen from Fig. 1 that when E_p curve is convex, the signs of E_p and ΔE_p slopes are the same. When E_p curve becomes concave, the signs are opposite. Therefore, if ΔE_p slope is added to or subtracted from E_p slope as appropriate, the network training is expected to speed up. This idea can be expressed as

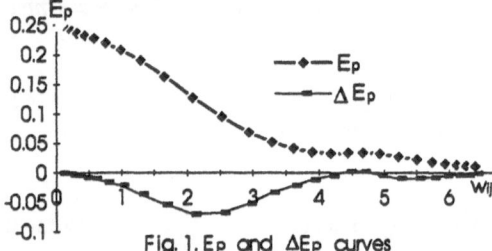

Fig. 1. E_p and ΔE_p curves

$$\Delta w_{ij} = -\eta \frac{\partial E_p}{\partial w_{ij}} + sign\left(\frac{\partial \Delta E_p}{\partial w_{ij}}\right) \cdot \xi \cdot \frac{\partial \Delta E_p}{\partial w_{ij}} \tag{1}$$

where ξ is a constant defined as an *accelerator*. For clarity, we define i, j and k as the unit in output, hidden and input layers. The derivatives of ΔE_p with respect to w_{ij} and w_{jk} are given as

$$\frac{\partial \Delta E_p}{\partial w_{ij}} = \left[\frac{\partial^2 o_{pi}}{\partial w_{ij}^2} \cdot \frac{\partial E_p}{\partial o_{pi}} + \left(\frac{\partial o_{pi}}{\partial w_{ij}}\right)^2\right]\Delta w_{ij} \tag{2}$$

$$\frac{\partial \Delta E_p}{\partial w_{jk}} = \sum_i \left[\left(\frac{\partial o_{pi}}{\partial w_{ij}} \cdot \frac{\partial o_{pi}}{\partial w_{jk}} + \frac{\partial E_p}{\partial o_{pi}} \cdot \frac{\partial^2 o_{pi}}{\partial w_{ij}\partial w_{jk}}\right)\Delta w_{ij}\right] \tag{3}$$

Simulations are carried out on the XOR, parity, encoding and symmetry problems. Results show that when compared to the conventional backpropagation (CBP) algorithm, MBP can reduce the training iterations by from 25% to 90%, depending on different problems.

Another attractive feature of MBP algorithm is that it is convergent when starting with equal or even zero initial weights. This eliminates the guesswork of finding proper initial weights for the network to avoid local minimum. Table 1 gives some of the MBP simulation results starting with zero initial weights. The CBP iterations listed in the table are the average of several training sessions with different random initial weights (η is the learning rate and α is momentum factor).

Table 1. MBP training with zero initial weights (compared to CBP training)

η	α	Training iterations in XOR problem		Training iterations in Symmetry problem	
		MBP (ξ=10)	CBP (average)	MBP (ξ=4.0)	CBP (average)
0.1	0.1	17380	57860	411120	532160
0.9	0.9	220	520	6080	8640

High-Order Boltzmann Machines for MAX-SAT and SAT

M.C. Hernández, F.X. Albizuri, A. d'Anjou, M. Graña, F. J. Torrealdea[#]
Facultad Informática UPV/EHU
Aptdo 649. 20080 San Sebastián. SPAIN
E-mail : ccpgrrom@si.ehu.es

We propose a mapping of the MAX-SAT problem, in the propositional calculus setting, into High-order Boltzmann Machines (HOBM). The approximate solution of MAX-SAT can be used as an approximate answer to the SAT question. An extensive experimental study of the behavior of HOBM for this problem has been conducted.

A HOBM can be defined as a pair (U, Λ), where U is the set of units, $U=\{u_1,...,u_N\}$, Λ is the set of connections and $\lambda \in \Lambda$ denotes a connection. The configuration or state of the BM with N units is defined by $x \in \{0,1\}^N$ and $x \in \{0,1\}$ is the state of the unit in the configuration x, a *connection* is a set of units $\lambda \in \Lambda, \lambda \subseteq U$ and $\lambda \neq 0$. The set of connections which contain the unit u_j, is denoted by $\Lambda^j = \{\lambda / u_j \in \lambda\}$. The connection strengths are given by a map $w:\Lambda \rightarrow R$. We denote w_λ the weight of the connection λ. The consensus or energy function associates a real value to every configuration $C : \{0,1\}^N \rightarrow \Re$, is defined as:

$$C(x) = \sum_{\lambda \in \Lambda} w_\lambda \prod_{i \in \lambda} x_i$$

We will state MAX-SAT as 0-1 nonlinear programing problem. Let $\{x_i\ i \in L\}$ be a set of $\{0,1\}$ variables. That is, we have a variable x_i for each literal, x_i takes value 0 when the literal is false, and 1 when true. MAX-SAT can be stated as the following maximization problem:

$$\max\left(\sum_{j=1}^{NC}(1 - \overline{c}_j) \right) \quad \text{where } \overline{c}_j = \prod_{i \in c_j}(1 - x_i) \quad \text{subject to the constraint } x_p = 1 - x_{\neg p} \ \forall p \in P$$

Note that each term in the addition will be either 0 (when the clause is false) or 1 (when the clause is true). The products, in fact, compute the truth value of the clause negation. Finally, the contraint implies that each literal and its negation must have opposite truth values. In other words, the values of the variables x_i can be interpreted as a truth assignemt to the propositions in P.

The formulation of MAX-SAT as a 0-1 programming problem leads to an inmediate mapping of the problem into HOBM. Given a propositional expression E defined over the set of propositions P, an HOBM=(U, Λ) for the approximate solution of MAX-SAT is built up as follows: The set of units is defined assigning an unit to each literal in the set of literals L, $U=\{u_i | i \in L\}$. The state x_i of each unit corresponds to the 0-1 variables. The set of connections is partitioned into two disjoint subsets $\Lambda = \Lambda^c \cup \Lambda^*$. The clauses are modeled by the connections in Λ^c. Those connections are constructed, following the formulation suggested above, as the clauses negations:

$$\Lambda^c = \left\{\lambda_j \text{ s.t. } u_{\neg i} \in \lambda_j \Leftrightarrow i \in c_j \quad j = 1..NC\right\}$$

Connections modelling clauses have a negative identical weight $w_\lambda = -\omega_c \ \lambda \in \Lambda^c$. The connections in Λ^* model the constraint that variables corresponding to opposite literals must have opposite 0-1 values. In its turn Λ^* is partitioned into bias and inhibitory connections:

$$\Lambda^* = \Lambda^i \cup \Lambda^b$$
$$\Lambda^b = \left\{\lambda_i = \{u_i\} \ \forall u_i \in U\right\} \qquad \Lambda^i = \left\{\lambda_p = \{u_p, u_{\neg p}\} \ \forall p \in P\right\}$$

Inhibitory connections have a negative value $-\omega_i$ to force opposite units to take opposite states, and the bias connections have a positive value ω_b to avoid the null configuration (all units in state 0) to be an optimum of the consensus function. This last condition may appear because all the weigths defined so far have negative values, our units are 0-1 variables and the assumed dynamic of the HOBM is the maximization of the consensus function.

[#] This work is being supported by grant PGV9220 of the Consejería de Educación del Gobierno Vasco

EBP Algorithm Can Work With Hard Limiters

Vojislav Kecman
Fulbright Visiting Scientist, MIT, Dept. of Mech. Eng.
Cambridge, MA 02139-4307, USA

The most important requirement for application of the error back-propagation (EBP) algorithm in multilayerd feed-forward artificial neural network (ANN) is that the neuron's activation function is a nonlinear, smooth and (sometime even nondecreasing) differentiable function. This is not valid anymore. This paper shows why and how simple treshold logic unit (TLU), or signum function, can be used instead of sigmoidal (or other different differentiable) activation function. The introduction of TLU is based on the understanding that the hidden layer is feeding the output layer with linearly separable patterns from its image space. To better understand the motivation for using TLU in multilayerd ANN and what is, actually, the hidden layer doing while facing nonseparable patterns of data, we use the famous XOR problem which, being two dimensional, can be explained in a graphical way, [3]. Thus, it is very important to realize that the output layer deals with linearly separable patterns, or that this layer is not separating data from input space but rather from internal, "invisible" representation or from, so called, image space and patterns in this space are separable.

Application of TLU in output layer to the XOR problem resulted in very fast convergence and without any error in just 12 learning cycles. Taking TLU for output layer nodes, equation for calculation of output deltas should be changed and standard algorithm for discrete perceptron was used without need for the value of the derivative of the activation function anymore. It is worthwhile to mention that the rest of the EBP algorithm remains the same. But there was still the unanswered question that could EBP algorithm have been more effective and maybe faster had the TLU been used in hidden layer, too. The answer was positive, but the solution in this case is now a little more complicated and diverges from theoreticaly 'clean' framework (we had in the previous first change of the EBP algorithm) to more heuristics. With that full-discrete ANN we solved XOR problem in just 5 cycles.

The paper has shown how TLU can be succesfully applied in ANN's at both output as well as the hidden layer. EBP algorithm is still applicable without global changes but the expressions for calculation of delta values should be accordingly changed. Significant reduction in number of iterative learning cycles is achieved making the algorithm faster and without any error. More practical investigation (primarely on larger scale problems) should be done in order to test robustness and applicability of proposed changes.

REFERENCES
[1] D. E. Rumelhart, J. L. McClelland, PDP, Vol. 1, MIT Press, 1986
[2] P. J. Werbos, Beyond regresion, Ph.D. dissertation, Harvard University, 1974
[3] J.M. Zurada, Introduction to Artificial Neural Systems,West Pub.Comp.,St. Paul,1992
[4] T. Poggio, F. Girosi, Networks for Approximation and Learning, Proc. of the IEEE, Sept., 1990

Stochastic Neural Networks

Keith Rennolls, Alan Soper, Phil Robbins, Ray Guthrie.

CASSM, University of Greenwich, London SE18 6PF.

ABSTRACT

The two layer feed-forward neural net (the perceptron) may be regarded as a generalization of a multiple logistic regression model with multivariate output, if the threshold function is interpreted as a generalised linear model link. The multi-layer perceptron can be, and has been, used as a highly structured and parameterised non-linear model of expectation behaviour. The usual back propagation rules for training a neural net are equivalent to fairly simple methods for minimizing the error sum of squares at the outputs, ie. OLS. There has been little work done which uses the multi-layer perceptron as a statistical trend model in which the error distributions at the output nodes are statistically specified. For such non-linear models, OLS will be inefficient as well as biased. However, an appropriate specification of the stochastic error structure will lead to either weighted least squares or maximum likelihood estimates, which may be achieved ether directly or by iteratively re-weighted least squares.

This general statistical approach allows the precision of trained weights to be estimated, as well as the correlation between these weights in the variance-covariance matrix. Such a feature is useful in examining the conjecture that different layers of the MLP have characterised different aspects of the training data set. Statistical methods can then also be used to prune the topology of the network using a step-wise approach.

Availability of the variance covariance matrix of the fitted weights allows the use of the trained MLP to make predictions. With simulated weights from the estimated distribution for the weight estimates prediction distributions may be obtained.

Steepest descent, and hence the back propagation algorithm, is one of the least efficient of optimisation algorithms available. The Fletcher Powell method is much more efficient for well behaved objective functions, whilst the Nelder-Mead algorithm is much faster, and more robust, for multi-modal functions. The use of simulated annealing is indicated for very badly behaved functions.

All of the above statistical features have been built in to a package which also implements the alternative training algorithms. A menu-based inter-face has also been developed for ease of usage. All features will be illustrated in the paper and in demonstration.

A Neurophysiologically Motivated Neural Network Model and its Application to the Superposition Problem

K. Eder[1], H. Geiger[1], W. Brauer[2]

[1]Kratzer Automatisierung GmbH, Carl v. Linde Str. 38, 8044 Unterschleißheim,
ederk@coco.informatik.tu-muenchen.de

[2]Fakultät für Informatik, Technische Universität München, Arcisstr. 21, 8000 München 2

Abstract

In this paper a new neural network model, the single spike model, is presented. This model is derived from neurophysiological knowledge about the membrane of neurons and synchronisation mechanisms recently found in the visual cortex. The model has two major advantages compared to the widespread frequency coded models: a temporal structure, microbursts, can be put to a sequence of spikes thus opening another information channel, where, for example, relations between superimposed patterns can be coded. In addition, the parameters used for simulating the model can be taken from neurophysiological experiments, thus showing the plausibility of the simulation. In an example, a solution to the superposition problem, some of the characteristics of the model are shown.

Introduction

The representation of objects and structures as a combination of features is an important task in language and image processing [1,2]. Using a model with mean firing rates in a superposition of such representations the identification of one object is not unambiguously possible any more. This is called the superposition catastrophe. A solution to this problem is to encode the relatedness of features by temporal correlation, as is proposed in [3]. Results from neurophysiology indicate a synchronisation of spikes or spike trains [4]. For coding geometrically related features the feature linking is proposed by [5,6], which uses the temporal correlation of single spikes. Another approach is an oscillator model, where not the temporal but the phase correlation of neural activity is used [7].

Here a single spike model is proposed which realizes temporal relations by correlation of spike trains, the so called microbursts. This opens another channel where information can be transmitted in the neural network. Information belonging to one input pattern can be coded in the spike frequency within the bursts and the duration of bursts, whereas the correlation of different superimposed input patterns is coded by the temporal structure of repetitive bursts [8]. Additionally amplitude information can be coded as mean burst frequency and information about signal confidence is coded in the standard deviation of the mean frequency within a burst. In our model the temporal distance between bursts adapts dynamically to the number of superimposed patterns. Thus the temporal resolution, which determines the quality of the feature correlation, is nearly independent of the number of competing patterns. Another advantage of the model compared to the phase correlation models is that one neuron can belong to an arbitrary number of superimposed patterns, thus firing simultaneously with all the patterns it belongs to.

The Membrane Model

Our model is derived from a very simplified membrane model, where the resulting membrane potential is calculated from a ground potential, an excitatory and an inhibitory subsynaptic potential [9]. The model is simple enough to be easily simulated and takes all the necessary effects into account which are needed for our applications. Connected to the membrane model is a system to create spikes. Here another potential (U_{Ref}) is added to the membrane potential (U_D) for simulating the refractory period of a neuron and for creating the postdepolarisation necessary for creating bursts. Another optional potential (U_R), which can be added for stochastic purposes, is deviated around zero with adjustable standard deviation. The resulting potential is compared to a threshold and if exceeding it, a spike is produced.

Simulation

An incoming spike influences the appropriate conductivity (S_{inh}/S_{ext}) dependent on w_i, the synaptic strength to the presynaptic neuron i. This influence can be defined as a function in time. In our application a spike sets the conductivity for one time interval to w_i. Other functions e.g. to widen the temporal effect of a Spike [10] or delays in transmission are investigated in current work.

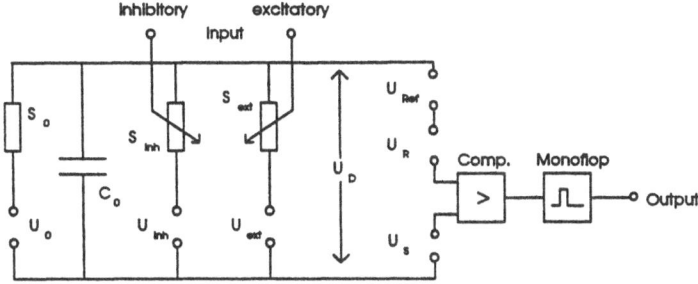

Circuit Diagram for the Model

Description:

U_0:	ground potential of membrane	U_{inh}:	inhibitory subsynaptic potential
U_{ext}:	excitatory subsynaptic potential	U_D:	membrane potential
U_S:	threshold potential	U_R:	stochastic potential with standard deviation σ_R
U_{Ref}:	refractory potential	S_0:	membrane conductivity
S_{inh}:	subsynaptic membrane conductivity caused by inhibitory input		
S_{ext}:	subsynaptic membrane conductivity caused by excitatory input		
C_0:	membrane capacity		

As we derived from the above circuit diagram the membrane potential is defined by:

$$S_0 \left[U_D(t) - U_0 \right] + C_0 \frac{dU_D(t)}{dt} + S_{ext}(t) \left[U_D(t) - U_{ext} \right] + S_{inh}(t) \left[U_D(t) - U_{inh} \right] = 0$$

$S_{inh}(t)$ and $S_{ext}(t)$ represent the sum of the incoming inhibitory and excitatory spikes at time t. When no spike arrives the conductivities are set to 0. For simulation purposes the above equation has to be discretized:

$$U_D(t) = \frac{S_0\, U_0 + S_{ext}(t)\, U_{ext}(t) + S_{inh}(t)\, U_{inh} + \frac{C_0}{\Delta t}\, U_D(t - \Delta t)}{S_0 + S_{ext}(t) + S_{inh}(t) + \frac{C_0}{\Delta t}}$$

During the simulation the delay between the production of a spike and its effect on the postsynaptic membrane has to be considered. But we were able to show that distances up to 100 simulation intervals had no significant effect on the application. The probability of producing a spike depends on the standard deviation σ_R of U_R and on U_{Ref}. $\sigma_R = 0$ is also allowed in order to get a deterministic behaviour of the model.

After producing a spike, the absolute refractory period is starting with a high negative refractory potential (U_{Ref}). So no further spikes can be produced. There is no influence of U_{Ref} to the membrane functionality. So incoming spikes can be processed even in the refractory period. Later U_{Ref} goes to a slightly negative value, thus reducing the firing probability, or even to a positive value allowing for the next spike of a burst.

Learning

A learning rule for our model can directly be derived from Hebbian learning [11]. Therefore the term simultaneity has to be adapted to our model. This is done by defining the change in the synaptic strength dependent on the temporal distance between the post- and presynaptic spike [12]. The following figure shows the weight change with regard to the temporal distance between the pre- and postsynaptic spike.

484

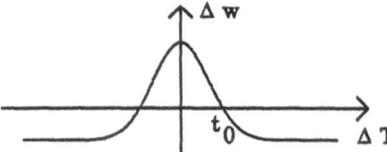

Δw:	weight change
ΔT:	$T_{post} - T_{pre}$
t_0:	time where sign of Δw changes
T_{post}:	time of postsynaptic spike
T_{pre}:	time of presynaptic spike

This function defines an interval, $[-t_0, +t_0]$, typically dependent on the mean duration of a burst, where the synaptic strength is growing and thus becomes more synchronizing. Outside of the defined interval the synaptic strength is becoming negative and thus desynchronizing. The above function is symmetric with regard to positive or negative temporal distances. With asymmetric learning functions, sequential processes can be trained. This is currently worked out [13].

An Application: The Superposition Catastrophe

We use the network structure proposed by Sejnowski [14]. This structure consists of pairs of neurons, one excitatory and one a synchronizing interneuron. This pair alone is able to produce bursts. The structure is now synchronized or desynchronized in comparison to other pairs by trained excitatory or inhibitory synapses.

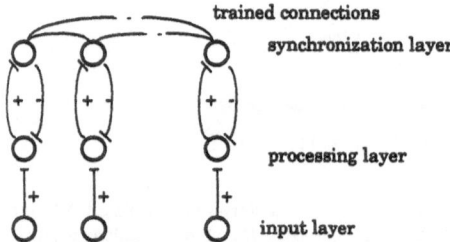

Network Structure for Solving the Superposition Problem

In the synchronization layer several patterns were trained one at a time. Then up to four patterns were superimposed on the trained network. In the following figure (next page) you can see spikes of one representative input neuron for each of the four superimposed patterns (neuron numbers 1-4) and the bursts produced by the corresponding synchronizing neurons (numbers 1a - 4a). The burst of the processing neurons are not shown. As you can see, the input starts with one pattern (time t_1), then one more pattern is superimposed (t_2) and the synchronizing neurons produce bursts alternately. After a while two more patterns are superimposed (t_3) and now all four neurons produce their bursts alternately.

Discussion

The proposed model results primarily from neurophysiological knowledge and modelling. Thus a lot of effects found in biological neural networks can be reproduced in our model. Nevertheless some recently found mechanisms had to be ignored for simplicity of the simulation.

A great advantage of the model is its robustness with regard to the adjustable parameters. The flexibility of the model is in no way exhausted by the shown example. In current work learning paradigms are derived, in particular a modification of the Delta-Rule. In addition the effect of asymmetric coupling between interneurons for training sequential processes is examined.

Of prime importance here is the ability to handle, store and retrieve time dependencies both on an absolute and relative timescale as well as event associations in time. Since these abilities must be present in addition to the ability of handling static information, learning spatial associations and so on, implementaions based on conventional state ANN´s become quite difficult. Time dependent models have, as has been pointed out, a second information channel, which can be used to represent different qualities at the same time within the same network.

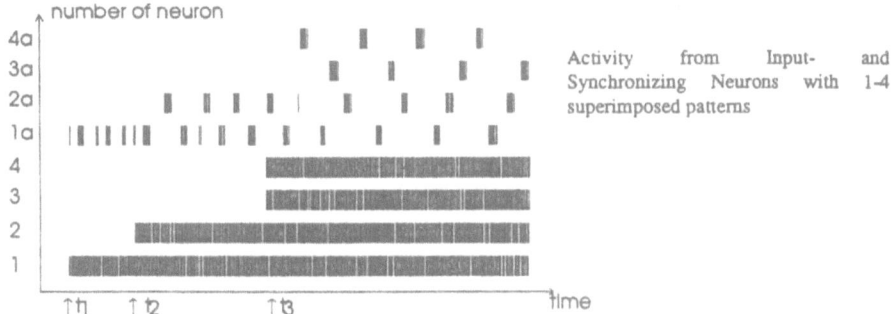

Activity from Input- and Synchronizing Neurons with 1-4 superimposed patterns

Applications are partially implemented and planned in the following fields:
- figure ground separation, a subtask of object recognition [15]
- associative trajectory memory for robotics and for speech and language recognition
- representation and selection of multiple hypotheses in a feedback system

References:

[1]   C. v. d. Malsburg: The Correlation Theory of Brain Function, Internal Report 81-2, Department of Neurobiology, Max-Planck-Institute for Biophysical Chemistry, December 1981.

[2]   K. Eder, R. Deffner, H. Geiger: Ein konnektionistisches System zur Satzerkennung am Beispiel der Übersetzung von Datenbankanfragen nach SQL. In: G. Görz (ed.) Proceedings KONVENS 92, Springer-Verlag 1992, pp 339-347.

[3]   C. v. d. Malsburg: Nervous Structures with Dynamic Links. In Ber. Bunsenges. Phys. Chem. 89, VCH-Verlagsgesellschaft mbH, Weinheim 1985, pp 703-710.

[4]   C. M. Gray, P. König, A. K. Engel, W. Singer: Oscillatory responses in cat visual cortex exhibit inter-columnar syncronization which reflects global stimulus properties. Nature, Vol. 338, 1989, pp 334-337.

[5]   G. Hartmann, S. Drüe: Self Organization of a Network Linking Features by Synchronization. In: R. Eckmiller et al. (eds.): Parallel Processing in Neural Systems and Computers, Elsevier North Holland, 1990, pp 361-364

[6]   R. Eckhorn, H. J. Reitboeck, P. Dicke, M. Arndt, W. Kruse: Feature Linking across Cortical Maps via Synchronisation. In R. Eckmiller et al. (eds.): Parallel Processing in Neural Systems and Computers, North Holland, 1990, pp 101-104.

[7]   P. König, B. Schillen: Segregation of Oscillatory Responses by Conflicting Stimuli - Desynchronizing Connections in Neural Oscillator Layers. In: R. Eckmiller et al. (eds.): Parallel Processing in Neural Systems and Computers, Elsevier North Holland, 1990, pp 117-120.

[8]   H. Geiger: Storing and Processing Information in Connectionist Systems, Proceedings of NSMS 90, Neuss, 1990

[9]   J. C. Eccles, P. Scheid: Physiologie der Nervenzelle und ihrer Synapsen. In: O. H. Gauer, et al.: Allgemeine Neurophysiologie, Urban & Schwarzenberg, München. Berlin, Wien, 1974, pp 108-158.

[10] O. D. Creutzfeld, W. Probst: Neurophysiologische Technik. In: O. H. Gauer, K. Kramer, R. Jung: Allgemeine Neurophysiologie, Urban & Schwarzenberg, München Berlin Wien 1974, pp 250-252.

[11] D. O. Hebb: The Organisation of Behaviour, Wiley, New York 1949

[12] A. Steiner: Vorverarbeitung digitalisierter Einzelwörter und Untersuchungen am Single-Spike-Modell in neuronalen Netzen, Diplomarbeit TU-München, 1991, pp 46-49

[13] D. Butz, K. Noack, W. Brauer: Representation of Real Valued Functions by a Three-Layered Artificial Neural Network with Topologically Ordered Output Units, accepted for proceedings of ICANN, 1993

[14] T. J. Sejnowski: Open Questions about Computation in Cerebral Cortex. In: J. L. McClelland, D. E. Rumelhart: Parallel Distributed Processing, MIT Press, Cambridge London 1986, pp 372-389.

[15] A. Nischwitz, H. Glünder, P. Klausner: Synchronization of Spikes in Populations of Laterally Coupled Model Neurons. In T. Kohonen et al. (eds.): Artificial Neural Networks, Elsevier North Holland, 1991, pp 1771-1774

A Symmetrical Lateral Inhibition Network
for PCA and Feature Decorrelation

Rüdiger W. Brause,
J.W. Goethe-University, FB20 NIPS,
D - 60 054 Frankfurt, Germany

Abstract

This paper introduces a new network model for data decorrelation and principal component analysis which relays on the biological plausible lateral inhibition. Different to already existing approaches, the assignment of the eigenvectors to the neuronal weights are not predefined but in the lateral inhibited network the weights evolve by the network dynamics alone to the eigenvectors of the input data crosscorrelation matrix.

The paper introduces the model and an objective function, presents the learning equations and computes the conditions for the parameters to assure the convergence of the weight vectors to different eigenvectors.

1 Introduction

The encoding of sensor information is a very important subject. Results are used in picture and music encoding and compression (video and audio transmission and storage), in the preprocessing for speech recognition or in tactile and position sensoring for robot control.

The encoding processe should consist of two stages: a linear transformation and a quantization of the output signals. In this paper we consider mainly the linear transformation. If we use the same number m of output channels as there are input lines, the $m=n$ output values y_i are just the projection of the input x on the vectors w_i or the coordinates of x in a new base system $\{w_i\}$. When the w_i are linear independent and complete then we do not loose information and a complete reconstruction of the input by $y = (y_1,...,y_m)$ is possible.

However, if we use with $m<n$ less output lines than input lines we will make a reconstruction error. For linear systems, it is well known that the mean square error is minimized by selecting only those base vectors (eigenvectors) with the biggest eigenvalues [12]. Thus, the eigenvector decomposition (*discrete Karhunen-Loéve transformation, principal component analysis (PCA)*) can be considered as an *optimal* transformation and should be preferred to all other current linear transformations as the discrete Walsh-Hadamard transformation, the descrete Fourier transformation or the discrete Cosinus transformation [6].

Since Oja's statement [7] that a linear, formal neuron using Hebb's learning rule and restricted weights will learn the eigenvector with the biggest eigenvalue several neural network architectures were proposed for a partial or complete eigenvector decomposition. Basically, they consist of two categories: networks which learns the eigenvectors sequentially ("asymmetric networks") which are based on the sequential Gram-Schmidt orthogonalization mechanism, and networks which learn them in parallel ("symmetric networks") and do not predetermine an order of the eigenvectors. The approaches use linear neurons, where each neural weight vector converges to one eigenvector.

Examples of the former architectures are the Sanger decomposition network [10], the lateral inhibition network of Rubner and Tavan [9] and the asymmetrical version of the lateral inhibition network of Földiák [4]. They use as a basic building block the linear correlation neuron which learns the input weights by a Hebb-rule, restricting the weights $w_1,..,w_n$. As Oja showed [7], this learning rule let the weight vector of the neuron converge to the eigenvector of the expected autocorrelation matrix C of the input patterns x with the biggest eigenvalue λ_{max}. The learning rule for one neuron can be generalized, yielding a network where the input is inhibited simultaniously by the projections of the input to all weight vectors. This corresponds to the symmetric network approach. The symmetrical Oja subspace network [8], the Williams subspace learning [13] and the symmetrical decorrelation network of Silva and Almeida [11] have the same property: They assume the propagation of weight values in the network which is not desirable neither from the biological point of view nor from the pathway restrictions of VLSI-implementations.

In fact, a fully symmetrical, stable network for eigenvector decomposition, construced by an objective function and implemented by a biological plausible and easily realizable network mechanism is still missing. Contrary to the opinion of Hornik and Kuan [5], who are not in favour of an symmetric PCA network due to convergence problems (e.g. the model of Földiák [4]), we will introduce a new symmetrical, stable model in this section which is not covered by their general convergence analysis of the PCA models mentioned above.

2 The symmetric network for eigenvector decomposition

Let us assume in a first step that we have m neurons which are laterally interconnected as shown in figure 1.

$$x = (\quad x_1, \qquad x_2, \qquad \cdots \qquad x_n)$$

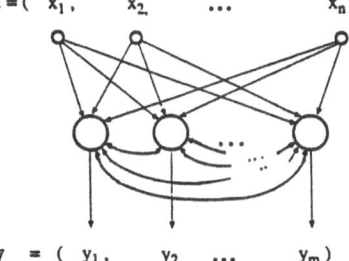

$$y \ = \ (\quad y_1, \qquad y_2, \qquad \cdots \qquad y_m)$$

Fig. 1 The symmetric, lateral interconnected network model

Each neuron i initially has a randomly chosen weight vector w_i. After we presented one input pattern x in parallel to each neuron of the linear system, the output of neuron i will result in

$$y_i = w_i^T x + T_i \qquad\qquad T_i = \Sigma_{j \neq i} u_{ij} y_j \qquad\qquad (2.1)$$

where T_i denotes the influence by the lateral connections which are weighted by the lateral weights u_{ij}. The input is assumed to be centered. If this is not the case, it can be made by introducing a special threshold weight learned with an Anti-Hebb-rule, see [2].

Although the model is quite linear, we have reactions for random input and weights due to the feedback lines which are difficult to analyze. Nevertheless, for the prediction of the system behaviour the analysis of the expected equilibrium states of the system is sufficient.

Let us assume that after an input pattern has been presented the system activity stabilizes. This is the case, when the feed-back does not induce additional oscillations, i.e. when all the eigenvalues of the feed-back matrix U are allways smaller than one [5].
Then the output for neuron i becomes with Eq. (2.1)

$$y_i = w_i^T x + \Sigma_{j \neq i} u_{ij} y_j \ = w_i^T x + u_i^T y - y_i \qquad\qquad u_{ii} = 1$$

and the output vector becomes $2y = Wx + Uy$ or $(2I-U)y = Wx$ with the identity matrix I.
Thus, the system output

$$y = (2I-U)^{-1} W \ x = A \ x \qquad\qquad A = (2I-U)^{-1} W \qquad\qquad (2.2)$$

depends again linearly on the input.

3 Learning the weights

The learning rule for the weights a_i is determined by the following three conditions

- The new features should be decorrelated $\langle y_i y_j \rangle = \langle y_i \rangle \langle y_j \rangle = 0$ (3.1)
- The variance of the features should be maximal $\sum_i \langle y_i^2 \rangle = \max$ (3.2)
- The decomposition should be neutral (no scaling) $|a_i| = 1$, i.e. $\det(A) = 1$ (3.3)

These conditions can be modelled by the minimum of deterministic objective function

$$R(a_1,...,a_m) = 1/4 \sum_i \sum_{j \neq i} (\langle y_i y_j \rangle)^2 - 1/2\, \alpha \sum_i \langle y_i^2 \rangle = R^1 + R^2 \qquad (3.4)$$

The first term R^1 ensures that the cross-correlation (3.1) is always counted positive. This results in a minimum of R(.) where the cross-correlation, becomes zero and $-R^2$, the sum of all variances, becomes maximal.

    The third condition (3.3) have to be additionally ensured during the learning process. This condition could also be integrated into the objective function. It was shown for one neuron [3] that this yields also the eigenvectors as solutions and can be compared to the approach using (3.2) to compute the unique maximum and minimum of the objective function [2]. It can be shown that the objective function R(a) takes its extrema when the a_i, the lines of the matrix A, are a subset of the eigenvectors of the autocorrelation matrix $C_{xx} = \langle xx^T \rangle$. Since C_{xx} is symmetric and real, the eigenvalues λ_i are real and the eigenvectors form an orthogonal base system. Here, the cross-correlations

$$\langle y_i y_j \rangle = a_i^T \langle xx^T \rangle a_j = a_i^T C_{xx} a_j = a_i^T \lambda_j a_j = 0 \qquad \forall\ i \neq j$$

become zero, and by (2.2), we have $U = I$ and $A = (2I-U)^{-1}W = W$. Thus, we can break up the weight vector a in two parts: the input weights w_i which should converge to the eigenvectors and the lateral inhibition weights u_{ij} which should become zero. The minimum of the objective function can therefore be approximated by a gradient search for the weight vectors w_i only where we assume the lateral inhibition to be a constant value at each learning step, learned seperately. The conditions (3.1), (3.2), (3.3) are ensured by using the objective function (3.4) for a=w which yields for w_i the eigenvectors as solution. The (t+1)-th iteration step is

$$w_i(t+1) = w_i(t) - \gamma(t)\, \nabla_w R(w_i) \qquad (3.5)$$

denoting the gradient by the Nabla-operator ∇_w. With the definition $u_{ij} = - \langle y_i y_j \rangle$ the learning rule is computed as

$$w_i(t+1) = w_i(t) + \gamma(t) \langle\, x\, (\alpha y_i + \sum_{j \neq i} u_{ij} y_j)\, \rangle \qquad (3.6)$$

The stochastic version is obtained by dropping the expectation brackets "$\langle\ \rangle$".

The lateral weights should be updated by a rule which let them become the expected cross-correlation. It can be shown that for the average value of $-y_i(t)y_j(t)$ at step t can be obtained by

$$u_{ij}(t) = u_{ij}(t-1) - 1/t\, (u_{ij}(t-1) + y_i(t)y_j(t)) \qquad (3.7)$$

This learning rule gets the average of the random variable $v = y_i y_j$. But, this is not the quantity we are looking for, because v is not stationary for changing w. Therefore, random initial values of the weights can disturb the average for a long period of simulation time. To get rid of these random values and to accelerate the convergence, we might use instead of the learning rule (3.7) the temporal floating average of N observed data.

4 Stability conditions for the learning fixpoints

It is well known that the sequential gradient descend algorithm (3.5) confirms a monotonic decrease of the quadratic function (3.4), because we have

$$\frac{\partial R(t)}{\partial t} = \frac{\partial R(a)}{\partial a}\frac{\partial a}{\partial t} = -\frac{\partial R(a)}{\partial a}\gamma(t)\frac{\partial R(a)}{\partial a} = -\gamma\left(\frac{\partial R}{\partial a}\right)^2 \le 0 \qquad (4.1)$$

Since the objective function R has a lower bound of $\min(-1/2\Sigma_i\langle y_i^2\rangle) = -1/2\ m\ \max(a_i^T\langle xx^T\rangle a_i) = -1/2\ m\ \max(a_i^T C_{xx} a_i) = -1/2\ m\lambda_{max}$ for linear systems, the objective function can be regarded as a Ljapunov function and the iteration will converge to the fixpoints.

It can be proven that all the fixpoints of the system are at the eigenvectors of the autocorrelation matrix. Note that this means only that the fixpoints of the system are eigenvectors, they have *not* to be necessarily *different* ones. To ensure different eigenvectors, we can compute the condition for the case when all weight vectors but one have converged to different eigenvectors. This leads to the condition (see [1]) for the autocorrelation

$$\alpha < \lambda_k^2/(\lambda_k-\lambda_l) \qquad (4.2)$$

for all eigenvalues λ_k and λ_l. Additional conditions are obtained by a local fixpoint analysis for α and for the learning rate γ, see [1]. Thus, the convergence region is limited by (4.2), but to ensure convergence at fixed values of α and γ for all different eigenvectors we should choose

$$\alpha < 4\lambda_{min}, \qquad \gamma < 2/\lambda^2_{max} \qquad (4.3)$$

which is valid for deterministic iterations of (3.6).

References

[1] R.Brause: Transform Coding by Lateral Inhibited Neural Nets; Internal Report (1993)

[2] R.Brause: The Minimum Entropy Network; Proc. IEEE Tools for AI TAI-92, Arlington (1992)

[3] Y.Chauvin: Principal Component Analysis by Gradient Descent on a Constrained Linear Hebbian Cell; IEEE Proc. Int. Conf. Neural Networks, pp. I/373-380 (1989).

[4] P.Földiák: Adaptive Network for Optimal Linear Feature Extraction; IEEE Proc. Int. Conf. Neural Networks; pp. I/401-405 (1989).

[5] K. Hornik, C.-M. Kuan: Convergence Analysis of Local Feature Extraction Algorithms, Neural Networks, Vol.5, pp.229-240, 1992

[6] A. Habibi, P. Wintz: Image Coding by Linear Transformation and Block Quantization; IEEE Trans. on Comm. Techn., Vol. COM-19, No 1, pp.50-62, 1971

[7] E. Oja: A Simplified Neuron Model as a Principal Component Analyzer, J. Math. Biol. 13: 267-273 (1982)

[8] E. Oja: Neural Networks, Principal Components, and subspaces, Int. J. Neural Systems, Vol 1/1 pp. 61-68 (1989)

[9] J. Rubner, P. Tavan: A Self-Organizing Network for Principal-Component Analysis, Europhys. Lett., 10(7), pp. 693-698 (1989).

[10] Sanger: Optimal unsupervised Learning in a Single-Layer Linear Feedforward Neural Network; Neural Networks Vol 2, pp.459-473 (1989)

[11] F. Silva, L. Almeida: A Distributed Solution for Data Orthonormalization; Proc. ICANN-91, Artificial Neural Networks, Elsevier Sc.Publ., pp.943-948 (1991)

[12] J.T. Tou, R.C. Gonzales: Pattern Recognition Principles; Addison-Wesley Publ. Comp., 1974

[13] R.Williams: Feature Discovery through Error-Correction Learning; ICS Report 8501, University of Cal. and San Diego, 1985

Minimizing the System Error in Feedforward Neural Networks with Evolution Strategy

Willfried Wienholt [1]

Ruhr–Universität Bochum
Institut für Neuroinformatik ND 04/584
44780 Bochum, Germany
ww@neuroinformatik.ruhr–uni–bochum.de

Abstract

In this paper evolution strategy is applied to minimize the system error in feedforward neural networks. The evolution strategy does not use externally tuned learning parameters. Moreover, it is not necessary to evaluate a gradient information as required by the backpropagation algorithm. Experimental results are presented and compared with the standard backpropagation technique.

1   Introduction

Several attempts have been made to improve the convergence properties of the backpropagation (BP) technique in feedforward neural networks (FFNN) [1, 2] introduced by Rummelhart and McClelland [3]. The problem is how to avoid getting stuck in suboptimal solutions caused by the steepest descent method and how to adjust the learning rate η and momentum α properly.

Baba [4] applies an alternative optimization method proposed by Solis & Wets [5]: The weights and thresholds of the FFNN are adjusted by Gaussian random numbers. The algorithm of Solis & Wets is based on a random optimization method introduced by Matyas [6]. Matyas proves that his optimization method ensures convergence to a global optimum with propability 1 on a compact set. This is an important advantage compared to the steepest descent method. Baba's results show a remarkable improvement of convergence speed compared with the BP but the variance of the Gaussian distribution needs to be adjusted by hand at the beginning of the optimization in order to guarantee *fast* convergence.

The evolution strategy (ES) first proposed by Rechenberg in the early 70th [7, 8] and expanded by Schwefel [9] is one kind of evolutionary algorithms. These algorithms imitate the rules of biological evolution for optimization purposes. An artifical population consists of a certain number of individuals. Each individual has a fitness which measures how well it is adapted to the system it lives in. The better the fitness the better is its chance for survival in the sense of Darwinian evolution.

In the following section the ES is explained and we show how it is applied to the minimization of the system error in FFNN. The system error E_S and the activation function used are defined as:

$$E_S = \frac{1}{\#P \cdot \#U} \cdot \sum_{i=1}^{\#P} E_i; \ E_i = \sum_{j=1}^{\#U} (o_{ij} - y_{ij})^2; \ f(In, \Theta, T) = \left(1 + \exp\left(-\frac{In + \Theta}{T}\right)\right)^{-1} \quad (1)$$

#P is the number of patterns to be learned, #U defines the number of units in the output layer.

---

[1]The author is supported under project 107 003 91 by the research Initiative *Parallel Computing* of Nordrhein–Westfalen, Land of the federal Republic of Germany

o_{ij} is the output produced by the jth component and the ith pattern, y_{ij} is the corresponding correct output given by the teacher. We use the sigmoid function as the activation function for each unit in the network with the input In, the threshold Θ, and the *temperature* T.

2 The Evolution Strategy

Following the notation of Schwefel [9] the different modes of ES are described by the abbreviation $(\mu + \lambda)$–ES and (μ, λ)–ES. μ is the number of parents of a generation and λ is the number of offsprings. In each generation, the λ children are breeded by the parents under a special selection/recombination mechanism. The ","; variant produces λ offsprings and the μ best offsprings are selected as parents for the next generation. The "+" variant selects the μ best out of the μ parents *and* λ offsprings for the next generation. The variables used are superscripted by "p" and "o" to differentiate between the parent and offspring population, respectively. The algorithm is as follows:

1. Code all weights and thresholds into a vector of real numbers $\underline{x} \in I\!R^n$, and assign a *strategy parameter $\underline{\sigma}$*. Thus, each individual of the population consists of a tupel $(\underline{x}, \underline{\sigma})$. The fitness of each individual is defined by its system error E_S.

2. Assign to each x_{ij}^p an equally distributed random number of the interval $[-0.5, 0.5]$ and set each σ_{ij}^p of the initial parent population to $\frac{1}{6}$. (i counts the individual and j counts the component.)

3. Breed an offspring population by the follwing recombination method: Repeat for each child i:

   (a) Define the *strategy parameter mutation factor* Γ by tossing a coin:
   $$\Gamma = 1.5 \text{ (head) or } \Gamma = 1/1.5 \text{ (tail)} \tag{2}$$

   (b) Repeat for each component j:
   - Breed a new strategy parameter σ_{ij}^o of an offspring by the so called *intermediate recombination* [9] of ρ randomly selected strategy parameters $\sigma_{\rho j}^p$ of the parent generation:
   $$\sigma_{ij}^o = \frac{1}{\rho} \cdot \sum_{k=1}^{\rho} \sigma_{kj}^p \tag{3}$$

   - Mutate the new strategy parameter by multiplication with the strategy parameter mutation factor Γ:
   $$\sigma_{ij}^o = \Gamma \cdot \sigma_{ij}^o \tag{4}$$

   - Breed a new offspring parameter by selecting randomly one parent k out of μ parents and mutate it with a Gaussian random number:
   $$x_{ij}^o = x_{kj}^p + \frac{\sigma_{ij}^o}{\sqrt{n}} \cdot z_{ij} \tag{5}$$

   z_{ij} is a N(0,1) normally distributed random number. n is the dimension of the parameter vector $\underline{x}$.

   (c) Calculate the system error E_S of the FFNN.

4. Select the best μ offsprings ($\rightarrow$ lowest E_S) and define them as parents for the next generation ($\rightarrow$ "," selection).

5. For terminating the algorithm check the termination criterions. If the conditions are fullfilled, then terminate the algorithm otherwise continue with step 3.

$(\sqrt{n})^{-1}$ in eq.5 is a factor which is derived from theory. It is important to guarantee good convergence [9]. The intermediate recombination (eq.3) and mutation (eq.4) of the strategy parameter is often called a *step width adaptation* because the strategy parameter defines the width of the Gaussian distribution mutating the offspring parameter (eq.5). The adaptation of the strategy parameters makes the ES perform better than other optimization techniques. Note that no gradient information needs to be evaluated.

3   Experiments

We have tested the "+" and "," variant of the ES. The test cases are shown in Tab.1. The optimization process is stopped if the system error E_S is less than 10^{-4} or the number of iterations exceeds 160000. We set the parameter ρ (eq.3) equal to μ. The number of iterations for the ES is the mean of the 10 best out of 20 runs, each. The parity problems are equal to the problems of Baba [4]. The # in the column BP stands for the number of complete training set presentations for a BP with learning rate and momentum. The # in the ES columns stands for the number of generations. In each generation, λ complete training pattern presentations are necessary because each individual must evaluate the complete training set. On a parallel computer architecture this could be done in parallel. The # columns are directly comparable between the BP and the $(1+2)$-ES. (Although the BP needs only one complete presentation of the pattern set, the error must be back-propagated through the FFNN. Furthermore, the gradient has to be evaluated by the BP. This is not the case in the ES.) The training patterns used are "1/0" patterns.

Test Case	#Parms		#Pat	(2,20)-ES		(1+2)-ES		BP
	nT	T		nT	T	nT	T	nT
: NN Structure				#	#	#	#	#
Encoder: 2-1-2	7	10	2	500	440	364	282	2638
Encoder: 4-2-4	22	28	4	3900	2780	3764	2408	2892
Encoder: 8-3-8	59	70	8	33300	15740	28782	13112	3120
Encoder: 16-4-16	148	168	16	154980	53020	111626	46246	–
Parity: 7-7-7-1	121	136	10	1920	1460	1800	1442	345
Parity: 6-6-6-1	91	104	64	75480	40120	–	–	–

Table 1:Results of the test cases. #Parms defines the number of parameters which must be coded into the real valued vector $\mathbf{x}$, #Pat defines the number of training patterns to be learned. T resp. nT means that the Temperature T in eq.1 is taken into account/not taken into account as an additional parameter in the optimization.

In the encoder problems the $(1+2)$-ES dominates over the $(2,20)$-ES. The number of generations needed to fall below $E_S = 10^{-4}$ is lower if the temperature T of the activation function (eq.1) is taken into account as an additionally parameter to be optimized. In this way, the number of generations needed is more than halved in the case of the $8-3-8$ and $16-4-16$ encoder. Compared to the BP, the $(1+2)$-ES shows a better performance except for the 8-3-8 encoder. It is an interesting fact that only the ES variants actually found solutions in the case of the 16-4-16 encoder.

The 7-7-7-1 parity problem has been investigated with a selection of 10 patterns out of 128 possible patterns. Both ES variants need approximately the same number of generations to satisfy the termination criterion but the BP performs best. The 6-6-6-1 parity problem is trained with the complete set of 64 training patterns. Only the $(2,20)$-ES finds solutions under the termination criterium used. Recent experiments show that a (30,200)-ES with $\rho = 3$ yields good results even for the 10 bit parity problem.

4   Conclusions

In this paper we presented evolution strategies as an optimization technique for optimizing the system error in FFNN. As opposed to the BP, the ES does not rely on external parameters (that have to be adjusted) and the evaluation of gradient information. The ES finds solutions which are conform with the boundary conditions even in search spaces of high dimensionality where we have not found any solution with BP. The method is based on a random optimization technique which guarantees to find the global optimum with probability one on a compact set. The only additional parameters used are the *strategy parameters* which are self-adaptive. We believe that the ES (especially the (μ, λ)–ES) is a robust algorithm concerning topology variations (number of layers and number of units) of the network because the algorithm is always the same and no external parameters have to be changed if the topology of the FFNN is modified. Additional parameters like the *temperature T* of the activation function used can easily be integrated into the algorithm. The ES would be more efficient on a parallel computer architecture because the individuals could be evaluated in parallel and the disadvantage of long convergence time on a sequential machine compared to the BP disappears. Further research in evolution strategy is concerned with its theoretical properties like the sizing of the population, global convergence, and the determination of the convergence rates [10]. Another parameter of the Gaussian distribution not taken into account yet is the momentum. [11] shows that an adaptation scheme for the momentum which is very similar to the self-adaptation of the strategy parameters improves the convergence speed in parameter optimization problems. The results are promising and we currently look into using the evolution strategy for a number of applications. One example is off-line learning of a FFNN as part of an obstacle avoidance system for road vehicles.

References

[1] Robert A. Jacobs. Increased rates of convergence through learning rate adaptation. *Neural Networks*, 1:295 – 307, 1988.

[2] Tom Tollenaere. SuperSAB: Fast adaptive back propagation with good scaling properties. *Neural Networks*, 3:561 – 573, 1990.

[3] D. E. Rummelhart, G. E. Hinton, and R. J. Williams. Learning internal representation by error propagation. In David E. Rummelhart and James L. McClelland, editors, *Parallel Distributed Processing, Volume 1: Foundations*, chapter 8, pages 318 – 362. MIT Press, 1986.

[4] Norio Baba. A new approach for finding the global minimum of error function of neural networks. *Neural Networks*, 2:367 – 373, 1989.

[5] Francisco J. Solis and Roger J–B. Wets. Minimzation by random search techniques. *Mathematics Of Operations Research*, 6:19 – 30, 1981.

[6] J. Matyas. Random optimization. *Automation and Remote Control*, 26:246 – 253, 1965.

[7] Ingo Rechenberg. Cybernetic solution paths of an experimental problem. Technical report, Royal Aircraft Establishment, Farnborough, Library Translation 1122, 1965.

[8] Ingo Rechenberg. The evolution strategy — a mathematical model for darwinian evolution. In F. Frehland, editor, *Synergetics — From Microscopic to Macroscopic Order*, pages 122 – 132, Berlin, 1984. Springer.

[9] Hans–Paul Schwefel. *Numerical Optimization of Computer Models*. John Wiley & Sons, Chichester, 1981.

[10] Thomas Bäck, Günter Rudolph, and Hans-Paul Schwefel. Evolutionary programming and evolution strategies: Similarities and differences. In D. B. Fogel and W. Atmar, editors, *Proceedings of the 2nd Annual Conference on Evolutionary Programming*, pages – , San Diego, CA, February 1993 (in print).

[11] Andreas Ostermeier. An evolution strategy with momentum adaptation of the random number distribution. In Reinhard Männer and Bernard Manderick, editors, *Parallel Problem Solving from Nature (PPSN II)*, pages 197 – 206, Elsevier Science Publishers, North–Holland, 1992.

Prove of Convergence of Extended Divide & Conquer Networks

Steve G. Romaniuk
Department of Information Systems and Computer Science
National University of Singapore
10 Kent Ridge Crescent
Singapore 0511
e-mail: stever@iscs.nus.sg

Abstract

The task of determining an effective architecture for multi-layer feed forward backpropagation like neural networks can be a time consuming effort. Over the past couple years several algorithms were proposed for dynamically constructing network architectures. Some of these algorithms have been shown to converge for binary data. Unfortunately, the results do not carry over for the non-binary input case. For classificatory problems no guarantees are provided, which would suggest otherwise. In this paper, we present an extension to the basic network growing algorithms, which allows constructing networks in bounded time. The algorithm guarantees convergence for any classificatory domain, as long as no contradictory training examples are present. The extension is described for the Divide & Conquer Networks (DCN) algorithm. The derived mathematical model can be readily incorporated into other network growing approaches to ensure convergence. The model is dependent on the usage of simple threshold cells, which can be applied to a variety of learning rules.

1 Introduction

In recent years network growing learning algorithms have received considerable attention, due to their ability to overcome some of the major shortcomings of more traditional neural network approaches. It is a well known fact, that in order to apply feed forward neural networks to learning requires making several choices, such as determining the number of hidden units, choosing a transfer function and fine tuning learning parameters. Several approaches have been proposed to dynamically create network structures, such as Divide & Conquer Networks [6], Cascade Correlation [3], Upstart Algorithm [4], and Extentron [2]. For several of these algorithms it was observed that after phases of rapid improvement (minimizing overall error) stages of no improvement would be entered, or even performance would decrease. These stages of stagnation would persist even though new hidden units would be introduced. Eventually, improvements would be obtained and solutions found. In some cases this resulted in exceedingly large network structures. For all cases only a limited number of patterns could not be recognized suggesting a general problem with picking at least a single example from the remaining set (not yet correctly recognized by the network) and learning it in conjunction with what is already known. In order to guarantee convergence and avoid unnecessary introduction of new hidden units (features) without improving the networks performance, it was deemed important to extent the DCN algorithm [6]. This extension allows picking at least one additional training pattern apart from those already correctly learned.

2 Divide & Conquer Networks

DCN dynamically creates network architectures based upon the domain examples presented, which allows more effective learning and does not require any preconceived guess on part of the human user. Training is done on the weights of the input links to one cell at any given time. Hence, backpropagation of error through hidden cells is not required. The architecture created may have hidden layers with multiple cells in them, as well as multiple hidden layers. Each cell introduced into the network has connections from all cells in the network that were previously introduced. One output at a time is trained in cell sharing mode, otherwise outputs are trained in parallel. The first cell introduced on a hidden layer is trained on all of the training examples. If it classifies the training set correctly, the cell will serve as an output (Conquer Phase). Succeeding cells on a layer are trained on subsets of the original train set with some (or all) examples correctly recognized omitted from the training set (Divide Phase). DCN attempts to economize on hidden layers to provide for more parallelism. The algorithm can effectively make use of various local learning rules, including quickprop, delta, and perceptron rule. Each cell in the network (other than the original input cells) serves as a form of feature detector. In introducing connections to new cells in the network a form of constructive induction (of features) takes place.

DCN differs from other approaches such as [1] in that it will add cells in multiple hidden layers. However, it can add multiple nodes on each layer unlike Cascade Correlation [3]. Furthermore, DCN does not require the use of any correlation measure. There is no dependence upon a tree structure as in [4]. Divide and Conquer Networks only require training examples that have not already been correctly classified during the divide phase.

3 Derivation of Extension Model

In this section we prove that given a feature F_K , which can correctly classify a subset of the original train set, a new feature F_{K+1} can be constructed that allows the recognition of at least one additional training pattern. Repeated application of this procedure - in case of failure of the incremental learning process - yields guaranteed convergence for any classificatory problem, unless contradictory training patterns are present in the train set. We state the above in the following theorem.

Theorem 1 *Let the current feature cell be $F_K = O(S_K)$, where $S_K = \sum_{i=1}^{K} F_{i-1} * W_{K,i}$ and the transfer function is the threshold function,*

$$O(a) = \begin{cases} 0, & a \leq 0 \\ 1, & a > 0 \end{cases}$$

Furthermore, let $T(I_l) \in \{0,1\}$ be the output for training example I_l. Let E be the set of all training examples and E_{F_K} be a true subset of E, $E_{F_K} \subset E$. Assume, $\forall I_l \in E_{F_K}$ $F_K = O(I_l * W_K) = T(I_l)$. Here, W_K represents the weight vector for feature F_K. Then, $\exists I_s \in E, I_s \ni E_{F_K}$ $O(I_s * W_K) \neq T(I_s)$. Given this condition the goal is to find a feature F_{K+1} such that: $O(I_s * W_{K+1}) = T(I_s) \wedge \forall I_l \in E_{F_K}$ $O(I_l * W_{K+1}) = T(I_l)$.

Proof Theorem 1 *The problem reduces to finding a weight vector W_{K+1} for feature F_{K+1}. The first case we investigate is for $T(I_s) = 0$. Given the initial constraints we can derive the following four conditions that have to be met in order to satisfy the earlier stated goal.*

$$\begin{aligned}
O(I_l * W_K) &= T(I_l), & \forall I_l \in E_{F_K} \ (a) \\
O(I_s * W_K) &= 1, & I_s \ni E_{F_K} \ (b) \\
O(I_l' * W_{K+1}) &= T(I_l), & \forall I_l \in E_{F_K} \ (c) \\
O(I_s' * W_{K+1}) &= 0, & I_s \ni E_{F_K} \ (d)
\end{aligned} \tag{1}$$

We define $I_l' = (I_l, F_K)$ and $I_s' = (I_s, F_K)$ as the composite of the original vectors I_l and I_s augmented by feature F_K. From Eq. (2d) it follows,

$$O(I_s' * W_{K+1}) = O(I_s * W_K' + F_K * W_{K+1,K+1}) = 0 \tag{2}$$

From the above condition we can derive the next set of conditions:

$$I_s * W_K' + F_K * W_{K+1,K+1} \leq 0 \Rightarrow F_K * W_{K+1,K+1} \leq -I_s * W_K' \tag{3}$$

Since $F_K = O(I_s * W_K) = 1$ (Eq. (1b)) we conclude from Eq. (3),

$$W_{K+1,K+1} \leq -I_s * W_K' \tag{4}$$

Using Eq. (2a) and (2c) we state,

$$O(I_l * W_K) = O(I_l' * W_{K+1}) = O(I_l * W_K' + F_K * W_{K+1,K+1}) \tag{5}$$

We consider two separate cases:
1. Case $T(I_l) = 1$:

$$O(I_l * W_K' + F_K * W_{K+1,K+1}) = 1 \Rightarrow I_l * W_K' + F_K * W_{K+1,K+1} > 0 \tag{6}$$

2. Case $T(I_l) = 0$:

$$\begin{aligned}
O(I_l * W_K) &= 0 \Rightarrow I_l * W_K \leq 0 \wedge \\
O(I_l' * W_{K+1}) &= O(I_l * W_K' + F_K * W_{K+1,K+1}) = 0 \Rightarrow \\
I_l * W_K' &+ F_K * W_{K+1,K+1} \leq 0
\end{aligned} \tag{7}$$

The above result leads to the following two conditions which need to be met for the case $T(I_s) = 0$.

$$W_{K+1,K+1} \leq -I_s * W_K' \qquad (a)$$
$$F_K * W_{K+1,K+1} > -I_l * W_K' \vee F_K * W_{K+1,K+1} \leq -I_l * W_K' \quad (b)$$
$$(8)$$

Since F_K is either,

$$F_K = O(I_l * W_K) = 1 \vee F_K = O(I_l * W_K) = 0 \qquad (9)$$

we obtain the following set of conditions,

$$W_{K+1,K+1} \leq -I_s * W_K' \qquad (a)$$
$$W_{K+1,K+1} > -I_l * W_K' \vee 0 \leq -I_l * W_K' \quad (b)$$
$$(10)$$

From Eq. (11a) and (11b) we derive:

$$-I_l * W_K' < W_{K+1,K+1} \leq -I_s * W_K' \Rightarrow I_l * W_K' > I_s * W_K' \qquad (11)$$

Finally, we can state for $T(I_s) = 0$,

$$(I_l - I_s) * W_K' > 0 \wedge W_{K+1,K+1} \leq -I_s * W_K', T(I_l) = 1$$
$$W_{K+1,K+1} \leq -I_s * W_K' \wedge I_l * W_K' \leq 0, T(I_l) = 0$$
$$(12)$$

For the second case we consider $T(I_s) = 1$. From Eq. (1) and (2) we obtain,

$$O(I_s' * W_{K+1}) = O(I_s * W_{K+1}' + F_K * W_{K+1}K + 1) = 1 \Rightarrow$$
$$I_s * W_K' + F_K * W_{K+1,K+1} > 0$$
$$(13)$$

Note, that the outputs of Eq. (1a) and (1c) have been reversed. Now, since $F_K = O(I_l * W_K) = 0$, it follows, $I_s * WK \leq 0$. By setting Eq. (1a) and (1c) equal we get,

$$O(I_l * W_K) = O(I_l' * W_{K+1}) = T(I_l) \qquad (14)$$

We again investigate two cases separately.
1. Case $T(I_l) = 0$:

$$O(I_l' * W_{K+1}) = 0 \Rightarrow I_l' * W_{K+1} \leq 0 \Rightarrow I_l * W_K' + F_K * W_{K+1,K+1} \leq 0 \qquad (15)$$

2. Case $T(I_l) = 1$:

$$O(I_l' * W_{K+1}) = 1 \Rightarrow I_l' * W_{K+1} > 0 \Rightarrow I_l * W_K' + F_K * W_{K+1,K+1} > 0 \Rightarrow I_l * W_K' > 0 \qquad (16)$$

This leads to the final set of conditions for $T(I_s) = 1$:

$$I_s * W_K' > 0 \wedge I_l * W_K' \leq 0, T(I_l) = 0$$
$$I_s * W_K' > 0 \wedge I_l * W_K' > 0, T(I_l) = 1$$
$$(17)$$

From Eq. (17) it follows that weight $W_{K+1,K+1}$ is not required. This of course further implies that the original feature F_K is also not required.
Q.E.D.

4 Application of Extension Model to DCN

A summary of the necessary conditions for implementing the decision strategy of the extension algorithm is displayed in Table 1. Solutions to this problem may be obtained by using results from operations research. The problem requires determining the vector W_{K+1}' with respect to the conditions imposed on the product between said vector and training patterns $I_l \in E_{F_K}$ and $I_s \ni E_{F_K}$. This can be accomplished using algorithms for solving linear programming problems [5]. Note, that the conditions for $T(I_s) = 1$ do not require the feature F_K, which implies the cell that implements this feature is redundant and can therefore be pruned from the network.

Table 1. Decision Strategy for Extension Model

$T(I_l)/T(I_s)$	0	1
0	$W_{K+1,K+1} \leq -I_s * W_K' \wedge I_l * W_K' \leq 0$	$I_s * W_K' > 0 \wedge I_l * W_K' \leq 0$
1	$W_{K+1,K+1} \leq -I_s * W_K' \wedge (I_l - I_s) * W_K' > 0$	$I_s * W_K' > 0 \wedge I_l * W_K' > 0$

(1) Let $ErrorBits = card(E_{F_K}) \wedge K = 0$
(2) While $ErrorBits > 0$ do
    (2.1) Perform DCN
    (2.2) Determine $NewErrorBits$
    (2.3) If $NewErrorBits \geq ErrorBits$ then
        (2.3.1) Select $I_s \in E \wedge I_s \ni E_{F_K}$
            (2.3.1.1)For all $I_l \in E_{F_K}$ do
                (2.3.1.1.1) Given $T(I_s) \wedge T(I_l)$ satisfy conditions in Table 1
    (2.4) $ErrorBits = NewErrorBits$
    (2.5) Create New Feature F_K
(3) Halt

Figure 1: High Level Description of Extended DCN Algorithm

We state the Extended Divide & Conquer algorithm in Figure 1. Note, that by the very nature of the algorithm convergence for classificatory domains (continuous inputs and binary outputs) is guaranteed. If the DCN algorithm can not improve on the ErrorBits parameter (Number of Training Patterns incorrectly classified by current feature F_K), then the extension part of the algorithm is executed. It guarantees that any training patterns $I_s \in E \wedge I_s \ni E_{F_K}$ will be recognized by the new feature F_{K+1}. Under worst case conditions K such features are generated and the condition in step (2.3) is true.

5 Summary

In this paper an extension for dynamic network growing algorithms is presented, which not only proves convergence of algorithms utilizing this method (for all classificatory domains), but also supports deriving simpler network architectures in cases, were no improvement in reduction of number of incorrectly classified training patterns is possible during installation of a new hidden unit. Derivation of the extension model is motivated by observing that during several experiments using Divide & Conquer Networks (as well as other methods), extended stages of no improvement (no reduction of error bits) were encountered. Extension of other network growing approaches is paramount, due to a similar inability to appropriately cope with this problem. Picking up at least a single (not yet correctly recognized) example pattern during training by a new feature cell is not guaranteed, - at least within some pre-determined number of epochs. Applying the extension algorithm in such a case, can guarantee improvement by at least one additional training pattern and eventual convergence by the network growing algorithm.

References

[1] Ash, T. (1989). *Dynamic Node Creation in Backpropagation Networks.* (Tech. Report ICS Report 8901), Inst. for Cognitive Science, University of California, San Diego, La Jolla, Ca.

[2] Baffes, P.T. and Zelle, J.M (1992). Growing Layers of Perceptrons: Introducing the Extentron Algorithm, *Proceedings of the 1992 International Joint Conference on Neural Networks* (pp. II-392- II-397), Baltimore, MD., June.

[3] Fahlman, S.E. and Lebiere, C. (1990). The Cascade-Correlation Learning Architecture, In D. Touretzky (Ed.), *Advances in Neural Information Processing Systems 2* (pp. 524-532). San Mateo, CA.: Morgan Kaufmann.

[4] Frean, M. (1991). The Upstart Algorithm: A Method for Constructing and Training FeedForward Neural Networks, *Neural Computation*, 2, 198-209.

[5] Phillips, D.T., Ravindran, A., Solberg, J. (1976) Operations Research: Principles and Practice, John Wiley & Sons, Inc.

[6] Romaniuk, S.G., Hall, L.O. (1993) Divide and Conquer Networks. To appear Neural Networks.

Monotonic Incrementation of Backpropagation Networks

Ingo Glöckner

Arbeitsbereich CL&KI, Sedanstr. 4, Universität Osnabrück, Germany

Abstract. One of the most challenging problems in neural network research is finding methods which increase approximation power but maintain generalization capabilities of simple nets. The approach pursued here employs methods of *monotonic network incrementation*, i.e. network modifications which locally change error surfaces, but retain network outputs. A moderate splitting algorithm plucked onto standard backpropagation is presented which detects units stuck in local minima and replaces them monotonically in order to speed up learning.

The case for monotonic incrementation. Modifications of neural networks will be called *monotonic* if, for arbitrary input patterns, network outputs are invariant with respect to the modification. They will be called *error-local* if backpropagated error signals (initially) remain the same for all non-modified units. Due to these properties, monotonic network incrementations are minimally disturbing in that they affect only a small fraction of the net and retain global network behavior. Generalizations of the smaller net are likely to be maintained in the further course of training as noise is introduced to the network only locally.

Monotonic unit splitting. Assume some layer C in which a unit c is to be replaced by n new units c_k monotonically. Let $C_{mod} = C \setminus \{c\} \cup C_{add}$ be the modified layer and, for each new unit c_k in $C_{add} = \{c_k \mid k = 1, ..., n\}$, let $w_{c_k}^{recp} = w_c^{recp}$. Now, in order for the modification to be monotonic, the projective weights $w_{c_k}^{proj}$ are initialized randomly such that they sum up to w_c^{proj}, i.e. [A] holds for each s in the 'successor layer' S. Using a random function rnd which provides uniformly distributed numbers in $[0,1]$, $w_{s,k}$ can be computed in turn $(k = 1, ..., n)$ by the formulae [B] and [C].

$$[A] \sum_k w_{s,c_k} = w_{s,c} \qquad [B] \ w_{s,c_k} = w_{s,c} \cdot X_{s,k} \qquad [C] \ X_{s,k} = (1 - \sum_{\nu=1}^{k-1} X_{s,\nu}) \cdot [rnd]^{n-k}$$

It is easy to check that the modification is both monotonic and error-local in the sense stated above. Apparently, this method can also be used to duplicate a whole layer monotonically without duplicating error surfaces. Adding pairs of new units $c^\bullet$, c° to a non-output layer can be treated as a special case of monotonic splitting where a 'virtual' unit c with $w_c^{proj} = 0$ is split into $c^\bullet$, c°. In this case, $w_{c^\bullet}^{recp} = w_{c^\circ}^{recp}$ and $w_{c^\bullet}^{proj}$ can be chosen arbitrarily, and $w_{c^\circ}^{proj} := -w_{c^\bullet}^{proj}$ ensures monotonicity. The method can also be applied to simple recurrent networks (SRNs). Splitting SRN hidden units simply requires to split the corresponding context units as well.

Escaping from local minima. In the MouseProp[1] algorithm, monotonic unit splitting is combined with *splitting strategies* aimed at detecting units which are stuck in local minima. MouseProp can be characterized as some kind of *meta algorithm* for backpropagation training which, in periodic control cycles, traces network training in order to choose a 'worst' unit for monotonic splitting. By intuition, a unit performs bad if its weights are strongly oscillating while, at the same time, it operates on a high level of error. The notions of 'oscillation' and 'level of error' have been formalized as follows:

$$[D] \ osc(c) = 1 - \frac{\sum_s |\sum_\tau \Delta w_{s,c}^\tau|}{\sum_\tau \sum_s |\Delta w_{s,c}^\tau|} \qquad [E] \ le_\blacktriangle(c) = \frac{\sum_\tau \delta_{\tau,c}^2}{\sum_{c' \varepsilon C} \sum_\tau \delta_{\tau,c'}^2} \qquad [F] \ le_\Diamond(c) = \frac{\sum_\tau \sum_s |\Delta w_{s,c}^\tau|}{\sum_{c' \varepsilon C} \sum_\tau \sum_s |\Delta w_{s,c'}^\tau|}$$

The 'level of error' is related either to error signals [E], or to weight changes [F]. τ is a time index which ranges over the current epoch only. I have tested several splitting strategies which detect the 'worst' unit, among others:

(**Proj.2Stp**) Of those units for which $le(c)$ is higher than average, split that with maximum oscillation.
(**Proj.Num**) Choose the unit for splitting which maximizes $split(c) = le(c) \cdot osc(c)$. If $le_\Diamond(c)$ is used, this is the unit maximizing $\sum_\tau \sum_s |\Delta w_{s,c}^\tau| - \sum_s |\sum_\tau \Delta w_{s,c}^\tau|$.

Analog methods can be obtained if $osc(c)$ and $le(c)$ are defined in terms of receptive weights.

MouseProp goes parity. A preliminary evaluation of MouseProp on 11-bit parity has put evidence that the extra effort posed by the splitting strategies pays off. MouseProp turned out to be superior to standard BP both in terms of speed and accuracy. Particularly, non-growing large nets performed worse than networks which reached the same size in a process of monotonic incrementation.

---

[1]MouseProp is an acronym for '<u>M</u>onotonic <u>u</u>nit <u>s</u>plitting in <u>e</u>rror <u>p</u>ro<u>p</u>agation'

A Multi-Layer Extension of a Bayesian Neural Network

Anders Holst
aho@sans.kth.se

Anders Lansner
ala@sans.kth.se

SANS, NADA, Royal Institute of Technology, S-100 44 Stockholm, Sweden

A Bayesian neural network [2] can be seen as implementing a Bayesian classifier, which is optimal for the case of independent evidence [3]. However, the performance of a standard one-layer Bayesian network is degraded when the underlying independence assumptions are violated. One way to solve this is to turn to a multi-layer architecture, where dependencies among the inputs can be removed or, in practice, at least reduced in the internal representation.

The normal way to treat strong dependencies among the evidence is to add an intermediate layer between the input (evidence) and the output (classes) [1]. This intermediate level consists of complex units, *i. e.* units that combine information from the input layer. If for example the units a and b are correlated, we produce a complex unit ab that is active when both a and b are active. The classes will now depend on a, b and ab, but these three units are still not independent, and the Bayesian model can not be used directly.

Instead we propose the following procedure. We start from a coding where each event is represented with one positive and negative unit. If we now find that two of these events are dependent, we merge the two events into one, with all possible combinations of the outcomes of the old events. For example, if the two primary events $A = \{a, \bar{a}\}$ and $B = \{b, \bar{b}\}$ are not independent, we insert in their place the composite event $AB = \{ab, a\bar{b}, \bar{a}b, \bar{a}\bar{b}\}$. We call this kind of group, with all possible combinations of some primary evidence, a *complex column*. If we manage to divide the primary events into independent groups, we can create a complex column from each such group. The Bayesian model is then directly applicable on these complex columns.

There are reasons to try to keep the order of the complex columns low. One is that the number of complex units increases rapidly with the order. Another reason is that when the columns grow larger, the generalization capability tends to decrease. This is the well known trade-off between probabilistic accuracy and generalization. Although the estimation of probabilities by the network gets better with larger columns, more training data are required to estimate the more detailed distribution.

Above it is assumed that each primary event occurs in only one complex column. If some primary event is a member of more than one column, then these will not be independent, and the estimation of probabilities will be distorted. However, if all primary events occur in the same number of complex columns, say m, it is possible to compensate for this by dividing all weights by m. We call such a coding m-covering.

There are cases in which a multiple covering will allow a lower order of the columns. Consider as a simple example the case of a 2×2 lattice with all patterns consisting of two active units, divided into three classes: vertical (▯▮ or ▮▯), horizontal (▭▭ or ▬▬) and diagonal (▯▮ or ▮▯). These classes are not linearly separable. If only a one-covering is used, a complex column of order four is required. However, if we allow a two-covering, we can solve it with only second order columns. In such situations this mechanism will thus save complexity, and at the same time increase generalization in the network.

[1] A. Lansner and Ö. Ekeberg. An associative network solving the "4-Bit ADDER problem". *Proc. of the IEEE First Annual Int. Conf. on Neural Networks*, 2:549–556, 1987. San Diego, USA.

[2] A. Lansner and Ö. Ekeberg. A one-layer feedback, artificial neural network with a Bayesian learning rule. *Int. J. Neural Systems*, 1(1):77–87, 1989.

[3] M. L. Minsky and S. A. Papert. *Perceptrons*. MIT Press, 1988.

YPROP: Yet Another Accelerating Technique for the Back Propagation

D.Anguita, M.Pampolini, G.Parodi, and R.Zunino

Department of Biophysical and Electronic Engineering - University of Genova
Via all'Opera Pia 11a, 16145 Genova, ITALY

Abstract

Many algorithms have been proposed to train a multi-layer feed-forward network, but most of them are extension of the classical generalized delta-rule: we will refer to these methods as Heuristical Adaptive Algorithms (HAAs). All HAAs are based on heuristics to compute a *good* learning step. Here we propose a new HAA that is an improvement of the VOGL method [1].

Let us consider the following updating rule:

$$\Delta w_{ij}(t) = -\eta(t)\sum_{p=1}^{N}\frac{\partial E_p}{\partial w_{ij}} + \alpha(t)\Delta w_{ij}(t-1) \tag{1}$$

where t is the index of the number of iteration in the learning, E_p is a typical cost function for pattern p, $\alpha(t)$ is the momentum term, and $\eta(t)$ is the adaptive learning step.

The VOGL technique can be briefly summarized as follows:

1) $\eta(0) = \eta_0$; $\alpha(0) = \alpha_0$

2) if $E_{tot}(t) < E_{tot}(t-1)$ then $\eta(t) = \phi \cdot \eta(t-1)$; $\alpha(t) = \alpha_0$

   if $E_{tot}(t) < (1+\varepsilon)E_{tot}(t-1)$ then $\eta(t) = \beta \cdot \eta(t-1)$; $\alpha(t) = 0$

   if $E_{tot}(t) > E_{tot}(t-1)$ then discard the last step; $\eta(t) = \beta \cdot \eta(t-1)$; $\alpha(t) = 0$

where $\phi > 1$ is the acceleration factor, $\beta < 1$ is the deceleration factor, E_{tot} is the total error, and $0.01 \leq \varepsilon \leq 0.05$.

The idea behind YPROP is very simple: instead of setting the acceleration and deceleration factors to fixed values, we change them at every learning step:

$$\phi(t) = 1 + \frac{K_a}{K_a + \eta(t-1)} \qquad \text{and} \qquad \beta(t) = \frac{K_d}{K_d + \eta(t-1)}$$

where K_a and K_d are two constants.

In the acceleration phase, if $\eta(t-1)$ is small ($\eta(t-1) << K_a$) then $\eta(t) \cong 2 \cdot \eta(t-1)$, if instead $\eta(t-1) >> K_a$ then $\eta(t) \cong \eta(t-1)$. The heuristic behind this is the following: we want η to increase rapidly if it is very small (like VOGL), but we want to be cautious if η is very large. The deceleration phase shows opposite behavior: if $\eta(t-1) >> K_d$ then $\eta(t) \cong K_d$; in other words the learning step is immediately (and drastically) shortened. If $\eta(t-1) << K_d$ then $\eta(t) \cong \eta(t-1)$; in this case, we are probably stuck in a local minimum (because the total error has increased even though η is very small), so letting the step almost unchanged can help to escape from it.

In order to test YPROP, we applied it to two well-known 'toy' problems and one real application: the parity problem, the encoder problem, and an automated diagnosys problem. YPROP showed good robustness to different learning parameters and starting learning step. Furthermore it showed better performances than the method proposed by Vogl et al., hence being far more efficient than standard back propagation.

References

[1]   T.P.Vogl, J.K.Mangis, A.K.Rigler, W.T.Zink, and D.L.Alkon - *Accelerating the Convergence of the Back-Propagation Method*. Biological Cybernetics, Vol.59, 1988, pp. 257-263.

Automatic Construction of Multilayer Networks for Non Linear Regression*

T. Denœux and R. Lengellé

University of Compiègne – U.R.A. CNRS 817
BP 649 - F-60206 Compiègne cedex - France
email: tdenoeux@hds.univ-compiegne.fr

Abstract

Although the back-propagation algorithm has proved very efficient in learning powerful internal representations, it leaves open the question of the optimal dimensionality of the space spanned by the hidden units. In [1], a novel approach to this problem has been proposed. This approach, applicable to discrimination problems, is based on direct optimization of an *objective function* for internal representations, which can be computed given a set of examples without specifying the network's outputs. Coupled with a strategy for recruiting units during the learning process, this concept provides a scheme for training a multilayer network layer by layer, until a simple correspondence can be found between the final, highest-level representations and a simple target coding scheme. The objective function introduced in [1] was a measure of class separability that resulted from the analysis of the relationships between discriminant analysis and multilayer neural networks. In this paper, this approach is transposed to the approximation of continuous functions. In that case, a useful strategy consists in increasing the *linear dependency* between the activations in the hidden layer and the target outputs. As a common measure of linearity, the sample coefficient of multiple determination ρ^2 is therefore a good candidate for an objective function.

The constructive algorithm is basically the same as described in [1]. The initial architecture consists of N_0 units in a single layer. A new randomly initialized unit is then added to the layer, and all the weights in that layer are iteratively updated so as to increase the value of the objective function, starting from the previous configuration. The layer is expanded until the addition of a new unit fails to increase the value of the objective function by a significant amount. The process can then be repeated with a new layer, until again no further improvement can be gained. The hidden-to-output weights can be computed in one step using a pseudo-inverse approach.

Numerical experiments show that this procedure is quite robust, with very small variability from one learning curve to the next, starting from different initial conditions. This robustness, which avoids doing many trials to reach a good solution, somehow compensates for the computational burden of inverting a matrix of size equal to the number of hidden units at each iteration. Future research effort will aim at extending the search space of the algorithm to different architectures and different types of hidden nodes.

References

[1] R. Lengellé and T. Denœux. Optimizing multilayer networks layer per layer without back-propagation. In Igor Aleksander and John Taylor, editors, *Artificial Neural Networks II*, pages 995–998. North-Holland, Amsterdam, 1992.

*This work has been supported by EEC funded Esprit II project nr. 5433 (NEUFODI); partners: BIKIT, ARIAI, Elorduy Sancho y Cia, LABEIN, Lyonnaise des Eaux-Dumez; Associated partner: RHEA S.A.

Generalization of a Parametric Learning Rule*

Samy Bengio   Yoshua Bengio   Jocelyn Cloutier   Jan Gecsei

Université de Montréal, Département IRO
Case Postale 6128, Succ. "A", Montréal, QC, Canada, H3C 3J7
e-mail: bengio@iro.umontreal.ca

Introduction

We proposed in previous work ([1, 2]) a method to find new learning rules for neural networks, considering them as parametric functions and using any standard optimization method (such as genetic algorithms, gradient descent, and simulated annealing) to select the parameters.

A parametric learning rule is a function of the form $f(x_1, x_2, ...x_n; \theta)$ where the x_i are local variables which can influence the synaptic efficiency, such as the presynaptic and postsynaptic activities, the synaptic weight and the activity of facilitatory (or modulatory) neurons, and θ is a set of parameters, which can be optimized using a cost measuring the performance of the rule over a test set. The question addressed in the next section is whether or not we can find a rule that will be able to learn tasks not used to select the parameters θ.

Capacity of a Parametric Learning Rule

In order for a learning rule obtained through optimization to be useful, it must be successfully applicable in training networks for new tasks. This property of a learning rule is a form of *generalization*, and it can be described using the same formalism used to derive the generalization property of learning systems, based on the notion of *capacity* (see [3] for an introduction).

The capacity h of a parametric learning rule $F(\theta)$ can be intuitively seen as a measure of the cardinality of the set of learning rules it can approximate. It is related to the average generalization ϵ of the rule over new tasks, and the number of training tasks N in the following way. For a fixed number of tasks N, starting from $h = 0$ and increasing it, one finds generalization ϵ to improve (decrease) until a critical value of the capacity is reached. After this point, increasing h makes generalization deteriorate (ϵ increases). For a fixed capacity, increasing the number of training tasks N improves generalization (ϵ asymptotes to a value that depends on h).

We performed experiments on classification problems to study the variation of N, h and the complexity of the tasks, over the learning rule's generalization property (ϵ). We concluded the following from these experiments:

1. The rules found generalized better over similar or simpler tasks (as expected from theory).

2. The generalization (ϵ) improves with the number of tasks (N) used to optimize the rule.

3. For a particular set of tasks, a rule constrained to 7 parameters using a-priori knowledge was better overall than a rule with 16 parameters, in terms of generalization performance and optimization time (as predicted, when we increase the capacity, the generalization can decrease).

References

[1] S. BENGIO, Y. BENGIO, J. CLOUTIER, AND J. GECSEI, *Aspects théoriques de l'optimisation d'une règle d'apprentissage*, in Actes de la conférence Neuro-Nimes 1992, Nimes, France, 1992.

[2] Y. BENGIO AND S. BENGIO, *Learning a synaptic learning rule*, Tech. Rep. 751, Département d'Informatique et de Recherche Opérationnelle, Université de Montréal, Montréal, QC, CANADA, 1990.

[3] V. N. VAPNIK, *Estimation of Dependencies Based on Empirical Data*, Springer-Verlag, New-York, NY, USA, 1982.

---

*A longer version of this paper is available by ftp in the neuroprose archive, file bengio.general.ps.Z.

Supervised Learning for Decorrelated Gaussian Networks

D. Obradovic and G. Deco
Siemens AG, Corporate R&D, Otto-Hahn-Ring 6, 8000 Munich 83, Germany

This paper presents a new two-stage learning paradigm that utilizes localization properties of Gaussian neurons. In the first stage, a single layer of Gaussian function is trained in a novel unsupervised fashion to model the distribution of the network input. The input model is obtained by minimizing a cost function whose first term can be seen as an implementation of the standard Hebbian learning law. The second term of the cost function has an "anti-Hebbian" effect which reinforces the competitive learning. In the second stage of the learning paradigm, the previously obtained receptive field distribution is further used for function approximation. For comparison, a standard single hidden-layer Gaussian network is optimized with the initial centers corresponding to the first stage learning.

Let a layer of Gaussian-neurons with centers w_i and widths σ_i has normalized outputs given by $O_i(\xi)$,where ξ is the input vector. Furthermore, let us define a cost function H'_i as follows:

$$H' = -\sum_i \left(\int d\xi P(\xi) O_i(\xi) \right) + \alpha \sum_i \sum_{j \neq i} \int d\xi P(\xi) O_i(\xi) O_j(\xi) \qquad \text{(EQ 1)}$$

where $P(\xi)$ is the probability density of the input patterns. Unsupervised learning is defined as minimization of the cost function presented in equation (EQ 1) with respect to the centers w_i. The first term on the r.h.s. of (EQ 1) induces attraction of all neurons toward the region of the input space were input data exist. The second term of the cost function penalizes overlapping of the outputs of the neurons with respect to each pattern. We refer to this learning procedure as the Decorrelated Hebbian Learning (DHL). We have shown that the value of scaling factor α is irrelevant due to the normalization of the Gaussian outputs.

Once when the centers of the Gaussians are determined, the output of the neurons are used to obtain a weighted piece-wise linear approximation of the single-output function by minimizing the following cost-function:

$$J = \frac{1}{N} \sum_{j=1}^{N} \left\{ z_j - \sum_{i=1}^{m} [a_i + \xi_j b_i] O_i(\xi_j) \right\}^2 = \frac{1}{N} \sum_{j=1}^{N} (error_j)^2 \qquad \text{(EQ 2)}$$

where z_j is the "j-th" pattern of the function output, N is the number of patterns and m is the number of the Gaussian neurons. The parameters to be adjusted are a_i and the vector b_i whose dimension is equal to the number of function inputs. Hence, the total number of parameters to be determined in the second learning stage is $m*[1+(number\ of\ function\ inputs)]$.

As an example we use the standard benchmark Mackey-Glass time series. The DHL is used as unsupervised learning method for positioning the centers of the Gaussians neurons to fit the Mackey-Glass chaotic attractor. The test set consists of 500 points not contained in the training set of 1000 data points. Once when the centers are determined, a piece-wise linear approximation of the function is obtained by minimizing the cost function defined in (EQ 2).

Approximation Type	Number of Parameters	Normalized Error
DHL + Optimal Linear Output Layer	401+100	0.2127
DHL + Piece-Wise Linear Approximation	401+500	0.056
Quasi-Newton (DHL for initialization)	901	0.0376
Locally Tuned Gaussian	900	0.23

Associative Memories that can form Hypotheses: Phase coded Network Architectures

Niels Kunstmann, Claus Hillermeier, Paul Tavan
Institut für Medizinische Optik, Ludwig-Maximilians-Universität München
Theresienstraße 37, W-8000 München 2, FRG

Abstract

Nonlinear associative memories as realized, e. g., by Hopfield nets are characterized by attractor type dynamics. When fed with a starting pattern they converge to *exactly one* of the stored patterns which is supposed to be the most similar one. These systems cannot render *hypotheses of classification*, i.e. render *several possible* answers to a given classification problem. Inspired by C. von der Malsburg's correlation theory of brain function we extend conventional neural network architectures by introducing additional dynamical variables, the so-called phases, one for each formal neuron in the net. The phases measure detailed correlations of neural activities neglected in conventional neural network architectures. Using simple selforganizing networks based on feature map algorithms we present an associative memory that actually is capable of forming hypotheses of classification.

1 Introduction

The conventional architecture of neural networks is characterized by the following features: (i) The state of the network at some time T is uniquely defined by the real-valued vector of the neural activities $\vec{a}(T)$. (ii) During a training phase the network acquires knowledge by means of a learning process. This knowledge is represented by matrix elements $C_{rr'}$ describing the synaptic influence of a neuron r' on a neuron r. (iii) In the working phase the matrix C is kept constant and the system starting with an initial activity distribution $\vec{a}(T = 0)$ evolves according to an updating equation

$$\vec{a}(T + 1) = \vec{F}(C\vec{a}(T)), \tag{1}$$

where $\vec{F}$ denotes some in general nonlinear function.

Apart from the chaotic range there are two possibilities of stable long time behavior of a nonlinear recurrent dynamics like in eq. 1. The trajectories in configuration space may either end up in a limit cycle or run towards a stable fixed point. In the latter case of a nonlinear attractor dynamics, the fixed points with their basins of attraction tesselate the input space of possible starting patterns and are prototypes classifying their respective tesselation volumes. However, such a dynamics discards any information about the initial activity distribution, especially whether it was located near a decision boundary, an instance that may render its classification by a unique prototype doubtful. If such a system is used as an early module in a hierarchical classification system, this loss of information strongly hampers the quality of the complete system. Speech recognition systems for example cannot renounce the formation of hypotheses in phoneme classification tasks.

In the conventional neural network architecture a set of classification hypotheses could only be represented in the form of a limit cycle leading past all pattern prototypes to which the input pattern exhibits similarities. However, the particular form of eq. 1 then disables any fixed point attractor dynamics towards the pattern prototypes, since these are elements of limit cycles. Thus, in conventional neural architecture a distinction between safe and doubtful classifications is impossible, i.e. a simultaneous representation of a prototype as fixed point within some small basin of attraction and as a point in configuration space visited regularly by limit cycles cannot be achieved.

This criticism leads us to the ideas presented in [1] by C. v.d. Malsburg, where he states that some fundamental abilities of the brain, e.g. figure-ground separation or the segmentation of images into features and the reintegration of these features to single objects, demand additional properties in the neural network architecture. C. v.d. Malsburg argues that the time correlations of neural firings carry information and must not be neglected. Furthermore he rejects the concept of rigid synaptic connections by introducing a short time scale synaptic plasticity or synaptic modulation [2].

In this paper we present a new kind of neural network based on these concepts and demonstrate its abilities by a simple example.

2   The phase coded network architecture

In the following we construct an associative memory which is actually capable to render hypotheses of classification. Inspired by the ideas presented in [1], we attribute a second real-valued variable, the phase $\phi_r \in [0, 2\pi]$, to every formal neuron r. The neural activity may then be viewed as a pointer in the plane of complex numbers. This approach incorporates the consideration of time correlations, because in case the neural firing is oscillatory with a fundamental frequency Ω, time correlations of neural activities can be expressed as phase differences.

First we give a brief survey of the autoassociative memory model which serves as the starting point for our work. The feature space of incoming signals $\vec{x}$ is discretized in a Kohonen-net of formal neurons r. The signals are sparsely coded through a distance dependent activation of neurons with

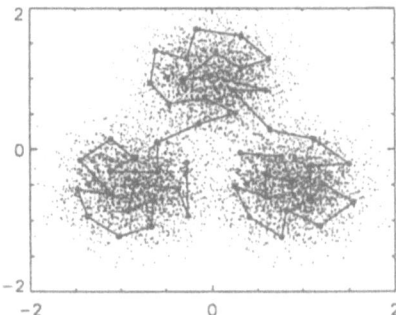

Figure 1: Distribution of stimuli in feature space and their representation by a one-dimensional Kohonen-map

neighboring codebook vectors. By competitive Hebbian learning lateral connections $C_{rr'}$ between the neurons are trained assuring that local maxima of the probability distribution $P(\vec{x})$ are fixed points of the diffusive-competitive dynamics

$$a_r(T + 1) = \sum_{r'} C_{rr'} a_{r'}(T) + k \sum_{r'} \frac{a_r(T)\{a_r(T) - a_{r'}(T)\}}{A^2} a_{r'}(T). \qquad (2)$$

The columns of the matrix C sum up to unity in order to preserve the total neural activity $A = \sum a_r$, and the parameter k controls the width of the neural activity distribution in feature space. Further details of this new associative memory can be found in [5, 7].

The guidelines for the extension of this architecture towards a phase coded network were: (i) The form of the activity dynamics should be preserved for neurons with equal phases. For strongly decorrelated neurons mutual flow of activity should vanish. (ii) The phase dynamics should tend to correlate neurons with strong connections. Decorrelations between non-interacting neurons should be enhanced.

To define a measure for phase differences we introduce a 2π-periodic function $g[\Delta\phi; \sigma]$ having well located maxima with extension σ at points $\Delta\phi = 2\pi l$, where l is some integer. The coupled

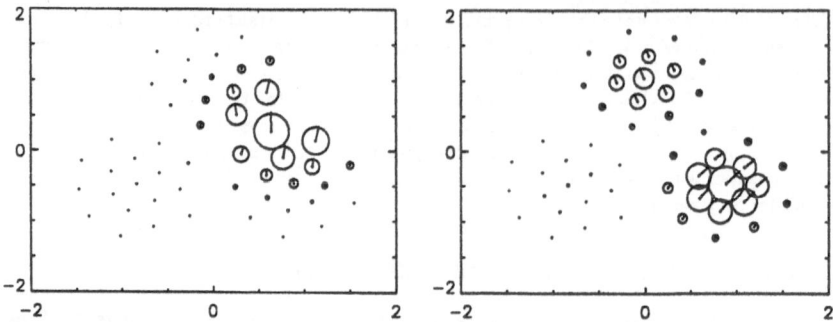

Figure 2: Initial and final state of the network caused by stimulus incident near a decision boundary

dynamics of activities and phases takes the form

$$
a_r(T+1) = \sum_{r'} \left(\tilde{C}_{rr'}[\vec{\phi}(T)] + k\frac{a_r(T)\{a_r(T) - a_{r'}(T)\}}{A_r(T)A_{r'}(T)} \right) g[\phi_r(T) - \phi_{r'}(T); \sigma] a_{r'}(T) \quad (3)
$$

$$
\phi_r(T+1) = \phi_r(T) + \epsilon \sum_{r'} \left(\frac{C_{rr'}\,a_{r'}(T)}{a_r(T) + \gamma} \frac{\partial}{\partial \phi_r} g[\phi_r(T) - \phi_{r'}(T); \sigma] + \delta\xi_{rr'} \right).
$$

The activity dynamics in eq. 3 resembles the simple dynamics from eq. 2. The main difference lies in the weight factor $g[\phi_r - \phi_{r'}]$ weakening the interaction between neuron r and neuron r'. Diffusion and competition as described in [5, 7] only occur between neurons with similar phases. In the nonlinear term the normalization is now carried out with respect to a phase dependent activity $A_r(T) = \sum_{r'} a_{r'}(T) g[\phi_r(T) - \phi_{r'}(T); \sigma]$ and, similarly, the axontree $C_{rr'}$ of neuron r' is normalized taking into account phase dependent quantities $\tilde{C}_{rr'}(\vec{\phi}) = C_{rr'}/\sum_{r''} C_{r''r'} g[\phi_{r''} - \phi_{r'}; \sigma]$. The resulting effective synaptic weights are thus determined by the phases of the interacting neurons and, therefore, are *implicitly* time dependent.

Obviously the original activity dynamics is retained, if all neurons fire correlated, i.e. $\phi_r \equiv \phi_0$. Large phase differences between two neurons inhibit any further exchange of activity. The dynamics governing the phases tries to enlarge correlations of neural firings by a simple gradient descent on a correlation product whereas the unspecific decorrelation term $\xi_{rr'}$ avoids a network state with global correlation of all neurons. The behavior of the phase dynamics thus implements the spirit of the propositions made in [1] in a most simple form.

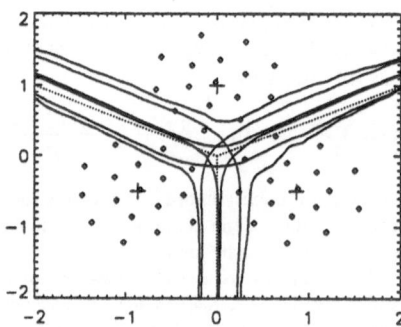

Figure 3: Contour lines (probability 20%, 50% and 80%) of the weights of classification hypotheses together with Bayesian decision boundaries in feature space

3   An example

We now demonstrate the capabilities of the phase coded network by an example. Fig. 1 shows a discretization of a two-dimensional feature space by a Kohonen-chain with 50 neurons. The probability distribution of incoming stimuli codes three pattern prototypes in terms of univariate Gaussian distributions of identical variance and amplitude. The set of training vectors for the quantization of the input space is indicated by the patterns of black pixels.

In Fig. 2 the initial and final state of the network are shown for an initial distribution centered near a decision boundary between two prototypes. Activity and phase of every neuron is depicted as a clockface whose diameter measures the neural activity and whose hand indicates the phase. In the stable long time limit, the neural activity is nearly equally distributed between the two prototypes and the respective phases are well separated.

Fig. 3 shows contour lines for the weights of the classification hypotheses as a function of the initial activity center. Small diamonds symbolize the virtual positions of the neurons in feature space, and the maxima of the probability density are indicated by small crosses. The dotted lines are the Bayesian decision boundaries. All prototypes are surrounded by a basin of attraction within which an incoming signal is uniquely classified. Initial patterns centered near decision boundaries are eventually mapped to two or three prototypes.

4   Summary

We have developed an associative memory capable of forming hypotheses of classification. The new architecture is based on a conventional associative memory with recurrent dynamics in population coded feature maps. Extensions towards high-dimensional feature spaces, large nets and to network models which render hypotheses of prediction appear straightforward.

References

[1] C. v. d. Malsburg. The correlation theory of brain function. Technical Report 81-2, Department für Neurobiologie, Max-Planck-Institut für Biophysikalische Chemie, Göttingen, 1981.

[2] C. v. d. Malsburg. Am I thinking assemblies? In G. Palm and A. Aertsen, editors, Proceedings of the Trieste Meeting on Brain Theory, October 1984, pages 161–176, Berlin, 1986. ICTP, Springer.

[3] B. Rabus. Hypothesenbildung in phasenkodierten Assoziativspeichern. Diplomarbeit, Physik-Department T30, Technische Universität München, April 1992.

[4] C. Hillermeier, N. Kunstmann, P. Tavan   Ein hypothesenbildender Assoziativspeicher in phasenkodierter Netzwerkarchitektur. Statusseminar Neuroinformatik 1992, Maurach, 1992

[5] C. Hillermeier, B. Rabus, P. Tavan. manuscript in preparation

[6] N. Kunstmann, B. Rabus, P. Tavan. manuscript in preparation

[7] C. Hillermeier, N. Kunstmann, P. Tavan   Population Dynamics on the Basis of Vector Quantization: A Method for Auto-Association and Classification   Proceedings of the ICANN 93, Amsterdam, 1993

A Formal Link between Multilayer Perceptrons and a Generalization of Linear Discriminant Analysis

E. Mousset [†‡], A. Faraj [†]

† Institut Français du Pétrole
1&4 avenue de Bois-Préau, BP311
92506 Rueil-Malmaison, France

‡ Compagnie Générale de Géophysique
1 rue Léon Migaux
91341 Massy, France

Abstract

We are interested here in studying MLPs' behavior for classification, in the case of overlapping classes, i.e. when Huygens's theorem (1) no longer holds.

$$C_{XX} = B_{XX} + W_{XX} \qquad (1)$$

(Let $X=(X_{ij})_{1 \leq i \leq n, 1 \leq j \leq p}$ be the data matrix. The p columns $X_{.j}$ are assimilated with variables and the n lines $X_{i.}$ with individuals. Let C_{XX}, B_{XX} and W_{XX} respectively be the total, between and within variance/covariance matrices.)

First, we describe a data analysis method - *contiguity analysis* (introduced by Lebart in 1969) - which aims to analyse data at local scales, induced by neighboring relations between individuals.

The key concept introduced in this method is the local variance/covariance matrix Γ_{XX}. By adding a new hypothese about the metric, we derive an extension of Huygens's theorem (2) to the case of overlapping classes (this result is due to Faraj, 1993).

$$C_{XX} = T_{XX} + \Gamma_{XX} \qquad (2)$$

For q disjoints groups, the new metric leads to the recovery in the T_{XX} and Γ_{XX} formulations, of the between- and within-class variance/covariance matrices. We derive from this result an extension of LDA - *generalized linear discriminant analysis* - to classification problems where classes overlap.

Second, we review a previous theoretical result in which MLP has been demonstrated to be equivalent to LDA (introduced by Gallinari et al. in 1990). We then propose conditions under which this result can be generalized and state a formal equivalence between linear MLP and generalized linear discriminant analysis (GLDA):

The first layer of a linear MLP with two weights layers performs calculations equivalent to GLDA, provided that the desired output matrix Y be equal to the *contiguity matrix* $G = (g_{ii'})_{1 \leq i, i' \leq n}$,

where $g_{ii'} = g_{i'i} = 1$ if $X_{i.}$ and $X_{i'.}$ are assumed to be neighbors (in other words, if they belong to the same group) and 0 otherwise. These are the only constraints imposed on G, and neighborhoods can be either user defined or data induced.

This results help us to go more deeply in the understanding of MLPs' behavior when they are used in classification problems. Moreover, they allow MLP to take into consideration neighboring relations between examples.

ADAPTIVE CRITIC FOR PROBABILISTIC LOGIC NETS

R.S.Neville and T.J.Stonham

Department of Electrical Engineering, Brunel University, Uxbridge UB8 3PH, U.K.

Abstract
In this paper we study the use of an Adaptive Critic to augment Probabilistic Logic nets' performance. The paper proposes a new Quantised Adaptive Critic Element for use with Probabilistic Logic Nodes(PLN). The Probabilistic units are logical nodes which respond to their input patterns in addressable locations; the locations then define the probability of the output being a logical "1". The technique uses an Unsupervised Adaptive Critic Element to predict a Reward signal for the net. The prediction is based on the standard scalar Reward, and the past and present predicted values of the reinforcement signal. The Quantised Adaptive Critic methodology may be utilised for problems which involve a large number of input variables(i.e. control). A new structure is proposed to deal with this situation and enable the Adaptive Critic to be Hardware realisable.

1. Quantised Adaptive Critic

The Quantised Adaptive Critic Element (QACE) stores the variables that it utilises in the same manner as the PLN, Neville [1993]. The QACE has a site address v which is addressed by an i bit vector $\{x_1, x_2, ... x_i\}$, this addresses three sites in parallel $P_{v(t)}$, $P_{v(t-1)} \in [+P_n, 0]$ and $\bar{x}_{v(t)} \in [+\bar{x}_n, 0]$, that are q bit numbers. The QACE uses the standard external reward signal $r = 1 - e_o$ where $r \in [+1, 0]$ is a binary scalar value. The reinforcement is then scaled $r' = (2 * r) - 1$ to obtain the reinforcement used by Barto et al.[1983], where $r' \in [0, -1]$. The scaled reward signal is then used to derive an improved or internal reinforcement signal, given by: $\hat{r}_{(t)} = r_{(t)}' + \gamma P_{v(t)} - P_{v(t-1)}$ (1)

where $P_{v(t)}$ is the present prediction and $P_{v(t-1)}$ is the past prediction. One should note that this is not the same as Barto's work, where he uses the prediction values $P_{(t)}$ and $P_{(t-1)}$. We use the present and previous prediction values for the given site address v. The Prediction value is updated by: $\Delta P_{v(t+1)} = \beta \hat{r} \bar{x}_{v(t)}$ (2)

where β is a positive constant determining the rate of change of P_v . All the input eligibility traces are updated by: $\bar{x}_{v(t+1)} = \lambda \bar{x}_{v(t)} + (1-\lambda) x_v$ (3)

where λ, $0 < \lambda < 1$ determines the eligibility trace decay rate. The binary scalar value x_v is a trigger for the eligibility trace, when the site v is addressed $x_v = 1$ and all other non-addressed traces are updated with $x(\bar{v}) = 0$. The internal reinforcement $\hat{r}$ is then re-scaled $r^* = (\hat{r} + 1.0)/2.0$ which denote a quantised real valued reinforcement $0.0 < r^* < 1.0$. The addressed sites are then adapted using: $\Delta S_{\mu} = r^* \alpha [Y - \sigma(S_{\mu})] + (1 - r^*) \alpha \lambda [1 - Y - \sigma(S_{\mu})]$ (4)

2. Concluding Remarks

In this work we have put forward a Quantised Adaptive Critic methodology for augmenting Probabilistic Logic Nets. The unsupervised Critic provides advisory information in the form of a quantised internal reinforcement, which promotes learning efficiency. The standard Adaptive Critic methodology has a single Critic as advisor, while we find it advantageous to have one Critic per layer to augment the training of the supervised net. An alternative approach to obtain the internal reinforcement was the use of the multicube structure. The use of several cubes bypasses the problem of exponentially growing resources with increased fan-in. The quantisation of the prediction and eligibility values enables the Quantised Adaptive Critic structure to be hardware realisable. One should note that in this study we have used the Critic to predict the present internal reinforcement $r_{(t)}$, the Critic is also able to predict the future reinforcement at $\hat{r}_{(t+1)}$, which may be more useful in temporal or control situations. In order to predict $\hat{r}_{(t+1)}$ one would utilise the future prediction $P_{v(t+1)}$ and the present prediction $P_{v(t)}$.

3. References

Barto, A.G. , Sutton, R.S., Anderson, C.W. (1983). Neuron like Adaptive Elements that can solve difficult learning control problems. IEEE Transactions on, Man, and Cybernetics, Volume SMC-13, Number 5, September/October. pp834-846.

Neville, R.S. (1993). Adaptive Critic for Probabilistic Logic Nets. World Congress on Neural Networks (WCNN'93 Portland), July 11-15, Portland, Oregon, USA.

DEFANET2 - ADVANCEMENTS OF A
DETERMINISTIC FUNCTION APPROXIMATOR

Wolfgang J. Daunicht

Neurologisches Therapiecentrum, Hohensandweg 37, D-4000 Düsseldorf, Germany

The approximation of an arbitrary continuous function by a neural network is, although possible in principle, an uncertain and time-consuming undertaking. First of all, a training procedure usually requires an unknown and generally high number of steps. Second, it is difficult to assess a size of a network, that is capable to learn the desired function; therefore it may be necessary to iterate the learning sequence to find a network of suitable size. Third, even if the network is capable of approximating the desired function, the learning procedure – depending on the initialization – may end up in a local minimum far from the global optimum. Again, the solution may be to repeat the training with different initializations.

One possible solution to such problems might be to remove part of the uncertainties in the topology of the network and in the values of the synaptic weights, so that the learning procedure is characterized by only one global minimum and it becomes possible to calculate the synaptic weights using a finite algorithm. Such a concept using a special three-layered architecture called DEFAnet has been suggested and it has been proven, that a DEFAnet is in fact a universal approximator [1].

In this advanced version DEFAnet2 a fast algorithm is presented, that allows to determine the synaptic weights so that the network function g is differentiable and matches the desired function f at all inputs $\underline{x}$ on the grid intersections indexed with the vector index $\underline{\lambda}$.

The basic idea to reduce the computational burden is to transform the problem into simpler subproblems. In DEFAnet2 the problem is broken down in two steps. First, it is transformed into the inversion of a set of 2^n matrices W, then the inversion of these matrices is reduced to the inversion of n (by orders of magnitude smaller) matrices W^*, n being the input dimension. Each matrix W is a multiple Kronecker product (denoted by $\otimes$) of some subset of the matrices W^*. The inverse of such a matrix W can be composed of the inverses of the factor matrices W_i by Kronecker multiplication

$$W^{-1} = \bigotimes_{i=1}^{j} (W_i^*)^{-1} \tag{1}$$

This formula assures considerable computational load reduction, even if the factor matrices have to be solved by means of a general inversion algorithm. However, it turns out that the factor matrices W^* are not necessarily of a general type, instead they can be made easier to invert, when mild restrictions are obeyed. For example, when all activation functions of those second first layer neurons receiving input from the same i-th neuron in the zeroth layer are identical, then the respective factor matrix W_i^* is a Toeplitz matrix. Toeplitz matrices, which are symmetrical and very regular, can be inverted quite effectively [2].

By holding to another restriction, i.e. using sigmoidal activation functions in the first layer that saturate exactly at 0 and 1, instead of approaching these values asymptotically, the factor matrices become symmetrical 'ribbon'-shaped, i.e. the elements are zero except for a narrow ribbon along the diagonal. Such matrices can be inverted even more effectively by an algorithm based on Cramer's rule.

Thus DEFAnet2 provides precise differentiable function approximator networks with efficient algorithms for weight calculation.

References

[1] W.J. Daunicht. DEFAnet - a deterministic neural network concept for function approximation. *Neural Networks*, 4:839–845, 1991.

[2] E.A. Robinson and S. Treitel. The Toeplitz recursion. In *Geophysical signal analysis*, pages 163–169. Englewood Cliffs, N.J., Prentice Hall, 1980.

Acknowledgements: This work was supported by DFG grant Da 199/2-1.

Document Retrieval and Protein Sequence Matching using a Neural Network

Björn Levin   Anders Lansner

blevin@sans.kth.se   ala@sans.kth.se

SANS, NADA, Royal Institute of Technology, S-100 44 Stockholm, Sweden

During the development of algorithms for the generation of higher order complex units in neural networks we have come to be interested in free text document retrieval and protein sequence matching. Document retrieval, as it is percieved here, consists of returning a list of the documents in the library that are the most relevant according to a description of a subject, sorted according to descending probability of relevance. Obviously, the key is having a reasonable measure of similarity between the individual documents and a given concept, a measure here provided by the neural network. [1] contains an overview of similar work.

Protein sequence matching is regarded as a special case of document retrieval. What makes it particularly interesting is the existence of generally accepted automatic (but resource consuming) procedures for determining the degree of similarity between sequences or parts of sequences. We have compared our results with those obtained using a system also running on the CM-200 at KTH, based on editing distance without swaps on a database containing 20 000 sequences [2].

In all cases our similarity measures are based on using binary sensors that indicate the existence of short sequences of characters in a document, i.e. feature detectors. These sensors are selected according to quality measures of the type in the table below. The similarity measure is then built up as a weighted sum over the sensors when applied to the search seed. These weights in turn, are computed during a learning phase when the documents themselves are fed to the sensors.

The evaluation of the sensor quality measures has been done on the 8 K Connection Machine CM-200 at KTH. These evaluations, done both by comparing rankings of protein sequences with the rankings of the editing distance based system as well as manual examination and testing using plain text documents, showed that especially measure #6 below performed quite well in returning an adequate ranking list of documents or protein sequences.

The ranking is also quite fast. Matching a sequence of 600 characters against a data base of 20 000 documents takes 3.6 s. The matching is then performed using a network that has been trained previously and has been read from the data vault. The training, i.e. selection of sensor units and setting up of the weights of the network, takes approximately 1.5 hours. It should be pointed out that after the training has been completed any number of matchings can be done. It should also be pointed out that a major part of the time is used selecting the sensors, a task that only requires a representative subset of the data. It is also possible to include new documents in the data base very quickly if the material is relatively homogeneous. More details can be found in [3].

#1	$\dfrac{P_s(\text{ABC}\ldots)}{P_s(\text{A})\cdot P_s(\text{B})\cdot P_s(\text{C})\cdots}$	#4	$\sqrt[n]{P_s(\text{ABC}\ldots)} - \sqrt[n]{P_s(\text{A})\cdot P_s(\text{B})\cdot P_s(\text{C})\cdots}$
#2	$P_s(\text{ABC}\ldots) - P_s(\text{A})\cdot P_s(\text{B})\cdot P_s(\text{C})\cdots$	#5	$\dfrac{P_s(\text{ABC}\ldots)}{\max\left(\max\limits_{X} P_s(X\text{ABC}\ldots),\ \max\limits_{X} P_s(\text{ABC}\ldots X)\right)}$
#3	$\dfrac{(P_s(\text{ABC}\ldots))^n}{P_s(\text{A})\cdot P_s(\text{B})\cdot P_s(\text{C})\cdots}$	#6	$\dfrac{\sqrt{2}\cdot P_s(\text{ABC}\ldots)}{\sqrt{\sum\limits_{X} P_s(X\text{ABC}\ldots)^2 + \sum\limits_{X} P_s(\text{ABC}\ldots X)^2}}$

Some of the quality measures tried for selecting sensors. $P_s(\text{ABC})$ stands for the probability that the sensor will be active at a random position while n is its size.

[1] Doszkocs T., Reggia J., and Lin X.: Connectionist Models and Information Retrieval; *Annual Review of Information Science and Technology (ARIST)*, Vol. 25 pp 209 - 260 (1990).

[2] Wallin E.: Optimized Sequence Matching on the CM-2; Masters Thesis, Royal Inst. of Technology, Sweden (1992).

[3] Levin, B. & Lansner, A.: Document Retrieval, Protein Sequence Matching and Sensor Selection Methods using a Neural Network; Royal Inst. of Technology, Tech. rep. TRITA-NA-P9238 (1992).

A Fuzzy Neural Architecture for Supervised Learning and Classification of Temporal Sequences

J.M. Cano Izquierdo, Y.A. Dimitriadis, and J. López Coronado

Department of Automatic Control and Systems
School of Industrial Engineering, University of Valladolid
Paseo del Cauce S/N, 47011, Valladolid, Spain
e-mail: isaetsii@cpd.uva.es

Abstract

The problem of learning and classifying ordered sequences is approached in this paper and a new architecture is proposed that incorporates elements from the Adaptive Resonance Theory (ART) and the theory of Fuzzy Sets.

The proposed architecture is composed of two stages. The first one is formed by a Fuzzy ART module that takes care of the unsupervised classification of the sequence components, and by a STORE module [1] that converts the sequence of category nodes provided by the Fuzzy ART to a spatial pattern. The second stage consists of another Fuzzy ART module and an associative memory.

The supervision is accomplished at the second stage, where a label is assigned to the global temporal sequence, that is introduced to the system. On the other hand, the spatial pattern produced by the first level is compared to the adaptive weights attached to the associative map, through a local distance that is based on the concept of fuzzy subset. This comparison permits the propagation of the supervision information to the first level, while taking advantage of the special form of the STORE patterns.

The proposed architecture extends the recently proposed Fuzzy ARTMAP [2] to the processing of temporal sequences, incorporating also an error checking mechanism at the level of an individual sequence component.

Experimental results are provided for the problem of on-line recognition of handwritten symbols, where each symbol is considered as a sequence of components. The obtained results compare favorably to those of a neural hierarchy [3] composed by ART2, ARTMAP and STORE modules, where the supervision information cannot be exploited at the symbol component level.

References

[1] G. Bradski, G.A. Carpenter and S. Grossberg, "Working memory networks for learning temporal order, with application to 3-D visual object recognition", *Neural Computation*, vol. 4, pp. 270-276, 1992.

[2] G.A. Carpenter, S. Grossberg, N. Markuzon, J.H. Reynolds and D.B. Rosen, "Fuzzy ARTMAP: A neural network architecture for incremental supervised learning of analog multidimensional maps", *IEEE Trans. on Neural Networks*, vol. 3, pp. 698-713, 1992.

[3] Y.A. Dimitriadis, J. López Coronado, C. García Moreno and J.M. Cano Izquierdo, "On-line handwritten symbol recognition, using an ART based neural network hierarchy", *Proc. of the 1993 IEEE Conf. on Neural Networks, ICNN'93*, vol. II, pp. 944-949, March 28 - April 1, 1993, San Francisco.

Hierarchical Reinforcement Learning

Magnus Borga
Computer Vision Laboratory
Linköping University
S-581 83 Linköping Sweden
tel. +46 13 28 27 83
fax. +46 13 13 85 26
email: magnus@isy.liu.se

Abstract

A response generating system can be seen as a mapping from a set of external states (inputs) to a set of actions (outputs). This mapping can be done in principally different ways. One method is to divide the state space into a set of discrete states and store the optimal response for each state. This is denominated a memory mapping system. Another method is to approximate continuous functions from the input space to the output space. I denominate this method projective mapping, although the function does not have to be linear. The latter method is the most common one in feed forward neural networks, where an input vector is projected on a "weight vector".

In reality this mapping is piecewise continuous. Consider the task of moving from one point to another. There is often an infinite number of solutions, and if two different paths are selected, it is often possible to interpolate these paths and get a new path that lies somewhere in between. On the other hand, if you for instance are passing a tree, you can choose to walk at the right side or the left side of the tree, but there is no success in trying to interpolate these two choices. There *are* at the same time an infinite number of ways to move on either side of the tree. This implies that there are two different kinds of responses, one that can be interpolated and another that can not.

In this paper the discrete set of alternative responses where there is no meaning in interpolation is defined as a set of *strategies*, and the continuous possibilities of outputs within each strategy are defined as *actions*. In the previous example the choice of which side of the tree to pass on is the choice between the two strategies, left and right, and the decision of which specific path to walk is a decision of a sequence of actions.

The input space is divided into a finite number of regions, where for each region there is one best choice of strategy and the input-output transition function varies in a smooth way. The total input-output transition function of the system is a set of continuous input-output transition functions, one for each strategy.

Now the problem of learning the total input-output transition function can be divided into several learning problems at two different levels. At the first level, there are a number of smooth and continuous functions to be learned, one for each strategy. For a given strategy an "ordinary" reinforcement learning problem is faced, where projective mapping could be used. At the second level, the best choice among the set of strategies is the learning task. The number of strategies are, however, finite, so at this level a memory mapping method could be efficient.

This paper describes how adaptive critic methods could be used to handle a two-level reinforcement learning problem. In adaptive critic methods predictions of future reinforcement are made, and these predictions are used to select strategy. For each strategy, a prediction of future reinforcement and an action are calculated. Then the strategy with the highest prediction is selected and the corresponding action is used as output. The update of the elements in the reinforcement association vectors and the action association vectors is made only for the chosen strategy and hence the internal reinforcement signal depends on which strategy was chosen. In this way the structural credit assignment problem is reduced considerably.

An example of an algorithm that solves a two-level learning task for a one-dimensional dynamic problem is presented. The system uses two strategies, each containing a linear action function. The choice of strategy and the proper actions are learned simultaneously by reinforcement learning.

Connectivity Maximization of Layered Neural Networks for Supervised Learning

E. Fiesler

IDIAP, C.P. 609, CH-1920 Martigny, Switzerland

Abstract

One of the main problems in current artificial neural network engineering is the lack of design rules for layered neural network topologies, namely how many hidden layers and how many neurons per hidden layer to choose for the network. This paper offers a theoretical basis for approaching this problem. Formally proven theories are developed which maximize the connectivity of layered neural networks, which use supervised learning, allowing for a maximum potential storage capacity. Due to space limitations, only the results, which are learning rule independent, are published here.

Since neural networks process and store information, their (storage) capacity is a key design issue, and since the interconnection strengths (or *weights*) contain the information of the neural network, the potential (storage) capacity is proportional to the number of weights in the network. Without a-priori knowledge, all the weights of a neural network are considered to have the same potential, that is, the range of their possible values is identical. A fully connected neural network will therefore have a higher potential capacity than any other interconnection scheme. However, the maximum number of connections depends on the topology of the neural network.

Most neural networks are trained using a supervised learning rule. For these networks, the number of neurons of the input and output layer is determined by the application and are therefore considered constants for the calculations of this publication. The total number of neurons in the network (N) is, without loss of generality, also considered a constant, in order to enable comparisons between different interconnection schemes (networks having interlayer, interlayer plus intralayer, interlayer plus supralayer connections, and plenary neural networks (which contain all three kinds)).

For each of the different interconnection schemes, the maximum connectivity topology can be stated as a bounded multivariate non-linear integer maximization problem and is solved using an extension of multidimensional Lagrange multipliers. The maximum connectivity topology and maximum number of weights for all layered neural networks with given (fixed) input and output layer sizes are given in the table below, where L_{min} represents the minimum number of layers, and L_{max} the maximum number of layers. For a description of neural network topologies and related terminology see [1], and for the proofs of all the theorems plus a more extensive bibliography, see [2]. Results for the case of variable input and output layer sizes can be found in [3].

interconn. scheme	maximum connectivity topology			
	L_{min}	L_{max}	# neurons per layer (N_l)	maximum number of weights (W_{max})
inter ($L=3$)	3	3	$N_2 = N - N_1 - N_L$	$(N - N_1 - N_L)(N_1 + N_L)$
inter ($L=4$)	4	4	$N_2 = \frac{2N}{L} - N_L,\ N_3 = \frac{2N}{L} - N_1$	$\frac{N^2}{L} - N_1 N_L$
inter+intra	1	2	fixed distribution	$\frac{N(N\pm1)}{2}$
inter+supra	3	none	$\forall l : N_l = \frac{N - N_1 - N_L}{L-2}$	$\frac{(L-3)N^2 + 2N(N_1+N_L) + (1-L)(N_1^2+N_L^2) - 2N_1 N_L}{2(L-2)}$
plenary	1	none	any distribution	$\frac{N(N\pm1)}{2}$

References

[1] E. Fiesler, Neural Network Formalization, IDIAP Technical report 92-01, IDIAP, Martigny, Switzerland, July 1992; extended version submitted to *Computer Standards & Interfaces*, special issue on Neural Network Standards, John Fulcher, editor, North-Holland/Elsevier, Amsterdam, The Netherlands.

[2] E. Fiesler, H. J. Caulfield, A. Choudry, and J. P. Ryan, **Maximal Interconnection Topologies for Neural Networks**, submitted to *IEEE Transactions on Neural Networks*.

[3] E. Fiesler et al., **Maximum Capacity Topologies for Fully Connected Layered Neural Networks with Bidirectional Connections**, *Proc. of the International Joint Conference on Neural Networks, San Diego*, volume 1, pp. 827–831, IEEE Neural Networks Council; Edward Brothers, Ann Arbor, MI, 1990.

THE OVERLAPPED TESSELLATION
A SUPERVISED NEURAL RULE

Ch. Molina, P. Baylou, M. Najim

Equipe Signal et Image and GDR 143 - CNRS
ENSERB. Université de Bordeaux I
351, Cours de la libération. 33405 Talence cedex. FRANCE
Tel. 56 84 61 40   Fax. 56 84 84 06

EXTENDED ABSTRACT

A three-layer Perceptron employing the *nearest neighbour* pattern classification rule was presented independently by O. Murphy [1] and P. Martin [2]. The main advantages claimed by the authors are a complete design and training process carried out in polynomial time in terms of n, the number of training patterns.

Unfortunately, as set out by P. Martin in (Martin, 92), this rule proves unworkable because of time computation $O(n^{[(d+1)]/2]})$ and storage $O(n^{[d/2]})$, which depend on the dimension d of the input vectors. Most of this time is employed to calculate the facets of the Voronoï polytopes, which make up the *Voronoï Tessellation*. Some facets discriminate patterns from opposite classes and, in terms of classification, these are essential; the rest separate patterns from the same class, which is unnecessary.

In order to reduce time computation and storage, we propose a new classification rule derived from the *Voronoï Tessellation*. This time, the new rule computes only those facets discriminating patterns from opposite classes. Voronoï polytopes are constructed in this way and optimized at a second stage, in order to obtain polytopes with the minimum possible number of facets. For a given pattern X, considered as a point in d-dimensional space, its polytope should be computed as follows. Construct a line segment from X to every other training pattern belonging to the opposite classes, and compute the hyperplane that bisects each line segment. This operation is computed in linear time. The intersection domain of all these half-planes is the Voronoï polytope of X. In terms of classification, we do not need all these facets to discriminate the pattern. We suggest a method which eliminates the useless facets of the polytope. A polytope facet is useless if, after its elimination, patterns from opposite classes are not included [3][4]. This operation is also computed in linear time. Computed polytopes usually overlap and they may contain several patterns, but only patterns from the same class, which improves the generalization capacity of the Perceptron. The process stops, with no classification error, when all patterns are included in the constructed polytopes.

This new classification rule, called Overlapped Tessellation (OT), designs and trains the three-layer Perceptron in linear time, and discriminates classes with far fewer neurons than the Voronoï Tessellation (VT).

REFERENCES
[1] Owen J. Murphy, October 1990, "Nearest Neighbor Pattern Classification Perceptrons," in IEEE proceedings, pp. 1595-1598.
[2] Philippe Martin, April 1992, "Réseaux de Neurones artificiels : Application à la Reconnaissance Optique de Partitions Musicales," Thèse de Doctorat de l'Université JOSEPH FOURIER.
[3] C. Molina, P. Baylou et al, 1993, "Evaluation of the '*Gift-Wrapping*' Neural Network in pattern classification," in press, APII, n. 2, 1993.
[4] C. Molina, P. Baylou et al, "New approach for the multi-perceptron architecture construction applied to the edge detection problem," pp. 1219-1222, EUSIPCO 92, 24-27 August 1992, Brussels-Belgium.

IABP: Interval Arithmetic Backpropagation

C.A. Hernández[1], J. Espí[1], K. Nakayama[2], M. Fernández[1]

[1] Department of Computers and
Electronics. Valencia University.
Doctor Moliner, 50.
46100 Burjassot (Valencia).
SPAIN.
E-mail:
HERNANDC@EVALUN11.BITNET

[2] Department of Electrical and Computer
Engineering. Kanazawa University.
Kodatsuno 2-20-40.
Kanazawa-shi. Ishikawa-ken. 920
JAPAN.
E-mail:
Nakayama@haspnn1.ec.kanazawa-u.ac.jp

Abstract

We present in this paper a new generalization of the BP (Backpropagation) algorithm by using interval arithmetic, this new algorithm permits the use of training samples and targets which can be indistinctly points and intervals. An interval vector in the input is translated into an interval vector in the output. It was already proposed an extension of BP to interval arithmetic [1], but it has the severe limitation that it can be only used for two class classification problems. This new algorithm can use any number of classification classes and represents a generalization of BP because every equation reduces to the normal BP equations in the case of point input vectors. This is due to a new definition of the error function:

$$E_P = \frac{1}{4} \cdot \sum_{K=1}^{Noutputs} \{ (t_{P,k}^L - O_{P,k}^L)^2 + (t_{P,k}^U - O_{P,k}^U)^2 \}$$

where t is the target, O the output, P denotes a pattern, L the lower limit of the interval and U the upper limit.

The algorithm can be used to integrate expert's knowledge and training samples. The kind of expert's knowledge which can be integrated is "if ... then" rules of the type: if $X_{P,1} \subset [A_1, B_1]$ ... and $X_{P,n} \subset [A_n, B_n]$ then $X_P \in G_K$, where X_P is a pattern vector and G_K its classification class, the intervals $[A_i, B_i]$ can be easily codified. Several two-dimensional simulation examples were presented.

It can be also used to efficiently represent "don't care attributes" [2]. We can codify a "don't care attribute" D by using an interval $[d_{min}, d_{max}]$, where d_{min} represents the minimum of all the possible values of the attribute and d_{max} the maximum. This codification overcomes an exponential increase in the training set which is needed in the case of normal BP and can be applied to discrete and continuous attributes. The same examples presented in [2] were investigated.

In general, the algorithm will add flexibility to the codification of inputs and targets. For example, in the case we have a strong subjectivity and imprecision (e.g., the codification of symptoms in a medical diagnosis problem) the use of intervals in the codification may reduce this subjectivity and imprecision, an imprecise or subjective input can be codified with a wider interval instead of a point.

References

[1]   H. Ishibuchi and H. Tanaka. "An extension of the BP-Algorithm to Interval Input Vectors - Learning from Numerical Data and Expert's Knowledge-". IJCNN-91. Singapoure. pp.1588-1593.

[2]   H.M. Lee and C. Hsu. "The handling of Don't Care Attributes". IJCNN-91. Singapoure. pp. 1085-1091.

Architecture of Associative Memory with Reduced Cross Talk and Its Performance Formulation

Yukio KUMAGAI[†] Joarder KAMRUZZAMAN[††] Jose L. PEREZ[†]

[†]Dept. of Comp. Science & Syst. Engg. [††] Dept. of Electrical & Electronic Engg.
Muroran Inst. of Tech. Bangladesh Univ. of Engg. & Tech.
27-1 Mizumoto-cho, Muroran 050, Japan Dhaka 1000, Bangladesh

Abstract - The extended conventional high order associative memory, including how to apply the input key and to obtain the recalled output, is represented as follows:

$z = F \{ \Sigma_{i=1}^{v} (y^{(i)} \otimes x_{(\alpha)}^{(i)}) \cdot x_{(\alpha)} \}$, where $\otimes$ denotes the outer product operator, F denotes the signum

function that operates componentwise and $x^{(i)}{}_{(\alpha)}$ denotes the i-th extended high order memorized key vector of $\Sigma_{k=0}^{\alpha}$ nCk - length whose components consist of mutually disjoint high order cross product up to α-th order. $x_{(\alpha)}$ is an extended high order input vector constructed under the same condition. Previously, we proposed a new architecture with reduced cross talk, which completely eliminated cross talk due to the memorizing keys having odd number Hamming distance from the input key and furthermore, a relatively large part of the remaining cross talk due to even number Hamming distance from the input key by introducing (N,n) Hamming code into all the memorizing keys and input key, i.e.,

$z = F [\Sigma_{i=1}^{v} (1/2) \{ (y^{(i)} \otimes \overline{x}_{(\alpha)}^{(i)}) \cdot \overline{x}_{(\alpha)} + (y^{(i)} \otimes \widetilde{x}_{(N-\alpha)}^{(i)}) \cdot \widetilde{x}_{(N-\alpha)} \}]$, where $\overline{x}_{(N-\alpha)}^{(i)}$, $\widetilde{x}_{(N-\alpha)}$ denote

the i-th high order memorized keys and input key extended from N to (N-α)-th order, encoded as (N,n) Hamming Code, respectively. The basic idea of the construction of this architecture is based upon the finding that the recalled output z of conventional associative memory, extended to high order correlation, can be represented in an explicit functional form in terms of the Hamming distance d between the memorized keys and input key as follows:

$z = F \{ \sum_{d=0}^{n} (\sum_{i_v(d) \in I^{(d)}} y^{(i_v(d))}) H(d,\alpha; n) \}$, where $H(d,\alpha;n) = \Sigma_{k=0}^{\alpha} h(d,k;n)$, $h(d,k:n) = \Sigma_{\zeta=0}^{d}$

$(-1)^\zeta \cdot dC_\zeta \cdot (n-d)C_{(k-\zeta)}$, and $I^{(d)} = \{ i_v(d) : i_v(d) \in I, v=1,...,v_d$ and $(1/2) \cdot \Sigma_{j=0}^{n} | x_j^{(i_v(d))} - x_j | = d \}$ is a partition

of the index set of data $I = \{ i : y^{(i)}, i=1,...,v \}$. In order to investigate the performance characteristics of conventional and the proposed architecture, computer experiments were performed by using randomly generated bipolar data and keys. To make the comparison on cost performance basis, the architectures considered use a nearly equal number of cross products though the order of correlation is different in each case. The proposed architectures require less order of correlation compared to the conventional architecture and this makes the construction of the proposed architecture less tedious. In this paper, to investigate quantitatively the ability of exact recall of the proposed architecture, we formularize the probability of exact data recall as the functional form of Hamming distance between the memorized keys and the input key under the assumption that all the memorizing keys and data are subjected to be mutually statistically independent. When the desired datum to be recalled exactly is $y^{(i^*)}$, the computational result is as shown below:

$$\text{Prob}(z=y^{(i^*)}) = \{ \frac{1}{\sqrt{2\pi}} \int_{-\infty}^{H(0,\alpha;N)/\sqrt{\sum_{d \in \{d : d \equiv 0 \text{ mod.2}\}} \frac{(v-1)}{2^n} \omega(d,N) H(d,\alpha;N)^2}} \exp(-\frac{u^2}{2}) du \}^m,$$

where $\omega(d,N) = {}_N C_d + \Sigma_{\gamma=0}^{d} (-1)^\gamma \cdot {}_{(N-1)/2)} C_{(d-\gamma)} \cdot {}_{((N+1)/2)} C_\gamma$. Experimental results show that the theoretical estimation of the probability of exact data recall is in well agreement with the experimental results even though the parameters n, N, m and v used in the experiments are not so large. In this paper, the percentage of data recall by the architecture was considered on all the exactly recalled data and also those data in which all the components except for only one is correctly recalled by using (15,11) Hamming Code for the memorized data. In this experiment, the proposed architecture achieved excellent recall (98.9 % by theoretical estimation and 99.2% by experimental result with 500 memorized pairs), and signified that, by employing error correcting code in all data of length 11, almost all the memorized data can be exactly recalled. Both the theoretical and experimental results strongly demonstrate the superior ability of the proposed architecture, especially the one with Hamming Code and substantiate the effective reduction of cross talk.

518

AUGMENTATION OF GENERALISATION IN PROBABILISTIC LOGIC NETS

R.S.Neville and T.J.Stonham

Department of Electrical Engineering, Brunel University, Uxbridge UB8 3PH, UK.

Abstract

In this paper we study Generalisation in feedforward nets with particular reference to Enhancing the net's performance. The paper studies several novel Methodologies for the training of probabilistic logic nodes. The nets are hardware realisable and the units are logical nodes that respond to their input patterns in addressable locations; the locations then define the probability of the output being a logical "1". Four methods are discussed; spreading, bit-streams, noise and a gaussian-weighted input layer. The new techniques use information that is readily available to the network as it is trained. The techniques augment existing training methodologies. The most important factor portrayed in this paper is that by using either a gaussian-weighted pre-processing layer or bit streams the generalisation of these units was enhanced.

1. Bit-Stream Training Theory

The basic work on bit streams was initiated by Gurney[1989]. The bit-stream structure, is said to automatically produce noise and hence promote generalisation, Neville [1993]. The bit-stream works by producing a noisy representation of the input. At each input, new output bits are obtained from the data in the stream, by interpreting its contents as a unary representation of the probability of outputting a 1.

2. Gaussian-Weighted Node Theory

The Gaussian-Weighted Node (GWN), was developed from work by Burt[1981] on Image processing. The GWN utilises gaussian weights on a spatially oriented receptive field. These are summed and Thresholded. The Threshold empowers the node with the ability to blur or de-blur its output. Blurring causes the diffusion of a sharp image into a wide or diffuse image[Koenderink, J. J. and Van Doorn, A. J. (1987)]. De-blurring is the reverse of this operation. Extra layers of the gaussian weights on a spatially oriented receptive field may be added in order to blur/de-blur to a larger extent. The spatially mapped input lines are centered on the pixel position, that was originally allocated to the input line of the PLN, and four other neighboring pixels. The output of the GWN was then connected to the input line of the PLN.

3. Concluding Remarks

The bit-streams promoted generalisation by inducing clusters of like valued sites on the hypercubes. The activation-output conversion unit on the output of the streams, gave rise to large clusters when a linear transfer function was used. While a sigmoidal transfer function advocates smaller clusters and by utilising a 'dead band' one induces tightly clustered like valued site values on the cubes. The Gaussian-weighted Node per-processing layer performs a gaussian convolution and thresholding operation on the input image. With a single layer GWN one may promote generalisation be blurring the image. The generalisation was augmented when a second GWN layer was added to de-blur the output of the first GWN layer. The GWN performs a gaussian normalisation function which reduces the images spurious resolution, which may induce augments the generalisation of the Probabilistic Nodes.

4. References

Burt, P. J. (1981). Fast Filter Transforms for Image Processing, Computer Graphics and Image Processing, 16, pp 20-51.

Gurney, K. (1992). Training Nets of Hardware Realisable Sigma-Pi Units, Neural Networks, 5, pp 289-303.

Koenderink, J. J. and Van Doorn, A. J. (1987). Representation of Local Geometry in the Visual System, Biological Cybernetics, 55, pp 367-375.

Neville, R. (1993). Augmentation of Generalisation in Probabilistic Logic Nets. World Congress on Neural Networks (WCNN'93 Portland), July 11-15, Portland, Oregon, USA.

Acknowledgement: I would like to thank SERC for my funding (Grant No. 90311957).

FUZZY EXPERT NETWORKS

H.C. FU, J.J. SHANN
Department of Computer Science and Information Engineering
National Chiao-Tung University
Hsinchu, Taiwan 300, R.O.C.

Abstract

The proposed fuzzy expert network is an event-driven, acyclic neural network designed for fine knowledge learning of a fuzzy expert system. The coarse knowledge, i.e. fuzzy rules, of a fuzzy expert system can be constructed in the structure of the network. The fuzzy expert network contains five types of nodes: Input, Membership-Function, AND, OR, and Defuzzification Nodes. Each input node of the fuzzy expert network represents an input variable of the fuzzy expert system, and is used as a buffer to broadcast the input to its membership-function nodes. Each membership-function node represents one of the membership functions associated with a particular variable. We define a modified Quadratic Sigmoid function [1] to approximate a trapezoidal normalized membership function. Each AND node represents the IF-part of some fuzzy rules. We define a parametric operation, called Fuzzy-MIN, which combines the upper bound (the min operator) and the lower bound (the drastic product) [2] of fuzzy intersections for the AND nodes. Each OR node represents the THEN-part of some fuzzy rules. Therefore, the operation performed in an OR node is to integrate the rules of the same consequence. We define another parametric operation, called Fuzzy-MAX, which combines the lower bound (the max operator) and the upper bound (the drastic sum) [2] of fuzzy unions for the OR nodes. Each defuzzification node represents either an intermediate variable or an output variable, and performs the defuzzification of all the related membership functions.

The backpropagation-like learning used in the network is focused on the learning of fine knowledge including the certainty factors of fuzzy rules, and the parameters of the fuzzy-MIN and fuzzy-MAX operations. The learning rules for the adjustment of certainty factors, the parameters of the operations of AND nodes, and the parameters of the operations of OR nodes are based on the gradient descent search. The evaluation of the gradients for adjustment of the learnable items of the network were derived.

A general purpose simulator of the proposed fuzzy expert network has been implemented in a Sun SPARC station. An exemplar fuzzy expert system was converted to a fuzzy expert network for observing the behavior of the network. After the fine knowledge learning, the fuzzy expert network contains precise knowledge such that the output can be produced in a more accurate manner than that of the original fuzzy system. In the near future, we plan to further study on the learnability of the parameters of membership functions.

References

[1] C.C. Chiang, H.C. Fu, "A Variant of Second-Order Multilayer Perceptron and Its Application to Function Approximations," *Proc. of IJCNN '92*, Baltimore, Vol. III, pp. 887-892, 1992.

[2] H.J. Zimmermann, *Fuzzy Set Theory and Its Application*, 2nd Ed., Kluwer Academic Publishers, Boston/Dordrecht/London, 1991.

Stochastic dynamical systems
—oral contributions

Using Boltzmann Machines for probability estimation

Bert Kappen

Department of Medical Physics and Biophysics, University of Nijmegen, Geert Grooteplein 21, 6525 EZ Nijmegen, The Netherlands.

1 Introduction

In most neural network applications, the generic idea is to find some complex non-linear mapping between an input domain and an output domain. This mapping can be deterministic but is in general probabilistic. Such a probabilistic map is formally described by the conditional probability $p(\vec{y}|\vec{x})$ to observe an output $\vec{y}$ given an input $\vec{x}$. Examples of deterministic mappings are Multi-Layered Perceptrons (MLPs) [1], which minimizes a quadratic error between the output of the network and the desired output. The output of such a network has in general no clear interpretation in terms of conditional probability.

A description of probabilistic mappings in terms of conditional probabilities can be obtained by deriving learning rules from a Log-likelihood or Kulback-Leibler error rate [2]. In the context of neural networks, this has been done by Solla et al. [3] and Baum and Wilczek [4] for MLPs, and by Hopfield [5] and Kappen [6] for BMs. In [6] also a comparison of these methods was given.

There are many important applications for which a priori probability is needed in addition to the conditional probability. Here we list some generic examples:

Classification with confidence

One of the well known attractive features of neural networks is that they generalize well to data that were not part of the training set. Nevertheless, when data are presented that are far from the training set in the input domain, the network is unlikely to give correct classifications. Thus, what is needed is an indication of the confidence of the output that the network produces. Such a confidence measure depends on two things:

1. whether the input pattern belongs to a part of the input space that was well sampled during training, and

2. if so, whether in that part of the input space a clear distinction between classes can be made.

The former is given by the a priori probability and the latter by the conditional probability.

Database repairing

In most practical data bases the majority of records can be easily learned by the network, and some records may be more difficult to learn. This may be because of errors in the database. For example, Brunak et al. [7] discovered in this way 7 typing errors in the EMBL nucleotide sequence databank. This can be understood by saying that the conditional probability as given by the majority of the database entries, was unlikely for these 7 patterns. Checking these DNA sequences in the original publications revealed the errors. In many databases, such an independent verification is not available. It is therefore important to have quantitative methods to decide whether or not records are erroneous. Since such errors may occur in the 'input' as well as in the 'output' fields, a correct model of the *conditional* probability is not sufficient to answer this question, and knowledge of the a priori probability distribution is required. The same applies to records with missing values: current neural networks can fill in the missing output values, but can not give the probability that the value that they provide is correct.

Inverse modeling

After a network has been succesfully trained, the question often arises to find the set of inputs that maximize the output of the network, or that correspond to another desired output value. Applications exist in for instance direct mailing, where a company wants to reach a group of potential customers with the largest expected interest in their products, or process control, where one wants to know all possible combinations of machine set points and raw material characteristics that will yield a product with the desired specifications. A good model of the a priori probabilities is needed to constrain this search task to probable input values.

2 Probability estimation with Boltzmann Machines

Thus, the general problem is to estimate a probability density $q(\vec{x}, \vec{y})$ of joint occurence of an input $\vec{x}$ and an output $\vec{y}$. If $\vec{x}$ denotes input patterns and $\vec{y}$ output patterns, this probability density can be written as

$$q(\vec{x}, \vec{y}) = q(\vec{y}|\vec{x})q(\vec{x}),$$

i.e. the joint probability is a product of the a priori probability $q(\vec{x})$ to observe an input $\vec{x}$ and a conditional probability $q(\vec{y}|\vec{x})$ to observe an output $\vec{y}$ given an input $\vec{x}$.

However, for joint probability estimation, the distinction between inputs and outputs disappears. We will therefore consider training patterns $(\vec{x}, \vec{y})$, with $x_i \in R, i = 1, \ldots, n$ and $y_k = \pm 1, k = 1, \ldots, m$. x_i and y_k can denote either input or output. Once $q(\vec{x}, \vec{y})$ is estimated, other probabilities, such as $q(\vec{x})$ or $q(\vec{y}|\vec{x})$, can be easily expressed in terms of of $q(\vec{x}, \vec{y})$.

Let us consider a network with h hidden units: $s_j = 0$ or $1, j = 1, \ldots, h$. When no continuous valued neurons are present, $n = 0$, the equilibrium distribution of the Boltzmann Machine (BM) under Glauber dynamics is given by

$$p(\vec{s}, \vec{y}) = \frac{1}{Z} \exp\{\beta(\sum_{j=0}^{h}\sum_{k=0}^{m} v_{jk}s_j y_k + \sum_{j,j' \neq j}^{h} a_{jj'}s_j s_{j'} + \sum_{k,k' \neq k}^{h} b_{kk'}y_k y_{k'}\} \tag{1}$$

$$= \frac{1}{Z} \exp\{\beta f(\vec{y}, \vec{s})\} \tag{2}$$

with Z such that $\sum_{\vec{s}, \vec{y}} p(\vec{s}, \vec{y}) = 1$. We have included thresholds $v_{0k}, k = 1, \ldots, m$ and $v_{j0}, j = 1, \ldots, h$ in the $\vec{y}$ and $\vec{s}$ units respectively, by assuming $s_0 = y_0 = 1$ and $v_{00} = 0$. The probability $p(\vec{y})$ to observe $\vec{y}$ on the visible units is $p(\vec{y}) = \sum_{\vec{s}} p(\vec{s}, \vec{y})$. Thus

$$p(\vec{y}) = \frac{\sum_{\vec{s}} \exp\{\beta f(\vec{y}, \vec{s})\}}{\sum_{\vec{s}, \vec{y}} \exp\{\beta f(\vec{y}, \vec{s})\}} \tag{3}$$

The BM minimizes the Kullback distance between $q(\vec{y})$ and $p(\vec{y})$:

$$d(p, q) = \sum_{\vec{y}} q(\vec{y}) \log\left(\frac{q(\vec{y})}{p(\vec{y})}\right). \tag{4}$$

The BM learning rule consists of gradient descent on d [8]. Note, that we can choose $\beta = 1$ without loss of generality: any other value of β will be compensated for by the learning process. The partial derivatives of $p(\vec{y})$ with respect to v_{jk}, $a_{jj'}$ and $b_{kk'}$ are difficult to calculate: One must either calculate the contribution from 2^{m+h} terms in Eq. (3), or use Glauber dynamics to measure $< s_j y_k >$, $< s_j s_{j'} >$ and $< y_k y_{k'} >$, respectively, after the system has reached equilibrium [8]. As a result, the Ackley Hinton learning rule for BMs is too slow to use in practice. The learning rule can be significantly accelerated by using the so-called mean field annealing [9], but this has been mainly used for optimization problems and has limited applicability for probability estimation [10]. This is because the mean field approximation assumes that $< y_k y_{k'} > \approx < y_k >< y_{k'} >$. It is easy to construct probabilities $q(\vec{y})$ for which this is not true. Consider for instance the AND problem, where y_1 and y_2 are inputs and y_3 is the output. Then $< y_1 > = 0$, $< y_3 > = \frac{1}{2}$ and $\frac{1}{2} = < y_1 y_3 > \neq < y_1 >< y_3 > = 0$.

With inclusion of the continuous neuron states $\vec{x}$, we propose the following special architecture:

- add inhibitory connections between the units $\vec{s}$ and thresholds at the units $\vec{s}$, such that the local field contribution at s_j from the other hidden units is $-J(\sum_{j' \neq j}^{h} s_{j'} - 1)$

- remove the adaptive connections among the units $\vec{x}$ and $\vec{y}$

- remove the adaptive connections among the units $\vec{s}$. This is not strictly necessary, but will not be pursued here.

The Boltzmann distribution becomes

$$p(\vec{x}, \vec{y}, \vec{s}) = \frac{1}{Z} \exp\{\sum_{i=0}^{n} \sum_{j=0}^{h} w_{ij} x_i s_j + \sum_{j=0}^{h} \sum_{k=0}^{m} v_{jk} s_j y_k - J(\sum_{j=1}^{h} s_j - 1)^2\} \tag{5}$$

The effect of these connections is that for large J, only *permissible states* $(\vec{y}, \vec{s}_j)$ with $(\vec{s}_j)_{j'} = \delta_{jj'}$ have a finite probabilty of occurence. It can also be shown that for any h, m, the contribution of all non-permissible states $(\vec{y}, \vec{s})$ to the partition function can be made arbitrarily small for sufficiently large J [10].

However, this probability is not normalizable: Z involves an integration over R^n, which is infinite. We therefore propose to change the local field contribution from $\vec{x}$ to unit s_j from

$$\sum_{i=0}^{n} w_{ij} x_i \rightarrow -\beta \|\vec{w}_j - \vec{x}\|^2. \tag{6}$$

Note, that we have reintroduced temperature, because with this change it can no longer be scaled away. $(\vec{w}_j)_i = w_{ij}$.

The probabilities on the visible units become:

$$p(\vec{x}, \vec{y}) = \frac{1}{Z} \sum_{j=1}^{h} \exp\{-\beta \|\vec{w}_j - \vec{x}\|^2 + \sum_{k=0}^{m} v_{jk} y_k\}$$

$$= \frac{1}{Z} \sum_{j=1}^{h} \exp\{H_j(\vec{x}, \vec{y})\} \tag{7}$$

with

$$Z = \int d\vec{x} \sum_{\vec{y}} \sum_{j=1}^{h} \exp\{-\beta \|\vec{w}_j - \vec{x}\|^2 + \sum_{k=0}^{m} v_{jk} y_k\} \tag{8}$$

$$= \left(\frac{\pi}{\beta}\right)^{n/2} \sum_{j=1}^{h} \exp(v_{j0}) \Pi_{k=1}^{m} 2\cosh(v_{jk}) \tag{9}$$

Note that Z is independent of $\vec{w}_j$.

It is easy to calculate the learning rules for w_{ij}, v_{jk} and v_{j0}:

$$\Delta w_{ij} = -\epsilon \frac{\partial d}{\partial w_{ij}}$$

$$= \epsilon \int d\vec{x} \sum_{\vec{y}} q(\vec{x}, \vec{y}) \frac{2\beta(x_i - w_{ij}) \exp\{H_j(\vec{x}, \vec{y})\}}{\sum_{j'=1}^{h} \exp\{H_{j'}(\vec{x}, \vec{y})\}} \tag{10}$$

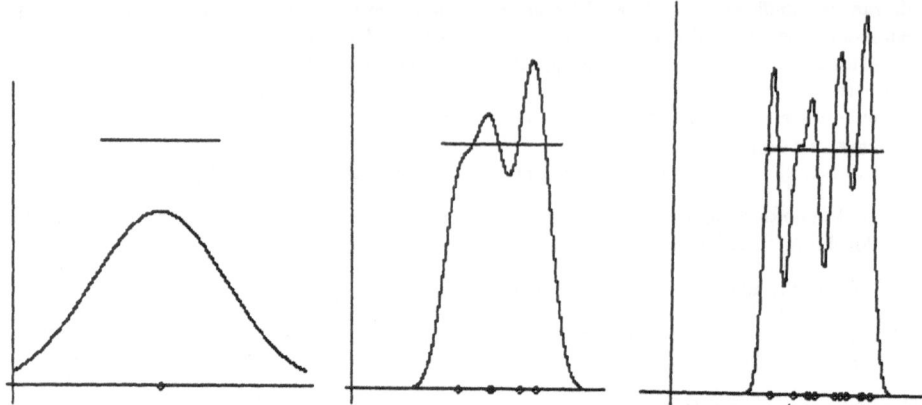

Figure 1: The probability densities $q(x)$ and $p(x)$ as a function of x for 16 hidden units and different values of $\beta(=b$ in the figure). The dots on the x-axis depict the locations of the Gaussian centers. a) $\beta = 0.1$, $d = 2.545$, b) $\beta = 2.0$, $d = 2.144$, c) $\beta = 10.0$, $d = 2.067$

$$(11)$$

$$= \epsilon \int d\vec{x} \sum_{\vec{y}} q(\vec{x}, \vec{y}) 2\beta(x_i - w_{ij}) p(\vec{s}_j | \vec{x}, \vec{y}) \tag{12}$$

$$\Delta v_{jk} = -\epsilon \frac{\partial d}{\partial v_{jk}}$$

$$= \epsilon \int d\vec{x} \sum_{\vec{y}} q(\vec{x}, \vec{y}) \frac{y_k \exp\{H_j(\vec{x}, \vec{y})\}}{\sum_{j'=1}^{h} \exp\{H_{j'}(\vec{x}, \vec{y})\}}$$

$$- \frac{\epsilon}{Z} \left(\frac{\pi}{\beta}\right)^{n/2} 2\sinh(v_{jk}) \exp(v_{j0}) \Pi_{k' \neq k}^{m} 2\cosh(v_{jk'}) \tag{13}$$

$$= \epsilon \int d\vec{x} \sum_{\vec{y}} y_k p(\vec{s}_j | \vec{x}, \vec{y})\{q(\vec{x}, \vec{y}) - p(\vec{x}, \vec{y})\} \tag{14}$$

$$\Delta v_{j0} = -\epsilon \frac{\partial d}{\partial v_{j0}} \tag{15}$$

$$= \epsilon \left(\int d\vec{x} \sum_{\vec{y}} q(\vec{x}, \vec{y}) \frac{\exp\{H_j(\vec{x}, \vec{y})\}}{\sum_{j'=1}^{h} \exp\{H_{j'}(\vec{x}, \vec{y})\}} - \frac{1}{Z} \left(\frac{\pi}{\beta}\right)^{n/2} \exp(v_{j0}) \Pi_{k=1}^{m} 2\cosh(v_{jk}) \right) \tag{16}$$

$$= \epsilon \int d\vec{x} \sum_{\vec{y}} p(\vec{s}_j | \vec{x}, \vec{y})\{q(\vec{x}, \vec{y}) - p(\vec{x}, \vec{y})\} \tag{17}$$

Eqs. (13) and (16) are learning rules with a computational complexity of $O(h(n + m))$ for single pattern presentation. This allows this BM to be used for practical applications. Eqs. (14) and (17) are just reinterpretation of these explicit expressions in terms of probabilities.

3   Clustering and symmetry breaking

An interesting symmetry breaking phenomenon occurs in the hidden layer of the BM, which can be illustrated on a simple example with one continuous valued input. The probability $q(x) = 1$ for $0 < x < 1$ and zero elsewhere. The network consists of 16 hidden units, and we set $v_{j0} = 0$. The training set consists of 100 patterns which are drawn at random from $q(x)$. We employed batch-mode learning for 1000 epochs. The results are given in Fig. 1 for different values of β. For high temperature, the optimal solution is such that all hidden units center on the average value of $q(x)$. At lower temperature, symmetry breaking occurs resulting in specialization of the hidden units. It

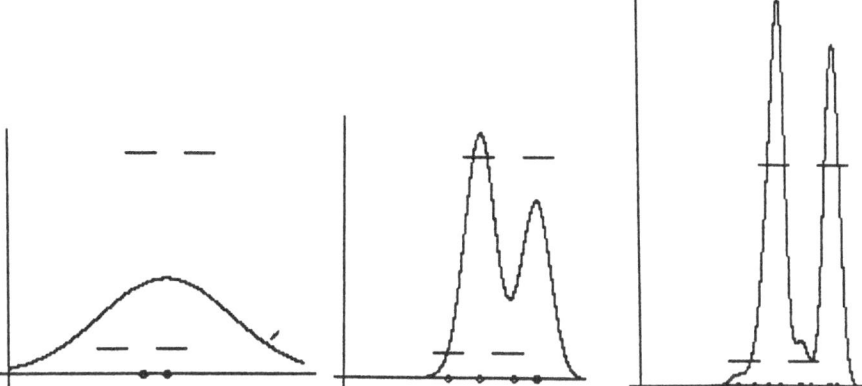

Figure 2: The probability densities $q(x, -1)$ and $p(x, -1)$ as a function of x for 16 hidden units and different values of β(=b in the figure). The dots on the x-axis depict the locations of the Gaussian centers. a) $\beta = 0.1$, d=2.525, b) $\beta = 2.0$,d=1.819, c) $\beta = 10.0$, d=1.637

can be shown in general that for the BM as discussed in this paper in the absence of discrete units $\tilde{y}$, the Kulback distance is identical to the Helmholtz free energy. The phenomena of symmetry breaking in the free energy was previously observed by Rose et al. [11]. The critical temperature for first symmetry breaking can be easily calculated and is $\beta = 0.38$ for this example.

When both continuous and discrete units are present the symmetry breaking phenomenon persists. We illustrate this with a system consisting of one continuous unit and one discrete unit. The probability $q(x, -1)$ is depicted in Fig. 2. $q(x, 1)$ is such that $q(x) = q(x, 1) + q(x, -1)$ is as above. We see, that even for high temperature the 16 units specialize with respect to the classes $y = \pm 1$, with 8 units for each class. Within each class no specialization occurs: all units center on the expectation value of $q(x, 1)$ and $q(x, -1)$ respectively. At lower temperature, the different constituents of the classes are discovered.

4 Discussion

In this paper a beginning was made with the design and study of Boltzmann Machines for probability density estimation. By using lateral inhibitory connections, it was possible to obtain fast learning rules. It was shown, that the hidden units specialize, depending on the temperature. As a result, the quality of the solution becomes largely *independent* of the number of hidden units. This opens the possibility of controling the generalization performance of the network by the temperature, instead of the number of hidden units.

Acknowledgements

This work was supported by the Dutch Foundation for Neural Networks (SNN).

References

[1] D. Rumelhart, G. Hinton, and R. Williams. Learning representations by back-propagating errors. *Nature*, 323:533–536, 1986.

[2] S. Kullback. *Information theory and statistics*. Wiley, New York, 1959.

[3] S. A. Solla, E. Levin, and M. Fleisher. Accelerated learning in layered neural networks. *Complex Systems*, 2:625–640, 1988.

[4] E. B. Baum and F. Wilczek. Supervised learning of probability distributions by neural networks. *Proc. IEEE NIPS*, 1987.

[5] Hopfield J.J. Learning algorithms and probability distributions in feed-forward and feed-back networks. *Proc. Natl. Acad. Sci. USA Vol. 84*, pages 8429–8433, 1987.

[6] H.J. Kappen. Using boltzmann machines as perceptrons. *Submitted to IEEE Trans. Neural Networks*, 1993.

[7] S. Brunak, J. Engelbrecht, and S. Knudsen. Cleaning up gene databases. *Nature*, 343:123, 1990.

[8] D. Ackley, G. Hinton, and T. Sejnowski. A learning algorithm for Boltzmann machines. *Cognitive Science*, 9:147–169, 1985.

[9] Peterson C. and Hartman E. Explorations of the mean field theory learning algorithm. *Neural Networks, Vol. 2*, pages 475–494, 1989.

[10] H.J. Kappen. Comment on the usefulness of mean-field approximation for probability estimation with boltzmann machines. Unpublished, 1993.

[11] K. Rose, E. Gurewitz, and G. Fox. Statistical mechanics of phase transitions in clustering. *Physical Review Letters*, 65:945–948, 1990.

Brownian Motion Updating of Multi-layered Perceptrons

Thorsteinn S. Rögnvaldsson
Department of Theoretical Physics, Sölvegatan 14 A, S-223 62 Lund, Sweden
email: denni@thep.lu.se

Abstract
The effect of adding noise during training is investigated for the Multilayer Perceptron. The Langevin updating, where noise is added directly to the weights, is advocated. It constitutes a natural extension of standard backpropagation learning and is easily controlled. It generates weight configurations within the Boltzmann ensemble, enabling the network to diffuse out of local extrema. Furthermore, "Manhattan" updating is shown to be similar to Langevin updating, explaining some of its successes in shortening the convergence time. The argument is illustrated on four different problems; separating overlapping Gaussian distributions, the parity problem, the Mackey-Glass time series, and a medical classification task.

1 Introduction

The presence of noise often improves the performance of a neural network: Noise in the weights can significantly improve the convergence and generalization of a Multilayer Perceptron (**MLP**) [1, 2]. Noisy training patterns improve the retrieval dynamics of Hopfield networks [3] and the generalization in linear Perceptrons [4]. So-called "on-line" back-propagation (**BP**) updating, which constitutes an indirect way of adding noise into the updating, usually shortens the convergence times considerably as compared to standard "off-line" BP updating. This paper investigates the effect of applying noise directly to different parts of an MLP during the learning process. Three different ways are studied; noise on the input nodes, noise on all nodes, and noise on the weights. The latter procedure, Langevin updating (**LV**), is advocated. It is easily controlled and a very natural extension to standard BP updating. The reason why noisy training shortens the convergence time is also discussed, which leads to a "rule of thumb" on when to use it.

2 Langevin updating

In Langevin updating a stochastic noise term is added to the BP updating equations

$$\Delta\omega(t) = -\eta\frac{\partial E(t)}{\partial \omega} + \sigma(t) \tag{1}$$

where ω is the weight, η the learning rate, E the error function, and σ a Gaussian noise. This updating corresponds to placing the network in a "heat bath", generating weight configurations within the Boltzmann ensemble. The network performs a Brownian motion in weight space and can diffuse out of plateaus and local minima in the error landscape.

A virtue of LV updating is that the noise term in eq. (1) is independent of the input patterns. This avoids the problem with the so-called "flat spot" when the output node is saturated and the derivative, which is propagated backwards, is approximately zero.

The Langevin updating is quite different from stochastic on-line BP updating, where noise is introduced in the error gradient by updating over randomly chosen subsets of training patterns. In that case local minima also become metastable states from which the network can escape [5]. This procedure, however, does not avoid the "flat spot" problem. Furthermore, the noise level here is difficult to control, since it depends on differences in the input patterns and the procedure how these are selected for updating. The maximum noise level in on-line BP is achieved by updating after each pattern presentation, which is time consuming. It is preferable to instead use a "block" update scheme with LV updating. This decreases the number of operations needed to converge considerably.

Relation to "Manhattan" updating:
In "Manhattan" (MH) updating [6] only the sign of the gradient is used

$$\Delta w(t) = -\eta_M \text{ sign} \left[\frac{\partial E(t)}{\partial w} \right] \tag{2}$$

which corresponds to a discretization of weight space into cells of size η_M. The learning rate η_M is decreased as training proceeds, allowing the network to converge to a good solution.

It is reasonable to assume that the distribution of standard BP weight updatings, at a specific time during the learning, is symmetric with two humps centred around $\pm \Delta w_{BP,0}$ [7]. If $\eta_M = \Delta w_{BP,0}$ then the "correction", $\Delta w_{BP} - \Delta w_M$, will be approximately random with a mean value of zero and a standard deviation of the same size as that of the Δw_{BP} distribution. Consequently, if the Manhattan learning rate η_M is properly tuned, MH updating will essentially imitate LV updating.

3 Noise Added to the Nodes

If noise is added to all nodes in an MLP with one hidden layer with tanh-sigmoids the differences in weight updatings can be written [7]

$$\delta w_{ij}(t) = -\eta \left(\frac{\partial E'}{\partial w_{ij}} - \frac{\partial E}{\partial w_{ij}} \right) \approx \sigma(t)\hat{o}_i(1 + \sum_k w_{jk}) \tag{3}$$

$$\delta w_{jk}(t) = -\eta \left(\frac{\partial E'}{\partial w_{jk}} - \frac{\partial E}{\partial w_{jk}} \right) \approx \sigma(t)\sum_i \hat{o}_i w_{ij} \tag{4}$$

where it has been assumed that the network is in the initial stages of training and the weight values are small. The error function calculated from the noisy nodes is denoted E' and $\hat{o}$ are the target values for the outputs. The indices ij and jk refer to weights connecting to the output and the hidden layer respectively. This kind of noise is much harder to control than the Langevin method since the noise gets propagated down through the weights, affecting lower layers with different amounts of noise. The noise will hardly affect weights in lower layers at all if the weight values are very small. Similarly, noise added only to the inputs will mostly affect weights connecting from the input layer, if the weights are small. Both these methods are hence inferior alternatives to the Langevin one, since the amount of noise added to each weight layer is very difficult to control during the learning.

4 Simulations

The effect of adding noise to different parts of the MLP during the learning process was studied experimentally for four different problems. The noise level σ was decreased geometrically during the training, in order for the network to converge towards the standard BP updating. The learning rate η was controlled using a "bold driver" technique, where η was increased if the error E was decreasing and decreased otherwise (the learning rate was set to decay slower than the noise level). Ensembles of 100 networks were trained with "block" BP (10 patterns per update), MH, LV, input noise, and node noise updating. The average performances were then compared for the different methods. The weights were initialized with small values around the origin, except for the medical classification, where another procedure was used in order to compare with previous results. All simulations were performed with the program JETNET 2.2 [8].

The first problem is to **separate two overlapping 10-dimensional Gaussian distributions** with the same mean value but different standard deviations [2]. MLP networks with 10 inputs, 20 hidden and 1 output unit were used. Each network was trained for 500 epochs, corresponding to $5 \cdot 10^5$ pattern presentations. The ensemble average errors and classification performances are tabulated in table 1a. This problem displays something similar to a local minima close to the origin and BP has trouble escaping from it if $\eta \leq 0.1$ (see the discussion section). However, Langevin updating takes the network out of the local minima for a wide spectrum of noise levels ($\sigma \in [10^{-5}, 10^{-2}]$ all give the same result). Manhattan also gives good results, if the learning rate is scaled down, but noise applied to the nodes of the network does not help at all.

The second problem, **5-parity** [9], is to determine whether the number of zeroes in a 5-dimensional binary vector is odd or even. MLP networks with 5 inputs, 8 hidden, and 1 output unit were used and each network was trained for 10^4 epochs or until convergence (corresponding to 100% correct classification). Table 1b shows the average convergence time, in epochs, for those networks that converged. It is evident that noise does not improve things.

Overlap. Gauss.			5-Parity		
Updating	$\langle E \rangle$	$\langle Class. \rangle$	Updating	$\langle \tau \rangle$	converged
BP, $\eta = 0.1$	0.102	0.652	BP, $\eta = 0.5$	2202	83%
BP, $\eta = 0.2$	0.033	0.912	BP, $\eta = 0.8$	2041	87%
MH, $\eta = 0.01$	0.046	0.890	MH, $\eta = 0.7$	390	4%
MH, $\eta = 0.005$	0.036	0.906			
LV, $\eta = 0.1$	0.029	0.922	LV, $\eta = 0.5$	2196	87%
NN, $\eta = 0.1$	0.089	0.721	NN, $\eta = 0.5$	2620	88%
IN, $\eta = 0.1$	0.102	0.662	IN, $\eta = 0.5$	2256	88%

Table 1: (a) Average error and classification on the overlapping Gaussian distributions. BP = Backpropagation, MH = Manhattan, LV = Langevin, NN = Node noise, IN = Input noise. The Bayes limit for this problem is 0.934. (b) Average convergence time, in epochs, for the 5-parity problem and the fraction of networks that converged within 10,000 epochs.

The third problem consists in **predicting the Mackey-Glass time series** [10] at time $x(t + 85)$ given $x(t), x(t - 6), x(t - 12)$ and $x(t - 18)$. An MLP with 4 inputs, two hidden layers with 10 and 5 units, and one output used. Figure 1 shows typical convergence

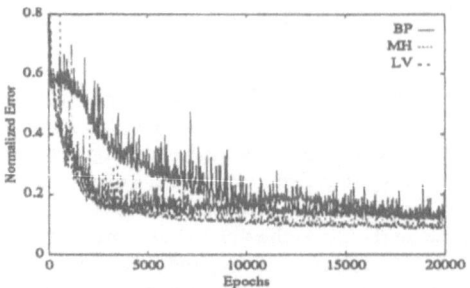

Figure 1: Convergence curves for BP, MH, and LV updating on the Mackey-Glass time series. One epoch corresponds to 500 training patterns.

curves using BP, MH, and LV updating. Adding noise significantly shortens the convergence time. The LV and MH updatings quickly move the network out of an initial plateau that keeps the network "trapped" in the BP case.

The fourth problem is to **determine whether a patient is hypothyroid or not** [11]. The data consists of 17200 patterns; 3772 used for learning and 3428 for testing the generalization ability. Each pattern has 21 input variables; 15 binary and 6 continuous, and 3 binary output variables; coding whether the patient has normal, hyper-thyroid, or subnormal functioning. The three different classes occur with probabilities 92.5%, 5.1%, and 2.4% respectively. This data set has recently been used as "a very hard practical classification task" in a benchmark test of different ANN training algorithms [12]. The same network architecture, one hidden layer with 10 units, and weight initialization was used as in ref. [12]. However, in contrast to [12], the inputs were centred and normalized to unit standard deviation in order to remove "stiffness" from the problem. This normalization procedure shortens the convergence time with an order of magnitude as compared to the results reported in [12]. Table 2 shows the average training and generalization errors for an ensemble of networks that were trained 1000 epochs each. Standard BP updating easily learns the problem (the learning takes about 30 minutes on a DECstation 3100) and LV does not shorten the convergence time.

Thyroid classification with MLP						
	Training set		Generalization			
Updating	$\langle E \rangle$	E_{min}	$\langle E \rangle$	E_{min}	$\langle$Class.$\rangle$	Class.$_{best}$
BP, $\eta = 15.0$	11.2	2.8	108.9	85.8	98.2%	98.6%
BP, $\eta = 20.0$	11.5	2.0	108.8	88.0	98.2%	98.6%
LV, $\eta = 20.0$	15.6	5.4	110.7	89.5	98.2%	98.7%
[12] (a)	-	21.5	-	96.6	-	98.4%
[12] (b)	-	0.82	-	101.4	-	98.5%

Table 2: Performance of MLP on classification of hypothyroid functioning. The lower two rows show the best results reported in ref. [12]; (a) using a normal MLP, and (b) using a cascade-correlation network.

5   Discussion

It is clear from the simulations that LV updating shortens the convergence time for some problems but not for others. The reason for this is the shape of the error surface, which is seen from examining the Hessian for the problems [7].

When the weights are initialized, one usually chooses small values to ensure that the gradient will be large. The Hessian typically has one large positive eigenvalue and several almost zero eigenvalues if the weights are small. The eigenvector belonging to the large positive eigenvalue corresponds to changes made to weights connecting to the output unit(s). The initial error surface is consequently sharply bent upwards in one direction, and practically "flat" in all directions perpendicular to this one direction. The network will thus spend the first epochs oscillating back and forth along the dominant eigendirection, making very little progress in the perpendicular subspace. Choosing an initial learning rate will hence be very tricky. A too large learning rate makes the network overshoot along the dominant eigendirection and get stuck in a completely flat spot, with a larger error than it started out with. A too small learning rate will not take the network out of the flat region within reasonable time. A possible way around this dilemma is to choose different learning rates (which has been done in the simulations), which also is quite tricky. It is not uncommon that the Hessian eigenvalues vary within several orders of magnitude and using learning rates of $\mathcal{O}(10^3)$ is risky business! The LV updating is much more robust and benevolent. The noise term in eq. (1) makes the network perform a Brownian motion in the flat subspace, searching for channels to converge along, while the gradient term ensures that it does not wander into regions with larger errors.

The error surface in the central flat subspace resembles a high-dimensional plateau, with more or less narrow valleys and ridges going out from its edges. It is the size of this plateau, and the size of the valleys leaving it, that determines how hard a time BP will have leaving it. This size can be experimentally estimated by monitoring the eigenvalue spectrum for the Hessian during the training; significant non-zero eigenvalues signal the edge of the plateau and/or the beginning of a valley. Doing this for the four problems mentioned above gives [7]: The (almost zero) eigenvalues for the overlapping Gaussians are grouped into three groups; $|\lambda| \sim 10^{-2}$, $|\lambda| \sim 10^{-4}$, and $|\lambda| \sim 10^{-8}$. The lower values move very slowly towards the higher ones if BP updating is used, whereas they quickly approach $|\lambda| \sim 10^{-(2-3)}$ if LV updating is used. For the Mackey-Glass problem, the eigenvalues are grouped into two groups; $|\lambda| \sim 10^{-2}$, and $|\lambda| \sim 10^{-9}$, where LV updating also quickly takes the lower values into the upper region. The Hessian eigenvalues for the parity problem are all $|\lambda| \sim 10^{-2}$, and for the Thyroid classification problem $|\lambda| \sim 10^{-4}$, which does not change when LV is used. There is subsequently a difference in Hessian eigenvalue spectrum between those problems where LV updating is efficient and those where it does not make any difference.

The conclusion is that LV updating is efficient whenever a large fraction of the Hessian eigenvalues, evaluated at the initial weight values, are significantly zero ($\leq 10^{-5}$). If the MLP has many layers and the initial weights are small, then the Hessian will have many small eigenvalues. This means that LV updating, or MH, will be especially efficient for networks with many hidden layers. Furthermore, since symmetric problems also tend to have smaller eigenvalues, LV updating also would be more efficient if the problem exhibits many symmetries.

6 Conclusion

The results show that Langevin updating (or Brownian motion) is an easily controlled and robust way of using noise to improve MLP performance. It is also a benevolent method; the performance is either improved or unaffected, as compared to standard BP, provided that a reasonable amount of noise is added. By analyzing the origin for the LV speedup, I have also given a "rule of thumb" on how to determine from the Hessian eigenvalue spectrum if LV updating will be beneficial or not. Furthermore, a connection was made between Manhattan updating and the Langevin method, which partly explains the success of MH updating schemes where the learning rate is cleverly tuned as in ref. [6].

References

[1] A. Murray, "Multilayer Perceptron Learning Optimized for On-Chip Implementation: A Noise-Robust System", *Neur. Comp.* **4**, 366 (1992)

[2] T. Rögnvaldsson, "Pattern Discrimination using Feed-forward Networks", *Neur. Comp.* **5**, (1993)

[3] H. Yau and D. Wallace, "Enlarging the Attractor Basins of Neural Networks with Noisy External Fields", *J. Phys. A* **24**, 5639 (1991)

[4] A. Krogh "Learning with Noise in a Linear Perceptron", *J. Phys. A* **25**, 1119 (1992); A. Krogh and J. Hertz, "Generalization in a Linear Perceptron in the Presence of Noise", *J. Phys. A* **25**, 1135 (1992)

[5] T. Heskes, E. Slijpen, and B. Kappen, "Learning in Neural Networks with Local Minima", *Phys. Rev. A* **46**, 5221 (1992)

[6] C. Peterson and E. Hartman, "Explorations of the Mean Field Theory Learning Algorithm", *Neural Networks* **2**, 475-494 (1989)

[7] T. Rögnvaldsson, to be published elsewhere.

[8] L. Lönnblad, C. Peterson, and T. Rögnvaldsson, to be published elsewhere.

[9] D. Rumelhart and J. McLelland (eds.) *Parallel Distributed Processing* (Vol. 1), MIT Press (1986)

[10] M. Mackey and L. Glass, "Oscillations and Chaos in Physiological Control Systems", *Science* **197**, 287 (1977)

[11] J.R. Quinlan , "Simplifying decision trees", *Int. J. Man-Machine Studies*, 221-234, (1987)

[12] W. Schiffmann, M. Joost, and R. Werner, "Comparison of optimized backpropagation algorithms", *Proc. ESANN'93* Brussels (1993)

Guaranteed Convergence of Learning in Neural Networks

Tom M. Heskes

Department of Medical Physics and Biophysics
University of Nijmegen Geert Grooteplein 21
6525 EZ Nijmegen The Netherlands
e-mail: tom@mbfys.kun.nl

Abstract

This paper describes schedules for the learning parameter that guarantee convergence to the optimal solution. It focuses on the difference between local and global optimization, i.e., learning in the presence of just one minimum and learning in the presence of several minima. In case of one minimum, the fastest possible cooling is an algebraic function of the number of learning steps, whereas in case of several minima the cooling must be "exponentially slow".

1 Scope

We will discuss on-line learning processes where at each learning step *one* of the examples is presented to the network. This is in contrast with batch-mode learning where a weight change takes place on account of the *whole* training set. Since the examples are drawn *at random*, learning becomes a stochastic process. The stochasticity leads to fluctuations in the network's representation. The larger the learning parameter, the larger the fluctuations. On the other hand, the smaller the learning parameter, the slower the convergence. In this paper, we will describe how to choose the learning parameter as a function of the number of learning steps. We will see that there is a fundamental difference in learning with just one minimum and learning with several minima. In case of just one minimum, fluctuations are cumbersome. One would like to choose a very small learning parameter, but one must be careful that it is large enough, i.e., such that it is still possible to reach the minimum. With several minima, fluctuations are a necessary evil to escape from a local minimum. The learning parameter must therefore decrease much more slowly. Learning in the presence of just one minimum is discussed in Section 2, learning in the presence of several minima in Section 3 (for details and a more thorough analysis see [1, 2] and [3, 4], respectively, and references therein).

Let us first introduce the notation. The network state is determined by the weight vector **w**. At learning step t, a training pattern $\vec{x}$ is drawn at random from the environment Ω. The weight vector changes from **w** to $\mathbf{w} + \Delta\mathbf{w}$, obeying

$$\Delta\mathbf{w} = \eta(t)\,\mathbf{f}(\mathbf{w}, \vec{x}) . \tag{1}$$

Here f is the learning rule, a function of the current network state w and the particular training pattern $\vec{x}$. $\eta(t)$ is the learning parameter at time t, it sets the typical magnitude of the weight changes. Most learning rules in neural-network literature can be written in the form (1), e.g., backpropagation [5], Kohonen learning [6], and so on. We will restrict ourselves to learning rules that can be written as the gradient of some energy function or error potential, i.e., such that

$$\langle \mathbf{f}(\mathbf{w}, \vec{x}) \rangle_\Omega = -\nabla E(\mathbf{w}),$$

where $\langle \ldots \rangle_\Omega$ stands for the average with respect to the training set Ω, ∇ denotes the gradient with respect to the weight vector w, and $E(\mathbf{w})$ is the error potential. This restriction is not necessary for the results in this paper, but simplifies the language and the notations. We now have a well-defined global measure of the network performance (the lower $E(\mathbf{w})$, the better w), the global minimum is the optimal network state, and local minima are undesired (meta)stable points.

The learning process described by (1) is a stochastic process, governed by the master equation

$$\frac{\partial P(\mathbf{w}, t)}{\partial t} = \int d^N w \left[T(\mathbf{w}'|\mathbf{w}, t) P(\mathbf{w}, t) - T(\mathbf{w}|\mathbf{w}', t) P(\mathbf{w}', t) \right], \tag{2}$$

with $P(\mathbf{w}, t)$ the probability that the network is at network state w at time t, and $T(\mathbf{w}'|\mathbf{w}, t)$ the transition probability

$$T(\mathbf{w}'|\mathbf{w}, t) = \left\langle \delta^N(\mathbf{w}' - \mathbf{w} - \eta(t) \mathbf{f}(\mathbf{w}, \vec{x})) \right\rangle_\Omega.$$

Let us assume that the error potential has only one global minimum (generalization to several global minima is trivial), located at $\mathbf{w}^*$ and let us scale the error such that $E(\mathbf{w}^*) = 0$. The goal is to find schedules for the learning parameter that guarantee convergence to this global minimum, i.e., that are such that the asymptotic solution of the master equation (2) obeys

$$\lim_{t \to \infty} P(\mathbf{w}, t) = \delta^N(\mathbf{w} - \mathbf{w}^*). \tag{3}$$

2 One minimum

Suppose that the error potential $E(\mathbf{w})$ has one global minimum at $\mathbf{w}^*$ and no other local minima. For nonzero learning parameters, the learning process (2) tends to this minimum. The question is how to get rid of the undesired fluctuations. In the neighborhood of the minimum, we may expand

$$\langle E(\mathbf{w}) \rangle_{\Xi(t)} \approx \frac{1}{2} \left\langle (\mathbf{w} - \mathbf{w}^*)^T H (\mathbf{w} - \mathbf{w}^*) \right\rangle_{\Xi(t)} = \mathrm{Tr}\,[HL(t)],$$

where $\langle \ldots \rangle_{\Xi(t)}$ denotes an average over a whole ensemble Ξ of neural networks at time t, i.e., weighted by $P(\mathbf{w}, t)$, and $L(t)$ and H are matrices with components

$$L_{ij}(t) \equiv \left\langle (w_i - w_i^*)(w_j - w_j^*) \right\rangle_{\Xi(t)}; \quad H_{ij} \equiv \left. \frac{\partial^2 E(\mathbf{w})}{\partial w_i \partial w_j} \right|_{\mathbf{w} = \mathbf{w}^*}.$$

H is the curvature of the error potential at the minimum, L is called the "misadjustment" [7]. Both are positive definite matrices, so, the requirement

$$\lim_{t \to \infty} \langle E(\mathbf{w}) \rangle_{\Xi(t)} = 0,$$

[recall the definition $E(\mathbf{w}^*) = 0$ and the goal (3)], may as well be written

$$\lim_{t \to \infty} L(t) = 0 .$$

Using Van Kampen's systems size expansion for small learning parameters $\eta(t)$, it is straightforward to show that the misadjustment obeys [1]

$$\frac{dL(t)}{dt} = -\eta(t)[H\,L(t) + L(t)\,H] + \eta^2(t)\,D , \qquad (4)$$

with D the diffusion matrix containing the local fluctuations in the learning rule

$$D_{ij} \equiv \langle f_i(\mathbf{w}^*, \vec{x})\, f_j(\mathbf{w}^*, \vec{x}) \rangle_\Omega .$$

The slowest convergence is in the direction of the eigenvector of H with the smallest eigenvalue $\bar{H}$ (we will use the notation $\bar{A} \equiv e^T A e$ where e stands for the normalized eigenvector of the Hessian H corresponding to the smallest eigenvalue). To get the fastest possible convergence along this direction, we must optimize the righthandside of (4) projected on the eigenvector e, i.e., we must choose

$$\eta(t) = \bar{H}\,\frac{\bar{L}(t)}{\bar{D}} . \qquad (5)$$

Substitution into (4) leads to the differential equation

$$\frac{d\bar{L}(t)}{dt} = -\frac{\bar{H}^2}{\bar{D}}\bar{L}^2(t) ,$$

with solution

$$\bar{L}(t) = \frac{\bar{D}}{\bar{H}^2 t} .$$

Substitution of this result into (5) yields finally

$$\eta(t) = \frac{1}{\bar{H}t} .$$

This is the optimal (asymptotic) behavior of the learning parameter in case of just one minimum [2, 7, 8].

3 Several minima

If the error potential has several minima, the problem of finding the global minimum is definitely much more difficult and very different from the case with only one minimum. Fluctuations are necessary to enable transitions from local minima to the global minimum. In this case, we will consider the "occupation numbers" at the various minima defined by

$$n_\alpha(t) \equiv \int_{I_\alpha} d^N w\, P(\mathbf{w}, t) ,$$

with I_α the attraction region of the minimum α, i.e., for all initial $\mathbf{z}(0) \in I_\alpha$, the deterministic differential equation

$$\frac{d\mathbf{z}(t)}{dt} = -\nabla E(\mathbf{z}(t)) ,$$

converges to the local minimum α. Suppose that the learning parameter is so small that the error difference due to the local fluctuations in the vicinity of the minima (this difference is of the order of the learning parameter) can be neglected if compared with the error differences between the various minima. Then the expectation value of the error is approximately given by

$$\langle E(\mathbf{w}) \rangle_{\Xi(t)} \approx \sum_{\alpha} n_{\alpha}(t) \, E_{\alpha} \,,$$

with E_{α} the value of the error potential at minimum α. By definition, we choose $E_1 = 0$, i.e., "1" denotes the global minimum. Now, the goal (3) can be written

$$\lim_{t \to \infty} n_{\alpha}(t) = \delta_{\alpha 1} \,. \tag{6}$$

Using the master equation (2) and some reasonable assumptions, frequently used in the theory on unstable stochastic processes [9], it can be shown that the evolution of the occupation numbers $n_{\alpha}(t)$ is governed by the set of linear differential equations [3]

$$\frac{dn_{\alpha}(t)}{dt} = -\sum_{\beta} \Gamma_{\alpha\beta}(\eta(t)) \, n_{\beta}(t) \,.$$

$\Gamma(\eta)$ is called the transition matrix. The reciprocal values of its elements are the transition times. These transition times scale exponentially with the learning parameter (for small learning parameters η), i.e., $\tau_{\alpha\beta}$, the transition time from minimum β to minimum α, reads [3]

$$\tau_{\alpha\beta}(\eta) \equiv \frac{1}{\Gamma_{\alpha\beta}(\eta)} \sim \exp \left[\frac{\bar{\eta}_{\alpha\beta}}{\eta} \right] \,. \tag{7}$$

$\bar{\eta}_{\alpha\beta}$ is called the reference learning parameter for the transition from minimum β to α. For learning parameters η much smaller than this reference learning parameter $\bar{\eta}_{\beta\alpha}$, the transition from minimum β to minimum α becomes very improbable. In [3] we give a scheme how to calculate or at least estimate this reference learning parameter. The reference learning parameter is similar to the Arrhenius factor in chemical reaction theory [9].

Let us assume that the matrix $\Gamma(\eta)$ is irreducible (if not then we should first decompose the original matrix and treat the remaining submatrices separately). Then, since $1 - \Gamma(\eta)$ is a stochastic matrix, exactly one of the eigenvalues of $\Gamma(\eta)$ is zero and all other eigenvalues are positive. Using the property (7), it can be shown that the smallest nonzero eigenvalue is of the form [4]

$$\bar{\Gamma}(\eta) \sim \exp \left[-\frac{\bar{\eta}}{\eta} \right] \,.$$

The constant $\bar{\eta}$ depends on the reference learning parameters for the various transitions. The component $\bar{n}(t)$ in the direction of the eigenvector corresponding to this eigenvalue obeys

$$\frac{d\bar{n}(t)}{dt} = -\bar{\Gamma}(\eta(t)) \, \bar{n}(t) \,.$$

So, in order to get (6), we must have

$$\int^{\infty} dt \, \bar{\Gamma}(\eta(t)) = \infty \,.$$

The fastest possible (asymptotic) decay of the learning parameter that fulfills this requirement is

$$\eta(t) \sim \frac{\bar{\eta}}{\ln t} \,.$$

This "exponentially slow" decay is typical for global optimization processes where only small steps are allowed (see also [10]). The learning parameter η appears to be similar to the temperature in simulated annealing processes [11]. In these processes the constant $\bar{\eta}$ is called "the critical depth". It is the depth (suitably defined) of the deepest local minimum which is not a global minimum state [12]. The important difference between on-line learning processes and simulated annealing processes is the nature of the noise. For simulated annealing the noise is artificial and homogeneous, i.e., the same in the whole state space. On the other hand, the noise in on-line learning processes is intrinsic, i.e., a direct consequence of the random presentation of patterns, and inhomogeneous. In general we have that the higher the error potential, the more there is to learn, the larger the fluctuations in the learning rule, so the easier to escape. Therefore, the inhomogenity of the noise is an advantage rather than a disadvantage. An example of this can be found in [13].

4 Conclusions

We discussed optimization of learning in neural networks. Both in the case of one minimum and in the case of several minima, there exists a typical relaxation time τ which is a function of the learning parameter η. In order to guarantee optimal convergence, this relaxation time has to fulfill the following two requirements.

$$\lim_{t\to\infty} \frac{1}{\tau(\eta(t))} = 0 \; ; \quad \lim_{t\to\infty} \int^t dt' \, \frac{1}{\tau(\eta(t'))} = \infty \, . \tag{8}$$

The first requirement is necessary to get rid of undesired asymptotic fluctuations. To fulfil this requirement the learning parameter must vanish. The second requirement gives the fastest possible cooling of the learning parameter without freezing the system in a non-optimal state. Equation (8) implies that the fastest possible cooling is such that

$$\tau(\eta(t)) \sim t \, .$$

In case of just one minimum, the only problem is to get rid of the undesired *local* fluctuations. The typical relaxation time of these fluctuations is inversely proportional to the learning parameter [1]:

$$\tau_{\text{one}}(\eta) \sim \frac{1}{\eta} \implies \eta_{\text{one}}(t) \sim \frac{1}{t} \, .$$

So, in case of just one minimum, the cooling of the learning parameter can be algebraic.

However, in case of several minima, the problem is much more difficult since the local fluctuations are necessary to enable transitions between the different minima. With algebraic cooling of the learning parameter the system may get stuck in a local minimum. The typical *global* relaxation time to the equilibrium distribution scales exponentially with some constant divided by the learning parameter [4]:

$$\tau_{\text{several}}(\eta) \sim \exp\left[\frac{\bar{\eta}}{\eta}\right] \implies \eta_{\text{several}}(t) \sim \frac{\bar{\eta}}{\ln t} \, .$$

So, in case of several minima, there is no way to escape from exponentially slow cooling if one really wants to guarantee convergence to the optimal solution.

This important distinction between *local* optimization and *global* optimization of learning in neural networks must be carefully noticed in making claims about the convergence of learning processes.

Acknowledgments

This work was partly supported by the Dutch Foundation for Neural Networks (SNN). I would like to thank Bert Kappen and Eddy Slijpen for stimulating discussions and Tonnie van Moorsel for carefully reading a previous version of this manuscript.

References

[1] T. Heskes and B. Kappen. Learning processes in neural networks. *Physical Review A*, 44:2718–2726, 1991.

[2] T. Heskes and B. Kappen. Learning-parameter adjustment in neural networks. *Physical Review A*, 45:8885–8893, 1992.

[3] T. Heskes, E. Slijpen, and B. Kappen. Learning in neural networks with local minima. *Physical Review A*, 46:5221–5231, 1992.

[4] T. Heskes, E. Slijpen, and B. Kappen. Cooling schedules for learning in neural networks. *Physical Review E*, 1993.

[5] D. Rumelhart, G. Hinton, and R. Williams. Learning representations by back-propagating errors. *Nature*, 323:533–536, 1986.

[6] T. Kohonen. Self-organized formation of topologically correct feature maps. *Biological Cybernetics*, 43:59–69, 1982.

[7] C. Darken and J. Moody. Note on learning rate schedules for stochastic optimization. In R. Lippmann, J. Moody, and D. Touretzky, editors, *Advances in Neural Information Processing Systems 3*, pages 832–838, San Mateo, 1990. Morgan Kaufmann.

[8] Y. Kasbashima and S. Shinomoto. Learning a decision boundary from stochastic examples: incremental algorithms with and without queries. *Preprint Kyoto University*, 1992.

[9] N. van Kampen. *Stochastic processes in physics and chemistry*. North-Holland, Amsterdam, 1981.

[10] H. Kushner. Asymptotic global behavior for stochastic approximation and diffusions with slowly decreasing noise effects: global minimization via Monte Carlo. *SIAM Journal of Applied Mathematics*, 47:169–185, 1987.

[11] S. Kirkpatrick, C. Gelatt, and M. Vecchi. Optimization by simulated annealing. *Science*, 220:671–680, 1983.

[12] B. Hajek. Cooling schedules for optimal annealing. *Mathematics of Operations Research*, 13:311–329, 1988.

[13] T. Heskes and B. Kappen. On-line learning processes in artificial neural networks. In J. Taylor, editor, *Mathematical Foundations of Neural Networks*. Elsevier, Amsterdam, 1993.

Activity-Conserving Dynamics for Neural Networks

Eric O. Postma, H. Jaap van den Herik, and Patrick T.W. Hudson
University of Limburg
Computer Science Department
P.O.Box 616   6200 MD Maastricht
The Netherlands

abstract

A new activation rule is proposed whose dynamic operation conserves the total activation in a neural network. Analogously to the standard stochastic dynamics that are based on *spin-flip* Glauber dynamics [1], activity-conserving dynamics are based on *spin-exchange* Kawasaki dynamics [2]. Through simulation studies, we show that stochastic activity-conserving dynamics perform satisfactorily on an optimization task in which the number of elements remains constant.

1   Introduction

Without a controlling mechanism, neural networks have a tendency for the total level of activation to rise out of bounds. The question now reads: what mechanism or structure guarantees that the activation level remains limited? The traditional approach uses inhibitory connections among the processing elements to ensure that the total activation does not rise without limit. Grossberg [3] showed such a network to converge to a state where activity is normalized. An alternative approach, pursued in this contribution, is to fix the total activity from the onset and only to allow for changes in the distribution of activation. This reduces the space of possible network states by one dimension and has been shown to lead to improved performance on optimization tasks [4,5].

2   Molecular Configuration Problem

As an example we consider the problem of finding the optimal *Molecular Configuration* (MC). This problem entails finding the set of coordinates of W atoms that minimizes, for each pair of atoms, the *Lennard-Jones* pair potential (see, e.g., [6])

$$LJ(i,j) = 4\left(d_{ij}^{-12} - d_{ij}^{-6}\right),\tag{1}$$

with d_{ij} representing the Euclidian distance between atoms i and j. The Lennard-Jones potential models interaction forces of real atoms. As shown in Figure 1, minimization of (1)

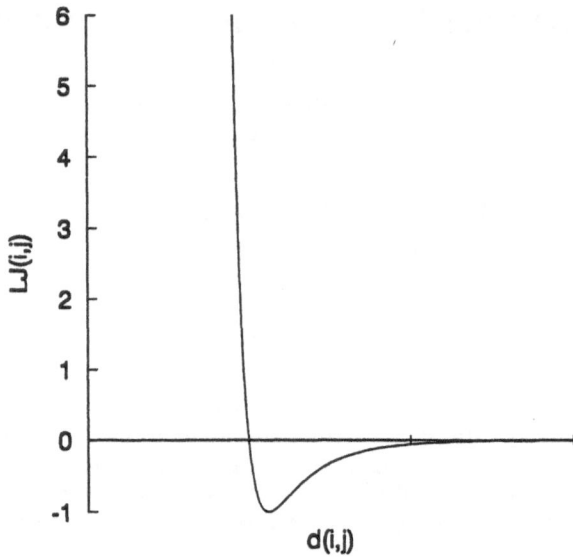

Figure 1: *Lennard-Jones potential $L(i,j)$ as a function of distance d_{ij}.*

implies strong repulsion for $d_{ij} < 1$, and attraction for $d_{ij} > 1$. (The latter is maximal at $d = 2^{\frac{1}{6}}$ and decays asymptotically with increasing distance.)

We define a neural-network structure appropriate for solving the MC problem in two dimensions: atom coordinates are represented by neurons organized in a two-dimensional grid. Network states are represented by the state vector n, the elements of which are the neuron variables $n_i \in \{0,1\}$ ($1 \leq i \leq N$). An active or inactive neuron i represents the presence ($n_i = 1$) or absence ($n_i = 0$), respectively, of an atom at position i. An appropriate energy function incorporating the constraints of the MC problem is

$$E(\mathbf{n})_{MC} = \frac{1}{2} \sum_i^N \sum_{j \neq i}^N LJ(i,j)\, n_i\, n_j + \frac{\alpha}{2} \left(W - \sum_i^N n_i \right)^2. \tag{2}$$

The first right-hand term is smallest when the Lennard-Jones potentials of all atom pairs are minimal. The last term incorporates the constraint that the molecule contains W atoms and is smallest (i.e., zero) when there are exactly W active neurons. (The parameter $\alpha > 0$ sets the strength of this constraint, relative to the first constraint.) By specifying neuron dynamics that minimize $E(\mathbf{n})_{MC}$ a solution of the MC problem is obtained.

3 Glauber Dynamics

Given an energy function, Glauber's update rule (cf. [1]) is commonly used to specify activation dynamics that minimize the energy (see, e.g., [7]). For our energy function (2) the appropriate Glauber rule reads

$$P(n_i = 0 \rightarrow n_i = 1) = \frac{1}{2}\left[\tanh\left(\frac{1}{T}\sum_{j\neq i}w_{ij}n_j + \frac{\alpha}{T}(W - \sum_i n_i)\right) + 1\right].\qquad(3)$$

In this equation $P(n_i = 0 \rightarrow n_i = 1)$ is the probability that the i-th neuron "flips" from an inactive to an active state. $(P(n_i = 1 \rightarrow n_i = 0) = 1 - P(n_i = 0 \rightarrow n_i = 1).)$ T is a "temperature" parameter that scales the argument of the sigmoid function. The weight of the connection between neurons i and j, w_{ij} $(= w_{ji})$ incorporates the atomic interaction forces and is defined as $w_{ij} = -LJ(i,j)$.

By applying the update rule (3) to N (randomly selected) neurons in each iteration, the network preferentially enters states that minimize (2). At equilibrium, the (average) network state represents a solution to the MC problem (i.e., a minimum of $E(n)_{MC}$).

4 Kawasaki Dynamics

Glauber dynamics constitute only one example of a (neural) updating rule that minimizes an energy function. An attractive alternative for problems where the total activation (i.e., number of active neurons) has to be conserved is Kawasaki's *exchange dynamics* [2]. Instead of "flipping" one neuron at a time, Kawasaki dynamics update neighbouring pairs of neurons by exchanging their states with a probability that is weighted by the change in energy. For our MC problem this leads to

$$P(n_i = 0; n_j = 1 \rightarrow n_i = 1; n_j = 0) = \frac{1}{2}\left[\tanh\left(\frac{1}{T}\sum_{k\neq i,j}(w_{ik} - w_{jk})n_k\right) + 1\right]\qquad(4)$$

where $P(n_i = 0, n_j = 1 \rightarrow n_i = 1, n_j = 0)$ is the probability that the active j-th neuron and neighbouring inactive i-th neuron exchange their states.

A stochastic network with Kawasaki dynamics is restricted to the part of the state space where the number of active neurons is fixed. This means that the activity-conserving dynamics (4) always satisfy the constraint expressed as the last term of (2). As a result, problem complexity is reduced considerably, leading to an improved solution quality and decreased relaxation times (cf. [4]).

5 Simulations

We tested our activity-conserving dynamics on a $W = 50$ MC problem. For this particular problem the energy values obtained with parallel global optimization algorithms are available [6]. In our simulations, an *update* is defined as the random selection of one of the W active neurons together with a (randomly selected) nearest neighbour, and exchanging their states with a probability defined by (4). A single iteration is defined as $W - 1$ updates. A square lattice of 1001^2 neurons was used to approximate two-dimensional space. the unit of distance (i.e., the distance between two nearest-neighbouring neurons) was scaled to ≈ 0.016. In all simulations, T was initially set to 10 and gradually lowered. At each iteration k, the temperature was set to $T(k) = 0.999T(k - 1)$ (i.e., simulated annealing [8]). Initially, $W - 1$ atom positions were randomly distributed in a small circular area with a radius of 25 neurons. One atom was fixed in the center of the lattice and not updated. A second atom was restricted to move along one direction only.

Figure 2 shows the results of our simulations. The left graph plots $E(n)_{MC}$ as a function of the number of iterations for a single simulation run. The right graph displays the results of four simulations for iterations > 5000. The grey horizontal bar indicates the *goal range*, i.e., the energy values obtained with algorithmic methods (obtained from [6]).

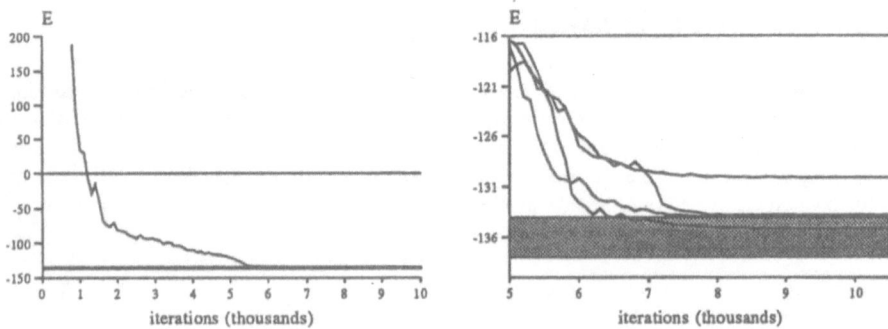

Figure 2: *MC energy as a function of the number of iterations.*

Figure 3 shows several snapshots of one simulation run (the circles represent atoms). An initial configuration is shown in the left panel. The middle panel depicts the configuration after 500 iterations. A typical final configuration is displayed in the right panel ($E(n)_{MC} \approx -133$). As is evident from the graphs, our neural-network approach leads to energy values near or within the goal range. Therefore, activity-conserving dynamics are competitive in performance with traditional algorithmic approaches. For all simulations, the lowest energy value was reached after approximately 8000 iterations. Long relaxation times represent the main disadvantage of simulated annealing methods. Implementation of our network on an asynchronous parallel computer may alleviate this problem to a certain extent.

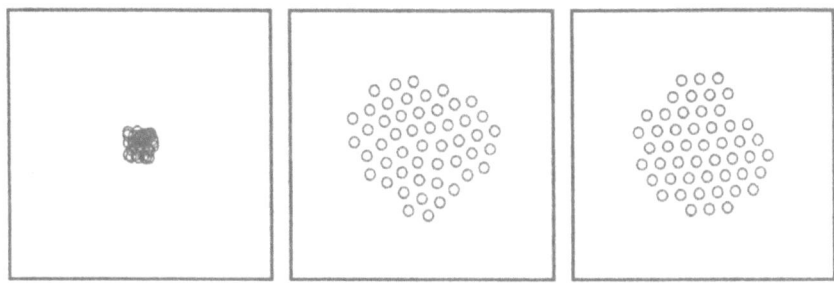

Figure 3: *Snapshots from a single simulation run.*

6   Conclusions and Future Work

Activity-conserving dynamics offer a viable alternative to the standard (Glauber) approach in neural networks. For optimization problems involving a conservation constraint, the proposed dynamics reduce problem complexity considerably. This leads to good solution quality as evidenced by the simulation results. Activity conservation may also be applied in associative memories to keep the total activity within bounds and to facilitate retrieval and storage.

We are currently studying the *mean-field approximation* to exchange dynamics analogously to the mean-field approximation to flip dynamics by Hopfield [9]. In the context of Ising models, Penrose [10] derived the following mean-field approximation of Kawasaki dynamics:

$$\frac{d}{dt}U_i = \frac{1}{2} \sum_{j \in A(i)} \left[(U_j - U_i) + (1 - U_i U_j) \tanh \left(\frac{1}{T} \sum_{k \neq i,j} (w_{ik} - w_{jk}) U_k \right) \right], \tag{5}$$

where $U_i \in [-1, +1]$ denotes the continuous activation value of the i-th neuron and $A(i)$ represents the set of nearest neighbours of i. There exists a Lyapunov function for (5) provided that all weights of the connections between nearest neighbours are equal [10]. We expect a further improvement in solution quality and relaxation speed when applying this mean-field approximation in neural networks.

Acknowledgements

The authors thank Bart Oldenkamp for helpful discussions regarding the molecular-configuration problem. IBM is acknowledged for their hardware support under the Joint Study Agreement DAEDALOS (#289651).

References

[1] Glauber, R.J. (1963). Time-dependent statistics of the Ising model. Journal of Physics, 4, 294-307.

[2] Kawasaki, K. (1972). Kinetics of Ising models. In C. Domb and M.S. Green (Eds.), *Phase transitions and critical phenomena, Volume 2*, pp. 443-501. London: Academic Press.

[3] Grossberg, S. (1973) Contour enhancement, short term memory, and constancies in reverberating neural networks. *Studies in Applied Mathematics*, LII, 213-257.

[4] Peterson, C. & Söderberg, B. (1989). A new method for mapping optimization problems onto neural networks. *International Journal of Neural Systems*, 1, 3-22.

[5] Yuille, A.L. (1990). Generalized deformable models, statistical physics, and matching problems. *Neural Computation*, 2, 1-24.

[6] Byrd, R.H., Eskow, E., Schnabel, R.B., & Smith, S.L. (1991). Parallel global optimization: numerical methods, dynamics scheduling methods, and application to molecular configuration. *Technical Report CU-CS-553-91*, Computer Science Department, University of Colorado, Boulder.

[7] Hertz, J., Krogh, A., & Palmer, R.G. (1991). *Introduction to the theory of neural computation*. Reading, MA: Addison-Wesley.

[8] Kirkpatrick, S., Gelatt, C.D. Jr., & Vecchi, M.P. (1983). Optimization by simulated annealing. *Science*, 220, 671-680.

[9] Hopfield, J.J. (1984). Neurons with graded response have collective computational properties like those of a two-state neuron. *Proceedings of the National Academy of Sciences, U.S.A.*, 81, 3088-3092.

[10] Penrose, O. (1991). A mean-field equation of motion for the dynamic Ising model. *Journal of Mathematical Physics*, 63, 975-986.

Stochastic dynamical systems
—poster contributions

The Lower Bound of the Capacity for a Neural Network with Multiple Hidden Layers

Masami YAMASAKI

Advanced Research Laboratory, Hitachi Ltd.
Hatoyama, Saitama 350-03, JAPAN
E-mail: yamasaki@harℓ.hitachi.co.jp

Abstract. We show the lower bound of the capacity of a hierarchical neural network, having multiple hidden layers whose node unit takes the value of a real number between zero and one as the output of a sigmoid function. It is shown that $n \cdot \lceil \frac{h_1}{2} \rceil + \lfloor \frac{h_1}{2} \rfloor \cdot \lceil \frac{h_2}{2} - 1 \rceil + \cdots + \lfloor \frac{h_{N-1}}{2} \rfloor \cdot \lceil \frac{h_N}{2} - 1 \rceil$ examples in the general position (i.e. no subset of n or less input vectors degenerate) can be memorized by the network which has n input units in the input layer, h_ℓ hidden units in the ℓ-th layer of N hidden layers, and a single output unit in the output layer.

1. Introduction

One question about artificial neural networks that is simple but difficult to answer is how many examples a network can memorize. If the network consists of just one threshold gate with n inputs, we can always find a set of weights and a threshold for any set of $n + 1$ examples in the general position (the examples do not share a hyperplane in n-dimensional Euclidean space) and their assigned outputs. The problem of extending this to multi-layer networks was left untouched for a long time. Mitchison and Durbin[1] and Akaho and Amari [2] recently gave order estimates of the upper and lower bounds of n-h-1 network capacity. The upper bound is $O(nh\log h)$ and the lower is $2n\lfloor h/2 \rfloor$ [3]. Sakurai [4] [5] determined conclusively that the capacity of n-h-1 networks with threshold elements is exactly $n \cdot h + 1$.

These studies all consider only threshold-gate networks, so the output is always binary. It is quite natural to ask what happens if the network outputs sigmoid functions. A lot of work has been done on the possibility of constructing networks that conform to a given input/output relation from the function approximation point of view (see e.g. [6] and its references), but their main concern is whether some networks can approximate any give function to any given accuracy. The results were positive, but do not say how many hidden gates are necessary for approximations of a certain accuracy.

Yamasaki[7] gave the lower bound of the capacity of a neural network consisting of n inputs, h hidden units in a single hidden layer, and one output unit. This paper will show the lower bound of the capacity of a network with multiple hidden layers. The lower bound of the capacity is given as minimum number of examples that can be memorized. We will provide some lemmas and a theorem without details (i.e. no proofs) in the following sections because of the limitation on paper length.

2. Notations

Consider the multiple layered feedforward network in Fig.1 . The network has n input units, h_ℓ hidden units in the ℓ-th layer of N hidden layers, one bias unit in each hidden layer, and one output unit.

Each unit excluding the bias units takes a real number between $(0, 1)$. The activation value of the bias unit is denoted as H_0^ℓ whose value is $f(0)$; the function f is a sigmoid function with parameter $\beta\ (\in R^+)$, i.e. $f(x) = (\exp(-\beta x) + 1)^{-1}$.

The value of the j-th hidden unit of the ℓ-th hidden layer H_j^ℓ is related to the value of the hidden units in the $(\ell-1)$-th hidden layer by $H_j^{(\ell)} = f(\sum_{i=0}^{h_{\ell-1}} w_{ji}^{(\ell-1)} H_i^{(\ell-1)})$, where $w_{ji}^{(\ell-1)}$ is the weight factor between the j-th unit in the ℓ-th hidden layer and the i-th unit in the $(\ell-1)$-th hidden layer.

To uniformly treat the input layer, the output layer and the hidden layers in the above expressions, we denote the input layer and the output layer as the 0-th hidden layer and the $(N+1)$-th hidden layer, respectively.

The μ-th output/input vector pair to be learned by the network is denoted as $< O_\mu, (I_\mu^1, \cdots, I_\mu^n) >$. We call the set of integer numbers between one and L the L-index set. A subset of the L-index set with m elements is called an m-index subset. We assume that the input vectors are in the general position, namely,

$$\forall S = \{\mu_1, \ldots, \mu_n\}; \quad \mathrm{rank} \begin{pmatrix} I_{\mu_1}^1 & \cdots & I_{\mu_1}^n \\ \vdots & \ddots & \vdots \\ I_{\mu_n}^1 & \cdots & I_{\mu_n}^n \end{pmatrix} = n, \tag{1}$$

where S is an m-index subset of the L-index set.

If an L-dimensional vector v and an m-index subset $S = \{\mu_1, \ldots, \mu_m\}$ are given, we can define an m-dimensional vector $(v_{\mu_1}, \ldots, v_{\mu_m})$ according to S. This m-dimensional vector is represented as v_S.

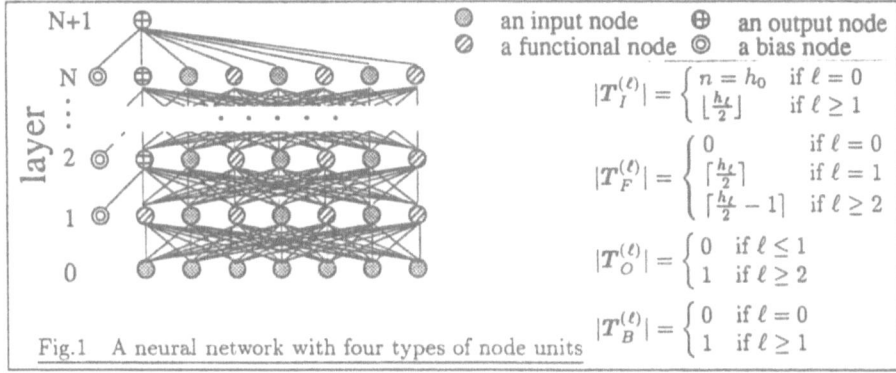

Fig.1   A neural network with four types of node units

3.   A Neural Network with a Special Configuration of Weights

We will show the lower bound of the capacity of a neural network with a special configuration of weight factors. To define the neural network, we first separate the node units in each layer of the network into the four categories (Fig.1). The subset of indices for the input node units, the functional node units, the output node units and the bias node unit are represented as $T_I^{(\ell)}$, $T_F^{(\ell)}$, $T_O^{(\ell)}$ and $T_B^{(\ell)}$, respectively.

We consider the neural network whose weight factor between node units are defined as follows,

$$w_{ji}^{(\ell)} = \begin{cases} t_j^\ell \sigma_i^\ell & \text{if } j \in T_I^{\ell+1} \text{ and } i \in T_I^\ell \\ t_j^\ell \sigma_i^\ell + x_{ji}^\ell & \text{if } j \in T_F^{\ell+1} \text{ and } i \in T_I^\ell \\ -\zeta^\ell & \text{if } j \in T_O^{\ell+1} \text{ and } i \in T_I^\ell \cup T_B^\ell \\ \zeta^\ell & \text{if } j \in T_O^{\ell+1} \text{ and } i \in T_F^\ell \cup T_O^\ell \\ 0 & \text{other cases.} \end{cases} \tag{2}$$

where t_j^ℓ, σ_i^ℓ and ζ^ℓ are real numbers.

To memorize L output/input data pairs, $\{< O^\mu, (I_1^\mu, \cdots, I_n^\mu) > | \mu \in \{1, 2, \ldots, L\}\}$, we can define the following map F from R^M into a domain D of R^L where $M = n \cdot \lceil \frac{h_1}{2} \rceil + \lfloor \frac{h_1}{2} \rfloor \cdot \lceil \frac{h_2}{2} - 1 \rceil + \cdots + \lfloor \frac{h_{N-1}}{2} \rfloor \cdot \lceil \frac{h_N}{2} - 1 \rceil$. The arguments of F are M variables $\boldsymbol{x} = \{x_{ji}^\ell \in R \mid j \in T_F^{\ell+1}, i \in T_I^\ell \text{ and } 0 \le \ell \le N\}$. And the μ-th component of the value $F(\boldsymbol{x})$ is $F_\mu = f^{-1}(H_1^{N+1}|_{(H_1^0, \cdots, H_1^0)=(I_1^\mu, \cdots, I_n^\mu)})/\zeta^N$.

We will show that we can always select appropriate values of t_j^ℓ and σ_i^ℓ to insure a non-zero Jacobian of the function F at $\boldsymbol{0}$, if the assumption on input data (Eq. (1)) is satisfied and $M = L$.

If a non-zero Jacobian at $\boldsymbol{x} = \boldsymbol{0}$ can be proved, by using the "Inverse Mapping Theorem", we can show that the following theorem holds.

Theorem 1.

$$\forall \boldsymbol{v} \in \boldsymbol{R}^L, \quad \exists \zeta^N \in \boldsymbol{R}, \quad \exists \boldsymbol{x} \in \boldsymbol{R}^M; \qquad \boldsymbol{v} = \zeta^N F(\boldsymbol{x}) \tag{3}$$

Theorem 1 shows that the relation, $O^\mu = H_1^{N+1}|_{(H_1^0, \cdots, H_1^0)=(I_1^\mu, \cdots, I_n^\mu)}$, holds for L arbitrary output/input data.

4. Linearly Independent Tangent Vectors

In order to prove the non-zero Jacobian of the map F at $\boldsymbol{0}$, we must show the linear independence of the L-dimensional vectors $\{e^{\ell j i} \mid j \in T_F^{\ell+1}, i \in T_I^\ell \text{ and } 0 \le \ell \le N\}$ which are defined as follows,

$$e_\mu^{\ell j i} = \frac{\partial H_j^{\ell+1}}{\partial x_{ji}^\ell}\bigg|_{(H_1^0, \cdots, H_1^0)=(I_1^\mu, \cdots, I_n^\mu), \, \boldsymbol{x}=\boldsymbol{0}} \tag{4}$$

By using the general position assumption on input data (Eq. (1)), we can construct an L-dimensional vector $(\eta_\mu^0) = (\sum_{i=1}^n \sigma_i I_i^\mu)$, such as $\eta_\mu^0 > 0$ for any μ and $\eta_\mu^0 \ne \eta_\nu^0$ for $\mu \ne \nu$. Yamasaki [7] introduced that the following lemma by using the property of (η_μ).

Lemma 2. Suppose there exist a kn-index subset $\boldsymbol{S}_{kn}$ and an m-index subset $\boldsymbol{S}_m$ such that

$$\boldsymbol{S}_{kn} \cap \boldsymbol{S}_m = \phi \tag{5}$$
$$\mu \in \boldsymbol{S}_{kn}, \nu \in \boldsymbol{S}_m \implies \eta_\mu > \eta_\nu \tag{6}$$

where $k \ge 1$, $m \le n$ and the following kn kn-dimensional vectors according to $\boldsymbol{S}_{kn}$,

$$\{e_{S_{kn}}^{011}, \ldots e_{S_{kn}}^{01n}, \ldots, e_{S_{kn}}^{0k1}, \ldots, e_{S_{kn}}^{0kn}, \} \tag{7}$$

are mutually linearly independent.
In this case, there exists an m-index subset $\boldsymbol{K} = \{i_1, \ldots, i_m\}$ of the n-index set. And the following $(kn + m)$ $(kn + m)$-dimensional vectors according to the $(kn + m)$-index subset $\boldsymbol{S} = \boldsymbol{S}_{kn} \cup \boldsymbol{S}_m$,

$$\{e_S^{011}, \ldots, e_S^{01n}, \ldots, e_S^{0k1}, \ldots, e_S^{0kn}, e_S^{0(k+1)i_1}, \ldots, e_S^{0(k+1)i_m}\} \tag{8}$$

are mutually linearly independent.

From Lemma 2, we can say that $n \cdot \lceil h_1/2 \rceil$ vectors of M tangent vectors $e^{\ell ji}$ are mutually linearly independent. The linear independence of the remaining tangent vectors of $e^{\ell ji}$ can be proved if a proposition similar to Lemma 2 holds for $e^{\ell ji}$ where $\ell \geq 1$. In order to prove a proposition similar to Lemma 2 for $e^{\ell ji}$, we must be able to prove the general position relations like Eq. (1) and the following "Order Conservation".

Lemma 3.

$$\eta_\mu^\ell = \sum_{i \in T_I^\ell} \sigma_i^\ell H_i^\ell |_{(H_1^0, \cdots, H_1^0) = (I_1^\mu, \cdots, I_n^\mu)}$$

$$\eta_\mu^\ell > \eta_\nu^\ell \quad \text{if} \quad \eta_\mu^0 > \eta_\nu^0 \tag{9}$$

Lemma 3 can be proved by using the monotonicity of sigmoid function f. General position relations like Eq. (1) hold for input nodes in each layer by assigning t_j^ℓ an appropriate value, since the following lemma can be proved.

Lemma 4. There exist $x_0 > 0$ such that the following equation for the variables x,

$$\alpha_1 f(\gamma_1 x) + \alpha_2 f(\gamma_2 x) + \cdots + \alpha_n f(\gamma_n x) = 0 \tag{10}$$

has at most $n - 1$ different solutions for any constants α_i if $x > x_0$ and $0 < \gamma_1 < \gamma_2 < \cdots < \gamma_n$.

With the use of Lemma 3 and Lemma 4, we can prove a proposition similar to Lemma 2 for $e^{\ell ji}$. Hence, we can confirm the linear independence of the M L-dimensional vectors $\{ e^{\ell ji} \mid j \in T_F^{\ell+1}, i \in T_I^\ell$ and $0 \leq \ell \leq N \}$. Consequently, we can show that there is a non-zero Jacobian of the map F at 0 if $M = L$, and that Theorem 1 is true.

5. Conclusion

We have shown that $n \cdot \lceil \frac{h_1}{2} \rceil + \lfloor \frac{h_1}{2} \rfloor \cdot \lceil \frac{h_2}{2} - 1 \rceil + \cdots + \lfloor \frac{h_{N-1}}{2} \rfloor \cdot \lceil \frac{h_N}{2} - 1 \rceil$ examples can be memorized exactly in a neural network with multiple hidden layers.

6. References

1   Mitchison, G. J. and Durbin R. M., "Bounds on the Learning Capacity of Some Multi-Layer Networks", Biological Cybernetics, vol. 60, pp 345-356 (1989).

2   Akaho, S. and Amari, S., "On the Capacity of Three-Layer Networks", Proc. IJCNN 90, vol. 3,pp 1-6 (1990).

3   Baum, E.B., "On the capabilities of multilayer perceptrons", Journal of Complexity, vol. 4, 193-215 (1988).

4   Sakurai, A., "n-h-1 Networks Store No Less n·h+1 Examples but Sometimes No More", Proc. IJCNN 92, III-936 - III-941 (June 1992).

5   Sakurai, A., and M. Yamasaki: "On the capacity of n-h-1 networks", Artificial Neural Networks,2, Elsevier Science Publishers, 237-240 (1992).

6   Funahashi, K., "On the Approximate Realization of Continuous Mappings by Neural Networks", Neural Networks, vol. 2, pp 183-192 (1989).

7   Yamasaki, M. and Sakurai, A., "On the Capacity of n-h-1 Networks with Sigmoidal Functions", Artificial Neural Networks,2, Elsevier Science Publishers, 229-232 (1992).

A method for finding the optimal number of learning samples and hidden units for function approximation with a feedforward network

Vytautas Vyšniauskas[*] Frans C.A. Groen, Ben J.A. Kröse

Faculty of Mathematics and Computer Science, University of Amsterdam
Kruislaan 403, 1098 SJ Amsterdam, The Netherlands

Abstract: This paper presents a methodology to estimate the optimal number of learning samples and hidden units needed to obtain a desired accuracy of a function approximation by a feedforward network. A model of the approximation error is derived of which the parameters can be determined experimentally. Given the computational complexity of the learning rule an optimal learning set size and number of hiden units can be found resulting in minimum computation time for a given desired precision of the approximation. This approach was successfully applied to optimize the learning of a function, which performs camera-robot mapping of a visually guided robot arm.

1 Introduction

Results of recent research established multilayer feedforward networks as a class of universal approximators ([2], [3], [4]). However, failures to approximate a function with the desired accuracy can be attributed to inadequate architecture of a network, inadequate number of of learning samples, the limited precision of a computer or to an inadequate learning procedure.

In this paper the attention is devoted to the problem how many hidden units must be used in a feedforward network and how many learning samples provide enough information to construct the mapping with the desired accuracy. The basic idea is to express the approximation error as a function of the number of learning samples and the number of the network weights. For each particular application it is necessary to estimate experimentally only a few parameters. Given the computational complexity of the learning procedure we can evaluate immediately the optimal size of the learning set and suitable architecture of a feedforward network in order to obtain the desired accuracy with minimal computational resources. The last section of this paper describes the application of this method for optimizing the learning of a function,which performs the camera-robot mapping of a visually guided robot arm.

2 Definitions

The input–output function for a feedforward network can be denoted as $g_w(x)$ where w is a set of the network adjustable parameters. A feedforward network with one hidden layer and output units without bias has $q = dim(w) = (\nu + \mu + 1)h$ adjustable parameters where ν, μ and h are respectively input, output and hidden layer dimensions. Let f be a function of multidimensional mapping from ν-dimensional input space to μ-dimensional output space $f : \mathbf{R}^\nu \to \mathbf{R}^\mu$. It was shown by Hornik et.al. [4] that there exists a *best approximation* g_{w_0} of the function f approximation for a given architecture of the feedforward network [1]. In practice, the goal is to create from a given set

---

[*] the current address is: the Institute of Mathematics and Informatics, Department of Neuroinformatics, Akademijos 4, 2600 Vilnius, Lithuania

[1] It is an evident fact, that a feedforward network has multiple solutions due to the weights permutation and sign flips, but after the proper rearrangement of the network weights we can consider this as one, unique solution.

of samples Z_N the *optimal approximation* $g_{w^o|Z_N}$. This optimal approximation can never be better than the best approximation evaluated from the infinite learning set.

We define the general error measure (approximation error in general) as the difference between an arbitrary solution w and the function f as $E_a(w) = \int_X \|g_w(x) - f(x)\|^2 dx$.

Representation error. The difference between the best available approximation g_{w^o} and the function f is defined as a representation error E_r which informs us how accurately can a given network *represent* a given function when we have perfect knowledge (noiseless, infinitive learning set) about the function. According to the main result in [4], the representation error E_r of the approximation can be arbitrarily small if a sufficient number of the hidden units is available

$$E_r(w^o) = E_a(w^o) = \int_X \|g_{w^o}(x) - f(x)\|^2 dx, \qquad \lim_{h \to \infty} E_r = 0. \tag{1}$$

Generalization error. Since we always have a finite learning set, an extention of this learning set domain (generalization) yields an additional error

$$E_g(w^o|Z_N) = E_a(w^o|Z_N) - E_r(w^o) \tag{2}$$

named as the generalization error which goes to zero when the learning set increases.

Optimization error. Unfortunately, we are not able even to find $g_{w^o|Z_N}$ since no theory at all exists how the knowledge about the function can be explicitly transformed into the weights of a suitable neural network. The vehicle to set the weights optimally $w \to w^o$ is a numerical procedure of nonlinear optimization, which produces the *actual approximation* $g_{w^*|Z_N}$ $(w^*|Z_N \in W^* \subset W)$, different from the $g_{w^o|Z_N}$ due to unperfect optimization. We define the optimization error as follows

$$E_{opt}(w^*|Z_N) = \int_X \|g_{w^*|Z_N}(x) - g_{w^o|Z_N}(x)\|^2 dx, \tag{3}$$

which is a measure of the difference between the actual and the optimal solution evaluated over the whole domain of X. In general, the total error of the approximation $E_a(w^*|Z_N)$ involves implicitly the optimization error. It can be shown that

$$E_a(w^*|Z_N) \le E_r(w^o) + E_g(w^o|Z_N) + E_{opt}(w^*|Z_N) \tag{4}$$

The equality holds only if $E_{opt} = 0$ $(w^* \equiv w^o)$, so E_a can never be lower than the representation error.

Likelihood of valid generalization. The errors we defined above are characteristic for an *individual realization* of the approximation, rather than a given network architecture and a given number of samples. Only the representation error E_r is unique for a given network architecture. We can make estimates in sense of average over all possible learning sets Z_N and random initializations of the network by introducing the average optimization, approximation and generalization errors ε_{opt}, ε_a^*, ε_g as functions of h and N. Now we derive immediately from (4) a relationship between the average errors

$$\varepsilon_a^* \le \varepsilon_g + E_r + \varepsilon_{opt} \tag{5}$$

A necessary condition for valid generalization is *uniform convergence* of E_g when N becomes large, otherwise a single realization from W^o may have a poor generalization (large E_g value). In other words, we want the probability that there is some $w^o \in W^o$ such that $E_g(w^o|Z_N)$ differs significantly from ε_g be very small

$$Pr\left[\sup_{w^o \in W^o} |E_g(w^o|Z_N) - \varepsilon_g| > \epsilon\right] \le \delta(N), \qquad \lim_{N \to \infty} \delta(N) = 0. \tag{6}$$

The problem was solved conceptually by Vapnik and Chervonenkis [5] by introducing the concept of Vapnik–Chervonenkis (VC) dimension, a measure how fast the convergence is achieved (likelihood of generalization). This is of practical importance (especially in the approximation of the multidimensional mapping, when $\nu > 1, \mu > 1$) because this determines the number of examples and suitable architecture of a network needed to guarantee generalization within given tolerance parameters. As was shown by Baum and Hausler [1], in the case of neural networks VC dimension is closely related to the number of weights in the architecture.

3 The approach

As the basis of our approach we explored the relationship (5). Without loss of generality we suppose that the optimization is perfect ($\varepsilon_{opt} = 0$).

Asymptotic expansion. Since the goal is to create an accurate function approximation we are interested in the behaviour of ε_a for $N >> 1$ and $h >> 1$. In such a case we can expand ε_a at the point (N_*, h_*) for any $N << N_*$ and $h << h_*$ as follows

$$\varepsilon_a(N, h) = \sum_p \sum_{\alpha+\beta=p} \gamma_{\alpha\beta} (\frac{1}{N} - \frac{1}{N_*})^\alpha (\frac{1}{h} - \frac{1}{h_*})^\beta \approx \sum_p \sum_{\alpha+\beta=p} \frac{\gamma_{\alpha\beta}}{N^\alpha h^\beta} = \sum_p \varepsilon_a^{(p)}(N, h), \qquad (7)$$

where α, β, p are natural numbers and $\gamma_{\alpha\beta}$ are unknown parameters of the expansion, which satisfies the asymptotic conditions from the relationships (1–2). As we want to represent the asymptotic behavior of ε_a for $N >> 1$ and $h >> 1$, only few terms of order $p_{min}, p_{min+1}, \cdots, p_{max}$ can be included. This truncation describes the asymptotical model of the error function (AMEF), different values of p_{min} and p_{max} yield different solutions.

Minimization of computational resources. In practice the knowledge about ε_a is of extreme importance, providing the possibility to find the optimal strategy so that the learning time would be minimal to to obtain the desired accuracy. If we know additionally the computational complexity of the learning procedure which can be expressed as $r \sim Nh^k$ as a function of N and h, where k is the order of complexity, it is possible to find an unique pair (N_o, h_o) resulting in the minimal computation time for a given precision of the approximation. Note that $k = 1$ for methods like *conjugate gradient* and *backpropagation*. The solution (N_o, h_o) is the minimum of r under the condition $\varepsilon_a = \varepsilon^o$.

Methodology to estimate γ parameters. The proper AMEF can be chosen by using the least–squares criterion. Note, the estimation of the average approximation error must be done by averaging over all different learning sets and over all random initializations of the network. Estimated parameters are needed to determine (N_o, h_o).

4 Results

We applied the approach to approximate the camera-robot mapping for the adaptive robot control, for which the number of inputs $\nu = 5$ and the number of outputs $\mu = 3$. We estimated the approximation error $\hat{\varepsilon}_a$ at 64 points on the 8×8 grid in (N, h) framework ($50 < N < 400$, $5 < h < 40$) from an independed test set consisted from 4000 samples, and the network learning was limited to 5000 epochs. At each point the average from 10 realizations was computed. We used *conjugate gradient* procedure for the network weights adjusting.

We obtained a very good fitting for the solution

$$\varepsilon_a^{(2)}(N, h) = \frac{\gamma_{20}}{N^2} + \frac{\gamma_{11}}{Nh} + \frac{\gamma_{02}}{h^2} \qquad (8)$$

(see Figure 1), where only the second term of the expansion was used. More detailed analysis showed that the approximation error can be described more exactly with the solution ($p_{min} = 1, p_{max} = 2$), and the relative contribution of these terms was respectively 26 and 74 percent. We found no practical gain to use more terms of the expansion, because the mean–square error decreases slowly with increasing of the regression order. These results also suggested that the optimal number of learning samples for these particular applications is very close to the number of the network weights.

For the solution $\varepsilon_a^{(2)}$ it can be shown that there exits a lower boundary of the number of learning samples and hidden units which are necessary to obtain the desired accuracy of approximation

$$N > \sqrt{\frac{\gamma_{20}}{\varepsilon^o}}, \qquad h > \sqrt{\frac{\gamma_{02}}{\varepsilon^o}}. \qquad (9)$$

where ε^o is the desired accuracy of approximation. Similar boundaries also exist for other solutions also. Probably, failures to approximate a function can be particularly explained as a lack of information about these boundaries.

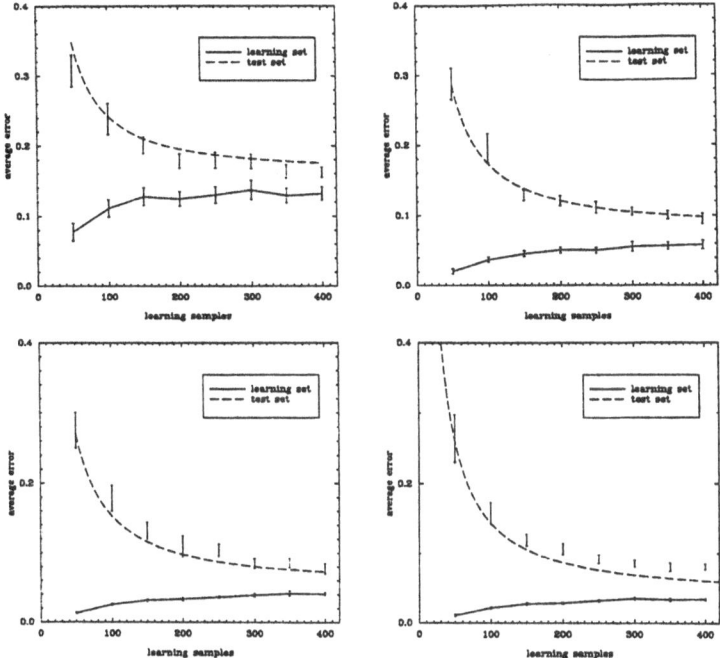

Figure 1: Comparison of the experimental results with the theoretical estimation (dashed line) for the network with 10, 20, 30, 40 hidden units. The learning error is represented by the solid line, error bars are indicators of a dispersion of the average value.

Concluding remarks. Future research will focus on a number of issues concerning the model and applying it to a number of different problems. We did not include in our approach the factor of the limited number of iterations needed to obtain the desired accuracy, because this was in our case not the most prominent source of error. The optimization error can be incorporated by adding extra-term as a function of N and h. A study will be carried out on the relationship between the architecture of a network and the number of terms in the expansion (7). Also the influence of the noise in the learning samples in the model needs some further attention.

References

[1] Baum E.B., Haussler D. "What Size Net Gives Valid Generalization?", *Neural Computation* 1,1989, pp. 151-160.

[2] Cybenko G. "Approximation by Superpositions of a Sigmoidal Function", *Math. Control Signals Systems*, **2**, 1989, pp. 303-314.

[3] Funahashi K. "On the Approximate Realization of Continuous Mappings by Neural Networks", *Neural Networks*, vol.2, 1989, pp. 183-192.

[4] Hornik K., Stinchcombe M., White, H. "Multilayer Feedforward Networks are Universal Approximators", *Neural Networks*, vol.2, 1989, pp.359-366.

[5] Vapnik V.N.,Chervonenkis A.YA. "On the uniform convergence of relative frequencies of events to their probabilities", *Theory of Probability and its Applications*, vol XVI, no.2, 1971, pp.264-280.

The N-2-N Encoder: A Matter of Representation

Paul Bakker, Steven Phillips†, Janet Wiles

Depts. of Computer Science & Psychology
†Dept. of Computer Science
The University of Queensland, QLD 4072 Australia
email: *bakker@cs.uq.oz.au*

Abstract Kruglyak [4] demonstrated that weights exist to implement the N-2-N encoder for all finite N, but it is known that Backpropagation (BP) is unusually poor at learning such solutions for N > 8 ([6]). We show that the learning problem lies not with BP, but with the pattern representation typically used in the encoder task. With an appropriate representation, we demonstrate that BP can learn encoders in approximately linear time with N as large as 100. This underlines yet again the crucial importance of pattern representation in neural network learning.

1 Introduction

The N-M-N encoder is a simple task that has been used to study the training properties of the Backpropagation learning algorithm [1, 2, 7]. A three-layer network consisting of N input and output units, and M hidden units, must learn to compress a set of N input patterns into M hidden units, and reconstruct them on the output units. If M is less than log_2 N, the encoder is termed a *tight* encoder [2].

The most popular encoder in the current literature is the N-2-N. This is the tightest encoder possible (for N > 2), and is known to present BP with a formidable learning task [5]. Kruglyak [4] proved geometrically that weights exist to implement the N-2-N encoder for *any* finite N. It was noted, however, that the existence of these weights does not imply that BP can learn them; he reported that he was only aware of encoders as large as 8-2-8 being learned successfully by a "modified" version of BP. The N-2-N was hence proposed as a challenging task for testing new learning algorithms [4].

The N-2-N is a good test of a learning algorithm because it requires that all input patterns be compressed into 2-dimensional hidden unit space. The efficient use of hidden unit space is vital in neural net learning tasks, where the number of hidden units is usually minimized to improve generalization.

2 Solving the N-2-N encoder task

Kruglyak [4] plotted a theoretical solution of the N-2-N encoder; in Figure 1 we show the hidden unit space of an *actual* solution[1] to the 19-2-19 encoder, similar to that found by Lister [5]. The activation range of hidden unit 1 is on the abscissa, and that of hidden unit 2 is on the ordinate. The 19 data points in Figure 1 correspond to the internal representations of the input patterns, and the line segments represent the 19 hyperplanes determined by the weights to each output unit. With respect to a particular output unit, the solution requires all hyperplanes to be arranged so that each point is on the *low side* of all but one of the hyperplanes; this corresponds to an output response of 0 for all patterns except one. When this constraint is extended to all output units, the solution requires the points to be evenly distributed to form an approximate circle with each hyperplane uniquely partitioning off one point [4].

The difficulty of the task is due to the critical positioning of the points and hyperplanes. As N increases, the triangular region which isolates a single point becomes smaller. Consequently, finding a solution requires greater precision and hence smaller steps across the error surface. Empirical results

---

[1]This 19-2-19 encoder was learned by standard BP in 38.3 million pattern presentations

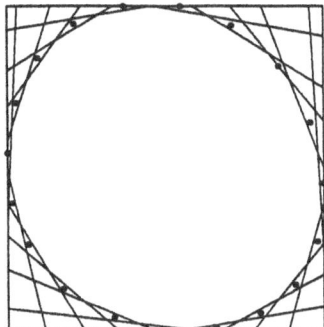

Figure 1: Solution of the 19-2-19 encoder with local encoding on the target vectors.

bear this out: Lister's [6] convergence figures demonstrate that learning time grows exponentially with N.

3   Would a Different Representation Help?

Representation is a crucial factor in neural network learning [3]. In studying the learning dynamics of the N-2-N problem, we noticed that the input and output representation traditionally used greatly handicaps Backpropagation. The standard encoding is *local*, with one unit on and all others off. For example, in the 6-2-6 encoder, the input (and target) set is:

$$100000 \ 010000 \ 001000 \ 000100 \ 000010 \ 000001$$

Note that for every output unit, the target is 0 for (N-1) patterns, and 1 for just one pattern. When training commences, we have observed that BP always sets all output values to near-zero values in the first few epochs. This is because a null output vector is a very close approximation of *all* of the target vectors shown above [6]. This initial action, though understandable, is actually a step *away* from the final solution. It leads to all weights in the output layer becoming negative and saturates the output units, making further learning very slow (the *flat-spot* effect; [2]).

Obviously, this problem is directly related to the preponderance of zeroes in the local pattern encoding. This prompted us to speculate that the difficulty that BP has in learning the N-2-N task may not be due so much to the demands of the task itself, but due to the output representation typically used.

We decided to experiment with different representations to test this hypothesis. The N-2-N task remains logically the same (information must be compressed into a bottleneck and reconstructed), but we used a pattern encoding that has equal (or near equal) quantities of 1's and 0's in the target patterns. This will be referred to as N/2 *block* encoding.

For example, in the 6-2-6 encoder, the target vectors become:

$$111000 \ 011100 \ 001110 \ 000111 \ 100011 \ 110001$$

It was expected that the use of block encoding on the target vectors would stop the counterproductive behaviour of Backpropagation in the initial stages of training, and would lead to faster convergence on N-2-N encoders, as well as the ability to learn larger encoders.

4   Simulations and Results

Standard Backpropagation [7] was used to learn N-2-N encoders for increasing N. For each N and each target pattern encoding (local and block), ten trials were run using a set of matched starting weights. Starting weights were matched to control for their influence on convergence times. A learning rate of 0.01, epoch update and a momentum factor of 0.9 were used in all runs; a training run was considered to have converged if all outputs were within 0.5 of their target values.

Table 1: Median number of pattern presentations required to achieve convergence.

	Median number of pattern presentations	
Encoder	Local Encoding	Block Encoding
4-2-4	5416	2208
9-2-9	743058	4077
14-2-14	8950718	4844
19-2-19	35583922	6973
50-2-50	—	12600
100-2-100	—	30300

With a learning rate of 0.01 and a local encoding scheme on the target vectors, BP was only able to learn up to N = 19. When the block representation was used, however, BP was able to learn all encoders faster, and was able to learn the 100-2-100 encoder consistently in just a few hundred epochs.

5 Analysis and Discussion

The use of N/2 block encoding on the target vectors resulted in significantly accelerated learning and the ability to learn much larger encoders than in the local case. This effect can be explained both in terms of learning dynamics, and in the organization of hyperplanes required to implement the solution.

Firstly, the equalization of the number of 1's and 0's presented to the output units as targets changed the learning behaviour of BP; instead of all output activations moving to 0 and saturating in the first few epochs, we now find all output units moving to an intermediate activation value of 0.5. Although significant, this does not completely explain the magnitude of the speedup reported. In examining the hidden unit space of the solution (see Figure 2), we found that the use of N/2 block encoding fundamentally changes the arrangement of hyperplanes required to solve the encoder. Instead of each hyperplane being required to carve off one unique data point, it now only has to carve off N/2 points; that is, in each case, half of the output units must be "on", and the other half "off". This requires far less precision than in the local encoding case, as each point can now be isolated within a larger region of hidden unit activation space.

In fact, with a block encoding, learning time (number of pattern presentations) grew approximately linearly with the size of the encoder (N). This is a dramatic improvement over the local encoding case where, with the same learning algorithm (BP) and parameters, learning time grew exponentially with N.

6 Conclusion

Because of its simplicity, the N-2-N encoder/decoder is a useful and interesting problem for characterizing the learning properties of Backpropagation. We found that BP learned much faster when a desired response was distributed over half the output units (block encoding), than when confined to a single output unit (local encoding), which is the more common practice. This reiterates the important point that learnability is determined not only by the search technique, but also by the representation of the problem [3]. We agree with Lister's [6] conclusions that the poor performance of BP on the encoder task is due to the error surface being ill-conditioned for gradient descent. The block representation used here creates an error surface more conducive to gradient descent search.

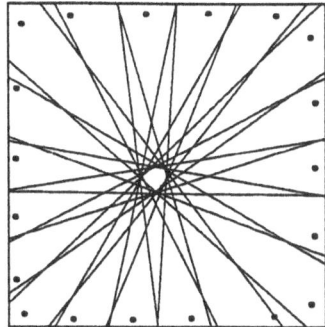

Figure2: Solution of the 19-2-19 encoder with block encoding on the target vectors.

In further work, we are trying to establish the limits of block encoding. We have been able to consistently solve 500-2-500 encoders, and are currently attempting the 1000-2-1000. We anticipate that, subject to the precision of the simulator, any size encoder can be tractably learned.

Acknowledgements

The authors would like to thank Ray Lister for insightful discussions on the N-2-N problem. We would also like to thank Anthony Bloesch for his help, and the rest of the Connectionists group at The University of Queensland for encouragement and discussions.

References

[1] Ackley, D.H., Hinton, G.E., & Sejnowski, T.J. (1985). A learning algorithm for Boltzmann Machines. *Cognitive Science, 9*, 147-169.

[2] Fahlman, S.E. (1989). Faster-learning variations on back-propagation: An empirical study. In D. Touretzky, G. Hinton & T. Sejnowski (Eds.), *Proceedings of the 1988 Connectionist Models Summer School* (pp. 38-51). San Mateo, CA: Morgan Kaufmann.

[3] Hertz, J., Krogh, K., & Palmer, R.G. (1991). *Introduction to the theory of neural computation* (p. 144). Redwood City, CA: Addison Wesley.

[4] Kruglyak, L. (1990). How to solve the N bit encoder problem with just two hidden units. *Neural Computation, 2*(4), 399-401.

[5] Lister, R. (1992). *On making the right moves: Neural networks, gradient descent, and simulated annealing.* Unpublished doctoral dissertation, The University of Sydney, Australia.

[6] Lister, R. (1993). Visualizing weight dynamics in the N-2-N encoder. In *IEEE International Conference on Neural Networks* (pp. 684-689). IEEE.

[7] Rumelhart, D. E., Hinton, G. E., & Williams, R. J. (1986). Learning internal representations by error propagation. In D. E. Rumelhart and J. L. McClelland (eds.) *Parallel Distributed Processing: Explorations in the Microstructures of Cognition*, Vol. 1. Cambridge, MA: The MIT Press.

Optimizing the Architecture of Multi-Layer Perceptrons for One-Dimensional Classification

Wim Wiegerinck and Bert Kappen

Department of Medical Physics and Biophysics, University of Nijmegen, Geert Grooteplein 21, 6525 EZ Nijmegen, The Netherlands

Abstract

We consider the problem of finding the optimal architecture of a multi-layer perceptron (MLP) for classification given a fixed training set. We restrict ourselves to problems with a one-dimensional input and a binary output. We propose a simple method to estimate the optimal number of hyperplanes from the training curve. We test two criteria and compare them with methods from the literature. We conclude that our method is superior to existing methods for our one-dimensional example, and may be useful for higher dimensional problems.

1 Neural Networks for Classification

Consider the problem of neural networks for classification where one has a network f which tries to learn to assign to an input $x \in \mathbb{R}^d$ the most probable output $y \in \{-1, 1\}$ from a *given* set of examples $\{x_i, y_i\}$, $i = 1 \ldots L$ drawn from an unknown underlying joint probability distribution $P(x, y)$. Such problems can not be classified perfectly, because of the intrinsic noise present in the data. The network f which chooses the most probable output y is the one which minimizes the generalization error,

$$\epsilon_g = \sum_y \int_x d(y, f(x)) P(x, y) dx \,,$$

with $d(y, -y) = 1$ and $d(y, y) = 0$. Usually, one tries to minimize generalization error by minimizing the training error

$$\epsilon_t = \frac{1}{L} \sum_{i=1}^{L} d(y_i, f(x_i))$$

This is correct for an infinite amount of training data. However, for a finite training set, minimizing training error does not necessarily minimize generalization error. The optimal value of the training and generalization error depends on the number of hidden units. We are interested in the number of hidden units that minimizes the generalization error. Too few hidden units yields high training error and thus high generalization error. Too many hidden units yields low training error, but overfitting may occur due to overspecialization on the training set, which yields also high generalization error.

In this paper we propose to study optimal generalization, by assuming that the noise in the training set has different statistical properties that the data. We illustrate the method for a one-dimensional example. This is not an essential limitation of the method, although extension to higher dimensions may be more difficult. In addition, we compare the method with Barron's complexity regularization [1] and Vapnik's structural minimization of risk [2] [3].

2 A Model for the Trained Networks

In one-dimension, a binary valued multi-layer perceptron (MLP) is of the form

$$f(x) = \text{sgn}(\sum_{n=1}^{H} w_n \, \text{sgn}(x - \theta_n)) \,. \tag{1}$$

Network (1) is described by H real parameters (the positions of the hyperplanes) and one binary parameter (the value of f at infinity).

For all H, let us call f_H the network with H hyperplanes which minimizes the training error ϵ_t. Let us denote H^* the number of hyperplanes that minimizes the generalization error. To determine this optimal number of hyperplanes exactly, one needs, of course, $P(x,y)$ which is not known.

In this one dimensional binary classification problem learning can be pictured in the following way: The training set consists of an array of values ± 1. This array consists of large areas of predominantly 1's or -1's, describing 'data'. Due to the probabilistic nature of the problem, small clusters with the opposite sign exist in these areas, which we call 'noise'. The H hyperplanes are placed such that the 1's and -1's are separated optimally. If too many hyperplanes are used, the noise clusters are classified as if they were data, and overfitting occurs. Optimal generalization in this context can be achieved by defining a criterion to ignore small clusters.

We assume that the noise is independent for each location, and that no nested noise clusters occur. The latter is approximately true for low noise values and/or for small training sets. Let us denote by N_k the number of clusters of length k in the training set and let us number the lengths of the clusters

$$k_1 > k_2 > \ldots > k_n > 0 .$$

For $H \leq H^*$, $\epsilon_t(H) \approx \epsilon_g(H)$ (Figure 2). For $H > H^*$, the dependence of the training error on H is fixed by the numbers k_j and N_{k_j},

$$\epsilon_t(H^* + 2\{\sum_{j=1}^{l} N_{k_j} + h\}) = \epsilon_g(H^*) - \frac{1}{L}\{\sum_{j=1}^{l} k_j N_{k_j} + k_{l+1}h\} , \qquad 0 \leq h < N_{k_{l+1}} . \qquad (2)$$

3 Optimal Number of Hyperplanes

The procedure to find the optimal number of hyperplanes is described in the following way: We chose a large value H_0 and make the hypothesis $H^* = H_0$. We determine the numbers $\{k_j, N_{k_j}\}$ from (2) and estimate the probability that $\{k_j, N_{k_j}\}$ occurs due to noise. If this probability is acceptable according to some criterion, then H_0 is decreased by one. The procedure stops if the hypothesis is not accepted, and H^* is determined by the previous hypothesis.

Instead of calculating the probability for the sequence $\{k_j, N_{k_j}\}$ exactly, we will calculate upper bounds for the probability that certain aspects of $\{k_j, N_{k_j}\}$ are due to noise. Then we will infer criteria by rejecting probabilities less than 0.5 .

For the first criterion we infer that, given the noise model, the probability that at any location in the data there occurs a cluster of length k is bounded by $L2^{-k}$. This probability is smallest for the the largest cluster k_1 in the training set. Therefore

$$\text{Criterion 1:} \qquad \text{If} \quad k_1 > \frac{\log(L)}{\log(2)} + 1 \qquad \text{then reject} \quad H^* = H_0 . \qquad (3)$$

For the second, more subtle criterion we calculate an upper bound for the probability $p(l)$ that there is some k such that no clusters of length $k, k+1, \ldots, k+l-1$, occurs, while some clusters of length $k+l$ occur.

Let us call $u(k)$ the probability that $N_k = 0$ given the noise model, then $p(l)$ is given by

$$p(l) = 1 - \prod_k (1 - u(k)u(k+1)\ldots u(k+l-1)(1 - u(k+l))) .$$

Define $q(l)$ as the probability for the case of a uniform distribution $P(x,1) = P(x,-1)$ in the limit $L \to \infty$. After some algebra this yields the bound

$$p(l) < q(l) \approx 1 - \exp\left(-\frac{1}{\ln(2)}\int_0^1 \frac{\ln(1 - x^{2^{l+1}-2}(1-x))}{x\ln(x)}dx\right)$$

From $q(1) \approx 0.5$ we infer

$$\text{Criterion 2:} \qquad \text{If} \quad k_1 - k_2 > 1 \qquad \text{then reject} \quad H^* = H_0 . \qquad (4)$$

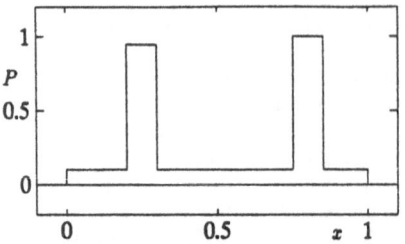 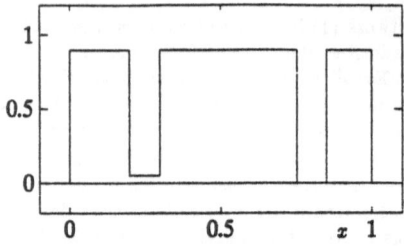

Figure 1: The probability distribution $P(x, 1)$ and $P(x, -1)$

4 Experiments

We tested our two criteria and two methods of the literature to estimate the optimal number of hyperplanes on artificial data sets drawn from a known probability distribution $P(x, y)$ (Figure 1). The first method is Barron's complexity regularization [1] which gives an estimate H_B for the optimal number of hyperplanes by minimizing

$$F_B(H) = \epsilon_t(H) + \sqrt{\frac{H}{2L}\left(\frac{\log L}{2} + 1\right)}.$$

The second method is Vapnik's method of structural minimization of risk [2] [3], which uses the well-known VC-dimension. This method estimates the optimal number of hyperplanes by the minimum H_V of $F_V(H)$. For our problem this function is

$$F_V(H) = \epsilon_t(H) + \Psi\left(1 + \sqrt{1 + \frac{\epsilon_t(H)}{\Psi}}\right),$$

with

$$\Psi = \frac{1}{L}\left[\left(\log(\frac{2L}{H+1}) + 1\right)(H + 1) + 1\right].$$

For each training set of size L ($L = 20, 30, \ldots, 100$), we computed the training curve $\epsilon_t(H)$, $H = 0, \ldots, H_{max}$, where H_{max} is defined as the smallest number of hyperplanes for which the training error vanishes identically. The networks f_H were found by exhaustive search in the hyperplane space. Then the optimal numbers of hyperplanes H_1 and H_2, according to the criteria (3) and (4), as well as H_B and H_V were determined. The performances of these estimations were measured by the generalization errors $\epsilon_g(H_1)$, $\epsilon_g(H_2)$, $\epsilon_g(H_B)$ and $\epsilon_g(H_V)$, which were calculated explicitly using $P(x, y)$. For each training set size L, 100 training sets were drawn. The averages $\bar{\epsilon}_g$ of the generalization errors are shown in Figure 3. As a reference, the averages of the generalization errors $\epsilon_g(H^*)$ and $\epsilon_g(H_{max})$ are drawn as well.

For small L, Figure 3 shows bad performance of all methods, even worse than the performance of the network with the maximal number of hyperplanes. Closer analysis indicates that all methods tend to underestimate the optimal number of hyperplanes (Figure 2). For large L, the performances of all methods converge to the optimal performance $\bar{\epsilon}_g(H^*)$ whereas $\bar{\epsilon}_g(H_{max})$ does not. For intermediate training set sizes, H_2 seems the best estimate.

5 Discussion

Based on a simple model for trained networks we inferred criteria to estimate the optimal number of hyperplanes for a one-dimensional MLP for classification from the shape of the training curve. The experiments show that even a very crude criterion gives reasonable generalization, and that a more subtle criterion, using more information about the shape of the training curve, outperforms existing methods.

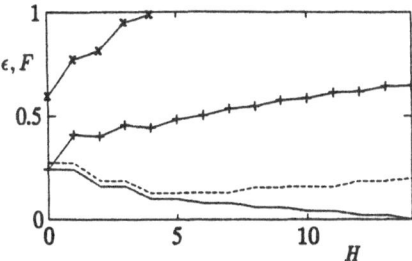

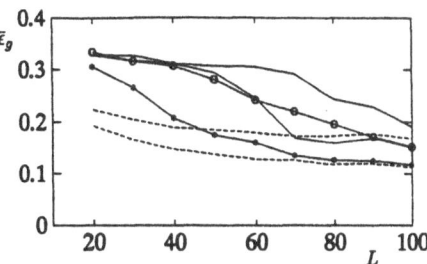

Figure 2: training error $\epsilon_t(H)$ (drawn), generalization error $\epsilon_g(H)$ (dashed), Barron's function $F_B(H)$ ('+') and Vapnik's function $F_V(H)$ ('x') as functions of the number of hyperplanes H for a random training set with $L = 50$ training samples. Note that $H^* = 4$ in this example.

Figure 3: averaged generalization errors $\bar{\epsilon}_g(H_1)$ ('o'), $\bar{\epsilon}_g(H_2)$ ('*') , $\bar{\epsilon}_g(H_B)$ (lower drawn line), $\bar{\epsilon}_g(H_V)$ (upper drawn line), $\bar{\epsilon}_g(H^*)$ (lower dashed line) and $\bar{\epsilon}_g(H_{max})$ (upper dashed line) as functions of the number of data points L. The average is over 100 data sets.

The main draw-back of our model is that its generalization to higher dimensions is not clear. However, this is also not a trivial matter for the two methods of the literature. Nevertheless, this study suggests that extension of this method to higher dimensions may be very fruitful.

6   Acknowledgements

This work was sponsored by the Dutch Foundation for Neural Networks. We thank Ole Winther for pointing us to reference [3].

References

[1] A. R. Barron, "Complexity Regularization with Applications to Artificial Neural Networks", in *Nonparametric Functional Estimation and Related Topics*, G. Roussas, Ed. Kluwer Academic Publ., Dordrecht, 1991.

[2] V. Vapnik, *Estimation of Dependences Based on Empirical Data*, Springer Verlag, New York, 1982.

[3] S. A. Solla, "Capacity Control in Classifiers for Pattern Recognition", in : *Neural Networks for Signal Processing 2 - Proceedings of the 1992 IEEE-SP Workshop.* S.Y.Kung, F.Fallside, J.A.Sørensen, and C.A.Kamm, Ed. IEEE Service Center,Piscataway NJ,255-266, 1992.

Neural Networks and Genetic Algorithms:
Improving the Fault Tolerance Capabilities

Hamed Elsimary*, Samia Mashali*, and Samir Shaheen**
* Electronics Research Institute, Cairo, Egypt
El-Tahrir St, Dokki, Cairo, Egypt
** Faculty of Engineering, Cairo University, Egypt
Email : ERI@EGFRCUVX.BITNET

Abstract:

Artificial neural networks, has shown great performance in many fields, and has shown its abilities in solving problems difficult to be solved by conventional methods of computation, this make it important to study and insure the reliability of its function under different working conditions. One of the advantages of ANNs is that the processing is distributed among many nodes, this provides a good degree of fault tolerance even if some of its nodes or interconnections fails to function properly. But this features needs to be emphasized, improved, and maximized through different techniques. In this work we are presenting a method for training ANNs to improve its fault tolerance capabilities using genetic algorithms.

Our measure for fault tolerance is the deviation of the network's output, when the interconnection weights are perturbed, from that output without perturbation. In our method we try to maintain that deviation as minimum as possible, i.e. the network should produce correct results even with the existence of error in the network structure.

In this work we show that, depending on how neural networks is trained, the sensitivity of its output to perturbation in interconnection weights can vary dramatically. The perturbation in interconnection weights (network parameters) can reflect the errors that arise when such network implemented in hardware. The fault tolerance of ANNs has been formulated as a nonlinear optimization problem with multivariants and nonlinear constraints, then solved by genetic algorithms.

Genetic algorithm has been proven a successful method for training, and optimizing the structure of neural networks. Genetic algorithms are a robust method based on biological principles. A population of strings representing possible problem solution is maintained. Search proceeds by recombining strings in the population. The theoretical foundations of genetic algorithms are based on the notion that selective reproduction and recombination of binary strings change the sampling rate of hyperplanes in the search space so as to reflect the average fitness of strings that reside in any particular hyperplane. In this work we have used genetic algorithms to train feed forward neural networks subject to maximize its fault tolerance capabilities and this has been presented, explained by examples, and supported by computer simulation results.

Entropy of Perceptrons

H.S. Toh

Department of Medical Physics and Biophysics,
University of Nijmegen, Geert Grooteplein Noord 21,
6525EZ Nijmegen, The Netherlands

We predict that, as the temperature is raised, the 2 layer perceptron undergoes a series of phase transitions from a greater number of first layer hyperplanes involved with the partitioning of the input space to a smaller number. This series of multiple phase transitions, in terms of the number of first layer hyperplanes, is not found in the 1 layer perceptron.

We obtain, via standard statistical mechanics, the training free energy :

$$f_t(\varphi, X_m) = E_t(\varphi, X_m) + T S_P(\varphi) \tag{1}$$

where $S_P(\varphi)$ is the entropy associated with the function φ.

$$S_P(\varphi) = -\log V(\varphi) \tag{2}$$

$V(\varphi)$ = fraction of weight space that realises the same perceptron function φ. X_m is the particular training set used. E_t is the training error. The temperature is the variance of the stochastic training noise η:

$$2T\delta_{ij}\delta(t - t') = <\eta_i(t)\eta_j(t') > \tag{3}$$

We restrict ourselves to the case when the inputs, hidden units and output unit are all Boolean. Our 2 layer perceptron has N input units, H hidden units, and 1 output unit. We state the following result without proof (The proof is given elsewhere [1]) :

Provided N is large enough relative to the number of hidden units ($NlnN >> lnH$), the entropy of the function mapped by 2 layer perceptrons, $S_P(\varphi)$, is of $O(KN)$. K is the number of first layer hyperplanes used to map the function. Thus the entropy of the 2 layer perceptron increases linearly with the number of hidden units used.

A function using fewer hidden units incurs a greater training error, but has a larger entropy. At some critical temperature T_c, the entropy term $T_c S_P$ will outweigh the error term in the free energy, causing a function using fewer hidden units to have a lower free energy than a function using more hidden units. At this critical temperature T_c, a phase transition occurs from the function using more hidden units to the function using fewer hidden units. T_c is given by

$$T_c = \frac{\Delta E_t}{\Delta S_P} \tag{4}$$

The discrete jump in E_t and S_P makes the transition first order. At the transition, there is a change in the number of hidden units used, so there is a discontinuous change in the value of the second layer weights - at least 1 second layer weight suddenly becomes weak.

The different functions occupy disjoint regions in weight space. This causes the phase transition to depend on the specific learning trajectory. This dependence upon specific learning history manifests itself tangibly upon the variation of the transition temperature T_c with different learning trajectories.

Acknowledgements

The financial support of the Dutch foundation for neural networks (SNN) is acknowledged.

References

[1] H.S. Toh. Entropy of perceptrons. *preprint.*

564

Assessing Generalization by 2–D Receptive Field Visualization

Michael K. Arras and Peter W. Protzel

Bavarian Research Center for Knowledge–Based Systems (FORWISS)
Am Weiselgarten 7, 91058 Erlangen, Germany
e-mail: arras@forwiss.uni-erlangen.de

Abstract A performance comparison of different neural networks for classification tasks has to take the convergence speed on the training data as well as the generalization capability on unseen test data into account. We present visualization results for two two–dimensional benchmark problems by scanning the receptive fields of different networks. This allows for the assessment of generalization and shows the evolution of the networks during training. The figure below shows the final receptive fields for three different networks that learn to distinguish between two classes arranged as an 8x8 checkerboard a), b) and c) and two intertwining spirals d), e) and f). The training points of the two classes are indicated by a cross and a square. The networks evaluated are a) a 2-12-10-1 MLP trained with backpropagation, d) a 2-5-5-5-1 MLP with fully connected "cross–cut" connections trained with backpropagation, b) and e) a standard Cascade–Correlation network and c) and f) a Cascade–Correlation network that uses a mixture of sine, cosine and sigmoid activation functions. Note that the number of epochs it took to solve the 8x8 checkerboard problem with a similar generalization differs by almost two orders of magnitude between a) and c). In contrast to c), while the use of mixed activation functions offers a similar speed advantage, the resulting generalization in f) is very poor. Our results show that convergence speed, network size and generalization performance do not necessarily go hand in hand, and that visualization is a valuable aid which should be part of any benchmark that assesses the performance of different network architectures.

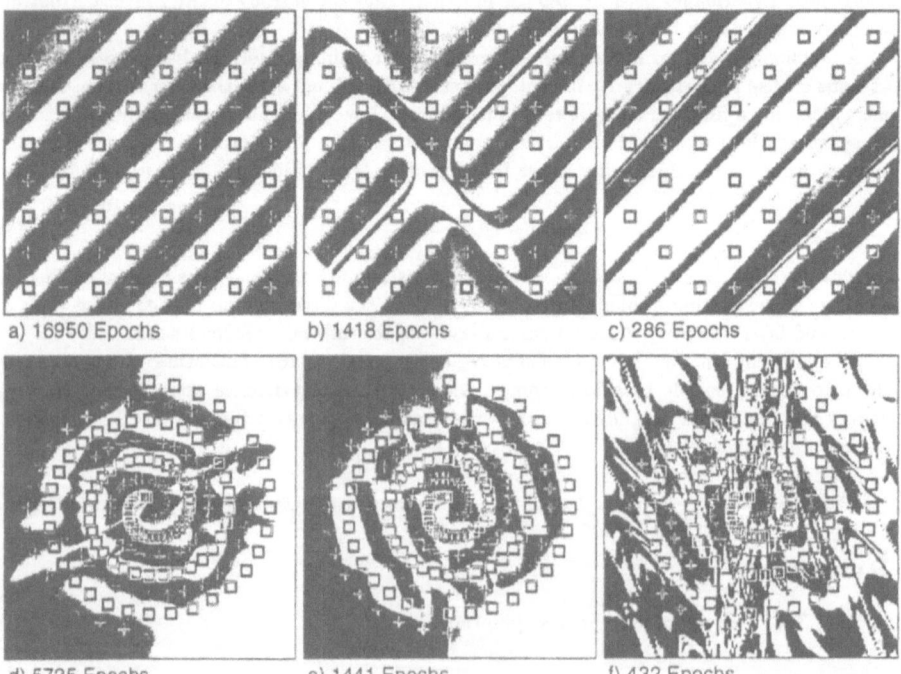

a) 16950 Epochs   b) 1418 Epochs   c) 286 Epochs

d) 5725 Epochs   e) 1441 Epochs   f) 432 Epochs

A Fast Training Algorithm for Feedforward Neural Networks

Y F Yam, T W S Chow
Department of Electronic Engineering, City Polytechnic of Hong Kong,
83 Tat Chee Avenue, Kowloon, Hong Kong.

The backpropagation algorithm for training feedforward neural networks is extensively used to solve pattern recognition, signal processing and control problems. However, the time required before convergence is long even for medium sized network problems. The choice of learning rate η and momentum coefficient α also have significant effect on the rate of convergence. In this abstract, a fast training algorithm for feedforward neural networks is briefed. This algorithm is based on the adaptation of learning rate by correlation coefficient. The algorithm has been extensively tested. Results show that this algorithm can significantly reduce the time required for convergence.

The η and α should be adapted according to the topography of error function. In this newly developed algorithm, the learning rate adaptation is simply based on the correlation coefficient r^t between the local downhill gradient $-\nabla E^t$ and the past weight update ΔW^{t-1}. The following function is suggested to alter the learning rate:

$$\eta^t = \eta^{t-1}(1 + \frac{1}{2}r^t)$$

However, the convergence rate is not optimized with a fixed momentum coefficient α. The α should be proportional to the η. Let

$$\alpha^t = \lambda^t \eta^t \quad \text{with} \quad \lambda^t = \lambda^0 |-\nabla E^t|/|\Delta W^{t-1}|$$

If $0 \leq \lambda^0 < 1$, the magnitude of momentum term is always shorter than that of negative gradient term. In order to prevent the neurons from being saturated in the very beginning stage, the η^t is reduced by half if the present mean error is greater than the previous one by 0.01.

To compare the convergence performance of this algorithm with the conventional one, we applied both algorithms to exclusive-or, 8-3-8 encoder/decoder and nonlinear function approximation problems. Each problem consisted of 250 runs with 10 different initial weight settings and 25 sets of η and α combinations. The termination criteria for XOR and 8-3-8 encoder/decoder were 0.0002 and 0.04 mean squared error respectively. For the nonlinear mapping problem, one hundred random samples of the following function were used as the training set.

$$y = f(x_1, x_2, x_3) = g(e^{x_1} - (x_2 - 0.5)^2 - 3\sin(\pi x_3))$$

where $g(z) = [1 - e(z)]/[1 + e(z)]$. The network structure was 3-10-1. The convergence criterion was 0.04 mean squared error. The results obtained by using the conventional backpropagation algorithm and the fast algorithm was shown in Table 1. For the XOR problem, the mean time required by the new algorithm to converge is only 1.71% of that by the conventional algorithm. For the 8-3-8 encoder/decoder and nonlinear function approximation problems, the mean time required were 40.2% and 4.89% of that required by the backpropagation algorithm respectively. Apart from the reduction in convergence time, it is noticed that the new algorithm is rather insensitive to the selection of η and α.

Problem	Mean Time (s) (Fast Algorithm) [1]	Mean Time (s) (Conventional Algorithm) [2]	Ratio [1]/[2]x100%
XOR	0.578	33.8	1.71
8-3-8	7.64	19.0	40.2
Function Approximation	22.1	452.5	4.88

Table 1: Comparison of the convergence performance of the fast and conventional algorithm

Improvement of the Covergence of the Learning Using the Modified Back-propagation Method

Masato SEKI[†] and Shinsaku MORI[††]

† TOYOTA MOTOR CORPORATION
1 Toyota-cho, Toyota, Aichi, 471 JAPAN
Phone: +81 565 28 2121
†† Dept.of Electrical Engineeriing, Keio University
3-14-1 Hiyoshi, Kouho, Yokohama, 223 JAPAN
Phone: +81 45 563 1141 Ext. 3319 FAX: +81 45 563 2773
email: mori@mori.elec.keio.ac.jp

abstract

The modified Back-Propagation(BP) method for improving the convergence of the learning is proposed. Our modification consists of two changes. 1) For accelerating the learning near convergence, we use a new type of activation function instead of a sigmoid function. 2) In order to accelerate the learning speed more effectively, we consider the role of momentum term and a propose a method of setting up the momentum rate. Though the momentum term is the heuristic method for the acceleration of the convergence and is customary used, there are few discussions for the momentum term. Using the proposed method, we carry out the computer simulations for two example probrems. We confirm that both of the convergence speed and convergence rate are improved.

Actuvation function

Generally, a continuous and nonlinear activation function is needed for the BP, and the sigmoid function has been used. But when the condition of network is near convergence, the characteristic of sigmoid function leads the small error signal and the slow learning speed. To recover the defect of the characteristic of the sigmoid function, we used the following activation function.

$$y=f(x)=\begin{cases} 0 & (x\leq-a) \\ k(x+a)^2 & (-a<x\leq0) \\ -k(x-a)^2+1 & (0<x\leq a) \\ 1 & (x>a)(2ka^2=1) \end{cases}$$

Momentum term

For setting up the momentum rate, the basic concept of the BP that is the start point of the learning is output layer is one of the important factors. Here we proposed the method that the momentum rate at the output layer is set up larger than that of hidden layer. That is we try to urge accelerating at the output layer. This will connect to accelerate at the hidden layer. When the error surface seems to be long slope without V-shaped varrey, the larger step size for weight change is the more effective for accelerating. On the other and the learning in the oscillation, the momentum term operates to reduce it. Then we propose the following method.

Case1: When the signs of the learning and momentum term are the same, we use the larger α.

Case2: When the signs of the learning and momentum term are different, we use the larger α.

Conlusions

From the simulation results, we recognized the following facts. The proposed activation function accelerates the learning near convergence. The method of case1 for the moentum rate is effective at the middle stage of learning. The method of case2 is also effective at the early stage. As such we succeedin reducing the number of the learning steps at all of the stages. In addition, the convergence rate of the network is also getting better.

References

[1] D.E.Rumelhart, et al.:Parallel Distributed Processing, vol1, pp.318-362, MA: MIT Press, 1986.

[2] T.P.Vogl, et al.: "Accelerating the convergence of the Back-Propagation Method", Biol. Cybern. 59, pp. 257-263, 1988.

[3] T.Yoshikawa, Y.Kawaguchi: "A High Speed Method for the Back-Propagation Rules in Neural Network", Trans. of the IEICE, vol. J75-DII, no.4, pp. 837-840, April, 1992.

Selforganization—oral contributions

Parametrized Self-Organizing Maps[1]

Helge Ritter

Department of Information Science
Bielefeld University , D-4800 Bielefeld, FRG
email: helge@techfak.uni-bielefeld.de

Abstract

The Self-organizing Map creates a dimension reducing mapping from a feature space onto a non-linear map manifold. In the basic mapping algorithm, this mapping is discretized and for higher-dimensional map manifolds a very large number of reference vectors must be used to obtain a good resolution. To overcome this limitation, we present an approach in which the map manifold is built from a small set of *basis manifolds*. The resulting *parametrized self-organizing maps* ("PSOM") require only very few reference vectors for their construction. They can be considered as recurrent networks with a continuous attractor manifold. We illustrate the approach with the construction of a PSOM for the six-dimensional configuration manifold of a puma manipulator, embedded in a 15-dimensional variable space.

1 Introduction

The construction of a good data representation is often the decisive step in the solution of a problem. This is particularly true for learning tasks, where the capability of *generalizing* from a limited set of examples to novel instances forms the central aim. To support this objective, a representation must serve two usually conflicting goals: it must provide a representation of the data that provides the *similarity relationships* among data elements as faithfully as possible. Yet, it also should provide a *compression* onto the essential variables and project out spurious or unimportant information.

The *Karhunen-Loeve transform* or *principal component analysis* is the well-known answer that results when looking for an optimal *linear* solution to these requirements [1]. Its basic idea is to construct a low-dimensional, linear subspace that is oriented in such a way in the original space that it captures as much of the data variation as possible. The wide-spread use of this method indicates the importance of constructing good data representations.

The Kohonen *self-organizing map* (SOM) can be viewed as a non-linear extension of principal component analysis [4, 5, 9]. It replaces the linear subspace of PCA by a *nonlinear manifold* that can represent even very non-linear data distributions. The manifold is constructed by an iterative learning procedure and can be viewed as a non-linear, "topology-conserving map" of the original data space. The great success of this method results from its capability to produce representations that yield a very good compromise between efficient data compression and a sufficient faithful preservation of the underlying "data topology" to support good generalization.

In the basic SOM-algorithm, the nonlinear map manifold is represented by a discrete approximation, using a lattice of reference vectors w_s [4, 5, 9]. Originally, this approach

---

[1]This work was supported by the German Ministry of Research and Technology (BMFT), Grant No. ITN9104AO. Any responsibility for the contents of this publication is with the author.

has been motivated by the desire to model the self-organization of topographic maps in populations of discrete neurons in the brain, and it also has the virtue of being very general. However, the discrete nature of the standard SOM can be a limitation when the construction of smooth, higher-dimensional map manifolds is desired. Since the number of nodes grows exponentially with the number of map dimensions, manageably sized lattices with, say, more than three dimensions (for an application of a 3d-SOM, see e.g. [9]) admit only very few nodes along each axis direction and can, therefore, be not sufficiently smooth for many purposes where continuity is very important, as e.g. in control tasks or in robotics.

This motivates the subject of the present paper: How can one construct a smooth map manifold that approximates the map manifold that would be obtained by a standard SOM-algorithm with sufficiently many nodes, but that uses only a much smaller number of topologically ordered reference vectors?

Mathematically, this constitutes an interpolation task in a high dimensional space, using the given reference vectors as support points. Below, we will suggest a solution that can be considered as a *parametrized self-organizing map* ("PSOM"): In a PSOM, the given reference vectors no longer represent the map directly. Instead, the "localist" representation of the SOM is replaced by a more distributed scheme, in which the parametrized map is built from a set of *basis* or *prototype manifolds*, the contribution of each basis manifold to the map being controlled by one of the reference vectors. In this way, smooth mappings can be constructed from very few data samples. The resulting PSOM can be considered as a recurrent network with a continuous attractor that is given by the map manifold and with the capability to transform different variable subsets into each other. In addition, it can be adaptively modified as additional data points become available, in a manner that is very similar to the adaptation of a standard SOM.

2 Parametrized Self-organizing Maps

Each point on a d-dimensional self-organizing map can be identified by a d-tuple $\mathbf{s}$ of "map coordinates" $\mathbf{s} = (s_1, s_2 \ldots s_d)$. In the case of a standard SOM the allowed values of $\mathbf{s}$ are restricted to a discrete set $\tilde{A}$ of lattice points, and each lattice point $\mathbf{s}$ is assigned a reference vector $\mathbf{w_s} \in V$ of the data ("feature") space V. Geometrically, this can be thought of as an embedding of the discrete lattice $\tilde{A}$ in V. Each point $\mathbf{x} \in V$ is then mapped to a lattice point $\mathbf{s}(\mathbf{x}) \in \tilde{A}$ by minimizing the distance $d(\mathbf{x}, \mathbf{w_s}) = \|\mathbf{x} - \mathbf{w_s}\|$, considered as a function of $\mathbf{s}$.

A *parametrized self-organizing map* generalizes this scheme to a manifold A on which the map coordinates $\mathbf{s}$ may vary continuously. As a consequence, the discrete assignment of reference vectors $\mathbf{w_s}$ is replaced by a continuous, vector-valued function $\mathbf{w} : A \mapsto V$, $\mathbf{s} \mapsto \mathbf{w}(\mathbf{s}) \in V$, and the map image $\mathbf{s}(\mathbf{x})$ of a point $\mathbf{x} \in V$ is found by minimizing the distance $d(\mathbf{x}, \mathbf{w}(\mathbf{s})) = \|\mathbf{x}, \mathbf{w}(\mathbf{s})\|$ with respect to the now continuous variable $\mathbf{s}$ (see below).

Since we want to construct a PSOM from an associated, discrete SOM with lattice $\tilde{A}$, we will require that $\mathbf{w}(\mathbf{s}) = \mathbf{w_s}$ on the set $\tilde{A} \subset A$ of lattice locations of the underlying SOM. For all intermediate positions $\mathbf{s}$, $\mathbf{w}(\mathbf{s})$ shall interpolate smoothly between these values. These conditions will not specify $\mathbf{w}(\cdot)$ uniquely, but a convenient, sufficiently general form results with the ansatz

$$\mathbf{w}(\mathbf{s}) = \sum_{\mathbf{a} \in \tilde{A}} H(\mathbf{s}, \mathbf{a}) \mathbf{w_s}. \tag{1}$$

Here, $H(\mathbf{s}, \mathbf{a})$ is a family of basis functions, one for each lattice location $\mathbf{a} \in \tilde{A}$, that must meet the orthogonality requirement

$$H(\mathbf{a}, \mathbf{a}') = \delta_{\mathbf{a}, \mathbf{a}'} \quad \forall \mathbf{a}, \mathbf{a}' \in \tilde{A} \tag{2}$$

on the subset of lattice locations $\tilde{A}$. This will guarantee the property $\mathbf{w}(\mathbf{s}) = \mathbf{w}_\mathbf{s}$ on $\tilde{A}$. Suitable families of basis functions can be constructed in several ways, and one possibility is given in Appendix I. It is based on polynomials and has the additional property that $\sum_{\mathbf{a} \in \tilde{A}} H(\mathbf{s}, \mathbf{a}) = 1$ (i.e. the functions $H(., \mathbf{a})$ form a "partition of unity"), which ensures that the shape of the map manifold remains invariant if we change all reference vectors $\mathbf{w}_\mathbf{s}$ by a constant translation.

Geometrically, Eq. 1 can be interpreted as a construction of the map manifold A from a discrete family of "basis manifolds" given by the functions $H(\mathbf{s}, \mathbf{a})$. Fig. 1 shows three such "basis manifolds" (using the functions specified in Appendix I) for the case of a two-dimensional PSOM with an underlying 3×3-lattice $\tilde{A}$ (there will be six further basis manifolds in this case, which are obtained from 90^0-rotations of the first two manifolds in Fig.1).

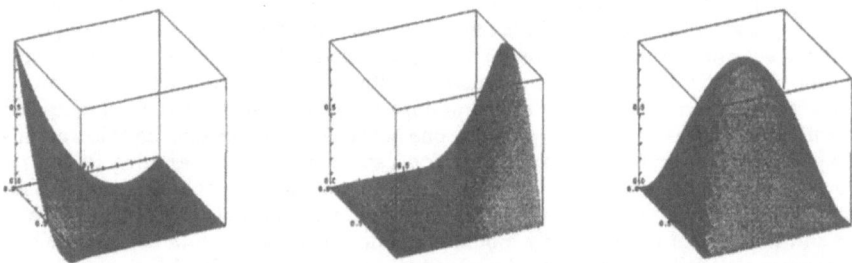

Figure 1: Three of the nine basis manifolds for a 3x3-PSOM. The remaining six basis manifolds are obtained by rotating the left two manifolds by multiples of 90^0 about a vertical axis through the center of the base square.

To determine the image $\mathbf{s}(\mathbf{x})$ of an input vector $\mathbf{x}$ in the PSOM requires minimization of the distance $d(\mathbf{x}, \mathbf{w}(\mathbf{s}))$. An explicit solution of this minimization task is not possible in general. Instead, one has to resort to an iterative procedure. The simplest approach is to use gradient descent in the variables $\mathbf{s}$, with the value $\mathbf{s}^*$ obtained from a prior best-match search on the discrete lattice $\tilde{A}$ as a starting value $\mathbf{s}(t = 0)$. Successive approximations $\mathbf{s}(t)$ to the minimizing value $\mathbf{s}(\mathbf{x})$ can then be obtained from

$$\mathbf{s}(t + 1) = \mathbf{s}(t) + \gamma(t)\mathbf{J}(\mathbf{s})^T \mathbf{P}(\mathbf{x} - \mathbf{w}(\mathbf{s}(t))) \tag{3}$$

Here, $\gamma(t) > 0$ is a step size parameter, $\mathbf{J}(\mathbf{s})$ is the Jacobian $\partial \mathbf{w}(\mathbf{s})/\partial \mathbf{s}$ evaluated at the current point, and $\mathbf{P}$ is an (optional) projection matrix (see below). Eq.3 can be viewed as a *network dynamics* for a recurrent network, however, with node activities represented parametrically by the map coordinate vector $\mathbf{s}$.

Besides mapping input vectors $\mathbf{x}$ to (usually lower dimensional) map coordinates $\mathbf{s}$, the network dynamics Eq.3 enables the PSOM also *to complete partial data vectors*. In this case, the minimized distance is taken only in the subspace V_S of specified data components (in Eq.3 this requires to set $\mathbf{P}$ to the projection matrix for the subspace V_S), and the specified components of the input vector act as a constraint for the determination of a map location $\mathbf{s}$, from which then the remaining data components are obtained by virtue of the associated vector $\mathbf{w}(\mathbf{s})$. Usually, if at least d components of $\mathbf{x}$ are specified, this constraint will be sufficient to constrain the set of compatible map locations to a single point, and the associated vector $\mathbf{w}(\mathbf{s})$ can be considered as the completion of the partial inputs[2].

---

[2]more precisely. in the non-degenerate case d specified values will constrain the possible solutions

This property is particularly useful if the map manifold represents the *graph of a function*, since it then allows to arbitrarily split the n-dimensional variable set of V into a d-dimensional subset of "independent" ("input") and $n - d$ remaining, "dependent" ("output") variables, the values of which are obtained from the value of $\mathbf{w}(\mathbf{s})$ at the fixed point $\mathbf{s}$ of Eq.3.

A final note concerns the possibility to adapt a PSOM as additional data vectors $\mathbf{x}$ become available. Stochastic gradient descent for the reference vectors $\mathbf{w_s}$ yields the adaptation rule

$$\Delta \mathbf{w_a} = \epsilon H(\mathbf{s}, \mathbf{a})(\mathbf{x} - \mathbf{w_a}), \quad \forall \mathbf{a} \in \tilde{\mathbf{A}} \tag{4}$$

where ϵ is a learning rate parameter and $\mathbf{s} = \mathbf{s}(\mathbf{x})$ is the map location computed from the network dynamics Eq.3 in response to an input $\mathbf{x}$. Eq.4 is the PSOM-equivalent for the Kohonen adaptation rule, however, with the role of the neighborhood function now taken by the basis functions $H(\mathbf{s}, \mathbf{a})$, which, in general, will vary with the lattice site $\mathbf{a}$.

3 Simulation Examples

As a first illustration, consider a 2-dimensional data manifold in $I\!R^3$ that is given by the portion of the unit sphere in the octant $x_i > 0$ ($i = 1, 2, 3$). Fig. 2, *left*, shows the "virtual net" of a SOM that tries to approximate this manifold with a 3×3-mesh. While the number of nodes could be easily increased to obtain a better approximation for the two-dimensional manifold of this example, this remedy becomes impractical for higher dimensional manifolds and the coarse approximation that results from having only three nodes along each manifold dimension is typical for these cases. However, we can use the nine reference vectors and Eq. 1 to construct a PSOM that provides a much better, fully continuous representation of the underlying manifold. Moreover, to determine the image of a data vector $\mathbf{x}$ on the map manifold, we need not specify all components of $\mathbf{x}$. Instead, in the present example the dynamics Eq.3 (with the corresponding projection matrix $\mathbf{P}$) will generate a value for any of the three variables x_i from values of the other two variables. For example, the central diagram in Fig.2 depicts the function $x_3(x_1, x_2)$ obtained from the PSOM when only x_1 and x_2 are specified (i.e. $\mathbf{P} = \mathbf{1} - \hat{e}_3 \hat{e}_3^T$ in Eq. 3), while the right diagram depicts the function $x_1(x_2, x_3)$ obtained when x_2 and x_3 are specified instead.

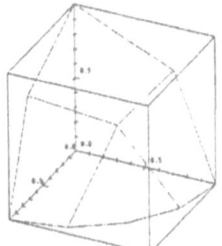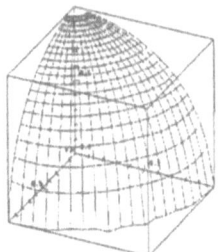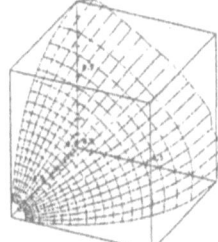

Figure 2: *Left:* Reference vectors of 3x3 SOM for part of sphere surface, *center:* mapping $x_3 = x_3(x_1, x_2)$ obtained from corresponding PSOM, *right:* same PSOM, but used for mapping $x_1 = x_1(x_2, x_3)$. Three of the nine basis manifolds used to represent the underlying PSOM are depicted in Fig.1.

This flexibility to use the mapping in different "directions" is useful in many contexts. For instance, in robotics a positioning constraint can be formulated in joint, cartesian or,

---

$\mathbf{s}$ to a zero-dimensional subset of the map manifold. This subset may contain one or several discrete alternatives

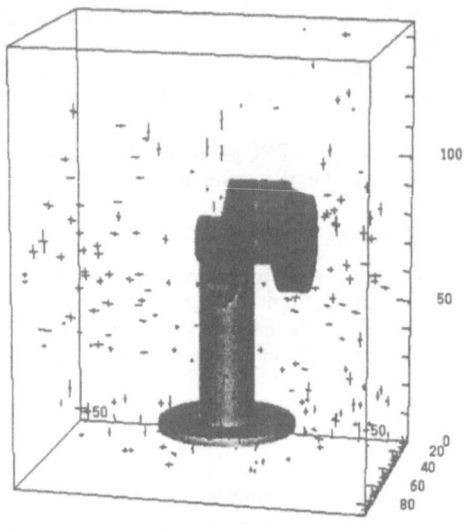

Figure 3: Positioning errors of PUMA robot arm with six degrees of freedom when the inverse kinematics transform is computed with a six-dimensional PSOM with 3^6 nodes, representing the configuration manifold of the manipulator embedded in a 15-dimensional $\vec{r}, \vec{a}, \vec{n}, \vec{\theta}$-space (for details, see text).

more general, in mixed variables, and one may need to know the respective complementary coordinate representation, requiring the direct and the inverse kinematics in the first two cases, and a mixed transform in the third case. If one knows the required cases beforehand, one can construct a separate mapping for each. However, it is clearly more desireable to work with a single network that can complete different sets of missing variable values from different sets of known values.

Of course the PSOM can yield only an approximation to the exact values. For the present sphere example, we found the RMS-errors for the reconstruction of x_i from the other two variables to be 0.011 0.039 and 0.034, for $i = 1, 2, 3$. The largest errors occur for those points where the surface is very steep (e.g., when obtaining x_3 from x_1, x_2 the slope of the sphere becomes arbitrarily large in the vicinity of the circle $x_1^2 + x_2^2 = 1$), and the error for the first case differs from that of the other two cases because the distribution of the reference vectors is not symmetric with respect to cyclic permutations of the x_i. To demonstrate the capabilities of the PSOM approach with a more realistic example, we apply it to construct an approximation to the kinematics of a PUMA 560 robot arm with six degrees of freedom. As embedding ("feature") space we use the 15-dimensional space V spanned by the variables $\mathbf{x} = (x, y, z, a_x, a_y, a_z, n_x, n_y, n_z, \theta_1 \ldots \theta_6)$. Here, x, y, z are the cartesian positions of the end effector, $\vec{a}$ and $\vec{n}$ denote the approach vector and the vector normal to the hand plane, and $\theta_1 \ldots \theta_6$ are the joint angles (for a more detailed description of these variables, see [3]). In this space, we must construct the six-dimensional map manifold that represents the configuration manifold of the robot. With three nodes per axis direction we require $3^6 = 729$ reference vectors $\mathbf{w_r} \in V$. The distribution of these vectors might have been found with a SOM, however, for the present demonstration we generate the values for the $\mathbf{w_r}$ by augmenting 729 joint angle vectors on a regular $3 \times 3 \times 3 \times 3 \times 3 \times 3$-grid in joint space with the missing $x, y, z, a_x, a_y, a_z, n_x, n_y, n_z$-components, using the forward kinematics transform equations. We then test the PSOM based on these 729 points in the inverse mapping direction. To this end, we specify desired cartesian position and orientation values $x, y, z, a_x, a_y, a_z, n_x, n_y, n_z$ at 200 randomly chosen, intermediate test points and use the PSOM to obtain the missing joint angles θ_i. The error is then computed from the difference between the specified cartesian location/orientation-values and those resulting from the angles computed with the PSOM. In Fig.3 , the resulting

xyz-positioning errors are indicated by error crosses (each error cross is centered at a tested location and has twice the x,y,z-positioning error as its respective "diameter" in the x,y and z-direction). We found a mean root square error of 3.3 cm in the position part. Values for the orientation errors and the normalized error values are summarized in Table 1.

	cartesian position	approach vector $\vec{a}$	normal vector $\vec{n}$
RMS-error:	3.3 cm	0.15	0.07
NRMS-error:	0.06	0.21	0.08

Table 1

The full 6-dimensional kinematics problem is already a rather demanding task. Most neural network applications to the robot kinematics problem have considered lower dimensional transforms, for instance [6] ($d = 5$), [9] ($d = 3$ and $d = 5$), a more recent investigation based on the backpropagation algorithm can be found in [10] ($d = 3$), and use on the order of several thousand training samples. To put the present approach in perspective with these results, we also report the application of a PSOM to the position problem alone, i.e. we consider the same PUMA robot, but with the last three joints fixed. Then we can work with a 6-dimensional embedding space of variables $\mathbf{x} = (x, y, z, \theta_1, \theta_2, \theta_3)$. Again using three nodes per axis direction, we now require *only 27 reference points* to specify the PSOM. Using the same joint ranges as in the previous six-dimensional case, we obtain a MRS-positioning error of 2.6 cm (corresponding to a NMRSE of 0.05). By using five instead of three nodes per axis (requiring 125 data samples), the error can be reduced to 0.5cm (NMRSE of 0.0097).

4 Discussion

The discrete representation that underlies the standard SOM algorithm poses a problem for applications where higher dimensional map manifolds together with continuity of the mapping are required. The proposed "parametrized self-organizing map" may offer a way to overcome this difficulty. It offers two interesting benefits:

First, it requires only a very small set of topologically ordered reference vectors, from which it constructs a map manifold in parametric form that is a good approximation to the manifold that would be obtained with the standard SOM, provided a sufficiently huge number of reference vectors could be used in that case. This can be viewed as a form of *learning from very few examples*: if for a small number of data samples their topological ordering is known, the full PSOM can be constructed. We have demonstrated this for the case of the kinematics mapping of a robot arm with six and with three degrees of freedom. In the latter case, only 27 samples are required to construct the entire mapping with an accuracy in the percent range. Regarding the dimensionality of the space, even the 729 points for the six-jointed robot arm must be considered to be a small number: when learning a general transform with d parameters for a hypercube region, one can hardly avoid to require at least one data sample to specify the transform at each of the 2^d extreme corners of its range. To capture the proper "curvature" of the transform within its definition range, one should have at least on further, intermediate point per axis direction. This raises the minimal number of points to 3^d, thus reaching the value used for the 6d-kinematics transform.

Second, the resulting mapping offers the flexibility to declare any subset of variables of the embedding space that is sufficient to uniquely specify a point on the map manifold as the independent "input" variable set, from which the values of the remaining variables can be constructed via Eqs.3 and 1. This can be viewed as a kind of nonlinear, "associative" completion of a partial variable set and distinguishes a PSOM from other, "feedforward-type"-approaches, such as orthogonal series expansions (which would result if an equation of the form Eq.1 could be written directly for the input variables of the problem at hand[3]), or RBF- and Spline-Networks [8, 2]. Instead, the map manifold of the PSOM plays the role of a smooth *attractor manifold* and a *dynamics* for the pattern completion is given by the gradient rule Eq.3. This strongly resembles the situation in *associative memory networks*. However, there is an important difference: most associative memory networks are intrinsically based on a *discrete set of point attractors* and do not allow the specification of attractors that are arbitrary smooth manifolds, as they commonly occur in robotics or in control applications.

Of course, these benefits also come at a price, namely the increased computational costs associated with the attractor dynamics specified by Eq.3. This is a common disadvantage of all recurrent networks, if compared to the simpler but less flexible feed forward nets. For large numbers of nodes, this may become a problem on a sequential computer, however, the problem is much less severe when parallel hardware is used, since the required computations are fully parallel in the number of nodes. A second point concerns the proper choice of the step size parameter $\epsilon(t)$. If the values $\gamma(t)$ are chosen too large, the dynamics may diverge, if they are too small, unnecessarily many steps are consumed. Appendix II specifies a step scaling procedure that works well if the data have been scaled such that they have comparable variances along all dimensions.

To judge the merits of the PSOM-approach will require its exploration in a wider range of contexts, a task to which the present contribution can only be a first step. In particular, the small number of required data samples to construct a mapping makes the use of PSOMs in larger network architectures an interesting option. For instance, using PSOMs in cascaded networks [7] might yield a particularly promising combination of the abilities of both schemes to provide good generalization from extremely few data samples.

Appendix I: Choice of Basis Functions

Below, we specify a suitable family of polynomial basis functions $H(\mathbf{s}, \mathbf{a})$ that can be constructed whenever the SOM-lattice $\tilde{A}$ is rectangular, i.e. a cartesian product $\tilde{A} = A_1 \times A_2 \times \ldots A_d$ of d sets $A_i = \{a_{i1} \ldots a_{in_i}\}$ of n_i *knot points* a_{ij} each. In this case, suitable functions $H(\cdot, \mathbf{a})$ are given by

$$H(\mathbf{s}, \mathbf{a}) = \prod_{i=1}^{d} f_{ik}(s_i) \tag{5}$$

where each f_{ij} is a polynomial of degree $n_i - 1$ of a single variable s obeying

$$f_{ij}(a_{ik}) = \delta_{jk}, \tag{6}$$

and is given by Lagrange's formula

$$f_{ij}(s) = \prod_{k, k \neq j} \frac{s - a_{ik}}{a_{ij} - a_{ik}}. \tag{7}$$

NOTE: This choice of polynomials is suitable for small numbers n_i of knot points. Larger values of n_i will require modifications (e.g., introduction of exponential damping factors) that, however, are beyond the scope of the present paper.

---

[3]Note, however, that there is a special situation in which the PSOM reduces to such a case, namely when the map coordinates s can be directly interpreted as the input variables of the task at hand.

Appendix II: Simulation Parameters

All simulations reported in Sec.3 were based on the polynomials $H(\mathbf{s}, \mathbf{a})$ for $d = 2$, $n_1 = n_2 = 3$ and equidistant knot points $a_{ij} = (j-1)/2$. The step size parameter $\gamma(t)$ was chosen such that $\|\Delta\mathbf{s}\| = \alpha d(t)/d(0)$, where $d(t) = \|\mathbf{Px} - \mathbf{Pw}(\mathbf{s}(t))\|$ is the euclidean distance between the current manifold point and the input vector, restricted to the subspace $\mathbf{P}V$ of specified input variables (for $\mathbf{x} = \mathbf{w_r}$ we have $d(0) = 0$, but in this case $\mathbf{s} = \mathbf{r}$ and no gradient descent needs to be performed). Termination condition was either $d(t) < 0.001$ or $t = 250$, whichever occurred earlier.

For the sphere example, a value of $\alpha = 1$ was used. The reference vectors were $\hat{e}_3, (\hat{e}_1 + \hat{e}_3)/\sqrt{2}$, $\hat{e}_1, (\hat{e}_1 + \hat{e}_2 + \hat{e}_3)/\sqrt{3}, \hat{e}_1 \cos 22.5^0 + \hat{e}_2 \sin 22.5^0$, $(\hat{e}_1 + \hat{e}_2)/\sqrt{2}$ plus those obtained by interchanging $\hat{e}_1$ and $\hat{e}_2$.

For the robot kinematics, the value $\alpha = 0.1$ and the puma geometry data in [3] were used. Joint angle ranges were $\theta_1 = 90^0 \pm 45^0$, $\theta_2 = -15^0 \pm 45^0$, $\theta_3 = 135 \pm 45^0$ and $\theta_i = \pm 45^0$ for $i = 4\text{-}6$. Test vectors were chosen randomly with uniform distribution from the resulting joint space hypercube.

References

[1] Devijver P., Kittler J., *Statistical Pattern Recognition*. Prentice Hall 1982.

[2] Friedman J.H. (1992) *Adaptive Spline Networks* In: *Advances in Neural Information Processing Systems 3*, eds. R.P.Lippmann at al., pp.675-683, Morgan Kaufman Publishers, San Mateo, CA.

[3] Fu K.S., Gonzalez R.C., Lee C.S.G. (1987), *Robotics*, McGraw-Hill, New York.

[4] Kohonen, T. (1984), *Self-Organization and Associative Memory*, Springer Series in Information Sciences 8, Springer, Heidelberg.

[5] Kohonen, T. (1990), *The Self-Organizing Map*, in *Proc. IEEE* **78**, pp. 1464–1480.

[6] Kuperstein, M. (1988) *Neural Model of Adaptive Hand-Eye Coordination for Single Postures*. Science 239:1308-1311.

[7] Littmann, E., Ritter, H. (1993), *Generalization Abilities of Cascade Network Architectures*, in *Advances in Neural Information Processing Systems 5*, eds. C.L. Giles, S.J. Hanson, J.D. Cowan, Morgan Kaufman Publishers, San Mateo, CA.

[8] Moody, J., Darken, C. (1988). *Learning with Localized Receptive Fields*, in *Proc. of the 1988 Connectionist Models Summer School*, Pittsburg, pp. 133-143, Morgan Kaufman Publishers, San Mateo, CA.

[9] Ritter, H., Martinetz, T., Schulten, K. (1992). *Neural Computation and Self-organizing Maps*, Addison-Wesley, Reading, MA.

[10] Yeung, D., Bekey, G. (1993) *On Reducing Learning Time in Context Dependent Mappings*. IEEE Transactions on Neural Networks, Vol.4, pp.31-42.

Population Dynamics on the Basis of Vector Quantization: A Method for Auto-Association and Classification

Claus Hillermeier, Niels Kunstmann, Paul Tavan
Institut für Medizinische Optik, Ludwig-Maximilians-Universität München
Theresienstraße 37, W-8000 München 2, FRG

Abstract

Recurrent, nonlinear, auto-associative memories like, e.g., Hopfield nets form a classic paradigm of neural network architecture. They are characterized by attractor type dynamics. Using vector quantizers as building blocks we present in this paper a most simple recurrent network of that type. However, due to its simplicity our model enables a thorough qualitative as well as analytical understanding. We show that the network is actually capable of performing a cluster analysis and hierarchical classification of data. Thus it also qualifies as a tool for unbiased statistical data analysis.

1 Introduction

Vector quantizers (VQ) are self-organizing tools for data compression. Given a set of data vectors $\vec{x}$ statistically distributed according to a probability density $P(\vec{x})$, a VQ distributes a chosen number N of reference vectors $\vec{w}_r$ in data (feature) space. The resulting point density $D(\vec{w}_r)$ is a slightly deformed version of $P(\vec{x})$. An incoming data vector $\vec{x}$ is then represented by the best-fitting reference vector $\vec{w}_r$.

As is suggested by the term 'k-means-clustering' for the Linde-Buzo-Gray VQ, VQ are also deemed to be clustering methods. Clustering denotes a self-organized classification scheme of data when the data-generating mechanism and, therefore, the a priori distributions of data classes C_j and conditional probabilities are unknown. Considering vector quantization as a clustering method would imply to interpret the reference vectors $\vec{w}_r$ as class prototypes.

The clustering performance of VQ like the LBG-algorithm suffers, however, from severe drawbacks. (i) The number of clusters does not adapt itself to the structure of the $P(\vec{x})$-distribution, but is pre-determined by the number N of chosen reference vectors $\vec{w}_r$. (ii) Well separated but weakly populated data clusters are often neglected by VQ. (iii) Clusters which are not compact clouds in feature space but have complicated shapes cannot be classified correctly by VQ.

Nevertheless, one can devise a classification procedure employing a VQ which meets the requirements of clustering in statistical data analysis. Instead of attempting to make the VQ directly cluster the data, this approach uses a VQ just as the scene on which the actual classification process takes place in the form of a recurrent dynamics leading to auto-association.

In a first step, the feature space is discretized by a VQ on a fine scale, i.e. employing a "large" number N of reference vectors $\vec{w}_r$ (which has to be small, of course, as compared to the number of data vectors $\vec{x}$). The vectors $\vec{w}_r$ are interpreted as "virtual positions" of neurons r in feature space, i.e., as centers of their respective gaussian-shaped receptive fields. The actual classification of an incoming feature vector $\vec{x}$ is performed by recurrently processing the activities $a_r(T)$ of the neurons

r starting with a pattern $a_r(T = 0)$ which has been induced by $\vec{x}$. Tavan et al. [2] earlier proposed a related concept and realized it by a two-layer network.

In the following we present a one-layer neural network as an improved and simpler realization of the VQ-based classification concept [3, 4]. During the training period, the new network not only learns the virtual positions $\vec{w}_r$ of the neurons r according to a VQ-algorithm, but also lateral synaptic connections $C_{rr'}$ from neuron r' to neuron r. In the classification process the initial activity distribution $a_r(T = 0)$ triggered by the incoming data vector $\vec{x}$ is subsequently updated according to the equation

$$\vec{a}(T + 1) = \vec{F}(\mathbf{C}\vec{a}(T)), \tag{1}$$

where $\vec{a}$ denotes the activity vector composed of the a_r and $\vec{F}$ is a nonlinear function. The learning rule for the weight matrix $\mathbf{C}$ and the nonlinearity of $\vec{F}$ guarantee that the activity population remains *sparse* during the dynamical process. By considering the virtual positions $\vec{w}_r$ of the neurons r in feature space, the activities a_r on the neural layer can be depicted as activities of "neurons" at virtual positions $\vec{w}_r$ belonging to the grid $W = \{\vec{w}_r | r = 1, \ldots, N\}$ in feature space. The construction of $\mathbf{C}$ and $\vec{F}$ also ensures that the cloud of appreciably active "neurons" at $\vec{w}_r$ remains compact and is attracted towards the closest local maximum of the $P(\vec{x})$-distribution. After this cloud has reached its stable end configuration, the prototype vector $\vec{x}_p$ which the system assigns to the incoming data vector $\vec{x}$ and which is identical to the closest $P(\vec{x})$-maximum can be read out by taking the average of the virtual positions $\vec{w}_r$ weighted by the activities a_r.

Thus, we obtain a partition of feature space into association classes each of which consists of the surroundings of a local maximum of the $P(\vec{x})$-distribution. The graining of this partition and therefore the scale up to which the $P(\vec{x})$-structure is resolved is controlled by a parameter. The limit of the resolution is given by the discretization density $D(\vec{w}_r)$ of the underlying VQ. Hence, the described classification procedure meets the requirements of unbiased statistical data analysis.

2 Auto-associative memory using population coding

The construction of the learning rules for the lateral $C_{rr'}$-connections was guided by the following requirements: (i) The rule should generate a connection structure which is local with respect to the distance between the *virtual* positions $\vec{w}_r$ and $\vec{w}_{r'}$ of the neurons in feature space. (ii) The connections emanating from neuron r' should direct the virtual flow of neural activity in feature space towards the nearest $P(\vec{x})$-maximum.

The lateral learning triggered by a training signal $\vec{x}$ is based on a gaussian activity response $a_r(\vec{x}) = \exp(-(\vec{x} - \vec{w}_r)^2/2\rho^2)$. The underlying rule reads

$$\delta C_{rr'}(\vec{x}) \sim \left[a_r(\vec{x})a_{r'}(\vec{x}) - C_{rr'}a_{r'}(\vec{x})\left(\sum_{r''}a_{r''}(\vec{x})\right)\right] \tag{2}$$

and consists of a Hebbian term responsible for the locality and a growth-limiting term $\sim -C_{rr'}$ which gives rise to the required asymmetry of the $C_{rr'}$-connections. From the stationary point of the learning dynamics,

$$C_{rr'} = \frac{\langle a_r a_{r'} \rangle_{P(\vec{x})}}{\sum_{r''} \langle a_{r''} a_{r'} \rangle_{P(\vec{x})}}, \tag{3}$$

one can infer that $\sum_r C_{rr'} = 1$ holds, i.e. the $C_{rr'}$-connections are normalized with respect to the axon-tree.

On the basis of the learnt lateral connections the initial activity $a_r(T = 0) \equiv a_r(\vec{x})$ shall be transformed dynamically. The governing rule consists of two parts:

a) Activity diffusion: Because of the locality of the $C_{rr'}$-connections, the linear matrix dynamics

$$a_r(T + 1) = \sum_{r'} C_{rr'} a_{r'}(T) \tag{4}$$

results in a diffusion process whose temporal and spatial continuum limit is described by a Fokker-Planck equation in feature space [1] . The potential landscape $U(\vec{w})$ where the diffusion takes place is determined by the $P(\vec{x})$-distribution through $U(\vec{w}) \sim -\ln P^{2\alpha+1}(\vec{w})$ (α denotes the distortion exponent of the discretization density $D(\vec{w}) = P^{\alpha}(\vec{w})$ of the VQ). The axon-tree normalization guarantees conservation of the total activity $A(T) = \sum_r a_r(T)$.

b) Nonlinear, competitive focussing of the activity: Due to the activity diffusion the linear dynamics results in a stationary activity distribution which is independent of the initial stimulus (data vector) $\vec{x}$. During the relaxation towards the equilibrium state, however, the center of gravity of the activity distribution performs a gradient descent in the potential $U(\vec{w})$ as long as this initially narrow distribution remains narrow as compared to the structures of $P(\vec{x})$. Hence, in order to enable association one has to add a nonlinear term which limits the diffusive broadening of the activity distribution and therefore obviates the transition into the global equilibrium state of the linear dynamics.

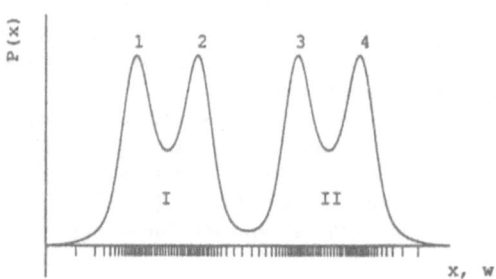

Figure 1: Probability density $P(x)$ consisting of four hills with maxima at $x_1 = 0.22$, $x_2 = 0.38$ and $x_3 = 0.62$, $x_4 = 0.78$ and its representation by a VQ $\{w_r | r = 1, \ldots, 100\}$

In order to enable an analytical understanding of the system this additional term has been chosen to

$$\Delta a_r^{nonlin} = k \sum_{r'} \frac{a_r [a_r - a_{r'}]}{A^2} a_{r'} . \tag{5}$$

Without destroying conservation of the total activity A this prevents the diffusive spreading of activity into the whole net. The activity distribution remains localized in feature space (population coding as a special form of sparse coding). After relaxation in the local potential, the prototype vector $\vec{x}_p$ encoded by the stationary distribution can be read out according to $\vec{x} = \sum_r \vec{w}_r a_r / \sum_r a_r$. By means of the parameter k governing the relative strength between diffusive broadening and nonlinear narrowing of the activity distribution the width of the final stationary distribution in feature space and therefore the resolution of the $P(\vec{x})$-structures can be controlled.

For a one-dimensional example, namely the four hill $P(x)$-distribution depicted in Figure 1, Figure 2 shows the stationary activity distributions in feature space which have been obtained by numerical simulations starting with a neural activation evoked by a stimulus at $x = 0.45$. The center of gravity of the activity distribution for the different values of the parameter k marks the associated prototypes. In case of very weak nonlinearities, i.e. small k-values, one obtains a four hill activity distribution which can easily be identified as the equilibrium distribution of a Brownian particle diffusing in a potential given by $-\ln P(x)$. The center of gravity of the activity distribution is here the average value $< x >_{P(x)}$. For growing k the diffusion is restricted to region I of the probability

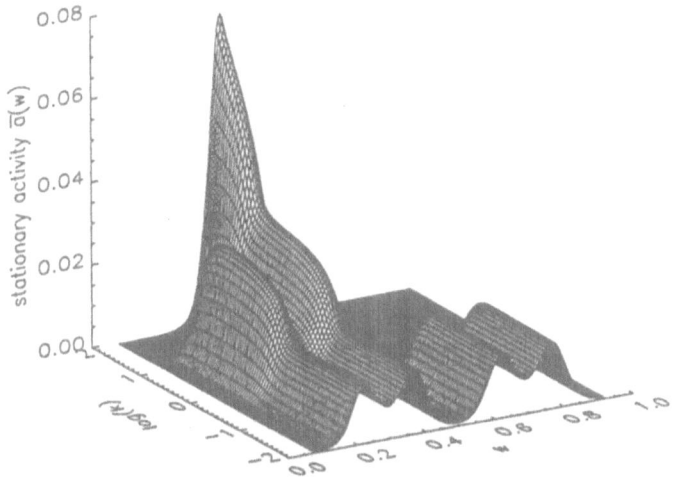

Figure 2: Stationary activity distributions $\bar{a}(w)$ for the nonlinear dynamics in their dependence on the nonlinearity parameter k; the underlying $P(x)$-distribution was that of Figure 1; $x = 0.45$ was used as initial stimulus.

distribution. The associated prototype is the mean value $\frac{1}{2}(x_1+x_2)$. Finally, in the presence of strong nonlinearities only the prototype x_2 next to the initial stimulus is associated. This k-dependent resolution of the $P(x)$-clusters demonstrates the network's capability of hierarchical classification of the initial stimulus [4].

References

[1] B. Rabus. Hypothesenbildung in phasenkodierten Assoziativspeichern. Diplomarbeit, Physik-Department T30, Technische Universität München, April 1992.

[2] P. Tavan, H. Grubmüller, H. Kühnel. Self-organization of assoziative memory and pattern classification: Recurrent signal processing on topological feature maps. *Biological Cybernetics*, 64:95–105, 1990.

[3] C. Hillermeier, N. Kunstmann, P. Tavan Ein hypothesenbildender Assoziativspeicher in phasenkodierter Netzwerkarchitektur. *Statusseminar Neuroinformatik 1992*, Maurach, 1992

[4] C. Hillermeier, B. Rabus, P. Tavan. manuscript in preparation

Vector Quantization with a Growing and Splitting Elastic Net

Bernd Fritzke

International Computer Science Institute
1947 Center Street, Suite 600
Berkeley, CA 94704-1105
USA
fritzke@icsi.berkeley.edu

Abstract

A new vector quantization method is proposed which generates codebooks incrementally. New vectors are inserted in areas of the input vector space where the quantization error is especially high until the desired number of codebook vectors is reached. A one-dimensional topological neighborhood makes it possible to interpolate new vectors from existing ones. Vectors not contributing to error minimization are removed. After the desired number of vectors is reached, a stochastic approximation phase fine tunes the codebook. The final quality of the codebooks is exceptional. A comparison with two well-known methods for vector quantization was performed by solving an image compression problem. The results indicate that the new method is significantly better than both other approaches.

1 Introduction

The huge amount of data in many technical applications (e. g. HDTV) makes it necessary to compress the original data before sending them over limited bandwidth channels. Usually correlations and structure in the data are removed to express the contained information using less storage. *Lossless* compression techniques allow a perfect reconstruction of the original information, but are rather limited in their ability to reduce the amount of data. *Lossy* compression techniques are much more effective but lead to a *reconstruction error* when the information is "decompressed".

One specific method of lossy data compression is *vector quantization* (VQ). The object of VQ are data which can be described as n-dimensional vectors. A limited set of *representative vectors*, the so-called *codebook*, is used to achieve compression. A given data vector is mapped onto the nearest codebook vector. Only the *number* of that codebook vector is transmitted. Assuming that the codebook is known to the receiver, he can reconstruct the original information to some degree by choosing the corresponding codebook vector.

The main problem in VQ is to find a codebook such that the loss of information caused by coding and reconstruction of the data is small. Equivalently, the mean distance between data vector and nearest codebook vector – the so-called *quantization error* (which *is* the reconstruction error in this case) – has to be minimized.

Many methods have been proposed to find a suitable codebook. The most well-known is the LBG-algorithm from Linde, Buzo, and Gray [6]. Recently also Kohonen feature maps have been used for VQ [7]. In this contribution a new method is introduced which is related to a recently proposed self-organizing network [2] and outperforms the LBG-algorithm as well as the Kohonen feature map w.r.t. the quality of the generated codebooks.

2 Principle

We assume from structured data in the sense that our data items are not distributed evenly over the n-dimensional space but rather concentrate in certain areas of R^n. This kind of distribution is sometimes denoted as *mixture distribution*. It is evident that for uniformly distributed data VQ techniques invariably lead to relatively large quantization errors.

The model we propose consists of a set A of formal units. Every unit $c \in A$ has an associated n-dimensional *representative vector* $w_c \in R^n$. The set W of all representative vectors is the current codebook. On the set of units we define a symmetric binary *neighborhood* relation $N \subset A \times A$ having the property that every unit has at most two neighbors:

$$(card\{x \in A|(x,c) \in N\} \in \{0,1,2\})(\forall c \in A). \tag{1}$$

In other words the set A is organized as a collection of one or more chains each of which consisting of one or more units.

The general idea of our method is to construct the codebook incrementally by *interpolating* new codebook vectors from existing ones. Interpolation is always done among topologically neighboring units, i.e. units a and b with $(a,b) \in N$. After each interpolation the current codebook is adapted with a fixed number of vectors from the original data.

3 Adaptation

One adaptation step can be described as follows:

1. Choose an input vector ξ from the original data.

2. Determine the best-matching unit (bmu) s with the nearest representative vector by

$$\|\xi - w_s\| = \min_{c \in A} \|\xi - w_c\| \tag{2}$$

3. Update the local quantization error of the bmu:

$$\Delta \tau_s = \|\xi - w_s\|^2. \tag{3}$$

4. Adapt the bmu and its direct neighbors:

$$\Delta w_s = \varepsilon_b(\xi - w_s) \tag{4}$$
$$\Delta w_n = \varepsilon_n(\xi - w_n) \quad (\forall n, (n,s) \in N) \tag{5}$$

Thereby ε_b and ε_n are constant parameters and $\varepsilon_b \gg \varepsilon_n$ holds. Furthermore for a unit c we denote with τ_c a local variable used for summation of quantization error. Throughout the paper the notation $\Delta x = y$ stands for $x^{\text{new}} = x^{\text{old}} + y$.

Since not only the best-matching unit, but also its direct neighbors are adapted, topological neighbors in the structure tend to have similar representative vectors. At the same time the reference vectors approach regions with many input signal.

4 Insertion

Always after a fixed number λ of adaptation steps a new unit is inserted. The local error variables of the units indicate whether the corresponding representative vectors lie in regions of R^n with large mean quantization error. It seems to be a promising strategy to insert new units in those regions. Therefore, the unit q with maximum error variable is determined by

$$\tau_q = \max_{c \in A} \tau_c. \tag{6}$$

Two possible cases can occur: the unit q may have topological neighbors or it may be single due to previous deletion of its neighbors (see below).

If q has one or two direct neighbors, the direct neighbor f with the largest error variable is determined. A new unit r is inserted into the chain between q and f. The representative vector of r is interpolated from the vectors of q and f according to

$$w_r = w_q + \gamma(w_f - w_q) \qquad \text{whereby} \qquad \gamma = \frac{\tau_f}{\tau_q + \tau_f}. \tag{7}$$

By interpolating among topological neighboring units chances are good that the new representative vectors also lie in regions with many data vectors and do further reduce the total quantization error. The purpose of the formula for γ is to position the new unit more closely (in input vector space) to the unit with the larger error variable. If both q and f have the same error value, γ has the value 0.5. If the error value of f approaches zero, then so does γ leading in the extreme case to identical representative vectors of q and r.

The new unit r has a local error variable, too. This is initialized with a certain non zero value. The idea behind this is that r, if it had existed already since the beginning, would probably have got a certain number of those signals which have actually been mapped onto q and f. Therefore, it is justified to reduce the error variables of q and f by a certain amount and initialize the error variable of r with the sum of these amounts. The strength of the reduction depends on the distance of the representative vectors and thus on γ. It is defined through

$$\Delta\tau_q = -\frac{(1 - \gamma)}{2}\tau_q, \qquad \Delta\tau_f = -\frac{\gamma}{2}\tau_f, \tag{8}$$

and the new cell r gets an initial error value according to

$$\tau_r^{\text{initial}} = -(\Delta\tau_q + \Delta\tau_f). \tag{9}$$

so that the total sum of all error variables stays equal. One should note that the amount which is subtracted from the error variables of q and f is indirectly proportional to the distance of the new unit in input vector space and can be at most 50%.

If the unit is single – i.e. has no topological neighbors – a new unit r is created and made a topological neighbor of q. The representative vector of r is copied from q. Half of the error value of q is subtracted and given to r as an initial error value.

5   Deletion

We have assumed a non-uniform distribution of data vectors. It can thus occasionally happen that through the interpolation method described above representative vectors are created with values very far away from all possibly occurring input signals. These vectors do not contribute to the reduction of quantization error and, therefore, the corresponding units are removed. The problem is to decide whether a given vector fulfills this condition. In the case of a finite set of possible data vectors (e.g. in image compression) the following strategy suffices:

- The input signals are chosen cyclically in a fixed order.
- Each unit which has not been best-matching unit for a complete cycle is removed.

This is easily realizable by numbering the input signals and marking at every adaptation step the best-matching unit with the current signal number. To be independent of any order of the input signals, it seems appropriate to generate a fixed *random* sequence of all input signals and use this sequence cyclically.

The removal of units splits the chain into subchains which will eventually also split. However, if the desired codebook size is small compared with the total number of training signals (and otherwise vector quantization does not make sense), insertions of units are much more frequent than deletions. The advantage of deletions lies in the fact that at the end of the process every unit is really used, i.e. has some input signals which are mapped onto it.

6 Stochastic Approximation

Once a codebook of the desired size has been generated it is fine tuned by *stochastic approximation* [8]. This denotes a number of adaptation steps with a strength α decaying

$$\text{slowly:} \quad \sum_{t=1}^{\infty} \alpha(t) = \infty, \quad \text{but not too slowly:} \quad \sum_{t=1}^{\infty} \alpha^2(t) < \infty. \quad (10)$$

One specific sequence of parameters fulfilling the above conditions is the harmonic series

$$\alpha(t) = 1/t \quad \text{or its more general form} \quad \alpha(t) = (c+1)/(c+t) \quad (11)$$

for a non-negative constant c.

For the stochastic approximation phase of our algorithm the neighborhood relations are completely ignored. Every adaptation step consists thus of:

1. Generation of an input signal ξ

2. Determination of the best-matching unit s

3. Adaptation of the best-matching unit by

$$\Delta w_s = \alpha(t)\varepsilon_b(\xi - w_s) \quad (12)$$

7 Simulation Results

To investigate the quality of the new algorithm, a large number of simulations have been performed. The new method has been compared with two other VQ techniques, the LBG-algorithm [6] and the Kohonen feature map [5].

The test problem was to find a suitable codebook for the well-known "Lenna"-image. We used a version of this picture consisting of 480×480 pixels each one coded by an 8 bit gray value.

To achieve a good overall compression ratio, we did not apply the VQ-methods directly to the test image. Instead we followed the approach of Schweitzer e.a. [7] and used a multi-layer perceptron (MLP) with linear activation functions to pre-process the picture. We divided the picture in square blocks of 8×8 pixels and trained an MLP with 64 input units, 16 hidden units, and 64 output units to reproduce these blocks at the output units when they were presented to the input units.

The small number of hidden units forces the network to find an efficient coding of the information in order to be able to reproduce it at the output layer. Baldi and Hornik [1] have proved that in such cases an MLP with linear activation functions performs a transformation equivalent to the optimal Karhunen-Loeve transform. By representing each of the 3600 blocks by the 16-dimensional vector of hidden unit activations, the dimensionality of the original data can be reduced by a factor four.

The resulting vector set was given as input data to the three VQ methods. With every method codebooks of size 256 were determined. Every codebook vector can be identified by an 8 bit number. The whole image thus can be transmitted with only 0.125 bits per pixel (bpp) compared to 8 bits/pixel for the original image.

For both the Kohonen method and the new approach a large number of parameters has been tested in preliminary runs. The final test runs were then made with those parameters giving the best results in the preliminary runs. Different random order of the chosen input vectors provided variations in the results.

The LBG method was always run until convergence. The initial codebook vectors were chosen at random from the input vectors.

Once the codebooks were determined the trained MLP was used to reconstruct the image. By comparison with the original image mean square error (MSE) and signal-to-noise ratio (SNR) were obtained. Moreover, subjective image quality was rated.

The results of our simulations are shown in the following table:

	# of runs	MSE	SNR (dB)	bits per pixel
MLP only	1	29.79	31.32	≥ 2.000[1]
LBG	100	$156.68 \pm 2.2\%$	$24.11 \pm 0.4\%$	0.125
Kohonen map	200	$151.09 \pm 3.7\%$	$24.27 \pm 0.7\%$	0.125
growing elastic net	200	$138.72 \pm 2.1\%$	$24.64 \pm 0.4\%$	0.125

It can be seen that Kohonen's model produced slightly better results than the LBG method. The growing and splitting elastic net, however, reliably produces codebooks with a quantization error more than 10 percent below the error of Kohonen's model. The running time of the two self-organizing networks was about equal in our experiments whereas the LBG-algorithm took considerably longer (about three times). This has, however, to be investigated further.

Good values for the simulation parameters of our model were $\varepsilon_b = 0.24$, $\varepsilon_n = 0.014$, $\lambda = 100$. The stochastic approximation was performed with two to four cycles through all 3600 input signals. The constant c (see eqn. 11) always had the value 2000.

The subjective quality of the reconstructed images correlates often rather well with the quantitative error measures as can be seen in fig. 1. This is, however, not always the case since humans tend to find errors in certain regions (e.g. faces) more disturbing than in others.

8 Discussion

We presented a new vector quantization method which incrementally builds up a codebook through interpolation. The new method produces significantly better results for an image compression problem than two other methods it was compared with.

The incremental character of the new method also has the important advantage that the size of the codebook can be increased until the quantization error falls below a given bound. For the other methods a complete restart with a larger codebook would be necessary if the final error was still above that bound.

The presented network is a variant of a more general class of algorithms the underlying principle of which is a growth process controlled by some insertion criterion. In this paper the insertion criterion has been based on the quantization error. Another interesting possibility is to locally count the input signals which leads to networks which can estimate the probability density of the input vectors (see [3]). Recently also a combination of a growing network with a radial basis function network has been proposed using the classification error to guide the growth process [4].

9 References

[1] Baldi, P. & K. Hornik, "Neural Networks and Principal Component Analysis: Learning from Examples Without Local Minima," *Neural Networks* 2 (1989), 53–58.

[2] Fritzke, B., "Growing cell structures – a self-organizing network in k dimensions," in *Artificial Neural Networks II*, I. Aleksander & J. Taylor, eds., North-Holland, Amsterdam, 1992, 1051–1056.

[3] Fritzke, B., "Kohonen feature maps and growing cell structures – a performance comparison," in *Advances in Neural Information Processing 5*, L. Giles, S. Hanson & J. Cowan, eds., Morgan Kaufmann Publishers, San Mateo, CA, 1993.

[4] Fritzke, B., "Growing Cell Structures – a self-organizing network for unsupervised and supervised learning," International Computer Science Institute, TR-93-026, Berkeley, 1993.

[5] Kohonen, T., "Self-Organized Formation of Topologically Correct Feature Maps," *Biological Cybernetics* 43 (1982), 59–69.

---

[1]This depends on the number of bits used to encode a hidden unit activation. If 8 bits are used, the rate is 2 bpp.

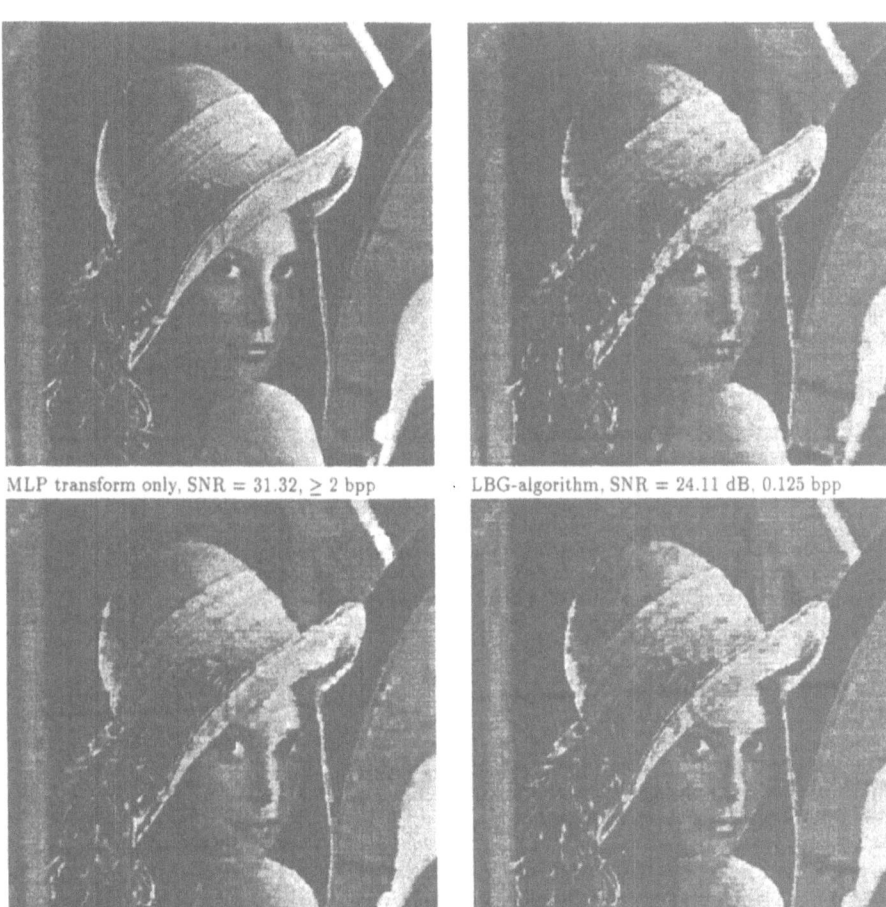

MLP transform only, SNR = 31.32, ≥ 2 bpp   LBG-algorithm, SNR = 24.11 dB, 0.125 bpp

Kohonen feature map, SNR = 24.28 dB, 0.125 bpp   Growing elastic net, SNR = 24.61 dB, 0.125 bpp

Figure 1: Typical reconstructed pictures produced by the various algorithms. The MLP-transform gives an image nearly indistinguishable from the original. The VQ-methods differ in their ability to properly encode this this image at a given bit rate (0.125 bpp).

[6] Linde, Y., A. Buzo & R.M. Gray, "An algorithm for vector quantizer design," *IEEE Transactions on Communication* COM-28 (1980), 84–95.

[7] Schweizer, L., G. Parladori, G.L. Sicuranza & S. Marsi, "A fully neural approach to image compression," in *Artificial Neural Networks*, T. Kohonen, K. Mäkisara, O. Simula & J. Kangas, eds., North-Holland, Amsterdam. 1991. 815–820.

[8] Wasan, M.T., *Stochastic Approximation*, Cambridge Univ. Press, 1969.

Learning Topology-Preserving Maps Using Self-Supervised Backpropagation

Arnfried Ossen
Department of Computer Science
Technical University of Berlin
Germany
<ao@cs.tu-berlin.de>.

Abstract

Self-supervised backpropagation is an unsupervised learning procedure for feedforward networks, where the desired output vector is identical with the input vector. For backpropagation, we are able to use powerful simulators running on parallel machines. Topology-preserving maps, on the other hand, can be developed by a variant of the competitive learning procedure. However, in a degenerate case, self-supervised backpropagation is a version of competitive learning. A simple extension of the cost function of backpropagation leads to a competitive version of self-supervised backpropagation, which can be used to produce topographic maps. We demonstrate the approach applied to the Traveling Salesman Problem (TSP).

1. Introduction

Topology-preserving maps can be developed by a variant of the competitive learning procedure. Self-supervised backpropagation is an unsupervised learning procedure for feedforward networks, in which the desired output vector is identical with the input vector. For backpropagation we are able to use powerful simulators running on parallel machines. In a degenerate case, self-supervised backpropagation is a version of competitive learning. This relationship between competitive learning and self-supervised backpropagation has been pointed out by Hinton [1].

In the simplest version of competitive learning, an input vector is represented by the weight vector of a single winning unit. The learning procedure then moves the weight vector in the direction of the input vector. The amount of change is proportional to the distance between weight and input vector. For small enough changes, this is equivalent to performing gradient descent in the sum-squared difference of weight and input vectors.

To see the relationship, we describe a degenerate case of self-supervised backpropagation. Suppose we add a layer of linear units with the same number of units as the input layer on top of the layer of competing units. We use the input vector activations as the desired output activations. We constrain the weights from the input units to the competing units to be identical to the corresponding weights from the competing units to the output layer. We keep the "winner-take-all" nonlinearity for the competing units and use backpropagation to determine the weight changes for the weights from the output layer to the competing layer. Then all error derivatives for the weights of non-winning units are zero and the derivatives of the winning unit will be proportional to the difference between the input/output and weight vectors. Thus, the above degenerate case of self-supervised backpropagation is equivalent to the simple competitive learning procedure.

2. Topology-Preserving Self-Supervised Backpropagation

One important version of competitive learning is concerned with learning topographic maps or feature mapping [3]. In this variant, there is some geometrical arrangement of the competing units. Usually they are placed on a line or in a plane. We then call a mapping between inputs and competitive units a *feature mapping* if it preserves neighborhood relations, i.e. if nearby input vectors are mapped to competitive units with nearby locations on the line (or in the plane).

A feature mapping can be learned by a competitive learning procedure if the winner-take-all non-linearity is changed to a "winner-take-more" rule. In this case, not only does the winning unit get its weights changed, but so do units close to the winner. The amount of change is usually defined as a Gaussian function of the distance between the winning unit and the unit in question.

The winner-take-more rule cannot be realized using self-supervised backpropagation with a winner-take-all nonlinearity at the hidden units. Since there is still a single winning unit, the error derivatives for the weights of non-winning units remain zero and non-winning units will never get an update of their weights. The neighborhood preserving feature mapping cannot develop. Instead, a smoothed version of the winner-take-all rule is needed, in which the winning patterns of activation will have a single peak at the winning unit and decreasing activation at neighboring units. We would also like to lift the equality constraints on the weights. One way to accomplish this is to view the winner-take-all nonlinearity as an extension of the usual backpropagation cost function.

For standard backpropagation we have:

$$E = \sum_i (t_i - o_i)^2 \qquad (1)$$

where t_i is the desired activation (target) for output unit i and o_i the activation of output unit i.[1]

Instead of using a winner-take-all nonlinearity at the hidden units, we simply define the needed winner-take-more patterns as targets t_j for the *hidden units*. We can now use standard backpropagation, but get an extended error function:

$$E = \frac{1}{2}\sum_i (t_i - o_i)^2 + \frac{1}{2}\sum_j (t_j - o_j)^2 \qquad (2)$$

The resulting error derivatives (assuming on-line mode and a sigmoidal nonlinearity) are then:

$$\frac{\partial w_{ij}}{\partial t} \propto -\frac{\partial E}{\partial w_{ij}} = o_i(1 - o_i)(t_i - o_i)o_j \equiv \delta_i o_j \qquad (3)$$

$$\frac{\partial w_{jk}}{\partial t} \propto -\frac{\partial E}{\partial w_{jk}} = o_j(1 - o_j)[(\sum_i \delta_i w_{ij}) + (t_j - o_j)]o_k \qquad (4)$$

Now, we can run the system on a standard backpropagation simulator. The only extension necessary is to find an appropriate winner-take-more pattern that can be used as a target pattern at the hidden units. Following the competitive learning procedure, we choose the closest (in squared error) winner-take-more pattern from a set of possible target patterns.

2.1. Teacher Forcing

A potential shortcut in the forward propagation phase is to use *teacher forcing*. Williams and Zipser [6] have shown its value for recurrent networks. Since a non-recurrent net is just a special case of a recurrent net, teacher forcing can be applied here, too.

---

[1] For notational convenience, index i is always used for output units, j for hidden units and k for input units

The idea in teacher forcing is to replace the activation value of any unit by its target value if the target value is available. This must be done *after* the error derivatives for the units in question have been calculated. In the backpropagation phase, the partial derivatives for any forced unit have to be set to zero. In the above case, the equations simplify considerably.

$$\frac{\partial w_{ij}}{\partial t} \propto -\frac{\partial E}{\partial w_{ij}} = o_i(1 - o_i)(t_i - o_i)t_j \tag{5}$$

$$\frac{\partial w_{jk}}{\partial t} \propto -\frac{\partial E}{\partial w_{jk}} = o_j(1 - o_j)(t_j - o_j)o_k \tag{6}$$

The resulting equations are exactly those of two one-layer feedforward nets, where the targets of the first are the inputs of the second net. However, the targets of the first net are a result of a competition: the closest winner-take-more pattern is chosen.

3. Application to the Traveling Salesman Problem

The Traveling Salesman Problem is an *NP-hard* problem, and even very restricted versions of the TSP are *hard* problems. One way to cope with the intractability of the general TSP is to approximate the solution. Many global or local heuristics are known [2]. More recently, neural network approaches were suggested for the TSP. A summary can be found in [5]. We compare our approach to these algorithms.

In order to apply the self-supervised backpropagation feature maps to the TSP, we use a strategy known from research on the Kohonen feature map. We restrict the problem to a random uniform distribution of city coordinates in a unit square and place the competing units on a ring. For N cities, the Kohonen approach needs far more competing units than there are cities. In our approach, the number of hidden units can actually be less than the number of cities. Only the number of competing patterns must be large enough. Empirically, it is sufficient to use only $N/2$ hidden units. We generate $4N$ winner-take-more patterns using a bell-shaped function. We then use backpropagation to minimize the error function.

Figures 1 to 4 demonstrate a typical run. In a first phase the patterns organize into a ring. In a second phase, the local irregularities become integrated. After about 100 epochs, a good approximation has developed.

In general, it is difficult to make precise statements about the quality of a TSP approximation. Since the solution is usually not known, we have to compute a lower bound on the tour length. The goodness of an algorithm is then its average behavior in comparison to the lower bound. Our approach has not yet been tested against a lower bound. Instead, we compare it to other Neural Network algorithms for the TSP. A detailed description of a benchmark study for the TSP can be found in [5]. This testbed consists of 50-, 100- and 200-city TSPs in the unit square. Table 1 shows the average tour lengths for *elastic nets (EN)*, *genetic algorithms (GA)* and *simulated annealing (SA)* from the benchmark study, and the results for the self-supervised backpropagation feature map (FM).

N	EN	GA	SA	FM
50	5.62	5.58	6.80	5.80
100	7.69	7.43	8.68	8.05
200	11.14	10.49	12.79	11.60

Table 1: Average tour length for several neural net algorithms

4. Discussion

The search for "good" internal representations is an important element for neural network learning procedures. Usually it is an indirect search, in which internal representations are defined by connection weights. However, it is possible to make this search more explicit. Krogh et al. [4] have suggested an extended error function which is an explicit function of the internal representations. The learning procedure then uses gradient descent for the two layers of connections *and* the hidden unit activations. The hidden unit activations "relax" to their optimal values. This approach achieves improved results on simple mapping problems.

We have shown how the search for internal representations can be used to realize a variant of competitive learning called feature mapping. Our approach also uses an extended cost function. But in contrast, we define targets for the hidden unit activations. An appropriate target is found using *competition*.

The use of backpropagation as learning procedure made it possible to take full advantage of existing parallel simulators in terms of execution time and implementation effort.

5. Acknowledgements

I would like to thank Jerome Feldman, Phil Kohn and Steve Omohundro for helpful comments.

References

[1] Geoffrey E. Hinton. Connectionist learning procedures. Technical Report CMU-CS-87-115, Carnegie-Mellon University, Pittsburgh PA 15213, 1987.

[2] D. S. Johnson. Local optimization and the traveling salesman problem. In *Proceedings 17th Colloquium on Automata, Languages and Programming*, pages 446–461, 1990.

[3] Teuvo Kohonen. *Self-Organization and Associative Memory*. Springer Verlag, 1989.

[4] Anders Krogh, G. I. Thorbergsson, and John A. Hertz. A cost function for internal representations. In David S. Touretzky, editor, *Advances in Neural Information Processing Systems II*, pages 733–740. Morgan Kaufmann Publishers, San Mateo, California, 1990.

[5] Carsten Peterson. Parallel distributed approaches to combinatorial optimization: Benchmark studies on traveling salesman problem. *Neural Computation*, 2(3):261–269, 1990.

[6] Ronald J. Williams and David Zipser. Experimental analysis of the real-time recurrent learning algorithm. *Neural Computation*, 1(1):87–111, 1989.

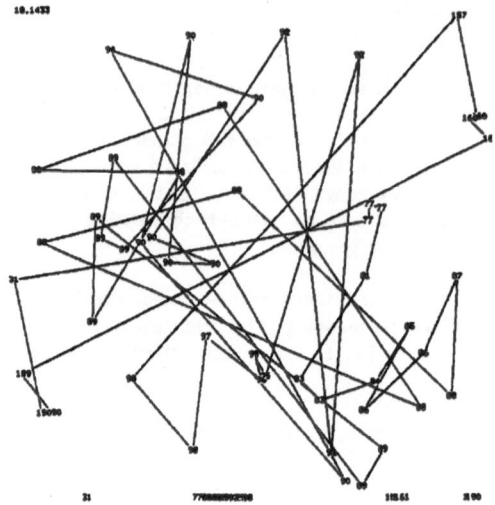

Figure 1: 50 city TSP, after 0 epochs

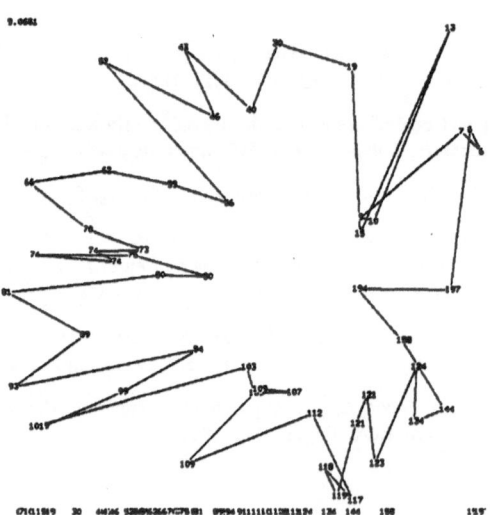

Figure 2: 50 city TSP, after 10 epochs

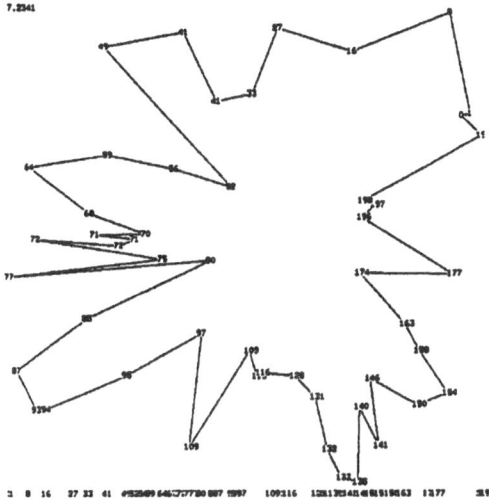

Figure 3: 50 city TSP, after 20 epochs

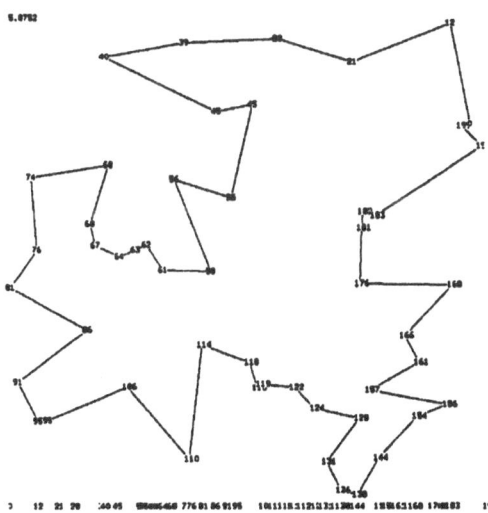

Figure 4: 50 city TSP, after 100 epochs

Selforganization
—poster contributions

A Multiassociative Memory for Control

Knut Möller
University of Bonn
Department of Computer Science
Römerstr. 164
D-5300 Bonn, FRG

Abstract

In control often many alternatives have to be considered in a specific state to reach a certain goal. Connectionist systems, that are currently being investigated in manipulator control ignore the issue of generating multiple outputs for a single input pattern. Current pattern association models have a principle problem with indeterminate many-to-many mappings. During training the learning system faces inconsistent data and has no possibility to resolve conflicting information in a satisfying manner.

In this paper we propose a multiassociative memory and discuss its utility in robot control applications.

INTRODUCTION

Complex real world tasks such as manipulator control and motion planning [16, 3] require powerful tools providing for adaptivity and selforganization. Despite some encouraging results [4, 10] current connectionist learning systems are limited in applicability to 1. small and 2. well behaved tasks. In earlier papers we proposed [1, 12, 10] special techniques realizing a selforganizing decomposition of a task. The resulting partitions are small and therefore easier to learn. Although using a similar approach in this paper we want to focus on the second point. By well behaved tasks we think of many-to-one mappings where associative memory models such as linear associators [17], Hopfield networks[2] or feed-forward networks [14] are applicable. For noisy or partial retrieval they proved to be useful. But many other tasks exist e.g. in motor learning indescribable within the (classical) associative framework. The problem of indeterminacy arises e.g. in the context of kinematic redundancy and in the domain of trajectory generation, but as well in many other domains like natural language processing or game playing where words are to be associated with its possible meanings or positions should be mapped to possible next moves.

Traditional symbolic or numerical approaches yield explicit representations of alternatives via stored links or implicitly through enumerative algorithms. But often it is not possible to find a closed mathematical formulation or the enumeration of solutions might be too expensive. On the other hand, linked lists in random-access memory do not scale well and due to missing generalization a possibly very large amount of memory is needed.

Connectionist attempts (restricted to discrete ranges) addressing this issue rely on a special coding scheme. The *power set approach* [9] captures any possible subset of the range set by allocating one output node to any range element. This localist approach is limited to discrete spaces and is again storage intensive.

To bypass this storage problem in [6] an enumerative approach using recurrent networks is proposed. But again it is limited to finite[1], discrete spaces and, more important in real time applications, it is by far not clear in which order items should be enumerated.

In our former control experiments [13, 16] we used an iterative technique introduced by [18, 7] of inverting a differentiable connectionist many-to-one mappings. But this process is cumbersome, time-consuming and there exists no systematic recall mechanism.

In an attempt to address this issue we propose a multiassociativ memory (MAM) capable of responding with one or more outputs from a set of possible outcomes to an input.

---

[1]From a practical point of view limited to very few items.

This paper starts with an introduction to the general architectural idea, identifying several weaknesses of the approach. In the following we present a number of interesting improvements, yielding a powerful tool.

LEARNING MANY-TO-MANY MAPPINGS

Associative memories implement mappings from each pattern of an input domain (e.g. $[0,1]^n$) to exactly one output pattern in a specified range (e.g. $[0,1]^m)^2$. They realize a one-to-one or a many-to-one mapping.

For obvious reasons associative memories fail to store many-to-many mappings. But these mappings appear quite frequently in learning control where a learning systems needs to capture the underlying functionality from uncontrolled observations. Current approaches simply focus on one solution[3] ignoring others. Thereby flexibility is lost, e.g. under different constraints a complete, new learning cycle has to be performed.

In our approach the complexity of many-to-many mappings is reduced, by learning finitely many unique versions instead, which all together approach the underlying distribution of the ambiguities (cf. Fig. 2). The general architecture of the MAM (cf. Fig.1) presented here is based on Kohonen's feature map (FM)[5, 15]. Feature maps separate the input domain into convex regions (Voronoi tesselation). In our proposed MAM the FM expands in the cross product of input domain ($[0,1]^n$) and output range ($[0,1]^m$), i.e. weightspace is of dimension $\tilde{n} = n + m$. In

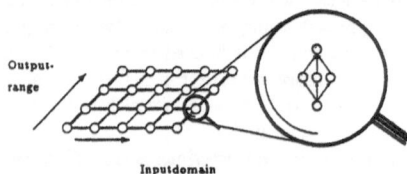

Output-range

Inputdomain

Figure 1: Architecture of the MAM. The FM unfolds in the crossproduct of input domain and range.

order to use this partitioning for supervised learning, each FM-unit has an associated (small) backpropagation net – still mapping from the input domain[3] to output range. Thus the task of each

---

[2]Consider the task to be learned as an unknown mapping $f : [0,1]^n \longrightarrow [0,1]^m$.

[3]which is restricted to the projection of the corresponding receptive field to the input domain.

FM-unit is to gate the input to its specific backpropagation network.

The system is now trained in the following way:
1. Each pattern pair (p,t) consisting of input p and target t is shifted by the winning FM unit to the corresponding backpropagation net which is then trained as usual.
2. The FM unfolds roughly proportional to pattern density in the extended weight space by using the usual FM-weight-update rule[15].

Now learning complexity depends on two facts: the degree of ambiguity and the complexity of the underlying functions. Given a fixed number of resources, there is a fundamental conflict of distributing them: Should the accuracy of the approximation be improved or the number of versions increased?

Error-driven decomposition [1, 10] is now used for *shifting resources between these two entities* – The backpropagation networks are moved due to both sources of error. In Fig.2(a) an artificial many-to-many mapping is shown. Fig.2(b) gives a typical distribution of FM-units after training. It is quite obvious, that the a priori topology of the map does not allow for an optimal distibution. We therefore incorporated the Neural gas learning rule [8] whose results[4] with and without error-driven decomposition are shown in Fig.3 respectively. The improvements are obvious; especially at the easy parts on the left and right the usual FM update rule concentrates too many ressources. These are missing for approximating the ambiguous part of the mapping (in the middle). But there is even a much more important benefit: the underlying topology can be approximated in a staightforward manner by counting patterns falling between two units. This topological information is extremly valuable during recall of patterns allowing for a fast search to solve the "reachability" question and a controlled interpolation even in the output dimension.

During learning both input and output are available. Thus, both FM- and backpropagation training is possible. In the recall phase of the MAM there is only input available. It is not straightforward to recall learned information, which is a consequence of the nondeterministic behavior of the underlying mapping. In principle, for a given input we want to find those ap-

---

[4]Topology learning is already included.

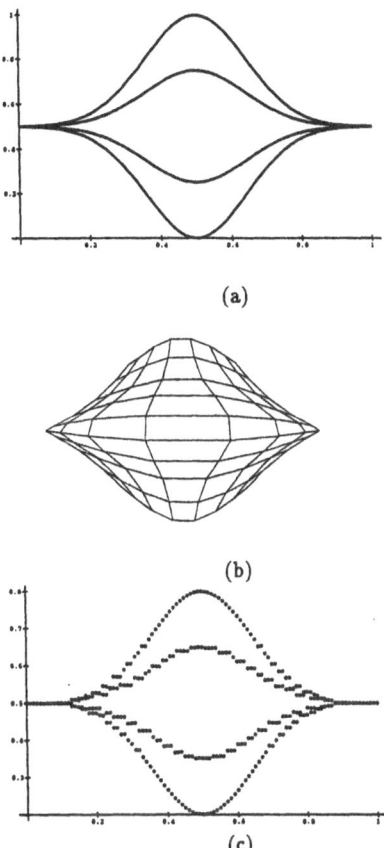

(a)

(b)

(c)

Figure 2: (a) Discrete problem space with 4 solutions; (b) Distribution of FM-units in $\mathcal{I} \times \mathcal{O}$-space; after 10000 pattern presentations;(c) resulting approximation

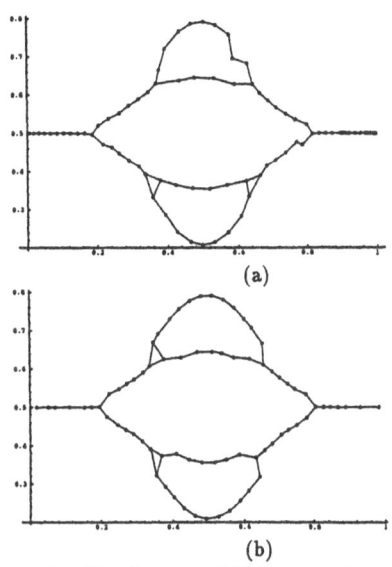

(a)

(b)

Figure 3: Distribution of FM-units using the Neural gas approach with topology learning (a) without (b) with error-driven decomposition;

proximators whose corresponding receptive fields intersect with the input-hyperplane – these are possible candidates for valid outputs.

Although a hierarchy of possibilities has been developed [10] we just want to describe an approach successfully used in our experiments.

Given an input each of the N backpropagation nets computes its output. Now the FM is evaluated for each of the resulting N input/output-pairs. If a FM-unit wins the competion whose associated bp-net produced the output the output is a valid solution. Otherwise it is removed from the candidate set. However, this is a heuristic since it might happen that FM units whose receptive fields have a non empty intersection with the input hyperplane get lost due to a bad approximation of the associated BP-network. Using an

ordering of the I/O-pairs e.g. activity level of FM unit the search for a "sufficient" solution is fast. The evaluation of a FM of size N is performed in $O(N)$ on a serial machine and in $O(1)$ on a parallel computer. The compution of a BP-net depends on the number of weights which are of the order[5] $O(n^2 + h^2 + m^2)$ (serial) and on the number of layers ($O(1)$) on a highly parallel machine. In the worst case we thus get a complexity $O(N^2(n + h + m)^2)$ (serial) which reduces – using the above mentioned ordering with the observation that on average a constant number of evaluations is neccessary – to $O(N+(n+h+m)^2)$.

DISCUSSION

We presented a novel scheme for multiassociative storage and recall. The crossproduct of input domain and output range of a function is decomposed into convex regions, each of which is learned by a separate backpropagation network. The idea is to produce finitely many versions covering the ambiguities best. Resources are shifted between the degree of ambiguity and the complexity of the functions. This view of many-to-many mappings is restricted in the sense that it

---

[5]$n, m =$ Input/Outputdimension. $h =$ number of hidden units

does not allow to generate infinitely many but finitely many outputs to each input. The benefits of this method were visualized by learning an artificial many-to-many mapping.

Subsequently, MAM was extended by the Neural gas method. The creation of a problem dependant topology now allows under certain restrictions to generate infinitely many outputs by interpolation. Further extensions include a growing number of approximators that are placed where they are needed most. This method and other experiments with more detailed descriptions (learning curves etc.) are included in a technical report [11].

Despite of the benefits demonstrated in the text the following disadvantages are to be mentioned: a) dimensionality of the FM might get large. b) recall of the system is slower than with usual associative memories[6]. The delay is depending on the special recall mechanism used. c) The system might not behave stable unless the learning rate is faded. d) As usual with gradient descent there is the danger of getting stuck in local minima.

Acknowledgments: The author wishes to thank D. Fox, V. Heinze for implementing part of the simulation system used for the experiments. M.Contzen, M. Eichhardt, M. Faßbender, S. Thrun and the research group on "connectionist systems and robotics" at Bonn University and the GMD contributed to this work with discussions and help with the robot experiments.

References

[1] D. Fox, V. Heinze, K. Möller, S.B. Thrun and G. Veenker. Learning by error-driven decomposition. In O. Simula, editor, *Proceedings of International Conference on Artificial Neural Networks (to appear)*, Amsterdam, 1991. Elsevier Publisher.

[2] J. J. Hopfield. Neural networks and physical systems with emergent collectiv computational abilities. *Proceeding of the National Acadamy of Science (USA)*, 79:2554–2558, April 1982.

[3] M. I. Jordan. Generic constraints on unspecified target constraints. In *Proceedings of the First International Joint Conference on Neural Networks, Washington, DC*, pages 217–226, San Diego, 1989. IEEE, IEEE TAB Neural Network Committee.

[4] M. Kawato. Computational schemes and neural network models for formation and control of multijoint arm trajectory. In W.T. Miller, R.S. Sutton, and P.J. Werbos, editors, *Neural Networks for Control*, pages 197 – 228. The MIT Press, Cambridge, MA, 1990.

[5] T Kohonen. *Self-Organization and Associative Memory*. Springer, Berlin New York, 1984.

[6] J.F. Kolen and J.B. Pollack Multiassociative Memory. In *Proceedings of the 13. Annual Conference of the Cognitive Science Society*, , 1991.

[7] Linden, A. and Kindermann, J. Inversion of Multilayer Nets. In *Proceedings of the First International Joint Conference on Neural Networks, Washington, DC*, San Diego, 1989. IEEE, IEEE TAB Neural Network Committee.

[8] T. Martinetz. *Selbstorganisierende neuronale Netwerkmodelle.* DISKI Bd.14, infix–Verlag, St.Augustin, 1992.

[9] J. L. McClelland and D. E. Rumelhart. *Explorations in the Microstructure of Cognition. Parallel Distributed Processing. Vol. II.* MIT Press, 1986.

[10] K. Möller. *Adaptive Roboterkontrolle mit konnektionistischen Systemen.* DISKI Bd.7, infix–Verlag, St.Augustin, 1991.

[11] K. Möller. *A Demand-driven Multiassociative Memory.* Tech.Report, Inst.f.Informatik, April 1993.

[12] K. Möller and S. Thrun. Task modularization by network modulation. In J. Rault, editor, *Neuro-Nimes '90*, pages 419–432, 1990.

[13] K. Möller and S. Thrun. Adaptive Roboterkontrolle. *Wirtschaftsinformatik*, 33(5),408–419, 1991.

[14] D. E. Rumelhart, G. E. Hinton, and R. J. Williams. Learning internal representations by error propagation. In D. E. Rumelhart and J. L. McClelland, editors, *Parallel Distributed Processing. Vol. I + II.* MIT Press, 1986.

[15] H. Ritter, T. Martinez, and K. Schulten. Topology-conserving maps for learning visuo-motor-coordination. *Neural Networks*, 2(3):159–168, 1989.

[16] S. Thrun and K. Möller. *Active exploration in Dynamic Environments.* In J. Moody, S. Hanson and R. Lippmann, editors, *Advances in Neural Information Processing Systems 4* , San Mateo, November 1991. Morgan Kaufmann.

[17] G. Widrow and M. E. Hoff. Adaptive switching circuits. Institute of Radio Engineers, Western Electronic Show and Convention, Convention Record, Part4, 1960.

[18] R. J. Williams. Inverting a connectionist network mapping by backpropagation of error. In *8th Annual Conference of the Cognitive Science Society*, Hillsdale, NJ, 1986. Lawrence Erlbaum.

---

[6]From a cognitive point of view it is quite natural, that recall is slower than recognition as in our MAM

Phase Transitions in Self–Organized Feature Maps

R. Der   M. Herrmann

Universität Leipzig, Inst. f. Informatik, D–O7010 Leipzig, Germany,

der, mherrman@informatik.uni–leipzig.dbp.de

Abstract

We present an analytical study of symmetry–breaking phenomena in Kohonen's self–organized feature map (SOFM) that occur due to a topological mismatch between the input space and the neuron setup. We give a microscopic derivation for the time dependent Ginzburg–Landau equations describing the behavior of the order parameter close to the critical point where the phase transition takes place.

1 Introduction

The Kohonen algorithm [?] provides a simple model of the self–organized generation of an internal representation of the environment in living systems. [?] For practical applications it may be used as a nonlinear generalization of principal component analysis which plays an important role in many problems of data analysis, above all for data in high dimensional data spaces. Considering from the point of view of physics the synaptic vectors as coordinates of a set of hypothetical particles with a complex dynamics created by the Kohonen algorithm. The complexity of this system is reflected by phenomena like spontaneous symmetry breaking (analogous to nonequilibrium phase transitions) and the occurrence of metastable states with partial ordering or the emergence of criticality under certain conditions [?]. From the point of view of systems theory Kohonen's algorithm models a system with competition and cooperation which may serve as a generic though degenerate example of the emergence of collective ordering phenomena in more complicated systems.

Concerning practical applications these effects may well interfere with the intentions of using the feature maps as a reliable tool for topology preserving mapping. On the other hand, if better understood they might be exploited for more sophisticated applications. In this sense, there is a great lack of theoretical study into the details of the map evolution. Pioneering work on this task has been performed by Ritter, Martinetz & Schulten [?]. They predicted a phase transition to occur due to dimensional conflicts between input space and net topology and calculated within the linear theory the critical value for the corresponding control parameter which measures the strength of the dimensional conflict.

In the present paper we present by the means of the time dependent Ginzburg–Landau formalism a general approach for the evaluation of order parameters in the SOFM. In particular, by including nonlinear terms we describe the behavior around the critical point where the phase transition takes place. The critical index and a modulating prefactor were obtained. The Ginzburg-Landau approach is quite general but we study here the simple case of mapping a two–dimensional input space onto a one–dimensional neuron chain. In input space the data points are assumed to be distributed homogeneously in a box of height s which plays the role of the control parameter measuring the strength of the dimensional mismatch.

2 Langevin picture of Kohonen's algorithm

Let us consider a set of N neurons situated at sites $\vec{r} \in \mathcal{N} \subseteq \mathcal{Z}^d$ ($\mathrm{card}\{\mathcal{N}\} = N$) of a d–dimensional lattice with lattice constant unity. Each neuron is connected with the input units by synaptic connectivities $w_{\vec{r}} \in \mathcal{R}^n$. The inputs to the network are given by random stimuli $\vec{v} \in \mathcal{R}^n$. The vectors of synaptic weights $\vec{w}_{\vec{r}}$ may be viewed as 'pointers' into the input space. Kohonen's learning rule is given by

$$\Delta \vec{w}_{\vec{r}}(t) = \epsilon h_{\vec{r},\vec{s}}(\vec{v} - \vec{w}_{\vec{r}}) \tag{1}$$

where $\vec{s}$ denotes the winner neuron defined as the one with the best match (in the sense of the Euclidian distance) between input $\vec{v}$ and synaptic vector $w_{\vec{r}}$, and $h_{\vec{r},\vec{s}}$ is a Gaussian neighborhood function measuring the degree by which the neurons in the vicinity of the winner may participate in the learning step. Eq. (??) represents a data driven algorithm since the $\vec{w}$ are incremented by the presentation of a datum (or input stimulus) $\vec{v}$.

Our approach is based on rewriting (??) into the form of a generalized Langevin equation

$$\Delta \vec{w}_{\vec{r}} = \epsilon \vec{\Phi}_{\vec{r}}(\mathbf{w}) + f_{\vec{r}}(\vec{v}), \tag{2}$$

where $\mathbf{w}$ is the combined vector $(w_{\vec{r}} | \vec{r} \in \mathcal{N})$, $\vec{\Phi}_{\vec{r}}(\mathbf{w}) = -\langle h_{\vec{r},\vec{s}}(\vec{w}_{\vec{r}} - \vec{v})\rangle$, the noise term is $f_{\vec{r}}(\vec{v}) = \epsilon h_{\vec{r},\vec{s}}(\vec{v} - \vec{w}_{\vec{r}}) - \epsilon \vec{\Phi}_{\vec{r}}(\mathbf{w})$, and $\langle \cdots \rangle = \int d^d v \, P(\vec{v}) \cdots$.

Manageable expressions are obtained close to the stationary state of the (Markovian) stochastic process $\mathbf{w}(t)$ by Taylor expanding $\vec{\Phi}$ in terms of the deviations $\vec{u}_{\vec{r}} = \vec{w}_{\vec{r}} - \vec{w}_{\vec{r}}^0$, where $\vec{w}_{\vec{r}}^0$ denotes the average over the stationary state distribution. In the symmetry breaking phenomenon to be discussed below the stationary state is not unique. Then $\vec{w}_{\vec{r}}^0$ is understood as the unstable asymptotic state average the symmetry of which is broken. By introducing the Fourier transforms (mode amplitudes)

$$u_{\vec{k}} = \frac{1}{\sqrt{N}^d} \sum_{\vec{r}} \exp(i\vec{k}\vec{r}) u_{\vec{r}}$$

where N is the number of neurons and the Taylor expansion of Φ we rewrite eq. ?? as

$$\Delta u_{\vec{k}m} = \epsilon \sum_{m'} B_{mm'}(\vec{k}) u_{\vec{k}m'} + \epsilon \sum_{\vec{q},m',m''} C_{mm'm''}(\vec{k},\vec{q}) u_{\vec{k}-\vec{q}m'} u_{\vec{q}m''} + f_m(\vec{k},\vec{v}) \tag{3}$$

We dropped terms of order higher than two. A more detailed investigation [?] yields that terms of third order are much smaller than the present terms. Explicit expressions for the coefficients $C_{ijk}(k,q)$ are given in the Appendix. The nonlinear coupling of the amplitudes of all modes appeared as the essential mechanism of the self–organized structure formation.

3 Ginzburg–Landau theory of feature maps

For a discussion of eq. (??) we study the case of the mapping of a two–dimensional rectangle onto a one–dimensional chain of neurons, where $\vec{k}, \vec{r}$ now are scalars k, r and the matrix $\mathbf{B}$ is two-dimensional, cf. Ref. [?]. If the height s of the rectangle is sufficiently small ($s \ll \sigma$), i.e. if the input space is pseudo one–dimensional, the chain is mapped onto the median which cuts the rectangle symmetrically into two parts and the virtual neuron sites are $\vec{w}_r^0$. For $s > s_c$ however, this situation is not stable any longer, i.e. the symmetric mapping is broken spontaneously and the chain begins to fold into the input space.

In the case considered here the matrix $\mathbf{B}$ is diagonal with elements $B_{11} = \lambda_{\parallel}(k), B_{22} = \lambda_{\perp}(k)$ [?] From the explicit expressions, $\lambda_{\perp}(k)$ is seen to change sign at $s = s_c \cong \sqrt{\frac{2}{3}}e \cong 2.02\sigma$ at a critical value of $k = k_0 \cong \sqrt{2}/\sigma$. Hence the mode with wave numbers k_0 or $-k_0$ would grow unrestricted for $s > s_c$ if the nonlinear term would be missing in eq. (??). For $s > s_c$ there is a band of k–values corresponding to unstable modes. Due to the nonlinearities we get a competition between these modes. From the derivations given below we justify a single mode ansatz as a solution for eq. (??), i. e. we assume that there is just one mode winning the competition which means

$$|u_{k_0 2}| \gg |u_{ki}| \tag{4}$$

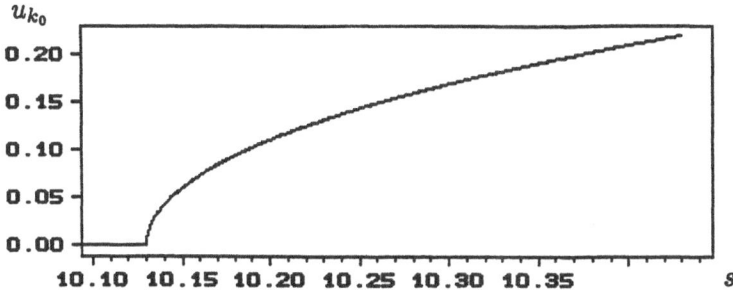

Fig. 1: *The order parameter $|u_{k_0 2}|$ corresponding to the amplitude of the winning mode vs. the control parameter s (the height of the rectangular input space) measuring the strength of the dimensional conflict.*

valid for $i = 1, 2$ if $k \neq k_0$ and for $i = 1$ if $k = k_0$, and k_0 being the wave number of the winner mode. Using (??) and the properties of C (see Appendix) we find from (??) the equation of motion for the amplitude of the winning mode in leading order [?]

$$\Delta u_{k_0 2} = -\epsilon \lambda_\perp (k_0) u_{k_0 2} + \epsilon \gamma (k_0) u_{2k_0, 1} u_{k_0 2}^* + f(k_0, \vec{v}) \qquad (5)$$

where the relations $u_{2\pi - q} = u_{-q} = u_q^*$ were used and

$$\gamma = C_{212} (k_0, 2\pi - k_0) + C_{221} (k_0, 2k_0)$$

Moreover we assumed that the contribution to the coupling term involving the $k = 0$ mode can be neglected. This is justified since $u_{k,1}$ represents for $k = 0$ just the longitudinal motion of the "center of gravity" of the neurons which may be assumed to be in rest.

By means of the stationary state solution of the accompanying equation for $u_{2k_0, 1}$ we obtain in leading order

$$\Delta u_{k_0 2} = -\epsilon \lambda_\perp (k_0) u_{k_0 2} + \epsilon g (k_0) |u_{k_0 2}|^2 u_{k_0 2} + F(k_0, \vec{v}) \qquad (6)$$

where

$$g (k_0) = \frac{C_{122} (2k_0, k_0) \gamma (k_0)}{\lambda_\| (2k_0)}$$

and $F(k_0, \vec{v})$ is a modified noise term. Eq. (??) can be given a standard form by assuming ϵ to be small so that a continuous time description is appropriate. Moreover, introducing a free energy as

$$\mathcal{H} (q) = \frac{1}{2} \lambda_\perp (k_0) |q|^2 - \frac{1}{4} g (k_0) |q|^4 \qquad (7)$$

where $q = u_{k_0 2}$ we may write eq. ?? as

$$\dot{q} (t) = -\frac{\partial}{\partial q^*} \mathcal{H} (q) + F(t) \qquad (8)$$

which in fact may be viewed as the standard form of the time dependent Ginzburg–Landau equation describing the time evolution of the amplitude of the winning mode which serves as the order parameter in the symmetry breaking phenomenon.

This equation is valid for values of $s \approx s_c$, i.e. just about the critical value of s where the dimensional conflict begins to make itself felt by a spontaneous symmetry breaking. For these values of s we find for the eigenvalue $\lambda_\perp$ of the transversal mode (cf. [?])

$$\lambda_\perp = (s - s_c) Q (k_0) \qquad (9)$$

where

$$Q(k_0) = -\frac{\sqrt{8\pi}\epsilon\sigma}{3N}s_c(1-\cos k_0)\exp(-k_0^2\sigma^2/2) \tag{10}$$

We obtain for the amplitude of the mode by means of eqs. (??), (??), and (??) the following expression

$$|u_{k_0 2}| = \quad 0 \qquad \text{for } s \leq s_c \tag{11}$$
$$|u_{k_0 2}| = \quad \alpha\sqrt{s-s_c} \quad \text{for } s > s_c, \tag{12}$$

where

$$\alpha = \sqrt{\frac{Q(k_0)}{g(k_0)}} = \sqrt{\frac{Q(k_0)\lambda_\parallel(2k_0)}{C_{122}(2k_0,k_0)\gamma(k_0)}} \tag{13}$$

The above theoretical expressions are typical for a second order phase transition from a disordered state ($q=0$) into an ordered one ($q \neq 0$),. We can establish a qualitative agreement with preliminary numerical results for the behavior of the order parameter close to the critical point as obtained from computer simulations for a chain of $N=256$ neurons and $\sigma=5$. Due to the large scale fluctuations the computer simulations need very long runs. Moreover, in order to reduce finite size effects the number of neurons should still be increased. We hope to present reliable numerical results in the near future [?].

In conclusion we remark that we have developed a theoretical description of the phase transition taking place if the dimensional conflict between the input space and net topology exceeds a certain value. By means of a single mode ansatz we could derive the time dependent Ginzburg–Landau equations describing the behavior of the order parameter close to the critical point where the phase transition takes place.

Acknowledgement: The reported results are based on work done in the project LADY sponsored by the German Federal Ministry of Research and Technology under grant 01 IN 106B/3.

References

[1] T. Kohonen (1984): Self–Organization and Associative Memory. *Springer Series in Information Science 8*, Springer: Berlin, Heidelberg.

[2] H. Ritter, T. Martinetz, K. Schulten (1992): Neural Computation and Self-Organizing Maps. Addison Wesley: Reading, Mass.

[3] R. Der, M. Herrmann, Th. Villmann: Dynamics of Self–Organized Feature Mapping. To be submitted.

[4] I. Csabai, T. Geszti, G. Vattay (1992): Criticality in the One–dimensional Kohonen Neural Map. *Physical Review A 46*, R6181.

Appendix

The coefficients C in eq. ?? are displayed in the case $n=1$, $d=2$, $\vec{v} \in [0,N) \times [-s,s]$, $P(\vec{v}) = \frac{1}{2N}$ subject to the symmetry relations $C_{mm'm''}((k,q) = C_{mm''m'}(k,k-q)$.

$$C_{112}(k,q) = C_{211}(k,q) = C_{222}(k,q) = 0$$

$$C_{111}(k,q) = i\left(\frac{1}{2}(\hat{h}(k-q)\sin(k-q) + \hat{h}(q)\sin(q)) - \frac{1}{4}\hat{h}(k)(\sin(k) + \sin(k-q) + \sin(q))\right)$$

$$C_{122}(k,q) = i\left(\frac{\partial}{\partial k}\hat{h}(k)(\cos(k)-1) + \frac{1}{2}\hat{h}(k)\sin(k) + \frac{s^2}{3}\hat{h}(k)(\sin(k) + \sin(k-q) + \sin(q))\right)$$

$$C_{221}(k,q) = i\left(\frac{1}{4}(\hat{h}(k-q)\sin(k-q) + \hat{h}(q)\sin(q)) - \frac{s^2}{3}\hat{h}(k)(\sin(k) - \sin(k-q) - \sin(q))\right)$$

UNSUPERVISED EXTRACTION OF PREDICTABLE ABSTRACT FEATURES

J. Schmidhuber and D. Prelinger
Institut für Informatik
Technische Univ. München
Arcisstr. 21, 8000 München 40, Germany

ABSTRACT. *We consider the problem of determining whether certain input patterns share any non-trivial abstract features. We present a neural system that invents non-trivial distributed representations of input patterns such that patterns having something in common are represented by the same distinct and informative output pattern. The system consists of two modules that learn in an unsupervised manner (no teacher provides target outputs): At a given time, each module sees an input pattern. Each module tries to represent different input patterns by different output patterns, but at the same time both modules try to emit output patterns that match. This procedure tends to create informative representations of non-trivial abstract features shared by both patterns. The approach can be related to the IMAX method of Hinton, Becker and Zemel (1989, 1991). Experiments include a stereo task proposed by Becker and Hinton, which can be solved more easily by our system.*

MOTIVATION AND BASIC IDEA

Consider the following stereo task proposed by Becker and Hinton [2]: There are two images called the 'left' image and the 'right' image. Each image consists of two 'strips' – each strip being a binary vector. The right image is purely random. The left image is generated from the right image by choosing, at random, a single global shift to be applied to each strip of the right image. An input pattern is generated by concatenating a strip from the right image with the corresponding strip from the left image. The input can be interpreted as a fronto-parallel surface at an integer depth. The only local property that is invariant across space is the stereoscopic depth.

Given a pair of input patterns, the first pattern can tell us something (but not everything) about the second pattern. Likewise, the second pattern can tell us something (but not everything) about the first pattern. Let us assume that with a given pair of different input patterns, an unsupervised learning system is told only that both patterns share some abstract feature. It is not told anything about the nature of the feature or about the concept of stereoscopic depth. The system's task is to classify each input pattern such that patterns with the same abstract feature (the ones with the same depth – but the system does not know that in advance) are represented by the same activation pattern. This activation pattern should be different from activation patterns representing input patterns with different depths (belonging to different classes). Thus, after the training phase (after exposure of the unsupervised system to a set of pairs of input patterns), different output patterns should correspond to different depths (the only non-trivial common properties of both elements of a pair of input patterns). In other words, the system's task is to discover the concept of stereoscopic depth by seeing positive training examples only.

Our approach to unsupervised extraction of abstract features common to different inputs (from positive training examples only) is described in [5]. The approach is based on two neural networks called T_1 and T_2. Both can be implemented as standard back-prop networks [7]. With a given pair of input patterns, T_1 sees the first pattern, T_2 sees the second pattern. We force each network to convey information about its input – under the constraint that each network has to emit the *same* output in response to the two (in general) different input patterns of each pair. Thus the output of both networks can be regarded as a representation of whatever non-trivial properties are common to both patterns of a pair.

Both networks have q output units. Let $p \in \{1, \ldots, m\}$ index the input patterns. T_1 produces as an output the representation $y^{p,1} \in [0, \ldots, 1]^q$ in response to an input vector $x^{p,1}$. T_2 produces as an output the representation $y^{p,2} \in [0, \ldots, 1]^q$ in response to an input vector $x^{p,2}$. The conflicting goals are: (A) $y^{p,1}$ should convey information about $x^{p,1}$, and $y^{p,2}$ should convey information about

$x^{p,2}$. (B) But $y^{p,1}$ and $y^{p,2}$ also should match.

We express the trade-off between (A) and (B) by means of two opposing costs.

(B) is expressed by an error term M (for 'Match'):

$$M = \sum_{p=1}^{m} \|y^{p,1} - y^{p,2}\|^2. \tag{1}$$

Here $\|v\|$ denotes the Euclidean norm.

(A) is enforced by additional error terms D_l ($l = \{1,2\}$) (for 'Discrimination'). D_l will be designed to encourage significant Euclidean distance between representations of different input patterns. As shown by Schmidhuber and Prelinger [5], D_l can be defined in more than one reasonable way. The various alternative definitions of D_l have mutual advantages and disadvantages – in the context of a given problem, the most appropriate definition of D_l can be plugged into equation (2) below. Due to limited space, however, we will limit ourselves to a technique called 'predictability minimization' recently introduced by Schmidhuber [4] (see next section). Both $T_l, l = 1, 2$ minimize

$$\epsilon M + (1 - \epsilon)D_l. \tag{2}$$

The objective function is minimized by gradient descent. The procedure is *unsupervised* in the sense that no teacher is required to tell the feature extractors how to represent their inputs.

DEFINING D_l BY PREDICTABILITY MINIMIZATION

As shown by Schmidhuber [4], D_l can be defined with the help of *intra-representational* adaptive predictors that try to predict each output unit of some T_l from its remaining output units, while each output unit in turn tries to extract properties of the environment that allow it to *escape* predictability. This was called the *principle of predictability minimization*. This principle encourages the output units to convey maximal information about the input patterns. Furthermore, each output unit of T_l is encouraged to represent environmental properties that are statistically independent from environmental properties represented by the remaining output units. The procedure aims at generating binary factorial codes [1]. Let us define

$$\bar{D}_l = -\frac{1}{2} \sum_i (s_i^{p,l} - y_i^{p,l})^2, \tag{3}$$

where the $s_i^{p,l}$ are the outputs of S_l^i, the i-th additional so-called *intra-representational* predictor network of T_l (one such additional predictor network is required for each output unit of T_l). The S_l^i are trained to predict the expected value of $y_i^{p,l}$ from $\{y_k^{p,l}, \ k \neq i\}$ by maximizing $\bar{D}_l$.

To encourage even distributions in output space, we slightly modify $\bar{D}_l$ and obtain the discriminating error term that goes into equation (2):

$$D_l = -\frac{1}{2} \sum_i (s_i^{p,l} - y_i^{p,l})^2 + \frac{\lambda}{2} \sum_i (0.5 - \bar{y}_i^{l})^2. \tag{4}$$

RELATION TO PREVIOUS WORK

Becker and Hinton [2] solve the stereo problem by maximizing the mutual information between the outputs of T_1 and T_2. This corresponds to the notion of finding mutually predictable yet informative input transformations. The method was called IMAX.

The nice thing about IMAX is that it expresses the goal of finding mutually predictable yet informative input transformations in a principled way (in terms of a single objective function). In contrast, our approach involves two separate objective functions that have to be combined using a relative weight factor. An interesting feature of our approach is that it conceptually separates two issues: (A) the desire for information preserving mappings from input to representation, and (B) the desire for mutually predictable representations. There are many different approaches (with mutual advantages and disadvantages) for satisfying (A). As mentioned above, in the context of a given problem, the most appropriate alternative approach can be 'plugged into' our basic architecture.

Another difference between IMAX and our approach is that our approach does not only enforce mutual predictability but also equality of $y^{p,1}$ and $y^{p,2}$. This does not affect the generality of our system, however. In fact, one advantage of our simple approach is that it makes it trivial to decide whether the outputs of both feature extractors essentially represent the same thing. With IMAX, this is in general more complicated.

Finally, it turns out that certain problems can be solved more easily using our approach instead of IMAX. See next section.

A COMPARISON WITH IMAX: THE STEREO EXPERIMENT

We conducted a number of successful experiments with systems based on the first section [5]. Due to space limitations, let us focus on an experiment that compares IMAX to our approach.

All networks used below were trained by Werbos' back-propagation algorithm [7]. In all cases we used the activation dynamics of Rumelhart et al. [3], as well as 'on-line' learning: Weight changes took place immediately after each presentation of some randomly chosen input pattern. Approximations of mean values $\bar{y}_i^l$ were updated by the formula

$$\hat{y}_i^l \leftarrow 0.95\hat{y}_i^l + 0.05y_i^l,$$

where $\hat{y}_i^l$ is the approximation of $\bar{y}_i^l$ after observing the current input pattern y^l. $\hat{y}_i^l$ was initially set to 0.5.

Details of the task. There are two binary images called the 'left' image and the 'right' image. Each image consists of 2 'strips' – each strip being a binary input vector with 4 components. There are two feature extractors with single output units and non-overlapping inputs: Each feature extractor has 8 input units and 'sees' an 8-dimensional input vector consisting of a strip from the right image and a corresponding strip from the left image generated as follows: The right image is purely random. The left image is generated from the right image by choosing, at random, a single global shift to be applied to each strip of the right image. The shift can be either one bit to the right or one bit to the left – 'overflow bits' generated by shifting some bit of a strip taken from the right image beyond the strip boundaries reappear on the opposite side of the corresponding 'shifted' strip of the left image ('wraparound'). Ambiguous shifts are excluded. The input may be interpreted as a fronto-parallel surface at an integer stereoscopic depth. Since the right image is random, the only common non-trivial property of both feature extractor inputs is the stereoscopic depth or shift [2]. The goal is to classify each input pattern such that patterns from the same class (the ones with the same depth – but the system is not told anything about depth) are represented by the same activation pattern. This activation pattern should be different from activation patterns representing patterns from different classes (patterns with different depths). In other words, the only information about the input of the second feature extractor that is embedded in the input of the first feature extractor (and vice versa) is the information about the depth. The goal is to extract this information.

Since the feature to be extracted is one-dimensional, only one predictor per feature extractor was necessary to predict the single output unit from a bias unit with constant activation (see the second section). It should be noted that for single output units predictability minimization degenerates to the procedure of maximizing the variance of the unit, which (in the binary case) is equivalent to maximizing the entropy of the unit.

The intra-representational predictors and the feature extractors learned simultaneously. Each of the two feature extractors T_1 and T_2 had 12 hidden units – the predictors had none. The learning rate of the predictors was 1.0, the feature extractor's learning rate was 0.5. Parameter settings were $\epsilon = 0.5$, $\lambda = 1.0$. The task was considered to be solved (the depth was considered to be extracted; the patterns were considered to be classified correctly) if (1) the outputs of both feature extractors were always equal (with an error margin of 0.1) and (2) each feature extractor emitted different binary outputs (again with an error margin of 0.1) in response to input patterns with different depths. This corresponds to 1 bit of mutual information between the outputs and the stereoscopic depth.

With a first experiment, we employed a separate set of weights for each feature extractor. With ten test runs involving 100,000 training patterns, the feature extractors always learned to extract the stereoscopic depth.

Becker and Hinton report that their system (based on binary probabilistic units) was able to extract the depth only if IMAX was applied in successive layer by layer 'bootstrap' stages. In addition, they heuristically tuned the learning rate during learning. Finally they introduced a maximal weight change for each weight during gradient ascent.

In contrast, our method (based on continuous-valued units) does not rely on successive training stages, bootstrap learning, or learning rate adjustments. Once the learning phase is started, no external mechanism influences the behavior of the system. The performance of our system, however, is comparable to the performance of Becker's and Hinton's bootstrapped system. (It should be noted that Becker and Hinton also devised learning procedures for continuous-valued units and for real-valued depths. In this paper, however, we do not attempt to apply our technique to the real-valued case.)

With a second experiment, we used only one set of feature extractor weights shared by both feature extractors (this leads to a reduction of free parameters). The result was a significant decrease of learning time – with ten test runs the system needed only between 20,000 and 50,000 training patterns to learn to extract the stereoscopic depth. No systematic attempt was made to optimize learning speed.

CONCLUSION

Our method tends to be simpler than IMAX. It does not require sequential layer by layer 'bootstrapping' or learning rate adjustments. In the binary case, Becker's and Hinton's stereo task can be solved more readily by our system. The abstract properties extracted by our networks are easier to analyze. It remains to be seen how well the method of this paper scales to larger problems.

ACKNOWLEDGEMENTS

Thanks to Mike Mozer for fruitful discussions. Thanks to Sue Becker and Rich Zemel for helpful comments. Parts of this paper are based on a previous publication [6]. This research was supported in part by a DFG fellowship to J. Schmidhuber, as well as by NSF award IRI–9058450, grant 90–21 from the James S. McDonnell Foundation.

References

[1] H. B. Barlow, T. P. Kaushal, and G. J. Mitchison. Finding minimum entropy codes. *Neural Computation*, 1(3):412–423, 1989.

[2] S. Becker and G. E. Hinton. Spatial coherence as an internal teacher for a neural network. Technical Report CRG-TR-89-7, Department of Computer Science, University of Toronto, Ontario, 1989.

[3] D. E. Rumelhart, G. E. Hinton, and R. J. Williams. Learning internal representations by error propagation. In *Parallel Distributed Processing*, volume 1, pages 318–362. MIT Press, 1986.

[4] J. H. Schmidhuber. Learning factorial codes by predictability minimization. *Neural Computation*, 4(6):863–879, 1992.

[5] J. H. Schmidhuber and D. Prelinger. Discovering predictable classifications. *Accepted by Neural Computation*, 1993.

[6] J. H. Schmidhuber and D. Prelinger. A novel unsupervised classification method. In *Proc. of the Intl. Conf. on Artificial Neural Networks, Brighton*, pages 91–96. IEE, 1993.

[7] P. J. Werbos. *Beyond Regression: New Tools for Prediction and Analysis in the Behavioral Sciences*. PhD thesis, Harvard University, 1974.

[8] R. S. Zemel and G. E. Hinton. Discovering viewpoint-invariant relationships that characterize objects. In D. S. Lippman, J. E. Moody, and D. S. Touretzky, editors, *Advances in Neural Information Processing Systems 3*, pages 299–305. San Mateo, CA: Morgan Kaufmann, 1991.

Genetic Algorithm with Migration on Topology Conserving Maps

Gábor J. Tóth[1] and András Lőrincz[2]
[1] Eötvös Loránd University,
[2]Institute of Isotopes, P.O.B. 77, Budapest, Hungary, H-1525
E-mails: jtoth@obelix.iki.kfki.hu, lorincz@obelix.iki.kfki.hu

Optimization problems depending on external variables (parameters) are treated with the help of a Kohonen network extended by a genetic algorithm (GA). The optimal solution is assumed to have continuous dependence on the external variables. The GA was generalized to organize individuals into subpopulations, which were allocated in the space of the external variables in an optimal fashion by Kohonen digitization. Individuals were allowed to breed within their own subpopulations and in neighboring ones (migration). To illustrate the strength of the modified GA the optimal control of a simulated robot-arm is treated: a falling ping-pong ball has to be caught by a bat without bouncing. It is shown that the simultaneous optimization problem (for different values of the external parameter) can be solved successfully, and that migration can considerably reduce computation time.

1. Introduction

The genetic algorithm (GA) is a general search technique based on biological examples that is widely used in many fields including a wide range of optimization and control problems (see e.g. [1]). The strength of the method lies mainly in its robustness and self-organizing nature.

The genetic algorithm tries to copy the laws of nature that govern natural selection. Conventionally it comprises a number of so called individuals, each of them coding – usually in a string of bits – a possible solution of the optimization problem. In each 'generation' all the individuals are tested or, as is usually said, their fitness value is evaluated. Subsequently the individuals are allowed to reproduce. A pair of individuals is chosen, giving individuals of higher fitness value more chance. Then a cross-over takes places with some probability, i.e. the strings associated with the individuals are cut eat some point and the half-strings exchanged. This may be followed by mutation, i.e. the changing of single bits. After this a new pair is chosen until the number of offsprings reaches the original number of individuals. Then the process continues with the testing of the new generation. Optimization stops if the performance is sufficiently good or after a given number of generations.

This form of GA may be applied to a wide range of optimization problems. Examples range from robotics applications [1] to the optimal control of molecules [2]. There is, however, a group of problems, for which the original GA approach seems inefficient. Consider, for example, table tennis. Most of the motions of the players are very similar: in a hitting action the player moves the bat forward, meets the ball, continues the motion, rotates the bat depending on the direction and the spin he intends to give to the ball. The whole procedure depends on the intentions of the player, the position, the speed and the spin of the ball, and the position and the speed of the player and the bat. There is a large variety of motions but assuming two cases when the intentions are the same and the positions, speeds and spin – let us call them external variables – are very similar, then the motion will also be very similar.

Let us give an other example: from acoustics [3]. The problem is the focusing and steering of ultrasonic waves generated by an array of local photothermal sources on the surface of some material. In this case computed delays between the local photothermal sources are applied in order to focus the acoustic energy to different points on the surface of the material. The situation is similar to the problem of table tennis from the point of view of continuity: a small change in the position of the aimed focal point means a small change in the delay parameters.

In both cases the problem may be solved by direct programming, as in the case of acoustic energy focusing, or through some adaptive techniques. There are numerous advantages in using adaptive techniques, such as (1) conditions (wind, elasticity of the ball, laser strength, etc.) can change but adaptivity may preserve performance, (2) the method can be used even if the material parameters are not known a priori, moreover (3) the solution of the adaptive technique may be used to determine the material parameters with the help of an inverse formulation of the problem.

Fortunately, we can easily find a matching biological example that can be used as a basis for our modified GA. In nature many species (plants, animals – e.g. the humans) have very large habitats within which the living conditions may vary considerably. Here the external variable is the location where a given individual lives. The example is similar to the mentioned problems: individuals living close to each other will not be very different from each other. Since nature solves this optimization problem successfully we try to modify the GA so as to incorporate the essence of this biological example.

Our goal is to develop a GA method that takes advantage of our assumption, that all optimal solutions involving slight changes in the external variables are very similar. This statement may be formulated in a more exact way: we assume that there is a list of external variables and that the optimal solution is a continuous function of these variables. By this we mean that (1) there are metrics in the spaces of both the external variables and the optimal solutions, (2) the problem in question defines a mapping of the space of external variables into the space of optimal solutions, (3) that mapping is continuous in the metrics. In the following we design a GA method that takes advantage of this continuity. First we digitize the problem, i.e. we take representative examples in the space of external variables. This should be done in an optimal fashion; we used the Kohonen algorithm. These representative examples shall be called 'sites'. Then we allow the 'individuals' to migrate to (visit) neighboring (close) sites and mate at those sites. Migration takes advantage of the assumed continuity. A successful individual of one site may also be successful at close neighboring sites. It will be shown that migration considerably speeds up adaptivity.

It may be worth mentioning that the method is not limited to continuous mappings. If the problem in question has continuous subdomains then the method is suitable. There is no need for a priori knowledge of the continuity regions: if continuity applies, then migration will take advantage of it. If not, then the new individuals will not survive and there will be no gain in adaptivity. The measure of success of the migrating individuals may provide an insight into the local properties of the mapping.

Genetic algorithms with migration have been extensively studied with the motivation of studying evolutionary systems (see e.g. [4]). Our study is mainly focused on the question of simultaneous optimization of many similar problems with the help of self-organizing digitization.

2. Description of the model

First of all we need to outline what our biological example is like, since different species behave differently. Our example species lives in small groups (subpopulations). Within each subpopulation individuals mate in the regular way: randomly but accordingly to their fitness. The main novelty is that we allow individuals to mate with others from *neighboring* subpopulations. This is essential because migration allows very small subpopulations to reproduce without a critical loss of biodiversity; this is why we call our model *genetic algorithm with migration (GAM)*. It is important to note, however, that the validity of this method relies on the assumption that a small change in the environment (external variable) will not dramatically change the optimal solution, the best individuals at close locations are similar to each other. It should also be noted that migration does not mean a permanent change of subpopulation, only that at mating time individuals may temporarily be found in neighboring subpopulations, too.

Having outlined the sample species let us construct the model. We need a structure embedding the subpopulations and a metric defined on it. We need an algorithm to locate the subpopulations in the external space (the space of the external variables) in a topology-conserving way, i.e. ensuring that subpopulations near to each other in the metric defined between them will be located at close points of the external space. Since in many cases we do not know in advance the likely values of the external variables, the algorithm must be capable of discovering these values (the inhabitable regions) and allocate subpopulations only there. And, finally, we need a modification of the breeding algorithm that incorporates migration.

In many cases the inhabitable region is – more or less – a (hyper)parallelepiped that is topologically equivalent to a (hyper)cube; for one external variable this is always the case. In such cases the structure embedding the subpopulation can be a (hyper)cube.

For the algorithm locating the subpopulations (still assuming the described inhabitable region) we used the Kohonen algorithm. During Kohonen learning, random samples from the external space are presented, one at a time. The location of each subpopulation is changed in accordance with the following equation:

$$x_r \to x_r + \epsilon G[\rho(r,s)](x - x_s), \tag{1}$$

where x is the sample presented, x_r denotes the location of an arbitrary subpopulation, x_s the location of subpopulation closest to the example, ϵ is a scaling factor, ρ is the metric defined among the subpopulations, and G is the distance dependence of the modification rate. In the usual case ρ is a Euclidean metric and G is an exponential function:

$$G(x) = e^{-x/\lambda}, \tag{2}$$

where λ characterizes the cooperativeness of the learning. This iteration is repeated n_{Koh} times.

If the inhabitable region is not so regular, similar but more complex techniques can be used (see e.g. [5]).

Once the subpopulations are fixed GAM can start to work. As mentioned before, some modifications to the algorithm are necessary to incorporate migration.

The first such modification is related to the testing. During the testing phase samples are taken from the external space in the same way as during Kohonen learning. The subpopulation closest to the sample is chosen and one individual of that subpopulation is tested. If all the individuals of the given subpopulation have already been tested then no testing takes place. Then a new sample is taken until each individual has been tested.

The other modification is made in the way of breeding. We assume that each individual is present in the mating pool of its own subpopulation with 100% probability, and with p_{mig} probability in the mating pool of each neighboring subpopulation. Let each individual in the mating pool have a relative fitness value such that the sum of all relative fitness values be one. Let us now define the relation between the fitness value f and the relative fitness value f^R so that the relative fitness can be interpreted as the probability of choosing an individual for breeding. For the sake of simplicity let us confine ourselves to linear functional dependence of the form

$$f^R = af + b \tag{3}$$

for individuals from the local subpopulation, and

$$f^R = \gamma(af + b) \tag{4}$$

for individuals from neighboring subpopulations. Since f^R is to be used as the probability of breeding, γ should equal p_{mig}. The value of a and b can easily be calculated (see e.g. [1]).

Once the relative fitness values are calculated, breeding can go on in the usual way: a pair of individuals is chosen with probabilities f_i^R, they are crossed over with probability p_x, and the offsprings undergo mutation with probability p_m. Then a new pair is chosen until enough offsprings are generated. The process then continues with the breeding of the next subpopulation. If all subpopulations have reproduced the testing of the new generation can begin.

3. Sample problem

To illustrate the capabilities of GAM we chose a simple, but not trivial problem. A robot arm had to catch a ping-pong ball falling from varying heights with a bat so that (1) the ball did not rebound, (2) at the end of the time interval the bat was not moving, (3) at the beginning and

608

at the end it was at the origin. A further constraint was that (4) the bat remained within a finite region (the robot arm has a finite length). The external variable in this case was the height from which a ball was dropped. Constraints (3) and (4) were automaticly satisfied by the decoding while the other two constraints were taken into account via the fitness function.

We allowed the dropping height to vary between 1 and 1.5 m. We found that a height variation of 5-10 cm did not change the optimal solution seriously. However, the uncertainty introduced by the variation of the height during testing could easily mislead the algorithm. So we set the number of subpopulations to 50. With large number of individuals per subpopulation (200) the performance was good regardless of the value of p_{mig}. However, we found that with $p_{mig} = 0.7$ the number of subpopulations could be decreased by a factor of 20, without any change in final performance, while with $p_{mig} = 0$ any substantial decrease led to severe fall of performance. The error as a function of generation for these cases is plotted in Fig. 1: curve (a) shows the GAM containing 200 individuals per subpopulations, (b) and (c) GAM's with 10 individuals per subpopulation, with $p_{mig} = 0.7$ and $p_{mig} = 0$ accordingly. An example of the final performance corresponding to curve (b) of Fig. 1 is shown in Fig. 2.

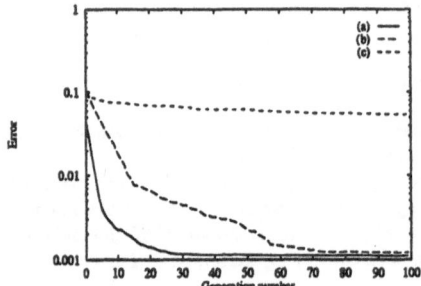 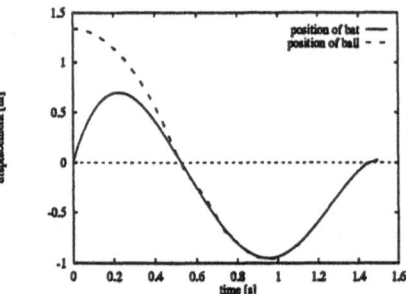

Figure 1: Error of GAM vs. generation (a) 200 individuals per subpopulation; (b) 10 individuals per subpopulation, $p_{mig} = 0.7$. (c) 10 individuals per subpopulation, $p_{mig} = 0.0$.

Figure 2: Position of bat and ball vs. time.

4. Conclusion

An extension of the GA was outlined to incorporate external variable(s) with the help of topology conserving Kohonen digitization. A sample problem was treated in detail. It was shown that the extended model could successfully solve the simultaneous optimization problem in only slightly increased time. It is worth mentioning that the algorithm can be run on a truly parallel computer thereby further decreasing the computation time.

References

1 D. E. Goldberg, *Genetic algorithms in search, optimization and machine learning* (Addison Wesley: Reading, Mass., 1989) and references therein.

2 R. S. Judson, H. Rabitz, Phys. Rev. Let. **68**, 1500 (1992)

3 M.-H. Noroy, D. Royer, and M. Fink, in *Proc. of SPIE at San Diego* (July 1992), p. 1.

4 H. Mühlenbein, in *Toward a Practice of Autonomous Systems*, edited by F. J. Varela and P. Bourgine, p. 236 (MIT Press: Cambridge, Mass., 1992).

5 (a) T. Martinetz and K. Schulten, in *Artificial Neural Networks*, edited by T. Kohonen, K. Mäkisara, O. Simula and J. Kangas, p. 397 (Elsevier: Amsterdam, 1991); (b) B. Fritzke, ibid., p. 403.

Analyzing Kohonen Maps With Geometry

Stéphane Zrehen
Laboratoire de Microinformatique
EPFL-DI, CH-1015 LAUSANNE

Abstract

An organization measure based on geometrical criteria has already been proposed for Kohonen maps in the two-dimensional case [3]. This measure is shown to be generalizable to all network topologies and input dimensions. It can be used for demonstrations on the convergence of the learning algorithm.

Introduction

Kohonen proposed in 1982 [1] a self-organizing neural algorithm. It has since then been widely used, mostly for data processing, when topological information needs to be conserved as much as possible [2]. However, a problem that remains is to measure the degree of reliability of the projection when the input space dimension is higher than three. Despite a few attempts [4, 5], the question about what "organized" should mean is still open.

An organization measure for the 2-D case was already proposed, based only on geometrical considerations about the map. The aim was to define an objective function that would explicit intuition when representation of the net in its weight-space is possible.

In this paper, I will show that the philosophy of this measure remains valid for all dimensions and topologies, and that the means to compute it remain the same.

Principles of the 2-D Geometrical Organization Measure

I will not give the details of the learning algorithm as they are not necessary for the definition of the organization measure. I will just recall that in the map neighborhood relations between cells are chosen in advance according to a given topology, usually linear, square or hexagonal. A discrete topological distance is associated to the map.

The justifications of the geometric organization measure rely on the utilization phase (after learning is over): an input vector is presented to the net; the activation of each cell is then defined as the distance between the weight of this cell and the input vector. The cell whose weight is closest to the input is called the winner cell. Learning consists in modifying the weights of the winner and some of its neighbors in function of the presented input.

It is common to use Euclidean distance, and our geometric measure is precisely based on Euclidean geometry.

When the input space has less than three dimensions, it is easy to represent the map in its weight space. It is usually said to be organized if there is an apparent ordering of the weights according to the position of the corresponding cells on the net. Unfortunately, there is no order relation in spaces of dimension higher than one that respects topology. Therefore some formal criterion needs to be found.

As exposed in [3], if the dimension of the topology of the map is smaller than that of the input space, the weight representation of the net folds itself, when the input distribution is uniform (Figure 1). Therefore the organization criterion proposed by Kohonen which states that two close inputs should activate close cells can not be verified in those cases. Indeed the measures proposed in [4] and [5] both state that such a network is highly disorganized. However one can expect the opposite property: *two close cells should only be activated by close input vectors*. Thus, at first sight, in the absence of crossing links (Figure 1) the network should intuitively be defined as organized.

Furthermore, the main feature expected from the organization in Kohonen maps is the continuous preservation of proximity properties of the input: it should be possible to move from one cell to another and keep track of the input represented. The simplest motion on a discrete graph is a straight line between neighbors.

Figure 1: "Peano curve": weight-space representation of a linear network in a 2-D uniform distribution

Following this idea, the geometrical organization measure gives the degree of correctness of the following statement:

(*) *A pair of neighbor cells A and B is locally organized if the straight line joining their weight vectors W_A and W_B contains points which are closer either to W_A or to W_B than they are to any other.*

Cells whose weight's Voronoi set intersect the straight line joining two other weight vectors are called intruders.

If the net shows no local disorganization, then it is possible to make a Manhattan motion between any two cells and guarantee that the trajectory never falls under the influence of intruders.

To determine the local disorganization of a pair of neighbors (A,B), one should count the number of intruders K which give rise to a local disorganization as defined in (*). With Euclidean distance, this can be done in several ways: a cell K≠A≠B disorganizes pair (A,B) (See Figure 2)

- (i)   if the scalar product
        $W_K W_A . W_K W_B < 0$.
- (ii) if W_K is inside the disk of diameter
        $[W_A W_B]$.
- (iii) if the angle $(W_A W_B W_K)$ is obtuse.
- (iv) if the center Ω of the circumcircle of
        triangle $W_A W_B W_K$ is outside the triangle.

Those four propositions are equivalent in the Euclidean plane as was proved in [3].

The disorganization D of the network is defined as the sum on all pairs of neighbors of the number of

intruder cells. It can be normalized with a factor that depends only on the size and topology of the network. That makes possible a comparison between nets of different sizes.

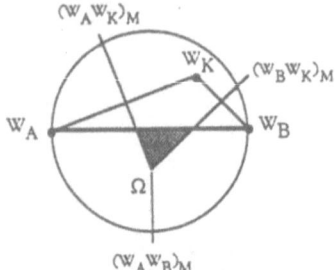

Figure 2 : Disorganization of pair AB by K

Although it is usual to define the network as organized if all its meshes are convex, the geometric point of view needs more (see Figures 3 & 4): some angles may not be obtuse in square meshes.

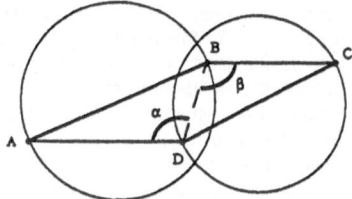

Figure 3: Disorganized square mesh: angles α and β are obtuse

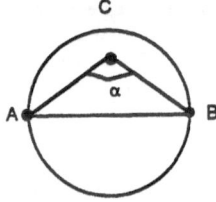

Figure 4: Disorganized triangular mesh

Extension to Higher Dimensions of Input Spaces

The basic philosophy of the geometric disorganization, expressed by (*) can be kept for all dimensions. Whatever the dimension of the input space, one should be able to monitor any Manhattan motion between two cells.

As the determination of the Voronoi tessellation of a set of points grows exponentially with dimension

[7], a crude application of that principle should prove too costly. We thus need to find a simpler way to determine the intruder cells, as we did in two dimensions.

Let $W_A=(a_i)$, $W_B=(b_i)$, $W_K=(k_i)$ the coordinates of the weight vectors of cells A, B and C in the N-dimensional space.

If $P=(p_i) \in [W_A W_B]$, then $\exists \mu \in [0,1]$ such that:
$p_i = \mu a_i + (1-\mu)b_i$.

(*) translates as:

K is an intruder to pair [AB] iff: one can find a point $P= (\mu a_i + (1-\mu)b_i)$ such that

$$\begin{cases} \sum_{i=1}^{N}(a_i - b_i) > \sum_{i=1}^{N}(k_i - p_i) \\ \sum_{i=1}^{N}(b_i - p_i) > \sum_{i=1}^{N}(k_i - p_i) \end{cases} \text{ with } \mu \in [0,1].$$

In two dimensions, we proved in [3] that it is equivalent to (i). Here is a sketch of a proof in any dimension:

The three points A, B and K form a two-dimensional plane in any space. Therefore, we can choose vectors $(W_A W_B)$ and $(W_A W_K)$ as the first vectors of the space base. In that base, all points of plane $(W_A W_B W_K)$ have zero coordinates except their first two. The above system of inequalities will thus reduce to the same system with $N=2$ which is known to be equivalent to (i).

The intersection of the hyper-sphere of diameter $[W_A W_B]$ with plane $(W_A W_B W_K)$ is the circle of diameter $[W_A W_B]$. This states the equivalence between (i) and (ii) for all dimensions. The equivalence of (ii) with (iii) and (iv) is immediate.

Organization and Learning

The disorganization of the net can be measured at any stage of learning. For usual smooth distributions, the net does organize almost perfectly (Figure 5), The border effects can alter the organization, but one may choose to exclude them from the computation of disorganization.

If one applies the simplified learning algorithm [2], it decreases in average for most distributions (Figure 6), although not monotonically, to a certain value that is not always null.

The learning algorithm seems to minimize the geometric disorganization in any case. A simple experiment shows that: if one takes the net after completion of learning, any permutation of weight vectors will lead to a higher disorganization

(Figure 7), except in pathological cases such as the "butterfly" (Figure 8).

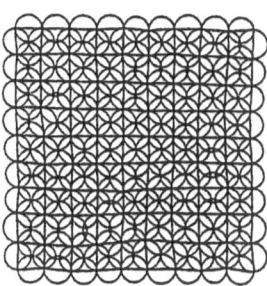

Figure 5: a perfectly organized 10x10 net

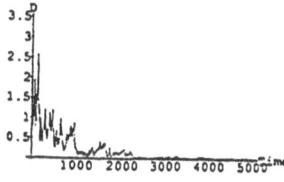

Figure 6: 10x10 Network. 15-d input space
Evolution of D measure through time

Figure 7: permutation of the weights of a 4x4 net

This is an important feature of the geometrical measure that helps to separate the two different operations performed by Kohonen networks: quantization of the input distribution and self-organization. Two nets whose weights have been permuted lead to the same quantization, but to a totally different organization.

612

One should note that for most applications,the input space is of higher dimension than the net topology. Thus, topological information is lost anyway. There is no a priori reason to choose a planar topology for the net rather than linear. Actually, linear nets usually get organized much better than planar ones.

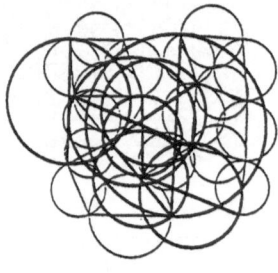

Figure 8: A butterfly. Thick circles indicate local disorganization

Conclusion

The geometric organization measure has been shown to be valid for all Kohonen networks, whatever their topology, and the dimension of their inputs.

It appears that the combination of certain input distributions with certain sizes of net do not allow a perfect organization. The value of the disorganization after learning can be used as a tool to dimension the network in function of the application. Unfortunately, no systematic method can yet be proposed.

Convergence properties of Kohonen nets remain a problem, mostly because there is no formal definition of an organized state [7]. The geometric organization measure provides a numerical tool for defining the state of the net. It seems to be minimized by the learning algorithm in most cases.

Future work should focus on defining a "matching function" between the geometry of the support of the input distribution and the size of the net. Such a result should give a lower bound to the disorganization of a net after learning.

References

[1] Kohonen T., (1982): *Self-Organization of Topologically Correct Feature Maps*, Biol. Cyb.43, 59-69

[2] Kohonen T., (1984): *Self-Organization and Associative Memory* (2nd ed.) Berlin, Springer Verlag.

[3] Zrehen S., Blayo F. (1992) : *A Geometric Organization Measure for Self-Organizing Kohonen Maps*, Proceedings of Neuro-Nîmes Conference 1992,p. 603-610, Paris: EC2.

[4] Demartines P. (1992): *Mesures d'Organisation du Réseau de Kohonen.* Congrès Satellite du Congrès Européen de Mathématiques, Paris.

[5] Bauer H-U, Pawelczik K (1992), *Quantifying the Neighborhood Preservation of Self-Organizing Maps*, IEEE Transactions on Neural Networks, 1992.

[6] Cottrell M. Fort J.C, (1987) : *Etude d'un processus d'auto-organisation.* Annales de l'Institut Henri Poincaré 23, n° 1, pp. 1-20.

[7] Cottrell M., Fort J.C, (1986): *A Stochastic Model of Retinotopoy: A Self-Organizing Process*, Biol. Cyb. 53, 405-411.

A Comparison Between Classical Unsupervised Classifiers and ART3 Neural Networks

Michael J. Berchtold

Carl Zeiss Germany, Postfach 1380, D-73444 Oberkochen, FRG
e-mail: berchtold@cae007.v-ppb.zeiss.de

Abstract

This contribution reports a comparative study concerning connectionist and classical unsupervised classifiers. In order to understand the main concepts behind the ART3 model, a mathematical analysis is carried out that makes it possible to categorise the investigated connectionist model in the framework of well-established pattern recognition methods. We concentrate our attention on the clustering algorithm and the search mechanism used in ART3 neural networks. New results on two topics will be presented: (1) There exists a formal equivalence between ART3 and the well-known cosine classifier with respect to their clustering algorithm. (2) In the batch adaptation mode, the ART3 model implements a gradient descent on an energy function.

1 Introduction

One of the most complex neural models is ART3, proposed by Carpenter & Grossberg ([1]). We analyse this model mathematically to understand its basic concepts. Our attention is focused on the clustering algorithm and the search mechanism, to be able to categorise ART3 with respect to classical clustering algorithms, especially the so-called cosine classifier (CC, [2]).

2 Mathematical Analysis

We want to describe mathematically two major results of our research on two layer ART3-networks: (1) There exists a formal equivalence between the ART3 clustering algorithm and the well-known classical unsupervised CC. (2) In the batch adaptation mode, ART3 implements a gradient descent on an energy function, like many other neural and classical classifiers.

2.1 Equivalence of clustering algorithms

We show that the ART3 similarity measure s^{ART3} is equivalent to the genuine cosine measure $\cos \alpha$. α is the angle between an active bottom-up prototype ψ_J (or a top-down prototype λ_J) and the current input vector i. As usual, we will write vectors in lower-case bold-faced type and matrices in upper-case bold-faced type. Let c and d denote two positive scalar parameters. η is the learning rate and $\mu = 1 - \eta$ the decay rate. Carpenter & Grossberg proposed a learning rate $\eta = 0.9$. $q_{[2]}$ denotes the second computation of the activity vector q, and $y_{\{2\}}$ represents the binary output vector of the F_2-layer. After bottom-up and top-down propagation of the input vector i through the network, one finally gets a normalised activity vector:

$$q_{[2]} = \frac{p_{[1]} + c\,\Lambda\,y_{\{2\}}}{\|p_{[1]} + c\,\Lambda\,y_{\{2\}}\|} = \frac{p_{[1]} + c\,\lambda_J}{\|p_{[1]} + c\,\lambda_J\|}. \tag{1}$$

The similarity measure s^{ART3} is given by the L_2-norm of the vector:

$$r = \frac{u_{[1]} + d\,q_{[2]}}{\|u_{[1]}\| + d\,\|q_{[2]}\|}.$$ (2)

Applying the identity $u_{[1]}\,\lambda_J = \|\lambda_J\|\cos\alpha$ and simple arithmetics one gets:

$$
\begin{aligned}
s^{ART3}(\cos\alpha, \|\lambda_J\|) \;=\; \|r\| \;&=\; \frac{\left[\sum_{i=1}^{K}\left(u_{[1]i} + d\,q_{[2]i}\right)^2\right]^{\frac{1}{2}}}{\|u_{[1]}\| + d\,\|q_{[2]}\|} \\[2mm]
&=\; \frac{1}{1+d}\left[1 + 2d\,\frac{1 + c\,\|\lambda_J\|\cos\alpha}{(1 + 2c\,\|\lambda_J\|\cos\alpha + c^2\,\|\lambda_J\|^2)^{\frac{1}{2}}} + d^2\right]^{\frac{1}{2}}.
\end{aligned}
$$ (3)

The following *Rule* expresses the condition for which two similarity measures are equivalent. Equivalence here means that one can get the same clustering result with two similar clustering devices 1 and 2, which differ only in their similarity measures s_1 and s_2.

Rule: (**Equivalence of similarity measures**) Two similarity measures s_1 and s_2 are equivalent, if s_2 is a strictly monotonically increasing function of s_1.

Therefore, we have to prove that the similarity measure of equation (3) is a strictly monotonically increasing function of the genuine cosine measure $\cos\alpha$. For this reason, we calculate the first partial derivation of s^{ART3} with respect to $\cos\alpha$. In equation (4), we substituted $\cos\alpha$ by x and $\|\lambda_J\|$ by y, respectively.

$$\frac{\partial s^{ART3}(x,y)}{\partial x} = \frac{\frac{c\,d\,y}{(1+2cxy+c^2y^2)^{\frac{1}{2}}}\left(1 - \frac{1+cxy}{1+2cxy+c^2y^2}\right)}{(1+d)\left[1 + d^2 + \frac{2d(1+cxy)}{(1+2cxy+c^2y^2)^{\frac{1}{2}}}\right]^{\frac{1}{2}}} \;>\; 0$$ (4)

Equation (4) satisfies the above *Rule*, because all parameters are larger than zero and $x > 0$ for $0 < \alpha < \pi/2$. This guarantees the equivalence of the two similarity measures. We now examine the adaptation rules for both methods. The ART3 adaptation rule for the bottom-up weight-matrix Ψ is given for the discrete case and for the k-th input presentation by:

$$\Delta\psi_J^{(k)} = \psi_J^{(k+1)} - \psi_J^{(k)} = \eta\,u^{(k+1)} - \mu\,\psi_J^{(k)}.$$ (5)

This can be easily transformed to:

$$\psi_J^{(k)} = (1-\mu)^k\,\psi_J^{(0)} + \eta\sum_{i=1}^{k}(1-\mu)^{k-i}\,u^{(i)} \;\stackrel{(k\to\infty)}{\longrightarrow}\; \eta\sum_{i=1}^{k}(1-\mu)^{k-i}\,u^{(i)}.$$ (6)

For large k equation (6) becomes $\eta\sum_{i=1}^{k}(1-\mu)^{k-i}\,u^{(i)}$, because in this case $(1-\mu)^k$ vanishes for $0 < \mu < 1$. Note, that the same result holds for λ_J, and that the vector norm of all ψ_J and λ_J asymptotically increases towards $g = 1/(1-\eta)$. For the proposed parameter values, we get a g-value of 10. The fact that not all the prototypes are equally long during the adaptation phase causes some difficulties in the search for the most similar prototype to the current input vector ([3]).

The CC adaptation rule for normalised input vectors x and learning rate $\eta_J = 1/n_J$ is given by (n_J is the number of assigned input vectors):

$$\psi_J^{(k+1)} = \left(1 - \eta_J^{(k+1)}\right)\psi_J^{(k)} + \eta_J^{(k+1)}\,x^{(k+1)}.$$ (7)

This leads for a learning set size N to final normalised weight vectors ψ_J:

$$\psi_J^{(N)} = \frac{1}{N} \sum_{k=1}^{N} x^{(k)} = \overline{x}. \tag{8}$$

Equations (6) and (8) both result in a mean of the previously presented input vectors. The ART3 case leads to a weighted mean, where the later presented vectors have got more influence on the resulting prototype ψ_J than earlier ones. This is the only difference between the two clustering results of ART3 and CC. In the case, where $\eta \to 1$ and $\mu \to 0$ the clustering results of ART3 and CC are the same, except a constant factor. Hence, our previously mentioned claim of the equivalence of both clustering methods holds indeed.

2.2 Batch adaptation mode

For the iterative adaptation mode no Ljapunow-function has been found until now. We will show, that in the batch adaptation mode, ART2 and ART3 neural networks implement a gradient descent on an energy function. The adaptation rules for the bottom-up and top-down weights are equivalent. Therefore, we will concentrate on the procedure for the bottom-up prototypes ψ_J:

$$\Delta\psi_J^{(k)} = \eta\, u^{(k+1)} - \mu\, \psi_J^{(k)}. \tag{9}$$

We can write the adaptation rule of equation (9) for prototypes ψ_J in a modified form, so that it is possible to use a similar Ljapunow-function as Ritter & Schulten ([4]) did for self-organizing maps.

$$\Delta\psi_{ij}^{(k)} = \mu\, o_j^{(k+1)} \left(\tilde{u}_i^{(k+1)} - \psi_{ij}^{(k)} \right), \quad \text{where } \tilde{u}_i^{(k+1)} = \frac{u_i^{(k+1)}}{1-\eta} = \frac{\eta}{\mu}\, u_i^{(k+1)} \tag{10}$$

Equation (11) as the Ljapunow-function of equation (10) has got the typical properties of energy functions like $V > 0$ and $\Delta V > 0$ for all N, Ψ, and $\tilde{u}$:

$$V(\Psi, N) = \frac{1}{2} \sum_{j=1}^{L} \sum_{k=1}^{N} \| \tilde{u}^{(k)} - \psi_j \|^2. \tag{11}$$

Here, ψ_j is not the currently active prototype like ψ_J in subsection 2.1, but the final prototype after batch adaptation.

The gradient descent for a single prototype ψ_q on function V results in:

$$-\frac{\mu}{N} \frac{\partial V}{\partial \psi_q} = \frac{1}{N} \sum_{j=1}^{L} \sum_{k=1}^{N} \mu \left(\tilde{u}^{(k)} - \psi_j \right) \delta_{jq} = \overline{\Delta\psi_q}. \tag{12}$$

Note that we used the ART3 adaptation rule of equation (9) to get (12). For any weight component ψ_{rs} we have to use the chain rule for multivariate functions:

$$
\begin{aligned}
-\frac{\mu}{N} \frac{\partial V}{\partial \psi_{rs}} &= -\frac{\mu}{N} \sum_{q=1}^{L} \left(\frac{\partial V}{\partial \psi_q} \right)^T \frac{\partial \psi_q}{\partial \psi_{rs}} \\
&= \sum_{q=1}^{L} \left[\frac{1}{N} \sum_{j=1}^{L} \sum_{k=1}^{N} \mu \left(\tilde{u}^{(k)} - \psi_j \right) \delta_{jq} \right]^T \frac{\partial \psi_q}{\partial \psi_{rs}} \\
&= \sum_{q=1}^{L} \left(\overline{\Delta\psi_q} \right)^T \frac{\partial \psi_q}{\partial \psi_{rs}} = \sum_{r=1}^{K} \overline{\Delta\psi_{rs}}.
\end{aligned}
\tag{13}
$$

A necessary condition for the stability of the above result is given by:

$$\sum_{r=1}^{K} \overline{\Delta \psi_{r_s}} \overset{!}{=} 0. \tag{14}$$

The consequence of equation (14) is that in the batch adaptation mode the prototypes are stable, if the gradient descent of equation (13) has been applied. This means that the final prototypes of the ART2 and ART3 adaptation rule are point attractors of the discrete Ljapunow-function V in the batch adaptation mode. This result includes two major drawbacks: First, it is possible that the procedure gets stuck in local minima of the function V ([5]). Second, the batch adaptation mode is not capable of processing the input data in real time with their frequency of presentation.

3 Conclusion

The theoretical analysis of the ART3 model of Carpenter & Grossberg ([1]) in this paper, in conjunction with established methods in the fields of connectionism and classical unsupervised clustering has caused various new insights towards a categorisation of ART3 in the arsenal of pattern recognition methods. In general, it will not be possible to make an a priori decision, which kind of classifier will work best for a given task.

As an application we used an ART3 neural network for a low-level image analysis task, the recognition of characteristic intensity variations. We achieved promising results, according to the recognition of gray value corners ([6]).

Acknowledgements

The author wishs to thank Carl Zeiss Germany and the Bundesministerium für Forschung und Technologie for the valuable funding of this work. J. Geissler, K. Rohr, and J. Zhang contributed many helpful suggestions to the paper.

References

[1] G. Carpenter, S. Grossberg, "ART3: Hierarchical Search Using Chemical Transmitters in Self-organizing Pattern Recognition Architectures," *Neural Networks*, vol. 3, pp. 129–152, Mar.-Apr. 1990.

[2] R. Duda, P. Hart, *Pattern Classification and Scene Analysis*, New York, NY: John Wiley & Sons, 1973.

[3] M. Berchtold, "Adaptive Resonance Theory and Classical Clustering Algorithms – A Mathematical Analysis," paper submitted to *IEEE Trans. on Neural Networks*, Dec. 92.

[4] H. Ritter, K. Schulten, "Kohonen's Self-organizing Maps: Exploring their Computational Capabilities," in *Proc. IEEE Int. Conf. on Neural Networks*, San Diego, CA, Jun. 1988, vol. 2, pp. 109–116.

[5] J. Hertz, A. Krogh, R. Palmer, *Introduction to the Theory of Neural Computation*, Redwood City, CA: Addison-Wesley Publishing Company, 1991.

[6] M. Berchtold, *Erkennung charakteristischer Grauwertübergänge mit ART-Netzwerken*, PhD thesis, Department of Computer Science, University of Ulm, Germany, 1993 (in German).

A Dynamic Procedure for Neural Network Design. Anna Maria Fanelli, Fabio Abbattista, Claudio Mangia and Nicola Abbattista, Department of Computer Science, University of Bari, Via Amendola, 173 - 70126 Bari, Italy

Introduction. We propose a procedure able to find the optimal size for an initially overdimensioned Multi-Layer Perceptron by pruning off units at the same time as the network learns. Our approach is based on a rule that changes only the threshold of every unit in the network, instead of imposing constraints on all weights [1].

Method description. The following rule for the weights and thresholds updating of Multi-Layer Perceptron networks has been derived:

$$\Delta w_{ij} = \eta \delta_j o_i$$

$$\Delta \vartheta_j = \eta \delta_j + sign(\vartheta_j) \varepsilon_j$$

where ε_j is a positive term representing the adaptative controlling parameter of pruning process. Its value is kept constant during each epoch of the training process and it is changed only at the end of a epoch, for each unit, according to the "history" of its activity. More precisely:

$$\varepsilon_j^{new} = a \cdot \varepsilon_j^{old} \qquad (\forall p \in S: O_j > 0.5) \vee (\forall p \in S: O_j < 0.5) \qquad \text{(concordant activity)}$$

$$\varepsilon_j^{new} = b \cdot \varepsilon_j^{old} \qquad \text{otherwise} \qquad \text{(discordant activity)}$$

where S is the training set, while a and b are constant values (a >1 and 0< b < 1). The elimination of a unit occurs when $|\vartheta_j| \geq \vartheta_{MAX}$ where ϑ_{MAX} is an appropriate constant value. The main feature of this algorithm is that the $sign(\vartheta_j)$ term creates the positive feedback effect of making the threshold values higher in those units that present a concordant activity and then must be pruned off; the positive feedback effect can be slowed down at will, in the initial phase, by a suitable choice of ε_j. Consequently, our procedure allows: a) the delay of the pruning process to minimize the training time by taking advantage of the high number of starting units; b) an extremely fast elimination phase; c) the disappearance of the pruning process, in the final training phase, thus permitting only the network learning model to act.

Experimental results. We have evaluated the behaviour of the previously described procedure in front of some well known problems and we have compared the results with those obtained from the "weight-decay" procedure [1]. As an example, the behaviour of the error function, as well as the number of hidden units during learning process of the 4-bit parity problem, is shown in the figure for the decay method (D) and our method (M).

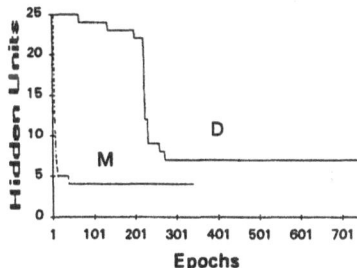
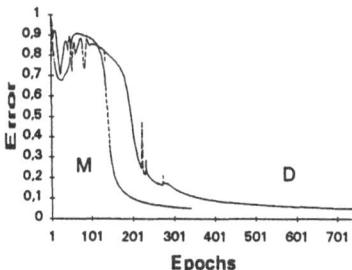

The figure clearly shows that our method is more effective than weight-decay algorithm in reducing the hidden units of the network as well as the learning epochs needed to reach convergence.

Reference.
[1] J. Hertz, A. Krogh, and R. Palmer, "Introduction to Theory of Neural Computation", Redwood City, CA: Addison-Wesley, 1991.

PCA in a Network with Full Lateral Connections

(Abstract)

Bernd Freisleben

Department of Computer Science (FB 20), University of Darmstadt, Alexanderstr. 10,
D–6100 Darmstadt, Germany, E-Mail: freisleb@isa.informatik.th-darmstadt.de

A two–layered network of linear neurons, with full connections between the two layers and full lateral connections among the output layer, is presented that organizes itself to perform *Principal Component Analysis (PCA)* of the set of input patterns. The output units compute their activations from the signals received from both the input units and the remaining output units, with the activation $y_i(t)$ of output unit i, $1 \leq i \leq M$, at time t given by $y_i(t) = \sum_{j=1}^{N} x_j(t) \cdot w_{ij}(t) + \sum_{k=1 \atop k \neq i}^{M} y_k(t) \cdot q_{ik}(t)$, where $x_j(t)$ is the j–th component of the input vector presented to input unit j, $1 \leq j \leq N$, at time t, $w_{ij}(t)$ is the weight of the connection between output unit i and input unit j, and $q_{ik}(t)$ is the weight of the lateral connection originating at output unit k and terminating at output unit i at time t. The lateral connection weights are assumed to be symmetric, i.e. $q_{ik}(t) = q_{ki}(t) \; \forall \, i, k,\; 1 \leq i, k \leq M$. The lateral weights q_{ik} are updated by $q_{ik}(t+1) = q_{ik}(t) - \beta(t) \cdot (q_{ik}(t) + y_i(t) \cdot y_k(t))$ and the weights w_{ij} are modified according to $w_{ij}(t+1) = w_{ij}(t) + \alpha(t) \cdot [x_j(t) \cdot (y_i(t) + C \cdot \sum_{k=1 \atop k \neq i}^{M} y_k(t) \cdot q_{ik}(t+1)) - y_i^2(t) \cdot w_{ij}(t)]$, where $\alpha(t) > 0$ and $\beta(t) > 0$ are decreasing learning parameters (proportional to $1/t$) and $C > 1$ is a coupling constant to strengthen the signals carried on the lateral connections.

Provided that the w_{ij}'s are initially set to random values, the q_{ik}'s are initialized to 0, and $\alpha(t)$, $\beta(t)$ and C are appropriately set, the $M < N$ weight vectors of the output units converge to the eigenvectors belonging to the M largest eigenvalues of the covariance matrix of the set of input vectors presented to the network.

The learning rules are purely local, since all the information necessary to update the connection weights is available at the corresponding output unit. The network is symmetric in the sense that all output units perform exactly the same computations, and it operates in a fully parallel mode, because all weight vectors converge simultaneously to the M leading eigenvectors. The learning rules provide more stability and have better convergence properties than other proposals for the same network topology.

Figure 1, (a) for $N = 16, M = 4$ and (b) for $N = 64, N = 4$, shows the convergence behaviour of the network when a set of 256 input vectors, with zero–mean Gaussian distributions along each component, but different variances for each distribution (to obtain different eigenvalues), is repeatedly presented to the network. The x-axis indicates the number of simulation steps (a simulation step is equivalent to updating the set of weights in response to the presentation of an input vector), and the y-axis indicates the average deviation (the mean of the Euclidean distances) of the weight vectors from the eigenvectors resulting after each simulation step.

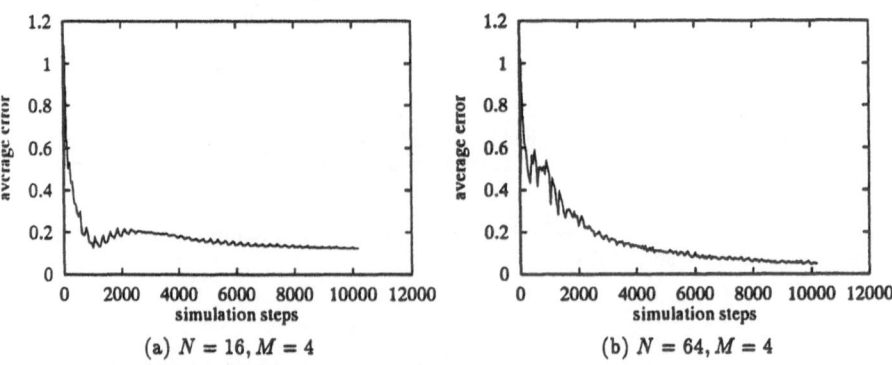

(a) $N = 16, M = 4$ (b) $N = 64, M = 4$

Figure 1: Convergence Behaviour of the Learning Rules

NON-UNIFORM CELLULAR AUTOMATA

Marifi GÜLER[1] and Hürevren KILIÇ[2]

[1]Department of Computer Engineering, Middle East Technical University,06531 Ankara, Türkiye
[2]Bilkent Computer Center, Bilkent University, Bilkent,06533 Ankara, Türkiye

Abstract- A model called Non-uniform Cellular Automata (NCA) is introduced and its capability to do mapping and computation is studied. The suggested model is similar to the original Cellular Automata, with its local neighborhood property, but neighborhood definition of cells do change in space and time (non-uniform) depending on the input-output template pairs introduced during the training phase.

Neighborhood degree of a cell can be defined as the number of neighbors that this cell uses to specify its next state. Only the cells having the same neighborhood degree have to evolve according to the same transition function rules. In other words, different neighborhood degrees may cause different transition functions.

NCA has two modes of operation similar to neural networks: training and mapping. At the beginning of the training mode, neighborhood of a cell is only itself. Neighborhood degree and transition functions for each cell gradually develop in accordance with the input patterns and the corresponding output bit desired to be mapped. At the end of the training, ultimate neighborhoods and cell functions are defined and one can introduce any input template to the system and get its intended output template. NCA can simply be considered as a parallel mapping system which makes its decision in one step of evolution.

In this paper, we have considered one-dimensional circularly connected cells. The length of an input template is the same as the length of an output template and is equal to the number of cells in the system. The algorithm used allows a cell to increase its neighborhood until all the templates are properly mapped. If the graph of the trained system has a strong component then the system will do computation.

The look-up table search and the lack of generalization are drawbacks to NCA. However, among some other possible ways of overcoming those drawbacks, relating the model to neural networks is discussed. Considering cells as neurons is one suggestion.

SUSOM
„SUpervised" Self-Organizing Maps

Harald Bayer
University of Stuttgart, IPVR
Breitwiesenstrasse 20-22, 7000 Stuttgart 80, Germany
email: bayer@informatik.uni-stuttgart.de

Self-organizing maps are very useful if the training vectors naturally fall into clusters. This paper presents an algorithm which will find a topology preserving mapping by combining the SOM algorithm with the use of available class information. While standard SOMs do not use classes, SUSOM makes use of them by adding a representation of the classes to the training vectors which are then used to train a map, thus improving the separation between the classes. When the training is complete, the weight vectors of the map are projected onto the original input space. Standard SOMs are a special case of the SUSOM algorithm.

Self-organizing maps (SOMs) form a topology preserving map of an input space. A map is defined to be topology preserving if it meets two conditions: first, neighboring neurons are activated by inputs that are close to each other in the input space, and second, two points which are close in the input space are mapped to the same or neighboring neurons. Not all authors agree upon the second condition as it holds only if the dimensionalities of the map and the subspace of the input space occupied by the training vectors are the same. If this is not the case the map folds and crincles, thus close points in the input space can be mapped to very distant neurons.

If additional information about the training vectors is available in form of class information, the trained map can be labeled accordingly. Each neuron is assigned the class which activated it most often. This implies an additional quality measure: Do the labels form clusters on the trained map? This will only be the case if the classes fall into natural clusters in the input space.

So if class information is available for the training vectors, the notion of topology preservation can be extended: A labeled map is topology preserving iff the map without labels satisfies the first condition of topology preservation and the labels form clusters on the map. The formation of label clusters replaces the second condition of topology preservation by substituting proximity of input vectors in the input space with semantic proximity. This also improves separation between close or overlapping classes.

SUSOM now replaces the original training vectors with vectors having more elements: The vector elements of the training vectors are maintained and some elements representing information about the classes are added. These additional elements are constructed to have the following properties: 1. They carry information about the distances between the classes. As a full distance matrix would require too many elements and is normally unknown a priori, it is a good guess to code the elements with constant distances between the classes. 2. They carry information about the „importance" of the class information. SOMs tend to concentrate on the part of the input space with the largest variances. So it is possible to increase or decrease the importance of an input vector element by scaling its range (and thus its variance) accordingly. If all elements of all input vectors are bounded within the interval $[0;1]$ and the elements are choosen as a binary 1-out-of-n vector, their variance is garanteed to be at least as large as the variance of the original vector elements. In this case the map concentrates mainly on separating the classes. Structure within classes is then of secondary importance. Note: Reducing the variance of the additional elements to 0 also reduces the importance of the class information to 0 resulting in the behavior of standard SOMs. These vectors are then used to train a map in the usual way. When this is done, the additional vector elements can be removed from the weight vectors.

Dynamical systems
—oral contributions

Synchrony in Integrate-and-Fire Networks

M. Tsodyks, I. Mitkov, and H. Sompolinsky
Racah Institute of Physics and Center for Neural Computation
Hebrew University, Jerusalem 91904, Israel

May 5, 1993

Abstract

It is shown that a network of globally coupled integrate-and-fire neurons with pulse interaction possesses a variety of dynamical states with different patterns of synchronization. In
the case of homogeneous external input, the network falls into the state of full synchrony in
which all oscillators are phase-locked. In a network of excitatory neurons, this state is unstable
to weak inhomogeneity: the system breaks into two subpopulations, one that is phase-locked
and another that consists of aperiodic oscillators, the overall network activity being a periodic
function of time. In the limit of vanishing inhomogeneity, the fraction of the unlocked population remains finite. Increasing the inhomogeneity quickly enters the network into the incoherent
state. Adding a population of inhibitory neurons stabilizes the synchronization substantially and
extends the dynamical variability of the system. Depending on the values of the parameters,
the system can display periodic activity with several subpopulations, or synchronized aperiodic
activity.

Whether a system of interacting oscillators will exhibit collective synchronized oscillations is
an important question from the dynamic point of view, as well as for applications to biology, and
in particular to the behavior of neural systems. Of particular interest is the question of whether a
phase-locked state is stable to inhomogeneity or noise in the local frequencies. This question has
been studied mostly in models in which the interaction between oscillators was taken to depend
smoothly on their state variables (see e.g. [1, 2, 3]). It was found that the phase-locked state is
stable to weak disorder or noise. In this paper we examine this issue in a system of oscillators with
pulse interactions, i.e., interactions that depend discontinuously on the state variables of oscillators.
Such interactions are of special interest for neural systems, where synaptic potentials are triggered
by the spiking of the presynaptic neurons. We first consider the network of excitatory neurons. We
show that for a small degree of inhomogeneity the neurons are divided into two populations, one of
which, consisting of the neurons with lower internal frequency, is phase-locked and characterized by
periodic spiking, whereas the other consists of unlocked neurons with aperiodic pattern of spiking,
which are only partially synchronized. Furthermore, the fraction of neurons in the phase-locked
population decreases as the inverse of the logarithm of the disorder that signals the rapid destruction of synchronization by the inhomogeneities. Despite the presence of two populations, the overall
activity of the network is a periodic function of time. These results are derived analytically, by
solving a set of self-consistent mean-field equations in the limit of weak disorder. Numerical results
supporting the analytical prediction are presented. We also present some results of simulations on
networks with a population of inhibitory neurons. In this case, the synchronization is much more
robust to inhomogeneities. Depending on the values of parameters, characterizing the relative effect of excitation and inhibition, and the distribution of external currents, the network can display
various patterns of synchronization. For small inhomogeneities in the excitatory population, the
network contains several populations of phase-locked neurons, which have spiking rates commensurate with each others', and a population of unlocked neurons with aperiodic spiking. Increasing
the values of inhomogeneity, the ratios between the rates of phase-locked populations jumps from
one rationale to another. The global activity of the network in all these states remains a periodic
function of time. Finally, adding inhomogeneity also in the inhibitory population leads to the new
dynamical regime of aperiodic synchronized activity.

We consider a network of fully connected *integrate-and-fire neurons*. The dynamics of neuron $i(i = 1, ..., N)$ are described by

$$\frac{dV_i}{dt} = -V_i + I + I_i^0 \tag{1}$$

$$\frac{dI(t)}{dt} = -\frac{I(t)}{\tau} + K\rho(t) \tag{2}$$

where

$$\rho(t) = \frac{1}{N} \sum_j \delta(t - t_j) \ . \tag{3}$$

Here $V_i(t)$ represents the membrane potential of the i-th neuron, and I_i^0 is an external, time-independent input to the neuron i. The current $I(t)$ is the mean-field synaptic current generated by the spikes of all the neurons in the system, and K is the normalized strength of the interaction. The normalized density of spikes is $\rho(t)$. The times t_j are the spiking times of the neuron j. These equations are supplemented by the condition that each time the potential V_i equals a threshold value, the neuron i emits a spike, and the value of its potential is set instantaneously at zero. Finally, τ is the integration-time constant of the synaptic current relative to that of V_i.

For the spatially homogeneous network with $I_i^0 = I_0 > 1$, simulations show that the phase-locked state is stable and is asymptotically reached from random initial conditions. In the phase-locked state, all the neurons fire simultaneously. The solution of equations (1) and (2) with $V_i = V$ is a periodic state with $\rho(t) = \sim_n \delta(t - nT)$. The period of oscillations T is determined from a simple self-consistency equation requiring that $V(t \to T^-) = 1$.

We focus on the effect of nonuniform local external currents on the phase-locked state. Let us assume that

$$I_i^0 = I_0 + \delta_i \ , \quad \Delta < \delta_i < \Delta \tag{4}$$

where Δ is the width of inhomogeneity, and we take for the sake of simplicity a uniform distribution of δ_i. The results of numerical simulation of such a network with $\Delta = 10^{-3}$ are shown in Fig. 1. The displayed spike density shows a substantial departure from full phase-locking even for such a small dispersion of local frequencies. Furthermore, the results indicate that the spike density consists of two contributions with substantially different widths. Further analysis shows that the two components originate from two distinct subpopulations. The narrow component represents neurons that are frequency-locked and phase-locked to the current I with fixed small phase shifts which depend on the local value of I_i. The rest of the neurons are not locked with $I(t)$ and exhibit aperiodic pattern of firing. Nevertheless, as is evident from the form of ρ that they do retain substantial degree of synchrony with I. In fact, they fire for a long time in close synchrony with $I(t)$ but undergo phase-slips which occur on a time scale that is long relative to the periodicity of I. The separation of the network into these two populations is demonstrated in Fig. 1, where the number of spikes fired by the different neurons in a fixed time window is displayed. It is seen that the phase-locked neurons consist of a slow population, i.e., $I_i^0 < I_c$, whereas the faster neurons with $I_i^0 > I_c$ are unlocked.

We explain these surprising results by finding a self-consistent solution of Eqs. ((1) and (2). Our basic assumption is that the self-consistent mean-field current is periodic. By studying the behavior of a single neuron in the periodic current, we then close the solution by demanding that I is consistent with the spike density ρ, as required by Eq. (2). The behavior of the neurons is best described by computing their phase, which is the time of their firing relative to the periodicity of the external current. These times are governed by a map $t_n \to t_{n+1} = \phi(t_n; I^0)$, where t_n is the phase of the neuron at the n-th period of $I(t)$, [4], defined as its time of spiking relative to the n-th peak of $I(t)$. This map can be easily computed from Eq. (1) (see [5] for more details). The map ϕ has a stable fixed point $t^0 = \phi(t^0; I^0)$ for the values of I^0 below some threshold value I_c. Corresponding neurons are phase-locked with the external current. Neurons with $I_i^0 > I_c$ are not phase-locked and have a spiking rate bigger than $1/T$.

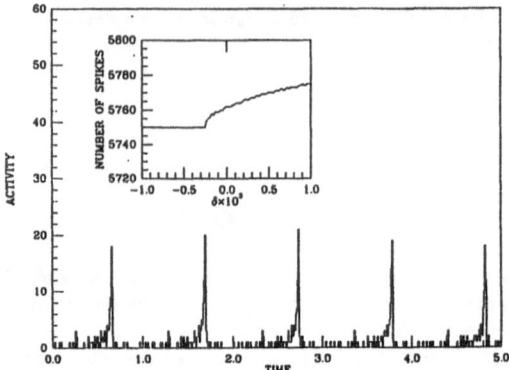

Figure 1: Simulated activity of the inhomogeneous network of 100 neurons, with $\Delta = 10^{-3}$. Parameters are $\tau = 0.5$, $K = 0.1$, and $I_0 = 1.5$. The number of spikes emitted by the network in the time interval $dt = 0.01$ is plotted vs. time. Inset - the number of spikes emitted by a neuron in a time window of length 6000, against the deviation of its external current from I_0.

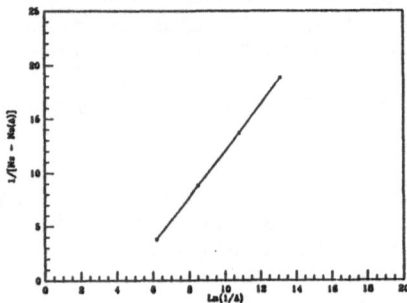

Figure 2: Simulation results of the fraction of phase-locked neurons for various values of Δ. N_s is obtained as the relative width of the plateau in Fig. 1.

As a result of the self-consistency condition we can obtain an asymptotic expression for a relative fraction of phase-locked population [5]:

$$N_s(\Delta) = N_s - k/|ln(\Delta)|. \tag{5}$$

The value of N_s is calculated as a function of parameters of the network. Surprisingly, the fraction of the population of unlocked neurons remains finite in the limit $\Delta \to 0$. However, these neurons become more and more synchronized with the first populations; thus the fully synchronized state is recovered in this limit. On the other hand, when Δ exceeds some threshold value (~ 0.01 for our choice of parameters), the fraction of phase-locked neurons shrinks to zero, and at the same point the network goes to a steady state, in which spike density and mean-field current are independent of time-constant values. This state is characterized by the absence of any synchronization between activities of different neurons.

Figure 2 presents simulation results for N_s vs $|\ln(\Delta)|$ with the same parameters as in Fig. 2. A linear relationship is obtained for $N_s \approx 0.65$. This value is in reasonable agreement with the analytical calculations, giving $N_s = 0.64$ for these parameters.

We now present some results for the integrate-and-fire network with a population of inhibitory neurons. The evolution of a neuron's membrane potential is described by the Eq. 1, and the mean-field current is given by a difference of two components, corresponding to excitatory and inhibitory populations: $I(t) = I^{ex}(t) - I^{inh}(t)$. These components are governed by the equations

$$\frac{dI^{ex,in}(t)}{dt} = -\frac{I^{ex,in}(t)}{\tau^{ex,in}} + K^{ex,in}\rho^{ex,in}(t) \ . \tag{6}$$

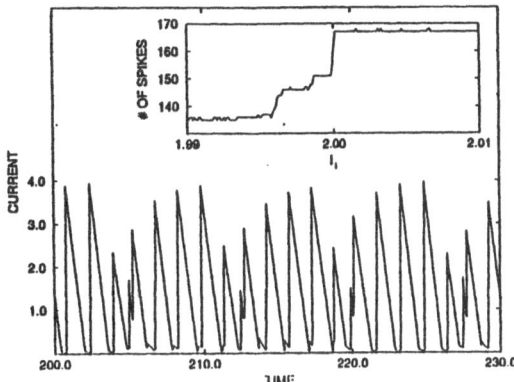

Figure 3: The mean-field current of excitatory population $I^{ext}(t)$. Inset - number of spikes emitted by an excitatory neuron as a function of its external current. Parameters are $\tau^{ext} = 0.3$, $\tau^{inh} = 0.8$, $J^{ext} = 4$, $J^{ext} = 2.8$, $I^0 = 2$, $\Delta^{ext} = 0.01$, $\Delta^{in} = 0$.

where $\rho^{ex}(t)(\rho^{in}(t))$ is the spike density of excitatory (inhibitory) neural population respectively:

$$\rho^{ex,in}(t) = \frac{1}{N^{ex,in}} \sum_j \delta(t - t_j^{ex,in}) \quad . \tag{7}$$

K^{ex} (K^{in}) is the normalized strength of excitatory (inhibitory) synapses. Finally, τ^{ex}, τ^{in} are the correspondent integration time constants of the synaptic currents.

Our simulations show that a network with inhibition has a variety of patterns of synchronization, depending on the relative values of parameters $K^{ex,in}$, $\tau^{ex,in}$ and $\Delta^{ex,in}$. Here we present some results for a network of 200 excitatory and 50 inhibitory neurons with $\tau^{ex} < \tau^{in}$, i.e., inhibition being slow relative to the excitation. In the case of homogeneous external input, i.e., $I_0^{ext} = I_0^{in}$ and $\Delta^{ex} = \Delta^{in} = 0$, the situation is the same as in the purely excitatory network: a fully synchronized state in which all the neurons are phase-locked is globally stable and is reached from random initital conditions. Introducing the inhomogenety in the external activation of the excitatory population leads to the division of the network into several populations, some containing neurons that are phase-locked and the others containing neurons that fire aperiodically. In this state the spiking rates of the different phase-locked populations are commensurate to each other, and the global activity of the network is periodic. Each phase-locked population gives rise to a plateau in the dependence of the number of spikes emitted by the neurons as a function of its external current. An example is shown in Fig. 3. In this case the ratio of the spiking rates of two main phase-locked populations is 4 : 5.

Another example, with greater inhomogeneity, is shown in Fig. 4. In this case, the network displays a regime with two phase-locked populations, with the period of spiking of the second population being twice as big as that of the first population. These populations manifest themselves in the two plaeaus in the dependence of the neuron's spiking rate on its external current. Note that the second plateau leads to the period doubling of the global activity of the network and of the mean-field current, which is shown in Fig. 4.

Finally, when inhomogeneities are added also in the external currents received by the inhibitory neurons, the network acquires the new dynamical regime, in which there are no phase-locked neurons, and the spiking rate of a neuron is a monotonically increasing function of its external current. However, in this regime the mean-field current is not a constant, as in the case of an excitatory network, but fluctuates irregularly with time. This behavior is illustrated in Fig. 5, where the mean-field current $I(t)$ and its power spectrum are plotted for some values of parameters. Nontrivial dependence of the mean-field current with time means that the activities of different neurons are correlated, because all the neurons are influenced by the same driving force. Thus, in this state the aperiodic activity is synchronized. Additional study is needed to check the robustness of this state in the networks of increasing sizes.

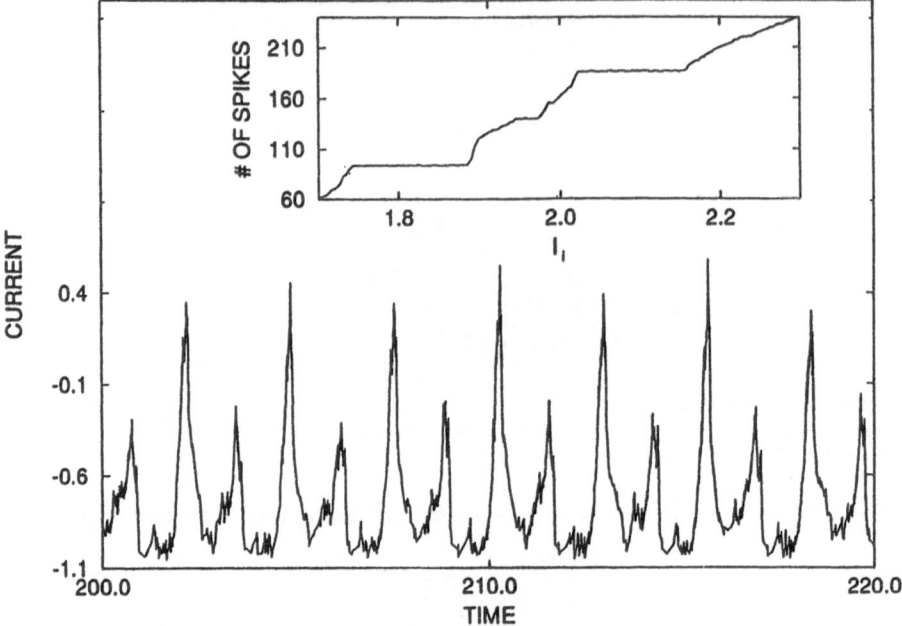

Figure 4: The mean-field current $I(t)$. Inset - number of spikes of an excitatory neuron as a function of its external current. Parameters are the same as in Fig. 3 but $\Delta^{ext} = 0.3$.

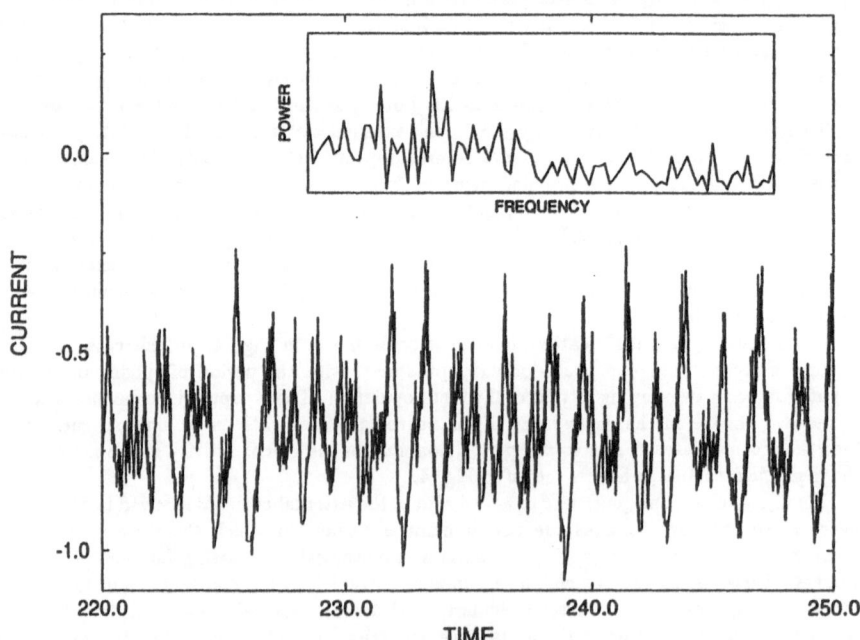

Figure 5: Mean-field current of excitatory population and its power spectrum. Parameters are the same as in Fig. 3 with $\Delta^{ex} = \Delta^{in} = 0.3$

In conclusion, we analyzed some possible patterns of synchronization in networks of integrate-and-fire neurons. In the case of excitatory networks, the synchronization is not robust to the inhomogeneities, and the network quickly enters into the regime with incoherent activity through the state with two distinct populations. Inhibition plays an important role in stabilizing the synchronization and extends the repertoire of possible dynamical regimes of the network. It is interesting that the effect of inhibition in chaotic synchronization was also observed in the simulations of a network of bursting neurons [6, 7]. Our study shows that a similar effect is obtained in the network of simple integrate-and-fire neurons. Given that integrate-and-fire model captures some basic features of neuronal activity, it gives added confidence in the biological relevance of these observations.

Acknowledgements

Most helpful discussions with I. Aranson, D. Hansel, and H. Mato are acknowledged. The work of MT is partially supported by the Israeli Ministry of Science and Technology and by a fellowship from Intel Electronics Ltd. HS is partially supported by the US-ISRAEL Binational Science Foundation.

References

[1] A. T. Winfree, J. Theor. Biol. **16**, 15 (1967).

[2] Y. Kuramoto, *Chemical Oscillations, Waves, and Turbulence* (Springer, Berlin, 1984).

[3] P. C. Matthews and S. H. Strogatz, Phys. Rev. Lett. **65**, 1701 (1990).

[4] J. P. Keener, F. C. Hoppenstead, and J.Rinzel, SIAM J. Appl. Math. **41**, 503 (1981).

[5] M. Tsodyks, I. Mitkov, and H. Sompolinsky, submitted to Phys. Rev. Lett.

[6] P.C. Bush and R. J. Douglas, Neural Comput. **3**, 19 (1991).

[7] D. Hansel and H. Sompolinsky, unpublished.

A NEURAL NETWORK FOR MOTION DETECTION

J.H. van Deemter and H.A.K. Mastebroek
Dept. of Biophysics, Physics Laboratory
Groningen State University
Nijenborgh 4 - 9747 AG Groningen
The Netherlands

Abstract- A neural network is presented to solve the motion correspondence problem. Solutions of this problem aim to maintain the identities of individuated elements as they move. In a preprocessing stage of image processing features are extracted from 2 snapshots of a moving scene. Each feature can be described by some attribute values. To learn which features in both frames are truely matching, a network is set up. This network consists of units divided up in 3 pools: one central pool and 2 attribute pools. Each central unit represents one possibly matching pair of features and each attribute unit represents a fixed difference or ratio in attribute value of a pair of features. The units are connected by fixed inhibitive and excitative weights specified by the 3 constraints 'group motion', 'one-to-one mapping' and 'minimal mapping'. After updating the activations of all units several times, the network finds a solution for the motion correspondence problem.

I. INTRODUCTION

Motion analysis provides a visual system with information that is used to detect moving objects and to get information about egomotion relative to the surroundings. It also contributes to tracking of objects, solving the structure-from-motion problem (i.e. recovering the 3D-structure of objects from their relative motion in 2D-images) and segmenting the optical image of the environment. Because of the variation in computational effort required for each of these tasks we assume that several different motion detection subsystems will cooperate in a more or less parallel manner.

Marr and Hildreth [1,2] suggested to compute a primitive description of the greylevel changes present in an image, the so-called primal sketch. This sketch consists of many sorts of changes (features), like blobs, lines, edges, shading edges, etc. To each of these sorts belong some parameters (attributes) like position, orientation, contrast, size, etc..

The problem of feature matching is to find the best correspondence between features in a frame at time t and features in a frame at time t+ Δt. Ullman [3] developed his "Minimal Mapping Theory" (M.M.T.) to perform this task. Minimal mapping is the process by which features in a given frame are matched to features in another (subsequent) frame such that the sum of the distances "travelled" is minimal. Ullman defined a sort of cost function which he tried to minimize by a linear programming method. Although this method always converges correctly, it is very slow (Grzywacz and Yuille [4]). To calculate the minimum of the cost function Grzywacz and Yuille used a neural network based on the network that Hopfield and Tank [5] used for solving the "traveling salesman problem". The time (t_c) needed for convergence on the network seemed to depend on the number of features (N): $t_c \approx C\sqrt{N}$ (C ≡ constant). This is much shorter than the best serial algorithms, which solved comparable problems in a time proportional to N^3 (Dinic and Kronrod [6]). A disadvantage of this network approach is that it suffers from the same problems as the Hopfield and Tank network, i.e. the solutions become worse when the number of features increases (N>30).

Schuling [7] proposed an alternative Minimal Mapping Theory that behaves properly with large numbers of features (calculation time proportional to N^2). Like Marr and Ullman he computed a primal sketch of the image consisting of N features, each described by a vector of M attributes. Instead of minimizing the total cost function he calculated translation-histograms for all attributes and multiplied these histograms. The most probable translation is indicated in the resulting histograms by a peak. Schuling's method is specialized in the detection of translations, but is unsuited for the detection of motiontypes other than translations. We developed a statistical correlation technique which is an improvement of Schuling's method (Van Deemter and Mastebroek [8,9]). This technique has the same properties concerning the computation time but is able to detect rotating and converging/ diverging movements as well. The main difference in strategy with Schuling's method is that we make a selection from the set of features (in this case we only use linepieces) and make use of their properties under rotation and divergence to detect 2D-motion. The statistical correlation technique is a base for the neural network which we will describe in the next paragraph.

II. A NEURAL NETWORK SOLUTION OF THE CORRESPONDENCE PROBLEM

As a basis for our network we used an interactive activation and competition network (McClelland and Rumelhart [10]). Our network contains units which are divided in 3 pools: one central pool and 2 attribute pools.

The central pool consists of units which each represent 2 possibly matching linepieces in both frames (N_1xN_2 units). The 2 attribute pools contain units which each correspond to one bin of one attribute value of 2 linepieces, viz. orientation difference and size ratio of 2 linepieces. The number of units in these attribute pools can be found by dividing the range of possible attribute values by the bin-size in the pool.

In the central pool every unit is excitatory connected by the same constant weight (w+) to those units in the other pools which represent the properties of its two linepieces. A central unit is only inhibitory connected to those other central units which represent the same linepiece in the first or the second frame. (e.g. If frame 1 contains linepieces A and B and frame 2 contains linepieces a and b, then the central pool contains units Aa, Ab, Ba and Bb. Of these 4 units the pairs (Aa,Ab), (Ba,Bb), (Aa,Ba) and (Ab,Bb) are inhibitory connected with the same strength). In each attribute pool all units are inhibitory connected to the other units within the same pool with the same constant weight

Scheme of the Network

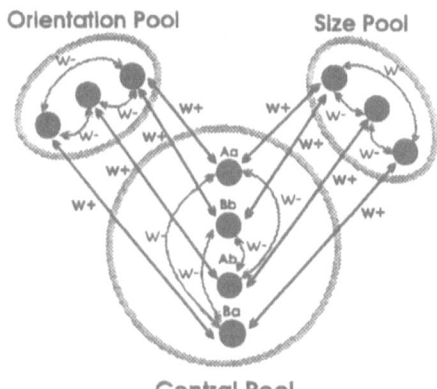

Fig.1. *A Scheme of the network is shown in this figure. It can be seen clearly that connections in every pool are inhibitory, while connections between different pools are excitatory.*

(w-). The inhibitive connections as well as the excitative connections are bidirectional (fig.1).

The activations of the units are restricted to continuous values between an upper bound (*max*) and a lower bound (*min*). If the activation of a unit i (a_i) is positive the output (*output$_i$*) of this unit is equal to its activation, otherwise the output is zero. The activations of all units change according to the following rules:

$$net_i = \sum_j w_{ij}.output_j \qquad (1)$$

in which net_i is the netto input of unit i and w_{ij} is the weight between unit i and unit j.

If ($net_i > 0$) then

$$\Delta a_i = (max - a_i).net_i - \text{decay}(d_i).(a_i - rest)$$

Otherwise $\qquad (2)$

$$\Delta a_i = (a_i - min).net_i - \text{decay}(d_i).(a_i - rest)$$

in which Δa_i is the change in activation of unit i, d_i is the distance between the 2 linepieces (which combination is represented by unit i), and *rest* is an activation value which a_i reaches if its input stays zero. Decay(d_i) is defined as:

$$\text{decay}(d_i) = \alpha.\exp(\beta.dd \,/\, ddmax) \qquad (3)$$

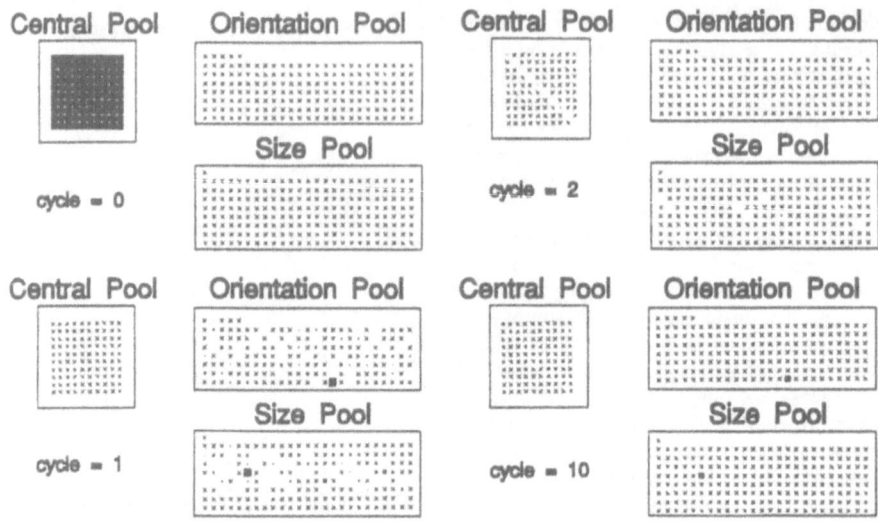

Fig.2. *The correspondence problem for linepieces which have rotated over 15° and diverged by factor 1.12 is put in the neural network. In this figure the activations of units are shown after different numbers of cycles. At the top of the figure a small scale for the activations is shown: The cross depicts a negative activation, while the row of black squares represents the activations of 0, 0.1, 0.2 .. 1.0. The central pool consists of 10x10 units, each of which represents one possible matching of 2 features in both frames. The Orientation pool consists of 180 units each representing a fixed orientation difference of 2 features (one unit per degree). The Size pool consists of 101 units which each depict a value $100 \times \log(L_2/L_1)$, in which L_2 and L_1 are the lengths of two features in frame 2 and frame 1 respectively. Initially the activations of the central units are all equal to max, while the activations in the other pools are all set to rest (upper left figure). After one cycle every unit in the central pool has received so much inhibitive input, that its activation has decreased below zero. At the same time units in the other pools have received excitative input from the central pool and some of their units representing the most common attribute differences have received the highest activations (lower left figure). After the next cycle central units possitively connected to the most strongly activated units in the attribute pools grow above zero, while in the attribute pools inhibition weakens the activations (upper right figure). When 10 cycles are completed, in each attribute pool only the unit with the highest activation has kept a positive activation, while the rest of the activations are below 0. In the central pool only the units representing the truely matching features have possitive activations (lower right figure), which is the solution of the correspondence problem.*

in which α, β and *ddmax* are constants.

Initially the units of the central pool all have the same activation value, because a priori it is unknown which features of both frames match. The activations in the 2 other pools are all set to zero. Every cycle all activations of the units in the network are changed according to formulas 1 and 2: First for every unit the input is calculated and depending on the height of this input the activation rizes towards $max = 1$ (high input) or decreases down to $min = -0.2$ (low input).

To get a better understanding of why the network is capable to solve the motion correspondence problem, it is useful to have a look at the constraints which underly the processing of the network. The first constraint (group motion) requests that central units which represent the most frequent orientation difference and size ratio of 2 linepieces are favoured above the other central units. The second constraint (one-to-one mapping) implies that units representing a similar linepiece in the first or second frame are suppressing each other. The last constraint (minimal mapping) favours units representing nearby lying linepieces above units representing linepieces which are lying farther apart.

As an example we rotated and diverged 10 randomly generated linepieces from frame 1 to frame 2 with an angle and factor of 15° and 1.12 respectively. To illustrate how this network solves the correspondence problem in time we show the positive activations of units after several updating cycles in fig.2. From the figure it can be seen that within a few cyles the correspondence problem is solved and the solution becomes more stable if we continue updating for some more cycles.

III. DISCUSSION

In order to get an impression of the performance of our neural network compared to the linear programming method of Ullman and the neural network of Grzywacz and Yuille, we tested our network for up to 200 features per frame. Although the values of excitative and inhibitive weights became more critical, the network still solved the problem within a few cycles. On the serial computer which we used the calculation times were excessive, but the network could be implemented on a parallel system to obtain the solution in a very short time. Compared to the methods of Ullman and Grzywacz and Yuille the computation time respectively maximal number of features have improved in our network.

The network shares many of the properties of the statistical correlation technique but it is a less roundabout method and solves the correspondence problem in a short time. An advantage of the network with respect to the statistical correlation technique is its smaller sensitivity to noise. When a correct peak in a distribution function is smaller than a peak due to noise, the statistical correlation technique is not able to overcome this problem, while the network combines activations from many units in the attribute pools so that it has a larger chance to overcome this pitfall.

Dawson [11] used psychophysical evidence for the constraints in his autoassociative network. Two of these constraints, minimal mapping and one-to-one mapping, are also used by Ullman, Grzywacz and Yuille and in our network. The third constraint which Dawson uses is the 'relative velocity' constraint, which favours nearby lying features with similar velocities above nearby lying features with different velocities. In one way this third constraint can be compared to our group motion because similar moving features favour each other. However, in the case of the relative velocity constraint this only happens locally while our group motion is a global constraint. A restriction of our neural network which is also found by Dawson in his network is that for every different scene a network with a different number of central units and different excitative connections has to be built. This means that the network can't be made hardwired and applicated to every moving scene. At this moment we are working on a solution of this problem by a network which contains structures like the network which we described here, but which does not depend on the number of linepieces in the 2 frames. The promising test results which we obtained with both networks are a stimulus for further research.

REFERENCES

[1] *D. Marr*, "Early Processing of Visual Information," Phil. Trans. R. Soc. London B, vol. 275, pp. 483-519, 1976

[2] *D. Marr and E. Hildreth*, "Theory of Edge Detection," Phil. Trans. R. Soc. London B, vol. 207, pp. 187-217, 1980

[3] *S. Ullman*, The Interpretation of Visual Motion, The MIT Press Cambridge, London and Massachusetts, England, 1979

[4] *N. M. Grzywacz and A. L. Yuille*, "Massively Parallel Implementations of Theories For Apparent Motion," Spatial Vision, vol.3, no.1, pp. 15-44, 1988

[5] *J. J. Hopfield, D. W. Tank*, "'Neural' Computations of Decisions in Optimization Problems," Biol. Cybern., vol.52, pp. 141- 152, 1985

[6] *E. A. Dinic and M. A. Kronrod*, "An Algorithm for the Solution of the Assignment Problem," Dokl. Akad. Nauk SSSR, Tom 189, no.1, pp. 1324-1326, 1969

[7] *F. H. Schuling, P. Altena, H.A.K. Mastebroek*, "The Computational Measurement of Apparent Motion: A Recurrent Pattern Recognition Strategy as an Approach to Solve the Correspondence Problem," Biol. Cybern., vol.62, pp.463-473, 1990

[8] *J. H. van Deemter and H. A. K. Mastebroek*, "A Neural Network Solution for the Correspondence Problem", Proceedings IEEE Int. Conf. on Neural Networks 1993, pp. 813-819

[9] *J. H. van Deemter and H. A. K. Mastebroek*, "A Statistical Correlation Technique and a Neural Network for the Motion Correspondence Problem", to be published in Biol. Cybern., 1993

[10] *J. L. McClelland and D. E. Rumelhart*, Ex- plorations in Parallel Distributed Processing, Chapter 2, The MIT Press, London, Massachusetts, England, 1988

[11] *M. R. W. Dawson*, "The How and Why of What Went Where in Apparent Motion: Modeling Solutions to the Motion Correspondence Problem," Psychological Review, vol. 98, no.4, pp.569-603, 1991

Spikes or Rates? – Stationary, Oscillatory, and Spatio-temporal States in an Associative Network of Spiking Neurons

Wulfram Gerstner and J. Leo van Hemmen
Physik-Department der TU München
D-85748 Garching bei München

Abstract: What happens if 'analog' or 'graded-response' neurons are replaced by *spiking* neurons? An analytical study of an associative network of spiking neurons, the 'spike response model', yields the following results: (i) In a stationary retrieval state, the network of spiking neurons can be reduced to a network of graded-response neurons. (ii) In a state of collective oscillations, however, the *spiking* of neuronal assemblies becomes important. (iii) Furthermore, there exist spatio-temporal states which cannot be described by firing rates or global activities. Here, the full information of the neuronal spike pattern is needed.

1 Introduction

So far, most theoretical studies of neural nets have concentrated on networks of 'analog' or 'graded-response' neurons. A graded-response neuron is defined by the gain function $f(I)$ where the output f is interpreted as the mean firing rate of the neuron. This *Ansatz* is based on the assumption that only firing *rates* are important and the time structure of neuronal *spike trains* can be neglected. Recent results based on a careful analysis of experimental data, however, suggest that the time structure of neuronal signals and even single spikes can be relevant in neuronal signal processing [1, 2]. Similarly it is known from information theory that coding by spikes or interspike intervals yields a much higher information capacity than a code which is based on the mean firing rate only [3, 4]. In view of these results the need for a network model with *spiking neurons* arises.

2 Model network

The neurons of our network are described by the Spike Response Model (SRM) which is characterized by two important response functions [5]: First, incoming spikes evoke an (excitatory or inhibitory) postsynaptic potential described by the response kernel $\epsilon(s)$. Second, spike emission causes a spike afterpotential (or increased threshold) which is described by the refractory function $\eta^{ref}(s)$. To be more precise, spike generation is modelled by a threshold process combined with a reset mechanism. If the internal variable h_i of neuron i crosses the threshold θ at time t_i^f, a spike is fired which travels along the axon to the synaptic connections. At the same time a negative contribution $\eta^{ref}(t - t_i^f)$ is added to the internal 'voltage' h_i

$$h_i(t) = h_i^{syn}(t) + \sum_{f=1}^{F} \eta^{ref}(t - t_i^f). \tag{1}$$

where h_i^{syn} describes the synaptic potential due to incoming spikes from other neurons. The sum is taken over all recent spikes, $1 \leq f \leq F$, of neuron i. The refractory function $\eta^{ref}(s)$ accounts for the reduced excitability (or higher threshold) of a neuron immediately after a first spike. It is set to $-\infty$ for $0 \leq s \leq \tau_{ref}$ where $\tau_{ref} = 4\ ms$ is the absolute refractory time and it decays to 0 afterwards, e.g., $\eta^{ref}(s) = -\eta_0/(s - \tau_{ref})$ for $s > \tau_{ref}$. It vanishes for $s < 0$.

Noise is modelled by the firing probability

$$P_F(h) = \frac{\delta t}{\tau(h)} \tag{2}$$

where δt is an infinitesimal time interval and $\tau(h)$ is a voltage dependent time constant $\tau(h) = \tau_0 \exp[-\beta(h - \theta)]$. For $\beta \to \infty$ we recover the noise free case, i.e., the neuron never fires, if $h < \theta$ ($\tau \to \infty$), and fires immediately, if $h > \theta$ ($\tau \to 0$).

Figure 1

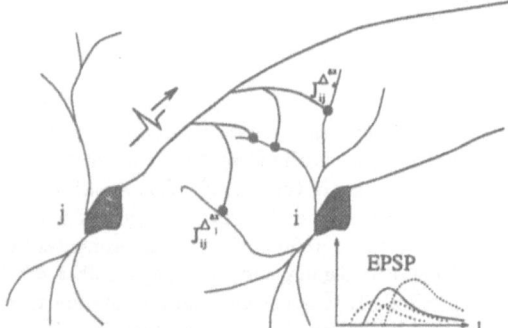

Neuronal signal transmission. Neurons are coupled through several synapses with variable delays. If neuron j fires a spike, the latter is transmitted to the synapses and evokes several postsynaptic potentials at the soma of neuron i. The EPSPs are delayed due to the axonal transmission time Δ^{ax}.

Every pair of neurons in the net is connected by one or more directed synaptic connections (Fig. 1). If neuron j fires, a spike is transmitted along the axon to the synapses. Consequently, one or several (excitatory) postsynaptic potentials (EPSPs) are evoked at the soma of the postsynaptic neuron. This is modelled by the response function $\epsilon(s) = [(s - \Delta^{ax})/\tau_s^2] \exp[-(s - \Delta^{ax})/\tau_s]$ for $s > \Delta^{ax}$ and 0 for $s < \Delta^{ax}$ with the parameter $\tau_s = 2\ ms$. This function is indicated in the inset on the lower right. It describes the smearing out of the sharp presynaptic signal caused by synaptic transmission and dendritic time constants. Due to the axonal transmission time, the EPSP is delayed by a time Δ^{ax} which is an important parameter of our model; cf. Fig. 3.

The EPSPs arriving from different synapses are weighted with the respective efficacy $J_{ij}^{\Delta^{ax}}$. If we add the contributions of all neurons and synapses, we find the total postsynaptic potential at the soma of neuron i,

$$h_i^{syn}(t) = \sum_{j=1}^{N} \sum_{\Delta^{ax}=\Delta_{min}}^{\Delta_{max}} J_{ij}^{\Delta^{ax}} \sum_{f=1}^{\infty} \epsilon(t - t_j^f). \tag{3}$$

Equations (1)-(3) define the network dynamics of the spike response model.

3 Results

Analytic solution

Patterns can be learnt by adjusting the synaptic weights $J_{ij}^{\Delta^{ax}}$ according to a local Hebbian learning rule [6, 9]. In the case of *stationary* random patterns, i.e., $\xi_i^\mu = \pm 1$ for $1 \le i \le N$ and $1 \le \mu \le q$ with equal probability, learning yields the standard efficacy matrix $J_{ij}^{\Delta^{ax}} = 2J_0 N^{-1} \sum_{\mu=1}^q \xi_i^\mu \xi_j^\mu$, independent of Δ^{ax}. In the following we set $F = 1$ and assume only a single axonal delay time, $\Delta_{min} = \Delta_{max}$. In this case an analytical solution can be given which is exact in the limit of an infinitely large network ($N \to \infty$)

$$A(\mathbf{x}, t) = \int_0^\infty ds\, A(\mathbf{x}, t - s)\tau^{-1}[h(\mathbf{x}, s, t)] \exp\left\{ -\int_0^s \tau^{-1}[h(\mathbf{x}, s', t - s + s')]ds' \right\}. \quad (4)$$

Here $A(\mathbf{x})$ is the activity of a *sublattice* $L(\mathbf{x})$ which defines an *assembly of equivalent neurons* [5, 6], i.e., $L(\mathbf{x}) = \{i | \xi_i^\mu = x^\mu\}$ where x^μ is the μth component of the sublattice vector $\mathbf{x} \in \{\pm 1\}^q$.

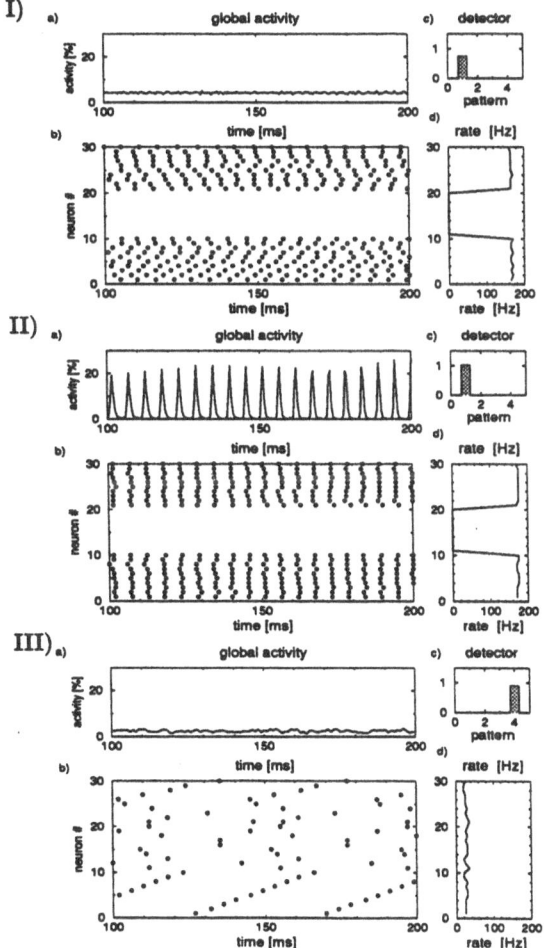

Figure 2

Neuronal activity patterns. I – stationary acivity or incoherent firing. II – oscillatory activity or coherent spiking. III – spatio-temporal acitvity pattern. In the *spike raster* b) each spike of a neuron $1 \le i \le 30$ (y-axis) is denoted by a dot at the corresponding time t (x-axis). If we count spikes along a horizontal line and divide by the total time, we find the *mean firing rate d)*. If spikes are counted vertically, this yields the *global activity a)*. Different spike patterns can be classified with the help of the *correlation detector c)*. Note that the incoherent state (I) can be characterized by the mean firing rate d) of the neurons, i.e., some neurons fire, others do not. The global activity a) can be used to distinguish the incoherent state from the coherent state (II) which shows large-amplitude oscillations of the activity. In the spatio-temporal state (III), neither firing rates nor spatial activity show any significant structure. Nevertheless a classification of the spike pattern based on the spatio-temporal correlations in the spike raster b) is possible. Note that both firing rate and spatial activity are low, as found in biological systems.

The integration kernel in (4), i.e.,

$$P_F^{(2)}(t|t-s) = \tau^{-1}[h(t)]\exp\left\{-\int_{t-s}^t \tau^{-1}[h(t')]dt'\right\},\qquad(5)$$

is the conditional probability that a neuron which has fired at time $t-s$ and has been subject to a potential $h(t')$ thereafter ($t-s \le t' \le t$) will fire again at time t. Here, the potential $h(t')$ has two components

$$h(\mathbf{x},s,t) = h^{syn}(\mathbf{x},t) + \eta^{ref}(s)\qquad(6)$$

where $\eta^{ref}(s)$ is the refractory potential due to the spike at $t-s$ and $h^{syn}(\mathbf{x},t)$ is the total postsynaptic potential evoked by the firing of neurons on other sublattices, viz.,

$$h^{syn}(\mathbf{x},t) = 2J_0 \sum_{\mu=1}^q x^\mu \int_0^\infty \epsilon(s')\sum_\mathbf{z} z^\mu p(\mathbf{z})A(\mathbf{z},t-s')ds'\qquad(7)$$

where $p(\mathbf{z})$ denotes the portion of neurons belonging to sublattice $L(\mathbf{z})$. Thus the network behaviour at time t has an intricate dependence on the spiking history as formalized by three different integrations over time; cf. Eqs. (4), (5), and (7) For a better understanding of the general structure of the solution we now consider two special cases, i.e., stationary and oscillatory solutions.

Universality of the stationary solution

A stationary solution can be defined by the condition that $A(\mathbf{x},t) \equiv A(\mathbf{x})$. This implies that all active neurons fire *incoherently*, see Fig. 2-I. In this case, the activity $A(\mathbf{x})$ of the assembly is equivalent to the mean firing rate of the neurons in this assembly, viz.,

$$A(\mathbf{x}) = f[h^{syn}(\mathbf{x})]\qquad(8)$$

where $h^{syn}(\mathbf{x}) = 2J_0\sum_{\mu=1}^q x^\mu \sum_\mathbf{z} z^\mu p(\mathbf{z})A(\mathbf{z})$ is the total postsynaptic potential of neurons in sublattice $L(\mathbf{x})$. At a given potential h^{syn} the mean firing rate f can be found from

$$f[h^{syn}] = \left\{\int_0^\infty ds\exp\left\{-\int_0^s \tau^{-1}[\eta^{ref}(s') + h^{syn}]ds'\right\}\right\}^{-1}.\qquad(9)$$

It can be shown that f as defined by (9) is the inverse of the mean interspike interval in a spike train of a neuron driven by the synaptic potential h^{syn} [5].

It follows from (8) that our network of spiking neurons can be reduced to a network of graded-response neurons which are defined by *mean firing rates* only. Thus, in a stationary state, a description by mean firing rates is sufficient [7]. This statement implies that all properties of standard associative nets can be translated into properties of spiking networks *in a stationary state*.

Coherent firing in an oscillatory state

Stationary states, however, are not necessarily stable. More strongly, it can be shown in the low-noise limit ($\beta \to \infty$) that the stationary state is always *unstable*, independent of the delay Δ^{ax} [10]. Noise increases the stability and allows stationary solutions, if the axonal delay is in a regime $\Delta_c^1 \le \Delta^{ax} \le \Delta_c^2$. The critical delays $\Delta_c^{1,2}$ depend on β and on the parameters of $\eta^{ref}(s)$ and $\epsilon(s)$. If the solution is unstable, an oscillation of the global activity evolves. This can be either a small-amplitude oscillation around the stationary state or a large-amplitude 'locked' oscillation – which is a completely different solution.

A collective or 'locked' oscillation can be defined by *coherent* spiking of all active neurons (Fig. 2-II). In the noiseless case, firing is exactly synchronous and is repeated periodically.

For the sake of simplicity we assume in the following that only a single pattern ν is active. If the collective oscillation has persisted for all $t \leq 0$, the activity is $A(\mathbf{x}, t) = \sum_{n=0}^{\infty} \delta(t + nT_{osc})$, if $x^{\nu} = +1$, and $A(\mathbf{x}, t) = 0$, if $x^{\nu} = -1$. The period T_{osc} can be derived selfconsistently from the threshold condition $h(\mathbf{x}, t) = \theta$ which must be fulfilled again at $t = T_{osc}$. Thus,

$$T_{osc} = \inf_s \left\{ J_0 \sum_{n=1}^{\infty} \epsilon(ns) + \sum_{n=1}^{F} \eta^{ref}(ns) = \theta \right\}. \tag{10}$$

If the contribution with $n = 1$ is dominant, then a simple graphical interpretation can be given, cf. Fig. 3. The first intersection of the effective threshold $\theta - \eta^{ref}(s)$ with the weighted EPSP $J_0\epsilon(s)$ yields the oscillation period. An analytical argument shows that locking is stable only if $\frac{d}{ds}\epsilon|_{T_{osc}} > 0$ [8].

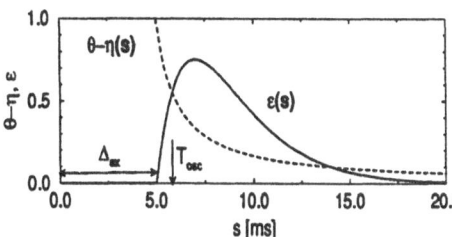

Figure 3
Coherent oscillations. The period T_{osc} of a coherent oscillation can be found from the intersection of the response function $J_0\epsilon(s)$ (solid) with the effective threshold $\theta - \eta^{ref}(s)$ (dashed) where $\eta^{ref}(s)$ is the refractory function. Locking is stable only if the slope of $\epsilon(s)$ is positive at the intersection. This poses a condition on the delay Δ^{ax}.

From the above considerations it is obvious that in an oscillatory state various details of spike transmission, viz., the delay Δ^{ax}, the shape of the EPSP, and the refractory behaviour of spiking neurons, become relevant. A rate description is therefore insufficient.

Spatio-temporal spike patterns

If the axonal delay time varies in a biologically plausible regime, say $1 \leq \Delta^{ax} \leq 4\ ms$, several arbitrary spatio-temporal spike patterns can be stored in the same network. In contrast to the standard efficacy matrix which we have used before, the synaptic weights $J_{ij}^{\Delta^{ax}}$ depend now explicitly on the delay Δ^{ax} [9]. Figure 2-III shows an example of a typical spatio-temporal spike pattern. Here the first ten neurons have learnt 'diagonal stripes' which are repeated every 40 ms. The other neurons fire in a random pattern which is also repeated. Note that neither the global activity Fig. 2-IIIa nor the mean firing rates d are sufficient to describe the pattern, but that the full information of the *spike raster b* is needed. This clearly shows the importance of a description by spikes.

4   Conclusion

It is only in the special case of *stationary* activity that neuronal spiking can be reduced to a firing-rate description. In the case of collective oscillations or general spatio-temporal spike patterns the *time structure* of neuronal signal transmission becomes important. In these cases, a theoretical Ansatz with *spiking* model neurons is essential.

Acknowledgements: This work has been supported by the Deutsche Forschungsgemeinschaft (DFG) under grant No. He 1729/2-1.

References

[1] Optican LM and Richmond BJ (1987) Temporal encoding of two-dimensional patterns by single units in primate inferior cortex. III. Information theoretic analysis. J. Neurophysiol. 57:162-178

[2] Bialek W, Rieke F, Ruyter van Stevenick RR, and Warland D (1991) Reading a neural code. Science 252:1854-1857

[3] MacKay DM, McCulloch WS (1952) The limiting information capacity of a neuronal link. Bull. of Mathm. Biophysics 14:127-135

[4] Stein RB (1967) The information capacity of nerve cells using a frequency code. Biophys. J. 7:797-826

[5] Gerstner W and van Hemmen JL (1992) Associative memory in a network of 'spiking' neurons. Network 3:139-164

[6] Herz AVM, Sulzer B, Kühn R, and van Hemmen JL (1988) The Hebb rule: Storing static and dynamic objects in an associative neural network. Europhys. Lett. 7:663-669

[7] Gerstner W and van Hemmen JL (1992) Universality in neural networks: The importance of the mean firing rate. Biol. Cybern. 67:195-205

[8] Gerstner W, Ritz R, and van Hemmen JL (1993) A biologically motivated and analytically soluble model of collective oscillations in the cortex: I. Theory of weak locking. Biol. Cybern. 68:363-374

[9] Gerstner W, Ritz R, and van Hemmen JL (1993) Why spikes? Hebbian learning and retrieval of time-resolved excitation patterns. Submitted to Biol. Cybern.

[10] Gerstner W and van Hemmen JL (1993) Coherence and incoherence in a globally coupled ensemble of pulse-emitting units. Preprint.

Cooperative Stochastic Effects In Globally Coupled Bistable Elements

A..R. Bulsara

NCCOSC-RDT&E Division, Materials Research Branch, San Diego, CA 92152-5000

ABSTRACT

We consider a network of N globally coupled bistable overdamped oscillators subject to Langevin noise and a weak periodic modulation. Assuming that one of the oscillators relaxes to its steady state on a time scale far slower than the remaining oscillators in the network, we extract its dynamics from the coupled stochastic differential equations describing the system, via adiabatic elimination. The bifurcation properties of this "reduced oscillator" model are discussed, together with cooperative stochastic effects (e.g. "stochastic resonance") that result from the interplay between the noise and modulation. Under suitably chosen operating conditions, this system is representative of globally coupled Hopfield-like circuit elements. It has also been proposed as a characterization of certain classes of neuro-dendritic interactions.

Recently, there has been an upsurge of interest in cooperative effects arising in networks of nonlinear oscillators interacting via mean-field type couplings [1]. The interest has spread to applications in biophysics and neural networks where a recent interest in single or few neuron dynamics has lead to the derivation of an effective single neuron model [2] starting from the continuum version of the connectionist neural network model of Hopfield [3], in the presence of Langevin and multiplicative fluctuations. In this work, we consider the influence of a large number of weakly nonlinear oscillators on the dynamics of a single (reference) nonlinear oscillator. The elements of the noisy network are assumed to include *a priori*, self-coupling terms as well as a weak, low-frequency periodic modulation:

$$C_i \, \dot{u}_i = \sum_{j=1}^{N} J_{ij} \tanh u_j - \frac{u_i}{R_i} + F_i(t) + q \sin \omega t . \tag{1}$$

An equation of this form describes a set of N nonlinearly coupled bistable oscillators. The $i=1$ index denotes the reference oscillator and the indices $i=2...N$ (where N is large) denote the "bath" oscillators. Systems of the form (1) have been used to describe connectionist-type electronic neural networks [4]. In such networks, u_i denotes the activation function (analogous to the membrane potential in neurophysiology) of the $i^{th.}$ element, C_i and R_i denoting the input capacitance and trans-membrane resistance. The coupling coefficients J_{ij} may be determined via a learning rule in an analog neural network framework. We assume the noise $F_i(t)$ to be Gaussian, delta-correlated with zero mean and variance σ_i^2 (the noise sources for different indices i are assumed to be uncorrelated).

We now assume that the time scale for relaxation of the reference oscillator is much longer than that for the bath:

$$C_i R_i \ll C_1 R_1 \quad (i > 1). \tag{2}$$

We may then adiabatically eliminate [2] the bath variables from (1). Specifically, an N-body Fokker Planck equation (FPE) is constructed from (1). Haken's slaving principle

[5] then permits us to decouple the N-body FPE into a FPE for the probability density function of u_1 (which contains the bath variables $u_{i>1}$) and another FPE for the bath variables. The latter is solved in the long-time limit, after invoking a local equilibrium assumption for the bath variables (this is tantamount to a quasi-linearization assumption for the bath dynamics). We are ultimately lead to a closed FPE for the slow variable u_1 whence a stochastic differential equation may be readily written down by inspection [6]:

$$\dot{u}_1 = -\alpha u_1 + \beta \tanh u_1 + \delta \sin \omega t + \sqrt{\sigma_e^2} F(t),$$ (3)

where,

$$\alpha = (R_1 C_1)^{-1}; \quad \beta = C_1^{-1}\left[J_{11} + \sum_{i>1} R_i G_i^{-1} J_{1i} J_{i1}\left(1 - \frac{\sigma_i^2 R_i}{2C_i}\right)\right];$$

$$\delta = \frac{q}{C_1}\left[1 + \sum_{i>1} R_i G_i^{-1} J_{1i}\left(1 - \frac{\sigma_i^2 R_i}{2C_i}\right)\right]; \quad G_i \equiv 1 - J_{ii} R_i; \quad \sigma_e^2 \equiv \sigma_1^2/C_1,$$ (4)

and $F(t)$ is now Gaussian delta-correlated noise having zero mean and unit variance. We have assumed further that the modulation frequency is smaller than the Kramers rate of the unmodulated system (the adiabatic assumption). Further, we assume that

$$\sigma_i^2 R_i < 2C_i \quad (i > 1).$$ (5)

This assumption [2,6] guarantees the convergence of the steepest descent techniques used to evaluate the coefficient β in (3) and places an upper limit on the noise strengths (with very large amounts of noise, the interesting cooperative behavior is lost).

The bifurcation properties of the reduced system (3) may be studied (in the absence of the noise and modulation terms) via the potential function,

$$U(u_1) = \frac{\alpha}{2} u_1^2 - \beta \ln \cosh u_1.$$ (6)

For positive α and β, the potential is bimodal (for $\beta/\alpha > 1$) with minima located at $c \approx \beta/\alpha \tanh(\beta/\alpha)$; in fact, we have global stability of the reduced dynamics (3) for $\alpha \geq 0$ with $U(u_1)$ being a Liapounov function. The transition to bimodality is accompanied by a pitchfork bifurcation in the most probable value of the activation u_1 with the two stable states (attractors) corresponding roughly to the quiescent and firing states (for the case when (3) is used to describe the reduced dynamics of an ensemble of neurons or a single exciteable cell coupled to a dendritic "bath" [6]). For this case the flow, given by the first two terms on the rhs of (3), exhibits the characteristic N-shaped characteristic known to exist in excitable cells [7]. The reduced dynamics (3) reproduces the long-time properties of the coupled system (1) very accurately when the inequality (2) is satisfied (in practice, one obtains reasonably good agreement when the inequality on (2) is far less stringent than predicated by theoretical considerations); this has been demonstrated in [6] via a comparison of the steady state probability density functions obtained from a numerical simulation of the coupled system (1) and from the reduced dynamics (3). In our subsequent analysis we assume that $\sigma_i^2 \equiv \sigma_2^2$ and $R_i \equiv R_2$ for the bath variables $i > 1$; also, we set $C_i = 1$ for all i. Further, the elements of the coupling matrix $\mathbf{J}$ are drawn from a Gaussian set having a specified mean and variance. Since, in general, N can be quite large, we must scale the coupling matrix by N to assure that the second and third terms in (3) do not become inordinately large. Henceforth, all the elements J_{ij} will be taken to be the scaled quantities J_{ij}/N. It is instructive to consider the effects of the bath dynamics on the transition to bimodality in the potential (6). From the expressions (4) we may obtain, approximately, the threshold value of R_2 at which the potential becomes bimodal (all other parameters being fixed). In the presence of a preponderance of excitatory (characterized by positive coupling coefficients) couplings the ratio β/α increases (for nonzero σ_2^2) upto a maximum value after which it decreases. The opposite effect occurs for the case of a mix of excitatory and inhibitory (characterized by negative coefficients) couplings. If the potential is monomodal in the absence of any coupling to the bath (this can be achieved by setting $J_{11}R_1 < 1$), then increasing R_2 leads to a transition to bimodality *only* for the case in which the sum $\sum_{i>1} J_{1i}J_{i1}$ is positive (keeping in mind the constraint imposed by the inequality (5)). This may be realized by imposing the same sign on the vast majority of the off-diagonal elements J_{1i} and J_{i1}. The coupling to the bath may thus actually introduce a phase-transition-like behavior, characterized by a *coupling-induced bimodality* in the effective neuron dynamics. The opposite effect can also occur: depending on the magnitude and sign of each element J_{ij}, a potential that is bistable in the absence of the bath coupling, can be rendered monostable by the bath.

We now consider the effects of the deterministic modulation, specifically stochastic resonance wherein a small amount of noise can introduce correlated switching events between the two stable states of the potential (6). An adiabatic theory, valid for very low frequency ω and weak amplitude q (such that there is no switching in the absence of noise) has been developed by McNamara and Wiesenfeld [8]. The central result of this theory is that if one computes the signal-to-noise-ratio (SNR) of a bistable system of the form (3) as a function of the noise variance, then the SNR passes through a maximum at a noise variance approximately equal to the potential barrier height. In the current context, we define the deterministic switching threshold as the critical value δ_c of the scaled modulation amplitude δ, above which one would obtain deterministic switching in the $\sigma_1^2 = 0 = \sigma_2^2$ case. Then, in order to satisfactorily explain stochastic resonance using adiabatic theory we must ensure that $\delta < \delta_c$ and $\omega < \omega_K$, the Kramers rate for the unmodulated system. The adiabatic conditions can be satisfied in the reduced dynamics (3) if we ensure that there is no deterministic switching in the isolated ($J_{1j} = 0$) case and we operate within the realm of validity (defined by (5)) of the theory.

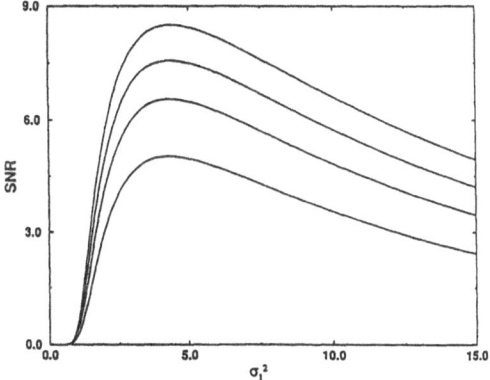

Fig 1. SNR vs. variance σ_1^2 for $(J_{11}, R_1, R_2, q, \omega, N) = (1, 10, 0.6, 0.1, 0.1, 100)$. $\bar{J}_{1i} = 1 = -\bar{J}_{i1}, J_{i1}^2 = 1 = J_{1i}^2$. $\bar{J}_{ii} = 0, J_{ii}^2 = 1, (i > 1)$. Bottom curve: $J_{1i} = 0$ (isolated case). Remaining curves: $\sigma_2^2 = 0$ (top), 1 (middle), and 2 (lower).

In figure 1 we show the SNR, obtained via the adiabatic theory [8], for the reduced equation (3). We take the elements of **J** to be a mix of excitatory and inhibitory couplings. The bottom curve shows the SNR that would be obtained for the isolated case ($J_{1i} = 0, i > 1$) with the remaining curves showing the effects of including the bath coupling with different values of the bath noise strength σ_2^2. The maximum enhancement is seen to occur for $\sigma_2^2 = 0$; increasing σ_2^2 degrades this enhancement. The important result is that the coupling to the bath enhances the SNR *even in the presence of noise* (recall that the inequality (5) imposes an upper limit on the noise). The enhancement of the SNR may be explained by observing that increasing R_2 from zero causes the ratio β/α (and therefore, the potential barrier height U_0), for this configuration of $\mathbf{J}$, to initially decrease and then increase. The renormalized modulation amplitude δ, however, can only increase since we have taken the set J_{1i} to be mainly excitatory in nature. Hence, one obtains a marked increase in the SNR as the potential barrier height decreases. Past the extremum of β/α the opposite effect occurs. A similar but smaller enhancement of the SNR occurs for the case of most of the off-diagonal elements of $\mathbf{J}$ being excitatory ($\bar{J}_{1i}=1=\bar{J}_{i1}$), or inhibitory. In both these cases, the barrier height increases; however the renormalized signal amplitude δ increases in the first case and decreases in the second case so that there are fewer switching events. Figure 2 shows the peak SNR (normalized to its value for the isolated, i.e., $J_{1j}=0$ case) as a function of R_2 for different bath noise strengths. This figure clearly shows that increasing the bath noise leads to a lower enhancement in the SNR. However, it is also important to point out that the bath may actually *induce* stochastic resonance through the coupling-induced bimodality (for the case when the slow oscillator is monostable in the absence of the bath) described earlier. The above analysis underscores the importance of the magnitudes as well as the *signs* of the interactions J_{ij}. In neurophysiological terms, we could argue that having a statistical mix of excitatory and inhibitory couplings provides superior performance to having all the couplings of the same sign.

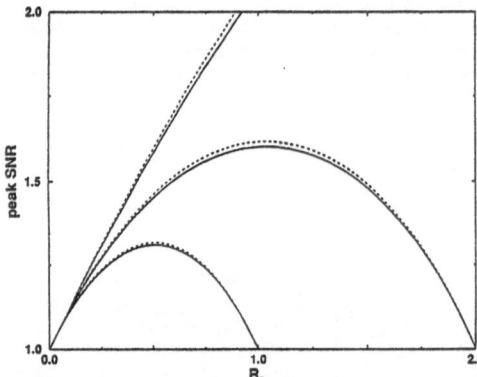

Fig 2. Peak SNR (normalized to its value for $J_{1i}=0$) vs. R_2 for $(J_{11},R_1,q,\omega,N)=(1,10,0.1,0.1,100)$ and $\sigma_2^2=0$ (top curve), 1 (middle curve), 2 (bottom curve). $\bar{J}_{ii}=0,\bar{J}_{ii}^2=1,(i>1)$. Solid curves: $\bar{J}_{1i}=1=\bar{J}_{i1},\bar{J}_{1i}^2=1=\bar{J}_{i1}^2$. Dotted curves: $\bar{J}_{1i}=1=-\bar{J}_{i1},\bar{J}_{1i}^2=1=\bar{J}_{i1}^2$.

The foregoing treatment clearly points up the benefits (from a signal/information processing standpoint) of coupling the reference oscillator to the bath with a faster time constant. The coupling clearly enhances the information flow (measured by the SNR) to the system output. Clearly, in order to achieve the best possible SNR (in the context of the adiabatic theory employed here) the potential barrier height in the effective dynamics (3) should be decreased and the effective modulation δ increased via the coupling to the bath. If the noise strength σ_2^2 is approximately known, then the optimum SNR can be achieved by setting $R_{2c}\approx\sigma_2^{-2}(1+h)$ where h satisfies the quadratic (taking the negative square-root),

$$(1-\sigma_2^{-2}H_1+\frac{5}{2}\sigma_2^{-4}H_2)h^2+(2\sigma_2^{-4}H_2-1)h+\frac{1}{4}\sigma_2^{-4}H_2+\frac{1}{2}\sigma_2^{-2}H_0=0$$

with the definitions $H_1\equiv\sum_{i>1}J_{1i}J_{i1}J_{ii}$, $H_2\equiv\sum_{i>1}J_{1i}J_{i1}J_{ii}^2$ and $H_0\equiv\sum_{i>1}J_{1i}J_{i1}$. If the external noise σ_1^2 and/or the signal amplitude q are approximately known then further optimization can be achieved by adjusting the coupling parameters J_{ij} such that the potential barrier height is roughly equal to σ_f^2 and, the effective signal amplitude δ, in the reduced system (3), is very close to the deterministic switching threshold δ_c. When suitably optimized, a network of oscillators such as (3) may well provide signal processing/detection capabilities that are far outside the purview of a single (i.e. isolated) oscillator.

We conclude with some comments regarding the role of noise in the processing of information by sensory neurons. It is important to point out, at the outset, that the inequality (2) which plays a central role in the reduction of the system (1) may hold for only a limited class of biological neural networks (or models of neuro-dendritic interaction). However, neurons are known to exhibit bistability at least over limited ranges of parameter values. Even the very simple integrate-fire models can be regarded as bistable in the sense that they treat neurons as driven two-state systems which reset themselves after each crossing of a boundary; all such systems are not, however, characterized by a bistable potential of the form (6). Stochastic resonance has been investigated [9] for a single (isolated) element of the form (3) with arbitrary α and β (the effects of multiplicative noise, i.e. a fluctuating β have also been investigated in this work) and has been observed [10], using a weak applied signal in *external* noise, in single-hair mechanoreceptors of the crayfish, *Procambarus clarkii*. Since noise is ubiquitous in the central nervous system (often several orders of magnitude larger than predicated by equilibrium statistical mechanics), one might expect that it plays a central role in the encoding of sensory information, exemplified for example by the Inter-Spike-Interval Histogram (ISIH). In a recent study, Longtin, Bulsara and Moss [11] demonstrated how experimental ISIHs could be explained via model noise driven bistable dynamics. This theory provides one of the simplest interpretations of the ISIHs. It explains [12] most of their salient features and is able to (via analog simulation) provide exceptionally good fits to experimental data by adjusting only one parameter (the signal or the noise strength) assuming the other to be comparable to the height of the potential barrier in the bistable model. The effects of using colored Gaussian noise, characterized by an exponentially decaying autocorrelation function, in place of delta-correlated noise have also been discussed in this work. It is important, however, to bear in mind the following caveat: while the ISIHs are not, by themselves, indicative of the presence of stochastic resonance as an underlying cooperative effect in neuroscience, many of their features lend themselves to an interpretation based on stochastic resonance [13]. Perhaps the most intrigueing feature is the fact that the ISIHs *cannot exist in the absence of noise*. This leads one to speculate whether noise is involved in the processing and transmission of sensory information. It is in fact tantalizing to speculate that neurophysiological systems constantly adjust the coupling coefficients J_{ij} as well as the remaining (internal) parameters in the dynamics (1) such that the effective potential function (3) admits of more than one minimum. Further, the height U_0 of the potential barrier is internally adjusted in response to the stimulus and background noise characteristics. This is an important point since, for given stimulus and noise, one obtains well-defined multi-peaked ISIHs for only a small range of U_0. The network then operates as a bistable switching element and actually can use the background noise so that its response (measured by the SNR or, equivalently, through the ISIH) is optimized. This implies that the network operates close to the maxima of the stochastic resonance curves of figure 1, while simultaneously obtaining other information about the stimulus (e.g. amplitude and frequency) via the ISIH. In effect, our construction and interpretation of the ISIH (together with the remarkable ability to explain most of the features of experimentally obtained ISIHs) as a natural outcome of our modelling the neuron as a noisy bistable switching element implies that the central nervous may measure the stimulus intensity by comparing it to the background noise, using the (internally adjusted) potential barrier height to optimize the measurement [F. Moss, 1991-92, private discussions]. Throughout this work, the underlying thread has been the positive role of noise; this, in fact, was recognized earlier by Buhmann and Schulten [14] who found that noise, deliberately added to the deterministic equations governing individual neurons in an artificial network, significantly enhanced the network's performance. They concluded that *"...the noise...is an essential feature of the information processing capabilities of the neural network, and not a mere source of disturbance, better suppressed..."*

Bibliography

1. See e.g. R. Desai and R. Zwanzig, J. Stat. Phys. 67, 313 (1978); J. Brey, J. Casado and M. Morillo, Physica 128A", 497 (1984); M. Shiino, Phys. Rev. A36, 2393 (1987); H. Sakaguchi, Progr. Theor. Phys. 79, 39 (1988); S. Strogatz and R. Mirollo, J. Stat. Phys. 63, 613 (1991); K. Wiesenfeld, Phys. Rev. B45, 431 (1992).
2. W. Schieve, A. Bulsara and G. Davis, Phys. Rev. A43, 2613 (1991)
3. J. Hopfield, Proc. Nat. Acad. Sci. 79, 2554 (1982); 81, 3088 (1984).
4. A. Maren, C. Harston and R. Pap, "Handbook of neural computing applications" (Academic Press, NY 1990).
5. H. Haken, "Synergetics" (Springer Verlag, Berlin, 1977).
6. A. Bulsara, A. Maren and G. Schmera, Biol Cyb. preprint.
7. J. Rinzel and B. Ermentrout; in "Methods of neuronal modelling," eds. C. Koch and I. Segev (MIT Press, Cambridge, MA 1989).
8. B. McNamara and K. Wiesenfeld, Phys. Rev. A39, 4854 (1989).
9. A. Bulsara, E. Jacobs, T. Zhou, F. Moss and L. Kiss, J. Theor. Biol. 152, 531 (1991).
10. J. Douglass, F. Moss and A. Longtin; in "Advances in neural information processing systems 4" (Morgan Kauffmann, 1993)
11. A. Longtin, A. Bulsara and F. Moss; Phys. Rev. Lett. 67, 656 (1991); see also Nature 352, 469 (1991).
12. A. Longtin, A. Bulsara, D. Pierson and F. Moss; Biol. Cyb. preprint. A. Longtin; J. Stat. Phys. 70, 309 (1993).
13. T. Zhou, F. Moss and P. Jung; Phys. Rev. A42, 3161 (1991).
14. J. Buhmann and K. Schulten; in "Neural networks for computing," ed. J. Denker (AIP, New York 1986); Biol. Cyb. 61, 313 (1987).

Dynamical systems
—poster contributions

Biologically Inspired Neural Network for

Trajectory Formation and Obstacle

Avoidance

R. Glasius , A. Komoda , S. Gielen

Department of Medical Physics and Biophysics, University of Nijmegen, Geert Grooteplein 21, 6525 EZ Nijmegen, The Netherlands

Abstract: We propose a biologically plausible neural network for trajectory formation and obstacle avoidance. Our model is a two-layer network of analog neurons with feed-forward connections between layers. The first layer, a sensory map, is a topologically organized network with short range excitatory connections. The second layer, a motor map, is a network with short-range excitatory and long-range inhibitory connections. The external input is assumed to activate a target neuron, corresponding to the target position, and to specify obstacles in the sensory map. The trajectory follows then from the neural network dynamics and is visualized by a neural activity cluster which travels through the motor map.

1 Introduction

Scientific investigation of the brain has resulted in models for neural information processing, underlying perception and action in animal and men. Examples are visual control of eye movement[1], identification and localization of acoustic sources[2], control of body, arm and hand motion[3]. Essential aspects of neural information processing are highly parallel execution of computation as well as performance which is robust against localized damages.

In this paper we present a biologically plausible model for robotic systems which can realize goal-oriented behavior in a complex environment and which captures most of the above mentioned properties. Beyond its functional realism, the model of interconnected maps (fig.1) captures the known physical structure of laminar cortex or midbrain superior colliculus [4].

2 The model

Our neural network is a collection of a large number of identical processing units called neurons. These neurons are arranged in a two-layer structure with a feed-forward connections between them. The first layer, called the sensory map, is a discrete topologically ordered representation of the external world. The strength of the synaptic connections in this map from neuron i to neuron j is represented by T_{ij}. They are excitatory ($T_{ij} > 0$), symmetric ($T_{ij} = T_{ji}$) and short range:

$$T_{ij} = \begin{cases} 1 & \text{if } 0 < \|i - j\| < r \\ 0 & \text{otherwise} \end{cases}$$

where r is a positive number and $\|i - j\|$ stands for the Euclidean distance between neuron i and j. The second layer is called the motor map. Here the connections are represented by

W_{ij} which is a sum of short-range interactions of the same type as in the sensory map, and of long-range symmetric and inhibitory connections J_{ij}

$$J_{ij} = \begin{cases} J_0 & \text{if } i \neq j \,,\, J_0 < 0 \\ 0 & \text{otherwise} \end{cases}$$

The feed-forward connections are such that only one neuron in the sensory map is connected to only one neuron in the motor map.

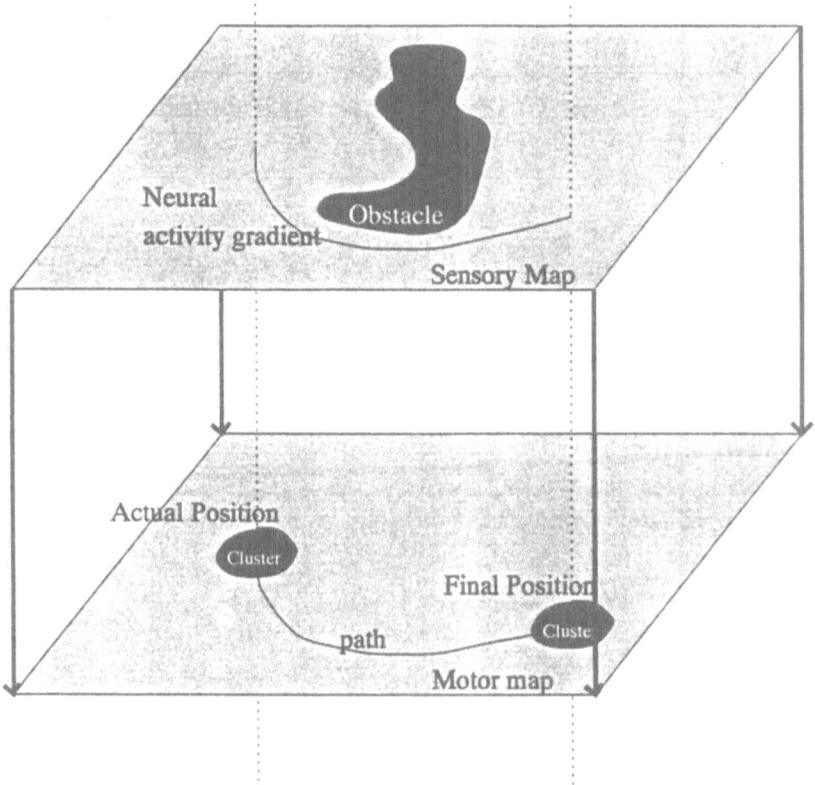

Figure 1. A schematic representation of the two-dimensional sensory and motor maps as a two-layer neural network.

The activities of the neurons are characterized by real-valued variables which are elements of the closed interval [-1,1]. A set $\mathcal{S}$ of all possible states of the network is called a phase space of the system. In our model $\mathcal{S} = [-1,1]^N$, where $[-1,1]^N$ stands for a hypercube in $\mathcal{R}^N$ of side [-1,1].

The external input to the sensory map is supposed to clamp the activities of neurons which correspond to the position of objects in the external world to the value "zero". The activity of the neuron corresponding to the target position, is clamped to the value "one".

3 Dynamics

We will now describe the dynamics of the network. The state variables σ_i, $i = 1 \ldots N$, of the neurons can change due to inputs from other neurons in the network and from an external input. The continuous-time evolution of activity of the neurons is given by the set of nonlinear differential equations. In the sensory map

$$\frac{d\sigma_i^s(t)}{dt} = \sum_j^N T_{ij}\sigma_j^s(t) + I_i^s(t) - \sigma_i^s(t) \tag{1}$$

where $g(x)$ is a monotonously increasing input-output transfer function, and where I_i^s is the external input to the neuron i in the sensory map. It is easy to verify that for this system the function

$$L_s(\sigma) = -\frac{1}{2}\sum_{i,j} T_{ij}\sigma_i^s\sigma_j^s - \sum_i I_i^s\sigma_i^s + \sum_i G(\sigma_i^s), \tag{2}$$

with $G(\sigma_i) = \int_0^{\sigma_i} g^{-1}(x)dx$, is a global Liapunov function. The existence of a Liapunov function guarantees that the dynamics of the network always converges to a fixed point. In the motor map

$$\frac{d\sigma_i^m(t)}{dt} = \sum_j^N W_{ij}\sigma_j^m(t) + J_0\mu + I_i^m - \sigma_i^m(t), \tag{3}$$

where $I_i^m = \alpha(1 - \sigma_i^s)$ is an input from the sensory map to the neuron i in the motor map and $\alpha < 0$. The parameter $\mu \in [-1, 1]$ defines the network activity level and the inhibitory interactions facilitate the stabilization of that level. The short-range interactions T_{ij} guarantee that all active neurons will form a connected cluster. The Liapunov function for the motor map reads

$$L_m(\sigma) = -\frac{1}{2}\sum_{i,j} W_{ij}\sigma_i^m\sigma_j^m - J_0\mu\sum_i \sigma_i^m - \sum_i I_i\sigma_i^m + \sum_i G(\sigma_i^m), \tag{4}$$

The set of equations (1) and (3) is a particular case of the more general equations studied by Cohen and Grossberg[5] and have the form of equations proposed by Hopfield[6].

At the initial time, $t = 0$, the activity of all neurons is set to "zero" except for a small cluster of neurons in the motor map corresponding to the initial position of the actuator. Under an external input to the sensory map, which clamps the activity of neurons, which correspond to the position of objects in the external world, to the value "zero" and the activity of the neuron corresponding to the target position to the value "one, the state of the system begins to change according to the dynamics of the network described by (1) and (3). The evolution of the network can be seen in the phase space S as a motion of a point along the curve on which the Liapunov function decreases. This down-hill motion ends when the network reaches an equilibrium state which is a local or global minimum of L. During this evolution the cluster of active neurons in the motor map travels with constant velocity to a new stable position. This motion stops when the center of mass of the cluster reachs the target position.

The result of numerical computer simulation demonstrate that this network always provides a feasible pathway from the initial position to the goal position without collision with obstacles of any shape. This is true as long as a collision free trajectory exist, irrespective of the complexity of the working space.

4 Conclusions

We proposed a model for trajectory formation and obstacles avoidance based on an analog neural network. When the network receives an external input, the initial state of the network starts to change towards a specific state which is a minimum of the Liapunov function. The direction of the motor response, the trajectory, is determined by the neural activity gradient in the sensory map.

Three points are worth noting about the functional properties of the network. First, it will remain functional despite local damages. The quality of the path will be somewhat degraded, but it will be still a feasible path. Secondly, continuous neurons eliminate oscillations related to the parallel dynamics of discrete neurons[8]. Third, the parallel architecture and dynamics of the model gives it a large speed needed for real-time processing of sensory information since all of the neurons simultaneously and continuously change their analog states. All three of these functional properties are biologically realistic.

Since the basic idea of the network can be expressed by electrical circuits, modern technologies should give possibilities to implement the network with a large number of processing elements in hardware.

Acknowlegdments
This work was partly supported by the Dutch Foundation for Neural Networks.

References

[1] R. H. S. Carpenter, Movement of the eyes. *Pion Lmtd., London (1988)*

[2] J. Blauert, Spatial hearing. The Psychophysics of human sound localization., *MIT Press, Cambridge, USA*

[3] Y. Burnod, P. Grandguillaume, I. Otto, S. Ferraina, P.B. Johnson, R. Caminiti, Visuomotor Transformations underlying arm movements toward visual targets: a neural network model of cerebral cortical operations., *The J. of Neuroscience 12 (1992) 1435*

[4] D. L. Sparks, Translation of sensory signals into commands or control of saccadic eye movements: role of primate superior colliculus. *Physiol. Rev. 66 (1983) 118*

[5] M. A. Cohen, S. Grossberg, Absolute Stability of Global Pattern Formation and Parallel Memory Storage by Competitive Neural Networks, *Transactions on Systems,Man, and Cybernetics 13 (1983) 815*

[6] J. J. Hopfield, Neurons with graded response have collective computational properties like those of two-state neurons, *Proc. Natl. Acad. Sci. USA 81 (1994) 81*

[7] R. Camaniti, P. B. Johnson, Y. Burnod, C. Galli, S. Ferraina, Shifts of preferred directions of premotor cortical cells with arm movements performed across the work-space, *Exp. Brain Res. 83 (1990) 228*

[8] C. M. Marcus, F. R. Waugh, and R. M. Westervelt, Associative memory in an analog iterated-map neural network, *Phys. Rev. A 41 (1990) 3355*

Catastrophic Phase Transitions in Exact ART Networks

Maartje E.J. Raijmakers and Peter C.M. Molenaar
Department of Developmental Psychology
University of Amsterdam, Amsterdam, The Netherlands
E-mail: op_raijmakers@macmail.psy.uva.nl

Abstract

To study the occurrence of sudden transitions in code development, bifurcation analyses of ART networks were carried out. In the interest of biological plausibility, we attempted to implement each ART network completely, including all regulatory and logical functions, as a system of differential equations capable of stand-alone running in real time. In particular, transient network behaviour thus remains intact because no asymptotic approximations were used. The most important functions of Exact ART are emergent properties of the network. Preliminary results of bifurcation analyses are presented. In closing, alternative connectionistic analyses of phase transitions are criticized and it is concluded that these analyses fall short on several accounts compared to Exact ART networks.

Summary

Theoretical background

Piagetian and neo-Piagetian research is traditionally concerned with stage transitions in human cognitive development. Recently, a catastrophe theoretical analysis of cognitive stage transitions was carried out and the ubiquity of such transitions was established by means of phenomenological methods drawn from topological singularity theory [1]. In order to arrive at a more specific analysis of these transitions, a set of biologically plausible ART networks, which are capable of generating code development, were simulated in real time.

The basic experimental paradigm in a bifurcation analysis involves determination of the manifold of stable equilibrium states in the phase space of network behaviour under continuous variation of a few network parameters. The parameter plane is partitioned in regions within which the network carries out equivalent equilibrium behaviour. A phase diagram can represent the behaviour of the system within a region. For example, bifurcation analysis is carried out to map the behaviour of so called neural oscillators [2,3].

In these oscillatory systems, the presence of distinct behavioural regimes fulfils a function in feature dependent synchronization (see also [4]). We suggest that it is possible to establish a connection between qualitatively different dynamic regimes in a network and stagewise cognitive development. The boundaries between distinct behavioural domains demarcate regions of instability between qualitative different dynamic regimes. The transition from one region to another is a discontinuous change of equilibrium behaviour of the system although the systems implementation (i.e. procedure) is continuous. This can be initially modelled by catastrophe theory. Thom [5] has proven that a large class of dynamical systems (involving up to four control variables) showing discontinuous behaviour can be classified into seven prototypical forms by means of smooth transformations of the system variables. Beside providing general topological models of discontinuous behaviour, catastrophe theory can also be applied to distinguish discontinuous transitions from continuous accelerations [6]. A system in the neighbourhood of a transition shows several specific characteristics, so called catastrophe flags. Among these flags are hysteresis and bimodality. A discontinuous transition, according to catastrophe theory, can be identified by establishing the presence of the flags without conducting further mathematical analyses of the system.

A stand-alone continuous ART network: Exact ART

ART-networks make up a class of neural networks which learn pattern recognition categories in response to arbitrary sequences of either binary (ART1, 2 & 3) or analog (ART2 & 3) patterns [7,8,9]. Learning takes place without supervision and in real time. Being able to handle the stability-plasticity dilemma is an important quality of ART that is essential for a biologically plausible, stand-alone network. Using the ART2 network as a point of departure, we implemented a completely continuous defined and stand-alone ART network, Exact ART. Although the dynamics of the original network are defined by a system of differential equations, the simulation procedure is commonly simplified (some aspects have been discontinuously or globally regulated, or have been approximated by making use of asymptotic dynamics). Carpenter and Grossberg [8] emphasized the efficiency of their implementation. We, on the other hand, concentrate on biological plausibility and employ ideas in Grossberg [10] to overcome the discontinuous aspects of the simulation procedure.

Firstly the multi-layer F1-structure is replaced by three layers of biological plausible networks, which are connected in a feedforward way (Input→ L1→ L2→ L3). L1 and L2 are feedforward competitive networks (A12, pg 46 of [11]) and L3 is a recurrent shunting network with linear signal function (A18, pg 48 of [11]). L1 enhances the contrast of the input, suppresses noise and prevents saturation of the system. L2 normalizes the input pattern regarding the summation of the positive activity (proofs in [11,12]). By approximation, the activity of L3 ($p = (p_1,..,p_i,..p_n)$) results in a normalized weighted sum of a bottom-up ($x = (x_1,..,x_i,..x_n)$ in L2) and a top-down (F2 * TD-weights = $z_J = (z_{J1},..,z_{Ji},..z_{Jn})$) pattern (i.e. $p_i = (1-\alpha\Sigma z_{Ji}) x_i + \alpha z_{Ji}$, α is constant; proof in [18]). In contrast to L1 and L2, L3 is sensitive to initial values and therefore must be reset in case of a mismatch (the latter also applies to F1 in Art2 [9]).

Secondly, the asymptotic reset procedure commonly used in simulation experiments of ART has been replaced by a real-time procedure in which L2 and L3 are compared online. A mismatch is detected when the networks expectation about the input is wrong or when the input changes suddenly. Detection of a mismatch is based on the sum of the elements of a vector r which depends on α, Σz_{Ji} and $\Sigma|x_i - z_{Ji}|$ ($Z_{Ji} = z_{Ji} /\Sigma z_{Ji}$). Σr_i has the following properties: 1) $\Sigma r_i = 0$, maximal match, if no learning occurred or top-down and bottom-up patterns match perfectly. 2) Σr_i becomes more sensitive for differences between x and z as Σz_{Ji} increases, i.e. during learning. 3) If Σz_{Ji} is maximal and x is orthogonal to Z_J, Σr_i is maximal, i.e. maximal mismatch. 4) Besides above mentioned factors, Σr_i only depends on the decay-rate of r_i, Σx_i and Σp_i which are all constants. At the moment a mismatch is detected, a subsystem (which is called 'orienting sub-system' in [8]) should produce a nonspecific inhibitory signal that both resets L3 and F2 until L3 can represent a new top-down pattern. We propose a system with two stable equilibria, a low activity and a high activity equilibrium, which jumps from the first to the second equilibrium if input (Σr_i) increases and jumps back if input decreases again. If, in addition, this process shows hysteresis, the subsystem is an appropriate orienting subsystem. A neural oscillator (extensively analysed in [3]) shows this typical behaviour for specific ranges of parameters.

Implementing this extended version of ART2 yields a network which is controlled completely by a set of differential equations. Exact ART meets the essential properties of an ART-network, mentioned by Carpenter and Grossberg [9]. The network generates a stable and consistent category structure. This means e.g. that a superset bottom-up input pattern cannot recode a subset top-down expectation. The 2/3 rule [8,9] is a intrinsic property of L3. The advantages of the exact implementation of ART are firstly that it can run stand-alone, secondly that most functions of the network including all the observed discontinuities in the behaviour are emergent properties of the system. Differential equations with which we implemented the dynamics of above described subsystems, analytical proofs of the claimed properties and simulation experiments are reported in [13].

Preliminary simulation results

Simulation experiments with ART should give insight in the functioning of the network in those areas of the parameter space that are biological plausible. Carpenter and Grossberg [11] mainly used analytical methods to analyse the subsystems and reported a few simulations with parameter values covering only part of the parameter space of interest. A completely continuous version of ART, however, is harder to analyse analytically because the behaviour of the total system does not only depend on equilibrium behaviour of the subsystems. Transient behaviour of the subsystems should be taken in consideration. We will apply bifurcation analysis and detection of catastrophe flags as alternative methods to analyse the behaviour of the network. Carpenter & Grossberg [10] reported on hysteresis in ART3. Their simulation experiments show that the classification of a

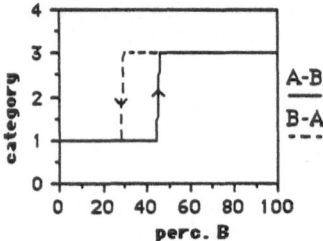

Figure 1: hysteresis in ART2. After learning patterns A and B, the network is presented pattern A which is classified in category 1. The input pattern A is slowly transformed into pattern B. If the input consists 46% of B a reset occurs and the pattern is classified as B. In a second test, the network is presented B which is slowly transformed into pattern A. The reset occurs now when the input pattern consisting 30% of B is presented. Naturally, the moments of reset and the difference between them depend on the vigilance parameter ρ.

slowly changing input, from one learned pattern (A) to an other pattern (B) belonging to a different class, results in a sudden shift of classification. However, according to catastrophe theory, hysteresis implies, in addition to a sudden jump, different shifting points with different sequences of the two stimuli (from A to B versus from B to A). Simulation experiments with Exact ART indeed revealed different shifting points (figure 1). This is an important indication that the network is able to perform saltatory behaviour, as hysteresis is a unique property of systems with discontinuous transitions. Two other catastrophe flags can easily be detected: bimodality and critical slowing down. Although hysteresis has been found at several times in neural networks and this example does not indicate a transition in the development of the system, it distinguishes ART from the PDP-models that are proposed as a model for stage-wise development.

Feedforward PDP-models of stagewise development

Among others, McClelland and Jenkins presented PDP-models which learn to solve, the well known developmental psychological balance-scale problems [14]. Although the feedforward backpropagation model of McClelland replicates some results of empirical studies, we have some comments on the model, mainly with respect to the claimed stage-wise character of the learning process. McClelland and Jenkins end with the following conclusion ([14], pg 70):

"This developmental progression seems to resolve the apparent paradoxical relation between observed stage-like behavioral development and assumed continuity of learning. To us this is the most impressive achievement of the model; it provides a simple, explicit alternative to maturational accounts of stage-like progression in development."

Firstly, replication experiments with the PDP-model made clear that the learning curve of the model (i.e. number of items correct evolving in time) is a power-function. If the learning behaviour of the network really is saltatory, a step-function instead of a smooth power-curve is expected. The learning curves of the individual item types, however, contain continuous accelerations. The stagewise character of the rule-application curve is the result of the categorization of the models performance.

Secondly, we mentioned above that discontinuous development has, according to catastrophe theory, a number of specific characteristics, e.g. jumps, bimodality and critical slowing down. Unfortunately, replication experiments indicate that the criteria for discontinuous transitions derived from catastrophe theory, do not hold for the backpropagation model. The backpropagation model learns to solve the balance scale items in a specific sequence found in empirical studies indeed, yet the presented models do not make qualitative jumps in their development. Beside this, as the authors say themselves, the model doesn't succeed to solve all problems properly and is in addition, biological implausible because of various properties.

Conclusion

Implementations of ART-networks [8,9,10] include some discontinuous and globally regulated procedures. Replacement of these discontinuities by continuous alternatives is feasible. The result is a stand-alone ART-network which is completely defined by a set of differential equations and runs in real time: Exact ART. The most important properties of this network, which meets the requirements stated by Carpenter and Grossberg [9], emerge from the dynamics of the subsystems and their interactions. In addition, we can remark that the observed discontinuities in the behaviour of the network, e.g. by gradual variation of control parameters or input, must be intrinsic properties of the system. ART-networks are, in contrast to connectionist models, biological plausible, self-organizing, unsupervised and overcome the stability-plasticity dilemma. Therefore a continuous ART is the most suitable neural network for studying stagewise development.

References

1. Van der Maas, H. L. J., Molenaar, P. C. M (1992). *Stagewise Development: An Application of Catastrophe Theory*. Psychological Review, vol. 99, no. 3, 395-417
2. Schuster, H. G., Wagner, P. (1990). *A model of neural oscillations in the visual cortex.* Biological Cybernetics, 64, 77-82.
3. Borisyuk, R. M., Kirillov, A. B. (1992). *Bifurcation analysis of a neural network model.* Biological Cybernetics, 66, 319-325.
4. Grossberg, S., Somers, D. (1991). *Synchronized Oscillations During Cooperative Feature Linking in a Cortical Model of Visual Perception.* Neural Networks, vol. 4, 453-466.
5. Thom, R. (1975). *Structural stability and morphogenesis.* Reading, MA: Benjamin.
6. Gilmore, R. (1981). *Catastrophe theory for scientists and engineers.* New York: Wiley.
7. Carpenter, G. A., Grossberg, G. (1987a). *A Massively Parallel Architecture for a Self-Organizing Neural Pattern Recognition Machine.* Comput. Vision Graphics Image Process., 37, 54.
8. Carpenter, G. A., Grossberg, G. (1987b). *ART 2: selforganization of stable category recognition codes for analog input patterns.* Applied Optics, vol. 26, no. 23, 4919-4930.
9. Carpenter, G. A., Grossberg, G. (1990). *ART 3: Hierarchical Search Using Chemical Transmitters in Self-Organizing Pattern Recognition Architectures.* Neural Networks, vol. 3, 129-152.
10. Grossberg, S. (1980). *How Does the Brain Build a Cognitive Code?* Psychological Review, 87, 2-51
11. Grossberg, S. (1973). *Contour Enhancement, Short Term Memory, and Constancies in reverberating neural networks.* Studies in Applied Mathematics, vol LII, no. 3.
12. McClelland, J. L., Jenkins, E. (1991). Nature, Nurture, and Connections: Implications of Connectionist Models for Cognitive Development. In: *Architectures for Intelligence: The Twenty-second Carnegie Mellon Symposion on Cognition*, (pp. 41-73). Lawrence Erlbaum Ass., Inc., Hillsdale, New York.
13 Raijmakers, M.E.J., Molenaar, P.C.M. (1993): *A continuous alternative for ART2.* Internal Report, Department of Developmental Psychology, University of Amsterdam.

654

Analysis of Chaotic Behaviour in Dynamical Systems Using Analog Neural Networks

M. Conti, S. Orcioni, C. Turchetti
Dipartimento di Elettronica, Univ. di Ancona, v. B. Bianche, 60131 Ancona, ITALY

Abstract

The analysis of chaotic behaviour requires, other than mathematical proof and computer simulation, experimental confirmation. Nonlinear electrical circuits are natural tools to study this phenomenon. In this paper a general approach, based on Approximate Identity Neural Networks, to the synthesis of nonlinear systems and then for the experimental analysis of chaotic behaviour is proposed.

1. Approximate Identity Neural Networks

A class of neural networks, named Approximate Identity Neural Networks (AINN's) particularly suitable for the implementation of nonlinear functions has recently been proposed in [1]. The networks are based on some results of the "theory of approximation". These can be summarized as follows. Let us given a function $f(\xi)$, and let us define the summation

$$\Lambda(\xi; \vartheta) = \sum_{j=1}^{v} \alpha_j \, \Omega(\xi - \lambda_j; \, n_j, \sigma_j) \;\; = \sum_{j=1}^{v} \alpha_j \, \omega(\xi_1 - \lambda_{1j}; n_{1j}, \sigma_{1j}) \, ... \, \omega(\xi_m - \lambda_{mj}; n_{mj}, \sigma_{mj}) \quad (1)$$

where

$$\omega(\xi; n, \sigma) = \tanh(\frac{n\xi + \sigma}{2}) - \tanh(\frac{n\xi - \sigma}{2}), \quad \xi \in \mathbb{R}, \quad \sigma \in \mathbb{R}^+, \quad n = 1, 2, ..; \quad (2)$$

and $\vartheta = (\alpha_j, n_{ij}, \sigma_{1j}, \lambda_{1j})$ $i=1, ... m$, and $j=1, ...v$ is a vector of parameters to be determined with a suitable algorithm which minimize the error between Λ and f. It can be shown that $\Lambda(\xi; \vartheta)$ converges to $f(\xi)$. This means that the functions (1) are able to approximate a given function to any degree of accuracy. The architecture of the neural network corresponding to (1) is made up of three-layers and it is shown for the 3-dimensional case in Fig. 1. The hidden layer is formed by elements implementing the single functions $\Omega(\xi)$, which are obtained as product of unidimensional functions $\omega(\xi)$.

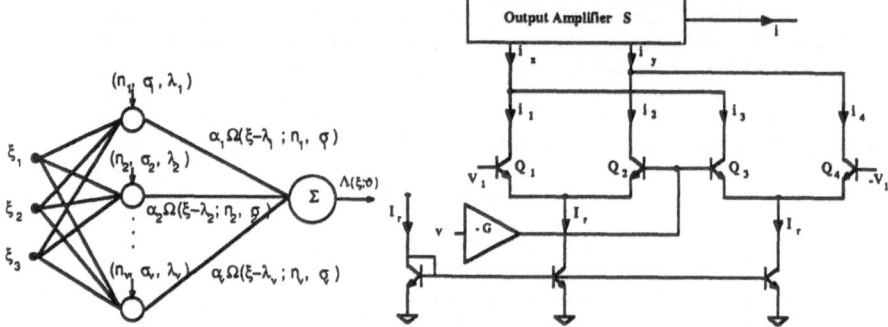

Fig. 1 AINN Architecture   Fig. 2 Circuit implementing the $\omega(\xi - \lambda; n, \sigma)$ term

Let us consider the circuit of Fig. 2, from an analysis of the diagram and by assuming that the output amplifier behaves linearly in the range of interest, we have

$$i = S \, I_r \, \omega(\xi - \lambda; n, \sigma). \quad (3)$$

Thus the generic unidimensional term $\omega(\xi - \lambda; n, \sigma)$ of the summation (1) can be implemented with the circuit of Fig. 2. On the basis of this fundamental block it can be shown that a nonlinear single-output multi-input transconductance amplifier, can be defined where the output current is given by

$$i(v; \vartheta) = \Lambda(v; \vartheta) = \sum_{j=1}^{v} \alpha_j \, \Omega(v - \lambda_j; n_j, \sigma_j) . \quad (4)$$

Hence, the amplifier represented with the symbol of Fig. 3 is able to approximate any given nonlinear function with an arbitrary error provided that the parameters ϑ are suitably chosen. In this sense can be considered as a universal approximator.

A large class of dynamical systems can be represented by a set of nonlinear first-order differential equations

$$\dot{x}(t) = f(x(t), u(t)), \qquad x(t_0) = x_0 \qquad (5)$$

and an algebraic vector equation of the form

$$y(t) = g(x(t), u(t)) \qquad (6)$$

in which $x(t)$ is the state vector, $u(t)$ the input-vector and $y(t)$ the output vector.

Let us consider the network shown in Fig. 4 consisting of p linear capacitors whose capacitances are C_i and p ideal nonlinear transconductance amplifiers whose output currents i_k are controlled by the voltages v_1, ..., v_p, u_1, ..., u_q. The circuit is described by the following equation

$$C\dot{v}(t) = i(w(t); \vartheta) = \Lambda(v(t), u(t); \vartheta) \qquad (7)$$

where v is the vector of voltages across the capacitors, $w = (v, u)$ the vector of voltages at the input of transconductance amplifiers, i the vector of currents at the output of transconductance amplifiers, and C a diagonal matrix of element c_i.

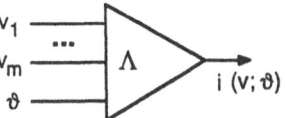

Fig. 3 Transconductance amplifier

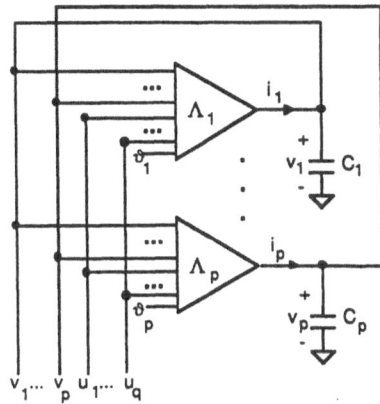

Fig. 4 Network corresponding to eq. (5)

Although eqs. (5) and (7) have the same structure, they differ for the two functions $f(.)$ and $i(.)$. However, for what said above the function $\Lambda(.)$ is able to approximate any function to any degree of accuracy. By defining $x(t) = C v(t)$ and by choosing appropriately the initial conditions, the following approximation holds

$$\dot{x}(t) = i(C^{-1}x(t), u(t); \vartheta) \approx f(x(t), u(t)), \quad x(t_0) = x_0. \qquad (8)$$

Thus eq. (7) map, apart an error which can be reduced to any degree of accuracy, into the network of Fig. 4.

2. Duffing's equation

The AINN's are fairly general since can be used to implement a large category of nonlinear systems and thus are particularly suited to experimental observation of chaotic behaviour [2-4]. As an application of the approach described the chaotic behaviour of the well known Duffing's equation has been analysed experimentally. This is a very interesting case since, in spite of its simplicity, many important effects such as quasi-harmonic resonance, subharmonic resonance and chaotic dynamics, can be studied by means of such an equation. The equation in its most general form is given by

$$\ddot{x} = -b\,\dot{x} - \alpha x - x^3 + A\cos t \qquad (9)$$

where b, α and A are constant nonnegative parameters. As you can see this equation represents a nonlinear nonautonomous systems due to the presence of the sinusoidal forcing term

$$u(t) = A\cos t \qquad (10)$$

Equivalently it can be rewritten in normal form as

$$\dot{x}_1 = \omega\sigma\, x_2$$

$$\dot{x}_2 = -b\,\omega x_2 - \alpha\frac{\omega}{\sigma}x_1 - v^3\frac{\omega}{\sigma}x_1^3 + \frac{A\omega}{\sigma v}\cos\omega\tau, \qquad \sigma, v, \omega > 0 \qquad (11)$$

where the new variables $x_1 = x/v$, $x_2 = \dot{x}/(\sigma v)$ and $t = \omega\tau$ have been defined, ω being the forcing frequency. The schematics of the circuit used to synthesize eq.(9) is reported in Fig. 5.

656

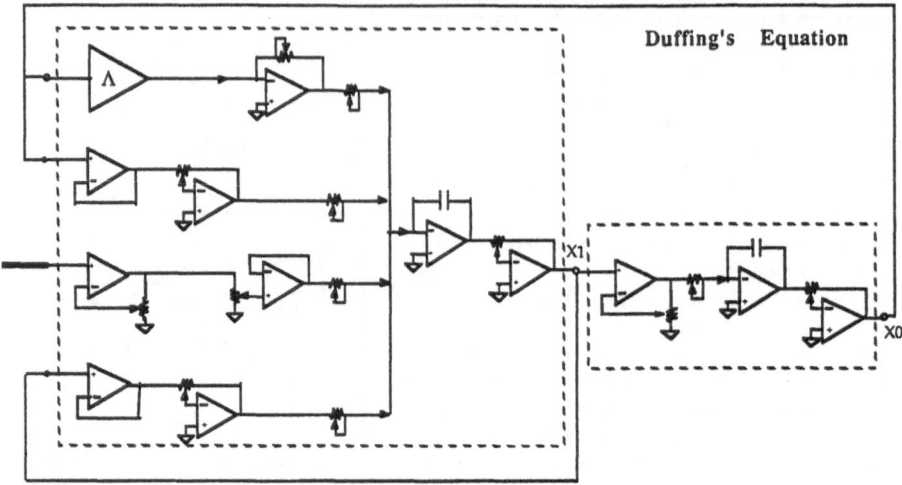

Fig. 5 Circuit implementing Duffing's Equation.

As you can see it is made-up by one nonlinear Λ–block alone, which realizes the cubic term, and several linear blocks, i.e. amplifiers and integrators. In order to obtain experimental results, a breadborded circuit based on this schematic has been built up. The cubic non linearity obtained from the Λ-block is plotted in the Fig.6. The agreement with exact curve is good being the average error of 0.65 %. It is worth noting that in this case only two elementary ω-blocks suffice to get the desired error. Obviously the more elementary blocks are used the less the error will be, as predicted from the theory. Duffing's equation was extensively studied by Ueda [5] for a wide range of the parameters b, A and with α=0. The results are of great interest since the equation behaves very differently depending on the values assumed for the two parameters. The circuit (and correspondingly Duffing's equation) exhibits, in some bounded regions a chaotic behaviour. In the first experimental condition analyzed we have chosen A=5.55, b=0.0393 and ω=40.59 rad/sec. Figs. 7a) and b), show the trajectory in the phase plain observed experimentally during a time interval of 1 sec. and 10 sec., respectively. It is evident that the trajectory, although is always bounded, cannot be considered as periodic. This is typical of chaotic behaviour. Such a result is confirmed by Figs. 8 which show the shapes of $x(\tau)$ and $u(\tau)$ during a time interval $\Delta\tau = 2$ sec (Fig. 8a), and the shapes of $x_1(\tau)$ and $x_2(\tau)$ during a time interval $\Delta\tau = 5$ sec. A further confirmation of this behaviour, and of goodness of the approach proposed for nonlinear synthesis, has been obtained by solving numerically eq. (9). In particular Fig. 9 and Figs. 10 a), b), c) show the trajectory and the shapes of $u(\tau)$, $x_1(\tau)$ and $x_2(\tau)$ derived from numerical analysis, for the same pair of values (A, b) previously used in the circuit. As you can see, the behaviour predicted numerically is very close (obviously not the same owing to the inevitable uncertainty in the starting condition and physical parameters) to the one observed experimentally. As another experimental condition we have chosen A=5.55, b=0.0393 and ω= 3171 rad/sec. Also for this example the trajectory and the shapes of x_1 and x_2 are reported in Figs. 11 and 12 respectively. In addition, a further confirmation of chaotic behaviour is given in this case by the Poincarè section, reported in Fig. 13, which clearly shows a non-periodic behaviour.

REFERENCES
[1]   C. Turchetti, M. Conti, "A new class of neural networks based on approximate identities for approximation and learning", 1992 IEEE ISCAS, San Diego, May 1992.
[2]   M. J. Hasler, "Electrical circuits with chaotic behavior", Proc. of the IEEE, Vol.75, n. 8, Aug. '87.
[3]   T. Matsumoto, "Chaos in electronic circuits", Proc. of the IEEE, Vol. 75, n. 8, August 1987.
[4]   J. Thompson, H. B. Stewart, "Nonlinear dynamics and chaos", J. Wiley and Sons, 1989, USA.
[5]   Y. Ueda, "Steady motions exhibited by Duffing's equation: a picture book of regular and chaotic motions", pp. 311-322, SIAM: Philadelphia, 1980.

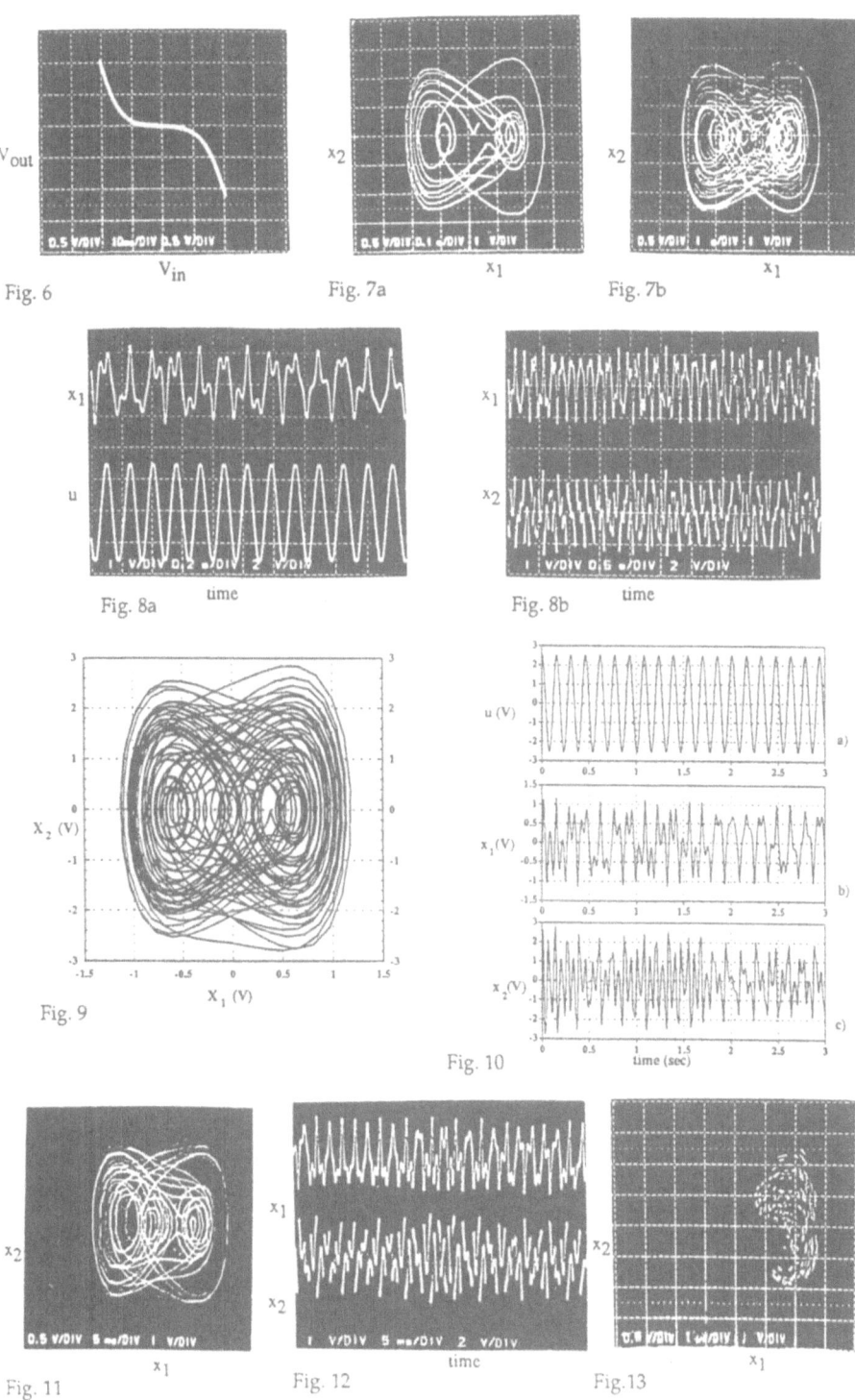

Fig. 6

Fig. 7a

Fig. 7b

Fig. 8a

Fig. 8b

Fig. 9

Fig. 10

Fig. 11

Fig. 12

Fig.13

A DYNAMICALLY GENERALISING WEIGHTLESS NEURAL ELEMENT

P. Ntourntoufis

Neural Systems Engineering Laboratory, Department of Electrical and Electronic Engineering, Imperial College of Science, Technology and Medicine, London SW7 2BT, U.K.

Abstract
The paper introduces the concept of the *Dynamically Generalising* weightless Neuron (DGN). The DGN is derived from the Generalising Random Access Memory [1], in which the *training* and *spreading* operations are replaced by a single *learning* phase. The DGN is able to store and spread patterns, through a dynamical process involving interactions between each memory location and its immediate neighbours, and external signals. The DGN exhibits very desirable properties. First, after the initial trained patterns have spread throughout the memory space, additional patterns can still be stored in the DGN. Secondly, it is possible to distinguish between trained and spread patterns. And finally a trained pattern and its associated spread patterns can be removed without affecting the rest of the stored patterns.

1. INTRODUCTION

The basic component of a weightless neural system [2] is the Random Access Memory (RAM). The inputs to a RAM node consist of N lines whose Boolean values form addresses to which correspond memory locations. The contents of a memory location is seen an the *internal state* in which the location can be. A memory location can be in one of two states, 1 or 0.

When a RAM learns a new pattern, during a *write* operation, the state transition of a memory location can be expressed as:

$$S' = S \cdot (\overline{w} + \overline{a} \cdot w) + a \cdot w [x \cdot 1 + \overline{x} \cdot 0], \tag{1}$$

where S and S' are the states of a memory location before and after the *write* operation, respectively; w is a Boolean variable whose value is 1 during a write operation and 0 otherwise; x represents the Boolean data-in value; and a is a Boolean variable whose value is 1 if the memory location is addressed and 0 otherwise. Memory locations not addressed are not affected by the write operation. This will no longer be the case for the DGN in which the state of a memory location is dynamically updated through interactions with neighbouring locations.

During a *read* operation, the RAM outputs 0 if a **0** state is addressed and 1 if a **1** state is addressed. We write [0] = 0 and [1] = 1, with [**X**] representing the value output by the RAM when the memory location addressed is in state **X**.

Similarly, in a Probabilistic Logic Node (PLN)[3], a memory location can be in one of the states **0**, **1** or **U**. When these states are addressed, during the *use* phase, the corresponding output values are [0] = 0, [1] = 1 and [U] = 0 or 1 chosen randomly with equal probability. For the *train* phase, the equation expressing the state transitions takes a form similar to that of (1):

$$S' = S \cdot (\overline{w} + \overline{a} \cdot w) + a \cdot w [x (\overline{s_0} \cdot 1 + s_0 \cdot U) + \overline{x} (\overline{s_1} \cdot 0 + s_1 \cdot U)],$$

$$\text{with} \quad s_X = (S = X) \quad \text{and} \quad \overline{s_X} = (S \neq X). \tag{2}$$

If the state of an addressed location is **1** and the data-in value is 0 then the state of this location is reset to **U**. State **1** is said to be *disrupted*, or to be in *contradiction* with state **0** [4]. Similarly, if the state of the addressed location is **0** and the data-in value is 1, then the state of this location is reset to **U**.

In a Generalising Random Access Memory (GRAM) [1] the *train* and *use* phases are identical to those in a PLN. There is an additional phase, called *spreading* [2], during which memory locations not addressed during the training phase, are affected by the use of a *diffusion algorithm* [5][2]. The *spreading* phase only affects memory locations which are in state U.

In order to characterise the influence on a memory location, of its neighbours, a variable g is introduced, which takes one of the values 0, 1, or U: 0, if at least one of the neighbouring locations is in the state 0 and the others are in states 0 or U; 1, if at least one of the neighbouring locations is in the state 1 and the others are in states 1 or U; U, otherwise. For the state transition of a memory location during the spreading phase, using the notation (2) and with $g_j = (g = j)$, we write:

$$S' = \bar{\sigma} \cdot S + \sigma [1 \cdot (s_1 + g_1 \cdot s_U) + 0 \cdot (s_0 + g_0 \cdot s_U) + U \cdot g_U \cdot s_U], \tag{3}$$

with σ being a Boolean variable that takes the value 1 during spreading and the value 0 at other times. (3) is applied synchronously to all memory locations and is repeated a number Γ of times corresponding to a desired *degree of generalisation*.

It is interesting to notice that the GRAM does not have some properties that might be desirable: all the training patterns have to be stored first, prior to any spreading; moreover, a *trained* pattern cannot be distinguished from a *spread* pattern; therefore it is not possible to remove a trained pattern together with its associated spread patterns.

The following section introduces a new weightless node called Dynamically Generalising Neuron (DGN) which operates in two phases. During the first phase, called *learning*, at each time step, each memory location in the DGN updates its state according to its own state, the states of its neighbours and external signals. This dynamical process enables trained patterns to spread automatically throughout the DGN memory space. The second phase is the use phase.

2. THE DYNAMICALLY GENERALISING NEURON

2.1. The states of a memory location

In order to learn additional patterns, after previous patterns have spread their states throughout the DGN, it is necessary to be able to distinguish between states associated with different *spreading levels*. The state of a memory location is said to be at a spreading level i when the memory location is at a Hamming distance i from the memory location which determined its state through spreading. Therefore the state of a memory location will be labelled by adding a suffix i, whose value is the spreading level of the state. A memory location can be in one of the states U, 1_i, 0_i, U_i or R_i, with $i = 0, \ldots, N$. U is the initial state of a memory location; its associated output is $[U] = 0/1$. The 1_i are states resulting from the spreading from an addressed memory location trained with a 1 pattern; their associated output value is $[1_i] = 1$. The 0_i are states resulting from the spreading from an addressed memory location trained with a 0 pattern; their associated output value is $[0_i] = 0$. The U_i are states resulting from a contradiction between states 1_i and 0_i; their associated output value is $[U_i] = 0/1$. The R_i are states assigned to memory locations from which a pattern is being removed; their associated output value is $[R_i] = 0/1$.

2.2. Write and remove operations

These operations affect an addressed memory location ($a = 1$) when the external *write* signal w has the value 1. There is an external *remove* signal r, whose value determines whether the operation is write ($r = 0$) or remove ($r = 1$). An addressed memory location in which a pattern 0 or 1 is written is set to state 0_0 or state 1_0, respectively. An addressed memory location from which a pattern 0 or 1 is removed is set to state R_0. The next state S'_{ext} of the memory location, resulting from the interaction with the external signals a, x, w and r can then be written:

$$S'_{ext} = \bar{r} \cdot x (s_{0_0} \cdot 1_0 + s_{0_0} \cdot U_0) + \bar{r} \cdot \bar{x} (s_{1_0} \cdot 0_0 + s_{1_0} \cdot U_0) + r \cdot R_0, \tag{4}$$

$$\text{with} \quad s_{X_i} = (S = X_i) \quad \text{and} \quad \overline{s_{X_i}} = (S \neq X_i). \tag{5}$$

When a pattern is stored at, or removed from a location, by external addressing, the dynamical system formed by all the memory locations of the DGN is no longer in a stable state. It is the interactions between neighbouring memory locations that enable the system to settle in a new stable state for which the node has the correct generalisation with respect to all the trained patterns.

2.3. Interactions with neighbouring memory locations

The influence of neighbouring memory locations (at Hamming distance 1) is taken into account by a variable g, associated with the considered memory location. First, are considered all the memory locations at distance 1 from the location to be updated, whose states have the lowest spreading level. The value of this spreading level is defined as $k - 1$. The states of the chosen memory locations form a set denoted M_{k-1}. The initial state U is defined as having the highest spreading level, say $N + 1$. g takes one of the values 0_k, 1_k, R_k or U_k: 0_k, if at least one element of M_{k-1} has the value 0_{k-1} and the others have values 0_{k-1} or U_{k-1}; 1_k, if at least one element of M_{k-1} has the value 1_{k-1} and the others have values 1_{k-1} or U_{k-1}; R_k, if at least one element of M_{k-1} has the value R_{k-1}; and U_k otherwise.

Let the considered memory location be in a state of spreading level i and the value of g characterising the influence of the neighbours be of level k. The next state S'_{dyn} of the memory location, resulting from the dynamical interactions between neighbouring locations, can then be written:

$$
\begin{aligned}
S'_{dyn} = {} & s_{R_i} \cdot U_{N+1} \\
& + (i < k) \cdot \overline{s_{R_i}} \cdot S + (i > k) \cdot [\overline{s_{R_i}} \cdot [G_{0_k} + G_{1_k} + G_{U_k}] + \overline{s_{U_{N+1}}} \cdot G_{R_k} + s_{R_i} \cdot S_{U_{N+1}}] \\
& + (i = k) \cdot [(g_{0_i} + g_{U_i}) \cdot S_{0_i} + (g_{1_i} + g_{U_i}) \cdot S_{1_i} + g_{U_i} \cdot S_{U_i} + (g_{0_i} \cdot s_{1_i} + g_{1_i} \cdot s_{0_i}) \cdot U_i \\
& \quad + s_{U_i} \cdot [G_{0_i} + G_{1_i}] + (s_{1_i} + s_{0_i} + s_{U_i}) \cdot G_{R_i}],
\end{aligned}
\tag{6}
$$

$$
\text{with} \quad g_{X_j} = (g = X_j), \quad S_{X_j} = s_{X_j} \cdot X_j \quad \text{and} \quad G_{X_j} = g_{X_j} \cdot X_j.
\tag{7}
$$

2.4. General update equation

The general update equation for a memory location in a DGN is obtained by combining the effects of write/remove operations (Eq. (4) and (5)) with those resulting from the interactions between neighbouring locations (Eq. (6) and (7)). We write:

$$
S' = (\overline{w} + \overline{a} \cdot w) \cdot S'_{dyn} + a \cdot w \cdot S'_{ext}.
\tag{8}
$$

2.5. Example

Table 1 shows the evolution through time of the contents of characteristic memory locations of a 5-input DGN during the storage of 2 patterns, followed by the removal of one of the stored patterns. The contents of memory locations with addresses 10000, 00000, 01000, 01100, 01110 and 01111 is shown.

3. CONCLUSIONS

In this paper a fundamental viewpoint was adopted in the description of weightless neural nodes, that a memory location contents is no longer considered as the value that is output when a memory location is addressed, but as an *internal state* in which a memory location can be. This has led to the definition of the DGN weightless node.

The DGN was derived from the GRAM, in which the training and spreading operations are performed in a single phase. The DGN is able to store patterns and spread them, via a dynamical process, governed by equations (4) to (8) above, involving interactions between each memory location and its neighbours and external signals. The DGN possesses certain advantages over the GRAM. First, it is possible to store

Time	10000	00000	01000	01100	01110	01111	
$T_1 - 1$	U	U	U	U	U	U	stable
T_1	0_0	U	U	U	U	U	unstable
$T_1 + 1$	0_0	0_1	U	U	U	U	
$T_1 + 2$	0_0	0_1	0_2	U	U	U	
$T_1 + 3$	0_0	0_1	0_2	0_3	U	U	
$T_1 + 4$	0_0	0_1	0_2	0_3	0_4	U	
$T_1 + 5$	0_0	0_1	0_2	0_3	0_4	0_5	stable
$\cdots$			PATTERN 1				
$T_2 - 1$	0_0	0_1	0_2	0_3	0_4	0_5	
T_2	0_0	0_1	0_2	1_0	0_4	0_5	unstable
$T_2 + 1$	0_0	0_1	1_1	1_0	1_1	0_5	
$T_2 + 2$	0_0	0_1	1_1	1_0	1_1	1_2	stable
$\cdots$	$\cdots$	$\cdots$	$\cdots$	$\cdots$	$\cdots$	$\cdots$	
$T_R - 1$	0_0	0_1	1_1	1_0	1_1	1_2	
T_R	R_0	0_1	1_1	1_0	1_1	1_2	unstable
$T_R + 1$	U	R_1	1_1	1_0	1_1	1_2	
$T_R + 2$	U	U	1_1	1_0	1_1	1_2	
$T_R + 3$	U	1_2	1_1	1_0	1_1	1_2	
$T_R + 4$	1_3	1_2	1_1	1_0	1_1	1_2	stable
$\cdots$			PATTERN 2				

Table 1: Evolution through time of the contents of characteristic memory locations of a 5-input DGN. First two patterns are stored, followed by the removal of one of the stored patterns.

additional patterns even after the spreading of previous patterns. Secondly, it is possible to distinguish between trained and spread patterns. And finally it is possible to remove trained patterns and their associated spread patterns without affecting the remaining stored patterns.

ACKNOWLEDGEMENTS

The author thanks C. Ioannou for useful comments.

REFERENCES

1   Aleksander, I., "Ideal Neurons for Neural Computers", *Proc. Int. Conf. on Parallel Processing in Neural Systems and Computers*, Düsseldorf, Springer Verlag, 1990.

2   Aleksander, I., "Weightless Neural Tools: Towards Cognitive Macrostructures", *The CAIP Neural Network Workshop*, Rutgers University, New Jersey, October 1990.

3   Kan, W. K., and Aleksander, I., "A Probabilistic Logic Neuron Network for Associative Learning", in Neural Computing Architectures, Ed. I. Aleksander, Kogan Page, London, MIT Press, Boston, 1989.

4   Ntourntoufis, P., "Storage Capacity and Retrieval Properties of an Auto-associative General Neural Unit", *Proc. IJCNN-91*, Seattle, 1991.

5   Wong, K. Y. M., and Sherrington, D., "Theory of Associative Memory in Randomly Connected Boolean Neural Networks", *J. Phys. A: Math. Gen.*, 22, pp. 2233-63, 1989.

Vector Quantization by Neuro-Dynamical System

Yoshinori Uesaka

Department of Information Sciences, Faculty of Science and Technology,
Science University of Tokyo
2641 Yamazaki, Noda-City 278 JAPAN

Abstract   A new type of objective function is proposed for solving vector quantization problems by means of neural networks of Hopfield type. The extension of the domain of the function gives an energy function, by which a neuro-dynamical system is introduced. It is proved that any minimal point of the objective function is an equilibrium and asymptotically stable point of the system. Hence, the state of the system will approach to an optimal solution of the vector quantization problem when the initial state is set in the basin of a minimum point of the objective function.

§ 1. Introduction : Problem to be Solved

Let $\Re$ be the set of all real numbers and $\Re^n$ a Cartesian product of n $\Re$'s. We call $\Re^n$ the set of *patterns*. For every pair of patterns $a = (a_1,...,a_n), b = (b_1,...,b_n) \in \Re^n$, a similarity between them is assumed to be defined as

(1) $d(a,b) \equiv \sqrt{(a_1 - b_1)^2 + \cdots + (a_n - b_n)^2}$.

Let $S \equiv \{a_1,...,a_N\} \subseteq \Re^n$ be a set of *sample patterns*. A family $\{S_1,...,S_M\}$ of subsets of S is said to be a *partition* iff $S_1 \cup \cdots \cup S_M = S$, $S_j \neq \varnothing$, and $j \neq k \Rightarrow S_j \cap S_k = \varnothing$ for $j,k = 1,...,M$. The *distortion* of S with a partition $\{S_1,...,S_M\}$ is assumed to be defined as

(2) $D(S_1,...,S_M) \equiv \sum_{j=1}^{M} D(S_j)$, where $D(S_j) \equiv \min\{ \sum_{a \in S_j} d(a,y)^2 \mid y \in \Re^n\}$.

We call $D(S_j)$ the *distortion of a cluster* S_j. Then, consider the following problem :

Problem 1. *For a given set of sample patterns*, $S \equiv \{a_1,...,a_N\} \subseteq \Re^n$, *find a partition of* S, $\{S_1^*,...,S_M^*\}$, *which gives the minimum distortion of* S, *that is, find a partition such that*

(3) $\{S_1^*,...,S_M^*\} = \arg\min\{D(S_1,...,S_M) \mid \{S_1,...,S_M\} : \text{partition of } S\}$.

The pattern giving $D(S_j^*)$, that is,

(4) $m_j \equiv \arg\min\{ \sum_{a \in S_j^*} d(a,y)^2 \mid y \in \Re^n\}$

is regarded as a representative of the cluster S_j^*, and may be called a *quantized vector* for S_j^*. So, to solve Problem 1 means the vector quantization procedure for a given set of sample patterns. Two algorithms have been reported so far [1], [2].

In the present paper another algorithm for Problem 1 will given by using a dynamical system which may be implemented by means of a mutually connected neural network.

§ 2. Main Results

In order to solve Problem 1, we consider a function F from $\{0,1\}^{NM} \times \Re^{nM}$ into $\Re$:

(5) $F(x,y) \equiv \sum_{i=1}^{N} \sum_{j=1}^{M} x_{ij} d(a_i,y_j)^2$,

where

(6) $x \equiv (x_{11},...,x_{1M},...,x_{N1},...,x_{NM}) \in \{0,1\}^{NM}$,

(7) $y \equiv (y_1,...,y_M) \in \Re^{nM}$ and $y_j \equiv (y_{j1},...,y_{jn}) \in \Re^n$ for $j = 1,...,M$.

Furthermore, let X be a subset of $\{0,1\}^{NM}$:

(8) $X \equiv \{x \mid x = (x_{11},...,x_{1M},...,x_{N1},...,x_{NM}) \in \{0,1\}^{NM}, \sum_{j=1}^{M} x_{ij} = 1$ for $i = 1,...,N\}$.

Now, consider the following problem:

Problem 2. *For a given set of sample patterns, $S \equiv \{a_1,...,a_N\} \subseteq \Re^n$, find a minimum point (x^*,y^*) of F under the constraint of $(x,y) \in X \times \Re^{nM}$, that is, find*

(9) $(x^*,y^*) = \arg\min\{F(x,y) \mid x \in X, \; y \in \Re^{nM}\}$.

Then, Problem 1 is reduced to Problem 2 as shown below (See Section 3 for proof):

Theorem 1 *If (x^*,y^*) is a minimum solution for Problem 2, then so is $\{S_1^*,...,S_M^*\}$ for Problem 1, where the partition $\{S_1^*,...,S_M^*\}$ is determined by x^* as follows:*

(10) $S_j^* = \{a_i \mid \exists i = 1,...,N : x_{ij}^* = 1\}$ *for* $j = 1,...,M$.

In view of Theorem 1, it is sufficient to consider Problem 2. But it has unfortunately the constraint (8). To remove the constraint we consider a function $H : \{0,1\}^{NM} \times \Re^{nM} \to \Re$:

(11) $H(x,y) \equiv F(x,y) + AG(x)$,

where A is a positive constant, and

(12) $G(x) \equiv \sum_{i=1}^{N} \sum_{j=1}^{M} (\sum_{j=1}^{M} x_{ij} - 1)^2$.

Note that $G(x)$ is positive iff x is not in X. So it is easily proved as in [4] that

Theorem 2 *There exist an $A > 0$ such that if (x^*,y^*) is a minimum point of H in $\{0,1\}^{NM} \times \Re^{nM}$, then so is (x^*,y^*) for F in $X \times \Re^{nM}$.*

Hence Problem 2 is reduced to the minimization problem of H without constraint. Next, according to [3-5], we construct a dynamical system by which this unconstrained optimization problem can be solved. To do so we rewrite H, by straightforward calculation, in *multi-linear form* with respect to x_{ij}:

(13) $H(x,y) \equiv F(x,y) + AG(x) = -\frac{1}{2} \sum_{i=1}^{N} \sum_{j=1}^{M} \sum_{k=1}^{N} \sum_{l=1}^{M} A_{ijkl} x_{ij} x_{kl} - \sum_{i=1}^{N} \sum_{j=1}^{M} B_{ij}(y) x_{ij} + C$,

where

(14) $A_{ijkl} \equiv -2A\delta_{ik}(1 - \delta_{jl})$, $B_{ij}(y) \equiv A - d(a_i,y_j)^2$, $C \equiv AN$,

δ_{ij} being Kronecker's delta. It should be noted that $A_{ijij} = 0$ and $A_{ijkl} = A_{klij}$.

By regarding that x_{ij} in H takes all real numbers, we may extend the domain of H from $\{0,1\}^{NM} \times \Re^{nM}$ to $\Re^{NM} \times \Re^{nM}$. The function obtained by such an extension is denoted as E and called an *energy function* induced from H. Obviously $E(x,y)$ coincides with $H(x,y)$ for all (x,y) in $\{0,1\}^{NM} \times \Re^{nM}$:

Now consider the following dynamical system with the potential energy E:

(15) $\begin{cases} \dfrac{dx_{ij}}{dt} = -x_{ij}(1 - x_{ij})\dfrac{\partial E}{\partial x_{ij}} \equiv f_{ij}(x,y), & i = 1,...,N, \; j = 1,...,M; \\[2mm] \dfrac{dy_{jp}}{dt} = -\dfrac{\partial E}{\partial y_{jp}} \equiv g_{jp}(x,y), & j = 1,...,M, \; p = 1,...,n. \end{cases}$

Then it will be shown that the stable point of the system has a tight relevance to the solution of Problem 2.

Before giving the relevance precisely we need a definition on minimal point of H. Let $x =$

$(x_{11},...,x_{1M},...,x_{N1},...,x_{NM})$ be in $\{0,1\}^{NM}$. By $x[k,l]$ we mean a vector $(x_{11},...,1-x_{kl},$ $...,x_{NM})$. Then, (x^*,y^*) is said to be a *minimal point* of H iff the inequality $H(x^*,y^*) <$ $H(x^*[k,l],y^*)$ holds for $k=1,...,N$ and $l=1,...,M$, and there exists $\varepsilon > 0$ such that $H(x^*,y^*) < H(x^*,y)$ for all y in the ε-neighborhood of y^* except y^*. With this definition on minimal point, the following holds (See Section 4 for proof) :

Theorem 3 *There exists $A > 0$ such that if (x^*,y^*) is a minimal point of H, then (x^*,y^*) is an asymptotically stable point of the dynamical system (15).*

This result shows that when the initial state is set in the basin of minimum point (x^*,y^*) of H, the state of the system will approach to (x^*,y^*) as $t \to \infty$. Thus we can get x^* as a solution of Problem 2, and hence $\{S_1^*,...,S_M^*\}$ as a solution of Problem 1 in view of Theorem 1.

From a neural network point of view, the dynamical system (15) may be realized by a mutually connected network consisting of the following neural elements of two types. In the first and second type, the inner potential u and the output z are governed by the input $x_1,...,x_n$ and $y_1,...,y_m$(or y), respectively, as follows :

(16) $\dfrac{du}{dt} = w_0 + \sum\limits_{i=1}^{n} w_i x_{kl} + \sum\limits_{j=1}^{m} (v_j - y_j)^2$, $z = \dfrac{1}{2}(\tanh u + 1)$; w_i, v_j : constants,

(17) $\dfrac{du}{dt} = \sum\limits_{i=1}^{n} x_i(v_i' - y)$, $z = u$; v_i' : constants.

Outlines of Proof for Theorem 1 and 3 are devoted to the following sections.

§3. Proof of Theorem 1

For any partition $\{S_1,...,S_M\}$ of S, we define

(18) $\hat{y}(S_j) \equiv \arg\min\{ \sum\limits_{a \in S_j} d(a,y)^2 \mid y \in \mathfrak{R}^n \}$ for $j=1,...,M$.

Then, the distortion of S_j is given by

(19) $D(S_j) \equiv \min\{ \sum\limits_{a \in S_j} d(a,y)^2 \mid y \in \mathfrak{R}^n \} = \sum\limits_{a \in S_j} d(a,\hat{y}(S_j))^2 = \sum\limits_{i=1}^{N} x_{ij} d(a_i,\hat{y}(S_j))^2$,

where $x_{ij} \equiv 1$ if $a_i \in S_j$ and $x_{ij} \equiv 0$ if $a_i \notin S_j$. Summing up (19) with respect to j and noting that x is in X and (x^*,y^*) is a minimum point of F under the constraint $(x,y) \in X \times \mathfrak{R}^{nM}$, we have

(20) $D(S_1,...,S_M) = F(x,\hat{y}(S_j)) \geq F(x^*,y^*)$.

On the other hand, for any $z \in \mathfrak{R}^n$

(21) $D(S_j^*) = \min\{ \sum\limits_{a \in S_j^*} d(a,y)^2 \mid y \in \mathfrak{R}^n \} = \sum\limits_{a \in S_j^*} d(a,\hat{y}(S_j^*))^2 \leq \sum\limits_{a \in S_j^*} d(a,z)^2 = \sum\limits_{i=1}^{N} x_{ij}^* d(a_i,z)^2$.

Taking y_j^* as z, we have that $D(S_j^*) \leq \sum_{i=1}^{N} x_{ij}^* d(a_i,y_j^*)^2$. Summing up with respect to j,

(22) $D(S_1^*,...,S_M^*) = \sum\limits_{j=1}^{M} D(S_j^*) \leq \sum\limits_{j=1}^{M} \sum\limits_{i=1}^{N} x_{ij}^* d(a_i,y_j^*)^2 = F(x^*,y^*)$.

Now, let $(\hat{S}_1,...,\hat{S}_M)$ be $\arg\min\{D(S_1,...,S_M) \mid \{S_1,...,S_M\}$: partition of $S\}$. Then, in view of (22) and (20), we have that $D(\hat{S}_1,...,\hat{S}_M) \leq D(S_1^*,...,S_M^*) \leq F(x^*,y^*) \leq D(S_1,...,S_M)$. Since $\{S_1,...,S_M\}$ is arbitrary, we may take $\{\hat{S}_1,...,\hat{S}_M\}$ as $\{S_1,...,S_M\}$, having $D(S_1^*,...,S_M^*) = D(\hat{S}_1,...,\hat{S}_M)$ which completes the proof of Theorem 1. (Q.E.D.)

§ 4. Outline of Proof for Theorem 3

We take the constant A as in Theorem 2. First, it is easily shown [4],[5] that if (x^*, y^*) is a minimum point of H, then it is an equilibrium point of the dynamical system (15). Second, it holds for any $(x, y) \in \{0,1\}^{NM} \times \Re^{nM}$ that

(23) $\quad H(x,y) - H(x[i,j],y) = (2x_{ij} - 1)\dfrac{\partial E}{\partial x_{ij}}(x,y)$ for $i = 1,\dots,N$ and $j = 1,\dots,M$.

In fact, since the energy function E is of multi-linear form with respect to x_{ij}, there exist functions E_1, E_2 of variables $x_{11},\dots,x_{NM}$ except x_{ij} and y such that

(24) $\quad E(x_{11},\dots,x_{ij},\dots,x_{NM},y) = E_1(x_{11},\dots,\circ,\dots,x_{NM},y) + x_{ij}E_2(x_{11},\dots,\circ,\dots,x_{NM},y)$.

Changing x_{ij} into $1 - x_{ij}$, we have

(25) $\quad E(x_{11},\dots,1 - x_{ij},\dots,x_{NM},y) = E_1(x_{11},\dots,\circ,\dots,x_{NM},y) + (1 - x_{ij})E_2(x_{11},\dots,\circ,\dots,x_{NM},y)$.

Subtracting (24) from (25), we have

(26) $\quad E(x,y) - E(x[i,j],y) = (2x_{ij} - 1)E_2(x_{11},\dots,\circ,\dots,x_{NM},y)$.

On the other hand, differentiating (24) with respect to x_{ij}, we have

(27) $\quad \dfrac{\partial E}{\partial x_{ij}}(x,y) = E_2(x_{11},\dots,\circ,\dots,x_{NM},y)$.

Inserting this into (26) and noting that $E(x,y)$ coincides with $H(x,y)$ on $\{0,1\}^{NM} \times \Re^{nM}$, we have (23).

Third, the straightforward differentiation of the right hand side of (15) with attention to (23) gives the following :

(28) $\quad \dfrac{\partial f_{ij}}{\partial x_{kl}}(x^*, y^*) = [H(x^*, y^*) - H(x^*[i,j], y^*)]\delta_{ik}\delta_{jl}$,

(29) $\quad \dfrac{\partial f_{ij}}{\partial y_{lq}}(x^*, y^*) = 0$, and $\dfrac{\partial g_{jp}}{\partial y_{lq}}(x^*, y^*) = -[\sum\limits_{i=1}^{N} x^*_{ij}]\delta_{jl}\delta_{pq}$.

Therefore the Jacobian matrix of (15) is triangle, and its eigen values are strictly given by

(30) $\quad \lambda_{ij} = H(x^*, y^*) - H(x^*[i,j], y^*); \quad \mu_{jp} = -\sum\limits_{i=1}^{N} x^*_{ij}$.

Since (x^*, y^*) is a minimum point of H, λ_{ij}'s are negative. Since x^* is in X and S^*_j is not empty, μ_{jp}'s are negative. Thus, the real parts of the eigenvalues of the Jacobian matrix at (x^*, y^*) are all negative. Hence, (x^*, y^*) is asymptotically stable as well known in the theory of dynamics [6]. (Q.E.D.)

References

[1] Linde,Y., Buzo,A. and Gray,R.M. (1980) : An algorithm for vector quantizer design, IEEE Trans. on Communication, COM-28, 84-95.

[2] Kohonen,T. (1989) : Self-Organization and Associative Memory, Third Edition, Springer-Verlag, Chap.7.

[3] Hopfield,J.J. and Tank,D.W. (1985) : "Neural" computation of decisions in optimization problems, Biological Cybernetics, 52, 141-152.

[4] Uesaka,Y. (1991) : Mathematical aspects of neuro-dynamics for combinatorial optimization, Proc. of ICANN-91, North-Holland, 1011-1014.

[5] Uesaka,Y. (to appear) : Mathematical basis of neuro-dynamics for combinatorial optimization, Optoelectronics.

[6] Hirsch,M.W. and Smale,S. (1974) : Differential equations, dynamical systems, and linear algebra, Academic Press.

The Effect of Synaptic Time Constants on Firing Patterns in Populations of Spiking Neurons.

C.A. van Vreeswijk and L.F. Abbott

Physics Department and Center for Complex Systems
Brandeis University, Waltham, MA 02254-9110

Abstract

We examine the conditions under which a population of spiking neurons with global excitatory coupling can fire asynchronously. Synapses with time constants larger than computed limits assure the stability of an asynchronous firing state even in the absense of inhibition. For smaller time constants the population starts to synchronize. Though the population then fires periodically, the individual neurons are quasiperiodic.

Model

We model a system of a large number N of tonically firing, identical integrate and fire neurons, with a global excitatory coupling. The time course of the voltage x_i of neuron i is given by

$$\frac{dx_i}{dt} = F(x_i) + G(x_i)E(t) \qquad (1.1)$$

with $F(x) > 0$ and $G(x) > 0$ for $0 \le x \le 1$. (The threshold potential is set at $x = 1$ and the reset at $x = 0$). $E(t)$ is given by

$$E(t) = \frac{\alpha^2}{N} \sum_{j,\kappa} (t - t_{j,\kappa}) e^{-\alpha(t - t_{j,\kappa})} \theta(t - t_{j,\kappa})$$

Here $t_{j,\kappa}$ is the time when neuron j fires for the κth time. One can make a transformation from x_i to $\tilde{x}_i$ so that $G(\tilde{x})$ is independent of $\tilde{x}$, so without loss of generality one can take $G(x) = g$.

Asynchronous State

For $g < 1$ there is a solution with constant firing rate E^0. We replace x_i with a phase ϕ_i so that

$$\frac{d\phi_i}{dt} = E^0 + \Gamma(\phi_i)e(t)$$

with $\Gamma(\phi) = \dfrac{gE^0}{F(x) + gE^0}$ and $e(t) = E(t) - E^0$.

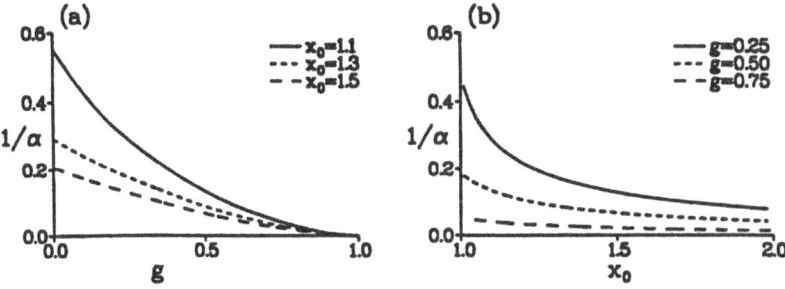

Figure 1: The value of α above which the asynchronous state is stable is plotted in (a) against g for different values of x^0 and in (b) against x^0 for different values of g.

With the density $\rho(\phi,t) \equiv \frac{1}{N}\sum_i \delta(\phi - \phi_i(t))$ and the current $J(\phi,t) \equiv (E^0 + \Gamma(\phi)e(t))\rho(\phi,t)$ the system obeys the equations [1-5]

$$\frac{\partial}{\partial t}\rho(\phi,t) = -\frac{\partial}{\partial\phi}J(\phi,t) \qquad (2.1)$$

$$\left(\frac{\partial}{\partial t} + \alpha\right)^2 E(t) = \alpha^2 J(1^-,t) \qquad (2.2)$$

With boundary conditions

$$J(0^+,t) = J(1^-,t). \qquad (2.3)$$

These equations have a static solution $\rho^0(\phi) = 1$ and $J^0(\phi) = E^0$.

Stability

If one expands around the static solution $J(\phi,t) = J^0 + j(\phi,t)$, $E(t) = E^0 + e(t)$, etc. one has after linearisation $j(\phi,t) = j(\phi)\exp(\lambda t)$ and $e(t) = \varepsilon\exp(\lambda t)$. λ obeys the eigenvalue equation [1,5]

$$(\lambda + \alpha)^2(e^{\lambda/E^0} - 1) = \frac{\alpha^2\lambda}{E^0}\int_0^1 d\phi\, \Gamma(\phi)e^{\lambda\phi/E^0} \qquad (3.1)$$

The static solution is stable if all solutions of this equation have a negative real part.

The eigenvalue equation has infinitely many solutions. In the uncoupled case, $g = 0$, one has solutions $\lambda_0^0 = -\alpha$, and $\lambda_n^0 = 2\pi n i E^0$ for $n \neq 0$, and the static solution is marginally stable. For small g, λ can be determined by expanding around λ_n^0. One finds that there is no stable solution if $\Gamma(0) > \Gamma(1)$. If $d\Gamma/d\phi > 0$ there is an $\alpha_{cr} > 0$, so that for $\alpha < \alpha_{cr}$ the asynchronous state is locally stable and for $\alpha > \alpha_{cr}$ unstable. This is also true for larger g, but in that case α_{cr} has to be determined numerically.

We calculated the critical value of α for $F(x) = x^0 - x$. The results are plotted in Fig. 1. In Fig. 1(a) α^{-1} is plotted against g for different values of x^0, in Fig. 1(b) α^{-1} is plotted against x^0 for fixed g.

Other Stable States

If $\alpha > \alpha_{cr}$ a periodic solution of eqns (2.1-3) becomes stable. This solution comes closer to complete synchronization with increasing α. A peculiarity of this state is that, while

$\rho(\phi, t)$, $J(\phi, t)$ and and $E(t)$ are periodic, $x_i(t)$ is quasi-periodic. Fig. 2(a) shows the firing rate $J(1^-, t)$, after the system has settled into a stable pattern, for a simulation with 100 neurons, with $x_i(0)$ chosen randomly between 0 and 1, $F(x) = 1.3 - x$ and $g = 0.4$. With Eqn. (3.1) one finds $\alpha_{cr} = 8.32\ldots$. The simulations show that the asynchronous state is *globally* stable for $\alpha \leq 8.32$ while as α increases the system synchronizes more and more. Fig 2(b) shows the distribution of the inter spike intervals for $\alpha = 9.0$. In Fig. 2(c) the inter spike interval is plotted against the previous inter spike interval, showing that the neurons fire quasi-periodically.

References

[1] D. Golomb, D. Hansel, B. Shraim and H. Somoplinsky, Phys. Rev. A45, 3516(1992).
[2] S.H. Strogatz and R.E. Morollo, Physica D31, 143(1988).
[3] S.H. Strogatz and R.E. Mirollo, J. Stat. Phys. 63, 613(1991).
[4] A. Treves, Network (in press).
[5] L.F. Abbott and C.A. van Vreeswijk, Phys. Ref. E (in press).

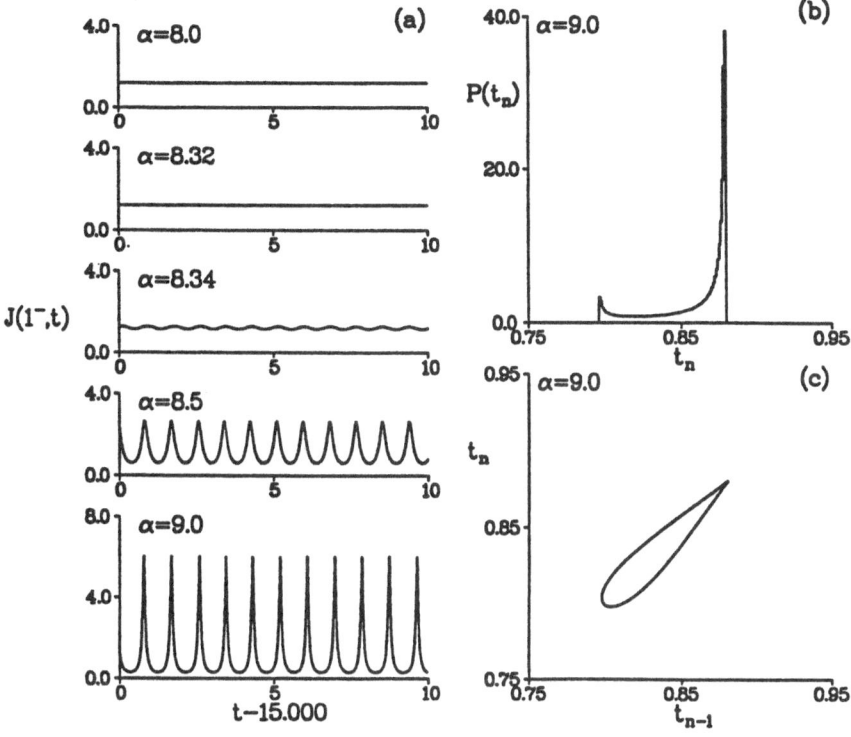

Figure 2: Results for simulations with 100 neurons, using $F(x) = 1.3 - x$ and $g = 0.4$. In (a) the firing-rate $J(1^-, t)$ is plotted against t after the system has settled into a stable pattern. In (b) the distribution of inter spike intervals t_n for one neuron is plotted. In (c) the inter spike interval t_n is plotted against the previous inter spike interval t_{n-1}

Information Processing by Spatio-temporal Chaotic Networks

A. Babloyantz and J.-A. Sepulchre

Service de Chimie-Physique
Université Libre de Bruxelles, CP 231 - Campus Plaine,
Boulevard du Triomphe, B-1050 Bruxelles, Belgium.
fax:32-2-6505767, email: ababloy@ulb.ac.be, jasepulc@ulb.ac.be

A chaotic categorizer made of two layered networks of interconnected oscillators is considered. Each layer shows chaotic dynamics. The latter are an infinite reservoir of unstable periodic orbits. Such orbits may be stabilized by minute changes in system variables or parameters, thus conveying a state of "attentiveness" to the network. These networks discriminate between different inputs. Due to the fact that the "attentive" network has an almost infinite variety of unstable cycles, the information processing capability of even small networks may be dramatically increased.

Introduction

A wide class of neural networks, loci of nonlinear processes, in a large range of parameter values, may exhibit spatio-temporal coherent activity of irregular nature, hereafter called spatio-temporal chaos.

Recent experimental and theoretical studies suggest that this is the case in the cortical activity of human brain where each behavioral state is characterized by a specific spatio-temporal chaos [1]. Thus the investigation of spatio-temporal chaos may be of importance for understanding of information processing of the cortex.

Such phenomena could also be of great importance in artificial neural networks. One may wish to suppress their onset if they appear as nuisance in network activity. Another possibility would be to think of these chaotic states, to convey to a neural network a more efficient way of operation. In this context one may note that most of the early neural networks correspond to dynamical systems possessing only fixed point attractors, i.e., asymptotic stationary states. This is the case for instance for the Hopfield networks [2], the feedforward networks [3] and the Kohonen maps [4].

The advantage of chaotic dynamics resides in their flexibility. Indeed, due to the sensitivity to initial conditions, chaotic dynamics may respond differently to external fluctuations, although keeping the same values of the parameters of the system. On the other hand, the great sensitivity to initial conditions makes a chaotic system difficult to predict and control. Thus a prerequisite to the use of a chaotic system, in a useful manner, is the possibility of controlling chaos.

A chaotic system is an infinite reservoir of unstable periodic orbits which, if stabilized, may serve as coding devices. Ott, Grebogi and Yorke [5] have proposed a method to achieve such a stabilization in systems with few degrees of freedom.

In this paper we show that the OGY method could be extended for stabilization of periodic orbits in larger systems. Moreover, in a multi-layer chaotic network, these stabilized orbits may be used to create an "attentive state" in the first layer of the network. Such attentive states are able to discriminate between two different inputs as recorded in the second layer.

The model

Figure 1 illustrates the schema of our device. The system is composed of two networks, each constituted of $N \times N$ oscillating units. In the absence of external input the dynamics of each network is assumed to be chaotic. An external pacemaker P sends micro-kicks to the network I such that an unstable cycle may be stabilized by the OGY procedure (see next section). In our model, this pacemaker plays the role of attentiveness which makes the system ready to process external information. The network II is not submitted to the OGY procedure, but is entrained by the network I. The entraining is achieved by $N \times N$ vertical links which join every unit of the network I to a unit of the network II, in a "one-to-one" correspondence. The vertical links are only activated in function of the input pattern.

For simplicity, the network is made of linearly coupled oscillators. In the future we intend to extend this model to more realistic neurons in a larger network and subject to nonlinear interactions.

The dynamics of such a system may be described by the following equations:

$$\frac{dZ_j}{dt} = Z_j(1 - (1 + i\beta)|Z_j|^2) + (1 + i\alpha)D\sum_k (Z_k - Z_j) + p_j(t)$$

$$\frac{dW_j}{dt} = W_j(1 - (1 + i\beta)|W_j|^2) + (1 + i\alpha)D\sum_k (W_k - W_j) + \gamma I_j(Z_j - W_j)$$

$$(j = 1, \ldots, N^2)$$

These equations are related to the complex Ginzburg Landau equation, which constitutes a generic model for oscillatory networks [6]. In Eqs. (1), each oscillator of the network I is described by a complex variable Z_j, whereas the W_j represent the variables of oscillators in network II. The summation $\sum k$ is extended over the first neighbours of each unit. As the network forms a square lattice, except on the boundaries of the network, there are four neighbouring units for each oscillator. The boundary conditions are of the zero-flux type.

The parameters α, β, γ and D are real quantities which are kept constant. These parameters are chosen such as, in the absence of the influence of the pacemaker P the system exhibits chaotic activity. A way to achieve chaotic dynamics is to consider

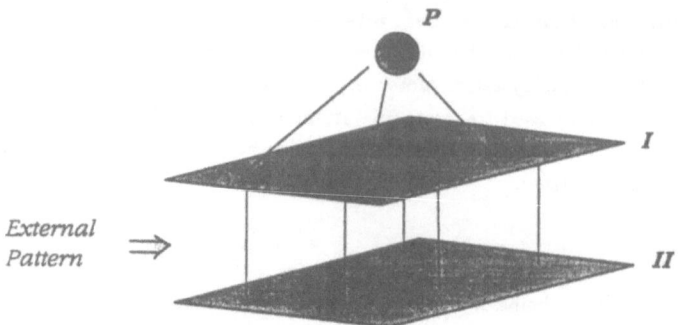

Figure 1: The pacemaker P sends micro-kicks to the network I and stabilizes periodic orbits which receive input patterns. The relevant connections are activated between the networks I and II and a coherent response is seen in II.

parameters for which the uniform state is unstable. We can show that this situation occurs if the following condition is fulfilled:

$$1 + \alpha\beta < D(1 + \alpha^2)(\cos \frac{\pi}{N-1} - 1) \tag{2}$$

The terms $p_j(t)$ in Eqs. (1), acting only on the network I represent the action of the pacemaker P which models whatever extremely small action is needed on the network to induce a state of attentiveness. This pacemaker sends micro-kicks to the network I in order to stabilize an unstable cycle of the network I. The stabilization of a periodic orbit is achieved through the OGY method. When an orbit is stabilized, we say that the system is in an *attentive* state. The network is thus ready to receive input from the external world.

The latter acts on the system by sending to the networks an input pattern associated to a matrix I_j. The external input activates a subset of links between the network I and II and is represented by the term $\gamma I_j(Z_j - W_j)$ in Eqs. (1). Each pattern induces characteristic output dynamics in the network II.

Stabilization of periodic orbits

We illustrate the dynamics of the system by the following simple example. We consider two networks of 9×9 elements, connected in a two-layered system, as depicted in Fig. 1.

Spatio-temporal chaos may be induced in the network I in the following manner. For fixed values of the parameters $\alpha = -10$ and $\beta = 2$ of Eq. (1), parameter D is decreased such that Eq. (2) is satisfied. By decreasing D from the critical value $D = 2.47$, the uniform state of the network looses its stability and various non-uniform dynamical regimes appear. We first observe periodic regime, then quasi-periodic dynamics. Finally chaos settles in for $D = 1.9$. In our simulations we fix the value of $D = 1.3$. The parameter γ will be chosen as $\gamma = 2.0$.

An infinite number of unstable periodic orbits are imbedded in this chaotic attractor. These unstable periodic orbits may be determined by numerical methods [7]. The basic recipe is to investigate a Poincaré section in the phase space. The recurrence of the flow may be followed on the Poincaré section until the approximate location of an unstable cycle is discovered. Then an iterative method is used to locate the precise position of the cycle. In this way, various unstable periodic orbits may be identified.

In our system we construct a Poincaré section corresponding to the hyperplane $\Re e\, Z_{(3,3)} = 0$, with the condition $\Re e\, \dot{Z}_{(3,3)} < 0$, where $(3,3)$ defines the position of an oscillator in the 9×9 network. With this Poincaré section, we found an unstable cycle $C_1(t)$ with a period $T_1 = 2.61$. Other unstable orbits C_2 and C_3 were also found, the respective periods were $T_2 = 2.55$ and $T_3 = 1.31$ [7].

Once the unstable periodic motions are identified, they can be stabilized by the OGY method. The OGY method uses the Poincaré map associated with the Poincaré section. Let $\mathcal{P}_\mu(\xi^{(n)})$ be this map, with $\xi^{(n)}$ representing a point of the Poincaré section and μ a control parameter. Consider also the linear map $M = \partial_\xi \mathcal{P}_\mu(\xi_s)$ and assume there is one unstable direction around the fixed point ξ_s, corresponding to the vector f_u and the eigenvalue λ_u of M^T (transpose of M). Then the prescription of OGY, derived in ref. [5], is to perform small fluctuations $\Delta\mu$ around μ according to the control feedback law:

$$\Delta\mu^{(n)} = -\lambda_u \frac{< f_u, \xi^{(n)} - \xi_s >}{< f_u, w >} \tag{3}$$

In this expression $w = \partial_\mu \mathcal{P}_\mu(\xi_s)$ and the bracket $<,>$ designates the scalar product. The feedback law (3) is applied only when $|\xi^{(n)} - \xi_s| < \delta$, where δ is function of the maximum allowed variation $\Delta\mu_{max}$ of μ (see ref. [5]). The effect of the OGY procedure is to pull the state $\xi^{(n)}$ towards the stable manifold of the fixed point ξ_s.

Equation (3) assumes that there exists only one unstable direction. If this is not the case, it can be easily shown that m unstable directions will require m independent variations of different parameters and a straightforward generalization of the Eq. (3) may be derived. In this paper, the unstable periodic motion used for information processing has two unstable directions associated with the eigenvalues $\lambda_1 = 7.3$ and $\lambda_2 = 6.4$ of the matrix M.

In OGY method, the small perturbations are applied to the parameters of the system. We extended the OGY method to include the perturbations of the *variables* of the system. This choice simplifies the computation without affecting the basic principles of the method, and enlarges the domain of the applicability of the method. Indeed let us assume that there exist m independant vectors v_j along which we may perturb the state of the dynamical system. Then we construct a new Poincaré map $\mathcal{P}_{\mu,\varepsilon}$ by introducing m new parameters ε_j such as:

$$\mathcal{P}_{\mu,\varepsilon}(\xi) = \mathcal{P}_\mu(\xi + \varepsilon_1 v_1 + \varepsilon_2 v_2 + \cdots + \varepsilon_m v_m)$$

In the case where the new parameters $\varepsilon_j = 0$. we recover the original Poincaré section.

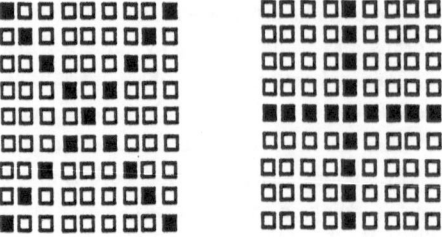

Figure 2: Two input patterns. The black squares represent the activated nodes, $I_j = 1$.

Consequently, fluctuation of the parameters ε_j is equivalent to perturbation of the variables of the dynamical system. As these perturbations are very small and occur whenever the dynamical flow intersects the Poincaré section, they may be regarded as micro-kicks periodically applied to the network.

Chaotic categorizer

Presently in the device of Fig. 1 the small fluctuation have stabilized a periodic orbit, say C_1, in network I. Thus the device is in an attentive state A_1. Let us note that the stabilization of orbits C_2, C_3 gives rise to states of attentiveness A_2, A_3.

Now, we consider the two different input patterns "×" and "+" represented on Fig. 2. When the pattern "×" is presented to the system in an attentive state, the corresponding links between network I and network II are activated. Network I then entrains network II which was in a chaotic state and coherent dynamics appear which are specific to the input pattern. Figure 3(a) shows the activity of a unit of network II in the absence of input. Whereas Fig. 3(b) depicts the coherent activity seen as a result of "×" input. On the other hand, if the pattern "+" is presented to the system, the dynamics of the network II exhibits a different coherent behaviour (see Fig. 3(c)).

We have thus a system, which is in a spatio-temporal chaotic state in the absence of input. If in an attentive state, whenever different patterns are presented to the system, the activity becomes coherent, and different signals are observed in function of the input patterns. This is a categorizer which does use the richness of chaotic dynamics.

Due to the fact that "attentive" network has an almost infinite variety of unstable cycles, the information processing capability of the network could be dramatically enhanced. In some way, the network I made of $N \times N$ units, behaves as if it was made of a very large number of units, or of multiple layers, each layer corresponding to one attentive state of the network.

The fact that the network I may attain an attentive state under the influence of extremely small parameter or variable perturbations may be of help in understanding human or animal attentiveness. One may speculate that minute changes in the state

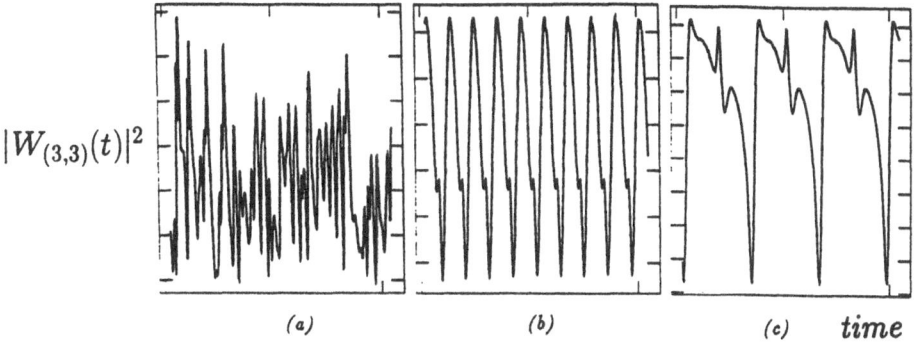

Figure 3: (a) The amplitude of the oscillator $(3,3)$ in a chaotic regime. (b) The response of network II to input pattern "×". (c) The response to the pattern "+".

of the system may bring about attentive states rapidly and without much physiological cost. Due to the almost infinite number of unstable periodic orbits, the information processing capabilities of even small regions of the cortex becomes practically limitless.

Acknowledgment

This research was supported by the Belgian Government (IMPULSE project RFO AI 10) and by the E.C.C. (ESPRIT, Basic Research, project 3234).

References

[1] A. Destexhe & A. Babloyantz, Pacemaker-induced coherence in cortical networks, *Neural Comp.* **3** (1991) 145-154.

[2] J.J. Hopfield, Neural networks and physical systems with emergent collective computational abilities, *Proc. Natl. Acad. Sci. U.S.A.* **79** (1982) 2554-2558.

[3] D.E. Rumelhart, G.E. Hinton & R.J. Williams, Learning representations by back-propagating errors, *Nature* **323** (1986) 533-536.

[4] T. Kohonen, Self-Organized Formation of Topologically Correct Feature Maps, *Biol. Cybernetics* **43** (1982) 59-69.

[5] E. Ott, C. Grebogi & J. A. Yorke, Controlling Chaos. *Phys. Rev. Lett.* **64** (1990) 1196-1199.

[6] Y. Kuramoto, *Chemical Oscillations, Waves, and Turbulence*, Springer-Verlag, Berlin Heidelberg New York Tokyo (1984)

[7] J.A. Sepulchre and A. Babloyantz, Controlling Chaos in a Network of Oscillators, submitted to Phys. Rev. E.

Hysteresis Phenomena and Bifurcation of Periodic Solutions in a Mathematical Model of Cortical Dynamics

Fotios Giannakopoulos
Mathematisches Institut der Universität zu Köln, Weyertal 86-90, D-5000 Köln, FRG
email: fotis@sun.mi.uni-koeln.de

In this paper we investigate the following system of two nonlinear differential equations

$$(1) \qquad \tau \dot{u}_1 = -u_1 + q_{11} f(u_1) - q_{12} u_2 + e_1 \quad , \quad \tau \dot{u}_2 = -u_2 + q_{21} f(u_1) + e_2 .$$

(1) describes the temporal behaviour of the spatially homogeneous solutions of a mathematical model for the dynamics of a cortically organized neural network. The network considered consists of an excitatory and an inhibitory cell population presented and discussed more detailed in [2,3]. Here, $u_1, u_2 : R \longrightarrow R$ denote the total potentials of the excitatory and inhibitory neurons. τ is a positive time constant describing the synaptic lowpasses and $f : R \longrightarrow R_+$ is a monotonically increasing, bounded function describing the nonlinear transformation of the total potential u_1 of the excitatory neurons into the according activity. For the sake of simplicity, we choose: $f(u_1) = 1 / (1 + \exp\{-4 a (u_1 - \theta)\})$, where $a > 0$ and $\theta \in R$. Carrying on with Eq. (1), q_{11}, q_{12}, q_{21} are appropriate positive constants. q_{ik} describes the connection strength on the line from neuron i to the neuron k. $e_1, e_2 \in R$ are external stimuli.

It has been shown analytically that the System (1) possesses characteristical nonlinear properties such as hysteresis of steady states and bifurcation of periodic solutions (Hopf bifurcation) [1,2]. These properties are controlled by just two parameters, i.e., positive q_{11} and negative feedback $q_{12}q_{21}$ (see left Figure). The most interesting case is the following. For $q_{12}q_{21} + \frac{1}{a} < q_{11} < 2q_{12}q_{21}$, we can prove the coexistence of hysteresis and Hopf bifurcation. On the upper and the lower branche of the hysteresis curve, there exist two Hopf bifurcation points (see right Figure). At these points periodic solutions bifurcate from the according steady states.

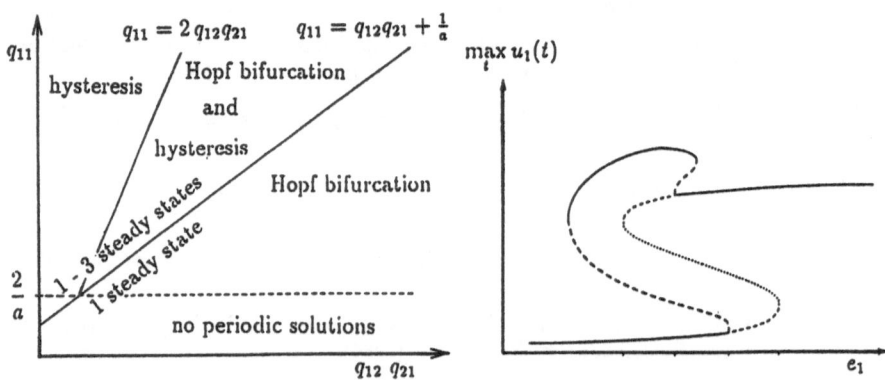

References

[1] F. Giannakopoulos, Dissertation, FB Mathematik, J. Gutenberg Universität, Mainz, 1989.
[2] H.A. Mallot and F. Giannakopoulos, in: J.G. Taylor, E.R. Cainiello, R.M.J. Cotterill and I.W. Clark (eds), Neural Network Dynamics, Springer, pp. 341-355 (1992).
[3] W. von Seelen, H.A. Mallot and F. Giannakopoulos, Biol. Cybern., 56, pp. 37-49 (1987) .

Computing Complexity of Symmetric Quadratic Neural Networks [1]

Eric Goles

Universidad de Chile, Facultad de Ciencias

Físicas y Matemá ticas, Departamento de Ingeniería

Matemá tica. Casilla 170-correo 3 Santiago, Chile

Martín Matamala[**]

Laboratoire de l'Informatique du Parallélllisme

Ecole Normale Supérieure de Lyon

46, Allée d'Italie, 69364 Lyon Cedex 07, France

Extended Abstract

A neural network of order $p \in I\!N$ (p-PNN in the sequel) [1, 4] is defined as a set of n neurons taking values in $\{0,1\}$ and interconnected by a tuple $(A^{(p)}, A^{(p-1)}, ..., A^{(0)})$ where $A^{(p)} = (a^{(p)}_{i_0,i_1,...,i_p}) \in I\!R^{(p+1)n}, ..., A^{(0)} = (a^{(0)}_{i_0}) \in I\!R^n$. The update function is as follows: For $1 \le i_0 \le n$

$$y_{i_0} = 1\!\!1\left(\frac{1}{p!}\sum_{i_1,...,i_p=1}^{n} a^{(p)}_{i_0 i_1...i_p} x_{i_1} \cdots x_{i_p} + ... + \frac{1}{2}\sum_{i_1,i_2=1}^{n} a^{(2)}_{i_0 i_1 i_2} x_{i_1} x_{i_2} + \sum_{i_1=1}^{n} a^{(1)}_{i_0 i_1} x_{i_1} + a^{(0)}_{i_0}\right)$$

where $1\!\!1(u) = 1$ iff $u \ge 0$ (0 otherwise).

The most usual dynamics for the previous network are the sequential and the parallel iteration. In the former case all cells are updated one by one in a prescribed order. The parallel iteration consists to update all the cells synchronously. We introduce also some hypothesis on the arrays $(A^{(p)}, ..., A^{(0)})$; Symmetry: $\forall 1 \le s \le p$, for any permutation, σ, of the indices : $i_0, ..., i_s \in \{1, ..., n\}$ $a^{(s)}_{i_0 i_1...i_s} = a^{(s)}_{\sigma(i_0)\sigma(i_1)...\sigma(i_s)}$ and Non-negativity: $\forall 1 \le s \le p$, if at least two indices have the same value; i.e. when $|\{i_0, i_1, ..., i_s\}| \le s$, then $a^{(s)}_{i_0,...,i_s} \ge 0$. Under these hypothesis one gets, for $p = 1$, the usual Hopfield model [3] of a linear symmetric neural networks: $y_i = 1\!\!1(\sum_{j=1}^{n} a_{ij} x_j - \theta_i)$, where $A = (a_{ij})$ is symmetric and $\text{diag}(A) \ge 0$.

Here, we present a simulation of arbitrary (non necessarily symmetric) linear neural networks by symmetric neural networks with quadratic ($q = 2$) arguments (2-PSNN). In order to make the simulation, we introduce a special unit: the Δ-unit which consists of two neurons connected by a quadratic weight. This unit can orientate the information and then it can simulate the interaction a_{ij} between two neurons i and j in the linear network.

Our general result is given by the following theorem whose proof can be seen in [2].

Theorem . Let (A, θ) be a linear neural network of size n, then there exists a 2-PSNN with $2(n+1)$ Δ-units and a morphism $\varphi : \{0,1\}^n \rightarrow \{0,1\}^{2(n+1)}$ such that: $\forall x \in \{0,1\}^n$ $\varphi(F(x)) = G^2(\varphi(x))$, where F and G are the global update functions of the linear and quadratic network respectively.

References

[1] Chen, H.H., Lee, Y.C., Maxwell, T., Sun, G.Z., Lee, H.Y., & Giles, C.L., (1986) 'High order correlation model for associative memory'. AIP Conference Proceeding, in J.S. Denker (Eds), Neural networks for computing, pp 86-92.

[2] Goles, E., Matamala, M. (1992) 'Dynamical and Complexity Results for High Order Neural Networks' Research Report, Dep. Ingeniería Matemática, Facultad Ciencias Físicas y Matemáticas, Univ. de Chile.

[3] Hopfield, J.J. (1982), 'Neural Networks and Physical Systems with Emergent Collective Computational Abilities', Proc. Natl. Acad. Sci. U.S.A. 79, 2554-2558, 1982.

[4] Psaltis, D., Park, C.H., (1986), 'Nonlinear Discriminant Functions and Associative Memories', J. Denker (Ed), AIP Conference Procc, N. York, Am. Inst. of Phys., p. 370.

[1] Partially supported by DTI, U. Chile, FONDECYT 91-1211 (EG) and FONDECYT 0047/90 (MM)

[**] Dept. Ingeniería Matemática, U. de Chile, from December 1993

Topology Learning Solved by Extended Objects: A Neural Network Model

Csaba Szepesvári[1] and András Lőrincz
Insitute of Isotopes, Budapest, P.O.B. 77, Hungary, H-1525
[1]Attila József University of Szeged, Szeged, Hungary, H-6720

We describe a self-organizing artificial neural network architecture that simultaneously creates (1) local filters with competitive learning, (2) a representation of the topology of the external world with Hebbian learning, and introduces (3) Kohonen type cooperative neighbour training through the self-developed connections. Such a network is capable of building up a 3-dimensional topology from two 2-dimensional images - similar to the working of the human eye - and it shows increased adaptivity.

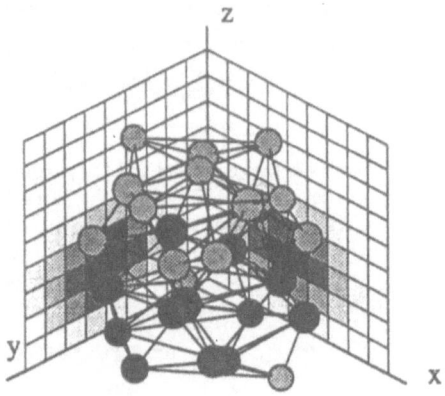

The system is based on a competitive network that builds up spatial filters of the external world in a self-organizing fashion. Inputs may overlap with a filter or several filters. The overlap allows the network to self-develop Hebbian connections that correspond to the topology of the external world. The self-developed connections then make self-developing Kohonen-type cooperative neighbour training possible. Neighbour training increases the adaptivity of the network considerably.

Figure 1. Input vector made of two 2-dimensional retinas are shown by the (y,z) and (z,x) planes. The self-developed receptive field of the black neuron is depicted on these planes. The self-developed wiring is given and the neighbours of the black neuron are coloured to grey.

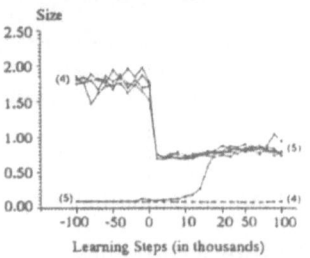 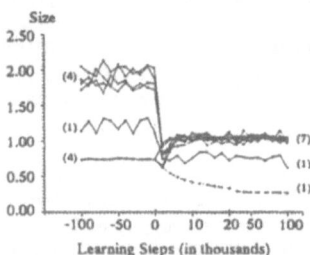

Figure 2. Test of adaptivity with and without neighbour training. Graphs on the left hand side show the behaviour of the competitive network: neurons of small role play very small role and after a sudden change in the external world (the object size was changed at step-number 0) it takes a long time for those to recover. Graphs on the right hand side show the behaviour of the network with self-developed Kohonen type neighbour training. Low role neurons have reasonable activity and the sudden change is compensated very quickly. In both cases the oscillations in the size of the neural receptive fields result from the ongoing competition. Because the network has no time dependent parameter it retains its adaptivity forever.

Higher Order Neural Networks in a Unified Learning Scheme

Arnd Bischoff[1] and Bernd Schürmann[1,2]

[1]Institut für Theoretische Physik, J.W. Goethe – Universität, D-6000 Frankfurt/M., Germany
[2]Corporate Research & Development, Siemens AG, D-8000 München, Germany

We propose a global approach to the derivation of learning rules for networks with arbitrary order of connectivity, and show how the most important rules for supervised learning can be handled in a unified fashion and discuss several special cases. For that we make use of the fact that learning usually takes place while the regarded system is in a fixed point state. The properties of different learning rules in feedforward and recurrent systems of first and higher orders are compared qualitatively by means of numerical simulations.

1 General Derivation

Neural systems that are trained *by* minimizing an *error function* for the output units are usually treated as being entirely unrelated to those that are trained *while* minimizing some *energy function* that also ensures stability of the activation state dynamics.

We now search for a more global point of view, which can be of advantage for the construction of new hybrid learning schemes as well as for the understanding of already widely known and used algorithms.

To start with, we assume a typical fixed point equation for signals in a network containing connections of order ω, reading

$$S_n = \Theta_n (Z_n) = \Theta_n \left(\sum_{\nu \in I_n} w_{n\nu} \prod_{j \in \nu} S_j \right) . \tag{1}$$

For recurrent systems, where fixed point states can not be found by a simple signal *propagation*, (1) is transformed into differential equations like

$$\frac{dS_n}{dt}\bigg|_t = -S_n(t) + \Theta_n (Z_n(t)) , \tag{2}$$

which tend to settle down into solutions of (1). $\Theta_n(Z_n)$ is a differentiable squashing function of the activation Z_n. The set of all connections influencing unit n is specified by I_n. $\nu^{(\omega)} = \{\nu_1, \ldots, \nu_\omega\}$, $n \notin \nu^{(\omega)}$ is a set of indices (of all units addressing unit n through the weight $w_{n\nu}$). Thereby we express the symmetry of all connections with regard to the "sending" units, which is an important consequence of the common definition of higher order connectivity used in eq. (1).

The gradient learning rule for an arbitrary objective function E has for all $\mu \in I_m$ the form

$$\frac{dw_{m\mu}}{dt} = -g \left[\left(\frac{dE}{dw_{m\mu}} \right)_{\text{impl.}} + \left(\frac{\partial E}{\partial w_{m\mu}} \right)_{\text{expl.}} \right] , \quad \left(\frac{dE}{dw_{m\mu}} \right)_{\text{impl.}} = \sum_n \Omega_n \frac{\partial E}{\partial S_n} \frac{dS_n}{dw_{m\mu}} . \tag{3}$$

Here g is the learning parameter, and $\Omega_n = 1$, if n is an output unit, $\Omega_n = 0$ otherwise. To proceed from this simple statement, we require that eq. (1) be valid. In other words, when *adapting the weights*, the network is assumed to be in a *fixed point state* (this can be regarded as introducing different time scales for signals and weights). In order to obtain a general learning rule, essentially the quantities $dS_n/dw_{m\mu}$ have to be calculated. Differentiation of (1) and several further mathematical operations (we assume $\mu \in I_m$) lead to the convenient equation

$$\left(\frac{dE}{dw_{m\mu}}\right)_{\text{impl.}} = \Delta_m \prod_{j\in\mu} S_j \,. \tag{4}$$

The "error signals" obey

$$\Delta_i = \Theta'_n(Z_i)\left[\Omega_i\frac{\partial E}{\partial S_i} + \sum_n\left(\sum_{\nu\in I_n:\nu\ni i} w_{n\nu}\prod_{j\in\nu\setminus\{i\}} S_j\right)\Delta_n\right] \,. \tag{5}$$

These equations can, according to the "Recurrent Backpropagation" algorithm introduced by Pineda [1] and Almeida [2] (for first order), be solved by use of the corresponding differential equations, similar to eq. (2). For that the $\Delta_n(t)$ are initialized with zero values, $\Delta_n(0) = 0$. The iterations appear to have the characteristics of a relaxation in a *linear first order* network.

We emphasize the importance of the "sender symmetry", which in this formulation can be handled easily. Pineda [1] ignored this point and obtained an unnecessarily complicated system of equations for his learning algorithm.

Using (4), the learning rule (3) can now be written as

$$\frac{dw_{m\mu}}{dt} = -g\left[\Delta_m\prod_{j\in\mu} S_j + \left(\frac{\partial E}{\partial w_{m\mu}}\right)_{\text{expl.}}\right] \,. \tag{6}$$

The entire learning and dynamics of our higher order network now just obeys the three equations (1), (5) and (6), in generalization to the unified treatment of the first order case [3]. For training sets with more than one pattern, a Δ has to be calculated for each of them. If batch learning is used, the bracket in (6) (at least the left part) must be replaced by the corresponding sum. For incremental learning, the equations are used unmodified.

The character of the regarded network is, apart from the choice of connections, determined by its objective function. In the following sections we will study some special cases.

2 Backpropagation Networks

For the well-known Backpropagation learning rule, a typical objective function is the error function

$$E = \frac{1}{2}\sum_n\left(S_n - S_n^{\text{target}}\right)^2 \,, \tag{7}$$

where for a single input pattern, S_n and S_n^{target} are the actual and the required output values, respectively. After inserting this into (5), the Delta Rule takes the form

$$\Delta_i = \Theta'_i(Z_i)\left[\Omega_i\left(S_i - S_i^{\text{target}}\right) + \sum_n\left(\sum_{\nu\in I_n:\nu\ni i} w_{n\nu}\prod_{j\in\nu\setminus\{i\}} S_j\right)\Delta_n\right] \,. \tag{8}$$

This is appropriate for Recurrent Backpropagation. We show how the feedforward case can be obtained quite easily from this.

In the output layer of a feedforward network, the rightmost term of (8) vanishes, because there are no connections to which the signal of unit i contributes. So, recalling the meaning of Ω_i, we write

$$\Delta_{i^{\text{out}}} = \Theta'_{i^{\text{out}}}(Z_{i^{\text{out}}})\left(S_{i^{\text{out}}} - S_{i^{\text{out}}}^{\text{target}}\right) \,. \tag{9}$$

For all units that belong to a "hidden layer", we get

$$\Delta_{i^{\text{hid}}} = \Theta'_{i^{\text{hid}}}(Z_{i^{\text{hid}}})\sum_n\Delta_n\sum_{\nu\in I_n:\nu\ni i^{\text{hid}}} w_{n\nu}\prod_{j\in\nu\setminus\{i^{\text{hid}}\}} S_j \,. \tag{10}$$

We have assumed that there is no feedback in our network. Therefore all $\Delta_{i\mathrm{hid}}$ only depend on the error signals of the following layer and thus can be determined successively (no relaxation method is needed). Because the input units do not receive any signals, there is no need for calculating corresponding error signals. Since the error function (7) does not explicitly depend on the connection weights, the learning rule just becomes

$$\frac{dw_{m\mu}}{dt} = -g\,\Delta_m \prod_{j\in\mu} S_j \tag{11}$$

where the adequate Δ_m have to be inserted. This has the form of the feedforward Backpropagation rule for arbitrary order of connectivity.

3 Networks Using Hebbian Learning

For networks with learning rules of Hebbian type, the objective function mainly consists of a parity term for the signals and connection weights. Including a forgetting term, we assume

$$E = -\sum_n \int_0^{Z_n} \varphi_n(\tilde{Z}_n)\,\Theta'_n(\tilde{Z}_n)\,d\tilde{Z}_n + \frac{1}{2}\sum_{\nu\in I}\lambda_\nu\,w_\nu^2 - \sum_{\nu\in I} w_\nu \prod_{j\in\nu} S_j \; . \tag{12}$$

Because of the total connection symmetry, all units involved in the same connection receive signals through the same weight. For simplicity, we have adapted our index notation to this special case, so that I is the set of all sets of indices that specify an existing connection. φ_n is a function of the corresponding unit's activation state, usually including a decay term (e.g. $\varphi_n(Z_n) = -Z_n + P_n$, where P_n is an input pattern component). The $\lambda_\nu > 0$ are forgetting constants that limit the weight values.

An important feature of this function is that the units' activation states as well as the weights can be updated according to a gradient descent rule,

$$\frac{dZ_n}{dt} = -\frac{\partial E}{\partial S_n} = \varphi_n + \sum_{\nu\in I\,:\,n\in\nu} w_\nu \prod_{j\in\nu\setminus\{n\}} S_j \; , \tag{13}$$

and

$$\frac{dw_\mu}{dt} = -g\left(\frac{\partial E}{\partial w_\mu}\right)_{\mathrm{expl.}} = -g\left(\lambda_\mu\,w_\mu - \prod_{j\in\mu} S_j\right) \; . \tag{14}$$

We recall that learning takes place while the network's activation state resembles a fixed point (e.g. with strong inputs forcing the system to assume a state according to a learning pattern), and therefore, according to eq. (12) and (13), $dZ_n/dt = -\partial E/\partial S_n = 0$. Thus eq. (3) holds for these systems, too, since

$$\left(\frac{dE}{dw_{m\mu}}\right)_{\mathrm{impl.}} = \sum_n \Omega_n \frac{\partial E}{\partial S_n}\frac{dS_n}{dw_{m\mu}} = 0 \; . \tag{15}$$

Eq. (5) for the error signals here allows for the trivial solution, $\Delta_i = 0 \;\forall\, i$. Thereby the general learning rule (6) also is reduced to eq. (14).

4 Simulation Results

For the test of higher order networks trained by Backpropagation rules, we chose a handwritten pattern recognition task and several small "benchmarks" like the four bit parity problem or the 8-3-8 encoder-decoder test. Similar to our studies for higher order Hebbian learning described in [4], we always compared networks of different orders *with equal numbers of degrees of freedom*, i.e. we *randomly pruned* the higher order networks to meet the complexity of the first order system.

Due to numerical effects, the higher order networks in some cases exhibited a somewhat slower convergence during learning. For these classification tasks, introducing recurrency did not generally improve performance.

Though most recurrent systems went through a phase of rapid training convergence, especially before the right signs for the output signals were reached, a quicker overall convergence of recurrent systems (by means of output accuracy and learning cycles) could only be found for the parity problem. Nonetheless, adding recurrent connections in the right places could speed up learning with regard to the signs of the output units. In addition, there still are various problems where recurrency is strictly needed.

Whenever emphasis was put on generalization ability, the higher order systems performed worse with increasing order of connectivity, independent of recurrency or the number of hidden units. On the other hand, when tested with the training set, i.e. when required to approximate the training data as exactly as possible, higher order networks appeared to be superior. The reason for this lies in the more complex shapes of decision regions formed by higher order networks, which can be of advantage for data approximation, but may cut off some generalization ability. This behaviour became obvious when we applied Perceptrons (no hidden units) of first to third order with equal numbers of connections to character recognition. Here after learning of 910 handwritten digits the recognition rates on the test set of 8050 patterns were 84.9%, 82.4%, and 71.3%, respectively. The reproduction of the test data yielded success rates of 97.8%, 99.1%, and 99.3%.

To summarize, the increased noise sensitivity but better resolution of correlations in randomly pruned higher order systems with Hebbian learning [4] has its equivalent in corresponding Backpropagation-trained networks.

An important means of improvement for the higher order systems will be achieved by network optimization. Here each connection can be selected from a greater variety of unit combinations than in a network of first order, and hence there will be more room for improvement. Concerning the generalization ability, a better procedure than random pruning *before* training will be to start training with a less reduced network and *then* optimizing by applying the well-known methods already used for first order networks (cf., e.g., [5]). Some of these methods (like stopped training, weight penalty, statistical significance test) may be taken over straightforwardly, others (like Optimal Brain Damage) have to be suitably generalized first.

For small, noisy input data sets, where optimization techniques are indispensable, even the size of a *fully connected* higher order network may be such that the number of weights is still manageable for adaptation. Moreover, restricting to a second order net may be sufficient in many cases.

Investigations along these lines are underway.

Acknowledgement

A. Bischoff is grateful to the Siemens AG for financial assistance by an E. v. Siemens fellowship.

References

[1] Pineda, F.J. (1989): *Generalization of Back-Propagation to recurrent and higher order neural networks*. In Touretzky, D., ed., Neural Information Processing Systems, vol. I, Denver 1988, pp. 602–11.

[2] Almeida, L.B. (1987): *A learning rule for asynchronous Perceptrons with feedback in a combinatorial environment*. In Proc. IEEE First Int. Conf. Neural Networks, vol. II, San Diego 1987, pp. 609–18.

[3] Ramacher, U., Schürmann, B. (1990): *Unified description of neural algorithms for time-independent pattern recognition*. In Ramacher, U., Rückert, U., ed., VLSI Design of Neural Networks, Kluwer Academic Publishers, pp. 255-70.

[4] Bischoff, A., et al. (1992): *Characteristic properties of stable recurrent higher order neural networks*. In Aleksander, I., Taylor, J., ed., Artificial Neural Networks, vol. 2, Brighton 1992, pp. 325–8.

[5] Hergert, F., Finnoff, W., Zimmermann, H.G.: *Improving model selection by dynamic regularization methods*. To be published in Neural Networks (1993).

On a Simple Hysteresis Network

Kenya JIN'NO   *Toshimichi SAITO*

Electrical Engineering Department, HOSEI University, Tokyo, 184, JAPAN
E-mail : jinno@toshi.hosei.ac.jp

Abstract

In order to approach the explication of rich dynamics from artificial neural networks (ab. ANN), this article analyses a simple continuous-time hysteresis network including only two parameters. The objective system is described by the following equation:

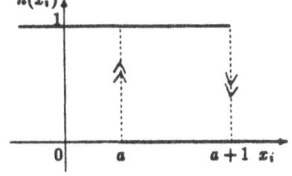

Fig.1 Hysteresis Function

$$\begin{cases} \dot{x}_i = -x_i + \sum_{j=1}^{m} y_j + T\, y_i, \\ y_i = h(x_i) \begin{cases} 0 & \text{, if } x_i > a \\ 1 & \text{, if } x_i < a+1, \end{cases} \end{cases} \quad (1)$$

where $x \equiv (x_1, \ldots, x_m)^t$, $x_i \in R$, is a state variable vector, $y \equiv (y_1, \ldots, y_m)^t$, $y_i \in \{0, 1\}$, is an output vector and m is the number of the cell. T is a self feedback parameter and $h(x_i)$ is a binary hysteresis function as shown in Fig.1. That is , y_i is switched from 0 to 1 if x_i hits the left threshold a and vice versa. This network is a simplified version of our original hysteresis neural network[1] and an implementation example is shown in [2]. The net includes only two parameters (a , T), but the system behavior is very interesting. Then we give the following results.

1. If the parameters satisfy $0 < K - a < 1$ and $0 < (T + l) + (K - a) < 1$, where K and l are non-negative intergers, then all attractors of the net are stable equilibrium points such that the number of "1" in corresponding output vector y is $K + n$, $n = 0 \sim l$. For any K and l, we can completely clarify the number of attractors and the their domain of attraction.

2. If the parameters satisfy $0 < K - a < 1$ and $1 < T + (K - a) < 2$, where K is a non-negative integer, then all attractors are stable periodic orbits such that the number of "1" in corresponding output vector y vibrates between K and $K - 1$. For any K , we can clarify the number of attractors and the their local domain of attraction. We can rigorously calculate all periodic orbits.

References

[1] K.Jin'no and T.Saito : *Trans. IEICE , J75-A , 3* (1992) *552*

[2] T.Saito and M.Oikawa : *IEEE Trans. Neural Networks , 4 , 1* (1993) *43*

[3] T.Saito , K.Jin'no and K.Kishiro : *in Proc. of IEEE/ISCAS'92 San Diego* (1992) *2769*

[4] M.Morita , S.Yoshizawa and K.Nakano : *Trans. IEICE , J73-DII* (1990) *232*

[5] J.A.Farrell and A.N.Michel : *IEEE Trans. CAS-37 , 5* (1990) *877*

[6] J.Hao and J.Vandewalle : *in Proc. of IJCNN'92 Baltimore , 2* (1992) *166*

[7] C.M.Marcus , F.R.Waugh and R.M.Westervelt : *Physica D , 51* (1991) *243*

Switching the Vector Field According to the Input of an Oscillatory Neural Network

Yukio Hayashi

ATR Human Information Processing Research Laboratories
2-2, Hikaridai, Seika-cho, Soraku-gun, Kyoto 619-02, Japan

Abstract

We propose a new concept of spatiotemporal pattern processing by artificial neural networks that is based on drastically changing the vector fields as attractors according to the input.

In the state of the art, most neural network models treat the mapping between static patterns with statistical distributions. This idea is based on space partition. However, recurrent neural networks can treat the mapping between spatiotemporal or time-sequential patterns (see Fig. 1).

As a special case of a recurrent neural network, the author has proposed the oscillatory neural network with simple connections between excitatory $\{x_i\}$ and inhibitory $\{y_i\}$ cells, which produces a limit cycle or chaos from a constant input bias.

The dynamic equations are $dx_i(t)/dt = -x_i(t) + G\left(\sum_{j=1}^{N} W_{ij}x_j(t) - K_{EI}y_i(t) + I_i\right)$, $dy_i(t)/dt = -y_i(t) + G\left(K_{IE}x_i(t)\right)$. The autonomous system with 2N variables x_i, y_i has an attractor independent of the 2N initial states $x_i(0), y_i(0)$ in the basin. However, the shadow variable y_i can be converted to $y_i(t) = \exp^{-t}\left[\int_0^t \exp^s G(K_{IE}x_i(s))ds + y_i(0)\right]$ as the hysteresis of x_i.

By the conversion, the crossing orbits can exist in the N-dimensional space of the non-autonomous system. Since the superimposed state space has crossed or linked trajectories from different vector fields, the orbits can approach the respective correct memories from the initial states of conflicting positions. For example, the two crossing orbits for an ambiguous letter such as A or H (e.g. / − \) can be discriminated adaptively, if the input is controlled by the higher-level context of CAT or THE. Such a drastic change is impossible in the conventional discrimination methods based on space partition in the low-level state space.

We will show an example of a drastically changing the vector field of an oscillatory neural network, whose attractors depend not only on the input biases but also on the initial states or past hysteresis. As a special case of a recurrent neural network, the characteristics are shown clearly.

< Conventional NN : Multi-layer Model, Hopfield Model etc. > < Recurrent NN >

Static Mapping: Initial State -> Prototype (SP)   Dynamic Mapping: Input ->SP, LC, or Chaos Attractor

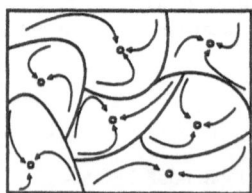

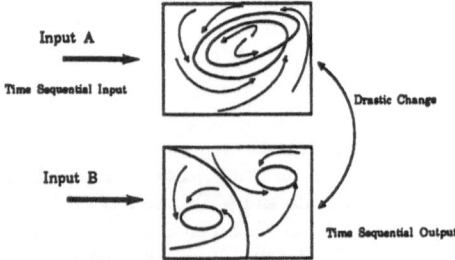

The weight parameters maintain global space partition
(Point Attractor & Basin, or Discriminant Function)

The weight parameters maintain each vector field

Figure 1: Static & dynamic mapping of neural networks

PROCESSING OF INFORMATION ENCODED IN COUPLED ONE-DIMENSIONAL MAPS

Alexander Yu. Loskutov and Valery M. Tereshko

Chair of Low-Temperature Physics,
Physics Faculty,
Moscow State University,
Moscow 119899,
RUSSIA
e-mail:khokh@mch.chem.msu.su

ABSTRACT

We propose an analytical method of ciphering and deciphering various information by means of creation and destruction of new stable periodic orbits respectively, in parametrically coupled one-dimensional maps. This method is based on the following theorem.

Denote the set of values of the parameter corresponding to the chaotic behavior of map, by C.

THEOREM [1]. There exists a subset D of the set C such that chaotic map with a cyclic parametric transformation generates stable cycles of a finite period.

The proof of this theorem is reduced to construction of multiple transformation maps and detection of certain parameter values at which iterations of such maps have stable periodic points.

Thus, parametric coupling of maps leads to creation of new periodic orbits of finite periods. Therefore, in a network of parametrically coupled maps a new information can be created. If the parameter values of maps belong to the chaotic set C, but they are outside D, then all periodic orbits are "mixed", and information corresponding such orbits is hidden. However, when parameters are distributed in the subset D, the necessary information can be extracted. One may say that in this case the cipher key is the subset D. This subset D has extremely complex structure but, apparently, its Lebesgue measure is positive. Moreover, if in network the connection weights are small or approach zero then information given in initially created periodic orbits is destroyed.

REFERENCES

[1] Loskutov A.Yu. and Shishmarev A.I. (1992), Preprint N236 of The Max-Planck Society, Munich, Germany.

Feedback in Single Continuous Neurons

J. Segovia, J. Rios, M. Lerma and D. Barrios
Facultad de Informática
Universidad Politécnica de Madrid
Campus de Montegancedo, s/n
Boadilla del Monte, 28660, Madrid
SPAIN

Abstract

In this paper we study the effect of the self-connections in the dynamic of isolated continuous neurons. A local study of the dynamic of self-connected neurons gives us an important knowledge of their behavior in response to a stream of pulses. It is known that a discrete single neuron with a self connection is capable of information latching, but it is also known that a discrete network with only local feedback is uncapable of performing a simple task such as implementing a counter. In this paper we study the dynamic of continuous neurons with a self-connection and a bias. Under certain circunstances, when the self-connection overruns a threshold value which depends on the activation function used, two stable states are created. These stable states modify the dynamic behavior of the neuron when reacting against a stream of pulses: the neuron is affected in different ways depending on the frequency, number, amplitude and shape of the stream of pulses. This property can be used when designing an architecture, providing layers with these neurons which will analyse the different characteristics of the input. This will let us implement, with only self-connections, systems such as counters or frequency meters.

A Neural Network for Decision Making
in Dynamic Environments

Abderrahim LABBI

LIFIA, Institut IMAG
46, avenue Félix Viallet,
38031 Grenoble cedex, France

Abstract

When designing a decision system (a pattern classifier, recognizer, or associator), we usually suppose that the information being processed by the system to compute a decision (output) remain unchanged (static) during the whole processing period called the System Autonomy Period (SAP). These systems are usually unable to process continuously dynamic information emerging from a changing environment since the internal states of such systems can not be forced to take into account sudden information changes in the environment. In this abstract is presented a competitive neural network which has the ability to process continuously dynamic information emerging from an unexpectedly changing environment. The network dynamics is described by a system of differential equations with external input E,

$$\frac{dx_i}{dt} = -c_i.x_i + (1-x_i).(\sum_{k=1}^{k=m} w_{ik}.f_k(e_k)) \quad - \quad (1+x_i).(\sum_{j=1}^{n} d_{ij}.g_j(x_j)); \quad i = 1,\ldots,n \quad (1)$$

To run a network represented by eq.(1), both E and $X(0)$ must be specified. The input E specifies the dynamics of the network, while $X(0)$ specifies the initial state of a trajectory obeying such dynamics. Each input vector E induces a set of attractors $\Gamma(E)$ when the system (1) is supposed to be convergent. When $\Gamma(E)$ contains more than one attractor, the dynamics of a same E may converge to different attractors depending on $X(0)$. Therefore, the network can not define a mapping from the environment to the attractor (decision) space. However, if for each E, $\Gamma(E)$ contains only one attractor (the system is saied to be globally convergent), $X(0)$ does not matter, and the network can work as a pattern classifier or associator. If in addition, such attractor is stable, the system is saied to be globally asymptotically stable. By assuming that, $w_{ik} \geq 0$, $d_{ii} = 0 \leq d_{ij} = d_{ji} \leq \alpha$, and $g_j(x) = \frac{1}{1+\exp(-x)}$, we have the following,

Theorem 1. The system (1) is convergent for any clamped input E.
Proof: By means of the variable change $z_i = 1 + x_i$, the system (1) can be written in the Cohen & Grossberg form.□
Theorem 2 . The system (1) is globally asymptotically stable for any clamped input vector E, provided, $c_i \geq (n-1).\alpha$; $i = 1,\ldots,n$
Proof: available from the author.□
Concerning learning, the network can be trained to classify a set of input data into classes by assigning a neuron in the decision layer to each class. The weights w_{ik} can then be determined by a competitive learning rule. The network runs continuously having regard for the environment changes, and may recant and change its output without resetting its state variables when tending towards a first decision and new external changes occur.
A natural extension of the analysis of the dynamics (1) is to explore other conditions on the network parameters under which more complex behaviour occurs and generates more complex attractors such as limit cycles. Such study is being developed in the framework of bifurcation theory.

Chaos in Neural Networks at Nonlinear Synapses

Algis Garliauskas, Remigijus Andžius

Department of Neuroinformatics
Institute of Mathematics and Informatics
Akademijos 4, 2600, Vilnius, Lithuania

Neural networks as massive parallel connectionist structures with nonlinearities are able to different complex phenomena as a self-organizing, associativity, bifurcation, and chaos.

It is well known an universal theory of behavior in nonlinear systems based on Los Alamos laboratory in 1971. For the large scale functions at increasing of the parameter a distruction of stable cycle is happened and a cycle of period-doubling is occured. The doubling of a period is continued until the endlessness. In this paper we pay an attention to different nonlinearities which take place on the interconnections between neurons, i.e., the functions of an activity on the input of neuron which rely through synaptic strength to output of another neuron. Using the Hebb's rule of learning, a mathematical model has been constructed on behalf slow and fast equations with different modifications of nonlinear synapses as follows:

The slow equation:
$$w_{ij}(k+1) = (1 - D_w)w_{ij}(k) + \eta a_i(k)a_j(k), \tag{1}$$
where $w_{ij}(k)$ is the weight (synaptical strength) between i and j neurons for the kth recursion step, a_i, a_j are potentials of i and j neurons, D_w is a decay parameter of the weights, η notates the rate of learning.

The fast equation:
$$a_i(s+1) = (1 - D_a)a_i(s) + Ea_\Sigma(s)\big(1 \pm a_i(s)\big), \tag{2}$$
where $a_\Sigma(s) = \sum_l [a_{il}(s)w_{il}(s)] + I_i(s)$, $I_i(s)$ is the external input of neuron i for sth iteration step, D_a is the decay parameter of potentials, E is the excitatory parameter, $a_{il}(s) = q\big(a_l(s)\big)$, i.e., there is nonlinear potential postsynaptic function from the potential of output (presynaptic potential) of the lth neuron. In the last member of Equation 2 the sign plus is taken if $a_\Sigma(s) \leqslant 0$, and minus if $a_\Sigma(s) > 0$. These conditions are expressed a bipolar sigmoid function of neuron.

The synaptic function $g\big(a_l(s)\big)$ (further $g(x)$) reflects the mutual activity among presynaptic and postsynaptic potentials or a result of complex synapse-dendrite activation.

A membrane characteristic of a dendrite is a nonlinear (with two stable areas and one unstable) N-shaped form. The approximation was taken:
$$g(a) = \big[x^3 - 3x^2 + 3x(1 - c^2)\big] \cdot \big[3(1 - c^2)\big]^{-1} \tag{3}$$
under the conditions:
$$g = \begin{cases} 1 & \text{if } g(x) > 1 \\ g(x) \\ -1 & \text{if } g(x) < -1 \end{cases} \tag{4}$$
where $[h_0, h_1]$, $[h_2 \leqslant x]$, $[h_1, h_2]$, h_0, h_1, and h_2 are restrictions of stable and unstable areas.

It was shown that the coupling strength nonlinearities between neurons in a neural network allow to obtain rich and complex bifurcation diagrams given some opportunities in recognizing of images.

Stability Conditions for
Nonlinear Continuous Random Neural Networks

Jie TIAN, Juwei TAI, Xinsheng ZHANG

National Laboratory of Pattern Recognition
Institute of Automation
Chinese Academy of Sciences
Beijing 100080, P. R. of China

Keywords: Stability condition, random neural networks, equilibrium distribution, white noise, diffusion processes

ABSTRACT

In this paper nonlinear conctinuous random nearal networks with asymmetric connection weights are considered. Some sufficient conditions for a given network to have a globally asymptotically stability equilibrium distribution for arbitrary input are derived. Finally, future research issues for these neuaral networks and briefly discussed.
The term neural network originally referred to a network of interconnected neurons. Today the neural network has come to mean any computing architection that consists of massing parallel inferconnection of simple "neural" processors. It is a network that performs some computional task (recognition, association, ect.) on a given pattern via interaction between a number of interconnected units having simple functions. Many types of neural networks are introduced and studied for various purpose since 1960, especially in the last decade. Recently, K. Matsuoka deal with the following type of neural networks which he called as nonlinear continuons neural network with asymmetric connection weights. The dynamics of the network is represented by

$$\begin{cases} \tau \frac{dx_i(t)}{d\tau} = -x_i(t) + \sum_j \bar{w}_{ij} y_j(t) + s_i \\ y_i(t) = \bar{g}_i(x_i(t)) \end{cases} \tag{1}$$

where $x_i(t)$ is the variable that represents a state of unit neuron i at time t $(i = 1, 2, \cdots, N)$. s_i a constant input form the outside of the network, and $y_i(t)$ an output at time t. $\bar{w}_{ij}$ in the strength of connection from unit j to unit i, τ is a time constant governing the rate of change of each unit's state. He obtained some stability condictions for these neural networks. In this paper, we consider the stable problem for system (1) with by white noise i.e. We deal with the following type of neural networks, which dynamics in represented by

$$\begin{cases} \tau \frac{dx_i(t)}{d\tau} = -x_i(t) + \sum_j \bar{w}_{ij} y_j(t) + s_i + \sum_j c_{ij} dB_j(t) \\ y_i(t) = \bar{g}_i(x_i(t)) \end{cases} \tag{2}$$

where $B_j(t), 1 \leq j \leq m$ are independent white noise, and $(c_{ij})_{1 \leq i \leq N, 1 \leq j \leq m}$ is constant matrix. We called system (2) as nonlinear continuons random neural networks with asymetric connection weights. Some stability conditions are obtained for these neural networks in this paper. In particular, when $(c_{ij}) = 0$, our results are same as K.Matsuoka's.

Attractor neural networks
—oral contributions

Optimal Classification with Multilayer Networks

T. L. H. Watkin
St. John's College
Cambridge CB2 1TP, U.K.

K. Y. M. Wong
Hong Kong University of Science and Technology
Clear Water Bay, Kowloon, Hong Kong

A. Rau
Theoretical Physics, Department of Physics
1, Keble Road, Oxford, OX1 3NP, U.K.

May 5, 1993

Abstract

We study how a neural network can optimally learn the best theoretical model of a classification task, the *proximity problem*. For any network we can obtain general scaling results by mapping learning a task from a finite number of examples to learning a harder task from an infinite number. For the perceptron we make exact predictions supported by numerical simulations.

1 Introduction

The application of neural networks to difficult computational problems, such as the recognition of spoken words, is a subject of increasing economic importance. Statistical mechanics is useful (as discussed extensively by [1]) because it gives direct insight into how to teach such a network. Unfortunately, the algebraic complexity of the statistical mechanics approach has meant that until recently most studies have dealt with "benchmark" problems in which simple networks, *perceptrons*, learn computational tasks, *rules*, for which they are particularly well suited. An important topic of present research is to see how much such insights help in more realistic situations.

This paper considers the best theoretical model of a classification task, and demonstrates the optimal way in which any network can learn the problem. It proves exact results which apply to any network, in particular a mapping exists between learning from a finite number of examples and from an infinite number. Thus we extract scaling and asymptotic results. We make theoretical predictions for learning the rule with a perceptron in limits more general than those previously studied and support them with simulations.

2 Background

A neural network is a computer whose processors, *neurons*, cooperate to process N digits of input data, S_i, $i = 1, \ldots, N$, which can also be written as the N-vector *question* S. S is normalised (i.e. $S \cdot S = N$) and the function the network performs is written $\mathcal{N}(S)$. One simple network, the *perceptron*, only has one processor whose output, S_0, is ± 1 according to whether S has a positive or negative overlap with a fixed N-vector $\mathbf{J}$. That is,

$$S_0 \equiv \mathcal{N}(S) = g\left[\mathbf{J} \cdot S\right]. \tag{1}$$

The normalised vector $\mathbf{J}$ defines the function this network performs on the input data. The only type of perceptron we will consider here has $g(x) \equiv \mathrm{sgn}(x)$.

More complicated networks can perform more complicated functions. In *multilayer networks*, each of one set of neurons, the *first hidden layer*, reacts to the inputs by a function like (1), then another set of neurons, the *second hidden layer*, perform a function similar to (1) on the first hidden layer, and so on, up to the output neuron. In some networks information flows from the input neurons to the output by exactly one path, in other networks it can flow in several ways. An example of the first case is *a committee machine with non–overlapping receptive fields* (NRF), whose function can be written

$$S_0 = \mathrm{sgn}\left[\sum_{k=1}^{K} \mathrm{sgn}\left(\sum_{i=N(k-1)/K}^{Nk/K} J_i^k S_i\right)\right], \tag{2}$$

for some $\{J_i^k\}$, while an example of the second is a *committee machine with overlapping receptive fields*, whose function can be written

$$S_0 = \mathrm{sgn}\left[\sum_{k=1}^{K} \mathrm{sgn}\left(\sum_{i=1}^{N} J_i^k S_i\right)\right]. \tag{3}$$

In both cases there is only one hidden layer, containing K neurons, and the output is just the sign of the sum of those units. It has been shown that any binary function of S can be written in the form (3), if there are enough hidden units.

Here we are concerned only with large networks (for which $N \to \infty$), and our results are accurate to order $1/\sqrt{N}$. For any given architecture of network, we will characterise the network's variable parameters by the vector $\mathbf{J}$. By "any network" we will mean any neural system, possibly with many layers, in which every neuron in the first hidden layer performs *synaptic emulation* (i.e. a function of the form (1), for some set of weights), and in which the number of neurons in the first hidden layer is much smaller than N. This definition is wide enough to include almost anything commonly called a neural network.

By adjusting the connections inside the network, we want to make $\mathcal{N}(S)$ close to some function $V(S)$, i.e. we want $\mathcal{N}(S) = V(S)$ for as many questions S as possible. The measure of how well we have done, if we knew what V was, would be the *generalisation function* $\epsilon_f(\mathcal{N}, V)$, which is the average over all questions of some measure of the disagreement between $\mathcal{N}$ and V:

$$\epsilon_f(\mathcal{N}, V) = \langle e\left(\mathcal{N}(S), V(S)\right)\rangle_S. \tag{4}$$

If e were chosen to be 1 for $\mathcal{N}(S) \neq V(S)$, and zero otherwise, then ϵ_f would just be the proportion of questions on which $\mathcal{N}$ and V disagree.

For real problems V is unknown, but when we choose to learn V with a network we are making a guess about its complexity. We assume that V is an element of a *rule space*. Now let us assume that we have a number of *example* inputs and outputs which are known to be connected by V. This is called the *training set*. Each example tells us more about where in rule space the true rule lies. The elements of rule space consistent with all the examples together constitute the *version space*, $\mathcal{V}$.

One strategy for designing a network is to choose an $\mathcal{N}$ which at least correctly reproduces all the examples of V. This is the idea behind most learning strategies, and we will call it here *Gibbs learning*. Gibbs learning is not necessarily optimal because it only selects a network which learns the training set, not which *generalises* best.

We know that V lies inside the version space, therefore the *expectation value* of the generalisation error of a network $\mathcal{N}$ is the average of $\epsilon_f(\mathcal{N}, V)$ over choices of V from $\mathcal{V}$. This is called the *network error*.

$$\epsilon_n(\mathcal{N}) = \langle \epsilon_f(\mathcal{N}, V)\rangle_{\mathcal{V}}. \tag{5}$$

Watkin [2] defined *optimal learning* as finding the $\mathcal{N}$ which minimises $\epsilon_n(\mathcal{N})$, because *by definition* this network has minimal expectation generalisation error. Intuitively, we see that this $\mathcal{N}$ agrees with as many rules as possible consistent with the data. Optimal learning is an example of decision theory.

It was possible to show that an optimally–taught perceptron learns the benchmark *linearly separable rule* as well as any possible network [2], and optimal learning may be easily implemented by an algorithm which samples version space. Often optimal learning has better asymptotic behaviour than Gibbs learning, and is easier to simulate.

Linearly separable rules, however, are a poor model of classification problems such as the recognition of spoken words mentioned above. A better model was proposed by [3, 4]. Questions, S, are clustered, so that they have overlap $S \cdot \boldsymbol{\eta}^\nu = mN$ ($0 < m < 1$) with one of p_0 normalised, N-vector *prototypes*, $\boldsymbol{\eta}^\nu$, $\nu = 1, \ldots, p_0$. In this paper we study the case of p_0 scaling with N. Each prototype ν has an *answer* η_0^ν which we will take to be ± 1. The right answer to question S is η_0^ν for the prototype such that $S \cdot \boldsymbol{\eta}^\nu = mN$. Thus the rule is: find the prototype ν closest to the question, and output η_0^ν.

V is defined by m, the $\{\boldsymbol{\eta}^\nu\}$ and $\{\eta_0^\nu\}$. Therefore, the generalisation function is $\epsilon_f(\mathcal{N}, \{\boldsymbol{\eta}^\nu\}, \{\eta_0^\nu\}, m)$. This function is easy to calculate, and a few examples will be given elsewhere [8]. It can always be written as $\epsilon_f(\mathcal{N}, \{\boldsymbol{\eta}^\nu\}, \{\eta_0^\nu\}, m) = \sum_\nu f(\mathcal{N}, \boldsymbol{\eta}^\nu, \eta_0^\nu, m)$, where f is a smooth function. In reality, the $\{\eta_0^\nu\}$ are known, and m can be deduced from a couple of examples; the problem is that the $\{\eta^\nu\}$ are not given. Instead, for each prototype η^ν we are given p normalised examples $\{\xi^{\mu\nu}\}$, $\mu = 1, \ldots, p$, such that $\xi^{\mu\nu} \cdot \eta^\nu = mN$.

Gibbs learning of this problem would be to find the network which gives the right answer to as many as possible of the questions in the training set. If the network is a perceptron this is equivalent to minimising a *training energy*,

$$E_t(\mathbf{J}) = \frac{1}{pp_0} \sum_{\mu,\nu} \Theta\left(-\eta_0^\nu \mathbf{J} \cdot \xi^{\nu\mu}\right). \tag{6}$$

3 Optimal learning of the Proximity Problem

3.1 General Results

Let us consider what the examples tell us about the prototypes. After we have p examples, we know that η^ν obeys the constraint $\eta^\nu \cdot \xi^{\nu\mu} = mN$ for all the μ and the normalisation constraint $\eta^\nu \cdot \eta^\nu = N$. That is, this prototype lies on the intersection of p hyperplanes with an N–sphere, which is an $(N - p)$–dimensional sphere offset from the origin.

Version space is all rules defined by a set of prototypes consistent with the examples. In the thermodynamic limit, it may be shown algebraically that the set of all questions consistent with at least one rule in the version space is equal to the set of questions consistent with *a single, noisier rule*: to be precise a rule defined by the same $\{\eta_0^\nu\}$, but with a different set of p_0 prototypes — $\{\frac{1}{\gamma} \sum_\mu \xi^{\nu\mu}\}$ where γ is a normalisation factor — and a noise parameter $\tilde{m} = m/\sqrt{1 + 1/\tilde{p}}$, where $\tilde{p} \equiv pm^2/(1 - m^2)$.

We can therefore write the key result that

$$\epsilon_n(\{\mathbf{J}\}) \equiv \langle \epsilon_f(\{\mathbf{J}\}, V) \rangle_V = \sum_\nu f(\mathcal{N}, \frac{1}{\gamma} \sum_\mu \xi^{\nu\mu}, \eta_0^\nu, \tilde{m}) \tag{7}$$

$$= \epsilon_f(\{\mathbf{J}\}, \{\frac{1}{\gamma} \sum_\mu \xi^{\nu\mu}\}, \{\eta_0^\nu\}, \tilde{m}). \tag{8}$$

Thus the optimal way to train any network is to minimise (7), which is a sum over prototypes of a function only of the sum of examples of the prototype.

Also, if the prototypes are distributed randomly in space, then so are the vectors $\frac{1}{\gamma} \sum_\mu \xi^{\nu\mu}$. Therefore, for any network, the ability to generalise an unknown rule from a certain number of examples (i.e. $\epsilon_g^{opt} \equiv \min_{\{\mathbf{J}\}} \epsilon_n(\{\mathbf{J}\})$) is equal to the ability to generalise a noisier rule in which all the prototypes are given. We call this result *the principle of compensation* : for all networks, noise and the number of examples combine so that the optimal generalisation ability is just a function of $m/\sqrt{1 + 1/\tilde{p}}$. This is particularly useful for analysis, because it is considerably easier to analyse a problem in which the only quenched disorder is in the rule, than one with a second quenched disorder in the examples. For example, optimal learning can then be mapped to the work of Wong and Sherrington [6] who considered memory storage using noisy examples.

Also, since the minimal error that a certain type of network can make on the rule (that is $\min_{\{J\}} \epsilon_f(\{J\}, \eta^\nu, \{\eta_0^\nu\}, m)$) is a smooth, increasing function of m, we can expand Eqn. (8) about $\tilde{p} = \infty$ to obtain the asymptotic optimal generalisation behaviour.

$$\epsilon_g^{opt}(\tilde{p}) - \epsilon_g^{opt}(\infty) = \frac{1}{2\tilde{p}}\left[\frac{\partial}{\partial m}\epsilon_g^{opt}(\infty)\right] + O(1/\tilde{p}^2). \tag{9}$$

3.2 Numerical Results for the Perceptron

Gibbs learning of the proximity problem with a perceptron has been studied by [3, 4]. [4] studied the limit in which $m \to 0$ but the rescaled variables $\tilde{p}_0 \equiv p_0(1 - m^2)/m^2$ and $\tilde{p} \equiv pm^2/(1 - m^2)$ remain of order 1. That is, a limit in which the noise in the problem becomes high but this is compensated for by having more examples of each prototype and fewer prototypes to learn. We will henceforth refer to this as the HS–limit.

Line 1 on Fig. 1 is taken from [4], and shows the generalisation error of Gibbs learning as a function of $\tilde{p}$ for $\tilde{p}_0 = 1.6N$. Line 2 is the training energy (6) resulting from Gibbs learning. It can be made to be zero for $\tilde{p}$ less than a critical value $\tilde{p}_c$, but afterwards is monotonously increasing. Note that the generalisation error of Gibbs learning rises as $\tilde{p} \to \tilde{p}_c$, which is called *overfitting*. [4] found that a better strategy was in fact to generate networks by a *Gibbs dynamics* at a *finite temperature*, using (6) as an energy; in this way the generalisation error may be reduced to line 3. Below $\tilde{p}_c$ simulations of zero temperature Gibbs learning are straightforward, since many algorithms are known which produce a perceptron with zero training error if one exists (for example, the ADATRON algorithm [7]); however, for $\tilde{p} > \tilde{p}_c$, the training energy must be minimised directly, possibly by simulated annealing. This turns out to be difficult, even at non-zero temperature (David Hansel, private communication, 1992), because the energy landscape has many local minima in which a gradient descent algorithm may become stuck.

Optimal learning is much easier in the HS-limit. We write the optimal J as $J^{opt} = \sum_\nu c_\nu(\frac{1}{\gamma}\sum_\mu \xi^{\nu\mu}) + c\mathbf{b}$, where $\mathbf{b}$ is a vector normal to the sums of examples, and insert this expression into Eqn. (7). Minimising Eqn. (7) with the constraint that J is normalised easily gives that $c = 0$ and that all the $\{c_\nu\}$ are the same. Thus J is simply the normalised sum of all the examples: the *Hebb rule*. The result that the Hebb rule was optimal in this limit was stated without proof by some of the present authors in a previous paper [5].

Line 4 on Fig. 1 (taken from [5]) shows the optimal generalisation error, for the values of p_0 and p chosen in [4]. Optimal learning is between one and two orders of magnitude more efficient in examples than Gibbs learning (the horizontal scale is logarithmic). The training error produced by optimal learning is shown as line 5. Line 5 is higher than line 2 as expected: optimal learning is the learning scheme which minimises generalisation error, while zero temperature Gibbs learning minimises the training error. Simulations of optimal learning are shown as crosses; each is the average of 25 runs with $N = 100$ and $m = 0.1$. The whole simulation took a few minutes on a SPARC station.

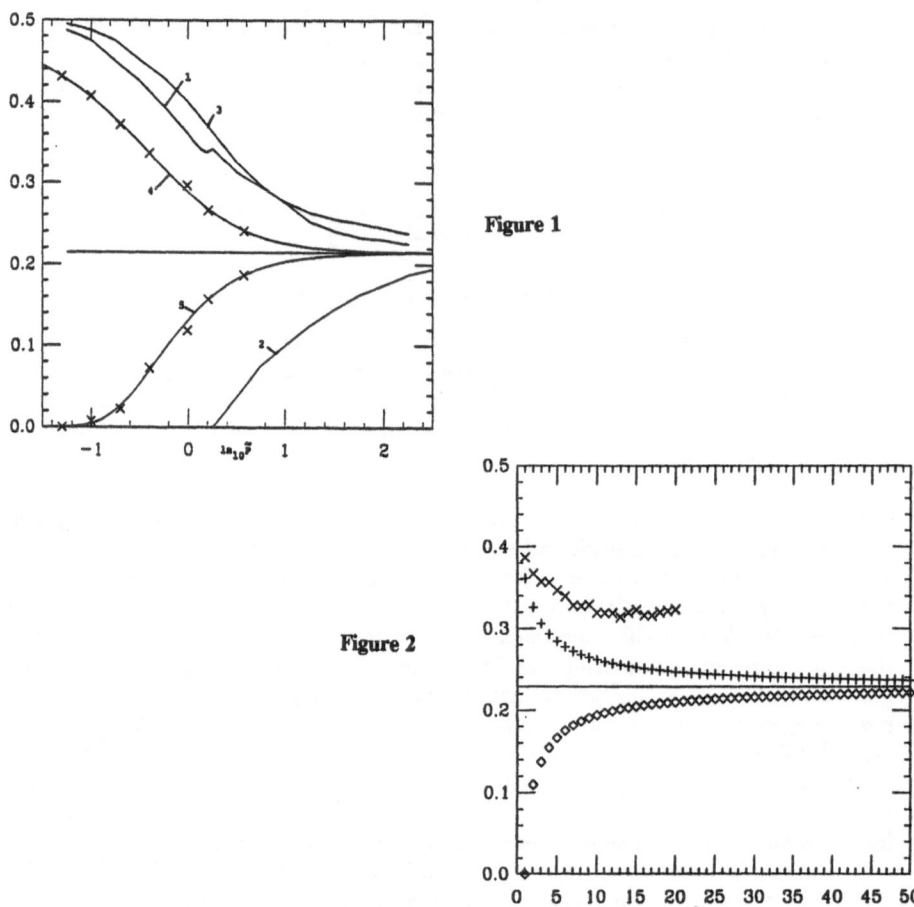

Figure 1

Figure 2

Figure 1: the HS limit for $\tilde{p}_0 = 1.6N$, as described in the text; line 4 shows the optimal generalisation error. Figure 2: the low-noise limit; the + symbols mark the optimal generalisation error against number of examples per prototype

Another interesting limit is that of "low noise", $m \sim 1$. For this limit the results of Gibbs learning are not known. However, for $\tilde{p} < \tilde{p}_c$ we simulate zero–temperature Gibbs learning using the ADATRON algorithm, and for $p > p_c$ we approximate the result of Gibbs learning by the state the ADATRON algorithm converges to. The crosses on Fig. 2 show the generalisation error of Gibbs learning averaged over 20 runs with $N = 50$ for the case of $m = 0.5$, and $p_0 = 0.5N$ (the horizontal axis is p).

In this limit too, optimal learning is easier to analyse than Gibbs learning, because it can be mapped to the results of Wong and Sherrington (see, for example, [6]). The optimal generalisation error can be painlessly deduced, and is shown as the + symbols on Fig.2. For interest, the ◇ symbols mark the training error produced by the optimal learning scheme. Preliminary numerical results seem to verify the theory.

4 Conclusion

We have studied optimal learning of a realistic classification problem and shown that insight into the rule lets us intelligently design energy functions to maximise the information extracted from examples. Despite the complexity of the problem, we can map learning a noisy rule from a finite number of examples to learning a noisier rule from an infinite number. This allows us to extract exact scaling and asymptotic results.

We have also studied particular cases of networks learning this problem optimally. For the perceptron we have shown that optimal learning can be more than an order of magnitude more efficient than Gibbs learning, and easier to implement.

This problem also shows that care must be taken in calling a machine powerful. In the HS limit, it may be shown that for any α_0 and m, $\epsilon_g^{opt}(\infty)$ is lower for the perceptron than for a NRF committee machine with any number of hidden units (David Hansel, private communication), thus the perceptron's whole optimal learning curve is better. This is despite the fact that an NRF committee machine is "more powerful" in the sense of being able to store more random input-output pairs [1]. The same does not apply to committee machines with overlapping receptive fields, since in the limit of $k \to \infty$ such a network can learn this rule (and any other rule, [1]) perfectly, $\epsilon_g^{opt} \to 0$. We are performing optimal learning in large networks, and our results will be presented elsewhere [8].

Two of us (T.L.H.W. and A. R.) are very grateful to the University of Science and Technology, Hong Kong, for their hospitality while this project developed and to David Hansel and Jean-Pierre Nadal for fruitful discussions. T.L.H.W. completed the project at the Laboratoire de Physique Statistique de L'Ecole Normale Supérieure, Paris.

References

[1] T.L.H. Watkin, A.Rau and M. Biehl, *The Statistical Mechanics of Learning a Rule*, Rev. Mod. Phys. (in press)

[2] T.L.H. Watkin, Europhys. Letts., **21**, 871 (1993)

[3] P. Del Guidice , S. Franz and M.A. Virasoro, J. Phys. I France, **50**, 121 (1989)

[4] D. Hansel and H. Sompolinsky, Europhys. Lett **11**, 687 (1990)

[5] T.L.H. Watkin, A. Rau, D. Bollé and J. van Mourik, J. Phys. I France, **2**, 167 (1992)

[6] K.Y.M. Wong and D. Sherrington, *Neural networks optimally trained with noisy data*, Phys. Rev. E (in press) (1993)

[7] J.K. Anlauf, M. Biehl, Europhys. Lett. **10**, 687 (1989)

[8] T.L.H. Watkin, K.Y.M. Wong and A. Rau (in preparation)

AN ATTRACTOR NETWORK MODEL FOR THE GENERATION OF EVENT-RELATED POTENTIALS USING INTEGRATIVE SYNAPSES

E.Ventouras[+] , C.Papageorgiou[++] , N.K.Uzunoglu[+] , A.Rabavilas[++] , and C.Stefanis[++]

+: Dept. of Electrical and Computer Engineering, National Technical University of Athens, Athens, 10682, Greece
++:Dept. of Psychiatry, Eginiteion Hospital, University of Athens, Athens 10431, Greece

Abstract

In this paper the generation of components of Event-Related Potentials (ERP) is achieved using a neural network model which activates synapses possessing integrative properties, acting on a target neural assembly. The summation of low frequency current sources created by the synaptic action produces an electric field detectable on the head surface as an ERP component. The behavior of the model is examined through simulations checking the network's temporal spiking rates and the low-pass filtering properties of the integrative synapses.

1. Introduction

During the last decade ERP, i.e. potentials measured on the head surface when a subject is exposed to stimuli possessing psychological importance, have gained wide interest [1]. The relation of ERP parameters, such as N_{100}, P_{200}, N_{200}, P_{300} and N_{400} components, to processes concerning attention, recent memory and decision making, sujests that ERP are capable to contribute in the elucidation of the mechanisms underlying the information process activities of the Central Nervous System [2],[3].

In the present work the functions of neural assemblies producing ERP are studied, using an attractor neural network model chosen so as to describe information processing properties in a way mathematically tractable and physiologically plausible. It is shown that, when the network reaches an attractor associated with the specific information content of the psychologically critical stimuli, the activation of efferent integrative synapses is able to create current sources, which in turn generate scalp-recorded ERP components.

2. Architecture and functions of the model

The model proposed for the simulation of ERP generation consists of three distinct units, hierarchically positioned, communicating through feed-forward connections (Fig.1) : a) Input Unit (IU), b) Attractor Neural Network (ANN) and c) Target Neural Assembly (TNA).

IU consists of a network which projects input patterns to the ANN, through stable synaptic connections of strength P_{ik} connecting neuron k of IU to neuron i of ANN. It might be considered as a preprocessing area for information related to the external stimuli, but its internal dynamics are neglected since it serves only as an input module.

The second unit is a highly recurrent attractor network, based on the network proposed in [4], being able of associative recall. The communication between neurons is dominated by excitatory and inhibitory, randomly diluted synapses, S_{ij} , which possess the ability of generalized Hebbian learning, including homosynaptic potentiation and homosynaptic/heterosynaptic depression, using time dependent, local and interactive mechanisms [5]. Probabilistic behavior may be added, slightly perturbating IU projection synchronization and ANN's neural firing homogenity. The equations describing the evolution of membrane potentials $U_i(t)$, are

$$\dot{U_i}(t) = k_1 U_i(t) + k_2 f(\Delta t_i) g(A_i(t)) \quad , \quad i=1,....,N \quad , \quad U_i(t) \in [U_F, U_T] \quad (1)$$

where N is the total number of neurons of unit 2, U_F the refractory potential, U_T the threshold potential and k_1, k_2 constants. $f(\Delta t_i)$ is a function representing the variable sensitivity of neuron i to external stimuli due to the refractory periods. Δt_i is the time elapsed since neuron i last fired, and g is the activation function response to incoming stimuli represented by

$$A_i(t) = \sum_j (S_{ij} \exp(-\Delta t_j / \tau)) + \sum_k (P_{ik} \exp(-\Delta t_k / \tau)) \quad (2)$$

where τ is constant. In case U_i reaches U_T the neuron fires and its potential is set to U_F.

The ANN used in [4] presents the serious problem of high temporal spiking rates, even when noise is added [6]. We surpassed this obstacle using steep sigmoidal activation functions instead of linear ones, making neurons more sensitive to projector inputs while not augmenting their firing rate. As a result, patterns projected from IU with realistically low repetition frequencies (15-30 Hz) produced a correspondingly acceptable neural activation frequency, inducing stabilized embedded pattern responses and demonstrating the associative memory capacities of the network.

It is assumed that the arrival of the ANN to an attractor state does not justify by itself the generation of ERP, as can be examplified in the

700

Fig.1: Architecture of the Model. IU: input unit, ANN: attractor neural network
IS: integrative synapses, TNA: target neural assembly

oddball paradigm, where only target stimuli create a P_{300} component,
though the categorization process of the stimuli sujests that a cognitively
meaningfull result has been reached in both target and non-target case [1].
We propose that the neurons active in an attractor state corresponding to a
cognitive condition related with an ERP component, project to the third unit,
a target neural assembly (TNA), exerting excitatory synaptic action through
synapses having integrative properties. This synaptic characteristic transforms
the spiked presynaptic inputs to reduced, graded and relatively smooth
extracellular electric currents in the synaptic cleft, effectively producing low-pass
filtering. Integrative synaptic action could be closely related to modulatory
mechanisms exerting modifications of rather long latency and duration to
the electrophysiological properties of the neuronal membrane[7],[8], especially
in connection with Short-Term Potentiation [9] and N-methyl-D-aspartate
(NMDA) receptors [10]. Since current flow in the active synaptic vicinity is
compensated by currents from more distant parts of the cell [11], a time
varying current dipole is created with frequency and amplitude properties
appropriate for the generation of Event-Related electric fields and potentials.
The neuronal geometric arrangement of the TNA should be sufficiently parallel
in order to enable the spatial summation of discrete elementary current dipoles,
therefore creating detectable far-field potentials, as is the case for cortex
pyramidal neurons.

The potential $V_{out}(\mathbf{r},t_i)$ measured on the head surface, produced by an
intracranial current source distribution, at time t_i, $\rho_J(\mathbf{r}',t_i)$ can be computed
solving a direct electromagnetic problem [12] using a three-layered concentric
spherical human head model, with layers corresponding to the brain, skull
and skin media.

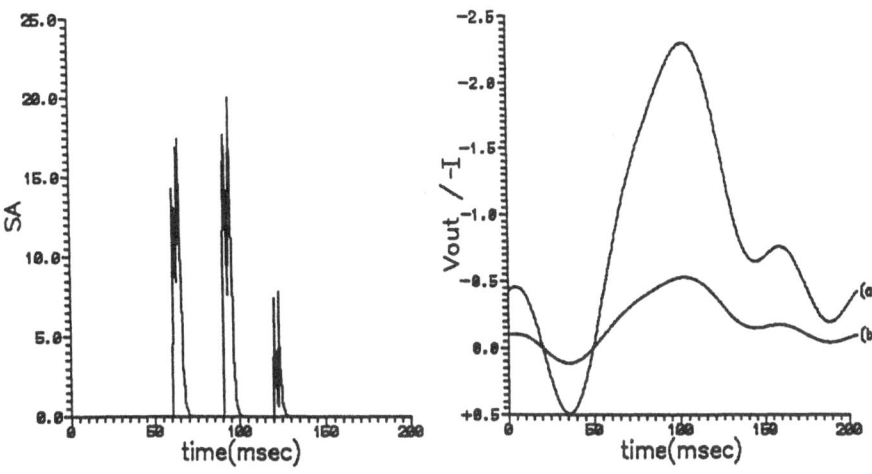

Fig.2: SA—summed activity in μV Fig.3:-I (curve a ,in μA) ,Vout (curve b ,in μV)

3. Simulations results and discussion

The simulations performed in order to evaluate the proposed model, comply to the following method: The summed input activity

$$SA(t) = \sum_k \lambda_k \exp(-\Delta t_k/\tau) \quad (3)$$

impressed on the TNA is computed, when the ANN, being stimulated by IU, recalls a previously learned input pattern which is considered to correspond to a condition related to an ERP component. In (3) summation is over the neurons active in the recalled pattern. $SA(t)$ is transformed by the integrative synapses to an electric current $I(t) = F(SA(t))$, where F is a low-pass filter function with cut-off frequency $f_c < 20Hz$. A dipole of strength $I(t)$ is consequently created, generating a potential computed on the external surface of the head model, whose similarity to real-data ERP can be checked.

Results are shown when an ANN with N=100 neurons was used, with parameters $k_1 = (2.5ms^{-1})$, $k_2 = 0.34mV/ms$, $\tau = 1.0ms$, $U_F = -15mV$, $U_T = 30mV$, $\lambda_k = \lambda = 10^{-5}V$. Two distinct patterns were memorized, each using about 10% of the networks' neurons. Pattern projection through IU, at 1/0.025 Hz frequency, was noiseless during the learning sessions, which lasted several hundred milliseconds for each pattern. Only one of the memories, (A), was considered to correspond to ERP generation and the neurons belonging to it possessed the potentiated specific efferent synaptic connections to the TNA, developed through a mechanism which has not been assessed in this study. In Fig.2 $SA(t)$ is represented when pattern A was recalled by projecting noise-corrupted incomplete inputs, starting at t=60ms, for 75ms, with 1/0.03 Hz repetition rate. A low-frequency current $I(t)$ resulted , F having cut-off frequency $f_c = 19.5Hz$. The head model adopted had layer boundaries at $r_1 = 9.0$cm, $r_2 = 9.8$cm, $r_3 = 10.0$cm and shell conductivities $\sigma_1 = \sigma_3 = 0.33$S/m, $\sigma_2 = 0.0042$S/m. Supposing a dipole of strength $I(t)$ is created, with source and sink at points having spherical coordinates $(r = 8.3$cm, $\vartheta = 30°$, $\varphi = 22.5°)$ and $(r = 8.5$cm, $\vartheta = 30°$, $\varphi = 22.5°)$ respectively, the sink current is represented in curve (a) of Fig.3. The ERP $= V_{out}(t)$, measured at head surface point $(r = 10.0$cm, $\vartheta = 30°$, $\varphi = 330°)$ is represented in curve (b), showing strong similarity to the N_{100} component.

Similar results were obtained for other ERP waveforms. The model can be extended to include waiting potentials such as CNV (Contingent Negative Variation) and PINV (Post Imperative Negative Variation) [1]. By modifying the parameters of unit 2 artificial neurons and the integrative synapses' properties, insight should be provided concerning the mechanisms of the generation of ERP obtained in both physiological and psychopathological conditions.

Further development of the model includes an ANN whose neurons obey Dale's Hypothesis [3], that is they only excitate or inhibit other neurons. Neurons in the cortex are believed to comply to this hypothesis. Explicit separation of excitatory and inhibitory neurons of unit 2 will further clarify conceptually the structure of the model, especially concerning the excitatory projections to TNA. Preliminary results indicate that the model retains and even improves its associative memory and low spiking frequency characteristics, therefore being able to simulate ERP generation under increased neurophysiological constraints.

References

[1] T.W.Picton (Ed.), *Human Event-Related Potentials, EEG Handbook (revised series, vol.3)*, Amsterdam, The Netherlands: Elsevier, 1988.

[2] J.R.Jennings and M.Coles (Eds.), *Handbook of Cognitive Psychophysiology, Central and Autonomic Nervous System Approaches*, New-York : Wiley, 1991.

[3] J.Cacioppo and L.Tassinary (Eds.), *Principles of Psychophysiology*, Cambridge: Cambridge University Press, 1990.

[4] J.Buhmann and K.Schulten, "Associative Recognition and Storage in a Model Network of Physiological Neurons," *Biol.Cybern.*, vol.54, pp.319-335, 1986.

[5] T.H.Brown, E.N.Kairriss, and C.L.Keenan, "Hebbian Synapses : Biophysical Mechanisms and Algorithms," *Annu.Rev.Neurosci.*, vol.13, pp.475-511, 1990.

[6] J.Buhmann and K.Schulten, "Influence of Noise on the Function of a "Physiological" Neural Network," *Biol.Cybern.*, vol.56, pp.313-327, 1987.

[7] R.M.Harris-Warrick and E.Marder, "Modulation of Neural Network for Behavior," *Annu.Rev.Neurosci.*, vol.14, pp.39-57, 1991.

[8] J.E.Desmedt and J.Debecker, "Wave Form and Neural Mechanism of the Decision P_{350} Elicited without Pre-Stimulus CNV or Readiness Potential

in Random Sequences of Near-Threshold Auditory Clicks and Finger Stimuli." *Electroenceph.Clin.Neurophysiol.*, vol.47, pp.648-670, 1979.

[9] L.Shen, "Neural Integration by Short-Term Potentiation," *Biol.Cybern.*, vol.61, pp.319-325, 1989.

[10] A.G.Brown, *Nerve Cells and Nervous Systems : An introduction to Neuroscience*, London : Springer, 1991.

[11] T.Allison, C.C.Wood, and G.McCarthy, "The Central Nervous System," in *Psychophysiology : Systems , Processes and Applications*, M.Coles,E.Donchin, and S.Porges (Eds.), Amsterdam, The Netherlands : Elsevier, 1986.

[12] N.K.Uzunoglu, E.Ventouras, C.Papageorgiou, A.Rabavilas, and C.Stefanis, "Inversion of Simulated Evoked Potentials to Charge Distribution Inside the Human Brain Using an Algebraic Reconstruction Technique," *IEEE Trans.Med.Imaging*, vol.10, pp.479-484, 1991.

[13] J.C.Eccles, *The Physiology of Synapses*, Berlin : Springer, 1964.

A Novel Liapunov Function for Additive Neural Networks.

Alexander N. Jourjine

Advanced Technology
Wang Laboratories
One Industrial Ave
Lowell, MA 01851
e-mail: alex.jourjine@office.wang.com

ABSTRACT

Using the properties of nonlinear integral equations of the Hammerstein type, it is proven that a new Liapunov function for additive neural networks exists, provided two conditions on symmetry and positivity of the weight matrix hold. The function does not require monotonicity of the transfer function.

Additive neural networks are dynamical systems on R^n with activation evolution given by

$$\dot{x} = -cx + W \sigma x + I \tag{1}$$

where $x = \{ x_i(t) \}$ are the activations, $W = \{w_{ij}\}$, $i,j = 1,2, \ldots ,n$, are the weights, c is the decay operator: $c = \text{diag}(c_i)$, $c_i > 0$, $\sigma = \{\sigma_j(x_j)\}$ are the transfer functions, and $I = \{I_i(t)\}$ are the external inputs. The Liapunov function (in the context of this paper, a continuous function which is non-decreasing on the orbits) for the additive neural networks with symmetric W and monotonic σ was found by Hopfield [1] and, for somewhat more general systems, by Cohen and Grossberg [2]

$$U(x) = (1/2) \text{Tr}(c \Psi x) - (1/2) (W \sigma x, \sigma x) - (I, x), \tag{2}$$

$$\Psi x = \int_0^{x_i} \xi \sigma' (\xi) d\xi ,$$

where Tr A is the trace of matrix A and $(x , y) \equiv \Sigma_i x_i \cdot y_i$. Here we construct an additional Liapunov function for additive networks. The idea for its construction comes from the results on the nonlinear integral equations of the Hammerstein type [3,4,5,6]. Further results on the connection between Hammerstein equation and additive neural networks with large number of nodes are given in [7].

First we note that for any matrix B with non-negative symmetric part, and any $U(x)$ such that

$$\dot{x} = - B \nabla_x U(x) \tag{3}$$

$U(x)$ will be non-decreasing on the orbits. Indeed

$$\dot{U} = (\nabla_x U, x) = - (\nabla_x U, B \nabla_x U) = - (\nabla_x U, B^S \nabla_x U) \leq 0. \tag{4}$$

We now can verify by direct computation that, provided $W^{-1} c$ is symmetric,

$$U = (1/2) (x, W^{-1} c x) - (e, \rho) - (W^{-1} I, x) \tag{5}$$

$$\rho_i(x_i) = \int_0^{x_i} \sigma_i(u) du , e = (1, 1, \ldots, 1)$$

is non-decreasing because we can take $B = W$ in (3). We, therefore conclude that $U(x)$ in (5) will be a Liapunov function if the symmetric part of W is non-negative definite. Obviously $U(x)$ will decrease on the orbits even if σ is not monotonic.

We should noted that the results of Cohen and Grossberg [2] apply to somewhat more general dynamical systems where the linear diagonal decay operator c is replaced by a arbitrary diagonal nonlinear operator. For such systems the approach of this paper in its present form fails.

It should also be pointed out that in the mathematical literature a Liapunov function is usually required to be bounded from below in the domain on which a dynamical system, such as (1), is defined. If the function is not bounded from below one cannot actually guarantee the convergence of orbits to some fixed points inside the domain of definition. For example the Liapunov function

706

in [1] as defined by Hopfield is not bounded (although its negative is) and as a result any trajectory which begins inside the hypercube on which the Hopfield system is defined eventually tends to its corners, which, in fact, are not fixed but boundary points of the hypercube. Boundedness of the functions presented here is actually connected with the value of the first derivative of the transfer function and to the spectral properties of the weight operator. This question is considered in more detail in [7]. The connection was also established in [8] for a particular subclass of additive neural networks where the decay operator is an identity.

REFERENCES

[1] Hopfield, J. J., *Neurons with Collective Response Have Collective Computational Properties Like Those of Two-State Neurons*, Proc Nat Acad Sci, *USA*, **81**, (1984), pp. 3088-3092;
[2] Cohen, M. A., Grossberg, S., *Absolute Stability of Global Pattern Formation and Parallel Memory Storage by Competitive Neural Networks*, IEEE Trans on Systems, Man and Cyb., **13**, (1983), pp. 815-826;
[3] Hammerstein, A., *Nichtlineare Integralgleichungen nebst Anwendungen*, Acta Math. **54**, (1930), pp. 117-176;
[4] Krasnosel'skii, M., *Topological Methods in the Theory of Nonlinear Integral Equations*, Pergamon Press, Oxford, 1965;
[5] Rall, L. B., *Variational Methods for Nonlinear Integral Equations*, in Anselone, P.M., Ed, Nonlinear Integral Equations; Proceedings of an Advanced Seminar at the University of Wisconsin, University of Wisconsin Press, Madison, 1964, pp. 155-189;
[6] Tricomi, F. G., *Integral Equations*, Chap 4, John Wiley/Interscience, New York, 1957;
[7] Jourjine, A. N., *Large n Limit of Additive Neural Networks*, Proceedings of Neuro-Nimes '92, Nimes, France, November 1992, pp. 635-643;
[8] Kelly, D. G. , *Stability in Contractive Nonlinear Neural Networks*, IEEE Trans Biomed Eng **37**, (1990), p.p. 231-242.

Capacity and Error Correction Ability of Sparsely Encoded Associative Memory with Forgetting Process

Shotaro Akaho

Electrotechnical Laboratory,
Mathematical Informatics Section,
1–1–4 Umezono, Tsukuba-shi, Ibaraki 305 Japan

Abstract Associative memory model of neural networks can not store items more than its memory capacity. When new items are given one after another, its connection weights should be decayed so that the number of stored items does not exceed the memory capacity (*forgetting process*). This paper analyzes the sparsely encoded associative memory, and presents the optimal decay rate that maximizes the number of stored items. The maximal number of stored items is given by $O(n/a \log n)$ when the decay rate is $1 - O(a \log n/n)$, where the network consists of n neurons with activity a.

1 Introduction

Neural network is an adaptive system that is trained with sample items given from outer environment, and it can store items up to its memory capacity.

In this paper, we consider the on-line type learning scheme where the learning is proceeded every time when a new item is provided (cf. batch type learning). This learning has the advantage that a little memory is needed (not necessary to memorize all patterns) and the network can adapts itself well to the change of the environment. However, when many new items are given one after another, it cannot be judged whether the number of stored items is more than its capacity or not, because the stored items are memorized implicitly on connection weights and the network does not memorize the number of items. One method to avoid exceeding the capacity is decaying connection weights to remove older items. If we want to store items as many as possible, connection weights should be decayed as slowly as possible, but if the decay is too slow, old items affects the recall of newer items. Thus there is an optimal decay rate that maximizes the number of stored items. We analyze the associative memory model[3] as an example.

Recently the associative memory model has attracted more attention because of *sparse coding scheme*, where most of components of pattern vectors are 0 and only a small ratio of those are 1. If patterns are sparsely encoded, the capacity of network becomes very large[2]. Sparse coding scheme is also supposed to act an important function in the brain memory such as hippocampus, and some important experimental results are coming out.

2 Associative Memory with Forgetting Process

We consider the autocorrelation associative memory with n input units and n output units. It is trained by a simple Hebbian rule as follows.

$$w_{ij}(t + 1) = (1 - \varepsilon)w_{ij}(t) + cs_i(t)s_j(t), \qquad i \neq j, \tag{1}$$

$$0 < \varepsilon < 1, \quad 0 < c,$$

where w_{ij} is a connection weight from j-th input to i-th output units, $s_i(t)$ is i-th component of pattern learned at time t, ε and c are time constants (we assume $c = 1$ without loss of generality) and $1 - \varepsilon$ denotes the decay rate. By the learning scheme above, connection weights become

$$w_{ij}(t) = \sum_{\mu=0}^{\infty} (1 - \varepsilon)^{\mu} s_i(t - \mu)s_j(t - \mu), \qquad i \neq j. \tag{2}$$

Each pattern component $s_i(t)$ takes binary value and there are two possible models. One is 0-1 model and the other is ± 1 model. Amari has shown that 0-1 model is superior in the sparse case and ± 1 model is superior in the non-sparse case[2]. In order to treat both cases, we shall encode patterns as follows.

$$s_i(t) = \begin{cases} 1-a & \text{Prob. } a \\ -a & \text{Prob. } 1-a \end{cases} \tag{3}$$

where a is called the *activity*. Each $s_i(t)$ takes a binary value independently according to the probability above. This coding works as 0-1 model in sparse case and ± 1 model in non-sparse case, and it also makes mathematical analysis easier.

Output x for input s is given by

$$x_i = 1_a\Big(\frac{1}{n}\sum_{j=0}^{n} w_{ij} s_j - h\Big), \tag{4}$$

where h is a threshold value and 1_a is a binary threshold function

$$1_a(u) = \begin{cases} 1-a & u \geq 0 \\ -a & u < 0 \end{cases} \tag{5}$$

3 Optimal Decay Rate and the Capacity

Let us define the capacity of the model described in the previous section. For some given item $s(t - \mu)$, if it is recalled correctly, namely, if

$$s_i = 1_a\Big(\frac{1}{n}\sum_{j=0}^{n} w_{ij} s_j - h\Big), \tag{6}$$

holds, pattern $s(t - \mu)$ is said to be stored. Memory capacity $M(\varepsilon; n, a)$ for given n, a and ε is defined as the maximal number of m, such that most recently learned m items $s(t), s(t - 1), \ldots, s(t - m + 1)$ can be recalled correctly. Since the patterns are randomly generated, we consider the case that items can be stored with probability 1 asymptotically for sufficiently large n.

Theorem 1 *The optimal decay rate of an associative memory with n neurons is given asymptotically by*

$$1 - \varepsilon_{\text{opt}} = 1 - \frac{8e(2+d)a(1-a)\log n}{n}, \tag{7}$$

where $d = -\log_n a$ and the memory capacity is given by

$$M_{\text{opt}} = \frac{1}{2\varepsilon_{\text{opt}}} = \frac{n}{16e(2+d)a(1-a)\log n}. \tag{8}$$

When $a = 1/2$ (non-sparse case), the capacity is $n/(8e\log n)$, while it is $n/(4\log n)$ in the case that the network learns only the finite number of patterns without decay (batch type learning). In general, the capacity of this model is $1/2e$ times the capacity in the batch type learning.

Next, we investigate the error correction ability of the learning. Consider the noisy version of m-th pattern as follows. Asymptotically, we can say that na components of m-th pattern are $1 - a$ and the others are $-a$. To make the noisy version of the pattern with keeping the activity a, we pick up randomly $na\xi$ components of $1 - a$ and flip them into $-a$, and similarly flip $na\xi$ components of $-a$ into $1 - a$. If $\xi = 0$, the pattern includes no noise.

$$\overbrace{1-a \;\; 1-a \;\; \cdots \;\; 1-a}^{na} \quad \overbrace{-a \;\; -a \;\; \cdots \;\; -a}^{n(1-a)} \quad : \text{original pattern}$$

$$\underbrace{\quad}_{na\xi} \qquad\qquad\qquad \underbrace{\quad}_{na\xi}$$
$$\Downarrow \qquad \Downarrow \qquad\qquad \Downarrow \qquad\qquad \Downarrow$$
$$-a \;\; -a \quad 1-a \qquad 1-a \;\; 1-a \quad -a \quad : \text{noisy pattern} \tag{9}$$

Theorem 2 *The maximal number of item patterns whose ξ noisy pattern is corrected is given by*

$$m(\xi) = \frac{(1 - a - \xi)^2}{(1 - a)^2} M_{opt},\tag{10}$$

when the decay rate is taken as

$$\varepsilon(\xi) = \frac{(1 - a)^2}{(1 - a - \xi)^2} \varepsilon_{opt},\tag{11}$$

where M_{opt} and ε_{opt} are defined in theorem 1.

Theorem 1 and 2 are proved basically by a similar technique to Amari's[2]. In the analysis, we only pay attention that the probability that the pattern is recalled correctly is different for each pattern. On the detail of the proof of theorems, see [1].

4 Simulation Results

We show some simulation results.

The following figure shows capacities for some values of ε normalized so that the optimal one becomes 1.0, where we gave the network ten times patterns as many as the theoretical capacity. The capacity is measured as the most recent pattern incorrectly recalled. We can see that the decay rate that maximizes the capacity is close to the theoretical value for all cases.

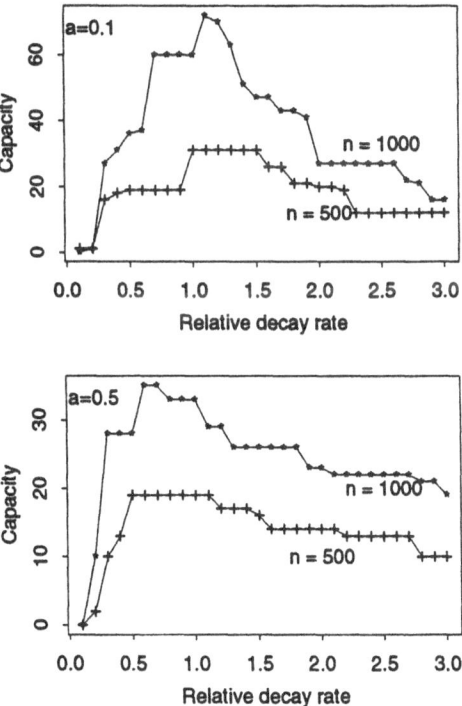

Figure: Capacity versus decay rate (ε). Decay rate is normalized by the optimal one. Up: a=0.1 (sparse case); Down: a=0.5 (non-sparse case)

5 Concluding Remarks

First of all, we give some remarks for the preceded sections.

We analyzed the case of autocorrelation associative memory for the simplicity, but the result is the same for the crosscorrelation case. In the later case, the activity can be different between input patterns and output patterns. We can show that the capacity and the decay rate does not depend on the activity of input. Thus only if output is sparsely encoded, the network has a large capacity, whether the input is sparsely encoded or not (but it is necessary that the activity of input is fixed). Moreover, the input does not need to be a binary pattern if the activity is fixed.

Some paper says that the capacity of associative memory is $O(n/\log n)$ and some paper says that it is $O(n)$ (the present paper is the former one; cf. [4]). This difference is caused by the difference of the definition of the capacity. While the former permits just $O(1)$ error components among n units, the later permits $O(n)$ error components. They are almost the same for a practical size of n.

Next, let us refer to the future problems. We have given attention only to the capacity as the measure of the network ability. However, there are two other aspects of the network as a learning machine. One is the adaptation ability for the change of environments (e.g. $f(x) = y_1$ in the past but $f(x) = y_2$ ($\neq y_1$) at present), which the on-line type learning is expected to have. The other is the generalization ability from the finite number of samples, i.e. the capability of describing unknown data only from the given sample data. (e.g. a sample indicates $f(x_1) = y$, hence $f(x_2)$ may be also y, where $x_2 \simeq x_1$).

Those three aspects are related to each other. Generalization theory of learning (such as VC dimension[5]) teaches us that the generalization ability of the network becomes higher, as the number of samples increases and the capacity of the network decreases. However it is the result induced under the situation that the environment does not change. If the environment changes gradually with time, old samples are not so reliable as newer items. Thus we should assign weights according to the "age" of each sample to ignore old samples, which corresponds to the forgetting process in this paper.

Thus new problem can be stated as follows:

Future Problem   How should we assign the optimal weights for samples so that the network can achieve the desired generalization ability?

In order to solve this problem, we will have to synthesize and develop the generalization theory of learning and the adaptive control theory.

References

[1] S. Akaho: Optimal decay rate of connection weights in covariance learning. Technical Report 92-37, Electrotechnical Laboratory, 1992.

[2] S. Amari: Characteristics of sparsely encoded associative memory. *Neural Networks*, Vol. 2, No. 6, pp. 451–457, 1989.

[3] J.A. Anderson: A simple neural network generating interactive memory. *Mathematical Biosciences*, Vol. 14, pp. 197–220, 1972.

[4] M. Mézard, J.P. Nadal, and G. Toulouse: Solvable models of working memories. *J. Physique*, Vol. 47, pp. 1457–1462, 1986.

[5] V.A. Vapnik: *Estimation of Dependences Based on Empirical Data*. Springer-Verlag, 1984.

Attractor neural networks
—poster contributions

Defining the Attractor of a Recurrent Neural Network by Boolean Expressions

H. Nikolaus Schaller[1], Klaus Ehrenberger[2]

Lehrstuhl für Datenverarbeitung
Technische Universität München
D-80290 München
Germany

Abstract

An autoassociator neural network can be operated to solve a computation problem with a high degree of parallelism. The set of stable states (solutions of the problem) that build up the attractor of such a recurrent network are determined by the feedback weights and biases. This set can be constructed by using the k-out-of-n design rule. It is shown how to convert arbitrary boolean expressions into a list of k-out-of-n constraints. Finally, a compiler for generating the network structure is described.

1. Introduction

The main idea of the k-out-of-n design rule introduced by Page and Tagliarini [1], [2], [3], [4] is to impose the constraint on a set of n neurons that exactly k neurons out of these n become active in a stable state (attractor, set of solutions). Several sets with individual k-values can be considered together in a common neural structure resulting in the conjunction of all these constraints.

As shown in [1], [3], [5], a set of k-out-of-n constraints

$$k_i = \sum_j C_{ij} z_j \tag{1}$$

can be converted into a recurrent network structure of Hopfield type [6], [7], [8]. Here, $0 \le z_j \le 1$ are the states of the neurons (Fig. 1), i is the constraint index, j the neuron index, k_i the k-value of the i-th constraint and $C_{ij} = 1$ if z_j belongs to the set constrained by constraint i and 0 otherwise. The connection weights $T_{\alpha\beta}$ and the constant biases I_α of the feedback structure are [5]

$$I_\alpha = \sum_i (2k_i - 1) C_{i\alpha} \tag{2}$$

$$T_{\alpha\beta} = -2 \sum_i C_{i\alpha} C_{i\beta} \text{ for } \alpha \ne \beta. \tag{3}$$

The activation (net input) of neuron α is

$$x_\alpha = \sum_{\beta=1}^{N} T_{\alpha\beta} z_\beta + I_\alpha. \tag{4}$$

Regarding this activation vector as the negative gradient of a potential

$$x_\alpha = -\text{grad}_\alpha E = -\frac{\partial E}{\partial z_\alpha} \tag{5}$$

---

1. hns@ldv.e-technik.tu-muenchen.de
2. now at: keb@munich.munich.ingr.com

results in the well known Hopfield energy [6], [7], [8] of the network

$$E = -\frac{1}{2} \sum_{\alpha=1}^{N} \sum_{\beta=1}^{N} T_{\alpha\beta} z_\alpha z_\beta - \sum_{\alpha=1}^{N} I_\alpha z_\alpha + E_0. \tag{6}$$

This energy has global minima for all binary states (state variables z_α either 0 or 1) fulfilling all constraints.

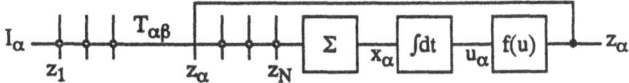

Fig. 1   Signal definition of a single neuron

In addition, the activity (degree of infeasibility) of a neuron can be defined as

$$a_\alpha = \left| x_\alpha - (2z_\alpha - 1) \sum_i C_{i\alpha} \right|. \tag{7}$$

If all constraints for neuron α are fulfilled, this implies that a_α becomes zero so that it needn't change its state. And only if all constraints are fulfilled, all $a_\alpha = 0$. This allows the detection of solution states.

Some neural dynamics like Hopfield's nonlinear differential equations

$$z_\alpha = 1/(1 + \exp(-u_\alpha/u_0)), \, du_\alpha/dt + u_\alpha/\tau = x_\alpha \tag{8}$$

will then turn an initial neuron state into one with minimal energy which is stable and an element of the attractor. But there is the problem, that other stable states besides the solutions may exist: the well known local minima of Hopfield networks. But by incorporating the activity as defined by (7) into the neural dynamic, as done in the SDNN [3], [4] for example, a distinction between local and global minima can be made to avoid the local minima. Other approaches like the Boltzmann machine or a random restart procedure are possible for finding global minima.

In this application, the neural network is operated as an autoassociator. But note, that contrary to the usual autoassociator, the stable states are neither stored nor learned patterns but solutions emerging from the set of given constraints. A combination with a trained pattern completion network with the same feedback and neuron structure is possible by superimposing the energy functions.

2.   Mapping boolean expressions to a recurrent neural network structure

Since the attractor of the network is determined by the conjunction (logical AND) of several constraints, the procedure to define the network structure by boolean expressions (axioms) is straightforward. The boolean variables are represented by neuron states ($z_\alpha = 1$: TRUE, $z_\alpha = 0$: FALSE). The boolean expression is firstly to be converted into a conjunctive form (CF: conjunction of disjunctions of variables and its negates). Then the negated variables have to be provided and at last, the disjunctions (logical OR) have to be represented by k-out-of-n constraints. This can be done as follows:

- Negated variables:

    Negating a variable is a 1-out-of-2 constraint for the original variable and a newly introduced negated variable [9], similar to the inverted output of a flip-flop.

- Disjunctions:

    A disjunction of some variables that is TRUE means to enforce at least one active variable. This is an at-least-1-out-of-n constraint, that can be converted into a n-out-of-$(2n-1)$ constraint imposed on the original and $n-1$ new slack variables [9].

714

- Conjunctions:

  The conjunction of disjunctions comes for free, since all constraints are to be fulfilled together if a state is a solution.

An example is shown in Fig. 2 where each oval represents a k-out-of-n constraint.

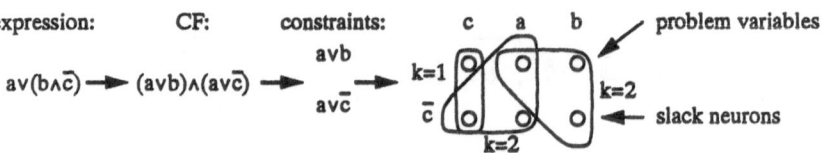

Fig. 2   Constraining the stable states by $a \vee (b \wedge \bar{c})$

A boolean function $y = f(a, b, c)$ (logic gate) can be represented by introducing a variable y for the function value and explicitly stating the equivalence constraint

$$f(a, b, c) \leftrightarrow y. \tag{9}$$

This expression can be translated into a conjunctive expression and mapped onto the network as shown before.

3.   Efficient minimization algorithms and a network compiler

The expressions should be minimized initially, to reduce the size of the resulting neural network. There are three major approaches to convert a given arbitrary boolean expression into a minimized disjunctive form:

- writing down the truth table as a Karnaugh map and finding irrelevant variables and/or disjunctions
- the algorithm of McCluskey
- the algorithms of Shannon and Brayton [10].

The last one is best suited for the neural network mapping problem, since it deals with symbolic expressions and not with an exponentially growing truth table and it is rather efficient because it can be programmed recursively. Using this algorithm and keeping in mind that a conjunctive form is required results in the following procedure for a network compiler

1. Read the expression; convert implications and equivalences to AND, OR, NOT; multiply parenthesized subexpressions to derive a (non-minimal) disjunctive form.

2. Find the minimized complement using Brayton's algorithm; the result is a disjunctive form.

3. Complement the result using De Morgan's rules; this results in a minimized conjunctive representation of the original expression.

4. Generate negated variables and 1-out-of-2 constraints for them and at-least-1-out-of-n constraints for all disjunctions.

A compiler [11] has been written in C language using the lex and yacc tools found on a Workstation with UNIX. It uses a library of procedures for manipulating terms of boolean variables. Symbolic variable names (like a, b, c as in the example in Fig. 2) are stored in a symbol table and assigned to neuron numbers. The input language (high level language) are boolean expressions including AND, OR, NOT, parenthesized subexpressions, implications, equivalences and antivalences. The output of the compiler is a list of neuron numbers tagged as input/output neurons and slack neurons and a list of k-out-of-n-constraints. From this, the network weights are calculated by a post-processor using (2) and (3). The weights can be regarded as the low level machine language of a neural computer.

4. Conclusions

There is a different approach for the same problem, the Neural Logic Gates (NLG) [3], [12] defined by Nakagawa. They allow a direct conversion of boolean expressions by building up a parse tree for the expression and generating a NLG for every node of the tree. This results in a larger network but has the benefit of eliminating the need for a logic minimization algorithm [11].

To make the difference to the well known neural implementation of boolean functions by McCulloch-Pitts neurons in a multilayer perceptron clear, it should be noted, that an approach using recurrent and not feedforward structures is presented here. Another difference is, that it is not intended to calculate the truth value of a boolean expression but to constrain a set of variables by boolean constraints. This allows to estimate input values for a set of given output variables and vice versa without generating a different network. This becomes possible due to the idea of autoassociative pattern completion where a given logical relationship between parts of the patterns is imposed by a connection structure.

5. References

[1] G. Tagliarini, J. Christ, E. Page, Optimization Using Neural Networks, IEEE Tr. on Computers, Vol. 40, No. 12, Dec. 1991, 1347-1358

[2] E. Page, G. Tagliarini, Algorithm development for neural networks, Proc. IEEE SPIE, Vol. 880, High Speed Computing, 1988, 11-19

[3] T. Nakagawa, E. Page, G. Tagliarini, SDNN: A Computation Model for Strictly Digital Neural Networks and its Application, Proc. 5th AAAIC'89, ACM/SIGART Dayton, OH, 1989

[4] T. Nakagawa, H. Kitagawa, E. Page, G. Tagliarini, SDNN-3: A Simple Processor Architecture for O(1) Parallel Processing in Combinatorial Optimization with Strictly Digital Neural Networks, IJCNN'91, Singapore Nov. 1991, 2444-2449

[5] H. N. Schaller, On the Problem of Systematically Designing Energy Functions for Neural Expert Systems Based on Combinatorial Optimization Networks, Neuro-Nîmes'92, EC2, Nîmes, 1992, 648-653

[6] D. Tank, J. Hopfield, Collective Computation in Neuronlike Circuits, Scientific American, Dec. 1987, 62-70

[7] D. Tank, J. Hopfield, Simple "Neural" Optimization Networks: An A/D Converter, Signal Decision Circuit, and a Linear Programming Circuit, IEEE Tr. on CAS, Vol. CAS-33, No. 5, May 1986, 533-541

[8] J. Hopfield, D. Tank, "Neural" Computation of Decisions in Optimization Problems, Biol. Cybernetics, Springer, Vol. 52, 1985, 141-152

[9] H. N. Schaller, A collection of constraint design rules for neural optimization networks, in Artificial Neural Networks II: Proc. ICANN'92, Brighton, (Eds.: I. Aleksander, J. Taylor), Elsevier, 1992, 1039-1042

[10] R. K. Brayton, G. D. Hachtel, et. al., Logic Minimization Algorithms for VLSI Synthesis, Kluwer Academic Publishers, 1984

[11] K. Ehrenberger, Automatische Umsetzung von booleschen Funktionen in eine neuronale Struktur, Diplomarbeit, Lehrstuhl für Datenverarbeitung, Technische Universität München, 1992

[12] M. Arai, T. Nakagawa, H. Kitagawa, An Approach to Automatic Test Pattern Generation Using Strictly Digital Neural Networks, IJCNN'92, Baltimore, June 1992, IV-474-479

USING REDUCE FOR REPLICA CALCULATIONS

Paul Lukowicz
Institut für Programmstrukturen
und Datenorganisation
Universität Karlsruhe
D-7500 Karlsruhe 1
Germany
lukowicz@ira.uka.de

Klaus-Robert Müller
GMD FIRST

Rudower Chaussee 5
D-1199 Berlin
Germany
klaus@first.gmd.de

Werner M. Seiler
Institut für Algorithmen und
Kognitive Systeme
Universität Karlsruhe
D-7500 Karlsruhe 1
Germany
kg04@dkauni2.bitnet

Abstract The REDUCE package REPPACK for performing replica calculations in the context of neural networks is presented. REPPACK provides the user with several functions needed to calculate partition functions and saddle point equations interactively. The application of REPPACK is demonstrated for the toy example of finitely many patterns ($p/N \to 0$) for the Hopfield model. The Gardner calculation with REPPACK is briefly outlined.

1. INTRODUCTION

Symmetric Hopfield networks [8] are employed as associative memories. They can store a set of patterns and perform fault-tolerant retrieval even from distorted input. The properties of Hopfield networks have been examined through large scale simulations determining important parameters like storage capacity and basins of attraction [4, 7, 10, 5, 18].

Amit et al. have found a way to calculate the storage capacity analytically for hebbian couplings [1] using statistical mechanics and the replica trick [15]. They take the spins s_i as dynamical variables to compute the free energy. Gardner has calculated the storage capacity for a fully connected network, where the patterns are embedded with a certain embedding strength [6]. In her analysis which also uses the replica trick the couplings w_{ij} are the dynamical variables.

Both of these so called replica calculations can be done for different pattern distributions [6, 9], for different types of models, e.g. Q-state, analog or diluted models [17, 11, 2, 19] or for generalisation problems [16]. They imply technical difficulties in the analysis and numerical problems in the evaluation of the saddle-point equations.

Of course we can not overcome the numerical difficulties with a computer algebra system. Our goal is to simplify the calculations leading to the saddle-point equations. They consist in every specific application of similar operations. REPPACK provides the user with functions supporting the computation of the partition function and the saddle-point equations in a replica (symmetric) calculation. In order to maintain a certain universality in our ansatz, the user has to do his calculations interactively. Thus it should be possible to perform different – even non-standard – calculations semi-automatically, i.e. the user expresses the individual steps by calls of procedures provided by REPPACK. REPPACK itself is a package developed in the computer algebra system REDUCE [14].

The next section presents the functionality of REPPACK; in the third section we demonstrate the application of our package to the toy example of $p/N \to 0$ and look at the standard Gardner calculation. Finally, some conclusions are given.

2. REPPACK

Replica calculations contain several typical operations. In REPPACK we have implemented general types of algebraic manipulations (cf. table and see also [12, 13]). The package is written in the symbolic mode of REDUCE. All procedures are nevertheless accessible from the algebraic mode. It supports the handling of vectors, matrices and tensors with variable (possibly infinite) dimensions. Elements of those structures can be accessed and modified by specifying an access pattern (e.g. diagonal elements or elements belonging to a certain block) and complex operations like determinant or inversion, rules can be specified.

The package allows flexible, easy-to-use access to parts of complicated expressions and simple alteration of their representation. This is especially important since a human user can immediately tell e.g. over which terms an averaging should take place or whether some terms in the exponentials factorize over a certain index. REPPACK solves such tasks according to certain matching or consistency rules suitable for replica calculations or rules explicitly given by the user.

REPPACK can perform Gauß linearisation and provides procedures to transform θ and δ functions into their integral representations. It contains a Gauß integrator which includes an automatic consistency check. For a given set of differentation variables, saddle-point equations are computed automatically. To enhance the readability of the results, we adapted the REDUCE-TEX-interface to our needs. The saddle-point equations can be written directly into a FORTRAN file through the FORTRAN interface of REDUCE. Then they can as usually be solved numerically.

nvec, nmat, nstruct	definition of a vector, matrix or tensor with variable dimension sizes
contr	reduces the dimensionality of an object (e.g. contracts a matrix to a vector)
nprod, nsum	new operators for sums and products with variable bounds
dint	new operator for definite, multidimensional integration with variable bounds
sav, strace	new operators for the average and the trace over system configurations
grpxxx, grpx	selects all factors and operators from a specified operator. xxx can be any REPPACK operator name
to!*xxx, tox	changes an exponential function over sums to a product of exponential functions and vice versa. xx can be nsum, nprod, sum, prod.
op!*, op!*part	selects the specified operator or its argument in a complex expression. op!*, op!*part can be used both on the left and right side of an assigmnent
gaussaux	introduces auxiliary variables using the gauss trick
thetaauxaux, deltaaux, oneaux	introduces auxiliary variables using the integral representation of θ, δ or unity
saddle!*eqs	computes the saddle point equations
gauss!*int	performs a gaussian integration (multidimensional) over a specified variable

Table 1: REPPACK contains about 50 different functions and operators, some of them are described in this table.

3. CALCULATIONS WITH REPPACK

3.1 A Toy Example

In this section we demonstrate the use of REPPACK for a toy example, namely the $p/N \rightarrow 0$

limit of the Hopfield model [1, 7]. In this limit, a Hopfield network is trained for p patterns σ_i^μ ($\mu = 1 \ldots p$, $i = \ldots N$) with the Hebb rule

$$w_{ij} = \frac{1}{N} \sum_\mu \sigma_i^\mu \sigma_j^\mu \, . \tag{1}$$

We look at the system in the thermodynamic limit $N \to \infty$, where the number of patterns will be kept fixed such that $\alpha = p/N$ becomes arbitrarily small. To evaluate the partition function Z, we start with some preliminary definitions[1]:

```
H   := NSUM(i,1,NN,beta*sigma(mu,i)*s(i));
```

$$h := \sum_{i=1}^{N} (s_i \cdot \sigma_{\mu,i} \cdot \beta) \tag{2}$$

and define the partition function as

```
Z   := EXP(-1/(2*NN*beta)*NSUM(mu,1,p,H**2));
```

$$z := e^{-\left(\left(\frac{1}{2}\right) \cdot \sum_{\mu=1}^{p} \left(\sum_{i=1}^{N} (s_i \cdot \sigma_{\mu,i} \cdot \beta)^2 \right) \cdot N^{-1} \cdot \beta^{-1} \right)} \tag{3}$$

where we omitted the diagonal terms $\exp(-\frac{1}{2}\beta p)$ that do not contribute to the saddle-point equations. Note that in the TEX representation of REDUCE capital letters become non-capital ones, if capital letters are necessary, we have to write e.g. NN in the REDUCE code to obtain N in the TEX representation. In order to average over the states s_i, we introduce the auxiliary variables m_μ linearising the quadratic term in s_i via a Gauß linearisation.

```
Z1  := GAUSSAUX(Z,H,m(mu));
```

$$z_1 := \prod_{\mu=1}^{p} \left(\sqrt{\beta} \cdot \pi^{-\left(\frac{1}{2}\right)} \cdot \sqrt{N} \cdot \int_{-\infty}^{\infty} \left(e^{-\left(\frac{1}{2}m_\mu^2 \cdot N \cdot \beta\right) + m_\mu \cdot \sum_{i=1}^{N} (s_i \cdot \sigma_{\mu,i} \cdot \beta)} \right) \, dm_\mu \right) \tag{4}$$

The next line is more tricky

```
Z3  := op!*(dint,m(mu),1,Z1) := fixrules(grpstrace(s,{-1,1},
           strace(s,{-1,1},op!*(dint,m(mu),1,to!*nprod(Z1,i))),{}) where sigtr);
```

We will explain it step by step. First, in the innermost command *strace*, we average over all $s_i \in \{\pm 1\}$ under the integral over m_μ. The command *grpstrace* tells REDUCE to perform this trace operation only where it is necessary, i.e. over $\exp(m_\mu \sum_i s_i \cdot \sigma_{\mu,i} \cdot \beta)$. Again we want to emphasize that a human being can easily see which term should be averaged over, but this is highly non-trivial for a computer. The expression *sigtr* makes the system use the exp ln 2 cosh representation of $2\cosh^2$ in its output, and the *fixrules* command forces REDUCE to use exp ln not only in its output but also in its internal representation. $op! * (dint, m(mu), 1, Z1) := \ldots$ tells REDUCE to substitute the result of the trace operation for the integral.

$$z_3 := \prod_{\mu=1}^{p} \left(\sqrt{\beta} \cdot \pi^{-\left(\frac{1}{2}\right)} \cdot \sqrt{N} \cdot \int_{-\infty}^{\infty} \left(e^{-\left(\frac{1}{2}m_\mu^2 \cdot N \cdot \beta\right)} \cdot \prod_{i=1}^{N} \left(e^{\ln(2\cosh(m_\mu \cdot \sigma_{\mu,i} \cdot \beta))} \right) \right) \, dm_\mu \right) \tag{5}$$

We finally arrive at the point, where we can differentiate with respect to m_μ

```
saddle!*eqs(op!*part(expt,e,1,to!*nsum(Z3,i))/-(N*beta),{m(mu)});
```

---

[1] Note, that multiple indices are separated by commata in REDUCE e.g. $\sigma_{\mu,i} \sim \sigma_i^\mu$.

[2] This is of course contraproductive for a computer algebra system, since it usually tends to simplify formulae instead of making them more complicated, whereas for humans the exp ln representation is just a matter of convenience.

and get the saddle-point equations

$$\left\{ -\left(\sum_{i=1}^{N} (tanh\,(m_\mu \cdot \sigma_{\mu,i} \cdot \beta) \cdot \sigma_{\mu,i}) \cdot N^{-1} \right) + m_\mu = 0 \right\} \tag{6}$$

This example should have made clear that even such a simple calculation can pose unsolvable problems to a computer, if the matching rules and the routines for accessing terms and altering representations are not worked out properly.

3.2 The Gardner Calculation

To show that the application of our package is not restricted to toy problems, we will briefly outline a Gardner type calculation in REPPACK. Technical details would go beyond the scope of this paper and can be found elsewhere [12, 13]. The full programm code we used is shown in the appendix. We only show some typical steps. As in the other example the REPPACK operators have to be defined first: the embedding strength γ_i^μ measures how good the patterns are learned

`gamma(a,mu,i) := nsum(j,1,NN,sigma(mu,i)*sigma(mu,j)*omega(a,i,j))/NN**(1/2);`

$$\gamma_{a,\mu,i} := \sum_{j=1}^{N} (\sigma_{\mu,i} \cdot \sigma_{\mu,j} \cdot \omega_{a,i,j}) \cdot N^{-1/2}, \tag{7}$$

and the replicated fractional phase space volume $\Omega_i = \langle V^n \rangle$ in the deterministic limit $T \to 0$

`Om(i):= sav(sigma,{-1,1},eta*nprod(a,1,n,dint(-inf,inf,rho(a)*`
`nprod(mu,1,p,theta(gamma(a,mu,i)-k)),nprod(j,1,NN,omega(a,i,j)))));`

$$\Omega_i := \left\langle \left(\prod_{a=1}^{n} \int_{-\infty}^{\infty} \left(\rho_a \cdot \prod_{\mu=1}^{p} \theta \left(\sum_{j=1}^{N} (\sigma_{\mu,i} \cdot \sigma_{\mu j} \cdot \omega_{a,i,j}) \cdot N^{-1/2} - k \right) \right) d \prod_{j=1}^{N} \omega_{a,i,j} \cdot \eta \right) \right\rangle \tag{8}$$

ρ^a is the integration measure for the spherical model, η is the normalisation constant. In the TEX notation of REPPACK $d \prod_{j=1}^{N} \omega_{a,i,j}$ stands for $\prod_{j=1}^{N} d\omega_{a,i,j}$. Multiple indices are separated by commata, e.g. $\sigma_{\mu,i} \sim \sigma_i^\mu$ and k is the embedding stability of the patterns usually denoted by κ. Now the θ or step function is replaced by its integral representation using the *thetaaux* function and the averaging is moved inside the integral (several lines of code in the appendix).

$$\Omega_i := \int_{-\infty}^{\infty} \int_{k}^{\infty} \int_{-\infty}^{\infty} \left(\prod_{j=1}^{N} \prod_{\mu=1}^{p} \left\langle \left(e^{-(\sigma_{\mu,i} \cdot \sigma_{\mu,j} \cdot \sum_{a=1}^{n} (x_{a,\mu} \cdot \omega_{a,i,j}) \cdot i \cdot N^{-1/2})} \right) \right\rangle \right) \cdot \prod_{\mu=1}^{p} \left(e^{\sum_{a=1}^{n} (x_{a,\mu} \cdot \lambda_{a,\mu}) \cdot i} \right)$$

$$\cdot d \prod_{\mu=1}^{p} \prod_{a=1}^{n} x_{a,\mu} \cdot d \prod_{\mu=1}^{p} \prod_{a=1}^{n} \frac{\lambda_{a,\mu}}{2\pi} \cdot \prod_{a=1}^{n} \rho_a \, d \prod_{j=1}^{N} \prod_{a=1}^{n} \omega_{a,i,j} \cdot \eta \tag{9}$$

After averaging and rearranging terms we consider $G_2(E_a, F_{ab})$ similar to Gardner [6].

$$g_2 := \ln \left(\int_{-\infty}^{\infty} \left(e^{\sum_{a=1}^{n} (-(i \cdot E)) + \sum_{a=1}^{n} \sum_{b=1}^{n} ((\frac{1}{2}) \cdot \omega_a \cdot \omega_b \cdot \delta_{a,b} \cdot i \cdot F + \omega_a \cdot \omega_b \cdot \delta_{a,b} \cdot i \cdot E - (\frac{1}{2}) \cdot \omega_a \cdot \omega_b \cdot i \cdot F)} \right) d \prod_{a=1}^{n} \omega_a \right) \tag{10}$$

For a replica symmetric ansatz we perform a Gauß integration

`on mcd; on div; h5 := gauss!*int(g2(EE,FF),11);`

$$h_5 := \ln \left(\frac{\pi^{\frac{n}{2}} \cdot e^{\sum_{a=1}^{n} (-(i \cdot E))} \cdot \sqrt{F + 2 \cdot E} \cdot 2^{\frac{n}{2}}}{\sqrt{-((-(i \cdot F) - 2 \cdot i \cdot E)^n \cdot n \cdot F) + (-(i \cdot F) - 2 \cdot i \cdot E)^n \cdot F + 2 \cdot (-(i \cdot F) - 2 \cdot i \cdot E)^n \cdot E}} \right)$$

and obtain the saddle-point equations for F and E in the limit $n \to 0$ and $q \to 1$,

```
sa:=saddle!*eqs(fix(h5 where !*dosum)+N*(N-1)*FF*q(a,b),{FF,EE})$
```

which completes our second example.

4. CONCLUSIONS

Our goal was to provide with REPPACK a modular package which is able to grow according to the needs of users. After some further testing on non-trivial problems, it will become generally available. An important advantage of using a computer algebra system is much higher probability for correct results. Many of the operations used in the replica trick are highly error-prone in hand calculations. The reported results may seem somewhat clumsy to readers not acquainted with computer algebra, but we hope that REPPACK will enable non-specialists to try methods of statistical mechanics for their problems and specialists familiar with the system to speed up their calculation considerably. Our future interest will be dedicated to a further simplification of the calculation and the consideration of replica symmetry breaking schemes within REPPACK.

APPENDIX

```
NMAT    sigma(p,N), lam(NN,p), x(NN,p), q(NN,NN),FF(a,b)$
NSTRUCT omega({NN,N,NN}),gamma({N,p,NN})$
NVEC s(N), rho(NN), Om(NN),EE(NN)$

gamma(a,mu,i) := nsum(j,1,NN,sigma(mu,i)*sigma(mu,j)*omega(a,i,j))/NN**(1/2);
Om(i):= sav(sigma,{-1,1},eta*nprod(a,1,n,dint(-inf,inf,rho(a)*nprod(mu,1,p,
                theta(gamma(a,mu,i)-k)),nprod(j,1,NN,omega(a,i,j)))));
Om(i):=(op!*part(nprod,mu,1,Om(i)):=
                thetaaux(op!*part(nprod,mu,1,Om(i)),lam(a,mu),x(a,mu)));
Om(i) := grpx(sv,{a,sigma},tox(s,grpx(pp,{mu,a},tox(p,Om(i),j)),a));
Om(i) :=(op!*(sav,sigma(j,mu),1,Om(i)):=e**(-1/2*nsum(a,1,N,nsum(b,1,N,
        x(a,mu)*omega(a,i,j)*x(b,mu)*omega(b,i,j)/NN))));
Om(i) := tox(p,fix(grpx(s,j,tox(ss,Om(i),{mu,j})) where insq),mu);
Om(i) := grpx(ii,{lam(a,mu),x(a,mu)},Om(i));
HGG := op!*(nprod,mu,1,Om(i));
Om(i) := tosum((op!*(nprod,mu,1,Om(i)) := exp(NN*alpha*gg(q(a,b)))));
rho(a) := deltaaux(delta(nsum(j,1,NN,omega(a,i,j)**2-NN)),EE(a));
H0:=nprod(a,1,N,nprod(b,1,a-1,
    oneaux((1/NN*nsum(j,1,NN,omega(a,i,j)*omega(b,i,j))),q(a,b),FF(a,b)*NN)));
Om(i) := grpx(ppi,{b,a,omega(a,i,j)},
                (op!*(nprod,a,1,Om(i)) := op!*(nprod,a,1,Om(i))*H0));
HG20 := op!*(dint,nprod(j,1,NN,nprod(a,1,N,omega(a,i,j))),1,Om(i))$
hg2  := (op!*part(nsum,j,1,hg20) := omega(a,i,j)**2-1);
Om(i):=(op!*(dint,nprod(j,1,NN,nprod(A,1,N,omega(a,i,j))),
                1,Om(i)):=e**(g2(EE,FF)));
h3 := grpx(i,{omega(a,i,j)},tox(ss,grpx(pp,{b,a},tox(p,hg2,j)),{b,a}));
h4 := fix(contr({EE,FF,FF,omega,omega},h3,{1,2,1,3,2}) where {!*npf,!*mrs});
off mcd; g2(EE,FF) := fix(ln(stm(a,b,omega(a),h4)) where !*lnr);
on mcd; on div; h5 := gauss!*int(g2(EE,FF),11);
sa:=saddle!*eqs(fix(h5 where !*dosum)+N*(N-1)*FF*q(a,b),{FF,EE})$
s1 := first(sa);s2 := second(sa);
off mcd; g1(x,lam):= fix(ln(stm(a,b,x(a),contr({q,q,x,lam},hgg,{2,2,2,2}))))
                        where {!*npf,!*lnr});
on mcd; h8 :=gauss!*int(g1(x,lam),ppq)$
```

```
h9 := fix(h8 where n*q = 0);
h10 := mts(a,b,lam(a),si(fix(h8 where n*q = 0),{2*n->0,4*n->0})); off mcd;
h11 :=tox(p,gaussaux(h10,nsum(a,1,n,lam(a)),(-q**(1/2)/(1-q)),z),a);
on mcd; on div; h11; h12 := fix(contr(lam,grpx(i,{lam(a)},h11),1) where !*npf);
```

References

[1] Amit, D.J., Gutfreund H., Sompolinsky, H., Phys. Rev. Lett. 55, 1530 (1985), Phys. Rev. A32,1007, (1985), Ann. Phys., NY 173, 30 (1987)

[2] Bouten, M., Engel, A., Komoda, A., Serneels, R., J. Phys. A:Math. Gen. 23, 4643 (1990)

[3] Buhmann, J., Divko, R., Schulten, K., Phys. Rev. A39, 2689 (1989)

[4] Forrest, B.M., J. Phys. A:Math. Gen. 21, 245 (1988)

[5] Forrest, B.M., Wallace, D.J., in: Models of Neural Networks, ed. Domany, E., van Hemmen, J.L. and Schulten K., Springer Heidelberg, 121 (1991)

[6] Gardner, E., J. Phys. A:Math. Gen. 21, 257 (1988)

[7] Hertz, J., Krogh, A., Palmer, R.G., Introduction to the Theory of Neural Computation, Addison-Wesley Redwood City (1991)

[8] Hopfield, J.J., Proc. Natl. Acad. Sci. USA, 79, 2554 (1982)

[9] Horner, H., Z. Phys. B 75, 133 (1989)

[10] Kohring, G.A., J. Stat. Phys. 59, 1077 (1990)

[11] Kühn, R., Bös, S., van Hemmen, J.L., Phys. Rev. A 43, 2084 (1991)

[12] Lukowicz, P., Anwendung von Computeralgebra auf Probleme Statistischer Physik neuronaler Netz, diploma thesis (in german), Universität Karlsruhe, Fakultät f. Physik (1993)

[13] Lukowicz, P., Müller, K.-R., Seiler, W.M., Application of Computeralgebra for Replica Calculations in Statistical Physics of Neural Networks, in preparation

[14] MacCallum, M.A.H., Wright, F.J., Algebraic Computing with REDUCE, Clarendon Press Oxford (1991)

[15] Mézard, M., Parisi, G., Virasoro, M.A., Spin Glass Theory and Beyond, World Scientific, Singapore, (1987)

[16] Opper, M., Kinzel, W., Kleinz, J., Nehl, R., J.Phys. A:Math.Gen. 23, L581 (1990)

[17] Rieger, H., J. Phys. A:Math. Gen. 23 L1273 (1990)

[18] Stiefvater, T., Müller, K.-R., J.Phys. A:Math.Gen. 25, 5919 (1992)

[19] van Hemmen, J.L., Kühn, R., in Models of Neural Networks, ed. Domany, E., van Hemmen, J.L. und Schulten K. (ed.), Springer Heidelberg, 1 (1991)

Equilibrium Statistical Mechanics of Non-Symmetric Neural Networks

A.C.C. Coolen   D. Sherrington

Dept. of Physics - Theoretical Physics, University of Oxford

1 Keble Road, Oxford OX1 3NP, U.K.

Abstract

We calculate the equilibrium probability distribution for a class of stochastic processes which need *not* obey detailed balance. For Ising spin neural network models this distribution takes the form of a high-temperature series. Our result implies that one can perform equilibrium statistical mechanical studies of *non-symmetric* neural networks.

1   Stochastic Systems in Equilibrium

We study finite discrete stochastic systems, the state of which at time t is described by a vector $|p_t\rangle$ in a D-dimensional Hilbert space. Each component $\langle n|p_t\rangle$, with respect to the orthogonal basis $\{|n\rangle\}$, represents the probability of finding the system at time t in microscopic configuration n. We restrict our analysis to systems with constant transition probabilities, in the form of a continuous-time master equation or a discrete-time Markov process:

$$\frac{d}{dt}|p_t\rangle = U[\beta]|p_t\rangle \quad \text{or} \quad |p_{t+1}\rangle = \{U[\beta]+1\}\,|p_t\rangle$$

The level of stochastic noise is assumed to decrease monotonically with the parameter β. *Equilibrium* probability distributions are state vectors $|p_\infty\rangle$ which, in addition to meeting certain physical requirements, are in the kernel of the operator $U[\beta]$. The uniform state is $\frac{1}{\sqrt{D}}|u\rangle$, with $|u\rangle \equiv D^{-\frac{1}{2}}\sum_n |n\rangle$. We restrict our analysis to operators with the properties:

(1) $U[\beta]$ is analytical in β: $U[\beta] = \sum_{n\geq 0} \beta^n U_n$

(2) $U[0]$ is Hermitian: $U_0^\dagger = U_0$

(3) $U[0]|\psi\rangle = 0 \;\Rightarrow\; |\psi\rangle = \lambda|u\rangle$

We can define the pseudo-inverse Q of U_0 in terms of its eigenvectors.

We assume the equilibrium probability distribution to depend analytically on β, insertion of a power series for $|p_\infty\rangle$ gives:

$$|p_\infty\rangle = \frac{1}{\sqrt{D}} \sum_{n \geq 0} \beta^n \sum_{m=0}^{n} \lambda_{n-m} \bar{S}_m |u\rangle \tag{1}$$

$$\bar{S}_0 \equiv 1, \quad \bar{S}_{n>0} \equiv \sum_{k=1}^{n} \sum_{m_1=1}^{n} \cdots \sum_{m_k=1}^{n} \delta_{n, m_1 + \cdots + m_k} a_{m_1} \cdots a_{m_k}, \quad a_l \equiv -QU_l$$

The coefficients $\{\lambda_k\}$ reflect the freedom to multiply the equilibrium distribution by a β-dependent scalar. Summation of the series (1) yields additional formal expressions. For the standard normalisation $\vec{\lambda} = (1, 0, 0, \ldots)$ we obtain:

$$\sqrt{D}|p_\infty\rangle = \sum_{k \geq 0} \{-|u\rangle\langle u| - QU[\beta]\}^k |u\rangle$$

$$\sqrt{D}|p_\infty\rangle = |u\rangle - \sum_{k \geq 0} [1 - QU[\beta]]^k QU[\beta]|u\rangle$$

$$\sqrt{D}|p_\infty\rangle = \lim_{m \to \infty} \{1 - QU[\beta]\}^m |u\rangle$$

We apply these results to systems of Ising spins $s_i \in \{-1, 1\}$, $i = 1, \ldots, N$. The microscopic N-spin configuration is denoted by the vector $\vec{s}$; the probability to find the system at time t in configuration $\vec{s}$ by $p_t(\vec{s})$. We study stochastic alignment to local fields $h_i(\vec{s})$. First we define sequential evolution in time of $p_t(\vec{s})$ by a continuous-time master equation:

$$\frac{d}{dt} p_t(\vec{s}) = \sum_{j=1}^{N} \{p_t(F_j\vec{s}) w_{j\beta}(F_j\vec{s}) - p_t(\vec{s}) w_{j\beta}(\vec{s})\} \quad w_{j\beta}(\vec{s}) \equiv \frac{1}{2}[1 - f_\beta(s_j h_j(\vec{s}))]$$

where F_j is the spin-flip operator, i.e. $F_j \Phi(s_1, .., s_N) \equiv \Phi(s_1, .., -s_j, .., s_N)$ and where the function $f_\beta(x)$ obeys:

$$f_\beta(x) \in [-1, 1] \qquad f_\beta(-x) = -f_\beta(x) \qquad \frac{d}{dx} f_\beta(x) \geq 0 \qquad f_0(x) = 0$$

$f_\beta(x)$ is also required to depend analytically on β. We now write the master equation in standard form

$$\frac{d}{dt}|p_t\rangle = U[\beta]|p_t\rangle \qquad U[\beta] \equiv \frac{1}{2} \sum_{j=1}^{N} \left[\sigma_x^j - 1\right][1 - \phi_{j\beta}]$$

$$\phi_{j\beta} \equiv \sum_{\vec{s}} |\vec{s}\rangle f_\beta(s_j h_j(\vec{s})) \langle \vec{s}| \qquad \phi_{jk} \equiv \frac{1}{k!} \frac{d^k \phi_{j\beta}}{d\beta^k} |_{\beta=0}$$

The σ_x^j are Pauli spin operators. The basis $\{|\vec{s}\rangle\}$ are the eigenstates of the Pauli operators $\{\sigma_z^j\}$. $U[\beta]$ is in the relevant class. We can now calculate the relevant operators in (1):

$$a_{2k} = 0 \qquad \langle \vec{s}|a_{2k+1}|\vec{s}'\rangle = 2^{-N} \sum_{j=1}^N s_j s_j' \phi_{jk}(\vec{s}') \, g_N \left[\sum_{l \neq j} s_l s_l' \right]$$

in which

$$g_N[m] \equiv \int_0^1 dt \, (1+t)^{\frac{1}{2}[N-1+m]} (1-t)^{\frac{1}{2}[N-1-m]}$$

We are done: the equilibrium probability distribution for the sequential process is given by the series (1), with the operators a_k as defined above. The second type of dynamics considered is parallel stochastic local field alignment in the form of a discrete-time Markov process:

$$p_{t+1}(\vec{s}) \equiv \sum_{\vec{s}'} w_\beta(\vec{s}' \to \vec{s}) p_t(\vec{s}') \qquad w_\beta(\vec{s}' \to \vec{s}) \equiv \prod_{j=1}^N \frac{1}{2} [1 + f_\beta(s_j h_j(\vec{s}'))]$$

We can write the Markov process in standard form:

$$|p_{t+1}\rangle = \left\{ \tilde{U}[\beta] + 1 \right\} |p_t\rangle \qquad \tilde{U}[\beta] \equiv 2^{-\frac{1}{2}N} \sum_{\vec{s}} |\vec{s}\rangle\langle u| \prod_{j=1}^N \left[1 + s_j \sigma_x^j \phi_{j\beta} \right] - 1$$

$\tilde{U}[\beta]$ is in the relevant class. We can now calculate the operators in (1):

$$\langle \vec{s}|\tilde{a}_k|\vec{s}'\rangle = 2^{-N} \left\{ \frac{1}{k!} \frac{d^k}{d\beta^k} \prod_{j=1}^N \left[1 + \sum_{m \geq 0} \beta^{2m+1} s_j s_j' \phi_{jm}(\vec{s}') \right] \right\}_{\beta=0}$$

We are done: the equilibrium probability distribution for the parallel process is given by the series (1), with the operators $\tilde{a}_k$ as defined above.

2 Non-Symmetric Neural Networks

We obtain the standard transition rules of stochastic Ising spin neural network models (for a review see [1]) upon making the following choices:

$$f_\beta(x) \equiv \tanh[\beta x] \qquad h_i(\vec{s}) \equiv \sum_{j=1}^N J_{ij} s_j + \theta_i$$

Detailed balance is violated as soon as the matrix $\{J_{ij}\}$ is non-symmetric. We can now calculate the sequential dynamics operators $\{a_k\}$ and the parallel dynamics operators $\{\tilde{a}_k\}$ for such models:

$$\langle \vec{s}|a_{2k}|\vec{s}'\rangle = 0, \qquad \langle \vec{s}|a_{2k+1}|\vec{s}'\rangle = \tau_k 2^{-N} \sum_{j=1}^{N} s_j \left[h_j(\vec{s}')\right]^{2k+1} g_N\left[\sum_{l\neq j} s_l s_l'\right]$$

$$\langle \vec{s}|\tilde{a}_k|\vec{s}'\rangle = 2^{-N}\left\{\frac{1}{k!}\frac{d^k}{d\beta^k}\prod_{j=1}^{N}\left[1+\sum_{m\geq 0}\tau_m \beta^{2m+1} s_j \left[h_j(\vec{s}')\right]^{2m+1}\right]\right\}_{\beta=0}$$

The $\{\tau_k\}$ are the expansion coefficients of tanh, $\tanh(x) \equiv \sum_{k\geq 0}\tau_k x^{2k+1}$ [2].

For the first few orders in the sequential case, for example, we find:

$$2^N p_\infty^{seq} = 1 + \beta\left[\frac{1}{2}\sum_{k\neq l}s_k J_{kl} s_l + \sum_k \theta_k s_k\right]$$

$$+\frac{1}{2}\beta^2\left[\frac{1}{2}\sum_{k\neq l}s_k s_l J_{kl} + \sum_k \theta_k s_k\right]^2 - \frac{1}{8}\beta^2\sum_{i|k\neq l}s_k[J_{kl}+J_{lk}][J_{li}-J_{il}]s_i$$

$$+\beta^2\left[\frac{1}{4}\sum_{k\neq l}s_k s_l J_{kk}[J_{kl}+J_{lk}] + \frac{1}{3}\sum_k \theta_k s_k J_{kk} - \frac{1}{2}\sum_{k\neq l}\theta_k[J_{kl}-J_{lk}]s_l + C\right]+\mathcal{O}(\beta^3)$$

The constant C simply follows from $\sum_{\vec{s}}p_\infty(\vec{s}) = 1$. For parallel dynamics the first few orders are found to be:

$$2^N p_\infty^{par} = 1 + \beta\sum_k \theta_k s_k + \beta^2\sum_{kl}s_k J_{kl}\theta_l + \frac{1}{2}\beta^2\sum_{k\neq l}s_k s_l\left[\theta_k \theta_l + \sum_i J_{ki}J_{li}\right]+\mathcal{O}(\beta^3)$$

References

[1] Coolen A.C.C. and Sherrington D. 1992 *Preprint OUTP-92-49S* University of Oxford

[2] Gradshteyn I.S. and Ryzhik I.M. 1980 *Table of Integrals, Series, and Products* (San Diego: Academic Press)

Storage of Words by Coupling Hopfield Nets

Patricio Perez and Giovanni Salini

Departamento de Fisica, Universidad de Santiago de Chile, Casilla 307,Correo 2, Santiago, Chile.

Abstract

We show that by coupling several Hopfield nets we can store correlated patterns. The model may be related to the human ability to store and retrieve words. The interaction term considers n-neuron synapses, where n is the number of letters in a word. Given that in general the number of neurons coding for single letters is much larger than the number of letters in a word, the connectivity is kept at a reasonable level. Numerical simulations on the storage of three-letter words show that for an appropriate strength of coupling the retrieval is very good.

It is well known that the Hopfield model[1] with bistable neurons fails in the storage of correlated patterns. Several alternative models for the synaptic coefficients have been proposed which permit the storage of an important amount of correlated patterns. A well known model is the pseudoinverse solution, which is based in the inversion of a $p \times N$ matrix [2]. In several models, in order to ensure stability, the synaptic matrix is built after a repeated presentation of the prescribed patterns [3,4].

In what follows we show a way to store correlated patterns in which the inconvenients of inverting a very large matrix or possible divergences of an iterative algorithm are not present. Let us assume that we have a Hopfield network with N Ising spins or two state neurons and let us write the states $\vec{S}$ as a sum of k terms or segments: $\vec{S} = \vec{S}^{(1)} + \vec{S}^{(2)} + \ldots + \vec{S}^{(k)}$, where each $\vec{S}^{(i)}$ has N_i consecutive components different from zero starting at position $N_1 + N_2 + \ldots + N_{i-1} + 1$. We can understand this as a partition of the network in k subnets, each of them with N_i neurons. The local field acting on neuron i which belongs to sub-net i_1 can be written as:

$$h_i^{(i_1)} = \sum_{i_2=1}^{k} \sum_{j=1, j \neq i}^{N} J_{ij}^{(i_1 i_2)} S_j^{(i_2)} \tag{1}$$

where

$$J_{ij}^{(i_1 i_2)} = \frac{1}{N} \sum_{\mu=1}^{p} \xi_i^{(\mu, i_1)} \xi_j^{(\mu, i_2)} \tag{2}$$

and $\xi_i^{(\mu, i_1)}$ is the i-component of the prescribed state μ if i is within sub-net i_1 and zero otherwise.

Equation (1) can be written as:

$$h_i^{(i_1)} = \frac{1}{N} \sum_{j=1, j \neq i}^{N} \sum_{\mu=1}^{p} \xi_i^{(\mu, i_1)} \xi_j^{(\mu, i_1)} S_j^{(i_1)} + \frac{1}{N} \sum_{i_2=1, i_2 \neq i_1}^{k} \sum_{j=1, j \neq i}^{N} \sum_{\mu=1}^{p} \xi_i^{(\mu, i_1)} \xi_j^{(\mu, i_2)} S_j^{(i_2)}. \tag{3}$$

If the second term were not present and N_{i_1} were large enough, we could store $p_{i_1} = \alpha_c N_{i_1}$ uncorrelated patterns in this sub-net ($\alpha_c = 0.144$). In the complete net we would have then $(\alpha_c)^k \prod_{i=1}^{k} N_i$ predetermined fixed points. These N-component patterns will be correlated but will not show properties of associative memory due to the decoupling of the sub-nets. Following an idea of U. Krey and G. Pöppel [5], we could include the interaction between sub-nets by introducing a coupling parameter and slightly modifying the second term of equation (3) such that:

$$h_i^{(i_1)I} = \frac{1}{N} \sum_{i_2=1, i_2 \neq i_1}^{k} \sum_{j=1, j \neq i}^{N} \sum_{\mu_{i_1}=1}^{p_{i_1}} \sum_{\mu_{i_2}=1}^{p_{i_2}} \lambda_{\mu_{i_1}, \mu_{i_2}}^{i_1 i_2} \xi_i^{(\mu_{i_1}, i_1)} \xi_j^{(\mu_{i_2}, i_2)} S_j^{(i_2)} \qquad (4)$$

where the superscript I denotes interaction and the parameter $\lambda_{\mu_{i_1}, \mu_{i_2}}^{i_1 i_2}$ allows for the choice of some of the possible combinations of segments of prescribed patterns from different sub-nets. From here on, patterns in the sub-nets will be called "letters" and the preferred combinations, "words". In order to have associative memory in the complete net, knowledge of a fraction of a word that includes the contribution of some of the sub-nets should induce the retrieval of the appropriate letters in the remaining sub-nets. However, for two-neuron interactions as is the case here, we cannot generate a letter of a stored word from the knowledge of more than one other letter of the word. Then, the interesting case can be reduced to two sub-nets or two-letter words. Besides, if we want that by knowing one letter of the word it be retrieved uniquely, a letter in a given sub-net, can at the most be associated with only one letter in the other sub-net. Reference [5] shows analytical results for this system, as for example the phase diagram relating storage capacity with the magnitude of the coupling parameter λ and temperature. However, since letters within sub-nets are uncorrelated, and due to the requirement of uniqueness, the patterns or words in the complete net will be also uncorrelated.

In the same scheme of the ideas described above, if we want to store and be able to retrieve n-letter words by requiring that knowledge of $n-1$ letters determines unambiguosly the word, we must have n-neuron interactions or synapses in the network. Due to the requirement mentioned above and in order to have stability, two stored words should differ in at least two sites. Then, if in each of the n sub-nets we store the same amount p of letters, we can build a maximum of p^{n-1} words. Now, a letter stored in the complete net will be correlated with several others, due to the repetition of letters in a given site. A practical implementation of a net with n-neuron synapses gets expensive in several grounds with increasing n due to connectivity. In order to aliviate this problem, we could use two-neuron synapses for the interactions within the sub-nets and leave the n-neuron only for the internet terms. Since in the context of real words the number of sub-nets will be much smaller than the number of neurons in a sub-net, this will represent an important simplification. For three-letter words the local field acting on a neuron in sub-net 1 will be reduced to:

$$h_i^{(1)} = \sum_j J_{ij}^{(11)} S_j^{(1)} + \sum_{j,k} J_{ijk}^{(123)} S_j^{(2)} S_k^{(3)} \qquad (5)$$

where the $J_{ij}^{(11)}$'s are the usual Hopfield coefficients for patterns within subnet 1 and

$$J_{ijk}^{(123)} = \frac{1}{N} \sum_{\mu_1=1}^{p} \sum_{\mu_2=1}^{p} \sum_{\mu_3=1}^{p} \lambda_{\mu_1 \mu_2 \mu_3} \xi_i^{\mu_1} \xi_j^{\mu_2} \xi_k^{\mu_3} \qquad (6)$$

with i within subnet 1, j within subnet 2 and k within subnet 3. Similar expressions apply for subnets 2 and 3, after a cyclic change of indices. In this situation, the number of words differing in at least two letters that we could in principle store are $p_{tot} = (p_c)^2 = (\alpha_c N/3)^2$, with $\alpha_c = 0.144$ in the limit $\lambda \to 0$. As an obvious extension, for n-letter words the storage capacity will be $(\alpha_c N/n)^{n-1}$. It is interesting to notice that this number is of the same order of magnitude as the storage capacity of the n-order fully connected network built as a generalization of the Hopfield model [6,7].

We have performed some numerical simulations on a system described by equations (5) and (6). We have divided a net with $N = 60$ neurons in three subnets of the same size. For a better visualization, our prescribed patterns were chosen to be real letters of the english alphabet. For each subnet we built 3 letters in which $+1$ corresponds to an 'x' and -1 to a blank in a two dimensional array. Care was taken so there were nearly the same amount of $+1$'s and -1's per letter and dot products close to zero for letters within nets and also between nets. These letters were: U, C, J, for the first sub-net, O, K, I, in the second subnet, and A, X, S, in the third. From the 27 possible combinations we chose 7 out of the maximum of 9 that differ in at least two sites between each other. In first place we checked that these words were stable fixed points of the network. In order to test the ability of the net to retrieve the prescribed words we did the following calculation: with the initial state corresponding to a pattern in which two of the letters of one of the seven words were present and the third site totally random, the fraction of times that the complete word was retrieved was plotted against the magnitude of the coupling parameter λ (assumed to be the same for all the non-zero values). The results after averaging over several trials are displayed in Figure 1. We observe that when only the two neuron interaction is present ($\lambda = 0$), there is no retrieval. For small values of λ, between 0.05 and 0.1, the three neuron term has a positive effect, allowing a very good recognition. However, when λ increases beyond 0.15, retrieval ability is lowered remaining stable in a value around 0.6. These results suggest that in Equation (5), the two terms on the right side are important and have a cooperative behaviour. The two neuron term stabilizes the single letters and the three neuron term is encharged of giving meaning to the word. The extension to the case of words with more letters is straightforward.

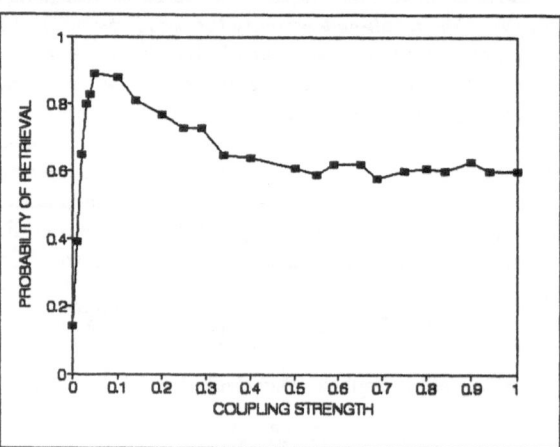

Figure 1. Fraction of times that a three letter word is retrieved when two letters are known and the third is random, as a function of coupling between the sub-nets.

We thank the support of Fondo Nacional de Desarrollo Cientifico y Tecnologico de Chile (FONDECYT) and Departamento de Investigaciones Cientificas y Tecnologicas, Universidad de Santiago de Chile (DICYT).

References

1) J.J. Hopfield, Proc. Natl. Acad. Sci., U.S.A. **79**, (1982) 2554.

2) L. Personnaz, I. Guyon and G. Dreyfus, J. Physique Lett. **46**, (1985) L-359.

3) S. Diederich and M. Opper, Phys. Rev. Lett. **58**, (1987) 949.

4) M.G. Blatt and E.G. Vergini, Phys. Rev. Lett. **66**, (1991) 1793.

5) U. Krey and G. Pöppel, Z. Phys. B **76**, (1989) 513.

6) E. Gardner, J. Phys. A: Math Gen. **20**, (1987) 3453.

7) I.J. Matus and P. Pérez, Phys. Rev. A **43**, (1991) 5683.

CONSTRAINTS ON LEARNING IN DYNAMIC SYNAPSES

Daniel J. Amit[1] and Stefano Fusi
INFN, Sezione di Roma, Istituto di Fisica
Università di Roma, La Sapienza, P.le Aldo Moro, Roma

Abstract

Some constraints intrinsic to unsupervised learning in attractor neural networks (ANN) are discussed. Hebbian type learning is discussed in a network whose synapses are analog, dynamic variables, with a fixed finite number of states that are stable on long time scales. It is shown that if the patterns to be learned are random words of ± 1 bits then in the limit of slow presentation, the network can learn at most $O(\ln N)$ patterns in N neurons. Going beyond the logarithmic contraint requires stochastic learning of patterns and low coding rate.

Introduction

We discuss constraints intrinsic to unsupervised learning in attractor neural networks (ANN). We have come across these constraints in connection with the design of an organically learning ANN, implemented in silicon [2]. The Hebbian framework provides a mechanism for affecting such learning [3]: if an afferent activates selectively certain sets of neurons in an 'assembly', then modifying each synapse by the correlation of the two neurons connected by it will create a corresponding attractor.

To arrive at a Hopfield matrix, synapses should have a temporal stability for the multiplicity of values generated in the course of one shot learning, both for the intermediate values as for the final ones. The problem of the analog depth of the synaptic values was treated by Sompolinsky [5]. The performance of the network was found to be midly reduced compared with performance of the intact network.

The practical problem, electronic or biological, is more difficult. It may be the case that the synapses do not have the analog depth to maintain the intermediate values at any time. It may also be that, while the synapse may be even analog, as would be a condenser, its acquired values cannot be stabilized in time (See also e.g. [6, 7]). What is common to both cases is that the synapses must be clipped, or refreshed, all along the learning process.

One could assume that a synapse has an electronic, or biochemical mechanism which pushes the synaptic value towards saturation, whenever that value is above some threshold, and allows the value of the synapse to decay to zero, if it is below the threshold. In this case information theory, or statistical mechanics, give a grossly inaccurate upper bound for storage capacities. On the other hand, the simple arguments introduced here, show several ways in which things can be improved.

Schematization of some typical learning situations

The performance and the properties of the network strongly depend on the kind of refresh mechanism, on the dynamics of the learning process and on the structure of the random patterns.

---

[1] On leave of absence from Racah Institute of Physics

Learning scenarios

There two typical learning scenarios:

1. **Low presentation rate**: random patterns are presented at very low rate so that clipping intervenes after each presentation. In this case the process of learning can be schematized giving a table containing all the possible transitions from one of the stable states J_m ($m = 0, ..., M - 1$) to the one which becomes stable after the presentation of a new pattern. These transitions depend only on the contribution that a synapse receives from each pattern presented: $\alpha_{ij}^\mu = \xi_i^\mu \xi_j^\mu$. This is probably the worst possible case because there is no possibility of using the advantages coming from the good analog depth which the synapses may reveal on short time scales.

2. **Fast presentation rate**: when frequency of patterns presentation gets higher then clipping intervenes after a certain number of presented patterns. In this case the whole set of the patterns to be learned is divided in subgroups. At each refresh cycle the all n patterns of the subgroup are presented. Neglecting a potential decay we have that each synapse sees, after the presentation of each subgroup, a contribution $A = 1/n \sum_{\mu=1}^n \xi_i^\mu \xi_j^\mu$.

Refresh mechanism

The refresh mechanism can be deterministic or stochastic. In the first case clipping intervenes in a deterministic way after each presentation of a single pattern or of a subgroup of patterns. The new stable value of the synapse is determined uniquely by the previous value of the synapse and by the contribution of the arriving input, i.e. α_{ij} for low presentation rate and A for fast presentation rate.

Stochastic refresh modifies each synapse with a probability q. At each presentation, only a fraction q of synapses is changed following the same rules of the deterministic case. Many presentations of the same pattern may be needed in order to make some noticeable modification on the synaptic structure.

Structure of the patterns to be learned

In what follows we shall consider two kinds of random patterns which differ in the way they are coded and in their coding level (i.e. the fraction of active neurons):

1. Hopfield like patterns: they are coded by ± 1 states for the neurons. Each pattern ξ_i^μ is a random word of ± 1 bits chosen with equal probability.

2. 0–1 patterns with low coding rate: the activity of the neuron is a binary variable ($\eta_i^\mu = 0, 1$) and the fraction f of non-zero bits is small.

Storage capacity of a network with 3-state synapses

The capacity of a network with 3-state synapses can be expressed in terms of a single parameter [1]: the fraction of synapses which retain a dependence on every pattern in the sequence of learned pattern. In fact, if d is the fraction of synapses which depend on the contribution of the oldest pattern, then the signal to noise ratio is:

$$\frac{S^2}{R^2} = const \cdot N d(p)^2 \gg 1$$

For the three learning situations described in the previous section we can sum up the main results:

1. Slow pattern presentation:

$$p_c < \frac{1}{2} \log_2 N$$

2. Fast pattern presentation (n patterns for each refresh cycle)

$$p_c < \frac{n}{-2\ln\left(1 - \frac{c}{2}\right)} \ln\left(\frac{N}{n-1}\right)$$

where c is the fraction of synapses changed by the presentation of a single subgroup.

3. Stochastic learning. The storage capacity depends on q and is expressed by:

$$p \simeq \frac{1}{-2\ln\left(1 - \frac{q}{2}\right)} \ln N$$

If q decreases, storage increases. q must not be $< \mathcal{O}(N^{-1/2})$ [1]. For the optimal choice $q = \mathcal{O}(N^{-1/2})$ we find: $p_c \simeq \sqrt{N}$. For a more detailed analysis of the three learning scenarios, the derivation of the three estimates of storage capacity and simulations see e.g. [1], [2].

Logarithmic limitation for multi-state synapses

The result can be extended to a very wide class of networks with dynamic synapses. The logarithmic restriction on the number of patterns that can be stored in a slow presentation scenario persists for networks with arbitrarily many stable states for the synapses, provided that the number of stable states is kept fixed as the number of neurons N grows.

Consider a network with synapses that have n possible stable states. We test the possibility of retrieval for the oldest of p patterns learned. When the pattern is presented the synaptic feed-back in the network creates an input 'post synaptic potential' at each neuron which is

$$h_i^p(\xi^1) = \sum_j J_{ij}^p \xi_j^1, \tag{1}$$

where the superscript p indicates the value of either h or J following the presentation of p patterns. The signal can be defined as:

$$S^p \stackrel{\text{def}}{=} \langle h_i^p(+1) - h_i^p(-1) \rangle. \tag{2}$$

If this difference is significantly greater than the noise: [1]:

$$R^2 \stackrel{\text{def}}{=} \langle (\xi_i^1 h_i^p - \langle \xi_i^1 h_i^p \rangle)^2 \rangle = \langle \frac{1}{N^2} \sum_j (J_{ij}^p)^2 \rangle \tag{3}$$

then a threshold can be found which will separate correctly the two outcomes to reproduce the retrieved pattern.

In the process of the presentation of random pattern a synapse undergoes a random walk within two reflecting barriers. If p is sufficiently large the signal can be approximated by:

$$S^p \simeq \lambda_{max}^{p-1} \frac{8q}{M^2} \sum_{m=0}^{M-1} J_m \cos\left(\frac{\pi m}{M-1}\right) \tag{4}$$

where M stands for the total number of stable synaptic states J_m, and λ_{max} is the maximal eigen-value lower than 1 of the stochastic transition matrix of the process:

$$\lambda_{max} = 1 - 2q \sin^2 \frac{\pi}{2M}$$

The noise is mainly contributed by the asymptotic part of the distribution:

$$\langle R^2 \rangle \simeq \frac{1}{MN} \sum_{m=0}^{M-1} (J_m)^2$$

so it does not depend on p. Consequently, when p is large enough, the signal-to-noise ratio goes as $\lambda_{max}^{p-1} NC$ where C is a constant which does not depend on p and N. This means that the number of storable patterns p cannot surpass $\log(CN)/\log(1/\lambda_{max})$ unless some other parameters as q or M which affect C and λ_{max} change with N.

Going beyond the logarithmic constraint

1. The network has M-state synapses, the patterns to be learned are Hopfield like, and the refresh mechanism is stochastic with a small transition probability q. In this case:

$$p_c < \frac{M^2}{2q} \ln \frac{Nq^2}{M^2}$$

So, even for an optimal choice of q value, the number of storable patterns can grow at most as $\sqrt{N/\ln N}$.

2. The network has 2-state synapses ($J = 1, 0$), the patterns are coded by 0–1 states for the neurons with coding rate f Refresh mechanism is deterministic or stochastic with fixed transition probabilities. In this case if f goes to zero as $\ln N/N$, $p_c = \mathcal{O}(N/\ln N)$.

3. Imposing a stochastic refresh mechanism with probability for a transition $1 \to 0$ for J_{ij} proportional to f. If $f = \mathcal{O}(\ln N/N)$ then the number of storable patterns reaches its absolute maximum $\mathcal{O}(N^2/\ln^2 N)$.

References

[1] Amit D.J., Fusi S., Constraints on learning in dynamic synapses, *Network* 3, (1992)

[2] Amit D.J., Fusi S., Genovese S., Badoni D., Riccardi R. e Salina G. 1992, LANN: Learning attractor neural network, Model and hardware implementation, Preprint ROM2F/92/51, 1992 (in italian)

[3] Hebb D.1949 *The Organization of Behavior* (Wiley, NY)

[4] Hopfield J.J. 1982 Neural networks and physical systems with emergent selective computational abilities, *Proc. Natl. Acad. Sci. USA* **79**, 2554

[5] Sompolinsky H., The theory of neural networks: The Hebb rule and beyond, in L. van Hemmen and I. Morgenstern eds. *Heidelberg Colloquium on Glassy Dynamics* (Springer-Verlag, Heidelberg, 1987)

[6] Shinomoto S., Memory-maintenance in neural networks, *J. Phys.*, **A20**, L1305 (1987)

[7] Dong DW and Hopfield JJ Dynamic properties of neural networks with adapting synapses, *Network* **3**, 267-284, (1992).

Recursive Construction of Neural Networks with Long Periodic Behavior[1]

Martín Matamala

Laboratoire de l'Informatique du Parallélisme, Ecole Normale Supérieure de Lyon

46, Allée d'Italie, 69364 Lyon Cedex 07, France.

From December 93: Departamento de Ingeniería Matemática, U. de Chile

Casilla 170-correo 3 Santiago, Chile

Extended Abstract

A neural network of size n, is a discrete dynamical system acting on $\Omega_n = \{-1,1\}^n$, whose transition function, F_A, is given in term of an nxn real matrix $A = (a_{ij})$ as follows:

$$F_A(x) = \overline{sgn}(Ax); \quad (Ax)_i = \sum_{j=1}^{n} a_{ij} x_j \quad i = 1, ..., n$$

$$\overline{sgn} : \mathbb{R}^n \to \{-1,1\}^n, \quad \overline{sgn}(y)_i = sgn(y_i) \; i = 1, ..., n \quad sgn(u) = \begin{cases} 1 & u \geq 0 \\ -1 & u < 0 \end{cases}$$

We are interested in the reverberation neural networks, i.e. neural networks where each state of the system, after a finite number of steps become to himself (hypercube permutations). We study the question: *how many reverberation neural netwoks have really different dynamical behavior ?*. A related question is asked in [2] where it is defined an equivalence relation whose equivalence class are characterized and the number of elements in any class is proved to be $2^n n!$ where n is the size of the neural network, but nothing is done concerning the number of different class.

Our approach consists in to look the neural networks , in particular reverberation neural networks, as dynamical systems. In order to give a partial answer to our question, we define an equivalence relation which is as the equivalence relation of dynamical systems and we proved, by building recursively 2^n non equivalent neural networks, that there exists at least 2^n different class when one considers this relation in the set of the reverberation neural networks of size n.

We give a way to build from a threshold function $f : \Omega_n \to \Omega_1$ another threshold function $g : \Omega_{n+1} \to \Omega_1$ such that g acts over a vector $(x, u) \in \Omega_{n+1}$, excepte for four vector y_1, y_2, y_3, y_4, as if only the n first coordinates, x, were considered, and then giving the value $f(x)$, moreover, the values $g(y_i), i = 1, 2, 3, 4$ are fixed by the construction. So, one can describe the dynamical evolution of g easily in term of those of f. Moreover, we show how to obtain threshold functions having an *a priori* desired behavior.

From these two constructions we give a recursive way for the construction of matrices , i.e., given a matrix A of size n satisfying some hyphotesis we build two matrices B and C of size $n + 1$, satisfying also those hyphotesis .

Later, we define an equivalence relation on P_n, the set of biyective function on Ω_n and we build a function η associating to each element in P_n a vector of size 2^n. We prove that two function F and G are equivalent iff $\eta(F) = \eta(G)$. Hence, B and C given by the construction define non equivalent neural networks because $\eta(F_B)$ and $\eta(F_C)$ are differents. Later, we prove that given two non equivalent neural networks A and A' their extentions are also non equivalent. This fact says that growing the size of the neural networks by one unit, one can double the non equivalent neural netwok number. That explains because we find 2^n non equivalent neural networks.

Like a corollary, we build a neural network A of size n which has only a cycle of period 2^n.

Reference

[1] M.Matamala, (1992)

"Recursive Contructions of Periodic Steady State Sequences for Neural Networks". Research Report, Dep. Ingeniería Matemática, Facultad Ciencias Físicas y Matemáticas, Univ. de Chile.

[2] M.Sato. C. Tanaka, (1982)

"Characterization and constructions of Reverberating Networks". Mathematical Biosciences 62:201-217.

---

[1]Partially supported by FONDECYT 0047/90

Phase-Space Gardening in the Binary-Couplings Memory Network

Norman Hendrich
Universität Hamburg · Fachbereich Informatik
Troplowitzstraße 7 · D-2000 Hamburg 54
e-mail: hendrich@informatik.uni-hamburg.de

Abstract

An iterative learning rule for the binary-couplings Hopfield/Gardner associative memory network is presented that allows to adjust the stabilities for each pattern at each neuron individually.

Simulations of the resulting networks show that the basins of attraction of the patterns can be shaped as desired. The dependence of the size of the basins of attraction, m_c, on the distribution $\rho(\lambda_{i\mu})$ of stabilities is shown.

Summary

A simple learning rule for the binary-couplings attractor neural network [1] is presented that allows to adjust the desired stabilities $\lambda_{i\mu}$ for each pattern and neuron independently, up to a storage ratio $\alpha \approx 0.4$.

The learning algorithm is a variant of the cost-function minimization algorithms introduced in [3]. The energy function used is $E(\lambda_{i\mu}, \kappa_{i\mu}) = \sum_{\mu=1}^{P} \Theta(\lambda_{i\mu} - \kappa_{i\mu}) \cdot (\lambda_{i\mu} - \kappa_{i\mu})$, where $\lambda_{i\mu}$ is the desired stability of pattern μ at neuron S_i and $\kappa_{i\mu} = ||J_{ij}||^{-1} \xi_i^\mu \sum_j J_{ij} \xi_j^\mu$ is its local field. For details and a full list of references see [2].

The dynamics of the resulting networks is studied in detail for two different choices of the distribution of desired stabilities. First, if $\lambda_{i\mu} \in \{\lambda_1, \lambda_2, \ldots, \lambda_m\}$, the size of the basins of attraction for the individual patterns depends on $\lambda_i(\mu)$ only. The dependence of m_c, the critical initial overlap to ensure pattern recall, on the values of $\lambda_i(\mu)$ is shown. Second, the effects of a linear distribution, $\lambda_{i\mu} = \mu \cdot \lambda_{max}/N$ are studied.

The generalization to correlated patterns and more special distributions of stabilities is obvious and allows anisotropic basins of attraction.

The possibility to shape the basins of attraction of the stored information patterns makes the networks more plausible as models of human associative memory and it may enable new applications of neural networks, as conventional RAM-based associative memories lack a storage-efficient means to adjust the recall of individual patterns.

References

[1] E. Gardner, B. Derrida, *Journal of Physics* A 21, 271–284 (1988).

[2] N. Hendrich, Phase-space gardening in attractor networks, Univ. Hamburg, FB Informatik, report 235/93 (1993).

[3] H. Koehler, S. Diederich, W. Kinzel and M. Opper, *Z. Physik B* 78, 333 (1992).

The Relationship Between Choice Of Representation, Network Structure And Performance In Harmony Theory Networks

T.Tambouratzis, Inst. of Informatics and Telecommunications, NCRPS "Demokritos", Aghia Paraskevi 153 10, Attiki, Greece.
D.Tambouratzis, Dept. of Mathematics, Agricultural University of Athens, Iera Odos 75, Athens 118 55, Greece.

Harmony Theory (HT) networks [1] are capable of solving constraint-satisfaction tasks in parallel via simulated annealing, whereas a global maximum (peak) of the internal-consistency landscape of the problem-space is found. Due to their capacity of directly encoding constraints, connectivity and the weight vector need not be formulated through a training phase but can be mathematically defined once the choice of representation (what part of the problem each network node and connection stands for) has been decided upon. This piece of research investigates the effect of the choice of representation and of the ensuing representation-dependent structure of a HT network on the internal-consistency landscape and subsequently on performance. This is of special interest in order to:
(1) Establish how the landscape is "sculpted" by the choice of representation.
HT network implementations of a given problem employing different representations [2] vary in terms of: a) the explicit versus implicit coding of the constraints of the problem [2], b) the weights [1] (values depend on the cardinality of the connections emanating from a node) and c) the strengths [1] (expressing the priority of a node in being satisfied). These determine the internal-consistency landscape describing the problem-space. Each point of the landscape corresponds to a possible state of the network, its "height" specifying its internal consistency. During simulated annealing [1], gradual ascent converges to a peak of the landscape, i.e. a solution to the problem.
(2) Determine the characteristics of the "optimal" landscape which allow for the best behavior in order to ascertain the attributes of the representation which can offer that.
The states of the network receive "heights" which vary according to the representation employed. Consequently, the form of the landscape (relative "height" and distance of the peaks and other points, configuration of valleys and plateaus, steepness and smoothness of the slopes) is representation-dependent and affects network performance in terms of: i) the power of convergence (the ability to reach a solution, the relative frequency with which each state is settled upon and the manner in which gradual crystallization becomes evident), ii) the accuracy of settling (the occurrence of probabilistic errors and of symmetry-breaking failures) and iii) the speed of settling, (the number of iterations necessary for settling, i.e. extreme values of temperature, decrements and critical values).

1. Smolensky, P., (1986). Foundations Of Harmony Theory, pp. 194-281, in "Parallel Distributed Processing: Foundations".
2. Tambouratzis, T., (1991). Harmony Theory Networks For Scene Analysis, pp. 1743-1746, in "Artificial Neural Networks".

Learning and generalization
—oral contributions

On The Power of Linearly Weighted Neural Networks

Martin Anthony
Mathematics Department
London School of Economics
Houghton Street
London WC2A 2AE, UK
anthony@vax.lse.ac.uk

Sean B. Holden
Engineering Department
University of Cambridge
Trumpington Street
Cambridge CB2 1PZ, UK
sbh@eng.cam.ac.uk

Abstract: The Vapnik-Chervonenkis dimension has proven to be of great use in the theoretical study of generalisation in artificial neural networks, providing bounds on the size of the sample one should use to achieve prescribed levels of generalisation in the 'probably approximately correct' model of generalisation. We investigate the VC dimension of certain types of linearly weighted neural networks. We first examine radial basis function networks (RBFNs). Using results of Micchelli and others on the interpolation properties of such networks, we obtain bounds on the VC dimensions of RBFNs with certain standard basis functions. We then calculate the VC dimension of polynomial discriminant functions (PDFs) defined over both real and binary-valued inputs.

1 Introduction

A *Linearly weighted neural network* (LWNN) computes a function $f_{\mathbf{w}} : \mathbf{R}^n \to \{0,1\}$ given by

$$f_{\mathbf{w}}(\mathbf{x}) = \rho[w_1\phi_1(\mathbf{x}) + \cdots + w_m\phi_m(\mathbf{x})],$$

where $\mathbf{w} = (w_1, \ldots, w_m)$ is a vector of weights, the $\phi_i : \mathbf{R}^n \to \mathbf{R}$ are fixed *basis* functions and ρ is the step function given by $\rho(y) = 1$ if $y > 0$ and $\rho(y) = 0$ if $y \leq 0$. The class $\mathcal{F}_n^{\phi}$ of functions computed by the network is $\mathcal{F}_n^{\phi} = \{f_{\mathbf{w}} \mid \mathbf{w} \in \mathbf{R}^m\}$. Networks of this general form have been studied extensively since the early 1960s; see [2] and the references therein.

In the case of PDFs the basis functions are formed as products of elements of the input vector $\mathbf{x}$; for example, $\phi_i(\mathbf{x}) = x_1^2 x_2^2 x_{n-1}^5$. For RBFNs, we use a set of m basis functions of the form $\phi_i(\mathbf{x}) = \phi(\|\mathbf{x} - \mathbf{y}_i\|)$ where $\mathbf{y}_i \in \mathbf{R}^n$ is a fixed *centre*, $\|.\|$ is the Euclidean norm and $\phi : \mathbf{R}^+ \cup \{0\} \to \mathbf{R}$ is a fixed function.

There is a simple and convenient interpretation of the way in which LWNNs operate [5]. Input vectors are mapped into an *extended space* using the basis functions; *extended vectors* in the new space are of the form $\Phi(\mathbf{x}) = (\phi_1(\mathbf{x}), \phi_2(\mathbf{x}), \ldots, \phi_m(\mathbf{x}))$. Clearly, $f_{\mathbf{w}}(\mathbf{x}) = 1$ if and only if the inner product $\langle \Phi(\mathbf{x}), \mathbf{w} \rangle$ is positive. Thus, in the extended space, the extended positive examples of f and the extended negative examples of f are linearly separable.

2 'PAC' learning and the VC dimension

In the basic 'probably approximately correct' (pac) model of neural learning, it is assumed that the neural network receives a stream of inputs, each input labelled with the value of a particular *target function* on that input. The target function is to be thought of as the function which is being 'learned'. A fundamental assumption of this model is that these inputs are presented independently and at random according to some probability distribution on the set of all inputs. How should generalisation be quantified? Some notation is required. Suppose that the set of inputs is $X = \mathbf{R}^n$ or $X = \{0,1\}^n$ where n is the number of input nodes to the network. Suppose that the target function t can be computed by the neural network in some state. A *training sample for t* of size m is a vector $\mathbf{x} = (x_1, x_2, \ldots, x_m) \in X^m$ of m inputs together with the values $t(x_1), \ldots, t(x_m)$. The learning algorithm (back-propogation, perceptron learning, linear programming, whatever it may be) accepts the training sample and changes the weights of the network in some way in response to the information provided by the sample. It is to be hoped that the resulting state is such that the function then computed by the network is in some sense 'close' to the target function. If $L(\mathbf{x})$ is the function computed by the network after training sample $\mathbf{x}$ has been presented and learning algorithm L has been applied, one way in which to assess the success of the learning process is to measure how close $L(\mathbf{x})$ is to t. Since there is assumed to be some probability distribution, P, on the set of all inputs, we may define the *error* of a function h (with respect to t) to be the P-probability that a further randomly chosen input is classified incorrectly by h. In other words, $\mathrm{er}_P(h)$ is $P(\{x \in X : h(x) \neq t(x)\})$. Now, we should hope that the error of $L(\mathbf{x})$ is 'usually' 'small'. Since each of the m inputs in the training sample is drawn randomly and independently according to P, the sample vector $\mathbf{x}$ is drawn randomly from X^m according to the product probability distribution P^m. Thus, more formally, we want it to be true that with high P^m-probability the sample $\mathbf{x}$ is such that the function $L(\mathbf{x})$ computed after training has small error with respect to t. This leads us to the following formal definition of PAC learning [11]: The learning algorithm L is a *PAC-learning algorithm* if *for any* given $\epsilon, \delta > 0$ there is a sample size $m(\epsilon, \delta)$ such that *for all* target functions computable by the network and *for all* probability distributions P on the set of inputs, we have

$$m \geq m(\epsilon, \delta) \Rightarrow P^m(\{\mathbf{x} : \mathrm{er}_P(L(\mathbf{x})) > \epsilon\}) < \delta.$$

In other words, provided the sample has size at least $m(\epsilon, \delta)$ then it is 'probably' the case that after training on that sample, the function computed by the network is 'approximately' correct. A number of points should be made about this definition. Note that the probability distribution P occurs twice in the definition: explicitly in the requirement that the P^m-probability of a misleading sample be small and implicitly through the fact that the error of $L(\mathbf{x})$ is measured with reference to P. One striking feature of the definition is that we require that the sample size $m(\epsilon, \delta)$ be independent of P and of t.

It is not difficult to show that if the network can only compute a finite number N of different functions (for example, if the weights are only allowed to take integer values in a certain bounded range) then any learning algorithm L such that $L(\mathbf{x})$ achieves the desired classification of the inputs in $\mathbf{x}$ is a PAC-learning algorithm. Such an algorithm is said to be *consistent*. In this case, a suitable $m(\epsilon, \delta)$ is $\epsilon^{-1} \ln(N/\delta)$. (In practice determining N may not be very easy.) However, it is by no means obvious that the definition can be satisfied if the network can compute infinitely many different functions (for example, if the network is the simple real perceptron). That this can be so was shown by Blumer *et al.* [3]. Their results show that the ability to PAC-learn and the sufficient sample size $m(\epsilon, \delta)$ are linked with a combinatorial parameter known as the *Vapnik-Chervonenkis dimension* [12] (or VC dimension).

Suppose $\mathcal{N}$ is a neural network with one output, which outputs 0 or 1 and suppose that $\mathcal{N}$ accepts inputs from a set X (for example, $X = \mathbf{R}^n$ where n is the number of input nodes). We say that a set T of inputs is *shattered* by $\mathcal{N}$ if for each of the $2^{|T|}$ possible ways of dividing T into two disjoint sets T_1 and T_0, there is some function f computable by $\mathcal{N}$ such that $f(x) = 1$ if $x \in T_1$ and $f(x) = 0$ if $x \in T_0$. The VC dimension [12] of $\mathcal{N}$, denoted $\mathcal{V}(\mathcal{N})$, is defined to be the largest size of a set of inputs shattered by $\mathcal{N}$. This parameter describes the 'expressive power' of the network. For example, the VC dimension of the simple real n-perceptron (that is, the single linear threshold neuron with n real inputs) is exactly $n + 1$ [3, 1].

The theory of Vapnik and Chervonenkis [12] and its modifications in [3] can be used to bound the sufficient sample size $m(\epsilon, \delta)$ in terms of the VC dimension. Specifically, Blumer *et al.* [3] showed that there is $K > 0$ such that if d is the VC dimension of the set of functions computable by a given network then a sufficient sample size is $K\epsilon^{-1}(d\log(\epsilon^{-1}) + \log(\delta^{-1}))$.

3   Radial basis function networks

Radial basis function networks in their most general form compute $\{0, 1\}$-valued functions on $\mathbf{R}^n$ of the form $f_{\mathbf{w}} = \rho[\bar{f}_{\mathbf{w}}(\mathbf{x})]$, where $\bar{f}_{\mathbf{w}} : \mathbf{R}^n \to \mathbf{R}$ is defined as follows:

$$\bar{f}_{\mathbf{w}}(\mathbf{x}) = \sum_{i=1}^{p} \lambda_i \phi(\|\mathbf{x} - \mathbf{y}_i\|) + \sum_{i=1}^{q} \theta_i \psi_i(\mathbf{x})$$

where $q \le p$, and where $\mathbf{w} = (\lambda_1, \lambda_2, \ldots, \lambda_p, \theta_1, \theta_2, \ldots, \theta_q)$ is a vector of weights, $\mathbf{y}_i \in \mathbf{R}^n$ are *centres* of the basis functions, $\phi : \mathbf{R}^+ \cup \{0\} \to \mathbf{R}$, $\|.\|$ is the Euclidean norm, and $\{\psi_i \mid i = 1, \ldots, q\}$ is a basis of the vector space $\pi_{d-1}(\mathbf{R}^n)$ of multinomials from $\mathbf{R}^n$ to $\mathbf{R}$ of degree at most $(d-1)$ for some specified d. (It follows from results in the next section that $q = \binom{n+d-1}{d-1}$.)

Broomhead and Lowe [4] introduced RBFNs on the basis that functions of the form of $\bar{f}_{\mathbf{w}}$ had previously proved very useful in the theory of multivariable interpolation (a review is given by Powell [9, 10]; see also [8] on which our review is based).

It is usual in practice not to include the polynomial terms in the network so that the network computes functions $f_{\mathbf{w}}(\mathbf{x}) = \rho[\sum_{i=1}^{p} \lambda_i \phi(\|\mathbf{x} - \mathbf{y}_i\|)]$. A single constant offset term λ_0 is often added to the summation, but is omitted here.

In this section we investigate the VC dimension of this class of networks, using various standard choices for the basis function ϕ.

We say that a set $T = \{(\mathbf{x}_1, o_1), \ldots, (\mathbf{x}_k, o_k)\}$ of points in $\mathbf{R}^{n+1}$ is *interpolated* by a class $\mathcal{G}$ of functions from $\mathbf{R}^n$ to $\mathbf{R}$ if there is $g \in \mathcal{G}$ such that for $1 \le i \le k$, $g(\mathbf{x}_i) = o_i$. The functions $\bar{f}_{\mathbf{w}}$ are useful because it is always possible to interpolate any k such points using a function of the form

$$\bar{f}_{\mathbf{w}}(\mathbf{x}) = \sum_{i=1}^{k} \lambda_i \phi(\|\mathbf{x} - \mathbf{x}_i\|) + \sum_{i=1}^{q} \theta_i \psi_i(\mathbf{x})$$

where $q \le k$, provided ϕ satisfies some simple conditions which we discuss below. It is clear that, in this case, the set $\{\mathbf{x}_1, \mathbf{x}_2, \ldots, \mathbf{x}_k\}$ is shattered by the class of functions computed by the network. Notice that the centres of $\bar{f}_{\mathbf{w}}$ have been taken to be the points $\mathbf{x}_i$. The sufficient condition on ϕ for the existence of an interpolating function of the required form is that $\phi \in \mathbf{P}_d(\mathbf{R}^n)$ where $\mathbf{P}_d(\mathbf{R}^n)$ is the set of *strictly conditionally positive definite (SCPD)* functions of order d.

Definition 1 *Let ϕ be a continuous function on $[0, \infty)$. This function is strictly conditionally positive definite of order $d \ge 1$ on $\mathbf{R}^n$ if for any k distinct points $\mathbf{x}_1, \ldots, \mathbf{x}_k$ in $\mathbf{R}^n$ and $c_1, \ldots, c_k \in \mathbf{R}$ (not all 0) where $\sum_{i=1}^{k} c_i p(\mathbf{x}_i) = 0$ for all $p \in \pi_{d-1}(\mathbf{R}^n)$, the quadratic form $\sum_{i=1}^{k} \sum_{j=1}^{k} c_i c_j \phi(\|\mathbf{x}_i - \mathbf{x}_j\|)$ is positive. We say that ϕ is SCPD of order 0 if the form $\sum_{i=1}^{k} \sum_{j=1}^{k} c_i c_j \phi(\|\mathbf{x}_i - \mathbf{x}_j\|)$ is positive definite.*

Let $\mathbf{P}_d$ be the set of functions which are in $\mathbf{P}_d(\mathbf{R}^n)$ over any $\mathbf{R}^n$, $\mathbf{P}_d = \bigcap_{n \ge 1} \mathbf{P}_d(\mathbf{R}^n)$. Note that for all d, $\mathbf{P}_d \subseteq \mathbf{P}_{d+1}$. An important theorem due to Micchelli provides us with a simple means of

Form of basis function	Type of basis function
$\phi_{LIN}(r) = r$	Linear
$\phi_{CUB}(r) = r^3$	Cubic
$\phi_{TPS}(r) = r^2 \ln r$	Thin plate spline
$\phi_{MQ}(r) = (r^2 + c^2)^{\frac{1}{2}}, c \in \mathbf{R}^+$	Multiquadric
$\phi_{IMQ}(r) = (r^2 + c^2)^{-\frac{1}{2}}, c \in \mathbf{R}^+$	Inverse Multiquadric
$\phi_{GAUSS}(r) = \exp\left[-(r/c)^2\right], c \in \mathbf{R}^+$	Gaussian

Table 1: Standard basis functions used in radial basis function networks.

determining whether some function ϕ is in $\mathbf{P}_d$, and hence whether it is a suitable basis function for use in forming $\overline{f}_\mathbf{w}$. We say that a function ζ is *completely monotonic* on $(0, \infty)$ if it is in $C^\infty(0, \infty)$ and its sequence of derivatives is such that $(-1)^i \zeta^{(i)}(x) \geq 0$ for $x \in (0, \infty)$ and $i \geq 0$.

Theorem 2 (Micchelli [7]) *If a function $\xi(x)$ is continuous on $[0, \infty)$, $\xi(x^2) \in C^\infty(0, \infty) \cap C[0, \infty)$ and $(-1)^d \xi^{(d)}$ is completely monotonic on $(0, \infty)$ but not constant then ϕ defined by $\phi(r) = \xi(r^2)$ is in $\mathbf{P}_d$.*

Now, consider the special case in which we attempt to interpolate the data using

$$\overline{f}_\mathbf{w}(\mathbf{x}) = \sum_{i=1}^k \lambda_i \phi(\|\mathbf{x} - \mathbf{x}_i\|).$$

The interpolation is possible provided we can find a solution to the matrix equation $\phi\lambda = o$, where $\phi_{ij} = \phi(\|\mathbf{x}_i - \mathbf{x}_j\|)$ and $o^T = (o_1, o_2, \ldots, o_k)$. It is possible to show (see Powell [10]) that ϕ is nonsingular if ϕ is SCPD of order 0, or if ϕ is SCPD of order 1 and $\phi(0) \leq 0$. Thus, in some cases theorem 2 will tell us whether a particular ϕ can be used successfully. An alternative sufficient condition also exists for this special case, again proved by Micchelli.

Theorem 3 (Micchelli [7]) *If ξ is continuous on $[0, \infty)$, positive on $(0, \infty)$, and has a first derivative that is completely monotonic but not constant on $(0, \infty)$, then for any n and any set of k vectors $\mathbf{x}_i \in \mathbf{R}^n$, $(-1)^{k-1} \det \xi(\|\mathbf{x}_i - \mathbf{x}_j\|^2) > 0$.*

Now clearly if we choose a suitable function ϕ such that $\xi(x) = \phi(\sqrt{x})$ satisfies the conditions in theorem 3, ϕ must be nonsingular and consequently there is a suitable weight vector λ regardless of the values for o_i.

In summary, provided we use a basis function ϕ chosen using the relevant conditions given in theorem 2 or theorem 3, then our radial basis function network having fixed centres shatters the set of p centres.

Table 1 summarizes some of the usual basis functions ϕ used in RBFNs. Note that the parameter c is fixed — it is not adapted during training. We immediately obtain the following two corollaries.

Corollary 4 *Consider the simple RBFNs of the form, $f_\mathbf{w}(\mathbf{x}) = \rho\left[\sum_{i=1}^p \lambda_i \phi(\|\mathbf{x} - \mathbf{y}_i\|)\right]$ where the centres $\mathbf{y}_i$ are fixed and distinct. If ϕ is one of the functions ϕ_{LIN}, ϕ_{GAUSS}, ϕ_{MQ} or ϕ_{IMQ} then the VC dimension $\mathcal{V}(\mathcal{F})$ of the network is exactly p. This result also applies if ϕ is one of the* *following two functions, which are more general forms of the multiquadric and inverse multiquadric respectively.*

$$\phi^*_{MQ}(r) = (r^2 + c^2)^\beta \text{ where } 0 < \beta < 1, \quad \phi^*_{IMQ}(r) = (r^2 + c^2)^{-\alpha} \text{ where } \alpha > 0.$$

Clearly ϕ_{MQ} and ϕ_{IMQ} are special cases of ϕ^*_{MQ} and ϕ^*_{IMQ} respectively.

Proof: The functions ϕ_{GAUSS} and ϕ^*_{IMQ} are in P_0 by theorem 2 and the functions $\sqrt{r}$ and $(r + c^2)^\beta$ where $0 < \beta < 1$ satisfy the conditions in theorem 3. This means that by the arguments given above $\mathcal{V}(\mathcal{F}) \geq p$ for all four cases. Also, we know from standard results that $\mathcal{V}(\mathcal{F}) \leq p$, and consequently we must have $\mathcal{V}(\mathcal{F}) = p$.

Corollary 5 *Consider the RBFNs of the form, $f_\mathbf{w}(\mathbf{x}) = \rho\left[\sum_{i=1}^p \lambda_i \phi(\|\mathbf{x} - \mathbf{y}_i\|) + p(\theta, \mathbf{x})\right]$ where $p(\theta, \mathbf{x})$ is the degree 1 polynomial, $p(\theta, \mathbf{x}) = \theta_0 + \theta_1 x_1 + \theta_2 x_2 + \cdots + \theta_n x_n$ and x_i are the entries of $\mathbf{x}$. Again, the centres $\mathbf{y}_i$ are fixed and distinct. If ϕ is the function ϕ_{CUB} or ϕ_{TPS} then the VC dimension of the network obeys $p \leq \mathcal{V}(\mathcal{F}) \leq p + n + 1$.*

Proof: By theorem 2 both ϕ_{CUB} and ϕ_{TPS} are in P_2 and hence $\mathcal{V}(\mathcal{F}) \geq p$. By standard results, $\mathcal{V}(\mathcal{F}) \leq p + n + 1$ and the results follows.

4 Polynomial Discriminant Functions

A PDF is said to be of order at most k when f can be realised as a PDF formed from basis functions of degree at most k. We shall denote by $\mathcal{P}(n, k)$ the class of PDFs of order at most k defined on $\mathbf{R}^n$. Further, we shall denote by $\mathcal{P}_\mathbf{B}(n, k)$ the class of *boolean PDFs* obtained by restricting $\mathcal{P}(n, k)$ to binary-valued inputs; i.e., to $\{0, 1\}^n$. We now introduce some notation. Let $[n]$ be the set $\{1, 2, \ldots, n\}$. We shall denote the set of all subsets of at most k objects from $[n]$ by $[n]^{(k)}$ and we denote by $[n]^k$ the set of all selections, in which repetition is allowed, of at most k objects from $[n]$. Thus, $[n]^k$ may be thought of as a collection of 'multi-sets'. For example, $[3]^{(2)}$ consists of the sets $\emptyset, \{1\}, \{2\}, \{3\}, \{1, 2\}, \{1, 3\}, \{2, 3\}$, while $[3]^2$ consists of the multisets $\emptyset, \{1\}, \{1, 1\}, \{2\}, \{2, 2\}, \{3\}, \{3, 3\}, \{1, 2\}, \{1, 3\}, \{2, 3\}$. Note that $[n]^{(k)}$ consists of $\sum_{i=0}^k \binom{n}{i}$ sets, and $[n]^k$ consists of $\binom{n+k}{k}$ multisets. With a slight abuse of mathematical notation, $[n]^{(k)} \subseteq [n]^k$. For each $\emptyset \neq S \in [n]^k$, and for any $\mathbf{x} = (x_1, x_2, \ldots, x_n) \in \mathbf{R}^n$, $\mathbf{x}_S$ denotes the product of the x_i for $i \in S$ (with repetitions as required). For example, $\mathbf{x}_{\{1,2,3\}} = x_1 x_2 x_3$ and $\mathbf{x}_{\{1,1,2\}} = x_1^2 x_2$. We interpret $\mathbf{x}_\emptyset$ to be 1. It follows that f defined on $\mathbf{R}^n$ is a PDF of order at most k if and only if there are constants w_S, one for each $S \in [n]^k$, such that $f(\mathbf{x}) = \rho\left[\sum_{S \in [n]^k} w_S \mathbf{x}_S\right]$. Further, because for $b = 0$ or 1, $b^r = b$ for all r, a function $f : \{0, 1\}^n \to \{0, 1\}$ is in $\mathcal{P}_\mathbf{B}(n, k)$ if and only if there are constants w_S, one for each $S \in [n]^{(k)}$, such that $f(\mathbf{x}) = \rho\left[\sum_{S \in [n]^{(k)}} w_S \mathbf{x}_S\right]$.

Dudley [6] showed that if $\mathcal{F}$ is a real vector space of real-valued functions defined on a set X, of vector-space dimension d, and if we define, for $f \in \mathcal{F}$, $f_+ : X \to \{0, 1\}$ by $f_+(\mathbf{x}) = 1 \iff f(\mathbf{x}) > 0$, then the VC dimension of $\text{pos}(\mathcal{F}) = \{f_+ : f \in \mathcal{F}\}$ is d. Along these lines, we have the following result.

Proposition 6 *For all n and k, let $B(n, k) = \{\mathbf{x}_S : S \in [n]^k\}$, regarded as a set of real functions on $\mathbf{R}^n$. Then $B(n, k)$ is a linearly independent set. Further, for $k \leq n$, let $C(n, k) = \{\mathbf{x}_S : S \in [n]^{(k)}\}$, regarded as a set of real functions on domain $\{0, 1\}^n$. Then $C(n, k)$ is a linearly independent set.*

Proof: The proof that $B(n,k)$ is linearly independent is omitted here. Consider $C(n,k)$. Let $n \geq 1$ and suppose $A(\mathbf{x}) = \sum_{S \in [n]^{(k)}} \alpha_S x_S = 0$ for all $\mathbf{x} \in \{0,1\}^n$, Set $\mathbf{x}$ to be the all-0 vector to deduce that $\alpha_\emptyset = 0$. Let $1 \leq l \leq k$ and assume, inductively, that $\alpha_S = 0$ for all $S \subseteq [n]$ with $|S| < l$. Let $S \subseteq [n]$ with $|S| = l$. Setting $x_i = 1 \Leftrightarrow i \in S$, we deduce $\alpha_S = A(\mathbf{x}) = 0$. Thus for all S of cardinality k, $\alpha_S = 0$. The result follows.

If V is the vector space of functions from $\mathbf{R}^n$ to $\mathbf{R}$ spanned by the functions in $B(n,k)$, then $\mathcal{P}(n,k) = \text{pos}(V)$. Similarly, if W is the vector space of functions from $\{0,1\}^n$ to $\mathbf{R}$ spanned by the functions in $C(n,k)$, then $\mathcal{P}_{\mathbf{B}}(n,k) = \text{pos}(W)$. The previous result shows that the (vector-space) dimension of V is $\binom{n+k}{k}$ and that the (vector-space) dimension of W is $\sum_{i=0}^{k} \binom{n}{i}$. Therefore, from Dudley's result, we have the following.

Corollary 7 *For all* n, k,

$$\mathcal{V}(\mathcal{P}(n,k)) = \binom{n+k}{k} \quad \text{and, for } k \leq n, \quad \mathcal{V}(\mathcal{P}_{\mathbf{B}}(n,k)) = \sum_{i=0}^{k} \binom{n}{i}.$$

References

[1] M. Anthony and N. Biggs, *Computational Learning Theory: An Introduction*, Cambridge Tracts in Theoretical Computer Science, Cambridge University Press, Cambridge, UK, 1992.

[2] M. Anthony and S.B. Holden, Quantifying generalization in linearly weighted neural networks, LSE Mathematics Preprint 42 and CUED report CUED/F-INFENG/TR.113 1993.

[3] A. Blumer, A. Ehrenfeucht, D. Haussler and M. Warmuth, Learnability and the Vapnik-Chervonenkis Dimension. *Journal of the ACM*, 36(4), 1989: 929–965.

[4] D.S. Broomhead and D. Lowe, Multivariable functional interpolation and adaptive networks, *Complex Systems*, 2, 1988: 321–355.

[5] T.M. Cover, Geometrical and statistical properties of systems of linear inequalities with applications in pattern recognition, *IEEE Trans. Electronic Computers* 14, 1965: 326–334.

[6] R.M. Dudley, Central limit theorems for empirical measures, *Ann. Probability* 6, 1978: 899–929.

[7] C.A. Micchelli. Interpolation of scattered data: Distance matrices and conditionally positive definite functions. *Constructive Approximation*, 2, 1986: 11–22.

[8] T. Poggio and F. Girosi, Networks for approximation and learning, *Proceedings of the IEEE*, 78(9), 1990: 1481–1497.

[9] M.J.D. Powell, Radial basis functions for multivariable interpolation: A review. Technical Report #DAMTP 1985/NA12, Department of Applied Mathematics and Theoretical Physics, University of Cambridge, October 1985.

[10] M.J.D.Powell, The Theory of Radial Basis Function Approximation in 1990. Technical report #DAMTP 1990/NA11, Department of Applied Mathematics and Theoretical Physics, University of Cambridge, December 1990.

[11] L.G. Valiant, A theory of the learnable, *Communications of the ACM, 27, 1984: 1134-1142.*

[12] V.N. Vapnik and A. Ya. Chervonenkis, On the uniform convergence of relative frequencies of events to their probabilities. *Theory of Probability and its Applications*, 16(2), 1971: 264-280.

Elimination of Overtraining by a Mutual Information Network

G. Deco, W. Finnoff and H.G. Zimmermann
Siemens AG, Corporate Research and Development, ZFE ST SN 41
Otto-Hahn-Ring 6, 8000 Munich 83, Germany

Abstract

The presented learning paradigm uses supervised back-propagation and introduces an extra penalty term in the cost function which controls the complexity and the internal representation of the hidden neurons in an unsupervised form. This term is the mutual information that punishes the learning of noise. This learning algorithm was applied to predict German interest rates by using real world data of the past. Excellent results are obtained. The effect of overtraining was eliminated, allowing implementation which finds the solution automatically without interactive strategies such as stopped training and pruning.

1.0 Introduction

Two principal approaches exist for connectionist learning paradigms: supervised and unsupervised. In the case of supervised learning the neural network learns with examples of input and desired output patterns to adjust the strength of its synapses in order to compute the presented patterns. In the unsupervised learning paradigm the neural network learns to optimize some task-independent cost function of the representation given by the outputs of its neurons. Barlow [1] defines the goal of unsupervised learning as to find the internal representation of the input data which make it easy to form new association between data and encoding. Linsker [2] formulated his theory of maximum information transfer as the goal of self-organization and applied it to a multilayer feedforward network of linear neurons for modeling the visual-system. Several authors [3-4] have afterwards tried to use the concept of "mutual information" in order to extend the principle of maximum information to non-linear units. We propose to use the mutual information for control of the diversity of the internal representation of the hidden units in a feed-forward network which learns by back-propagation. We introduce this term in the cost function but in the inverse sense of maximum information principle. We try with this term to reduce the amount of information contained in the hidden neuron and to reduce the complexity of the network. In real world data, the noise in the inputs variables, that contain the information is normally the cause of overtraining. So by forcing the internal representation in an unsupervised way to minimize the information transferred from the data, we obtain very good results without overtraining and produce a network with reduced complexity.

2.0 Theoretical Framework of the Model

In this section we formulate the architecture of the neural network. The first layer is used to represent the input data ξ_i of dimension n. The second layer is a layer of m hidden neurons with activation function given by the logistic function $\Phi(\ )$. An extra layer is used in order to normalize the outputs of the hidden units. These normalized outputs are interpreted as probability that the presented input contains the "hidden" feature described by the corresponding hidden neuron. The output layer is also given by a set of T neurons with activation function given by $\sigma(\ ) = \tanh()$. The outputs of the network are then given by,

$$O_t = \sigma\left(\sum_j W_{tj} y_j\right) ; \ y_j = \frac{\Phi\left(\sum_i \omega_{ji}\xi_i\right)}{\sum_k \Phi\left(\sum_i \omega_{ki}\xi_i\right)} \tag{EQ 1}$$

Let us relate the outputs of the hidden layer y_j^a, when the pattern a is presented, with the conditional probability $p(j|a)$. Then the cost function E involves two terms: the usual quadratic error and the mutual information. That is,

$$E = \sum_a \sum_t (T_t - O_t)^2 + \lambda\left(-\sum_j \bar{y}_j \log(\bar{y}_j) + \left(\frac{1}{N}\right)\sum_a \sum_j y_j^a \log(y_j^a)\right) \tag{EQ 2}$$

where N is the number of pattern in the training set.

The mutual information as used in equation (3) is the difference between the entropy of the average of the conditional probability and the average of the entropy of the conditional probability between two random process. The mutual information is minimized by including it in the cost function with a positive sign. In other words, we penalize the learning of the noise. We can see the hidden layer as an non-linear filter of the input layer. Two terms are competing in the learning of the weights between the input and hidden layer. The first one, supervised, is given by back-propagation and it tends to learn perfectly all the input-output mapping including the noise. The second term given in an unsupervised way by the mutual information between the input pattern and the internal representation (probability distribution given by the outputs of the hidden layer) tend to filter the noise and to represent just the principal structure of the data. After some algebra, the weight update equations for the usual gradient method are as follows,

$$\Delta W_{tj} = \eta \sum_a (T_t^a - O_t^a)\dot{\sigma}_t(\ )y_j^a \tag{EQ 3}$$

$$\Delta\omega_{ji} = \eta\sum_a \delta_j \xi_i^a (1 - y_j^a) -$$

$$\eta\frac{\lambda}{N}\sum_a (y_j^a(1 - \Phi_j)\xi_i^a(\log(y_j^a) - \log(\bar{y}_j) - \sum_k y_k^a \log(y_k^a) + \sum_k y_k^a \log(\bar{y}_k))) \tag{EQ 4}$$

3.0 Complexity Penalty Terms

Assume that a finite data set $\wp = \{(x_t, y_t) \mid t = 1, ..., T\}$ consists of inputs $\{x_t \mid t = 1, ..., T\}$ and targets $\{y_t \mid t = 1, ..., T\}$. Although the data set contains all the information about the data generating process available, the problem that one wants to solve is to choose a network activation function f^* which minimizes the expected error $R = E((y - f^*(x))^2)$ on future exemplars (y, x) drawn from the same distribution as the original data. Here E denotes the mathematical expectation. Since $R(f)$ is unobservable, one must approximate it with the 'empirical error', which is defined for a network activation function f by setting

$$L(f) = \frac{1}{T} \sum_{t=1}^{T} (y_t - f(x_t))^2 \qquad \text{(EQ 5)}$$

This approximation only becomes exact for $T \to \infty$. Network training involves choosing a specific network $f_\wp$ from a (generally restricted class) $\Im$, so that $f_\wp \approx f^*$ using the information contained in the data set. As such, this always produces models $f_\wp$ that, since they depend on the data, must be viewed as a random elements, taking values in a function space (estimator). The expected error of using such an estimator is the given by the 'statistical risk' $ER(f_\wp)$. There are two contributions to the risk of a network found by optimization of the empirical error: specifically the approximation error between f^* and $f_\wp$ on the one hand, and the estimation error between the expected and empirical error on the other. The approximation error can generally only be reduced by choosing $f_\wp$ from a large, complex class. To bound the second source of error it will be necessary place restrictions on the complexity of the class from which $f_\wp$ is chosen. In the method of structural risk minimization developed by Vapnik one attempts to minimize the statistical risk. To achieve this one defines the complexity of a class of functions by a 'capacity'. Examples for such capacities are the VC dimension of a set of classifiers, or the metric entropy of a class of continuous functions. Within this class of functions one then defines a structure consisting of a nested family of subsets $(F_\gamma)_{\gamma \in \Gamma}$, where Γ is some ordered index set, and $F_\gamma \subset F_{\bar{\gamma}}$ for $\gamma \leq \bar{\gamma}$ and $F_\gamma \subset \Im$ for every $\gamma \in \Gamma$. By this construction one insures that the capacity C_γ of the subset F_γ is less than that of $F_{\bar{\gamma}}$. The objective of structural risk minimization then consists of finding the subset F_{γ^*} so that if $f_\wp \in F_{\gamma^*}$ is chosen to minimize the empirical error, this will also yield the best generalization performance when compared to the alternative functions taken from other subsets using the same criterion.

A complexity penalty term added to the empirical error can be interpreted in the following fashion: Assume that for every $f \in \Im$ a penalty term $P(f) \geq 0$ is given. One then defines the nested family by setting for $t \geq 0$, $F_t = \{f \in \Im \mid P(f) < t\}$. If a weighted penalty term $\lambda P(f)$ is added to the empirical error term and the function $f^\lambda \in \Im$ is chosen to minimize this extended criterion, it follows that for $\lambda \leq \bar{\lambda}$, the corresponding functions $f^\lambda, f^{\bar{\lambda}}$, have the property $P(f^\lambda) \geq P(f^{\bar{\lambda}})$. Therefore, by varying λ across a wide range of values, $\lambda_1 < \lambda_2 < \lambda_3 < ...$ one generates a sequence of functions $f^{\lambda_1}, f^{\lambda_2}, f^{\lambda_3}, ...$ with $P(f^{\lambda_1}) \geq P(f^{\lambda_2}) \geq P(f^{\lambda_3}) \geq ...$ as required by the principles of structural risk minimization.

There are essentially two problems in the implementation of this program. The first is estimating the generalization performance, usually achieved using a validation set of data over the different subsets. The second and more critical problem, is designing the structure so that a decrease in the capacity can be achieved with the least possible increase in approximation error. Most of the complexity penalty terms suggested in the literature depend only on the weights of the network activation function to which they are applied. Two essential factors are not taken into account by this type of penalty term. First is the effect of correlation between hidden units. A network activation function will only be able to approximate complex functions if the various hidden units in the network perform distinct functions. If the outputs of the hidden units in a network activation function are highly correlated this won't be possible. A penalty term that only depends on weight size may indicate a lower level of complexity for a network with smaller or fewer weights than a second one in which the hidden units are all producing essentially the same output across the same entire data set. Obviously, the ability of a network of the first type to overfit the data will be much better than a network of the second type.

A second issue that is not addressed by the usual penalty terms is the problem of dependence on the distribution of the input data. To see how this plays a role, one should consider the situation where the data for two or more of the input data are highly correlated. In this case, the genuine complexity of the class of functions should be measured as though it were defined on a lower dimensional input space. Penalty terms that don't take this into account will overestimate the complexity increase induced by weights connected to the correlated inputs, while underestimating that from the uncorrelated inputs.

Interestingly, the maximum mutual information criterion between inputs and network outputs, when used to steer unsupervised learning as suggested by Linsker [2] has the effect of producing a network that will try to utilize the computational resources of the network as an optimal channel to code the data. As such, the individual hidden units of the network will be forced to become feature detectors, taking different roles which have the least possible functional overlap.

Further, the network will be modified so that the computational resources are placed on those portions of the input space where the data is concentrated to produce the most effective representation. In the network architecture that we propose, in which the hidden units are normed so that the outputs always sum to one, if the network is to fit the data, it must find a representation of the input data in the hidden units with the smallest loss of information. During the training process, there is a continuous increase in this mutual information until the process converges (see Figure 1). On the other hand, by introducing the mutual information as complexity penalty term, it performs essentially the reverse function as that usually intended in unsupervised learning. In contrast to more commonly used penalty terms, it will resist the decorrelation and separation of the functionality of the different hidden units, preferring to reduce the information transfer of the network to the fewest features possible. Further, it will ignore any 'spurious complexity' which the network might produce on regions of the input space where little or no data is found. Therefore, to reduce the empirical error during training the network will have to find the few most relevant features in the input data needed to explain the variance of the targets. The effectiveness of this combination of architecture and penalty term in extracting the structure from a small noisy data set is demonstrated in the following empirical results.

4.0 Applications and Results on Economical Series Predictions

We applied our model to predict german interest rates by using high dimensional real world data.

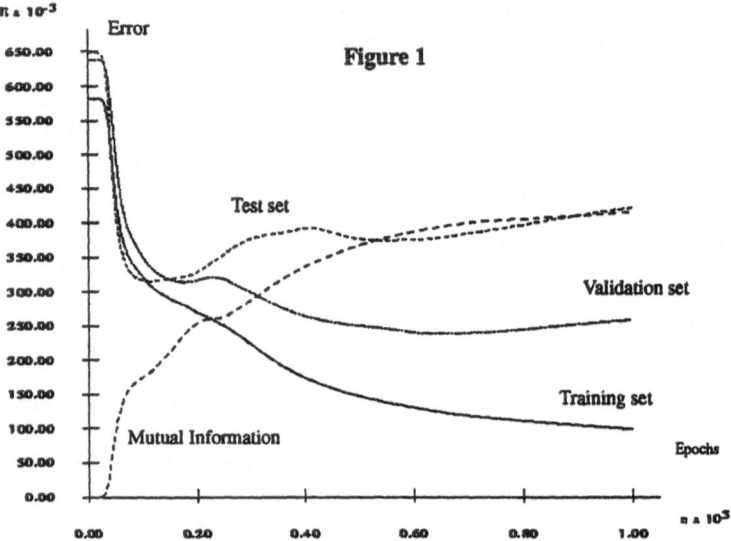

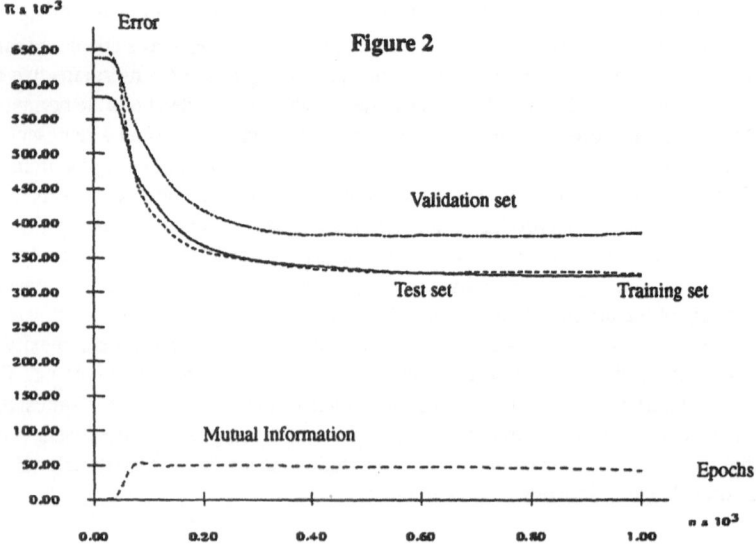

The dimension of the inputs is 14 and the dimension of the output vector is 9. The input represents the monthly development of economic time series (most of them are fundamentals,e.g. the income of private households or the amount of german investments on banks or foreign countries) between 1972 and 1991, whereas the first three outputs gives the qualitative tendency of the interest rate in 3,6,12 months respectively. The other outputs represent a continuous representation of the same information. It can be shown, that it is useful, not to take the time series itself but the difference between two succeeding measurements - this applies when the underlying time series shows only small changes relative to its absolute values. As a training set we used 132 pattern. The validation set contains randomly selected 44 pattern in the same period. The generalization ability was measured on a different test set of 45 pattern in the period from 1986 - 1991. One should have in mind that this generalization test is a complicated task, because the reunification of Germany occurred in this time period. The value of the used constants are: $\eta = 0.02$, $\lambda = 0$ or 1. The number of hidden units were 10. Figure1 shows the results obtained with the network here formulated but without the mutual information term in the cost function, e.g. $\lambda = 0$. The mutual information between the input data and the representation given by the hidden neurons is also presented in this figure. It is important to see that the increase of mutual information in the first 70 epochs correspond to real structure to be learned. The change of pendent and the abrupt increase of mutual information after 70 epochs correspond to the information gained by the learning of noise which is the cause of overtraining reflected in the increase of the error in the validation and test set. Figure 2 shows the result obtained when the mutual information was introduced in the cost function. The overtraining was eliminated in complete accord with the stabilization of the mutual information after the 70 epochs. The first amount of mutual information gained by back-propagation were not eliminated by the penalty term but just stabilized.

The table 1 shows the percent of correct tendency prediction measured on the test set for the 3,6 and12 months.

TABLE 1.

Model	3 Months	6 Months	12 Months
Back-Prop	57%	64%	84%
Mutual Inf.	60%	80%	84%

5.0 References

- [1] Barlow H., 1989, "Unsupervised Learning", Neural Computation 1, 295-311.

- [2] Linsker R., 1989, "How to generate ordered maps by maximizing the mutual information between input and output signals", Neural Computation 1, 402-411.

- [3] Linsker R., 1992, "Local Synaptic Learning Rules Suffice to Maximize Mutual Information in a Linear Network", Neural Computation 4, 691-702.

- [4] Becker S., 1992, "An Information-theoretic Unsupervised Learning Algorithm for Neural Networks", Ph.D. Thesis, Univ. of Toronto.

CASCADE CORRELATION: AN INCREMENTAL TOOL FOR FUNCTION APPROXIMATION

Gian Paolo Drago

Istituto per i Circuiti Elettronici - CNR

via Opera Pia, 11 - 16145 Genova, Italy

Sandro Ridella

DIBE - Universita' di Genova

via Opera Pia, 11A - 16145 Genova, Italy

abstract- In this paper we show the capability of the *Cascade Correlation* (CC) algorithm of finding, through an *incremental* procedure, a sum of weighted hyperbolic tangents, which approximate every function of practical interest to any desired degree of accuracy. The incremental algorithm works only on one-layer perceptron and it is, then, a way of solving the *credit assignment problem*. We show that the integrated squared error has a speed of convergence of order $O(1/n_h)$, where n_h is the number of hidden neurons: numerical results, through computer simulation, agree with the theory. Our analysis shows that CC represents an efficient implementation of the *Projection Pursuit* algorithm.

1 Incremental Constructive Approximations

Neural networks, using a feed-forward *Multilayer Perceptron* (MLP) topology with one layer of sigmoidal non-linearities, are able to approximate any function of practical interest to any desired degree of accuracy [1]. These networks implement function of the form

$$y = v_0 + \sum_{j=1}^{n_h} v_j \tanh(\sum_{i=1}^{n_i} w_{ji} x_i + w_{j0}) \qquad (1)$$

when connections between hidden neurons are not allowed and $tanh(.)$ is used as the activation function. Barron [1] has shown that an *Integrated Squared Error* (ISE) of order $O(1/n_h)$ is achieved for smooth functions, more precisely, those which are bounded on the first moment of the magnitude distribution of the Fourier transform. In contrast, under the same hypothesis, Barron [1] has shown that for series expansions with n_h terms, in which only the parameters of linear combination are adjusted, the ISE can not be made uniformly smaller than order $O[(1/n_h)^{2/n_i}]$ where n_i is the number of inputs. The approximation rate and the parsimony of the parameterization of the neural network are strongly advantageous in high-dimensional settings.

Our research started from trying to understand how much Barron's result may be used in helping the design of neural networks: in doing so we need finding the values of the weights in eq. 1. This problem is more difficult with respect to that of finding the coefficients of the linear combination of known functions. Finding unknown parameters in eq. 1 is a non-linear process: this difficulty may be hardly eliminated since it is originated by the shape of the sigmoid. On the contrary, finding *all together* the weights in eq. 1 with an optimization algorithm may be substituted with an *incremental* procedure, similar to the one used for computing the coefficients of a Fourier series, where they may be found one by one without any need of a global optimization.

Cascade Correlation (CC) [6] is an incremental algorithm for training MLP: it is a neural network whose topology is growing in order to increase its capability of approximating the sample targets and where the weights connected at the inputs of every new hidden neuron are found as soon as it is connected in the network.

Considering convergence problem, CC looks very similar to *Projection Pursuit* (PP) [7] [8], a well known incremental algorithm for data processing in statistical analysis, which approximates a function with a sum of *ridge* functions. The parameters estimation process starts by computing the residuals of the approximations (determined by the first $(j-1)$ terms) and then it minimizes them by finding the new j_{th} direction and a smooth function. The process is iterated until the residual error is less than a predetermined quantity. CC suggests to solve the technical difficulty relative to the choice of the direction and the bandwidth of the smooth function by maximizing the magnitude of the covariance of the error (at the output of the neural network) and the output of the new hidden neuron. Clearly, the successive approximations, through which eq. 1 is obtained, both in PP and in CC, need not be the best possible for n_h summands. PP improves the fit by various versions of backfitting [7], while training in CC is performed in two steps: in the first the weights at the inputs of the new hidden unit are found, while in the second step the weights connecting both the hidden units and the network inputs to the the output units are optimized. When both connections between hidden neurons and between inputs and output neurons are activated, one has a *Fully Feed-forward Perceptron* (FFP) and eq. 1 must be substituited with

$$y = v_0 + \sum_{j=1}^{n_h} v_j z_j + \sum_{i=1}^{n_i} a_i x_i \tag{2}$$

$$z_j = \tanh(\sum_{i=1}^{n_i} w_{ji} x_i + w_{j0} + \sum_{k=1}^{j-1} b_{jk} z_k) \tag{3}$$

where v_0, v_j, a_i, w_{ji}, w_{j0} and b_{jk} are weights.

Inserting this connections may be wise for solving efficiently some problems: we will discuss this point when doing computer simulation.

We were able to show [5] that the CC algorithm is able to converge with performance similar to Barron's iterative sequence of approximations [2]. In more details, being n_p the number of training samples, one can define the correlation coefficient

$$\rho^2 = \frac{R^2}{[\frac{1}{n_p} \sum_{l=1}^{n_p} (y_l - \bar{y})^2][\frac{1}{n_p} \sum_{l=1}^{n_p} (\delta_l - \bar{\delta})^2]} = \frac{R^2}{V_y V_\delta} \tag{4}$$

where V_y and V_δ are the estimated variances of y (the output of the new hidden neuron) and δ (the error at the output neuron) respectively. The squared covariance is:

$$R^2 = \{\frac{1}{n_p} \sum_{l=1}^{n_p} [(y_l - \bar{y})^2 (\delta_l - \bar{\delta})^2]\}^2 \tag{5}$$

Straightforward computation concerning the minimization of the ISE with respect to the weight connecting the new hidden neuron to the output neuron and its bias, gives the ISE before and after the insertion of the new hidden neuron

$$C_{new} = C_{old}(1 - \rho^2) \qquad (6)$$

The main difference between PP and CC is that PP makes a simultaneous optimization of all the parameters relative to the smooth function (the new hidden unit in our case), while CC splits the search into two stages. That means that PP makes ρ^2 maximum, while CC maximizes R^2: that leads to

$$C_{new}^{PP} \le C_{new}^{CC} < C_{old} \qquad (7)$$

With regard to the speed of convergence, under the same Barron's hypotheses [2], one obtains

$$\rho^2 = \frac{C_{old}}{C_{old} + b_f{}^2} \qquad (8)$$

where $b_f{}^2$ depends on the square of the norm of the contribution of the new hidden neuron to the activation of the output neuron and on C_{old} [2].

This guarantees that the convergence of CC is of the order of $1/n_h$, as in [2].

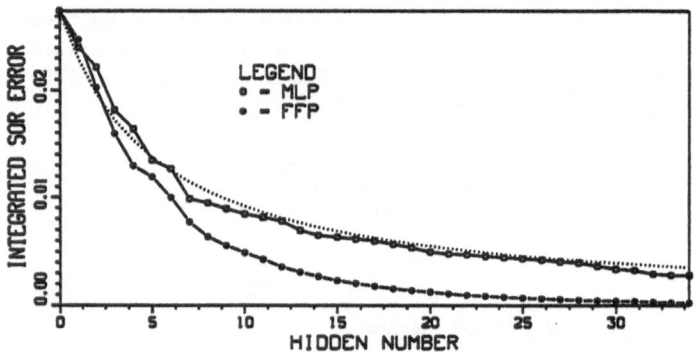

Figure 1: Capabilities of CC in approximating a two dimensional function: the dotted line is the prediction from theory for the MLP case.

2 Computer simulation

In this paper an improved CC code [4] is used to test its capabilities for function approximation. One of the main improvements is in using SCAWI procedure [3]. As benchmark we use the the two-dimensional functions used by Widrow [9], which is a surface described by

$$d(x_1, x_2) = 0.5 \sin(\pi x_1^2) \sin(2\pi x_2) \; ; \quad -1 \le x_1 \le 1, \; -1 \le x_2 \le 1.$$

The first result is that the convergence follows the theoretical behavior without inter-hidden connections, as shown in Fig.1.

With inter-hidden connections, convergence is better both for one- and for two-dimensional functions (fig.1). In this configuration however the number of weights increases quadratically with n_h, instead that linearly. With regard to the computation time, in order to

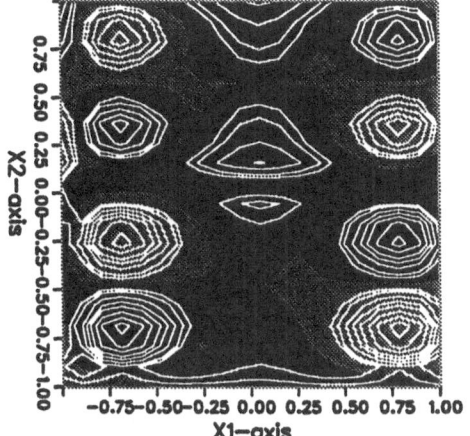

Figure 2: The level contour curves represent the error at the output neuron when two hidden units have been fully trained and connected (with inter-hidden connections), while the dark (light) gray shaded areas represent regions where the new (third) hidden unit forecasts a reduction (increase) of the previous error.

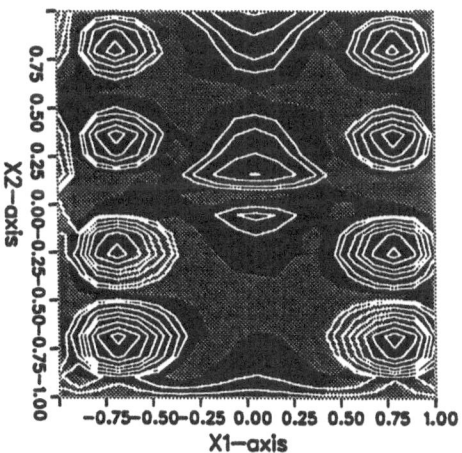

Figure 3: The level contour curves are the same as in fig.2, while the dark (light) gray shaded areas represent regions where C_{new} is less (greater) than C_{old}, when the third hidden unit has been fully trained and connected.

reach the same ISE ≤ 0.0005, one needs 1692 seconds and 1450 seconds with and without inter-hidden connections respectively. In the first case $n_h = 30$, which corresponds to 558 weights, while in the second case $n_h = 85$ and 343 weights.

Fig.2 shows the regions where a new hidden units forecasts a reduction of the previous error at the output neuron, while fig.3 shows the regions where this reduction is really achieved. The purpose of fig.2 and Fig.3 is to show how the maximization of R^2 is able not only to reduce the ISE (as shown in eq.6), but also to forecast, with good approximation, on which samples there will be a reduction or an increase of the error.

Similar results have been obtained for the one-dimensional function used by Jones [8].

As conclusion, Cascade Correlation represents a very efficient incremental tool for function approximation, both with regard the quality of the approximation itself and with regard the speed of convergence.

Acknowledgments: we thank Scott Fahlman for helpful suggestions and for making available reports and software.

References

[1] Barron A.R., "Universal approximation bounds for superpositions of a sigmoidal function", to appear in *IEEE Trans. information Theory*.

[2] Barron A.R., "Universal approximation bounds for superpositions of a sigmoidal function", *technical Rep. No.58*, Univ. Illinois, 4 march 1991.

[3] Drago G.P. and Ridella S., "Statistically Controlled Activation Weight Initialization", IEEE Transactions on Neural Networks, Vol.3, No. 4, July 1992, pp.627-631.

[4] Drago G.P. and Ridella S., "Convergence proprerties of Cascade Correlation in function approximation", to be submitted to *IEEE Trans. on Neural Networks*

[5] Drago G.P. and Ridella S., "A comparison between a SCAWI Backpropagation and an improved Cascade Correlation", submitted to 6^0 Italian Workshop on Neural Nets, Wirn Vietri, 12-14 Maggio 1993.

[6] Fahlman S.E. and Lebiere C., "The Cascade-Correlation Learning Architecture", *CMU-CS-90-100*, February 14, 1990.

[7] Huber P.J., "Projection Pursuit", *Ann. Stat. 1965*, vol.13, No.2, pp.435-475.

[8] Jones L.K., "Constructive approximations for neural networks by sigmoidal functions", *PIEEE*, vol.78, No.10, oct.90, pp.1586-1589.

[9] Nguyen D. and Widrow B., "Improving the learning speed of 2-layer neural networks by choosing initial values of the adaptive weights", *IJCNN*, 1990, pp.III/21-26.

Learning and generalization
—poster contributions

Bounds on the Complexity of Testing and Loading Neurons

Martin Anthony
Mathematics
London School of Economics
Houghton Street
London WC2A 2AE, UK
anthony@vax.lse.ac.uk

John Shawe-Taylor
Computer Science
Royal Holloway
Egham Hill
Egham, Surrey TW20 0EX, UK
john@dcs.rhbnc.ac.uk

Abstract: In this paper we investigate two distinct aspects of the learning theory of the boolean linear threshold neuron and the sigmoid neuron. First, we present results on the number of examples needed for *testing* a given boolean linear threshold function. Then, using this theory, we analyse the time some well-known learning algorithms require to *load* a given training sample onto a simple boolean perceptron and a single sigmoid neuron.

1. Introduction and Definitions

A boolean function t defined on $\{0,1\}^n$ is *linearly separable* if there are $\alpha \in \mathbf{R}^n$ and $\theta \in \mathbf{R}$ such that $t(x) = 1$ if $\langle \alpha, x \rangle \geq \theta$ and $t(x) = 0$ otherwise. Here, $\langle \alpha, x \rangle$ is the standard inner product of α and x. Given such α and θ, we say that t is represented by $[\alpha, \theta]$ and we write $t \leftarrow [\alpha, \theta]$. The vector α is known as the *weight-vector*, and θ is known as the *threshold*. This class of functions is the set of boolean functions computable by the *linear threshold node* or *simple boolean perceptron* (see [9,10,7]). The set of all linearly separable boolean functions defined on $\{0,1\}^n$ will be denoted by BP_n.

A *sigmoid neuron* has output $\sigma(\langle \alpha, x \rangle - \theta)$, where $\sigma(z) = 1/(1+e^{-z})$. Following Maass *et al.* [6], we will say that a sigmoid network $\mathcal{N}$ computes a boolean function $F : \{0,1\}^n \to \{0,1\}$ with *separation* ϵ if the output of $\mathcal{N}$ is greater than $0.5 + \epsilon$ whenever F gives output 1 and is less than $0.5 - \epsilon$ otherwise.

For $t \in BP_n$, a given linearly separable boolean function, we say that a set $T \subseteq \{0,1\}^n$ of inputs is a *test set* for t if when $h \in BP_n$ and h classifies the inputs in T in the same way as t does, then h is necessarily equal to t. In other words, T is a test set for t if the inputs in T serve to specify uniquely the function t. This notion is clearly of use in *testing*: suppose we have a simple perceptron with an unknown weight assignment which is supposed to compute the function t on binary inputs. We may check whether this is the case by providing as inputs a test set for t. If the outputs are as they should be, then we may deduce that the perceptron is in one of the set of possible states which realise t and that every other possible input in $\{0,1\}^n$ will be classified correctly. Another interesting application of the idea of test sets is to *teaching*; we refer the reader to [4,2]. In this paper, we present results on the size of test sets for linearly separable boolean functions and apply these results to analysis of the time required for certain learning algorithms to 'load' a training sample (as defined below).

A *training sample* for $t \in BP_n$ is a sequence $\mathbf{x} = (x_1, x_2, \ldots, x_m)$ of inputs from $\{0,1\}^n$ *together with* the corresponding classifications $t(x_1), t(x_2), \ldots, t(x_m)$. Thus, a training sample is a labelled set of examples. A *learning algorithm* $\mathcal{A}$ takes as input training samples and by some means adjusts the weight-vector correspondingly. We say that the learning algorithm *loads* the training sample if it produces a weight-vector α and threshold θ such that the function h then computed by the perceptron (that is, the function represented by $[\alpha, \theta]$) agrees with the target function t on the inputs $x_1, x_2, \ldots, x_m$ of the training sample. In other words, $\mathcal{A}$ loads the training sample if the resulting function computed by the perceptron is *consistent* with the training sample. It is of key importance that loading take place in a computation time polynomial in n and the size m of the training sample: unless such polynomiality holds, the learning algorithm may be thought of as inefficient and its use may be completely impractical for anything other than small-scale problems. The field of Computational Learning Theory places much emphasis on such computational issues (see [1,3]).

2. The Complexity of Testing

For $t \in BP_n$, denote by $\sigma(t)$ the cardinality of the smallest test set for t; this is the least number of examples required to test that a boolean perceptron computes the function t. We call the parameter $\sigma(t)$ the *testing number* of t. This parameter clearly quantifies in a strong sense the complexity of testing for t. Examples may be found in [2].

Theorem 1 ([2]) *Suppose $t \in BP_n$ and suppose that $k \geq 1$ is such that any weight-vector realising t has at least k non-zero weights and that there is a weight-vector realising t which has exactly k non-zero weights. Then*

$$2^{n-k}(k+1) \leq \sigma(t) \leq 2^{n-k}\binom{k+1}{\lfloor \frac{k+1}{2} \rfloor},$$

and equality is possible in both of these inequalities. □

Despite the fact that the testing number can be exponential, we have shown [2] that the average, or expected, testing number of a function in BP_n is at most n^2.

Fixing attention for the moment on the case $k = n$ above, it has been shown in [2] that there is a large family of linearly separable boolean functions — the *nested* functions — each having testing number $n + 1$. Employing the standard formula notation for decribing boolean functions on $\{0,1\}^n$, in terms of *literals* $u_1, u_2, \ldots, u_n$ (and their negations) and the 'OR' operation $\vee$ and 'AND' operation $\wedge$, examples of nested functions include the functions with formulae $u_1 \wedge u_2 \wedge \ldots \wedge u_n$, $u_1 \vee u_2 \vee \ldots \vee u_n$, and the function $f_n = u_n \wedge (u_{n-1} \vee (u_{n-2} \wedge (u_{n-3} \vee (\ldots u_1) \ldots)$. This latter function has proven to be of interest in the context of the perceptron learning algorithm [5]

Consider the function

$$f_{2n} = u_{2n} \wedge (u_{2n-1} \vee (u_{2n-2} \wedge (u_{2n-3} \vee (\ldots (u_2 \wedge u_1)) \ldots)).$$

As mentioned above, f_{2n} lies in BP_{2n} and has testing number $2n + 1$. For this particular function, we can construct explicitly a test set T_n for f_{2n} of minimal size, as follows.

Theorem 2 *Define $T_n \subseteq \{0,1\}^{2n}$ inductively, as follows: $T_1 = \{(0,1),(1,0),(1,1)\}$, and for $n \geq 1$,*

$$T_{n+1} = \{x01 : x \in S_n\} \cup \{(11\ldots10),(00\ldots011)\}.$$

Then T_n is a test set for f_{2n}.

Proof: We prove by induction that the only $h \in BP_n$ consistent with f_{2n} on T_n is f_{2n} itself. This is easily seen to be true for $n = 1$. Suppose it is true for $n \geq 1$ and suppose that $h \leftarrow [\alpha, \theta]$ is consistent with f_{2n} on T_{n+1}. The examples $x01$, for $x \in T_n$, are in T_{n+1} and so h agrees with f_{2n+2} on all such examples. But, since $f_{2n+2}(y01) = f_{2n}(y)$ and T_n is a test set for f_{2n}, the function g defined by $g(y) = h(y01)$ must equal f_{2n}. Therefore, $f_{2n} \leftarrow [(\alpha_1, \alpha_2, \ldots, \alpha_{2n}), \theta - \alpha_{2n+2}]$, h agrees with f_{2n+2} on all examples of the form $y01$, and $\alpha_1, \ldots, \alpha_{2n} > 0$. By this latter condition, $h((11\ldots10)) = f_{2n+2}((11\ldots10)) = 0$ implies $h(y10) = 0 = f_{2n+2}(y10)$ for all y and similarly, $h((00\ldots01)) = f_{2n+2}((00\ldots011)) = 1$ implies $h(y11) = 1 = f_{2n+2}(y11)$ for all y. Now let y be such that $f_{2n}(y) = 0$. Then $h(y01) = 0$ and $h(y11) = 1$, from which it follows that $\alpha_{2n+1} > 0$. Hence, additionally, for any y, $h(y00) = 0 = f_{2n+2}(y00)$. All cases are now covered, and $h = f_{2n+2}$. The result now follows. □

3. Loading by Means of the Perceptron Learning Algorithm

We now give a fleeting description of the perceptron learning algorithm, and refer to [7,1] for more details. For any *learning constant* $\nu > 0$, we have the *perceptron learning algorithm* L_ν, devised by Rosenblatt [9,10], which acts sequentially as follows. Let t be any function in BP_n, which may be thought of as the *target*. L_ν maintains at each stage a *current hypothesis*, which is updated (in order to converge to t) on the basis of an example in $\{0,1\}^n$, presented together with its classification $t(x)$. (The initial hypothesis is some fixed 'simple' hypothesis. We shall take the initial hypothesis to have the all-0 vector as weight-vector, and threshold 0.) Suppose the current hypothesis is $h \leftarrow [\alpha, \theta]$ and that an example x is presented. Then the new current hypothesis is

$h' \leftarrow [\alpha', \theta']$ where $\alpha' = \alpha + \nu(t(x) - h(x))x$, $\theta' = \theta - \nu(t(x) - h(x))$. The *Perceptron Convergence Theorem* [9,7] asserts that no matter how many examples are presented, the algorithm makes only a finite number of changes, or updates (provided ν, which can be a function of n, is small enough).

As indicated in [3], given $t \in BP_n$ and a sample $\mathbf{x} = (x_1, x_2, \ldots, x_m)$ of examples, we may use L_ν to load a training sample for t. This is achieved by cycling through x_1 to x_m in turn until no updates are made in a complete cycle. It is clear that the resulting weight vector is such that the linearly separable boolean function realised is consistent with the training sample. A natural question is whether this is an efficient means of finding a consistent function. In fact, it is not, in the sense that the number of complete cycles required can be exponential in m, the size of the sample. Here, we shall use the notion of test set to prove that when one uses the perceptron algorithm in the batch sense described above to load a given training sample, then the running time can be exponential *in the size of the sample*. (We remark that this result is very different from previous results of Minsky and Papert [7] and Hampson and Volper [5], in which the performance of the perceptron learning algorithm is considered: their results concern the difficulty of converging *exactly* to the target function (as opposed to merely loading a training sample), and they show that this can take exponential time with respect to the domain dimension n (as opposed to the sample-size, m). Note that there is a polynomial time method for loading a training sample on a simple boolean perceptron: linear programming. Thus the problem of finding a consistent hypothesis has no *intrinsic* computational difficulty.

The following result is easily proved [2]: Muroga [8] has proved similar results.

Proposition 3 *Let n be any positive integer. Suppose $f_{2n} \leftarrow [\alpha, \theta]$. Then for each $1 \leq k \leq n$, $\alpha_{2k} \geq \sqrt{3}^{k-1} \min(\alpha_1, \alpha_2)$. In particular, $\alpha_{2n} \geq \sqrt{3}^{n-1} \min(\alpha_1, \alpha_2)$.*

Combining this result with Theorem 2, we obtain the following result.

Theorem 4 *For any fixed $\nu > 0$, the loading algorithm using the perceptron learning algorithm L_ν is not a polynomial time algorithm. That is, it does not always run in time polynomial in the size of its input training sample. This result also holds if $\nu = \nu(n)$ is a function of n, bounded above by some constant.*

4. Loading Sigmoid Neurons

We now ask when learning algorithms for a single sigmoid neuron will load in polynomial time, achieving a reasonable separation. For simplicity, we will incorporate the threshold as an additional weight ($\alpha_0 = -\theta$) on an input line which is always on.

In the same way as the perceptron learning algorithm may be used cyclically to load a training sample of a linearly separable boolean function, we may use the standard back-propogation algorithm to load a sigmoid neuron. Our first observation is an indirect consequence of Proposition 3 and Theorem 4. By standard back-propagation we consider on-line application of the back-propagation algorithm (i.e. not batch mode). Loading will be considered successful if the sigmoid network computes the function with separation $1/P(n)$ for some fixed polynomial $P(n)$ in the input dimension n.

Proposition 5 *For any fixed update constant $\nu > 0$, the loading method for the sigmoid neuron based on the standard back-propagation algorithm does not always run in time polynomial in the size of its input.*

Proof: Assume there is a polynomial $Q(n, m)$ such that the loading time is bounded by $Q(n, m)$ for samples from $\{0, 1\}^n$ of size m. Consider the loading problem for f_{2n} used in the previous section and suppose we find a weight assignment which gives the positive examples an output greater than 0.5 and the negative examples an output less then 0.5. In this case we can use the weights for a perceptron which correctly classifies the inputs, by simply taking the threshold $\theta = -\alpha_0$. This follows from the fact that $\sigma(z) > 0.5 \iff z > 0$. Hence by Proposition 3 the weights must satisfy the inequalities $\alpha_{2k} \geq \sqrt{3}^{k-1} \min(\alpha_1, \alpha_2)$, $1 \leq k \leq n$. Each iteration of the algorithm causes a change of $\delta\alpha = \nu(t^p - o^p)(1 - o^p)o^p x^p$ in the weight vector (including the threshold weight α_0), where t^p is the required output on the p-th input x^p, and o^p is the actual output. We wish to round the value $\nu(t^p - o^p)(1 - o^p)o^p$ to a multiple of $1/nP(n)Q(n,m)$. This will cause a change $\delta\alpha'$, whose entries satisfy $|\delta\alpha_i'| < 1/nP(n)Q(n,m)$. They cause a change in the final output of an input x of at most $1/P(n)$ (in fact a quarter of this amount since the derivative of $\sigma(z)$ is bounded by 0.25) and hence will still leave the weight vector satisfying the condition described above, if loading was successful with separation $1/P(n)$. We may therefore assume that the non-zero components of individual updates satisfy $\delta\alpha_i = u/nP(n)Q(n,m)$, for some integer u bounded by a polynomial in n and m. Again rounding the initial hypothesis to have entries which are multiples of $1/nP(n)Q(n,m)$ and assuming that there is not an exponential ratio between initial individual weights, the $Q(n,m)$ iterations will not be enough to generate the required exponentially large ratio between α_{2n} and $\min(\alpha_1, \alpha_2)$. □

We now consider an alternative algorithm for training multi-layer sigmoid networks [11] which relies on solving a linear programme for each iteration of the algorithm. The algorithm places additional constraints on the weight updates by requiring that no individual errors should be allowed to increase locally. This places linear contraints (involving the local derivatives) on the weight updates. Subject to these the overall error is reduced maximally. The solution of the linear programme delivers a direction in weight space. In the version of the algorithm considered here the weight change made is the maximum in the given direction provided the overall error is reducing and individual errors remain bounded by 0.9 and those which are reducing are still greater than $c2^{-n}/n$, for some fixed constant $c < 0.001$. In order to be able to guarantee errors are bounded by 0.9, initial weights should be chosen with $|\alpha_j| < 1/n$ and iterations should always keep errors bounded away from 0.9. We have the following result for the linear programming algorithm, the proof of which is omitted in this abstract.

Proposition 6 *The loading method for the sigmoid neuron based on the linear programming algorithm for multi-layer sigmoid networks runs in polynomial time. In particular the required number of cycles through the training sample is at most m, the size of the training sample.*

References

[1] M. Anthony and N. Biggs, *Computational Learning Theory: An Introduction*, Cambridge University Press: Cambridge, UK, 1992.

[2] M. Anthony, G. Brightwell and J. Shawe-Taylor, *On specifying boolean functions by labelled examples*, LSE Mathematics Preprint Series, LSE-MPS-32, September 1992.

[3] A. Blumer, A. Ehrenfeucht, D. Haussler and M. Warmuth, Learnability and the Vapnik-Chervonenkis Dimension. *Journal of the ACM*, 36(4), 1989: 929–965.

[4] S.A. Goldman and M.J. Kearns, On the complexity of teaching, In *COLT'91, Proceedings of the Fourth Annual Workshop on Computational Learning Theory*, 1991: Morgan Kaufmann, San Mateo, CA.

[5] S.E. Hampson and D.J. Volper, Linear function neurons: structure and training. *Biological Cybernetics* 53, 1986: 203–217.

[6] W. Maass, G. Schnitger and E.D. Sontag, On the Computational Power of Sigmoid versus Boolean Threshold Circuits, Proceedings FOCS, 91.

[7] M. Minsky and S. Papert, *Perceptrons*. MIT Press, Cambridge, MA., 1969. (Expanded edition 1988.)

[8] S. Muroga, Lower bounds of the number of threshold functions and a maximum weight. *IEEE Transactions on Electronic Computers*, 14, 1965: 136–148.

[9] F. Rosenblatt, Two theorems of statistical separability in the perceptron. In *Mechanisation of Thought Processes: Proceedings of a Symposium Held at the National Physical Laboratory, November 1958. Vol. 1.* HM Stationery Office, London, 1959.

[10] F. Rosenblatt, *Principles of Neurodynamics*. Spartan, New York, 1962.

[11] J. Shawe-Taylor and D. Cohen, The Linear Programming Algorithm for Neural Networks, Neural Networks, 3 (1990) 575–582.

Principal Hidden Unit Analysis with Minimum Entropy Method

Ryotaro Kamimura
Information Science Laboratory
Tokai University
1117 Kitakaname Hiratsuka Kanagawa 259-12, Japan

Abstract- In the present paper, a principal hidden unit analysis with entropy minimization is proposed. By this principal hidden unit analysis, an oversized network can be reduced to a simple principal network with several principal hidden units. To obtain principal hidden units, an entropy function of the hidden unit activity must be minimized. In a state of low entropy, only a small number of hidden units are activated and all the other units are turned off for all the input patterns. In this case, the variance of connections into activated hidden units becomes larger than that of other connections. Thus, major hidden units with large variances, called *principal hidden units* can be selected and used to construct principal networks. These principal networks can produce targets as correctly as original oversized networks and enable us to infer the mechanism of original networks. Applied to a symmetry problem and an autoencoder, it was confirmed that by using entropy method, we could obtain principal hidden units with large variances. With these principal hidden units, a principal network was constructed, producing targets almost perfectly.

1 Introduction

In the present paper, we propose a principal hidden unit analysis with entropy minimization. This analysis can be used to determine a principal network with several hidden units, called *principal hidden units*, by which the majority of the network mechanism can be explained. To generate a principal network from an oversized network, we must select some hidden units, which greatly contribute to the production of outputs. Then, by using these principal hidden units, we can construct a principal network. If this principal network can produce target outputs almost perfectly, an original network can now be reduced to a principal network. In this case, the interpretation of the internal representation of the principal network is much easier because the number of principal hidden units is expected to be much smaller than that of an original network.

To produce principal networks, we have employed an entropy minimization method[1]. By minimizing an entropy defined with respect to the activity of hidden units, we can obtain a small number of principal hidden units. Because in a state of minimum entropy, only one hidden unit is turned on, while all the other units are turned off. On the other hand, in a state of maximum entropy, all the hidden units are equally activated. In addition, it has been observed that hidden units activated by entropy minimization, usually have input-hidden connections with large variance. Thus, principal hidden units can be considered to be units with input-hidden connections with larger variance.

2 Theory and Computational Methods

Let us formulate a method of entropy minimization, applied to the standard back-propagation. Suppose that a network is composed of three layers: input, hidden and output layer. A hidden unit (v_i) can produce an activity

$$v_i = f(u_i),$$

where

$$u_i = \sum_{j=1}^{K} w_{ij} \xi_j.$$

where w_{ij} is an input-hidden connections, ξ_i is a ith element of an input pattern and K is the number of elements in the pattern. An entropy function at hidden layer is defined by

$$H = -\sum_{i=1}^{M} p_i \log p_i,$$

where

$$p_i = \frac{v_i}{\sum_{r=1}^{M} v_r},$$

and M is the number of hidden units.

An update rule is formulated as follows. First, for hidden-output connections, only delta rule must be used. For input-hidden connections, total function (F) to be minimized is defined by

$$F = \alpha H + \beta E,$$

where E is a quadratic error function and α and β are parameters. To minimize this function, in addition to delta rule, phi rule must be incorporated as

$$\begin{aligned}
\Delta w_{ij} &= -\frac{\partial F}{\partial w_{ij}} \\
&= -\alpha \frac{\partial H}{\partial w_{ij}} - \beta \frac{\partial E}{\partial w_{ij}} \\
&= \alpha \phi_i \xi_j + \beta \delta_i \xi_j,
\end{aligned}$$

where δ is the ordinary delta and ϕ is obtained by differentiating the entropy function and formulated as:

$$\phi_i = (\log p_i + 1) \frac{\sum_r v_r - v_i}{(\sum_r v_r)^2} f'(u_i).$$

By using this update rule, entropy is decreased and simultaneously the error is decreased.

3   Results and Discussion

3.1   Application to Symmetry Problem

We applied our method to the so-called *symmetry problem*. Because we have already known its suitable network size, that is, a network with two hidden units [2]. In addition, it has been well known that typical symmetric connections appear in input-hidden connections.

First, by using entropy method, we decreased entropy as much as possible. A left figure of Figure 1 shows entropy as a function of the parameter α. Entropy decreases gradually as the parameter increases. When the parameter is 0.0005, entropy reaches the lowest point of 0.068. In this state with a low entropy, we computed the variance of connections into ith hidden unit by the following equation:

$$s_i^2 = \frac{1}{M-1} \sum_{j=1}^{M} (w_{ij} - \overline{w_i})^2, \tag{1}$$

where $\overline{w_i}$ is an average over all connections into ith hidden units. The variance of a right figure in Figure 1 was obtained by dividing the variance s_i^2 by the total variance $\sum_j s_j^2$. As can be seen in the figure, we can immediately detect two principal hidden units. Connections into the first and sixth hidden units have larger variance, compared with other connections. Thus, a principal network can be constructed with these two hidden units. Figure 2 shows a principal network generated by minimum entropy method. As can clearly be seen in the figure, connections symmetric about the

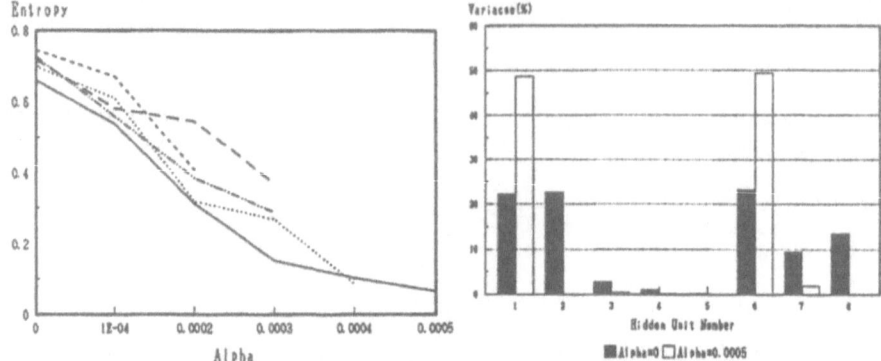

Figure 1: Entropy as a function of the parameter α(left) and the variance of connections into six hidden units (right) when α was zero(black) and 0.0005(white). The values of entropy were averaged over all input patterns and normalized, ranging between zero (minimum) and one(maximum).

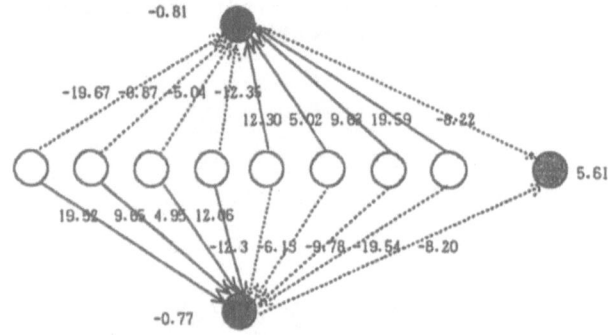

Figure 2: A principal network with two hidden units obtained from an original network with eight hidden units. Numbers in the figure shows values of biases and connections. Solid and dotted lines show positive and negative connections respectively. The parameter α was 0.0005.

middle are equal in magnitude and opposite in sign, as described in [2]. As the reconstruction error is zero, that is, this principal network can produce outputs perfectly, only connections into and from principal hidden units can be taken into consideration for the interpretation of the internal representation.

3.2 Application to Autoencoder

We applied the method to a larger network in which 35 input, hidden and output units were employed. The network must exactly reproduce 26 alphabet letters at output units.

Figure 3 shows the variance of all the hidden units. As can be seen in the figure, several principal hidden units can immediately be pointed out. For example, 2nd, 10th, 16th, 27th, 29th and 31st hidden units can perfectly be considered to be principal hidden units, because their variances are considerably higher than the variances of other hidden units. A principal network can be composed of only six principal hidden units.

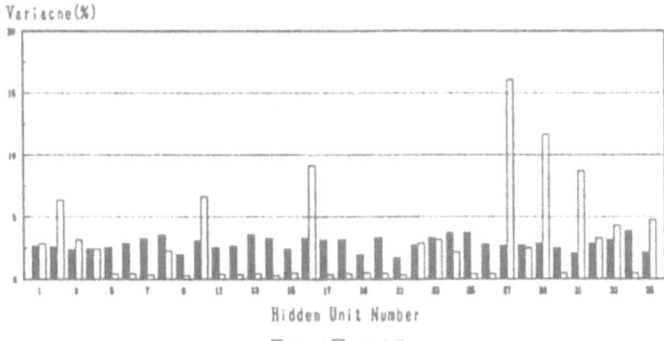

Figure 3: Variance of connections from input units to 35 hidden units, when α was zero(black) and 0.07(white).

4 Conclusion

In this paper, we have proposed a principal hidden unit analysis with entropy minimization, used to produce a simplified principal network from an original oversized network. One of the difficult problems in neural networks is concerned with the interpretation of hidden unit or the internal representation. As the information is distributed over many hidden units, the interpretation of the internal representation is extremely difficult. A principal network is composed of a smaller number of hidden units. The interpretation of the internal representation can be much easier than an original oversized network. Then, our method enables a network itself to determine the minimum number of hidden units for the learning, that is, the automatic determination of a suitable size of a network can be possible. Finally, we think that the principal hidden unit analysis with entropy minimization can be extended to the unsupervised learning and can be used to extract some features from redundant input patterns.

References

[1] R. Kamimura, "Minimum entropy method in neural networks," to appear in *Proceeding of 1993 IEEE International Conference on Neural Networks*.

[2] D. E. Rumelhart, G. E. Hinton and R. J. Williams, "Learning internal representation by error propagation," in *Parallel Distributed Processing*, D. E. Rumelhart, J. L. McClelland, and the PDP Research Group, Cambridge, Massachusetts: the MIT Press, Vol.1, pp.318-362, 1986.

Empirical Criteria to Compare
The Performance of Neuro Algorithms

Stéphane Canu*

Université de Technologie de Compiègne – U.R.A. 817 du CNRS
BP 649 - F-60206 Compiègne cedex - France - E-mail : scanu@hds.univ-compiegne.fr
Lyonnaise des Eaux - Dumez / LIAC. Rue du Fonds Pernant. 60471 Compiègne cedex- France

Abstract

When dealing with benchmarking a major task is to determine the criteria used for comparing algorithms. Instead of taking theoretical comparison criterion some empirical ones have been proposed. These criteria are the generalization of previous ones such as *efficiency* or *information*, only defined for classification applications. These criteria were built to empirically measure both the computational and sample complexity of an algorithm. Other criteria try to characterize the influence of the problem type on the results such as the relative information score. Some preliminary experiments have been done but the efficiency of such criteria can only be given experimentally.

Introduction

Neural networks inspire many different training algorithms and it is hard to be able to compare all of them in order to choose one for a given application. There are two ways to tackle such a problem : the theoretical study of the algorithm or an empirical one. From the theoretical point of view two kinds of complexities permit to compare learning algorithms : computing complexity and sample complexity. It was shown that the loading problem, that is the task of memorization of some given data by feedforward neural network with a fixed architecture, is $\mathcal{NP}$-complete. But in practice back propagation algorithms are used and provide relevant solutions. In order to compare different loading algorithms an empirical computing time measure has to be settled. That is to say that many trials have to be performed and an empirical mean of some relevant criterion has to be taken. The problem lies in the definition of such a criterion. Regarding the sample complexity of the loading problem, it is known to be polynomial. But the bounds provided by the theory are not useful in practice because they are too large. Here also an empirical characterization of the complexity may be useful. The main goal of this paper is to define simple an useful statistical criteria for comparing different learning algorithms. Most "neural" algorithms are stochastic and depend on initial and computational parameters. In order to be able to compare different algorithms, many trials are necessary, varying the conditions of the algorithm. Reviewing these different fields we are going to give some empirical criteria for comparing different algorithms. Note that once an empirical measure is settled this measure will depend on the problem used for benchmarking. In order to avoid such a drawback some measures have to be proposed to characterize the different kinds of problems.

Benchmarking computational complexity

This problem was studied in [1]. First the author proposes to take into account the empirical mean $\hat{\theta}_a$ of the computing time t_i on N trials. But since some trials may not converge (reach some success criteerion), this measure may be infinite. In order to avoid such a drawback, [1] has proposed to take into account the number of successful trials S (e.g. those that have converged at time T). In this case, the computational time becomes a function depending on stopping time T. The complementary measure of this mean computational time is the success rate of the N trials.

*This work has been supported by the EEC funded Esprit II project nr. 5433 (NEUFODI). Neufodi's partners are : Austrian Research Institute for Artificial Intelligence. Babbage Institute for Knowledge and Information Technology, Elorduy, Sancho y CIA. S.A., LABEIN Technological Research Center, Lyonnaise des Eaux - Dumez

Under the assumption that the capacity of the multi layered perceptron is sufficient, the limit of $\hat{\lambda}(T)$ as $T \to \infty$ is one. Because of spurious local minima, this is no longer verified when taking into account the generalization performance for noisy data. Regarding this measure, [1] states that there are three important questions to be answered by empirical criteria : What is the optimal value for the stopping time T ? What happens as $T \to \infty$? Is it possible to build a measure independent of time T ? To answer these questions [1] introduced the efficiency $e(t)$ as a function of the stopping time.

$$e(t) = K \frac{S(t)}{\sum_{i=1}^{N} min(t_i, t)} \tag{1}$$

Where K is a scaling constant taken equal to 1000 in [1] and $S(t)$ the number of successful trials at time t. The efficiency permit to define an empirical measure independent from computing time : the optimal epoch limit T_e defined as the maximum of efficiency according to the stopping time. This measure is easy to interpret and gives an idea of a kind of "optimal" training stopping time : $T_e = \arg\max_t e(t)$ In conclusion, in the case where a success training criterion can be defined, the measure of the optimal epoch limit T_e is very interesting because it is independent from the stopping time and seems to be unbiased. But some questions may arise when taking into account the assumptions made by [1] : What is the influence of the capacity of the neural networks on the performance ? What happens when taking into acount the generalization performances instead of the training performance ? What is the influence of noise on the reliability of the measure ? In order to answer such questions a generalization of the criterion has to be built.

Unfortunately in presence of noise the success of a trial is hard to characterize. In the case of approximation application (e.g. forecasting) the definition of such a criterion is even harder. Sometime a convergence criterion can be given (for instance the stabilization of the synaptic weights). But in other cases cross validation data sets are used to stop the training phase. In this case the comparison criteria given in the previous section* have to be generalized. In order to do so, the efficiency can be defined as follow :

$$e_a(t) = \frac{\sum_{i=1;t_i<t}^{N} 1 - \frac{J_{emp}^i}{J_K}}{\sum_{i=1}^{N} min(t_i, t)} \tag{2}$$

where J_K is a scaling constant.

Problem and Algorithm characterization

The criterion described in [2] is defined for classification algorithms. The goal of this criterion is to permit to compare different algorithms on different problems taking into account a measure of the difficulty of the problem modeled by its entropy E_c. For a given classifier let $\mathbb{P}' \{C_i | x_j\}$ denote the a posteriori probability of having class C_i knowing the input x_j. Then the information provided by the algorithm for a given pattern x_j may be given by

$$\begin{cases} I(x_j) = -log_2\mathbb{P}\{C_i\} + log_2\mathbb{P}'\{C_i|x_j\} & \text{if } x_j \text{ is well classified} \\ I(x_j) = log_2(1 - \mathbb{P}\{C_i\}) - log_2(1 - \mathbb{P}'\{C_i|x_j\}) & \text{if } x_j \text{ is misclassified} \end{cases} \tag{3}$$

According to this information, the average information score I_C may be defined. The criterion proposed in [2] is the relative information score I_{rc} defined as follows :

$$I_{rc} = 100 * \frac{I_C}{E_c} \tag{4}$$

This criterion permit to compare the performances on different problems whose difficulty is represented by the entropy. The relative information score is a percentage built to be problem independent.

In the case of continuous output, the probability density of the data is unknown. One way to tackle this difficulty is to use the entropy definition based on the Fourier transform. Another way to define the entropy of a continuous function consists in transposing the continuous problem into a discrete one considering the histogram of the output data. In this case the entropy E_a can be defined by as :

$$E_c = \sum_{i=0}^{N_h} \frac{n_i}{\ell} log_2 \frac{n_i}{\ell} \tag{5}$$

where N_h denotes the number of classes of the histogram and n_i denotes the numer of data output values included in a single part of the histogram. Note that the classes containing few items (e. g. less than five) can be merged to the closest one. In this case the probability of belonging to this class has to be weighted by a class size factor. The information provided by a given example may be defined by :

$$\begin{cases} I_a(x_j) = -log_2 \frac{n_+}{\ell} + log_2 \frac{n_{i\ell}}{\ell_T} & \text{if } y_j\prime \text{ belongs to the same partition than } y_j \\ I_a(x_j) = log_2(1 - \frac{n_+}{\ell}) - log_2(1 - \frac{n_i\ell}{\ell_T}) & \text{if } y_j\prime \text{ does not belong to it} \end{cases} \tag{6}$$

where ℓ_T denotes the size of the test set, $y_j\prime$ the output computed by the algorithm when x_j in given as input and n_i the number of output elements classified in the same class than the target value y_i. According to this information, the average information score and the relative information score may be defined as in the classification application.

The $VCdim$ gives information about the capacity of a given class of models [3]. An empirical way of estimating this quantity is given in [4]. It consists first in splitting the training data into two subsets S_1 and S_2 and compute the maximum Δ_J of the empirical error over all the possible functions found in the given class of models. Then according to the theorical prediction for such maximal discrepancy the value of the $VCdim$ is given by solving the following equation :

$$\frac{\Phi_\epsilon(\Delta_J)}{N} = \frac{9}{2} \frac{(2\ell)^{VCdim}}{VCdim!} exp^{-\epsilon^2 \ell/4} \tag{7}$$

where $\Phi_\epsilon(\Delta_J)$ denotes the number of trial where $\Delta_J \leq \epsilon$ and N the total number of trials choosing randomly the partition of the examples.

Generalization ability : sample complexity

The main drawback of the computation complexity as defined above is that it does not take into account the generalization ability of the model according to both the size of the network and the size of the training set. One way to deal with such a problem is to replace the training error J by an estimation of the generalization error given by some cross validation set. But it is not enough and the empirical estimation of the relationship between these three quantities has to be done for at least two reasons. One is the search of some minimal training set size ℓ^* such that for any training set larger than ℓ^* a bound can be given on the expected error. The other one is the assymptotic study of the algorithm. In other words how many examples do we need to reach a certain level of performance ?

Some theorems give explicit relationships between the expected cost and the empirical cost but unfortunately they give unusable bounds for practical applications. Nevertheless, [3] proposes the following relationship between these two quantities both depending on the parameters w of the parametrized model :

$$J(w) \leq J_{emp}(w) + \mathcal{G}(VCdim, \sigma, \ell, \alpha)$$

where σ denotes the variance in the response variable , ℓ the size of the training set and α some confidence level. The identification of the function $\mathcal{G}$ allows to minimize the expected cost function. For a small J_{emp} The regret function is given by [3] :

$$\mathcal{G}_1 = c_1 \frac{VCdim(\log(2\ell/VCdim + 1)) - \alpha}{\ell} \tag{8}$$

where c_1 is a constant to be empiricaly determined. The regret function $\mathcal{G}$ is roughly proportional to the network capacity $VCdim$. In this case, the regret function is independant of the training process and has to be considered after having identified the synaptic weights. This is used to determine an "optimal" network architecture structurally avoiding over-training. Other results about the regret function were found such as Moody's one [5] based on the generalization of Akaike's AIC. For Moody the regret function has the following form :

$$\mathcal{G}_2 = 2\sigma^2 \frac{\|W\|}{\ell} \tag{9}$$

where $\|W\|$ denotes the effective number of parameters. The advantage of this second regret function $\mathcal{G}_2$ is that it is easy to identify making some experiment varying independently the noise of the data (for σ), the size of the neural network ($\|W\|$) and the number of examples taken into account for the learning phase. Its major drawback lays in its linearity with respect to $\sigma, \|W\|$ and $\frac{1}{\ell}$. The reality is known to be sometimes non linear.

The learning curve is defined to be the representation of the evolution of the expected performance of an algorithm when the size of the training set increases. This curve is an average over different training sets and initial conditions randomly chosen [6]. It may be empirically built using Monte Carlo experiments. Such an experiment can be found in [7]. But note that here also the generalization perfomance is estimated based on a given test set supposed to be independant of the training data. In other words the performance measure based on the test set is assumed to be representative of the expected cost.

Conclusion

Some preliminary experiments have been done but the efficiency of such criteria can only be given on "real world" applications. For implementing such criteria, each experiment has to be repeated at least 10 times starting from random initial conditions. the averages of the following criteria have to be reported : the minimum quadratic error on the test set, the average quadratic error on the test set and the standard deviation, the empirical efficiency, the information score, the Relative information score and the empirical $VCdim$. The learning curves may be given. They will provide an interesting reference for future applications. Note that this methodology can also be used for comparing any stochastic algorithm such as for instance the Kalman filter, simulated anealing of learning algorithms from machine learning.[1]

References

[1] L. G. C. Hamey, "Benchmarking fed-forward neural networks : models and measure," in *Advances in Neural Information Processing Systems 4* (D. Touretzky, ed.), pp. 1167–1174, San Mateo, CA: Morgan Kaufmann, 1992.

[2] I. Kononenko and Bratko, "Information-based evaluation criterion for classifier's performance," *Machine Learning*, vol. 6, pp. 67–80, 1991.

[3] V. Vapnik, *Estimation of Dependences Based on Empirical Data*. No. 82 in Statistic Series, New York: Springer–Verlag, 1982.

[4] I. Guyon, V. Vapnik, B. Boser, L. Bottou, and S. A. Solla, "Structural risk minimization for character recognition," in *Advances in Neural Information Processing Systems 4* (D. Touretzky, ed.), pp. 471–479, San Mateo, CA: Morgan Kaufmann, 1992.

[5] J. E. Moody, "The effective number of parameters : an analysis of generalization and regularization in nonlinear learning systems," in *Advances in Neural Information Processing Systems* (D. Touretzky, ed.), vol. 4, (San Mateo, CA), pp. 847–854, Morgan Kaufmann, 1992.

[6] Statlog, "Deliverable d2.5 : Criteria for the sequential evaluation of the test results," Tech. Rep. statlog/delivers/wp2/d2.5, Statlog, Esprit Project, 1992.

[7] E. B. Baum, "When are k-nearest and back propagation accurate for feasible sized sets of examples ?," in *Neural Networks: Proceeding of EURASIP workshop*, pp. 2–25, Springer-Verlag, 1990.

[1] Acknowledgements : The author would like to thank Thierry Denœux, Yves Grandvalet and Stéphane Boucheron for their very helpful comments on the draft of this article, their numerous suggestions for improvements and for stimulating discussion on these topics. Thanks also to the people involved in the Statlog Esprit project for useful results they provide me.

LS-BACKPROPAGATION ALGORITHM FOR TRAINING MULTILAYER PERCEPTRONS

E.D. Di Claudio, R. Parisi and G. Orlandi
INFOCOM Dept. - University of Roma "La Sapienza"
Via Eudossiana 18, 00184 Roma - Italy
Phone: +39-6-44585837, Fax: + 39-6-4873300

A new training algorithm is presented as a faster alternative to the backpropagation method. The new approach is based on the solution of a linear system at each step of the learning phase. The squared error at the output of each layer before the non linearity is minimised on the entire set of the learning patterns by a block LS algorithm. The optimal weights for each layer are then computed by using the SVD technique.
The simulation results have shown considerable improvements from the point of view of both the accuracy and the speed of convergence.

Introduction

The multilayer perceptron is one of the most commonly used types of feed-forward neural networks and it is utilised in a large number of applications of pattern recognition and classification. Its strength resides in its capacity of mapping arbitrarily complex non-linear functions by a convenient number of layers of sigmoidal non-linearities [1]. The determination of the optimal weights of the structure is made by repeated presentations of the patterns of the learning set. At each step the output error is computed as the difference between the desired and the actual outputs. The error minimisation is made during the learning procedure according to a pre-specified criterion.

The backpropagation algorithm [2] is still the most used learning algorithm; it consists in the minimisation of the Mean-Squared Error (MSE) at the network output performed by means of a gradient descent on the error surface in the space of weights.

Anyway the backpropagation algorithm suffers from a number of shortcomings; above all the relatively slow rate of convergence and the final misadjustment that can not guarantee the success of the training procedure in real applications. The choice of a different learning algorithm, based on a different minimisation criterion, can help to overcome these drawbacks. These considerations have led to introduce the LS-backpropagation training algorithm as a way to get considerable improvements from the point of view of both the accuracy and the speed of convergence.

The algorithm

In the backpropagation algorithm the output error is defined as:

$$E = \sum_p E_p \qquad (1)$$

where E_p is the output squared error for the p-th pattern.
The weights are updated by computing the derivatives of E according to the formula:

$$\Delta w_{ij}^{(k)} = - \eta \frac{\partial E}{\partial w_{ij}^{(k)}} \qquad (2)$$

where $w_{ij}^{(k)}$ is the weight from the i-th neuron in layer (k-1) to the j-th neuron in layer (k) and η is the learning rate.
The learning rule thus derived is :

$$w_{ij}^{(k)}(t+1) = w_{ij}^{(k)}(t) + \eta \, e_{pj}^{(k)} \, x_{pi}^{(k-1)} \qquad (3)$$

where $e_{pj}^{(k)}$ is the error signal for the j-th unit in layer (k) and $x_{pi}^{(k-1)}$ is the output of the i-th unit in layer (k-1), relatively to the p-th input pattern. The error signal is computed as

$$e_{pj}^{(L)} = f'(y_{pj}^{(L)}) \, (t_{pj} - x_{pj}^{(L+1)}) \qquad (4)$$

for the output layer, and as

$$e_{pi}^{(k)} = f'(y_{pi}^{(k)}) \sum_j e_{pj}^{(k+1)} \, w_{ij}^{(k+1)} \qquad (5)$$

for all the other layers. In these formulas t_{pj} is the j-th desired target output for the p-th pattern, $y_{pi}^{(k)}$ is the input to the generic non linearity, being $f()$ the non-linear activation function, typically the sigmoidal one.

The backpropagation algorithm is based on the use of a continuous, derivable function as the activation function. Its presence makes it difficult to apply to multilayer perceptrons the linear techniques successfully used in a large number of problems in the field of adaptive filtering. Anyway, since each layer can be separated in a linear part (the multiplication by the weights) and a non linear one (the sigmoidal functions), it is possible to use the LS method to update the weights of the network. Specifically the LS algorithm is applied to the linear part by introducing the error components backpropagated from the outputs.

In the algorithm description the following matrices are utilised:

$$X = \begin{pmatrix} x_1^T \\ .. \\ x_P^T \end{pmatrix} ; \; Y = \begin{pmatrix} y_1^T \\ .. \\ y_P^T \end{pmatrix} ; \; E = \begin{pmatrix} e_1^T \\ .. \\ e_P^T \end{pmatrix} \qquad (6)$$

where the layer index has been omitted. In these expressions x_i^T, y_i^T and e_i^T are the input, output and error vectors relatively to the linear part of the generic layer, for the i-th learning pattern.

The new algorithm works by epochs and consists of the following steps:
1) the weights are randomly initialised ;
2) each pattern of the learning set is presented in input to the network and forward propagated through it; during this phase the matrices X and Y for each layer are constructed;
3) for each pattern, the output of the network is compared to the desired output; the signal error thus obtained is used to compute the signal errors at the output of the linear part of each layer; in this phase it is constructed the perturbation matrix E for each layer;
4) after the presentation of all patterns, for each layer the following linear system is solved in the LS sense:

$$X \, W = Y + \eta \, E \qquad (7)$$

using the SVD technique [3] for computing the optimal set of weights in the sense of the minimal 2-norm of the weight solution matrix W.
5) if the output error with the new weights is still higher than a specified threshold the procedure is repeated from the point 2); otherwise the training has terminated successfully.

The SVD-LS method allows a complete control over the internal structure of matrix X and the regularity of the weight matrix W. The SVD can be replaced in most cases by a cheaper QR Decomposition (QRD) [3] in order to reduce the computation cost.

Experimental results

1) XOR
The algorithm has been tested at first with the XOR problem. A network with 2 hidden neurons is used , while the non linearity adopted is the sigmoid function $f(x)=1/(1+\exp(-x))$. Several

770

trials with different configurations of initial weights and different values of learning rate have been performed, showing convergence in about 90% of cases. Fig. 1 reports the MSE as a function of the number of iterations in a typical case; both the rapidity of convergence (about 30 iterations to get MSE<.01) and its depth (MSE<10^{-4} after 100 iterations) can be easily verified .

2) *Pattern recognition*

The considered problem is that of recognising a circle with a radius R=0.35 inside a square with side of length L=1, both centered in the origin of the coordinated axes. A network with eight hidden neurons is chosen; the tangentoid function f(x)=(1-exp(-ax))/(1+exp(-ax)) is used, with different choices for the parameter a. In the following examples the value a=2 is adopted. The training patterns are totally randomly selected inside the square; if the selected point is internal to the circle the chosen target is +1, if it is external the target is -1. One parameter of the new algorithm to be properly chosen is the number of points to present to the network at each epoch (block LS iteration); it can not be too low since in this case it could generate an excessive oscillation of the weights from an iteration to the next one, then compromising the stability of the learning procedure. At the other side it can not be too high because in this case the approximation provided by the solution of the system (7) may be not satisfactory. As a trade-off value it has been chosen P=50 points per epoch.

A comparison with the backpropagation algorithm has been realised. In both cases the learning parameters were selected to maximise the performance of the algorithms. In the case of the backpropagation slope a=0.2 and a learning rate η=2 were chosen; for the LS algorithm the adopted value was η=0.1.Fig.s 3 and 4 show the output of the two networks in a typical case at different steps of the learning phase, showing the higher speed of convergence of the new algorithm. Moreover, with the selected step size (optimizing the convergence rate), the classical backpropagation algorithm exhibited misadjustment in the converged solution, which caused a fluctuation of the circle radius. The behaviour of the proposed LS algorithm was instead regular, with a monotonic decrease of the error functional during the last phase of the learning procedure.

Acknowledgment: The work was supported in part by Italian CNR and in part by Italian MURST

References

[1] G. Cybenko, "Approximation by superpositions of a sigmoidal function", Signal and Systems, Springer-Verlag, New York, 1989, pp. 303-314.

[2] D.E. Rumelhart, G.E. Hinton, R.G. Williams, "Learning internal representations by error propagation", in D.E Rumelhart, J.L. McLelland (Eds.), "Parallel Distributed Processing: Exploration in the Microstructure of Cognition", Vol. 1, MIT Press/Bradford books, Cambridge, MA, 1986.

[3] G.H. Golub, C.F. van Loan, "Matrix computations", John Hopkins Universiy Press, Second edition, 1989.

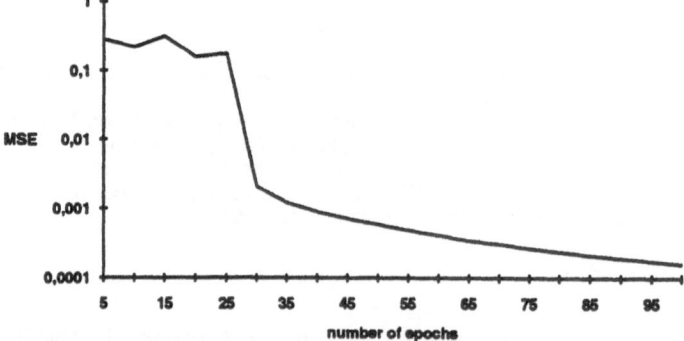

Fig. 1 : MSE versus number of epochs for XOR problem

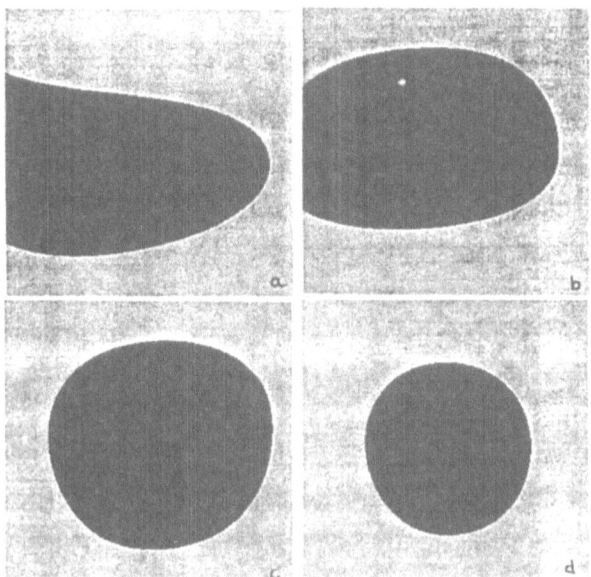

Fig. 2 : Backpropagation. (a) Iterations=900, MSE=.766. (b) Iter.=2800, MSE=.578. (c) Iter.=5300, MSE=.368. (d) Iter.=7800, MSE=.305.

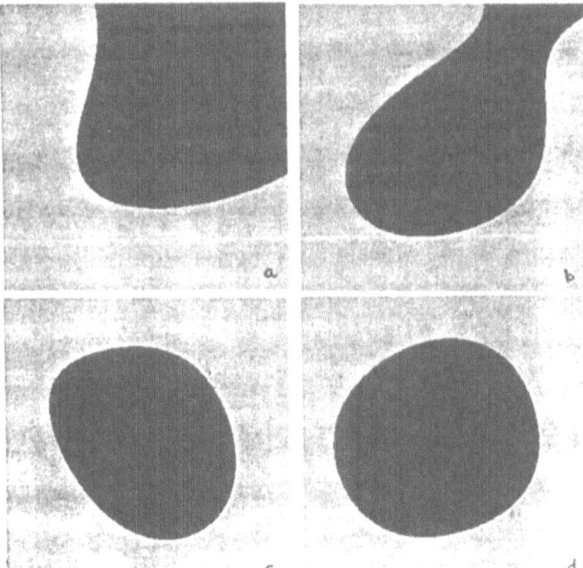

Fig. 3 : LS algorithm. (a) Iterations=16, MSE=.923. (b) Iter=18, MSE=.501. (c) Iter=20, MSE=.390. (d) Iter=24, MSE=.298.

Do Backpropagation Trained Neural Networks have Normal Weight Distributions ?

I. Bellido and E. Fiesler

Institut Dalle Molle d'Intelligence Artificielle Perceptive (IDIAP)
Case postale 609, CH-1920 Martigny, Switzerland

Abstract

Although artificial neural networks are employed in an ever growing variety of applications, their inner workings are still viewed as a black box, which is due to the complexity of the non-linear dynamics that govern neural network learning. The key parameters in this learning process are the so called interconnection strengths or weights of the connections between the neurons. Because of the lack of data, mathematical approaches for studying the 'inside' of neural networks have to resort to assumptions like a Normal distribution of the weight values. In order to better understand what goes on inside neural networks, a thorough study of the real probability distribution of the weight values is important. Besides this, knowledge about weight distributions is also a main ingredient for weight reduction schemes enabling the creation of partially connected neural networks and for network capacity calculations. This paper reports on the findings of an extensive empirical study of the distributions of weights in backpropagation neural networks, and tests formally whether the weights of a trained neural network have indeed a Normal distribution.

1 Introduction

When a large set of numbers is studied, it is usually assumed that its probability distribution is a Normal distribution. This is supported by the *Central Limit theorem*, which states that a large enough set of *independent* random variables tends to be Normally distributed. Although the distribution[1] of the weights in a trained neural network is often assumed to be Normal, the *Central Limit theorem* may not be valid for this case because of the mutual dependencies of the weights due to the learning process. Since the distribution of the weights in a trained neural network is often used in neural network analysis [1] and weight reduction methods [2, 3], it is important to examine how weights are distributed by the backpropagation learning process. For example, Hanson and Burr [1] observed "Normalness" with large kurtosis (peakedness) in the three large experiments they conducted.

To test whether trained weights are really Normally distributed, a set of widely known benchmark problems with a range of different topologies and complexities, including real-world problems, were selected: six encoder problems [4], the three Monks problems [5], the sonar identification problem [6], a genetic sequences identification of the promoters activity problem [7], and a Finnish vowel phoneme recognition problem [8].

2 Statistics

2.1 Comparison

Several statistical tests are available in order to formally test whether a distribution is a Normal distribution. The best known of these is the χ^2 test, which appeared to be too tough for this purpose, most likely due to the non-clustered nature of the weight data. Two goodness-of-fit techniques have been applied: the modified Anderson-Darling test (A^* test) and the D'Agostino's D test [9]. The value returned by either of these tests is inversely proportional to the goodness-of-fit of the weight distribution to the Normal distribution. This value, together with the number of elements of the data set determines the *level of significance*[2]. For example, a level of significance of 0.005 means that if the hypothesis is rejected, the probability that the distribution is Normal is less than or equal to

---

[1] In this publication, the words *distribution* and *probability distribution* are used interchangeably.
[2] The probability of rejection of the hypothesis by the test, when the hypothesis is actually true.

0.5%. In this publication, the hypotheses of Normality are rejected with a fixed level of significance. Non-rejected weight distributions are assumed to correspond to Normal distributions.

In order to provide further information, the third and fourth momentums have also been calculated. These statistics give information about the shape of the distribution with respect to the Normal distribution. If the third moment value is smaller than zero, the distribution is skewed to the right, if the value is larger than zero, the distribution is skewed to the left. If the fourth moment (*kurtosis*) value is larger than three, the distribution is more peaked, and if the value is smaller than three, it is less peaked than the Normal distribution.

2.2   Graphical Representation

The main problem in visualizing a weight distribution is how to transform one-dimensional (density) data into a two dimensional weight distribution graph. One solution is to create a histogram of the distribution, dividing the observed weights into several classes and counting the number of weights whose value lies inside those classes. This set of frequencies can be plotted to show the distribution. This method is sensitive to the parameters chosen; small changes in the width of the classes or in the choice of the centers often causes deviations in the obtained graphs. It also requires a large number of classes to obtain an accurate picture.

A second solution is to create as many classes as there are elements in the data set, and to count how many neighbors are inside the interval $[w_i - d, w_i + d]$, where d is a fixed distance and w_i is a weight. After all the elements of the distribution have been processed, a plot can be generated joining the obtained ordinates by straight lines. One problem of this representation is that in areas where the density is supposed to be zero, like large distances between two consecutive weights, the plotted density is not zero. To avoid this problem, the following heuristic can be applied: if the distance between two weight values is larger than $2d$, two zero density values will be introduced between them, one at $w_i + d$ and the other at $w_{i+1} - d$, assuming that the weights are sorted. The selection of d depends on the number of elements available in the data set. If a very small distance is chosen, a very noisy representation (spikes at each weight) is obtained. On the other hand, if a very large distance is chosen, the representation obtained approaches a uniform distribution.

3   The Benchmarks

Even though the field of Neural Networks has surged, the number of available benchmarks is surprisingly small. A substantial subset of these is used in this article and listed in table 1. Several key characteristics of these problems are also listed: the topology used (the number of layers and the number of units per layer), the number of weights used, and the way in which the inputs and outputs are encoded. The topology notation expresses the number of units per layer from input to output layers (left to right). W is the number of weights of the network (including biases) and P is the *number of input-output pairs used for training*. Biases are considered here as weights of connections from a constant unit, as is customary in backpropagation implementations. The normalization used for the Finnish vowel recognition problem consists of dividing each value of the input pattern by the largest absolute value in the pattern. This preserves the relations between these values.

4   The Results

The above benchmark problems listed in table 1 have been implemented, their weight distributions have been plotted, and goodness-of-fit techniques have been applied to them. Graphical results show differences in weight distributions depending on the number of weights of the network and the desired training accuracy. For most experiments the same convergence criterion has been used: a maximum error ($\epsilon = \Delta |t - o|$) of 0.1 for all the output neurons (this implies a large accuracy in learning that can lead to poor generalization). Only in the Finnish vowel recognition problem a maximum error (ϵ) of 0.4 is used. Each problem has been evaluated repeatedly with different initial conditions (weight values) to obtain different experiments.

| Benchmark | Data Types | | Topology | W | P |
Name	Input	Output			
Monks-1	Binary	Binary	17-3-1	58	124
Monks-2	Binary	Binary	17-2-1	39	169
Monks-3	Binary	Binary	17-3-1	58	122
8-bits Encoder	Binary	Binary	8-3-8	59	10
16-bits Encoder	Binary	Binary	16-4-16	148	18
16-bits Encoder	Binary	Binary	16-8-16	280	18
8-bits-Inverse Encoder	Binary	Binary	8-3-8	59	10
16-bits-Inverse Encoder	Binary	Binary	16-4-16	148	18
16-bits-Inverse Encoder	Binary	Binary	16-8-16	280	18
Sonar Signals Classification	Real	Binary	60-15-1	931	104
Gene promoters	Binary	Binary	228-1	229	106
Finnish vowels	Normalized Real	Binary	20-50-5	1305	300

Table 1: Benchmarks used.

The way the A^* and the D tests have been used (see section 2.1) implies the choice of a level of significance. The level of significance selected for this study is 0.005 for both tests, which implies a very small number of invalid rejections and a very generous acceptance criterion. Table 2 shows the percentage of rejected experiments for each benchmark. It also shows the percentage of experiments that were rejected by both tests. The values in the table for the skewness (third momentum) represent the percentage of cases for which the distribution is deviated to the left or to the right, and the values for kurtosis (fourth momentum) represent the percentage of the experiments with larger or smaller kurtosis (peakedness) as compared to the Normal distribution.

| | Number of | Rejected by | | | Skewness | | Kurtosis | |
	Experiments	A^* test	D test	Both	Left	Right	More	Less
Monks-1	42	95.24	97.62	95.24	57.14	42.86	100.00	0.00
Monks-2	40	62.50	30.00	30.00	52.50	47.50	97.50	2.50
Monks-3	42	14.29	23.81	11.90	59.52	40.48	97.62	2.38
E. 8-3-8	48	97.92	0.00	0.00	0.00	100.00	93.75	6.25
E. 16-4-16	48	33.33	22.92	6.25	2.08	97.92	0.00	100.00
E. 16-8-16	48	97.92	95.83	95.83	12.50	87.50	4.17	95.83
I. E. 8-3-8	48	97.92	12.50	12.50	100.00	0.00	93.75	6.25
I. E. 16-4-16	48	60.42	20.83	18.75	100.00	0.00	2.08	97.92
I. E. 16-8-16	48	97.92	95.83	95.83	37.50	62.50	4.17	95.83
Sonar	44	100.00	100.00	100.00	18.18	81.82	100.00	0.00
Gene promoters	14	7.14	14.29	7.14	78.57	21.43	14.29	85.71
Finnish vowels	10	100.00	100.00	100.00	0.00	100.00	100.00	0.00

In the left column: E. means Encoder problem and I.E. Inverse Encoder problem.

Table 2: Results obtained for each benchmark.

It can be observed from the table that for all the benchmarks but three, the A^* test rejects more than the 60% of the cases, and often more than 95%. The D test rejects a slightly smaller percentage of cases. The differences between the tests used may be caused by the differences in the power of each test. The A^* test is one of the most powerful tests for Normality, which implies that the rejections are actually valid with large probability. For the sonar identification and Finnish vowel problems the results are unanimous: the distribution of the weights is not Normal. Following the A^* results, only one benchmark, the gene promoters problem, passes this Normality test for more than 90% of the cases. This is coherent with the D test result. It has to be noted that the gene promoters problem is the only problem solved with a network without hidden layers.

Figure 1 shows an example of the distribution of weights in a monks-2 problem compared to its corresponding Normal distribution. Two large maxima can be observed in this case, which was reported as non-Normal by both the A^* and the D tests. Figure 2 corresponds to a distribution of weights for the sonar signal identification problem. The kurtosis given by the fourth momentum test is clearly seen. The third figure corresponds to the gene promoters problem, which yielded minimal rejection ratios.

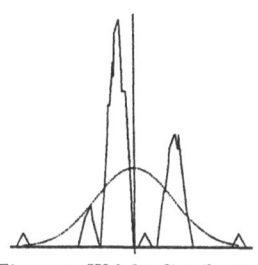 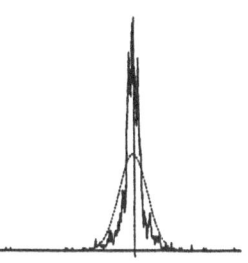 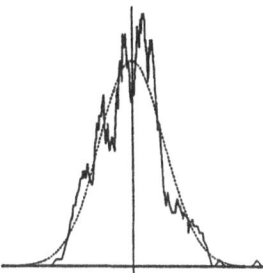

Figure 1: Weight distribution and corresponding Normal distribution for the Monks-2 problem.

Figure 2: Weight distribution and corresponding Normal distribution for the Sonar problem.

Figure 3: Weight distribution and corresponding Normal distribution for the Gene promoters problem.

5 Conclusions

Even while using a very small probability of invalid rejections (0.005), the majority of the 480 weight distributions were labeled not Normal by either of the statistical tests used. The results also show a strong problem dependency for the weight distributions.

References

[1] S. J. Hanson and D. J. Burr, "What Connectionists Models Learn: Learning and Representation in Connectionists Networks", *Behavioral and Brain Sciences*, Vol. 13, Nr. 3, pp. 471-518, 1990.

[2] S. J. Nowlan and G. E. Hinton, "Simplifying Neural Networks by Soft Weight-Sharing", *Neural Computation*, Vol. 4, Nr. 4, July 1992.

[3] W. Banzhaf et al.,"A Sparsely Connected Asymmetric Neural Network and its Possible Application to the Processing of Transient Spatio-Temporal Signals", *Proceedings of the International Neural Network Conference (INNC) 90*, Vol. 2, pp. 1005-1008, Kluwer Academic, 1990.

[4] D. E. Rumelhart and J. L. McClelland, *Parallel Distributed Processing: Explorations in the Microstructure of Cognition*, Vol. 1, Chap. 8, pp. 318-362, MIT Press, Cambridge, MA, 1986.

[5] S. B. Thrun et al., "The MONK's Problems: A Performance Comparison of Different Learning Algorithms", Carnegie Mellon University Technical Report CMU-CS-91-197, December 1991.

[6] R. P. Gorman and T. J. Sejnowsky, "Analysis of Hidden Units in a Layered Network Trained to Classify Sonar Targets", *Neural Networks*, Vol. 1, pp. 75-89, 1988.

[7] G. G. Towel et al.,"Refinement of Approximate Correct Domain Theories by Knowledge-Based Neural Networks", *Proceedings of the AAAI'90*, pp. 861-866, AAAI Press 1990.

[8] T. Kohonen, J. Kangas, J. Laaksonen, and K. Torkkola, "LVQ_PAK: A program package for the correct application of Learning Vector Quantization algorithms", *Proceedings of the International Joint Conference on Neural Networks*, Vol. I, pp. 725-730, Baltimore, June 1992.

[9] R. B. D'Agostino, "Tests for the Normal Distribution", in R. B. D'Agostino and M. A. Stephens Eds. *Goodness of Fit Techniques*, Chap. 9, Marcel Decker Inc. 1986.

A Constructive Algorithm for Binary Mapping

F. M. Frattale Mascioli and G. Martinelli

INFO-COM dept. - University of Rome

Via Eudossiana, 18 - 00184 Roma - Italy

I. INTRODUCTION: In the present contribution we propose a constructive algorithm useful for determining a neural network that implements any binary mapping. It yields a two layer perceptron-like neural network: the first layer is a *hidden layer*, the second layer is an *output neuron* that implements an OR operation, all the neurons are *linear threshold units*. The algorithm is characterised by directly controlling the separating hyperplanes of the desired *decision region*, taking account of both the training set and a *smoothing* generalisation rule for covering the unspecified part of the mapping. This result is obtained by relying on a topological approach, based on the representation of the mapping onto the *binary hypercube* of the input space: each pattern of the training set is in fact a vertex of the hypercube, labelled with the value of the corresponding desired output ('1', '0' or 'don't care'). Finally, we remark that, from a topological point of view, the hypercube is a *connected graph* where the *nodes* correspond to the vertices and the *arcs* to the edges.

II. THE CONSTRUCTIVE ALGORITHM: The proposed algorithm is based on Lemmas 1, 2 and 3 of [1], which state the *basic conditions* which guarantee for a set of vertices of the binary hypercube to be *linearly separable*, i.e. the existence of a hyperplane which separates it from the rest. A linearly separable set that contains only '1' or 'd' vertices is a *positive cut* of the hypercube. A convenient strategy for determining a positive cut is to construct step-by-step a set of vertices, initially composed by only one '1' vertex, by controlling that the growth meets the basic conditions. The algorithm achieves this goal by suitably visiting the vertices of the binary hypercube, considered as a connected graph. Each determined positive cut corresponds to a hidden neuron of the net to be constructed. The OR implemented by the output neuron shapes the desired decision region. The parameters of the hidden neuron are determined by means of simple operations based on geometrical considerations.

III. CONCLUSIONS: The constructive algorithm proposed in the present contribution is characterised by controlling directly the number of neurons required by the training set. Consequently, the resulting nets are simpler than those obtained with other methods. It is moreover important to note that the computational cost required by the algorithm is low and the construction of nets in general fast.

REFERENCE

[1] M.R. Emamy-Khansary: *"On the cuts and cut number of the 4-cube"*, J. Combinatorial Theory, Series A, Vol. 41, 1986, pp. 221-227.

BOXES Revisited

Shlomo Geva & Joaquin Sitte
Faculty of Information Technology
Queensland University of Technology
GPO Box 2434 Brisbane Q 4001 Australia
geva@sleet.qut.edu.au

In their BOXES experiment Michie and Chambers [1] demonstrated how the cart-pole problem can be solved in an unsupervised manner (ie. without a teacher) by a collection of co-operating independent processes, each responsible for providing a solution to a small part of the problem at hand, but all using an identical learning rule. The BOXES experiment, and its Cart-Pole formulation, inspired the subsequent development of the Adaptive Critic method of Barto, Sutton and Anderson [2]. The Cart-Pole experiment became a *de facto* benchmark for neural networks learning schemes. Michie and Chambers did not publish the exact parameter settings they have used in their experiments. Subsequently, Barto et al. have reproduced the experiment to provide a comparison with their adaptive critic learning scheme. Although they made it clear that their experiments did not provide a definitive statement on the relative performance of the methods, the BOXES performance was portrayed as poorer, a result that was never questioned.

The performance of the actual controller of neither method was analyzed other than with respect to survival time. It should be stressed that the primary motivation of the experiments was to explore learning methods rather than control strategies *per se*. Nevertheless, this information is crucial to the control engineer who may wish to use the methods in solving real problems.

In this paper we analyse the BOXES Cart-Pole experiment and show that with appropriate parameter setting the BOXES method performs at least as well as the Adaptive Critic reinforcement learning method. The paper shows that the original heuristic interpretation of the experiment is inaccurate. The BOXES scheme is shown to work as a directed random search. The mechanism described in this paper is more detailed and is not entirely consistent with the original heuristic explanation of the BOXES experiment.

Despite good performance in learning to avoid failure, Analysis of the quality of the evolved cart-pole controllers and a perofrmance comparison of both BOXES and the Adaptive Critic methods with that of a simple PD controller [3] are not favourable. Neither BOXES nor the Adaptive Critic methods are adequate *practical methods* for controller design as they stand. From the control engineering perspective the reinforcement learning formulation of the Cart-Pole benchmark is obviously too forgiving and should include, apart from balancing, the requirement of centring the cart-pole, as distinct from failure avoidance on a track of limited length.

References

[1] D. Michie and R. A. Chambers, "BOXES: An experiment in adaptive control," in *Machine Intelligence 2*, E. Dale and D. Michie, Eds. Edinburgh: Oliver and Boyd, (1968), pp. 137-152.

[2] A. G. Barto, R. Sutton, and C. W. Anderson, "Neuronlike Adaptive Elements That Can Solve Difficult Learning Control Problems," *IEEE transactions on systems, man, and cybernetics*, vol. SMC-13, NO. 5, (1983), pp. 834–846.

[3] S. Geva and J. Sitte, "Is the broom-balancer a useful test case for learning methods?," Proc. IEEE Intl. Workshop on Emerging Technologies for Factory Automation (1992), pp. 283–289.

Mathematical Properties Of Multi-Layer Adaptive Filters

M. IBN KAHLA, Z. FARAJ and F. CASTANIE

ENSEEIHT-GAPSE, 2 rue Camichel
31071 TOULOUSE CEDEX, FRANCE

This paper studies the mathematical properties of a scalar input multi-layer linear neural network (MLLNN) trained with back propagation (BP) (Fig. 1) [1] [3]. The network output is a linear function of the input (FIR filtering) : $y(n) = h(n)x(n)$, where the filter coefficient $h(n)$ is the product of all neuron weights : $h(n) = \prod_{i=1}^{L} h_i(n)$. The BP algorithm is designed to minimize the mean square error $e(n)$ which is the difference between the actual output $y(n)$ and the desired response $d(n)$. The weights are adjusted according to the equations : $h_i(n+1) = h_i(n) + 2\mu e(n)x(n) \prod_{j=1, j\neq i}^{L} h_j(n)$. We suppose that all weights are initialized with the same manner, the filter coefficient update is then given by : $h(n+1) = h(n)(1 + 2\mu e(n)x(n)h^{\alpha}(n))^{L}$, where $\alpha = \frac{L-2}{L}$. To determine the mean weight behaviour, we define a weight error variable at time n as : $v(n) = \ln(\frac{h(n)}{h_{opt}})$, where h_{opt} is the optimum Wiener coefficient: $h_{opt} = \frac{E(x(n)d(n))}{r_x}$. Let $m(n) = E(v(n))$, then $m(n)$ satisfies the following recursive relation : $m(n+1) = f(m(n))$ where $f(x) = ax + bx^2$. a and b are functions of different parameters : $a = 1 - 2\mu L h_{opt}^{1+\alpha} r_x$; $b = (4 - 3L)\mu h_{opt}^{1+\alpha} r_x$.

The stability conditions (SC) (In which case $m(n)$ converges to 0) can be studied in 2 steps : Find the fixed points of f and study Lipshitz conditions (They will not be detailed in this abstract). In [1] the case of $L = 2$ is analysed. We give here the necessary condition for choosing the step size μ : $0 < \mu < \frac{1}{Lh_{opt}^{\alpha+1} r_x}$.

The graph of f is a parabola with a positive vertex. The convergence can be established in very few iterations if the coefficient value at time 0 is near the value that annuls f. On the other hand, if we drow the functions f for different values of L, we see that the parabola vertexes decrease when the number of layers increases : So, if the filter coefficient is initialized with a value greater than the optimum one, then the greater the number of layers, the faster the convergence rate.

The steady state mean square error (SSMSE) is a function of the input signal variance, the optimum filter coefficient, the learning rate, and the number of layers : $E(\infty) = \frac{E_{min}}{1 - \mu L r_x h_{opt}^{\alpha+1}}$, where $E_{min} = E((d(n) - h_{opt}x(n))^2)$. A computational example is given in Fig. 2 in order to show the influence of the number of layers on the convergence speed (Different parameters are choosen such that the SSMSE are the same for all layers).

The generalization of this study to a p-Dim. input MLLNN can be found in [2]. It is particularly shown the influence of the number of layers on the mean weight behaviour, the SC, and the SSMSE.

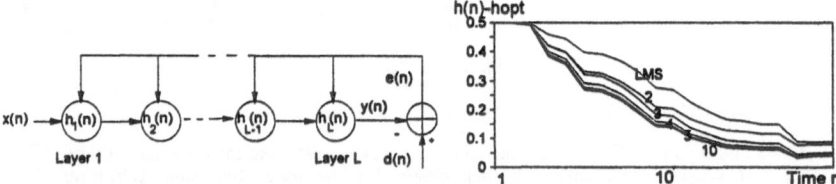

Fig. 1 : Multi-layer adaptive filter.   *Fig. 2* : Learning curves for different values of L.

REFERENCES :
[1] Z. FARAJ, M. IBN KAHLA and F. CASTANIE, "Multi-layer adaptive filtering", pro. of the 6th SP workshop on statistical signal and array processing, Victoria, Canada, pp. 58-61, October 1992.
[2] M. IBN KAHLA, F. CASTANIE and J. C. HOFFMANN, "Statistical analysis of a p-Dim. input multi-layer linear neural network trained with back propagation", to be published, 1993.
[3] R. P. LIPPMANN, "An introduction to computing with neural nets", IEEE ASSP magazine, pp. 4-22, April 1987.

Weight Zero Enhancement In Speech Synthesis Using Neural Networks

G.C. Cawley, M.I. Heywood, P.D. Noakes.

Neural and VLSI Systems Laboratory, Department of Electronic Systems Engineering,
University of Essex, Wivenhoe Park, Colchester, Essex, CO4 3SQ, United Kingdom.

Introduction.

In continuous speech the boundaries between allophones are not distinct, but are considerably blurred, an effect known as coarticulation, caused by the inertia of articulators (eg lips and tongue). Coarticulation carries little semantic content of an utterance, but we subconsciously expect to hear its effects in natural speech. This work is concerned with the training of neural networks for speech synthesis through generation of Linear Predictive Coding (LPC) parameters corresponding to the sequence of allophones. LPC attempts to find the coefficients A_k of an all pole filter with a transfer function $H(z)$, such that the spectral properties are optimally similar to that of a segment of sampled speech.

Data Pre - Processing

Unfortunately, LPC coefficients are not suitable for training neural networks as they are highly sensitive to error. Consequently, the LPC coefficients are transformed to an equivalent Line Spectral Pair (LSP) representation [1], where LSP schemes are known to offer excellent quantisation and interpolation properties. LSP coding records the frequency of the zeros of two polynomials $P(z)$ and $Q(z)$ which are related to the predictor polynomial $A(z)$. The zeros of $P(z)$ and $Q(z)$ lie on the unit circle in the z plane, this reduction in the search space allows efficient root finding methods to be employed. For the synthesis filter to be stable, the zeros of $P(z)$ alternate with the poles of $Q(z)$ around the unit circle. The overall spectral sensitivity of LSP parameters is less than that of PARCOR or log area ratio parameters. Furthermore, the spectral sensitivity of individual LSP parameters are uniform, whereas low order PARCOR parameters exhibit higher sensitivities.

Network Training

Heywood and Noakes [2] introduce a framework for performing pruning in feed forward neural networks, interactively with the standard back-propagation (BP) learning rule. To achieve this two extra learning parameters are introduced. One is a stability threshold which effectively identifies when the weights feeding a neuron represent 'useful' information. The second parameter indicates whether a stable weight is of significance at the neuron. For the purposes of this paper, the magnitude of 'stable' weights are merely compared to a Zero Weight threshold, and set to zero if they fall below this threshold [2]. However, there is no reason why a more stringent measure of weight contribution cannot be employed (eg product of weight and stimuli). Secondary controls to limit the number of weights introduced at any one weight update are also included.

Results

The main aim of these experiments is to determine the extent, if any, of weight redundancy and resulting generalisation improvements, when the performance is compared to that of the standard BP algorithm (MLP, single hidden layer). Simulation results indicate that significant weight redundancy is detected (25% minimum, 90% maximum). In the case of the more important phonetic classes of vowel, nasal and plosive, where the data set is larger, increases in generalisation ability (10% - 18%) are returned, in spite of employing a simple magnitude type zero weight criteria (pruning is facilitated by a correction path provided via the learning algorithm).

Acknowledgments.

The authors acknowledge the funding of the United Kingdom Science and Engineering Research Council.

References.

[1] Cawley G.C., Noakes P.D. "LSP Speech Synthesis Using Backpropagation Networks", in - press, ANN - 93 Brighton, May 1993.

[2] Heywood M.I., Noakes P.D., "Simple addition to Back-Propagation Learning for Dynamic Weight Pruning, Sparse Network Extraction and Faster Learning", IEEE ICNN 93, San Francisco, Vol 2, pp 620 - 625.

Biological Metaphors in Designing Modular Artificial Neural Networks

Egbert J.W. Boers[1],
Herman Kuiper[2],
Bart L.M. Happel[3],
Ida G. Sprinkhuizen-Kuyper[4]

[1]Department of Computer Science. Leiden University, P.O. Box 9512, 2300 RA Leiden, The Netherlands, email: boers@wi.LeidenUniv.nl, Tel: +31-71-277093.

[2]Currently a Research Engineer at the Knowledge Engineering Group of the National Aerospace Laboratory (NLR), Amsterdam, The Netherlands, email: kuiperh@nlr.nl.

[3]Department of Experimental and Theoretical Psychology, Leiden University, member of the Dutch Foundation for Neural Networks (SNN) and the Leiden Connectionist Group, email: happel@rulfsw.LeidenUniv.nl.

[4]Department of Computer Science, Leiden University, email: kuyper@wi.LeidenUniv.nl.

Abstract

This paper presents a method for designing artificial neural network architectures. The method is based upon a *reversed engineering* of the processes which resulted in the mammalian brain. The method extends the brain metaphor in neural network design with genetic algorithms and L-systems, modelling natural evolution and growth. It will be argued that a principle of *modularity*, which is inherent to the design method as well as the resulting network topologies, improves network performance as measured by learning speed and generalization capability.

The method uses *context sensitive* production rules, creating the possibility of architectures being self-similar in more than one way. The coding of the production rules in the chromosomes is inspired by DNA, using markers, allowing the chromosome to be read twelve times, which might increase the level of implicit parallelism.

Our simulation results indicate that the combined use of genetic algorithms and L-systems results in an efficient search, characterized by fast convergence towards a solution and better architectures. This might be explained by a number of theoretical advantages of the use of L-systems to code network topologies over 'blueprint representations' where the genetic algorithm has to specify every single connection: .

1. Coding can be sparse (less free parameters). A few production rules can produce already very complex architectures.
2. Scalability of solutions. Arbitrarily large architectures can easily evolve from small architectures by relatively minor changes in a fixed number of production rules. This significantly reduces the time needed for genetically optimizing large architectures.
3. Modularity of solutions. The same production rules can be applied many times in the growth process resulting in the multiple application of efficient pieces of architecture, and self-similar fractal-like architectures.
4. This method of encoding network topology is much related to the way biological architectures are genetically encoded and might therefore be expected to provide some efficient design principles.

Learning and Generalization Controlled by Contradiction[†]

Cory Barker and Tony Martinez

Computer Science Department, Brigham Young University, Provo, Utah 84602

SG (Specific to General) is a network that learns from a *training set* containing *examples*. Each example gives an *input* pattern along with the *output* that the network should produce for that input. The training set is a subset of the complete mapping from input to output. Therefore, the network should not only converge to a *representation* that contains the information given by the training set, but also *generalize* that information so that the network will respond well to inputs that it has not been trained on.

SG networks learn by monitoring *features* in the input space. A feature is some subset of network inputs along with their associated values. A feature is *matched* when the values on the network inputs that are part of the feature are equal to the values for those inputs as given in the feature. Inputs that are not part of the feature can be any value. A feature with few inputs is a *general* feature; it matches many points in the input space. A feature with many inputs is a *specific* feature; it matches few points in the input space. It is impractical to monitor all possible input features because the number of features is exponential in the number of inputs.

The network is made up of many simple nodes. Each node contains an input feature and a vector of counters, one counter for each possible output value. When the feature matches an example, the counter for the example's output value is incremented. The *strength* of an output value V for a given feature is determined by dividing the counter for V by the total of all the counters. If most examples that match a feature have the same output value, then the feature is a strong feature. If examples that match a feature have a near equal number of each of the possible output values, then the feature is a weak feature. The output of a node is the value with highest strength. The output of the network is the output of the strongest matching node.

When an example is given during learning, the network creates a fully specific node that matches only the new example, if such a node does not already exist. Existing nodes that output the same as the training example then intersect their current feature with the training example giving a new feature. The new feature is general since intersection drops some inputs from the original feature. The new feature encompasses the original feature and the input example as well as points nearby in the input space not defined by any example. If the new feature does not conflict with other nodes then a new node is created with the feature. Thus nodes effectively expand their features around the original training examples until other expanding nodes are encountered. The expansion process causes the network to cover the entire input space.

Because nodes are created incrementally, many times nodes are created that are not essential for representing the training set. Deletion of unnecessary nodes can improve generalization. Nodes determine if they should delete themselves by monitoring their own usage. All nodes keep a flag that is set whenever the node is used. After each pass over the training set, each node checks and resets its use flag. If the flag was not set during the training iteration, the node deletes itself.

SG was tested using Iris, Voting, Hepatitis, and Mushroom data from UC Irvine [1]. The results given here are averaged over ten runs. For each run half of the examples were randomly selected for the training set. The remaining examples were used as a test set to test generalization performance. The table shows percent correct on the training set, percent correct on the test set, and number of training iterations.

Iris			Voting			Hepatitis			Mushroom		
98.13	90.67	2.3	99.95	94.24	5.3	99.74	77.69	6.1	100.0	99.97	4.6

The SG network has a number of potential advantages over other learning methods. Evaluation of the model on four problems has shown that the network provides good generalization performance. The network converges within a small number of training passes. The network provides these benefits while automatically allocating and deleting nodes and without requiring user adjustment of any parameters. The network learns incrementally and operates in a parallel fashion.

Future research will focus on improved methods for creating general nodes, network architectures to support better parallelism, testing on additional data sets, and extensions to real-valued inputs.

[1] Murphy, P. M. & Aha, D. W. (1992). *UCI Repository of machine learning databases*. Irvine, CA: University of California, Department of Information and Computer Science.

[†] This research has been supported in part by the Air Force Office of Scientific Research.

EXTRACTION OF SYMBOLIC STATEMENTS
FROM SYNAPTIC WEIGHTS

W.J. Daunicht, R. Steiner, and H. Franz

Neurologisches Therapiecentrum, Hohensandweg 37, D-4000 Düsseldorf, Germany

In general, neural networks are 'opaque', i.e. it seems impossible to understand the meaning of their adaptive parameters such as synaptic weights. The difficulty to extract symbolic information or derive rules about the internal structure of a problem by interpretation of the synaptic weights after training is one of the fundamental problems in neural network theory.

The DEFAnet concept [1, 2], however, lends itself to such an analysis. A DEFAnet is a 3-layered feedforward network with n inputs where the input space is subdivided into an n-dimensional grid in the first layer. The outputs of the second layer form polynomials of first order within a grid cell. The output layer of the network is fully connected to the neurons of the second layer forming scalar products of the weights and the output of these neurons. Only these synapses are plastic and can be determined by calculation or training. Then, they can be used to derive symbolic information about the internal structure of the network function.

For each pair of inputs it is tested, whether all the synaptic weights corresponding to a product of the pair is smaller than a small positive ϵ. If so an additive relation between the two inputs has been detected and the network function can be considered as the sum of two – possibly nonlinear – functions of one of the inputs, as long as the rest of the arguments is constant. Thus, the desired function can be interpreted as the sum of the outputs of two separate networks, each omitting one of the inputs of the pair. A positive result of this test can be entered as 1, in a symmetric $n \times n$ incidence matrix M holding the relations between all possible pairs of inputs. If a row or column (except the diagonal element) is filled with 1s only, the contribution of that single input can be identified as a summand to the rest of the function.

Furthermore, the linearity of the function in an input can be tested. For this analysis, each input interval in the network used must be subdivided into at least 2 subintervals. Then, the synaptic weights corresponding to products differing only in the subinterval of the particular input are analysed. If all these weights differ by less than ϵ, the network function is linear in this input and the corresponding diagonal element of M is set to 1. If the weights are small themselves, the input does not contribute to the network function at all and the appropriate diagonal matrix element is set to -1; otherwise it remains 0.

Some special cases of the matrix M can be easily interpreted as shown in the table where f is the network function, f^* is a general nonlinear function, and a, b, c are parameters.

Matrix M	Statement
M = 0	$f(\underline{x}) = f^*(\underline{x})$
M = 1	$f(\underline{x}) = \sum_{\nu=1}^{n} a_\nu x_\nu + c$
M = I	$f(\underline{x}) = \prod_{\nu=1}^{n}(a_\nu x_\nu + b_\nu)$
M = 1 − I	$f(\underline{x}) = \sum_{\nu=1}^{n} f_\nu^*(x_n)$
M = 1 − 2I	$f(\underline{x}) = c$

The results can be used to tailor a DEFAnet in such a way as to minimize both network size and training time. In case of additivity between two inputs the corresponding second layer neurons can be omitted. In case of linearity, those neurons in the first layer covering the linear range can be replaced by single neurons thereby reducing the number of second layer neurons and plastic synapses.

In addition, the results of that analysis can be used for the evaluation of the influence of certain inputs and for the detection of some internal structure of an unknown function. It may also help to uncover architectures of unknown nonlinear systems or may detect regularities in discrete data. It is concluded, that the DEFAnet concept and the extraction of information from synaptic weights can be a useful tool for the analysis of unknown functions.

References

[1] W.J. Daunicht. DEFAnet - a deterministic neural network concept for function approximation. *Neural Networks*, 4:839–845, 1991.

[2] W.J. Daunicht. DEFAnet2 – advancements of a deterministic function approximator. In *this conference*, 1993.

Acknowledgements: This work was supported by DFG grant Da 199/2-1.

A NOVEL BACK PROPAGATION ALGORITHM WITH OPTIMAL NUMBER OF HIDDEN UNITS

Hiroaki Kurokawa, Chun-Ying Ho, Shinsaku Mori
Dept. of Electrical Engineering, Keio University, 3-14-1 Hiyoshi,
Kouhoku-ku, Yokohama, 223, JAPAN
Phone: +81 45 563 1141 Ext. 3319 FAX: +81 45 563 2773
email: kuro@mori.elec.keio.ac.jp

Abstract

A large fraction of recent work in artificial neural networks uses feed-forward networks trained with the back-propagation algorithm described by Rumelhart et.al. However, to decide the size of the network, how many layers and/or neurons, needed to solve a particular problem is a difficult question. This study discusses an approach to ease this problem based on hidden neuron's reduction method and shows an application for 3-layer networks. The network learned by proposed algorithm performs almost the same as the network with original back-propagation. By using this proposed method, the network structure (synaptic weight and number of neurons) can be decide simultaneously during a single learning.

summary

The multilayer feed-forward networks of interconnected of nonlinear neurons are widely used in adaptive problem [1]. One of the valuable characteristics of multilayer network is that the network can be trained. Back-propagation is a popular training algorithm and it applies to fields of technology like pattern recognition, classification etc. However, network performances are affected by the size of the network (layers and/or neurons). If training is started with too small network, no learning can occur and it needs to increase the size of the network and the network has to be retrained again. If the network is larger than that required, then learning is slowed, particularly in a conventional Von Neumann computer. The problem of the size of the network is a difficult problem to solve. Recent studies proposed some algorithms to avert these problems. The method based on information criteria was proposed in [2]. Although this method costs short time to determine the network size, the network that constructed by this method has no guarantee to converge to optimal solution. Also, the neuron reduction method was proposed in [3]. However, since the reduction of one neuron needs some epochs that are decided by human, a drastic time reduction is achieved for the determination of optimal number of hidden neurons. To solve this problem, we have proposed a new algorithm to optimize the size of the multilayer network under the learning with back-propagation in a single learning. This algorithm is based on the new character values C_j of hidden neurons which proposed here. These C_j s indicate neurons' stability and influence to the next layer. The value of C_j is given by,

$$C_j = \frac{2}{1 - x_j{}^2} \quad ,$$

where x_j is the output of the neuron. This equation is given by the differential form of the input-output function of the neuron (*i.e.* a sigmoid function which is bounded by (-1, 1)). This value C_j can be considered as how stable the output of the neuron is and how influential the output of the neuron is to the next layer. So it can be considered that neurons with large C_j are more important to the network than those with small C_j, then neurons which have a small C_j can be deleted. Since cutting down some neurons also means that the network loses its flexibility, these procedures are applied in the convergence stage of the learning. By cutting down neurons in the convergence stage, the influence of this matter can be almost negligible. We show simulation results of two standard test problems. First our proposed algorithm has been tested on artificial problem such as the XOR problem that is realized small size network. Then it has been tested on simple character classification as comparatively complex and large size network. These simulation results show that we can obtain an optimal size network even if this network size is not minimum size. And this optimized network performs as well as that initially structured to equal size network. This algorithm needs human's task. It is that we have to define the cutting line of neurons by the value of C_{j_i}. However this task is very easy to do, and is allowed some misjudgment. Therefore, since the number of final hidden neurons was concentrated around one value on both simulations XOR problem and character classification, it can be considered that the algorithm has flexibility for a complex problem.

References

[1]   D. E. Rumelhart, et al. : " Parallel Distributed Processing ", MIT Press (1986)

[2]   Takio Kurita. : " A Method to Determine the Number of Hidden Units of Three Layered Neural Networks by Information Criteria ", IEICE Trans. inf. & syst., Vol.J73-D-II, No.11, pp.1872-1878 (November 1990)

[3]   Masafumi Hagiwara. : " Back-Propagation with Artificial Selection — Reduction of the Number of Learning Times and That of Hidden Units — ", IEICE Trans. inf. & syst., Vol.J74-D-II, No.6, pp.812-818 (June 1991)

Two Neural Models for fast Category Learning – Neural Associative Memories and the Restricted Coulomb Energy Model*

A. König, A. Korn, F. Quint† M. Glesner
Darmstadt University of Technology
Institute for Microelectronic Systems
Karlstrasse 15, D-6100 Darmstadt, Germany

Abstract

In this work we provide a qualitative evaluation and comparison of two models for fast category learning. The first model is the Neural Associative Memory (NAM) model [1] and the second the Restricted Coulomb Energy (RCE) model [2]. The latter is embedded in a commercial product. Both models provide iterative training procedures that select prototypes from the initial training set. Both models allow supervised learning of pattern categories separated by arbitrary complex boundaries. NAM generates a piecewise linear approximation of the class boundaries, RCE a piecewise approximation with hyperspherical surfaces. Training is an incremental process, as prototypes are iteratively inserted or deleted from the network. Thus, the training process is much faster than for other comparable interpolative neural approaches, e.g. LVQ. No a priori knowledge of data distribution is necessary, as the correct amount of prototypes for each class is determined by the training procedures of NAM and RCE. Dynamic LVQ techniques also generate a sufficient number of prototypes for each class, but they do not optimize the number. All these vector quantizing approaches are very transparent and allow the easy interpretation of the weight vectors. RCE and NAM model both provide rapid training and recall, which makes them excellent candidates as classifiers in a generic inspection system for visual industrial quality control. In the context of the research project SIOB such a system is developed and implementations of NAM and RCE are tested as modules here. To assess the superior model qualitative and quantitative comparison is undertaken. Both models start with an initial empty set of prototypes. Driven by the objective to correctly classify the training patterns, prototypes are iteratively selected from these patterns. If the existing prototypes give correct classification no insertion takes place, else a new prototype is stored. There are two main differences between the models. The NAM conducts an unrestricted competition, whereas the RCE has an radius of influence for each prototype. The NAM eliminates redundant prototypes, whereas RCE only adjusts the respective radii of influence. In the case of misclassification both models store a new prototype, RCE adjusts the according radii of influence. The unrestricted competition of the NAM can cause an overfitting to the training set. Thus, in this work the NAM model was extended introducing thresholds of similarity for training and recall. This approach was validated with data from the SIOB project. Based on this qualitative comparison the authors will in future work elaborate a quantitative comparison using SIOB feature data obtained from an inspection line. The interest of the authors is to obtain a rapid classification mechanism for the visual quality inspection system supporting rapid installation and change of objects.

References

[1] W. Pöchmüller, A. König, M. Glesner, "Iterative Data Reduction Algorithms and their Application to Binary Associative Networks", in *Proceedings of the International Joint Conference on Neural Networks IJCNN'92*, Vol. II, pp. 3 - 9, Beijing, 1992

[2] D. L. Reilly, L. N. Cooper, C. Elbaum, "A Neural Model for Category Learning", Biological Cybernetics, 45, pp. 35 - 41, 1982

---

*This work was accomplished within the scope of a research project for visual object inspection in industrial quality control under grant of the German Federal Ministry of Research and Development (BMFT) grant number 01 IN 110 B/6 (SIOB)
†Dr. A. Korn and F. Quint are with Fraunhofer-Institut für Informations- und Datenverarbeitung IITB Karlsruhe

Storage Capacity Results for Decomposed Structures of Generalizing RAM Nodes

Janko Mrsic-Flögel
Neural Systems Engineering
Department of Electrical Engineering
Imperial College, London SW7 2BT (jmf@uk.ac.ic.ee)

Abstract

This paper presents novel storage capacity results for decomposed structures using generalizing RAM nodes [1].
Two storage capacity results are derived:
a) K_α - the number of number of random input patterns stored in a systems that are guaranteed not to overwrite each other and thus produce errors at output,
b) K_ω - the maximum number of selected patterns trained into the system that can be correctly retrieved.
Both storage capacity results can be improved by introducing the Overwrite Rule [4] into the system.The method involves making a distinction between RAM locations whose contents have been overwritten and other locations that have been trained only once. It uses the notion that any location that has its contents set more than once becomes less significant at Sigma Unit level. Results for decomposed systems using the Overwrite Rule are also presented.

References

[1] Aleksander I. (1990) Ideal Neurons for Neural Computers, Parallel Proc. on Neural Systems and Computers, Ed. R. Eckmiller/G.Hartmann/G.Hauske,pp.225-228, North-Holland
[2] Aleksander I.;Thomas W.;Bowden P.(1984) WISARD. *Sensor Review* 4(3):120-124.
[3] Redgers A. (1989) WIS system report. Internal Report. Imperial College, London
[4] Mrsic-Flogel J. (1992) Aspects of Planning with Neural Systems, PhD Thesis, Imperial College

APPLICATIONS

Industrial applications
—oral contributions

Novelty Detection and Neural Network Validation

C. M. Bishop

Applied Neurocomputing Centre
AEA Technology
Harwell Laboratory
Oxfordshire, U.K.

Abstract

One of the key factors limiting the use of neural networks in many industrial applications has been the difficulty of demonstrating that a trained network will continue to generate reliable outputs once it is in routine use. An important potential source of errors arises from input data which differs significantly from that used to train the network. In this paper we investigate the relation between the degree of *novelty* of input data and the corresponding reliability of the output data. We provide a quantitative procedure for measuring novelty, and we demonstrate its performance using an application involving the monitoring of oil flow in multi-phase pipelines.

1 Introduction

Neural networks have been shown to have a useful degree of performance in a wide range of industrial and medical applications. However, a key factor limiting the widespread implementation of neural network solutions has been the difficulty of demonstrating that the outputs generated by the network in the field are reliable. In general, the problem of network validation is a difficult one. Here, we consider the restricted problem of providing a measure of confidence associated with the response of a trained network to a new input vector.

Intuitively we expect that a network will generate reliable results when presented with data which is similar to that used during training, but that when substantially *novel* data is presented the network outputs will be prone to serious error. In Section 2 we investigate the relationship between novelty of the input data and validity of the network output, which can be used as the basis of a practical system for network validation. The approach is illustrated in Section 3 using an example from the monitoring of multiphase flows in oil pipelines.

2 Network Validation

Consider a feedforward network trained by minimising a sum-of-squares error function. In the limit of an infinite data set we can write the error in the form

$$E = \sum_{j=1}^{m} \int [y_j(\mathbf{x}; \mathbf{w}) - t_j]^2 \, p(\mathbf{x}, t_j) \, d\mathbf{x} \, dt_j \qquad (1)$$

where $j = 1, \ldots, m$ labels the output units, $\mathbf{x}$ is the input vector to the network, y_j denotes the output from unit j, and t_j is the target value for that unit. The functions $p(\mathbf{x}, t_j)$ represent the joint probability density functions for the training data. The network corresponds to a set of functional mappings $y_j(\mathbf{x}; \mathbf{w})$, parameterised by a set of weights and biases $\mathbf{w}$ whose values are found by minimising E.

We note that the joint density $p(\mathbf{x}, t_j)$ can be factored into the product of the unconditional density of the input data $p(\mathbf{x})$ and the conditional density of the target data $p(t_j \mid \mathbf{x})$. After some simple algebra, and dropping terms which are independent of the network weights $\mathbf{w}$, we can write the error (1) in the form

$$E = \sum_{j=1}^{m} \int [y_j(\mathbf{x}; \mathbf{w}) - \langle t_j \mid \mathbf{x} \rangle]^2 \, p(\mathbf{x}) \, d\mathbf{x} \qquad (2)$$

where we have defined the conditional average of the target data as

$$\langle t_j \mid \mathbf{x} \rangle \equiv \int t_j \, p(t_j \mid \mathbf{x}) \, dt_j \qquad (3)$$

Provided the functions $y_j(\mathbf{x}; \mathbf{w})$ have sufficient flexibility, (for instance if they have a sufficiently large number of hidden units) the minimum of this error function occurs when $y_j(\mathbf{x}; \mathbf{w}) = \langle t_j \mid \mathbf{x} \rangle$ so that the network outputs represent the *regression* of the target data conditioned on the input vector. In this sense the neural network solution can be regarded as optimal, since if the training data was generated from a deterministic function with superimposed zero-mean noise, then the network will average over the noise and learn the underlying function. Similarly, for classification problems in which the training data has a 1-of-N target coding scheme, the network outputs represent *a-posteriori* probabilities of the input vector $\mathbf{x}$ belonging to the corresponding classes, and so again can be regarded as optimal.

The key point to note is that the error function in equation (2) is weighted by $p(\mathbf{x})$ which represents the *unconditional* density of the input data. In a practical problem in which a finite set of training data is used we expect the network outputs to approximate the regression of the target data only for regions of input space for which this density is high. The unconditional density provides a quantitative measure of novelty with respect to the training data. If a new input vector falls in a region of input space for which the density $p(\mathbf{x})$ is high then the network must effectively interpolate between training data points and the network performance will generally be good. If the input vector falls in a region of input space for which $p(\mathbf{x})$ is low then the input data is essentially *novel* and the network could easily generate erroneous outputs.

We therefore arrive at the following procedure for validating network outputs. The data which is used to train the network is also used as the basis for an estimate $\hat{p}(\mathbf{x})$ of the (unknown) density $p(\mathbf{x})$. Standard cross-validation may be used at this point to optimise the network topology, and this can also serve to allow any smoothing parameters in the estimated density $\hat{p}(\mathbf{x})$ to be chosen. An independent test set is then used to confirm the performance of the trained network. Evaluation of $\hat{p}(\mathbf{x})$ using the test set also provides an indication of values of $\hat{p}(\mathbf{x})$ which are typical of data which is not to be regarded as novel. In this way a threshold can be set on $\hat{p}(\mathbf{x})$ if desired. When the network is in use, each new input is used to evaluate $\hat{p}(\mathbf{x})$ and this provides a quantitative measure of the degree of novelty of the input vector. Inputs which have relatively small values of $\hat{p}$ are those which are likely to generate spurious outputs. Many conventional techniques exist for estimating probability densities from finite samples [1] and various adaptive 'neural' approaches have also been suggested. In the next section we illustrate these ideas using one of the standard methods for density estimation.

3   An Example Application

In order to illustrate the concept of novelty detection as a form of network validation, we consider a specific industrial application of neural networks concerned with the determination of the oil fraction in a multiphase oil pipeline. Full details of this application can be found in ref [2].

In order to minimise costs, the oil industry makes use of multiphase pipelines to transfer mixtures of oil, water and gas directly from the off-shore production field to the on-shore facility. This leads to the need for an effective non-invasive method for monitoring the oil fraction in such pipes. A major difficulty arises from the fact that the multiphase flow can exhibit a wide variety of configurations, and numerical modelling of such flows is notoriously difficult.

One technique for determinine the phase fractions in the pipeline is based on multiple-beam dual-energy gamma densitometry [2]. The attenuation of a gamma beam passing through the pipe depends both on the particular material in the path of the beam and on the gamma energy (i.e. wavelength). By measuring the attenuation of a gamma beam at two energies along a narrow path through the pipe, and knowing the absorption coefficients of oil, water and gas, it is possible to evaluate the fractional path lengths in each of the three phases. If measurements could be made along many transverse paths through the pipe it would in principle be possible to perform tomographic reconstruction of the phase configuration and hence calculate the oil fraction. In a practical system only a few lines of sight will be available, and so alternative analysis methods need to be used.

We have considered a system with 3 vertical and 3 horizontal beam lines, and we have used neural network techniques to analyse the outputs from the corresponding densitometers. Each beam line generates two signals representing the attenuation at each of the two wavelengths, and these are first pre-processed to extract the fractional path lengths in the oil and water phases. (The fractional path length in the gas phase is not considered since this represents redundant information). The resulting 12 numbers are used as inputs to a multilayer perceptron having a single hidden layer of sigmoidal units followed by two linear output units whose activations represent the fractions of oil and water in the pipeline.

For the purposes of this study, synthetic data have been generated using configurations selected at random from the four model configurations shown in Figure 1. Note that these are not intended to be accurate representations of real flows, but are chosen mainly for computational simplicity. To generate a data point, a configuration is chosen at random, and the oil, water and gas fractions are then selected at random (with uniform distributions). The path lengths are then calculated, allowing for the effects of noise. In a practical system, the dominant contribution to the noise arises from the photon statistics due to the limited source strength of the gamma beams, and the limited integration time of the detectors, and has been accurately modelled with the correct Poisson statistics.

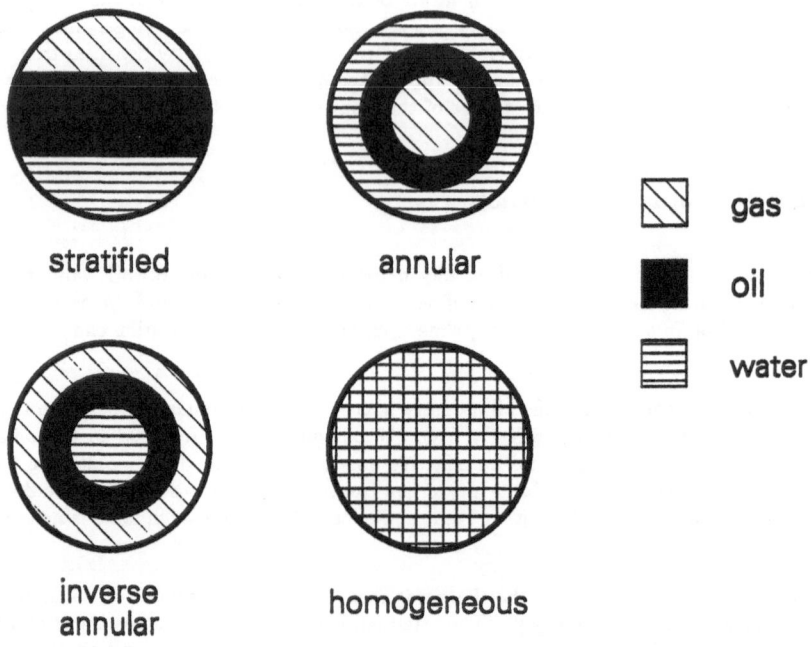

stratified annular

inverse annular homogeneous

gas

oil

water

Figure 1. Cross-sections of the pipe illustrating the four multiphase configurations used for network training.

For a training set with 1,000 points, a network with 8 hidden units is found to have the best performance on an independent test set (also of 1,000 points) generated by the same procedure as used to produce the training set. Figure 2 shows a plot of the predicted oil fraction versus the actual oil fraction for the test set data. The corresponding plot obtained using the optimal *linear* mapping is shown in Figure 3. It is clear that the neural network can determine the oil fraction to high accuracy, and that the non-linearity provided by the network gives a useful improvement in accuracy compared with the linear mapping (the average error is reduced by about a factor of 2). In a practical application of this technique the network would be trained using data from a multiphase flow test rig in which a variety of conditions, typical of those expected to occur in practice, could be produced.

In order to determine the degree of novelty of the input data we need a procedure for estimating the unconditional probability density corresponding to the training data set. For simplicity we have chosen a standard kernel based approach with gaussian kernel functions [1]. There is one kernel function for each point in the training set with its center given by the corresponding input vector. We therefore write our estimate of the density in the form

$$\hat{p}(\mathbf{x}) = \frac{1}{(2\pi)^{d/2}\sigma^d} \sum_{q=1}^{n} \exp\left\{-\frac{|\mathbf{x}-\mathbf{x}^q|^2}{2\sigma^2}\right\} \tag{4}$$

where $\mathbf{x}^q$ represents the qth data point from the training set, and d is the dimensionality of the input space (so that here $d = 12$). The smoothing parameter σ controls the degree of smoothness of the estimated density function, and its value must be neither too large (since this gives the estimate a large bias) nor too small (since this gives the estimate a large variance). We have chosen σ heuristically to be the average distance of the ten nearest neighbours, averaged over all points in the training data set.

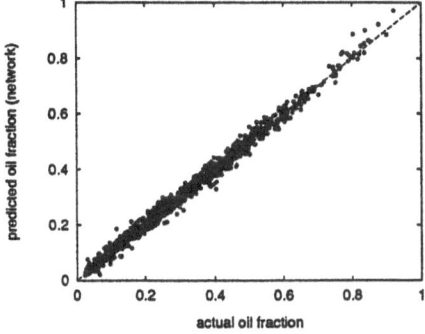
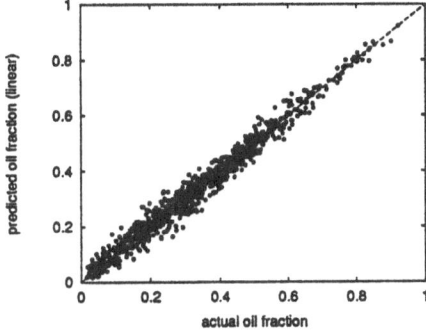

Figure 2. Scatter plot of the oil fraction predicted by the neural network versus the actual oil fraction, for all points in the test set.

Figure 3. As in Figure 2, but for the optimal linear mapping.

In order to test the performance of this novelty detector we have generated a further data set, consisting of 1,000 examples with randomly chosen oil and water fractions, corresponding to a 5th configuration referred to as 'inverted-stratified' which is obtained by inverting the stratified configuration of Figure 1.

When viewed as a function of $\mathbf{x}$, the density function $p(\mathbf{x})$ is generally called a likelihood. Figure 4 shows a plot of the magnitude of the error between the oil fraction predicted by the neural network and the true value obtained from the data set versus the log of the likelihood. Here the network with 8 hidden units described above was used, and the crosses show the 1,000 points from the test set as used to plot Figure 2. The circles show the 1,000 samples corresponding to the inverted-stratified configuration, and it is clear that the majority of these points have log likelihood values which are substantially smaller than those of the test set points, with a correspondingly larger range of oil fraction errors. We see that the network can indeed generate poor results when presented with data from this new configuration. Such data points can, however, easily be rejected on the basis of their log likelihood values. Setting a threshold anywhere between −5 and −10 would reject all data points having significant phase fraction errors. In practice, a suitable threshold can be chosen by examining the log likelihood values of the test set data.

It can also be seen from Figure 4 that there are some inverted-stratified points (circles) lying within the cluster of crosses. Examination of the phase fractions for these points shows that they represent configurations which could also be classified as stratified configurations. For instance, if the oil phase fraction is sufficiently large then the three horizontal beam lines pass through oil only, and there exist stratified and inverted-stratified configurations having the same phase fractions which give rise to the same 12 path length measurements. The novelty detector 'correctly' interprets these as being similar to the training data, and indeed the network predicts the phase fraction to high accuracy.

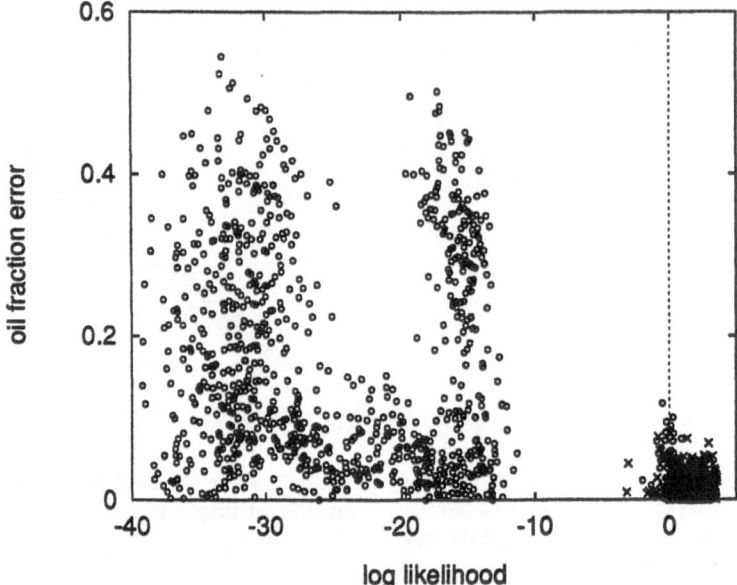

Figure 4. Oil fraction error of the neural network prediction versus the log likelihood from the novelty detector. Crosses correspond to the original test set while the circles correspond to novel phase configurations.

We have illustrated the estimation of the probability density of the input data using a simple technique based on a variable number of fixed kernel functions. One of the disadvantages of such an approach is that the complexity of the model grows with the size of the training set (since there is one kernel function for each data point). This difficulty can be overcome by using a fixed number of *adaptive* kernel functions in the form of a gaussian mixture model. The parameters of the gaussians can be determined by maximising the likelihood (for instance through re-estimation using the EM algorithm). Furthermore, the gaussian kernels can also form the basis functions for a radial basis function network, which can then replace the multilayer perceptron (to determine the oil and water fractions in the particular application considered here). The second-layer weights can be found by a fast linear supervised technique to give the least-squares solution. In this approach, the process of novelty detection and regression are thereby combined into a single network structure.

References

[1] Silverman B W (1986) *Density Estimation*, Chapman and Hall, New York.

[2] Bishop, C.M. and James, G.D. (1993) Monitoring of Multiphase Flows using Dual-Energy Gamma Densitometry and Neural Networks, *Nuclear Instruments and Methods in Physics Research* **A327** 580–593.

Estimating Material Properties for Process Optimization

Thomas Poppe and Thomas Martinetz
Siemens AG
Corporate Research and Development
Otto-Hahn-Ring 6
8000 München 83, Germany

A neural network approach to the problem of estimating physical properties of a material based on the material's chemical composition is presented. The network consists of sigmoidal hidden units and a linear output unit arranged in a feed-forward architecture. As a component of a process optimization system which is applied in production processes with a priori unknown and eventually drifting characteristics, robust and fast on-line adaptation of the network is required. Therefore, a permanently updated, stack-like organized training data set and a line-search procedure for adjusting the network weights is employed. A first application has been the estimation of the "relative yield stress" of different steel qualities, which is necessary for optimizing the rolling process at a hot line rolling mill. Compared to the current state-of-the-art method a reduction of the average estimation error of about 35% has been achieved.

1. Introduction

Process optimization requires knowledge about the relevant properties of the processed material. Depending on the material transformation process to be controlled, physical properties of the material like its heat capacity, its viscosity, its heat conductivity, or its hardness (just to mention a few) determine the optimal choice for the control parameter values. In most cases, however, the respective material property cannot be measured directly but must be estimated based on the thermodynamic state of the material, i.e., its chemical composition, its temperature, the given pressure, and eventually geometric quantities. The quality of the estimation result determines to a great extent the cost effectiveness and the product quality of the production process.

To be able to estimate material properties based on the thermodynamic state variables, the respective physical relationship has to be known. A common approach is to try to describe this relationship through physical models. However, in most cases the underlying physics is too intricate and/or not understood sufficiently to allow the design of feasible physical models which yield satisfying estimation results. In addition, the development of physical models is time consuming, requires precise knowledge about the usually very complex physical processes, and each model is specific for each material and each material transformation process.

To increase cost effectiveness and product quality also of intricate material transformation processes, an approach is necessary which *learns* the underlying physical relationship instead of modeling it based on specific prior knowledge. In addition, it would be highly desirable to have an approach which is generic and can be applied to a variety of materials and transformation processes. In the following we demonstrate that neural networks as adaptive modeling schemes

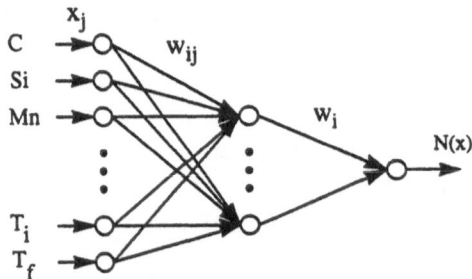

Figure 1: The architecture of the neural network.

have the desired capabilities. We describe the application of a neural network to the problem of estimating the *relative yield stress* (plasticity) of steel plates based on the steel plates' chemical composition, temperature, and shape. Knowledge about the relative yield stress is necessary for optimizing rolling processes, in our case the rolling of steel at a hot line rolling mill.

2. The Neural Network Architecture

The neural network has to model the relation

$$\alpha = F(C, Si, Mn, P, S, Al, N, Cu, Cr, Ni, Sn, V, Mo, Ti, Nb, B, d, b, T_i, T_f)$$

between the relative yield stress α of the steel plate and the concentrations of the sixteen chemical additives $C, Si, ..., B$, the steel plate's thickness d and its width b. T_i and T_f denote the temperature of the steel plate before and after the rolling, respectively. These two temperatures serve as a measure for the actual rolling temperature T, which cannot be determined explicitly. The concentration of the sixteen chemical additives $C, Si, ..., B$ is obtained from a material analysis during the steel cooking.

Figure 1 shows the neural network architecture, a three-layer feedforward network consisting of ten sigmoidal hidden units and one linear output unit. Each hidden unit receives the same twenty-dimensional input vector $\mathbf{x} = (C, Si, ..., Nb, B, d, b, T_i, T_f)$. The weights of the hidden units i, $i = 1, ..., 10$, are denoted by $\mathbf{w}_i = (w_{i1}, ..., w_{i20})$, and the weights of the linear output unit are denoted by $\mathbf{w} = (w_1, ..., w_{10})$. The thresholds of the hidden units and the output unit are denoted by θ_i and θ, respectively. Hence, when the network receives the input $\mathbf{x}$ which carries the information about the steel plate to be rolled, the network generates the output

$$\mathcal{N}_{\mathcal{W}}(\mathbf{x}) = -\theta + \sum_{i=1}^{10} w_i\, \sigma\!\left(\sum_{j=1}^{20} w_{ij} x_j - \theta_i\right)$$

as an estimation for the relative yield stress of the steel plate, with $\sigma(.) = 1/(\exp(-.) + 1)$ forming the sigmoidal output of the hidden units. The index $\mathcal{W}$ denotes the dependence of the network output $\mathcal{N}_{\mathcal{W}}(\mathbf{x})$ on the set $\mathcal{W} = (\mathbf{w}_i, \theta_i, \mathbf{w}, \theta)$ of all network weights and thresholds.

3. The Adaptation of the Network Weights

The estimation error of the network has to be minimized by selecting the right set of network weights $\mathcal{W} = (\mathbf{w}_i, \theta_i, \mathbf{w}, \theta)$. This is achieved through gradient descent on the mean square error

$$E(\mathcal{W}) = \frac{1}{2P} \sum_{\mu=1}^{P} (\alpha^\mu - \mathcal{N}_{\mathcal{W}}(\mathbf{x}^\mu))^2$$

of the last P estimation trials. $\mathbf{x}^\mu$ comprises the chemical composition, thickness, width, and temperature of the μ-th steel plate, the actual relative yield stress of which was α^μ.

The training data $(\mathbf{x}^\mu, \alpha^\mu)$, $\mu = 1, ..., P$ are accumulated on-line in a FIFO stack (first in first out). With each steel plate which is rolled a new data pair $(\mathbf{x}^\mu, \alpha^\mu)$ is available and put on top of the stack. The data pair at the bottom of the stack, i.e., the data pair which occurred P rolling processes ago, is removed from the stack. The stack size P, i.e., the size of the window on the incoming data stream, is determined by the available memory size and the computation time requirements.

With each update of the FIFO stack by a new data pair the network weights are adjusted through gradient descent on the cost function $E(\mathcal{W})$, which yields, by calculating

$$\Delta \mathcal{W} = -\eta_{dyn} \frac{\partial E(\mathcal{W})}{\partial \mathcal{W}}, \tag{1}$$

the backpropagation learning rules [1, 2]. The step size η_{dyn} is determined dynamically by a line-search procedure.

The need for fast on-line learning was the reason for choosing the stack-like organized training data set together with a line-search procedure for adjusting the weights. About 500 steel plates are rolled per day. Accumulating all data pairs $(\mathbf{x}^\mu, \alpha^\mu)$ would lead to a training data set which is much too large. Besides the huge memory and computation time requirement, this is also not desirable since the characteristics of the material transformation process might be drifting, and, therefore, very old data pairs might not be representative anymore. On the other hand, avoiding storage of training data completely by performing pattern by pattern learning makes sense only with a very small adaptation step size η_{dyn}. The corresponding slow-down of the adaptation procedure and the increased probability to get stuck in local minima, however, is not acceptable in the described application.

The line-search procedure for determining the optimal value of η_{dyn} looks for the minimum of $E(\mathcal{W})$ along the gradient of $E(\mathcal{W})$ at the current weight set $\mathcal{W}_0$, i.e., it looks for that $\eta_{dyn} = \eta_{opt}$ which minimizes

$$E\left(\mathcal{W}_0 - \eta_{dyn} \left.\frac{\partial E}{\partial \mathcal{W}}\right|_{\mathcal{W}_0}\right). \tag{2}$$

Determining η_{opt} based on a quadratic approximation of (2) is sufficient in our case. For this purpose we calculate two support points $(\mathcal{W}_1, E(\mathcal{W}_1))$, $(\mathcal{W}_2, E(\mathcal{W}_2))$ along the gradient in addition to $(\mathcal{W}_0, E(\mathcal{W}_0))$. These three support points define a parabola which forms a quadratic approximation of $E(\mathcal{W})$ along the gradient, i.e., a quadratic approximation of (2). As the size of adaptation step (1) we choose that $\eta_{dyn} \approx \eta_{opt}$ which minimizes the parabola.

4. The Performance

For testing the performance of the neural network approach and comparing it with the current state-of-the art method, 12000 data pairs $(\mathbf{x}^\mu, \alpha^\mu)$ from the rolling mill were made available by the steel manufacturer. 9000 data pairs formed the training set which was used for adapting the network, and the other 3000 data pairs formed the independent test set. The stack size P was chosen to be 500. The on-line training of the network was simulated by sequentially and randomly picking data pairs $(\mathbf{x}^\mu, \alpha^\mu)$ from the training set, putting the respective data pair on top of the FIFO stack, and performing an adaptation step (1). Already after having presented about 5000 samples, the network has converged from its randomly chosen initial to its final state. The achieved estimation performance is shown in Table 1.

<E_net>	<E_cur>	Δ	Worst_net	Worst_cur	Δ
34.9%	53.6%	35%	262.8%	302.2%	13%

Table 1: The RMS and the worst case estimation error of the neural network and the current state-of-the-art method, relative to standard deviation of the test data.

$\langle E_{net} \rangle$ denotes the root mean square (RMS) estimation error of the neural network on the data of the test set, relative to the standard deviation of the test data. $\langle E_{cur} \rangle$ denotes the RMS estimation error of the current state-of-the-art method on the test set, and Δ is the achieved improvement. Table 1 also shows the worst-case, i.e., the largest estimation error which occurred, for the neural network approach as well as for the state-of-the-art method. For the average estimation error the neural network approach yields an improvement of 35%, and for the worst-case the neural network approach yields an improvement of 13.2% over the current state-of-the-art method.

5. Discussion

The results obtained with the neural network approach are very promising. In the application described, the estimation of the relative yield stress of steel, the improvement of the estimation quality is so significant that the neural network approach will replace the current method and soon be a component of a commercially available process optimization system for rolling mills.

There are a couple of reasons for the favorable results with the neural network approach. The main reason is the on-line adaptation of the network. The network weights are permanently adjusted to the changing characteristics of the rolling mill and the drifts of the measuring devices for the chemical composition, thickness, width and temperature of the steel plate. Particularly the calibration of the measuring devices is not very reliable because of the very hazardous environment at a hot line rolling mill.

Another reason for the favorable results seems to be the superiority of feedforward neural networks with sigmoidal hidden units in modeling moderately complex, multivariate functions. To achieve a mean square error of ϵ, a feedforward neural network with $\mathcal{O}(\epsilon^{-1})$ hidden units is sufficient, whereas approximation through traditional trigonometric, spline or polynomial expansion requires at least $\mathcal{O}(\epsilon^{-D/2})$ terms [3]. D denotes the dimension of the input vector x, i.e., the number of input variables, and was 20 for estimating the relative yield stress. D is large in the described application domain since properties of a material depend on many variables.

References

[1] Werbos P (1974) "Beyond Regression: New Tools for Prediction and Analysis in the Behavioral Sciences." Ph.D. thesis, Harvard Univ. Committee on Applied Mathematics.

[2] Rumelhart DE, Hinton GE, Williams RJ (1986) "Learning Representations by Back-Propagating Errors." Nature, 323:533–536.

[3] Barron AR (1992) "Universal Approximation Bounds for Superpositions of a Sigmoidal Function." Technical Report #8, Department of Statistics, University of Illinois at Urbana-Champaign (to appear in IEEE Trans. Information Theory).

Hybrid Digital Signal Processing and Neural Networks for Automated Diagnostics Using Eddy Current Inspection

Wu Yan and Belle R. Upadhyaya
The University of Tennessee
Department of Nuclear Engineering
Knoxville, Tennessee 37996-2300, U. S. A.

ABSTRACT

The primary purpose of the current research is to develop an integrated approach by combining information compression methods and artificial neural networks for the monitoring of plant components using nondestructive examination (NDE) data. Specifically, data from eddy current inspection of heat exchanger tubing are utilized to develop this technology. The results of analysis show that for effective (low-error) artifact type classification and estimation of parameters, it is necessary to identify proper feature vectors using different data representation methods.

INTRODUCTION

A nuclear power plant is a complex system with the various components fulfilling the needs of process control and plant safety. Continued operation of these systems has both safety and economic implications. Near-term operation through proper maintenance and long-term plant life with consideration to component aging are important aspects of nuclear plant design and operation. Nondestructive examination (NDE) methods are being increasingly applied to the monitoring of critical components. Eddy current inspection, ultrasonic inspection, and thermographic inspection are the most commonly used methods.

The research being undertaken at the University of Tennessee focuses on the problem of automating NDE data analysis using a hybrid neural network and digital signal processing technique. In recent years research in neural networks has been advanced to the point where several real-world applications have been successfully demonstrated [3,10]. These include automated pattern classification, signal validation, nuclear plant monitoring, plant state identification during transients, estimation of performance related parameters, underwater acoustic signature classification and text recognition.

Eddy current inspection has been applied for crack and flaw detection; corrosion, thickness, and coating monitoring; and for monitoring changes in metallurgical properties. Three of the important areas of application are pressure vessel, steam generator tubes, and turbine-generators. Eddy current technique is used as the standard technique for steam generator tubing inspection [1,2]. The integration of neural networks and digital signal processing techniques for the automation of NDE signature analysis is a unique feature of this research. This research will also provide a technology base for the safety assessment of system and subsystem technologies used in nuclear power applications of neural networks.

EDDY CURRENT INSPECTION DATA

A large multi-frequency eddy current (EC) inspection database was acquired from the Metals and Ceramics Division of Oak Ridge National Laboratory. The data were obtained from

800

two series of measurements on an ASME Section XI standard specimen. A typical impedance plane (resistance versus inductive reactance) trajectory of data from an eddy current probe transducer is shown in Figure 1. The defect is in the tube support at a depth of 20% and the distance from the center of the defect to the artifact is 0.1 inch. The phase plane plot shows the trajectory corresponding to an AC source of frequency 60 kHz.

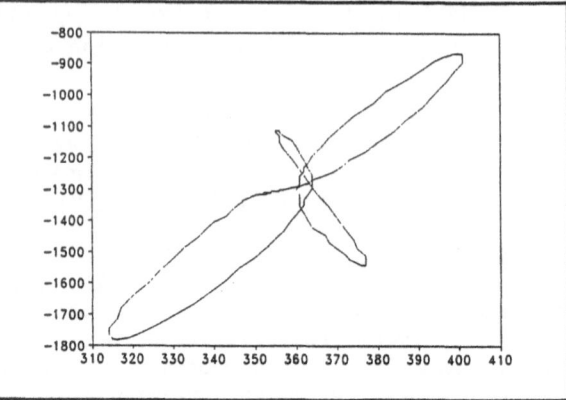

Figure 1 A typical impedance plane trajectory of data from an eddy current probe transducer.

SYSTEM DESCRIPTION

The general approach for automated artifact type classification and defect parameter estimation is shown in Figure 2. This integrates data representation and compression, and the development of artificial neural networks for anomaly detection and estimation. The key issues for the development of the automated system for NDE diagnostics include the following: (1) Digital data representation and information compression. (2) Development of robust neural networks with low probability of misclassification. (3) Correlation of neural networks results to failure or fault modes. (4) Provide guidelines for assessing neural networks technology as related to nuclear safety issues. The product of this research is envisioned to be an automated system for anomaly classification using NDE data and neural networks technology.

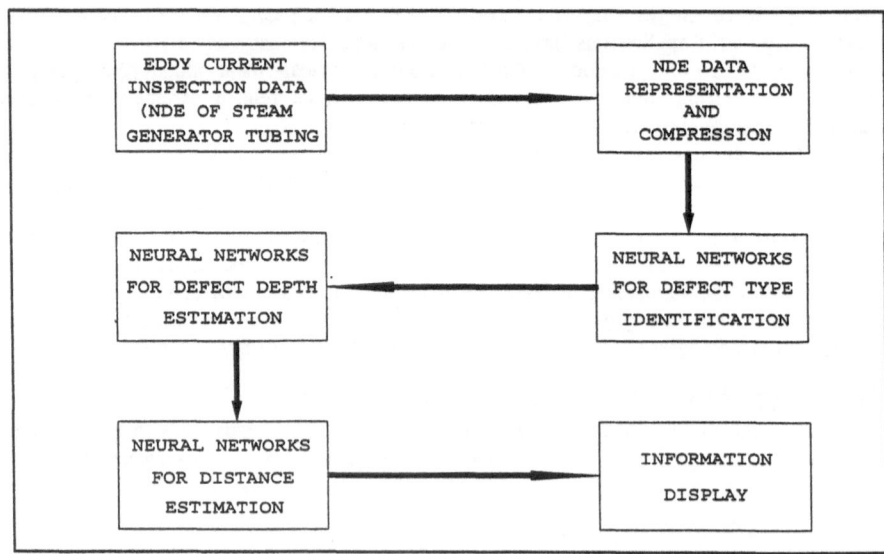

Figure 2 Schematic showing eddy current NDE data analysis using artificial neural networks.

DATA REPRESENTATION

For the neural network approach to be effective in defect type classification and defect parameter estimation, the information input to the network must have certain features. These are (a) size of data vector, (b) invariance to data scaling, (c) invariance to data orientation, and (d) sensitivity of the defect type and defect parameters to input signatures.

Data representation methods involve reorganizing the raw measurement data using (1) direct compression of raw data, (2) subtraction from a reference dada, (3) magnitude and phase of the raw data, (4) integral value of the raw data, (5) sequence of radii from the center of gravity to the closed contour of the shape, and (6) Fourier descriptors. Impedance plane integration and radii from the center of gravity have been found to be effective in data representation.

Compressed Integral Signal Representation (CINT). By observing the plots of integrated raw data, it is seen that the line integration of impedance data is sensitive to certain frequencies, and may be used to identify artifact type and estimate defect depth.

Radii From the Center of Gravity (CG). Since the defect parameters will influence the center of gravity of the complex impedance plot, a sequence of radii from the center of gravity to the contour of the shape is used to train neural networks to estimate defect parameters.

NEURAL NETWORKS FOR DEFECT TYPE CLASSIFICATION AND DEFECT PARAMETER ESTIMATION

Neural networks provide general mapping between two sets of information. This nonlinear mapping from data to data is very useful in associating information pairs where a clear mathematical relationship is not available. Neural networks have been applied to the problems of pattern classification, signal validation, plant monitoring, transient state identification in power plants, underwater acoustic signature recognition, and many others [12]. A general architecture of a four-layer feedforward network is shown in Figure 3. The input layer requires a signature vector from measured data. The network output may be in the form of a signature vector, or a pattern classification index. The number of processing elements in the intermediate layer is often determined experimentally. The back-propagation [11] algorithm is used to train the above neural network. Once the network is established it can be later used for diagnostic applications by providing measured signatures and tracking the network output.

The preprocessed eddy current impedance data are used as input feature to multi-layer neural networks, with the output map providing defect type and

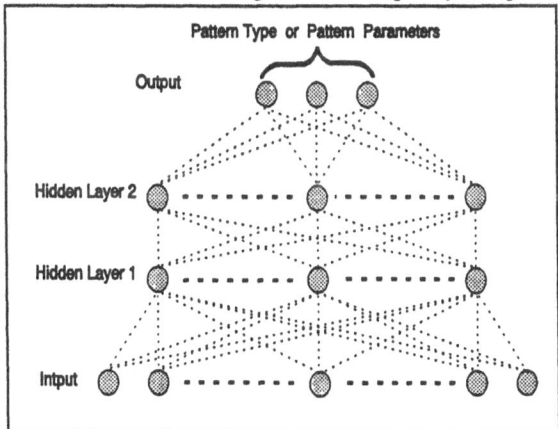

Figure 3 Architecture of a multi-layer neural network used in eddy current pattern classification and for defect parameters estimation.

estimates of defect parameters. Different data compression methods are implemented here and are ranked according to their effectiveness in producing proper mapping. Separate networks for artifact type classification and parameter estimation have been developed. A PC-based software called the NeuralWorks Professional II/Plus is used in most of the implementation.

RESULTS OF DEFECT DIAGNOSTICS USING EDDY CURRENT DATA
Estimation of Tube Defect Parameters

The defect parameters include defect depth and the distance from the center of the defect to the artifact. In order to make the networks perform effectively, separate networks are established to estimate either defect depth or distance. Currently, our efforts are focused on the estimation of the distance from the center of the defect to the artifact. Only one back propagation network was trained for depth estimation.

In the estimation of the distance from the center of the defect to the artifact, in order to compare the effects of the different data representation techniques, only the data corresponding to defect in tube-support, 20% depth, 0.1 to 2.0 inch sizes are used. In defect depth estimation, only the defect in tube-support, at 20% to 100% depths are used. The center of gravity radii signature are employed.

Generally, the mean-squared error (MSE) is used to evaluate the accuracy of the network. The MSE is defined as

$$MSE = \frac{1}{N} \sum_{k=1}^{N} [x_p(k) - x_m(k)]^2$$

where
N = Number of patterns.
$x_p(k)$ = the network predicted value for pattern k, and
$x_m(k)$ = the measured value for pattern k.

In parameter estimation, an Averaged Absolute Scale Error (AASR) is also used to judge the accuracy of the network. This is defined as

$$AARE(\%) = \frac{1}{N} \sum_{k=1}^{N} |\frac{100(x_p(k) - x_m(k))}{x_m(k)}|$$

where N, $x_p(k)$, and $x_m(k)$ are the same as above.

Different backpropagation networks (BPN) have been tried using different number of hidden layers, hidden-layer elements, number of iterations, and training coefficients. Table 1 lists the information for the best backpropagation networks obtained from training for the different data representation techniques. Figure 4 shows the recall result of the distance from the center of the defect to the artifact using the center of gravity (CG) training data.

Table 1. Networks for Distance Estimation

	CG	CRS	FD	FS
No. of Input Elements	50	100	8	8
No. of Hidden Layers	1	1	2	1
No. of Hidden Elements	50	50	15 10	20
Learning Coefficient	0.3	0.3	0.5	0.3
Transfer Function's Shape	0.7	0.7	1.0	0.7
Momentum Term	0.5	0.5	0.6	0.5
No. of Iterations	13501	30233	10000	228012
Mean Squared Error (MSE)	0.00323	0.00416	0.2128	0.0423
Avg. Abs. Scale Error (AARE) (%)	0.169	1.095	--	12.11

CG: center of gravity training data, CRS: compressed raw data subtraction training data,
FD: Fourier descriptors training data, FS: Fourier descriptors subtraction training data.

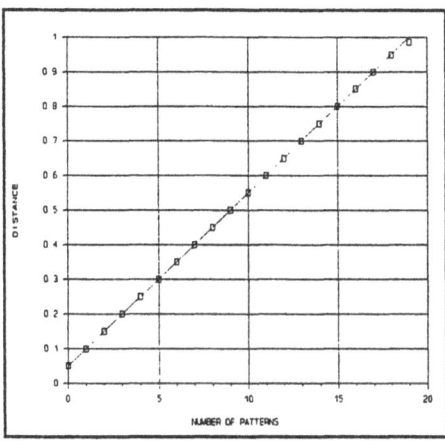 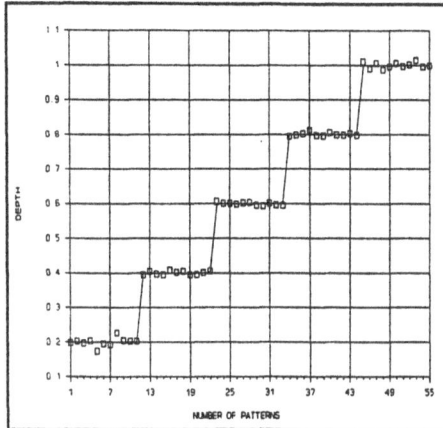

Figure 4   Recall result of BPN using CG training   **Figure 5**   Recall result of BPN using CG training
              data   data for depth estimation

From the above results, it can be seen that the AARE for the CRS data is 1.095%, and for the CG data is 0.169%. Therefore, both of them can be used to estimate the distance from the center of the defect to the artifact.

One network was trained for the tube-support defect depth estimation. Figure 5 shows the recall results of the BPN using center of gravity training data. It can be seen that the estimated depths are very close to the desired results. Therefore, center of gravity data representation method is also good for defect depth estimation. More study will be performed for depth estimation using other data representation methods.

EC Defect Type Identification

Table 2 lists the information for the best networks obtained for each of the Fourier descriptors (FD), compressed magnitude and phase (MP) and compressed integral (CINT) signatures.

Table 2. Networks for EC Defect Type Identification

	FD	MP	CINT
No. of Input Elements	8	100	100
No. of Hidden Layers	2	2	1
No. of Hidden Elements	15 10	10 10	50
Learning Coefficient	0.5	0.7	0.3
Transfer Function's Shape	1.0	1.0	0.7
Momentum Term	0.6	0.9	0.5
No. of Iterations	10000	50000	12561
Mean Squared Error (MSE) for Output 1	0.2687	0.0238	0.0369
Mean Squared Error (MSE) for Output 2	--	0.1928	0.1531
Mean Squared Error (MSE) for Output 3	--	0.1942	0.1914

From the current results, it can be concluded that the BPN network is able to identify the defect type very accurately.

CONCLUSIONS AND FUTURE WORK

The current results of analysis show that for effective (low-error) defect type classification and estimation of parameters, it is necessary to identify proper feature vectors using various data representation methods. "Center of gravity" and "data subtraction" have been found to be sensitive in defect parameter estimation. The magnitude and phase signature and the compressed integral signature are very effective in defect type identification.

The continuing research includes development of networks for defect type identification, defect parameter estimation, further study on feature representation relating NDE data to incipient changes in components behavior (relationship to aging), and evaluation of performance of neural networks using a large eddy current test database.

ACKNOWLEDGMENTS

This research and development work has been made possible by a University Grant sponsored by the U.S. Nuclear Regulatory Commission, Office of Nuclear Regulatory Research. The authors acknowledge the assistance of Dr. C. V. Dodd, Metals and Ceramics Division of ORNL, for technical discussion and providing eddy current tube inspection data.

REFERENCES

1. W. E. Deeds and C. V. Dodd, "Eddy Current Inspection of Steam Generator Tubing," Electromagnetic Methods of Nondestructive Testing, Vol. 3 of Nondestructive Testing Monographs and Tracts, W. Lord, Ed., Gordon and Breach, New York, 1985.
2. C. V. Dodd and W. E. Deeds, "In-Service Inspection of Steam-Generator Tubing Using Multiple-Frequency Eddy-Current Techniques," Special Technical Publication, American Society for Testing and Material, Philadelphia, PA., 1981.
3. L. Udpa and S.S. Udpa, "Neural Networks for the Classification of Nondestructive Evaluation Signals," IEE Proceedings-F, Vol. 138, No. 1, February 1991.
4. T. Pavlidis, "Algorithms for Shape Analysis of Contours and Waveforms," IEEE Transaction on Pattern Analysis and Machine Intelligence, Vol. PAMI-2, No. 4, July 1980.
5. "Eddy Current Nondestructive Testing," U.S. National Bureau of Standards Special Publication 589, 1981.
6. E. Persoon and K. S. Fu, "Shape Discrimination Using Fourier Descriptors," IEEE Trans., SMC-7, pp. 170-179, March 1977.
7. G. H. Granlund, "Fourier Preprocessing for Hand Print Character Recognition," IEEE Trans., C-21, pp. 195-201, February 1972.
8. "Neural Computing," NeuralWorks Professional II/PLUS and neural Works Explorer Software, NeuralWare, Inc., 1991.
9. R. C. McMaster, P. McIntire, and M. L. Mester (Eds), "Nondestructive Testing Handbook," Vol. 4 (Electromagnetic Testing), Am. Soc. for Nondestructive Testing, 1987.
10. B. R. Upadhyaya and E. Eryurek, "Application of Neural Networks for Sensor Validation and Plant Monitoring," Nuclear Technology, Vol. 27, No. 2, pp. 170-176, 1992.
11. D. Rumelhart and J. McClelland, Paralled Distributed Processing, Vol. 2, Bradford Books/MIT Press, Cambridge, MA, 1986.
12. R. P. Lippmann, "An Introduction to Computing with Neural Nets," IEEE ASSP Magazine, Vol. 4, No. 2, pp. 4-22, April 1987.

Industrial applications
—poster contributions

Self-Organizing Neural Network for Diagnosis

Pietro Morasso, Alberto Pareto, Stefano Pagliano, and Vittorio Sanguineti

Department of Informatics, Systems, and Telecommunications
University of Genova, Italy

Abstract. The paper describes an approach to diagnostic applications that uses a self-organizing classifier, capable of performing incremental learning and of dealing with noisy data, and allows to estimate the *distance* from pathological regions and the *time-to-failure*.

1. Introduction

Both in industrial and medical applications the need is emerging of diagnostic systems that, in addition to the classification of a system's state, can also provide some estimate, for example when the state is classified as *normal*, of the pathological condition which is more likely to develop and of the expected *time-to-failure*. Moreover, in many cases, the problem has a markedly dynamic character because the experimental data may become available incrementally, i.e. after the design phase, and/or pathological classes might slowly evolve with time. Neural network models are quite useful for establishing effective and robust solutions, particularly if the studied processes are complex enough to escape the possibility of a reliable formalization. Such models should be able to carry out incremental category formation, category tracking, and should provide an explicit representation of the decision surfaces. In the next sections, we outline a neural model, based on self-organizing concepts.

2. Incremental category formation: the probabilistic SOC model

For the type of applications outlined above, it is not appropriate to rigidly separate a learning phase, based on a pre-established database of examples, from a subsequent operational phase. On the contrary, it may be desirable to rapidly reach a minimum performance with a small training set, in order to bootstrap a useful behavior, and to continuously update the network, thus blurring the boundary between learning and recognition. This is the concept of *incremental category formation* (*ICF*), that implies a dynamic neural architecture without a pre-fixed number of neurons but self-adapting to the intrinsic complexity of the categorization problem. In general, adaptation of a network's size and connectivity can be achieved either by means of pruning down (a large network) or growing up a simple one. The methods of the former type, like "optimal brain damage" [1], are not very appropriate for *ICF* because they require a quite heavy bootstrap phase. As regards the growing methods, we may distinguish between networks that construct multiple layers of simple perceptrons and networks based on the tessellation of the feature space by means of prototypical feature vectors corresponding to the weight-vectors of the different neurons. The methods of the latter type are much better for our purpose because neurons have a localized receptive field that allows to limit each learning episod to a small part of the network. In particular, vector-quantization models such as RCE [2], GAL [3], or the growth model [4] attempt to cover classification regions by means of a set of prototypes whereas the SOC model [5] follows a different philosophy in the learning process, i.e. allocating the prototypes where they are more relevant (near the decision boundaries). We also used self-organization concepts proposed in LVQ/LVQ2 [6] and in the "neural gas" [7].

The first version of the SOC model was deterministic, in the sense that it took decisions affecting network's size and topology instantaneously, without any consideration of the probabilistic distribution of the classes. On the contrary, the nature of the problem that we are considering requires a sufficiently stable identification of decision boundaries, particularly when the noise in the data determines a significant overlap of the probability distributions of the different classes. This goal can be obtained by means of a probabilistic SOC model that exploits a local statistical representation.

In more detail, a probabilistic SOC model (figure 1) has a two-layer structure with a variable set of hidden neurons and a set of output neurons equal to the number of categories. The hidden neurons N_i's have the same input vector $\mathbf{x}$ (i.e. the input layer has full connectivity) and are identified by a weight vector $\mathbf{w}_i$ with the same dimensionality of the input vector and a class label C_i; their activation function is competitive (winner-take-all). The output neurons are only connected to the hidden neurons with the same category label and their activation function is simply an inclusive-or (the neurons perform a Voronoi tessellation of the input feature space).

Incremental learning is based on three rules (*birth*, *tuning*, and *death* rule), used for each new input vector, which may create new hidden neurons, delete old ones, or update the weight vectors of resonant neurons. For the purpose of learning, the winner-take-all activation function singles out both the first and the second best matching hidden neurons, matching being based on the euclidean distances between the input and the weight vectors.

The tuning rule is the basic self-organizing engine, quite in accordance with the LVQ2 model [6]. The rule is applied if the two winners are labelled differently and one of the labels is correct, which means that the input pattern is near the border between a class region and another one. Supposing that N_i and N_j are the two winners and $C=C_i=C_j$, then the rule can be written as follows:

$$\Delta\mathbf{w}_i=\eta(\mathbf{x}-\mathbf{w}_i) \qquad \Delta\mathbf{w}_j=\eta(\mathbf{w}_j-\mathbf{x}).$$

In addition to this mechanism, which is known to optimally distribute in the feature space a given set of vector prototypes (or weight vectors), the birth and death rules are meant to self-adapt the total number of prototypes and disseminate them to the different categories. In a previous deterministic SOC model [5] the decision to create/kill neurons was performed instantaneously, i.e. on the base of the current input vector. This worked very well for noiseless data or for noisy data with little overlap between probability distributions of different categories. For noisy data with substantial overlap, the instantaneous birth/death approach leads to irregular boundaries characterized by spurious neurons and undamped non-linear dynamics.

The problem is, in general, that birth/death of neurons in a neural network is a highly non-linear operation that must be dealt with care. A global approximation of the probability distribution of each class would certainly be useful as it is performed in some probabilistic networks [8] by means of gaussian mixtures, but this is excessive for the purpose of SOC, which can work with a simpler local statistics (and a smaller set of neurons). In particular, for each neuron of the network four counters are used that estimate the following probabilities: p_{11}, p_{10}, p_{01}, p_{00}. They are updated whenever the neuron resonates, also taking into account the second-best neuron, by detecting four possible case: (i) both neurons give a correct classification (p_{11}), (ii) only the first one is correct (p_{10}), (iii) only the second one is correct (p_{01}), (iv) both are wrong (p_{00}). A fifth probability (p_r) depends on how frequently the neuron actually resonates. The death rule can then be implemented by detecting three circumstances:

- If p_r is very small, i.e. if the neuron is located in a tail of the distribution or has been "left behind" in the case of time-varying distributions.

- If p_{11} is close to 1, which means that, very likely, the weight vector of that neuron corresponds to an "internal" region of the Voronoi tessellation and thus does not directly contribute to the decision surface.

- If the neuron is "near" the decision surface but it is likely beyond the local Bayesian class boundary (characterized by the fact that the posterior probabilities of the two neighboring classes are equal). This event can be detected if $p_{01} > p_{10}$.

As regards the birth rule, it has been modified, in the same line of reasoning, by conditioning the decision to the fact that p_{00} is close to 1. In the simple case of two classes, the algorithm finds just two neurons that well approximate the Bayes condition. Similar results were found in two dimensions with overlapping Gaussian distributions (figure 2), i.e. the probabilistic SOC model approximates the decision boundaries of the different classes with a self-adapted number of prototypes and, at the same time, can cope with substantial overlapping of the distributions.

3. Diagnostic functions

A SOC model provides an explicit representation of the decision surfaces, at the local level: the neurons perform a Voronoi tessellation of the input feature space (hyper-polyhedra centered around each weight vector) and the decision surface between a class and another one is piece-wise planar, i.e. it is a composition of hyper-faces of the boundary hyper-polyhedra. In particular, if x_i and x_j are two prototype vectors which are neighbors in the Voronoi tessellation with different class labels, then the local decision surface is a hyperplane through the mid-point ($q_{ij}=(x_i+x_j)/2$) with a normal directed as the line joining the two points ($n_{ij}=(x_i-x_j)/ \mid x_i-x_j \mid$) and is characterized by the following equation: $(x-q_{ij})^T n_{ij} = 0$. A typical diagnostic function is to estimate the "safety" of the current operating condition and the relative probability of occurrence of the different failures. This can be achieved by measuring the distance of the current state vector $x(t_k)$ from the boundaries of the different pathological classes: $d_1(t_k), d_2(t_k), \ldots$ We implemented this function, in the context of SOC, by means of a *diffusion metaphor*. This means that an array of probing directions is generated starting from the current state and for each direction the point is detected where the bondary between the "normal" region and an "abnormal" one is crossed. Therefore, the different pathological conditions can be ranked according to the values of the corresponding distance parameters. As regards the *time-to-failure*, it can be estimated by computing distance at different time instants ($d_\lambda(t_k)$) and performing a linear or non-linear extrapolation to the zero-crossing. Preliminary experiments have been performed with promising results.

References

[1] Le Cun, Y., Denker, J.S., and S.A. Solla (1990) Optimal brain damage. In "Advances in neural information processing systems" (D.S. Touretzky ed.), 2, Morgan Kaufman, 598-605.

[2] Reilly, D.L., Cooper, L.N., and C. Elbaum (1982) A neural model for category learning. Biological Cybernetcs, 45, 35-41.

[3] Alpaydin, E. (1991) Grow and learn: an incremental method for category learning. International Neural Network Conference, Paris, France.

[4] Fritzke, B. (1991) Let it grow. Self-organizing feature maps with problem dependent cell structure. In "Artificial Neural Networks" (T. Kohonen, K. Makisara, O. Simula, and J. Kangas, Eds.), 1, 403-408, North Holland, Amsterdam.

[5] Morasso, P., Pagliano, S., and A. Pareto (1992) Neural models for handwriting recognition. In "Proceedings of the Second International Workshop on Frontiers in Handwriting Recognition" (S. Impedovo and J.C. Simon, Editors), Elsevier, Amsterdam.

[6] Kohonen, T. (1989) Self-Organisation and Associative Memory (3rd ed.). Springer-Verlag Series in Information Sciences, Berlin.

[7] Martinetz, T. and K. Schulten (1991) A "neural gas" network learns topologies. In "Artificial Neural Networks" (T. Kohonen, K. Makisara, O. Simula, and J. Kangas, Eds.), 1, 397-402, North Holland, Amsterdam.

[8] Specht, D. F. (1990) Probabilistic neural networks, Neural Networks, 3, 109-118.

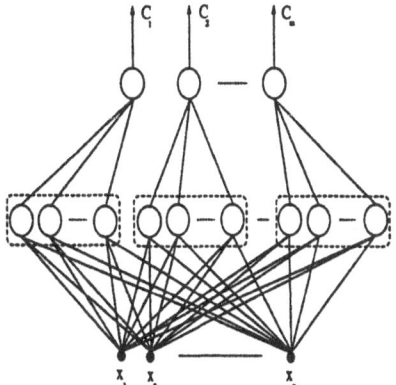

Figure 1: SOC-Net architecture

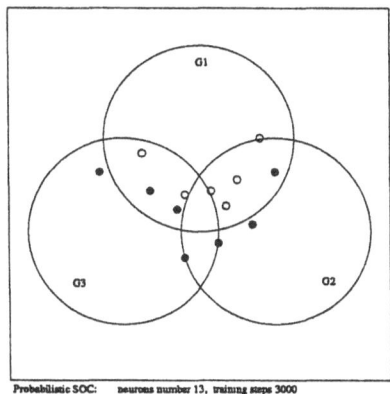

Figure 2: Gaussian overlapping distributions

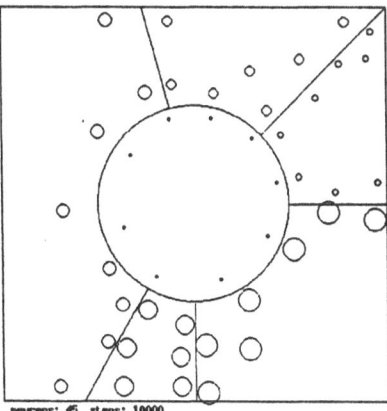

Figure 3: A central "normal" class surrounded by "pathological" classes

Limitations of Adaptive Critic Control Schemes

W.T.C. van Luenen[1], P.J. de Jager[1], J. van Amerongen[1] and H.M. Franken[2]

1: Control Laboratory and Mechatronics Research Centre Twente
2: Biomedical Engineering Division
Dept. of Electr. Eng., University of Twente P.O. Box 217, 7500 AE Enschede,
Netherlands, e-mail: lnn@rt.eltn.utwente.nl

Abstract. *Earlier publications about learning control systems used adaptive critic schemes. We argue that the results published about adaptive critic schemes in control problems appeared positive due to the choice of the inverted pendulum or cart-pole system as a test case. This conclusion was drawn after e*
xamination of a number of publications and after our experiments with the original adaptive critic scheme. In this paper we describe an adaptive critic scheme derived from a control point of view. This scheme allows us to indicate more clearly the limitations of the scheme in a control context. We show how an adaptive critic scheme fails to learn a set-point control for a simple second order non-linear process in a simulation.

1. Introduction

The adaptive critic algorithm as originally published by Barto et al. [2] may be regarded as a self-learning controller, and was applied successfully for learning to balance a pole. The attractive property of the adaptive critic algorithm is that it learns a control task without the use of a model of the process. It uses a criterion predictor, CP, instead of a model and it has been traced back to early literature [10]. Here we discuss an approach in which the CP and the controller are learning simultaneously. Many papers have confirmed the result in [2] and have elaborated on it ([1],[4],[5],[6]). To the best of our knowledge, these results have all been obtained by simulations. Application of this algorithm on an experimental set-up, an inverted pendulum, has resulted in rather poor behaviour [6], motivating the development of an improved algorithm. This paper is the result of tracing the fundamentals of the adaptive heuristic critic algorithm and experimental work in simulation as well as on practical systems.

Problems encountered with the original algorithm when applied to an experimental set-up were:
a) increased sensitivity for disturbances due to quantized state variables;
b) the excitation of dynamics caused by the binary control output;
c) the process states were not driven towards a particular set-point.

An algorithm published by Anderson [1] provides an approach which solves the problem of the quantized states using two neural networks. Werbos [10] has suggested a gradient learning approach and various algorithms to obtain a continuous real-valued controller output. This approach could solve the problem of the undesired binary control.

As far as we know, the problem of not reaching a set-point is overlooked in earlier publications ([3], [8],[9]). The example of the cart pole more or less hides this problem, because both the adaptive critic scheme and a conventional control algorithm (say an LQG state feedback controller) result in a balanced pole. However, careful examination of the original experiment shows that each time a learning sequence is completed (the pole is balanced) the cart is at a different (average) position along the track. This aspect becomes more obvious when a different process is taken as shown in our example later on.

In the original scheme, a binary reinforcement signal is used for learning. One might argue this is the reason that the system does not reach a set-point. To solve this, we have used a scheme with a real valued reinforcement signal (or criterion in control terms). This criterion has a minimum value for the desired zero state. If the adaptive critic scheme is appropriate, it will find the minimum and this will solve the set-point problem.

The temporal difference approach [9] used in [2] was maintained. The control aim is to learn an optimal controller mapping from state space to control space. This mapping is created by learning an optimal control action for each particular state as will be explained in the following.

A sequence of control actions is needed to bring a process from an initial state to a desired state (set-point) while minimizing a criterion (a reinforcement in terms of [1]). In previous publications ([1],[2],[4], [5]) the reinforcement simply indicated a desired area of state space. We used a more conventional (in terms of classical control) real-valued criterion in which states and control signal are taken into account because we aim to use a real-valued control signal and we desire to position the system at a set-point.

Normally a process model is used to calculate a control sequence, using e.g. a dynamic programming procedure. Because an optimal trajectory exists for each initial state, this results in a unique sequence of control actions and therefore in a unique control action for each (intermediate) state. Hence, there is a unique mapping from state space to control action space. Implementation of a closed loop controller requires a knowledge structure (e.g. a neural network) [7] in which a control action is stored for each state. Such an algorithm can operate in closed loop, because at each discrete time instant (sampling moment), the present state will result in a control action. As a result of this action and possible disturbances, a new state is entered on the next time instant for which the related optimal control action will be used.

Adaptive critic designs use a Criterion Predictor (CP) instead of a process model in order to adjust the controller mapping. In a normal dynamic programming procedure, the actual criterion outcome J is calculated from all states passed between the initial state and the set-point and the accompanying control signals. This requires a model of the process. In the adaptive critic algorithm, an approximation $\hat{J}$ is obtained by successive predictions of the criterion outcome. This prediction is based on the current state $\underline{x}$ and the current control signal $\underline{u}$. It therefore requires a mapping with $\underline{x}$ and $\underline{u}$ as input and $\hat{J}$ as output. This mapping, the predicted criterion surface, is unique for a particular process in combination with a particular criterion. An implementation of a CP can be obtained using various knowledge representations. Examples in literature use a table [2], a multi layer feed forward network [1] or a fuzzy rule base [4].

2. Adaptive critic learning control scheme

The following algorithm, based on the ideas mentioned before, has been used in our experiments. Given a non-linear process $\underline{x}_{k+1} = f(\underline{x}_k, \underline{u}_k)$ with discrete state equations (state $\underline{x}$, control $\underline{u}$, time k) and a criterion over an infinite time with exponential discounting to ensure boundedness:

$$J_k = \sum_{i=0}^{\infty} \gamma^i g_{k+i+1}(\underline{x}_{k+i+1}, \underline{u}_{k+i+1}) = \sum_{i=k}^{\infty} \gamma^{i-k} g_{i+1}(\underline{x}_{i+1}, \underline{u}_{i+1}) \tag{1}$$

where γ $(0 < \gamma \le 1)$ is a discount factor. The momentary state transition costs are:

$$g_i = \underline{x}_i^T Q \underline{x}_i + \underline{u}_i^T R \underline{u}_i \tag{2}$$

This is a more severe criterion than the constraints used in [2]. The learning algorithm uses the error between successive predictions of the criterion (temporal difference learning [9]) to indirectly (via the CP) adjust the controller. The relation between two successive values of J_k and J_{k+1} is

$$J_k = g_{k+1} + \sum_{i=k+1}^{\infty} \gamma^{i-k} g_{i+1} = g_{k+1} + \gamma \sum_{i=k+1}^{\infty} \gamma^{i-(k+1)} g_{i+1} = g_{k+1} + \gamma J_{k+1} \tag{3}$$

A prediction error resulting from an incorrect prediction can be written as

$$\varepsilon_k = \hat{J}_k - \left(g_{k+1} + \gamma \hat{J}_{k+1} \right) \quad or \quad \varepsilon_{k-1} = \hat{J}_{k-1} - \left(g_k + \gamma \hat{J}_k \right) \tag{4}$$

The control signal is calculated from the measured state while the prediction of the CP is calculated from this control signal and the measured state. Using the criterion, a gradient approach can be used to calculate the adjustments of the parameters of the CP ($\underline{w}_j$) and the controller ($\underline{w}_c$). In our case, layered neural networks and backpropagation [8] were used.

In some ways, the CP has a similar function to a process model which needs to be identified. As a result of this, it is preferable that the CP has learned before the controller is adjusted. However, within the commonly used adaptive critic algorithm, both controller and CP are learning simultaneously. Therefore, the learning factor of the CP will be given a higher value than the learning factor of the controller. The pseudo code for the control algorithm at time instant k is:

```
retrieve last state x_{k-1} and last control output u_{k-1}
compute estimated J_{k-1} using w_{j,k-1} from x_{k-1} and u_{k-1}
input x_k, compute u*_k, a suggestion for u_k using w_{c,k-1} and x_k
compute g_k and J_k using w_{j,k-1} from x_k and u*_k
compute prediction error ε_{k-1}
use ε_{k-1} to adjust weights in CP to w_{j,k}
once again compute J_k but now using w_{j,k}
compute gradient dJ_k/du_k
adjust control weights to w_{c,k} using gradient dJ/du
compute u_k but now using w_{c,k}
send u_k to process resulting in next state x_{k+1}
```

3. Analysis

We assume no a priori knowledge is available within the controller and the CP. Practically this means that the criterion surface is approximated by some appropriate knowledge structure with a large set of parameters, e.g a neural network. The initial parameter settings are random. This approach is similar to the approach taken by for instance [1] and [2]. As a result, the initial shape of the CP criterion surface is flat. During a (stochastic) learning procedure, this surface will develop. A flat area indicates that the predicted criterion outcome is relatively insensitive for changes in the values of its inputs: the state and the control. This causes the following sequence of events.

1) In the early stages of learning only a few parameters will be adjusted while the majority stays relatively small. This roughly divides the surface into regions of which some are flat, because they were not visited during learning.

2) Each time a flat area is visited, the CP tries to minimize the error

$$\varepsilon_{k-1} = \hat{J}_{k-1} - \left(g_k + \gamma \hat{J}_k\right) \approx \hat{J} - \left(g_k + \gamma \hat{J}\right) \tag{5}$$

because the difference between successive predictions in a flat region is small. This error can only become zero if $\hat{J}_{k-1} \approx \hat{J}_k = \hat{J} = g_k/(1-\gamma)$ which may be achieved by adjusting only a few parameters. As a result of this an entire (flat) region takes this value of $\hat{J}_k$ and the CP stops adjusting its weights in this region.

3) In a flat region in the CP surface, the gradient is negligible and the behaviour of the process is mainly determined by the dynamics of the process and the control signal u. The control signal will not be adjusted in a flat region because of the small gradient.

4) Depending on the process, the (incompletely learned) controller may either I) drive it out of the flat region, II) drive it into a limit cycle within the flat region or III) drive it to a steady state within the flat region. The learning strategy of trial and error may bring the process in the region in subsequential trials. If it enters the region and remains there, it will either be in a limit cycle or in a steady state. A limit cycle may contain the set-point but there is no reason to believe that a steady state is equal to the set-point. As a result, the complete system is at rest but not at the set-point.

4. Experimental results

The process model considered is a simple non-linear 2nd order model of a 1-DOF robotic arm. The equation is

$$\ddot{\varphi} = \frac{1}{J}\left(-F_v\dot{\varphi} + u + mgl\sin(\varphi)\right) \tag{6}$$

with φ [rad] the position of the arm, u [Nm] the input torque, the mass m = 1 [kg], the gravity constant g = 9.81 [m/s^2], the length l = 0.5 [m] and the viscous friction F_v = 0.05 [kgm^2/s]. The maximum torque U_{max} = 10 [Nm].

A state feedback controller with gravity compensation was designed to serve as a reference and to test CP learning:

$$u_k = K_p(\varphi_{sp} - \varphi_k) - K_d\dot{\varphi}_k - mgl\sin(\varphi_k) \qquad K_p = 10 \ and \ K_d = 2.4 \tag{7}$$

The linear part of this controller resulted from LQG theory and the solution of the Riccati equation for the criterion function (2). Controller inputs were the set-point and the measured position and the measured velocity. The CP also used these inputs and in addition the control signal generated by the controller. The velocity and control signal were scaled down by a factor 0.1 compared with the position signals [1].

The artificial neural networks for the controller and the CP were multi-layer feed forward networks with two layers of sigmoid units and a linear output unit. All networks had 40 units in the first layer, 20 in the second layer and one linear output unit. A constant learning factor was used for all weights.

For the discount factor (see (2)), a value of $\gamma = 0.98$ was chosen, which appeared to be a fair compromise between having a large enough horizon while maintaining a large enough influence of the current control signal on the output of the predicted criterion [3]. This value of γ was used in all subsequent experiments.

In a first experiment the conventional controller was used to test the temporal difference learning algorithm of the CP. This resulted in a criterion surface which had errors at the origin and the edges but which had a shape which resembled the real criterion surface as computed for each initial state.

In a second experiment the complete algorithm was evaluated. Because the weights in the controller and criterion predictor were initialized randomly, at first the controller failed to stabilize the process. After a number of trials (typically 200 or more) the behaviour improved. First the controller failed to stabilize the process, the state moved away from the origin immediately. Later the trajectories were starting to circle around the origin. Finally, after ± 600 trials equivalent to 1.5 hours simulation, the system converged to a steady state which appeared to be different for each sequence of trials. The steady state was not equal to the origin. Inspection revealed that this steady state corresponded to a global minimum in the CP surface [3]. Further trials from various initial states all resulted in convergence to the undesired steady state.

In a third experiment, gaussian noise was added to the control signal with a variance of 30% of the maximum torque. The results were similar to the ones shown without noise.

5. Conclusions

This paper indicates limitations of the adaptive critic scheme if regarded from a control point of view. The straightforward application of the original scheme on an experimental setup results in a number of problems. A modified scheme, designed to solve these problems, is used to explain the origin of these limitations. A second order non-linear example illustrates what goes wrong.

Literature

[1] C.W. Anderson, "Learning to control an inverted pendulum using neural networks", IEEE Control Systems Magazine, vol. 9, pp. 31-37, Apr. 1989
[2] A.G. Barto, R.S. Sutton, and C.W. Anderson, "Neuronlike adaptive elements that can solve difficult learning control problems", IEEE Tr. Syst. Man Cybern., SMC-13, pp. 834-846, 1983
[3] P.J. de Jager, "Approximating on-line predictive control using artificial neural networks", M.Sc. Thesis, report 92R080, Control lab, Dept. of El. Eng., University of Twente, Netherlands
[4] C.C. Lee, and H.R. Berenji, "An intelligent controller based on approximate reasoning and reinforcement learning", Proc. IEEE Int. Symp. on Intell. Control, pp. 200-205, 1989
[5] C.S. Lin, and H. Kim, "CMAC-based adaptive critic self-learning control", IEEE Tr. Neural Networks, vol. 2, no. 5, pp. 530-533, Sept. 1991
[6] W.T.C van Luenen, "Real time reinforcement learning control of dynamic systems applied to an inverted pendulum", In "Symbols versus Neurons", Eds. J. Stender, T. Addis, IOS Press, 1990
[7] W.T.C van Luenen, "Artificial neural networks used for on line modelling and control", internal report, report 91R110, Control lab, Dept. of El. Eng., University of Twente, Netherlands
[8] D.E. Rumelhart, G.E. Hinton, and R.J. Williams, " Learning internal representation by error propagation", in "Parallel Distributed Processing", vol 1, Cambridge MA: MIT Press, 1986
[9] R.S. Sutton, "Learning to predict by the methods of temporal differences", Machine Learning, vol. 3, pp. 9-44, 1988
[10] P.J. Werbos, "A menu of designs for reinforcement learning over time", in "Neural networks for control", Eds. Miller, Sutton, Werbos, Cambridge, MA: MIT Press, 1990

Periodic Disturbance Rejection: A Neural Network Approach

Alexander Medvedev   Gunnar Hillerström

Division of Automatic Control, Luleå University of Technology

S-951 87 Luleå, SWEDEN

Abstract

A feedback controller comprising a Luenberger observer and a Time–Delay Feedforward Neural Network is shown to possess a periodic disturbance rejection property for linear multivariable dynamic systems. Stability of the resulting closed–loop system is proved and the controller performance is exemplified by a simulation example.

1. Introduction

Model–based disturbance rejection can conventionally be done in two ways. The Internal Model Principle(IMP) (Francis and Wonham[2]) stipulates that a feedback controller should *incorporate* a disturbance model to possess a disturbance rejection property. In contrast, Tomizuka *et.al.*[8] introduced an External Model *outside* the feedback loop so that the model output cancels the disturbance signal.

The controllers which contain a model of a periodic disturbance/set–point are known as *repetitive controllers*, Tomizuka *et.al.*[7].

Recently, neural networks have received much attention from the point of view of their possible application to the control system area. The benefits expected from the neuromorphic architectures are: (1) simpler controller structures, (2) ability to handle nonlinearities, and (3) wide–range adaptation possibilities. Unfortunately, in practice, the lack of theoretical analysis and formal design procedures leaves the neuromorphic controllers behind their mainstream rivals. Very seldom a neural network–based control design overperforms the traditional approaches in the field of dynamic systems, in contrast with their definite success in solving static problems like handwritten character recognition, XOR–problem, etc.

The model–based control structures, such as state vector estimators and model reference controllers, pose a problem of training a neural network to simulate the desired or actual plant dynamics. To be used in control systems, a neural model should satisfy certain design criteria, such as modeling accuracy, robustness, filtering properties etc., which are hard to achieve simultaneously by training. Thus, it becomes rather important to find the neural network architectures which are akin to the well–known non-neural control structures and use these analogies for the basic design analysis.

In the present paper, we develop a neural network–based feedback controller to eliminate disturbance effects in the output signal of a multivariable continuous system.

The paper is composed as follows. Firstly, we describe the structure of the proposed controller and the nature of the underlying state estimation problem. A similarity between Time–Delay Neural Networks (TDNNs) and Continuous Dead-beat Observers (CDBO) allows to use the former for reconstructing disturbance from the output and input signal measurements of the plant.

Then, a combination of a conventional Luenberger observer and TDNN is exploited to achieve disturbance rejection in a feedback control system subjected to a periodic disturbance of a form unknown in advance. It is shown that TDNN's training can be performed without any impact on the closed–loop stability properties. Finally, an application of the proposed structure to a pump is described to demonstrate feasibility of the method.

2. Problem Statement

Consider a multivariable linear system

$$\dot{x}_p(t) = A_p x_p(t) + B_p u(t) + E_1 \omega(t)$$
$$y(t) = C_p x_p(t) + E_2 \omega(t) \qquad (1)$$

where $x_p(t) \in R^n$ is the state vector, $u(t) \in R^m$ is the control vector, $y(t) \in R^l$ is the observation vector, and $\omega(t) \in R^r$ is a disturbance signal. A_p, B_p, C_p are real matrices of appropriate dimensions. Moreover, assume that the disturbance signal can be described as the output of an autonomous system governed by the following equation

$$\dot{\omega}(t) = F\omega(t) \qquad (2)$$

where $F \in R^{r \times r}$.

Now, the disturbance rejection problem can be defined as designing a feedback control law $\varphi(\cdot)$

$$u(t) = \varphi(y(t)) \qquad (3)$$

to make the system output $y(t)$ independent of the disturbance ω.

3. Analytical solution

Consider a Luenberger observer estimating the states of (1)

$$\dot{\bar{x}}_p(t) = A_p \bar{x}_p(t) + B_p u(t) + G(y(t) - C_p \bar{x}_p(t))$$

where $G \in R^{n \times l}$ is the gain matrix. The estimation error

$$e(t) = x_p(t) - \bar{x}(t)$$

satisfies the differential equation

$$\dot{e}(t) = (A_p - GC_p)e(t) + (E_1 + GE_2)\omega(t)$$

Taking into account (2), we define an augmented system

$$\dot{x}(t) = Ax(t)$$
$$y_l(t) = Cx(t) \qquad (4)$$

where $x^T(t) = (\, e^T(t) \quad \omega^T(t)\,)$, and the matrices A, C are

$$A = \begin{pmatrix} A_p - GC_p & E_1 + GE_2 \\ 0 & F \end{pmatrix} ; \; C = (\, C_p \quad E_2\,)$$

Apparently, the vector $y_l(t)$ is the Luenberger observer's residual and thus can be easily evaluated.

Theorem 1: Provided that the matrix

$$\mathcal{W}_k = \sum_{i=0}^{k} \exp(-A^T \tau_i) C^T C \exp(-A\tau_i)$$

is positive definite, then the observer

$$\hat{x}_k(t) = \mathcal{W}_k^{-1} \sum_{i=0}^{k} \exp(-A^T \tau_i) C^T y_l(t - \tau_i) \quad (5)$$

has the following properties (i) finite memory, limited by the largest time-delay τ_k; (ii) deadbeat performance, i.e. $e(t) = x(t) - \hat{x}_k(t) = 0$; $t > \tau_k$ for any initial function $\phi_0 = y(t)$, $t = [-\tau_k, 0]$.

Proof: See Medvedev, Toivonen [5] ∎

The result of Theorem 1 shows that it is possible to estimate both the plant state variables and the disturbance with an estimation error vanishing deadbeatly. This reduces the original disturbance rejection problem to a simpler one assuming perfect knowledge of the state vector and the disturbance. Indeed, partitioning the deadbeat estimate in (5) as

$$\hat{x}_k^T = (\, \hat{e}^T \quad \hat{\omega}^T\,)$$

one can observe that the following equations hold

$$x_p(t) = \bar{x}_p(t) + \hat{e}(t)$$
$$\omega(t) = \hat{\omega}(t); \; t > \tau_k$$

Theorem 2: If there exist matrices $T \in R^{n \times r}$ and $K_g \in R^{m \times r}$ such as

$$C_p T = E_2$$
$$-A_p T + B_p K_g + TF = -E_1 \qquad (6)$$

then perfect disturbance rejection with respect to the output signal $y(t)$ in (1) can be achieved by the control signal

$$u(t) = g(t) \triangleq K_g \omega(t)$$

Proof: Consider a nonsingular state vector transformation

$$z = \begin{pmatrix} I & T \\ 0 & I \end{pmatrix} x$$

Being applied to the augmented system (4) and

taking into account (6), it gives the following state equations

$$\dot{z}(t) = \begin{pmatrix} A_p & 0 \\ 0 & F \end{pmatrix} z(t)$$
$$y(t) = (\, C_p \quad 0\,) z(t)$$

and the disturbance signal does not influence the system output. ∎

To be used in a closed-loop system, the observer (5) should not inflict instability when its estimate is being fed back through a linear controller.

Theorem 3: Suppose the feedback control law

$$u(t) = K_x(\bar{x}(t) + \hat{e}(t)) \qquad (7)$$

is used to stabilize the plant (1) and $\omega(t) = 0$. Then, the closed-loop system (1), (5), (7) is asymptotically stable iff $(A_p - GC_p)$ and $(A_p + B_p K_x)$ are Hurwitz matrices.

Proof: Straightforward computations show that the characteristic polynomial of the system above is

$$\det(sI - A_p + GC_p)\det(sI - A_p - B_p K_x) = 0$$

and, therefore, both necessity and sufficiency follow immediately. ∎

4. Neural Network based solution

The observer-based controller discussed above gives rise to a disturbance rejection technique exploiting the perfect (deadbeat) estimation of the disturbance signal. Unfortunately enough, such a structure can be seldom implemented in practice due to the lack of information about disturbance dynamics. The literature gives many examples of successful control implementations by means of neural networks under various kinds of uncertainties. A major obstacle on this way is the choice of an appropriate neural network architecture to be used in the control system. However, some control structures are closely related to their neural counterparts, which fact facilitates neural implementation.

Time Delay Neural Networks (TDNN) trained by the backpropagation algorithm have been applied successfully to speech recognition problems [4]. Being a concoction of the concepts of Adaptive Filtering (the structure) and Neural Networks (training algorithm), TDNN boast a number of properties which can be used for dealing with the wide range of dynamic problems in Control Science, Signal Processing[6], Pattern Recognition[1]. Generally, TDNN can be viewed upon as a set of static neural networks processing input information sequentially at the subsequent time instants.

In the sequel we consider a Neural Network-based control system (Fig. 1) that guarantees desired closed-loop stability and effectively lessens disturbance effects in the output signal.

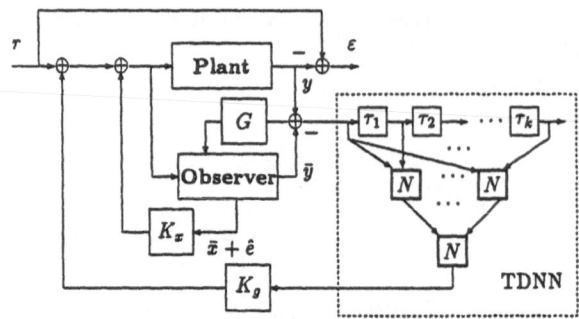

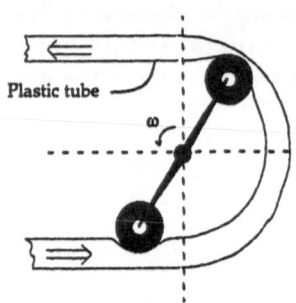

Fig. 1: *The ANN controller structure*

Fig. 2: *Illustration of the blood pump*

The controller comprises a conventional Luenberger observer estimating the system state variables and a TDNN trained to generate a control signal counteracting the disturbance. The observer residuals are fed into the TDNN, which fact exempts the time delays from influencing the closed–loop system characteristic polynomial. Both state–vector estimate and the output of the TDNN are processed by a linear feedback controller.

In can be shown, see e.g [6], that the observer (5) derived in the previous section inherently possess the structure which short of nonlinear activation function coincides with TDNN. This structural property justifies the use of TDNN instead of the CDBO.

To accomplish disturbance rejection, the TDNN in Fig. 1 is trained to minimize the criterion

$$\int_0^\infty (y(t) - r(t))^2 \, dt = \int_0^\infty \varepsilon^2 \, dt$$

where $r(t)$ is desired output trajectory.

A supervised learning method such as back-propagation can not be used for training the neural network because it requires $g(t)$ to be *known*. This minimization problem can be solved numerically by a global nonlinear optimization procedure.

5. Application

The experimental plant, a peristaltic pump intended for blood-pumping in a dialysis machine, is described in Fig. 2. The construction is chosen not to destroy the blood cells. The rollers are pushing the fluid around.

Every lap the tube must be compressed and released by each roller. This introduces a large periodic disturbance, and the axis may almost stop at low paces, which is undesirable.

The first order differential equations below describes the pump fairly well, θ is the axis angle

and ω is the angular velocity.

$$\dot{\omega}(t) = -\frac{1}{\tau}\omega(t) + \frac{k}{\tau}u(t - \tau_d) + \frac{k}{\tau}v(t)$$
$$\dot{\theta}(t) = \omega(t) \tag{8}$$

The disturbance v is coupled to the axis angle by some periodic function $f(\cdot)$, and thus a good model is $v(t) = f(\theta(t))$. This is obviously a nonlinear feedback(Hillerström[3]). The problem with this process model is that $f(\cdot)$ is large and *unknown*, and controller design can not be done using linear theory.

Since on–line training of the network requires much computer power, and involves repeated runs of the pump from stand still to the working point, an analytic model has been used for this purpose instead of the actual process. The plant parameters obtained from an identification experiment are $k = 1$ and $\tau = 0.04s$. The time delay for the continuous time model is small and therefore neglected.

The periodic disturbance is modeled as a process disturbance and is not explicitly affecting the output, i.e. $E_1 = 1$ and $E_2 = 0$. Because of sufficiently fast process dynamics no state feedback is used. The observers feed–back gain is chosen as $G = 0.1$.

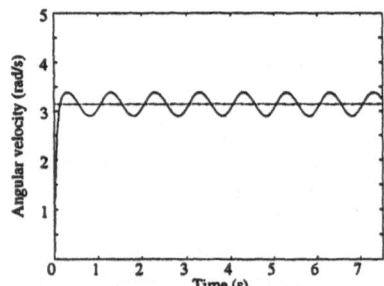

Fig. 3: *Plant output without ANN, $g(t) \equiv 0$*

The disturbance is heavily angular velocity dependent. The fundamental frequency will be

2π rad/s at a nominal speed of about $r =$ π rad/s, due to the fact that there are *two* equal rollers mounted on the pump axis. A simple disturbance model is $v(t) = 2sin(2\pi t)$. This model is used during the training. The magnitude of the fluctuations in the process output because of this disturbance is shown in Fig. 3.

Layer 1 of the TDNN consists of 5 tan-sigmoid neurons, while layer 2 is made up of 1 linear neuron. Such 2-layer networks (with one sigmoid and one linear layer) are capable of approximating almost any function with arbitrarily small error.

The closed loop system is simulated a number of times up to a fixed time instant $T_{final} \approx 7.5s$ and the loss function is minimized using unconstrained nonlinear optimization with the weights (including K_g defined in Fig. 1) as parameters.

$$loss = \int_0^{T_{final}} \varepsilon^2 \, dt$$

The loss function minimization is shown in Fig. 4. Note that after the initial oscillations the loss function gets stuck on a constant level. At this time the network manage to get rid of the biases in the output. After this the network learns the waveform of the disturbance and the loss decreases further.

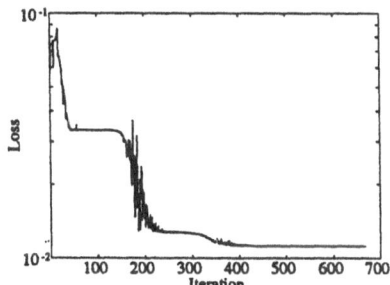

Fig. 4: *Loss function minimization*

The network can thus be trained to reject the periodic disturbance $v(t) = 2sin(2\pi t)$ and what is more surprising it rejects $v(t) = 4sin(2\pi t)$ as well. Inspired by this, the trained network is tried for periodic component rejection in the dynamic system with a nonlinear feedback, $v(t) = 2sin(2\theta(t))$. This disturbance also vanishes from the process output. Finally the nonlinear feedback model $v(t) = 4sin(2\theta(t))$ is tried. The controller performance is shown in Fig. 5. This shows the networks ability to generalize with respect to disturbance waveform.

6. Conclusions

A periodic disturbance rejection problem is solved by a control structure comprising a Luenberger observer and a Time–Delay Neural Network. It is demonstrated, that in case of known disturbance dynamics, both disturbance signal and the plant state variables can be reconstructed with the estimation error vanishing deadbeatly in continuous time. A linear feedback controller based on the proposed observer is shown to be asymptotically stable.

A neural–network implementation of the controller is suggested for the case of unmodeled disturbance dynamics.

The network is trained using unconstrained nonlinear optimization, and appears to generalize to more complex disturbance model such as nonlinear feedback through a periodic function. The generalization property of this kind of network will be further explored.

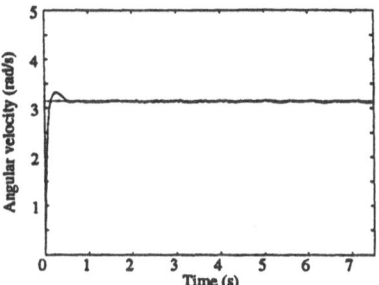

Fig. 5: *Plant output with ANN controller after training,* $f(\theta) = 4sin(2\theta) \Rightarrow v(t) = 4sin(2\theta(t))$

References

[1] A. Bulsari, A. Medvedev, and Saxén H. Sensor fault isolation in dynamic systems by a neural network hierarchy. *Preprints of the IFAC Symposium on Intelligent Components and Instruments for Control Applications*, pages 437–442, 1992.

[2] B.A. Francis and W.M. Wonham. The internal model principle of control theory. *Automatica*, 12(5):457–465, 1976.

[3] G. Hillerström. Rejection of periodic disturbances. Licentiate thesis 1992:21L, Luleå University of Technology, October 1992.

[4] K.J. Lang, A.H. Waibel, and G.E. Hinton. A time delay neural network architecture for isolated word recognition. *Neural Networks*, 3:23–43, 1990.

[5] A.V. Medvedev and H.T. Toivonen. A continuous finite–memory deadbeat observer. *American Control Conference*, pages 1800–1804, 1992.

[6] A.V. Medvedev and H.T. Toivonen. A systematic synthesis of a neural network–based smoother. *Proceedings of the 1992 IEEE International Symposium Intelligent Control*, 1992b.

[7] Masayoshi Tomizuka, Kok-Kia Chew, and Tsu-Chin Tsao. Discrete-time domain analysis and synthesis of repetitive controllers. *American Control Conference*, pages 860–866, 1988.

[8] Masayoshi Tomizuka, Kok-Kia Chew, and Wei-Chi Yang. Disturbance rejection through an external model. *ASME Journal of Dynamic Systems, Measurement, and Control*, 112:559–564, 1990.

REPRESENTATION OF REAL-VALUED FUNCTIONS BY A THREE-LAYERED ARTIFICIAL NEURAL NETWORK WITH TOPOLOGICALLY ORDERED INPUT AND OUTPUT UNITS

Dieter Butz[1], Klaus Noack[2], Wilfried Brauer[2]

[1] Kratzer Automatisierung GmbH, Carl-von-Linde-Straße 38, D-8044 Unterschleißheim
[2] Fakultät für Informatik, Technische Universität München, Arcisstr. 21, D-8000 München 2

ABSTRACT

Many technical tasks in control engineering require the representation of (quasi-)stationary functions by means of a limited set of given measurement values. In this contribution we present a connectionist method to store these values offline in a distributed associative memory to achieve interpolation in the recall phase. A Perceptron-like feed-forward-network with three layers is used. Initially the hidden layer is stochastically connected to the input layer and the connections are frozen; therefore only one layer of connections must be learnt. The ordering relations of real numbers are represented in a topologically oriented coding scheme imposed on the input and output processing units.

The network architecture is shown; simulation results with respect to a particular 2-dimensional function are presented.

INTRODUCTION

Many industrial processes (e.g. extrusion [1]) are difficult to describe by theoretical models, because their mathematical structure is unknown. Moreover experimental non-parametric identification methods such as frequency response, Fourier and correlation analysis etc. are applied; nevertheless process control often must be supported by human operators. It is the effort of many groups to utilize human knowledge by storing, representing and processing those informations in data processing machines.

Fuzzy logic is quite a new method to transform human experience and acting to a machine-storable knowledge base. Formulation of correct fuzzy logic rules in multi-dimensional input spaces, however, is a very sophisticated problem because number and complexity of rules to be stated increase exponentially with respect to the input dimension. In addition the rule bases are not adaptive.

To model complex processes with very low explicit knowledge, we first of all lay down some essential properties the applied method should exhibit:
* in general binary decision rules (on/off) are not sufficient, real-valued outputs should be producable;
* costs in memory and processing time with respect to input space dimension should increase only proportional or less than exponential;
* no particular knowledge of the structure of the process should be necessary;
* self-organizing adjustment of free parameters;
* associativity, i.e. interpolation based on a limited set of measurement points should be possible.

In our opinion, artificial neural networks could help solving such tasks. The remainder of this chapter is a short overview of neural networks in control. A more extensive description is given in [2].

Ritter et al. [3] used a self-organizing Kohonen-map to duplicate a pole-balance controller with two state-variables as input and replaced it gradually the controller.

Nguyen et al. [4] took a backpropagation network to get a model of the process (truck backer-upper), whereby as many various control outputs as possible should be generated. The finished network was placed as process emulator feeding back the control error to a second controller network. Schmidhuber speaks of a "sequential version" of adaptive control and proposes a "parallel version", by training both the model and control network at the same time in a non-Markovian environment [5].

Further structurally different possibilities for backpropagation network controllers are discussed by Psaltis et al. [6]. Essentially the networks try to map the inverse model of the processes. A simulation example converts polar to cartesian coordinates. Another realization of this can be found in [7], where the stationary behaviour of single-input-single-output control loops with respect to a sudden change of the desired output value is investigated.

Ersü et al. [8] place 2 networks (modified versions of Perceptrons) in a control system of the self-tuning-regulator-type (cf. [9]). One presents a predictive model of the unknown environment, the other is the controller itself. Incoming information is stored content-addressable (similar to hash-coding) and a certain generalization is achived by incorporating adjacent memory locations. The distributedly stored information can be recalled by building a weighted sum.

One of the authors accomplished simulations in a closed control loop, where a structured Perceptron-like network had to store static 1-dimensional variable relations online [10]. The aim was to optimize control of a mathematically construed input-output-system, which is in one of many possible states and produces a quality value as reaction on a controller output. No apriori knowledge was used in any part of the network. If a controller output (supplied with a stochastic portion) produced an improved quality, both new values had been stored in a distributed manner so that adjacent regions also took benefit of this action. Thus the controller's quality could be increased step by step.

This quite incomplete list shows, that different neural networks have already been combined in many ways together with processes to be controlled to realize adaptive and self-adaptive methods. Process identification plays a central part herein. In this contribution we present a neural network, different from the backpropagation type but also able to construct a model of non-linear unknown processes, which show a static "input-output-mapping of the stimulus-response-type" as in [8].

NETWORK ARCHITECTURE

The backpropagation algorithm [11] has some shortcomings, letting the extension of the input-space-dimension to let's say m>5 be unrealistic: a) network's convergence depends very sensible on the choice of the learning parameters; b) connections develop very slow; c) the number of hidden units must neither be too small nor too big. To overcome these, we propose a different approach.

Instead of mapping a value x on a single unit, we extend the number of units per variable to typically n_x=10-50 (local vs. distributed coding representation), depending on the desired accuracy. Each unit i is assigned a fixed place x^i on the coordinate axis; so the topological ordering of numeric values can be found in the alignment of the units in one dimension. A numeric value x is encoded by imprinting an

integer activity of a_i=rnd$(255 * \exp(-\frac{(x-x^i)^2}{2\sigma^2}))$, i=1...$n_x$, x^{i-1}<x^i<x^{i+1} , on each unit i (see fig.1a).

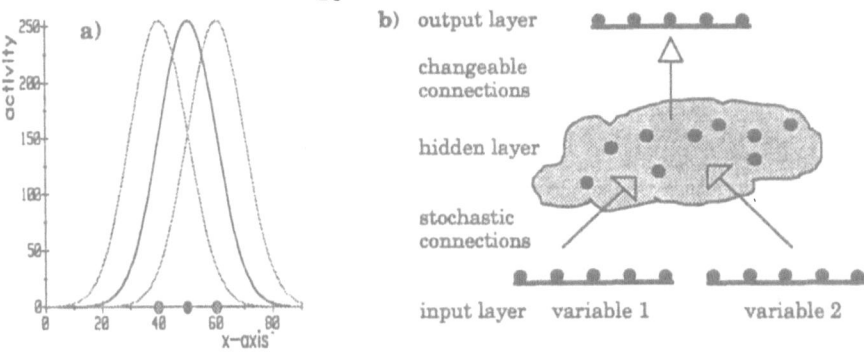

Fig. 1a) Three input units at x^{i-1}=40, x^i=50, x^{i+1}=60 in one dimension and the corresponding overlapping (Gaussian) transfer-functions for the activities are shown. σ is chosen of the size of the units' distance on the axis. This is called "topological encoding" of a numeric value, proposed by Geiger [12], and presents a decisive extension of Ballard's "interpolation coding" [13], where only a single value x, given by a number of noisy measurements, is represented by "votes" for adjacent "value units" to store invariants (e.g. constants of motion).

Fig. 1b) A network for two input and one output variables. All units are displayed as full circles. The visible units are alinged on coordinate axes. The hidden units are not arranged topologically in the numeric sense. This is symbolized by a cloud and irregular distribution. In all layers, units with integer activities in the range of [0,255] are used unexceptionally (continuous drawing in Fig. 1a only for simplicity). The hidden and output units are linear threshold units, the actual activity of which is given by the weighted sum and linear interpolation between a lower and upper threshold boundary. The plot is only schematic. Further explanations see text.

Extending this procedure to m variables, we instead place only of the order n*m units on the m coordinate axes and encode a m-dimensional vector $\underline{x}$=$(x_1...x_m)$ in m separate sets of units (fig. 1b: m=2 for input space). Of course relevant topological information is lost thus, because a particular unit does only carry numeric information of the corresponding variable - independent of the other variables. But exactly this would be essential for storing the function value y=f($x_1...x_m$) unambiguously. This resembles the XOR-problem to some respect (cf. [11]) from a single unit's and local learning algorithm's point of view, because information should be globally present and is only locally represented. Feldman et al. speak of "cross-talk in distributed representations" in this context [14]. For this reason we introduce one hidden layer, where all units are stochastically connected to all of the input layer. The connection strengths are chosen equally distributed out of a symmetric interval around 0. The hidden units possess a very steep threshold-function to perform non-linear processing in the saturation region on the one hand, but on the other hand to exhibit a better behaviour in representing neighbourhood in the hidden layer than pure threshold units with a step function. Roughly spoken, every hidden unit represents some kind of "coarse coding" (cf. [14]) of the input information, showing activity in stochastically shaped, but topologically continuous regions over the entire input space.

We proceed as follows: 1.) Encoding of the m input coordinates of a given measurement vector $\underline{x}$ in the m separated sets of input units. 2.) Updating the activity of the hidden units. 3.) Updating the activity of the output units to produce the actual output activity distribution. 4.) Producing the desired output activity distribution by encoding the target function value $y=f(\underline{x})$. 5.) Performing a learning step on the connections between all hidden and output units according to the familiar Widrow-Hoff-rule [15].

The idea behind this all is, to find a compromise in network architecture between unacceptable amount of a) computing time and b) memory together with an acceptable interpolation quality (as will be seen later).

Ad a): The feed-forward architecture and the integer activities of units are a simple, but straightforward extension of the Perceptron. For this reason, one can easily spend several hundred units in each layer on modern GP-hardware without losing control over computation time, because neither a single processing step nor the complete number of steps exceed an unreasonable limit.

Ad b): The distributed representation of numeric output information avoids a combinatoric increase in connections with respect to the number of input units, which would appear, if one follows e.g. the concept of Feldman et al. [14]. They used a table-look-up strategy ("function table"), realized with logic "conjunctive connections" for every possible grid location of the input space to store a function. The input space there is represented by non-overlapping units. The number of connections therefore increase exponentially with input space dimension.

SIMULATION RESULTS

The following 2-dimensional function, depicted in fig. 2a, was chosen:

$$f(x_1,x_2) = \frac{1}{2}\left(\sin(2\pi x_1) * \sin(2\pi x_2) + 1\right), \text{ with } x_1, x_2, f(x_1,x_2) \in [0,1] \pm 1/1000.$$

The input layer of the network consisted of $2 * 11$ units, sufficient to encode the numeric interval $[0,1]$ with the desired accuracy (in fact, there were some few more units to avoid boundary effects). In the hidden layer were 100 (fig. 2b) and 300 (fig. 2c) units and in the output layer again 11. Thus we had 133 units and 3300 connections (333 and 9900 respectively).

We took 100 random training pairs (x_1,x_2), which were unequally distributed over the input space, to make the task more difficult: In the 1st input square 70 pairs were located, in the 2nd, 3rd and 4th square 10 pairs each. Fig. 2b and 2c show results after 100 training cycles through the complete training set. It is clear, that the function is best represented in the 1st input square in both cases. As can additionally be seen, the interpolation quality increases with the number of hidden units. But simulations with 1200 hidden units (not shown here) did not really improve the results any more.

The whole method is very robust against the initial choice of network parameters. The learning rates were not really controlled, but only roughly adjusted. The number of units in either layer must only be chosen in a reasonable, widely spanned interval. Even scaling up the input space can be done without running into capacity problems. This is an important advantage of our method, compared to classical interpolation algorithms. A further advantage is, that new training points can easily be added without running the whole procedure from the beginning and having lost (or destroyed) previous experiences.

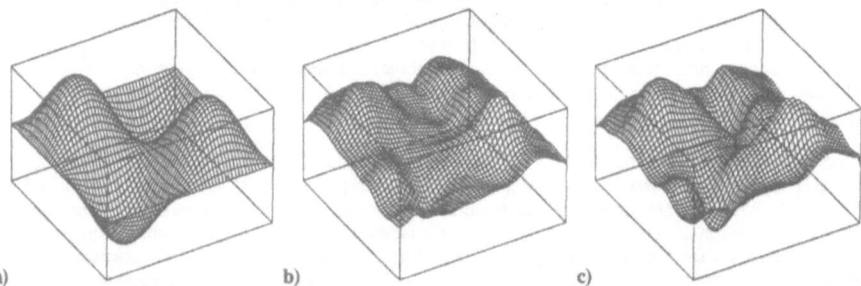

a)　　　　　　　　　b)　　　　　　　　　c)

Fig 2a: The function $y = f(x_1,x_2)$. The bottom of the cubes represent the input space (x_1,x_2). The function is plotted against the vertical axis. The origin is on the left back side of the cubes.

Fig. 2b and Fig. 2c: The network output after 100 learning cycles with 100 (2b) and 300 (2c) hidden units. The training pairs were unequally distributed over input space.

CONCLUSION

A three-layered network was presented, where a limited set of given points could be stored offline in order to accomplish interpolation. The learning algorithm touched only the output connection layer; the advances lie in a constant amount of memory and low computing time to store new measurement values. We speak of a continuous table-look-up (we used a similar storing method already *online* in a controller for car engines on the test bench [16]).

All network architectures described here use extern driven units with respect to time. Therefore modelling and control are only possible, when the process can be represented by a complete set of known and measurable state variables (as in [3] and [16], where besides the measured values also time-derivatives are calculated and used as inputs, which we consider another application of the "functional link" ansatz [17]). In other cases, where time correlations play a distinctive but unknown role, one has to extend the model of units explicitly with respect to time (see e.g. [18]). This is not the place to discuss the dynamic network aspects fully; for further reading see [5]. Nevertheless many cases exist, where pure input-output-relations are to be stored, that our network may present a helpful method.

REFERENCES

[1] Kulshreshtha, M.K., Zaror, C.A., Jukes, D.J., "Automatic Control of Food Extrusion: Problems and Perspectives", Food Control, April 1991, pp. 80-86.

[2] Miller, W.T., Sutton, R.S., Werbos, P.J. (eds.), "Neural Networks for Control", MIT Press, Cambridge, MA, 1990.

[3] Ritter, H., Schulten, K., "Topology Conserving Mappings for Learning Motor Tasks", in: Denker, J.S. (ed.), "Neural Networks for Computing", AIP, Snowbird, UT, 1986.

[4] Nguyen, D., Widrow, B., "The Truck Backer-Upper: An Example of Self-Learning in Neural Networks", in [2], pp. 287-300.

[5] Schmidhuber, J., "Dynamische neuronale Netze und das fundamentale raumzeitliche Lernproblem", thesis, Technische Universität München, 1990.

[6] Psaltis, D., Sideris, A., Yamamura, A., "Neural Controllers", in: Proc. of the IEEE First International Conference on Neural Networks, Caudill, M., Butler, Ch. (eds.), San Diego, 1987, vol. 4, pp. 551-558.

[7] Schiffmann, W.H., Geffers, W., "Adaptive Control of Dynamic Systems by Backpropagation Networks", to appear in: Neural Networks, Pergamon Press, 1992.

[8] Ersü, E., Tolle, H., "Learning Control Structures with Neuron-Like Associative Memory Systems", in: von Seelen, W., Shaw, G., Leinhos, U.M. (eds.), "Organization of Neural Networks. Structures and Models", VCH-Verlagsgesellschaft, Weinheim, 1988, pp. 416-440.

[9] Kraft, L.G., Campagna, D.P., "A Summary Comparison of CMAC Neural Network and Traditional Adaptive Control Systems", in: [2], pp. 143-171.

[10] Butz, D., "Selbstlernende nicht-algorithmische Prozeßregelung in einem strukturierten konnektionistischen System", in: Krönig, D., Lang, M. (eds.), "Physik und Informatik - Informatik und Physik", Informatik Fachberichte vol. 306, Springer, Berlin, 1991, pp. 81-84.

[11] Rumelhart, D.E., Hinton, G.E., Williams, R.J., "Learning Internal Representations by Error Propagation", in: Rumelhart, D.E., McClelland, J.L. and the PDP Research Group (eds.), "Parallel Distributed Processing", vol. 1, MIT Press, Cambridge, MA, 1986.

[12] Geiger, H., "Storing and Processing Information in Connectionist Systems", in: Eckmiller, R., (ed.), "Advanced Neural Computers", Elsevier Science Publishers, North Holland, 1990, pp. 271-277.

[13] Ballard, D.H., "Interpolation Coding: A Representation for Numbers in Neural Models", Biol. Cybern. 57, 389-402 (1987).

[14] Feldman, J.A., Ballard, D.H., "Connectionist Models and their Properties", Cogn. Sci. 6, 205-254 (1982).

[15] Widrow, B., Hoff, M.E., "Adaptive Switching Circuits", Western Electronic Show and Convention, Convention Record IV, 1960, pp. 96-104.

[16] Waschulzik, T., Böller, D., Butz, D., Geiger, H., Walter, H., "Neuronale Netze in der Automatisierungstechnik", in: Brauer, W., Hernández, D. (eds.): "Verteilte künstliche Intelligenz und kooperatives Arbeiten", Informatik-Fachberichte vol. 291, Springer, Berlin, 1991, pp. 486-497.

[17] Pao, Y.H., "Adaptive Pattern Recognition and Neural Networks", Addison-Wesley, Reading, MA, 1989.

[18] Eder, K., "A Neurophysiologically Motivated Neural Network Method and its Application to the Superposition Problem", accepted contribution for the ICANN'93 conference, Amsterdam, Springer, London.

Prediction of Reflectance Values: Towards the Integration of Neural and Conventional Colorimetry

Dr. J. M. Bishop[1] & Dr. S.Westland[2]

[1] Department of Cybernetics, University of Reading, Berkshire, UK.
email: shsbishp@uk.ac.reading

[2] Dept. of Communication & Neuroscience, University of Keele, Staffs, UK.
email: coa23@uk.ac.keele.seq1

ABSTRACT

Colour control systems based on spectrophotometers and microprocessors are finding increased use in production environments. One of the most important aspects of quality control in manufacturing processes is the maintenance of colour in the product. This involves selecting a recipe of appropriate dyes or pigments which when applied at a specific concentration to the product will render the required colour. This process is known as recipe prediction and is traditionally carried out by trained colourists who achieve a colour match via a combination of experience and trial-and-error. Instrumental recipe prediction was introduced commercially in the 1960's and has become one of the most important industrial applications of colorimetry. The model that is almost exclusively used is known as the Kubelka-Munk theory, however its operation in certain areas of coloration is such as to warrant an alternative approach. Previous papers [10,11] have demonstrated the feasibility of a neural network method of predicting dye recipes from a colour specification, however these techniques have not been directly applicable to commercial systems. This paper will describe new research at Keele that enables neural techniques to be simply embodied into current recipe prediction systems.

INTRODUCTION

An industrial colour control system will typically perform two primary functions relating to the problems encountered by the manufacturer of a coloured product. Firstly the manufacturer needs to find a means of producing a particular colour. This involves selecting a recipe of appropriate dyes or pigments which when applied at a specific concentration to the product in a particular way, will render the required colour. This process is known as recipe prediction and is traditionally carried out by trained colourists who achieve a colour match via a combination of experience and trial-and-error. The second function of a colour control system is the evaluation of colour difference between a batch of the coloured product and the standard on a pass/fail basis.

The first commercial computer for recipe prediction [1] was an analog device known as the COMIC (COlorant MIxture Computer) but all colour systems on the market today employ digital computers. A typical colour control system consists of a reflectance spectrophotometer connected to a PC-based machine with various peripherals and costs in the region of £20,000 - £50,000. All computer recipe prediction systems developed commercially to date are based on an optical model that relates the concentrations of individual colorants to some measurable property of the colorant in use (e.g. reflectance). The model must also describe how the colorants behave when used in mixtures with each other.

The model that is almost exclusively used is known as the Kubelka-Munk theory [2]. It relates measured reflectance values to colorant concentrations via two terms K and S, which are the Kubelka-Munk version of the absorption and scattering coefficients of the colorant. The Kubelka-Munk theory is a highly simplified version of rigorous radiative-transfer theory [3] whereby only two fluxes of radiation are considered. Attempts have been made to introduce more complex theories by allowing the use of three or more fluxes [4], but the application of these more complex theories is generally not practical [5]. The use of the exact theory of radiative transfer is not of practical interest to the coloration industry.

COMPUTER RECIPE PREDICTION

The operation of a typical recipe prediction system is as follows (see Figure 1):

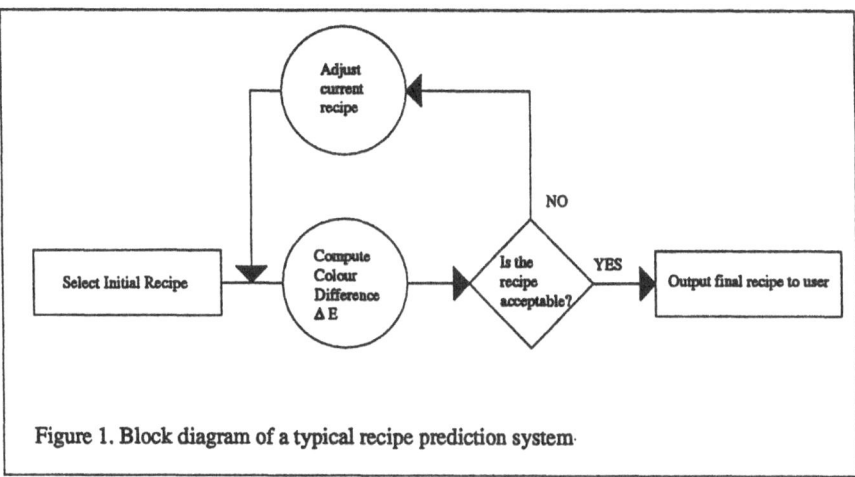

Figure 1. Block diagram of a typical recipe prediction system

i.   The computer makes an initial *guess* at a suitable recipe for the specified colour, based on historical data and the choice of dyes available.

ii.   The reflectance values for the given recipe are computed using Kubelka-Munk theory.

iii. The colour difference (ΔE) between the generated recipe and the desired colour is computed using the CMC 2:1 colour difference equation.

iv.   If the ΔE value is within specification the recipe is acceptable and is presented to the user. Otherwise, the recipe is modified (typically using a gradient descent optimisation technique) and the procedure repeated from step (ii).

However the use of the conventional two-flux Kubelka-Munk theory has attracted criticism [6]. The popularity of the Kubelka-Munk equations is undoubtedly due to their simplicity and ease of use. The equations give insight and can be used to predict recipes with reasonable accuracy in many cases. In addition the simple principles involved in the theory are easily understood by the non-specialist. However in order for the Kubelka-Munk approximation to be valid a number of restrictions are assumed [6].

There are many applications of the Kubelka-Munk theory in the coloration industry, where these assumptions are known to be false. In particular,the applications to thin layers of colorants, for example, lithographic printing inks [7] and fluorescent dyestuffs [8,9] have generally yielded poor results. For these reasons an alternative approach using a neural network to generate reflectance values from a specified recipe has been investigated.

NEURAL NETWORK METHODS

A data set was obtained that contains a list of measured reflectance values (at 5nm intervals between 400 and 700nm ie. 31 wavelengths) with their corresponding recipes (concentrations of colorants used to generate the samples). The total number of colorants in the set was 5, and recipes could be combinations of any number of these 5 dyes. Two of the five colorants are fluorescent, hence some reflectance values are greater than 1.0, and thus the data would present a considerable challenge to Kubelka-Munk theory.

A simple multi-layer perceptron architecture was chosen (see Figure 2). The network has 5 input units, three hidden units in a single layer, and 31 output units. The network is therefore designed to map recipe to reflectance. It is known that this mapping is a many-to-one mapping since there may be more than one way of generating a set of reflectance values from a large set of colorants.

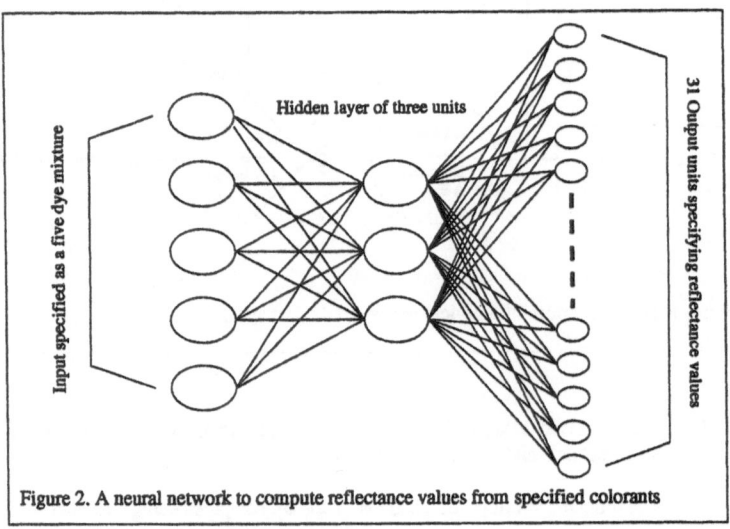

Figure 2. A neural network to compute reflectance values from specified colorants

Input and output data were scaled before presentation to the network and 200 samples were used for training the network. The network was trained using the standard back-propagation algorithm [10]. The learning rate (*initially 0.3*) and momentum (*initially 0.5*) terms were reduced during training at *epochs* (200,000; 400,000; 700,000) with the complete training lasting approximately 3,000,000 epochs (an epoch being defined as the presentation of a single sample chosen at random from the 200 sample data set).

RESULTS

After learning a measure of the effectiveness of the network was made using the test data. The output data ranges from 0 to 100 and the average output error (averaged over all 31 output units and all 200 training samples) is 3.41. This corresponds favourably with the performance of the best Kubelka-Munk model of the data even though it is known that the training data was not evenly spread though colour space, making successful modelling of the data harder. Analsysis of the weights in the trained network showed a trichromatic representation of reflectance data. One of the hidden units was responsible for the blue part of the spectrum, one for the red part, and one for the green part. A paper by Usui [12] showed that a network could be used to compress reflectance data into three hidden units; two units were insufficient and when four units were used there was a correlation between two of them.

CONCLUSION

The performance of the neural network on the specified reflectance prediction task is very promising - training on a second commercially secret data set being significantly better than that quoted above. Earlier attempts at using neural networks in colour recipe prediction concentrated on the inverse problem of predicting dye concentrations from a colour specification. Not only is this a more difficult task (it is a one to many mapping), but such a system, even if perfect, would have limited commercial appeal since the user needs to retain control of the colorants to be used. The method outlined in this paper enables for the first time a neural solution to be embodied as a black box replacement for the Kubelka-Munk stage in any recipe prediction system, and hence promises to have widespread use within the colour industry.

The significance of the network's hidden nodes arriving at a trichromatic representation of reflectance data is currently being investigated.

REFERENCES

1.   Davidson, H.R., Hemmendinger, H. & Landry, J.L.R., *A System of Instrumental Colour Control for the Textile Industry*, Journal of the Society of Dyers and Colourists, Vol.79, pp. 577, 1963.

2.   Judd, D.B. & Wyszecki, G., *Color in Business*, Science and Industry. 3rd ed., Wiley, New York, 1975, pp. 438-461, 1975.

3.   Chandrasekhar, S., *Radiative Transfer*. Clarendon Press, Oxford, 1950.

4.   Mudgett, P.S. & Richards, L.W., *Multiple Scattering Calculations for Technology*. Applied Optics, Vol.0, pp. 1485-1502, 1971.

5.   Mehta, K.T. & Shah, H.S., *Simplified Equations to Calculate MIE-Theory Parameters for use in Many-Flux Calculation for Predicting the Reflectance of Paint Films*. Color Research and Application, Vol.12, pp. 147-153, 1987.

6.   Nobbs, J.H., *Review of Progress in Coloration*. The Society of Dyers and Colourists, Bradford, 1986.

7.   Westland, S., *The Optical Properties of Printing Inks*. PhD Thesis, University of Leeds, (UK), 1988.

8.   Ganz, E., *Problems of Fluorescence in Colorant Formulation*. Colour Research and Application, Vol.2, pp. 81, 1977.

9.   McKay, D.B., *Practical Recipe Prediction Procedures including the use of Fluorescent Dyes*. PhD Thesis, University of Bradford (U.K), 1976.

10. Bishop, J.M., Bushnell, M.J. & Westland, S., *The Application of Neural Networks to Computer Recipe Prediction*. Color, Vol.16, No.1, pp.3-9, (USA), 1991.

11. Bishop, J.M., Bushnell, M.J. & Westland, S., *Computer Recipe Prediction Using Neural Networks*. Proc. Expert Systems '90. (London), 1990.

12. Usui S, Nakauhci S, Nakano M., *Reconstruction of Munsell Color Space by a Five-Layer Neural Network*. Journal of the Optical Society of America, Series A, Vol 9, no 4, 516-520, 1992.

Neural Network Modeling and Prediction of Multivariate Time Series Using Predictive MDL Principle

Mikko Lehtokangas[1], Jukka Saarinen[1], Pentti Huuhtanen[2] and Kimmo Kaski[1]

[1]Tampere University of Technology
Microelectronics Laboratory
P.O.Box 692, SF-33101 Tampere, Finland
{ mikkol, jukkas, kaski } @ee.tut.fi

[2]University of Tampere
Department of Mathematical Sciences
P.O.Box 607, SF-33101 Tampere, Finland
pvh@uta.fi

Abstract

A neural network approach for modeling and prediction on non-linear multivariate time series is presented. The main aim is to reduce the size and complexity of the network and use the least number of weights and nodes for any predictive mapping. The problem of selecting the number of input and hidden nodes is studied by the predictive minimum description length principle. We discuss comparatively the performance of neural networks and conventional methods in predicting non-linear time series. The neural network is found to yield better predictions than an optimum MARMA model.

1 Introduction

In recent years, primarily due to the work of Box and Jenkins [1], a class of ARMA models originally proposed by Yule (1927) and Slutsky (1937), has been found useful in representing the serially dependent relationship of many practical time series. Different non-linear models have also been proposed and tested with good results, *e.g.* TAR and ARCH models [2]. The common problem of time series is the identification of the model and the estimation of its parameters, so that a concise representation is constructed for the time series data.

In the past few years some artificial neural network techniques have been introduced for predicting time series [3,4]. An important but difficult problem is to determine the number of input and hidden nodes needed to model the system by using only input-output examples. In this paper, a method of the Predictive Minimum Description Length (PMDL) [5] has been used to solve this problem. This results in a network of lesser complexity in terms of its weights. Such networks reduce the risk of overfitting and have better generalization properties. The PMDL method provides a systematic procedure for determining the topology of the network for many applications. One purpose of this paper is also to show that neural networks are capable of making better predictions on non-linear time series data than an optimum MARMA-model.

2 Prediction with Neural Network

In the network prediction method, feedforward neural networks are trained to represent a given process as a time delay mapping:

$$\hat{x}_t = F[x_1, ..., x_{t-1}, W], \tag{1}$$

where $\hat{x}_t$ is the approximation at time t of a vector x_t and W is a matrix of network weights. The weights are chosen to minimize the error:

$$\min_W \sum [x_t - \hat{x}_t]^2. \tag{2}$$

The weights are calculated iteratively using a steepest descent-based optimization routine in order to minimize a given objective function. The standard architecture of the multilayer perceptron consists of an input layer, one or more hidden layers and an output layer. The input layer consists of simple distributing nodes, while the other two or more layers have nodes with sigmoidal non-linearities. Most of the knowledge representation is done in the hidden layer. Cybenko [6] has shown that any mapping from R^n to R^m could be achieved with two layers of hidden nodes. Later, Hornik *et al.* [7] showed that with one hidden layer and a sufficient number of nodes, any mapping could be achieved to an arbitrary degree of accuracy. In spite of this proof two or even three hidden layers with a large number of nodes are often used for a given mapping problem.

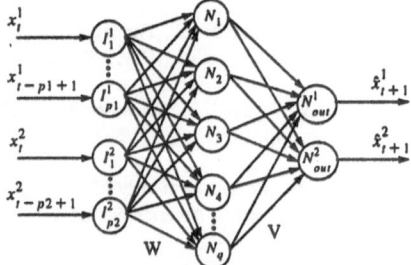

Figure 1. Architecture of neural network.

In this study a neural network (Fig. 1) is used to model two-variate time series. The architecture consists of three layers: one input, one hidden, and one output layer. Learning is achieved using the backpropagation method. The number of input nodes for first variable x_t^1 is $p1$ and for the second variable x_t^2 $p2$, and the number of hidden nodes is q. The number of output nodes is set to two. The output nodes represent the one-step ahead values for variables of the time series. The non-linearities are located in the activation function of the hidden and the output nodes. The *tanh* function is used for the activation function.

3 Model Selection Using Predictive MDL Principle

Currently, the backpropagation neural network is one widely used network paradigm for solving mapping problems. Despite its wide usage, there is no definite methodology for determining the network structure for a particular mapping application. The lack of a network model selection method has resulted in a tendency to use networks much larger than needed. Such models have excessive number of parameters or weights. Hence it is important to find networks which have the simplest possible structure, *i.e.* use the minimum number of weights and nodes.

There are many ways in which model selection for time series can be done and these have been discussed for example by Shibata [8]. A variety of statistical tests have been developed for testing different models. One approach is to find a criterion which balances the overfitting and underfitting characteristics of the model. Such criteria are for example, the Akaike information criterion [9], Schwartz criterion [10], Rissanen criterion [11] and Hannan-Quinn criterion [12]. In the following we use the PMDL principle created and presented by Rissanen [5,11,13] for model selection.

Based on the algorithmic notion of complexity as well as Akaike's work on the model selection, Rissanen proposed the shortest code length for the observed data as a criterion for model selection [11]. This evolved gradually into stochastic complexity [14]. We may also regard the stochastic complexity as a generalization of Shannon's information or complexity. Indeed, for a given distribution $P(x)$ its negative logarithm $I_s(x) = -\log P(x)$, evaluated at the observed data x, may be called Shannon's complexity of the data relative to a given distribution, or the singleton model class $M = \{P(x)\}$.

For applications a most important coding system is obtained from a class of parametric probability models:

$$M = \{f(x|\theta), \pi(\theta) | \theta \in \Omega^k, k = 1, 2, ...\}, \tag{3}$$

where Ω^k is a subset of the k-dimensional Euclidian space with non-empty interior. Hence, there are k "free" parameters. The stochastic complexity of x, relative to the model class M is now according to Rissanen [14]:

$$I(x|M) = -\log f(x|M), \quad \text{where} \quad f(x|M) = \int_{\theta \in \Omega^k} f(x|\theta) \, d\pi(\theta). \tag{4}$$

Although the model class M includes the so-called "prior" distribution π, its role is not the same as in Bayesian inference. In fact, we need not select the distribution π at all, since we can construct a distribution $\pi(\theta|x)$ proportional to $f(x|\theta)$ [11]. The stochastic complexity represents now the shortest code length attainable by the given model class. Frequently, for example in curve fitting and related problems, the models are not primarily in terms of a distribution. Rather we are given a parametric predictor $\hat{x}_{t+1} = F(x|\theta)$ as in the case of neural networks, where $x = (x_{t-p}, ..., x_t)$ is the input and θ denotes the array of all the weights as parameters.

In addition, there is a distance function $\delta(\varepsilon_t)$ to measure the prediction error $\varepsilon_t = x_t - \hat{x}_t$. Such a prediction model, however, can immediately be reduced to a probabilistic model in which optimization of one causes the optimization of the other. In this case we define the conditional r-dimensional multivariate gaussian distribution

$$f(x_{t+1} | x^t, \theta, \Sigma) = (2\pi)^{-r/2} |\Sigma|^{-1/2} exp\left(-\frac{1}{2}\varepsilon_{t+1}^T \Sigma^{-1} \varepsilon_{t+1}\right), \tag{5}$$

where $x^t = x_1, ..., x_t$ and $|\Sigma|$ is the determinant of the covariance matrix Σ of the errorvector ε_t. The density (5) is extended to sequences by multiplication with the result of the total code length

$$-lnf(x^n | \theta, \Sigma) = \frac{1}{2}\sum_{t=0}^{n-1} \varepsilon_{t+1}^T \Sigma^{-1} \varepsilon_{t+1} + \frac{n}{2}ln|\Sigma| + \frac{nr}{2}ln(2\pi). \tag{6}$$

After having fixed the model class, we have the problem of how to estimate the shortest code length obtainable with this class of models. Let $\hat{\theta}(x^t)$, $\hat{\Sigma}(x^t)$ be written briefly as $\hat{\theta}_t$ and $\hat{\Sigma}_t$. They are the maximum likelihood estimates, *i.e.* the parameter values which minimize the code length for the past data, in particular

$$\hat{\Sigma}_t = \frac{1}{t}\sum_{i=1}^{t} \varepsilon_i \varepsilon_i^T. \tag{7}$$

The predictive code length for the data is now given by

$$-lnf(x^n | k) = \frac{1}{2}\sum_{t=0}^{n-1} [\varepsilon_{t+1}^T \hat{\Sigma}_t^{-1} \varepsilon_{t+1} + ln|\hat{\Sigma}_t|] + \frac{nr}{2}ln(2\pi), \tag{8}$$

in which a suitable initial value for $\hat{\Sigma}_0$ is picked. In this form the model cost, *i.e.* the code length needed to encode the model, appears only implicitly. The MDL principles have been described in detail by Rissanen [15], where an example of the time series prediction is also given. In this study we have not used exactly the general form (8) of code length but its approximative version, where the covariance Σ has been assumed as constant. Therefore we need not estimate it. More details of the practical predictive MDL-algorithm for uni-variate case has been explained in [16].

4 Experiments

In this section we present two examples showing the effectiveness of using neural networks in predicting non-linear time series. Each training set consists of $n=190$ successive time-delayed values of the series.

Example 1. The first example time series is defined as:

$$x_1(t+1) = (\alpha_1 + \alpha_2 e^{\alpha_3 x_1^2(t)}) x_1(t) + \beta_1 x_1(t) |x_2(t-1)|^{\beta_2},$$

$$x_2(t+1) = (\alpha_4 + \alpha_5 \pi^{\alpha_6 x_2^2(t-1)}) x_2(t) + \beta_3 x_2(t-1) |x_1(t)|^{\beta_4}, \tag{9}$$

where the parameter values were $\alpha_1=-0.2$, $\alpha_2=4.9$, $\beta_2=0.2$, $\alpha_5=3.0$, $\beta_4=0.4$ and $\alpha_3=\beta_1=\alpha_4=\alpha_6=\beta_3=-0.3$. The starting values were $x_1(-1)=0$, $x_2(-1)=-0.5$ and $x_1(0)=x_2(0)=0.5$. The variance of x_1 was $\sigma_1^2=15.00$ and for x_2 the variance was $\sigma_2^2=17.38$. No noise was added to the time series data. The generated time series data was centered around zero and scaled by a factor of two. In the learning phase the dimension of $p1$, $p2$ and q were varied. The optimal values for $p1$, $p2$ and q were found to be $p1=1$, $p2=2$ and $q=5$. The prediction error R is 0.04.

In order to assess the predictive ability of the neural network against a standard statistical model we fit the same time series data using an optimum MAR and MARMA process. Employing the mean square error the optimum MAR model was found to be 4th order and it's prediction error is 0.69. Optimum MARMA model was MARMA(2,2,1) which gave the prediction error 0.67.

After training the optimal neural network and the optimal MARMA-model were used to predict 60 new values of the future $x(t)$. Results of the neural network and the MARMA-model in predicting one step ahead compared with the actual values are given in Figs. 2 and 3, respectively. We can see that the predicted values of the neural network correspond quite well with actual values and clearly better than the values of the MARMA-model.

Example 2. For the second example a more non-linear time series was selected with additive noise terms. We choose to consider the time series generated by the following equations:

$$x_1(t+1) = (\alpha_1 + \alpha_2 e^{\alpha_3 x_1^2(t)}) x_1(t) + \beta_1 x_1(t) |x_2(t-1)|^{\beta_2} + \varepsilon_1(t+1) - x_1(t)\varepsilon_2(t+1),$$

$$x_2(t+1) = (\alpha_4 + \alpha_5 \pi^{\alpha_6 x_2^2(t-1)}) x_2(t) + \beta_3 x_2(t-1) |x_1(t)|^{\beta_4} + x_2(t-1)\varepsilon_1(t+1) - \varepsilon_2(t+1). \tag{10}$$

The parameter values were $\alpha_1=-0.2$, $\alpha_2=4.9$, $\beta_2=0.2$, $\alpha_5=3.0$, $\beta_4=0.4$ and $\alpha_3=\beta_1=\alpha_4=\alpha_6=\beta_3=-0.3$. The noise was a gaussian sequence with zero mean and 0.1 variance. The starting values were $x_1(-1)=0$, $x_2(-1)=-0.5$ and $x_1(0)=x_2(0)=0.5$. The generated time sequence data was centered around zero and scaled by a factor of 1.5. The variance of x_1 was $\sigma_1^2=16.39$ and for x_2 the variance was $\sigma_2^2=21.37$. The optimum values were $p1=1$, $p2=3$ and $q=8$ with the prediction error 0.19.

The best MAR-model was the 3th order model with the prediction error 0.92. The prediction error for the best MARMA model, MARMA(2,2,1), was also 0.92. The results for the neural network and MARMA-models predictions are demonstrated in Figs. 4 and 5, respectively. Clearly, the neural network is capable of capturing the underlying non-linear dynamics of the time series far more better than the linear models.

5 Conclusions

In this study we have used the MDL approach for the prediction problem based on a flexible feedforward neural network. We investigated the problem of determining the optimal number of neural network parameters from the statistical point of view. The predictive MDL method provides an opportunity to search and construct an optimal model based on input-output observations. The PMDL approach demonstrates potential usefulness of modeling a non-linear multivariate time series. The neural network predictions were found to be significantly better than the linear models in our experiments.

References

[1] G. E. P. Box and G. M. Jenkins, Time Series Analysis, Forecasting and Control, Holden-Day, 1970.
[2] H. Tong, Non-Linear Time Series, A Dynamical System Approach, Oxford University Press, 1990.
[3] A. S. Weigend, B. A. Huberman and D. R. Rumelhart, "Predicting the future: a connectionist approach," International Journal of Neural Systems, Vol. 1, No. 3, 1990, pp. 193-209.
[4] A. S. Lapades and R. M. Farber, "Nonlinear signal processing using neural networks: prediction and system modelling," Technical Report LA-UR-87-2662, Los Alamos National Laboratory, 1987.
[5] J. Rissanen, Stochastic Complexity in Statistical Inquiry, Series in Computer Science, Vol. 15, World Scientific Publishing Co., Singapore, 1989.

[6] G. Cybenko, "Approximations by superpositions of a sigmoidal function," Mathematics of Control, Signals, and Systems, Vol. 2, 1989, pp. 303-314.

[7] K. Hornik, M. Stinchcombe and H. White, "Multi-layered feed-forward neural networks are universal approximations," Neural Networks, Vol. 2, 1990, pp. 359-366.

[8] R. Shibata, "Various model selection techniques in time series analysis," Handbook of Statistics, E. J. Hannan, P. R. Krishnaiah and M. M. Rao (Eds.), Vol. 5, 1985, pp. 179-187.

[9] H. Akaike, "A new look at the statistical model identification," IEEE Trans. Autom. Control, Vol. AC-19, No. 6, Dec. 1974, pp. 716-723.

[10] G. Schwarz, "Estimating the dimension of a model," Ann. Stat., Vol. 6, 1978, pp. 461-464.

[11] J. Rissanen, "Modelling by shortest data description," Automatica, Vol. 14, 1978, pp. 465-471.

[12] E. J. Hannan and B. G. Quinn, "The determination of the order of an auto-regression," J. R. Statistic. Soc. Ser., B41, 1979, pp. 190-195.

[13] J. Rissanen, "A universal prior for integers and estimation by minimum description length," Annals of Statistics, Vol. 11, No. 2, 1983, pp. 416-431.

[14] J. Rissanen, "Stochastic complexity," The Journal of the Royal Statistical Society, Series B, Vol. 49, No. 3, 1987, pp. 223-239 and 252-265.

[15] J. Rissanen, "Information theory and neural nets," submitted for publication, 50 pages.

[16] M. Lehtokangas, J. Saarinen, P. Huuhtanen and K. Kaski, "Neural network prediction of non-linear time series using predictive MDL principle," Proceedings of IEEE Winter Workshop on Nonlinear Digital Signal Processing, January 17-20 1993, Tampere, Finland, pp. 7.2-2.1 - 7.2-2.6.

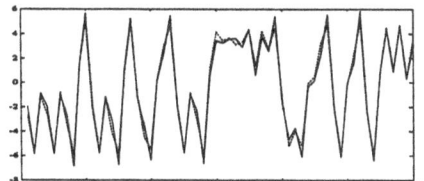 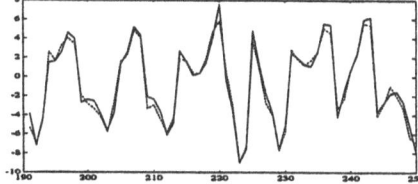

Figure 2. Comparison of the actual time series (solid line) with neural network prediction (dashed line) for first time series: $x_1(t)$ on the left side and $x_2(t)$ on the right side.

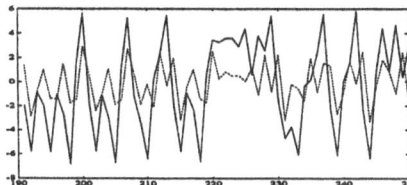 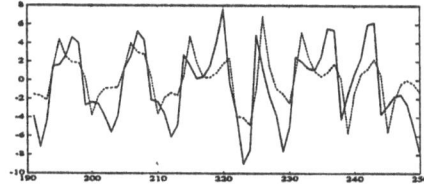

Figure 3. Comparison of the actual time series (solid line) with MARMA-model prediction (dashed line) for first time series: $x_1(t)$ on the left side and $x_2(t)$ on the right side.

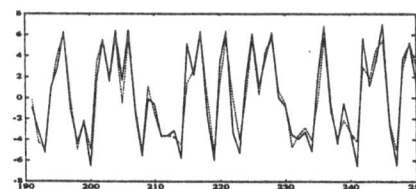 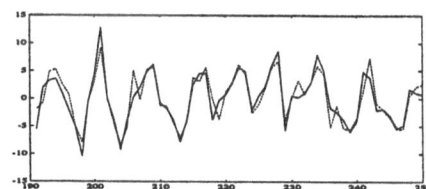

Figure 4. Comparison of the actual time series (solid line) with neural network prediction (dashed line) for second time series: $x_1(t)$ on the left side and $x_2(t)$ on the right side.

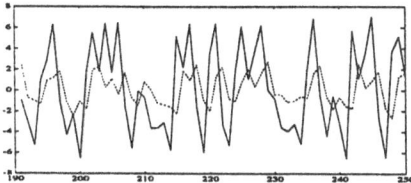 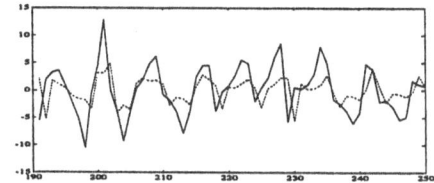

Figure 5. Comparison of the actual time series (solid line) with MARMA-model prediction (dashed line) for second time series: $x_1(t)$ on the left side and $x_2(t)$ on the right side.

Dynamics of a Neural Network-based Financial Market

Sergio Margarita[+] and Andrea Beltratti[*]

(+) Istituto di Matematica Finanziaria, Università di Torino, Piazza Arbarello 8, 10122 Torino, Italia; Tel. +39 11 546805; Fax +39 11 544004
(*) Istituto di Economia Politica G. Prato, Università di Torino, Via della Cittadella 10/E, 10122 Torino, Italia; Tel. +39 11 540900; Fax +39 11 541497

Abstract: we study a financial market populated by neural network-based agents who build autonomously their own behavioral rules through a learning process. Two populations of agents interact in the market: technical traders make their decisions on the basis of an information set including past prices, while fundamental traders make their decisions on the basis of an information set including the past level of dividends paid by the share. Their interactions determine every day the market price and the volume of transactions. We analyze heterogeneity and its effects on the market volume and on the volatility of prices and we study the behavior of the market as a function of the proportions of traders in the two populations.

1. INTRODUCTION

The powerful computational abilities of artificial neural networks (ANNs) make them ideal to help understand the evolution of complex systems of which agents have an imperfect knowledge. This is especially true for economic and financial systems. In this paper we focus on the price and volume dynamics in a financial market populated by heterogeneous learning agents who evaluate the best action by means of ANNs.

Our paper considers two different interacting populations of agents: (i) technical traders making their decisions on the basis of an information set including past prices, and (ii) fundamental traders making their decisions on the basis of an information set including the past level of dividends paid by the share. The two kinds of agents learn over time some features of the other agents' models of the world, affect the environment with their own actions and evaluate assets according to different methodologies. Even more importantly, our agents do not behave in order to maximize a well-specified time-invariant utility function, but learn to decide the valuation of the share, the quantity and the direction (purchase or sale) of the trade with the purpose of increasing their wealth over time. In this sense our model has some behavioristic features of the bounded rationality hypothesis of Simon [1], Nelson and Winter [2] and others. Arthur [3] has recently shown the relevance for economic behavior of learning methods similar in spirit to the one we use in this paper, while Holland and Miller [4] argue for the importance of modelling artificial adaptive markets. This paper continues the line of research of Beltratti and Margarita [5]. We pursue our goals by using ANNs, both as algorithms that may simulate human behavior as they learn from the environment and as a black-box device that allows the building of models that do not assume homogeneous utility-maximizing agents. Agents create their own procedures for deciding the action that is appropriate for any given information set. They trade pair-wise and a market model determines endogenously both the price and the volume of transactions.

The structure of the paper is the following: section 2 describes the economic behavior of agents; section 3 considers the structure of ANNs. Section 4 describes the marketplace and section 5 the learning technique. Section 6 presents the results and section 7 concludes.

2. DESCRIPTION OF THE AGENTS

Agent i enters the early morning of day t with a given amount of money $M_{i,t}$, a stock of shares $S_{i,t}$ and an information set $I_{i,t}$. She has two choices for carrying her wealth to the following morning: money and shares, about which she will have to make a decision at noon of day t. Each day t, shares can be bought or sold in a stock market at a price $p_{ij,t}$ depending on the value assigned to shares by the agent j who will be (randomly) met in the marketplace (see section 4); Money is a riskless asset; investing \$1 in money at t means having \$1 of money at t+1. Buying shares is risky because there is uncertainty about the price which will prevail at t+1, and about the dividend that will be paid at the end of period t, denoted with D_t. Each transaction involves buying or selling a number of shares which depends on the choices of the two agents who meet in the market (see section 4). The actions of the traders depend on the specific structure, described in the next sections.

3. THE ECONOMIC BEHAVIOR OF AGENTS

3.1 Technical traders

Each technical trader is modelled as a network that forms an expectation about the price at which the share may be sold in period t+1 and about the dividend that will be paid at the end of period t, D_t. Agent i does not know the identity of agent j with whom she will transact, so we denote the expectation of agent i for the price with the symbol $E_{i,t}P_{t+1}$, while the expectation of the dividend is $E_{i,t}D_t$. These expectations are formed on the basis of an information set containing the previous history of market prices, the market volume and the value of the trade relating to the previous period (Figure 1).

3.2 Fundamental traders

Fundamental traders adopt a long-run perspective of the market, basing their action on the estimated fundamental value of the asset. The main differences between the technical and fundamental traders lie in the information sets and in the interpretation of the first output (Figure 2).

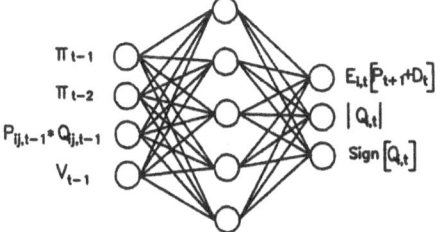

Figure 1. The technical traders' ANN

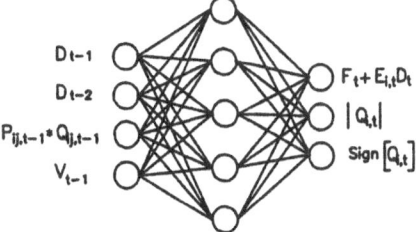

Figure 2. The fundamental traders' ANN

As figure 2 shows, the information set contains two past values of dividends rather than two past prices. This is done in order to estimate the fundamental value of the asset at time t, that we define $F_{i,t}$. The sum of such value and the expectation of the dividend is the first output of the agent, and also in this case this sum is an important reference point. We assume that such fundamental value differs from agent to agent; we let agent j learn over time a fundamental value equal to:

$$F_{jt} = g_j\, m + 0.5[(1+\exp(-\sum_{i=0}^{4} D_{t-i}))^{-1} - 0.5]$$
(1)

where g_j is extracted from a uniform distribution $U(g_{min}, g_{max})$. Therefore each trader recognizes an estimate of the fundamental value of the asset that is a smooth function of the realization of the past values of dividends. This is the reason why two values of dividends are incorporated into the information set: they can help the trader to learn more efficiently the relationship between dividends and fundamental value.

3.3. The neural network representation of an agent

As we have seen in figures 1 and 2, an agent is represented by an ANN, with a given information set. The net has 3 outputs, a price, a quantity (absolute value) and its sign. Given the information set for a technical trader we can therefore write (with an obvious similar extension for the fundamental traders):

$$E_{i,t}P_{t+1} = g_{i1t}(\Pi_{t-1}, \Pi_{t-2}, p_{ij,t-1}Q_{ij,t-1}, V_{t-1})$$
(2a)

$$Q_{i,t} = g_{i2t}(\Pi_{t-1}, \Pi_{t-2}, p_{ij,t-1}Q_{ij,t-1}, V_{t-1})$$
(2b)

$$T_{it} = g_{i3t}(\Pi_{t-1}, \Pi_{t-2}, p_{ij,t-1}Q_{ij,t-1}, V_{t-1})$$
(2c)

where Π_t is the market price at time t, $p_{ij,t-1}$ is the price for the transaction between agents i and j, $Q_{ij,t}$ is the amount of the transaction between agents i and j, $p_{ij,t-1}Q_{ij,t-1}$ is the value of the transaction made by the agent during the last period of time, V_{t-1} is the market volume at t-1 (these variables will be defined more precisely in section 4), T_{it} is the sign of the transaction chosen, a purchase (sale) if the sign is positive (negative). The functional form g_{ijt} in (2) describes the function mapping the inputs of the i-th network at time t to the j-th output at time t. Such form is equal for each network, but the weights vary from one to the other and from one period to the other. For example $T_{it} = 1$ ($T_{it} = -1$), $Q_{i,t} = 3$ and $E_{i,t}p_{t+1} = 100$ means that agent i wants to buy (sell) 3 shares at a price equal to or lower than (equal to or greater than) 100. Of course agents face a budget constraint. To take it into account we check whether the planned purchase is affordable on the basis of available money, that is whether $Q_{i,t} < (M_{i,t}/E_{i,t}p_{t+1})$. The actual planned quantity is therefore equal to $Q^*_{i,t} = \min(Q_{i,t}, (M_{i,t}/E_{i,t}p_{t+1}))$. In case of selling we check that the sale is compatible with previous holdings, so that the actual quantity is $Q^*_{i,t} = \min(Q_{i,t}, S_{i,t})$

4. THE MARKETPLACE

After forming their price expectations agents meet randomly in the market. From the population of J agents we randomly select two agents and match them. Given the assumptions we have made, a transaction may or may not occur. There is no transaction for example if the two agents are both seller or both buyers; in this case we just put them again in the sample of potential traders and extract other two. If a trading possibility exists we organize a transaction which is dominated by the short side of the market. An example is the best way to describe the transaction technology of the model. If two technical traders meet in the marketplace with the following outputs $Q^*_{i,t} = 10$, $T_{it} = 1$, $Q^*_{j,t} = 5$, $T_{jt} = -1$; $E_{i,t}p_{t+1} = 97$, $E_{j,t}p_{t+1} = 92$, then the actual transaction is $Q_{ij,t} = 5$ at a price $p_{ij,t}$ equal to the weighted average between the expected prices of the two agents; the weights are such as to push the price in the direction of excess demand. In our example the price is larger than the simple average 95, to take into account the fact that the demand of i is larger than the supply of j. After the transaction is completed we exclude agents i and j from the sample. When N_t transactions, where N_t is the random number of meetings that at time t satisfy the necessary conditions for a successful conclusion of trade, have been performed we have a set of prices whose simple average is considered for determining the market price at time t, Π_t. Such market price is therefore a weighted average of the expected prices of the single traders. We also compute the market volume V_t by summing the change in the holdings of shares of each single agent.

5. LEARNING

At the end of the day agents know the market price Π_t, and can therefore calculate the value of their wealth and update their view of the world. As to wealth dynamics, for an agent i who bought during the day, wealth (evaluated at market values) is equal to $W_{i,t+1} = (M_{i,t}-Q_{ij,t}p_{ij,t})+\Pi_t(S_{i,t}+Q_{ij,t})$; if the agent was instead a seller we have $W_{i,t+1} = (M_{i,t}+Q_{ij,t}p_{ij,t})+\Pi_t(S_{i,t} - Q_{ij,t})$. Agents can also update their "view of the world" (the weights of the network); this gives rise to two sets of problems that will be discussed separately. As to quantity, in order to understand how the agent learns we need to know what she would like to have done at t-1 if she had known the actual price at t. From the point of view of agent i we can say that if $p_{ij,t}>p_{il,t-1}$ then i would have liked to buy more shares at t-1. The target we compute for back-propagation is equal to the quantity she could have bought had all the money been spent in shares, that is $M_{i,t-1}/p_{ij,t-1}$. As a consequence we measure the error by comparing the quantity output at t-1, $Q_{i,t-1}$ with $M_{i,t-1}/p_{ij,t-1}$. On the contrary, if $p_{ij,t}<p_{il,t-1}$ then i would have liked to sell more shares at t-1, regretting not keeping all wealth in the form of money. She compares $Q_{i,t-1}$ with $S_{i,t-1}$, the maximum quantity she could have sold at t. As to the price, note that the forecast of the price and of the dividends made at the beginning of day t by traders cannot be compared to anything known at the end of period t, since such forecast is about variables which will be known during period t+1. What the network can do at the end of day t is to compare the forecast which was made at time t-1 with the new information available on day t. We compare the forecast of the net at time t-1 with the actual transaction price at t plus the dividend paid at t. As to fundamental traders, we compare the price and dividend output of the network with the sum of the fundamental value we have defined in (1) and the actual dividend.

6. RESULTS

Figures 3 to 5 contain the market price at each time t and the volume that is exchanged on the market under different hypothesis in the proportions of traders. The dividend is stochastic and uniformly distributed between 0.01 and 0.04, while the parameter g_i is uniformly distributed between 0.9 and 1.1. In figure 3 there are 50 technicals and 50 fundamentals, and the price path is reasonably stable, even in the face of much variability in transactions. Transactions do not tend to disappear in 300 periods, contrary to what happens in figure 4, where all the agents are fundamental traders. The importance of divergences of opinions for the volume has been noted among the others by Tauchen and Pitts [7]. The divergence in the opinions about the fundamental value of the shares is not enough to generate trading in the long period. The same is true of figure 5, where everybody is a technical, and where a new feature appears, that is the increased volatility in price. It seems therefore that technicals reach a consensus opinion by trading and by destabilizing the price. The extrapolative nature of their behavior may in fact increase volatility with respect to what is done by long-term traders looking at dividends.

7. CONCLUSIONS

We have presented a model of a financial market where agents learn to determine their own behavioral rules in order to maximize their wealth. The rules consist of a choice of price and quantity that meets a certain budget constraint, and that is proposed for trade with another agent. We have provided some evidence that increasing the heterogeneity of the trading strategies affect most of all the volume of trades, while increasing the proportion of agent following a technical strategy increases the volatility of prices.

There are many possible extensions of this model, that we hope to pursue in our future work. We plan to experiment more with the current version, in order to analyze the consequences of different learning methods for accumulation of wealth of the various agents. Also, we want to study the volume and price outputs of the simulations of the model to see whether they present some of the statistical features (e.g. autocorrelations, cross-correlations, and so on) that is usually possible to find in real data. Finally, we plan to elaborate on this version to create agents who follow predetermined trading strategies, in order to see whether one strategy takes over the others.

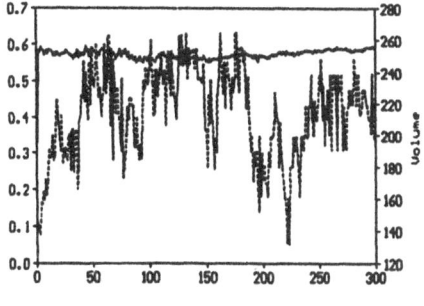

Figure 3. 50 technical and
50 fundamental traders

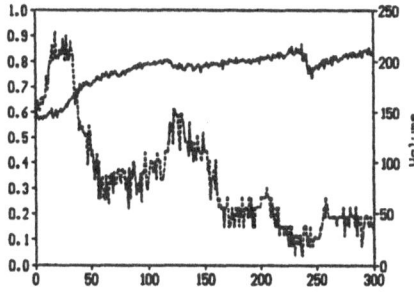

Figure 5. 100 technical traders

References
[1] Simon, H.A., 1979, Rational decision making in business organizations, American Economic Review, 69
[2] Nelson R.N. and S. Winter, 1982, An evolutionary theory of economic change, Harvard University Press, CA
[3] Arthur W.B., 1991, Designing economic agents that act like human agents: a behavioral approach to bounded rationality, American Economic Review, 81, 2
[4] Holland, J.H. and J.H. Miller, 1991, Artificial adaptive agents in economic theory, American Economic Review, 81
[5] Beltratti A. and S. Margarita, 1992, Evolution of trading strategies among heterogeneous artificial economic agents, in J.A. Meyer, H.L. Roitblat, S.W. Wilson (eds.), From animals to animats 2, Cambridge, Mass., MIT Press (forthcoming)
[6] Tauchen G.E. and M. Pitts, 1979, The price variability-volume relationship on speculative markets, Econometrica, 51.

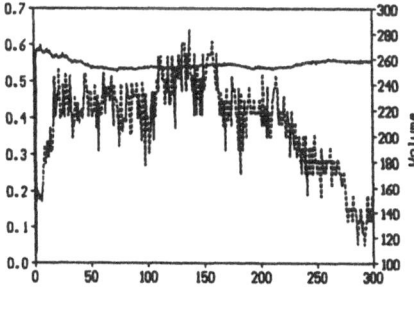

Figure 4. 100 fundamental traders

Evolving Neurocontrollers for Pole Balancing.

Dipankar Dasgupta and Douglas R. McGregor
Department of Computer Science
University of Strathclyde
Glasgow G1 1XH, U. K.

Abstract

The paper describes the application of a Structured Genetic Algorithm (sGA) for the automatic design of neurocontrollers using genetic reinforcement learning. The novelty of the method is that it can design and train neural nets for a specific task in a single evolutionary process. It is possible because of its multi-level chromosomal representation which can define both topology and weight-bias space of neural nets. The preliminary results reported here demonstrate that the evolved nets with sGA can efficiently control a typical dynamic system - a simulated pole cart system.

1   Introduction

When building a controller for a dynamic system, traditional control theory requires a mathematical model to predict the behavior of the system. Many systems are either too complicated to model accurately or sufficient information is not available about the system environment. The pole balancing problem is one such inherently unstable classical control problem. The complexity of the task is significant enough to make the problem interesting while still being simple enough to make it computationally tractable.

The problem has often been used as an exercise in the control of dynamic systems and has been studied extensively by researchers in different fields. Other than classical control theory approaches, it has been solved mostly using different techniques of Artificial Intelligence (AI). They include machine learning [9], fuzzy logic [3], qualitative modelling [7], neural networks [1, 2, 6], genetic algorithms [8, 11] etc.

Among all these AI methods, neural-based methods are widely used. The advantage of using neural networks is twofold: versatile mapping capabilities from input to output and its learning ability. It is widely recognized that the architecture of a neural network can have significant a impact on the network's function and processing capability. But in most cases, predefined architectures are used for performing tasks with neural nets. Genetic algorithms can replace the effort of human designers in determining the network structures and also can be used for training neural nets.

This paper uses a different genetic model, called the Structured Genetic Algorithm (sGA) for both designing and training of neural-based controllers in a population-based evolutionary process.

2   Problem description

A rigid pole is hinged on the top of a cart which is free to move along a bounded straight track; the task is to balance the system such that the pole must not fall beyond a predefined vertical angle and the cart must not go off the ends of the track limits. This is to be achieved by applying a force of fixed magnitude (a bang-bang force) to the left or right of the cart. The simulation equations are used to map the system's state and the control action into a linear acceleration of the carriage and an angular acceleration of the pole, which are then integrated

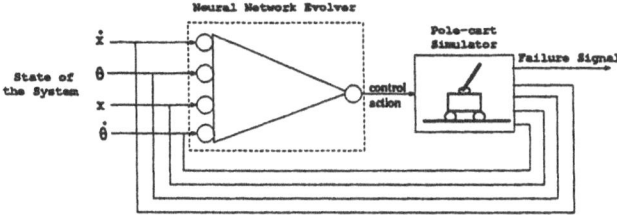

Figure 1: Genetic design of neurocontrollers.

using a simple version of Euler's approach to generate the simulated pole-cart system's state vector (as formulated by Anderson [1]) for next time step.

The state space can be regarded as a four-dimensional space and a state variable defines each dimension. The sampling rate of the system's state variables are the same as the rate of application of the control force. The initial starting position of the cart is randomly set between ±0.1 metres, the starting pole angle between ±6°; the cart velocity and pole's angular velocity are set to 0.0 at the start of each training phase [8, 10]. These values are considered as the initial inputs for each individual neural net at every generation. The input state vector is normalized so that the values lie in the range 0 and 1. The algorithm terminates if at least one evolved net holds the pole for 50,000 steps or the allowed number of iterations are used up.

3 Structured Genetic Algorithms

The central feature of the Structured Genetic Algorithm (sGA) [4] is its use of genetic redundancy and quasi-hierarchical structure in its genotype. The primary mechanism for eliminating the conflict of redundancy is through regulatory genes (as in biological systems) which act as switching operators to turn genes *on* and *off* respectively. The model also uses conventional genetic operators and the *survival of the fittest* criterion to evolve increasingly fit individual offspring.

This special representation allows the model to define the network configuration and its connection weights in its chromosome, and these parameter sets can be optimized in a single unified process. So in each generation, while some members of the population are engaged in searching for the feasible topology, others are searching for a set of optimal weights if they have feasible structures; the process continues until a fully trained network evolves which can solve the pole balancing task. The detail of the model and its application for full designing of neural nets are not included here because of space restriction, but were explained in our previous work [4, 5].

3.1 Neurocontroller design using sGA

Figure 1 shows the different functional blocks of genetically designed neural network controllers. The approach considers two black boxes communicating with each other without knowing their internal dynamics. One is designing the controller to adapt the environment of the other through an evolutionary process, while the other is responding to the control action of the first and feedbacking the system state at each time step. So the only information for evaluating the performance is a failure signal indicating that the pole-cart system is out of control.

For this empirical study, we have considered a two-level Structured Genetic Algorithm for the description of complete neural networks. Each individual is composed of two substrings which represent the two-level of genomic structure. The higher level defines the network configurations, while the lower level encodes the connection weights and biases. So the high level of sGA searches the connectivity space of N units (to evolve a minimal network structure), while the low-level searches for an optimal (near-optimal) set of weights for the network which can solve the task. The fitness of each individual is determined by the combined performance of these two components, i.e. each individual is treated as a single complete network [4, 5]. A set of individuals (population) is generated randomly to initialize the evolution-learning process.

4   Fitness measure

We have used a criterion that rewards the net which is feasible, having fewer nodes and connections, while learning quickly and accurately. In every generation, each chromosome is decoded into its phenotype (a network structure with its weights), and the fitness is evaluated by taking into account the feasibility, complexity (number of nodes and their connectivities) and learnability of the evolved network. The higher the fitness of an individual, greater the probability of its being selected as a parent in the subsequent generations. In our fitness criteria (details in [4]), if an individual decodes to a feasible structure, it is rewarded such that its high level portion remains stable (no change is allowed to occur), while only the weight-bias space is explored through learning. However, while training a feasible net if no improvement is noticed in balancing the pole for 50 successive generations, the net then loses its structural stability. Also the feasible individuals which have fewer nodes and links get an advantage for reproduction relative to the competing feasible individuals with more complex structures. Each feasible net (individual) is trained through reinforcement learning [12], where the learning process is also an object of evolution. The learning process starts with providing the initial state of the system to the net and the net's output response is applied to the simulated system. The output of the net is either 0.0 (push left) or 1.0 (push right) representing the direction of a bang-bang control force. The output of the system is a new state vector which is then reintroduced as new input to the net. This continues until failure occurs or successful control is performed for the prescribed maximal period of time.

5   Experimental Results

In this experiment we used a mixed encoding technique, where the high level portion of the chromosome is binary-coded representing the network topology. The low level is real-valued encoding the weight-bias space in the range of -1.0 to +1.0 and crossover is allowed to occur between the weights. A simple bit mutation is used on high level and a floating point mutation is used in low level where a random value (between -0.1 to +0.1) is added to the existing weight rather than replacing it. The reported results used a population size of 80 and a two-point crossover operator with a probability of 80%. The size of connectivity matrix used is 10 by 10 along with a logistic transfer function for all the nodes. We used a ranking selection scheme in which an individual receives an expected number of offspring which is based on the rank of performance and not on its magnitude.

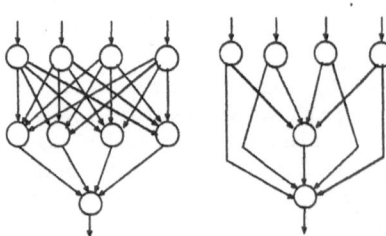
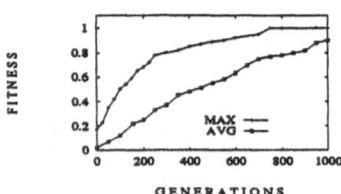

Figure 2: The evolved neural networks.   Figure 3: Pop convergence in evolving net.

It is observed that different structures evolved during trial runs, some were irregular and some having direct connections from inputs to the output node. Figure 2 shows the networks which evolved in two independent runs when given selection privilege to the regular structures. The details of these experiments will be reported elsewhere. In most test runs, the algorithm takes 800 generations (in average) to evolve neural net controllers that successfully balanced the pole-cart system for the specified period of time and the success rate is 75% of the total number of runs. Figure 3 shows the graphs for evolving a successful net and population convergence in a typical run.

6 Conclusions

The paper presented a different genetic algorithm for evolving neurocontrollers which can learn a mapping between a dynamic system's state space and the space of possible actions. The Structured GA approach offers the following advantages: 1) it can evolve network structures and their weights in a single evolutionary process; 2) each individual net can be trained using genetic reinforcement learning; 3) the method does not require partitioning of the state space for the problem; 4) no supervisory training data is required for performing the balancing task; 5) it uses global search rather than local search; 6) it can be implemented in parallel to improve the speed of convergence.

We conclude that there are many domains where this approach can make a unique contribution in fully automatizing design and training processes of application-specific neural nets.

The first author gratefully acknowledges the support of the Government of Assam (India) for awarding a State Overseas Scholarship.

References

1 C. W. Anderson. Strategy learning with multilayer connectionist representations. In *Proceedings of the Fourth International Workshop on Machine Learning*, pages 103–114. Morgan Kaufmann, Los Altos, 1987.

2 Andrew G. Barto, Richard S. Sutton, and Charles W. Anderson. Neuronlike adaptive elements that can solve difficult learning control problems. *IEEE Transactions on Systems, Man and Cybernetics*, SMC-13(5):834–846, Sept/Oct 1983.

3 Hamid R. Berenji and Pratap Khedkar. Learning and tuning fuzzy logic controllers through reinforcements. *IEEE Transaction on Neural Networks*, 3(5):724–740, September 1992.

4 Dipankar Dasgupta and D. R. McGregor. Designing Application-Specific Neural Networks using the Structured Genetic Algorithm. In *Proceedings of COGANN-92*, pages 87–96, IEEE Computer Society Press, June 6, U.S.A 1992.

5 Dipankar Dasgupta and D. R. McGregor. Designing Neural Networks using the Structured Genetic Algorithm. In *Proceedings of ICANN-92*, pages 263–268, U K, 4-7 September 1992.

6 E. Grant and Bing Zhang. A neural-net approach to supervised learning of pole balancing. In *Proceedings of IEEE International Symposium on Intelligent Control*, pages 123–129, Albany, New York, 25-26 September 1989.

7 A. Makarovic. A qualitative way of solving the pole balancing problem, *Machine Intelligence*, volume 12, chapter 16, pages 241–258. Oxford University Press, 1988.

8 D. R. McGregor, M. O. Odeytayo, and D. Dasgupta. Adaptive control of a dynamic system using genetic-based methods. In *IEEE International Symposium on Intelligent Control*, August 11-13, UK, 1992.

9 D. Miche and R. A. Chambers. Boxes: An experiment in adaptive control. *Machine Intelligence*, 2:137–152, 1968.

10 M. O. Odetayo and D. R. McGregor. Genetic algorithm for control rules for a dynamic system. In *Proceedings of ICGA-89*, pages 177–181, 1989.

11 Dirk Thierens and Leo Vercauteren. A topology exploiting genetic algorithm to control dynamic systems. In *Lecture Notes in Computer Science*, pages 104–108, 1991.

12 D. Whitley, S. Dominic, and R. Das. Genetic reinforcement learning with multilayer neural networks. In *4th International Conference on Genetic Algorithms*, pages 562–569, 1991.

BACKPROPAGATION VECTOR QUANTIZATION FOR SATELLITE COVERAGE PLANS OPTIMIZATION

Pietro Burrascano
INFO-COM Dept. - Universita' di Roma "La Sapienza"
via Eudossiana, 18 - 00184 Roma - Italy

ABSTRACT
The use of the Backpropagation Vector Quantization neural network is proposed for the solution of the constrained optimization problem of satellite antenna coverage plans project. Network's structure and characteristics are described in order to clarify the proposed procedure. The effectiveness of the approach is demonstrated by means of several examples.

INTRODUCTION AND BACKGROUND

Resources available on board of the space segment of a satellite system are limited, by definition: consequently, a great attention is paid off to the dimensioning of all the equipments and components. To obtain the best performance without violating the given constraints, the architectural and system design phases usually rely on optimization processes. The problem that arises is that of complexity: as far as the only antenna coverage plan optimization problem is concerned, the variables and constraints involved are the main geographical factors (samples of the area to be covered, antenna pointing, propagation and orbit characteristics); their number can become large and a mathematical model that takes into account all them can be too complex to manage. Most of the today used optimization algorithms experience, in fact, an exponential increase in their computational complexity with the number of constraints and variables. Complexity increases even more in our case with the introduction of other categories of parameters, such as for example spot geometry, traffic volume and access method.

In the present paper a possible solution is proposed, based on the use of neural networks: the intrinsic parallel processing performed by neural networks let in fact their use to be suitable in all the applications dealing with a large amount of data. In this context, the problem becomes that of choosing a neural network paradigm which allows to model the particular constrained optimization problem at hand. A solution to this optimization problem, based on neural networks, was considered in [1]: it relies on the Constraint Satisfaction neural network. After an appropriate data pre-processing phase, very good results were obtained, at the expense of an affordable computational cost.

In the present paper a different solution to the same optimization problem is proposed, based on the Backpropagation Vector Quantization (BVQ) Neural Network [2]. The new approach, whose effectiveness will be demonstrated by means of several examples, leads to a neural structure much simpler than the former, and the computational cost is furthermore reduced. Moreover, the present approach does not require any pre-processing on the data and can be extended in a simple way in order to include other constraints, such as single station traffic volume and frequency and/or polarisation reuse.

THE OPTIMIZATION PROBLEM

The particular problem of satellite coverage plans optimization can be described as follows: given the area to be covered (the SErvice Area, SEA), described by means of its samples, the number of spots and the projection on the earth surface of each antenna spot, the problem is to define the positions of spot centres in order to obtain a global coverage of the SEA. The optimum solution is the one which both minimises the maximum excursion of the electric field within the SEA and the energy radiated outside the SEA. In the following, the antenna spots will be defined by their -4 dB coverage, assumed to be circular with diameter D and covered area SPA. Each spot corresponds to a feeder in the antenna system located on board: owing to technological constraints, the -4dB projections of antenna spots cannot overlap. This constraint must be introduced in the optimization procedure: in other words we are required to constraint the neural model to avoid that different SPA's overlap.

THE BACKPROPAGATION VECTOR QUANTIZATION

The Backpropagation Vector Quantization (BVQ) optimization procedure [2] allows an approximation of the probability density function from which the training data samples are drawn: it relies on the minimisation of a particular cost function defined in the L_1 norm. The neural implementation of the approximator proposed in [2] is a layered network whose hidden layer consists of a number of non linear units with a gaussian activation function. The output layer consists of a single, linear unit, as shown in Figure 1: the input-output function $\Phi(x,\omega)$ implemented by the network consists thus of the superposition of gaussian kernels. It can be shown that the minimisation of the L_1 norm cost function implies the approximation of the probability density function $p(x)$ from which the training data samples are drawn: the procedure allocates the gaussian kernels according to training data distribution [2]. At the end of the optimization procedure, kernel's centres can be regarded as the reference vectors defined by a vector quantization of the pattern space performed contextually to the pdf approximation. Owing to this characteristic, the proposed procedure has been denoted as Backpropagation Vector Quantization (BVQ).

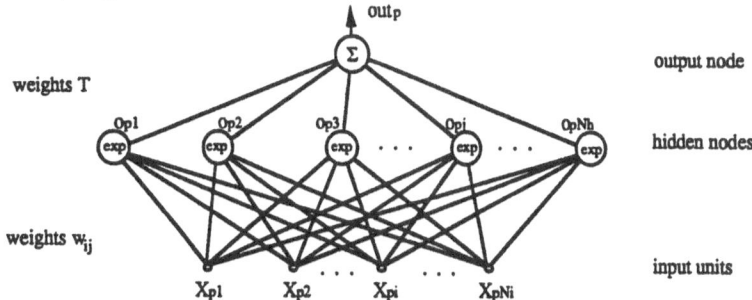

Figure 1: topological structure of the BVQ Neural Network

The optimization (training) procedure proposed in [2] is based on a particularisation of the generalised delta rule to the L_1 norm and to the topological structure described above; a point which is particularly interesting for the application we are interested to in the present paper, is that the training procedure implies that if two or more kernels excessively overlap during the training phase, the weight changes, while maintaining their amplitudes, tend to separate the overlapping kernels: this characteristic of the training procedure implies a sort of "repulsive potential" among the different gaussian kernels, which ensures that, at the end of training phase, the amount of overlap of the different kernels is controlled by the training parameters.

BVQ FOR SATELLITE COVERAGE PLANS OPTIMIZATION

Let's see now how the satellite plans optimization problem can be approached by means of the BVQ optimization procedure. The Service Area to be covered is represented by means of samples taken on a regular grid: they can be interpreted as samples drawn from an uniform distribution in a two dimensional feature space; the shape of the uniform distribution is the one of the Service Area. Under this interpretation, the satellite coverage plans optimization problem can be approached by means of the BVQ procedure if we identify the gaussian kernels of the approximating function with the spots of the satellite antenna and use the samples of the Service Area to train the network. The constraint that the -4dB projections of antenna spots cannot overlap is fulfilled if we associate to the centre of each gaussian kernel a Spot Area of appropriate diameter: the spots will not overlap because the BVQ training procedure implies a repulsive potential among kernels, such that they cannot overlap more than a pre defined amount.

Several experiments have been performed to validate the above procedure. They have been performed on a DEC VaxStation; the training phase for the most complex experiments, relative to the global coverage of western Europe (an old figure nowadays) sampled with about 500 points, took less than three minutes to run. Some of the results obtained are reported in the following: in each figure there are reported the spot positions at the end of the competitive phase and the corresponding global coverage contour plot (dB), obtained by assuming that each spot produces an electric field described by the function $E(r)=J_2(r)/r^2$. Spot radii are expressed in number of sampling units.

In order to give a quantitative evaluation of the coverage results obtained, the following figures of merit were considered:

$$\eta_1=\frac{\text{portion of the SEA covered by spots}}{\text{sum of spot areas}} \quad ; \quad \eta_2=\frac{\text{portion of the SEA covered by spots}}{\text{SEA}}$$

Moreover, procedure's stability has been tested by repeating each coverage experiment for twenty times and evaluating the average and variance of both figures of merit. The mean and variance of both η_1 and η_2 are reported for all experiments.

CONCLUSIONS

The use of the Backpropagation Vector Quantization neural network was proposed for the solution of the constrained optimization problem of satellite antenna coverage plans project. The proposed approach leads to a very simple neural structure, whose computational cost is extremely light. It improves a previously proposed neural approach: no pre-processing is needed, because no boundary effect is present and because the case of spot centre's optimal location *outside* the area to be covered, as for example in the case of gulfs, is directly included in the procedure. Moreover, the present approach can be extended in a simple way in order to include other constraints, such as single station traffic volume and frequency and/or polarisation reuse.

The effectiveness of the procedure was demonstrated by means of several examples. Further work is in progress to verify the possibility of associating traffic volume of an earth station to the probability of occurrence of the relative SEA sample. Moreover, the early experiments with kernels different from the gaussians seem to indicate that further improvements to the final solution are possible.

ACKNOWLEDGEMENTS

This research was financially supported by the italian "Ministero dell'Università e della Ricerca Scientifica" (MURST) and by the italian "Consiglio Nazionale delle Ricerche" (CNR).

REFERENCES

[1] P. Burrascano, M. Melis:"A Neural Technique for Satellite Coverage Plans Optimisation"; Proc. Int. Conf. on Artificial Neural Networks, ICANN-92, Brighton, U.K., Sept. 4-7 1992;

[2] P. Burrascano:"Backpropagation Vector Quantization Neural Network for Probability Density Functions Modelling"; Proc. Fifth Italian Workshop on Parallel architectures and Neural Networks, Vietri sul Mare, May 1992;

[3] P. Burrascano:"A Norm Selection Criterion for the Generalized Delta Rule"; IEEE Trans. on Neural Networks, vol 2, n. 1, Jan 1991, pp. 125-130;

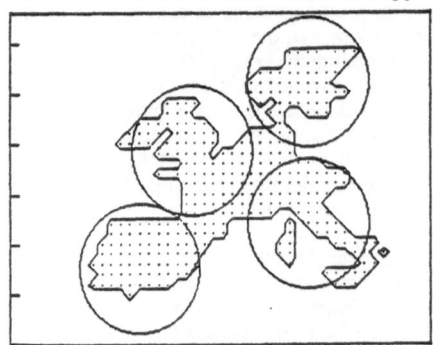

Figure 2: Europe coverage.
4 spots of radius 6.5
η_1 :avrg 44.68; st.dev 0.32
η_2 :avrg 83.57; st.dev 1.13

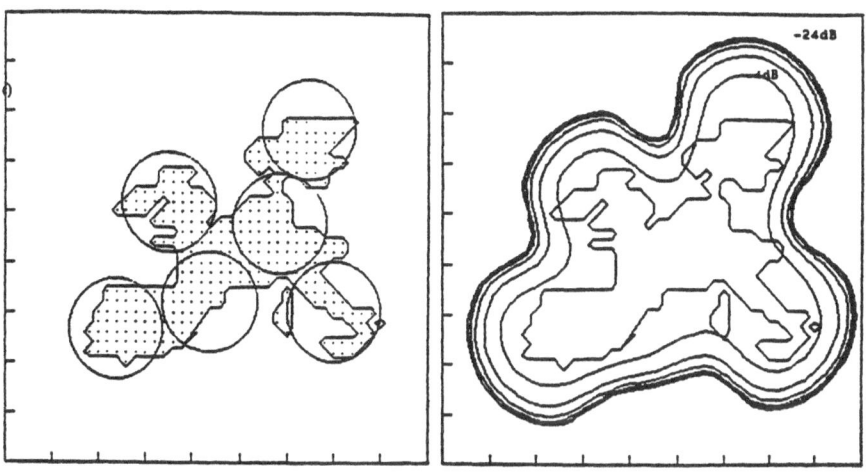

Figure 3: Europe coverage. 6 spots of radius 5
η_1 : average = 59.72; st. dev = 0.71 η_2 : average = 85.50; st. dev = 1.45

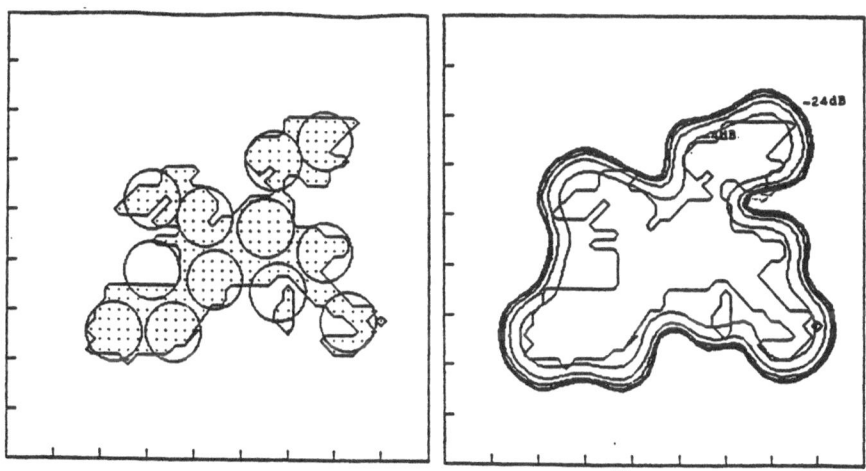

Figure 4: Europe coverage. 12 spots of radius 3
η_1 : average = 73.84; st. dev = 2.31 η_2 : average = 75.92; st. dev = 2.46

Sequential Self-Organization for the Traveling Salesman Problem

Shi-Chen Lee, Jiann-Ming Wu, and Cheng-Yuan Liou

Corresponding address: Cheng-Yuan Liou, Dept. of Computer Science and Information Engineering, National Taiwan University, Taipei, Taiwan, R.O.C. Tel.8862 3630231 ext. 3229, FAX. 8862 3628167

Abstract
In this work, by exploring geometrical feature, we modify the self-organization algorithm by Bernard et al to solve the traveling salesman problem (TSP) and obtain efficient computation in execution time and memory space with feasibility in computing with ring structure. The new algorithm is applied to the 532-city TSP problem in Padberg's paper. Numerical simulations show that the performance of the new algorithm is faster than that of the Bernard's approach in speed and better than that of the annealing with perturbing synapses by Lin et al in quality.

I Introduction
Hopfield and Tank [3] in 1985 pioneered a heuristic method based on the energy landscape to solve the TSP problem. A neural network based algorithm, called the elastic net method [4], gradually and non-uniformly elongates a circuit by using iterative procedure. The elastic net has been shown to reach a favor answer in short time. Following the same idea, Bernard et al proposed another analog algorithm on the basis of the self-organization principle for the TSP. Bernard's algorithm contains two major components, a mapping function from cities to nodes and an adaption method of updating the circuit. The simplicity in hardware structure and fast convergence with excellent quality in speed is attractive of their method [1]. Since there are too many unpredictable number of creations and deletions of the nodes in the algorithm, the unpredictable increases of nodes may much exceed the total number of cities and the well developed nodes may be deleted in later iteration. The unpredictable excess of nodes may cost the hardware load. The many deletions may neglect the accumulation informations during the early evolutions which are important for the cluster distribution of the cities. We modified their algorithm by keeping the total number of nodes as a constant, the same number as cities, during the evolution. And a ring structured hardware is well feasible for our algorithm. Except the distance criterion as in Bernard's algorithm, we explore another geometrical feature to the construction of the mapping function with efficient computation in both memory space and execution time. We present our algorithm in section II, and show performance comparisons between our approach, Bernard's algorithm, and the APS[2] in section III. We draw our conclusions in the final section.

II The New Algorithm
Given positions of N cities in a two-dimensional plane, the TSP problem asks to find a shortest path which visits each city once and returns to where it starts. We first review Bernard's algorithm briefly in following context. Let $v_i = (v_{xi}, v_{yi})$ denote the coordinate of city i. And let $u^t_i = (u^t_{xi}, u^t_{yi})$ denote the coordinate of node i at time step t. The algorithm is an iterative procedure. For each iteration, each city is selected once in a random sequence to update the circuit. The time step is increased by one for each city updating. At time step t, the updating for city i is

    a. by using the following mapping function ϕ to select a node on the circuit as the representative of city i.

$$\phi(i) = p \quad \text{where p satisfies } d_p = \underset{j}{\text{MIN}} \{d_j\} = \underset{j}{\text{MIN}} \{ \parallel v_i - u_j \parallel \}$$

    b. in according to the value $\phi(i)$ to elongate each node j on the circuit as following equations.

$$u^{t+1}_{xj} = u^t_{xj} + f(b(t),m_j) * (v_{xi} - u^t_{xj}) \; ; \; u^{t+1}_{yj} = u^t_{yj} + f(b(t),m_j) * (v_{yi} - u^t_{yj})$$

where $\quad f(b(t), m_j) = \frac{1}{\sqrt{2}} * \exp(-\frac{m_j^2}{b(t)^2})$ and $m_j = \text{MIN} \{ p-j(\text{mod } N), j-p \text{ (mod } N)\}$

b(t) is a non-increasing function which decays every iteration. In the algorithm, the function ϕ may map different cities to the same node during the same iteration. This results in improper adaption of the circuit toward cities. In Bernard's algorithm, an additional procedure is applied to deal with such improper adaption. If mapped by multi cities during an iteration, the node is duplicated one copy for each mapping city. If a node is not selected as the representative of any city for three continuous iterations, the procedure deletes the node. In above algorithm, the number of the nodes on the circuit is not fixed. The creations and eliminations of nodes may cause degrading quality and convergence speed.

We use another geometrical feature to modify the mapping function ϕ to minimum creations and eliminations. In the modified algorithm, the number of the nodes on the circuit is fixed as the number of the cities at all time. At the beginning of each iteration, the new algorithm sets the auxiliary variable s_i to 0, $1 \le i \le N$. Let $hr(k) = k-1 \pmod N$ and $hl(k) = k+1 \pmod N$ denote the two neighbors of the node k. The initial N nodes are located in a closed circle with very small radius. The rule for the selection of the representative of a city k and updating the circuit is as following.

    i. Setting q to $\phi(k)$.

    ii. If $s_q = 0$, increasing s_q by one and mapping q as the representative of city k,
        otherwise

        for case 1, both of $s_{hr(q)}$ and $s_{hl(q)}$ are not zero, neglecting q.

        for case 2, only one of $s_{hr(q)}$ and $s_{hl(q)}$ is zero, mapping the one with s value zero
           and increasing its s value by one.

        for case 3, both of $s_{hr(q)}$ and $s_{hl(q)}$ are zero, applying an angle bisector method to
           select one of two neighbors as mapping node and increasing its s value by one.

    iii. do as b. to elongate the mapped node q.

At the end of one iteration, the new algorithm deletes the nodes with s value zero, and if with s value greater than one, the node k is duplicated for $s_k - 1$ copies. At each iteration, the new algorithm maintains N nodes on the circuit. In the new mapping method, when one node has been mapped by city j and is later selected by another city i as the nearest node, as in case 3, we apply the following angle bisector method to decide helpfully which neighbor should be assigned to be the representative of city i.

    1. As shown in figure 1, the angle formed by node q, hr(q) and hl(q) is coppied to city j by moving the angle vertex from the node q to city j in parallel.

    2. Drawing the angle bisector for the copied angle and separating the plane into two sides, each corresponding to one neighbor of node q.

    3. Checking on into which side the city i is belonged to and assigning the corresponding neighbor of node q to be the representative of city i.

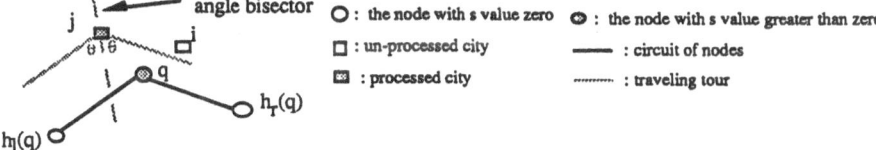

Figure 1. The angle bisector method first translates the vertex of the angle formed by node q, hl(q) and hr(q) to city j and draws the angle bisector of the copied angle. In this case, city i is mapped to node hr(q).

The figure 2 shows the advantages of the new mapping method. In figure 2a, node q has been mapped by city j and is the nearest node of the city i. If we use the distance as only criteri decide which neighbor should be mapped to, hr(q) will be selected and the obtained traveling tour is as in figure 2b. By using the angle bisector method, we obtain the traveling tour as in figure 2c, which is free of a twist.

This angle bisector method can be easily implemented. As shown in figure 3, we first find two points c for hr(q) and d for hl(q) on two edges of the copied angle, where two points have the same distance to the city j. And we can decide the representative of the city i by comparing the distance from city i to point c with the distance to point d. In figure 3, the city i is closer to point c than point d. The node hr(q) is selected as representative of the city i.

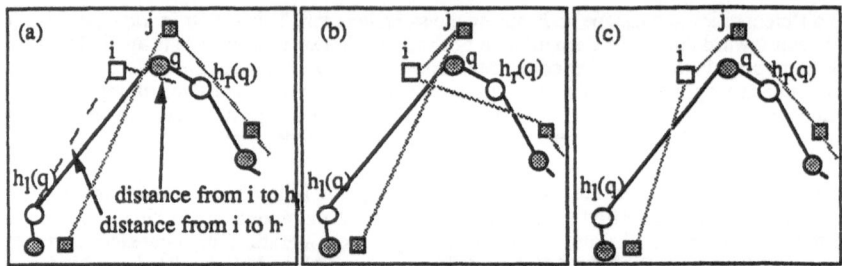

Figure 2. When the competition criterion is only the distance, in the case shown in figure 2a, node hr(q) is closer to city i than node hl(q). And the obtained traveling tour shown in figure 2b is a tour with a twist. By the angle bisector method, the obtained representative of city i is node hl(q), which corresponds to a tour free of a twist.

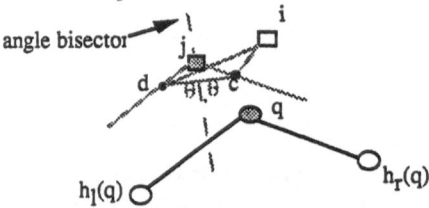

excution time average	30 cities	1000 cities
new algorithm	0.31 sec	3 min 8 sec
Bernard	0.44 sec	6 min 58 sec

Figure 3. By comparing the distance from i to point c and the distance from i to point d, the mapping can be deicded. In this case, c will map to i.

Table 1.Average execution time used on SPARC II workstation by the Bernard's algorithm and the new algorithm for the simulations.

III Numerical Simulations

In our simulations, we compare the performances of the Bernard's algorithm, APS method and our algorithm for 30-city and 1000-city TSP problems. The positions of the cities in the problems are randomly generated within a 1x1 square. The 30-city TSP problem is the example in [2]. For the Bernard's algorithm, the circuit is initialized as one node with random position in the plane and for the new algorithm, the circuit is initialized as N nodes with position on the center mass of the cities. All simulations are run on SUN SPARC 2 workstation. We run 2000 converges of each algorithm for 30-city TSP and 200 converges of Bernard's algorithm and the new algorithm for 1000-city TSP problem. For each converge, the algorithms are restarted with different initial neuron values for APS method, or with different sequence of city process for both the Bernard's algorithm and the new algorithm. For both of the Bernard's algorithm and the new algorithm, the b value is initialized at 40 and decays by rate 0.9 per iteration to the minimum value 0.1. Figure 4 shows the tour length distribution graphs of each algorithm for 30-city TSP problem. In a tour length distribution graph, the horizontal coordinate represents the tour length and the vertical coordinate represents the frequency of the execution converges occurring within the horizontal coordinate interval of tour length.We see that the new algorithm is much better than the APS method in solution quality. This is because the APS method is based on design of cost function and the new algorithm is based on design of a reward function. This example shows the effectiveness of the reward function. The Hopfield network [3] is also based on a designed cost function. Extensive examples show that designing a reward function is much better than using a cost function. The performances of the new algorithm are no less than those of the Bernard's algorithm in quality. The simulations for 1000-city TSP problem also display their comparable performance. The comparisons of the CPU time consumed by the Bernard's algorithm and the new algorithm in our simulations are shown on table 1. In the same computing environment, the new algorithm is faster than the Bernard's algorithm. The APS method consumes the CPU time much more than that listed on the tables.We test the new algorithm for another 532-city TSP problem [5], of which the length of the shortest path is known. We run the new algorithm for 400 converges and display the relative performance in figure 5. In the figure, the horizontal coordinate represents the relative ratio, which is defined as (obtained path length - shortest path length) / optimal path length.The vertical coordinate denotes the frequency of occurrence at each relative ratio. The best tour obtained within 400 converges is shown in figure 6 with relative ratio 0.0492.

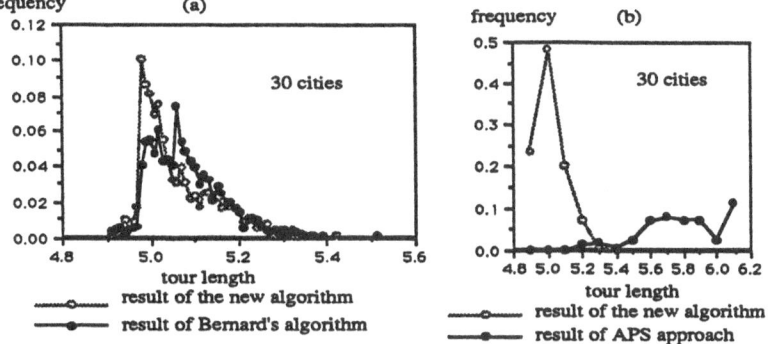

Figure 4. (a) The tour length distribution graphs of the Bernard's algorithm and the new algorithm for 30-city TSP problem. (b) The tour length distribution graphs of the APS approach and the new algorithm for 30-city TSP problem.

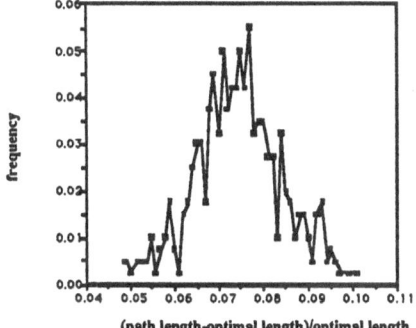

Figure 5. Relative performance of the new algorithm for the 532-city TSP problem with 400 convergence.

Figure 6. The best tour obtained from 400 convergence of the new agorithm in figure 5. The relative ratio is 0.0492.

IV Conclusions

By exploring the geometric feature, we present another self-organization based algorithm for the TSP problem, which obtains good solutions for the large scale TSP problems in short time. For the 1000-city TSP problem, a good solution is obtained with average execution time 3 min. 8 sec. on SUN SPARC 2 workstation. We show the self-organization based algorithm is capable of obtaining high quality solution in short time.

References:
[1] Bernard AngÈniol, GalÎde La Croix Vaubois and Jean-Yves Le Texier,"Self-Organizing Feature Maps and the Travelling Salesman Problem", Neural Network, Vol. 1, pp. 289-293, 1988
[2] Shiao-Lin Lin, Jiann-Ming Wu and Cheng-Yuan Liou, "Annealing by Perturbing Synapses", IEICE Trans. Inf.& Syst., Vol. E75-D, No. 2, pp. 210-218, March 1992.
[3] J.J. Hopfield and D.W. Tank, "Neural Computation of Decisions in Optimization Problem" Biological Cybernetics 52, pp.141-152,1985.
[4] Richard Durbin and David Willshaw, " A analogue approach to the travelling salesman problem using an elastic net method", Nature Vol. 326 16, pp. 689-691, April 1987.
[5] Padberg M. and Rinaldi G.: Optimization of a 532-city symmetric traveling salesman problem by branch and cut, Operations Research Letters, Vol 6, No. 1, 1-7(1987)

Invariant Process Control Using Neural Networks

Zahari Taha , Sigeru Omatu
Department of Information Science and Systems Engineering
Faculty of Engineering,
University of Tokushima,
Tokushima 770
Japan

Abstract

Using neural networks in direct inverse control is a promising approach mainly because of the simplicity of implementation in which the inverse model has immediate utility for control. This paper presents a study of the behaviour of such a controller. Specifically in the control of a water temperature control system it was discovered that if a NN is trained with its input being the rate of change of the plant output but instead, as a controller, the input of the NN is the plant error, the control loop is reduced to a first order system. However this is subject to the plant parameters being closely matched.by the network Thus the plant behaves as a first order system and whilst doing so becomes invariant to disturbances .

1.0 Introduction

One of the major methods in applying neural networks to control problems is direct inverse control . This method is based on supervised learning in which a network learns the inverse dynamics of the plant to be controlled. The training data consist of the vector X(t)- a set of observations of the plant's output and U(t)- a set of control actions governing the plant's ouput. The inverse dynamics is then used directly in the control loop. This method has been used in many applications both in its basic and modified form [1,2,3]. Theoretically if the network is the plant inverse then applying the desired plant output to the controller network will cause the plant input to reproduce the desired plant ouput. However this does not seem to be the general case particularly when the plant's inverse is not well defined. This is because the inverse mapping have been averaged over various targets and therefore is not neccessarily an inverse of the plant. Despite this, direct inverse control is still a promising approach mainly because of its simple configuration in which the inverse model has an immediate utility for control.

In this paper, a study of a neural network inverse controller applied to a water bath temperature control system is presented in which it was discovered that the network behaves as a first order system. Whilst doing so, it becomes invariant to disturbances. This results from the network being configured differently when in the learning phase and consequently as a controller. A linear neural network is used so that a study in continous time domain can be done.

2.0 Linear Network

A fully connected neural network can be represented by a structure as illustrated in figure 1 with the following logic [4]. The same notations are used as in [4].

$$x_i = X_i \qquad\qquad 1 \le i \le m$$

$$net_j = \sum_{j=1}^{i-1} W_{ij}x_j \qquad\qquad m < i \le N+n$$

$$x_i = s(net_i) \qquad\qquad m < i \le N+n$$

$$Y_i = x_{i+N} \qquad\qquad 1 \le i \le n$$

In this study, a linear neuron function is used, defined as

$$s(net_i) = 0.5net_i$$

with the constraints

$$s(net_i) = 1 \quad \text{if } net_i > 1$$
$$s(net_i) = -1 \quad \text{if } net_i < -1$$

A three layered network is used- an input layer, "hidden" layer, and an output layer with connection weights only between layers. The number of input neurons is m=2, the number of ouput neurons n=1 and the number of hidden neurons is 8 so that N=10.

3.0 Water Bath Temperature Control System

A schematic diagram of the water bath temperature control system is shown in figure 2. The heater is switched on and off according to the following constraints:

$$\text{if } u(kT) \leq 0.0, \text{ then } V_i = 0.0V$$
$$\text{if } u(kT) \geq 5.0, \text{ then } V_i = 5.0V$$
$$\text{else } V_i = u(kT)V$$

where u(kT) is the output of the controller, T denotes the sampling interval, k is the sampling number (k=0,1,2,3...), and V_i is the input voltage to the thyristor controlling the heater. The continous time model of the plant is given by

$$\frac{dy(t)}{dt} = \frac{f(t)}{C} + \frac{(Y_0 - y(t))}{RC} \tag{1}$$

where Y_0 is the room temperature, R is the thermal resistance, C is the thermal capacity, f(t) is the energy flow into the water bath and it is asummed that f(t)= Ku(t). However this model is not required by the neural controller but is useful in the ensuing discussions

4.0 Controller Implementation

The water bath is an open loop stable system that exhibits linear behaviour up to about 70° and then become nonlinear and saturates at about 80°. In the learning phase a ramp signal is applied to the water bath. This coupled with the corresponding output rate of change of temperature y(kT)-y((k-1)T) is used as training data. The objective is to obtain the model

$$u(kT) = \alpha(y(kT) - y((k-1)T)) + \beta y(kT) \tag{2}$$

where α and β are the plant parameters that can be extracted from the weights of the linear network. The network is then used as a controller with the input being the error, $y_d - y(kT)$ where y_d is the set point. The controller is thus of the form

$$u(kT) = \alpha(y_d - y(kT)) + \beta y(kT) \tag{3}$$

In the continous time domain , the Laplace transforms of equations (2) and (3) are given by

$$U(s) = (\alpha s + \beta)Y(s) \tag{4}$$

and

$$U(s) = \alpha E(s) + \beta Y(s) \qquad (5)$$

respectively. The neural network control loop can be represented in the familiar block diagram of figure 3. The second term of equation (1) which is $(Y_0 - y(t))/R$ is always negative since the temperature $y(t)$ is always above the room temperature. By analysing the weights of the linear network it was found that $\beta y(t)$ is positive and compensates for the second term of equation (1). The value of α can be seen as a proportional gain which has been tuned in the open loop during the learning phase. The CLTF can then be derived as

$$\frac{Y(s)}{Y_d} = \frac{1}{1 + \frac{C}{K\alpha}s} \qquad (6)$$

The whole system is thus reduced to a first order system. While the plant is behaving as a first order system it becomes invariant to disturbances. Figure 4 shows the plant performance with various setpoints. The transient obviously emulates a first order system. Figure 5 shows that the plant is also invariant to disturbances.

5.0 Conclusions

In this paper it has been discovered that by using a neural network as an inverse controller the plant is reduced to a first order system. The controller implementation is such that during learning the neural network input is the rate of change of plant output whilst as a controller the neural network input is the plant error. The plant in this case is open loop stable and the network has actually tuned the optimal gain of a proportional controller as well as identified unmodelled dynamics. Future work will include applying this approach to MIMO and nonlinear systems.

References

[1].   Widrow, B., Mc Cool, J., and Medoff, B (1978). Adaptive Control by inverse modelling . In Twelfth Asilomar Conference on Circuits, Systems and Computers.

[2].   Widrow, B. and Stearns, S.D. (1985). Adaptive Signal Processing, Englewood Cliffs, NJ: Prentice Hall.

[3]   Widrow, B. (1986). Adaptive inverse control. In Proceedings of the Second IFAC Workshop on Adaptive Systems in Control and Signal processing, 1-5, Lund, Sweden: Lund Institute of Technology.

[4]   Werbos, P.J, Backpropagation Through Time: What It Does and How to Do It. In Proceeding IEEE, Vol. 78, no. 10, Oct. 1990, pp. 1550-1560.

Acknowledgements

The first author is supported by a Hitachi Scholarship Foundation Fellowship as Visiting Researcher at the University of Tokushima.

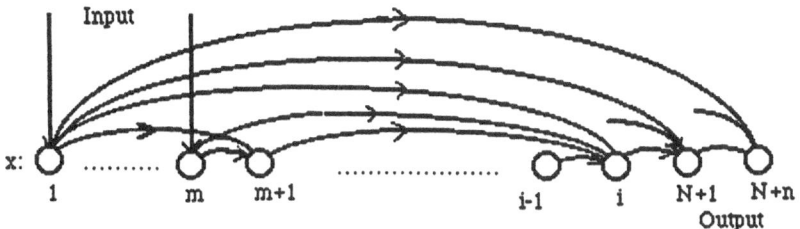

Figure 1. General Structure of Neural Network

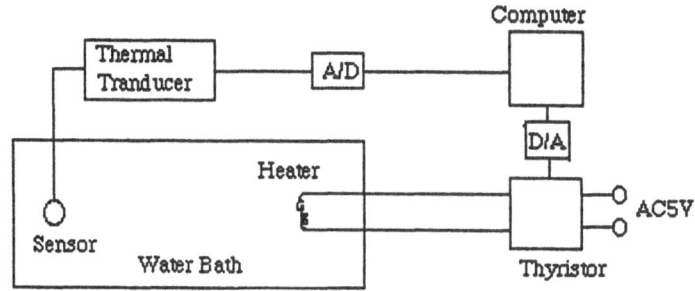

Figure 2. Schematic Diagram of Water Bath Temperature Control System

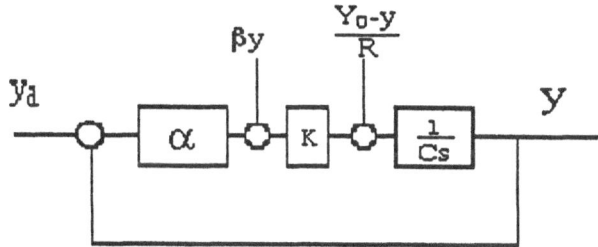

Figure 3. Block diagram of Neural Network Inverse Control Loop

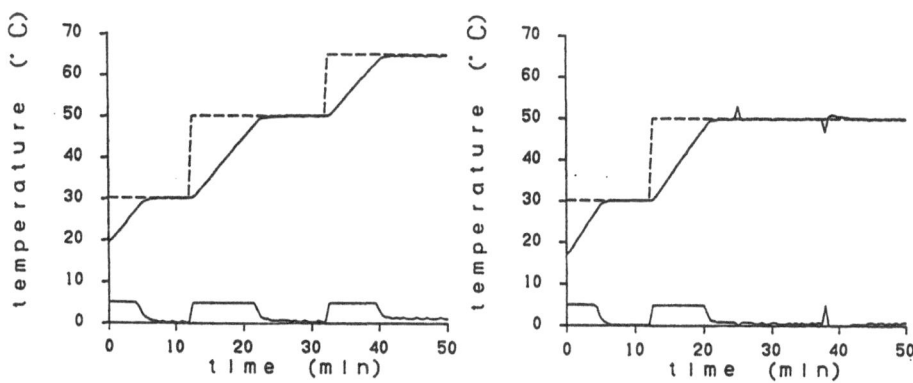

Figure 4. Tracking performance of Inverse controller

Figure 5. Performance of Inverse Controller under noise disturbance.

Optimal Control of Dynamic Systems
Using Self-Organizing Maps

Heikki Hyötyniemi
Helsinki University of Technology
Control Engineering Laboratory
Otakaari 5 A, SF-02150 Espoo, Finland

Abstract

Self-organising Kohonen maps can efficiently be used for process modeling purposes. The states of the closed loop system are used as input feature vectors to the neural net, and the map converges according to the distribution of the state variables. This map can be seen as a compressed version of the state space.

The map of the system states can be used for controlling of the dynamic system. In this paper, an approach to constructing a suboptimal control strategy utilising the self-organising net approach is presented. Simulation example is given to illustrate the potential of the method in optimization tasks.

1 Introduction

Control of complex non-linear processes is among the most active fields of research in control engineering. One of the most promising approaches is the use of *artificial neural networks* in process identification and control tasks ([6],[8],[7]). However, from the point of view of dynamic systems, the *de facto* standard neural networks approaches, where adaptation is based mainly on the back propagation of errors, have some less appealing features:

1. these methods only learn the input-output behavior of the system and the created process model is implicit, and

2. the dynamic nature of the processes is not taken into account, as the methods have been developed for modeling static mappings.

Another way of utilizing neural networks in control is the use of *self-organizing maps*. Usually, in the field of control engineering, these methods have only been used for fault diagnosis or measurement data preprocessing tasks. In this paper, an approach to controlling of dynamical systems based on *Kohonen networks* is presented. An application example is shown to illustrate the method.

2 Principles

Kohonen neural network adaptation algorithm [3] is the standard example of *unsupervised* learning. The network is composed of N independent computing elements, *nodes*, each of which contains a *feature vector* f_i, $1 \leq i \leq N$, containing the 'prototype' of vectors matching that node. The adaptation algorithm that is carried out for each input vector x consists of two steps:

1. **Matching.** Find node c with the feature vector f_c nearest to the input vector x

$$c = \operatorname{argmin}_{i \in [1,N]} \left\{ \| x - f_c \| \right\}. \tag{1}$$

2. **Updating.** For each node i in the neural net, modify the corresponding feature vector f_i according to the law

$$f_i \leftarrow f_i + \alpha(x - f_i). \tag{2}$$

In Step 2, the value of the parameter α is between 0 and 1, being larger if some specific neuron is to be updated more according to the input. The value of α is a decaying function of time, assuring that the network finally converges. Also, α determines the net topology, giving large values if nodes c and i are 'near' each other in the net and small values otherwise, causing the feature vectors of neighboring nodes to be pulled nearer to each other.

When modeling dynamic systems with a self-organizing net, the measured state vector samples are used as input vectors to the net. Hereby, 'typical dynamics', or the most probable states of the closed loop system, will be learned by the net. If the control signals are saved in the net together with the process states, the adapted neural net can be used as an associative memory for finding control signals, and the process can be controlled by the neural net.

One of the basic methods for optimizing the control laws is called dynamic programming (see, for example, [2]). The *principle of optimality* states that the optimal control law can be regarded as a series of independent control actions, and it allows one to calculate optimal paths in a recursive manner. This recursion is general and applies to non-linear systems just as well as for linear ones. Normally, dynamic programming has only theoretical interest, because, with some exceptions, the search space grows rapidly. This restriction is commonly called the *curse of dimensionality.*

In this paper, the traditional dynamic programming concept is made applicable, or the explosion of complexity is avoided, utilizing self-organization. During iterative optimization steps, the adaptation of the net keeps the net contents valid as the set of relevant state combinations is changed.

3 Theoretical treatment

In what follows, we concentrate on time-invariant processes, with state dimension n and control dimension m. The system is completely observable and controllable, and the state vector is assumed to be known through measurements. The structure as well as the parameters of the process are supposed to be known. The process is supposed to be discretized with a reasonably short sampling interval.

Let $u(k)$ be the control signal vector of length m to be applied to the process at time instant k, and let $x(k)$ be the state vector of length n. The system can be written in the form

$$x(k+1) = F(x(k), u(k)) \tag{3}$$

Suppose that the control signal is a smooth and continuous function of the state variables. Locally, linear approximations of these functions are valid to sufficient accuracy, and the system can be linearized about the state x. Suppose that the time steps are short enough to justify using the linearized process model between successive time points:

$$x(k+1) \approx F_x x(k) + G_x u(k) \tag{4}$$

The optimality criterion for the process is supposed to be quadratic:

$$J = \sum_{k=0}^{\infty} u^T(k)Ru(k) + x^T(k+1)Qx(k+1) \tag{5}$$

The weighting matrices R and Q, for control signal and the states, respectively, are symmetrical. Minimizing J is a time-invariant optimization problem, because no final time is given in (5). According to the dynamic programming principle, introduce the recursive *cost function* $C(x(k))$ for any state $x(k)$ as

$$C(x(k)) = \min_{u(k)}\{u^T(k)Ru(k) + x^T(k+1)Qx(k+1) + C(x(k+1))\} \tag{6}$$

This recursion has to result in a finite cost, so that the process has to be asymptotically stable. This means that there has to be some kind of a stabilizing controller available, before optimization process is started. Origin $x = 0$ is supposed to be the final state, so that $C(0) = 0$. In practice, for any state x represented in the net, the value of the cost function $C(x)$ in (6) is calculated utilizing the cost function values of the nearest neighbor nodes.

The control signal is calculated for each state that is represented in the net, minimizing the cost functional. Assuming that local linearization of the process and cost function is meaningful in state $x = x(k)$, minimizing (6) is approximately equal to minimizing

$$u_x^T \left(R + G_x^T Q G_x \right) u_x + \left(2x^T F_x^T Q G_x + C_x^T G_x \right) u_x \tag{7}$$

where the next state $x(k+1)$ has been expressed using $x(k) = x$ and $u(k) = u_x$, calculated from (4). Vector C_x represents the parameters of the linearized cost function hyperplane, so that $C(x') \approx C_x^T x'$ for any x' that is near x. In this case, for simpler notations, the hyperplane always goes through origin. This quadratic optimization problem can be solved analytically by differentiation, so that an explicit expression for u_x is found:

$$u_x = - \left(R + G_x^T Q G_x \right)^{-1} G_x^T \left(Q F_x x + \frac{1}{2} C_x \right) \tag{8}$$

The problem can also be formulated as a *linear complementarity problem* that can efficiently be solved using, for example, *quadratic programming*, which is a modification of the SIMPLEX algorithm [1]. Additional non-linearities, like constraints on states or control signals, can then be introduced in the optimization problem.

Even if a more optimal control signal could be calculated this way explicitly for all states that are represented in the net, the closed loop dynamics must not be changed abruptly. To smoothen the changes, the updating step in the standard Kohonen algorithm has to be modified. For each node i, where $1 \leq i \leq N$, both f_i and u_i are modified gradually as follows:

$$f_i \leftarrow f_i + \alpha(x - f_i)$$
$$u_i \leftarrow u_i + \lambda(u_x - u_i).$$

The *forgetting factor* λ is used to weigh also older signal values when calculating new controls. The value of λ is dependent of the same factors as the value of α is.

When the neural net controller is used in the closed loop control system, the output control signal is calculated using linear interpolation, determined by the nodes that are nearest to the actual state. Assuming that a linear interpolation is to be calculated using controls u_i in nodes $i \in S$, where S is some subset of all N nodes, the control signal is $u_x = U_S^T x$, where the parameter vector U_S of the approximating hyperplane is calculated using least squares method:

$$U_S = (\sum_{i \in S} f_i f_i^T)^{-1} \sum_{i \in S} f_i u_i^T. \tag{9}$$

4 Example

In this example, a simple and linear process model is used to show how optimization is carried out. The second order continuous system represents a mass point ($m = 1$) with a force acting on it. The continuous system has been discretized with sampling period T:

$$x(k+1) = \begin{pmatrix} 1 & T \\ 0 & 1 \end{pmatrix} x(t) + \begin{pmatrix} \frac{1}{2}T^2/m \\ T/m \end{pmatrix} u(t) \tag{10}$$

State x_1 is the cart location and x_2 is the velocity. The initial state of the system is $(-1\ 0)^T$ and the desired final state is $(0\ 0)^T$. The sampling interval in the experiments is $T = 1$. The weighting matrices used in the optimality criterion (5) are

$$R = (1) \qquad \text{and} \qquad Q = \begin{pmatrix} 1 & 0 \\ 0 & 1 \end{pmatrix} \tag{11}$$

The training and optimization consists of a succession of simulation runs, always starting from the initial state, the controller trying to drive the cart to the final state. First, some non-optimal stabilizing controller is used to drive the system to the final state. After some dozen of runs the two-dimensional net of 10×10 nodes has adapted enough to be used as the controller in the closed loop, and after that the behavior is gradually optimized (Fig. 1). The organization of the map after optimization is shown in Fig. 2.

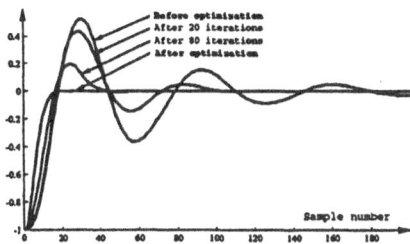

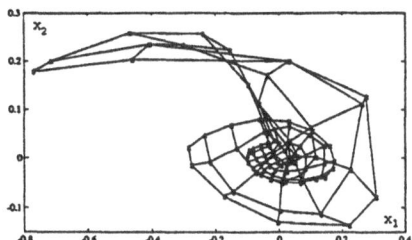

Figure 1. Simulation results as the optimization process continues

Figure 2. The converged map projected in
the state space

5 Comparison to existing methods

The heuristic *tabulation method,* where the whole process state space is tabulated and appropriate
control actions are calculated, is an old approach to handle very complex processes. As the dimension
of the dynamic system grows, the complexity of this method soon becomes intolerable. The approach
that is presented in this paper is a near relative to this method. However, now the explosion of
complexity is avoided, because the self-organizing net automatically allocates the available memory
in an optimal way.

Plenty of more or less qualitatively oriented approaches have been proposed to tackle with highly
complex processes. Usually a stabilizing control law is regarded as sufficient, no matter how poor
the process behavior actually is. These proposed approaches include genetic algorithms (see, for
example, [4]), fuzzy and linguistic methods, and other (for example, [5]). There is a very interesting
possibility that would deserve some attention: It would be possible to let some of these AI oriented
and qualitative methods to generate some kind of a stabilizing control strategy, and after that the
neural optimization could take place, finding a more useful control law.

Acknowledgement

During this work financial support has been provided by the Academy of Finland and by Jenny and
Antti Wihuri Foundation, and this support is gratefully acknowledged.

References

[1] Bazaraa, M. S. and Shetty, C. M.: *Nonlinear Programming: Theory and Algorithms.* Wiley, New York,
1979.

[2] Jacobs, O. L. R.: *Introduction to Control Theory.* Clarendon Press, Oxford, 1974.

[3] Kohonen, T.: *Self-Organization and Associative Memory.* Springer-Verlag, Berlin, 1984.

[4] Koza, J. R. and Keane, M. A.: *Cart Centering and Broom Balancing by Genetically Breeding Populations
of Control Strategy Programs.* IJCNN'90, Washington, VA, January 15–19, 1990.

[5] Mitchie, D. and Chambers, R. A.: *BOXES: An Experiment in Adaptive Control.* Machine Intelligence 2,
1968.

[6] Narendra, K. S. and Parthasarathy, K.: *Identification and Control of Dynamical Systems Using Neural
Networks.* IEEE Transactions on Neural Networks, Vol. 1, No. 1, March 1990.

[7] Warwick, K., Irwin, G.W., and Hunt, K.J. (eds.): *Neural networks for control and systems.* Peter Pere-
grinus, London, 1992.

[8] Willis, M. J., Di Massimo, C., Montague, G. A., Tham, M. T., and Morris, A. J.: *Artificial Neural
Networks in Process Engineering.* IEE Proceedings D, Vol. 138, No. 3, May 1991.

Monitoring a Control System with a Hybrid Neural Network Architecture

Thierry Catfolis

Expert Systems Applications Development Group
Department of Chemical Engineering
Katholieke Universiteit Leuven
de Croylaan 46, B-3001 Heverlee, Belgium

Abstract

In this paper we simulated four working conditions of a control system, one normal and three abnormal - hysteresis at a valve, systematic deviation on the output of the measurement device and an abnormal amount of noise on the measurement device output. The monitoring system was build to distinguish these four conditions. The information available for the system consists of the setpoint, the measurement and the controller output.

We constructed a network architecture with recurrent and feedforward layers for monitoring that control system. This design decomposes the spatio-temporal pattern recognition task in a temporal preprocessing part and a spatial analysing part. The first part of the system is a recurrent layer [1] that performs a conversion of the dynamic data stream into a static image. This image contains all information needed by the spatial part of the error detector to perform its monitoring task. The second part are backpropagation layers performing a pattern recognition task on the image created by the recurrent layer. The result of this second part is the response of the overall network to the data stream. The advantages of the combined architecture are the acceleration of the training by using smaller recurrent layers and the possibility to preprocess data without further analysis.

The entire system consists of 4 equivalent architectures - called error detectors -, one for each working condition of the control system. The advantages of dividing the problem in subproblems are that the recurrent networks are smaller (and thus trainable in a reasonable time) and that adding or deleting some error detectors is possible without retraining the entire system.

Because we trained each error detector to distinguish its problem from the normal conditions, the results of each subproblem were not always relevant when other problems were shown. To solve this problem we added a postprocessing module that analyses the result of each subproblem network and reports the detected problem. This module consists of rules generated from a set of data by OCCAM [2], a software package for rule induction. The addition of a rule-based postprocessing module that combines the results of the error detectors creates a hybridisation of the architecture and increases the application range of the entire system by the possibility of stressing the importance of some results.

For one of the best systems we found a misclassification error of 8.67%. About 55% of these misclassifications are due to transient errors - faults that already occur but can not yet be detected. By stressing the importance of some results during the generation of the rules we can lower the misclassification rate to 4.7%. This also lower the generalization capabilities of the entire system.

References

[1] Williams, R.J. & Zipser, D. (1989) Experimental Analysis of the Real-time Recurrent Learning Algorithm. *Connection Science*, 1, 87-111

[2] OCCAM, Optimizing Case-based Classifier for Analysis and Modelling ©1992 Information Technology Research vzw.

Paper Web Profile Analysis Using Neural Networks

Jukka Vanhala[1], Pekka Pakarinen[2], and Kimmo Kaski[1]

[1]Tampere University of Technology, Microelectronics Laboratory, P.O.Box 692, 33101 Tampere, Finland
[2]Technical Research Centre of Finland, Combustion and Thermal Engineering Laboratory, P.O.Box 221,
40101 Jyväskylä, Finland
jv@ee.tut.fi, PEKKA.PAKARINEN@vtt.fi, kaski@ee.tut.fi

In order to guarantee the uniform quality of paper across the whole cross direction of the paper machine, on-line measurements and control equipments related to different paper quality profiles are necessary. In modern paper machines only the basis weight, the moisture and the caliber profiles are controlled by an on-line control system. However, there are several important profiles that can not be easily measured or controlled directly. As an example, the fiber-orientation profile is difficult to measure on-line and in addition to this the control mechanism is not clear. This is mainly due to the insufficient knowledge about the complex relationship between the orientation profile and the other profiles. In this study, the neural network method is applied to the fiber-orientation profile analysis and its control. Similarly the same method can be used to control other profiles as well [1,2].

The cross-machine flow causes changes in the fibre orientation. The cross-machine flow can be due to *e.g.* adjustments to the slice opening or irregular flow from the turbulence generator [3]. The fiber-orientation profile can be adjusted to a near optimal shape but only on the cost of the basis weight and formation profiles. Since the latter are normally more important factors and also easier to control, the fiber orientation is not optimized [4]. When the paper machine is run on the drag side, *i.e.* the velocity of the wire is higher than the velocity of the jet, all the suspension will filtrate in a same manner and the cross-machine component of the flow will show up in the fiber-orientation profile. If the profile is adjusted in this situation the paper machine may be run also on rush side.

As a test case we have created a system which is intended to help the paper machine operator to control the fiber-orientation profile. The fiber-orientation profile is measured in the laboratory using the standard methods of ultra-sound velocity measurements. Using this profile the neural network system will give the required change to the basis-weight profile which will correct the fiber-orientation profile. The operator checks that the change to the basis-weight profile is acceptable and feeds the change into the control system of the paper machine. We have used a back-propagation type network with two hidden layers for profile data analysis. The case is a typical mapping task were the input and output are large vectors. We have used one hundred cross-machine points. The number of units in the hidden layers vary from test to test from ten to one hundred. The network is able to learn the training set and also to make accurate generalizations when given an input profile that has not been included in the training set, Fig 1.

[1] K.E.Smith, "Cross-direction control is still top process automation trend," Pulp and Paper, Feb. 1985.
[2] D.B.Brewster, W.I.Robinson, "How computers are controlling functions on the paper machine," Pulp and Paper, Nov. 1973.
[3] R.Turunen "Experiences of Sym-Flo headbox and automatic profile control," Papermakers conf., 1986.
[4] D.B.Brewster, "Interactions in CD control," Symposium on Systems in the pulp and paper industry, May 1986.

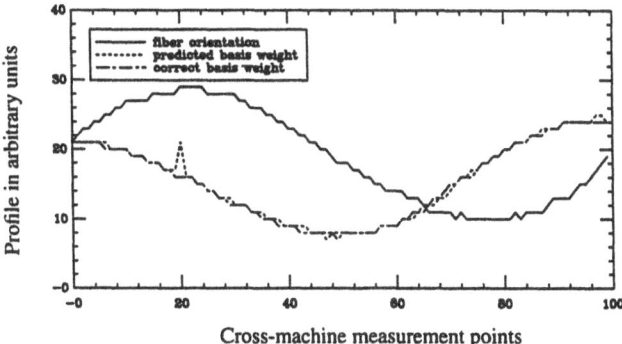

Figure 1. An example of the quality profile prediction.

Modelling of Quality Properties in Paper Drying with Multilayer Perceptron Network

P. Elo, J. Saarinen, K. Kaski

P. Pakarinen, H. Kiiskinen, S. Kaijaluoto, K. Edelmann

Tampere University of Technology,
Microelectronics Laboratory,
P.O.Box 692, SF–33101 Tampere, Finland

Technical Research Centre of Finland,
Combustion and Thermal Engineering Laboratory,
P.O.Box 221, SF–40101 Jyväskylä, Finland

Water in wet paper is located between fibers, inside of capillaries and even inside of fiber walls. Part of the water is also chemically bound as gels. This means that evaporation of water is a very complex process leading to physical and chemical changes in paper [1]. This is why the controlling of paper quality during drying is very difficult. In addition paper is also a porous material where the heat conductivity is strongly dependent on moisture.

The quality of paper is strongly dependent on drying conditions. It is difficult, if not impossible, to model the development of paper quality based on theoretical considerations. In this study neural networks have been applied to model the relationship between paper quality and drying conditions in a laboratory test dryer (Figure 1). By using trained neural networks it is possible to simulate behavior of the characteristics features of the paper quality without time-consuming test runs. The most important quality measures are related to strength, optical and printability properties.

The main drying parameters and quality properties measured are shown in Table 1. Three different backpropagation neural network were trained with different combination of data values generated by our laboratory drying apparatus.

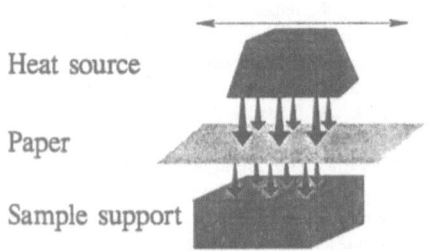

Figure 1. Laboratory drying system.

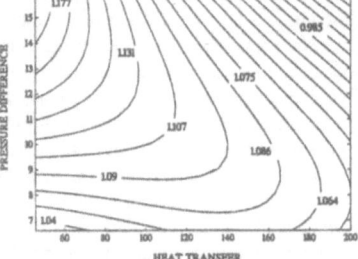

Figure 2. Simulation example.

Table 1. Contests of used datasets.

Drying parameters	Quality properties	
• heat transfer • pressure difference • drying time	• caliber • bottom side smoothness • top side smoothness • porosity	• optical properties • tensile properties in cross direction • tensile properties in machine direction • moisture of the paper • temperature of paper

The main aim of this study was to use neural networks to model and illustrate the developement of paper quality during drying section in our laboratory drying apparatus. It much easier to study the trends and the behaviour of a drying process by using a model than by doing repeated drying experiments. With a neural network we can fix all simulation parameters except one. By varying the last parameter we can analyse how the quality values behave as a function of this parameter. In Figure 2 an example of the simulations runs in arbitary units is given.

The neural network models can be used to interpolate new test points from original learning material and to visualize the relationships between quality values and operational parameters. The main benefit of this work is the reduction of time needed for the analysis and understanding of experimental data. In the future neural networks could e.g. be used as intelligent tools in paper mill laboratories. These simulation results have given new information and helped to understand the paper drying process more deeply. Encouraged by this project a new industrial application is beginning.

[1] A. S. Mujumdar (Ed.), Handbook of Industrial Drying, Marcel Dekker, New York, USA, 1987.

INTERPOLATION OF STATIONARY NON-LINEAR TIME SERIES BY AN OPTIMIZED NEURAL NETWORK

F.QENDRO, R.LENGELLE, T.DENOEUX, P.GAILLARD
University of Compiègne - U.R.A. CNRS 817
BP. 649 - F - 60206 Compiègne cedex FRANCE

Abstract

This paper presents a method for restoring samples in non-linear stationary discrete-time signals. Using an objective function for internal representation, the estimate of the unknown samples are obtained by a multilayer neural network with a criterion for optimising the number of units and the hidden layers.

In our approach the missing samples are in the between of known samples. No a priori information is assumed to be available. The strategy used for optimising the hidden units consists in increasing from one layer to the next, a measure of linear dependecy between the vector of internal representation and the desired outputs.

To test the performance of proposed method we have chosen an artificially generated time series (Mackey-Glass time series) and a natural stationary non-linear time series. A comparison with a linear model is presented too.

The numerical examples have shown promising results, indicating the potential use of the proposed interpolator for many stationary non-linear time series applications.

TWO-SENSOR NEURAL NETWORK MODELING FOR FAULT DETECTION

Robert E. Uhrig*, Israel Alguindigue, Anna Loskewicz-Buczak,
Andreas Ikonomopoulos, and Lefteri, Tsoukalas**
University of Tennessee, Knoxville, TN 37996-2300
* Also at Oak Ridge National Laboratory, Oak Ridge, TN 37831-6005
** Now with Japan Atomic Energy Research Institute, Tokai-Mura, Japan

ABSTRACT

The technique described here involves training a neural network to model the internal behavior of a component or system from vibration data taken from two sensors (accelerometers) located at different positions or mounted in different directions on the component or system. The power spectral density (PSD) of the sampled time-series from one accelerometer is used as the input to the neural network, and the PSD of the sampled time-series from the other accelerometer is the desired output of the neural network. The network is trained while the signals varies over the normal range of operation when the component or system is known to be operating properly. The trained neural network is then used in a monitoring mode to predict the output sensor PSD from the input PSD, and a comparison is made between the predicted and actual output PSDs. Significant deviations indicate that the interrelationship between the input and output signals has been modified due to a change (failure) in the component or system. The usefulness of this methodology has been demonstrated in the monitoring of the operability of check valves and a pump-motor bearing.

Check Valve Monitoring. The most common problems associated with check valve failures are due to system flow oscillations or system piping vibrations. These vibrations and oscillations induce measurable accelerations that produce check valve component wear and thus component failure. Analysis of time-records from piezoelectric accelerometers attached to check valves on a large nuclear power plant has been used to demonstrate this process. An autoassociative-like neural network in which the inputs and desired outputs are values of the PSDs of two related time-series representing vibration at two different positions on the valve was trained to produce a neural network model of the interrelationship when the valve is operating properly. During monitoring, the output PSD of one accelerometer is used to predict the output PSD of the other accelerometer. This predicted PSD is then compared with the actual PSD, and a significant deviation indicates failure of the check valve. Comparison of PSD spectra between identical 30-inch check valves (one broken and one normal), operating under identical conditions, demonstrated that this technique can identify the failed valve. Subsequent measurements taken on the broken valve after it was repaired further confirmed the validity of this technique.

Analysis of Vibrations. The two-sensor technique was also used to analyze the progressive failure of a large (950 HP) motor-pump bearing in a nuclear power plant. A series of measurements of horizontal and vertical components of acceleration for the motor-pump bearing were taken periodically throughout the operating lifetime of the bearing and as it began to fail. The PSDs of the horizontal and vertical components of acceleration on the bearing during the first four sets of measurements (when the bearing was known to be operating properly) were the input and desired output respectively of a neural network while it was being trained. For the next eight sets of measurements, while the bearing progressed towards failure, the predicted value of the vertical component of acceleration (obtained from the neural network using the horizontal component as the input) was compared with the actual value of the vertical component. The deviation between the predicted and actual vertical PSDs grew as the bearing progressed towards failure.

Modelling The Fed Batch Fermentation Process Using Artificial Neural Networks

N.A. Jalel and J.R. Leigh
Industrial Control Centre, University of Westminster, 115 New Cavendish St., London W1M 8JS, U.K.

Abstract
In this contribution an artificial neural network technique has been used for the on-line estimation of the important state variables of the fed batch fermentation process. The neural network tasks includes both modelling and state estimation of the residual nitrogen inside the fermenter.

Neural Network For Modelling The OTC Fermentation Process
A neural network trained on one batch of data was adopted for the on-line estimation of the residual nitrogen in the fed batch fermentation process with the neural network taking on the task of both modelling and state estimation. In this work, two layered, fully connected, feed-forward networks learning the fermentation process by standard back-propagation algorithm have been used to model the process. The networks consist of five processing elements in the input layer, eight in the hidden layer and one in the output layer. From the network results and by comparing with the off-line measurements it is clear that the network was able to provide a good estimation for the residual nitrogen.

Training The Network On More Than One Batch Of Data
It has been decided to train the network on more than one batch of data so that the network is able to learn different behaviour of the residual nitrogen inside the fermenter. Two approaches have been used to train the network. The first is based on training on each batch sequentially; this means that the network will be trained on the first batch, then the second and then continuously until the last batch. The results indicate that the network is able to provide a reasonably good estimation of the residual nitrogen. In the second approach, for each batch of data a neural network has been generated and trained. Since the aim is to use ten batches of data, ten neural networks have been generated and trained. The neural network which is considered to be representing the ten batches of data is found by averaging the weights of the ten networks. The results of the main network represents the mean average of the residual nitrogen of each batch which is not very satisfactory.

Sequential Neural Network
Since fermentation is a time varying process, a sequential neural network approach has been investigated in which the fermentation process is divided into phases according to the operation time of the process. From manual inspection of the data and from a physical understanding of the process together with some experimentation it has been decided to divide the process into three phases. Each phase has been treated separately and a neural network model for each has been derived. The unmeasurable state variables of the whole process have been taken to be the sequential combination of the three networks. The sequential modelling approach has achieved good modelling and representation of the process, presumably because this approach can accommodate the known time varying behaviour of the process.

Improving The Neural Network Ability By Adding Physical Understanding Of The Process
To improve the modelling ability of the neural network it has been decided to add some physical understanding of the process into the actual structure of the neural network. This structure involves a mixed neural network and integrator. A back propagation neural network has been trained to estimate the nitrogen uptake rate. The estimated value of the neural network is subtracted from the amount of nitrogen fed by the operator in order to calculate the level of the residual nitrogen. The predicted value of the residual nitrogen is calculated by simply integrating the level of the residual nitrogen inside the fermenter. Comparing with the previous results, a very good prediction has been achieved. Adding well-known physical understanding of the process in the form of a summation and integrator into the neural network has remarkably improved the neural network's ability to estimate the residual nitrogen.

Conclusion
Different neural network topologies for modelling the fed batch fermentation process have been investigated. In general, the neural network approach has been found to provide good estimation of the state variables and some of the variants give excellent results.

Identification of Car Body Steel
by an optical on line System and a Kohonen's self-organizing Map

W. Kessler[1], D. Ende[1], R. W. Kessler[1], W. Rosenstiel[2]
[1]Institut für Angewandte Forschung, FH Reutlingen
Alteburgst. 150, D-7410 Reutlingen, Germany
[2]Institut für Technische Informatik, Universität Tübingen,
Sand 13, D-7400 Tübingen, Germany

The corrosion of materials is a result of the superposition of complex reactions which depend on surface thickness, its porosity and chemical composition. By means of optical spectroscopy, especially diffuse reflectance spectroscopy, the surface of low carbon steel can be analysed within milliseconds. From these results it is possible to predict its future corrosion resistance[1].

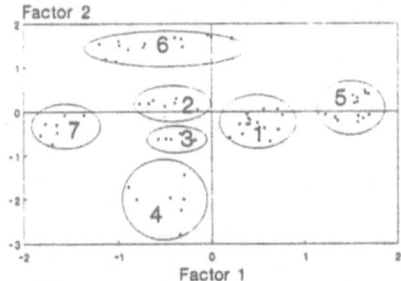

Figure 1: Classification of steel samples by the factor scores of the diffuse reflectance spectra

The pattern of the reflectance spectra in combination with a Kohonen's self-organizing map can be used to determine the corrosion behaviour of the steel sample as well as to typify the steel manufacturers. Steel samples from 7 different steel manufacturers (1 .. 7) in Europe were measured by the optical on line system. The reflectance spectra are analyzed by means of factor analysis and a Kohonen's self-organizing map.

Figure 1 shows the scores of the steel samples in the factor 1 versus factor 2 chart. The samples are classified by the factor scores. Factor 1 may be attributed to the chemical composition of the surface oxide (10 - 50 nm thickness) and factor 2 represents the porosity of the oxide. It is obvious that the 7 different manufacturers can be distinguished.

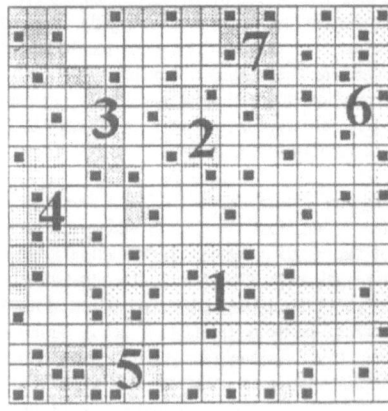

The Kohonen's self-organising map is as well used as a classifier. The map is trained with the same known input data , that were used for the factor analysis. An example of a classification with a Kohonen map is shown in figure.2

The advantage of using the Kohonen map for classification compared to the factor analysis is a higher differentiation of the spectra especially of samples with high resemblance. The advantage of the factor analysis is the shorter computing time and the possibility of a scientific interpretation of the results.

Figure 2: Classification of car body steel samples by a Kohonen self-organizing map

References

[1] R. W. Kessler et al., Werkstoffe und Korrosion 40, 539 - 544 (1989)

Simulation of Pulsed Laser Material Processing Controlled by an Extended Self-Organizing Kohonen Feature Map

Gábor J. Tóth[1], Tamás Szakács and András Lőrincz
Institute of Isotopes, P.O.B. 77, Budapest, Hungary H-1525,
[1] Eötvös Loránd University, Budapest, Hungary
E-mails: jtoth@obelix.iki.kfki.hu, lorincz@obelix.iki.kfki.hu

We have simulated the control of laser material processing. The process to control was surface polishing and milling. In the simulations we assumed pulsed ultraviolet laser beams of 100 nanosecond duration. The controller was an artificial neural network (ANN): an extended self-organizing Kohonen feature map. The controller was trained – tuned – on examples of laser evaporation of one dimensional 'surfaces' by using a Widrow-Hoff type Delta-rule for error correction. The strength of the approach lies in the speed of parallel computing and the adaptive properties of the controller in a changing environment, like drift in laser properties, optical components, etc. Results show more than an order of magnitude improvement in surface roughness after an ANN designed laser shot onto a large surface. Restrictions of the model are discussed.

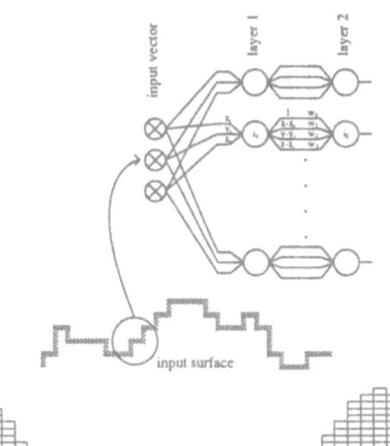

Figure 1: The architecture of the network

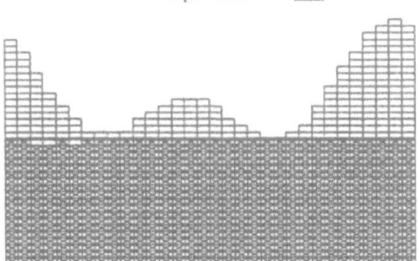

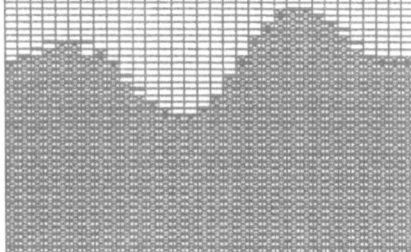

Figure 2: Smoothing surfaces: typical performance of the trained network. Empty boxes correspond to the original surface, filled boxes to surface after the shot. Wide line shows the target surface level. Box size is 200 nm horizontally and 100 nm vertically, material is aluminium.

Figure 3: Milling patterns: typical performance of the trained network. Empty boxes correspond to the original surface, filled boxes to surface after the shot. Wide line shows the target surface level. Box size is 200 nm horizontally and 100 nm vertically, material is aluminium.

Process Modelling Using Artificial Neural Networks

Roland Orre and Anders Lansner

NADA, Royal Institute of Technology, S-100 44 Stockholm, Sweden

E-mail: orre@sans.kth.se, ala@sans.kth.se FAX: +46-8-790 09 30

We model a part of a process in pulp to paper production using feed forward connected neural networks. A set of parameters related to paper quality is predicted from a set of process values. The predicted values are results from laboratory experiments which are time consuming. The number of training vectors were rather limited. Therefore, our work was focused on finding the relevant inputs for each signal and to find the architecture that was most efficient for each output. The output vector is separated into single values which are predicted on different architectures adapted to each output. A strategy that continuously adapts the process model seems to be useful. In this work the backprop learning algorithm has been used.

The general problem is to model a process P as vector transformations. P has as inputs a known parameter vector x, an unknown parameter vector y and a random disturbance vector d. The output vector of the process is thus $o = P(x, y, d)$. We want to predict the output vector o when only the input vector x is given. Due to measurement errors and manual input errors the vector x is also an estimate. The modeled vector o thus becomes $\hat{o} = \hat{P}_{model}(\hat{x})$. For simplicity we assume that the process only depends upon its instantaneous inputs and not its history. The specific problem in this study was to investigate how some measures of pulp quality, could be predicted. The laboratory measurements predicted were *csf, density, elongation, tear* and *tensile* with the emphasis on *tear* and *tensile*. After proper scaling all values were coded as single analogue neuron inputs and outputs.

As we deal with feed forward functional mapping networks, the output vector can be separated into single valued functions.

$$f_{ab}(x, y, z) = [f_a(x, y, z), f_b(x, y, z)]$$

There are good reasons for doing this separation when the training data set is small.

- The network complexity needed may differ for different outputs.
- All available inputs may not be relevant for all output values.

We searched for the best architecture for each of the separate output values using 1–3 layered networks of reasonable size [1] according to the number of data samples available. The architecture that showed the best average performance on the test set over different partitions of the training set, was considered the best one. We also tried to remove insignificant inputs by checking the performance when the inputs were "removed" one at a time. The reasons for removing insignificant inputs are twofold.

- Irrelevant input may disturb the training, thus increasing the demand on the amount of data.
- Further relationship analysis becomes less complex when only relevant inputs are present.

We used the standard deviation of the difference between the actual and the predicted signal as a quality measurement. The average prediction difference was used as an indication of systematic errors. The data was split into training and test parts to be able to measure the *generalization* capability of the network, *i.e.*, its performance on data which it has not been trained on. We used four ways of partitioning the data samples into training and test sets, *first, last, random* and *split*. In an ideal case, when data from a process is collected over a long time, the method of partitioning should not make a big difference. In reality, a process may show varying characteristics over time and the sampling period may be too short to give a well-represented distribution.

The results have indicated that quite good performance can be obtained even though the amount of data available for training and evaluation of the networks is limited. We did, however, not get a complete answer to how the artificial neural network technique compares with other methods. The learning strategy that showed the best result was the one that continously adapted the model to the training data and threw away old samples. Methods to extend the use of the neural network technique to on-line verification and filling in of data has been proposed earlier [2].

[1] Abu-Mostafa, Y. The Vapnik-Chervonenkis dimension: Information versus complexity in learning. *Neural Computation*, 1:312–317, 1989.

[2] Orre, R. and Lansner, A. A study of process modelling using artificial neural networks. Tech. Rep. TRITA-NA-P9239, Dept. of Numerical Analysis and Computing Science, Royal Institute of Technology, Stockholm, Sweden, 1992.

REAL-TIME NUCLEAR POWER PLANT MONITORING
WITH ADAPTIVELY TRAINED NEURAL NETWORK

K.Nabeshima, E. Türkcan, Ö. Ciftcioglu
Netherlands Energy Research Foundation ECN
P.O.Box 1 NL-1755 ZG,Petten, The Netherlands

EXTENDED SUMMARY

Real-Time nuclear power plant monitoring by feedforward neural network is described. For the training, standard backpropagation algorithm is used for an auto-associative network structure having 12 inputs/outputs. After initial training of the network, the real-time process signals are introduced to the network as input and plant status is monitored through the network's response at the output. At the same time the training of the network is performed in adaptive form so that the sensitivity of the network's failure detection sensitivity is enhanced in the course of the operation.The real-time monitoring system considered is primarily for the Borssele nuclear power plant (NPP) from which signals are available on-line with a sophisticated data acquisition system [1]. The neural network uses a three-layered feedforward structure with 7 nodes in the hidden layer. Standard backpropagation algorithm with sigmoidal nonlinearity is used for training and adaptive learning. All input and output data are so normalized that the range extends from 0.1 to 0.9. The program is coded in FORTRAN and executed at VAX-4200. For initial training 1850 patterns covering start-up, steady-state and shut-down are used, so that the wide-range operation of the plant is considered. After the initial learning, real-time monitoring and adaptive learning starts to progress. The fault severity level (ε_f) that is, measured and estimated signals difference normalized by related range, is taken to be 0.1. The schematic block diagram of the monitoring system is shown in Fig.1 where ε_{max} is the error bound used during the training as well as in real time. To illustrate the training and real-time monitoring system's performance, the results obtained from a process signal is given in Fig.2. The detailed description of this study is presented elsewhere [2].

The study demonstrates that a complex and dynamic system like a NPP can successfully be modelled by ANN, by the help of the measurement information only. Such a computational model can be used for monitoring of the system having several important advantages compared to conventional applied monitoring. For instance, the model comprises nonlinear dynamics which is approximated to linear system dynamics in most physical modeling applications due to the complexity of the problem.

REFERENCES

[1] Türkcan E.,Quaadvliet W.H.J.,Peeters T.T.J.M., Verhoef J.P.: "Operational Experiences on the Borssele NPP using Computer Based Surveillance and Diagnostics System On-Line" ECN-RX--91-057, The Netherlands, (June 1991).

[2] Nabeshima, K., Türkcan E., Ciftcioglu, Ö.: "Real-Time Nuclear Power Plant Monitoring with Adaptively Trained Neural Network", ECN-RX--93-146, The Netherlands (1993).

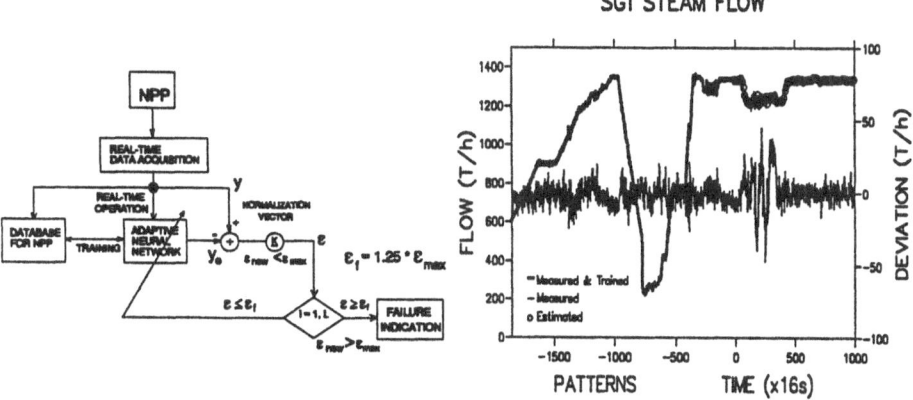

Fig.1. Block diagram of the monitoring system.　Fig.2. Training and real-time estimation.

Self Organized Feature Maps for Monitoring and Knowledge Aquisition of a Chemical Process.

Alfred Ultsch
FG Informatik
Philipps University of Marburg
Hans-Meerwein Str. 22
3500 Marburg/Lahnberge
Germany

An application of Self-Organizing Feature Maps to the problem of process control for a chemical process is described. Very few of the nature and structure of the process can be learned from the trained Feature Map itself. A set of methods called U-Matrix methods have been developed to enhance the representation of data on Feature Maps. This methods allows to discover structure in the process data and allows to judge the quality of the learned maps. It can be used to extract knowledge for expert systems from the Feature Maps. The extracted rules were able to control the chemical process as well as a human designed expert system. The methods presented here are in particular usefull to monitor critical processes and for the design of suitable human interfaces for process monitoring stands.

1. Introduction

Some processes, in particular in the chemical industry are hard to control. One of the reasons for that is that it may be impossible to observe or steer the process directly. This may be due to the general physics of the process or the difficultis of observation without interfering with the process. Other reasons are the nonlinearity and coupeling respectively interference of the control parameters. This means, that if some parameters are changed only slighly, some drastical and unwanted changes in the process may occur. In this paper we describe an application of self-organizing feature maps (SOFM) to the problem of process control for a chemical process [Wayand 92].

2. SOFM for Process Control

The state of a process at a certain time t can be described by a vector of parameters x(t). Components of x(t) are either measurements of some physical parameters or controling information.SOFMs may be trained with these vectors. We know that a SOFM adapts to the training data such that similar vectors are represented in close topological neigborhoods [Kohonen 82]. After enough learning steps the states of the process may be represented on a 2-dimensional SOFM as Figure 1 shows.

The SOFM can directly be used to steer the process. For a given set of measurements from the process the corresponding (most similar) location is found on the map and the settings of the steering parameters are taken from the state-vector. This approach has been followed, for example, by [Tryba/Goser 91].

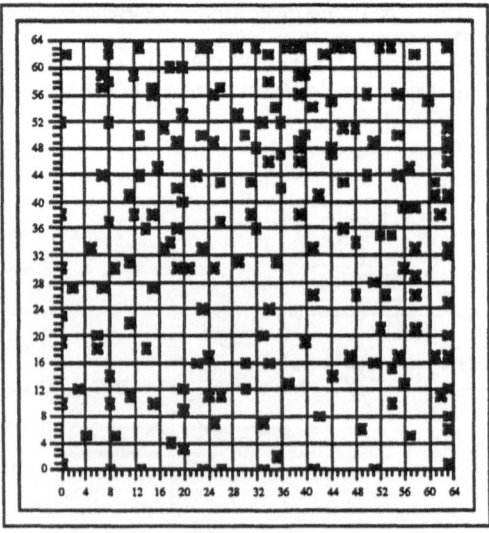

Figure 1: Representations of process states

If the process' representations on the map are observed in some time intervals, the following Figure arises. It shows the sequence of states at two different time intervals.

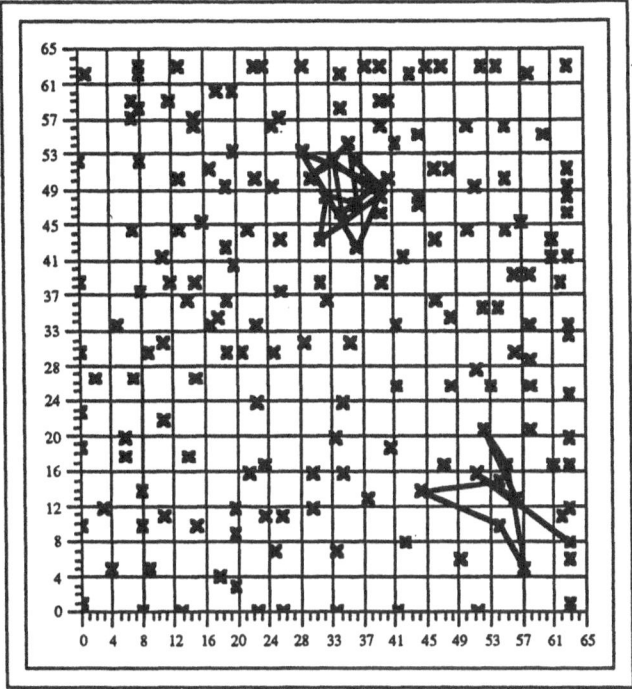

Figure 2: Representations of the process' dynamic

The consecutive process states form a line or route on the map. Note in particular, that the process stays in almost the same location on the map for the given interval.

Of the representation of Figure 1 and 2 very few can be learned about the process itself. With a creative eye some regions on the map may be detected, where the process states are more densely distributed than in other regions. These regions may be seen to be separated by gaps where no process state can be found. In this form the interpretation of the map is rather vaque. Looking at the distribution of the parameters on the map, as suggested in [Marks/Goser 88], may help. This approach is however limited to one or only a few parameters. It is clearly prohibitive, if the process can only be described by a large number of parameters i.e. a high dimensionality of the process vector is given.

3. U-matrix methods

Since several years we have been developing methods to detect structures in trained SOFM [Ultsch 91, Ultsch/Panda 91]. These methods we call unified distance matrix (U-matrix-method or UMM). The simplest UMM is to calculate at each map coordinate (X;Y) the sum of the distances of the weight vector at (X;Y) to its neighbouring weight vectors [Ultsch 92a]. The matrix of these distances can be displayed as 3-dimensional landscape on top of the positions of Figure 1 resp. 2.

The landscape may be interpreted as follows: if there are hills or walls, then the neighbouring weights are quite distant. I.e., the process states differ significantly. If a depression or valley can be seen in the U-matrix, then the process states are quite similar. A U-matrix representation of the process control problem can be seen in Figure 3.

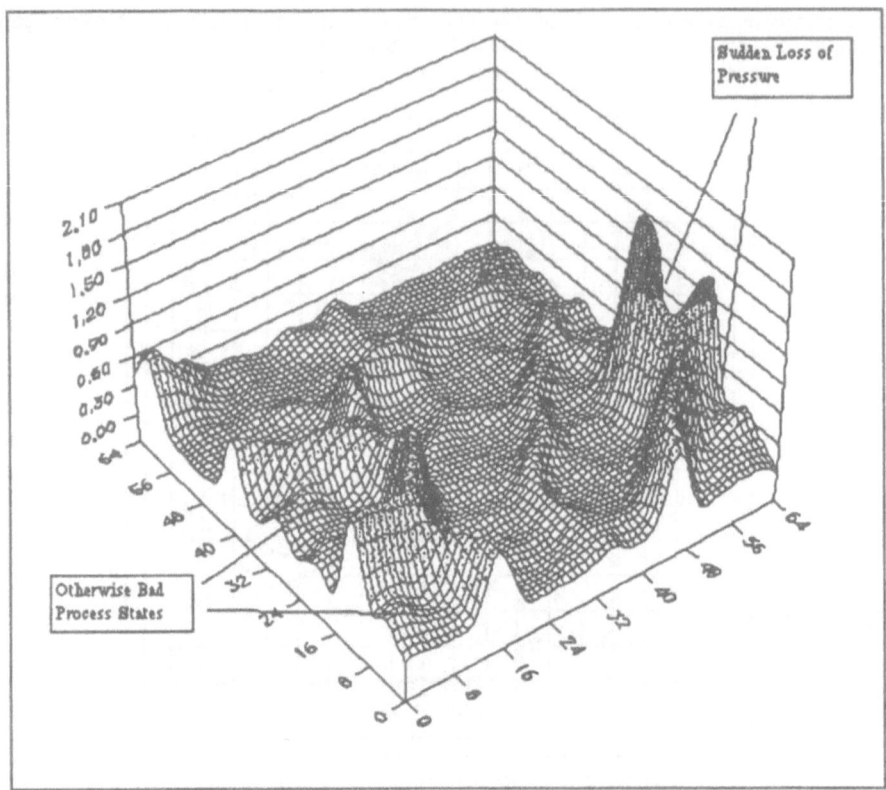

Figure 3: U-Matrix for a chemical process

In Figure 3 some definitive states can be distinguished. Some of these states may be recognized by people who are experts in the particular process. In Figure 3, for example, one of the states that could be identified is "loss of pressure" in the process.

4. Applications

SOFM with the UMM may be used in process documentation. Think of an online representation of the actual state of the process as a marker in the U-matrix of Figure 3. Process supervision is then the observation of the movements of a point in the landscape given by the U-matrix. This can be a significant improvement over the observation of a possible large number of gauges. If critical process states are identifieable in the U-matrix, the observer may quickly identify crucial points. The process runs normal as long as the process representation x(t) on the U-matrix stays in safe regions. If x(t) proceeds towards crucial areas, appropriate actions may be taken. This may be a contribution to make process observation more human-like and more safe.

Another application is to extract the structural information from SOFM that can be detected by UMM. This may be done by a rule extraction algorithm [Ultsch 91, Ultsch 92]. In this application the walls or hills in the U-matrix are thought to seperate different classes or clusters on the SOFM. From the weights of the SOFM that are classified with this method a special designed machine learning algorithm, called sig* derives abstract descriptions of the classes. These descriptions may be used on one hand to discuss the structure of the SOFM with an domain expert. This results also in a methods to judge the quality of the trained map. On the other hand the extracted descriptions may also be used in some expert system.

We have tried this approach successfully for several different applications [Ultsch/Halmans 91, Ultsch/Panda 91, Ultsch et al 91]. For the process control application presented here, the extracted rules were successfully to steer the chemical process [Wayand 92]. The extracted rules showed the same performance as human generated rules that have been build for an expert system for process control. The major cost factor of this expert system was the knowledge aquisition, i.e. the questioning of the experts and the correct design of the rules. This took almost two man-years. With the rules extracted from SOFM/UMM this process can be accelerated significantly.

5. Conclusion

SOFM enhanced with U-matrix methods may be used successfully in process control applications. The artificial landscapes that arise from the U-matrix methodes can be used to judge the quality of the trained map. More important, however, they reveal the intrinsic structure of the data. This means that humans can detect structures in a high-dimensional data space. For process monitoring stands this may result in a better and more safe human-interface. U-matrix methodes may also be used to extract symbolic knowledge from SOFM. With this costly knowledge aquisition steps in the realisation of expert systems may be efficiently accelerated.

Aknowledgements

The realisation of the programs and the field testing was done by M. Wayand. We thank in particular T. Soltysiak for the supply of data. H.-P. Siemon cosupervised the practical work and did some Macintosh adaptation of the graphics. This work has been supported in part by the Bennigsen-Foerde research price of Nordrhein-Westfalia and the BMFT research grant WINA.

Literatur

Kohonen 82   KOHONEN, T.: Clustering, Taxonomy, and Topological Maps of Patterns. In: Lang, M. (Eds), Proc.Confe on Pattern Recognition, Silver Spring, MD, IEEE Computer Society Press, 1982, pp 114-128

Marks/Goser 88   Marks,K., Goser,K.: Analysis of VLSI Process Data Based on Self-organizing Feature Maps; In: Proc. Neuro-Nimes 1988, pp.337-348

Tryba/Goser 91   Self-Organizing Feature Maps for Process control in Chemistry, Proc. ICANN, Helsinki 1991, pp 847-852.

Ultsch 91   Ultsch, A.: Konnectionist Models and their Integration with Knowledge Based Systems, Habilitationsschrift, Univ. Dortmund, (in german) 1991.

Ultsch 92   Ultsch, A., Self-Organizing Neural Networks for Knowledge Akquisition, Proc ECAI, Wien, 1992, pp 208-210

Ultsch 92a   Ultsch, A.: Self-Organizing Neural Networks for Visualisation and Classification, Proc. Conf. Soc. for Information and Classification, Dortmund, April 1992.

Ultsch/Halmans 91   Ultsch,A., Halmans,G.: Data Normalization with Self-Organizing Feature Maps, Proc. IJCNN, Seattle, July 1991, Vol I, pp 403 - 407.

Ultsch/Halmans 91a   Ultsch, A., Halmans, G.: Neuronal Networks for Environmental Research, Symp. Computer Science for Environmental Protection, Munich, Informatik Fachberichte 296, Springer, Dezember 1991.

Ultsch/Panda 91   Ultsch, A., Panda, PG.: Die Kopplung konnektionistischer Modelle mit wissensbasierten Systemen, Tagungsband Expertensystemtage Dortmund, Februar 1991 VDI Verlag, pp 74-94.

Ultsch et al 91   Ultsch, A., Palm, G., Rückert, U.: Wissensverarbeitung in neuro-naler Architektur, in: Brauer/Hernandez (Eds.): Verteilte künstliche Intelligenz und kooperatives Arbeiten, GI-Kongress, München, 1991, p 508-518.

Wayand 92   Kohonen Feature Maps for Process control, Diplomarbeit, Dept. of Computer Schience, University of Dortmund (in german), 1992.

Flow Regime Identification by a Self-Organising Neural Network

S. Cai, H. Toral, and J. Qiu

Department of Mineral Resources Engineering
Imperial College of Science, Technology and Medicine
Prince Concert Road, London, SW7 2BP, United Kingdom

Abstract

One of the prerequisites of effective operation of multiphase flow systems is the need to identify flow regimes. A number of flow regime classification models have been reported in the literature based on the subjective and ambiguous interpretation of visual observations. With the advent of microprocessors and recent advances in signal processing, pattern recognition and neural network techniques, we are now on the threshold of finding more objective and quantitative technique for flow regime identification and flowrate measurement in multiphase systems. This paper presents a technique for the classification of flow regimes in horizontal air-water two-phase flow lines by the application of the Kohonen self-organising feature map (KSOFM).

The principle of the technique rests on the characterisation and classification of turbulent pressure signals in relation with flow regimes. Characterisation means deriving stochastic features from time domain signals which are likely to be associated with specific physico-fluid dynamic groups based on space-time average values. Classification means defining the boundaries of the fluid regimes as represented by the fluid dynamic groups.

KSOFM is known to be capable of organising patterns from an arbitrary n-dimension space into lower dimension space and preserving the topological relationships among the features. In this case, a KSOFM with 8*8 neurons was selected and trained with stochastic features derived from the absolute pressure signals obtained from a range of flow regimes. The resulting feature map succeeded in organising samples from similar flow regimes into same or adjacent neurons and separating samples from different flow regimes into different areas.

The samples used in the tests were obtained in laboratory measurements conducted in a horizontal 2 inch diameter air-water flow line. KSOFM identified the flow regimes observed visually as bubbly, slug and wavy/stratified flow regimes as well as the more elusive transition regimes. For future work, it is proposed that the KSOFM offers a pre-processing capability in separating training regions into sub-regions where within each sub-region other supervised network models, such as the back propagation neural networks can be trained to identify primary quantities of interest such as average mass/volumetric flow rates of individual phases.

Functional Electrical Stimulation with Neural Network Controlled State Feedback

J. Beckmann, W.J. Daunicht, V. Hömberg

Neurologisches Therapiecentrum at the Heinrich-Heine-University Düsseldorf, FRG

Optimal linear control: As described earlier, we proposed the use of artificial neural networks for the control of free stance and simple movements in spinal cord injured people by Functional Electrical Stimulation (FES). For simulation purposes we presently use a planar patient model with 3 segments (angles θ) and 6 muscles [1].

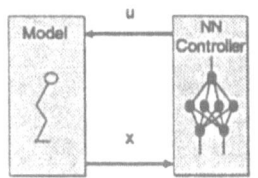

With the state vector x and the command vector u of stimulation currents the nonlinear dynamics of this model can be denoted as: $\dot{x} = f(x, u)$. In the theory of optimal linear control a *state feedback* approach, $u := K \cdot x + V \cdot \theta_{desired}$, is proposed for the linearized dynamics. K is determined in order to minimize the following performance index with positive definite cost-matrices Q, R [2]:

$$E := \int_0^\infty dt(e^{\alpha t} \cdot x^T \cdot Q \cdot x + u^T \cdot R \cdot u).$$

Results: A reference path $\theta_R(\tau)$ forming a continuous movement of standing up from a chair followed by kneeling down was chosen such that itself and its surroundings can be considered as a purposeful working range (cf. left figure below). Next, the state feedback matrices $K(\theta_R), V(\theta_R)$ were determined at different locations $\theta_R(\tau)$ of the reference path. The resulting data-set was then implemented into a neural network. Our network uses radial basis functions and a normalization procedure at the output [1]. In a simple controller test the points of the reference path were presented in a time sequence yielding a desired trajectory that should be tracked. No single state feedback controller with constant K, V - matrices was able to track the trajectory. In contrast, the neural network controller that comprises the knowledge of gain matrices along the whole reference path correctly tracked the path. A similar success of the network could be observed for other tasks.

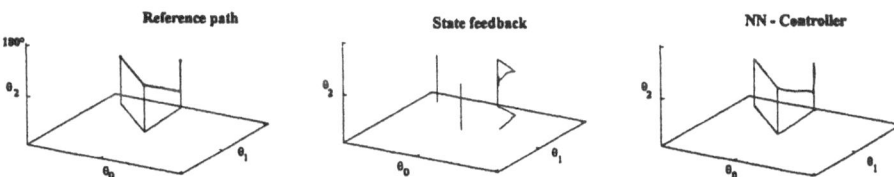

References

1. J. Beckmann, W.J. Daunicht and V. Hömberg, Control of a Paraplegic Patient Model by Neuroprothetic Networks, in: I. Aleksander, J.Taylor (eds.): *Artificial Neural Networks II*, pp.471-474 (1992).

2. T. Kailath, Linear Systems, Prentice-Hall (1980).

Acknowledgements: This work was supported by DFG grant Da 199/2-1.

Artificial Interacting Agents for Stock Market Experiments: the Cross-target Method

Pietro Terna

Università di Torino, Istituto di Economia Politica "G. Prato", Facoltà di Economia e Commercio, Piazza Arbarello 8, 10122 Torino, Italy.

Abstract

The purpose of the *cross-target* (CT) method is the development of consistency between two types of guesses made by an artificial neural subject: *i*) Guesses about actions that the subject wants to do; *ii*) Guesses about the effects of those actions. We present here an improvement of CTs, with the use of external objectives (EOs) to direct the guesses of the effects and with the introduction of external proposals (EPs) to influence the guesses about actions.

With CTs the targets for training are generated by the neural model itself while it learns and acts in a *crossed* way. Actual effects are estimated by environment rules on the basis of the guessed actions, taking account also of the consequences of interaction between agents; the results are used to train the mechanism that guesses the effects. The evaluations of the actions necessary to match guessed effects are, on the contrary, employed to train the decision mechanism (that guesses actions). EPs and EOs are external targets: EOs substitute the cross one to train specific output elements related to effect guesses; EPs represents one of the multiple targets - from which the highest is chosen - used to train the side of the model that guesses the actions.

Artificial interacting agents (AIA) can simulate, with CTs, complex economic systems, showing a short term or a long term rational behaviour. Our AIA buy and sell shares with or without self developed risk aversion, following cyclical behaviour (buying when the price rises and selling when the price diminishes) or developing anti cyclical strategies. A small quantity of randomness, introduced as EP in the training of the guess capability about actions, is sufficient to help rational behaviour to emerge.

The experiments are initially based on the independence of the agents. Without EOs (and EPs) we have all the agents acting to avoid risk: They sell all the shares and keep money, so greatly simplifying the task of developing consistence between the guesses of the actions and those of their effects. With an EO (to improve wealth at a daily fixed rate), and always without EPs, the agents act in the short term selling when the price diminishes and vice versa.

Introducing the interaction, with four AIA of two types (one population of two agents acting with EO and without EPs and the other without both EOs and EPs), we see the consequences of the introduction of market constraints due to the effect of the combined choices of any couple of matching agents. Agents with the EO (always that of improving wealth) are forced to keep shares, in the attempt of improving wealth: the other two agents confirm the risk averse attitude, obstructing the desired behaviour of the first population.

Maintaining interaction and with four AIA of two types (one population of two agents acting with EO and without EPs and the other without EOs, but with an EP -- intended as passive adhesion to the proposal of the counterpart, but only if it represents the max value between potential targets --, the original behaviour (to keep money) of the first population is restored, at least partially; we see here a new interesting behaviour of the second population, that is now acting (being an active counterpart of the first population) against the cycle, with apparent long term rationality.

We can also conduct the experiment introducing a random EP (as consequences of news, imitations, subjective choices) in the second population of agents. The results, although lessened, are very close to those of the previous run. We stress the importance of this experiment, explaining what apparently is the consequence of the *reason* as the result of small random shocks in a constrained environment. In other terms, *randomness vs. reason.*

NEURAL NETWORK TRAINING BY PARAMETER OPTIMIZATION APPROACH

Ö. Ciftcioglu and E. Türkcan
Netherlands Energy Research Foundation ECN
P.O.Box 1 NL-1755ZG Petten, The Netherlands

EXTENTED SUMMARY

A neural network training method with fast convergence is introduced. As the feedforward neural network training can be viewed as a special case of function minimization, powerful non-linear optimization algorithms with fast convergence are applied for synaptic weights estimation. Extensive experimental investigation was performed by means of the data from the Borssele nuclear power plant. The Borssele power plant is a two-loop pressure water reactor with nominal electric power output of 477 MWe. The on-line signal analysis system designed for the multi-level mode operation is capable of monitoring the plant states by tracking 32 DC and 32 AC signals simultaneously. The preparation of the data used in this study and results obtained by standard backpropagation were reported earlier [1].

For the experimental research, a comprehensive data set is considered. From this data set, 12 different signals from the Borssele power plant were used to estimate the generated electric power. Signals positions are indicated in the schematic representation of the power plant (Fig.1). The optimization algorithms used are quasi-Newton, congugate gradient and standard backpropagation. The detailed description of the present study is given elsewhere [2]. The comparison of the results with those obtained by standard backpropagation algorithm indicated that the approach to the global minimum is much improved by the optimization algorithms. Although the training time is shorter than that obtained by standard backpropagation, one may argue that the improvement in training time elapsed is not essential for the training problem since this is done only one time. However, if one considers that the ratio of the training times can extend to more than 500, the importance of the efficient training becomes conspicuous apart from the high risk of trappings by local optima encountered during standard backpropagation training approaches.

Since the statistical test for the decision on the selection of the number of hidden layer nodes and the training of the network are done at the same time, the utilization of the powerful non-linear optimization techniques for neural network is essential for efficient and effective training.

REFERENCESS
[1] Ciftcioglu, Ö. and E. Türkcan:"Selection of Hidden Layers Nodes in Neural Networks by Statistical Tests" EUSIPCO-92, Signal Processing VI: Theories and Applications. J. Vande Valle et al. (Edts.) Elsevier, Amsterdam, (1992).
[2] Ciftcioglu, Ö. and E. Türkcan:"Neural Network Training by Parameter Optimization Approach" ECN-RX--93-147, The Netherlands (1993).

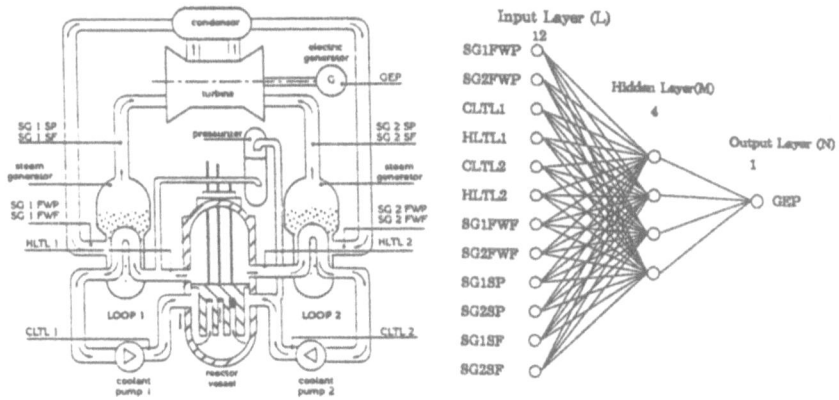

Fig.1. Schematic Representation of Borssele NPP, Signals and ANN Structure.

Neural Network Analysis of the Hungarian Party-State System

M. Csanádi* and A. Lőrincz**
* Institute for Economic Sciences, Hungarian Academy of Sciences, P.O.B. 262, H-1502 Budapest, Hungary
** Institute of Isotopes, Hungarian Academy of Sciences, P.O. B. 77, H-1525 Budapest, Hungary

The paper is based on empirical surveys in different spheres of the decision-making system on the last 20-25 years of the Hungarian party-state system. It treats the origin of the rigidity of the interlocked communist party-state system which failed to accomodate the world market. On the basis of the experiences we suggest a model that is also given mathematically. We describe the well-defined principles of the interlocking, interdependent entangled network of political, governmental and economic decision making of the Hungarian party-state system. We show that it was built on a self similar fashion at every level.

Our starting point is that the so called "planned economy" constantly failed in planning because the politically monopolized system was based on political rather than economic rationality of behaviors. This politically rational behavior running through the whole structure distorted the economic system led to a constant drive for growth, to unavoidable interventions, to consuming instead of investing resources, to constrained paths in distributing investments, to the absorption and neutralization of reforms, to the rigidity of the economic structure, to indefinite dependences and interest enforcement possibilities, to selectivity based on political criteria, and thus as an overall result to uncertainty in economic planning. In view of this there was no possibility to make comparison and judgement on an economic basis among goals, resources, outputs, economic power, performances, real needs, real expenses, etc. This lack of possibility was stabilized and strengthened by the constant aspirations towards autarchy. The final result was that these uncertainties - as we show - led to pattern recognition type decision making in Hungary.

We analyze the system as a recognizing machine based on political criteria, an architecture which under the influence of input patterns creates, self-organizes and self-stabilizes its recognition codes. Learning takes place by the feedback of behavioral success of the system. We show that the Hungarian party-state system had strong similarities and interesting dissimilarities with the ART model [1]. We follow how this quasi-ART system develops self-supporting mechanisms, and treat the concepts of stability, adaptability, etc. with the help of concepts developed in the ART model. Briefly, we discuss, why other party-state systems were different in spite of the self-similar structures. We establish the presence of resonance like decision-making in the Hungarian party-state system. We show, that the some of the basic ingredients of ART learning were present; (i) the system was a two level system, (ii) the appropriate hierarchy of short term memories and long term memories existed, and (iii) the system tried to learn with the help of a mismatch-mediated activity rearrangement procedure. The system's fate was determined by its deviances from the self-organizing and self-stabilizing architecture of ART.

We note, that ART model uses finely tuned mechanisms to solve the design problems. The existence of such mechanisms is questionable in the party-state system:

- Learning requires the fine and tunable distinction between input noise and input signals, i.e. there is a need for a mechanism of turning inputs considered as noise under less demanding conditions into features of new recognition categories under more demanding conditions, that is after punishing environmental feedbacks. However, the system was finely tuned by political instead of economic rationality, and filters out economic input patterns as noise and thus there was no economic control within the system.

- The producing stage could always excite the directing stage. This is called attentional priming in the ART model. Attentional priming may result in resonances without real inputs. In ART a special mechanism was designed to avoid the generation of such virtual resonances. This mechanism was not present in the quasi-ART system of the Hungarian party-state. The lack of that mechanism led to self-excitations.

- The arousal of the STM activity of the production stage should result in the stable learning of the new pattern. This seems not to be the case during the learning course of the Hungarian party-state system.

1   G.A. Carpenter and S. Grossberg, *Comp. Vision, Graphics, and Image Proc.*, **37**, 54-115 (1987).

Modelling Time-Varying Industrial Processes Using MLP Networks.

Edward J. Williams & Michael J. Denham

School Of Computing, University Of Plymouth.
Drake Circus, Plymouth, Devon. PL4 9QD.
United Kingdom.

Abstract.

The modelling of time-invariant systems using a multilayer perceptron (MLP) network has been studied. As MLPs have been demonstrated to be universal approximators - given sufficient hidden processing elements within the hidden layers and a suitable input vector - a time-invariant systems output can be predicted by an MLP trained to functionally approximate the system given current input and historic input and output information, thus

$$\hat{y} = f(y, u)$$

if the system is an infinite impulse response system.

Experimental modelling of the nonlinear startup regime of a real industrial process using this method met with poor results. One explanation for this failure is that the system is time-varying in that at any stage of operation, the output of the process is functionally dependant not only upon previous input and output measurements, but also on time. Thus an estimate for y will be

$$\hat{y} = g(y, u, t)$$

and an MLP will need to be provided with a representation for time in its input vector in order to approximate this function. However, many industrial processes are not smoothly dependant upon time as would be an MLP with time as an input. Study of the industrial process revealed several distinct phases of operation within its startup regime; stages characterised by the switching in and out of various process components and changes in control set points which alter the underlying operation of the process. As each stage of operation was time-invariant in isolation, the system can be described as being piecewise time-invariant overall. Thus the system is more disjointedly dependant upon time, and attempting to model it using an MLP with time as an input provided no greater success than before.

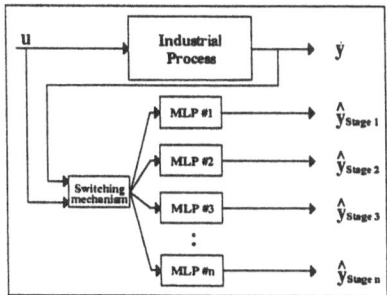

Figure 1 . A mechanism for modelling a time-varying process with n identifiable stages of operation.

An alternative, where it is possible to clearly distinguish between several stages of execution in the operation of a system, is to treat each stage as a function in its own right and attempt to model each with a separate MLP. This would result in a cascade of MLPs which it should be possible to switch between during the normal running of the process to provide a continuous input-output mapping (Figure 1). This method met with far greater success for the industrial process.

Pattern recognition I
—oral contributions

LITHOFACIES IDENTIFICATION FROM WIRELINE LOGS
- Bringing Neural Networks to Application -

Willem J.M. Epping, Sandra M. Oudshoff and Frances V. Abbots

Shell Research B.V.

Koninklijke/Shell Exploratie en Produktie Laboratorium

P.O. Box 60, NL-2280 AB Rijswijk, The Netherlands

Invited paper to be presented at the

International Conference on Artificial Neural Networks (ICANN '93)

Amsterdam, September 13-16, 1993

Correspondence to:

W.J.M. Epping

phone : 070-3112610

fax : 070-3113110

e-mail : eppingw@KSEPL.nl

LITHOFACIES IDENTIFICATION FROM WIRELINE LOGS
- Bringing Neural Networks to Application -

Willem J.M. Epping, Sandra M. Oudshoff* and Frances V. Abbots

Shell Research B.V.

Koninklijke/Shell Exploratie en Produktie Laboratorium

P.O. Box 60, NL-2280 AB Rijswijk, The Netherlands

Abstract

Neural-network classifiers, of the multi-layered perceptron type, were trained to identify lithofacies (rock types) from wireline logs in two ways: (1) by a semi-autonomous method based on pre-segmented log intervals and (2) by an autonomous method based on non-segmented data. Good results were achieved in reservoirs from two geologically different environments (siliciclastics and carbonates) without having to fine-tune the network parameters. The performance was substantially better than that of linear discriminant and Gaussian classifiers. The standard neural-network algorithm has been modified to increase its robustness e.g. to over-training. A user-friendly neural-network module has been embedded in an existing computing environment to facilitate the use of neural networks by formation analysts.

Introduction

Whether economically recoverable oil or gas is present in subsurface reservoirs is highly dependent on physical properties of the reservoir rock such as porosity and permeability. These properties can be determined directly from actual rock samples (cores) taken from wells. For economic reasons, however, cores tend to be available from only a limited number of wells. In most instances, formation analysts have to rely on measurements from wireline tools. These tools are lowered into a well to measure a suite of physical parameters such as electrical resistivity, sound velocity and gamma-radiation as a function of depth. The resulting data are recorded on wireline logs. These wireline logs are interpreted by a formation analyst to identify lithofacies which, in turn, are indicative of the reservoir properties. Lithofacies are the sum of the (physical, chemical and biological) rock characteristics that distinguish rock beds from each other.

The inherent variability of wireline-log responses due to gradations in rock characteristics, effects of data acquisition and statistical fluctuations in count rates of nuclear tools often requires a statistical approach to lithofacies identification. The conventional method of identifying lithofacies by cross-plotting pairs of wireline logs is adequate for identification of simple mineral mixtures, but fails in more complex and heterogeneous reservoirs. The statistical pattern-recognition techniques used thus far, whether based on unsupervised clustering [1] or on supervised classification [2,3], often assume Gaussian probability distributions and linear decision boundaries. The need for powerful statistical tools in the processing of wireline logs becomes even more urgent in the light of the steady increase in the data volumes requiring rapid processing.

Neural networks, which do not assume normality and linearity of data and yet are fairly easy to use, offer an alternative to the more traditional pattern-recognition approach. They have been shown to be more accurate than linear statistical techniques [4,5,6] for lithofacies identification.

---

* Present address: PTT Research, Groningen, The Netherlands

It is anticipated that neural networks will have the largest impact in the hands of end-users (in our case geologists and petrophysicists), who want to employ this technology for their own problems, rather than neural-net experts. The aim of this contribution is to describe the transfer of neural-network technology to a real-world application. Issues we will address are: (1) the performance in different geological settings (siliciclastic and carbonate); (2) the robustness to non-expert use and (3) the implementation of neural networks in an existing computing environment used by formation analysts.

Methods

Lithofacies classification procedure

Lithofacies identification was based on a suite of four wireline logs (Fig. 1a), which had been properly calibrated and depth-matched. For the siliciclastic environment, neutron, density, sonic and gamma-ray logs were used. For the carbonate environment, the gamma-ray log was replaced by a resistivity log.

Two lithofacies-identification procedures have been developed. The first, *"classification by layer"*, is semi-autonomous. The formation analyst has to specify the depths at which the lithofacies type changes, after which the neural network assigns the pre-segmented intervals to the correct classes. Each interval is characterised by 13 parameters, comprising the layer thickness and attributes derived from the logs such as average amplitudes, trends and separations between logs. These parameters, which serve as input for the neural network, have been scaled to a mean value of zero and a standard deviation of one.

The second procedure, *"classification by depth"*, operates on non-segmented log data and is therefore more challenging. The neural-network input is obtained by sampling the amplitudes of the four logs using a sliding window covering 9 depth samples. In this way, the neural network receives not only log information from the depth of interest, but also context information from neighbouring depth samples. Hence, the neural net receives 36 input parameters (9 depth samples times 4 logs) at each depth. The amplitudes are scaled to zero mean and unit standard deviation for each log. The number of depth samples used in this technique does not seem to be very critical.

A robust neural-network algorithm

Neural-network models of the feed-forward, multi-layered perceptron type, which are usually trained by an iterative gradient-descent procedure, are very powerful and can approximate accurately to any decent mathematical function. However, the basic procedure for specifying and training such networks, as described in many textbooks (e.g. [7]), is not suitable for use by end-users since the performance depends critically on too many parameters such as numbers of units and layers, learning rate and stopping criterion. End-users will generally not be in a position to become neural-net experts, or to spend a long time fine-tuning the parameters for optimum performance. Under these circumstances, neural networks are prone to converge only very slowly, probably to a sub-optimal solution, and perform poorly on new data.

The basic neural-network algorithm was modified in a number of respects, as listed below, to increase its robustness for use by end-users. Most of the modifications have been previously described in the literature - but rarely if ever combined in a single algorithm.

- *Weight initialisation.* At the start of learning, the connection weights are usually initialised to small random values taken from the range $[-u_o, u_o]$. To prevent premature saturation of the units, u_o is divided by the square root of the fan-in (number of incoming connections).

- *Transfer functions.* The hyperbolic tangent tanh(sx) was chosen as the transfer function of the hidden and output units. This function has a number of favourable properties; in particular, a value of 1.5 for the steepness factor s leads to a uniform spread of the error information over the various layers [8].

- *Error function.* The usual quadratic error criterion was replaced by a logarithmic cross-entropy criterion:

$$E = \sum_{n=1}^{N} \left\{ d_n \ln(d_n / o_n) + (1 - d_n) \ln[(1 - d_n) / (1 - o_n)] \right\}$$

where o_n and d_n are the output value and desired value of output unit n, respectively. For classification tasks, this entropy criterion induces a smoother error landscape and hence faster and better convergence [9].

- *Complexity regularisation.* To avoid over-training effects, thus making the choice of the numbers of hidden units and training iterations less critical, a cost term was added to the above entropy term. This cost term, which is related to the information-theoretic concept of minimum description length, penalises excessive weights (w_i) and hence excessive network complexity [10]:

$$E_{cost} = \lambda \sum_i (w_i / w_o)^2 / (1 + (w_i / w_o)^2);$$

where λ is a small constant that trades-off approximation error and complexity, and w_o usually has unit order.

- *Pattern presentation.* Input-output patterns were presented to the network in a random order. The weights were adapted after each single presentation (on-line learning). This approach introduces stochastic fluctuations that help the net to escape from local minima.

- *Learning rate schedule.* A good compromise between speed and accuracy of convergence is achieved by the "search-then-converge" schedule [11]: $\eta(t) = \eta_o/(1 + t/\eta_1)$. Here, the learning rate $\eta(t)$ first remains at a constant level η_o to explore the error surface for favourable regions. After a certain time, regulated by η_1, it starts to converge with an asymptotic speed of $1/t$. To guarantee convergence to the global minimum, an asymptotic $1/\log(t)$ behaviour would be needed. However, this would be prohibitively slow [12].

- *Addition of noise.* Gaussian noise of exponentially decreasing amplitude was added to the activation values of all units to improve generalisation properties and to help the network to escape from local minima (e.g. [12]):

$$g(t) = g_o \exp(-t/g_1)$$

The above modifications introduce several new parameters, which cannot be chosen freely if performance is to be optimal. The number of free parameters is reduced by establishing heuristic relations between parameter values known to be good, while default values appropriate for the given type of problem are provided for the remaining free parameters. In cases where a log had missing or corrupted data over some interval, we used a classification result obtained with the aid of another neural network trained on the intact logs only.

It has been shown that the outputs o_n of neural networks of the type described above represent accurate approximations to Bayesian *a posteriori* probabilities [13]:

$$(o_n+1)/2 \approx p(C_n|x) = p(x|C_n) \bullet p(C_n)/p(x);$$

where $p(x|C_n)$ is the likelihood of producing log response x given lithology type C_n, $p(C_n)$ is the *a priori* probability of lithology type C_n and $p(x)$ is the unconditional probability of observing response vector x. The most probable lithology at a certain depth is that corresponding to the output unit with maximum activity (Fig. 1b). The Bayesian framework permits compensation of different *a priori* class probabilities in training and test sets, and use of minimum-risk criteria that weight the various types of errors differently. Moreover, a confidence measure may be determined by checking whether the estimated *a posteriori* probabilities sum to one.

Software implementation

A prototype neural-network algorithm was designed and tested with the commercial simulator Neural Works Professional II+ (NeuralWare Inc., Pittsburgh). The optional facilities User-Defined Neuro Dynamics and Designer Pack allowed us to extend the functionality and to generate C code, respectively. The C modules obtained in this way have been incorporated in Shell's geological computing environment, in which the formation analyst can train and apply neural networks via a familiar, user-friendly graphical interface.

(a) Input representation

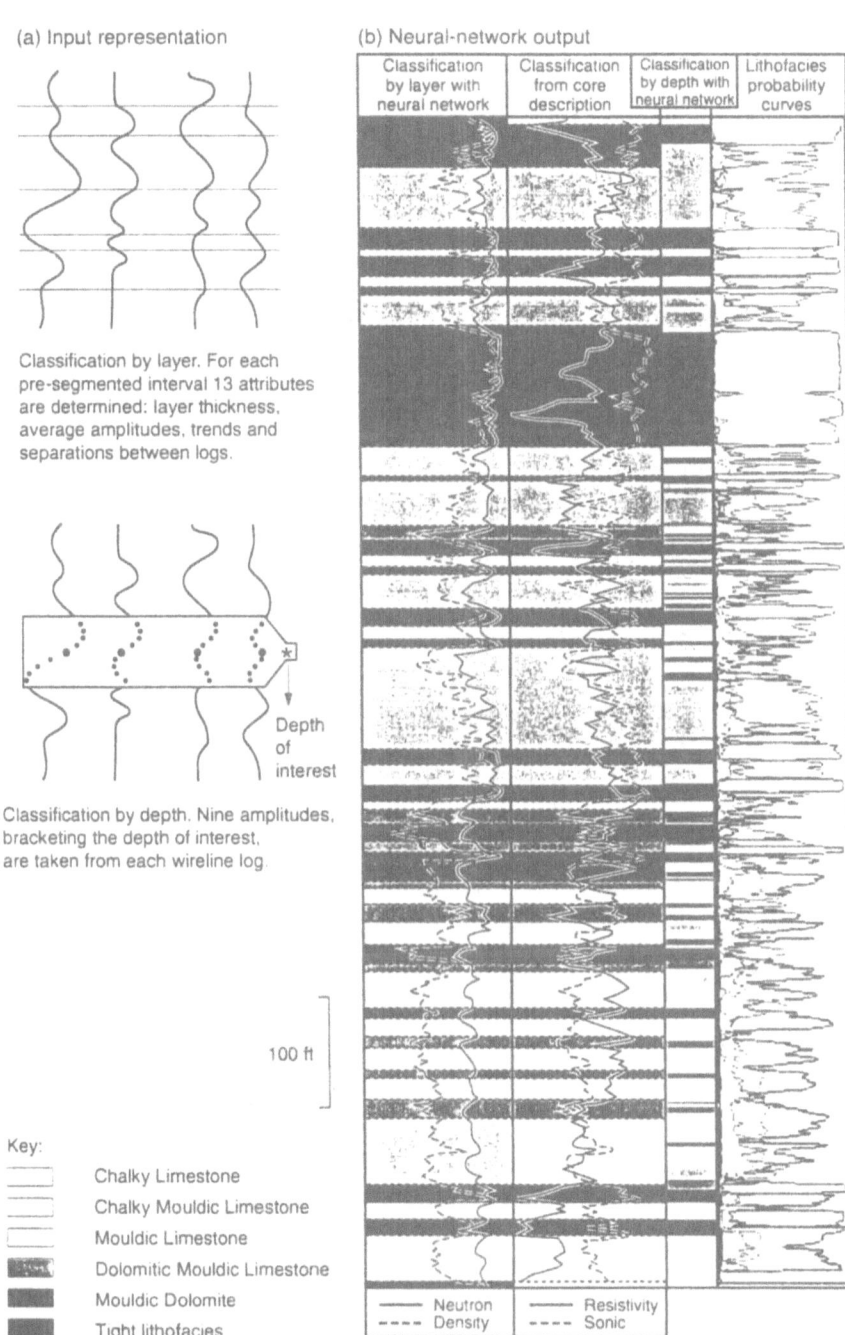

Classification by layer. For each pre-segmented interval 13 attributes are determined: layer thickness, average amplitudes, trends and separations between logs.

Classification by depth. Nine amplitudes, bracketing the depth of interest, are taken from each wireline log.

Depth of interest

100 ft

Key:

Chalky Limestone

Chalky Mouldic Limestone

Mouldic Limestone

Dolomitic Mouldic Limestone

Mouldic Dolomite

Tight lithofacies

(b) Neural-network output

| Classification by layer with neural network | Classification from core description | Classification by depth with neural network | Lithofacies probability curves |

Neutron — Resistivity
Density — Sonic

Fig. 1 Lithofacies identification by neural networks. The data shown are from a carbonate reservoir. The lithofacies probability curves represent the outputs of the classification-by-depth network.

Dr.no 104469

Results

The neural-network procedure has been used to identify lithofacies in hydrocarbon reservoirs from two, entirely different geological environments: siliciclastic rocks (sandstones and shales) and carbonate rocks (limestones and dolomites) .

Siliciclastic lithofacies

Five lithofacies classes had to be identified from wireline-log data recorded in a North-Sea Jurassic coastal plain reservoir: channel-fill, sheet-sand, mouthbar, coal and shale. Data from 12 wells were used for training and from another 6 wells for testing. A core description of the wells considered was not available to provide the 'true' classification. Instead, we relied on a more subjective classification made by an experienced geologist on the basis of the wireline logs.

For the classification-by-layer task, a neural network with a single hidden layer of eight units agreed with the geologist in 88% of the test cases (93% for training data). This accuracy is roughly equivalent to the degree of consensus among different geologists. Linear discriminant analysis performed some 10% worse.

For the classification-by-depth task, a neural network with ten hidden units achieved a score of 78% correct for test data (and 81% for the training set). In a number of cases where the neural network disagreed with its trainer, the neural-network answer was considered reasonable on second thought. Disagreement could often be ascribed to the gradational nature of lithofacies and to the presence of thin beds.

Carbonate lithofacies

Carbonate lithofacies identification from wireline logs is generally considered to be more difficult than the identification of siliciclastic lithofacies. The difficulty of the task is underscored by the failure of a Gaussian classifier to achieve satisfactory results in many instances. Two cored wells drilled in a South-East Asian Miocene build-up were available for training a neural network. Six lithofacies types were identified: one tight (poor reservoir) lithofacies class and five gradational porous types (chalky limestone, chalky mouldic limestone, mouldic limestone, dolomitic mouldic limestone and mouldic dolomite). The neural-network set-up used was almost the same as in the previous siliciclastic case. The only difference was an extra output unit and a larger hidden layer reflecting the increased difficulty of the task; all other parameters were left untouched.

A neural network with eight hidden units achieved a score of 93% on training data for the classification-by-layer task (Fig. 1b). For classification-by-depth, a neural network with fifteen hidden units classified 80% of the training data correctly (Fig. 1b). At the moment, not enough core data is available for an accurate estimate of the performance on new data. However, a more qualitative study in which the trained neural networks were applied - without retraining - to a well from a different Miocene carbonate reservoir in another province yielded very promising results. Analysis of microscopic sections and petrophysical measurements from sidewall samples showed that the neural nets had correctly distinguished tight, low-permeability and high-permeability lithofacies types.

Discussion and Conclusions

The neural classifier for the siliciclastic lithofacies was designed by neural-network researchers [5]. After demonstration of its potential, geologists and neural-network experts co-operated to extend its applicability to the more challenging carbonate data. Since the neural-network technology has been made available to geologists, they have found new applications and are now in a position to achieve promising results on their own.

The neural network was made robust to non-expert use. However, the formation analyst remains responsible for the proper selection of representative training data and an adequate wireline-log suite. In particular, a neural network should be applied with care to formations where the fluid composition is different from that of the training set. This is because the wireline logs can only be corrected approximately for different formation fluids.

The neural networks performed substantially better than commonly used conventional classifiers. In the realms of formation evaluation, neural networks may be applied to other problems than lithofacies identification as well. Applications to log inversion [14] and permeability estimation [15] have been described.

References

[1] Wolff, M. and Pelissier-Combescure, J.: "Faciolog - automatic electrofacies determination", SPWLA 23rd Annual Logging Symposium (1982) July 6-9, paper FF.

[2] Delfiner, P., Peyret, O. and Serra, O.: "Automatic determination of lithology from well logs", SPE Formation Evaluation (1987) September, 303-310.

[3] Bush, J.M., Fortney, W.G. and Berry, L.N.: "Determination of lithology from well logs by statistical analysis", SPE Formation Evaluation (1987) December, 412-418.

[4] Baldwin, J.L., Bateman, R.M. and Wheatley, C.L.: "Application of a neural network to the problem of mineral identification from well logs", The Log Analyst (1990) September-October, 279-293.

[5] Cardon, H.R.A., Van Hoogstraten, R. and Davies, P.: "A neural network application in geology: identification of genetic facies", Proceedings of the International Conference on Artificial Neural Networks, Espoo Finland, (1991) June 24-28, p. 809-813.

[6] Liu, R.L., Zhou, C.D. and Jin, Z.W.: "Lithofacies sequence recognition from well logs using time-delay neural networks", SPWLA 33rd Annual Logging Symposium (1992) June 14-17, paper L.

[7] Rumelhart, D.E., McClelland, J.L. and the PDP Research Group. Parallel distributed processing, explorations in the microstructure of cognition, Vol. 1: Foundations. MIT Press, Cambridge MA, 1986.

[8] Kalman, B.L. and Kwasny, S.C.: "Why tanh: choosing a sigmoidal function", Proceedings of the International Joint Conference on Neural Networks, Baltimore MD, (1992) June 7-11, Vol. 4, p. 578-581.

[9] Solla, S.A., Levin, E. and Fleisher, M.: "Accelerated learning in layered neural networks", Complex Systems (1988) 2, 625-640.

[10] Weigend, A.S., Rumelhart, D.E. and Huberman, B.A.: "Generalisation by weight-elimination with application to forecasting", Advances of Neural Information Processing Systems 3, Morgan Kaufmann, San Mateo CA, (1991), p. 875-882.

[11] Heskes, T.M., Slijpen, E.T.P. and Kappen, B.: "Cooling schedules for learning in neural networks", Physical Review E (accepted).

[12] Holmström, L. and Koistinen, P.: "Using additive noise in back-propagation training", IEEE trans. Neural Networks (1992) 3, 24-38.

[13] Richard, M.D. and Lippmann, R.P.: "Neural network classifiers estimate Bayesian a posteriori probabilities, Neural Computation (1991) 3, 461-483.

[14] Garcia, G. and Whitman, W.W.: "Inversion of a lateral log by using neural networks", SPE 7th Petroleum Computer Conference, Houston TX, (1992) July 19-22, paper 24454.

[15] Wiener, J.M., Rogers, J.A., Rogers, J.R. and Moll, R.F.: "Predicting carbonate permeabilities from wireline logs using a back-propagation neural network", SEG 61st Annual International Meeting, Houston TX, (1991) November 10-14, p. 285-288.

Using Selforganizing Feature Maps to Classify EEG Coherence Maps

Georg Dorffner[1], Peter Rappelsberger[2], Arthur Flexer[1]

[1]*Dept. of Medical Cybernetics and Artificial Intelligence, University of Vienna, and Austrian Research Institute for Artificial Intelligence*

[2]*Inst. of Neurophysiology, University of Vienna*

Abstract

In this work we have been applying self-organizing feature maps [3] to the problem of unsupervised classification of EEG data. The type of EEG used are so-called coherence maps based on 19 electrodes, which were derived during specific cognitive taks such as mental rotation. The goal was to exploit the network learning scheme as extractor for any task- (or other parameter-) related information in the data. In other words, we used the self-organizing feature maps to detect whether the EEG inputs can be classified acording to underlying parameters such as the type of task performed. This paper reports about the very promising results of the experiments.

The application

This paper describes the application of self-organizing feature maps for the unsupervised classification of EEG. The underlying data base contained EEG data of 78 students (39 females and 39 males) who had to perform spatial imagination tests. A cube rotation test (henceforth "WU") was performed by 19 female and 16 male students [1]. The same 19 female students also performed a rope figure ("SF") test [9]. Finally, a figure rotation ("FI") test was done by 20 (different) females and 23 males [8]. During the tests, lasting for a few minutes, EEG was recorded with 19 electrodes according to the 10/20 system against averaged ear lobe signals. Averaged spectral parameters such as amplitue ("AM"), local coherence ("LC"), and interhemispheric coherence ("IC") were computed. LC corresponds to the coherence between adjacent electrodes along the transversal and longitudinal electrode rows. IC is computed between corresponding sites of the two hemispheres. All in all, 19 AM, 30 LC, and 8 IC values were obtained per case. The computations were based on at least 20 artifact-free 2 second epochs for the six frequency bands theta, alpha, beta1, beta2, beta3, and delta. To obtain a reference for comparison, all parameters were also computed for each person with their eyes open at rest ("AA"). In contrast to related work (e.g. [2, 5, 10]) the resulting input for the network is a stationary pattern.

The network

The network employed for classification was a two-dimensional self-organizing feature map [3] with 4 by 4, or 5 by 5, units. The input layer consisted of 30, or 38, units directly encoding the difference of the LC or LC+IC values (which are all between −1 and 1) of the EEG during task performance, minus the corresponding values for the reference task ("AA"). We omitted the AM values (amplitudes), since previous results had shown that they contain relatively little information with respect to the given task. The weights between the input and the map layer were initialized randomly between −1 and 1 before each training run. For training, the learning rule in [7], p.58, was taken:

$$w_r^{new} = w_r + \epsilon\, h_{rr'}\, (v - w_r^{old})$$

where w_r denotes the vector of incoming weights of unit r, and v the vector of input values. The factor $h_{rr'}$ reflects the neighborhood among units in the map (r' stands for the winning unit) and was chosen according to an exponentially declining function with 0.8 for the four immediate vertical and horizontal neighbors in the two-dimensional grid ("near1"), and with 0.5 for the four diagonal neighbors ("near2", see fig.1). The decay factor ϵ started off with 0.5 and was piecewise-linearly decremented toward 0.0 at each training epoch. At each epoch all input patterns were presented in a fixed order. Fig. 1 shows the development of the product of both factors over time. Each training run con-

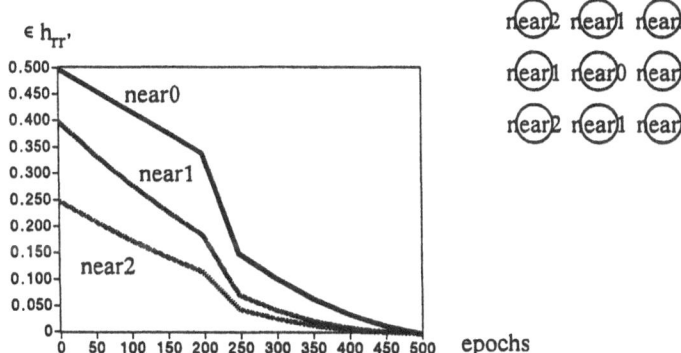

Fig. 1: The decay of the factors scaling the weight change over time, and the neighborhood of a unit (depicted as "near0" in the center).

sisted of 500 epochs.

The choice for this particular type of neural network was guided by two considerations. First, one of our ultimate goals is to exploit the network to extract not only information known a priori (such as the task performed). In other words, the network ultimately should find groupings of cases independent of the obvious given classifications. Secondly, since the input data themselves form a two-dimensional topological map, choosing the network by Kohonen seems adequate so as to lead to a method of "map transformation." There also exists similar work using feature maps (such as [2, 4]), albeit not all with the same success.

The interpretation of the resulting map

The great disadvantage of an unsupervised learning algorithm is, of course, the lack of an immediate means of interpretation of the results. Thus we have employed the following method to label the units in each resulting map. First a level of confidence c in percent was chosen. Then all training patterns were presented to the network. For each unit that became the winner in the resulting update the occurences of the known categories of the input ("FI", "WU", or "SF" in most of the experiments) were counted. If percentage of the number of occurences of one category was above the confidence value, this unit was labeled by the name of that category. If there were two above c, the higher one was taken. For instance, if a unit was the winner for patterns of category "FI" 70% of the time, but only 15% for "WU" and "SF", respectively, the unit was labeled "FI" if the chosen confidence value was not larger than 70%. Obviously, the confidence value should be larger than the probabilities in a random distribution (33% in this example). If no category reaches a percentage above the confidence, the unit was not labeled.

After this labeling, all the patterns were presented again. Now for each category, and for all categories together, the percentage of correct responses was computed, thus resulting in a measure for the overall performance of the network. A response was "correct" if the winner was labeled by the correct known category of the input pattern. This was done for several values of the confidence, resulting in several curves depicting the percentage of correct responses depending on the chosen value of confidence. Since there can be unlabeled winners, an anaologous curve for "no decisions" ("none") can be drawn.

The results

Most experiments were run with all training patterns and the three categories "FI", "WU", and "SF," for all frequency bands. The inputs were either the 30 LC values, or the 38 values consisting of LC+IC. The best results were obtained in the frequency bands beta1 (13–18 Hz) and theta (4–7.5 Hz). In contrast, the alpha band yielded very poor results. This confirmed previous observations on the relevance of each band in predicting underlying cognitive parameters. Fig. 2 shows the results of a 5 by 5 map trained with the full 38 dimensional input vectors. Fig 2a. depicts the overall results (in the way explained above), 2b through 2d depict the results for each category separately. Fig 3 shows the corresponding labeled map for a ·confidence value of 60%. The results show that the categories "FI" and "WU" can be predicted relatively well, at low confidences with up to 90%. Contrary to that, prediction of "SF" almost completely fails. One reason for this could be due to the fact that the training set was not homogenous. In particular, all "SF"-cases were performed by persons who also went through the "WU"-test, while all other persons performed only one test. In another run, the network was trained with "WU" and "FI" patterns alone, which improved their prediction rate even further. Fig. 2 further shows that all curves remain relatively constant up to a confidence value of 55–60%, where they decline relatively rapidly. The map in fig. 3 shows the topological compactness of each of the three areas. This underlines the good results, since a high prediction without topological closeness of units labeled by the same category would not be very informative (given the relatively low number of cases, as compared with the number of units in the map).

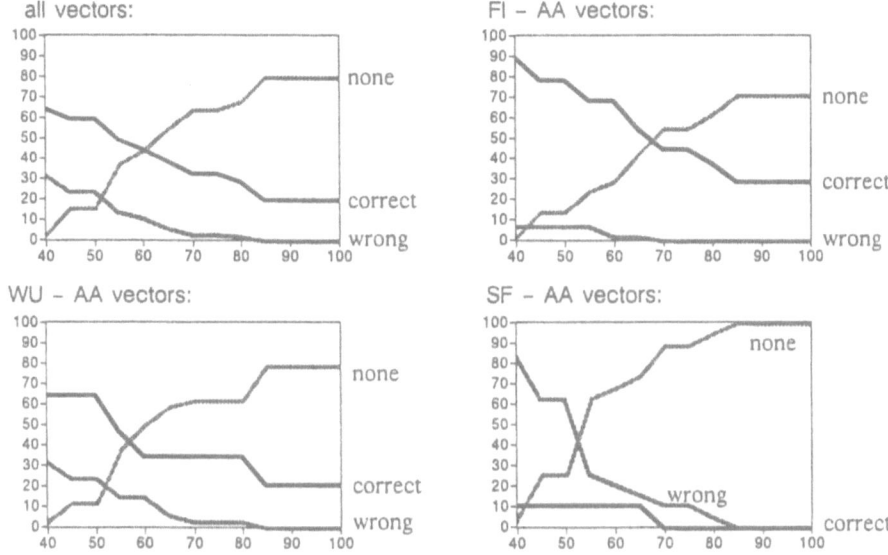

Fig. 2a (upper left) – 2d (lower right): The results of the classification of three imagery tasks.

```
   --  --  FI  FI  FI
WU  --  --  FI  FI
WU  --  --  --  FI
   -- WU  FI  --  FI
SF  FI  FI  FI  --
```

Fig. 3: The resulting map after labeling at a confidence of 60 %.

For further evaluation of the method, experiments with different interpretations of the output were performed. The first one concerned the information whether the test person was generally good or bad in imaging tasks. The second one concerned the test persons' sex. Both informations could be extracted from the maps relatively well. The percentage of prediction, however, was lower than in the previous experiments (Fig. 4, see also [6]).

Discussion

The results of these experiments demonstrate that neural networks can serve as a valuable tool in predicting cognitive parameters based on stationary EEG coherence maps. The outcome of the prediction concerning the cognitive tasks performed had not been equaled by any other method before. The outcomes of the second series of experiments, in particular the ones about predicting the person's sex, demonstrate the bandwidth of the method with respect to the variety of parameters to be predicted. Although performance was not satisfactory in all cases, most results are very encouraging for further investigations. In particular, we plan to combine the self-organizing feature map with more supervised learning schemes, and to find automatic methods of interpretation to perhaps extract information which is not known a priori. The long-term goal of this research is to develop this method into a routine tool for psychiatric clinical uses.

886

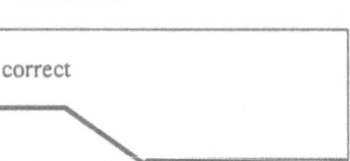

all vectors

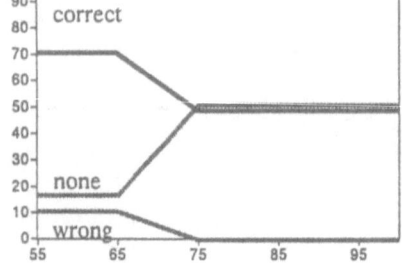

Fig. 4: The results of the classification according to the persons' sex.

An important aspect of the presented results is the constancy of classification over large areas of variation of the confidence value used during labeling. This is especially important for future clinical uses, where often it is vital to have a method, the performance of which can be poor percentage-wise but which predicts classes with a high probability of correctness. In the graphs in figures 2 and 4 those regions would be where the curve for "wrong" (inverse specificity) goes below a certain value (e.g. 10 %) or even approaches 0. The graphs show that often a large percentage (e.g. 45 % in fig. 2b) of classifications with a certainty of up to 100 % can be achieved if one accepts a large number of "none"-responses. In this sense, even curves below "chance-level" can be useful.

Acknowledgments

Part of this project has been supported by the Jubiläumsfonds der Oesterreichischen Nationalbank, project no. 4547. The Austrian Research Institute for Artificial Intelligence is supported by the Austrian Federal Ministry for Science and Research. We thank Prof. Robert Trappl and Prof. Giselher Guttmann for their support for this project.

References

[1] Amthauer R.: IST 70 – Intelligenz–Struktur–Test, Verlag Hogrefe, 1970.

[2] Elo P., J. Saarinen, A. Värri, H. Nieminen, K. Kaski: Classification of Epileptic EEG by Using Self-Organizing Maps, in Aleksander I., Taylor J.(eds.): Artificial Neural Networks 2, Elsevier Science Publishers, 1147–1150, 1992.

[3] Kohonen T.: Self–organized formation of topologically correct feature maps, Biological Cybernetics 43:59–69, 1982.

[4] Litscher G., Flotzinger D., Pfurtscheller G.: Die Verwendung neuronaler Netzwerke bei der Mustererkennung von Schlafprofilen; to appear in: Löffler (ed.): Central Nervous System Monitoring, Verlag Wilhelm Maudrich Wien-München-Bern, 1993.

[5] Lo P.–C., Principe J.C.: Dimensionality Analysis of EEG Segments: Experimental Considerations, in IEEE International Conference On Neural Networks, Washington D.C., IEEE, Volume I, pp.693–698, 1989.

[6] Rappelsberger P., Dorffner G., Flexer A.: Classification of EEG coherence maps of cognitive processes; to appear in: Löffler (ed.): Central Nervous System Monitoring, Verlag Wilhelm Maudrich Wien-München-Bern, 1993.

[7] Ritter H., Martinetz T., Schulten K.: Neuronale Netze, Addison– Wesley, Reading, MA, 1990.

[8] Shepard R.N., Cooper L.A.: Mental images and their transformations, MIT Press Cambridge, 1982.

[9] Stumpf H., Fay E.: Schlauchfiguren: Ein Test zur Beurteilung des räumlichen Vorstellungsvermögens, 1983.

[10] Wang G., Takigawa M., Miyazaki T., Takeishi T.: Analysis of EEG Changes Between Frontal and Occipital Area in Speaking Process, in Caudill M.(ed.), Proceedings of the International Joint Conference on Neural Networks (Winter Meeting), Washington D.C., Lawrence Erlbaum, Hillsdale, NJ, pp.27–30, 1990.

Building an Artificial Retina for
Distance- and Orientation-Invariant Pattern Recognition[1]

M. Busemann, J. Dunker, G. Hartmann, K. O. Kräuter, E. Seidenberg, H. Wiemers
Fachbereich 14 Elektrotechnik, Universität-GH Paderborn
Pohlweg 47-49, D 4790 Paderborn, Germany

Abstract

The present paper demonstrates how - due to a special construction principle for an artificial retina - orientation- and distance-invariant pattern recognition can be realized. Logarithmic retinae in their pure form can support invariant recognition, but they suffer from difficulties concerning an inhomogeneous representation of objects. Selection of ring shaped representations from a superposition of differently scaled logarithmic retinae helps to overcome these difficulties. On the basis of this retinal structure the orientation of objects can be identified with an angular resolution of $\pm 1.4°$. This orientation information combined with available distance information serves as input for mapping processes producing normalized object representations.

1. Introduction

The visual system demonstrates excellent performance in recognizing arbitrarily placed objects in a 3D environment. The underlying mechanisms are far from being understood, so they are not directly usable for the construction of artificial vision systems with comparable performance. Within a robot vision project we try to provide at least some of these features using neural network techniques. While common approaches rely upon model-based recognition working with previously entered object models, we prefer a subsymbolic approach using an one-step learning strategy. Parametric mappings are used to normalize the neural representation of objects before matching in an associative memory.

The normalization step is crucially dependent on the way our artificial retina takes up and represents object information. In order to provide orientation invariance object information can appropriately be represented in a polar coordinate system; to support distance invariance a logarithmic representation is suitable. Retinae of this type are already known (eg. [1], [2]), but suffer from difficulties concerning the exploding number of sampling points in the center provided that a sufficiently high resolution shall be obtained in the periphery. We will concentrate on a detailed description of our retina that overcomes these difficulties and we will demonstrate its usefulness in achieving invariances against distance and orientation by simple operations on the retinal representation. Additionally we will show how the orientation of objects can be extracted at an angular resolution of $\pm 1.4°$ due to the structure of our retina, though our binary detector neurons work with an angular resolution of $\pm 15°$.

Moreover our system offers abilities for reliable foveation of objects as well as for foreground/background separation and is tolerant against minor deviations in size, position, orientation and shape of the object representation (more information about the taken approach can be found in [3],[4]).

---

[1] Supported by the German Minister for Research and Technology (BMFT), ITN 910 506

2. The artificial retina

In our system the differently sized retinal images of an object at different distances are mapped to a normalized representation at a target layer of a given resolution. Under ideal conditions the number of pixels representing the retinal image of the object should correspond with the number of pixels or neurons of the target layer. Obviously, this could only be achieved if the resolution of the retina was changed with the object distance or if the appropriate retina was selected from a set of retinae with different resolutions (fig. 1a, 1c).

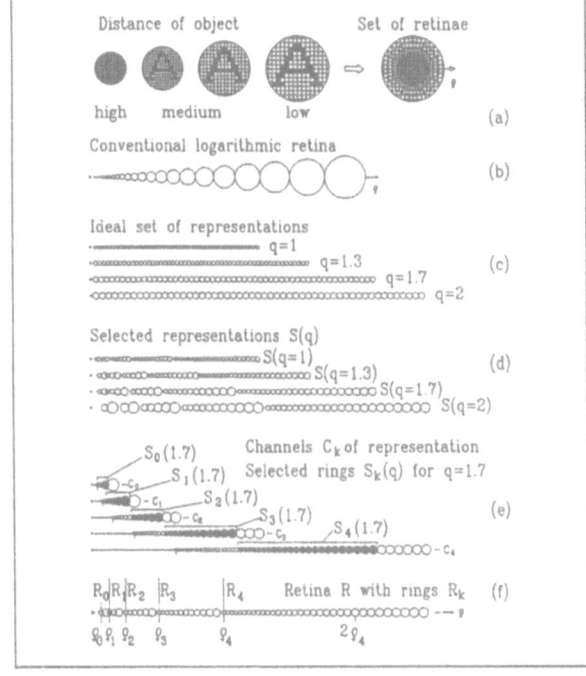

Fig. 1: Representations with variable resolution (a) can be provided by our retina (f), and by selective mappings from representations C_k (e). These selected representations S(q) (d) are good approximations to representations with homogeneous resolution (c), while the "logarithmic retina" (b) is extremely inhomogeneous.

A superposition of a set of retinae (fig. 1a right hand) shows that the required resolution decreases from the center to the periphery. This observation suggests that it should be possible to reconstruct a set of virtual retina from a single retina. This retina should have decreasing resolution from center to periphery. Low resolution in the central area could be achieved by smoothing and subsampling in the case of low object distance. The well known polar coordinate retina with equidistant pixels in an orthogonal $(\ln \rho/\rho_0, \alpha)$-plot seems appropriate at first sight (fig. 1b). There is, however, a crucial disadvantage concerning the exploding number of sampling points. If we start with a sufficient resolution in the periphery of a logarithmic retina, the number of sampling points per area unit increases exponentially approaching the center.

This problem can be overcome, extracting different regions from retinae with different resolution and combining them into an arrangement of rings R_k (fig. 1f, 2a) to provide approximately constant resolution within a radius $\rho \leq \rho_4$. Doing this, we are able to build a specific polar coordinate retina with approximately equidistant pixels allowing invariance by simply shifting the internal representation in an orthogonal $(\ln \rho/\rho_0, \alpha)$- plot.

The high symmetry of hexagonal pixel arrangements and the availability of well proven detector sets for this raster convinced us to use this pixel arrangement in our retina. So we start with the description of a hexagonal logarithmic retina, continue with the description of our composed retina and proceed with the reconstruction of a set of virtual retinae from this representation.

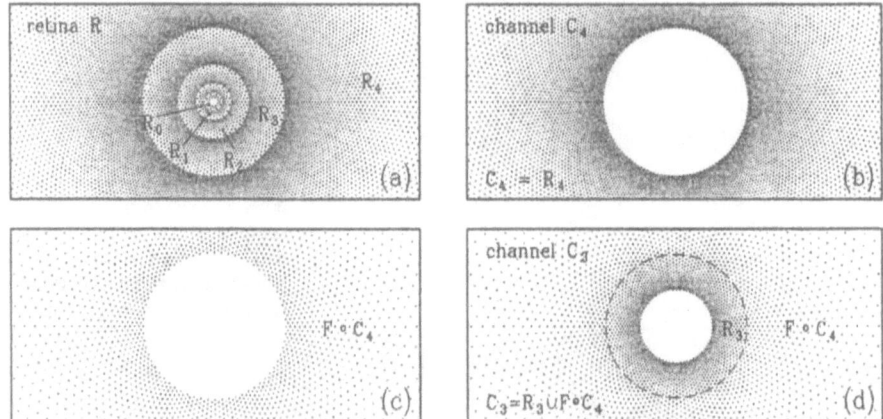

Fig. 2: The complete retina (a). A channel C_3 (d) can be constructed by taking C_4 (b), filtering it (c) and combining the result with R_3 (d).

In order to get a retina with equidistant sampling points in a $(\ln \rho/\rho_0, \alpha)$-plot one firstly has to exponentially expand the sampling raster in its radial direction. A single sampling circle shall consist of q sampling points. To provide a hexagonal structure neighbouring sampling circles are rotated against each other by π/q degrees. Let furthermore r(i) denote the radial distance of the i-th sampling circle, then the ratio $r(i+1)/r(i)$ is constant for an exponentially scaled retina $(i=1,...,n;$ with n being the number of retinal sampling circles). In the following this constant will be denoted by p. In order to get a symmetrical hexagonal arrangement of sampling points in an $(\ln \rho/\rho_0, \alpha)$-plot, one has to specify p in dependence on q as: $p = \exp(\pi/(q\sqrt{3}))$ (a detailed derivation can be found in [5]).

In order to construct our ring shaped retina with approximately constant resolution we proceed as follows (fig. 2a): beginning with a sampling raster (R_0) which provides a sufficiently high sampling density in the center, we switch to a new raster (R_1) as soon as the sampling density of R_0 falls beneath a critical value. Repeating this procedure, we add the rings R_2, R_3 and R_4 providing our composed retina (fig. 1f, 2a). The ring R_0 of our retina has $q_0 = 16$ sampling points on each of its sampling circles; therefore the corresponding p_0 computes to 1.12 which is approximately $\sqrt[6]{2}$. Working with this approximation, the radius of the innermost (the first) sampling circle is doubled at the seventh sampling circle. At this location the distance of two neighbouring points on a sampling circle - the inverse of which is the resolution - is doubled and we switch to sampling raster R_1 by doubling the number of sampling points on a sampling circle.

Looking at figure 3, one observers that as soon as the resolution in a specific ring R_i $(i = 0,...,3)$ falls beneath a critical value res_0, a new sampling raster (giving R_{i+1}) is selected and the original resolution res_0 can again be obtained.

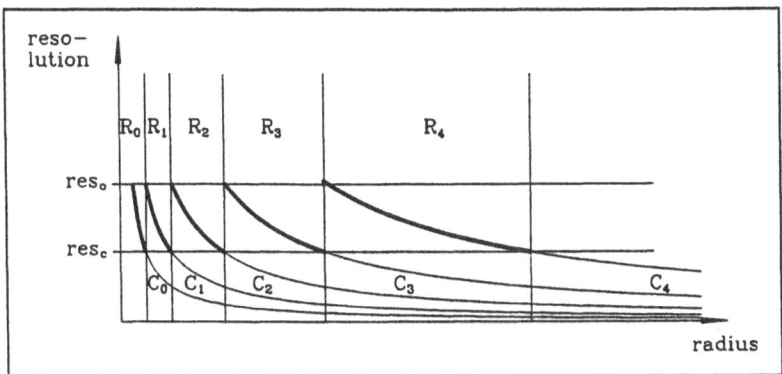

Fig. 3: Resolution falls with increasing radius in each of the rings R_0 - R_4.

3. Representations and Mappings

From this composed retina we are able to reconstruct a set of overlapping virtual retinae C_k (k=0,...,4) describing visual structures at different resolution (fig. 1e). For this reason we have called these representations C_k spatial frequency channels. The outermost ring of our retina R_4 is identical with the channel C_4 (fig. 1e, fig. 2b). Low pass filtering F, combining smoothing and subsampling, provides a representation $F \circ C_4$ of half the resolution (fig. 2c). This representation $F \circ C_4$ is combined with the retinal representation R_3 and $R_3 \cup F \circ C_4$ forms the channel C_3 (fig. 2d). In a similar way C_2, C_1 and C_0 are generated using $C_k = R_k \cup F \circ C_{k+1}$ (for k=0,...,3). The integration of the spatial frequency channels in the processing system can be seen in fig. 4 while the development of their resolution with increasing radius is shown in fig. 3.

The retinal representations can directly be used to identify objects by region based overlap with prototype objects. To support good shape discrimination, we additionally extract information about the orientation of contour elements and generate contour representations which are tolerant against unavoidable foveation and normalization errors.

In this short contribution we can only sketch the transformations between C_k and the corresponding normalized contour representation S (fig. 4), which is the input vector for the associative match. A center-periphery coupling between neighbouring neurons in C_k provides neurons similar to on-center and off-center neurons. These neurons are thresholded in our system and provide a binary representation (L_k^+, L_k^-) containing the information about bright/dark transitions. Local interconnections between neighbouring neurons of (L_k^-, L_k^+) provide neurons in D_k with oriented receptive fields encoding the contour information about the object. These neurons are similar to the complex cortical neurons and have extended receptive fields. Correspondingly, these neurons respond even if the contour is placed outside the center of their receptive field. Therefore contour elements are represented by clouds of neurons in D_k. The final object representation is composed of different areas of D_0,...,D_4. A parametric mapping controlled by the object distance determines the selection of those rings shaped areas which are chosen to make up the final representation S (fig. 1d). This enables our system to recognize an object independent of its distance from the camera.

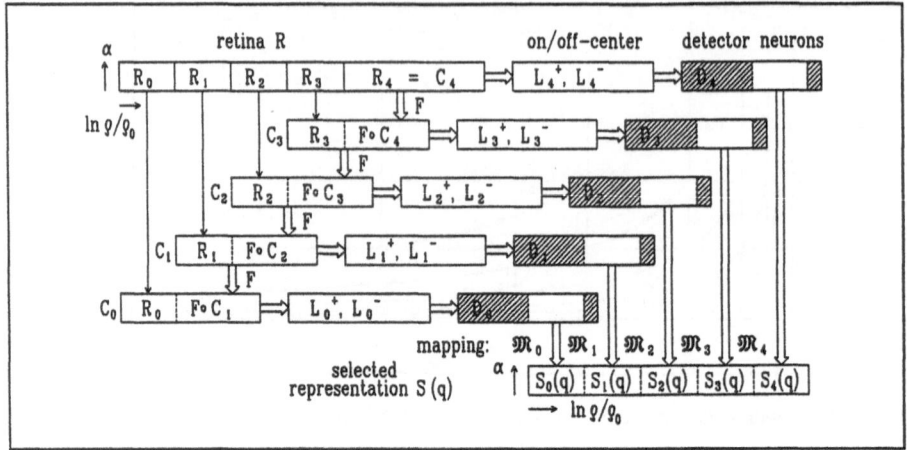

Fig. 4: Block diagramm of the complete processing, starting with the artificial retina (upper left side) to the normalized representation.

In order to achieve also orientation invariant recognition we firstly determine the angle between the actual object orientation and a predefined orientation. This angle is the input parameter for a further process mapping the object image onto its normalized representation. A neural network for the extraction of an object's orientation has been developed. 128 output neurons - each representing the total activity of all neurons with similarly oriented receptive fields - compete with each other. The winning neuron indicates the object orientation with an angular resolution of $1.4°$. Fig. 5 shows the activities of the 128 output units for two workpieces each presented in two different orientations.

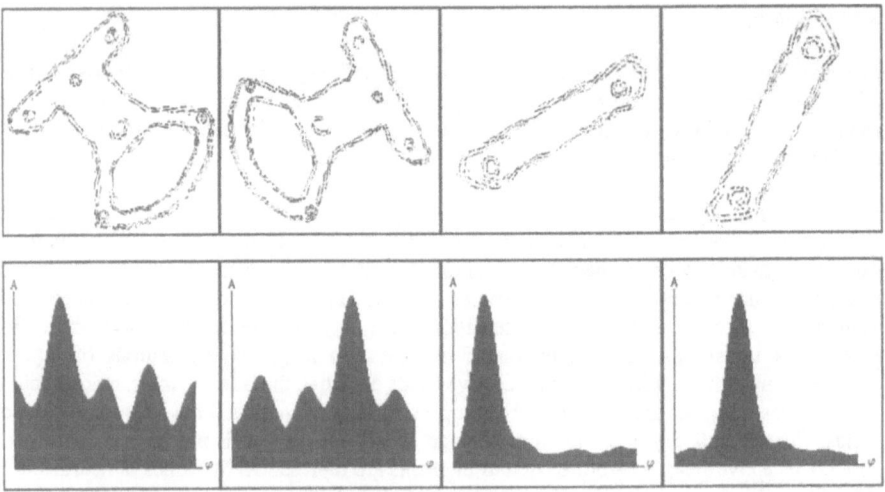

Fig. 5: Objects presented in different orientations (upper line) are rotated back according to the angle where the highest peak of the corresponding histogram (lower line) is found.

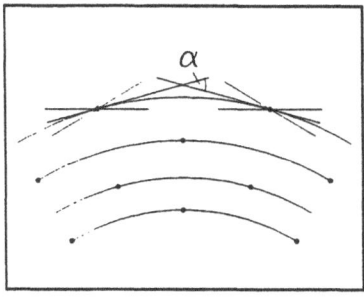

Fig. 6: Two neighbouring sampling points in R_4 and their detector orientation.

The high angular resolution of $1.4°$ can be obtained although our binary detector neurons work with a resolution of $\pm 15°$ only. Looking at two neighbouring sampling points in R_4 with identical radial distance (fig. 5) one can see that the local axis of corresponding detectors is shifted by α ($\alpha = 360/q_4$; $q_4 = 256$). Hence high angular resolutions can be realized on the basis of detectors with comparably rough resolution using ensemble coding in the pixel arrangement of our special retina.

4. Conclusion

It was shown that due to the structure of our artificial retina distance- and orientation-invariant pattern recognition can be achieved. This retina has been implemented as a front end of our robot vision system and provides good results in combination with the above mentioned invariance modules.

References:

[1] Reitboeck, H.J., Altmann, J.: A Model for Size- and Rotation-Invariant Pattern Processing in the Visual System. In: Biological Cybernetics, 51, 1984, 113-121.

[2] Schwarz, E.L. Cortical Anatomy, Size Invariance and Spatial Frequency Analysis. Perception, 10, 1981, 831-835.

[3] Busemann, M., Hartmann, G., Kräuter, K.O., Seidenberg, E., Wiemers, H.: Digit classification using an edge based hierarchical neural representation. In: Aleksander, I., Taylor, J.G. (eds.): Artificial Neural Networks (2), Elsevier Publishers 1992, 1579-1582

[4] Hartmann, G.: Hierarchical Neural Representation by Synchronized Activity: A Concept for Visual Pattern Recognition. In: Taylor, J.G. et al (eds.): Neural Network Dynamics (Springer) 1991, 356-370

[5] Hartmann, G., Drüe, S., Kräuter, K.O., Seidenberg, E.: Simulations with an Artificial Retina. Submitted to World Congress on Neural Networks 1993 (WCNN 93)

MLP-RBF: A COOPERATIVE MULTI-MODULAR NEURAL NETWORK APPLICATION IN HIGH-ENERGY PHYSICS

Joseph PRORIOL

Laboratoire de Physique Corpusculaire

Clermond-Ferrand

e-mail: proriol@frcpn11.in2p3.fr

ABSTRACT

We present a cooperative multi-modular neural network architecture: a Multi-Layer Perceptron (MLP), followed by a Radial Basis Function network (RBF). We show that, in the LEP experiment of electron-positron collision run at CERN, this architecture was able to outperform both a simple multi-layer perceptron, a multi-modular MLP+LVQ (LVQ: Learning Vector Quantization) and MLP+RBF trained sequentially and a conventional technique (Discriminant Analysis).

1- INTRODUCTION

Various researchers have recently advocated the use of multi-modular neural networks [1, 2, 3, 4, 5, 7]. Various networks were used: multi-layer perceptrons (MLP), Time Delay Neural Networks (TDNN), Learning Vector Quantization (LVQ) or Radial Basis Function nets (RBF). The most common applications have been character recognition [7] or speech [1, 2, 5]. The idea underlying these efforts is to get improved performances from the cooperation of networks: for example, a MLP is first used on the raw noisy data, then a LVQ or a RBF network is trained on the MLP-extracted features and will be able to perform a better classification than the last layer of the MLP.

Most often these architectures are trained sequentially, one module after the other. However, such a training procedure is sub-optimal: global optimality can be obtained by cooperatively training all modules together [4].

Our aim was to classify very noisy data coming from High Energy Physics (HEP) problems [6]: events produced by the collisions of particles (e^+e^-) at the LEP accelerator at CERN (Geneva). In work reported previously [9], we compared the performances of conventional techniques (cut on most discriminant variable, Discriminant Analysis) and various Neural Networks (MLP, LVQ). In this paper, we will introduce a MLP-RBF cooperative neural network; the first part of the processing is done by the first layers of a MLP. The final classification is done by a RBF. Instead of using the usual initialisation by a k-means method [7], we have initialized by small random numbers. Our model is also different from [8]: we first use a MLP and then a RBF structure.

In section 2, we will describe our model and in section 3 discuss our application.

2- THE MULTI-MODULAR COOPERATIVE ARCHITECTURE

A multi-modular architecture is made of various modules: in this paper, we will consider architectures with two modules MLP and another network, used in succession: the MLP generates features in its last hidden layer and these features are used by the next module as inputs. In sequential training, the two modules are trained one after the other. The global architecture is thus sub-optimal, since the features computed by the MLP are optimized for its own classification and not for that of the next module. We will present in section 3 results obtained with MLP+LVQ and MLP+RBF architectures trained sequentially. In this section, we indicate a method to cooperatively train such architectures [4].

The architecture is made up from two modules. First a MLP, with L layers of neurons; the MLP used for classification would have L+1 layers, but here we have stripped the last layer, and only retain the features computed by the MLP whih are used as inputs to the RBF. Then a RBF with 2 layers of neurons (the first one is a set of gaussian kernels and the other is a standard MLP layer: fig. 1). Let us denote $W_{ij}^{\lambda-1}$ and $T_i^{\lambda-1}$ the weights and bias of neuron i, in layer λ; x_i^λ its activation, y_i^λ its state and f_λ the transfer function, n_λ the number of neurons in layer λ.

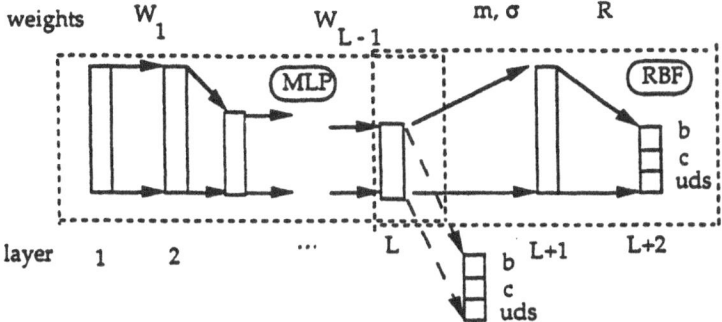

Figure 1: multi-modular architecture MLP+RBF. The MLP has L layers (if it were trained on its own, it should have one more layer, shown with dotted arrows); the RBF has 2 additional layers.

The feed-forward activations of the MLP are given by ($\lambda = 1 \ldots L$):

$$x_i^\lambda = \sum_{j=1}^{n_{\lambda-1}} W_{ij}^{\lambda-1} y_j^{\lambda-1} + T^{\lambda-1} \qquad y_i^\lambda = f_\lambda [x_i^\lambda] \tag{1}$$

The RBF network has an input layer which is the last layer of the MLP (layer L), then a hidden layer of n_{L+1} kernel neurons and an output layer with n_{L+2} neurons. We denote m_i the weight vector of neuron i in layer L+1 (the means of the kernels), σ_i the width of the kernel and R_{ik} the weights of neuron i in the output layer.

The states in RBF are computed by:

$$x_i^{L+1} = \exp\left(-\frac{\| m_i - y^L \|^2}{2\sigma_i^2} \right) \qquad y_i^{L+1} = \frac{x_i^{L+1}}{\sum_{j=1}^{n_{L+1}} x_j^{L+1}} \tag{2}$$

$$x_i^{L+2} = \sum_{k=1}^{n_{L+1}} R_{ik} y_k^{L+1} \qquad y_i^{L+2} = f_{L+2}(x_i^{L+2})$$

Now the weight modification is done in the following way [4]:
The RBF is trained to minimize the error $E(k) = \| t(k) - y^{L+2}(k) \|^2$ where $y^{L+2}(k)$ is the output computed on layer L+2, when example x(k) is presented on the input layer of MLP and t(k) is the associated target, ie x(k) class. Let us denote:

$$V_i = f'_{L+2}(x_i^{L+2}) \qquad U_i = \frac{\sum_{j \neq i} x_j^{L+1}}{\left(\sum_{j=1}^{n_{L+1}} x_j^{L+1} \right)^2} \tag{3}$$

$$\Delta R_{ik} = [t_i - y_i^{L+2}] y_k^{L+1} V_i$$

$$\Delta m_{kj} = \frac{U_k}{\sigma_k^2} \cdot [m_{kj} - y_j^L] \cdot \sum_i [t_i - y_i^{L+2}] V_i R_{ik}$$

$$\Delta \sigma_k = \frac{U_k \parallel m_k - y^L \parallel}{\sigma_k^3} \cdot \sum_i [\, t_i - y_i^L + 2\,] \; V_i \; R_{ik}$$

The MLP is trained as usual, with the Gradient-Back-Propagation rule [10], except that the error δ which is back-propagated comes from the RBF:

$$\delta_j^L \;\; = \;\; - f'_L(x_j^L) \cdot \sum_k \Delta m_{kj} \qquad\qquad (4)$$

$$\Delta W_{ji}^{L-1} = \delta_j^L \;\; y_i^{L-1} \qquad\qquad\qquad \Delta T_j^{L-1} \;\; = \;\; \delta_j^L$$

With this rule for weight modification, the two modules are cooperatively trained [4], the architecture is globally optimal. Note that the MLP is not trained to perform classification (its usual output layer has been stripped here), but chiefly to derive the optimal features for the RBF.

3- APPLICATION

We have used our multi-modular architecture to classify events generated from the LEP accelerator at CERN (Geneva). The physics of the events has been described by Denby [6]: electrons e^- and positrons e^+ are accelerated, at high energy (91.2Gev), in opposite directions in a ring and made to collide. Resulting from these collisions, various particles are emitted, namely quarks of different types (*flavours*). In the work reported here, we are interested in 3 such types: two heavy quarks (b: bottom and c: charm) and one class grouping all three light quarks (u: up, d: down and s: strange).

One common problem to use Neural Networks in High Energy Physics (HEP) is that experimental data can never be labelled directly: however, since very sophisticated simulators have been developed, one can use them to produce labelled data sets. Here we produced our data sets by a full simulation of the ALEPH detector, which has been demonstrated to produce events very similar to experimental ones.

The learning set was composed of 30 000 events, 10 000 per class; for each event we generated 150 variables. Careful analysis of the variables [9] allowed us to select the 20 most discriminant variables.

In real data, the three classes are not equally probable. The respective proportions have been estimated as: 21% b + 17 % c + 62 % uds. Training with a training set realistically balanced would lead to a somewhat poorer performance on the detection of the heavy quarks, which is our main concern. We thus trained with an equally balanced training set, but tested our classifiers on a realistically balanced test set, generated from the same ALEPH simulator: it has 73 376 events, 21% b (16086 events), 17 % c (12757) and 62 % uds (44533).

We compared the performances of Discriminant Analysis, MLP, MLP+LVQ (a full classifier MLP is trained, and then the LVQ is trained sequentially on the features extracted by the MLP) [9], MLP+RBF trained sequentially (as for the MLP+LVQ case) and our MLP+RBF architecture, cooperatively trained.

Our MLP had, in all 3 cases, 20 input units, a first hidden layer with 20 units, a second hidden layer with 8 units, and, in the case of MLP and MLP+LVQ, an output layer of 3 units. Both LVQ and RBF used 21 reference vectors. The weights of the MLP+RBF architecture were all initialized with small random numbers (that is in the same way as a MLP), except σ which was set at 0.75.

Training required 200 sweeps through the training set, which took about 2 hours on a 30-90.

Figure 2 shows the results on the test set, in terms of ratio of correctly classified events.

These results demonstrate that:

- neural networks improve the performances of the conventional Discriminant Analysis classifier.
- multi-modular architectures, both MLP+LVQ and MLP+RBF, are better than just the MLP alone.
- the cooperatively trained MLP+RBF architecture achieves the best performances: it improves upon the other classifiers, and in particular upon the MLP+RBF trained sequentially.

It is a matter of further research to analyse in details the results obtained by the various classifiers, in the context of the detection of the b-quark.

METHOD	TEST SET %
Discriminant Analysis	67.1
MLP	71.2
MLP + LVQ (sequential training)	72.8
MLP + RBF (sequential training)	73.3
MLP + RBF (cooperative training)	73.5

Figure 2: performances of various classifiers on the classification of b / c / uds quarks.

4- CONCLUSION

We have presented a cooperative neural network: after processing through a MLP, the classification is performed by a RBF network. We have indicated how to compute the error to be back-propagated through the two successive architectures, so as to cooperatively train them: the MLP computes the best features so that the task of the RBF is easiest. We have initialized the MLP-RBF architecture in the same way as a MLP.

On an application to a High Energy Physics experiment, we have found that the MLP-RBF architecture performed better than the MLP and other classifiers.

ACKNOWLEDGEMENTS

I would like to thank Professor F. Fogelman-Soulié who suggested to apply multi-modular networks to High Energy Physics problems. Several discussions on Neural Networks proved very stimulating. This paper is part of a work going on in the ALEPH collaboration: I would like to thank the physicists of the collaboration for their support.

5- REFERENCES

[1] BENGIO Y., DE MORI R.: Global optimization of a neural network-Hidden Markov Model hybrid. IEEE Trans. on Neural Networks, vol. 3, n° 2, pp252-259, (1992).

[2] BENNANI Y., GALLINARI P.: A modular connectionist architecture for text-independent talker identification. IJCNN'91, Seattle, vol. II, 857-860, (1991).

[3] BOLLIVIER M. de , GALLINARI P., THIRIA S.: Cooperation of neural nets and task decomposition. IJCNN'91, Seattle, vol. II, 573-576, (1991).

[4] BOTTOU L.Y., GALLINARI P.: A framework for the cooperation of learning algorithms. In "Neural Information Processing Systems", NIPS'90, R.P. Lippmann, J.E. Moody, D. S. Touretzky eds., Morgan Kaufmann, vol. 3, 781-788, (1991).

[5] BOURLARD H., MORGAN, N.: A continuous speech recognition system embedding MLP into HMM. NIPS'89, D. Touretsky ed., Morgan Kaufman, vol.2, 186-193 , (1990).

[6] DENBY B.: The use of neural networks in High Energy Physics. Neural Computation, (to appear).

[7] FOGELMAN SOULIE F., LAMY B., VIENNET E.: Multi-modular neural network architectures for pattern recognition: applications in optical character recognition and human face recognition. Int. Jl Pattern Recognition and Artificial Intelligence (to appear).

[8] LEE S.: In Neural Networks for Signal Processing, B. Kosko ed., Prentice Hall, Englewood Cliffs (1992).

[9] PRORIOL J., JOUSSET J., GUICHENEY C., FALVARD A., HENRARD P., PALLIN D., PERRET P.: Tagging b quark events in Aleph with neural networks. In Proceedings of the Elba workshop: Neural Networks: from Biology to High Energy Physics. O. Benhar, C. Bosio, P. del Giudice, E. Tabet, eds. 419-444, (1991).

[10] RUMELHART D.E., Mc CLELLAND J.L.: Parallel distributed processing: explorations in the microstructures of cognition, MIT Press, (1986).

Pattern recognition I
—poster contributions

OPERATIONAL CLOUD CLASSIFIER BASED ON
THE TOPOLOGICAL FEATURE MAP

Olli Simula, Ari Visa, and Kimmo Valkealahti
Helsinki University of Technology
Laboratory of Information and Computer Science
Rakentajanaukio 2 C, SF-02150 Espoo, Finland

ABSTRACT

Recently, an adaptive method to partition and interpret satellite images for cloud classification has been presented. In this application, texture measures are used as features and the segmentation and classification is based the self-organizing process. The cloud classification system has been in operational use since September 1991. In this paper, performance characteristics obtained so far are presented.

INTRODUCTION

The recognition and classification of clouds for better weather forecasting and cloud climatology research are important applications in meteorology. Automatic cloud classification has been studied since 1970s [1]. Several methods exist for cloud detection and classification from visible and infrared satellite measured radiances. Methods based partly on texture information have been introduced by Kittler [2]. Ebert [3] combined texture measures and a maximum likelihood decision rule to identify regions of various surface and cloud types at high latitudes. Multispectral threshold algorithms have been developed by Karlsson and Liljas [4], utilizing all five channels of the NOAA satellite and textural information from one infrared channel. The classification was suited to seasons of the year, to sun elevation, and to satellite. The classifier is in operational use but the number of thresholds needed in this method is very high, some thousands.

Recently some neural network based methods have been suggested [5, 6, 7]. Smottroff et.al. used artificial neural networks as trainable classifiers. They reported good classification results [5]. Visa et.al. used a self-organizing process to cluster features to a topological feature map. Because clouds form stochastic textures in satellite imagery the method proposed by Visa [8] is also applicable to cloud classification. The texture features are selected in such a way that principal part of classification is based on textures. Known samples were located on that map and the labelled map was used in interpretation of weather pictures. Welch et.al. [7] have tested several neural network methods on six AVHRR (Advanced Very High Resolution Radar) images. They promote the use of neural network as a classifier.

The purpose of this research project has been to develop an operational system that classifies images from NOAA weather satellites continuously from day to day and from year to year. In the present version, the images are classified into 22 classes. Four of them, i.e. land, open sea, ice/snow, and sunglints, are of limited interest in meteorology. Other recognized classes are pure cloud types (9 classes) or mixtures of clouds (9 classes). The pure cloud types are cirrus, cirrostratus, altostratus, altocumulus, stratus, stratocumulus, cumulus, cumulonimbus, nimbustratus. It is possible that there are up to three layers of clouds on each other at the same time. The system has been in operation since September 1991 and some performance results are given.

CLASSIFICATION TECHNIQUES

The NOAA satellite images consist of five channels, two visible and three infrared. For the first pure infrared channel, i.e. channel four, the texture information is extracted using the co-occurrence matrices. They are a kind of normalized histograms representing the probability that two gray level values occur at certain position to each other.

The presented classification method is based on a labelled topological feature map, called the texture map. First, the topological feature map is generated in an unsupervised way via the self-organizing process [9]. The interpretation of this feature map is then obtained by labelling the features using known samples. This results in forming the texture map [6]. However, only cloud pixels are classified by the texture map. A texture description is calculated for a cloud pixel from thermal information, i.e. channel four of the NOAA satellite. This description is fed into the map, which responses classification. The feature map is created by means of neural computation. Surface pixels are separated from clouds by comparing them with multispectral prototypes in the daytime, and by thresholding in the night.

The segmentation and classification method is illustrated in Figure 1. A window scans over the whole image from left to right and from top to bottom. For each pixel, the co-occurrence matrix is calculated for three different distances within the window centered at the pixel. The original 256x256 co-occurrence matrices calculated for one pixel are reduced to 16x16 matrices and concatenated to form a feature vector. This feature vector is then compared to the earlier created topological feature map. The comparison is done on the nearest neighbour basis. In this way, the vector element on the map mostly resembling the extracted feature vector is found resulting in image segmentation, shown in Figure 1 (a). Because the texture map is labelled, i.e. each element has an interpretation defining the physical meaning (low or high cloud etc.), the classification of each pixel can be done (Figure 1 (b)). This interpretation is then transferred to the corresponding image point resulting in classification of the segmented image. Thus, the closely related regions of the segmented image are combined to form cloud classes. In some cases, when the interpretation in the texture map is ambiguous extra measures from the other available channels, i.e. visible information, can be used.

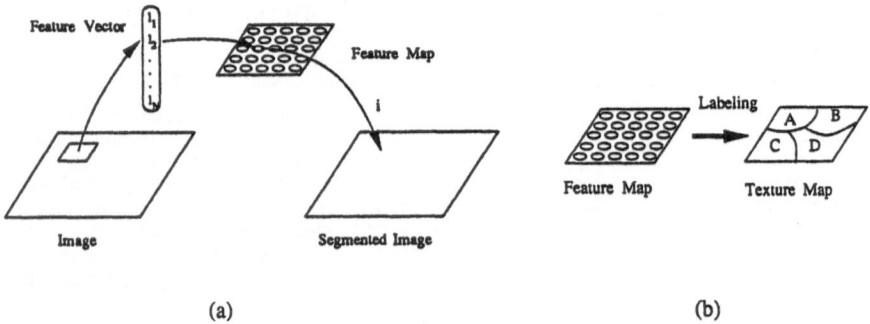

(a)   (b)

Figure 1. The segmentation method (a) and classification based on interpretation (b)

EXPERIMENTAL RESULTS

The verification of classification results is a difficult task. The main problem is the lack of suitable reference material. The reports from weather stations tell only the cloud type but neglect the shape and size information. Weather radars give us information about the shape and location of rain clouds. This information is useful but it is available only for raining clouds in certain parts of Finland. The classification rate of the system for raining clouds is about 95%.

The classified satellite images have also been compared with the existing Swedish system. The problem is that their system covers only a limited neighbourhood of Sweden. However, some comparison results have been obtained and presented in Figure 2. Situations on September 1, 1992, early in the morning (04:07) and at noon (12:21) are shown. The larger figures on the left are the classification results of our system. On these images the squares correspond to the location of the areas of the reference images shown on the right. One should note that the colour codes of the classified images are not the same.

In Figure 2 (a) it can be seen that some parts in central Norway and Sweden are clear (blue/gray on left versus blue/green on right). Some stratus and cumulus clouds are also seen (yellow versus yellow). Above the northern part of Sweden nimbustratus (magenta versus pink) and some stratus clouds or rain clouds (pink versus red) are seen. The classified images resemble each other quite well in this case.

1992-09-01-0407

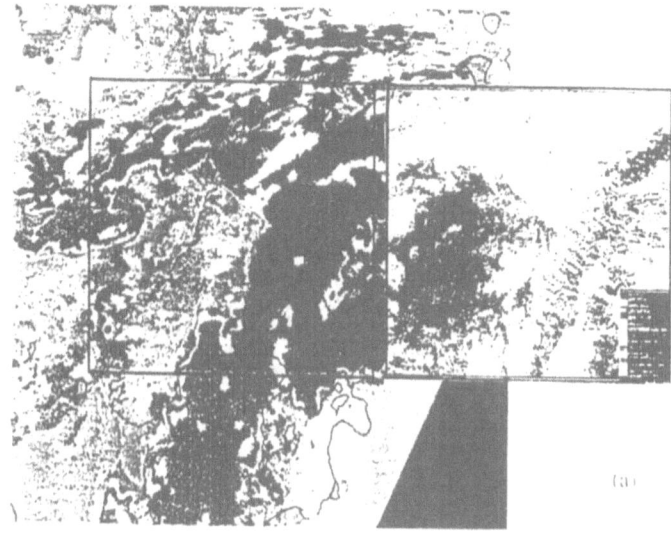

1992-09-01-1221

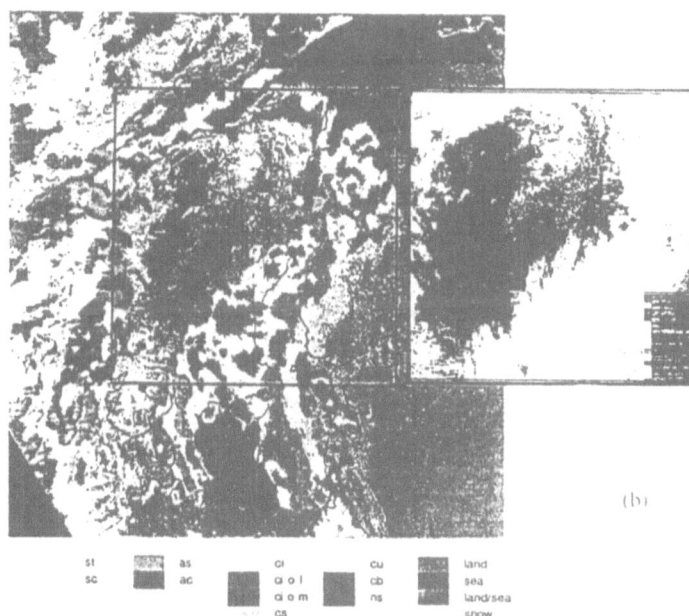

Figure 2. Classification results based on topological feature map (left) and classified reference images obtained from the Swedish Meteorological Institute (right); Northern and central parts of Sweden in the morning (04:07) (a) and at noon (12:21) (b) are shown

In Figure 2 (b) the situation in the same area, on the same day at noon, i.e. about eight hours later is shown. The unclouded area over Norway is bigger than in the morning (blue/gray versus blue/green). The raining clouds (magenta versus pink) are above Sweden moving east towards Finland. The similarity between these images is also quite good.

Based on the verification results obtained so far it is still too early to state anything about classification accuracy of small areas (about one pixel) or about the borders of clouds. The example images, however, depict that investigating the sequences of classified images the development and movements of cloud frontiers is clearly shown.

CONCLUSIONS

A segmentation and classification method for NOAA weather satellite images has been developed. The method is trainable by examples and hence it is adaptive to modified circumstances. The method has been tested on hundreds of images. The comparisons with other classification results have shown that the neural networks based classification performs well in real operational environment. The existing problem is to get more reliable reference results especially from difficult classification situations. The use of additional features, e.g. visible channel information, should also be investigated in order to improve classification accuracy.

The cloud classification method described above seems to be suitable for aerial photographs as well. Preliminary tests for monitoring the amount of trees in a forest surveying application have successfully been done. The method could be easily applied to other applications based on aerial photographs, too.

REFERENCES

[1] Anderson, R. K., Veltishchev, N.F., "The use of satellite pictures in weather analysis and forecasting, World Meteorological Organization", Technical Note No. 124, WMO-No. 333, 1973.
[2] Kittler, J., Pairman, D., "Contextual Pattern Recognition Applied to Cloud Detection and Identification", IEEE Transactions on geoscience and remote sensing, Vol. GE-23, November, 1985, pp. 855-863.
[3] Ebert, E. E., "A Pattern Recognition Technique for Distinguishing Surface and Cloud Types in the Polar Regions", Journal of Climate and Applied Meteorology, Vol. 26, pp.1412-1427, 1987.
[4] Karlsson, K., Liljas, E., "The SMHI Model for Cloud and Precipitation Analysis from Multispectral AVHRR Data", SMHI Promis Reports, Nr 10, August, 1990.
[5] Smottroff, I. G., Howells, T.P., Lehar, S., "Meteorological Classification of Satellite Imagery and ground Sensor Data Using Neural Network Data Fusion", Proc. of IJCNN90, San Diego, California, USA, June 17-21, Vol II , pp. 239-243, 1990
[6] Visa, A., Valkealahti, K., Simula, O., "Cloud Detection Based on Texture Segmentation by Neural Network Methods", Proc. of IEEE International Joint Conference on Neural Networks, Singapore, November 18-21, Vol. 2, pp. 1001-1006, 1991.
[7] Welch, R.M., Sengupta, S.K., Goroch, A.K., Rapindra, P., Rangaraj, N., Navar, M.S., "Polar Cloud and Surface Classification Using AVHRR Imagery: An Intercomparison of Methods", Journal of Applied Meteorology, Volume 31, May, pp. 405-419, 1992.
[8] Visa, A., "Texture Classification and Segmentation by Neural Network Methods", Doctoral Thesis, Helsinki University of Technology, Department of Information Technology, Espoo, Finland, 1990
[9] Kohonen, T., "The Self-Organizing Map", Proceedings of the IEEE, Vol. 78, No. 9, September 1990.

Image Segmentation Using a Self-Organising Logical Neural Network

G. Tambouratzis, D. Patel[1] and T. J. Stonham

Brunel University,
Electrical Engineering Department,
Uxbridge, UB8 3PH.
Middlesex, U.K..

[1] Royal Holloway College, University of London,
Department of Physics,
Egham, TW20 0EX.
Surrey,U.K..

Abstract

In this paper, we present a method for image segmentation via texture recognition and feature space clustering, using a logical neural network paradigm. The texture analysis strategy results in a transformation of the image into a Texture Co-occurrence Spectrum (TCS). The logical neural network clusters the TCS in feature space by operating in an unsupervised mode, which eliminates the need for external supervision during operation. The method as presented is applied to, and shown to be capable of segmenting natural texture images. The results are in good agreement with perceived ones.

1. Introduction.

The process of partitioning an image into its constituent parts is commonly referred to as segmentation. Image segmentation, whilst a complex task, is an essential low-level processing aspect in an image analysis hierarchy. The resulting fragmented image from this stage can be labelled on the basis of domain specific knowledge in the higher stages of the hierarchy. The better the results in the initial stage, the less effort is required in the refinement process. Image segmentation has been an area of considerable research [4]. The plethora of techniques have been partitioned into 3 principal approaches: (a) region-based methods, (b) edge-based methods and (c) feature-space clustering methods.

This paper is concerned with the segmentation of texture images via feature-space clustering. Texture provides essential structural information in regions of an image and is thus a valuable feature for segmentation. There exists a large number of texture analysis methods and a summary of the majority of these can be found in [9]. We use a statistical texture analysis technique which transforms the texture image data into a new Texture Co-occurrence Spectrum. A logical artificial neural network paradigm is used to assess the relation amongst the texture spectra by organising the spectra into groups or clusters, such that entities within a cluster are more closely associated to each other than to entities in other clusters. Visa [10], suggested a method based on an analogue self-organising neural network paradigm.

Unsupervised clustering groups the texture patterns without assuming any a-priori knowledge about the textures. The unsupervised approach may reveal clusters in the data which might otherwise remain unobserved. Also, in the most general case, there exists an infinite number of textures in the real-world and for segmentation of any input into labelled regions to be automatic, segmenting in an unsupervised mode is mandatory.

2. Texture Measure.

The texture analysis technique used operates on a novel transformation of the image data, calculating the frequency of occurrence of nth order RANK and strength-of-direction of grey level pixel values to generate a **Texture Co-occurrence Spectrum** (TCS). Four orientation masks at 0, 45, 135 and 180 degrees around each pixel are used to map n-points from the texture window. The window which scans across the image for feature extraction is overlapped to smooth out the transition from one textural region to another. These masks extract the textural information in terms of the occurrence of n-conjoint pixel values at varying spatial frequencies. For the image data analyzed in this research, only 4 of the immediate neighbours around the centre pixel of the mask are used. The 4-tuple extracted by the directional masks is sorted in descending order and this ordered state is referred to as the RANK of the intensity values. The occurrence of these RANKs across the texture window is recorded. We also introduce the "strength-of-direction" for the grey levels of each directional mask to be used in conjunction with the RANK statistics. The Variance is used to measure the dispersion of the n-tuple elements for each RANK. This gives an indication of the strength-of-direction of the grey level values in all directions around the pixel. The mask corresponding to the minimum dispersion is selected and the RANK for that n-tuple increased. The resulting 24x4 directional RANK-strength vector, which reveals the texture information is referred to as the Texture Co-occurrence Spectrum. A full description of the texture feature extraction process can be found in references [5,6]. The co-occurrence transformation maps the segmentation problem from the spatial domain to the feature space domain. In multi-dimensional feature space, the logical neural network is used to group these features, such that these groupings in feature space represent homogeneous regions on the image.

3. Self-Organising Network Description.

The logical neural network used to cluster the TCS is based on the discriminator network [1]. The discriminator consists of a number of logic functions, each of which samples a number of n pixels from the input image. Each function comprises 2^n memory locations which correspond to all possible combinations of the binary pixels sampled by the function. The memory locations have a capacity of a single bit, where a binary number reflecting the occurrence of the corresponding tuple combination is stored.

The basic discriminator structure has been modified to adapt it to unsupervised learning tasks [7]. The main change involves extending the memory locations of the discriminator functions so as to store a k-bit number (where k>1 and typically equal to 8). This number reflects the relative frequency of occurrence of each tuple combination and is constantly updated during training by applying an adaptation rule. This rule has been designed specifically for logic neural networks [7] and consists of shifting units of information towards the current input. For each function, the content of address a_1 which corresponds to the current input is increased by a given amount, while the content of address a_2 which is at a maximum Hamming distance from a_1 and has a non-zero content is decreased by the same amount.

The self-organising network consists of a number of discriminators, each of which can be used to represent a different pattern class. The discriminators are interconnected to form a one-dimensional (or higher-dimensional) structure as in Kohonen's self-organising maps [3]. During the learning phase, the interconnections are used to define a neighbourhood of discriminators, whose size is reduced as training progresses. All discriminators compete for each input, the winning discriminator together with its neighbours being adapted towards the input using the adaptation rule. The winning discriminator is the one which generates the highest response to the current input among all discriminators, provided that it is well matched to the input. This matching requirement is achieved by incorporating the **distribution constraint** in the selection process [8]. It ensures that the utilisation of system resources (i.e.

discriminators) is maximised and avoids the excessive clustering of patterns.

4. Segmentation Results.

The window which is scanned across the 256x256 resolution texture image generates 1444 texture spectra. The frequency of occurrence of each RANK-strength value is typically in the range 0-384. This range is clipped to a maximum of 96 before being input to the network. This clipping is justified as most of the information lies in the positions of the peaks in the vector rather than their relative heights, while very high peaks are extremely rare. Hence, the TCS presented to the self-organising network is a digitised 96x96 array.

The self-organising network consists of 8-tuple functions. The number of functions in each discriminator is sufficient to sample each pixel of the TCS once. Using all of the 1444 TCS, would make the self-organising task extremely time-consuming. Instead, only 16 training samples are selected in random for each texture class. The network initially classifies these and then by generalising in the space of the training samples segments the entire image. The patterns are presented in a fixed sequence for 100 iterations. The system comprises a relatively small number of discriminator nodes (less than 10 in the experiments presented in this paper). For such a small population, the notion of the neighbourhood becomes less important than in other applications and for a small number of distinct texture classes (2 or 3) the topological ordering is not possible. Hence, the size of the neighbourhood is permanently set to 1, the network operating effectively on a competitive learning principle.

The proposed unsupervised segmentation scheme is illustrated with the aid of natural image composites which are constructed from images of naturally occurring textures from Brodatz's album [2]. The textures in Brodatz's album have become a common benchmark in texture analysis research. The first natural texture composite shown in figure 1a is made up of two textures, both of Herringbone weave at different magnification. The purpose of this is to highlight the salient properties of operating in an unsupervised mode using the logical ANN paradigm. The neural network comprises 6 discriminators. The resulting segmentation is shown in figure 1b. Here the textures are segmented into subtextures each representing opposite orientations of the weave. By merging the discriminator labels, according to the classification of the training patterns, a segmentation of the image into 2 classes is obtained, as shown in figure 1c. The classification rate obtained is 96.16%. The second image composite used to illustrate the efficacy of the proposed method is a 3 texture class image, shown in figure 2a. A higher population of discriminators is used (8 discriminators). The resulting classification rate is 96.11% and the output image is shown in figure 2b.

5. Conclusions.

The proposed system has been shown to successfully segment the different textures. It is worth noting that due to using an unsupervised logical neural network, the system actually requires a minimum amount of external input. This is restricted to identifying the discriminator to which each training pattern is assigned in order to label the classification outputs. Also, as the system is based on a logical neural network, it is readily implementable and thus well-suited to real-time applications.

A major advantage of the proposed architecture is the ability to modify the system's sensitivity to textures in order to suit the segmentation task. This sensitivity is modified by simply raising or lowering the distribution constraint threshold. The only limitation of the method is that the output of the texture analysis method, viz. texture co-occurrence spectrum, is rather large and requires clipping to reduce computational overheads. Currently, methods for compressing this data, whilst preserving the discriminatory information are being evaluated.

906

References

[1]. Aleksander, I. & Morton, H. (1990) An Introduction To Neural Computing. Chapman and Hall, England.

[2]. Brodatz, P (1966) Textures: A Photographic Album, Dover, New York.

[3]. Kohonen, T. (1989) Self-Organisation and Associative Memory (3rd edition). Springer Verlag, Heidelberg.

[4]. Nevatia, R (1986) Image Segmentation. In Young, T.Y & Fu, K-S (eds.) Handbook of Pattern Recognition and Image Processing, pp. 215-232, Academic Press, New York.

[5]. Patel, D. & Stonham, T.J (1992a) Texture Image Classification and Segmentation using RANK-order Clustering. In Proc. of the 11th Int. Conf. on Pattern Recognition, The Hague, The Netherlands, Vol. 3, pp. 92-95.

[6]. Patel, D. & Stonham, T.J (1992b) Unsupervised/Supervised Texture Segmentation and its Application to real-world data. In SPIE Int. Conf. on Image Processing and Image Processing, Boston, U.S.A..

[7]. Tambouratzis, G. & Stonham, T.J. (1992a) Implementing Hard Self-Organisation Tasks Using Logical Neural Networks. In Aleksander, I., & Taylor, J. (eds.) Artificial Neural Networks-2, Vol. 1, pp. 643-646. North-Holland, Amsterdam.

[8]. Tambouratzis, G. & Stonham, T.J. (1992b) Evaluating the Topology-Preservation Capabilities of a Self-Organising Logical Neural Network. Pattern Recognition Letters (in print).

[9]. Van Gool, L, Dewaele, P & Oosterlink, A (1985) SURVEY - Texture Analysis Anno, 1983. Comp. Vis. Graph. Image Process., CVGIP-29, pp. 336-357.

[10]. Visa, A (1992) Unsupervised Image Segmentation based on a Self-Organising Feature Map and a Texture Measure. In Proc. of the 11th Int. Conf. on Pattern Recognition, The Hague, The Netherlands, Vol. 3, pp. 101-104.

 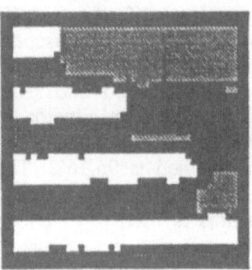 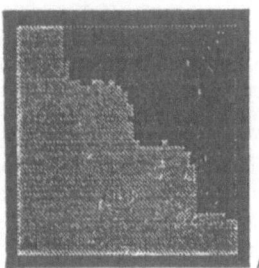

Figure 1: Segmentation of a two-texture composite.

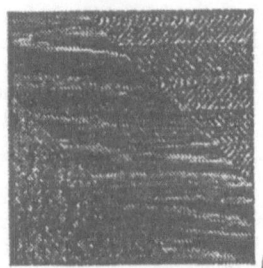 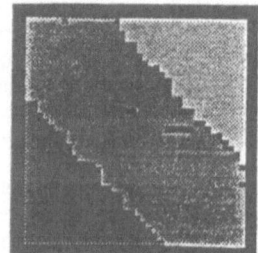

Figure 2: Segmentation of a three-texture composite.

High-resolution Classification of Papanicoloau Smear Cells
using Back-propagation Neural Networks

S J McKenna, A Y Cairns, I W Ricketts
The Microcentre, Department of Mathematics and Computer Science,
The University, Dundee DD1 4HN, Scotland

K A Hussein
Department of Pathology, Ninewells Hospital, Dundee, Scotland

ABSTRACT

Cervical cancer screening programmes based upon examination of the Papanicolaou smear are effective in reducing incidence of the disease. Millions of smears are generated by such programmes each year and must be analysed by trained cytologists. For nearly four decades the possibility of automating parts of the screening process has been investigated. In particular, the automatic removal of clearly healthy specimens from the cytologists' workload would enable them to concentrate their efforts on the diagnosis of suspicious smears. A dual-resolution image processing strategy has been adopted in the analysis of cervical smears. Once located by an initial, low-resolution search, suspicious objects of interest are analysed in detail at high-resolution.

This paper describes experiments in which neural networks were used to perform high-resolution classification of cellular objects as either normal or abnormal. Classification was based upon a set of features extracted from the frequency domain and originally derived for use with a parametric Bayesian classifier. Single-layer and multi-layer back-propagation networks were trained and tested on cells drawn from an expertly verified database of grey-scale images. Back-propagation networks improved upon the classification rates of the original Bayesian classifier. Preliminary experiments with a new image database indicated that increased spatial and grey-scale resolutions would result in even lower error rates.

INTRODUCTION

There is evidence that screening reduces the probability of a woman developing invasive cervical cancer by approximately 90% [1]. The U.K. screening programme produces approximately 4 million smears for analysis every year and one smear can contain as many as 200,000 cells. The possibility of automating the screening process has been investigated since the 1950's with the aims of removing the bulk of healthy specimens from laboratory workloads, enabling screening of a larger population and increasing accuracy. A review of previous developments in this area can be found in Banda-Gamboa *et al.* [2]. Descriptions of some of the more advanced systems under development for the analysis of cervical smears can be found elsewhere [3]. The complex, heterogeneous nature of cervical smear scenes and the resulting difficulties in segmenting clustered cells and discarding artifacts make image analysis of standard Papanicolaou smears a demanding task. The false-negative rate must be minimised whilst keeping the false-positive rate reasonably low. A dual-resolution strategy has been adopted in which suspicious objects are located by an initial low-resolution search. These objects are then analysed in detail at increased resolution and classified as either normal or suspect. Suspect objects are either passed to a further classification stage or subjected to human review. The number of abnormal cells found on cervical smears is relatively small and as a result the number of normal cells misclassified as suspect (false-positives) must be minimised in order to avoid swamping any genuinely abnormal cells. This paper describes experiments with a method which uses the frequency domain and neural network classifiers to perform high-resolution analysis of cellular objects.

METHOD

Initial experiments were conducted using 7-bit grey-scale images containing 256x256 pixels with a spatial resolution of 0.29 microns per pixel. Each image was of a single cell's nucleus along with part or all of its associated cytoplasm. In addition, artifacts and cytoplasmic material from other cells was also often present. Disjoint training and test sets were formed containing normal cells of varying maturity (superficial, intermediate and parabasal) and abnormal cells with varying degrees of abnormality (mildly, moderately and severely dyskaryotic).

In a preprocessing step, correction was made for shading effects due to uneven illumination. Images were then transformed to the frequency domain by application of the 2-dimensional discrete Fourier transform. A set of 80 features was then derived from each cell's Fourier spectrum using a ring-wedge arrangement. The feature extraction process has been described in detail elsewhere [4]. It is worth noting that features were obtained without recourse to high-resolution segmentation. Classification of cells as either normal or abnormal was performed using these 80 features.

The features were originally used with a parametric Bayes' classifier which calculated a linear discriminant function in order to minimise misclassifications under the assumptions that the features were drawn from a multivariate normal distribution and that both classes had identical covariance matrices [5]. The 80 features were subsequently used as input to feed-forward neural network classifiers with the aim of improving accuracy. Fully-connected single layer networks as well as multi-layer networks with one or two hidden layers were trained using error back-propagation [6]. These networks employed symmetric sigmoid activation functions and added 0.1 to the derivative of each unit's activation function in order to avoid derivatives of zero ('flat spots') [7]. Learning times were thus shortened substantially. Learning rate and momentum parameters were set to values of 0.003 and 0.9 respectively using the empirically derived rule of Eaton and Olivier [8]. These values compared favourably with others tested by the authors.

A second set of experiments formed a preliminary study into the use of images of higher quality. A total of 318 8-bit images were acquired at a spatial resolution of 0.15 microns per pixel (512x512 pixels). These images were of superficial, intermediate and abnormal (mildly, moderately and severely dyskaryotic) cells. In order to assess the relative merits of increased spatial and grey-scale resolutions, two further image sets were derived. These were formed by resampling to 256x256 pixels using pixel averaging and had 7-bit and 8-bit grey-scales. Test set classification rates obtained using the three image sets were compared using the Bayesian classifier. Neural network classifiers were also trained and tested on these images.

CLASSIFICATION RESULTS

Each of the networks was trained 10 times with different initial random weights. In the first set of experiments, the training set and test set each consisted of 702 feature vectors (351

Type of classifier	Epochs to train	Test set classification rates (%)
Parametric Bayes' classifier	-	91.3
Single-layer network (80-1)	1745 (54)	89.4 (0.1)
One hidden layer net. (80-6-1)	1965 (1051)	92.9 (0.3)
Two hidden layer net. (80-6-2-1)	2180 (1601)	92.7 (0.4)

Table 1. Comparison of Bayesian and back-propagation classifiers

		No. of test cases	Predicted class	
			Normal	Abnormal
Normal cells:	superficial	150	150	0
	intermediate	150	150	0
	parabasal	51	46	5
Abnormal cells:	mildly dyskaryotic	117	2	115
	moderatly dyskaryotic	117	9	108
	severely dyskaryotic	117	16	101

Table 2. Classification results for 80-6-1 network with output threshold adjusted

normal and 351 abnormal cells). After experimentation with various back-propagation networks, a single hidden layer of 6 units was found to give the best test-set performance. In particular, the addition of a second hidden layer was not of benefit. Table 1 shows the percentage of test set images correctly classified by the Bayesian classifier and the back-propagation networks with zero, one and two hidden layers. Standard deviations are given where appropriate and appear in parentheses.

The back-propagation networks with hidden layers were always more accurate than the Bayesian classifier. The most successful of the single-hidden layer networks trained obtained a test set classification rate of 93.4%. The misclassified cells consisted of 25 false-positives and 21 false-negatives. Sixteen of the 25 false-positive cases were parabasal cells. The specificity of the classifier was 92.9% and the sensitivity was 94.0%.

In order to reduce the false-positive rate, the output unit's threshold was adjusted to bias classification in favour of normal cells. Varying the threshold value allowed the response operating characteristic to be plotted. In particular, lowering the threshold from 0.0 to -0.2 resulted in the classification results shown in table 2. The specificity increased to 98.6% and the sensitivity decreased slightly to 92.6%. The only false-positive cases were parabasal cells.

Table 3 shows the classification results obtained using the new improved image database and the Bayesian classifier. Training and test sets comprised 70% and 30% of the 318 images respectively. A single-layer neural network was also able to classify the 8-bit 512x512 pixel images with perfect accuracy.

Spatial resolution	Grey-scale resolution	Classification rate (%)
256x256	7-bits	95
256x256	7-bits	99
512x512	8-bits	100

Table 3. The effect of increased spatial and grey-level resolutions (Bayesian classifier)

DISCUSSION

A back-propagation classifier was able to consistently outperform the parametric Bayesian classifier using the original image database. It should be noted that the set of 80 features used was arrived at after much experimentation with the Bayesian classifier. The features have been specially tailored to suit this particular classifier. Adjustment of the best network's output threshold resulted in only 5 false-positives all of which were parabasal cells. Parabasal cells are very rare and not critical to the screening process. In conjunction with this low false-positive rate, a large percentage (92.6%) of abnormal cells were detected. These results are very promising.

Both increased spatial and grey-scale resolutions resulted in improved classification accuracy. The new higher resolution image database is currently being expanded. Future work will concentrate on high-resolution analysis using neural networks and larger image sets. In addition, the possibility of using neural networks in the initial low-resolution search for suspicious objects will be investigated.

REFERENCES

1. Eddy D. M., 1990, Annals of internal medicine, 113, 214-225

2. Banda-Gamboa H., Ricketts I. W., Cairns A. Y., Hussein K. A., Husain O. A. N., 1992, Analyt Cell Path, 4, 25-48

3. Data on automated cytology systems as submitted by their developers 1991, Analyt Quant Cytol Histol, 13, 300-306

4. Banda-Gamboa H., 1990, Classification of cervical cells using computer vision and the frequency domain, Ph.D. Thesis, University of Dundee, Scotland

5. Duda R. O. and Hart P. E., 1973, Pattern classification and scene analysis, John Wiley & Sons

6. Rumelhart D. E., Hinton G. E., Williams R. J., 1986, Learning internal representation by error propagation, in: Parallel distributed processing vol. 1, Rumelhart D. E. and McClelland J. L. (eds.), M.I.T. Press, Cambridge, MA.

7. Fahlman S. E., 1988, Faster learning variations on back-propagation: an empirical study, Proc. 1988 connectionist models summer school, 38-51

8. Eaton H. A. C. and Olivier T. L., 1992, Neural networks, 5, 283-288

Artificial Neural Networks Detect Subtle Differences Between Anesthetics

Richard Watt, Gene Maslana, Mohammad Navabi, Steve Hill,
Dave Boujak, Amy Gale, and Ken Mylrea

Advanced Biotechnology Laboratory, Departments of Anesthesiology and Electrical and Computer Engineering, University of Arizona, Tucson, Arizona.

ABSTRACT

Artificial Neural Networks (ANN) have proven useful in a wide variety of pattern recognition tasks in anesthesia monitoring research. However, routine anesthesia monitoring during surgical procedures does not include sophisticated data analysis such as spectral content of electroencephalograms (EEG) or heart rate variability (HRV).

Sevoflurane is a new halogenated ether anesthetic, an isomer of the commonly used anesthetic, enflurane. In this study, ANN were used to analyze and compare dose-dependent EEG and HRV changes during sevoflurane and enflurane anesthesia. Conventional cardiovascular and neurologic variables showed no statistically significant difference between sevoflurane and enflurane. However, ANN trained on EEG spectral signatures were able to correctly classify 75% of the EEG samples as either sevoflurane or enflurane. ANN trained on heart rate variability spectral signatures were able to correctly classify HRV epochs as belonging to sevoflurane or enflurane 86% of the time. In this study, ANN's have been successfully applied to recognize subtle differences between EEG and HRV data patterns observed with sevoflurane and enflurane.

INTRODUCTION

Although the brain is the target organ of general anesthesia, brainwave activity (EEG) is not routinely monitored. Changes in the complex EEG waveform during anesthesia are difficult to interpret. The electrically noisy operating room environment makes acquisition of clean EEG data (10-100 microvolts) problematic. The multiplicity of confounding variables (variations in anesthetic depth, changes in cerebral profusion, temperature changes, etc.) results in ambiguous effects. Finally, inter-drug and inter-patient differences make generalization difficult. In recent studies, ANN have performed well in categorizing electroencephalographic data according to anesthetic depth [1], [2], [3], as well as enhancing other anesthesia monitoring modalities [4].

Routine cardiovascular monitoring during anesthesia usually includes visual observation of the electrocardiogram (ECG) and blood pressure (BP) waveforms, as well as tracking derived variables such as average heart rate (HR), systolic, mean and diastolic blood pressure (SBP, MBP, DBP). For anesthesia operating room monitors, HR is averaged over a window of 10 or more beats, so that beat-to-beat variability is obscured. However, analysis of beat-to-beat heart rate variability (HRV) has been shown to be a useful clinical research tool for diseases such as diabetes mellitus where impairment of the autonomic nervous system (ANS) exists [5]. A recent study suggests that HRV may provide useful clinical information on anesthetic depth, reflecting ANS mediation [6].

Sevoflurane is a new halogenated ether anesthetic, an isomer of the commonly used anesthetic, enflurane. Its low blood-gas coefficient of 0.6 and pleasant smell make it more desirable for use as an induction agent. In this study, ANN were used to analyze and compare dose-dependent EEG and HRV changes during sevoflurane and enflurane anesthesia.

METHODS

With human subjects committee approval, 14 healthy male volunteer subjects were randomly selected to receive either sevoflurane or enflurane general anesthesia (7 each) at a level of 1 to 1.2 minimum alveolar concentration (MAC) for a duration of 2.5 to 3 hours. In addition to routine cardiovascular monitoring, an additional set of ECG electrodes were connected to a modified Hewlett-Packard 7830A ECG monitor which triggered a digital input port of an IBM PC. At each R-wave, R-to-R intervals of heart rate were calculated and stored to record HRV. A moving window interpolation algorithm was employed to calculate R-to-R intervals at equal periods of .25 seconds, yielding data sets of 4096 points. Using a fast Fourier transform algorithm, the power spectrum for each set of heart rate values was determined.

Two channels of EEG waveform (left and right cerebral hemisphere) were recorded on FM tape. For each subject, ten 16-second epochs of EEG (chosen by visual inspection to be artifact-free) were digitized at 256 Hertz for FFT analysis to derive spectral signatures.

Each spectral signature for HRV and EEG was analyzed by a three-layer back propagation network (64 input nodes, 8 hidden nodes, 2 output nodes), which was trained using all but the spectral signature under test to attempt differentiation between sevoflurane and enflurane. Each ANN was trained using the generalized delta rule, with 15 different learning procedures consisting of various combinations of learning constant and momentum values. Conventional derived cardiovascular variables (HR, BP) and EEG derived variables (median frequency and RMS amplitude) were analyzed with descriptive statistics.

RESULTS

There was no statistically significant difference between sevoflurane and enflurane with respect to conventional cardiovascular variables (HR, SBP, MBP, and DBP). Artificial neural networks trained on heart rate variability spectral signatures were able to correctly classify HRV epochs as belonging to sevoflurane or enflurane 86% of the time. Further examination of HRV spectral signatures reveals that power in the low frequency band (0.02-0.06 Hz) of the sevoflurane subjects is less than that of the enflurane subjects.

There was no statistically significant difference between sevoflurane and enflurane using conventional EEG-derived variables (median frequency, RMS amplitude). Artificial neural networks trained with EEG spectral signatures were able to correctly classify 75% of the EEG samples as either sevoflurane or enflurane. Further examination of the EEG data reveals that it is possible to derive frequency based variables which show a statistically significant difference between sevoflurane and enflurane (Beta/Theta band power).

DISCUSSION

The results of this study indicate a difference in ANS mediation of heart rate between sevoflurane and enflurane as shown by ANN comparison. ANN analysis of EEG spectral signatures indicated a subtle difference between sevoflurane and enflurane. These results may suggest dissimilar mechanisms in sevoflurane and enflurane action. Our results show that ANNs may be used to augment conventional statistical analysis when searching for subtle differences among grouped variables. This exemplifies an application of ANNs in clinical research which capitalizes on their pattern recognition sensitivity. ANNs offer further advantages over classical statistical approaches such as discriminant analysis, because new data can be accommodated by re-training. Clinical research often suffers from inadequate patient databases. Although vast amounts of physiologic data is monitored on millions of patients every year during surgical/anesthetic procedures, such data is rarely collected and stored in a computer database suitable for analytical research. If ANN were implemented on operating room physiologic monitors, retraining could occur over years of data exposure which could provide an alternative to prohibitively large databases, thus enhancing clinical research. Furthermore, ANN may offer superior performance due to greater suitability for classifying non-linear processes [7].

REFERENCES

[1] Watt RC, Saumelson H, Navabi MJ & Mylrea K, Pattern classification of EEG spectral signatures during anesthesia. *Proc Ann Int Conf IEEE EMBS* 13:433-434, 1991.

[2] Veselis RA, Reinsel R, Sommer S & Carlon G, Use of neural network analysis to classify electroencephalographic patterns against depth of midazolam sedation in intensive care unit patients. *J Clin Mon* 7(3):259-267, 1991.

[3] Sharma A, Wilson SE & Roy RJ, EEG classification for estimating anesthetic depth during halothane anesthesia. *Proc 14th Ann Int Conf IEEE EMBS* 6:2409-2410, 1992.

[4] Orr JA, Kuck K, Farrell RM & Westenskow DR, Neural network breathing circuit alarms in an anesthesia workstation. *1991 Annual Meeting of the American Society of Anesthesiologists Abstracts* 75(3A):A1005, 1991.

[5] Lishner M, Akselrod S, Mor Avi V, Oz O, Divon M & Ravid M, Spectral analysis of heart rate fluctuations. A non-invasive, sensitive method for early diagnosis of autonomic neuropathy in diabetes mellitus. *Journal of the Autonomic Nervous System* 19:119-125, 1987.

[6] Kato M, Komatsu T, Kimura T, Sugiyama F, Nakashima K & Shimada Y, Spectral analysis of heart rate variability during isoflurane anesthesia. *Anesthesiology* 77:669-674, 1992.

[7] Gallinari P, Thiria S, Badran F & Fogelman-Soulie F, On the relation between discriminant analysis and multilayer perceptrons. *Neural Networks* 4:349-369, 1991.

Pattern Segmentation and Feature Linking
as Simultaneous Processes
in an Associative Network of Spiking Neurons

Raphael Ritz and J. Leo van Hemmen

Physik–Department der TU München, D–85748 Garching bei München

Abstract: Feature linking and segmentation of four stationary patterns are shown to be performed as *simultaneous* processes by a fully connected, auto–associative neural network of *spiking* neurons. The patterns have been learned through an asymmetric, Hebbian rule that can handle a varying low activity. In this case the total activity of the patterns ranges between 4 and 7%. The underlying model is the 'spike response model'. Spiking is achieved by an absolute refractory period ($1ms$) while an inhibitory delay loop prevents continuous firing. Reaching the synapse after some axonal delay each spike evokes an excitatory or inhibitory postsynaptic potential (EPSP or IPSP) with a realistic response at the receiving neuron. Each neuron sums up its input signals linearly and acts as a noisy threshold element for generating a new spike.

1 Introduction

Feature linking and pattern segmentation have been known since long as notorious problems in the field of pattern recognition. As to the visual system, a theoretical Ansatz to solve this kind of problem is the idea that *temporal* coincidence (and disjointness) of neural activity can code relevant information. In recent time some experimental evidence [1, 2] has supported this idea – at least for the feature linking. Here we propose a biologically motivated model [3], the spike response model, and show how it is able to simultaneously perform both associative reconstruction (feature linking) and pattern segmentation through *spiking* neurons.

2· Network

We discretize time ($\Delta t = 1ms$) and consider a fully connected network of $\{0,1\}$ Ising spins were the state $S_i = +1$ denotes a single spike. The basic time step is taken to be the duration of a single spike ($1ms$). The network elements are updated in parallel through a noisy threshold dynamics (*Glauber dynamics*).

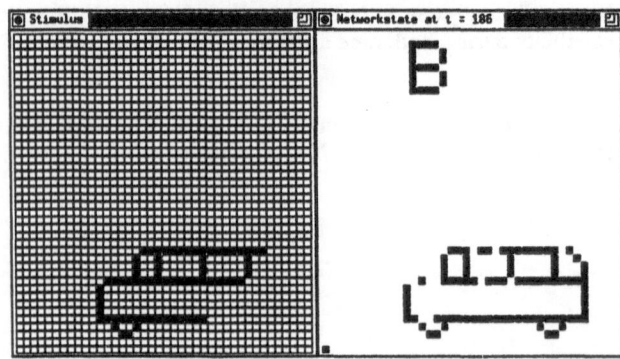

Figure 1:

Feature linking: Reconstruction of the bus (left: stimulus, right: momentary retrieval state) which has been 'learned' before. Associative retrieval and 'labeling' of learned patterns are performed by a network of 40x42 *spiking* neurons.

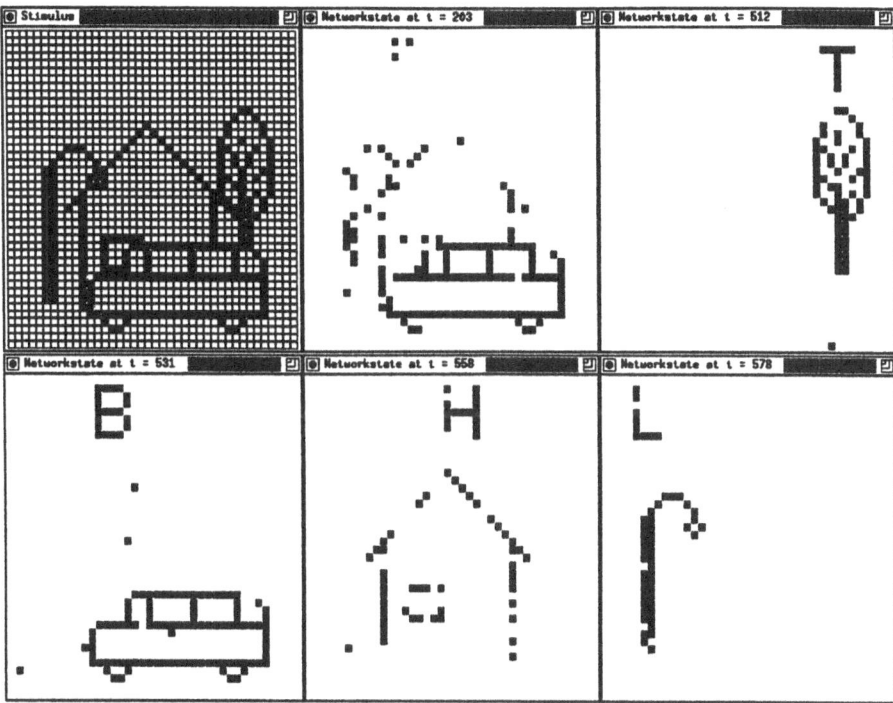

Figure 2: Pattern segmentation: Four patterns which have been taught to the network as separate entities are now presented in a superposition (top left). After an intermediate phase the network nicely completes and separates the original patterns (T-B-H-L) *through a coherent spiking* which is specific to each pattern.

The local field at neuron i is the sum of an external signal and the inputs from other neurons. It also includes an absolute refractory period so as to guarantee that a series of distinct spikes will be generated. That is,

$$h_i(t) = h_i^{syn}(t) + h_i^{ext}(t) + h_i^{ref}(t), \tag{1}$$

$$h_i^{syn}(t) = \sum_{j=1}^{N} J_{ij} \sum_{\tau=0}^{\infty} \epsilon(\tau) S_j(t{-}\tau{-}\Delta_i^{ax}) + \sum_{\tau=0}^{\tau_{max}} \eta(\tau) S_i(t{-}\tau{-}\Delta_i^{inh}), \tag{2}$$

$$h_i^{ref}(t) = -\infty \quad \text{for } t_F \le t \le t_F + \tau_{ref}, \text{ and } 0 \text{ otherwise.} \tag{3}$$

The first term in $h_i^{syn}(t)$ stems from the interaction with all other neurons of the network while the second one mimics a local inhibition to prevent continuous firing of a persistently excited neuron. The synaptic efficacies J_{ij} are chosen according to an asymmetric, Hebbian rule [3]. The axonal delays Δ^{ax} are equally distributed between 2 and 4ms and the inhibitory delays vary between $3 \le \Delta^{inh} \le 6ms$. To simulate a realistic time course of the incoming excitatory and inhibitory postsynaptic potentials, we have taken the memory kernel $\epsilon(t) = t/\tau_\epsilon^2 \exp(-t/\tau_\epsilon)$ and $\eta \propto \exp(-t/\tau_\eta)$. For a more detailed discussion of the model see [3].

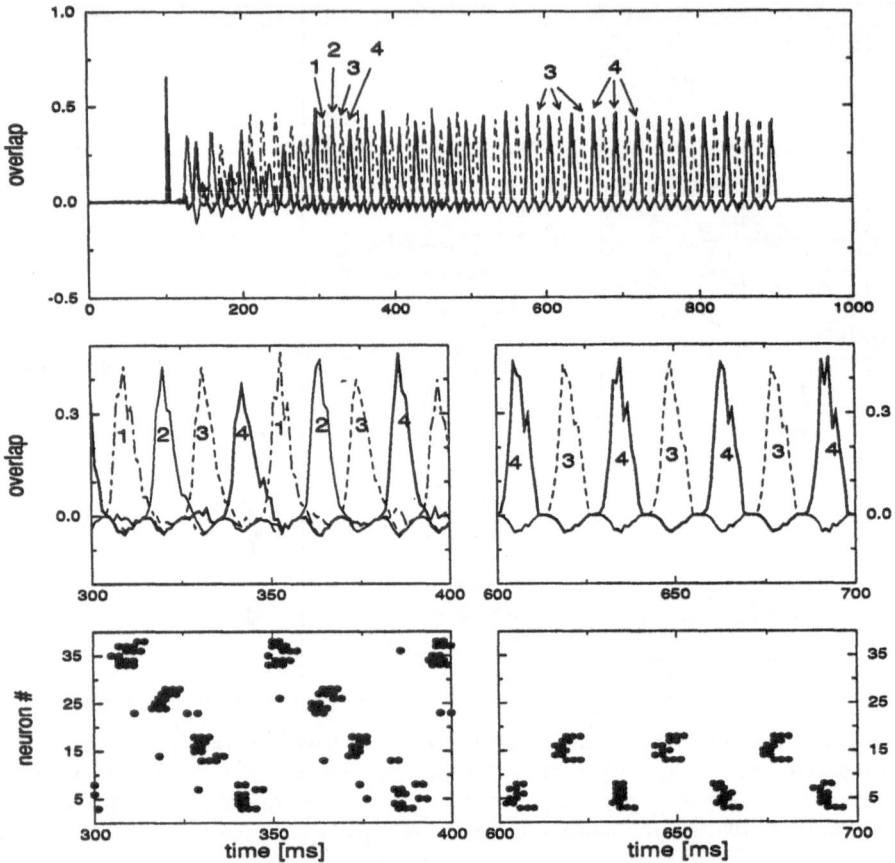

Figure 3: Overlap and spike raster (bottom) during a segmentation run. The network reacts immediately once patterns 1 and 2 have been removed at $t = 500ms$.

3 Simulation results

First of all, it is worth noticing that the associative capabilities of the Hopfield model can be found in this network of *spiking* neurons too. To show this we have simulated a network of 1680 neurons which has learned several (7) patterns. The neurons are represented by pixels to allow an easy interpretation to the human reader. This does not mean that the network has been arranged according to some two–dimensional topology. Neurons which have fired at least one spike during the last five time steps are coded by black pixels while the rest is white. Figure 1 shows on the left–hand side the stimulus which is switched on and kept constant thereafter. On the right–hand side one sees how the reconstruction is done and how even a sort of hetero–association can be accomplished by assigning a 'label', here a B, to a retrieved pattern. Note that this kind of pattern retrieval is time–dependent because, due to inhibition, the neurons are not able to fire all the time.

On the other hand, the intermittent firing opens the possibility to excite several patterns one after the other. This is exactly what is done when a superposition of *several* patterns is presented to the network. Figure 2 first shows the stimulus, then the network 100 time steps after the onset of the stimulus and finally the system as it has evolved to a locked oscillatory behavior where the four simultaneously presented patterns are active one after another. The frequency of this oscillation mainly depends on the strength and duration of the local inhibition. Note that parts of the patterns hidden in the stimulus are reconstructed. Due to noise, locking is not perfect in that some neurons drop out of the picture as they come a little to early or to late. It is fair to say, though, that the network can perform the segmentation of the presented scene.

The last picture (Fig. 3) shows the time structure of such a segmentation run in more detail. A macroscopic measure called overlap which compares the present network state with the learned patterns is plotted as well as a spike raster of some representative neurons from each of the stimulated patterns. All microscopic parameters are now in a biologically reasonable range. Here we see that the maximum number of patterns that can be distinguished at the same time is restricted to *four*. In principle this number is arbitrary and depends only on the strength and duration of the local inhibition.

In addition, the network is able to change its global behavior on a fast time scale (ms range) as can be seen in the upper part of Fig. 3. At $t = 500$ two of the four presented patterns are removed from the stimulus and the corresponding overlaps immediately stop oscillating. This can be observed once again at $t = 900$ when the stimulus is removed completely.

4 Discussion

Incorporating a *spiking* model neuron and a distribution of transmission delays into an associative neural network opens up the way to realizing both feature linking and pattern segmentation in the very same network. The number of patterns that can be segregated is in principle arbitrary and bounded only by the inverse activity of the patterns and strength as well as duration of the local inhibition. A biologically realistic set of parameters leads to the result that *up to four* patterns can be separated in a biological network.

References

[1] R. Eckhorn, R. Bauer, W. Jordan, M. Brosch, W. Kruse, M. Munk, and H. J. Reitboeck; *Coherent oscillations: A mechanism of feature linking in the visual cortex?* Biol. Cybern. **60**, 121-130 (1988)

[2] C. M. Gray, P. König, A. K. Engel, and W. Singer; *Oscillatory responses in cat visual cortex exhibit inter-columnar synchronization which reflects global stimulus properties*. Nature **338**, 334-337 (1989)

[3] W. Gerstner, R. Ritz, and J. L. van Hemmen; *A biologically motivated and analytically soluble model of collective oscillations in the cortex: I. Theory of weak locking*. Biol. Cybern. **68**, 363-374 (1993)

Spatial Topology Distance for Handprinted Character Recognition

Cheng-Yuan Liou and Hsin-Chang Yang
Department of Computer Science and Information Engineering
National Taiwan University

Correspondence address : Cheng-Yuan Liou, Department of Computer Science and Information Engineering, National Taiwan University, Taipei, Taiwan, 10764, R.O.C. Tel. 8862-3630231 ext. 3229

Keyword : neural network, self-organization, handprinted character recognition, perception, image recognition, distance measurement

Abstract

In this work we introduce new distance measurements for calculating spatial topology dissimilarity and apply them in recognition of handprinted characters. In many applications (e.g. recognition of deformed images, handprinted characters, signatures, biomedical and geophysical patterns), it is of interest to analyze and recognize phenomena occuring at varying deformations. The recently introduced neural network models provide a learning-by-training and collective recognition capability that offer the possibility of such analysis and recognition. At present, however, there is no corresponding effective framework to support the development of effective neural network for rigorous real applications, such as handprinted character recognition for large set of characters. In this paper we provide such a framework. The devision of analog topology preserving map leads naturally to order the deformed patterns to their standard forms. Vector fields can be obtained from such ordering. The cumulation of each individual vector in the vector field is then properly defined and used as the perception energy. Several empirical perception energies are introduced for recognition of handprinted characters. An automatic recognition of mail address system which contains isolated Chinese characters, ten digits, and all English alphabets is developed based on such framework. A rigorous test is performed on the ten handprinted digits using their ten standard forms as reference database. More than eighty percent of the test database which contains 5000 different deformed handprinted digits can be correctly recognized. To our knowledge, this reference database is the smallest database for the same performance.

I Introduction

The many various distance mesaurements (e.g. correlation, Hamming distance, ...) [1] devised so far can not be used successfully in handprinted character recognition. The investigation of topological representations of handprinted characters and the development of perception energies or spatial topological distances (STDs) have been and remain a topic of much interest in many contexts [2,3]. In some cases, such as in the use of elastic net models (elastic matching) for handprinted character and images recognitions [2,4,5] the motivation has directly been the fact that the phenomenon of interest exhibits patterns of importance at spatial topology similarity. A second motivation has been the possibility of developing accurate distance measurements of the perception energies (or inverse distances) based on such topology representations. Nine such perception energies are obtained empirically in this paper. A third motivation stems from the so-called "measurement fusion" problem in which one is interested in finding an emergent character by combining together all nine distance measurements.

One of the more recent areas of investigation in handprinted character recognition has been the development of a distance measurement for the abstracted key features of the handprinted character. This method has draw considerable attention in several disciplines including image recognitions, pattern recognitions, and robot visions because it provide a normal way to perform a critical decomposition of characters [3]. Essential all methods developed recently, including the elastic net involve overcoming nonlinear deformations. We will not pursue a thoroughly review of the handprinted character recognition subject in this paper. Interest readers may look into the review paper [3]. A full detailed recognition method is in the next and third sections.

II Spatial Topology Representations

We will depict the full detailed self-organization method plus the introduced 12 perception energies for the self-organization map in this section. We will use an example to illustrate the whole idea to save the contents. The well thinned skeletons of the standard 5401 chinese Kai-font characters

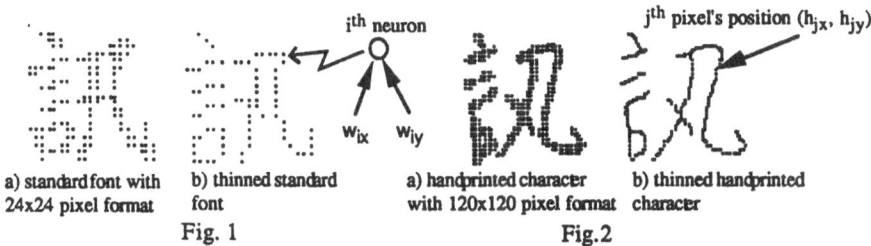

a) standard font with   b) thinned standard   a) handprinted character   b) thinned handprinted
24x24 pixel format   font   with 120x120 pixel format   character

Fig. 1   Fig.2

compose the standard skeleton set where each standard character has a square 24×24 pixel format. The appearance of the Kai-font character is very close to the handprinted character. The thinned skeletons can be obtained by the method in [6]. An example is shown in Fig. 1. And then they are improved by user program. Each standard character's skeleton is used as the object support and will be used as the support for the self-organizing map. The neurons are spread on the pixels of the thinned skeleton. Each neuron has two synapse weights. The geometry locations of the neurons are the same as the coordinates of the pixels on the skeleton. The size of the skeleton is normalized to an 1×1 square. The initial strengths for the two weights of a neuron are assigned to its X-Y coordinates on the normalized square with origin on the center of the square. The neurons plus their synapses compose the support for the self-organizing map for a standard character. Let $\omega_{ix}^k(0)$ denote the horizontal coordinate and $\omega_{iy}^k(0)$ denote the vertical coordinate of the ith neuron, $1 \leq i \leq N^k$, where k denote the kth character in the chinese character set, $1 \leq k \leq 5401$. N^k is the total number of pixels of the kth standard character's skeleton. The above process completes the preparation of the standard object support set. The coordinates, $\{(h_{jx}, h_{jy}), j = 1 \sim M\}$ of the pixels on the properly thinned and normalized skeleton of a handprinted character are used as the M training templates or patterns where h_{jx} and h_{jy} denote the horizontal and vertical coordinates of the jth pixel on the skeleton of the handprinted character, $1 \leq j \leq M$. M is the total number of pixels of the handprinted character's skeleton. A handprinted character caseis shown in Fig. 2. We use 120×120 pixel format for recording handprinted inputs in all our simulations. With the learning algorithm in [1] and the initial strengths, the kth object support's weights $(\omega_{ix}^k(t), \omega_{iy}^k(t))$ are trained using the template set $[(h_{jx}, h_{jy}), j = 1 \sim M]$. The training parameters and algorithm are the same as in [1]. In the algorithm the topological neighborhood $N_c(t)$ is linearly decayed from the two third of full size to a single pixel. The gain sequence $\alpha(t)$ is also linearly decayed from 0.4 to zero. The number of training cycles is properly scaled by experiments for efficiency. The number of cycles is fixed as 50, $t = 1 \sim 50$, for all training procedures to save computations with tolerable convergence accuracy in all simulations in this paper. A training result is shown in Fig. 3(a).

Then we define the topology distance, E_1, as the mean square distance between the kth object character support and the handprinted character support

$$E_1^k = \frac{1}{N^k} \sum_{i=1}^{N^k} D_i^k$$

$D_i^k = \left(\omega_{ix}^k(50) - \omega_{ix}^k(0)\right)^2 + \left(\omega_{iy}^k(50) - \omega_{iy}^k(0)\right)^2$ is the ith distance between a pixel in the object support and its corresponding location after self-organization using the handprinted character support. Fig. 3(b) shows the vector field of all locations of $(D_i^k)^{\frac{1}{2}}$. The average of the smallest 1/3 number of the N^k distances $\{D_i^k, i = 1 \sim N^k\}$ is defined as $E_2^k = \frac{1}{N^k/3} \sum^{SUM} \{D_i^k | D_i^k$ belong to the smallest one third number of all $D_i^k\}$. Accordingly, the average of the smallest 10% number of the N^k distances, $\{D_i^k | D_i^k$ belong to the smallest 10% number of $D_i^k\}$, is defined as E_3^k.

After self-organization the topology distances (E_1^k, E_2^k, E_3^k) can be further reduced by partial matching the disconnected parts of the object support separately. This partial matching is accomplished by shifting, scaling, and light rotating the disconnected parts to their corresponding parts on the handprinted character support. We will not use rotation in this partial matching. Because it cost computations. The corresponding part of each disconnected part is first identified through neuron

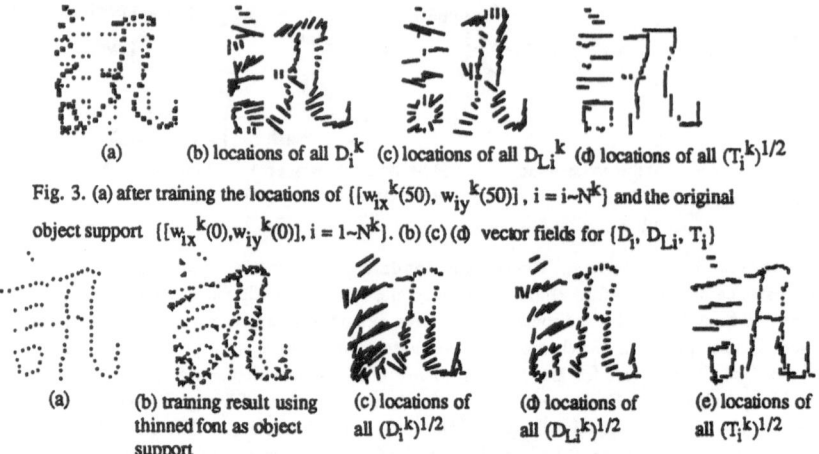

(a)   (b) locations of all D_i^k (c) locations of all D_{Li}^k (d) locations of all $(T_i^k)^{1/2}$

Fig. 3. (a) after training the locations of $\{[w_{ix}^k(50), w_{iy}^k(50)], i = i{\sim}N^k\}$ and the original object support $\{[w_{ix}^k(0), w_{iy}^k(0)], i = 1{\sim}N^k\}$. (b) (c) (d) vector fields for $\{D_i, D_{Li}, T_i\}$

(a)   (b) training result using thinned font as object support   (c) locations of all $(D_i^k)^{1/2}$ (d) locations of all $(D_{Li}^k)^{1/2}$ (e) locations of all $(T_i^k)^{1/2}$

Fig. 4 (a) An improved object support obtained by using the Voronoi thinning method on the 300x300 dpi standard font

excitations. The distances of a whole disconnected part can be reduced by scaling and centralizing of this disconnected part support to its correspondent. After this improvement new distances, D_{Li}^k can be obtained. The E_4^k, E_5^k and E_6^k follow the same definitions as for E_1^k, E_2^k and E_3^k respectively by replacing D_i^k with D_{Li}^k. Fig. 3(c) shows the vector field of $(D_{Li}^k)^{\frac{1}{2}}$.

E_7^k, E_8^k and E_9^k emphasize the neighborhood topology. The average of neighborhood topology E_{10}^k is obtained by averaging $\{T_i^k, i = 1 \sim N^k\}$, where $T_i^k = |\omega_{rx}^k(0) - \omega_{ix}^k(0)|^2 + |\omega_{ry}^k(0) - \omega_{iy}^k(0)|^2$ where the rth neuron is the representative of the template closest to the location $(\omega_{ix}^k(50), \omega_{iy}^k(50))$ on the handprinted support. The rth and i^{th} neurons are different neurons and r^{th} neuron is the representative of the closest neighbor of the position $(\omega_{ix}^k(50), \omega_{iy}^k(50))$. Fig. 3(d) shows the vector field of $(T_i^k)^{\frac{1}{2}}$. E_8^k, E_9^k is defined consequently as the average one third and 10% smallest of $\{T_i^k\}$. The empirically introduced quantities $\{E_1^k...E_9^k\}$ will be used to discriminate among 5401 object characters. The recognition is identified by the object character which has small value E_i^k.

We find most of the thinned skeleton supports of the 24 × 24 pixel standard character are poor in shape and in representation. This may cause topology confusions. We derive an improved skeleton supports by thinning the 300 × 300 dpi standard characters using the Voronoi method [7]. A character is depicted in Fig. 4 for comparison. To reduce computations, an improved sparse skeleton supports can be obtained by sampling the pixels along the lines of the skeleton every 10 pixels and keeping the joint pixels. A typical well prepared skeleton support is depicted in Fig. 4. Then we use the improved sparse skeleton supports as the object character representations. Fig. 4(b-e) shows the resulting vector fields using this improved sparse support. With the improved

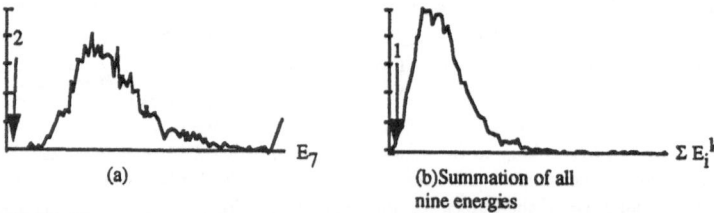

(a)   (b)Summation of all nine energies

Fig. 5. (a) An example of the statistics distribution on the E_7 distance measurement between the 5401 characters and the character in Fig. 2(b). (b)The statistics distribution of the summation of the nine distances for the 5401 characters. Correct recognition is obtained in this case.

sparse skeleton support as in Fig. 5, we always obtain the correct object character in all of our simulations through the 9 STDs.

The recognition is obtained by listing the character set which contains the intersection of the nine emergent character sets for each one of the nine measurements. Each emergent character set contains one thousand characters (or $\sim$ 10%) which have the minimum E_i values among the 12000 characters. The characters in the intersection contained in all the nine emergent character sets are easily identified. The character in the intersection which has the minimum of the summation of the nine distance measurements is the solution of our method. Fig. 5(a) shows a typical statistics distribution of a distance E_7^k for the chinese character in Fig. 2(b) which means "information". Among the selected nine distance measurements the E_4, E_5 and E_6 are most effective measurements for our purpose. Fig. 5(b) shows a typical statistics distribution of the values of summations for all the 5401 character set using the 24 × 24 pixel object supports. The correct recognition of the handprinted character as in Fig. 2(b) is clear in this figure.

III Conclusion

Self-organization maps were used to derive a new approach for the recognition of the handprinted chinese characters. The perception energies were obtained from the topological distances between the thinned handprinted character and the object standard character support, which was the thinned skeleton support of a standard chinese character. Spatial deformations of the handprinted character were then relieved under similar topology. These perception energies were employed to discriminate among the many object characters. Our method may be seen as new distance measurements derived for dissimilarity in terms of topology.

The self-organization alogorithm was used to suppress the spatial deformations of the handprinted character. Our method is totally different from all existing modern algorithms which use a vast number of features plus varieties of other distance measurements. Due to the large various deformations, the topology might be the major clue for recognition. This may be the primary reason for the success of our method based on topology distances. It was demonstrated through simulations that when the deformation is large, our method offers an improvement in all threshold values. When improved skeleton object supports can be obtained, additional improvements can be obtained.

The choice of the self-organization parameters as in our simulations gives the satisfactory results for our method in consideration of both saving computations and convergence accuracy experimentally. A prototype mail address recognition system is integrated using several modern DSP processors. The recognition speed of the system is approximately 1 second for a character in our laboratory and the system includes all isolated English alphabet characters, Arabic numerals and hundreds Chinese characters. This system contain hundreds of small subsets. Each subset contains a very limited set of characters. These subsets have a hierachical structure.

Since no quantitative analysis is available at this point to verify the results reported here for large character set, future work is directed towards a statistical analysis using volumn tests.

References

[1] Teuvo Kohonen, *Self-organization and Associative Memory second edition*, Springer-Verlag, 1988.

[2] D. J. Burr, "Elastic Matching of Line Drawings", *IEEE Trans. on Pattern Analysis and Machine Intelligence*, Vol. 3, no. 6, pp. 708-713, 1981.

[3] Ching Y. Suen, Marc Berthod and Shunji Mori, "Automatic Recognition of Handprinted Characters - the State of the Art", *Proc. IEEE*, Vol. 68, No. 4, pp. 469-487, April, 1980.

[4] R. Durbin, R. Szeliski and A. L. Yuille,"An Analysis of the Elastic Net Approach to the Travelling Salesman Problem", *Neural Computation*, Vol. 1, pp. 348-358, 1989.

[5] Y. Le Cun, B. Boser, J. Denker, D. Henderson, R. Howard, W. Hubbard and L. Jackel, "Handwritten Digit Recognition With A Back-propagation Network" in *Advances in Neural Information Processing Systems 2*, pp. 396-404, Morgan Kaufmann, 1990.

[6] T. Y. Zhang and C. Y. Suen, "A Fast Parallel Algorithm for Thinning Digital Patterns", *Communications of the ACM*, Vol. 27, No. 3, pp. 236-239, March 1984.

[7] D. T. Lee and R. L. Drysdale, "Generalization of Voronoi Diagrams in the Plane", *SIAM J. Comput.*, Vol. 10, pp. 73-87, Feb. 1981.

IDENTIFICATION OF UNDERWATER SONAR IMAGES
USING FUZZY-NEURAL ARCHITECTURE FuNe I

S. K. Halgamuge, W. Poechmueller, S. Ting, M. Hoehn, M. Glesner
Darmstadt University of Technology
Institute of Microelectronic Systems
Karlstr. 15, D-6100 Darmstadt, Germany
Tel.: ++49 6151 16-5136   Fax.: ++49 6151 16-4936
email: *saman@microelectronic.e-technik.th-darmstadt.de*

Abstract

FuNe I is a Classificator based on a Fuzzy-Neural Architecture. Physically interpretable fuzzy linguistic rules are generated from numerical sample data using supervised learning in the first phase and the antecedent membership functions are tuned in the second phase. The posteriori reduction of input features and the possibility of integrating partial apriori knowledge into the trained network are the special features of FuNe I.

Identification of underwater images in realistic situations is a tedious task. This paper describes the application of FuNe I for identification of underwater sonar images. The images are preprocessed into numerical data sets before classification. Authors also present results of conventional classifiers and a multilayer perceptron for comparison.

Introduction

The Fuzzy-Neural combined architectures are becoming increasingly popular among researchers in the fields of Fuzzy Logic and Neural Networks. Several methods for combining the both techniques are reported in literature [2, 6, 7]. Different learning strategies are used in those applications e.g. unsupervised learning [1], supervised learning [3] and differential competitive learning [6].

FuNe I is developed for application in numerical pattern classification problems. This paper illustrates a typical real world application for FuNe I demonstrating its abilities to generate physically interpretable rules and to tune initially created membership functions.

The resulting network after extraction of rules is a fuzzy system, which is completely structured and human interpretable. Therefore, existing partial apriori knowledge can be integrated into the system and superfluous input features can be removed, which makes FuNe I superior to a conventional neural and fuzzy approach.

After describing the application, authors outline the FuNe I technique briefly. The results of FuNe I are then compared with the results obtained from conventional classifiers and multilayer perceptron.

Underwater Sonar Images

The classification of underwater sonar signals is a very hard classification task. Already in the late eighties some neural network applications to such a type of problem became wellknown as e.g. Gorman's and Sejnowski's [4] classification of sonar targets (cylinders versus rocks). Another application is the detection and classification of fishes as proposed by Ramani et al. [5]. Unlike using sonar data created by artificial objects in an artificial environment (water tank), the classification with real underwater sonar data is tedious. If real data is used from objects lying on a sea bed, a terribly large amount of echos not coming from the relevant objects makes classification a really hard task. Furthermore, the problem will become especially hard if realistic objects are taken instead of using perfectly round balls or cylinders filled with air.

The echos we received from objects are generated by emitting gaussian shaped pulses with varying carrier frequency. Each object to be classified was hit by a total of 19 different pulses with carrier frequencies of 5kHz up to 14kHz with a step size of 0.5kHz. Furthermore, this measurement was done several times for each object. To obtain a considerable amount of data several objects were used, the position of each object was repeatedly changed as well as the position of the emitter. The objects were separated into two classes, namely stones and artificial objects (e.g. metal can). Figure 1 (left) shows a typical echo signal from a single pulse. One of the pulses found in the echo signal was emitted by a reflection from the object. This pulse is found by a run length measurement and then segmented (with a window width of 128 amplitudes) for further data preprocessing and classification.

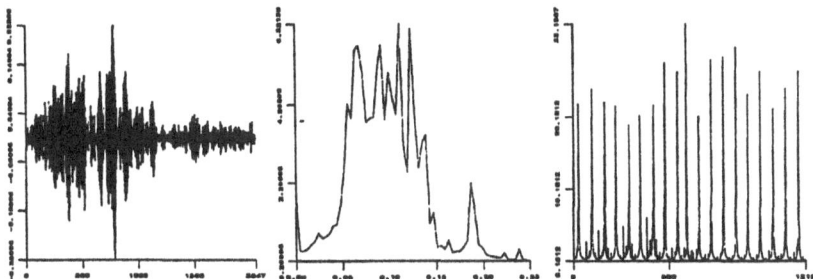

Figure 1: Received sonar echo (left), 19 concatenated spectral diagrams (right), and 19 added spectral diagrams (center)

Data preprocessing after segmentation of the echo signal is done in the following way. From the window of 128 object echo amplitudes we obtain 64 spectral coefficients from a Fast Fourier Transform (FFT). If we take these coefficients from all 19 measurements with different pulses by concatenation this would result in a 1216 element feature vector for classification. This is nearly impossible to classify with a multilayer Perceptron (backpropagation algorithm) since too many weights have to be optimized by the learning algorithm. Therefore, we added the spectral diagrams from the 19 different measurements of one object. In the right part of Figure 1, one can see 19 concatenated spectra whereas the center part shows the same 19 spectra, now added together.

The added spectra information was finally normalized and served as basis for classification. Different methods were used to classify this information. Conventional classifiers i.e. Euclidean Distance Classifier (EDC), City Block Classifier (CBC), Maximum Likelihood Classifier (MLC), Mahalanobis Classifier (MAC), Linear Polynomial Classifier (LPC), Quadratic Polynomial Classifier (QPC), and Nearest Neighbour Classifier (NNC) were tested as well as a standard multilayer Perceptron (MLP) and the FuNe I architecture.

To test the different classification approaches the sonar data set was separated into three sets. The first set is the training set. The second set, called RECALL 1 (soft recall set), consists of sonar data from the same objects used in the training set but the objects were in different positions during the echo measurement. A third set is RECALL 2 which contains only signal echos from objects not present in the training set (hard recall set).

The FuNe I Architecture

The conventional fuzzy model consists of three basic steps; fuzzification, inference process, and defuzzification. Crisp or non-fuzzy input data to the system is converted to fuzzy values using membership functions in the fuzzification section. The rule base contains knowledge acquired from experts or training data to evaluate the fuzzified inputs. Defuzzication is the process which transfers the fuzzy outputs to crisp outputs. In contrast to conventional models, the fuzzy model used in FuNe I contains an non-transparent defuzzification, which makes the implementation as a neural network easier.

The FuNe architecture contains 4 neuron blocks. The input block feeds preprocessed input signals to the fuzzification block where antecedent membership functions are realized. The rule block receives all the membership values, and rules are formed and presented to the output block which delivers crisp outputs directly after a post-processing step.

A set of bell shaped membership functions was created exploiting the possibilities of shifting, scaling and reflecting the sigmoid transfer function. Three (e.g. Low (L), Medium (M), and High (H)) or more user selectable membership functions can be formed according to the users preference for each input [7]. This semiautomatic procedure may be extended to generate triangular or other shapes for membership functions using an additional neuron layer and pre-training.

Three types of fuzzy rules are considered for extraction from sample data in FuNe I. These are simple rules (with premises consisting of a single fuzzy variable), conjunction rules and disjunction rules. A conjunction (disjunction) rule with premises containing more than two fuzzy variables can be approximated by a number of conjunction (disjunction) rules with only two fuzzy variables in their premises. The rule block of FuNe I contains 5 layers to extract rules during the first phase. The fuzzified inputs are analysed in such a way that three basic lists containing possible candidates (nodes) for simple, conjunction and disjunction rules can be created at the end of the first phase. During the second phase all layers in the rule block are deleted and two new layers are created for conjunction and disjunction rules. All possible conjunction and disjunction rules with two variables in their premises are created using the tables obtained during the first phase. Furthermore, a layer connecting all the extracted nodes from the first phase is also considered for simple rules. After a number of training steps the network can be pruned in order to remove weak rule nodes in all three rule layers. The extracted rules (presented in the subsequent section) are weighted due to the non-transparent defuzzification used. This also indicates how strong a rule can influence the decision.

The restriction in the fuzzification block (that the connection weights remain fixed) is automatically removed after a certain number of training steps, which can be specified by the user. This permits the tuning of antecedent membership functions. The desired number of rules to be extracted can also be regulated by the user. The aim is to get the best performance with the least possible number of rules. Automatic generation of physically interpretable rules is further illustrated in [8].

After both phases the generated rules can be analysed and superfluous inputs can be removed. From the rules extracted, we present those rules having strong weights:

W1:	IF Input 1 IS Low OR Input 35 IS Low	THEN Class 1
W2:	IF Input 24 IS Low OR Input 30 IS Low	THEN Class 1
W3:	IF Input 19 IS Low OR Input 31 IS Low	THEN Class 1
W4:	IF Input 13 IS Low AND Input 64 IS Low	THEN Class 2
W5:	IF Input 14 IS Low OR Input 29 IS High	THEN Class 2
W6:	IF Input 47 IS Medium OR Input 49 IS Medium	THEN Class 2

Comparison of Results

Table 1 gives the comparison of achieved classification accuracy of the different classification methods. The first column gives the used classification method whereas following columns report about the achieved classification accuracy. The classification results of "Class 1" (stones) and "Class 2" (artificial objects) are given. In addition to the two recall sets we also tested training success.

From the classical techniques, it seems that EAK and CBK show the best results. MLP shows relatively better results than conventional classifiers used. FuNe I also gives better results in addition to the advantage of feature reduction (34%).

Discussion

We believe that the sonar application problem is a difficult one and the classification can be improved by using more data, especially from a large number of stones and man made objects that are typical for a certain area in a sea bed. MLP and FuNe I proved to be the best approaches in this example. Both techniques provide better performance than conventional techniques in identifying man made objects in soft and hard recall data sets.

However, with FuNe I one can neglect the dependence on an expert, which is a handicap in conventional fuzzy systems and one can also avoid the "black box" nature, which is common to neural networks.

	TRAINING		RECALL 1		RECALL 2	
CLASSIF	class 1	class 2	class 1	class 2	class 1	class 2
EAK	81.54%	93.33%	83.67%	100%	73.75%	58.89%
CBK	76.92%	99.17%	83.33%	100%	62.50%	63.33%
MLK	100%	100%	100%	0.00%	100%	0.00%
MAK	99.23%	100%	80.33%	81.67%	64.17%	56.67%
LPK	99.23%	100%	83.00%	76.67%	64.17%	55.56%
QPK	100%	100%	59.00%	15.00%	80.00%	36.67%
NNC	100%	100%	77.33%	41.67%	84.58%	34.44%
MLP	100 %	90.77 %	84.17%	93.33 %	72.08%	65.56 %
FuNe I	100 %	98.33%	89.33%	98.33%	65.83 %	68.89 %

Table 1: Obtained classification results

Two versions of FuNe I were tested with popular IRIS classification data first. FuNe I version 1.0 generates only minimum rules while version 2.0 also extracts maximum and simple rules. Recent results clearly show that version 2 (99%) is superior to version 1(95%) [7]. In both cases 25% feature reduction was achieved. By applying version 2 on automatic classification of solder joint images (23 input features and 2 classes) [8] we obtain 99% classification rate with 43% feature reduction.

The work in progress shows that the resulting networks from FuNe I, which are fuzzy systems without hardware consuming defuzzification methods, can be easily implemented in commercially available fuzzy hardware.

Acknowledgement

We thank Dr. Guicking, K. Goerk, and H. Peine from the "Drittes Physikalisches Institut" of the University of Goettingen for providing us with all their sonar data files.

References

[1] T. Kohonen *Self-Organization and Associative Memory*, Springer Verlag, 1989.

[2] H. Surmann, B. Moeller, K. Goser, "A Distributed Self-Organizing Fuzzy Rule Based System", *Proceedings of the Neuro Nimes 92*, Nimes, France, November 1992.

[3] D.E. Rumelhart, J.L. McClelland, *Parallel Distributed Processing: Explorations in the Microstructure of Cognition*, Volume 1, The MIT Press, 1986.

[4] R.P. Gorman, T.J. Sejnowski, "Learned Classification of Sonar Targets Using a Massively Parallel Network", *IEEE Transactions on Acoustics, Speech, and Signal Processing*, Volume 36, No. 7, July 1988

[5] N. Ramani, P.H. Patrick, W.G. Hanson, H. Anderson, "Fish Detection and Classification Using a Neural-Network-Based Active Sonar System – Preliminary Results", *Proceedings of the International Joint Conference on Neural Networks IJCNN'90*, Volume II, pp. 527-530, Washington, U.S.A., 1990

[6] B. Kosko, *Neural Networks and Fuzzy Systems,* Prentice-Hall, 1992.

[7] S.K. Halgamuge, M. Glesner, "A fuzzy-neural approach for pattern classification with the generation of rules based on supervised learning", *Proceedings of the Neuro Nimes 92*, Nimes, France, November 1992.

[8] S.K. Halgamuge, W. Poechmueller, M. Glesner, "A Rule based Prototype System for Automatic Classification in Industrial Quality Control", *ICNN'93*, Volume I, pp. 238-243, San Fransisco, U.S.A., March/April 1993

Fault Detection in Multivariate Time Series with a Coding Approach

Peter Weierich

Bavarian Research Center For Knowledge-Based Systems (FORWISS)
Knowledge Processing Research Group
Am Weichselgarten 7, D-91058 Erlangen, Germany

email weierich@forwiss.uni-erlangen.de Phone: +49-9131-691-134, FAX -185

Abstract

In this paper a coding approach is presented, which is able to detect erreonous passages in multivariate time series. The field of application is the experimental stress analysis of cars during the development of new models. We use a coding approach to distinguish correct from suspicious signals. After simple multi-layer perceptrons (MLPs) had shown to cause too many false positive alarms, nets with recurrent links in Jordan and Elman style were investigated. The first results are very promising.

1   Field of Application

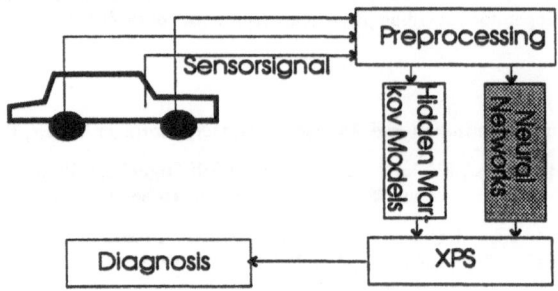

Figure 1: Overview

Neural Networks are investigated as one alternative to Hidden Markov Models to produce the input to an expert system (MESSPERT) for the automated analysis of multivariate measured data.

New assembly groups in car development afford costly methods to prove the fulfillment of requirements of the market [3]. Measured data acquired during test runs is characterized by a huge storage requirement. For instance, during one hour of measurement the 64 sensors provide about 100 MB of data. Though different methods have been developed to compress the data and to extract relevant features, it has to be assured that no faulty signal is used for evaluation. A typical "faulty" signal may contain spikes, shifts and drifts in the signal, caused by amplifiers, or the loss of one channel caused by broken wires.

On the other hand, real mechanical changes in the test model have to be recognized. These changes and the faulty signals should be automatically extracted as "suspicious" for further inspection. This inspection can be done by experts or with an expert system which is being developed at our institute. An overview is shown in figure 1.

2 Basic Idea

It is well known that neural networks with structures like the 8-3-8 coding net can produce a mapping like classical singular value decomposition of the target data [4] when linear neurons are used.

This property motivated the following approach to detect suspicious passages in the signal:

A neural net should be trained, which is able to reconstruct the input signal even if the number of hidden units is lower than the number of inputs. During training the net implicitly learns correlations between input channels.

After the training the neural net is expected to have two properties:

- Correct signals presented at the input layer are reconstructed completly at the output layer.

- Suspicious signals fail to be reconstructed completely.

3 Experiments

3.1 Methods

In first tests a simple 8-6-8 MLP was trained with a training set of 1100 vectors. Testing was done with the same data.

In a second attempt a network with 20 input and output units and 15 hidden units with a sigmoid activation function were used. Two different types of recurrent links were used: Elman style[1] and Jordan style[2] links. The total number of computed vectors was 4000, with identical training and testing set. The learning rate to provide convergence was 0.01, but had to be increased in experiments when 13000 vectors were used for training. The standard backpropagation algorithm was used.

For the automatic detection of suspicious signals two types of faults were simulated: Spikes which reached the maximal possible value and offsets (10 percent of the dynamic range) with short duration (500 milliseconds).

The signal was presented "as is", i.e. there was no feature extraction carried out as Fourier transform.

For the simulation of neural networks the SNNS (Stuttgarter Neuronale Netze Simulator)[5] and the FAST (FORWISS Artificial Neural Network Simulator), developed at the FORWISS by Michael Arras, were used.

3.2 Results

The simple coding net showed a poor behaviour, especially in passages where the signal amplitudes changed (see fig. 2). The lack of time dependent information was identified as a cause.

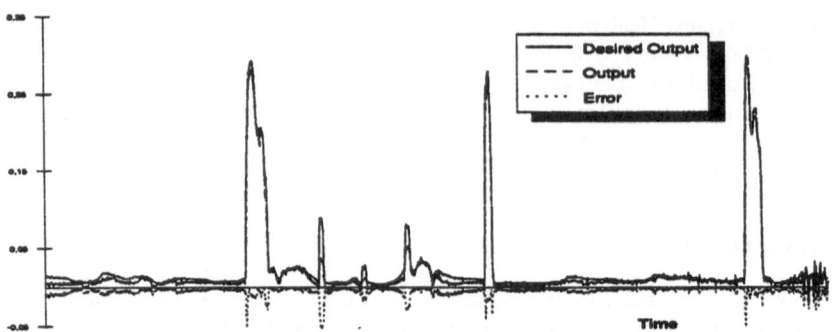

Figure 2: Results of a 8-6-8-Coding Net

Extreme deviations of the real from the desired output occur when signal amplitudes are changing. The absolute error is displayed negative.

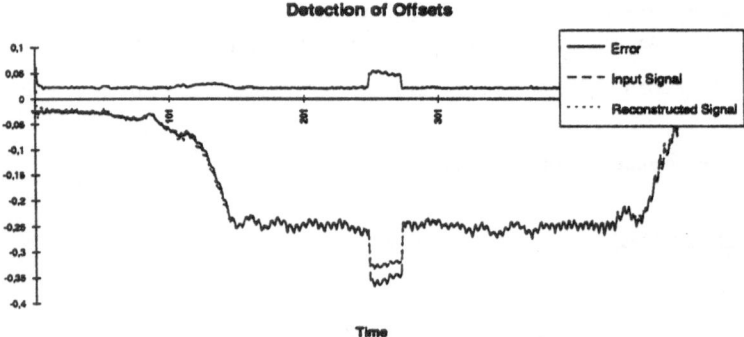

Figure 3: Detection of offsets in one channel

Displayed is one measured channel with a negative offset, the output activation and the absolute error plus a small constant (here 0.03 for display purposes). A clear decrease of the error signal is visible during the offset.

The second attempt brought better results with Jordan-style recurrent links by a factor of two. Networks with Elman links were even slightly worse than networks without recurrent links.

Small offsets in one channel led to a clearly visible error in the affected channel (see figure 3). During the occurrence of spikes not only the disturbed channel produced a high error output, but also other channels (figure 4).

3.3 Discussion

The presented coding approach is well suited for the detection of suspicious multivariate measured data. The side effects of deviations on one channel on the output values of others were expected and could improve the classification rates in the real application.

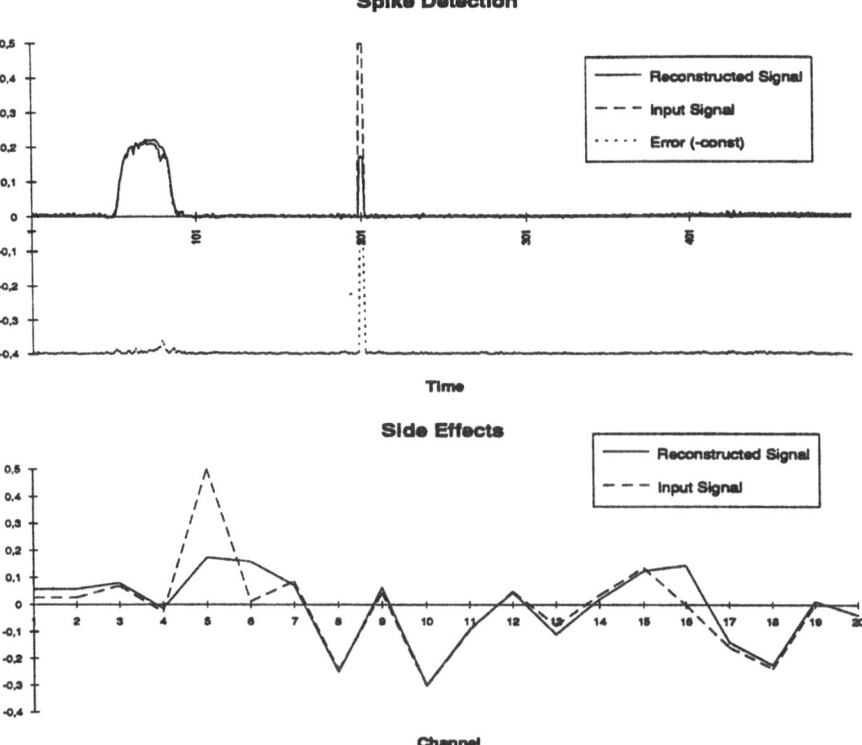

Figure 4: Spike detection and side effects to other channels
Above: Sharp error signal during a spike, as in figure 3. Below: The input and the reconstructed signal of the 20 channels are is displayed for one frame while channel 5 is distorted. Extrem side effects are visible on channels 6 and 16 besides smaller deviations in half of the remaining channels.

References

[1] Elman, Jeffrey L. Finding Structure in Time Cognitive Science, Volume 14, pp. 179 – 211, 1990

[2] Jordan, Michael I. Attractor dynamics and parallelism in a connectionist sequential machine. In Proc. of the 1986 Cognitive Science Conference, PP. 531 – 547, 1986

[3] Petersen, J., Tunker H. Vom Messaufnehmer zur Datenbank. Informatik Spektrum, Springer, Berlin, Heidelberg, New York, Vol. 14 (2), pp. 69 – 73, 1991.

[4] Webb, Andrew R., Lowe, David The Optimized Internal Representation of Multilayer Classifier Networks Performs Nonlinear Discriminant Analysis. Neural Networks, Volume 3, pp. 367 – 375, 1990

[5] Zell, A., Mache, N. Sommer, T., Korb, T. Recent Developments of the SNNS Neural Network Simulator. In Proc. Applications of Neural Networks Conference, SPIE, Vol. 1469, pp. 708–719, Orlando Florida, 1991. Aerospace Sensing Intl. Symposium.

Practical Implementation of a Radial Basis Function Network for Handwritten Digit Recognition

Bernard Lemarié - Service de Recherche Technique de la Poste
10, Rue de l'Île Mabon F-44038 Nantes Cedex, France
e-mail: lemarie@srtp.srt-poste.fr

Abstract : We present a practical implementation of a Radial Basis Function Network to handwritten digits recognition. Reduction of the number of hidden nodes which is an important and necessary step to obtain a computationally tractable network is made using an original technique. A comparison is made with the k-nearest neighbour method. Results appear better for the network at a much lower computational cost.

1. Introduction

In the field of neural networks Radial Basis Function models have been introduced in the context of the regularization theory [1]. Regularization models allow to assure the existence and also to obtain a global minimum of a function interpolation problem under some constraints of regularity. Therefore, when viewing layered Neural Networks as universal approximators, RBF models offer a theoretical framework.

For a classification task, the function to be approximated by the neural network is the bayesian probability. For this problem, non parametric statistical techniques have been investigated for many years and offer today good reference, especially under the aspect of bias/variance. We thus compare the network with the k-nearest neighbour method. Such a comparison have yet been done [3][4], but remains very dependant on experiment conditions like the database for learning and testing and also on the precise implementation of the network. We chose to realize most of parameter learning steps by a unique backpropagation algorithm. Another important point, may be the most relevant, is the method of reduction of centres which might considerably influence the performance of the recognition system.

We use here a database of about 18,000 handwritten digits from french postal code and introduce the network as a classifier after a process of morphological extraction of characteristics. We argue that for the present comparison with statistical methods, morphological or topological aspects of neural networks do not have to be considered and thus classification of characteristic vectors is convenient.

The organization of the paper is as follows. Design of the network on the basis of regularization theory and Parzen windows model is sketched in section 2. Section 3 presents the method of reduction of the number of centres. Section 4 presents the practical application and reports the comparative results. The last section discusses some of the remaining issues and extensions of the method.

2. Regularization network and Parzen windows estimator.

RBF networks are issued from regularization theory. In this model, one considers that «world is smooth»: Given $S = \{(x_i, y_i) \in R^d \times R, i = 1, 2, ..., N\}$ a set of points we try to build a function from $R^d \to R$ which minimizes the sum of Mean Square Error on the given points and a regularity constraint on the unknown function:

$$H(f) = \sum_{i=1}^{N} (y_i - f(x_i))^2 + \lambda \|P(f)\| \qquad (1)$$

The regularity constraint is represented by a differential operator on the unknown function f. As described in [1] for some form of regularity operator we can obtain and build the solution of (1). In particular for:

$$\|P(f)\| = \sum_{m=0}^{\infty} a_m \left(\int_{R^{n_{i_1 i_2 \cdots i_m}}} \sum \left(\frac{\partial^m}{\partial i_1 \partial i_2 \cdots \partial i_m} f(x) \right)^2 dx \right) \qquad a_m = \frac{\sigma^{2m}}{m! 2^m} \qquad (2)$$

the solution of (1) is given by:

$$f(x) = \left(\sum_{i=1}^{N} C_i \times \left(exp \left(-\frac{\|x - x_i\|^2}{2\sigma^2} \right) \right) \right) \qquad \forall x, x \in R^d \qquad \text{with} \qquad C_i = (y_i - f(x_i)) / \lambda$$

C_i can be found by solving the N equations system obtained with (3) applied to each (x_i, y_i) . As pointed out in [1][2], the formula (3) can easily be viewed as the output of a neural network with one hidden layer of

N neurons. The formalism is also easily extended to the approximation of functions from R^d to R^c and then can be applied to the estimation of a posteriori density of probability in a problem of classification.

Thus under a condition of regularity we can obtain a global minimum and solve the learning task. In practice however some problems appear. The first difficulty is to select the values of the regularization and kernel width parameters σ and λ. The second difficulty is due to the computational cost of the method because we have to compute the kernel activity for each element of the learning set. Also it's interesting to consider the regularization method under the aspect of neural networks and to try to apply a learning method for some parameters on a network. For a neural network a mean square error criteria is equivalent to (1) except the regularization coefficient. We would like to suppress this one but doing this with a complete network the trivial solution of (1) is σ = 0 (a Dirac function) and thus without any interest. However on a reduced network this solution disappears and it becomes relevant to minimize the mean square error. In this sense reduction of centres appear to counterbalance the omission of the regularization parameter avoiding thus a too important overfitting on the learning set.

We have to note that the formula (3) is closed to the expression of Parzen windows non parametric estimator with gaussian kernel. Given a n elements data set in a d-dimension space, one writes:

$$\hat{p}_n(x) = \left(\frac{1}{nh_n^d}\right)\sum_{i=1}^{n} exp\left(-\frac{\|x-x_i\|^2}{2h_n^2}\right) \qquad (4)$$

Let's recall that this estimator is consistent under the following conditions:

$$\lim_{n\to inf} h_n = 0 \qquad and \qquad \lim_{n\to inf} h_n^d \times n = \infty \qquad (5)$$

For a classification task we then estimate for each class i of n_i elements the a posteriori probabilities and get the bayesian probabilities by Bayes' rule:

$$\hat{p}(Ci|x) = \left(\frac{K}{h_{n_i}^d}\right)\sum_{j=1}^{n_i} exp\left(-\frac{\|x-x_{ij}\|^2}{2h_{n_i}^2}\right) \qquad with \qquad p(Ci) = \frac{n_i}{N} \qquad (6)$$

Here again such a formula can be interpreted as the dynamic of a neural network with one hidden layer. This relation have been pointed out in [10], but without any subsequent learning. Considering only the network at age zero, we can therefore affirm that if the learning set grows neural network first and second order moments will converge if the conditions (5) are respected. On the other part dealing with risk of local minimum, results of learning are very sensitive to the initial state of the networks. In this sense taking the Parzen estimator as the initial state of the network would ensure a convenient performance. Here again, the reduction of the references and the choice of the kernel radius are the main difficulties. As for the regularization reduction is required so as to decrease computational cost and also to avoid the trivial solution. Again, we propose the reduction as the first step following by the learning step of kernels width.

Finally, Radial Basis Function networks offer a new model of neural network architecture with theoretical justification and proof of existence [2]. The parallelism with the Parzen windows estimator gives us the initial state of the network. At the last step, using a learning algorithm like back propagation should help us to improve the performance of such a network with respect to the bayesian probabilities. However, this network presents too much hidden nodes to be tractable for any learning database and we have to try to reduce this number. Moreover, this reduction turns out to be required to avoid overfitting.

3. Method of reduction.

Clustering methods like K-means are the most used tools of reduction of data sets and have been applied to RBF-Networks [5]: Neighbour points in the data set are grouped into clusters which become the hidden neurons of the network. In this method however, one have to choose the number of final clusters. On the other part, for neural networks the number of hidden neurons is a fundamental parameter of the functional capacity of the networks to approximate with more or less consistency the unknown function. [12][13]. Next we note that the cluster positioning processing before the learning by the neural network could be included in the network learning with this time the criterion of the network instead of the clustering method's one. And at last as pointed out in [11]classical clustering does not take into account the classification problem in the sense that clustering is done on the whole data set. Thus, a reduction method introducing differentiation between classes would be welcomed.

Finally we have tried to concentrate on a reduction method according to the following guidelines: Number of elements, classification aspects and relation with consistency. The first step is to reduce each Parzen estimator of the a posteriori density for each class. Few solutions exist, but we can hope that grouping near elements of a given class should not affect too much the consistency of the non parametric estimator. The formula

(6) becomes:

$$\hat{p}_n(x) = \left(\frac{1}{nh^d_n}\right) \sum_{i=1}^{n} exp\left(-\frac{\|x - C(x_i)\|^2}{2h^2_n}\right)$$

where $C(x_i), i = 1, 2, n$ is the cluster assigned to xi and finally for our classification problem:

$$\hat{p}(Ci|x) = \left(\frac{K}{h^d_{ni}}\right) \sum_{j=1}^{nc_i} n_{ij} \times exp\left(-\frac{\|x - c_{ij}\|^2}{2h^2_{ni}}\right)$$

where nc_i is the number of clusters found for the class i and $Cij, j = 1, 2, nci$ the j-th centre of the class i containing n_{ij} points. This formula gives us the initial state of the network where h_{ni} is chosen so as to respect the condition (5).

For each class cluster list is built from a chosen radius r. Taking randomly one element x we pick up all the other elements within the same class which are included in the hypersphere centred at x and with a radius of r. If r is too large, not only Parzen estimation will be false but classification task won't be realized. So we try to choose the clustering radius for each class in a classification objective. For that, we plot for each pair of classes the average number of points of the two classes which distance is below r:

$$N(i, j, r) = \frac{1}{N_i \times N_j} \times \sum_{x \in C_i} \sum_{y \in C_j} H(x, y, r)$$

where H is the Heavyside function:

$$H(x, y, r) = 1 \quad if \quad d(x, y) \le r \quad and \quad H(x, y, r) = 0 \quad otherwise$$

The figure below shows these curves for class 0:

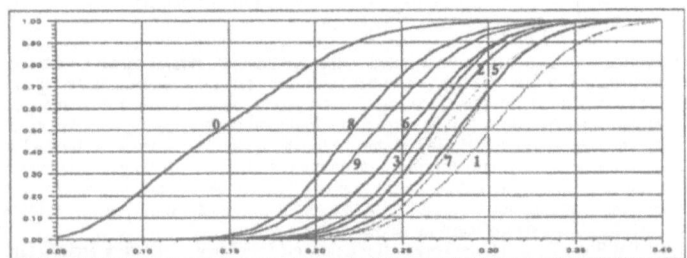

The curves are then used to select the radius of the clustering for each class. For example for class 0 we take the radius 0.175.

4. Application and Results: Building Network from learning set.

We now present the application of the RBF Networks to the recognition of handwritten digits issued from the segmentation of french postal codes. A base of 8783 examples is extracted for learning. A test base of 2197 elements is also used for testing and finally an evaluation base of 7398 elements is required for final comparison with other methods. The morphological preprocessing extracts a vector of length 138 from the bitmap image of the character. The method is inspired from «characteristic loci method» [8][9]: For each white point we consider the axis issued from the point and oriented in one of the eight main cardinal directions. We characterise the axis and therefore the point to cross the character (black pixel).

Curves of distance estimation as described in the preceding section have been plotted for each class and allow to get a set of 356 centres form the 8783 elements learning set. Next, the network adjusted the radius of each centre and the output weights via a stochastique learning backpropagation algorithm. The following chart gives the results for the networks and for the k-nearest neighbours, with for this latter the use of the 8783 learning set elements as references.

database	learn	test	validation
# elements	8783	2197	7394
3-NN	97.51/1.75	96.13/2.81	95.00/3.97
RBF-Network	98.14/1.86	97.04/2.96	96.29/3.71

For each case the first value represents the recognition rate RR and the second value the substitution rate SR. With the statistical method, the best equilibrium between recognition rate and generalisation on the test set is reached while considering the three nearest neighbor. Nevertheless, these results remain below the network's ones on the valid set.The network shows an interesting balance between performance and overfitting i.e. between bias and variance of the estimator. Backpropagation learning reduces mainly bias and that the measure of performance on the test set is generally used to prevent overfitting. With the RBF network the limitation of overfitting is reinforced by the structure of the network and by the reduction of centres. In fact at a local level presence of several examples of the same class near a centre forbids the reduction of this centre width leading thus to a local equilibrium closed to the variance of the set of the initial points included in the hypersphere around the centre and in counterpart presence of near elements of a different class leads to a reduction of the width. For centres with low probabilities i.e. centres with few near examples of the same class the variance should be low. But because this centre has a low probability, gradient updating is also rare and the initial width does not decrease too much.Thus for those points overfitting is reduced.

We have also to consider the computational cost of the two best methods in recognition mode. The most expensive operation is the computation of the distance between centres or references and the pattern to recognize, embedded in the 138 dimension space. k-NN method requires 8783 distance computations while RBF-Networks process only 356 ones. We get thus a cost reduction by a factor of 20.

5. Conclusion

The application of Radial Basis Function presented in this paper gives fairly good results.The comparison with statistical non parametric estimator reveals that in a practical situations the RBF network will be well suited because it gives better performance at a much lower computational cost. It appears that regularization networks if they do not solve the theoretical bias variance/dilemma [7] present at least a fruitful framework for non parametric estimation with practical applications. This mainly comes from the neural architecture which allows the integration of most parameters in an unique formalism.

This model should raise in performance by adding more free parameters in the network as reviewed in [1]: Moving centres or multidimensional gaussian functions. But finally we must recall the necessity to control the liberation of parameters first in order to avoid bad convergence with back propagation and secondly to limit the variance problem. Another required improvement of the method is the automation of the selection of the centres nodes: For this a criterion is needed for separability of classes against radius of clustering. Our approach is to try to integrate directly the selection of the centres in the learning step. This point is currently under investigation.

6. References

[1]   Poggio T., Girosi F., Networks for approximation and learning.Proceedings of the IEEE, Vol 78, No. 9, 1990.
[2]   Girosi F., Poggio T., Networks and the best approximation property. Biological Cybernetic 63, 169-176, 1990.
[3]   Richard M. D., Lippman R. P., Neural Networks Classifiers estimate a posteriori Probabilities. Neural Computation, 4,461-483, 1991.
[4]   Lee Y., Handwritten recognition using K Nearest-Neighbour, Radial Basis Function and Backpropagation Neural Networks. Neural Computation ,3,440-449, 1991.
[5]   Moody J., Darken C. J. , Fast learning in Networks of locally tuned processing units, Neural Computation, 1, 281-294, 1989.
[6]   Ng, Lipmann R.P. , A comparative study of the practical characteristics of neural networks and conventional pattern clasifiers, in Neural Information Processing Systems 3, 1991, D.S. Touretzky, ed. Morgan Kaufmann, San Mateo, Ca.
[7]   Geman S, Bienenstock E.,Boursat R., Neural Networks and the bias variance dilemma, Neural Computation, 4, 1-58,1992.
[8]   Gluksman H. A., Classification of Mixed Font alphabetics by characteristics loci., 1st annual IEEE Computer Conference, 138-141,1967.
[9]   Gaillat G., Berthod M., Panorama des techniques d'extraction de traits caractéristiques en lecture, optique des caractères , Revue Technique THOMSON-CSF, Vol 11, No 4, 1979.
[10] Specht D. F., Probabilistic Neural Networks, Neural Networks, vol.3, 109-118, 1990.
[11] Musavi M. T., Ahmed W.,Chand K. H., Faris K. B., Hummels D. M., On the Training of Radial Basis Function Classifiers, Neural Networks , Vol 5,595-605, 1992.
[12] Hausler D., Decision Theoretic Generalization of the PAC Model for Neural Net and Other Learning Applications, Technical Report, UCSC-CRL-91-02, 1991.
[13] White H, Conectionist Non Parametric Regression: Multilayer feedforward Networks can learn Arbitrary Mappings, Neural Networks, Vol.3, 535-549,1990.

934

An Efficient Method of Neural Network Application to Recognizing of Handwritten Digits in Zip Codes

Sławomir Skoneczny Rafał Foltyniewicz and Maciej Sitnik

Institute of Control & Industrial Electronics,
Warsaw University of Technology,
00–662 Warszawa, ul. Koszykowa 75, Poland
tel. (+48–2) 6280665, fax. (+48–2) 6256633, e–mail: skonecz@plwatu21.bitnet or
rfoltyn@plwatu21.bitnet

Abstract

We present some neural networks models applied to the recognition of hand-written digits for postal application. We apply both, multilayer perceptron and neocognitron network. We also propose a novel neural network structure built of Hopfield - like network connected with the multilayer perceptron. The advantage of such structure enables to achieve high rate of correctly classified digits. Both digits and feature vectors are given to the input of neural network. A comparison of the results for different network structures is given.

1 Introduction

For many years automatic recognition of handwritten digits is a serious problem with results not satisfying high postal requirements. Optical Character Recognition (OCR) machine performs well on typewritten mail but have difficulty with handwritten pieces. Mail pieces which are rejected by these machines must be currently manually sorted. Classical Zip Code recognition methods although still being developed and improved are far from such efficiency that would enable to be applied in postal offices. On the other hand neural networks and their computational properties have attracted the interest of researchers in the area of machine perception by presenting a complementary solution to symbolic processing paradigms. Obviously, there exist many different neural network structures, each of them having its own advantages and disadvantages, but what is exciting is the ability to learn from examples and to generalize some features difficult to be describe in precise way.

In our paper we describe different neural network models multilayer perceptron with backpropagation learning algorithm and neocognitron. In order to improve recognition rate and assure short training time we propose a new structure consisted of Hopfield network (for learning and recognizing and multilayer perceptron (for final classification).

2 Multilayer Perceptron with Backpropagation

When an input pattern p is applied to the network, the activation of each unit is dynamically determined using the function [3]:

$$o_{pj} = \frac{1}{1 + e^{-\left(\sum_i w_{ji} o_{pi} + \theta_j\right)}} \tag{1}$$

where o_{pj} is the activation of unit j as a result of the application of pattern p, w_{ji} is the weight from unit i to unit j, and θ_j is the bias for unit j. Backpropagation is then invoked to update all of the weights in the network according to the following rule:

$$\Delta w_{ji}(n+1) = \eta \cdot \delta_{pj} \cdot o_{pi} + \alpha \cdot \Delta w_{ji}(n), \tag{2}$$

where n is the presentation number i.e., the number of times the system has been presented a pattern, η is the learning rate, δ_{pj} is the error signal for unit j, and α is the momentum factor. The error signal δ_{pj} for and *output* unit j is calculated from the difference between the target value and the actual value for that unit:

$$\delta_{pj} = (t_{pj} - o_{pj}) \cdot o_{pj} \cdot (1 - o_{pj}). \tag{3}$$

The error signal δ_{pj} for a *hidden* unit j is a function of the error signals of those units in the next higher layer connected to unit j and the weights of those connections:

$$\delta_{pj} = o_{pj} \cdot (1 - o_{pj}) \cdot \sum_k \delta_{pk} w_{kj}. \tag{4}$$

3 Neocognitron

The neocognitron [2] is a multilayer network consisting of a cascade of many layers. Each layer consists of a two sub-layers: S-layer and C-layer (simple and complex). There are forward connections between cells in adjoining layers. The initial stage of the network is the input layer and consists of a 2D array of receptor cells. Each of the succeeding stages has a layer of "S-cells" followed by a layer of "C-cells". Thus, in the whole network, layers of S-cells are arranged sequentially. Incidentally, each S-layer contains subsidiary inhibitory cells, called V-cells. S-cells are feature extracting cells. Connections converging to feature-extracting S-cells are variable and are reinforced during learning process. When learning is finished S-cells with the aid of subsidiary V-cells, can extract features from the input pattern. The features which are extracted are determined during learning. Generally speaking in the lower stages, local features, such as lines, bends are detected. In higher stages, more global features, such as a part of the training pattern are extracted. The C-cells are inserted in the network to allow for the positional errors in the features of stimulus. Connections from S-cells to C-cells are fixed and invariable. Each C-cell receives a signal from a group of S-cells which extract the same feature, but from slightly different positions. The C-cell is activated when at least one of these S-cells is active. Therefore even if the stimulus is shifted in the position and another

S–cell is activated instead of the first one, the same C–cell keeps responding. Cells in a layer are divided into subgroups (cell planes) according to the kinds of feature to which they respond. The density of cells in each layer is designed to decrease with the order stage, because the cells in higher stages usually receive signals from larger areas of the input layer and the neighboring cells come to receive similar signals. Hence, in the highest stage, only one C–cell exists in each cell–plane. Thus, in the whole network, in which layers of S–cells and C–cells are arranged alternately, the process of feature extraction by S–cell and toleration of positional shift by C–cells are repeated. During this process local features extracted in the lower stages are gradually integrated into more global features. Finally, each C–cell of the highest stage integrates all the information of the input pattern, and responds only to one specific pattern.

Only connections converging to the S–cells are variable and are reinforced gradually in accordance with stimuli given to the network during the process of learning. Both processes, "learning–with–a–teacher" and "learning–without–a–teacher" can be used to train neocognitron to recognize patterns.

4 Hopfield Network

The Hopfield network [1] is constructed by connecting a large number of simple processing elements to each other. Those connections are described by a connection matrix T, which contains weighting factors for each input. The network output v is formed by nonlinear nondecreasing and bounded operator acting on sum of all weight weighted inputs. We use the simple synchronous discrete–time model:

$$v^{k+1} = sign(T v^k) \tag{5}$$

5 Experimental Results

The performance was measured both on the training set of digits as well as on the test set. For multilayered perceptron the training set to test set ratio was 25% (the whole character set was about 5000 digits, 500 samples per digit). All simulations were performed on IBM PC 386/486. A special software package was built in C language (Watcom 9.0 C) in order to perform experiments with different network structures.

The neocognitron was trained with "learning–with–a–teacher" procedure. For simulation we used 4 stage network (each consisting of 2 sub layers of S–cells and C–cells). Size of the cell planes and their number per layer were as follows: Us_1 19x19x12, Uc_1 21x21x8, Us_2 21x21x38, Uc_2 13x13x19, Us_3 13x13x35, Uc_3 7x7x23, Us_4 3x3x11, Uc_4 1x1x10. The size of the input layer was 19x19 pixels. Thus the total number of neurons in the whole network (including inhibitory cells for S layers) was 36 321. As a training set of digits for the neocognitron we regarded training patterns for layer Us_4 as they were pure digits while training patterns for the previous layers contained only certain geometrical features. Recognition rate for the training set was 100% while for the training set it reached

80%. It should be mentioned, however, that misclassified training patters were difficult to judge even for a human reader. The learning time was significantly shorter than for a multilayered perceptron with backpropagation while the recognition itself took longer.

In our model, the input layer consisted of 256 neurons, each taking information from binary 16x16 input image. The hidden layer included variable number of neurons, varying from 10 to 96. The output was built of 10 neurons. The network was fully connected and the initial weights were randomly set within a range from −0.5 to +0.5. Learning took 7000 lessons. Recognition rate for the training set was 100% while for the test one about 82%. The high recognition rate for training set was reached after only 300 lessons and than remained at the same level.

The Hopfield algorithm was used to train the Hopfield network. The problem, however, arises as there exist undesirable stable states in the Hopfield network called spurious states. We found some useful methods for reducing their number. The recognition rate on a training set was 100%, on the test set 88%. The 12% of misclasified digits results from some spurious states still remaining in the network. In order to increase the recognition rate we connected the Hopfield network with multilayered perceptron. Hence we had two stages recognition process. The recognizing process is firstly performed by the Hopfield network and than by the multilayered perceptron. The recognition rate of such structure reached 92%.

References

[1] S. Aiyer, M. Niranjan, and F. Fallside. A Theoretical Investigation into the Performance of the Hopfield Model. *IEEE Transactions on Neural Networks*, 1:204–219, 1990.

[2] K. Fukushima and N. Wake. Handwritten Alphanumeric Character Recognition by the Neocognitron. *IEEE Transaction on Neural Networks*, 2:355–365, 1991.

[3] D.E. Rumelhart and J.L. McClelland. *Parallel Distributed Processing*. MIT Press, 1986.

The Application of Average Gradient Matrices for Fingerprint Classification using Neural Networks

P.A. Hughes and P.D. Noakes

Neural and VLSI Systems Laboratory,
Department of Electronic Systems Engineering,
University of Essex,
Colchester, Essex, CO4 3SQ
UNITED KINGDOM

ABSTRACT

Currently the task of classifying fingerprints consumes much time and labour by trained fingerprint technicians. Manual fingerprinting storage and retrieval systems require a vast amount of human skill and expertise. Technicians who classify fingerprints have first to undergo several years training. Many law enforcement agencies use a fingerprint classification system known as the Henry Classification system in order to partition fingerprint databases. Using this system fingerprints are sub-divided into one of eight possible fingerprint pattern types. Numerous attempts have been made to automate such classification methods using conventional image processing techniques but very few have been embraced by law enforcement agencies due to their limited successes in solving the problem. The re-emergence of interest in neural networks (NNs) in recent years has caught the attention of those involved in the fingerprint field as they begin to recognise the potential advantages of a NN approach.

In this paper a method of classifying fingerprints using Back Propagation (BP) NNs and average gradient matrices is presented. For this research we assembled a database of fingerprint images made up of 512 by 512 pixels and having 256 grey levels. A dynamic grey level threshold transformation was used in order to obtain a 512 by 512 bi-level image. As such an image contains a high degree of redundant pictorial information which is not required in any input data set to a neural network, the image was reduced to a matrix of varying sizes. This matrix contained the average direction of the ridge lines within a certain pixel subregion. The average directions were estimated by counting the occurrences of five types of "micropattern" which may occur in any group of four neighbouring pixels.

Rather than having one large BP network for the task of classifying the fingerprints presented the network was decomposed. This means that rather than train a network to learn the classifications of perhaps seven classes of fingerprint, separate individual networks were trained to classify a particular class of fingerprint as opposed to another class or all other classes. The advantages of such decomposition are that the convergence-training time is greatly reduced and the results obtained are often superior to those for a fully composed network. Input to each BP network consisted of varying array sizes depending on the average direction matrix size chosen. Thus each neuron represented a fingerprint subregion average gradient. Each BP network contained one hidden layer which in each case had twenty neurons. The output of a particular network consisted of a single neuron. This neuron represented the degree to which an input fingerprint matched the 'learned' classification.

Results of the our research have been encouraging. Using one part of our fingerprint database for training exemplars and the remainder for test data, various experiments have been conducted using various average gradient matrix sizes. The results being both noteworthy and surprising. It was found that no one matrix size yielded the best classification accuracy across all the classes. In the case of training a BP neural network to classify a whorl type fingerprint as opposed to any other sort of fingerprint it was found that the use of an average gradient matrix of size 32 x 32 produced the highest accuracy. This was also true for arches, however for loop type fingerprints the best matrix size was found to be one of 16 x 16. The results of this research confirm that using the average gradient technique is a suitable method to use with neural networks for fingerprint classification. The percentage of correctly classified test fingerprints has been found to be in excess of 90%.

Acknowledgement - This research has been funded by SERC.

Neural Architectures for Motion Tracking

Lino D'Agnese, Andrea Ferro, Giancarlo Parodi, Rodolfo Zunino
DIBE - Dept. of Biophysical and Electronic Engineering - University of Genoa
Via all'Opera Pia 11a - 16145 GENOA - ITALY

Abstract

The goal of a visual tracking system is to evaluate the motion parameters of an object in a scene. The present research approaches this problem from a connectionist perspective, aiming at an effective methodology with the fewest assumptions on the observed domain. The research basic constraints are simplicity and flexibility, which aim at lowering computational loads, allowing real-time implementations, and ensuring general applicability.

The described architecture processes binary images containing one object, which translates in the image plane; two image frames must be available (so that motion can be detected), and the system is supposed to know the time difference between consequent frames. Tracking information is derived from consecutive frames in a sequence; the key idea is to map each image into a lower-dimensional representation (message) where the NN can induce motion parameters from time-consecutive differences.

To stress simplicity, the procedure to compute a message from an image (M-GEN) just sums pixel values along rows and columns; however, the approach allows any mapping that preserves positional information. By this representation a frame series maps into a message series. Then a standard feedforward network processes differences between contiguous messages in a sequence; the output is an analog estimate of the object-translation 2-D vector. As no assumption is made about either the object present in the scene or its scale factor, this schema does not assume previous object classification. This property found experimental confirmation, where test images differ from training ones in both shape and scale factor. Fig.1 presents a schema of the overall architecture.

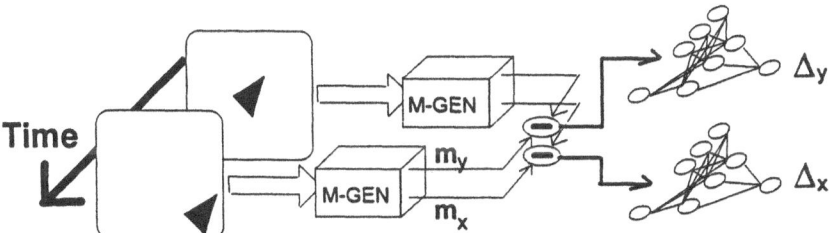

Fig.1 - A schema of the motion-tracking architecture

The testbed domain was given by five classes of object shapes (e.g., circles, squares, triangles, diamonds, L-shapes, etc.); each class included sixteen images at increasing scale factors. In training, three images per class (corresponding to three "central" scale factors) were shifted along different translation vectors to simulate a variable-speed motion sequence. The neural network was trained on message differences accordingly. System test included shape images different from training ones, covering the whole scale range for each test class. Results demonstrate how very accurate motion estimation occurs in a scale range close to (and wider than) the one used in training, performance degrades at the borders of the scale range (as expectable). Permuting classes in training and test aimed at avoiding biased result assessments.

A Multi-Agent Classifier
Using Associative Networks in Parallel

Bahram Moobed, Lionel Montoliu*,
Jean-Dominique Gascuel, Michel Weinfeld

Laboratoire d'Informatique (LIX)
Ecole Polytechnique, F-91128 Palaiseau Cedex, France
<moobed, ...>@lix.polytechnique.fr

*CEA C.E. Saclay DMT/SERMA/LETR
F-91191 Gif-sur-Yvette Cedex, France

The associative networks have rarely been used for classification. Here we propose to use binary associative networks (Hopfield Network) as agents in a *multi-agent* architecture for classification of binary patterns. In this approach there is a network dedicated to each possible class for recognizing the patterns belonging to that class. Each network has a *Parasite Detection Mechanism* [1] that allows it to detect if it has converged on a prototype or on a *parasite* (a spurious attractor not learned by network). This mechanism appends a detection code to the prototypes before learning and checks it after relaxation.

Each network learns some patterns (called *prototypes*) representing its corresponding class (*goal class*) and some other patterns representing all the other classes. The difference is that the patterns that represent other classes are learned with a false parasite detection code. So if a network converges on one of these patterns, it will detect it as a parasite. These patterns are called *Artificial Parasites*. For classifying a pattern, it is fed to all the networks and all the networks are relaxed simultaneously. Each network uses the parasite detection mechanism and provides a boolean recognition signal indicating if it has converged on a prototype of his goal class or not. The prototypes must cover roughly the space of the goal class and the artificial parasites must cover the space of all the other classes as well as possible. We use an LVQ network to choose the prototypes and artificial parasites for each network.

For some ambiguous input patterns, several networks may activate their recognition signals. In this case, an algorithm (separate from agents) called the *arbiter* is needed to take an appropriate action: either remove the ambiguity and determine the class of the pattern or start another process that gives more information about the competing classes. The arbiter that we use feeds some noised versions of the pattern as well as the original one to the networks. Then the network that activates most frequently his recognition signal is considered as the winner.

The parasite detection mechanism, as well as the noise generation mechanism and the learning and relaxation algorithms are implemented in a VLSI circuit [2]. This circuit can be used a building block for the hardware implementation of this classifier. In this approach, the modifications and optimizations of a particular network, or the addition of a new class, can be done without much disturbance on the other networks.

References:

[1] J.D.Gascuel, P.Y.Alla, J.Roman, M.Weinfeld, "Implementation of internal annealing and self-identification of successful prototype retrieval in a VLSI feedback neural network chip", International Workshop on Algorithms and Parallel VLSI Architectures, Pont-à-Mousson, Juin 1990, Published in "Algorithms and Parallel VLSI Architectures", E.F. Deprettere ed., Elsevier Science Publishing 1991.

[2] A.Johannet, L.Personnaz, G.Dreyfus, J.D.Gascuel, M.Weinfeld. "Specification and implementation of a digital Hopfield-type associative memory with on-chip training", IEEE Transactions on Neural Networks, Vol. 3, N°. 4, July 1992.

HANDWRITTEN ALPHABET AND DIGIT CHARACTER RECOGNITION USING SKELETON PATTERN MAPPING WITH STRUCTURAL CONSTRAINTS

Kenji NAKAYAMA   Osamu HASEGAWA   Carlos HERNANDEZ E.[+]

Dept. of Electrical & Computer Eng., Kanazawa Univ. Japan
+Computers & Electronics Dept., Valencia Univ. Spain

ABSTRACT

Handwritten character recognition is important application fields of neural networks. Many approaches, based on multilayer neural networks, modified back-propagation and self-organizing methods have been proposed [1]-[3]. However, rotation and high distortions, such as bold lines, non-uniform line width and blurred lines, prevent high recognition rates.

This paper presents a new character recognition method, which can be applied to handwritten alphabet and digit characters, having the above distortion. Basic strategy of the proposed follows human brain like processing, that is "mental distortion". Distorted patterns can be directly applied to the proposed system. The process is based on pattern mapping from standard to distorted patterns, while maintaining essential structure information. This process is similar to Kohonen's self-organizing feature map (SOM) [4]. However, several improvements are proposed in order to apply it to highly distorted pattern recognition.

Let the distorted pattern, standard pattern and its skeleton pattern be $Q(m)$, $P(n)$ and $S(n)$, respectively. First, $Q(m)$ is applied, and appropriate candidates for $Q(m)$ are selected from $P(n)$. Their skeleton pattern $S(n)$ are mapped onto $Q(m)$ following the modified SOM. A point of $Q(m)$, q_i, is randomly selected. The point of $S(n)$, s_j, which locates close to q_i, is selected. s_j is shifted toward q_i with the structure constraints. As the mapping makes progress, structural constraints are gradually relaxed. After the mapping, $Q(m)$ is recognized based on consistency between $Q(m)$ and the mapped $S(n)$. The consistency is measured using line lengths of $S(n)$, which stick out from $Q(m)$, and line lengths of $Q(m)$, which are not covered by $S(n)$. In order to avoid one-stroke pattern mapping onto another one-stroke pattern, $S(n)$ is restored to the standard pattern at some intervals. Furthermore, in order to compensate for line lengths of scaled and distorted patterns, redundant points are removed, and extra points are added in $S(n)$ during the mapping process.

Computer simulation using so many kinds of distorted patterns have been done. High recognition rates are obtained. Figure 1 shows examples.

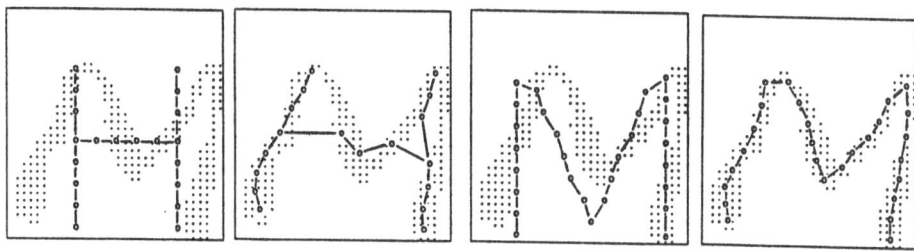

Fig.1 Examples of the mapping, using distorted 「M」, and standard 「H」 and 「M」.

REFERENCES

[1]K.Fukushima et al.,IEEE Trans. vol.SMC-13, pp.826-834, 1983.
[2]Y.Chen et al., Neural Network, vol.1, no.4, pp.541-551, 1989.
[3]K.Nakayama et al., Proc. IJCNN'92 Baltimore, pp.IV-235-239, June 1992.
[4]T.Kohonen, Proc. IEEE, vol.78, n0.9, pp.1464-1480, Sept. 1990.

Optimization of a Signature Verification System Using Neural Networks

Luan Ling Lee
DECOM-FEE-UNICAMP
C.P. 6101, Campinas, SP,
13081-970, Brazil

Toby Berger
School of Electrical Engineering
Cornell University
Ithaca, NY 14853, USA

Signature verification is a two-pattern classification problem or a hypothesis testing problem. A signature verification system for point-of-sales applications requires an extremely low false rejection rate of genuine signature due to the fact that the satisfaction of customers is a top priority [1]. On the other hand, a system for purpose of granting entry to a secure facility must operate at or near zero false acceptance of forgery in order to be assured of barring any intruder. This paper describe two neural network (NN) based approaches for optimization of a signature verification system which uses a majority classifier: 1) minimizing the total frequency of misclassification using the pocket perceptron learning algorithm [3], and 2) simultaneously minimizing the false rejection error and the total frequency of misclassification using a modification of the pocket algorithm.

A majority rule based signature verification system employs a set of 42 personalized parameter features consisting of both dynamic and static features extracted from sampled signatures by using a graphic tablet [2]. A majority classifier assigns weight one or zero to each feature according to a defined condition. The architecture of the networks is inspired by that of the single-cell model proposed in [3]. The NN classifier is a feed-forward network with one-hidden layer. The output unit computes a weighted sum of its binary inputs obtained from normalized feature values after hard limiters. The pocket algorithm and its modification, because of their positive feedback, are well-behaved even for nonseparable problems. The basic principle of the both algorithm is to keep currently the best set of weights in a "pocket" while training the network until a better set is found in the sense of a higher percentage of correct classification.

Two sets of experiments were carried out. Simulated and real signature data were used in the first set and second set of experiments, respectively. Neural network methods are particularly attractive when statistical properties of the feature set are unknown. In all experiments, the neural networks perform better than the majority classifier. The NN classifier trained by the pocket algorithm is considerably sensitive to the size of genuine and forgery training set. This fact is confirmed by the receiver operating characteristic (ROC) mathematical model in [4]. However, the classifier has the smallest number of misclassifications which is the real virtue of the pocket algorithm. The modified pocket algorithm is less sensitive to the size of training sets and is capable of providing better asymptotic performance than the NN trained by the pocket algorithm. We believe that the proposed learning algorithms and the NN classifiers can extent their applications to other practical problems of pattern classification.

References

[1] L. L. Lee and T. Berger, "Adaptive method and system for real time verification of dynamic human signature," *United States Letters Patent*, submitted on 8 November 1991, in review.

[2] L. L. Lee, "On-line systems for human signature verification," Ph.D. thesis, Cornell University, School of Electrical Engineering, January 1992.

[3] S. I. Gallant, "Perceptron-based Learning Algorithms," *IEEE Trans. Neural Networks*, vol. 1, NO. 2, pp. 179-191, June 1990.

[4] L. L. Lee and T. Berger, "On two-pattern classification using neural networks," *Int. Conf. on Signal Processing '93, Beijing, Oct. 1993*, to appear.

Incremental Case-based Pattern Classifier

Joo-Hwee LIM

Institute of Systems Science, National University of Singapore
Heng Mui Keng Terrace, Kent Ridge
Singapore 0511, Republic of Singapore
email: joohwee@iss.nus.sg

ABSTRACT

In this paper, we propose a new Incremental Case-based neural nEtwork (ICE) for pattern classification. We develop the framework upon a case-based learning rule. We show its simplicity, efficiency, and balance between storing specific instances and generalizing exemplars. It also allows nested exemplars to be formed and control of cluster fineness. Finally we compare it with other similar incremental learning algorithms by applying them on off-line handwritten numeral recognition. On the digit database obtained from the Centre for Pattern Recognition and Machine Intelligence, Concordia University, Canada, ICE has attained 87.6% recognition accuracy on the test set, which is at least as good as other learning algorithms: Instance-Based learning algorithm (85.6%), Instance-Averaging learning algorithm (87.6%), Restricted Coulomb Energy network learning algorithm (56.5%), Back-Propagation learning algorithm (85.8%). We also observe that the recognition accuracy of ICE improves as more instances are used in the training and even though the number of prototypes (ie. hidden nodes) increases as well, the rate of increment levels off.

Cepstral Blur Identification by Neural Network for Image Restoration Purpose

Sławomir Skoneczny and Rafał Foltyniewicz

Institute of Control & Industrial Electronics,
Warsaw University of Technology,
00–662 Warszawa, ul. Koszykowa 75, Poland
tel. (+48–2) 6280665, fax. (+48–2) 6256633, e–mail: skonecz@plwatu21.bitnet or
rfoltyn@plwatu21.bitnet

Abstract

A deterministic shift–invariant linear blur is a kind of degradation very often encountered in image processing as a serious problem we must deal with. Examples of such degradation are: camera–motion blur, out–of–focus blur and atmospheric turbulence blur. The exact knowledge of blur type is required for restoration purposes. After restoration we would like to have the perfect image, which is obviously impossible in practice. Instead of that, many sophisticated techniques must be applied in order to obtain satisfying version of the picture but still far from degraded original not available in most cases. The crucial thing is to identify blur type and its parameters. The more accurate we can describe blur phenomenon of degraded image the more precisely we can construct proper filters, which results in better restoration. We propose a method of identifying blur type and parameters by using neural networks which learns and recognizes power cepstra of sums of eigenimages obtained in Singular Value Decomposition. A comparison to learning and recognizing of power cepstra of degraded images without SVD by neural network is given. Theoretical investigations and practical results are presented.

Reduced Pattern Recognizing Neural Nets

Orly Yadid-Pecht and Moshe Gur

Julius Silver Institute and Department of Biomedical Engineering,

Technion – Israel Institute of Technology, Haifa 32000, Israel

ABSTRACT

A scheme that fits a reduced number of feature extractors to a particular recognition task is suggested. The main motivation of the work is to cope with the "curse of dimensionality" which exists in pattern classification systems.

A special scheme is proposed for grading and selecting the "best" features for specific recognition tasks. The "best" features for each class are then assembled to a reduced feature set. The size of the reduced set is determined by the performance quality required. The initial "bank" of feature extractors used is biologically inspired, but the same process can be implemented on any set of input features. Results show that feature reduction is drastic and that very compact nets, of the order of tens of neurons, can be used to classify patterns, even in a noisy environment.

Digit Recognition by The Random Neural Network using Supervised Learning

M. Mokhtari

E.H.E.I., Université René Descartes, 45 rue des Saints Pères, 75006 Paris, France

Abstract

The Random Neural Network model with positive and negative neurons is studied as an autoassociative memory for pattern recognition. A neuron can excite its neighbours if it emits positive signals or can inhibit them by emission of negative signals. There is accumulation of positive signals at positive neurons and negative signals at negative neurons. Let $\underline{k}(t)$ be the vector of neuron potentials at time t. If all the steady state excitation probabilities q_i are such that $0<q_i<1$, the stationary probability distribution of the network's state is

expressed by : $p(\underline{k}) = \prod^{n}_{i=1} (1-q_i) q_i^{k_i}$

where $\quad q_i = \gamma^+(i)/[r(i)+\gamma^-(i)]$ if $i \in$ **P**, $\quad q_i = \gamma^-(i)/[r(i)+\gamma^+(i)]$ if $i \in$ **N** $\qquad$ (1)

$$\gamma^+(i) = \Lambda(i) + \Sigma_{j \in P} \; q_j \; r(j) p^+(j,i) + \Sigma_{j \in N} \; q_j \; r(j) p^-(j,i) \qquad (2)$$

$$\gamma^-(i) = \lambda(i) + \Sigma_{j \in P} \; q_j \; r(j) p^-(j,i) + \Sigma_{j \in N} \; q_j \; r(j) p^+(j,i) \qquad (3)$$

with

P (respectively **N**) : set of positive (respectively negative) neurons

$\Lambda(i)$ (respectively $\lambda(i)$) : external arrival rate of positive (respectively negative) signals to neuron i

r(i): emission rate of signals by neuron i

$p^+(i,j)$ (respectively $p^-(i,j)$): probability that i emits a positive (respectively negative) signal to j, if $i \in$ **P**

$p^+(i,j)$ (respectively $p^-(i,j)$): probability that i emits a negative (respectively positive) signal to j, if $i \in$ **N**.

Let us write: $w^+(i,j)=r(i)p^+(i,j) \geq 0$, $w^-(i,j)=r(i)p^-(i,j) \geq 0$, $r(i)=\Sigma_j[w^+(i,j)+w^-(i,j)]$. E . Gelenbe has presented a learning method based on gradient descent for the recurrent (i.e. general) random network model. The algorithm lets the network learn <u>both</u> n by n matrices $W^{+k} = \{w^+(i,j)\}$ and $W^{-k} = \{w^-(i,j)\}$ by computing for each kth input, a new value W^{+k} and W^{-k} which minimizes the cost function $E^k = 1/2 \; \Sigma^n_{i=1} a_i \; (q_i^k-d_i^k)^2$ where a_i is the contribution of neuron and $d_i^k \in [0,1]$ correspond to its desired value. The rule for weight update is (4) : $w^k(u,v) = w^{k-1}(u,v) - \eta \; \Sigma^n_{i=1} a_i \; (q_i^k-d_i^k) \; [\partial q_i/\partial w(u,v)]^k$ where η is a constant and $w(u,v)$ is either $w^-(u,v)$ or $w^+(u,v)$.

We wanted to realize an auto-associative memory, which learns a set of K patterns and recognizes them even if they are noisy. We have taken 2 layers of n neurons. The output layer is a dupplication of the input layer. When a pattern x^k is applied as input, it characterizes network neurons as positive or negative :
if $x_i^k=-1$ then $i \in$ **N**, $\lambda(i)=\lambda^k(i)$, $\Lambda(i)=0$; if $x_i^k=1$ then $i \in$ **P**, $\Lambda(i)=\Lambda^k(i)$, $\lambda(i)=0$ where $i \in$ Layer 1.

Let us denote by α_0 the storage threshold once the learning stage is finished: $\alpha_0 = \min_{1 \leq k \leq K}(\min_{1 \leq i \leq n} q_i^k)$. The recognition of a pattern x is defined as:

(1) for $i \in$ Layer 1, if $x_i=1$ then $i \in$ **P** else $i \in$ **N**;

(2) for $i \in$ Layer 2, compute its q_i as a positive neuron (q_i^+) and a negative neuron (q_i^-) :

(3) compute the output y

$\qquad$ if $q_i^+ > q_i^-$ then $i \in$ **P** => $y_i=1$ and $q_i = q_i^+$

$\qquad$ if $q_i^- > q_i^+$ then $i \in$ **N** => $y_i=-1$ and $q_i = q_i^-$

$\qquad$ if $q_i^+ = q_i^-$ then $y_i=x_i$ and $q_i = q_i^+ = q_i^-$

(4) if $q_i \geq \alpha_0$ for $1 \leq i \leq n$, the recognition is finished meaning that y (=x) is a stored pattern.

(5) else, x is still noisy. Reinject this output y to the system which then becomes a new input x. Go to (1)

(6) we stop x injection when the reinjection number is larger than n by the maximal supposed noise rate, because at worst there is only 1 correction per pass.

The recognition has been applied to the 10 decimal numerals coded on 64 components We have corrupted the input patterns by a noise rate which reach 20%. The recognition performance is successful for 97%.

Detecting Abnormalities in MRI Images Using the Difference Method

Kwee Hiong Lee

Neural Systems Engineering Laboratory, Department of Electrical Engineering
Imperial College of Science, Technology and Medicine
Exhibition Road, London SW7 2BT

Abstract

With advances in MRI techniques and equipment, it is possible to obtain images of good contrast or high resolution. Taking one step further is to develop a system which processes these images and detects abnormalities within them with reasonable accuracy. This paper describes such an application using neural networks on MRI images of the human head.

As the two halves of the human brain are approximately mirror-image of the other [1], this symmetry property can be used. Hence, instead of operating directly on the image, the Difference Operation is applied on symmetrical areas of the image. Before this, a pre-processing step is needed to position the head (within the image) so that it occupies the top central position in the image. The Difference Operation involves evaluating the absolute difference in intensity between pixels taken from two regions that occupy symmetrical positions in the image. Pairs of pixels are taken with the first from one region and the other from the symmetrical position in the second region. The resulting difference map is operated on by the neural network.

The neural network is a discriminator-based system similar to WISARD [2]. Instead of using Random Access Memory (RAM) as nodes, Multi-valued Probabilistic Logic Nodes (MPLNs) [3] are used. The inputs to the nodes are taken randomly from the input image. As the input to the nodes are binary while the pixel intensities from the image are in 256 gray levels, an intensity coding module is needed. Each input is assigned a buffer with a certain number of memory locations in it. The specified intensity range is divided uniformly among these memory locations which are initially set to "0.5" .With training, the contents are increased to "1" or decreased to "0". This will form a define mapping from a gray scale input to a binary output for each input.

Three main parameters were varied in the simulations. For intensity coding, the number of intensity ranges and the number of gray levels in each range were varied. The product of these two parameters was used as the upper limit of the allowed intensity range. Thus, any pixel intensity above this limit would be set to it. The number of training iterations was also varied from 1 times to 16 times. From the results, it is clear that proposed system can be used to detect abnormalities. Using a small number of intensity ranges and a medium number of gray levels in each range gives the best detection. The detection rate increases with training but decreases after reaching an optimum number of training iterations.

References

[1] Springer, S.P. and Deutsch, G., "Left Brain, Right Brain", Freeman, San Francisco, 1981.

[2] Aleksander,I.;Thomas,W.V. and Bowden,P.A., "WISARD, a radical step forward in image recognition", Sensor Review,vol.4,no 3,1984.

[3] Aleksander,I., "The Logic of Connectionist Systems", in Aleksander,I., eds., "Neural Computing Architectures", Boston, MIT press, 1989.

Neural Networks for the Echographic Diagnosis of
Diffuse Liver Diseases

J.M. Thijssen[1], H.J. Huisman[1], M.S. klein Gebbink[1,2],
J.T.M. Verhoeven[1], T.E. Schouten[2]

[1]Biophysics Laboratory, Department of Ophthalmology, University Hospital and [2]Department of Real-Time Systems, Faculty of Mathematics and Computer Science, Catholic University, 6500 HB Nijmegen, The Netherlands

Three different methods were investigated to determine their ability to detect and classify various categories of diffuse liver disease. A statistical method, i.e. discriminant analysis [1,2], supervised and non-supervised neural networks [3] were examined. The differentiation of disease classes was performed by using a selected set of acoustospectrographic- and image texture parameters [4].

The limited numbers of patients in the various disease classes were successfully extended by generating additional but independent data with identical statistical properties. The generated data were used as the training and test sets. The final assessment of each of the methods was made with the original patient data as a validation set.

It is concluded that neural networks can significantly improve the diagnosis and, therefore, may become an attractive alternative to traditional statistical techniques when dealing with medical detection and classification tasks. Moreover, the use of generated data for the training of the networks and of the discriminant classifier was shown to be justified and profitable.

[1]   B.J. Oosterveld, J.M. Thijssen, P.C. Hartman, G.J.E. Rosenbusch. Detection of diffuse liver disease by quantitative echography: dependence on a priori choice of parameters. Ultrasound Med. Biol. 1993; 19: 21-25.
[2]   P.C. Hartman, B.J. Oosterveld, J.M. Thijssen, G.J.E. Rosenbusch. Detection and differentiation of diffuse liver disease by quantitative echography. Invest. Radiolog. 1993; 28: 1-6.
[3]   M.S. Klein Gebbink, J.T.M. Verhoeven, J.M. Thijssen, T.E. Schouten. Application of neural networks for the classification of diffuse liver disease by quantitative echography. Ultrasonic Imag. 1993; 15: in press.
[4]   J.M. Thijssen, B.J. Oosterveld, P.C. Hartman, G.J.F. Rosenbusch. Correlations between acoustic and texture parameters from RF- and B-mode liver echograms. Ultrasound Med. Biol. 1993; 19: 13-20.

Pattern recognition II
—oral contributions

Neural Networks and the Travelling Salesman Problem
(Extended Abstract)

E.H.L. Aarts[1,2] and H.P. Stehouwer[2]

[1] Philips Research Laboratories, P.O. Box 80000, NL-5600 JA Eindhoven, The Netherlands
[2] Eindhoven University of Technology, Box 513, NL-5600 MB Eindhoven, The Netherlands

Abstract. We briefly review the state of the art of applying neural networks to find approximate solutions to the travelling salesman problem. Furthermore, we present a new self-organizing feature map algorithm and demonstrate that this algorithm can find approximate solutions within a few percent from the optimal tourlength for TSP instances with several thousands of cities.

1. Introduction

In an instance of the travelling salesman problem (TSP) one is given a set of n cities and a distance d_{ij} between each pair i, j of cities. The problem is to find a permutation $\pi : \{1, \ldots, n\} \to \{1, \ldots, n\}$ of the n cities that minimizes the following quantity:

$$\sum_{i=1}^{n} d_{\pi(i),\pi(i+1)} + d_{\pi(n),\pi(1)}.$$

In this quantity, $\pi(i)$ denotes the city that is visited at the ith position in the tour and the quantity itself specifies the length of the tour a salesman would make if he visits the cities in the order specified by the permutation. In this paper we only consider instances of the TSP with symmetric distances satisfying the triangular inequality.

The TSP is probably the best known combinatorial optimization problem and it has served as a proving ground for many new algorithmic ideas; see for instance [16]. The TSP belongs to the class of *NP-hard problems* [11], and, consequently, it is unlikely that polynomial-time algorithms exist that solve each instance of the problem to optimality. So, roughly speaking, there are two options. Either one requires optimality of solutions, at the risk of very large, possibly impracticable running times, or one strives for more quickly obtainable solutions at the risk of sub-optimality. The first option corresponds to optimization, the second one to approximation.

Optimization. Successful optimization algorithms for the TSP are based on enumeration methods using *branch and bound* techniques in combination with sophisticated techniques for generating *cutting planes*. Padberg & Rinaldi [20] report optimal solutions of instances with up to 2392 cities, obtained within 27.3 hours on a Cyber-205. Recently, it was reported that Applegate, Bixby, Chvátal, & Cook obtained an optimal solution of a 3038-city problem [26]. The running time needed to find this solution was estimated to be one-and-a-half year, indicating that the available techniques have reached the limits of tractable instance sizes. Nevertheless, instances with sizes up to a few hundreds of cities are currently routinely solvable.

Approximation. Practice shows that there are many instances of much larger sizes that must be handled. For instance, in printed circuit board design [18] and X-ray crystallography [4] instances are known with sizes up to several tens of thousands of cities. This has raised the question of approximation algorithms, that can find high near-optimal solutions preferably in small running times. The most popular techniques are tour construction heuristics such as *nearest neighbour, nearest insertion, farthest insertion*, and *Christofides' algorithm.* Typical running times of these algorithms range from $O(n^2)$ to $O(n^3)$ [16]. Faster algorithms with running times equal to $O(n \log n)$ or $O(n)$ are obtained by using partitioning approaches; see for instance Karp [14] and Reinelt [23]. Though fast, the effectiveness of these algorithms is moderate; not better than 10% from optimal [13].

Other approaches use *local search* algorithms which are based on the exploration of neighbourhoods. Well-known examples for the TSP are the 2-exchange, 3-exchange, and the variable-depth search algorithms of Lin & Kernighan [17]. These algorithms can find solutions within a few percent from optimal but are relatively slow; straightforward implementations have time complexity functions equal to $\mathcal{O}(n^3)$ or more. Recently the performance of these algorithms was improved by Johnson and others by using sophisticated data structures and multi-level neighbourhoods [13]. Reinelt obtained similar improvements by using reduced neighbourhoods [23]. Both authors report results within a few percent from optimal obtained with running times slightly more than quadratic. Furthermore, instances could be handled with as many as a million cities!

2. Neural network solutions to the TSP

We now address the question of what neural networks can contribute to the search for better TSP algorithms. To this end we first briefly survey the achievements obtained so far. All known results relate to the problem of approximation and the applied techniques can be roughly divided into two classes: recurrent networks and adaptive networks.

Recurrent networks. The seminal paper in this area is due to Hopfield & Tank [12], who use a *Hopfield neural network* to find approximate solutions to the TSP. They introduce a neural network consisting of $\mathcal{O}(n^2)$ neurons and $\mathcal{O}(n^4)$ connections, and an energy function, derived from a quadratic assignment formulation, whose minimization leads to short tour lengths, provided the weights of the connections are carefully chosen.

Hopfield & Tank report simulations for a 10 cities and a 30 cities instance showing that results can be obtained within 5% from optimal. The simulations required a few hours of running time on a main frame. It should be mentioned that these running times are of little significance in this context since hardware implementations of the neural networks will speed up computations with several orders of magnitude. The sole purpose here is to investigate the question of feasibility.

Wilson & Pawley [30] report on a study in which they failed to reproduce the results published by Hopfield & Tank. They found that their network often converged to infeasible solutions and that the approach was very sensitive to the weights of the connections. Furthermore, attempts to scale the algorithm to larger problem instances failed, which they explain with a number of convincing arguments.

Despite its criticism, Wilson & Pawley's paper had a very stimulating effect. Since then, a series of papers appeared in the literature, all aiming at improving Hopfield & Tank's approach, with respect to one or more of the shortcomings pointed out by Wilson & Pawley. For instance, Aarts & Korst [1] present necessary and sufficient conditions for the connection strengths to obtain feasible solutions. Cuykendall & Reese [6] present heuristic scaling rules for the connection strengths. Aiyer, Niranjan & Fallside [2] use an eigenvalue analysis of the Hopfield model to derive theoretical scaling rules. Several authors propose to use modified cost functions; see for instance Van den Bout & Miller [29]. Xu & Tsai [31] present a hybrid approach that combines a classical Hopfield network with *subtour-elimination* heuristics [16]. Finally, Peterson & Söderberg [21] propose to replace the strict binary decision units in the Hopfield model by integer decision units, which is identical to adopting a *Potts glass model* rather than a *spin glass model* in the physical analogue.

These modifications gradually led to an improvement of the performance of the Hopfield & Tank approach. Solutions within 20% from optimal can now be found for instances up to a few hundreds of cities. However, despite all these efforts, we believe that this approach is not the neural network solution to the TSP, since the quadratic assignment formulation of the problem, which is essential to the approach, greatly hampers the network's convergence and thus the scaling of the approach to handle effectively instances larger than a few hundred cities.

Adaptive networks. A second line of neural network approaches to the TSP is based on self-organizing networks such as *elastic nets* and Kohonen's *self-organizing feature maps*. Durbin & Willshaw [7] present an *elastic net algorithm* for the construction of approximate TSP solutions as topologically ordered maps, based on the *tea-trade model* of Von der Malsburg [19]. Roughly speaking, the approach tries to find a topology preserving mapping of a two-dimensional map given by the cities in a plane, onto the nodes of a ring corresponding to a cyclic permutation of the cities.

Durbin & Willshaw's approach can be visualized as an elastic net, which is gradually elongated until it captures all the cities in the plane. A cost function can be associated with the elongation process, whose minimum corresponds to a minimum quadratic length tour. Simic [27] shows that this cost function yields only valid approximations in those cases where the number of units - nodes in the ring - is sufficiently large compared to the number of cities. Durbin & Willshaw present simulation results for the 30-cities instance of Hopfield & Tank [12] and report fast convergence to an optimal solution. For a 100-cities instance they report solutions within a few percent from optimal, which suggests promising scaling properties. The elastic net method was further refined by Burr [5] who considers several modifications which yield better convergence results.

The issue of self-organizing feature maps has been addressed by Kohonen [15] in a more general context. He modelled a self-organization process for the formation of topology preserving feature maps, which can be viewed as an extension Von der Malsburg's model [19]. Furthermore, Kohonen introduced a self-organizing feature map (SOFM) algorithm and showed that the algorithm can be implemented as a neural network with units whose functionality is similar to that of perceptrons [25].

Ritter & Schulten [24] derived a mathematical formalism of Kohonen's SOFM algorithm and applied it to the TSP in a way similar to that of the elastic net method of Durbin & Willshaw [7]. Ritter & Schulten present simulation results for a random 30-cities instance in the unit square and report convergence to an optimal solution. Fort [9] presents a modified SOFM algorithm with no *ad hoc* parameter choices. Furthermore, he discusses its implementation as a classical neural network. On the basis of simulations he claims satisfactory results for instances up to 400 cities.

A major breakthrough was obtained by Angéniol, De la Croix Vaubois & Le Texier [3], who extended Durbin & Willshaw's approach by including a unit creation and deletion mechanism, which greatly improves the network's convergence. They show results of simulations yielding solutions within 20% from optimal for instances up to 1000 cities, and claim similar quality results in all cases. Favata & Walker [8] present an algorithm based on the Kohonen's original SOFM algorithm [15] with some additional features. They present simulation results for instances up to 10000 cities and claim solutions within 20% from optimal. In a recent paper, Fritzke & Wilke [10] propose to use a hybrid approach that is partly based on Kohonen's algorithm. They replace the fixed-size ring structure by a dynamic structure that is initially small and gradually encreases to a full-size structure with n nodes. They claim a linear time complexity and present solutions within 9% from optimal for instances with up to 2392 cities.

3. A new SOFM algorithm for large TSP instances

While the recurrent network approach to the TSP is open to criticism, the results obtained with adaptive networks are quite promising. This motivates a study to fully exploit the capabilities of adaptation through self-organization to handle large TSP instances. In this section we propose a new SOFM algorithm for generating topology preserving bijective mappings. We present a stochastic gradient descent method which we conjecture to converge to mappings that are feasible and order preserving, thus corresponding to high-quality solutions of the TSP.

Network description. Given an Euclidean instance of the TSP with n cities. Each city has coordinates $z \in \mathbb{R}^2$ and the set of n city coordinates is denoted by C. The Euclidean distance is denoted by $d_E(\cdot, \cdot)$. The corresponding network consists of a fully connected network K of n units, labeled $1, 2, \ldots, n$, and two additional input units to which each unit in K is connected. The weights of the connections between the units are chosen as follows. For each unit $i \in K$, $w_i \in \mathbb{R}^2$ denotes the weights of the connections between i and the two input units, and $v_i \in \mathbb{R}^n$ denotes the weights of the connections between i and all units in K, including i. The output of a unit $i \in K$ is denoted by o_i and the output of all units in K is denoted by $o \in \mathbb{R}^n$.

The units are ordered on a ring which is represented by a *cycle* on n nodes. The underlying idea is similar to that used by Favata & Walker [8], Angéniol, De la Croix Vaubois & Le Texier [3], and Ritter & Schulten [24]. A tour is obtained by a bijective mapping of the city coordinates onto the nodes of the cycle in such a way, that distance relations are preserved "as much as possible", i.e., cities that are close with respect to their distance measure should be mapped onto nodes on the cycle that are close in some sense.

In the earlier work mentioned above, one uses the quadratic Euclidean distance measure. This measure may deviate arbitrarily far from the linear Euclidean distance measure and thus may lead to low quality solutions. Therefore, we consider here the linear Euclidean distance measure. The distance in the cycle is defined as the ordinary distance in a graph, i.e. the length of a shortest path between two vertices in terms of the number of edges of the path, which we denote by $\Delta(\cdot, \cdot)$.

Algorithm description. The algorithm we describe here is an iterative procedure. Each iteration contains two steps.

In the first step, the coordinates z of a randomly chosen city are presented to the input units of the network. The output of unit i at time t is equivalent to that described by Tolat [28] and is given by

$$o_i(t) = \sigma[\exp(-f(d_E(z, w_i)) + v_i^T o(t-1)], \tag{1}$$

where $f(\cdot)$ is a monotonically increasing function of the Euclidean distance between z and w_i and $\sigma(\cdot)$ is a sigmoidal real-valued function with a range between 0 and 1.

Kohonen [15] has shown that, for appropriate weights v_i, a network with the dynamics given by (1) exhibits a winner-takes-it-all performance, i.e., all activity is concentrated around the unit j whose weight vector w_j is closest to the input vector z. This unit j is called the *image* in the cycle of input vector z. The exact form of the final activity depends on the weights v_i, but is generally Gaussian-shaped. To speed up the simulations, we replace the output of unit i, determined by the network dynamics of (1), with a simple expression given by

$$o_i = h_\beta(\Delta(i, j)), \tag{2}$$

where $h_\beta(\cdot)$ is a Gaussian function of the form $h_\beta(x) = \exp(-x^2/2\beta^2)$ with $\beta \in \mathbb{R}^+$ and j denotes the unit for which $d_E(w(j), z)$ is minimal over $j \in K$.

In the second step, the weights w_i are updated according to the gradient descent rule given by

$$
\begin{aligned}
w_i(s+1) &= w_i(s) - \varepsilon(s) o_i(s) \nabla_w d_E(z, w_i(s)) \\
&= w_i(s) + \varepsilon(s) h_\beta(\Delta(i, j))(s) \frac{(z - w_i(s))}{d_E(z, w_i(s))},
\end{aligned} \tag{3}
$$

where the s's have been added to indicate the sth step in the iteration and $\varepsilon \in \mathbb{R}^+$ denotes the *step size*.

In pseudo code the algorithm can be written as follows.

```
INITIALIZE(s_max, ε, β, w_1, w_2, ..., w_n);
s := 0;
while s < s_max do
begin
    SELECT(z ∈ C');
    j :∈ K;
    for all k ∈ K do if d_E(z, w_k) < d_E(z, w_j) then j := k;
    for all i ∈ K do w_i := w_i + εh_β(Δ(i,j))(z − w_i)/d_E(z, w_i);
    ADJUST(ε, β);
    t := t + 1;
end;
```

Both ε and β are adjusted according to a predefined schedule and the weights w_i are randomly initialized.

After termination of the algorithm, the images of the cities in the cycle are calculated and a solution to the TSP is obtained by sorting the cities according to their images. This may lead to a situation in which two or more cities have the same image and thus the mapping is not a bijection representing a tour. To overcome this problem we enforce a bijective mapping by randomly ordering the cities that have the same image.

Simulation results. In all our simulations we used the following functions. **INITIALIZE** randomly computes initial values of the weights w_i from the smallest enveloping rectangle containing all the city coordinates. **SELECT** randomly selects a coordinate pair. **ADJUST** uses a linear decrement of ε from ε_i to ε_f and a geometric decrement of β from β_i to β_f. The value of β_i was chosen equal to the number of units and in all simulations β_f was set equal to 1.0. The values of ε_i and ε_f were adjusted to tune the convergence. We have performed simulations with

Table 1: Simulation results.

instance	n	l_{lwb}	s_{max}	ε_i	ε_f	d_{best}	d_{avg}	N
ATT532	532	27686	5+e5	50.0	5.0	4.9	6.5	5
PR1002	1002	259045	1+e6	100.0	5.0	6.0	7.6	5
U2319	2319	234256	1+e6	100.0	1.0	6.2	6.7	5
PR2392	2392	378032	5+e6	100.0	1.0	8.6	8.7	2
PCB3038	3038	137694	5+e6	50.0	0.1	10.0	10.1	4
FL3795	3795	28594	5+e6	10.0	0.4	12.2	12.4	2
RL5915	5915	560937	5+e6	55.0	0.1	18.7	18.8	2
RL5934	5934	552082	5+e6	100.0	0.9	15.8	16.5	4
RL11849	11849	913981	5+e6	100.0	0.9	17.4	17.4	2

instances from the Reinelt data base [22] ranging from 532 to 11849 cities, with known lower bounds. Table 1 shows the results obtained from a number of simulation runs for each instance using different initializations. d_{best} and d_{avg} denote the smallest and the average deviation, respectively, from the lower bound l_{lwb} obtained in N simulation runs. The lower bounds are taken from Reinelt [22].

4. Concluding remarks

We have discussed the significance of neural networks to the problem of handling effectively large instances of the TSP. Broadly speaking, it is unlikely that recurrent networks such as the Hopfield networks can contribute much to the solution of this problem. More promising are self-organizing networks such as the Kohonen feature maps. With these networks solutions with reasonable quality can be obtained for instances up to several thousands of cities. Nevertheless, there is still a large gap between the effectiveness of neural networks and that of the best-known approximation algorithm for the TSP.

References

[1] AARTS, E.H.L., AND J.H.M. KORST (1989), Boltzmann machines for travelling salesman problems, *European Journal of Operational Research* **39**, 79-95.

[2] AIYER, S.V.B., M. NIRANJAN, AND F. FALLSIDE (1990), A theoretical investigation into the performance of the Hopfield model, *IEEE Transactions on Neural Networks* **1**, 204-215.

[3] ANGÉNIOL, B., G. DE LA CROIX VAUBOIS, AND J.Y. LE TEXIER (1988), Self-organizing feature maps and the travelling salesman problem, *Neural Networks* **1**, 289-293.

[4] BLAND, R.G., AND D.F. SHALLCROSS (1989), Large traveling salesman problems arising from experiments in X-ray crystallography, a preliminary report on computation, *Operations Research Letters* **8**, 123-133.

[5] BURR, D.J. (1988), An improved elastic net method for the traveling salesman problem, *Proceedings of the IEEE Conference on Neural Networks*, San Diego, 69-76.

[6] CUYKENDALL, R., AND R. REESE (1989), Scaling the neural TSP algorithm, *Biological Cybernetics* **60**, 365-371.

[7] DURBIN, R., AND D. WILLSHAW (1987), An analogue approach to the travelling problem using an elastic net method, *Nature* **326**, 689-691.

[8] FAVATA, F., AND R. WALKER (1991), A study of the application of Kohonen type neural networks to the travelling salesman problem, *Biological Cybernetics* **64**, 463-468.

[9] FORT, J.C. (1988), Solving a combinatorial problem via self-organizing process: an application of the Kohonen algorithm to the traveling salesman problem, *Biological Cybernetics* **59**, 33-40.

[10] FRITZKE, B., AND P. WILKE (1992), FLEXMAP - a neural network for the traveling salesman problem with linear time and space constraints, *Proceedings of the IEEE Joint Conference on Neural Networks*, 929-934.

[11] GAREY, M.R., AND D.S. JOHNSON (1979), *Computers and Intractability: A Guide to the Theory of NP-Completeness*, W.H. Freeman and Co, San Francisco.

[12] HOPFIELD, J.J., AND D.W. TANK (1985), Neural computations of decisions in optimization problems, *Biological Cybernetics* **52**, 141-152.

[13] JOHNSON, D.S. (1990), Local optimization and the travelling salesman problem, *Springer Lecture Notes in Computer Science* **447**, 446-461.

[14] KARP, R.M. (1977), Probabilistic analysis of partitioning algorithms for the traveling-salesman in the plane, *Mathematics of Operations Research* **2**, 209-224.

[15] KOHONEN, T. (1988), *Self-Organization and Associative Memory*, 2nd ed., Springer-Verlag, Berlin.

[16] LAWLER, E.L., J.K. LENSTRA, A.H.G. RINNOOY KAN, AND D.B. SHMOYS (1985), *The Traveling Salesman Problem*, John Wiley & Sons, Chichester.

[17] LIN, S., AND B.W. KERNIGHAN (1973), An effective heuristic algorithm for the traveling salesman problem, *Operations Research* **21**, 498-516.

[18] LITKE, J.D. (1984), An improved solution to the traveling salesman problem with thousands of nodes, *Communications of the ACM* **27**, 1227-1236.

[19] MALSBURG, C. VON DER, AND D.J. WILLSHAW (1977), How to label nerve cells so that they can interconnect in an ordered fashion, *Proceeding of the National Academy of Sciences, USA* **74**, 5176-5178.

[20] PADBERG, M., AND G. RINALDI (1988), *A branch-and-cut algorithm for the resolution of large-scale symmetric traveling salesman problems*, Report R. 247, Instituto di Analisi dei Sistemi ed Informatica del CNR, Rome.

[21] PETERSON, C., AND B. SÖDERBERG (1989), A new method for mapping optimization problems onto neural networks, *International Journal of Neural Systems* **1**, 3-22.

[22] REINELT, G. (1991), TSPLIB, A traveling salesman problem library, *ORSA Journal on Computing* **3**, 376-384.

[23] REINELT, G. (1992), Fast heuristics for large geometric traveling salesman problems, *ORSA Journal on Computing* **4**, 206-223.

[24] RITTER, H., AND K. SCHULTEN (1988), Convergence properties of Kohonen's topology conserving maps: fluctuations, stability and dimension selection, *Biological Cybernetics* **60**, 59-71.

[25] ROSENBLATT, F. (1962), *Principles of Neurodynamics*, Spartan Books, Washington DC.

[26] SANGALLI, A. (1992), Short-circuiting the travelling salesman problem, *New Scientist*, June 1992, 16-17.

[27] SIMIC, P.D. (1990), Statistical mechanics as the underlying theory of 'elastic' and 'neural' optimisations, *Networks* **1**, 89-103.

[28] TOLAT, V.V. (1989), A self-organizing neural network for classifying sequences, *Proceedings of the IEEE Joint Conference on Neural Networks*, 561-566.

[29] VAN DEN BOUT, D.E., AND T.K. MILLER (1989), Improving the performance of the Hopfield-Tank neural network through normalization and annealing, *Biological Cybernetics* **62**, 129-139.

[30] WILSON, G.V., AND G.S. PAWLEY (1988), On the stability of the travelling salesman problem algorithm of Hopfield and Tank, *Biological Cybernetics* **58**, 63-70.

[31] XU, X., AND W.T. TSAI (1991), Effective neural algorithms for the traveling salesman problem, *Neural Networks* **4**, 193-205.

Automatically Structured Neural Networks For Handwritten Character And Word Recognition

Ulrich Bodenhausen and Stefan Manke
University of Karlsruhe, Computer Science Department
Postbox 6980, 7500 Karlsruhe 1, FRG

Abstract

Highly structured neural networks like the Time-Delay Neural Network (TDNN) can achieve very high recognition accuracies in real world applications like on-line handwritten character and speech recognition systems. Achieving the best possible performance greatly depends on the optimization of all structural parameters for the given task and amount of training data. We propose an Automatic Structure Optimization (ASO) algorithm that avoids time-consuming manual optimization and apply it to Multi State Time-Delay Neural Networks (MSTDNNs), a recent extension of the TDNN. We show that MSTDNNs are a very powerful approach to on-line handwritten character and word recognition and that the ASO algorithm can automatically structure this type of architecture efficiently in a single training run.

Introduction

Time-Delay Neural Networks (TDNN) [1] with shifted input windows have been successfully applied to speech recognition and on-line handwritten character recognition tasks [2,3,4]. The main feature of the TDNN architecture is the use of a highly structured connectivity between the units of the network. This structured connectivity reduces the number of trainable parameters and also ensures a translation invariant recognition. One reason for the introduction of structure to the network is the relationship between the number of trainable parameters, amount of training data and generalization [5]. Networks with too many trainable parameters for the given amount of training data learn well, but do not generalize well. This phenomenon is usually called overfitting. With too few trainable parameters, the network fails to learn the training data and performs very poorly on the testing data. Imposing structure into the network can increase the generalization performance by reducing the number of trainable parameters [1].

The use of a highly structured approach leads to the problem of finding the best possible structure for the given task and amount of training data. In order to achieve optimal performance without time-consuming manual optimization of the architecture, we propose an Automatic Structure Optimization (ASO) algorithm that automatically optimizes the structure and the total number of parameters in a single training run and also considers the current amount of training data. Rather than starting with a distributed internal representation, the structure of the network is constructed by adding units and connections in order to selectively improve certain parts of the network.

The ASO algorithm is applied to the optimization of Multi State Time-Delay Neural Networks (MSTDNNs), a recent extension of the TDNN [6, 7]. These networks have been originally proposed for continuous speech recognition, but we show that they can also be successfully applied to on-line handwritten character recognition tasks. In principle, they allow the recognition of sequences of ordered events that have to be observed jointly. For example, in their application to speech recognition, the recognition of words is decomposed into the recognition of sequences of phonemes or phoneme like units. In handwritten character recognition the recognition of characters can be decomposed into the joined recognition of characteristic strokes etc.. The approach can be extended to the recognition of handwritten cursive words by decomposing the recognition of words into the joined recognition of characteristic strokes.

The combination of the proposed ASO algorithm with the MSTDNN was applied successfully to on-line handwritten character recognition tasks with varying amounts of training data.

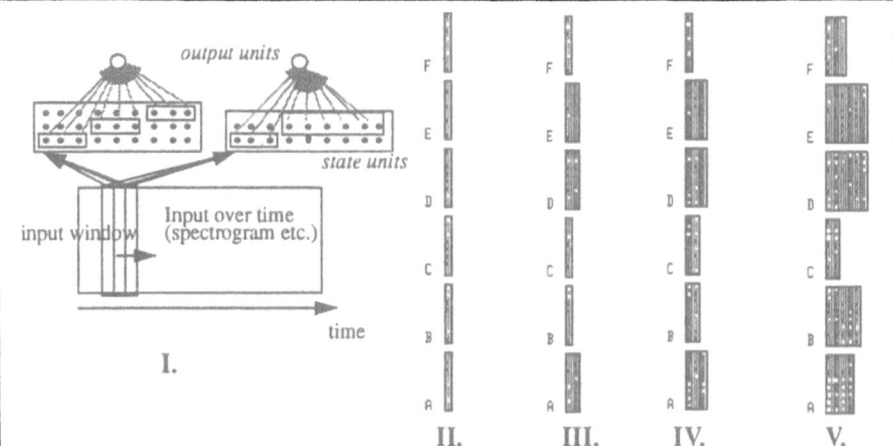

Fig. 1: I. *An example of a simple MSTDNN with an input layer, a state layer and an output layer (consisting of two ouput units). In this example the first output unit is connected with three state units and the second output unit is connected with two state units. **II.** The weights from the input units to the state units for the recognition of the capital letters A .. F after initialization. All windows have a size of one. Negative weights are displayed by white blobs, positive weights by black blobs. Eight input units are used to represent the features recorded from the touch sensitive tablet (see [2]). **III.** The weights after the first epoch. The input window for the characters "A", "D" and "E" were increased to a size of two. The input windows for the characters "B", "C" and "F" were not increased. **IV.** The input window for the characters "A", "D" and "E" were increased again. The windows for the characters "B" and "C" were increased to a size of two. **V.** The weights after the fifth epoch.*

The Automatic Structure Optimization Algorithm (ASO)

The ASO algorithm is based on four principles: Built-in invariances, task decomposition, confusion matrix dependent construction of the network and early constructive changes which are described in more detail in [3, 4]. Multi State Time-Delay Neural Networks (MSTDNNs), an extension of the TDNN, are a suitable candidate for the application of the ASO algorithm because they have been shown to be very powerful classifiers if the structure of the network is well adapted to the task [6, 7]. For best performance, the following architectural parameters are important:

- the size of the input windows (i.e. how much temporal context does a unit need)
- the number of hidden units between the input and the state units (i.e. how complex is the mapping from the input to the state units)
- the state sequence topologies (i.e. the number of states per output unit).

The ASO algorithm optimizes all of these parameters synergetically and does not require the computational effort to compute second derivatives like Optimal Brain Damage [9] or Optimal Brain Surgeon [10]. Other constructive algorithms like Cascade Correlation [8] only consider the optimization of the number of hidden units, which is only one relevant architectural parameter for the given application.

The ASO algorithm starts with the minimal configuration of a MSTDNN which consists of an input layer, a state layer and an output layer (see Fig. 1). Let us consider an on-line handwritten character recognition task where each output unit represents a character. The basic idea of our application of the MSTDNN is to recognize a character by recognizing a sequence of parts (e.g. strokes) of the character. Each state unit represents a certain stroke of the character.

The network is initialized with a window size of one (see Fig. 1.II.) and a single state unit per output unit. The net input of the output units is computed by integrating the weighted activity of the single or multiple state unit(s) over time.

The idea behind the ASO algorithm is that the confusion matrix C (see Table 1) is used to determine the size of the input windows and the number of states. The confusion-symmetry matrix S (see Table 2) is used

D 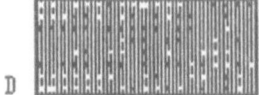   D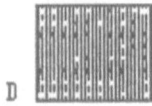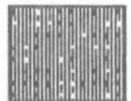

Fig. 2: The splitting of a state in the same training run as in Fig. 1. The modeling of the character "D" with one state and a big window is still insufficient in the 18th epoch (left side). A new state is added in the 19th epoch and the size of the old window is halfed (right side).

to allocate hidden units. At first, the size of the input window is increased for the units that make more mistakes than the average unit by adding one set of random connections (see Fig. 1, II - V). If the size of the input window of a state unit gets close to the average duration of a character and the corresponding output unit still makes more mistakes than the average unit, then a new state unit is added (see Fig. 2). The size of the input window of the 'old' (first) state unit is halved to avoid a dramatic increase of the number of trainable parameters. The 'new' (second) state unit receives input from an input window of the same size as the 'old' state unit, but it is installed with random connections. From now on, the output unit receives input from both state units.

TABLE 1. The confusion matrix on the training set for the recognition of capital letters after 100 epochs. The target outputs are shown on the horizontal axis. The actual output of the network is shown on the vertical axis. The numbers on the diagonal indicate the correct responses of the network. All other numbers show wrong responses and are used for resource allocation. The gray column shows the class with most mistakes.

	A	B	C	D	E	F	G	H	I	J	K	L	M	N	O	P	Q	R	S	T	U	V	W	X	Y	Z
A	74	1						4				1	2											1		
B		42		1	9		1								1		1	1	1							1
C			55	1		3							2							1						1
D		2		67											1	2	1									
E					60	2	1																			4
F		4			3	26	1	7					4	1				1		1				1	1	
G				2			53	2					1		2						1				2	
H	2							49	2												1			3		
I		2				1	1		38	2								3	5			2				
J									5	47										1		2		1	2	2
K						1					62					2					1					
L												58														1
M						1	2	2					59	3				1			1					
N	1					1		1					1	50						1	2	1				
O			2			1	1		1						42		6		1		1					
P		1		1		5										59		1	1							
Q			1		1										1		41	2								
R		6				7		1								1		50								
S				1					1										55					2		
T						1														48			1			
U		1											1								42	4	2		1	
V		2					1						2	1	1	1					6	46				
W							1			1	1										2	3	47			2
X		1				4		1							1				1	1				53	1	
Y			1				1	1	1											2	1	1			48	
Z				1			2	2																		46

The confusion-symmetry matrix is used to add hidden units between the input and the state units similar to the Cascade Correlation algorithm [8]. Fig. 3 shows the basic idea. If the network has no hidden units but the data requires a nonlinear decision boundary, the linear decision boundary will try to compromise and pairwise confusions appear (see Fig. 3). Hidden units are installed to solve the most important pairwise confusions. This is done by initializing the connections between the new hidden unit and the state units of the pairwise confused characters with weights with different signs. The inputs to the hidden units are ini-

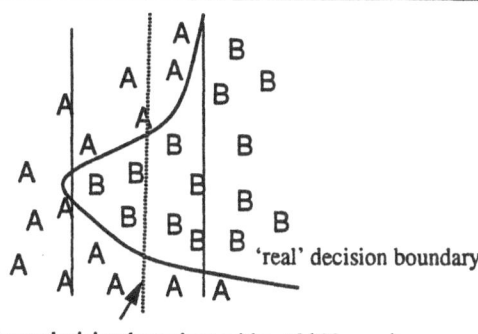

Fig. 3: An example where the classes 'A' and 'B' are confused because only linear decision boundaries can be learned without hidden units. Without hidden units, the decision boundary will be somewhere between the two straigt lines. One possible solution is shown by the dotted line. In this case the pairwise confusions ("A" with "B" and vice versa) will be very frequent.

'real' decision boundary

linear decision boundary without hidden units

tialized randomly. If the same pairwise confusions appear in the following epochs, the input window of the hidden unit is increased before new hidden units are allocated.

The ASO algorithm modifies the criterium for the allocation of new resources depending on the number of training patterns and the total number of parameters smoothly. The underlying idea is that adding more resources should be easy if the number of connections is small compared to the number of training patterns and should get harder with an increasing number of parameters. This avoids hard upper bounds for the network resources.

TABLE 2. The confusion-symmetry matrix for the recognition of capital letters after 100 training epochs. This matrix shows pairwise confusions (class "A" is confused with class "B" and vice versa). The elements s_{ij} are computed from the elements c_{ij} of the confusion matrix as $s_{ij} = c_{ij}c_{ji}$. Because $s_{ij} = s_{ji}$, only elements below the diagonal are shown. This particular confusion-symmetry matrix shows that "B" and "F" is the most frequent pairwise confusion. Hidden units are allocated to solve these pairwise confusions. (*: not relevant for the structuring process)

	A	B	C	D	E	F	G	H	I	J	K	L	M	N	O	P	Q	R	S	T	U	V	W	X	Y	Z
A	*	-	-	-	-	-	-	-	-	-	-	-	-	-	-	-	-	-	-	-	-	-	-	-	-	-
B		*	-	-	-	-	-	-	-	-	-	-	-	-	-	-	-	-	-	-	-	-	-	-	-	-
C			*	-	-	-	-	-	-	-	-	-	-	-	-	-	-	-	-	-	-	-	-	-	-	-
D				*	-	-	-	-	-	-	-	-	-	-	-	-	-	-	-	-	-	-	-	-	-	-
E					*	-	-	-	-	-	-	-	-	-	-	-	-	-	-	-	-	-	-	-	-	-
F		36		6		*	-	-	-	-	-	-	-	-	-	-	-	-	-	-	-	-	-	-	-	-
G				2			*	-	-	-	-	-	-	-	-	-	-	-	-	-	-	-	-	-	-	-
H	8							*	-	-	-	-	-	-	-	-	-	-	-	-	-	-	-	-	-	-
I		2				7			*	-	-	-	-	-	-	-	-	-	-	-	-	-	-	-	-	-
J								10		*	-	-	-	-	-	-	-	-	-	-	-	-	-	-	-	-
K											*	-	-	-	-	-	-	-	-	-	-	-	-	-	-	-
L												*	-	-	-	-	-	-	-	-	-	-	-	-	-	-
M					4								*	-	-	-	-	-	-	-	-	-	-	-	-	-
N	2												3	*	-	-	-	-	-	-	-	-	-	-	-	-
O			4			1									*	-	-	-	-	-	-	-	-	-	-	-
P			2													*	-	-	-	-	-	-	-	-	-	-
Q						2								6			*	-	-	-	-	-	-	-	-	-
R		6				7									1			*	-	-	-	-	-	-	-	-
S									1										*	-	-	-	-	-	-	-
T						1														*	-	-	-	-	-	-
U																					*	-	-	-	-	-
V														2							24	*	-	-	-	-
W																					4		*	-	-	-
X					12													1						*	-	-
Y								2													4	1			*	-
Z				4			4	4																		*

Most resources that are added to the network are initialized randomly. This reduces the risk that new resources disturb the learning process. A positive side-effect is that noise is added which prevents the network from getting stuck in local minima and also greatly reduces the risk of overfitting. This noise is reduced afterwards by gradient decent training.

In case of more than one state unit per output unit the inputs of the output units can be computed in three different ways: The simplest way is to give each state unit an equal share of the time slice that the output unit respresents. The second possibility is to use *Dynamic Time Warping* (DTW) [11] to find the best path through the activation matrix of the state units. The third possibility is to smooth the DTW path by Gaussian functions positioned according to the DTW segmentation [4]. Smoothing of the DTW path allows the states to model the transitions between two states more accurately.

Simulations

The ASO algorithm was tested with two on-line handwritten character recognition tasks:

- Recognition of the digits 0, 1, 2, ..., 9 written on a touch sensitive tablet (aprox. 1000 digits written by aprox. 70 writers, recorded as described in [2]: During writing, the position and the pressure of the pen are recorded from the tablet. Resampling is used to reduce the temporal variations of the digits. From these data points, the directions and the curvatures of the pen strokes are computed and are added to the data.

- Recognition of the capital letters A, B, ..., Z written on a touch sensitive tablet (aprox. 2500 capital letters written by aprox. 50 writers, recorded in the same way)

These tasks are small enough to allow a reasonable number of experiments with the algorithm and are large enough to be relevant for an application oriented algorithm. The databases were cut into training data, validation data and testing data. The validation data is used to determine the stopping criterium for the training phase. The results for both manually and automatically optimized architectures are summarized below (see Table 3 and 4). Results with different manually optimized architectures (single state TDNNs with hidden layer) are added for the handwritten digit recognition task for comparison.

Table 4 shows that the MSTDNN network optimized by the ASO algorithm can adapt to different amounts of training data. The handtuned architecture performed equally well for the amount of data that it was optimized for, but did not generalize as well for more data and failed to learn a small subset completely for various learning rates and momentums.

Conclusions

In this paper we have shown that the ASO algorithm can achieve equal or better recognition performances than manually tuned network architectures. It offers the flexibility to use a given amount of available training data without the need to manually adapt the architecture to this amount. The handtuned architectures performed equally well for the amount of training data they were optimized for, but did not completely make use of a larger training set and even failed to learn a very small subset of the training set. The ASO algorithm performs similar on a speech recognition task we have tested [3, 4].

Another advantage of the ASO algorithm is the reduced danger of overfitting. Many connections are added late in the training run when the error is already very low. These connections are never trained by large error derivatives and their weights remain very close to their random initialization. This explains our observation that ASO sometimes constructs rather large networks which generalize surprisingly well. Similar effects have been investigated by Moody who introduced the *effective* number of parameters as the relevant variable for generalization capability [5]. Our experimental results suggest that the ASO algorithm builds networks with a small number of *effective* parameters. The generalization performances of the algorithm seems to benefit from *unequal* training: Very few connections get a lot of training and most connections get very little training and remain close to their random initialization.

Compared to subtractive optimization techniques like Optimal Brain Damage [9] and Optimal Brain Surgeon [10] the algorithm is rather simple to implement because it does not require the computation of n^2 second derivatives for a network with n connections (which is expensive for typical speech and on-line character recognition applications with $n > 5000$). Another disadvantage of the subtractive methods is the inability to recover when an important connection was accidentally removed. The ASO algorithm behaves more smoothly in this respect: If a new connection is installed that was not necessary, only a slight loss in generalization performance can appear.

Our results show that the ASO algorithm is able to optimize MSTDNN architectures for spatio-temporal real world applications with varying amounts of training data effectively. The principles of the ASO algorithm should also be applicable for other architectures.

We have also shown that MSTDNN architectures can be used effectively for handwritten character recognition tasks. Work is in progress to extend our MSTDNN architectures for continuous (cursive) handwriting. In preliminary simulations we have obtained a 80.15% word accuracy for a 400 word vocabulary.

Table 3: Digit Recognition Performances

	training	testing
manually optimized MSTDNN architecture	98.3%	96.5%
automatic optimization of the window size and the number of state units	99.6%	98.0%
automatically optimized architecture with gaussian smoothing of the DTW path	100%	99.5%
TDNN architecture proposed by [2] on the same data	100%	95.5%
TDNN architecture manually optimized for the same data	100%	98.5%

Table 4: Capital Letter Recognition Performances depending on training set size

number of training patterns	TDNN architecture manually optimized for 1170 training patterns	automatically optimized MSTDNN architecture
520	no convergence	81.5%
1170	88.5%	88.5%
1560	90.5%	91.3%

Acknowledgements

The authors gratefully acknowledge the support of the McDonnel-Pew Foundation (Cognitive Neuroscience Program) and would like to thank Alex Waibel and Scott Fahlman for lots of helpful discussions.

References

[1] A. Waibel, T. Hanazawa, G. Hinton, K. Shiano, and K. Lang. Phoneme Recognition using Time-Delay Neural Networks. *IEEE Transactions on Acoustics, Speech and Signal Processing*, March 1989.

[2] I. Guyon, P. Albrecht, Y. Le Cun, J. Denker, and W. Hubbard. Design of a Neural Network Character Recognizer for a Touch Terminal. *Pattern Recognition*, 24(2), 1991.

[3] U. Bodenhausen and S. Manke. Connectionist Architectural Learning for High Performance Character and Speech Recognition. In: *Proceedings ICASSP-93*, Minneapolis, April 1993.

[4] U. Bodenhausen and A. Waibel. Application Oriented Automatic Structuring of Time-Delay Neural Networks for High Performance Character and Speech Recognition. In: *Proceedings ICNN 93*, San Francisco, March 1993.

[5] J. Moody. The Effective Number of Parameters: An Analysis of Generalization and Regularization in Nonlinear Learning Systems. In: *Advances in Neural Information Processing Systems 4*, 1991.

[6] P. Haffner, M. Franzini, and A. Waibel. Integrating Time Alignment and Neural Networks for High Performance Continuous Speech Recognition. In *Proceedings of the ICASSP-91*.

[7] P. Haffner and A. Waibel. Time-Delay Neural Networks Embedding Time Alignment: A Performance Analysis.In: *Proceedings Eurospeech 91*.

[8] S. E. Fahlman and C. Lebiere. The Cascade-Correlation Learning Architecture. In: *Advances in Neural Information Processing Systems 2*, 1989.

[9] Y. Le Cun, J. S. Denker, and S. A. Solla. Optimal Brain Damage.. In: *Advances in Neural Information Processing Systems 2*, 1989.

[10] B. Hassibi and D. G. Stork. Second Order Derivatives for Network Pruning: Optimal Brain Surgeon. In: *Advances in Neural Information Processing Systems 5*, 1993

[11] H. Sakoe and S. Chiba. Dynamic Programming Algorithm Optimization for Spoken Word Recognition. *IEEE Transactions on Acoustics, Speech and Signal Processing*, (26): 43-49, 1978.

Tracking Rain Cells in Radar Images using Multilayer Neural Networks*

X. Ding[1], T. Denœux[2] and F. Helloco[3]

[1] Lyonnaise des Eaux - Dumez / LIAC
Technopolis - 14 Rue du Fonds Pernant
F-60471 Compiègne cedex- France

[2] Université de Technologie de Compiègne – U.R.A. CNRS 817
BP 649 - F-60206 Compiègne cedex - France
email: tdenoeux@hds.univ-compiegne.fr

[3] RHEA - 11 rue du Vieux Pont F-92000 Nanterre - France

Abstract

Very short-term forecasting of rainfall amounts on small watersheds can be achieved by extrapolating the advection of rain cells perceived as "echoes" on weather radar images. Pattern recognition oriented approaches [3] to this problem are based on (1) echo definition; (2) echo description by features; (3) echo matching; and (4) forecast by advection extrapolation. In this paper, the most critical phase — echo matching — is tackled using multilayer neural networks. A comparison with decision trees is provided, and several ways of combining both approaches are considered.

1   Introduction

The utility of weather radars for short term rainfall forecasting has been recognized since the early 1950's. The first automatic forecasting systems have been based on the estimation of a mean advection vector for all reflectivity patterns in the whole image [1]. In order to take heterogeneous advection into account, recent research has focused on forecasting the advection individually for each rain cell, perceived on radar images as reflectivity patterns [3] (Figure 1). Pattern recognition oriented approaches have been proposed [3], based on (1) echo definition; (2) echo description by features; (3) echo matching; and (4) forecast by advection extrapolation. Because the variability of precipitation systems can be extremely high in some meteorological situations, the matching phase, involving

*This work has been supported by EEC funded Esprit II project nr. 5433 (NEUFODI); partners: BIKIT, ARIAI, Elorduy Sancho y Cia, LABEIN, Lyonnaise des Eaux-Dumez; Associated partner: RHEA S.A.

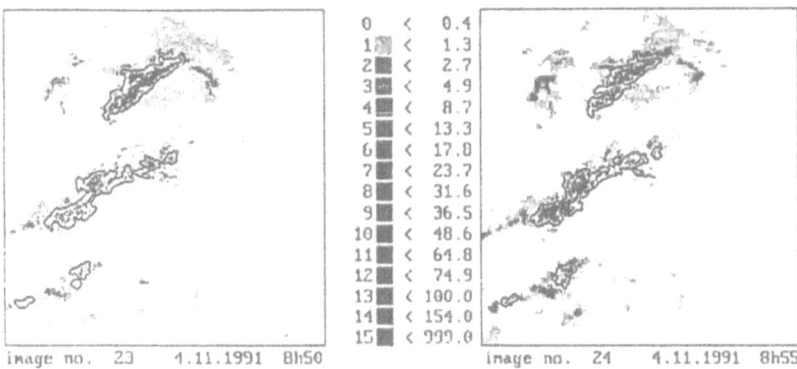

0	<	0.4
1	<	1.3
2	<	2.7
3	<	4.9
4	<	8.7
5	<	13.3
6	<	17.0
7	<	23.7
8	<	31.6
9	<	36.5
10	<	48.6
11	<	64.8
12	<	74.9
13	<	100.0
14	<	154.0
15	<	999.0

image no. 23   4.11.1991 8h50   image no. 24   4.11.1991 8h55

Figure 1: Images from the C-band radar in Trappes. Reflectivity values are represented in grey levels. The echoes considered in the matching procedure are contoured.

the identification of the same rain cell at different time steps, is the most challenging. The possibility for echoes to split or merge makes the problem very different from other conventional matching problems. Most of the cell matching methods proposed so far rely on a set of heuristic rules [3] based on cell characteristics and evolution. Recently, Neumann [6] has proposed to generate such rules automatically from examples of correct and incorrect matching using Machine Learning techniques.

A first attempt to apply connectionist algorithms to radar-based rainfall forecasting has been presented very recently in [4]. In this approach, a three-layer backpropagation network is presented with the most recent image, and trained to predict the rainfall field one hour ahead. However, this "brute-force" approach has been found to be very time-consuming, and would require the processing of several successive images in order to extract dynamic information. In this paper, we propose to restrict the application of multilayer networks to the matching phase, which can be formalized as a classification problem. Results obtained with the back propagation algorithm are compared with Neumann's decision tree [6], and mixed approaches involving the design of a neural network using a decision tree are experimented.

2   Description of the problem

The problem consists in identifying echoes in successive images as a unique rainfall generating system. "Simple" echos are first defined as sets of connected pixels with reflectivity value above some threshold. In order to cope with the splitting and merging of simple echoes from one image to the next, "structured" echos are then contructed from simple ones using the following hierarchical clustering algorithm [3]:

Considering a set of n simple echoes $\{e_1(t), \ldots, e_n(t)\}$ in one image at time t, a first structured echo $e^*(t)$ is defined by $e^*(t) = e_i(t) \cup e_j(t)$ with

$$d(e_i(t), e_j(t)) = \min_{k \neq l} d(e_k(t), e_l(t))$$

and

$$d(e_k(t), e_l(t)) = \frac{m_k(t) + m_l(t)}{m_k(t) m_l(t)} \|c_k(t) - c_l(t)\|$$

where $m_k(t)$ and $c_k(t)$ are, respectively, the mass (defined as the sum of the rainfall intensities in each pixel) and the centroid of echo $e_k(t)$, and $\|.\|$ denotes the euclidian norm.

For defining the next structured echo, $e^*(t)$ substitutes $e_i(t)$ and $e_j(t)$, and the algorithm is repeated until all of the $n-1$ possible structured echoes have been obtained.

Attached to each simple of structured echo is a list of attributes describing geometric characteristics of that echo (e.g. size, mass, principle moments of inertia, etc.) and, if that echo has previously been matched, evolution characteristics of the *sequence* of echoes to which it belongs (e.g mean speed and direction of advection, mean relative difference of size, mass and elongation, etc.).

The matching problem can be formulated as a classification task. A pair

$$(e_i(t), e_j(t+1))$$

of simple or structured echoes at times t and $t+1$ has to be classified as corresponding to the same rain cell, or not. This classification can be based on numerical features calculated from the attributes of each echo in the pair. Twenty six features have been defined, belonging to two main categories:

- *instantaneous* features, describing the differences in the characteristics of $e_i(t)$ and $e_j(t+1)$, e.g. $\Delta m_{i,j}(t+1) = (m_i(t) - m_j(t+1))/m_i(t)$

- *historic* features, describing the change in the characteristics of $e_i(t)$ and $e_j(t+1)$, compared to the mean change previously observed in the sequence

Note that these historic features are defined in such a way that their value is systematically unknown in some cases, namely when $e_i(t)$ corresponds to a rain cell that has just entered the scope of the radar. The treatment of such features in decision tree induction has been studied by Neumann [5]. A solution to that problem when using multilayer perceptrons is proposed in this paper (see Section 3.2).

3   Methodology

3.1   Classification techniques

The objective of the present study is to assess the performance of a neural network classifier in extracting matching rules from correct and incorrect matching examples. Since decision tree generation techniques have previously been applied to this problem [5], this also represents an interesting opportunity of comparing the performance of decision trees and neural networks on a difficult real-world problem, and of investigating ways of taking the best of both techniques. Three approaches have been implemented and tested against Neumann's decision tree: (1) using the decision tree to initialize a network; (2) using the decision tree as a feature selector; and (3) training back propagation networks from scratch, with different architectures.

In the first approach, the existing decision tree has been mapped onto a multilayer neural network, using the conversion rules proposed in [7]. The resulting network is composed of a partitionning layer, an AND layer and an OR layer. The initial weights of the network are set following the constraints defined in [2]. To allow further training of the network, the step activation function is replaced by a sigmoid function with a gain parameter α that controls the non-linearity, and the weights are updated using batch back propagation with adaptive learning rates.

Although the total number of features used is 26, the best decision tree obtained by Neumann [6] contains only 8 features, which indicates that the decision tree generation algorithm has the ability to select relevant features. Therefore, we have used it as a feature selector in the second approach. The number of units in each hidden layer has been determined empirically, several structures being tested in order to evaluate the sensitivity of the results to the network architecture.

3.2 Treatment of unknown values

The problem of unknown values is a classical one in Statistics. Typical approaches to this problem are either to discard the corresponding feature, or to replace each missing value by the sample mean of the feature, or alternatively by a random number with the same statistics. None of these approaches is strictly applicable in our case, because some values are *systematically* unknown, i.e. are simply not defined for some examples. Of the 26 features, 16 have occasionally unknown values, and the 26 values are known simultaneously for only 25 % of the training samples. Therefore, discarding the features with missing values would have a dramatic effect on the size of the training set. On the other hand, the assumption that missing values have the same statistics as known values simply does not make sense in that case.

A new approach to this problem has been experimented in this application. It consists in telling the network when to use a feature, by using an additional input unit that is on when that feature is present, and off when it is absent. Based on this information, the network can learn to take into account the value of a feature only when it is present. This approach has been compared to the replacement of missing values by random numbers which, although not suitable *a priori*, is a reasonable alternative.

3.3 Training data

The raw data used in this study are images from the weather radar in Trappes (near Paris), operated by the French National Weather Service. The images are characterized by a temporal resolution of 5 min and are composed of 256×256 pixels on a 800 m grid, digitized to 16 reflectivity levels. The training data are composed of echo matching examples from ten rainfall events chosen for their representativity of the climate in this area (large and narrow frontal rainbands, convective precipitation). Each pair of echoes has been labelled manually as "correct" or "incorrect" matching by visual examination of the images. All correctly matched pairs of echoes have been included. Examples of incorrect matching have been generated using the concept of "near-miss": only pairs of echoes which could "nearly" be matched, but nevertheless do not correspond to the same rain cell, have been selected [6]. The 4662 available examples have been partitionned randomly into a learning set (2/3) and a cross-validation set (1/3) that has been used to prevent overtraining. An independent test set containing 7010 correct and incorrect matching examples has been generated in the same way using ten different rainfall events.

4 Results

Table 1 shows the results obtained by the different methods on the learning set (L), the cross-validation set (C), and the test set (T). As can be seen, the decision tree is outperformed by approaches (2) and (3). Only slight differences in error rates can be noticed between the different structures, which is partly due to the use of crossvalidation

Table 1: Classification error rates on learning (L), cross-validation (C) and test (T) sets, for the decision tree (DT) and the three neural network (NN) based methods: (1) NN initialized with DT; (2) NN with same inputs as DT; (3) NN with random initial weights and 26 inputs. Variants for methods (2) and (3) are: (a) randomly generated unknown values and (b) additional inputs. The results of methods (2) and (3) are the best out of 5 trials

method	variant	architecture	Error (%)		
			L	C	T
decision tree		8-18-19-2	14.5	14.4	8.7
(1)		8-18-19-2	12.2	13.5	9.5
(2)	(a)	8-10-5-1	11.6	11.0	6.9
	(b)	12-10-5-1	11.3	10.0	5.9
	(b)	12-20-10-1	11.4	11.0	6.3
	(b)	12-30-1	11.1	10.5	6.0
	(b)	12-10-1	11.9	10.2	6.4
(3)	(a)	26-10-5-1	9.8	11.1	7.6
	(b)	42-20-10-1	9.0	11.0	7.2

to avoid the overtraining effect. The best performance of all has been obtained with the 12-10-5-1 network using only the 8 features present in the decision tree, and 4 additional input units to account for unknown values of historic features.

Using the first method, further training has surprisingly failed to improve the generalization performance of the decision tree, although some improvement has been obtained on the learning set. This result is somehow in contradiction with those presented in [7] for other problems (and a different learning algorithm), but more experiments would obviously be needed in order to draw any final conclusion. At least, these results suggest that the gain in training time obtained by initializing a neural network with a decision tree does not always guarantee optimal performance. Moreover, this method seems to be rather sensitive to the choice of the gain parameters controlling the non-linearity of the sigmoids, which makes it difficult to apply in an automatic manner.

5 Conclusions

The problem of tracking rain cells in radar images has been addressed using several connectionist and hybrid learning procedures. Among the various approaches that have been tested, the best performance has been obtained with a multilayer network presented with a limited number of features selected during the generation of the decision tree, and additional inputs to account for missing values. The lowest test error rate obtained is 5.9 %, which represents a significant improvement over the 8.7 % obtained with the decision tree alone. The networks constructed independently from the decision tree also showed relatively good classification performance, but with a greater number of hidden units. On the contrary, the performance obtained when initializing the neural network with the decision tree has not been as high as expected, perhaps because of the difficulty to find a good choice of the gain parameters.

Detailed analysis of the forecasts produced by our method reveals that its most severe limitation now resides in its inability to take into account the growth and decay of rain areas. Future research efforts in the context of the NEUFODI project will aim at predicting the evolution and deformation of rain cells by extending the application of connectionist techniques to the extrapolation of reflectivity patterns.

References

[1] G. L. Austin and A. Bellon. The use of digital weather radar records for short-term precipitation forecasting. *Quarterly Journal of the Royal Meteorological Society*, 100:658–664, 1974.

[2] T. Denœux. Generation of symbolic rules in back-propagation networks. In Igor Aleksander and John Taylor, editors, *Artificial Neural Networks II*, pages 711–714. North-Holland, Amsterdam, 1992.

[3] Thomas Einfalt, Thierry Denœux, and Guy Jacquet. A radar rainfall forecasting method designed for hydrological purposes. *Journal of Hydrology*, 114:229–244, 1990.

[4] Mark N. French, Witold F. Krajewski, and Robert R. Cuykendall. Rainfall forecasting in space and time using a neural network. *Journal of Hydrology*, 137:1–31, 1992.

[5] A. Neumann. Systematically unknown values in decision tree induction. In E. Diday and Y. Lechevallier, editors, *Symbolic-Numeric data analysis and learning*. NOVA Publishers, 1991.

[6] Andreas Neumann. *Introduction d'outils de l'Intelligence Artificielle dans la prévision de pluie par radar (In French)*. PhD thesis, Ecole Nationale des Ponts et Chaussées, Paris, 1991.

[7] I. K. Sethi. Entropy nets: From decision trees to neural networks. *Proceedings of the IEEE*, 78(10):1605–1613, 1990.

Neural Network Analysis of Plasma Spectra

Chuen-Tsai Sun
Department of Computer and Information Science
National Chiao Tung University, Hsinchu, Taiwan

Jyh-Shing Jang
Department of Electrical Engineering and Computer Sciences
University of California, Berkeley

Chi Yung Fu
Lawrence Livermore National Laboratory

Abstract

A neuro-fuzzy model of adaptive learning and feature detection is presented for plasma analysis. The inherent issues of a network with massive amount of input channels are analyzed based on a statistical argument. Remedial solutions are suggested. A fuzzy filtering architecture for learning efficiency and feature detection is described together with simulation results.

1 Introduction

The importance of plasma analysis has long been asserted by both the scientific and the engineering communities. For problems ranging from outer space physics to medical diagnosis, the success of the solutions depends highly on our understanding of the primative but complicated information: spectral signals emitted from plasmas. Take VLSI manufacturing as an example, it is desired to indicate endpoints of an etching process and to detect contamination of a chemical chamber. Existing recipes or rules of thumb are mostly heuristic and therefore not guaranteed for all the possible scenarios. Moreover, these approaches highly depend on the knowledge of which species are indicant to the process, they can not find the species on their own.

The most direct way of monitoring a chemical process should be using the full range of optical signals (the spectra), generated by an optical emission spectrometer, to tell us the actual status of the chemical reaction. The complexity and uncertainty involved in plasma analysis and the lack of human knowledge make fuzzy neural network a proper tool for this problem.

To serve our need of a multichannel spectrum analyzer, a neural network must be able to (1) drive learning error to a low value during supervised training, and (2) provide information of important species with regard to the monitored process. For an artificial network to learn, there are many issues to be considered, such as input data scaling, weight initialization, learning rate adaptation, to name a few. For spectrum analysis, we are dealing with massive amount of signal channels, the above considerations should be put in this context.

The complexity and limitation of traditional neural network architectures are largely due to the lack of an effective way of extracting meaningful information from the learned configuration. This problem becomes more intractable when the number of physical channels used for measurement increases.

Besides, due to drifting of equipment, clouding of the window, and variations in the plasma, any single wavelength detector would miss the specified signal; therefore, a certain kind of signal filtering is needed. Further, we know the number of essential species in the chamber is much less than the number of channels. Taking all these factors into consideration, we propose a mechanism of *fuzzy filtering* to cope with the complexity and uncertainty of plasma analysis.

Fuzzy filtering is the task of partitioning a massive amount of physical channels into much fewer *fuzzy channels*. These channels, adaptive during the training process, are employed for both noise filtering and feature detection. In the proposed neuro-fuzzy model, system parameters such as the *membership functions* defined for each fuzzy channel and the weights in the feedforward network are calibrated with backward error propagation. We also apply Kalman filter algorithm to improve the overall performance.

In the following sections, first we provide the simulation results of using a backpropagation neural network to analyze plasma spectra generated from an etching process. Issues such as input scaling and weight initialization are discussed briefly. Next, we introduce the use of fuzzy channels and the associated mechanism of learning. Simulation results are demonstrated for the validity of fuzzy filtering in identifying important chemical species.

2 Neural Network Training

The importance of data preparation in neural network training was discussed frequently in previous research, e.g., see [3]. For spectra, there are basically two ways of rescaling the data: (1) across-spectrum, or (2) across-channel. The first method is good in detecting the changes on a certain channel across spectra; contrastedly, the second emphasizes on the difference across channels within a certain spectrum. In this study, we found that the first method gives more meaningful results.

In the past, there were many suggestions of improving the learning efficiency and effectiveness of the backpropagation algorithm, such as changing the number of hidden units dynamically [1], rescaling of variables [5], initializing network weights [6], and using adaptive learning rates [8]. However, they did not take a large number of input variables into consideration.

In the following, the use of the logistic derivative in backward error propagation suggests one source of ill-conditioning to be the large number of input channels. We will suggest a way of weight initialization based on a statistical analysis as a remedy.

Let $I_k^{[s+1]}$ denote the weighted sum of node k at layer $s + 1$, i.e.,

$$I_k^{[s+1]} = \sum_j w_{kj}^{[s+1]} * x_j^{[s]}, \qquad (1)$$

where $x_j^{[s]}$ is the output of node j at layer s and $w_{kj}^{[s+1]}$ is the weight of the synapse between node j at layer s and node k at layer $s + 1$.

We have

$$\frac{\partial I_k^{[s+1]}}{\partial I_j^{[s]}} = \frac{\partial I_k^{[s+1]}}{\partial f(I_j^{[s]})} * \frac{\partial f(I_j^{[s]})}{\partial I_j^{[s]}}$$
$$= f'(I_j^{[s]}) * w_{kj}^{[s+1]}, \qquad (2)$$

where f is the transfer function.

Let $e_j^{[s]} = -\frac{\partial E}{\partial I_j^{[s]}}$, where E is the square error of the actual output for the current weight configuration with respect to the desired output. From Equation (1) and (2), we have

$$e_j^{[s]} = f'(I_j^{[s]}) * (\sum_k e_k^{[s+1]} * w_{kj}^{[s+1]}). \qquad (3)$$

The component step $\triangle w_{ji}^{[s]}$ is defined as follows:

$$\triangle w_{ji}^{[s]} = -l * \frac{\partial E}{\partial w_{ji}^{[s]}}, \qquad (4)$$

where l, a positive parameter, is the learning rate.

Thus, when we use the sigmoidal function as our transfer function, we have

$$\triangle w_{ji}^{[s]} = l * e_j^{[s]} * x_i^{[s-1]}$$
$$= l * x_j^{[s]} * (1 - x_j^{[s]}) * (\sum_k e_k^{[s+1]} * w_{kj}^{[s+1]}) * x_i^{[s-1]}. \qquad (5)$$

With the observation that the chain rule of differentiation introduces the factor $x_j^{[s]} * (1 - x_j^{[s]})$ and $x_j^{[s]} = f(I_j^{[s]})$, we know that when the magnitude of $I_j^{[s]}$ is large, the sigmoidal function f will produce a saturated value, 0 or 1. From Equation (5) we know the update amount will become zero and result in a no-learn situation for the network. When the number of inputs is large, this worst scenario is likely to happen unless we carefully set the initial weights.

By the *Central Limit Theorem*, we know that if $X_1, X_2, ..., X_n$ are independent, identically distributed random variables with mean μ and variance σ^2, then we have

$$\lim_{n \to \infty} P(\frac{S_n - n\mu}{\sigma \sqrt{n}} \leq x) = \Phi(x), \quad -\infty < x < \infty, \qquad (6)$$

where $S_n = X_1 + \cdots + X_n$ and Φ denotes the standard normal density. Therefore, when n is large, which is exactly the case in spectrum analysis, and input values are scaled within $[0, 1]$, we can find the proper range for randomizing weights initially.

Assume each weight is uniformly distributed in $[-a, a]$. Now $\mu = 0, \sigma^2 = \frac{a^2}{3}$. From 6 we have

$$P(\frac{S_n}{\frac{a}{\sqrt{3}}\sqrt{n}} \leq n_\rho) \doteq \Phi(n_\rho) = \rho. \qquad (7)$$

Given ρ, we can find n_ρ by table checking. Now, to avoid saturation, our goal is

$$P(S_n \leq K) = \rho. \qquad (8)$$

Thus, a can be calculated with the following formula:

$$a = \frac{\sqrt{3}K}{\sqrt{n}n_\rho}. \qquad (9)$$

This will prevent the hidden units from saturation at the early stages of learning. In the mean while, compensatory learning rates are set to match the weights so that fast convergence can still be achieved. Our computer-simulated network has a three-layer (731-8-4) feedforward architecture. It examines 731 optical channels and uses backpropagation to adjust the weights so that the inputs are associated to four

RMSE(%)

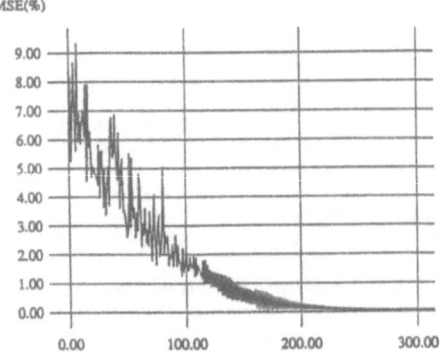

Figure 1: *Training error of a 731-8-4 network.*

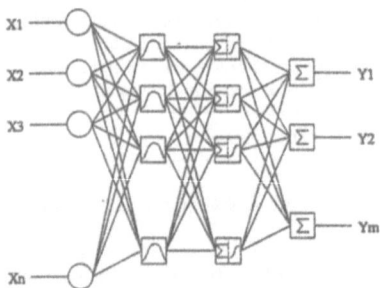

Figure 2: *A fuzzy-filtered neural network for plasma analysis.*

control variables of an oxide etching process: power, chamber pressure, and two gas flows (H_2, CF_3). The training data set is generated by *center cubic experimental design*. We record the plasma's spectral data when the etch is working well. The differences across spectra are so subtle that a process engineer can not detect them. However, the neural network effectively identified small signal changes in a very noisy environment After 200 training epochs on 30 spectra, the root-mean-square error was driven to below 0.3%, see Figure 1. The neural network successfully identified the underlying one-to-one correspondence between the plasma emission and the set of input conditions.

The above analysis can be applied to an on-line training situation. Here we initialize weights in $[0, 1]$ and treat inputs as random variables. We can determine the range of input scaling by Equation 9.

3 Fuzzy Channels for Feature Detection

Fuzzy filtering assume the boundary between two neighboring *meaningful* channels as a continuous, overlapping area in which a physical channel has partial membership in both fuzzy channels. A fuzzy channel defines a range of input wavelength characterized by an appropriate *membership function* [9]. The position and shape of this membership function is adjusted during the learning process so that the system error is minimized. At the end of training, these fuzzy channels are expected to hook on salient physical channels so as to provide a meaningful interpretation of the qualitative aspects of neural networks.

To implement the idea, we use a multi-layer feedforward *adaptive network* in which each node performs a particular function (*node function*) based on incom-

ing signals and a set of parameters pertaining to this node. Figure 2 demonstrates a fuzzy-filtered adaptive network, in which x_i's are inputs and y_j's are outputs. The nodes in the same layer have the same type of node function.

Layer 1 is the input layer. Each node at *Layer 2* is associated with a parameterized bell-shaped membership function represented as

$$\mu_A(x_i) = \frac{1}{1 + [(\frac{x_i - c_i}{a_i})^2]^{b_i}}, \qquad (10)$$

where x_i is one of the input variables, A is the linguistic term (e.g., ultra-purple) associated with this node function, and $\{a_i, b_i, c_i\}$ is the parameter set. The node output is a normalized weighted sum:

$$\frac{\sum_i \mu_A(x_i) f(x_i)}{\sum_i \mu_A(x_i)}, \qquad (11)$$

where $f(x_i)$ is the intensity of input channel x_i.

The initial values of the parameters are set in such a way that the membership functions satisfy ϵ *completeness* [4] ($\epsilon = 0.5$ in our case), *normality* and *convexity* [2]. Although these initial membership functions are set heuristically and subjectively, they do provide an easy interpretation parallel to human thinking. The parameters are then tuned with backpropagation in the learning process based on the training data set.

Each node at *Layer 3* performs as in a standard network: taking weighted sum of inputs and producing the transferred output through a sigmoidal function. *Layer 4* is similar except the neglection of the transfer function because our output values are also pre-scaled.

From Figure 2, it is observed that given the values of the membership parameters and P training data, we can form P linear equations in terms of the parameters in *Layer 4*. The equations can be solved on-line by using the *Kalman filter algorithm* so that the learning process is speeded up. Please refer to [7] for details.

Intensity/Membership

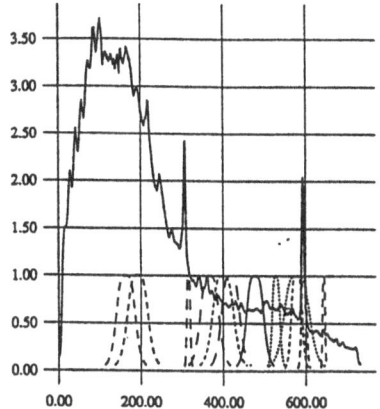

Figure 3: *A typical spectrum and fuzzy channels after training.* The three values associated with each membership function are a, b, c in Equation 10, respectively.

The fuzzy filtering mechanism proposed above simplifies the neural network architecture because much less system parameters need to be adjusted. The 731-8-4 network mentioned in the previous section has $731 \times 8 + 8 \times 4 = 5880$ weights to fine-tune; as a counterpart, a 731-8-8-4 fuzzy filtered network has only $3 \times 8 + 8 \times 8 + 8 \times 4 = 120$ parameters. This benefits learning efficiency. A more important point is that the fuzzy approach provides a meaningful interpretation for the training results so that we can obtain better understanding of the sophisticated nature of plasma emission, see Figure 3.

A spectrum is shown in Figure 3 with the y-axis representing the (scaled) intensity of the plasma emission signal and the x-axis representing the optical channel number, which corresponds to the wavelength of a certain species. We used a 731-15-15-4 network in handling the data. Three channels were lost in the learning process (driven out of the wavelength range); the left ones are shown in the same figure.

For an etching process, the important channels are typically those have the highest intensity, and change the most dramatically as the etch proceeds. We can match the wavelengths to the species that emit them. The fuzzy channels automatically identified two hydrogen peaks and a CO emission line, see Figure 3. The nitrogen lines indicate a possible leak in the vacuum system.

In brief, the network actually learns on its own from the training data and selects the most pertinent channels without any guidance from human experts of plasma diagnostics. The ability of feature detection is

one of the advantages that fuzzy filtering mechanism provides us. This capability is significant because the human experts can actually learn from the network and gain better understanding of the plasma discharge so as to produce better manufacturing recipes.

4 Concluding remarks

We investigated the use of fuzzy filtered neural networks to monitor a plasma environment through optical signals. Input scaling is considered in the context of multiple channel spectrum analysis. By a statistical argument, initial weights are set for avoiding premature saturation early in the learning phase. The neuro-fuzzy model we propose in this paper has not only the capability to learn and to adapt to changes in operating conditions, but the advantage of pertinent feature identification. We employed Kalman filters to improve the overall performance.

Simulations on experimental spectra substantiate the effectiveness of fuzzy channels. The location and shape of membership functions provide new insight for the complicated chemical reaction. This helps us to have an idea of what wavelengths are of the greatest interest. Once the critical wavelengths are identified, further automations are made possible, such as endpointing, contamination monitoring, and process control.

This technique can also be used for spectrum analysis in other fields such as medical diagnosis or global change, in which the explanation of the training results plays a role as important as the applicability of the working model. In conclusion, optical emission spectra are not self-explanatory, they require a good deal of interpretation. Fuzzy filtered neural networks successfully serve this purpose.

References

[1] Yoshio Hirose, Koichi Yamashita, and Shimpei Hijiya. Back-propagation algorithm which varies the number of hidden units. *Neural Networks*, 4:61–66, 1991.

[2] Arnold Kaufmann and Madan M. Gupta. *Introduction to Fuzzy Arithmetic: Theory and Applications.* Van Nostrand Reinhold Co., 1985.

[3] Jeannette Lawrence. Data preparation for a neural network. *AI Expert*, November 1991.

[4] C.C. Lee. Fuzzy logic in control systems: Fuzzy logic controller. *IEEE Trans. on Systems, Man, and Cybernetics*, 20(2):404–435, 1990.

[5] A. K. Rigler, J. M. Irvine, and T. P. Vogl. Rescaling of variables in back propagation learning. *Neural Networks*, 4:225–229, 1991.

[6] Peter H. Singer. Diagnosing plasmas and detecting endpoints. *Semiconductor International*, pages 66–70, August 1988.

[7] Chuen-Tsai Sun and Jyh-Shing Jang. Fuzzy modeling based on generalized neural networks and fuzzy clustering objective functions. In *Proceedings of the 30th IEEE Conference on Decision and Control*, 1991.

[8] Michael K. Weir. A method for self-determination of adaptive learning rates in back propagation. *Neural Networks*, 4:371–379, 1991.

[9] Lotfi A. Zadeh. Fuzzy sets. *Information and Control*, 8:338–353, 1965.

Pattern recognition II
—poster contributions

Monitoring EEG Signal with the Self-Organizing Map

Samuel Kaski* and Sirkka-Liisa Joutsiniemi**

*Helsinki University of Technology
Laboratory of Computer and Information Science
Rakentajanaukio 2 C, SF-02150 Espoo, Finland

**Tammiharju Hospital
Laboratory of clinical Neurophysiology
10600 Tammisaari, Finland

Abstract

We apply the Self-Organized Map to monitoring and analyzing EEG. The signal is visualized as a trajectory on a two-dimensional map. The shape and location of the trajectory give information about the variations of the state of the subject. We show that already a fairly simple and fast FFT-based feature extraction scheme is sufficient for identifying some overall states of the individual. This method may be used e.g. as diagnostic assistance.

Introduction

The Self-Organizing Map [4, 5] developed by Prof. Teuvo Kohonen is suitable for visualizing complicated processes and finding structures in multidimensional data (e.g. [3, 7, 10]). The features that need to be monitored in EEG are largely unknown, and there does not exist a usable model of the process that generates the signals. Moreover, the amount of data is enormous and must be reduced somehow — e.g. with a projection into the coordinates of a two-dimensional map.

In addition to the conventional methods (review: [8]), also neural networks have been applied to the analysis of EEG. These approaches (see e.g. [1, 2, 9]) have aimed at classifying the signals into sharp categories, however. We *visualize and describe* the signal with quite simple and clinically usable procedures for assistance in diagnosis and monitoring of patients.

Methods and feature extraction

Materials. Two EEG recordings were measured from a child with minor learning disabilities. At the time of the first recording he was 7 years old; the second EEG was measured 12 months later. During the 25-minute sessions the subject was lying awake mainly with eyes closed. At times he was asked to open the eyes for a moment and then close them again.

The EEG was recorded with 22 electrodes (channels) located according to the standard 10-20 system [8], using Cz as a common reference. The signal was low-pass filtered at 70 Hz and sampled at the rate of 200 Hz.

In order to quantify the ability of the Self-Organized Map to visualize the signal, 6 typical EEG-phenomena were defined :

'a': well organized alpha frequencies (8-12 Hz) dominating over the posterior part of the scalp

'b': well organized posterior alpha with clear, visually detectable, low-amplitude beta (13 - 30 Hz) activity in most channels

'f': low-amplitude, flat signal without clear rhythmic activity

'e': eye movement artifacts over anterior and/or lateral parts of the scalp

'm': muscle activity, mainly in anterior and lateral channels

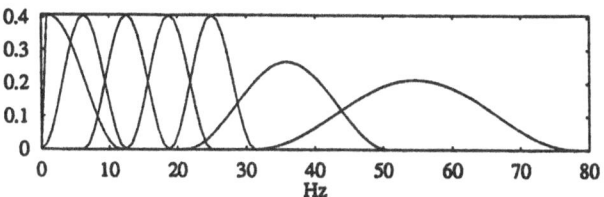

Figure 1: Passbands of the filters, the outputs of which were used as features. The filters were raised-sine shaped and the area below each curve was normalized to unity.

'g': artifacts due to big movements of the subject; in some channels the signal exceeds the cut-off limit of the amplifier

About 700 samples representing the phenomena 'a' — 'g' defined above were selected from each EEG recording and labeled accordingly.

Preprocessing. The feature extraction was based on short-time FFT power spectra. This approach was chosen because it is fast and already in clinical use.

The spectra were computed every half a second from each channel using a 256-point Hamming window, spanning about one second. The window was chosen fairly narrow in order to minimize the effects of the unstationarity of the signal.

Feature extraction. The features were computed in the frequency domain with filters, the passbands of which are shown in Fig. 1. The filters were chosen so that 1) they are mainly concentrated in the frequencies below 30 Hz, because in customary EEG recording higher frequencies are often attenuated, 2) the bands are wide and overlapping in order to make the system insensitive to small changes in the frequency content, and 3) some filters extend also above 30 Hz, so that artifacts caused by muscular activity can be taken into account.

The outputs of the seven filters of each channel were collected to a short-time feature vector, which thus consisted of 154 components. The total number of feature vectors in the two measurement sessions was 5340.

Teaching the maps. The maps were initialized, taught, and evaluated using the routines in the SOM_PAK program package developed by Prof. Kohonen et al. [6].

Evaluating the performance of the method. It is not obvious how to quantify the visualization capacity of the map. The interpretation of the trajectory must be learned by practice. We quantify the performance of the map by dividing the map into areas describing different classes, and thereafter measuring how often the trajectory goes to the right area *during samples representative of each category*. The areas were defined by projecting all classified samples onto the map and joining each unit of the map in the class from which it received the majority of samples.

Teaching of the maps and defining the areas was always done with a data set different from the one used in testing. First, one half of each measurement session was used in teaching and the other half in testing, and thereafter the roles of the sets were changed. The results reported below are averages of these two independent trials.

Results and discussion

When the successive feature vectors are projected on the map they form a trajectory which describes changes in the EEG-signal. An example is given in Fig. 2: the projection moves from the alpha-area ('a') via beta ('b') to the area denoting eye-movements ('e'), and further to the other beta-area ('b').

Samples belonging to most of the classes could clearly be detected by their area of projection on

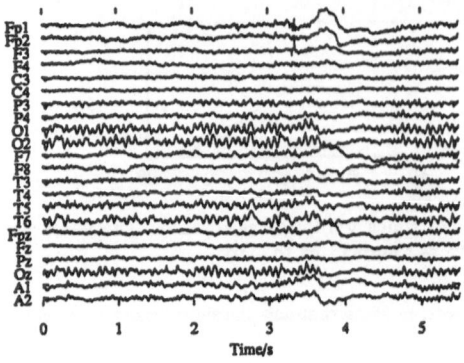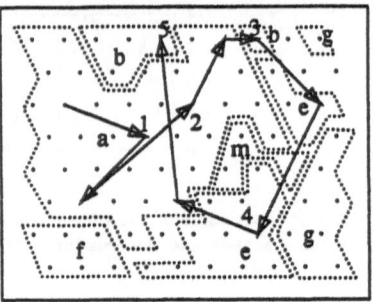

Figure 2: The EEG signal is visualized as a trajectory on the map. The raw signal from the 22 channels is shown on the left, and the corresponding trajectory on the right. The letters on the map denote the typical projection areas of each class, the numbers refer to the time instants of the raw recording.

Class	'a'	'f'	'e'	'g'	'm'	'b'	no label (tie)	Number of samples	Accuracy
'a'	858	3	7	0	0	14	4	886	97%
'f'	9	175	4	0	0	0	15	203	86%
'e'	11	14	165	7	0	2	2	201	82%
'g'	6	1	61	4	2	0	0	74	5%
'm'	3	0	3	1	55	0	8	70	79%
'b'	34	0	1	0	2	10	0	47	21%
Total								1481	86%

Table 1: Demonstrating the visualization capacity by classifying representative samples. On the left is the confusion matrix and on the right the classification results. It should be noted, that the system is not *optimized* for classification—it learns in a totally unsupervised manner, and these figures are presented only as a means of quantifying the visualization capacity of the map.

the map (Table 1). The samples belonging to class 'g' (messy data) were not projected onto the corresponding area. The quantization error, i.e. the distance between the feature vector and the closest model vector, was high for these samples, however (Fig. 3). As this observation suggests, detection of the distortions caused by the movement of the subject would require either different feature extraction methods or a more extensive data base.

Class 'b' was also quite poorly detected. Partial explanations of this result may be due to the small number of samples belonging to this class, and the overlap between the spectral content of classes 'a' and 'b'.

Conclusions

The organization of the maps reflected closely the visually defined EEG-phenomena. The samples, which clearly belonged to the classes studied, were projected to the corresponding areas on the map. However, it is not reasonable to expect all kinds of states to be describable with short-time spectra. This simple FFT-based method seems to be suitable for monitoring the overall state of the subject; in general the features must of course be chosen according to the phenomena to be studied.

A self-organized EEG map is fast enough for real-time monitoring of the EEG signal. By choosing

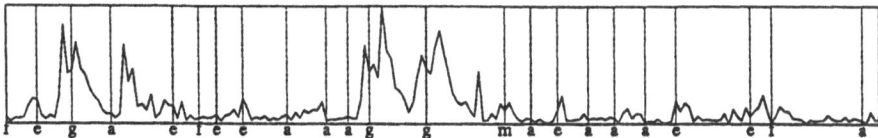

Figure 3: An example sequence of the quantization error corresponding to consecutive feature vectors. The classes of the samples are given below the figure. The quantization error is high around the messy measurements (label 'g').

the features according to a specific diagnostic problem it might be possible to reduce the dimension of the feature vector. Already a cheap PC would then be sufficient for some applications.

References

[1] R. C. Eberhart, R. W. Robbins, and W. R. S. Webber. EEG waveform analysis using CaseNet. In *Proceedings of the Annual International Conference of the IEEE Engineering in Medicine and Biology Society*, pages 2046–2047, New York, 1989.

[2] P. Elo, J. Saarinen, A. Värri, H. Nieminen, and K. Kaski. Classification of epileptic EEG by using Self-Organizing Maps. In I. Aleksander and J. Taylor, editors, *Artificial Neural Networks, 2. Proceedings of the 1992 International Conference on Artificial Neural Networks (ICANN-92)*, Brighton, United Kingdom, pages 1147–1150, Amsterdam, 1992. North-Holland.

[3] M. Kasslin, J. Kangas, and O. Simula. Process state monitoring using Self-Organizing Maps. In I. Aleksander and J. Taylor, editors, *Artificial Neural Networks, 2. Proceedings of the 1992 International Conference on Artificial Neural Networks (ICANN-92)*, Brighton, United Kingdom, pages 1531–1534, Amsterdam, 1992. North-Holland.

[4] T. Kohonen. Clustering, taxonomy, and topological maps of patterns. In *Proceedings of the Sixth International Conference on Pattern Recognition, Munich, Germany*, pages 114–128, Silver Spring, MD, 1982. IEEE Computer Society Press.

[5] T. Kohonen. The Self-Organizing Map. *Proceedings of the IEEE*, 78:1464–1480, 1990.

[6] T. Kohonen, J. Kangas, and J. Laaksonen. SOM_PAK: The Self-Organizing Map program package. Obtainable via anonymous ftp from the internet address "cochlea.hut.fi" (130.233.168.48), 1992.

[7] T. Kohonen, K. Mäkisara, and T. Saramäki. Phonotopic maps — insightful representation of phonological features for speech recognition. In *Proceedings of the 7th International Conference on Pattern Recognition (7th ICPR), Montreal, Canada*, pages 182–185, Silver Spring, MD, 1984. IEEE Computer Society Press.

[8] E. Niedermeyer and F. Lopes da Silva, editors. *Electroencephalography: Basic Principles, Clinical Applications and Related Fields*. Urban & Schwarzenberg, Baltimore-Munich, second edition, 1987.

[9] G. Pfurtscheller, D. Flotzinger, W. Mohl, and M. Peltoranta. Prediction of the side of hand movements from single trial multi-channel EEG data using neural networks. *Electroencephalography and clinical Neurophysiology*, 82:313–315, 1992.

[10] V. Tryba and K. Goser. Self-Organizing Feature Maps for process control in chemistry. In *Artificial Neural Networks: Proceedings of the International Conference on Artificial Neural Networks (ICANN-91), Espoo, Finland*, pages 847–852, Amsterdam, 1991. North-Holland.

Invariant Pattern Recognition with Recovery of Transformation Parameters

K. Jeschke, J. Reinhardt, J.A. Maruhn

Institut für Theoretische Physik, Universität Frankfurt, Germany

Abstract

In this paper a method is described for using a multilayered perceptron for recognizing a rotation and translation invariant pattern and for recovering the transformation parameters. This is achieved by employing an invariant preprocessing stage. By using the knowledge of the pattern class and a second preprocessed input the original rotation angle is determined. Simulation results are presented and discussed.

Introduction

Many attempts have been made for neural-net-based pattern recognition invariant against geometrical transformations like rotation or translation. Most of these employed multilayered perceptrons (MLP), for which backpropagation provides a practical and robust learning method [1].

Brute force learning of many transformed versions of the same pattern is quite inefficient so that interest lies in preprocessing of the net inputs for achieving the invariances. Many authors have used the Fourier or Hough transforms [2,3] or moment methods like regular or Zernike moments [4,5,6] and achieved good results, but these methods are noise sensitive or they rely on conventional first search of object position. In [7] a more direct model of using invariance classes for input was proposed and in [8] it was shown that this performs better than moment methods.

The invariant input for our net is the sum over all products of intensities of two points in the original image having a distance, within a certain interval. The result of the recognition process together with other rotation-sensitive inputs is fed into a special higher-order neural network to recover the rotation parameter.

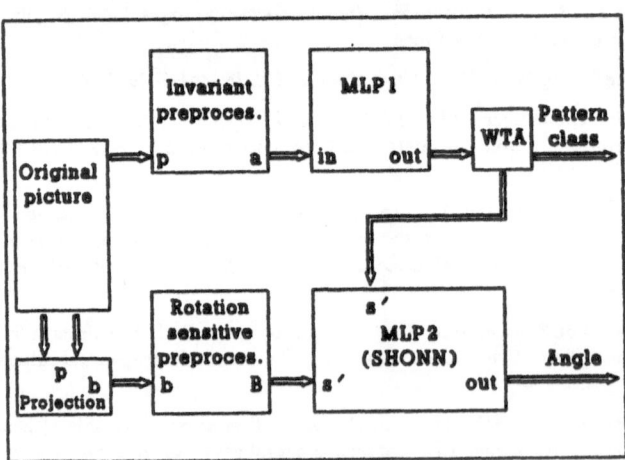

Fig. 1: Block diagram of the system.

Feature Extraction

A block diagram of the system is shown in Fig. 1 and explained in the following. There are two neural nets both fed with preprocessed data from the pixel input. The original input area has N^2 pixels which are indexed by two dimensional index vectors x in the index space. $p(x)$ is the intensity at point $\vec{x}$. The first network (MLP1) is used for pattern recognition and receives as rotation and translation invariant input the following autocorrelation function of the pattern

$$a_n'' = \sum_z \sum_x p_z p_{z+x} F_n(z, z+x) \,. \tag{1}$$

F represents the geometrical restriction function

$$F_n(x_1, x_2) = \begin{cases} 1 & \text{for } nr_d \leq d(x_1, x_2) < (n+1)r_d \\ 0 & \text{otherwise} \end{cases} \tag{2}$$

with r_d the resolution parameter and $n = 0, 1, .., N/(2r_d)$. For lower resolutions it was found advantageous to replace the step-function F_n by a piece-wise linear ramp function. Final input for MLP1 is a normalised version a_n derived from a_n''. It is first normalised by the weight of a class n multiplied with the class index

$$A_n = \sum_z \sum_x F_n(z, z+x) \quad , \quad a_n' = a_n''(n + \tfrac{1}{2})/A_n \tag{3}$$

followed by a second normalisation

$$a_n = N a_n' / \sum_n a_n' \,. \tag{4}$$

For the second net (MLP2) we need a different feature extraction method, which is translation invariant but rotation-sensitive. We can use projections of selected rotated versions of the picture onto its edge. Calling the rotated versions $r(\phi, p)$, with different angles $\phi(z) = \pi z/Z$ indexed by $z = 0, 1, ...Z - 1$, we project like:

$$b_{z,x} = \sum_y^N r_{x,y}(\phi, p) \quad , \quad x = 0, 1, \ldots, N \,. \tag{5}$$

These projections enter a direction sensitive function B' as follows:

$$B_{z,d}' = \sum_i^N g(b_{z,i} b_{z,i+d}) = \sum_i^N b_{z,i}^2 b_{z,i+d} \quad , \quad d = -N/2, \ldots, N/2 \,. \tag{6}$$

Various functions g were tried, but this version proved to be most efficient. g must be a nonlinear function of at least of one of its two arguments. For normalisation we used

$$B_{z,d} = B_{z,d}' / \sum_d B_{z,d}' \,. \tag{7}$$

As second input the results of MLP1 were used but modified by a binary winner takes all algorithm (WTA), which sets the most excited output line to one and the rest to zero. These values are denoted by s_k'.

Computation and Learning

A great advantage of this construction is its high degree of separability, leading to very short learning times. MLP1 is an ordinary MLP, which can be trained for its invariant input without considering the rest of the system. In normal working mode of the system the outputs of

MLP1 together with the projection data drive MLP2. But in the learning phase it is better to separate these two nets. MLP2 should be trained with correct MLP1-outputs and the projection data as yielded by (6,7). When enough projection directions Z are used, learning can be very fast and the results are good. Otherwise long learning times dont improve the results essentially. In practice $Z = 6$ was found to be satisfactory.

MLP2 is a special form of a higher order 3-layer neural net (Fig. 2). It has the lines s'_k as strobes. So the equation for the $(l+1)$th layer activations s_{il+1} is

$$s_{il+1} = S\Big(\sum_{j=0}^{N}\sum_{k=1}^{M} w_{ijkl}s_{jl}s'_k\Big) \quad , \quad i = 1,2,\ldots,N \tag{8}$$

where as usual the threshold is incorporated in the weight with $j = 0$ and the zeroth neuron of each layer is stuck at binary one. (S is the Fermi function). Combined with normal error back propagation (t_i is the correct answer vector) minimization of the cost function L

$$L = \frac{1}{2}\sum_{i}(t_i - s_{i2})^2 \quad , \quad dw_{ijkl} = -\eta\frac{\partial L}{\partial w_{ijkl}} \tag{9}$$

yields a learning rule ($s'^n_k = s'_k$ for $s' \in \{0,1\}$) :

$$dw_{ijkl} = \eta\delta^l_{ik}s_j\,s'_k \quad \text{with} \quad \delta^l_{ik} = \sum_n \delta^{l+1}_{nk}S'w_{nik(l+1)} \quad \text{and} \quad \delta^2_i = S'(t_i - s_{i2})\,. \tag{10}$$

and this is almost the same as the familiar backpropagation rule except for an s'_k in each term. This implies that learning can be performed separately for each value of k (i.e. for each class of patterns), enhancing the efficiency of the process considerably. Therefore this special net is called "separating higher order network" (SHONN). It is equivalent to a farm of toggled perceptrons but simpler because nodes are not duplicated.

The components of the output vector t_i are sensitive to restricted and overlapping ranges of the angular variable and produce a triangular shaped output function, see Fig. 3. From the two output signals with largest amplitude the angle is easily determined.

Results

The goal of this work was to find an efficient neural-net-based way to recognize translated and rotated patterns and recover their rotation angle. The inputs for the network have been 32×32 pixels pictures. Smaller input areas would cause a too large discretisation loss when translation by non-integer numbers of pixels or rotations about nontrivial angles are performed. Therefore we restricted the minimum structure thickness of the patterns to 3 pixels.

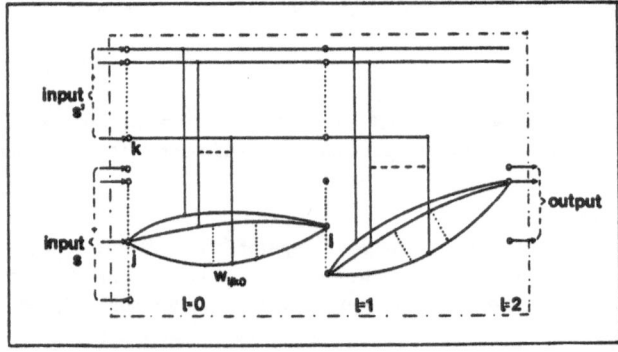

Fig. 2: Schematic drawing of the internal structure of SHONN.

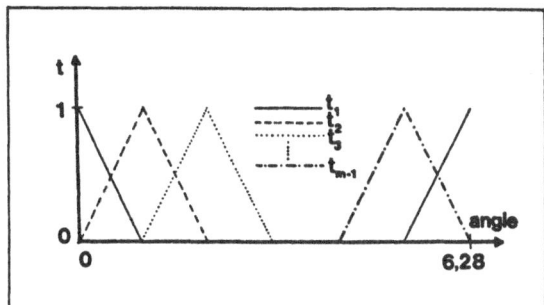

Fig. 3: The response of the output neurons t_i which is used to reconstruct the angle information.

For this case $r_d = 1.2$ was found to work best Smaller or larger r_d have been tried and degraded the results marginally. For reducing computing time larger values of r_d may be acceptable, especially when combined with a smoothed geometry function F_n.

The networks were trained for 5 sets of computer-generated patterns and tested for different versions translated and rotated by random parameters and with random noise added. The perturbed values were given by $\bar{s} = s + (r - 0.5)p$ with r an uniform random number in $[0, 1]$ and p the noise amplitude (given in %). The recognition results are shown in the Table. Note that the results of angle reconstruction refer only to such patterns, where class reconstruction was correct. In these cases angle detection was almost error-free.

noise	0%	75%	105%	135%
class correct	100%	99,5%	97.5%	91.2%
angle $\pm 10^0$	100%	100%	99.1%	96.7%

Table: Recognition rate of the pattern class and the rotation angle in dependence of the noise level.

We alternatively tried to work with a simple fully forward-connected MLP having the same inputs as SHONN. It works fairly well but not as efficiently as SHONN. Although the SHONN as a higer-order network has a larger number of connections than the MLP, at any given time in the recognition phase most of these are inactive because of the separablity and the computational expense becomes smaller on average. Similiarly, learning is also more efficient because the rotational properties are learned separately for each pattern class.

References

[1] D.E. Rumelhard, G.E. Hinton, R.J.Williams, "Learning Internal Representations by Error Propagation" PDP vol 1, Cambridge, MA: MIT Press, pp 318-362, 1986

[2] G. Elliman R.N. Banks, "Shift Invariant Neural Net for Machine Vision", Proc. IEEE, vol 137, pt 1, pp 183-187, 1990

[3] A.P. Reeves, R.J. Prokop, S.E. Andrews, F.P. Kuhl, "3D Shape Analysis Using Moment and Fourier Descriptors", IEEE Trans. Pattern Anal. Mach. Intell., vol 12, pp 487-497, 1990

[4] A. Khotanzad, J.H. Lu, "Distortion Invariant Character Recognition by a MLP and Back-Propagation Learning", Proc 2nd Intl. Conf. on NN vol I, pp 625-632, 1988.

[5] M.K. Hu, "Visual Pattern Recognition by Moment Invariants" IRE Trans. Inform. Theory, vol IT-8, pp 1028-1038, 1990

[6] A. Khotanzad, Y. H. Hong, "Invariant Image Recognition by Zernike Moments", IEEE Trans. Pattern Anal. Mach. Intell., vol 12, pp 489-497, 1990

[7] M.B.Reid, L. Spirkovska, E. Ochoa, "Rapid Training of Higher Order Neural Networks for Invariant Pattern Recognition", Conf. Proc. IJCNN 1989, vol I, pp 689-692

[8] P. Zimmerer and A.Zell, "Translationsinvariante Erkennung von Werkstücken mit Neuronalen Netzwerken", Informatik Fachberichte 290, pp 51-58, 1991

Applying Dynamic Link Matching to Object Recognition in Real World Images [*]

Wolfgang Konen,[†] Jan C. Vorbrüggen

Institut für Neuroinformatik, Ruhr-Universität Bochum, FRG

Abstract

We apply the dynamic link matching algorithm to object recognition in gray level images. The algorithm is able to map from one view of an object to different—e. g., translated, rotated, or mirror-reflected—views, being at the same time tolerant of small distortions. A sparse representation (10%) of the image data is used as a boundary condition for a self-organizing mechanism which performs the object match within a modest number of iterations ($\sim 10^2$). The mechanism can be derived from local neural dynamics [1].

Introduction. Invariant object or pattern recognition is considered to be one of the hardest problems in vision research. This is mainly due to the large variety of transformations which the image of an object can undergo while the object is still being perceived as the same by our visual system. Neural networks offer promising approaches to this problem since they can deal well with the variability of natural scenes as far as incomplete information or noise (e. g., associative memory [2,3]) or distortions (e. g., in handwritten digits [4,5]) are concerned. They cannot, however, easily perform geometric transformations of an object's image. Graph matching schemes [6,7] offer the ability to match transformed (e. g., translated, rotated, or stretched) patterns, since the quality of the match is only evaluated in terms of *relative* and not of absolute arrangement of features within the patterns. Graph matching has been applied successfully to the recognition of facial images [8].

An algorithm for finding the best graph match may, however, be hindered by energy barriers between the initial estimate and the globally optimal solution. Consider for example the case of mirror reflection, where the whole graph has to be flipped in order to achieve the best match. In this paper, we want to overcome these limitations by connecting the graph nodes with dynamic links which code for the probability that two nodes are matching counterparts. This frees the system from specifying an initial estimate. An appropriate self-organizing mechanism then allows it to search in parallel among many possible transformations, even those (like mirror reflection) which cannot be reached from the identity mapping by successive infinitesimal transformations. The principle of the dynamic link matching (DLM) algorithm and a possible implementation by neural dynamics have been reported previously [1,9] for the case of symmetry recognition in synthetic data. Here, we want to demonstrate its applicability to realistic gray level images.

Dynamic Link Matching (DLM). In order to allow for a flexible match between images, we will not work on the pixel level but use a sparse graph of nodes within each image. Each node carries local feature information obtained by preprocessing the image with localized filters (see below). For simplicity, graphs will be rectangular grids (see Fig. 1), but the algorithm could use arbitrary planar graphs as well.

In all object recognition tasks we have an image graph I and one or several model graphs M^i, $i = 1, \ldots, N$. All nodes $a \in I$ and $b^i \in M^i$ are connected by dynamic links $J_{ba}^i \in [0,1]$, where $a, b, \ldots$ specify two-dimensional locations in a planar graph.

A high value of J_{ba}^i denotes a high confidence that nodes a and b^i are matching counterparts, a low value denotes low confidence. The effective connectivity $J_{ba}^i T_{ba}^i$ between two nodes a and b^i is the product of the dynamic link and the local feature similarity $T_{ba}^i \in [0,1]$. Again, a high value of T_{ba}^i denotes strong similarity of the features attached to a and b^i, while a low value denotes dissimilarity.

Within this system architecture the task of object recognition can be formulated as a self-organizing mechanism selectively strengthening those links J^i which connect nodes in I with their corresponding nodes in M^i. This mechanism uses activations of subgraphs of nodes—termed *blobs* in this work—within each graph. In a truly neural system, the activation would be mediated by lateral connectivity within each graph layer [1]. For simplicity, we replace the lateral dynamics with an algorithm where nodes within a blob $B(a_c)$ are set to an active state. $B(a_c)$ is a square of $m \times m$ nodes with center at a_c. (If a part of the square crosses the border of the graph, it is truncated.)

---

[*]Supported by a grant from the German Federal Ministry for Science and Technology (413-5839-01 IN 101 B/9).
[†]Email address: wolfgang@neuroinformatik.ruhr-uni-bochum.de

The DLM algorithm. Initialize the dynamic links with $J_{ba}^i = T_{ba}^i / \sum_{a'} T_{ba'}^i$.

(i) Randomly choose a center a_c among the nodes of I and activate the nodes $a \in B(a_c)$. The effective connectivity from I to each of the model graphs M^i leads to an input activity at each node b^i:

$$I(b^i) = \sum_{a \in B(a_c)} J_{ba}^i T_{ba}^i. \tag{1}$$

(ii) Find the center position b_c^i at which the blob $B(b_c^i)$ will have its largest overlap with the input activity:

$$\sum_{b \in B(b_c^i)} I(b) = \max_{i, \beta^i \in M^i} \sum_{b \in B(\beta^i)} I(b) \tag{2}$$

and activate the nodes $b \in B(b_c^i)$.

(iii) Update the dynamic links between active nodes such that the total link strength converging on each cell b is kept constant:

$$J_{ba}^i = \frac{J_{ba}^i + \epsilon J_{ba}^i T_{ba}^i}{\sum_{a'}(J_{ba'}^i + \epsilon J_{ba'}^i T_{ba'}^i)} \qquad \forall a \in B(a_c), b \in B(b_c^i). \tag{3}$$

(iv) Reset all active nodes and proceed with step (i).

The algorithm has three free parameters: the update parameter ϵ and the sizes m_I and m_M of the blobs in I and M^i, respectively.

Image Preprocessing. As feature input to the DLM algorithm we use a simple preprocessing stage based on localized filtering of camera images. Two requirements have to be met by the filters: (i) The absolute placement of the rectangular grid on the image should be irrelevant, i.e., the feature information should be insensitive against a shift of the grid by at most half the distance of neighboring nodes. This implies sufficient spatial extent of the filters. (ii) Since we want to match rotated and reflected images, the local features must be invariant against local rotation and reflection.

We extract approximately rotationally invariant features $c_k(a)$ by convolving the image $\mathcal{I}$ at node location $\vec{r}_a$ with a family of Gabor wavelets $g_{k\phi}$ [8]:

$$c_k(a) = \sum_{\phi} |(g_{k\phi} * \mathcal{I})(\vec{r}_a)|, \quad g_{k\phi}(\vec{r}) = 2\pi G(\vec{r}, \sigma/k)(e^{i\vec{k}\vec{x}} - e^{-\sigma^2/2}) \tag{4}$$

with normalized Gaussians $G(\vec{r}, s)$ of width s and $\sigma = 2\pi$. The sum runs over 8 different orientations ϕ and we use 7 frequency levels with half-octave spacings starting from the lowest frequency $k = 3\pi/32$. We note in passing that we also tested DoG (Laplace) filters as an alternative to the Gabor wavelets and found qualitatively the same results.

Coding the image information in terms of the coefficients $c_k(a)$ instead of the original 128×128 pixels means a large information reduction due to the sparseness of the grid (down to 11% and 2.7% for a 16×16- or 8×8-grid, resp., using 7 frequency levels). From this sparse representation we compute the local feature similarity of nodes a and b:

$$T_{ba} = \exp\left[-\alpha \sum_k (c_k(a) - c_k(b))^2\right]. \tag{5}$$

Note that the DLM algorithm is independent of the specific kind of preprocessing used: Any kind of feature can be incorporated once a suitable similarity function T_{ba} has been defined.

Object recognition. We have tested the DLM algorithm with different recognition tasks:

Single object recognition: Given two different views of the same object (Fig. 1), the task is to find the best match from the graph of Fig. 1A onto Fig. 1B. After 100 iterations the dynamic links have established the correct transformation as can be seen from the smoothed maximum link map shown in Fig. 1C. Minor imperfections of the map are mainly due to the coarseness of the grids and to the use of filters with low spatial resolution. Note that node pixels in grid 1A in general do not have an exact correspondence among the node pixels of grid 1B.

Object discrimination and patch identification: The DLM algorithm is also able to discriminate an object I among a number of model choices $M^i, i = 1, 2, 3$ (Fig. 2), or to locate different patches

984

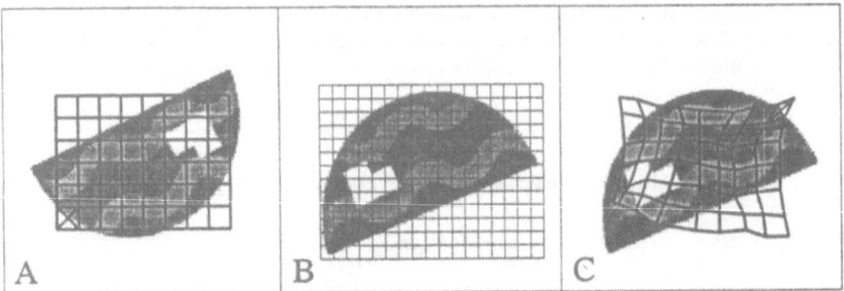

Figure 1: **Single object recognition.** The task is to find an estimate of the transformation (here: a 180°-rotation) which matches image A onto image B. Model graph M and image graph I are shown overlayed in A and B, resp. After 100 iterations of the DLM algorithm the dynamic links attached to M have grown into the correct regions of image B. This result is demonstrated in C as the *maximum link map* which is defined as follows: For each node $b \in M$ we mark the location of node $a \in I$ whose link J_{ba} is largest among all links converging on b. For better visualization the resulting map of locations is smoothed with a window of 3×3 nodes (the nearest neighbors of b) giving the map shown in C. One corresponding corner of graph and map is marked by an 'x' showing that the map is rotated correctly. Parameters are $\epsilon = 1.0, m_M = 4, m_I = 7, \alpha = 0.8$.

M^i of an object (e. g., the head region of the elephant in Fig. 3) within a different image of this object.

Conclusion. The DLM algorithm has sufficient flexibility to deal with the noisy data of gray level images. Various object matching tasks requiring different invariance transformations have been demonstrated. The advantage of the system is that the *same* algorithm can be used for searching in parallel among different image transformations, as well as solving different recognition tasks. Currently, objects are represented by planar graphs, limiting the system to map only views of an 3D-object from similar viewpoints. Object discrimination has been investigated only for a small number (3) of objects; it remains to be shown that the algorithm can also cope with a larger number of objects. Of course, this algorithm is only a simple sketch of a more mature system, because it uses only coarse image features on fixed rectangular grids which are by no means adapted to the object information. For a more realistic neural system the grids should be replaced by arbitrary planar graphs of salient nodes with activation spreading along the edges of these graphs. Further work will go into this direction.

Acknowledgements. We would like to thank C. v. d. Malsburg, L. Wiskott and R. Würtz for helpful discussions and K. Behrmann and A. Weitzenfeldt for valuable support in programming.

References

[1] W. Konen and C.v.d. Malsburg. Learning symmetries from single examples in the dynamic link architecture. *To appear in: Neural Computation*, 1993.

[2] J.J. Hopfield. Neural networks and physical systems with emergent collective computational abilities. *Proceedings of the National Academy of Sciences*, 79:2554–2558, 1982.

[3] T. Kohonen. *Associative Memory.* Springer, Berlin, 1977.

[4] Y. LeCun, B. Boser, J.S. Denker, D. Henderson, R.E. Howard, W. Hubbard and L.D. Jackel. Backpropagation applied to handwritten zip code recognition. *Neural Computation*, 1:541–551, 1989.

[5] K. Fukushima. Neocognitron: A self-organizing neural network model for a mechanism of pattern recognition. *Biological Cybernetics*, 36:193, 1980.

[6] C.v.d. Malsburg. Pattern recognition by labeled graph matching. *Neural Networks*, 1:141–148, 1988.

[7] E. Bienenstock and R. Doursat. Elastic matching and pattern recognition in neural networks. In L. Personnaz and G. Dreyfus, editors, *Neural Networks: From Models to Applications*. IDSET, Paris, 1989.

[8] M. Lades, J.C. Vorbrüggen, J. Buhmann, J. Lange, C.v.d. Malsburg, R.P. Würtz, and W. Konen. Distortion invariant object recognition in the dynamic link architecture. *IEEE Trans. Comp.*, March 1993.

[9] W. Konen and C.v.d. Malsburg. Unsupervised symmetry detection: A network which learns from single examples. In I. Aleksander, editor, *Proceedings of the International Conference on Artificial Neural Networks*, pages 121–125. North-Holland, Amsterdam, 1992.

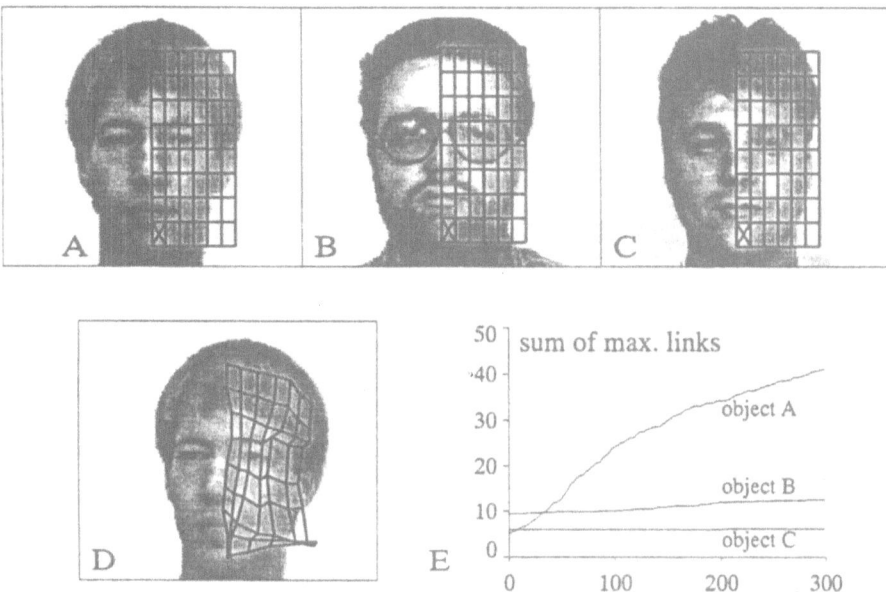

Figure 2: Object discrimination. The DLM algorithm allows fast discrimination to which stored model M^i (**A, B, C**) a different image D of object A belongs. **E:** The best matching M^i can be easily read off from the sum of the maximal links converging on objects **A, B** and **C**, resp. Note that initially object A has the lowest sum, i. e., the discrimination could not have been obtained from the similarity function $T^i_{b_a}$ alone. Parameters are $\alpha = \epsilon = 1.0, m_I = m_M = 4$ for the first 100 iterations and $m_I = m_M = 3$ thereafter.

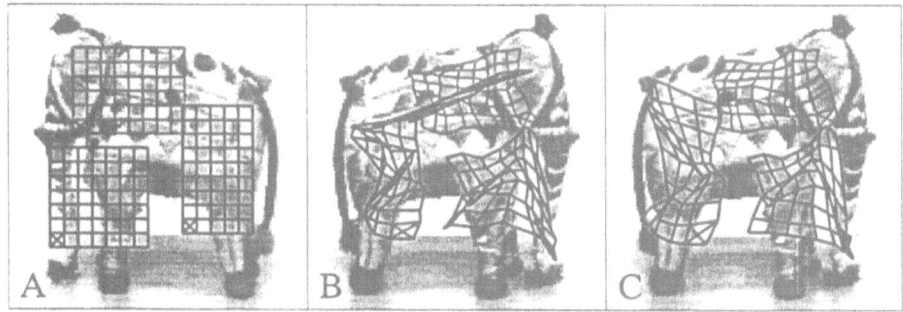

Figure 3: Patch identification. This figure demonstrates the ability of the system to correctly locate and orientate the three patches M^i defined by the grids in **A** within a mirror-reflected image I shown in **B**. Note that both images additionally differ in a slight rotation in depth of the elephant. Image I was covered by a 16×16-grid (not shown). **B** and **C** show the maximum link maps for the different patches obtained after 300 and 600 iterations, resp. Parameters are $m_I = m_M = 5, \alpha = 0.5$ and a varying update parameter ϵ, starting with 0.1 and doubled every 150 iterations.

Performance of the Backpropagation Neural Network for Recognition of Radio Signals Using Time-Domain Features

René Lamontagne
Defence Research Establishment Ottawa
Ottawa, Ontario
K1A 0K2, Canada
lamontag@crow.ewd.dreo.dnd.ca

Abstract

An approach to the problem of automatic modulation recognition is presented. The algorithm is based upon an exhaustive set of 21 time-domain features obtained from statistical analysis of the modulated signals. Traditionally, such patterns were classified using probabilistic classifiers. However, it is shown that radio signals produce non-parametric patterns in the decision space and that performance can be significantly improved by using more robust classifiers. Namely, performance of probabilistic, k-nearest neighbour, and backpropagation neural network classifiers were compared for different cases of increasing complexity. Computer simulation using 10 modulation types showed that neural nets are the clear winner for this application.

1 Introduction

With the increasing demand for radio communications, the task of monitoring electromagnetic signal transmissions in the RF spectrum has become a major one in electronic warfare and broadcasting control. Nowadays, a lot of effort is expended to develop automated systems that can enhance the effectiveness of human operators, allowing them to perform higher level functions in the monitoring process.

One of the major attributes of intercepted signals is the modulation type. Automatic Modulation Recognition (AMR) is essentially a pattern recognition problem, trying to classify a radio signal as belonging to one of several predefined classes. Numerous algorithms capable of estimating the modulation types of communication signals have been proposed [1]. However, they lack robustness when the signal-to-noise ratio (SNR) is low.

This paper proposes a robust approach based on an exhaustive feature set and a robust classification engine. We demonstrated that by using artificial neural networks, AMR can be achieved successfully on the test set of 10 modulation types [2]. These modulation schemes are described in the next section. The signal features, the classifiers, and the results are presented in sections 3, 4, and 5 respectively. Section 6 presents concluding remarks on the work herein and some recommendations for further work.

2 Input Signals

Baseband signals produced by various information sources are not suitable for direct transmission. These signals must be processed to allow radio transmission. The conversion process, known as *modulation*, consists of varying the parameters of a high frequency sinusoid, called a carrier, according to the baseband message. Therefore, the general model can be represented by a function of the form

$$s(t) = A_c A(t) \cos(2\pi f_c t + \varphi(t) + \theta_0),$$

where the signal envelope $A(t)$ and the zero phase $\varphi(t)$ are determined by the baseband message, A_c is related to the carrier power, f_c defines the carrier frequency, and θ_0 is the initial phase angle. Particular modulation types are obtained by encoding differently the baseband message into $A(t)$

and $\varphi(t)$. The modulation types considered are the following: amplitude modulation (AM), on-off keying (OOK), upper-side band (USB), lower-side band (LSB), frequency modulation (FM), bipolar phase-shift keying (BPSK), quadrature phase-shift keying (QPSK), non-coherent frequency-shift keying (FSK1), continuous-phase frequency-shift keying (FSK2), and continuous wave (CW).

Simulators were used to generate the various modulated signals. Based on hardware models, the simulators can reproduce specific signal characteristics (modulation index, frequency deviation, bandwidth) as well as some signal distortions (filtering, hard-limiting, additive noise, frequency synthesizer instability). For each experiment, the chosen simulator was set to generate a snapshot of the signal consisting of 8192 points. The sampling frequency was set at 75 kHz, which corresponds to a sample time of 109.23 ms. Other signal parameters were randomly specified and changed for each example. Using the simulator, four data banks of increasing complexity were generated. As shown in Table 1, the complexity was increased by adding features and modulation types, and by diminishing the SNR.

Case	Modulation Type	SNR (dB)	# of features	# of examples per class (Train/Test)
1	AM, FM, OOK, QPSK	8 to 25	17	257/63
2	AM, FM, OOK, QPSK, FSK1	7 to 25	21	726/148
3	AM, FM, OOK, QPSK, FSK1	5 to 25	21	817/200
4	all 10 modulations	5 to 25	21	999/200

Table 1: The data bank: four cases of increasing complexity.

3 Time-Domain Features

The feature set is obtained from signal parameters that are assumed to contain enough discriminative information about the modulation types. Five signal parameters were used: the signal envelope, the instantaneous frequency, the squared envelope, the envelope times the derivative, and the squared envelope times the instantaneous frequency.

Since the dimensionality of the parameters obtained precludes direct classification, feature extraction is prescribed. Specifically, a statistical approach based on the first moments was chosen to create 21 features. A detailed description of those features is provided in [2]. Each feature was normalized to fit within the range {-1, +1 }. Finally, the logarithm function was used to scale the features having a dynamic range greater than four octaves.

4 Classification Algorithm

We have limited the scope of our investigation to three classifiers: probabilistic, k-nearest neighbour (kNN), and artificial neural net (ANN) classifiers.

Traditionally, probabilistic linear classifiers have been used for AMR. They are fast and easy to implement. Based upon the Bayes expressions for conditional probabilities, the discriminant functions are expressed by

$$g_i = \log(p(\vec{x} \mid \omega_i)P(\omega_i)) = \log(p(\vec{x} \mid \omega_i)) + \log((P(\omega_i)),$$

where $\vec{x}$ is the input vector, $p(\vec{x} \mid \omega_i)$ is the probability density function (pdf) for class ω_i, and $P(\omega_i)$ is the *a priori* probability of occurrence of samples from class ω_i [3].

To estimate the pdf, a parametric assumption is made. Assuming multivariate normal densities of mean $\vec{\mu_i}$ and a covariance matrix Σ, i.e., $p(\vec{x} \mid \omega_i) \equiv \mathcal{N}(\vec{\mu_i}, \Sigma)$, the discriminant functions become

$$g_i = -\frac{1}{2}(\vec{x} - \vec{\mu_i})^t \Sigma^{-1}(\vec{x} - \vec{\mu_i}) + \log(P(\omega_i)).$$

The kNN method does not need a parametric assumption and can estimate arbitrary density functions. Given vector $\vec{x}$, its k-nearest neighbours are computed using any valid metric, such as Euclidean distance. For an M-class problem, let $k_1, k_2, ..., k_M$ be the respective number of nearest neighbours from each class, $\omega_1, \omega_2, ..., \omega_M$, such that $k_1 + k_2 + ... + k_M = k$. The new pdf is

$$p(\vec{x} \mid \omega_i) = \frac{k_i - 1}{N_i} \frac{1}{\mathcal{A}(k, N, \vec{x})},$$

where N_i is the number of training samples in the ith class, and $\mathcal{A}(k, N, \vec{x})$ is the volume of a hypersphere [4]. The decision rule becomes

$$\frac{N_i}{N} p(\vec{x} \mid \omega_i) = \max_i \left(\frac{N_i}{N} p(\vec{x} \mid \omega_i) \right) \longrightarrow \vec{x} \in \omega_i.$$

Feedforward nets trained with the backpropagation learning rule have proven their ability to solve complex pattern recognition problems. By using hidden units, they can define arbitrarily complex decision boundaries. Unfortunately, training this kind of classifier with the conventional momentum descent algorithm is a tedious task. To alleviate the problem, an alternate algorithm based on the conjugate gradient optimization technique was used.

First, the nets were initialized with small random values to avoid premature saturation. Then, using the conventional momentum descent algorithm, the nets were trained in on-line mode until preliminary convergence was achieved. Finally, the momentum descent algorithm was replaced by the conjugate gradient optimization method and the nets were further trained.

Conjugate gradient is an optimization technique used to determine a set of search directions that can lead efficiently to a function minimum [5]. In batch mode, the algorithm is known to converge better than the steepest gradient descent technique [6].

5 Experimental Results

Results are reported in Table 2 for Case 4 (from Table 1), and in Figure 1 for all 4 cases of increasing complexity. The probabilistic classifier is of particular interest because it has been traditionally used for AMR. It becomes apparent with these results that this classifier cannot handle the complexity of the noisy patterns. For example, many misclassifications occurred between "bauded" digital modulation types. These classes have multi-modal distributions, which are not supported by the parametric model. Moreover, FM and CW signals were confused. This might be caused by unmodulated segments inside the FM signals that created patterns hard to distinguish from those of CW.

The kNN classifier offered significant improvement over the probabilistic classifier. However, the performance broke down when the number of modulation types was increased (Case 4). Particular difficulty was encountered for the single-side-band modulation types (LSB and USB).

The ANN classifier showed very little degradation in performance as complexity was increased, indicating good robustness for noisy patterns. All patterns were identified with accuracies above 99%. The selected network topology (found experimentally) included 21 hidden units organized in a single-hidden-layer fashion.

Classifier	AM	FM	ASK	QPSK	FSK1	LSB	USB	FSK2	BPSK	CW	Overall
Probabilistic	97.0	74.0	98.5	73.0	75.0	91.0	94.0	78.5	88.0	69.5	83.85
kNN	99.0	100	99.5	97.0	100	61.0	43.0	100	90.5	100	89.00
ANN	100	100	100	100	99.0	99.0	95.5	100	99.5	100	99.30

Table 2: Performance of the various classifiers for Case 4. Results in percent.

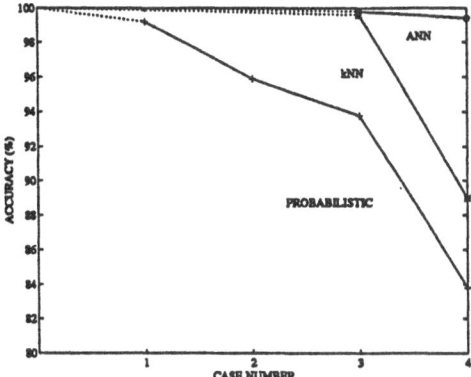

Figure 1: Performance of the classifiers for the 4 Cases of increasing complexity.

6   Conclusion

In this research work, backpropagation neural network classifiers have very successfully classified the 10 signal classes. Parametric statistical classifiers were not as successful indicating that the decision regions are fairly complex. Although the kNN classifier can provide arbitrarily complex decision functions, properly trained feedforward nets offered much better robustness and generalization. Moreover, computation time for classification is less. Preliminary work also indicated that data requirements (number of training examples) is less for ANN than kNN.

Future work on the automatic recognition problem is proposed. First, optimization of the feature set has started. Preliminary tests indicate that the number of features could be reduced without decreasing the accuracy. Second, other preprocessors will be investigated: namely, spectral-domain and cyclostationary features. Third, different neural networks, will be studied. Specifically, radial-basis-function, self-organizing-map, and fussy-ARTMAP networks are proposed. Finally, to model the problem more accurately, real radio signals will be used.

References

[1] Lamontagne, René. *Modulation Recognition: An Overview.* Technical Note 91-3, Defence Research Establishment Ottawa, 1991.

[2] Lamontagne, René. *An Approach To Automatic Modulation Recognition Using Time-Domain Features And Artificial Neural Networks.* Technical Report, Defence Research Establishment Ottawa, 1993.

[3] Duda, Richard O. and Peter E. Hart. *Pattern Classification and Scene Analysis.* John Wiley, 1973, Chap. 2.

[4] Fukunaga, K. *Introduction to Statistical Pattern Recognition.* Academic Press, London, 1972.

[5] Press, W. H., B. P. Flannery, S. A. Teukolsky, and W. T. Vetterling. *Numerical Recipes in C.* Cambridge University Press, 1988.

[6] Barnard, Etienne. *Optimization for Training Neural Nets.* IEEE Trans. on Neural Networks, Vol. 3, No. 2, March 1992.

Conceptual Fuzzy Sets Application to Facial Expression Recognition using Associative Memory System

Hirohide USHIDA, Tomohiro TAKAGI, and Toru YAMAGUCHI*

Laboratory for International Fuzzy Engineering Research,
SiberHegner Buil 3F, 89-1 Yamamashita-cho, Naka-ku, Yokohama-shi, 231, JAPAN

*System & Software Engineering Laboratory, Research & Development Center, Toshiba Corp.
70 Yanagi-cho, Saiwai-ku, Kawasaki-shi, 230, JAPAN

ABSTRACT

Real world consists of a very large number of events and continuous numeric values. The concepts used in logical thinking process are essentially vague, since they are derived from generalization of these instances and numeric values. We have constructed Fuzzy Associative Memory Organizing Units System (FAMOUS) as a software package tool implemented on a Unix-workstation. In this paper, we realize previously proposed Conceptual Fuzzy Sets[3] (CFS) to represent vague concepts on FAMOUS, which are capable of context sensitive recognition based on fusion of bottom-up and top-down processing. We also discuss the representation and recognition of facial expressions as an illustrative example.

1. INTRODUCTION

It is difficult to represent the meaning of a concept by using simple logic because the meaning depends on the context. For example, a human facial expression consists of hazy and weak patterns on different parts of the face rather than one conspicuous pattern on one part of the face[1]. Although eyes look angry, for instance, the whole face may seem to be laughing depending on the expression of the mouth or other parts. Also it is difficult to explicitly represent the concepts of facial expressions by using rule-based logic. For example, in the case of an angry face, it is easy to tell the degree of anger but not easy to explain why the face looks angry. Therefore, in the definition of a fuzzy set[2], a denotative description is generally used by means of the membership function, which associates each element with a grade of membership in the interval [0,1]. However the definition of the fuzzy set, described by instances "A, B, C", cannot determine the membership value of "D", which is a new instance. This means that a fuzzy set is not able to generalize knowledge from instances.

In order to solve all these problems, we proposed Associative Memory System called Conceptual Fuzzy Sets (CFS)[3]. In this Associative Memory System "CFS", the meaning of a concept is represented by the distribution of activations of labels that have concepts. Since the distribution changes, depending on the activated labels, to indicate a situation, the CFS can represent context dependent meanings.

CFS also carry out Multi-Layered Reasoning (called MLR in this paper) based on association that is driven by propagation of activation of labels[4]. MLR by means of CFS has the following features:
1. Capability of simultaneous symbolic and quantitative processing (semantic guideline)
2. Capability of simultaneous top-down and bottom-up processing (context sensitive processing)

In this paper, we realize CFS using FAMOUS (Fuzzy Associative Memory Organizing Units System) implemented on a Unix-Workstation and apply MLR by means of CFS to recognition of human facial expressions. In section 2, we discuss the general characteristics of CFS. In section 3, we show the FAMOUS which simulates CFS. In section 4, we propose a CFS network for recognizing human facial expressions and discuss the results of recognition experiments.

2. CONCEPTUAL FUZZY SETS AND MULTI-LAYERED REASONING[4]

The shape of a fuzzy set should be determined from the meaning of the label (word) depending on various situations. We can assign grades of activation showing compatibility degrees between different labels in CFS, so that the distribution of activation can represent the meaning of a label depending on context.

Generally in image processing, recognition is carried out using characteristic values that are already obtained by low level image processing. However, when the model or context of an object is known, image recognition is more efficient. In fact, human mechanisms simultaneously realize both image processing and recognition, by means of the effective fusion of bottom-up and top-down processing supported by simultaneous information exchange and parallel processing. CFS can realize parallel processing to support the fusion of bottom-up and top-down processing by combining the semantic information processing in the upper layer and local processing in the lower layer. For example, in image recognition, the upper layer describes the knowledge about a context while the lower layer describes primitive concepts or instances. The concepts in the upper layer are explained by the instances in the lower layer. The characteristic values extracted from an original image activate the corresponding nodes in the lower layer. This results in the activation of the concept in the upper layer. At the same time the context described in the upper layer depresses the contradictory patterns of distribution of activation and promotes the meaningful patterns of activation in the lower layer. Thus the nodes denoting instances become active so as to satisfy both the characteristic values and the context. This context sensitive processing provides us with an accurate result. It uses the context to eliminate vagueness which may come from noisy and vague data and which would otherwise cause misunderstandings.

3. CFS BASED ON FUZZY ASSOCIATIVE MEMORY ORGANIZING UNITS SYSTEM

CFS are realized using associative memories called FAMOUS (Fuzzy Associative Memory Organizing Units System). The FAMOUS is a Multi-layered network (Fig. 1), in which a node represents a concept and the strength of a link is determined by the strength of the relation between two connected concepts. Concepts are usually classified into a layer. Activations of nodes produce reverberation and system energy is stabilized to a local minimum. As a result, corresponding concepts are recollected.

In the FAMOUS, the recollections are realized by means of Bidirectional Associative Memories (BAMs)[5]. During the association in BAMs reverberation is carried out according to:

$$Y(t) = \phi\left(M \cdot X(t)\right), \ X(t+1) = \phi\left(M^T \cdot Y(t)\right) \tag{1}$$

where, $X(t) = [x1, x2, ..., xm]^T$, $Y(t) = [y1, y2, ..., yn]^T$ are activation vectors on x and y layers at the reverberation step t, and $\phi(\cdot)$ is a sigmoid function of each node. BAMs memorize corresponding pairs of elements at each layer in terms of a synaptic weight matrix M to memorize CFS. When the pairs of patterns (X_1, Y_1), (X_2, Y_2), ... , and (X_p, Y_p) are given, M is calculated from these pairs with coefficient β:

$$M = \sum_{i=1}^{p} \beta Y_i X_i^T, \ M^T = \sum_{i=1}^{p} \beta X_i Y_i^T \tag{2}$$

CFS can be inductively constructed using the Hebbian learning law[6]. The CFS is realized using associative memories in which a link represents the strength of the relation between two concepts. In the Hebbian learning law the strength m_{ij} of a link is modified by the product of the activation of two nodes x_i and y_j according to:

$$m_{ij} = -m_{ij} + x_i y_j \tag{3}$$

A complex CFS is realized by combining several pieces of associative memory structured individually. Such a combination of pieces of knowledge enables the CFS to realize context dependent representation.

CFS are realized using a FAMOUS implemented on a Unix-workstation. The FAMOUS is a CFS development and simulation system which is programmed in C-language. It has user-friendly tee-structured menu and the design of CFS network can be changed flexibly. The activation values of nodes are illustrated using figures on X-window system. When the program starts, it indicates menus so that the user can select a menu-item (Fig. 2). The operations are as follows. First, the user should select no. 1 item and fix the number of layers and nodes and specify the network parameters. Secondly, the training data should be loaded from a file system using menu no. 3. Next the user can get correlation matrix M in eq. (2) using menu no. 4. If the network should be trained according to eq. (3), the user can train the network using menu no. 5. The correlation matrix M can be saved to a file system using menu no. 7 and loaded from it using menu no. 8. Finally, The fuzzy reasoning is carried out by menu no. 6. We apply this system to recognition of facial expressions.

4. RECOGNITION OF FACIAL EXPRESSIONS USING CFS NETWORKS

There are generally six types of basic facial expressions: surprise, fear, disgust, anger, happiness, and sadness[1]. In this paper, we limit facial expressions to three categories: anger, happiness, and sadness, and propose a network model that recognizes these facial expressions. The network has links containing the knowledge that is constructed of characteristics extracted from instances (Fig. 5). The network consists of three stages and the lowest stage is composed of input layers. The network between the lowest stage and the middle stage is like an LVQ (Learning Vector Quantization) network. This network compares the knowledge in links with inputs by using the Kohonen algorithm[6] so that activations of nodes in the middle stage increase according to the degree of agreement. The network between the middle stage and the highest stage is composed of CFS networks and the distribution of activations of nodes in the network converges to satisfy contexts after the reverberation.

The coordinates of each facial part illustrated in Fig. 3 are used as the characteristic values and transformed into relative coordinates for inputting to the network. The input nodes of each facial part connect with the nodes of the corresponding part in the middle stage as shown in Fig. 5. The weight vector of the links between the input nodes and the instance node in the middle stage is the vector whose components are the inputted relative coordinates. Thus if the input nodes are activated by a new pattern, the activations are propagated to the middle stage and the instance nodes have activations proportional to the nearness between the link vector and the input vector. Therefore these activations represent the instance of the middle stage that is the nearest to the input pattern in each layer. The distribution of the activations of the instance nodes can represent the facial expression nearest to the input pattern in each part but cannot do this for the whole face. For example, when the input vector of the eyebrow is near to "Angry A" but those of the mouth are near to "Sad B", we cannot explain the facial expression of the input. CFS networks between the middle stage and the highest stage are bi-directional associative memories in order to solve such problems. The activation in the middle stage is propagated to the upper stage. The highest stage has three nodes and each node denotes a facial expression: anger, happiness, or sadness. The final distribution of activations in the highest stage represents the facial expression of the input. In this network, the contexts are standard patterns so a combination of characteristics similar to that of the standard patterns is promoted and the mismatch with the context is inhibited. Therefore the CFS network realizes context sensitive recognition.

In our experiment, 56 faces including three type of expressions were used. The data were the same as Kobayashi's[8]. Three facial expressions for three people, A, B, and C, were used to construct the knowledge of the network. The data were transformed into relative coordinates and normalized to make the length of the vector equal to 1. The Numbers of elements in the characteristic vectors of each facial part are 12 (eyebrows), 7(right eye), and 8(mouth). We used data only from the right eye, on the assumption that the face is symmetrical.

After training the network with 9 data, the network scored 100% correct faces about 9 training data. About 47 generalization testing data, the score was 78.7%. An example of a subject image is shown in Fig. 4 and the results of recognition are illustrated in Figs. 5 and 6. The face in Fig. 4 is a happy face and was not taught to the network. In Fig. 5, the activations of the nodes in the lowest stage denote values of characteristics of the face. The middle stage in the network shows the initial activation state (before reverberation). We cannot judge the facial expression of the object from the output of the middle stage. Figure 6 illustrates the state of the network after reverberation. Fuzzy entropies have decreased during cycles of the reverberation and the distribution of activation of nodes converged to the state representing "Happy". The experimental results show that the distribution of activations converges to a state that is meaningful to the context after reverberation though the initial state of activation is too vague to judge the facial expression.

5. CONCLUSION

We described CFS based on Associative Memory System "FAMOUS" and its application to recognition of facial expressions. CFS can be realized on the FAMOUS and it simulates Multi-layered reasoning. MLR by means of CFS can simultaneously process both numeric values and abstract concepts, and achieve recognition depending on context by bi-directional parallel processing.

A facial expression is represented by vague and weak patterns at different parts of the face instead of one clear pattern at one part. We showed that the proposed network matches characteristics of each part with the knowledge constructed from instances and achieves context sensitive recognition by decreasing fuzzy entropies by means of bi-directional processing.

ACKNOWLEDGMENTS

We express our deepest gratitude to Dr. Hara and Mr. Kobayashi, Science University of Tokyo, for useful advice and providing us with important data.

REFERENCES

[1] P.Ekman and W.V.Friesen: Unmasking The Face, Prentice-Hall, Inc., Englewood Cliffs, New Jersey (1975)
[2] L. A. Zadeh, Fuzzy sets, Inform. & Control, Vol. 8, pp. 338-353 (1965)
[3] T.Takagi, T.Yamaguchi and M.Sugeno: Conceptual Fuzzy Sets, International Fuzzy Engineering Symposium '91, PART II, pp. 261-272 (1991)
[4] T.Takagi, A.Imura, H.Ushida and T.Yamaguchi: Multi-layered Reasoning by Means of Conceptual Fuzzy Sets, Third International Workshop on Neural Networks and Fuzzy Logic '92, NASA Johnson Space Center (1992)
[5] B.Kosko: Adaptive Bidirectional Associative Memories, Applied Optics, Vol.26, No.23, pp. 4947-4960 (1987)
[6] T.Takagi, A.Imura, H.Ushida and T.Yamaguchi: Inductive learning of Conceptual Fuzzy Sets, 2nd International Conference on Fuzzy Logic and Neural Networks IIZUKA '92 (1992)
[7] T.Kohonen: The neural phonetic typewriter, IEEE Computer, 21, 3, pp. 11-22 (1988)
[8] H.Kobayashi and F.Hara: The Recognition of Basic Facial Expressions by Neural Network, Proc. of IJCNN '91, pp. 460-466 (1991)

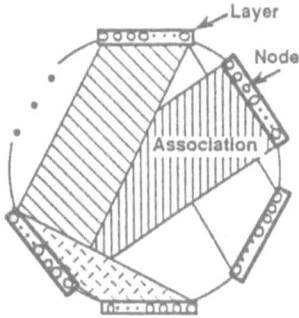

Fig.1 Configuration of FAMOUS

Fig. 2 An example of FAMOUS menus

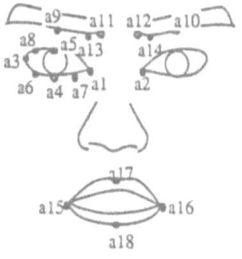

Fig. 3 Facial characteristic points

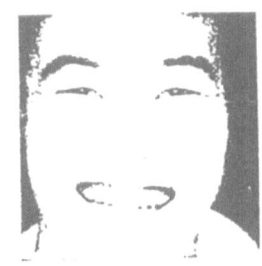

Fig. 4 Object image

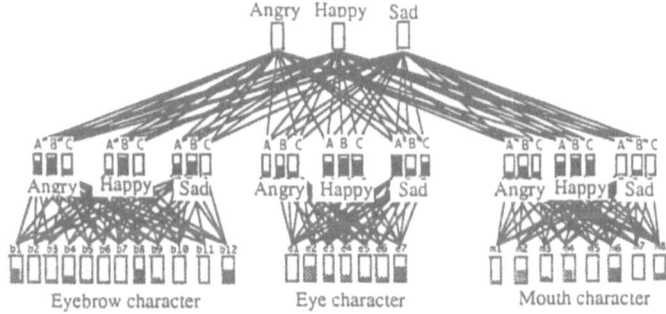

Fig.5 Recognition without context (before reverberation)

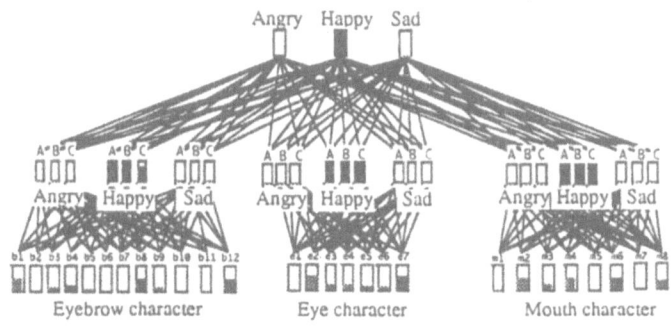

Fig.6 Recognition with context (after reverberation)

Neocognitron with Non-uniform Receptive Fields

Kunihiko Fukushima,   Masato Okada,   Kouichirou Yamauchi,
Michihiro Ohno   and   Kazuhito Hiroshige

Faculty of Engineering Science, Osaka University,   Toyonaka, Osaka 560, Japan

Abstract
The neocognitron is a hierarchical neural network model, which is capable of deformation-resistant pattern recognition. In the conventional neocognitron, each cell has a uniform receptive field. In other words, all parts within a receptive field have the same characteristics. Non-uniformity within a receptive field, however, often produces greater robustness in recognizing patterns. This paper discusses two kinds of non-uniformity: non-uniform blurring and non-uniform sensitivity within a receptive field. Several methods to produce such non-uniformity in a modified neocognitron are offered in the paper.

1   Introduction

The neocognitron is a hierarchical neural network model, which is capable of deformation-resistant pattern recognition [1],[2],[3]. Various experiments have demonstrated its powerful ability to recognize visual patterns.

In the conventional neocognitron, each cell has a uniform receptive field. In other words, all parts within a receptive field have the same characteristics.

Non-uniformity within a receptive field, however, often produces greater robustness in recognizing patterns. This paper discusses two kinds of such non-uniformity: non-uniform blurring and non-uniform sensitivity within a receptive field.

2   Non-uniform Blurring

In the hierarchical network of the neocognitron, feature extraction by the S-cells and a blurring operation by the C-cells are repeated. The ability of each S-cell to robustly extract deformed features is created by the blurring operation of the C-cells placed in front of the S-cell [4]. Positional errors of the local features extracted by the S-cells in the preceding stage can be tolerated as a result of the blurring operation of the C-cells.

Let an S-cell have previously been trained to extract a feature such as a corner or a cross point of two lines. If a deformed version of the training pattern is presented as a stimulus to the input layer, the discrepancy between the feature in the training pattern and that of the stimulus pattern usually becomes larger in the periphery than at the center of the receptive field of the S-cell. In the conventional neocognitron, however, the amount of blurring produced by the C-cells is uniform throughout the receptive field of the S-cell. It can be expected that the S-cell might accept a much larger deformation of the feature, if a non-uniform blurring could be produced in such a way that a larger blurring is generated in the periphery than at the center of the receptive field.

A straightforward way of realizing this non-uniform blurring operation is to prepare for each individual S-cell a set of C-cells of its own, and to make the C-cells at the center have a smaller blurring while the C-cells near the periphery have a larger blurring proportional to the distance from the center. This is not practical, however, because of the enormous number of C-cells that would be required by the network.

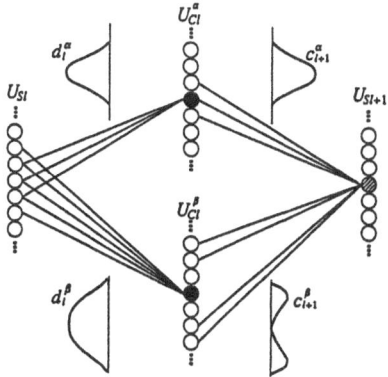

U_{Sl}: S-cell layer of the lth stage
U_{Cl}^{α}: small-blurring C-cell layer of the lth stage
U_{Cl}^{β}: large-blurring C-cell layer of the lth stage
d_l^{α}: connections from U_{Sl} to a cell of U_{Cl}^{α}
d_l^{β}: connections from U_{Sl} to a cell of U_{Cl}^{β}
c_{l+1}^{α}: connections* from U_{Cl}^{α} to a cell of U_{Sl+1}
c_{l+1}^{β}: connections* from U_{Cl}^{β} to a cell of U_{Sl+1}
* c_{l+1}^{α} and c_{l+1}^{β} are parameters corresponding to $c(\nu)$ in Figure 2.

Figure 1: One-dimensional cross section of the connections to and from a dualized layer of C-cells.

In order to get a similar effect much more economically, we propose a neocognitron with dual C-cell layers [5],[6]. Each layer of C-cells is divided into two sub-layers: one with a small blurring and the other with a larger blurring. Each S-cell in the succeeding layer is constructed so as to receive input connections from the small-blurring C-cell layer at the center of its connecting area. and also receive connections from the large-blurring C-cell layer in the periphery of its connecting area. This situation is illustrated in Figure 1.

A neocognitron with this new architecture has been simulated on a computer. The simulated neocognitron is a four-stage network (i.e., a nine-layered network), in which the first stage is devoted to line-extraction. The network has been trained to recognize handwritten numeric characters by supervised learning. In the new network, only the C-cell layer of the first stage has been dualized, because the effect of dualization is not as prominent in other stages. It has been shown that the new neocognitron recognizes characters more robustly than the conventional neocognitron: The error rate in recognizing test patterns has been halved compared to the conventional neocognitron.

3 Non-uniform Sensitivity

Non-uniform sensitivity within a receptive field is also useful for increasing the robustness of the neocognitron. By selectively reducing the sensitivity at some part of the receptive field, we can virtually cut away that part of the receptive field. Thus, the shape of the receptive field can be adaptively adjusted to the shape of the stimulus feature to be extracted. Conversely, if the sensitivity at some part of the receptive field is increased, the ability of the cell to discriminate the difference of stimuli presented there, thus, the ability to differentiate similar features, can be increased. In the following subsections, these two opposing types of sensitivity modulation are discussed.

3.1 Adjusting the Shape of Receptive Fields

The local features which S-cells are to extract are usually different in shape and size from each other. If we want to let an S-cell extract one particular local feature without being disturbed by the existence of other features around it, it is desirable to make the cell have a receptive field identical in shape and size to the feature to be extracted.

In the conventional neocognitron, the excitatory input connections to an S-cell are modifiable by learning and can be adjusted to match the shape of the stimulus feature. Smaller excitatory connections from a part of the receptive field, however, do not necessarily make the S-cell insensitive to stimuli given to that part of the receptive field. It simply makes that part of the receptive field more inhibitory. In other words, reducing the strength of the excitatory connections is not a sufficient

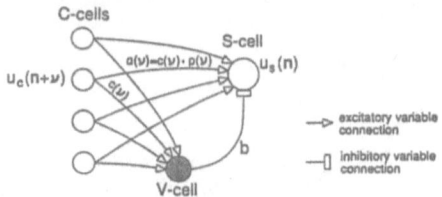

Figure 2: Connections converging to an S-cell

procedure to adjust the shape and the size of the receptive field. We will discuss this process in more detail below.

Figure 2 illustrates the signal paths converging to an S-cell. Let us consider the signal paths to the S-cell from an arbitrary C-cell in the preceding stage. The S-cell receives an excitatory signal through direct excitatory connection from the C-cell, and also an inhibitory signal via the V-cell which makes a pair with the S-cell. The effect of the C-cell on the response of the S-cell is determined depending on the relative strength of the direct excitatory path and the indirect inhibitory path. If the direct excitatory path is stronger than the indirect inhibitory path, that part of the receptive field becomes excitatory. If not, it becomes inhibitory.

In the conventional neocognitron [4], the input connections $c(\nu)$ to the intermediate V-cell are all fixed and unmodifiable. Only the excitatory connections $a(\nu)$ to the S-cell are modifiable. Hence, the shape and the size of the receptive field cannot be changed. Weakening of the excitatory connection does not make the S-cell insensitive to the signal from the C-cell, but simply makes the inhibitory signal from the C-cell more dominant.

In order to adjust the sensitivity within a receptive field of the S-cell, it is necessary to change the input connections to the V-cell in conjunction with the direct input connections to the S-cell. Therefore, we propose a neocognitron with this mechanism of modifying connections. Each input connection $a(\nu)$ to the S-cell is decomposed into two components, that is, $a(\nu) = c(\nu) \cdot p(\nu)$. The sensitivity to stimuli, which is determined by $c(\nu)$, is not uniform within the receptive field of an S-cell in the new neocognitron. It can be adaptively adjusted to the shape of a feature of a training pattern during the learning process. An unnecessary part of the receptive field can be virtually cut away by reducing the sensitivity of that part, and the shape of the receptive field can be adjusted to match the shape of the local feature which the cell is to extract.

Many kinds of training methods can be devised for this new neocognitron, which are currently being tested. In this subsection, we present a training method using an unsupervised learning procedure [7]. Similarly to the conventional unsupervised learning of the neocognitron [2],[8], "seed cells" are selected with a kind of winner-take-all procedure. Let us assume the S-cell $u_S(n)$ in Figure 2 has been selected as a seed cell. The input connections to the seed cell (S-cell) and the partner V-cell are modified in two steps.

Initially, the excitatory input connections to the seed S-cell are reinforced in a similar way as the conventional method, in which the input connections to the V-cells are not changed. To be more precice, the value of $a(\nu)$ is reinforced by the amount proportional to the output of C-cells $u_C(n + \nu)$.

Then, the sensitivity within the S-cell's receptive field is modulated by reducing the input connections to the S-cell in conjunction with the input connections to the partner V-cell. The value of $c(\nu)$ is reduced by the amount proportional to the degree of discrepancy between the spatial shape of $p(\nu)$ and that of the response of C-cells $u_C(n + \nu)$ generated by the the training pattern. If the discrepancy is small, however, the reduced value gradually recovers to its initial value.

During the learning process, each local feature generally appears together with other local features in a single training pattern. A different training pattern usually has a different combination of local features. This means that a relevant training feature is not always presented alone in isolation from other features in the receptive field of a seed cell. In the peripheral part of the receptive field of the seed cell, some fractions of other features often exist. If we suppose the combination of local

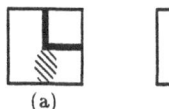

(a)   (b)

Figure 3: Similar features to be discriminated.

features in different training patterns be random, the fractions of irrelevant features in the periphery also changes randomly in different training patterns. Hence the sensitivity at the peripheral part of the receptive fields can be adaptively changed depending on the shape of the relevant feature. This automatic shaping of the receptive field by the proposed learning method was demonstrated by computer simulation [7].

3.2 Local Sensitization in a Receptive Field

Non-uniform sensitivity within a receptive field is also useful for increasing the ability to differentiate similar patterns. If the sensitivity in some part of a receptive field is selectively increased, the difference in shape of the stimuli presented there can be emphasized.

Let us consider a concrete example. When we want to discriminate the alphabetical character E from F, it is important to distinguish local feature (a) from (b) in Figure 3. Since the difference between these two local features is not large, they are apt to be confused if the selectivity of the S-cell is low. Too high a selectivity, however, is not desirable because it reduces the S-cell's ability to accept deformed versions of the stimulus feature.

The difference between the two features exists only in the place indicated by the hatching in the figure. Therefore, if the sensitivity of the S-cell is selectively increased in this place, the cell can acquire the ability to discriminate feature (a) from (b), without losing the ability to accept a deformed version of the feature.

In order to automatically create this kind of non-uniform sensitization of receptive field, we have trained the neocognitron by supervised learning using a backpropagation algorithm [9]. The simulated neocognitron is a four-stage (i.e., nine-layered) network.

To be more precise, the parameters $p(\nu)$ in Figure 2 are first reinforced by unsupervised learning as in the case of the conventional neocognitron. After finishing the initial training by unsupervised learning, the parameters $c(\nu)$ and $p(\nu)$ are modified by a backpropagation algorithm, in which the cost function is defined as the total sum of the squared errors of the outputs from individual cells in the recognition layer at the highest stage. Computer simulation has shown that the ability to differentiate similar patterns is increased without losing the ability to generalize for deformed patterns.

Acknowledgment This work was supported in part by Grant-in-Aid #02402035 for Scientific Research (A), and #04246105 for Scientific Research on Priority Areas on "Higher-Order Brain Functions", both from the Ministry of Education, Science and Culture of Japan, and also by the grant for Frontier Research Project in Telecommunications from the Ministry of Posts and Telecommunications of Japan.

References

[1] K. Fukushima: *Biological Cybernetics*, **36**[4], pp. 193–202 (April 1980).

[2] K. Fukushima: *Neural Networks*, **1**[2], pp. 119–130 (1988).

[3] K. Fukushima, N. Wake: *IEEE Trans. on Neural Networks*, **2**[3], pp. 355–365 (May 1991).

[4] K. Fukushima: *Neural Networks*, **2**[6], pp. 413–420 (1989).

[5] K. Hiroshige, M. Okada, K. Fukushima: *Technical Report, IEICE*, No. **NC91**-160 (Mar. 1992).

[6] K. Fukushima, M. Okada, K. Hiroshige: "Neocognitron with Dual C-cell Layers", submitted to *Neural Networks*.

[7] K. Yamauchi, K. Fukushima: "New learning methods for temporal-pattern recognition in the neocognitron", (in Japanese), submitted to *IEICE Trans. D-II*.

[8] K. Fukushima, S. Miyake: "Neocognitron: A *Pattern Recognition*, **15**[6], pp. 455–469 (1982).

[9] M. Ohno, M. Okada, K. Fukushima: *JNNS'92*, Osaka, Japan, No. P316 (Dec. 1992).

Hand-Written Character Recognition by a Structured Self-Growing Neural Network "CombNET-II"

Akira IWATA, Yoshihisa SUWA

Dept. of Electrical and Computer Eng., Nagoya Institute of Technology
Gokiso-cho, Showa-ku, Nagoya, 466, JAPAN
e-mail : iwata@mars.elcom.nitech.ac.jp

ABSTRACT

CombNET-II is a self-growing 4 layered neural network model which has a comb structure. The first layer constitutes a stem network which quantizes an input feature vector space into several sub-spaces and the following 2-4 layers constitutes branch network modules which classify input data in each sub-space into specified categories. CombNET-II employs a self-growing neural network learning procedure for training the stem network and Back Propagation for branch modules. The excellent performance of CombNET-II are demonstrated using hand-written digits (10 categories), alphabets (26 categories), and symbols (4 categories) using several databases.

1. INTRODUCTION

We have proposed a self-growing 4 layered neural network model, "CombNET-II"[1]. It has a comb structure. The first layer constitutes a stem network which quantizes an input feature vector space into several sub-spaces and the following 2-4 layers constitutes branch network modules which classify input data in each sub-space into specified categories. We have implemented a large scale neural network to classify 2965 printed Kanji characters with the recognition rate of 99.9% by CombNET-II[1]. CombNET-II is a general purpose neural network model for pattern classifications. We have applied CombNET-II to recognition of hand-written alpha-numeric characters[2] and hand-written Kanji characters[3].

Here we have investigated the performances of CombNET-II for hand-written character recognition using several databases. Because there are an enormous number of different characters and wide varieties of character patterns make very complicated discriminating boundaries in the input feature space, it is very difficult to implement a neural network with a simple 3 layered hierarchical network by BP algorithm. In those cases the learning procedure may often fall into the local minimum state and then cannot find the optimum solution. However, CombNET-II divides a input feature space into sub-spaces where a restricted number of categories are involved by a stem neuron so that each branch module has always a restricted number of output units and inter-connections. As increasing categories to be classified and varieties of input data, more stem neurons and branch modules are created so as to keep proper sizes of branch modules. Therefore, CombNET-II does not cause the local minimum state since the complexities of the problems to be solved for each branch module are restricted by the stem network.

This paper describes the network structure and the learning procedure of CombNET-II together with the experimental results of hand-written character recognition.

2. A SELF GROWING NEURAL NETWORK MODEL "CombNET-II"

CombNET-II consists of 4 layered network with a comb structure as shown in Fig.1. A vector quantizing network forms the first layer as a stem and many 3 layered network modules form 2-4 layer as branches. The number of branch network modules is as many as the number of neurons of the stem network. As an input data is given to the stem network, several (3-5) neurons which give higher matching scores between the input data vector and the synaptic weights are selected. Then the input data is led to the plural branch modules which are connected to the selected neurons in the stem network. Final scores are given by the following criterion

$$Z=(SM)^a(SB)^b \qquad (1)$$

where, SM: the matching score in the stem network

SB : the maximum output score in a branch network module

CombNET-II employs the self-growing neural network learning procedure, an original learning procedure [1,2], for training the stem network. The purpose of the stem network is to divide the input feature space into several sub-spaces in which a restricted numbers of the categories are involved. The self-glowing neural network learning procedure constructs such a vector quantizing network.

After learning process of the stem network by the procedure, the input feature space is partitioned into sub-spaces according to the best-matching criteria to the synaptic weight vectors of neurons. The synaptic weight vector becomes a template pattern which represent the common feature of an input data group.

After training the stem network, all input data are partitioned into category groups according to the best-matching criteria to the synaptic weight vectors of neurons. Then branch network modules are trained for every sub-space to make discriminating boundaries. Back propagation is utilized to train branch modules. Each branch neural network which is 3 layered hierarchical network has a restricted number of output neurons and inter-connections so that it is easy to train.

3. HAND-WRITTEN CHARACTER RECOGNITION

A neural network to recognize hand-written digits (10 categories), alphabets (capital letters, 26 categories), and symbols (4 categories, +,-,*,=) has been implemented by CombNET-II. Three databases has been used in this study. Two of them named ETL-6 and ETL-1 are the databases delivered from the Electro-Technical Laboratory (ETL), Ministry of International Trade and Industry. The database ETL-6 involves 1000 exemplars of each characters which were written carefully according to the sample patterns. So those character patterns in ETL-6 are typical and similar to each other in a category. The database ETL-1 involves also 1000 exemplars of each characters. But those were written in free style without seeing the sample patterns. So there are wide varieties of character patterns. The other database, POST involves also 1000 exemplars of each digits which are collected from ZIP code on post cards at the Institute for Posts and Telecommunications Policy, Ministry of Posts and Telecommunications. Those characters were written in free style with any kinds of pen including Japanese writing brushes. So, the database POST has the widest varieties of character patterns among three databases. The first 600 exemplars of each character in each databases made up the training data set and the remaining 400 exemplars went into the test data set.

After a character region was picked up, a character image was converted to an gray level image of 8x8 pixels. To extract features about local line directions of patterns, the input patterns of characters are preprocessed using the network[2]. The network provides 128 (4x4x8) features from a character pattern. Those features are used as the input data of CombNET-II.

The networks have been implemented respectively using each database as the training data. Two

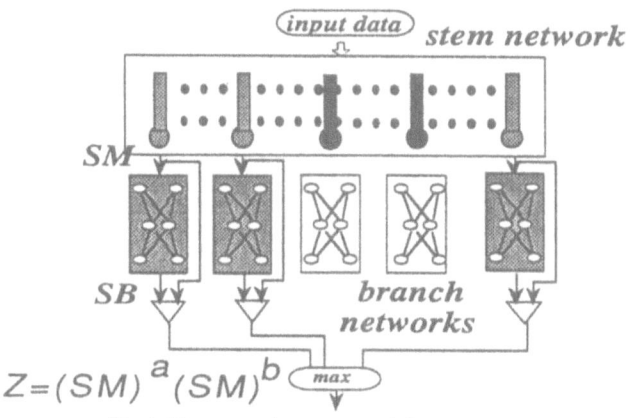

Fig.1 The network structure of CombNET-II

kinds of the networks have been implemented for two kinds of character category sets, digits (10 categories), or digits, alphabets and symbols (40 categories), for the databases ETL-6 and ETL-1. Using the database POST, a network for classifying digits have been implemented. The performances of the networks have been investigated using the remaining test data-set in each database. Table 1 shows the results. The recognition rates for digits were 99.40% to 96.67%. The recognition rates for digits, alphabets and symbols were 98.38% to 93.49%. Fig.2 shows examples of the recognition of character patterns in each database.

We compared the performance of CombNET-II with the nearest neighbor classifier. The results are shown in Table 2. In the nearest neighbor method, all training patterns (600 patterns for each character) are used as the templates. The recognition rates for digits were 99.38% to 96.83% by the nearest neighbor classifier.

Though the nearest neighbor classifier gives high performances, it uses all of the character patterns as templates which need large mount of memory to store templates. CombNET-II needs small amount of memory, which is only 6.4% (49.4k/768k) to 9.8% (75.6k/768k) of the amount needed for the nearest neighbor classifier as shown in Table 2. CombNET-II compresses the informations for classifying character patterns which are involeved in the databases and implies them as the value of the inter-connecting weights. Moreover, in the recognizing phase, all templates has to be used to compute the similarity measures in the nearest neighbor classifier. But CombNET-II needs only 2.88% (21.1k/768k) to 2.95% (22.7k/768k) of the amount of the nearest neighbor classifier since only several (3 - 5) branch network modules are used in the recognizing phase (Table 2). These are a big advantage to implement a system for practical use.

4. CONCLUSIONS

A self-growing neural network model, "CombNET-II" has been applied to the recognition of hand-written characters. The recognition rates of each network are as good as the nearest neighbor classifiers. The memory sizes of the networks are 6 to 10% of the nearest neighbor classifiers. The nearest neighbor classifiers provides the best performances which are determined by the databases. Therefore, CombNET-II can develop neural networks which provide the same performances as the nearest neighbor classifiers using compact memory sizes.

CombNET-II employs the self-growing neural network learning procedure, an original learning procedure, for training the stem network. The purpose of the stem network is to divide the input feature space into several sub-spaces in which a restricted numbers of categories of input data are involved. After training the stem network, all input data are partitioned into category groups. Then branch network modules are trained for every sub-space. Back propagation is utilized to train branch

Table 1 The recognition rates for classifying the hand-written characters by CombNET-II

Database	ETL-6	ETL-1	POST
digits (10 categ.)	99.40%	98.17%	96.67%
digits, alphabets, symbols (40 categ.)	98.38%	93.49%	-----

Table 2 The recognition rates for classifying the hand-written digits
and the amounts of the weights for CombNET-II and the nearest neighbor classifier.

Database	ETL-6		ETL-1		POST	
Classifier	CombNET-II	Nearest Neighbor	CombNET-II	Nearest Neighbor	CombNET-II	Nearest Neighbor
Recognition Rate	99.40%	99.38%	98.17%	98.20%	96.67%	96.83%
Amount of Weights	49.4k	768k	66.4k	768k	75.6k	768k
Amount of Weights used at Recognition	21.5k	768k	22.1k	768k	22.7k	768k

networks. Each branch module which is 3 layered hierarchical network has a restricted number of output neurons and inter-connections so that it is easy to train. Therefore CombNET-II does not cause the local minimum state since the complexities of the problems to be solved for each branch module are restricted by the stem network.

ACKNOWLEDGMENTS
The authors express their appreciation to the Electro-technical Laboratory, Ministry of International Trade and Industry and the Institute for Posts and Telecommunications Policy, Ministry of Posts and Telecommunications for providing the hand-written character database. The authors also express their appreciation to the partial support of this work by Grant in aid of the Ministry of Education.

REFERENCES
[1]HOTTA,K., IWATA,A., MATSUO,H. and SUZUMURA,N. : "Large Scale Neural Network "CombNET-II" ", Technical and Research Reports of IEICEJ, Vol.NC-90, No.34, pp29-36, 1990 (in Japanese).
[2]IWATA, A., SUWA, Y., INO, Y., SUZUMURA, N.:"Hand-written alpha-numeric recognition by a self-growing neural network "CombNET-II", Proc. of 1992 IEEE&INNS Int. Joint Conf. on Neural Networks, Baltimore, Vol.4,pp.228-234, 1992
[3]IWATA, A., INO, Y., HOTTA, K., SUZUMURA, N. : Hand-written Japanese Kanji character recognition by a structured neural network "CombNET-II", Proc. of 1992 Int. Conf. on Artificial Neural Networks, Brighton, pp.1189-1192, 1992.

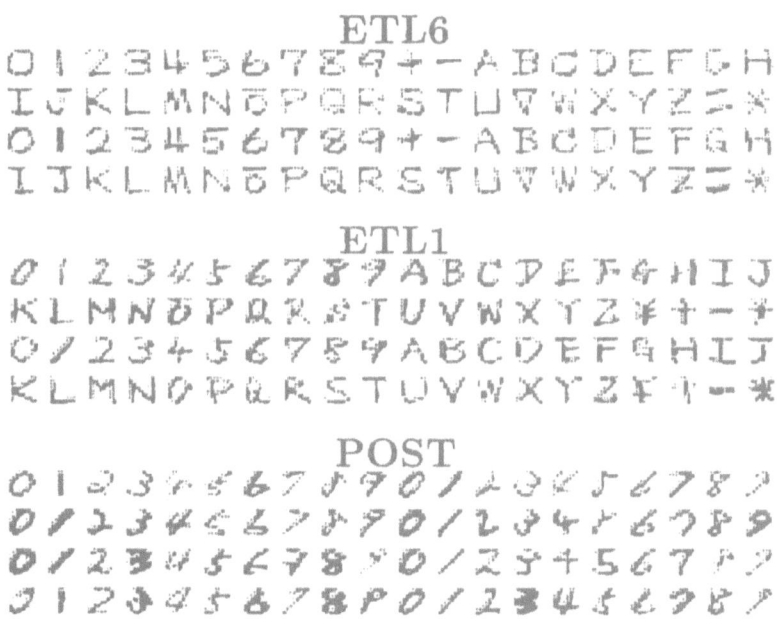

Fig.2 Examples of character patterns in the databases, ETL6, ETL1, and POST.

Segmentation Of Image Sequences Using Self-Organising Feature Maps

René Natowicz[a] - Fabrizio Bosio[b,a] - Serge Sean[c,a]

(a) E.S.I.E.E. Laboratoire de Traitement de l'Information et des Systèmes - Cité Descartes B.P. 99
93162 Noisy le Grand - France. phone/fax : (33)(1) 45 92 67 14/66 99
(b) Politecnico de Milano - Dipartimento di Elettronica e Informazione
20133 Milano - Italia
(c) Université de Paris XII - Génie Biologique et Médical
Bât. P2 Pièce 230 - 94010 Créteil cedex - France

Abstract
One can segment a single digital image in grey level after having quantized it by a self-organising map. Because of the great computing time needed by the quantization process, one cannot quantize each image of a sequence when real time processing is at aim. We propose a method in which 1.- one has a set of precomputed maps among which is chosen the most appropriate one to segment the current image. Quantization errors determine the most appropriate map. 2.- one lets the current map continuously adapt itself to new images as long as it is the most appropriate. Time complexity and possibilities of parallelisation of the deduced algorithm are considered.

1. Introduction

Kohonen's self-organising feature maps [1,2] have been used in image processing for compression purposes [3,4], texture based segmentation [5,6], and grey level based segmentation [7]. We propose an extension of this last method to define a process of image sequence segmentation using self-organising maps. The method for segmenting isolated images defined in [7] relies upon the "tonotopy" property of self-organising maps and consists in

1. quantifying the set of grey levels of the image using a one dimension chain connected self-organising map : the computed quantization defines a compression mapping from the set of image grey levels onto the set of map's cells

2. defining a spatial neighbourhood of every pixel on the image and a spatial neighbourhood of every cell on the map

3. computing the set of pixels for which tonotopy property does not spatially hold within the neighbourhoods defined.

When real time processing is aimed, one cannot segment an image sequence by computing a new quantizing map at any image occurrence because learning for the quantizing map (step 1.) is too costly in computer time.

The method we propose consists in

1. having a set of precomputed maps : segmentation of the first image uses one of these maps and one can at any image occurrence, replace the current map by a precomputed one which better quantizes the grey level statistic distribution of the new image

2. letting the current map continuously adapt itself to the images of the sequence.

Every image is segmented through the map among the precomputed ones and the adapted current one, that better quantizes its grey level distribution. The upper method relies on the segmentation of single images using self-organising maps which will be briefly described before describing sequence segmentation itself.

2. Single image segmentation using self-organising map

One considers an image I in grey level to be segmented by using a self-organising map. Let P be its set of pixels an G its set of grey levels. A pixel $p = (g, (x,y))$, $p \in P$, is characterised by its grey level

$g \in G$ and its position *(x,y)* in the image. One also considers a self-organising map K *of m* cells connected in chain, $K = \{k_1, ..., k_m\}$, quantizing the set G of image grey levels [3,4].

As the map K quantizes the grey level set G, it defines a compression mapping C, $C: G \to K, g \mapsto C(g)$, which links any grey level g to map's cell $C(g)$ having the grey level closest to g. "Tonotopy" property of map K insures that two close grey level values g and g' are mapped onto cells $C(g)$ and $C(g')$ spatially close on map K.

One considers the mapping C as acting on set of pixels P, i.e. $C: P \to K, p = (g, (x,y)) \mapsto C(g)$ and one defines a spatial neighbourhood on the image and a spatial neighbourhood on the map as follows.

Spatial neighbourhood on the image : any interior pixel p is assigned the pixel set $V_I(p)$, $V_I(p)$ being the 9-neighbourhood centred on pixel p.

Spatial neighbourhood on the map : any cell k is assigned the set of cells $V_K(k)$, which chain path length up to cell k is less or equal than r. Parameter r is the radius of the neighbourhood centred on cell k.

One defines the set S of pixel segmentation as the set of pixels within the neighbourhood of which the tonotopy property does not spatially hold :

$$\forall p \in P, \ p \in S \Leftrightarrow \exists p' \in V_I(p), \ C(p') \notin V_K(C(p))$$

The image neighbourhood here chosen is the classical 9-neighbourhood but any other one (e.g. 5-neighbourhood) would be convenient. The neighbourhood defined on the map is parametrized by radius r, and one obtains segmentations at varying levels of details depending upon radius values : S_r and $S_{r'}$ being the sets of pixel segmentation computed for respective radius r and r', one has $r \leq r' \Rightarrow S_{r'} \subseteq S_r$.

3. Segmenting image sequences using self-organising maps

As underlined in chapter 1., single image segmentation has a step of quantization learning which time computing cost prohibits to process it on every image of the sequence. By contrast, computation of segmentation pixel set is quick and further, can be processed in a highly parallel way. The extension to image sequence segmentation eliminates the quantization step by using a set of precomputed maps resulting from the quantization of various images. One can describe it by the following algorithm after what one will describe the the two main points of it, i.e. the selection of the best map for segmenting an image and the adaptation of the current map to a new image.

Algorithm for image sequence segmentation

Inputs : $I_s = (I_0, I_1, ...)$ non empty image sequence, $K_s = \{K_1, ..., K_n\}$ set of n precomputed self-organising maps of m chain connected cells.

Output : $S_s = (S_0, S_1, ...)$ segmented input image sequence.

Initialisation : $K^* \leftarrow bestMap(K_s, I_0)$; – select in K_s map K^* to segment I_0

$S \leftarrow segment(I_0, K^*)$;   – segment image I_0 with map K^*

$display(S)$;

$next(I_s)$;

```
While input sequence not empty
    I ← first(I_s);                              -- new image
    K ← adapt(K*, I);                            -- adapt current map to the new image
    K* ← bestMap(K_s ∪ {K}, I);                  -- map to use is the current adapted one or is in K_s
    S ← segment(I, K*);
    display(S);
    next(I_s)
end while

end algorithm
```

3.1 Best map selection

Let P be the set of pixels of the current image, G its set of grey levels and $p(g)$ the probability distribution over set G. For any map K let C_K be the compression mapping which assigns any grey level g of set G its representing map cell $C_K(g)$ and let us denote $gl(C_K(g))$ the grey level of cell $C_K(g)$.

The map K^* which is the best for segmenting the current image is the one which minimises the quantization mean error over the set of maps :

$$Err(K^*) = \min_{K \in K_s}(Err(K)), \text{ with } Err(K) = \sum_{g \in G}|g - gl(C_K(g))|.p(g)$$

Time complexity for selecting the best map K^* is an $O(|P|+|G|.|K|)$ on a sequential computer (probability distribution $p(g)$ is computed in $O(|P|)$; for any map, quantization mean error computation time is an $O(|G|)$; selecting the best map is processed "on the fly" when computing quantization errors).

Quantization errors can be computed in parallel for all the maps and one obtains a complexity in $O(|P|+|G|)$ in this case of full parallelism.

3.2 Map adaptation

When the grey level probability distribution of a new image is close to the previous' one, adapting the best current map K^* may be sufficient to obtain the map best suited for segmenting the new image. The adaptation process is nothing else than the competitive learning process described in [1,2]. The learning set is the grey level set G: grey level of every pixel is used one time and pixels are selected at random. The radius r of the spatial neighbourhood defined on the map (cf. 2.) is the initial radius used in the learning process.

Adapting the best current map to the new image can be done in parallel with the computing of its grey level probability distribution. This last computing is needed for selecting the best map in set of precomputed maps K_s.

4. Conclusion

We have set out a method for segmenting image sequences as an extension of grey level based segmentation of single images using Kohonen's self-organising maps. Single image segmentation asks for a quantization of the set of grey level values which is costly in computing time. The extension we proposed is to use a set of precomputed maps and to let the current map adapt itself to new images of the sequence. Complexity evaluation of the on line segmentation process and considerations on its parallel computation allow us to aim at segmenting sequences in real time. Effective implementation of the method on a parallel computer is under study.

References

[1] T. Kohonen, "Self-organisation and associative memory", Springer-Verlag Berlin, 1984.

[2] T. Kohonen, "The self-organising feature map", proc. of the I.E.E.E., vol. 78, n° 9, Sept,1990.

[3] N.M. Nasrabadi, Y. Feng, "Vector quantization of images based upon the Kohonen self-organising feature map", I.E.E.E. Int. Conf. on Neural Networks, pp. 101-108, San Diego Ca.,1988.

[4] E. le Bail, A. Mitchie, "Quantification vectorielle par le réseau neuronal de Kohonen", Traitement du Signal, vol. 6, n° 6, 1989.

[5] A. Visa, "Identification of stochastic textures with multiresolution features and self-organising maps", Int. Conf. on Pattern Recognition, pp. 518-522, Atlantic City, 1990.

[6] O. Simula, A. Visa, "Self-organising feature maps in texture classification and segmentation", I.C.A.N.N. 92, pp. 1621-1628, Brighton, 1992.

[7] R. Natowicz, R. Sokol, "self-organising feature maps for image segmentation", Int. Work. on Artificial Neural Networks, Barcelona, June 1993.

Combining Neural-Network and Statistical Methods in Seismic First-Arrival Picking

Guozhong An and Willem J. M. Epping
Shell Research B.V.
P.O. Box 60, 2280 AB Rijswijk
The Netherlands

Abstract

First-arrival picking is an important and time-consuming task in the processing of seismic data for oil and gas exploration purposes. It is basically a pattern-recognition problem. Here, we report on a neural-Bayesian solution to this problem. By interpreting the output of the neural network as Bayesian *a posteriori* probability, a Bayesian classification was achieved. In addition, a statistical post-processing procedure was developed that reduced the error rate by more than 50%.

Introduction

Oil and gas are often found trapped in the earth's structures, a few kilometer below the surface. The primary means of obtaining information about the earth's substructure is a seismic survey. During such a survey, sound waves are generated at the surface by explosion or vibration. The acoustic energy that propagates downward is reflected back to the surface whenever it encounters discontinuities along its path. The seismic signal originating from a single source and recorded at a single location is called a trace. Commonly used sources in land surveys are dynamite and mechanical vibrators. Seismic data produced by sources of the latter type are called vibroseis data. Normally, vibroseis data have a lower signal-to-noise ratio than dynamite data in the part containing the first arrivals. The first-arrival time of a seismic trace denotes the moment when the seismic signal first attains a non-zero value (the onset). Finding the first-arrival time of the seismic traces enables the correction of distortions in the seismic image of the earth caused by near-surface layers [1]. Similar picking procedures are required in the processing and analysis of seismic signals from earthquakes and underground nuclear explosions.

Manual first-arrival picking is a rather time-consuming procedure. Recently, this task has been approached from a pattern-recognition point of view with the aid of neural networks [2, 3]. In this paper, we shall report on a neural-network picking system that embraces Bayesian probability theory and incorporates multi-trace information.

Neural Bayesian pattern classification

Actual recorded traces consist of both signal and noise. Although, the onset of the signal is often masked by the noise, the magnitude and location of the peaks in the signal are relatively insensitive to noise. On the assumption that the lag between the onset of the signal and the subsequent peak is constant for all the traces in a survey, it suffices to consider only the peaks instead of all the trace samples as potential first arrivals.

Each peak of a trace is characterised by a feature vector $\mathbf{x}$, which serves as input to the neural network. The feature vector must contain sufficient information to permit discrimination of the first-arrival peak from the remaining noise and from other peaks. One way of forming the feature vector $\mathbf{x}$ is to take N trace samples centred about a peak. (Recall that a trace is a discrete time series.) A value of $N = 3$ or 5 as used in Refs. [2] and [3] was considered too small by us; the vector $\mathbf{x}$ does not carry enough contextual information in these cases. An appropriate value for N is that corresponding to about twice the time between the first-arrival peak and the next peak. For the data we tested, the value of N obtained in this way varies from 19 to 23. The poor results reported using this input representation in Refs. [2] and [3] are probably due to the smallness of N in

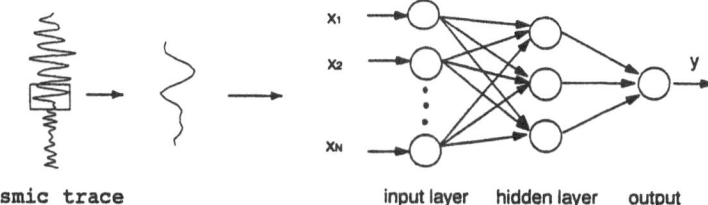

Figure 1: Schematic representation of one of the neural network configurations used along with an example seismic trace.

comparison with our values. much smaller than this. We have also investigated the use of alternative input representations, the results are reported in Ref. [4].

The neural network we used is the well-known feed-forward multilayer perceptron with one hidden layer. It approximates the relation between its input vector $\mathbf{x}$ and its output $\mathbf{y}$ by combining a number of simple $\tanh(x)$ functions. The output $\mathbf{y}$ is related to its input $\mathbf{x}$ as follows:

$$\begin{cases} f_j(\mathbf{x}) &= \sum_{i=1}^{n} a_i \tanh(\mathbf{x} \cdot \mathbf{w}_i + t_i) \\ y_j &= \tanh(f_j(\mathbf{x}) + g_j) \end{cases} \tag{1}$$

As it is usually done, the parameters of the neural network, $\{\mathbf{w}_i, t_i, a_i, g_j\}$, are determined by minimising an "error function" E in a training procedure. The function E is given by

$$E = \frac{1}{2}\frac{1}{M}\sum_{j,p}[y_j^p - d_j(\mathbf{x}_p)]^2, \tag{2}$$

where p is the index of the training example ($p \in \{1, 2, .., M\}$), j the index of the output units, and $d_j(\mathbf{x}_p)$ is the desired output for pattern $\mathbf{x}_p$ at output unit j. It can be shown [5] that the output of the neural network as specified above is related to the Bayesian probability by

$$y_j \approx \sum_k P(C_k|\mathbf{x})d_j(C_k), \tag{3}$$

where $P(C_k|\mathbf{x})$ is the conditional probability that $\mathbf{x}$ belongs to class C_k, and $d_j(C_k) \equiv d_j(\mathbf{x})$ for $x \in C_k$ is the desired output value for class C_k at output unit j. The desired output $d_j(\mathbf{x})$ associated with a training pattern is set to 0.9 (if $\mathbf{x} \in C_j$) and -0.9 (if $\mathbf{x} \notin C_j$). Using the implied values for $d_j(C_k)$ in Eq. (3) and the relation $\sum_k P(C_k|\mathbf{x}) = 1$, we have

$$y_j \approx 0.9[2P(C_j|\mathbf{x}) - 1]. \tag{4}$$

The problem of first-arrival picking can be viewed as a two-class classification problem, one class for all the first arrivals, and the other class for all the non-first arrivals. From a physical point of view, the input falls more naturally into three classes, i.e., the noise (pre-arrival), first-arrival and later-arrival classes. We investigated two- and three-class cases using separate neural-network configurations for each cases. The neural network used for a two-class classification has one output unit (Fig. 1). Its three-class counterpart has three output units. Equation 4 makes it possible to apply Bayesian classification theory [6]. According to this theory, an input $\mathbf{x}$ is classified as the first arrival if $P(A|\mathbf{x}) > P(B|\mathbf{x})$, where $A =\{$first arrivals$\}$ and $B =\{$the rest of the classes$\}$. For the two-class case, Bayesian classification leads to $y > 0$. Note that the interpretation of the neural-network output as a Bayesian *a posteriori* probability, automatically specifies the threshold that controls the class boundary; it would have to be set heuristically otherwise. Since a trace can have only one first arrival, it is natural to select among all the possible first-arrival patterns in a trace the one which has the highest y-value as the first arrival. If no first arrival is found for a trace, it is rejected.

Post-processing using regression

In the aforementioned (single-trace) method, first arrivals are determined trace by trace without explicit reference to neighbouring traces. Although contextual information along the time dimension around a potential first arrival is included in the input vector **x**, spatial contextual information from neighboring traces is not available to the neural network at all. For low-noise data, the single-trace operation gives very good results. However, for noisy data the information from a single trace is often insufficient to determine the first arrival. As a matter of fact, the coherency of the picks among neighbouring traces is a key criterion by which first arrivals are picked manually. For those traces in which the first arrival cannot be determined, a seismologist interpolates the first arrivals on the basis of contextual information from neighbouring traces. A post-processing procedure that captures some of the essence of this aspect of manual picking was therefore developed to supplement the neural network with spatial information. If the thickness and sound velocity of the earth's near-surface layer change relatively little over the source-receiver distance, the first-arrival time has an approximately linear dependence on the source-receiver distance. A linear regression model was constructed for every half of a " split-spread shot gather" (the set of traces originating from a common source in the middle of a line of receivers), to fit the arrival time and source-receiver distance based on picks provided by the single-trace neural network. Picks that deviate by more than an amount Δt from the arrival time of the regression model are replaced by alternative ones that are within Δt of the regression model. If alternative picks are not available, an interpolation is made on the basis of the model. Those picks that are within Δt of the model are not modified. We have found that a value of 16 ms for Δt is appropriate. The above procedure was iterated until converges was reached.

Application to real data

The number of input neurons is determined by the dimension of the input vector **x**. However, the number of hidden neurons has to be found experimentally. The neural networks were trained on a commercially available package (NeuralWorks Professional II) using the back-propagation algorithm. The training set (comprising about 1500 patterns) was generated from a manually picked shot gather comprising 150 traces. The input vectors were normalised per trace and per component.

Some quite serious problems were encountered with the initial training set: less than 5% of the training patterns were first arrivals. In this case, the neural network failed to recognise nearly all of the first arrivals even though the overall root-mean-square error was low. This problem was overcome by increasing the proportion of first-arrival patterns in the training set to about 50% (by copying the existing first-arrival patterns).

The Bayesian framework offers an explanation for the above empirical success. Recall that Bayes rule: $P(C|\mathbf{x}) = P(\mathbf{x}|C)P(C)/P(\mathbf{x})$, i.e., the *a posteriori* class probability equals the class likelihood times the *a priori* class probability divided by the *a priori* input probability. Neural networks estimate $P(C|\mathbf{x})$ for all the classes simultaneously during learning. The *a posteriori* class probabilities for those classes that have small $P(C)$ and $P(\mathbf{x})$ are therefore poorly estimated. The corrective measure alters $P(C)$ and $P(\mathbf{x})$ without changing $P(\mathbf{x}|C)$ and $P(C)/P(\mathbf{x})$, forcing the neural network to pay roughly equal attention to all the classes in learning $P(C|\mathbf{x})$.

The training procedure took about 2 to 10 minutes on a Sun-SPARC 2 workstation for a typical neural-network configuration. The picking of an entire seismic line, that consists of 22500 traces, took only 10 seconds. Our neural networks are trained and tested on real vibroseis data, which are considerablely more difficult than those used in previous investigations [2, 3].

Table 1: Picking performance on the testing data.

Architecture	Percentage		
	correctly picked	picked in error	rejected[a]
N19-10-3[b]	58 (77)[c]	31 (12)	11 (11)
N19-10-1	58 (76)	34 (16)	8 (8)
N19-15-1	56 (72)	31 (15)	13 (13)

[a] A trace is rejected if no first arrival is found.

[b] The notation Nm-n-l describes the numbers of neurons of a neural network. For example, N19-10-3 represents a neural network with 19 input units, 10 hidden units and 3 output unit.

[c] The numbers within parentheses are the results obtained using post-processing.

The percentage of correctly picked traces is used as the performance measure. A trace is considered to be correctly classified if the neural network picks the same first-arrival peak as a seismologist. The neural-picking system that classifies the patterns into three classes has a picking accuracy that is a few percent higher than its two-class counterpart (Table 1). The neural-network picking system operating in single-trace mode is able to classify about 57% of the traces correctly (Fig 2a). With our simplistic post-processing procedure, a substantial improvement is obtained. The percentage of correctly classified traces increased by almost 20%, and the percentage of misclassified traces is more than halfed. In Table 1, the performances of a number of neural-network configurations are listed.

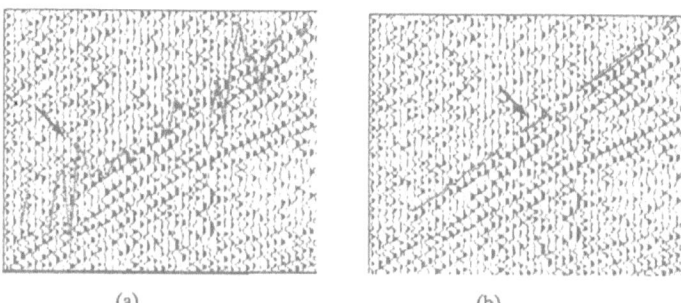

| (a) | (b) |

Figure 2: Part of a seismic shot-gather from a vibroseis survey used to test the neural picking system. (a) without post-processing; (b) with post-processing. The grey lines (pointed out by the arrow) represent the positions of the first arrivals as estimated by a neural network. The rejected traces are not marked.

Conclusions

We have adopted the Bayesian view of probability for neural-network classification throughout this paper. The threshold used for discriminating the input patterns into classes are set according to the minimum-risk classification theory [6]. It was found that classifying the input patterns as three classes (pre-arrival, first-arrival and later-arrival) gives a slightly better picking accuracy than classifying the input patterns as two classes (first-arrivals and non-first-arrivals). An important result of this paper is that a simple statistical post-processing procedure improves the picking accuracy substantially. For cases in which noisy data are accompanied by an unpredictable dependence of the first-arrival time on the source-receiver distance, a more sophisticated method of incorporating spatial information still needs to be developed.

References

[1] M.B. Dobrin and C.H. Savit. *Introduction to Geophysical Prospecting*, chapter 7,11, pages 230–232, 450–457. McGraw-Hill, New York, 4th edition, 1988.

[2] J. Veezhinathan and D. Wagner. A neural network approach to first break picking. In *Proc. of Int. Joint Conf. on Neural Networks*, pages 235–240, San Diego, 1990. IEEE Neural Council.

[3] M.E. Murat and A.J. Rudman. Automated first arrival picking: A neural network approach. *Geophysical Prospecting*, 40:587–604, 1992.

[4] Guozhong An and W.J.M. Epping. Seismic first-arrival picking using neural networks. *Submitted to 1993 World Congress On Neural Networks*, Portland, Oregon.

[5] M.D. Richard and R.P. Lippmann. Neural network classifiers estimate Bayesian *a posteriori* probabilities. *Neural Computation*, 3:461–483, 1991.

[6] R.O. Duda and P.E. Hart. *Pattern Classification and Scene Analysis*, pages 10–35. Jonh Wiley, New York, 1973.

A Study of Neural Network Input Data for Ground Cover Identification in Satellite Images[1]

Marijke F. Augusteijn, Kelly A. Shaw, and Rebecca J. Watson

University of Colorado at Colorado Springs, P.O. Box 7150, Colorado Springs, CO 80933, USA

Abstract: The performance is compared of several feature extraction methods for the classification of ground covers in satellite images with a neural network. These methods include first and second order statistics features, features derived from the Fourier transform, and Gabor filters. Some simple methods, allowing the neural network to extract its own features, are also investigated. When features of many spectral bands are combined, all methods are seen to perform within the same range. The simple methods generally lead to good results with reduced processing time.

1. Introduction

A large volume of satellite image data is currently available, and additional data are being generated on a daily basis. There is a strong interest in developing methods that automatically index images based on their contents. This paper addresses the classification of natural ground covers in satellite images. A supervised neural network architecture is used as the classifier. The Cascade-Correlation architecture [1] was selected for this application. This is a feedforward network which has been shown to be capable of much faster learning than backpropagation. It is a dynamic architecture which builds its internal structure during training. Thus, given some measure of accuracy, the network will allocate the number of hidden units necessary to solve the problem.

Supervised classification requires the existence of a set of training patterns and their corresponding target values. In the domain of satellite images, this translates into the availability of images and their associated ground truth. The ground truth specifies the actual ground cover on a pixel by pixel basis. Generating the ground truth for a particular image is a difficult task and, as a result, not many exist. This project had access to a fairly large (1784 x 1907) Thematic Mapper (TM) image of the central mountains of Colorado, USA, for which a ground truth had been generated in a previous study [2].

Most traditional classification methods require some preprocessing. First, a set of features is extracted from the data, and these features are used as a basis for classification. A variety of features has been developed for the categorization of image data. The various ground covers could be distinguished by their textural information. This requires the measurement of characteristics like homogeneity, contrast, correlation, and entropy. There are statistical methods that calculate these measures in the spatial domain of the image. They may be first or second order depending on the dimension of the probability density from which they are obtained. Alternatively, measures may be derived from the Fourier spectrum of the image. Recently, Gabor filters, which measure the spectrum locally, have also been used. All these feature extraction methods have the disadvantage of, sometimes extensive, computational overhead. This overhead may be reduced or eliminated by taking advantage of the feature extraction capabilities of neural networks. For example, the actual pixel values of an image segment could be given as input patterns. The network must then extract the pertinent features. Or, simple preprocessing may be used like the calculation of average pixel values and standard deviations. This study compares the performance of several standard preprocessing methods and the achievements of these simpler techniques when combined with a neural network classifier.

2. The Experiments

Image segments were extracted from the TM image in nine different categories: ponderosa pine, douglas fir, spruce/fir, mixed conifer, non-vegetated, water, riparian deciduous shrub, dry meadow, and alpine. Each category provided a data set of 100 segments of size 8 x 8 pixels. One half of these segments were used for training, and the other half formed the test set. This study used six of the seven frequency bands that are standard for TM images. Only the thermal band (band 6), which is recorded at a lower resolution, was not included. The relative performance of data extracted from different band combinations was also investigated. Each band is preprocessed separately, and features from different bands may be concatenated. Experiments were performed using the band 2 image (green), combining the three bands in the visible spectrum: 1, 2 and 3, merging these bands with the infrared band 7, and combining all available six bands.

---

[1]Project supported by funds from the Colorado Advanced Software Institute

The performance of the various feature sets, derived from the same data segments, is compared. The Cascade-Correlation neural network was always trained five times on the same input. The weights of each network were randomly initialized which resulted in slightly different performance of each trained network. Both the range of performances and the average value are reported for each experiment.

2.1 Second-Order Statistics Features

The second order gray level statistics method [3] requires the calculation of a set of co-occurrence matrices which measure the frequency of the simultaneous occurrence of two specified gray levels in designated relative positions in an image segment. In this study, four different matrices were used, each computing the frequency of gray-level co-occurrence at neighboring positions in four different directions (horizontal, vertical, and along the diagonal directions). Textural features were calculated from these matrices and averaged over the four directions. Elaborate tests were performed to determine which of the fourteen features defined in [3] provide the best texture representation in this application. The first experiment involved a set of four: homogeneity, contrast, correlation and entropy. These four features did not perform well as shown by the leftmost columns of Table 1. The performance improved significantly when the average gray level of each segment was added to these data (middle columns of Table 1). This subset was gradually expanded to include other features defined in [3] and the standard deviation of the pixel values in the segment. Because of the large number of experiments, only data from band 2 and the 4-band combination were used. None of the other features caused any major improvements. A set of nine features was determined to give small, but statistically significant, improvements over the former set of five. This set included the sum average, sum variance, and an information measure of correlation, as defined in [3], and the standard deviation. The performance of this set is shown in the rightmost columns of Table 1.

Table 1. Performance of Second-Order Statistical Features

Band(s)	Classification Range for 4 Features	Average Classification for 4 Features	Classification Range for 5 Features	Average Classification for 5 Features	Classification Range for 9 Features	Average Classification for 9 Features
2	53.8 - 59.8	55.8	82.2 - 86.4	84.2	84.7 - 87.8	86.1
1, 2, 3	78.2 - 81.6	80.0	92.4 - 95.3	93.8		
1 - 3, 7	84.0 - 86.9	85.7	97.6 - 98.4	98.1	98.7 - 99.3	99.0
1 - 5, 7	85.3 - 88.4	87.5	98.4 - 99.6	98.9		

2.2 First-Order Statistics Features

A different technique, also defined in the spatial domain of the image, is first order gray level statistics, which calculates textural features from a one-dimensional probability density. Two different sources of first order measures were used. One set was based on the probability density of the occurrence of differences between pairs of gray levels, as described in [4]. Four measures (contrast, homogeneity, entropy, and the mean of the probability density components) were computed from these distributions and combined with the mean and standard deviation of the pixel gray levels of the segment to form a six component feature vector. The performance of this feature set is shown in the leftmost columns of Table 2.

The second set consisted of seventeen "texture-tone" measures, defined in [5]. These measures include the four central moments and various pixel averages calculated over the image segments. A variety of experiments was performed to determine the optimal subset of these seventeen measures. Good performance was

Table 2. Performance of First-Order Statistical Features

Band(s)	Classification Range for 6 Features	Average Classification for 6 Features	Classification Range for 5 Features	Average Classification for 5 Features	Classification range for 11 Features	Average Classification for 11 Features
2	78.2 - 85.1	81.2	84.4 - 86.7	85.4	84.7 -86.9	86.0
1, 2, 3	93.3 - 94.7	93.9	88.9 - 93.1	91.5	91.8 - 93.8	92.4
1, - 3, 7	97.3 - 98.9	98.0	92.9 - 95.3	94.2	94.9 - 96.0	95.5
1, - 5, 7	96.9 - 98.9	98.2	97.1 - 98.9	98.2	98.2 - 99.1	98.7

observed for a set of five features consisting of the four central moments and the deviation of the mean. When this set was combined with the six average measures (measures 12 through 17 in [5]), a slight increase in performance was observed, but this difference was not statistically significant. The performances of these two feature sets are also shown in Table 2.

2.3 Features Derived from the Fourier Spectrum

A variety of measures was extracted from the Fourier transform of each image segment. These were all based on the spectral magnitude, no phase information was used. One set consisted of the rings and wedges defined in [4]. The 8 x 8 Fourier spectra were partitioned into four rings and five wedges, and the average amplitude in each ring and wedge was calculated. The performance of these measures is shown in the leftmost columns of Table 3.

A different set of Fourier measures can be defined as follows [6]. If $I(j,k)$ is the matrix containing the amplitudes of the spectrum, and N is the number of frequency components, then four useful measures are:

Maximum Magnitude $= Max[I(j,k) - I(0,0)],$

Average magnitude (A_M) $= \sum_{j,k} I(j,k) / N,$

Energy of Magnitude $= \left[\sum_{j,k} I(j,k)^2 \right]^{\frac{1}{2}},$

and Variance of Magnitude $= \sum_{j,k} [I(j,k) - A_M]^2 / N.$

These four measures were combined with the average gray level value of the image segments. The performance of this feature set is reported in the middle columns of Table 3.

During an investigation of all spectra generated by the various ground cover categories, it was observed that several frequencies were dominant (they consistently appeared with the highest amplitude values). The amplitudes of the seven most dominant frequencies were also used as a feature set. The performance of this set is shown in the rightmost columns of Table 3.

Table 3. Performance of Features Derived from the Fourier Spectrum

Band(s)	Classification Range for 9 Features	Average Classification for 9 Features	Classification Range for 5 Features	Average Classification for 5 Features	Classification Range for 7 Features	Average Classification for 7 Features
2	80.0 - 87.8	85.0	88.2 - 89.6	88.6	87.1 - 91.6	89.0
1; 2, 3	92.4 - 94.2	93.0	94.2 - 98.7	95.3	93.1 - 96.9	95.5
1, - 3, 7	97.3 - 98.0	97.7	98.2 - 99.3	98.9	96.4 - 98.0	97.4
1, - 5, 7	98.4 - 99.3	99.0	99.6 - 100	99.8	96.7 - 99.6	98.5

2.4 Gabor Filters

Gabor filters were calculated as defined in [7]. Each filter consisted of a two-dimensional, complex sinusoidal function modulated by a Gaussian distribution. Three parameters need to be selected for each filter: the Gaussian width, σ, and two spatial frequencies. A feature vector was constructed for each image segment through the application of a set of filters to the segment. Each experiment used seven filters with the same σ and a selected set of frequency pairs. One of these pairs was always the (0, 0) frequency corresponding to the average gray level value of the image segment. Two methods were used to generate the other pairs. The first one used the dominant frequencies of the Fourier spectrum. The best performance of these features was observed in combination with a small Gaussian width ($\sigma = 0.2$). The results

Table 4. Performance of Gabor Filters

Band(s)	Classification Range for dominant freq	Average Classification	Classification Range for eq-spaced freq.	Average Classification
2	84.2 - 88.2	86.8	85.3 - 87.1	86.5
1, 2, 3	91.8 - 93.3	92.6	91.1 - 93.1	91.8
1, - 3, 7	98.2 - 99.1	98.7	97.6 - 98.9	98.4
1, - 5, 7	99.3 - 99.8	99.6	99.6 - 99.8	99.6

are shown in the leftmost columns of Table 4. The second method used a set of frequencies that were equally distributed over the spectrum. Here, performance was better for a larger width. The results for $\sigma = 2.0$ are shown in the rightmost columns of Table 4.

2.5 Simple Preprocessing Methods

Three experiments were performed that required no or simple preprocessing. The first experiment involved the classification of a single pixel of each image segment. Many classification methods are based on single pixel values. In the extreme case, where ground covers are expected to vary from one pixel to the next, each pixel needs to be classified separately. Pixels were selected near the center of each image segment. Classification performances are shown in the leftmost columns of Table 5. The second experiment used a set of pixel values to represent each texture. Subsets of size 5 x 5 were selected from the original image segments, so that each band provided 25 pixel values. (A subset was selected to limit the size of the input vectors.) The results of this test are reported in the middle columns of Table 5. The third experiment used the single pixel values of the first experiment combined with the average values of its 3 x 3 and 5 x 5 neighborhoods, and the mean and standard deviation of all pixel values in the segment. Performances of this simple preprocessing method are shown in the rightmost columns of Table 5.

Table 5. Performance of Simple Features

Band(s)	Classification Range for Single Pixel	Average Classification for Single Pixel	Classification Range for 25 Pixels	Average Classification for 25 Pixels	Classification Range for 5 Features	Average Classification for 5 Features
2	76.8 - 78.2	77.5	80.9 - 83.1	82.5	84.2 - 87.3	85.3
1, 2, 3	84.4 - 86.2	85.3	88.0 - 91.3	89.1	95.1 - 95.6	95.3
1, - 3, 7	94.7 - 95.3	94.9	96.7 - 98.7	98.0	98.4 - 99.3	98.8
1, - 5, 7	98.7 - 99.3	98.8	98.4 - 99.1	98.8	99.1 - 99.6	99.3

3. Conclusion

The experiments show that a neural network can be a powerful instrument in the classification of ground covers in satellite images. The main objective of this study was to investigate various image preprocessing methods. It was found that when data from all six frequency bands were combined, most methods gave similar results. In this case, classification based on single pixel values is preferred, since it requires no preprocessing and does not depend on a neighborhood of pixels belonging to the same class. The differences in performance were greatest for the classification of single band data. Here, the set of Fourier measures defined in this paper or a set of dominant frequencies in the Fourier spectrum showed the best performance. The simple, neighborhood averaging technique can also be used. The last mentioned is slightly faster, but most methods used in these experiments (except for the second order statistics method) have comparable execution times. The second order statistics method was found to be a poor choice since it executed by far the slowest of all methods included in this study. The neural network was found capable of extracting useful features directly from pixel values or averages of pixel values.

References

[1] Fahlman, S. E., and Lebiere, C. (1990). The Cascade-Correlation Learning Architecture. In D. Touretzky (Ed.), Advances in Neural Information Processing Systems 2, Morgan Kaufmann, San Mateo, pp. 524-532.

[2] Huber, T. P., and Casler, K. E. (1990). Initial Analysis of Landsat TM Data for Elk Habitat Mapping. Int. Journal Remote Sensing, Vol. 11, pp. 907-912.

[3] Haralick, R. M., Shanmugam, K., and Dinstein, I. (1973). Texture Features for Image Classification. IEEE Transactions on Systems, Man and Cybernetics, Vol. 3, pp. 610-621.

[4] Weszka, J. S., Dyer, C. R., and Rosenfeld, A. (1976). A Comparative Study of Texture Measures for Terrain Classification. IEEE Transactions on Systems, Man, and Cybernetics, Vol. 6, pp. 269-274.

[5] Hsu, S. (1978). Texture-Tone Analysis for Automated Land-Use Mapping. Photogrammetic Engineering and Remote Sensing, Vol. 44, No. 11, pp. 1393-1404.

[6] Phillips, K. Unpublished notes.

[7] Lu, S., Hernandez, J. E., and Clark, G. A. (1991). Texture Segmentation by Clustering of Gabor Feature vectors. International Joint Conference on Neural Networks, Vol. I, pp. 683-687.

On Generalization Ability of Cascaded Neural Net Architecture

Joarder KAMRUZZAMAN[†] Yukio KUMAGAI[††] Hiromitsu HIKITA[†††]

[†]Dept. of Electrical & Electronic Engg.   [††]Dept. of Comp. Science & Syst. Engg.
Bangladesh Univ. of Engg. & Tech.   [†††]Dept. of Mechanical Syst. Engg.
Dhaka 1000, Bangladesh   Muroran Inst. of Tech.
27-1 Mizumoto-cho, Muroran 050, Japan

Abstract - We present construction of a 3-layer feedforward network called cascaded network and make a performance comparison with the conventional Backpropagation network [1] in terms of generalization ability of the trained networks. Cascaded networks are trained to realize category classification employing binary input vectors and locally represented target output vectors. Empirical study in [2] shows that when a neural network is trained to classify binary input vectors into locally represented binary target output vectors, better saturation of hidden outputs in response to the training set yields better robustness of the network, and in conventional Backpropagation network hidden outputs usually do not get saturated in response to the training set. Based on this observation a 3-layer cascaded network is constructed by cascading two 2-layer networks trained independently by delta rule [3] whose intermediate layer can be viewed as hidden layer and is trained to attain preassigned saturated outputs in response to the training set. Each input-output pair is associated with a preassigned code. Each 2-layer network can thus be considered as a separate module; the first module is trained with the training pattern as the binary input signal and the corresponding preassigned code as the teacher signal. The second module is trained with this preassigned code as the input signal and the desired binary output of the corresponding training pattern as the teacher signal. Thus the first module maps the training patterns onto the preassigned codes and the second maps these preassigned codes onto the desired outputs. After cascading, outputs of the first module become inputs to the second module and thus the intermediate layer of the resultant 3-layer cascaded network can be considered as the hidden layer which is trained to attain predefined saturated outputs as the internal representation of the training set. For linearly separable tasks cascaded network can be built straightforwardly as described earlier, and for nonlinearly separable task, cascaded network is constructed by employing high order cross product inputs at the input layer. These high order cross product inputs are implemented as the mutually disjoint cross products of the components of the binary input vectors received from the training patterns. Experimenting with character recognition problems we demonstrate by simulation results that, for both linearly and nonlinearly separable tasks with binary input and locally represented binary target out put vectors, cascaded network yields generalization ability far better than that of Backpropagation network. The better performance of cascaded network is duo to the attainment of saturated hidden outputs in response to the training set.

References
[1] D. E. Rumelhalt, J. L. McClelland and the PDP research Group, *Parallel Distributed Processing,* vol. 1, M.I.T. Press, 1986.
[2] M. Hamamoto, J. Kamruzzaman, Y. Kumagai and H. Hikita," Generalization ability of feedforward neural network trained by Fahlman and Lebiere's learning algorithm," *IEICE Trans. Fundamentals,* Japan, vol. E75-A, no. 11, pp. 1597-1601, Nov. 1992.
[3] B.Widrow and M. E. Hoff," Adaptive switching circuits," *IRE WESTCON Convention Record,* part 4, pp. 96-104, 1960.

Assessing the Latency of Peak Pa in Auditory Evoked Potentials using Neural Networks

M.J. van Gils and P.J.M. Cluitmans
Division of Medical Electrical Engineering
Eindhoven University of Technology
P.O. Box 513, 5600 MB Eindhoven, The Netherlands

The electrical response of the central nervous system to an auditory stimulus is called an Auditory Evoked Potential (AEP), it can be extracted from the electroencephalogram (EEG) by averaging a large number of epochs of the EEG that are time-locked to the auditory stimuli. There are strong indications that information on anesthetic depth is reflected in certain features of AEPs; the location as well as the amplitude of characteristic peaks in AEPs tend to change when the level of anesthetic depth changes [1][2]. Therefore, it is to be expected that AEPs can be used to serve in a monitor for anesthetic depth. The interpretation of AEPs and the extraction of these features is a task that is performed off-line by human experts who inspect the signals visually. If AEPs are to be used in an on-line monitor for anesthetic depth it is necessary that these features can be extracted automatically. Current research assesses the value of artificial neural networks for performing this task. The work described here concentrates on the recognition of one important peak, peak Pa, in AEPs.

The neural networks that were examined consist of a Kohonen self-organizing feature map [3] succeeded by an output layer. During the training phase, prototypes of AEPs are constructed in the Kohonen layer and the nodes in the output layer are trained to generate the expected latency of peak Pa for each prototype. The performance of the networks as function of the sizes of the Kohonen layer was tested. Also, the effect of applying a post-processing algorithm following the neural network was examined. This post-processing algorithm uses as its input the expected latency of the peak as estimated by the neural network. It calculates the local maximum in the AEP within a window around the expected latency and generates this local maximum as its output. In the experiments 150 AEPs were used to train and test the neural network and post-processor. The patterns that were used consisted of 200 samples that were measured between 10 and 50 ms after the auditory stimulus was applied. The performance was calculated using 10-fold cross-validation. The determination of the latency of peak Pa was considered correct if the network output and the desired output (as specified by a human expert) differed less than 2.0 ms. This "allowed error" of 2.0 ms comes forth from the variance that exists between the assessments of latencies of this peak if they are to be identified by two different human experts.

The results of the experiments show that a moderately sized neural network (containing a Kohonen layer of 4×4 or 5×5 nodes) combined with a suitable post-processing algorithm is capable of determining the latency of peak Pa correctly in (74.0 ± 3.5) % of the cases. The maximum absolute error made in a latency determination is smaller than 5.0 ms for all network sizes. The moderate size of the network and the fact that learning can be done very fast (compared to earlier investigated neural networks that use the backpropagation algorithm for learning) indicate the potential usefulness of this network for on-line recognition of peaks in auditory evoked potentials. Experiments with other networks and algorithms for this task will indicate whether this network will be used in the eventual implementation.

[1] P.J.M. Cluitmans, *Neurophysiological Monitoring of Anesthetic Depth*, Ph.D. thesis, Eindhoven University of Technology, Eindhoven, The Netherlands, June 1990.
[2] C. Thornton, *Assessment of graded changes in the Central Nervous System during general anaesthesia and surgery in man, using the auditory evoked response*, Ph.D. thesis, University of London, 1990.
[3] T. Kohonen, *Self-Organization and Associative Memory*, Springer-Verlag, New-York, 2nd ed., 1988.

A Self-organizing Network of Alterable Competitive Layer for Pattern Cluster

Zhiling Wang, G. Sylos Labini and Marco De Sario[†]
Center for Space Geodesy, Italian Space Agency, P.O. Box 11, 75100 Matera, Italy
†Dept. of Electric Engineering-University of Bari, Via Re David, 200-70125 Bari -Italy

Introduction

This paper apply ART1, only for binary input patterns, to perform the classification of image features. The category number of the competitive layer can be altered automatically to improve the net plasticity. There are two smoothing approaches used to reduce the effection of the noise on the extraction of image features. Fuzzy mapping functions are used to solve the un uncertainty/ambiquity problem of the image.

System Architecture

Layer 0: consist of a gray image. Its data can be used both to segment the image edges, textures etc. and to run the more complex segmentation algorithm when the learning fails.

Layer 1: include the neurons presenting different image features with binary 0/1.

Layer 2: the alterable competitive layer to provide the cluster of image features.

Feature Extraction

Smoothing operation: Average of of pixels in n frame images is used when the camera is static; average of pixels in a window of a frame image is taken when it is moving.

Extraction of Feature: Here the multi-region *fuzzy mapping function* is used to estimate the distributions of the sizes, heights, textures, colour, ratios of wide/length, and adjacent/above/left relationships of regions in the image. The membership values are translated to binary form for the ART1 network,

Neural Networks Training

The algorithm for growthing neurons of competitive layer in improved ART1 is:

Step 1: Allocate arrays w'_{ij}, w'_{ji}, and v'_i as same dimensions as w_{ij}, w_{ji}, and L_2 v_i.

Step 2: Copy all contents of the current arrays to the new medium arrays.

Step 3: Free the current arrays.

Step 4: The number of neurons in the L_2 is added by one, i.e. M+1.

Step 5: Allocate arrays w_{ij}, w_{ji}, and v_i again by the neuron number added (M+1).

Step 6: Copy the contents of the medium arrays to the first M items of w_{ij}, w_{ji}, and v_i.

Step 7: Initialize the weights of the (M+1)th neuron:

$$w_{N+1j} = \frac{1.}{1.+N}; w_{jN+1} = 1. \tag{1}$$

Step 8: Let the (M+1)th be a winning neuron and calculate its output.

Step 9: Free the medium arrays allocated in the step 1.

Step 10: Return to the improved ART1 algorithm to adjust weights of the winning neuron.

Experiments

Here an 512×512 image with 255 gray-level grades is considered. In plasticity test, the neuron number of competitive layer is supposed as 4, arbitrarily. The set of 1127 frame images with binary is adapted. There are other tests having been done to compare the learning time of alterable neuron number of competitive layer with the fixed number. The correction of pattern classification by the improved algorithm is 100%.

Conclusion

In this paper, two ways are used to smooth the image noise. The fuzzy mapping functions are applied to extract the image features. Because the category number of the competitive layer can be altered automatically by growing neuron number of the competitive layer, The network is able to work normally when the new patterns are inputed, i.e. good plasticity. The series of experiments have been completed. The all programs are written by C language and have been performed on a SUN-4/330 sparc-station system with an image board IT-150 and a CCD camera.

PREDICTION OF SECONDARY STRUCTURES OF PROTEINS:
Comparison of Neural Networks(FUZZY ARTMAP) and Statistical Techniques

By
Bhavin V. Mehta, Assistant Professor
Ravi Soni & Lokesh Vij, Research Assistant
Department of Mechanical Engineering
and
Luis C. Rabelo, Assistant Professor
Department of Industrial and Systems Engineering
Ohio University, Athens, Ohio 45701

Abstract

Artificial neural network techniques and Statistical techniques have been used for the prediction of secondary structures of proteins. Back-error propagation was used initially. However it has some inherent shortcomings in its implementation, one being a long convergence time for training and the other being the occurrence of local minima. To overcome the above mentioned drawbacks, a different algorithm called Fuzzy-Adaptive Resonance Theory Mapping (ARTMAP) was employed for our complex pattern mapping problem. Fuzzy ARTMAP is an incremental supervised learning algorithm which combines fuzzy logic and adaptive resonance theory neural network for the recognition of pattern categories.

The Chou-Fasman(C-F) algorithm is a statistical procedure based on assigning conformation potentials to all the amino acid residues. The conformational potentials, one for each confirmation state, are obtained from statistical analysis of proteins of known secondary structure. A commercial package microgene, which implements the Garnier's algorithm(statistical method) was also used in this research. Results of all the three methods are compared in this paper.

The following conclusions have been drawn from this study:
1. The artificial neural networks can be employed to successfully predict the 3-dimensional structures of the unknown proteins and provide an alternative to the experimental techniques such as X-ray crystallography.
2. The neural network based on the supervised adaptive resonance theory provides a system which can make 3D structure predictions, comparable in accuracy to the back- propagation algorithm and in a much shorter time.
3. The statistical technique based on Chou-Fasman rules also gave good prediction results. But with the neural network technique the network can be improved by adding more proteins to it. Also, the lower percentage prediction in neural networks was observed in proteins having alpha-helixes which means results should improve after adding another 20-30 proteins having alpha-helixes in them.

COGNITIVE GRAMMAR AND MAP DIGITIZATION

Eero Carlson

National Land Survey of Finland
P.O. Box 84
FIN-00521 Helsinki, Finland
carlson@mmh.mailnet.fi

The new conception of natural language structure introduced by Ronald Langacker is applied to graphic language. Symbolic structures reside in the relationship between a semantic and a phonological or graphical structure. The usage-based nature and the content requirement of the cognitive grammar corresponds to the self-organizing systems, where usage events are categorized and form regions in the given domains.

Meaningful structures are learned from examples and clustered around prototypes. The structure at a given level is elaborated by substructures. Primary and secondary participant - termed the trajector and the landmark - motivates the categorization of the usage events and the recognition of the figures from the ground.

Two or more component structures are integrated to form a composite structure. Components are categorized by the component Kohonen map and their relationships by the composite Kohonen map. Salient objects described as trajector/landmark pairs focus the attention to the entities with maximal meaning and other components can be obligatory or optional.

Map digitization consists of three phases. Firstly old originals are scanned, secondly raster data is vectorized to graphemes and thirdly the meanings of the symbolic structures are given or recognized.

In the test environment graphemes are formed as one to five vectors. These graphemes are self-organized on the component Kohonen map. Composite structures consist of one to five trajector/landmark pairs and are organized on the composite Kohonen map.

The recognition of the handwritten character "2" ('the bench under the tree') is tested and the recognition rate of over 90 % is achieved, depending on the chosen distance to the prototype. If all characters are recognized, then many wrong characters, which have the same components (8, 3, 5 and others) are also recognized and the ambiguities should be solved in some way.

On-line Learning With Learning Vector Quantization: A Case Study Of EEG Classification*

D. Flotzinger, J. Kalcher, G. Pfurtscheller

*Department for Medical Informatics, Institute of Biomedical Engineering
and Ludwig Boltzmann Institute of Medical Informatics and Neuroinformatics
Graz University of Technology, Brockmanngasse 41, A-8010 Graz / Austria / Europe*

Abstract

Real-time classification, e.g. of EEG, is one possible application of the Learning Vector Quantizer (LVQ) [1]. Its main advantage over other classifiers is its simplicity and speed but also the possibility for on-line learning. Usually, real-time EEG classifiers must be created off-line on the basis of a seperate recording. It would be preferable if the classifier could create itself on-line in a training session, i.e. start from a very sub-optimal initial state, e.g. using a classifier of another subject, and train itself on-line.

Data recorded in three subjects during sessions, where a cursor was controlled in real-time based on EEG classification (Graz Brain-Computer Interface, BCI) [2], are examined in off-line simulations. Each subject participated in 4-6 sessions and the subject-dependent LVQs were updated between sessions to improve their generalization ability. Training LVQs on these data sets with varying values of the learning parameter α, suitable parameter ranges are derived for performances which are comparable to those obtained during the recording. The simulation results show that LVQ1 is a learning algorithm which is well suited for on-line learning because of a number of facilities:

(1) LVQ is a very fast and well-understood classification method. Speed is an important factor when EEG classification is carried out in real-time.

(2) On-line learning can be incorporated without much additional expense: only one reference vector (the winner) must be updated (either drawn further towards the current input vector or pushed slightly away), therefore the additional calculation time only depends on the input vector dimension (in our case: 10) but not on the number of reference vectors used (usually about 8). Note that this would be different if we used a Multi-Layer Perceptron: the error has to be backpropagated through the whole network and therefore the time for calculation depends on the topology (the size) of the network.

(3) The amount of update can be controlled via the learning parameter α. As a rule of thumb, α should range between 0.001 and 0.01, depending on the performance in former sessions. For the first session, a big α is preferable to bring the LVQ, which can stem from some other subject, into the right general position of the current user. For the following sessions, α can be lowered to freeze the LVQ in its position.

Additional improvement can be expected if both learning during data processing and learning between sessions are combined. Although this study was based only on off-line examination of existing recordings, subjects in future experiments can also be expected to give improved performance: if even the first session gives more than random results they will experience the system as more trustworthy and will be more motivated in the following sessions.

References

[1] D. Flotzinger, J. Kalcher, G. Pfurtscheller: EEG Classification by Learning Vector Quantization. Biomedizinische Technik, 37, 12/1992, pp. 303-309.
[2] G. Pfurtscheller, D. Flotzinger, J. Kalcher: BCI - A New Communication Device for Handicapped Persons, Proceedings of the 3rd ICCHP-Conference on Computer for Handicapped Persons, Vienna 1992, pp. 409-415.

---

* Supported by a grant of the Austrian Federal Ministry of Science and Research, project GZ.45.167/3-27b 91, the "Lorenz-Böhler Foundation", project 7/92, and the "Fonds zur Förderung der wissenschaftlichen Forschung", project P9043.

Image Sequence Coding Using
a Neural
Vector Quantization

Sylvie S. Jumpertz- Eduardo J. Garcia
CCETT, Departement of Picture Coding and Psychovisual Studies
4, rue du clos Courtel - B.P. 59 - 35512 Cesson Sévigné Cedex - France
e-mail : jumpertz@ccett.fr - garcia@ccett.fr

Some image sequence coding systems are already defined, and some are even normalised for specific applications. Researchs always try to improve this type of system to reduce the transmission bit rate while maintaining a good reconstructed image quality. The study presented here is in this direction : we try to improve the "classical" image sequence coding system by introducing a specific neural box. We have introduced in a system near of the normalised ETSI (1) sequence coding system a vector quantization realised with a Kohonen Map in which modifications was done on the learning neighbourhood.

I VECTOR QUANTIZATION AND KOHONEN MAP

The main problem of the vector quantization is the codebook generation : the classical algorithm for this codebook generation is the LBG algorithm (2). Kohonen has proposed a topological map wich can be used to build optimal vector quantizers (3). We have used this type of neural network to define our codebook.This network type keeps the topological relations existing in the input signal on its activation map. For our application we have chosen to use a structure where each neuron represents a codeword. At the learning end, weigth connexion of a neuron represents vector components of the corresponding codeword. During the learning procedure, we present blocks randomly chosen in the image. At each iteration, the basic operation consistes in determining the neurone c with the closest matching weights to input vector x. When the "winner" neurone c is found, we increase activity of this neurone and of its neighbourhood.
We have tested different neighbourhoods and learning gain functions (linear, exponential, inversely proportional to t, etc.), but we have obtained better results using a learning function which decreases once with the learning iteration and with the neurone position in the neighbourhood.

II INTEGRATION IN A CODING SYSTEM AND RESULTS

Once the codebook defined for our application, using the Kohonen map with made on the neighbourhood function, we have introduce this vector quantization in the nomalised ETSI sequence coding system based on the DCT. Specific modifications have be done on this system to be able to integrate this new quantization. To reduce the bit rate, we don't transmit the label of the codeword choosed for the block buts its difference with the label found for the previous block.This method gives a better efficacity of the variable length code used at the end of the coding, by exploiting the topological capabilities of Kohonen map.

III RESULTS AND CONCLUSION

Simulations has been done on sequence TV images of 4:2:2 format. The learning phase, for the codebook generation, was done in each case on the first image of the sequence. We don't try to have a general purpose codebook but we propose to reguliarly refresh it. So, we increase the reconstructed image quality. We have try simulations with blocks 2x2, 4x4 and 8x8. The better results was obtained with 2x2 block size. For this size of block we had a compression rate of 0.6 bits per pel, and the signal to noise ratio was about 30dB.
Results presented here seems to be interessant in the fact that they are a direct application of neural network technique. Effectivly, the coding scheme used is really classical, only one "box", the quantization, is realised by using neural network technique. The coding system exploites the topological capabilities of the Kohonen map used to define the codebook. And this caracteristic of the codebook obtained by this method permits to reduce significantly the bit rate transmission. It's interessant to observe that the definition of the codebook using the Kohonen map is really simple and can easily be realised by hardware in opposition with the LBG method.

IV REFERENCES

(1) "Network aspect digital coding of component television signals for contribution quality applications in the range 34-45Mbits/s", Draft pr RTS 300174, ETSI, June 1991
(2) J. Linde, A. Buzo, R. Gray, "An Algorithm For Vector Quantizer Design", IEEE Trans. Comm. Vol. COM-28, pp 84-95, Jan 1980
(3) T. Kohonen, "Self Organisation and Associative Memory", Ed. Springer Verlag 1984

Minimum Distance Pattern Classifiers Based On A New Distance Metric

N. Gaitanis, G. Kapogianopoulos and D.A. Karras

Institute of Informatics and Telecommunications,National Research Center "Demokritos", 153 10 Aghia Paraskevi, Greece, Fax.:+30 1 6532175.

Abstract

A method for the design of minimum distance bipolar pattern classifiers based on a new distance metric between the bipolar patterns is described. The new distance metric is defined and its properties are demonstrated. Neural Networks with the Nearest-neighbor recall mechanism [2], trained according to this method, distinguish the images of a prototype pattern $\mathbf{X}^k$ from the images of the opponent prototype functions $\mathbf{X}^q$, taking into account not only the erroneous pattern elements but also their distinguishing ability. A Neural pattern classifier can be defined by R linear discriminant functions [1] $\{F_1(\mathbf{X}), .., F_R(\mathbf{X})\}$ with $F_k(\mathbf{X}) = \mathbf{X} * \mathbf{W}^k = x_1 * W_1^k + x_2 * W_2^k + \ldots + x_n * W_n^k$ where, x_i are bipolar variables $x_i = +1, -1$ and the weights W_i^k are real numbers and $-1 \leq W_i^k \leq +1$. In the proposed neural net architecture, the discriminant function $F_k(\mathbf{X})$ is then, followed by an output function $T_k(F)$ defined as

$$T_k(F) = \begin{cases} 1 & \text{if } F_k(\mathbf{X}) > 1 - mND(\mathbf{X}^k)/2 \\ -1 & \text{otherwise} \end{cases} \quad, \ k = 1, \ldots, R \text{ and } mND() \text{ is the new distance}$$

metric defined next. Let $\mathbf{X}^k = (x_1^k, x_2^k, \ldots, x_n^k)$ be a prototype pattern with $k = 1, 2, \ldots, R$ and $\mathbf{X} = (x_1, \ldots, x_n)$ be an unknown pattern. Their Hamming distance $D(\mathbf{X}^k, \mathbf{X})$ can then be obtained from the Hamming Discriminant Function $HF_k(X) = x_1 * x_1^k + x_2 * x_2^k + \ldots + x_n * x_n^k$ with $W_i^k = x_i^k$, $i = 1, \ldots, n$, as $D(\mathbf{X}^k, \mathbf{X}) = (HF_k(\mathbf{X}^k) - HF_k(\mathbf{X}))/2$. Then, in the case of a finite set of R prototype patterns $\mathbf{X}^k$, $k = 1, ., R$, a new Discriminant Function $F_k(\mathbf{X}) = x_1 * W_1^k + \ldots + x_n * W_n^k$ can be defined for every pattern $\mathbf{X}^k$, $k = 1, ., R$ with respect to its opponent patterns $\mathbf{X}^q$, $q = 1, ., R$ and $q \neq k$ as follows:

The weights W_i^k, $i = 1, \ldots, n$ of the Discriminant Function $F_k(\mathbf{X})$ are calculated according to the formula

$$W_i^k = (x_i^k) * \{\sum_{q \neq k} (1 - x_i^k * x_i^q)/(2 * R * D(\mathbf{X}^k, \mathbf{X}^q))\}.$$

From the above formula we can see that the minimum value of a weight W_i^k will be equal to $W_i^k = 0$, in the case where $x_i^k = x_i^q$ for every $q = 1, \ldots, R$ with $q \neq k$ and the maximum value of the weight W_i^k when $x_i^k = -x_i^q$ for every q, will be equal to $W_i^k = (2 * x_i^k) * \{\sum_{q \neq k} 1/(2 * R * D(\mathbf{X}^k, \mathbf{X}^q))\}$. The properties of the new Discriminant Function can be demonstrated using the above formula of the weights W_i^k written as $W_i^k = (x_i^k) * \{\sum_{q \neq k} 1/(2 * R * D(\mathbf{X}^k, \mathbf{X}^q))\} - \{\sum_{q \neq k} (x_i^q)/(2 * R * D(\mathbf{X}^k, \mathbf{X}^q))\}$.

According to the above, the output of the Discriminant Function $F_k(\mathbf{X}^k)$ for the pattern $\mathbf{X}^k$ will be equal to the maximum value $F_k(\mathbf{X}^k) = 1$. Also, the output of the Discriminant Function $F_k(\mathbf{X}^{k'})$ for the complement $\mathbf{X}^{k'}$ of the prototype pattern $\mathbf{X}^k$ will be equal to the minimum value $F_k(\mathbf{X}^{k'}) = -1$. Finally, the minimum New Distance $mND(\mathbf{X}^k)$ between pattern $\mathbf{X}^k$ and its opponent patterns $\mathbf{X}^q$, $q = 1, \ldots, R$, $q \neq k$ will be equal to $mND(\mathbf{X}^k) = 1 - \max\{F_k(\mathbf{X}^q)\}$, $q \neq k$.

References

[1] N. Nilsson. *Learning Machines*. McGraw-Hill, 1965.

[2] P. K. Simpson. *Artificial Neural Systems*. Pergamon Press, 1990.

Knowledge Extraction by Self Organising Maps.

R. Kohlus, M. Bottlinger
Deutsches Institut für Lebensmitteltechnik e. V.
D-4570 Quakenbrück / Germany

Abstract: At the example of particle shape analysis the application of selforganisung maps and feedforward backpropagation networks is shown. The main problem is the feature extraction out of a big amount of data. This is done by interpretation of the structure of the Kohonen-layer.

Introduction: In process engineering there are often complex signals of properties which have a deep influence on the process behavior. As an example the particle shape in respect of crystalisation can be mentioned. As it is difficult to discribe the particle shape in a relevant way its influence is often neglected. Some works however show, that it is possible to find shapefactors which gives a quanitative measurement of the shape.

By using Neural Nets it is easier to find the optimal shapedescribtor respectively to classify the shape of particles. The first attempt was made by using a classical feedforward backpropagation network to classify the contour of particles on the basis of small parts of the contour. In this investigation the Fourierdescriptors of the outer boundary were used.

Results: The overall classification results for the contours of quartz, limestone and coffee reached up to 92 %. and are more or less equal for the different nets. The interessting point is, that the different nets have a special ability to distinguish between special materials, e.g. the best backpropagation network works superb on coffee and limestone but very poor on quartz. This means that the best results would be obtain by using one net for each classification tasks of the type "coffee or not coffee".

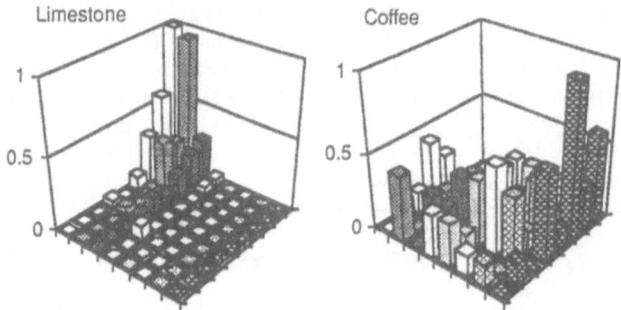

Besides different backpropagation and pruned backpropagation networks self organizing maps were used to classify the particle by their contour. Since the results of the classifation are equal or even better than the results whith the backpropation nets we had a look at the Kohonen Layer. In the Figure the relative frequency of the winning Neurons are shown for an 8x8 Kohonen Layer. As the classification results of the coffee particles are very good compared with those of the mineral particles one might expect a clear location on the Kohonen layer for the "coffeeneurons" but in opposite to that the location are very distributed. The locations of the minerals were more concentrated and laid nearer to each other. The location of coffee is very seperated from the minerals. These effects can be explained by the big differences between the minerals and coffee signals and the variations within the shape of coffee sample. Using self organizing maps gives us the opportunity to quantify a distance between the signals and to get an impression of the variations within a sample.

Conclusion: In techinical problems which are concerned with feature extraction neural nets might be an appropriate way to determine the relationship between data and effect. As shown by the example of particle shape analysis unsupervised learning networks as self organising maps can be very useful.

Acknowledgements: The project is supported by the Deutsche Forschungsgemeinschaft (DFG).

Application Of The Sensitivity Algorithm In Biological Fields.

M. Giacomini, T. Parisini, C. Ruggiero, R. Sacile

Dept. Communication, Computer and System Science University of Genova, Italy.

One of the greatest problem in using an artificial neural network is the determination of its most appropriate size. The advantages of the use of the smallest reliable network can be summarized as follows:
-a small net is more efficient both in forward computations and in learning;
-neural nets which are too large tend to have poor generalization ability (specially if they are trained with a small set of patterns);
-it may be possible that a smaller net will exhibit a behaviour that can be described by a simple set of rules.
The last feature above is very important in biology where artificial neural networks can give the opportunity of determining the most important features for classification in the considered field.
In this paper we present an application of the pruning algorithm by Karnin [1] for the determination of the main classification features in three biological fields: the determination of the protein secondary structure starting from NMR data, the determination of the same secondary structure starting from the primary structure, the classification of marine bacteria starting from their gas chromatographic profile.
The knowledge used in examining NMR raw data is often largely heuristic, reflecting both the incompleteness of the data available and the uncertainty as to the best method to interpret it. In this situation of absence of a well assessed method for the analysis of these data, we thought it would be worthwhile to use a neural network which learns from a certain number of examples the way in which it is possible to determine in which state an amminoacid can be, given four characterizing distances. We have already shown the possibility of such a net to determine the correct state of an amminoacid. [2]. Here we want to present the first results of the sensitivity check to determine if all these distances have the same importance or not. Our first results show that the variation in sensitivity of the four distances is not significant. This indicates that their importance for the determination of the secondary state is about the same.
The application of the sensitivity check to the problem of determining the secondary structure from the primary one [3] aims to determine the best width of the input window to avoid the huge number of calculations due to an overestimated value of this value and not to miss important information about the co-operation of nearby aminoacids. After some trials we can affirm that sensitivity is quite constant for the four aminoacids nearer to the central one (for both sides) and decreases considerably for farther aminoacids, becoming almost negligible for amminoacid farther than 6 places than the central one. So we conclude that the best width for the input window is of 13 aminoacids.
Gas liquid chromatography has been found to be very useful for a correct and speedy classification of the bacterial genus; in fact all the bacterial genus does not contain the same quantities of the different fatty acids. So we have trained a neural network with gas-chromatographic data and specifically using 113 different fatty acids [4]. In this case the application of the sensitivity check aims to determine the real importance of such a high number of indexes. Using this method we have found that the 46% of the input fatty acids are almost uninfluent for the classification. In fact, after pruning the corrections of these uninfluent input neurons we can classify correctly almost all the bacteria not considered in the learning set.

References

[1] Karnin, E. D., 1990, A simple procedure for pruning back-propagation tranied neural networks, *IEEE Transactions on Neural Networks* 2: pp. 239-242.
[2] Giacomini, M., Parisini, T., and Ruggiero, C., 1992, Secondary structure of proteins from NMR data by neural nets, *Artificial Neural Networks* 2 (I. Aleksander and J. Taylor eds) Elsevier.
[3] Ruggiero, C., Sacile, R., and Rauch, G. 1993, Peptides secondary structure prediction with neural networks: a criterion fro building appropriate learning sets, IEEE Transaction on Biomedical Engineering, in press.
[4] Giacomini, M., Ruggiero, C., Calegari, L., Bertone, S., Casareto, L., and Reina, S., 1993, Marine bacteria classification by neural nets, MIE93, in press.

Neural hardware and software
—oral contributions

Challenge of ANN to microelectronics

Karl F. Goser

University of Dortmund, Faculty of Electrical Engineering,
POB 500500, 44221 Dortmund, Germany

Abstract
The paper reviews the state of the art of integrated circuitry and neural networks. In the near future special technologies on silicon, as optoelectronic devices and analog memory cells, may improve the performance of integrated neural networks. For the far future a rough survey shows the potential of nano and molecular electronics which will offer a gigantic potential for neural networks.

Introduction

Artificial neural network (ANN) researchers hope that simplified functional models of nervous tissue can help us to design algorithms and machines that are better than conventional computers for difficult problems in machine perception and intelligence. The response and the characteristics of present models of ANNs are investigated primarily by simulation on vector computers, workstations, special coprocessors, or transputer arrays. The fundamental drawback of such simulators is that the spatiotemporal parallelism in the processing of information that is inherent to the neural net is lost entirely or partially and that the computing time of a simulated net, especially for large associations of neurons, increases to such order of magnitudes that a speedy acquisition of neural behaviour is hindered or made impossible. With specially designed neural hardware not only the processing time would drastically decrease, but also the smaller volume, the reduced power supply requirements, and the easy retraining possibilities would render miniature neurocomputers very attractive.

Despite the impressive developments of microelectronics during the last decades, there is no clear consensus on how to exploit these technological capabilities for massively parallel ANN algorithms. It is currently not possible to determine the best way to perform ANN calculations for any given application. This is one reason for the huge variety of approaches for ANN hardware implementation known in literature. In this paper, an overview of the major trends in ANN hardware implementation with an emphasis on integrated circuits (ICs) will be given by grouping the different approaches into few categories and by discussing the key features of each of these categories.

Standard ICs for Neurocomputer

Several institutions have developed board-level systems, so called add-on-boards or accelerator boards, built around a modern microprocessor (proprietary, DSP, or RISC(reduced instruction set computer)) that excel at the floating-point operations required by many ANNs. Accelerator boards speed up ANN processing by at least one order of magnitude albeit significantly higher costs. Most of them are designed for IBM PCs and simulate ANNs with supercomputer-like performance of up to 50MCPS (million connections per second/multiply&add per second, feed-forward net) and 20MCUPS (million connection updates per second, backpropagation net)[1].

The next step towards higher performance are multiprocessor neurocomputers which are very similar to general-purpose parallel computers. The connections between processors can either be implemented through a single high speed data path (bus-oriented) or via short point-to-point links. Literature gives performance figures of up to 500MCPS for multiprocessor neurocomputers.

Since the components for such neurocomputers already exist they are being improved as fast as the state of the art of technology will allow it. This is obviously a strong point for neurocomputers based on standard ICs. The challenge in ANN simulation lies in developing architectures to effectively use these standard components.

Neural ASICs

There has been much work in the design of VLSI ICs implementing ANN algorithms: they are digital or analog or a combination thereof. - Digital neural ASICs can be categorized into general-purpose approaches, for emulating different ANN models, and special-purpose approaches which are dedicated to a specific ANN model, however, the more general a neurocomputer is the slower it is. The synaptic weights can be stored on or off chip. The advantage of on-chip weight storage is the fast weight access, the disadvantage is the large chip area or number of chips. - The N6400 chip [1] is an example for a general purpose digital neural ASIC with on chip storage of weights. A single chip (25 MHz) can perform 1.6 GCPS (256 MCUPS, backprop) for 8 or 16 bit weights and 12.8 GCPS for 1 bit weights, respectively.

Another interesting example of this category is the WSI (wafer scale integration) neural net from Hitachi [6,7]. It comprises 576 digital neurons and 36KBytes weight memory. It performs 1.25 GCPS and 118MCUPS. - The MA16 neural signal processor from Siemens is an example of a general-purpose digital neural ASIC with off chip weight storage. The MA16 chip has a systolic architecture build by four processing units each with four 16bitx16bit multipliers. The chip performs about 800MCPS at a clock rate of 50MHz. A prototype of the Synapse-1 neurocomputer has eight MA16 neural signal processors, two MC68040 CISC processors for control purposes, and a 128MByte DRAM bank. The prototype performs 4.2GCPS and 330MCUPS [8,5].

These examples of digital neural ASICs show the potential for further improvement of ANN simulation speed compared with general-purpose parallel computers. A speed of up to two orders of magnitude higher seems to be possible with current technologies.

At analog neural ASICs the current approaches can be divided into continuous-time and discrete-time concepts. Some additional options arise relating to the connectivity (local/low or full connectivity) within the ANN and the transistor's mode of operation (weak inversion or strong inversion) [4]. Analog processing derives its main advantage when physical processes can be used to perform required computational functions. For example, the weighted sum of input signals (activation function) which incurs the largest computational load in the recall phase can be efficiently implemented in analog circuit technique by means of current or charge summing [4]. Most of the proposed analog neural circuits make use of current summing as, for example, the associative memory chips [4]. With current state-of-the-art of microelectronics simple neural associative memory chips with more than 1000 neurons and 1000 inputs can be integrated on a single chip performing about 100GCPS.

The associative memory chips mentioned above are programmable, but not trainable. Learning does require incremental adjustment of the synapses (weights) in small steps. The design of multivalued weights must balance the cell size and the resolution of the weight. Proposals for analog synapses include charge-coupled devices (CCDs), MNOS (metal nitride oxide silicon) transistors combined with CCDs, and concepts based on special materials like bismuth sesquioxide or a-silicon. Very promising concepts offer floating-gate transistors as used in EEPROMs [3] and analog storage cells with ferroelectric films,e.g. PZT [2,6].

Analog and programmable weights are a significant feature of an effective and flexible VLSI implementation of ANNs. At the moment, there is one such analog neural ASIC commercially available, which uses two floating-gate transistors as an adaptive weight [1]. It contains 64 neurons, 10240 synapses and performs about 2GCPS. Learning has to be done off chip.

A rather new analog design methodology is the so called neuromorphic approach, which can achieve significant improvements in computing hardware capability compared with conventional analog and digital techniques. This neuromorphic approach starts by identifying several structural levels in the nervous system, and then attempts to capture these organizing principles. At the lowest level, the computational primitives of the nervous system are identified and silicon analogies are designed by creatively harnessing the available physics of semiconductors. At the next level various ensembles of primitives can be organized to perform complex computational tasks, such as signal preprocessing. This novel methodology was inspired by the work of C. Mead [9] and has let to several interesting prototype chips, for example, a silicon retina [9], an electronic cochlea [9], or ear prothesis.

Analog circuits can boost the performance beyond that of digital designs. But in general they are special-purpose implementations of a selected ANN model for a specific application. In the next future analog techniques will make possible dynamic neurons, which will lead to networks with new features.

Emerging Technologies for ANNs

Optoelectronic devices and light wave guides integrated on silicon offer interesting aspects for ANN: First we get a flexible interconnectivity with high data rates, second we can process many data in parallel, and third we can store analog weights, in PZT films for example. Such concepts may be relevant for the electronic eye in the far future.

Much higher integration densities follow from nano and molecular electronics. For example the crystal Zeolith with its regular pipeline structure, where conducting polymers and semiconductor molecules are embedded, offers a very high density of weights, about 109 mm-3. The drawback is the very simple architecture with low interconnectivity, which is similar to cellular nets.

Molecular electronics offers a huge potential for ANN [10]. The tactile molecular processing unit with proteins has a relative low processing speed, but solves the interconnectivity in an elegant way: The information packets flow parallel in the cell liquid all searching their neuronal goals. Despite of such features the discussion about the technological way to large neural systems, in the long term to so-called artificial brains, is open. It might be the way of biological concepts with proteins or the physical way with nanostructured devices.

Final Remarks

One of the most important differences between ANN research today and what was possible 30 years ago is the huge improvement in the technological capabilities. State-of-the-art VLSI and the emerging ULSI technologies are able to integrate thousands of neurons on a single chip with clock rates reaching 1 GHz. In the far future even more computational power for ANN may be obtained by using optoelectronic and molecular electronics.

Acknowledgement

The author would like to thank the DFG (Deutsche Forschungsgemeinschaft), the BMFT (german ministry of research and technology), the European community (ESPRIT project NERVES), and the Siemens AG Munich for financial support of the work in his institute at the University of Dortmund.

References

[1]   Examples for VLSI-ANN-Chips are described in the following books:
      DAPRA, Neural Network Study, AFCEA Int. Press, Fairfax Verginia, 1988.
      Digest of the Intermational Solid State Circuit Conference
      Proc. of IJCNN

[2]   Goser, K., Hilleringmann, U., Rückert, U., Schumacher, K.: VLSI Technologies for Artificial Neural Networks, IEEE Micro, Vol. 9, No. 6, pp. 28-44.

[3]   Goser, K., Ramacher, U., Rückert, U.: Microelectronics for Neural Networks, Proceedings of the 1st Int. Workshop, University of Dortmund, 1990.

[4]   Ramacher, U., Rückert, U.: VLSI Design of Neural Networks, Kluwer Academics, Boston 1991.

[5]   Goser, K. and Ramacher, U.: Mikroelektronische Realisierung von künstlichen neuronalen Netzen, (in german) Informationstechnik it, Vol.34, No. 4, 241-247.

[6]   Ramacher, U., Rückert, U., Nossek, J.A., Microelctronics for Neural Networks, Proceedings of the 2nd Int. Workshop, Kyrill&Method Verlag, M?nchen 1991.

[7]   Yasunaga, M., et.al.: Design, Fabrication and Evaluation of a 5-inch Wafer Scale Neural Network LSI composed of 576 Digital Neurons, Proc. IJCNN, Vol. 2, June 1990, pp. 527-535.

[8]   Ramacher, U.: SYNAPSE-A Neurocomputer That Sysnthezis Neural Algorithms on a Parallel Systolic Engine, Journ. of Parallel and Distributed Computing 14, 1992, pp. 306-318.

[9]   Mead, C.: Neuromorphic Electronic Systems, Proc.IEEE Vol.78, No.10, 1629-1636.

[10] Conrad, M.: Molecular Computing, Computer, Vol.25,No.11, 6-81.

Implementation of Million Connections Neural Hardware with URAN-I

Subject Area : Neural Hardware

Il-Song Han*, Ki-Hwan Ahn**
Principal Member of Technical Staff*, Member of Technical Staff**
Korea Telecom Research Centre
17, Woomyun-dong, Suhcho-ku
Seoul, 137-140
KOREA

FAX: 82-2-526-5568
Telephone: 82-2-526-5048

Abstract

This paper describes the neural hardware build of the full custom neural VLSI. The newly developed URAN-I neuro-chips of 135,424 connections are used to construct with their speed of hundreds giga connections per second. The basic hardware module is suggested with 8 chips and digital computer interface to yield thousands giga connections per second in speed and more than one million connections in size.

I. Introduction/Overview

Recently, the neural network hardware with neuro-chip has been expected in respects of its relatively fast speed, huge network size and effective cost comparing software simulation. In this paper, the neural hardware implementation for a conventional computer is proposed, which is made of URAN(Universally Reconstructable Artificial Neural-network)-I. The URAN-I is implemented with the electronic synapse of the electrically-controllable MOSFET resistance and is of 135,424 in synaptic connection size. The URAN-I of Fig 1 is 13 X 13 mm^2 in chip-size using the standard 1.0μ digital CMOS technology. Computational speed of hundreds giga connections-per-second is achieved and fully asynchronous/dierct wired-OR expansion of chip-to-chip connection is also attained. These features make possible larger scale and higher speed neural network chips and system.

II. URAN of analog-digital hybrid neuro-chip

In general, most of digital, analog, or analog-digital mixed neuro-chips are constrained in accuracy, speed, size or flexibility. There has been made new advancement in those aspects with the suggested analog-digital hybrid neural network circuit. The accuracy is improved by the linear voltage-controlled MOSFET linear resistance circuit for the synapse emulation . The speed is increased by employing the post synaptic current switch controlled by the digital neural state. The general flexibility is realized by the inherent electrical characteristic of each synapse cell and the modula architecture of chip. Chip features are summarized in Table 1. As in Table 1, the chip performs under the flexible control, that is, the various mode of synaptic connection per neuron, the extendible weight accuracy and the unlimited asynchronous/direct interchip expansion in size and speed. In fact, 16 fully connected module is selected from external and independently – either one by one selection or all at once selection is possible. Additionally, the speed or the modularity of each module is improved by introducing the individual external weight input. The neural hardware of huge size and high speed is straightforwardly implementable with chips in the same way as the chip is with module.

Considering the operation of the circuit itself, all circuits over the chip except digital decoder unit are operated in analog. And as they are almost virtually static except switching transistor controlled by neural input, the computation speed is high and even can be improved substantially with the advance of memory production technology. As a basic cell, 9 transistors are used per cell including weight memory. The cell size including interconnection area for URAN-I is reduced to less than 40 μm in diameter. With the linear voltage-controlled bipolar current source of each synapse cell, the synaptic function of multiplication is done with the switching transistor, i.e. half-in-analog and half-in-digital. The use of bipolar pulse improved stray effect from switching. Pulses of neural input for switching are not limited in style, time and numbers, that is, they are fully independent from each other. The linearity is based on the compensated channel resistance of balanced configuration in the triode region, and proved to have more than 8 bit if necessary. The accuracy extendibility and flexible modularity are inherent in electrical wired-OR characteristics from each independent bipolar current source. No clocking or any synchronous operation is needed in this case, while it is indispensible in most of conventional digital neural hardware or analog-digital neural chip. Any size of network can be integrated or implemented by merely placing the cell in 2 dimensional array without considering the timing requirement of digital or the load effect of analog. In the case of URAN-I, the delay of control signal is considered in designing the logic interface due to the long path on the chip.

The neuron chip is currently designed for the pulse stream operation of multiple mode. For the easiness in interfacing to the conventional computer, simplified A-D, D-A devices are used with 8 neural chips for one million connections and multiplexed one thousand neurons.

III. Conclusion/Discussion

The test chip of Fig 1 demonstrates the extensible weight value accuracy with the cost of total logical network size due to the programmable architecture of parallel module selection. By changing the control bit of module select input, the number of synaptic connection is programmable from 92 synapse/neuron to 1,472 synapse/neuron internally. URAN-I has been fabricated in conventional 4 mega DRAM process(in GoldStar Electron Corp.).

The computational speed of synaptic connection is not degraded as the modular structure and extendible cell performance allow the multiplexed parallel access. The modular extendibility of URAN enables the construction of more than million connections neural hardware with 8 chips to yield the internal speed of 1,000 Giga connections per second or more.

IV. References

1. Il-Song Han, "Analogue Circuit for a Neural Network," ICANN'91, Vol.2, 1991, pp. 1577

2. Il-Song Han, "Fully Programmable Neural Chip," 3rd Korea-Japan Joint Seminar for Neural Network and VLSI implementation, 1990 pp. 255

3. M. Brownlow, L. Tarassenko, A. F. Murray, A. Hamilton, Il-Song Han, H. M. Reekie, "Pulse Firing Neural Chips Implementing Hundreds of Neurons," NIPS2, 1990, pp. 785

4. Il-Song Han, "Extensible Hybrid Neuro-chip and its Characteristics," JCEANF'92, 1992, pp. 516

5. M. S. Tomlinson, D. J. Walker, M. A. Sivilotti, "A Digital Neural Network Architecture for VLSI," IJCNN, 1990, Vol. II, pp. 545

6. G. Fahner, N. Goerke, R. Eckmiller, "Structural Adaptation of Boolean Higher Order Neurons," ICANN'92, Vol. 1, 1992, pp. 285

7. Il-Song Han et al, "Adaptable VLSI Neural Network of Tens of Thousand Connections," ICANN'92, Vol.2 1992, pp. 1423

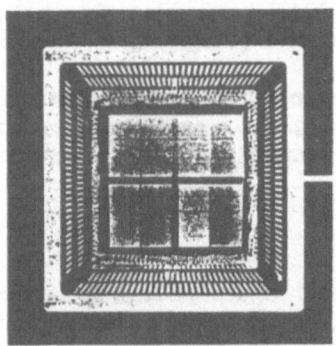

Fig 1. URAN-I of 135,424 connections

Table 1. URAN-I features

Speed	200×10^9 connections /s
Synaptic Connections	135,424
Weight Accuracy	8 bit
Organization (synapses/neuron)	$92 \times n$, n= 1 to 16 (electrically programmable)
Function of interchip expansion	Fully asynchronous and direct electrical wired-OR at output
Supply Voltage	3 V, -3 V
Chip Size	$13 \times 13 \text{ mm}^2$
Technology	1.0 µ digital CMOS
PGA package	257 pin

Multiprocessor and Memory Architecture of the Neurocomputer SYNAPSE-1

U. Ramacher, W. Raab, J. Anlauf, J. Beichter,
U. Hachmann, N. Brüls, M. Weßeling, E. Sicheneder

Siemens AG, Corp. R & D
Otto-Hahn-Ring 6
D-8000 Munich 83
Germany

R. Männer, J. Gläß, A. Wurz

Informatik V
University of Mannheim
A5
D-6800 Mannheim 1
Germany

ABSTRACT

A general purpose neurocomputer, SYNAPSE-1, is presented which exhibits a multi processor and memory architecture. It offers wide flexibility with respect to neural algorithms and a speed-up factor of several orders of magnitude -- including learning. The computational power is provided by a 2-dimensional systolic array of neural signal processors. Since the weights are stored outside these NSPs memory size and processing power can be adapted individually to the applicational needs. A neural Algorithms Programming Language, embedded in C++, has been defined for the user to cope with the neurocomputer. In a benchmark test the prototype of SYNAPSE-1 was 8000 times as fast as a standard workstation.

INTRODUCTION

The architecture of artificial neural nets differs greatly from the one of conventional computers. Instead of a sequential execution of operations by a complex processor, neural nets exhibit a massively parallel and non-linear processing by means of very simple processors, called neurons. This mismatch of architectures is the reason why simulation of neural nets on today's computers is extremely time consuming for larger networks. Therefore, the acquisition of 'neural' know how is hindered or even made impossible. For several important applicational domains, [1] reports the estimated sizes of neural nets along with the computational throughput.

In consideration of the resulting lack of neural engineering know-how as to neural architectures and learning rules, it is thus a matter of designing a neurocomputer architecture that

-- accelerates the neural algorithms and the learning algorithms as well by orders of magnitude,
-- exhibits a sufficient measure of algorithmic flexibility for coping with the known neural paradigms and those to come,
-- is configurable to the needs of the applications in terms of processing power and weights memory size,
-- and offers a user interface acceptable for a wide spectrum of users.

MULTI PROCESSOR AND MEMORY ARCHITECTURE

The requirements outlined above are fulfilled by the multi processor and memory architecture of the neurocomputer SYNAPSE-1 (fig. 1).

The compute-intensive operations (e.g. matrix multiplication) common to all neural algorithms including learning paradigms are executed on an array of neural signal processors MA16, each providing a peak performance of $640 \cdot 10^6$ conncections (16x16 bit) per second at 40 MHz [2]. The non-compute-intensive operations are performed on the 'Data Unit', while the 'Control Unit' cares for the concatenation of compute-intensive and non-compute-intensive operations as well as the communication with the host.

The neural algorithms have a common set of compute-intensive operations. They differ in the non-compute-intensive operations (e.g. evaluation of the transfer function of neurons) and the order of concatenation of compute-intensive and non-compute-intensive operations [3]. Both the Data Unit and the Control Unit with their standard CPUs are well prepared for a universal programming of their respective tasks.

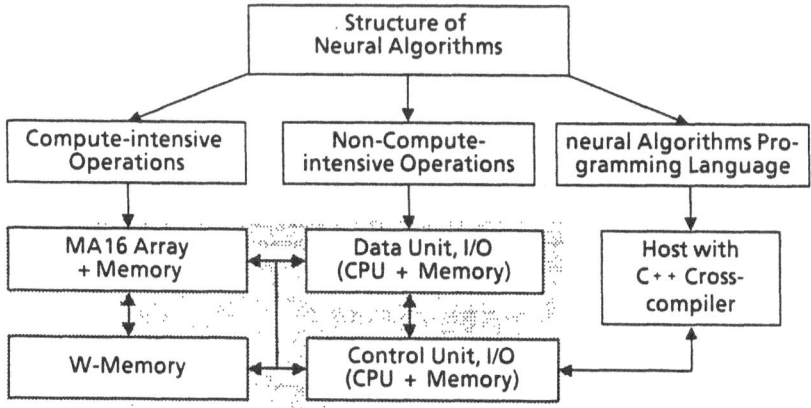

Figure 1   System concept of SYNAPSE-1. The neurocomputer consists of the coprocessor (shaded) and the host workstation.

The synaptic weights are stored outside the MA16 array in a DRAM bank 'W-Memory'. The number of MA16 boards and memory boards in a system can be chosen according to the requirements of an application.

The syntax of the neural Algorithms Programming Language (nAPL) reflects the multi processor and memory architecture of SYNAPSE-1. It is embedded in C++ and provides -- by means of a class library -- the elementary operations which are supported by the MA16s.

MA16 ARRAY   The processor board hosts a 2-dimensional systolic array of neural signal processors, arranged in two rows by four columns. Each MA16 is accompanied by a local memory for intermediate data (Z-Memory). Both processor rows are connected to the same weights bus, thus executing the same neural net operations for different input patterns. The MA16s of a row form a linear systolic array where input data as well as (intermediate) results propagate from the Data Unit through the MA16s back to the Data Unit.

Each MA16 itself contains a linear systolic array of 4 processing modules PM (fig. 2). The data paths and operations of a PM are set up such that both neural algorithms and classical signal processing algorithms (e.g. DCT, correlation, etc.) can be performed.

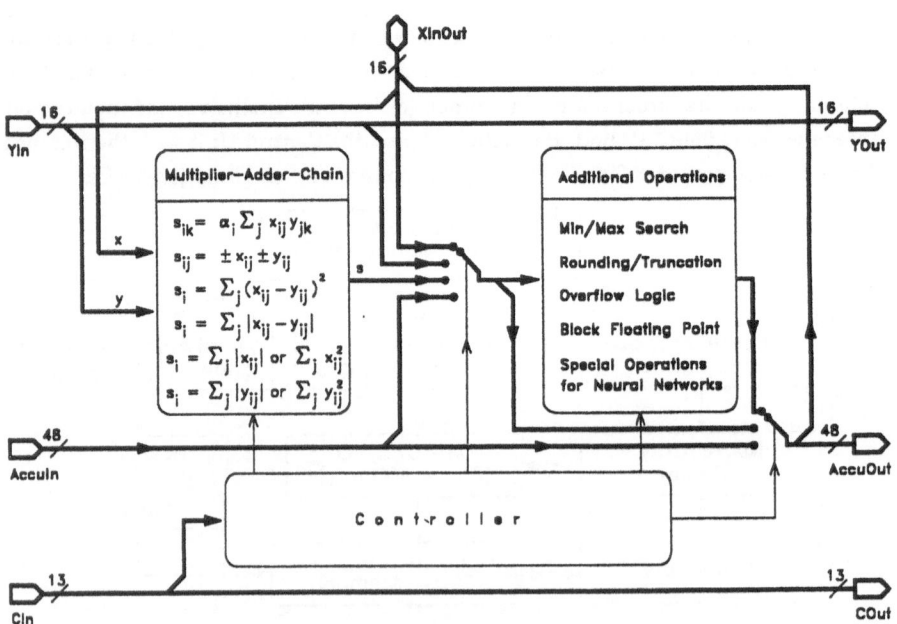

Figure 2   Block diagram of a processor module PM

Each MA16 features a total bandwith of 10.9 Gbit/s, the MA16 array and bus frequency being 40 MHz.

W-MEMORY   The W-Memory provides the weights for the MA16 array. The data are transferred over the system backplane on a special purpose synchronous bus with 40 MHz. To achieve this frequency two DRAM banks operating in fast page mode are interleaved. The memory supports parity check and generation for surveillance of data transfers.

Besides the connection to the backplane the W-Memory offers a second port to the Control Unit. Thus single weights can be communicated without transferring whole matrices through the processor array. This feature enhances both speed and flexibility of inspecting or modifying weights.

DATA UNIT   The Data Unit is responsible for performing the non-compute-intensive operations. For highest flexibility a CPU (MC68040) takes the data coming from the MA16s (AccuOut) via FIFOs, executes the required operations, and stores them into a memory reserved for neuron states/activations and I/O data (Y-Memory). To speed up standard operations like barrel shifting, addition, scalar multiplication, table look up, the CPU can be bypassed using a pipelined special purpose hardware and a DMA channel to the Y-Memory. The latter is the standard processing path for neural algorithms.

Through a second port of the Y-Memory data are transferred to the MA16 input (YIn). These transfers are performed by a DMA controller which receives the instructions from an address generator on the Control Unit.

CONTROL UNIT   The Control Unit is divided into 2 main groups:

A MC68040 CPU takes over the communication with the host and the Data Unit CPU. It also controls the other parts of the board, thereby synthesizing the complete algorithm to be performed (SYNAPSE = synthesis of neural algorithms on a parallel systolic engine).

A microprogram sequencer controls the MA16 array along with all components which must operate synchronously to the array. These are the W-Memory, the Z-Memory, FIFOs, and the Backplane Transceivers. Therefore a microprogram generally corresponds to the compute-intensive part of an elementary operation of SYNAPSE-1.

The Contol Unit is completed by the Y-Address Generator which provides the addresses of data to be transferred to the YIn port of the MA16s, and an interface to the W-Memory.

PERFORMANCE OF SYNAPSE-1

SYNAPSE-1 in its standard configuration is depicted in fig. 3. With the 8 MA16s on the processor board running at 40 MHz, the peak performance of SYNAPSE-1 is $5.1 \cdot 10^9$ connections per second. A first benchmark test consists of the computation of one layer of neurons. In a contest with a SPARCstation2, SYNAPSE-1 proved to be 8000 times faster than the workstation. Here the clock frequency was still kept at 25 MHz, and the microprograms were not optimized. Figure 4 shows a photograph of SYNAPSE-1.

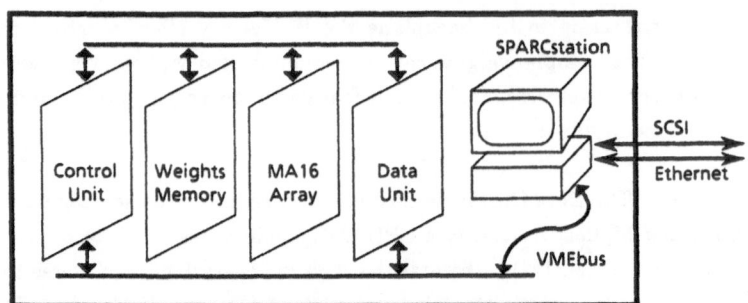

Figure 3   Configuration of SYNAPSE-1

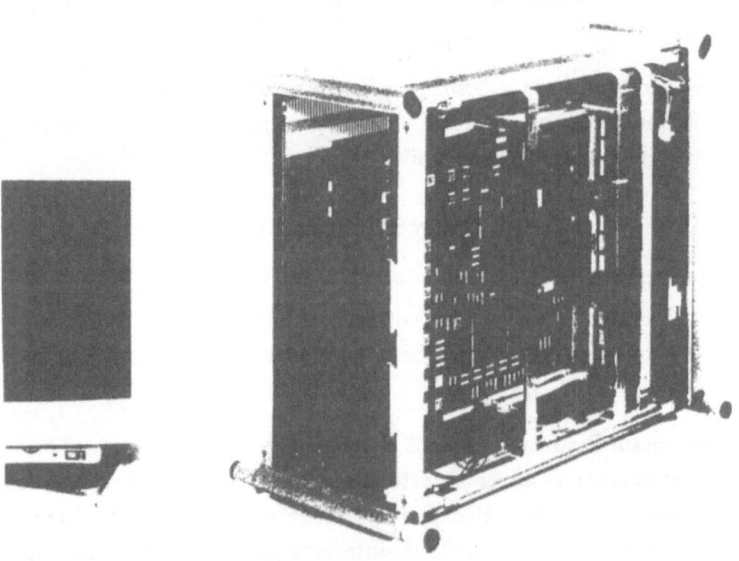

Figure 4   SYNAPSE-1 --- 8000 times as fast as a SPARCstation2

REFERENCES

[1]   DARPA Neural Network Study, AFCEA International Press, Nov. 1988, p. 34 (figure 2.14-15), p. 330 (figure 28.5).

[2]   Beichter, J., et al., A VLSI Array Processor For Neural Network Algorithms, Proc. Custom Integrated Circuit Conference, San Diego, May 1993, to be published.

[3]   Ramacher, U., SYNAPSE -- A Neurocomputer That Synthesizes Neural Algorithms on a Parallel Systolic Engine, Journal of Parallel and Distributed Computing, Vol. 14, pp. 306-318 (1992).

COKOS: A COprocessor for KOhonen's Selforganizing Map

H. Speckmann, P. Thole, W. Rosenstiel
Wilhelm-Schickard-Institut f. Informatik
University of Tübingen
7400 Tübingen, Sand 13, Germany

Abstract

In this paper we present a system which enables easy and fast computation of Kohonen's selforganizing map (SOM). It is a hardware supported system consisting of different parts. A neural coprocessor (COKOS) is connected to a personal computer (PC) by a special, asynchronous interface. The neural coprocessor works vector-component-parallel and processing-element-serial and speeds up the performance of Kohonen's algorithm of the selforganizing map by magnitude. An userfriendly interface supports the preprocessing of the input data and the control and assessment of the learning process.

1 Introduction

Large CPU-times for training large data sets to neural networks and the impossibility of real-time evaluation constitute a serious obstacle for practical applications of neural networks, especially software simulations on MIMD-computers, e.g. transputer networks [8], or on SIMD-computers (we are using MasPar with the language Parallaxis [1]), bring up the performance. For some general concepts which have many applications specialized hardware may further improve the performance. One of these concepts is the selforganizing map which has been introduced by T. Kohonen [3]. This unsupervised learning network is used with many applications, like impulse test fault diagnosis on power transformers (by classifying digital signals obtained during transformer impulse testing) [5], speech and character recognition [4], or data analysis of chemical gas-sensors what we are currently doing [2]. The different applications require different architectures of the map concerning the topology of the net and the number of components and processing-units (PUs). Second there is no satisfying theory saying which parameters are to choose for which application. So a system implementing the SOM should include tools for verifying the learning results.

2 Short description of Kohonen's selforganizing map algorithm

For our implementations we use a modified version of Kohonen's algorithm [7, 3]. Generally the SOM consists of a two-dimensional array of identical processor units PU_{ij} with each processor unit storing a single vector w_{ij} with the components w_{ijk}, i and j denote the position of the processor unit in the array. When the map is trained, each processor unit computes the euclidian distance D_{ij} of the input vector $S = (s_1,..,s_n)$ and the stored vector w_{ij} according to (1).

$$D_{ij} = \sum_{k=1}^{k=n} (w_{ijk} - s_k)^2 \qquad (1)$$

After computing the distance for all PE_{ij}, the map searches the processor unit which stores the most similar vector with the minimum distance D_{ijmin} to the input vector of the array, and then the processor unit and its neighbourhood are adapted to the input vector according to (2).

$$w_{ijk}(t+1) = (1 - e_{ij}(t)) * w_{ijk}(t) + e_{ij}(t) * s_k \qquad (2)$$

where $e_{ij}(t)$ is the adaptation function which indicates the degree of adaptation of the processor units towards the input vector ($0 \le e_{ij}(t) \le 1$, $e_{ij}(t) \to 0$ towards the training time t). For the adaptation function e_{ij} we use the Gaussian function, which decreases linearly during training time in width and height.

The concept of the SOM is to map a high dimensional space to a lower dimensional space while preserving the topology of the high dimensional input space. This describes the SOM's algorithm for 2-dimensional output space, but higher dimensional output spaces are possible. Our hardware realisation is not restricted to any dimension of the input or the output space because the structure of the map is not hardwired.

3 The system configuration

3.1 Overview

A neural coprocessor is connected to a personal computer by a special interface (figure 1). The neural coprocessor calculates the parallel parts of the SOM's algorithm. In addition to the realization of the asynchronous communication the interface has the ability to interprete complex commands and controls the neural coprocessor to simplify the programming of the neural coprocessor. The PC holds the software implementation of the serial parts of the SOM's algorithm, controls the neural coprocessor and delivers a software package for analyzing the learning results.

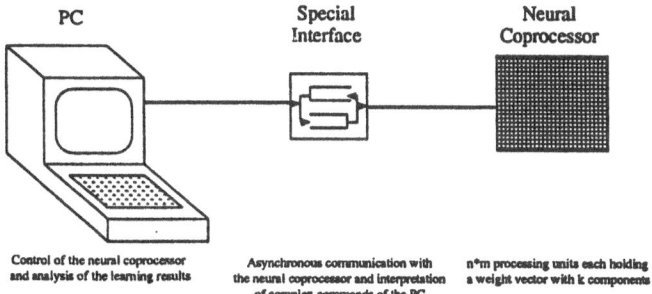

Figure 1: Structure of the whole system

The PC determines randomly an input vector from the input data set and delivers it to the neural coprocessor which finds rapidly the position of the processing unit (PU) with the minimum euclidean distance to the input vector. With the coordinates of the nearest PU the PC calculates the excitation matrix for the map holding the adaptation factor for each PU and the coprocessor calculates the adaption of each PU.

3.2 COKOS: A COprocessor for KOhonen's Selforganizing map

In [6] we discussed the two possibilities of parallizing the SOM, neuron parallelism, leading to the architecture of an array processor and synapse parallelism, leading to the architecture of the vector processor. Weighing up the pros and cons of the two concepts we choose the last concept because it is easier to control, has less effort of hardware and the search for the minimum distance is avoided.

As you see in figure 2, the neural coprocessor consists of the following parts, n so called Memory-and-Arithmetic-Boards (MAB), $\lceil \log n \rceil$ stage adder tree and the controller of the coprocessor. In our prototype we use eight MABs and a three stage adder. Each MAB holds the arithmetic and memory necessary for learning one input vector component. So you can process one input vector for one PU in parallel and store the weights of the whole SOM with the neural coprocessor. The arithmetic of the MAB is pipelined.

To determine the PU with the minimum euclidean distance to the input vector you calculate the partial distances in the MAB arithmetic SUB-MULT pipelines. The outputs of the pipelines are added to the whole distance by the adder tree, which is organized as a three stage pipeline. For adaption the adaptation factor is loaded into every MAB and according to equation (2) we calculate adaptation using the MAB's SUB-MULT-ADD pipeline.

For many applications eight components are not sufficient. But with the coprocessor's ability of time multiplexing every number of vector components can be handled. The only restriction is the finite memory size of the MAB's. The restriction is that the number of PU's multiplied by the degree of time multiplex must not exceed 2^{16}.

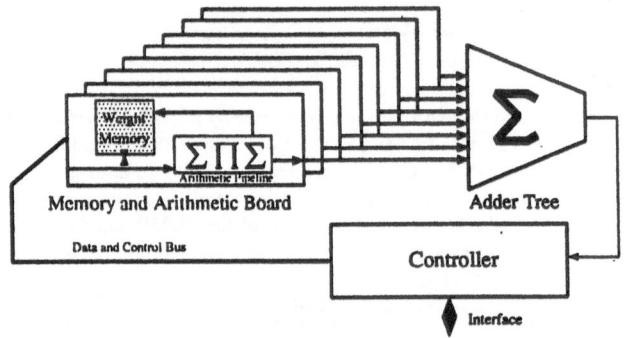

Figure 2: The architecture of COKOS.

3.3 The asynchronous interface

The asynchronous interface reduces the communication requirement as follows. Data is buffered in 2 FIFOs, one for reading data from the PC and one for the opposite direction. The flow of control is reduced by connecting the neural coprocessor to a command interpreter interpreting complex PC commands to easier commands understandable by the coprocessor. The interface is connected to the PC-EISA bus (IEEE B996). As you can see in figure 3 the interface consists of 5 main parts. The command interpreter demultiplexes the 16 bit data from the PC to 45 bit data for the neural coprocessor. The read and write FIFO consist of a RAM where the adress is generated by a counter. The decode unit and the register unit control the communication of PC and interface, like connection to the PC boards and interrupts.

4 The PC task

The PC holds two software packages for working with the neural coprocessor. First it holds the serial parts of the SOM's algorithm embedded in a userfriendly system enabling the user to choose easily the different learning parameters like size of the SOM and heights and width of the learning function. Different forms of learning functions are choosable and input data can be normalized.

Second a software package, ULTIKOS, for analyzing the learning results is available. Additionally graphical analysis like component cards, spanning trees, U-matrix, D-matrix etc., and different quality measures can be calculated. Input data can be normalized and different forms of learning functions can be selected.

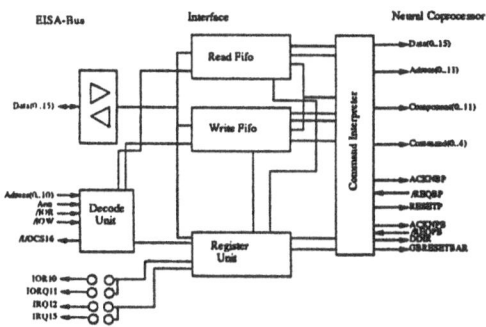

Figure 3: The architecture of the asynchronous interface

4.1 Learning

For one learning step the PC chooses an input vector from the set of the learning data set, determines a set of PU's for learning and send these information to COKOS. COKOS calculates the location of the PU of the given set with minimum euclidean distance according to (1) and send this location back to the PC.

Then the PC selects a set of PU's in the neighbourhood of the winning PU for adaptation and the corresponding adaptation factors and send these data to COKOS. COKOS adapts these PU's performing according to (2).

Because of the interface's FIFOs for each direction neither the PC nor COKOS has to wait for the other part. The PC puts several tasks for minimum calculation into the FIFO until COKOS has determined the location of the PU holding the minimum distance. Then the PC sends the adaptation information to the FIFO meanwhile COKOS calculates the next minimum locations.

Considering the memory in the MAB's it is possible to store a limited input data set in COKOS, too. This reduces the data transfer.

4.2 Evaluation

To evaluate the SOM after finishing the learning you have to read out the weights of the whole SOM or of a part of it. Especially for online evaluation it is important to read only a part of the SOM.

During online evaluation the PC normalizes the input data and transfers them to the input FIFO of the interface. While COKOS calculates the location of the winning PU, the PC can put the next data to the FIFO. In the FIFO for the opposite direction the PC can get the locations of the PUs holding the most similar vector in the same order as the input vectors occured. The PC can controll the filling of the two pipelines.

5 Hardware implementation

For the implementation of the coprocessor and the interface we use off-the-shelf elements and FPGAs. Off-the-shelf elements were used for building fast arithmetic like multiplier and the addertree. Irregular structures like a subtractor with different datapathes for calculating the euclidean distance and adaptation, controller and command interpreter, were realized with XILINX FPGAs. This concept yields to a fast and flexible prototype.

6 Results and conclusion

We compared our hardware implementation to a software implementation running on a SISD-computer (SUN 4). The virtual selforganizing map consists of 60x60 PUs learning 10000 vectors with 8 components. The algorithm written in C achieves a learning rate of nearly 1 MCUPS. We define one CUPS as a complete learning step for one vector component and one PU including the calculation of the minimum distance and adaptation. Our hardware implementation achieves about 16 MCUPS for the same application running with a clock rate of 10 MHz.

Therefore COKOS is useful for learning and absolutely necessary for fast, online evaluation. The hardware implementation with programmable gate arrays ensures the system's flexibility to different requirements of the various applications.

References

[1] I. Barth, T. Bräunl, S. Engelhardt, und F. Sembach. Parallaxis version 2 user manual. Technical Report 2/91, University of Stuttgart, 1991.

[2] J. Göppert, H. Speckmann, W. Rosenstiel, W. Kessler, G. Kraus, und G. Gauglitz. Evaluation of Spectra in Chemistry and Physics with Kohonen's Selforganizing Feature Map. In *Neuronimes 92*, 1992.

[3] T. Kohonen. *Self-Organization and Associative Memory*. Springer Verlag Heidelberg New York Tokyo, 1984.

[4] T. Kohonen. The Neural Phonetic Typewriter. *IEEE Computer*, Seiten 11–22, March 1988.

[5] D. Niebur und A. J. Germond. Power System Static Security Assessment Using the Kohonen Neural Network Classifier. In *IEEE Power Industry Computer Applications Conference*, 1991.

[6] H. Speckmann, P. Thole, und W. Rosenstiel. Hardware Implementations of Kohonen's Selforganizing Feature Map. In *IJCNN 92, Bejing, China*, Seiten III 183–187, 1992.

[7] V. Tryba. *Selbstorganisierende Karten: Theorie, Anwendung und VLSI-Implementierung*. VDI Verlag, 1992.

[8] A. Ultsch und H. P. Siemon. Exploratory Data Analysis: Using Kohonen Networks on Transputers. Technical report, Universität Dortmund, 1989.

Neural hardware and software
—poster contributions

Hardware Implementation of Kohonen's Feature Map by Scalar and SIMD-Array Processors [*]

A. König, X. Geng, M. Glesner
Darmstadt University of Technology
Institute for Microelectronic Systems
Karlstrasse 15, D-6100 Darmstadt, Germany

Abstract

In this work we present two implementations of Kohonen's self-organizing feature map. First by scalar processor slice for learning and recall, that can also be cascaded to a SIMD-array computer. Second by a SIMD-array computer with bit-serial processing elements that is also applicable as a fast parallel multi-reference or nearest-neighbour classifier. Prior to the implementation we have studied the accuracy requirements and the tolerable algorithmic simplifications of the Kohonen algorithm. The investigation methodology and the results for accuracy and simplification will be presented. Based on the implemented recall architecture of the bit-serial processor we present an extended architecture for an on-line learning implementation. We target on full-custom VLSI-implementation of our processors and envision further integration by multi-chip modules.

1 Introduction

Various neural network algorithms are more or less inspired by biological evidence. Models like backpropagation are purely mathematical models and only have a principal relation to actual neural structures. In contrast, Kohonen' s Self-Organizing feature Map (SOM) algorithm was developed according to observations with regard to the organization of neurons in the cortex of the human brain. Neighbouring neurons are sensitive to stimuli that are close or similar to each other in vector space of the stimuli.

During learning of the feature map the neuron's weight vectors are distributed in the feature space according to the underlying probability of the features (input stimuli). Thus, a process of vector quantization of the input data takes place, similar to well known algorithms as the LBG-algorithm [1]. The decisive difference is, that the SOM also computes a mapping from $L \rightarrow N$ dimensions, preserving the topological order of the input stimuli on the map surface. Commonly N is chosen N=2. This choice makes the SOM excellent candidate for data inspection and analysis. By means of dedicated inspection tools, e.g. grey value output of weight vectors, mesh display, or labeling of the neurons the SOM can be applied for problems that otherwise remain obscure. For instance, the separability of data for a chosen feature in classification can be assessed or numbers of classes and labeling of prototypes can be defined according to the clusters and structure found in the feature data. Another application of the SOM is in image coding for vector quantization techniques [2]. In image coding real-time constraints promote the application of fast dedicated hardware, but on-line learning is not mandatory. Thus, a very simplified architecture can be used for SOM implementation in this case. As the recall phase of a group of networks like LVQ, SOM, Neural Associative Memories (NAM) are identical in simplified form, such an architecture can be generically employed. By a slight enhancement, also the principle of the RCE-model

---

[*]This work was accomplished within the context of a research cooperation for hardware implementation of neural networks with German Telecom Forschungs- und Technologiezentrum Darmstadt FI 16

[4] can be implemented this way, providing a hardware platform both for classification and vector quantization. We have implemented a small prototype of such a generic system that is used as a NAM classifier system [2]. In this paper we will present the enhancement of this basic architecture to a distributed parallel learning system that implements a simplified SOM algorithm. Such a system could be used for adaptive vector quantization in image coding and fast classification. For data analysis and classification we pursue the development of a fixed-point SOM processor.

2 Accuracy Requirements and Algorithmic Simplification

The simplified Kohonen algorithm [5] was used in our work, limiting our investigation to orthogonal matrix topology, euclidean or manhattan distance, and the learn rule $w_i(t+1) = w_i(t) + \alpha(t, r) \times [x_k - w_i(t)]$ in vector notation. Most investigations and applications of SOMs base on emulation by standard computers or hardware accelerators using floating point computation. This seems not feasible for a massive parallel implementation. Instead a most simple fixed point computation appears to be mandatory. Also implementation can be alleviated by reducing the complexity of the algorithm, e.g. choice of metric or neighbourhood function. Of course, these modifications of the algorithm must be carefully checked, so that proper behaviour is retained.

For this reason we designed a powerful SOM simulator in C++ that provides all the interesting variants of the algorithm. The simulator was applied for image coding and in a quality control research project. By operator overloading we can replace floating point by a user defined fixed point computation for all calculations. In the range of 31 bits the user can define dynamic and resolution of signed numbers. Over/underflow is prevented by automatic saturation. As just one data-type is not enough for the various dynamic and resolution requirements within the SOM algorithm, we introduced for our work four data-types, e.g. input, weight, learn, and response type. Type casting uses truncation and saturation, which is easy to implement in VLSI. Usually, due to enormous simulation times, only very simple benchmarks are used for the verification of the network function with reduced accuracy. For verification we also took a very simple benchmark, by training the network with equally distributed random numbers. By displaying the mesh of weight vectors, ordering of the map could be assessed. Additionally, this can be compared visually and numerically to floating point computation results. For the real tests we used the application of vector quantization as benchmark and the achieved image coding quality in dB versus fixed point bit number as criterion for the network functionality. The ordering of the map for high dimensional data is of course also a criterion, and by employing the sammons mapping [3] a mesh display for high dimensional data was generated. Subject to the criteria mentioned above, we found that SOM can be simplified to using a box function as neighbourhood function N_c and manhattan distance as metric. Further, we found that the multiplication in the learning phase may be replaced by appropriate shift operations, thus considerably reducing the area requirement for a neuron.

The necessary accuracy depends on the accuracy of the problem data. Several parameters of the four data-types can be preset according to the accuracy of input data and vector length. Based on vector quantization of grey-value video phone images featuring 8 bit grey-values and $8 \times 8 = 64$ pixel image blocks we have the following presetting:

Type	Dynamic	Resolution
Input type	9	0
Weight type	9	X = 6
Response type	15	X = 5
Learn rate type	2	X = 6

The table entries marked by an X were determined by simulations. The values represent the minimum values achievable according to the quality criterion. Quality sharply deteriorated to an unacceptable value for further accuracy reduction of any data-type. As indicated above, to achieve a simplified learning unit we replaced the multiplication in the learning by shift right operation of the difference controlled by α, i.e. the learn rule takes the form:

$$w_{ij}(t+1) = w_{ij}(t) + rightshift((x_i - w_{ij}(t), \alpha) \tag{1}$$

By appropriate mapping of intervals of α to shift steps we have achieved encouraging results for the simple benchmark.

3 Hardware Implementation

Currently, we pursue two tracks of implementation. On the one hand we target on implementing a scalar processor architecture for small and low cost implementation. This processor features a *Subtract-Multiply-Accumulate (SUMA)* structure. The data path is under design as a highly pipelined full-custom approach. We provide accuracy rounded to the next power of 2 from our minimum results to achieve a regular structure. Euclidean distance will be used but the form of learn and neighbourhood functions are arbitrary. The architecture supports cascading of several of those scalar units. We target on integrating several units in a full-custom chip design. The simplified block diagram of this processor is displayed in Fig. 1. On the other hand we develop a massively parallel implementation

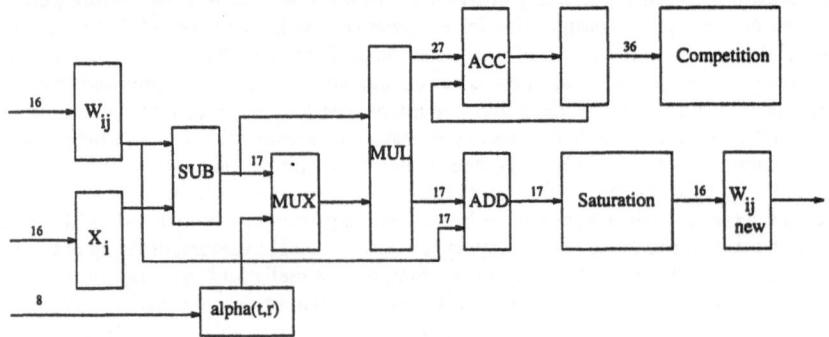

Figure 1: SOM scalar processor with SUMA structure

of the SOM based on very simple computation units. In prior work we have developed a bit-serial neural unit that implements the SOM recall phase using manhattan distance [2]. This structure can be interpreted as SOM or as NAM implementation. We have developed and implemented a very efficient bit-serial competition mechanism that finds the winner in 25 (response type + 1) processing cycles. The system is cascadable, so that networks of arbitrary size can be assembled. For the application of this network as a classifier system a standard cell design has been carried out and will be manufactured in the end of 1993. Especially for learning the computation accuracy of 8 bit implemented in [2] is not sufficient. Thus the concept was enlarged. We started work on a full-custom design of this concept. The basic cell of both the recall and the training architecture is a counter with several count inputs. (Fig. 2). The counter is used for the manhattan distance calculation in recall. In the current chip design the counter provides 24 bits with 16 inputs. The counter cell was optimized and pipelined and thus runs according to simulations with 80 MHz. In learning we intend to use the counter for calculating $(x_i - w_{ij})$ and transfer this to a shift register, shift dependent on α and accumulate this δw_{ij} to w_{ij}. The learn rate $\alpha(t)$ will be globally computed and broadcasted. Based on the box function neighbourhood the spatial learning rate $\alpha(r)$ is determined by the distance to the winners X and Y coordinates and the current radius. Figure 2 illustrates the neuron cell of this architecture with X and Y registers. Targeting on a large system for both classification and adaptive vector quantization, we

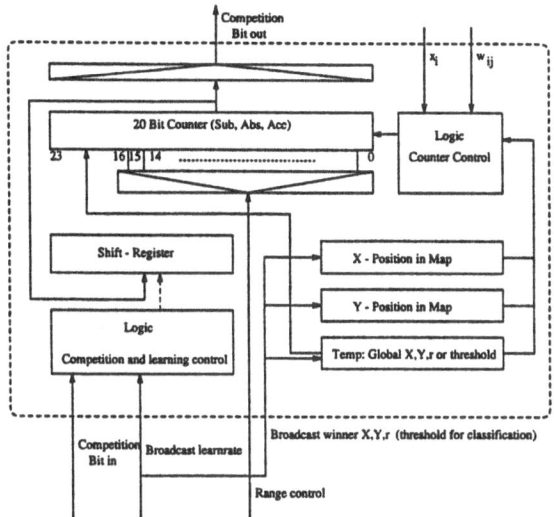

Figure 2: SOM neuron for on-chip learning array implementation

use a full-custom design approach accomplishing a considerable amount of units on a single die and also a high clocking rate (0.8μ process and 40 to 50 MHz). A multi-chip module approach is envisioned for the compact integration of the system on one or two extension boards to a general purpose host.

4 Conclusion and Future Work

We have presented two architectures for the implementation of the SOM algorithm. The array implementation is also applicable as a parallel multi-reference classifier. A prototype of the recall architecture is available. Fast recall chip and a SUMA-processor are under design. In future work we will elaborate and implement the bit-serial SIMD array processor with learning capability, the two full-custom designs will be manufactured, and system prototypes will be developed.

References

[1] Y. Linde, A. Buzo, R. M. Gray, "An Algorithm for Vector Quantizer Design ", *IEEE Transactions on Communications*, Vol.28, No.1 , January 1980

[2] A. König, M. Glesner, "VLSI-Implementation of Associative Memory Systems for Neural Information Processing", *3rd International Workshop on VLSI for Artificial Intelligence and Neural Networks*, Oxford, 1992

[3] D. L. Reilly, L. N. Cooper, C. Elbaum, "A Neural Model for Category Learning", Biological Cybernetics, 45, pp. 35 - 41, 1982

[4] T. Kohonen, *Self-Organization and Associative Memory*, Third Edition, Springer 1989

[5] J. W. Sammon, "A Nonlinear Mapping for Data Structure Analysis", *IEEE Transactions on Computers*, Vol. 18, pp. 401-409, 1969

A Nonlinear Electronic Layer for Distributed Neural Nets

S.B.Colak, F.P.Widdershoven

Philips Research Laboratories, P.O.Box 80.000, 5600 JA Eindhoven, the Netherlands

Abstract

A new distributed approach to obtain neural network operation using a nonlinear inhomogeneous electronic layer is described. For this purpose, the electronic transport through the two dimensional inversion layer of a multi-terminal field effect semiconductor device is studied by numerical modeling. Information extraction and operations such as inhibition and signal crossing are handled in such a layer by defining difference or relative signals at the output. The results of calculations give examples for nontrivial vector mapping abilities including the XOR operation. In addition, continuous input to output mapping is demonstrated and compared with feedforward neural networks.

1. Introduction

Despite the large variation in their architectures, the basic operation done by neural networks is a nonlinear mapping between the input and output signals in a parallel and adaptive fashion. Conventionally, neural networks has been implemented by using individual and distinct network elements called synapses and neurons performing separately multiplication, summation and sigmoid functions. Here, our purpose is to investigate different options for such mapping techniques using distributed and collective properties of nonlinear devices and materials. The origins of this aim can be found in the analogies between the Hopfield network and spin glass material systems [1]. Similar to that analogy, we expect that some other device and materials systems can achieve nonlinear mapping in a fully distributed fashion without the necessity of clearly defined areas of neurons and synapses. To demonstrate this, we study the mapping abilities of a simple two-dimensional (2D) electronic transport layer composed of distibuted nonlinear conduction paths between an array of input and output contacts. Simplest form of such transport systems can be realized by using subthreshold conduction in large-area multi-terminal field effect devices and conducting granular layers including organics or superconductors. We show that such layers can be used in vector mapping applications and also as the front processing layers of high-order neural networks [2].

2. Multi-Terminal Nonlinear Conducting Layer

We now consider a nonlinear multi-port inhomogeneous electronic layer and its connections as sketched in Fig.1. The basic property needed is nonlinear transport between areas of varying conductivity as indicated schematically by contoured regions in the figure. This layer is used to transform a set of input voltages (V_i) into a set of output currents (I_j). We assume a fully planar device with no negative differential resistance effects in order to be able to show the vector mapping abilities of a nonlinear electronic layer in its simplest form. In such a conducting system, the general form of the output current can be given by,

$$I_j^q = \sum_{i=1}^{n} G_{ji}^q(V) \ V_i^q$$

where $G_{ji}(V)$ are the nonlinear conductances which depend on the (q)th input vector of bias voltages. The response of such a layer to increasing input bias is monotonously increasing. In order to obtain any type of useful mapping with such a system we consider the difference or relative values of output signals with respect to other varying signals from the layer. For example, the output information can be defined as $O = I_k - I_l$ or $O = I_k / I_l$. Using such definitions, it can be shown that inhibition and signal crossing functions can be achieved even within such a simple system [3].

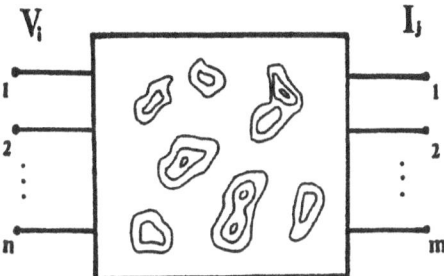

FIGURE 1. Multi-terminal inhomogeneous 2D nonlinear electronic layer configuration.

3. Numerical Model and Mapping Examples

To give an example for the vector mapping abilities of a nonlinear electronic layer, we have modeled the electronic transport through a 2D conducting plane representing approximately the inhomogeneous surface inversion layer of a field effect semiconductor device. This layer is assumed to consist of a 2D array of electron puddles with differing electron densities due to the differences in the surface potential distribution. The electron conduction between individual puddles of such an array provides the highly nonlinear effects needed. All other effects related to bipolar doping and surface gates which could be included in a real devices, and other output vectors in terms of capacitances or charges were ignored. Such added features at later stages can only enhance the effects. The device concept studied here may be considered to be similar in nature to the experimental Neuro-MOS [4] study. However, in sharp contrast to that device which serves only as synaptic connections of a single neuron, our nonlinear electronic layer is utilized as a distributed neural transformation system in between multiple input and multiple output terminals.

The schematic drawing of the modeled device topology is shown in the inset of Fig.2 as an inhomogeneous array of 5x5 dots representing the location of free electron puddles. Such areas can be formed by trapped charges within an insulator at the surface of the semiconductor. These electron puddle regions are coupled to each other by nonlinear capacitances and conductances. The input/output signals are provided at the boundaries by a set of linear resistors as shown in Fig.2. In principle, input/output contacts do not always have to be at the boundary, but can also be distributed within the surface . The output currents in the layer are shown in Fig.2 for one varying (V_1) and

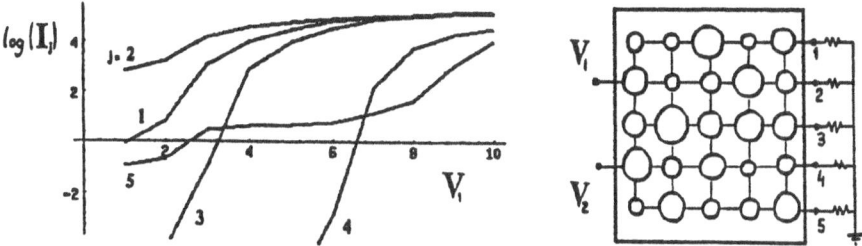

FIGURE 2. The output currents as a function of input voltage 1 in an electron puddle array.

one fixed (V_2) input. These characteristics are similar to subthreshold response of a single MOS-FET. In this case however, we have a set of floating subthreshold MOSFETs connected similar to a cellular array. These results can also be interpreted as a bias dependent "percolation" through the puddles.

Fig.3 gives results of the 5x5 electron puddle array with its input signal vector defined by the existence (1) or absence (0) of a connection to a common voltage supply with a bias value of V. Fig.3a shows the output node current values normalized to the total current for the array with two input and three output nodes. By putting proper thresholds for normalized current values in Fig.3a, we can obtain various combinations of vector mappings made possible by the nonlinearity in this specific device. Note that it is possible to obtain XOR operation from the output node (1) since the level of normalized current output for input (1,1) is lower than those for inputs (0,1) and (1,0). A threshold placed at 0.65 at output node (1) will result in XOR operation with proper leakage for the input (0,0). This result is obtained for a fixed but random distribution of the number of electrons in the puddles. Alternatively, the conducting array can be put into learning, by for example a suitable learning method such as stochastic learning, in order to optimize XOR or other functions.

Fig.3b gives the actual output current values for another puddle array with 3 input and 5 output nodes. This figure demonstrates another method of vector mapping defined as the difference currents between two pairs of output nodes. The difference operation can be implemented by resistors between the terminal pairs. The mapping is achieved by comparing the difference in the current values of different output nodes as indicated in the figure by the arrows. This shows how the response at output "B" changes from positive to negative, or in other words provides inversion, as the system receives more "on" inputs . Output "A" on the other hand demonstrates another example of XOR as the input vectors (100),(010),(110),(000) generate (1),(1),(0),(0) respectively with respect to a threshold of 3. In this case, the functionality of the nonlinear electronic layer is equivalent to a two layer feedforward network with the exception of one final threshold/limiting operation. It is worthwhile to mention that if the output coding for the example of Fig.3b were the output responses normalized to the total current rather than differences, the pairs of vectors mapped into each other would obviously change. This case would result in vector

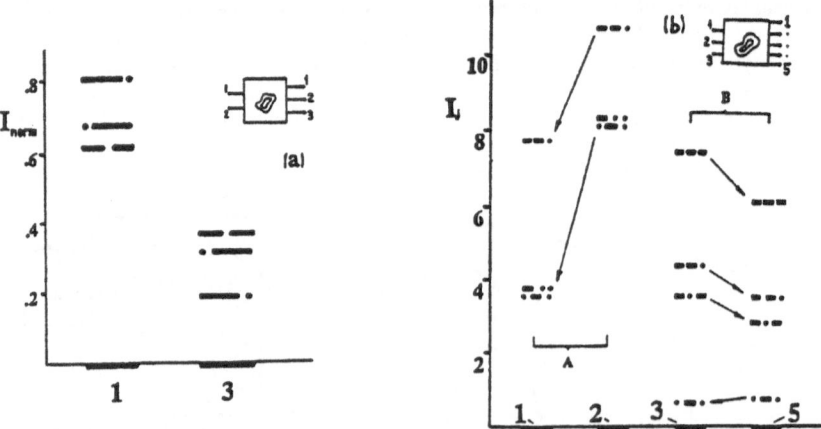

FIGURE 3. (a) Normalized output currents in an electron puddle array with two inputs and three outputs. (b) Actual output currents and the difference currents (arrows) between two pairs of output nodes for an electron puddle array with 3 inputs and 5 outputs. The current values are coded with dashed and dotted patterns meaning a (1) and a (0) input connection to a bias of V.

mappings such as (011) --> (11101) and (111)-->(10101) demonstrating inversion and signal crossing simultaneously.

Our final example demonstrates a continuous mapping achieved by an array with 5 output and 2 input contacts (fig.1). In this case the bias level to each input contact is varied independently of the other. These results are obtained by biasing one input with a constant value (V_2) and changing the other (V_1) continuously. If we than extract the information contained in these output responses in terms of difference or normalized values between different output terminals, we get the results shown in Fig.4 . Fig.4a shows that an input can be mapped nonlinearly onto another function as done by a two layer feedforward network with the exception of simple output resistors.

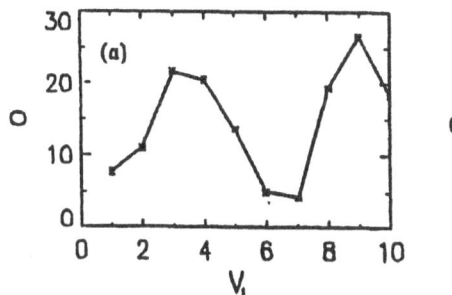 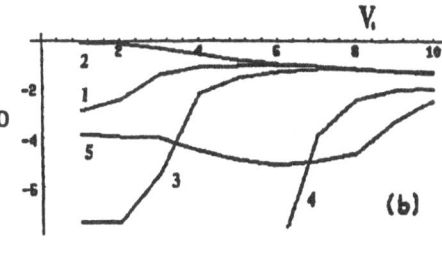

FIGURE 4. Examples of continuous mapping possible with a 2 input/5 output electronic layer with one of the input biases fixed. The outputs (O) are defined as differences (a) or normalized (b) values of output currents. Difference output is obtained from four terminals (achievable by resistors).

4. Conclusions

The results obtained in this study demonstrate that the electronic transport characteristics in a multi-terminal 2D nonlinear conducting layer can provide nontrivial vector mapping and continuous nonlinear mapping between input and output signals to help in designing alternative forms of neural networks. Such mappings are done by parallel and collective processing and by involving "memorized" patterns of the distributed conductivities. Even in its simplest from, a distributed nonlinear 2D electronic layer can be utilized to replace parts of a neural network in order to convert it to a high-order neural net optimized for specific applications. Larger scale systems can be achieved by adding area to the system as done in cellular neural networks. At some point this will require sparse but long range interconnections [5] to avoid screening problems. The learning or adaptation issues have not been addressed in this study. For the type of distributed physical system which we have at hand, stochastic learning techniques may be the most advisable ones.

5. References

[1]   J.J.Hopfield, Proc.Nat.Acad.Sci.USA, vol.79, p.2554 (1982).
[2]   C.Lee Giles, T.Maxwell, Appl.Opt. vol.26, p.4972 (1987).
[3]   S.Colak, Philips Tech.Report, (Oct.1992), Unpublished.
[4]   T.Shibata, T.Ohmi, IEEE Trans.Electron Devices, vol. 39, p. 1444 (1992).
[5]   A.J.Noest, Phys.Rev.Lett., vol.63, p.1739 (1989).

HOW TO FIND A NEAR OPTIMAL MAPPING OF NEURAL NETWORKS ONTO MESSAGE PASSING MULTICOMPUTERS

B. Kreimeier, M. Schöne, R. Steiner and R. Eckmiller

Department of Computer Science VI – Neuroinformatics
University of Bonn,
Römerstr. 164, W-5300 Bonn 1, Germany
e–mail: bernd@nero.uni-bonn.de

Abstract

Concerning the simulation of neural networks with arbitrary topology on distributed memory multiprocessor systems, we introduce our approach to an automatic determination of a near-optimal mapping of neural networks onto a given multiprocessor. Our approach is based on stochastic local search. We propose a decomposition of the mapping into a network partitioning step followed by a placement of partitions onto processors.

1 Introduction

Typical implementations of neural networks on message passing multicomputers do not map each neuron or synapse to a single hardware component but rather simulate clusters of neurons and synapses as network partitions on a coarse-grain [2] or fine-grain [10] multiprocessor system. In our case, the multiprocessor system consists of a cubic lattice of processing units with six bidirectional serial links to adjacent processing units [10]. The message routing mechanism is an autonomous component of the processing units, which generates direct paths to a demanded target by local routing decisions between possible steps.

Following the network decomposition approach for the distributed implementation of neural networks on message passing multicomputers, the problem of efficiently mapping neuron processes to processing units has to be solved. This well known problem [1] is very similar to the placement problem encountered in VLSI implementations of neural networks. We propose a decomposition of the mapping operation into a k-partitioning of the neural network process graph followed by a placement of k partitions onto k processors.

2 Process Graph Representation

To represent sparsely connected neural networks, we use an adjacency list representation, which is structured by a tree of partitions. A neural network topology with N_{pe} processing elements (neurons) and N_{ce} connecting elements (synapses) is represented as an edge-weighted, node-weighted directed multi-graph $(\mathbf{N}_{pe}, \mathbf{N}_{ce})$, with each node $i \in \mathbf{N}_{pe}$ and each edge $(i,j) \in \mathbf{N}_{ce}$:

$$\mathbf{N}_{pe} = \{1, \ldots, N_{pe}\} \qquad \text{and} \qquad \mathbf{N}_{ce} \subseteq \{1, \ldots, N_{pe}\} \times \{1, \ldots, N_{pe}\}$$

Before partitioning all self-edges are removed. Multi-edges are substituted by a single edge with summed edge weight. Thus N_{ce} varies between $N_{pe} - 1$ and $N_{pe}(N_{pe} - 1)$.

In the case of synchronous neural network models, the edge weights w_{ij} are normalized to unity, which is not always possible in the case of asynchronous updates. The neuron process is subdivided into three steps: processing of input messages (i.e. synaptic events), computation of the neuron's state and output, and generation of output messages (i.e. neuron output). Each of these steps requires a number of instructions depending on the details of the message encoding and the complexity of the neuron model:

$$w_+ \quad : \quad \text{\#instructions required to handle input messages}$$
$$w_- \quad : \quad \text{\#instructions required to generate ouput messages}$$
$$w_o \quad : \quad \text{\#instructions required to calculate neuron state}$$

The total amount of process load due to a single neuron process depends on the degree d_i of the process node (i.e. the number of connections). In the case of neuron processes with different complexity, the node weights are given by:

$$w_i = w_{+,i} d_{+,i} + w_{o,i} + w_{-,i} d_{-,i} \qquad \text{with} \qquad d_i = d_{+,i} + d_{-,i}$$

For small variations in $w_{o,i}$, $w_{+,i}$, $w_{-,i}$ and in $d_{+,i}$, $d_{-,i}$, the node weights w_i are normalized to unity, which allows for estimation of lower bounds and accelerated calculation of cost function changes.

3 Decomposition of the Mapping

In our approach we restrict ourselves to a static process graph representation. The process-to-processor-mapping problem is given by finding a mapping $\Pi_{map} : \{1, .., N_{pe}\} \mapsto \{1, .., N_{pu}\}$ with N_{pe} neuron processes and N_{pu} processing units. If the number of processes exceeds the number of processors (which is not the case for VLSI cell placement), the mapping problem includes finding an appropriate partition of the process graph. Therefore the mapping operation is decomposed into a graph partitioning operation followed by a placement of partitions onto processors:

$$\Pi_{map} = \Pi_{place} \, \Pi_{part} \qquad \text{with} \qquad \begin{aligned} \Pi_{part} &: \{1, .., N_{pe}\} \mapsto \{1, .., N_{pu}\} \\ \Pi_{place} &: \{1, .., N_{pu}\} \mapsto \{1, .., N_{pu}\} \end{aligned}$$

Graph partitioning as well as placement belongs to the class of NP-complete combinatorial optimization problems. Various heuristics [5, 9] have been suggested to automatically achieve near-optimal solutions of such problems, including neural network approaches [3, 4] and simulated annealing [6, 8]. Our approach is based on stochastic local search and iterative improvement, where the improvements are estimated by properly chosen cost functions.

4 Cost Function for Process Graph Partitioning

Using constraint penalization, both displacement of process load and total amount of communication load contribute to the total cost of a given partition. The cost function is:

$$f_{cost} = (1 - \gamma) f_{displace} + \gamma f_{comm} \qquad \text{with} \qquad 0 \leq \gamma \leq 1$$

The solution depends on the value of γ, which must be derived from the processing time per operation versus the communication time per operands for given processing unit and communication network. The displacement is measured by:

$$f_{displace} = \sum_{i=1}^{N_{pu}} |s_i - \bar{s}| \qquad \text{with average partition size} \qquad \bar{s} = \frac{1}{N_{pu}} \sum_{i=1}^{N_{pe}} w_i$$

The remaining part of f_{cost} is f_{comm} given by the total amount of communication load, which is proportional to the sum over all edge weights in the process graph:

$$f_{comm} = \sum_{i=1}^{N_{pe}} \sum_{j=1}^{N_{pe}} c_{ij} \qquad \text{with} \qquad c_{ij} = \left\{ \begin{array}{lll} 0 & : & \text{if } i,j \text{ same partition,} \\ w_{ij} & : & \text{weight of } (i,j) \text{ otherwise} \end{array} \right.$$

If all edge weights are equal and normalized to unity, this cost function is equivalent to the max-cut or minimum-edge separator solution for the graph partitioning problem.

5 Cost Function for Partition Placement

A given near optimal partition of the process graph is simplified by removing all edges within the partitions and by contracting all nodes of each partition to a single node with summed node weight w_i^*. Subsequently multi-edges are substituted by single edges with summed edge weight w_{ij}^*. The cost function has to take into account the dilation due to the message routing through the multiprocessor communication network. In general, the dilation has to be estimated by determination of the path length through the multiprocessor, e.g. by Dijkstra's shortest path algorithm or by an appropriate metric. In our case, the dilation is equal for all admissible paths and given by the Manhattan metric.

$$f_{dilate} = \sum_{i=1}^{N_{pu}} \sum_{j=1}^{N_{pu}} c_{ij}^* l_{ij} \qquad \text{with} \qquad \begin{array}{ll} c_{ij}^* : & \text{sum over all } c_{ij} \text{ of edges } (i,j) \\ & \text{from nodes on source processor} \\ & \text{to nodes on destination processor} \\ l_{ij} : & \text{\#hops from source to destination processor} \end{array}$$

A similar cost function is used in simulated annealing approaches without decomposition [8]. Note that in our approach the problem size for a 2-opt local search decreases by $\frac{1}{s^2}$, if the partition size s is the number of nodes in each partition.

6 Iterative Improvement with 2-opt strategy

To achieve near-optimal partitioning and placement, we use a 2-opt strategy, i.e. the pairwise exchange of nodes between partitions and the pairwise exchange of partition places on the multiprocessor. If there are clusters of adjacent nodes, a 2-opt iterative improvement approach based on dilation costs will stick in local minima, which is usually avoided by use of simulated annealing [8] and other approaches [5, 9]. Our simulations have shown that in many cases the possible partitions of neural network process graphs have similar costs [7], and that the possible placements of a near-optimal partition do not differ in terms of dilation costs.

7 Conclusions

An efficient distributed implementation of neural networks on message passing multicomputers requires near optimal mapping. The mapping problem can be solved by applying stochastic local search. Using a decomposition of the mapping, we are able to restrict ourselves to iterative improvement instead of using simulated annealing to avoid local minima.

Acknowledgments

The authors wish to thank Th. Reski and R. Lüling for valuable discussions. This work was supported by the Bundesminister für Forschung und Technologie under grant 01 IN 105 A/O (SENROB), and by the Minister für Wissenschaft und Forschung in Nordrhein-Westfalen.

References

[1] S. H. Bokhari. On the mapping problem. *IEEE Trans. Comp.*, 30(3):207–214, March 1981.

[2] T. Bossomaier and A. Loeff. Coherent oscillation in neural networks: Its transputer simulation. In P. Welch et. al., editor, *Transputing '91*, pages 676–686. IOS Press, 1991.

[3] D. E. van den Bout and T. K. Miller, III. Graph partitioning using annealed neural networks. *IEEE Trans. Neural Networks*, 1(2):192–203, June 1990.

[4] L. Hérault and J.-J. Niez. Neural networks & combinatorial optimization: A study of NP-complete graph problems. In E. Gelenbe, editor, *Neural Networks: Advances and Applications*, pages 165–213. Elsevier, Amsterdam, 1991.

[5] B. W. Kernighan and S. Lin. An efficient heuristic procedure for partitioning graphs. *The Bell System Technical Journal*, 49:291–307, Feb. 1970.

[6] S. Kirkpatrick, C.D. Gelatt jr., and M.P. Vecchi. Optimization by simulated annealing. *Science*, 220:671–680, 1983.

[7] B. Kreimeier, M. Schöne, and R. Eckmiller. Communication load reduction for neural network implementations on message passing multicomputers. In I. Aleksander and J. Taylor, editors, *Artificial Neural Networks 2*, volume 2, pages 1655–1658. Elsevier, Amsterdam, 1992.

[8] C. Lee and L. Bic. On the mapping problem using simulated annealing. In *8th Int. Conf. on Computers and Communications*, pages 40–44. IEEE Computer Soc. Press, 1989.

[9] J. E. Savage and M. G. Wloka. Parallelism in graph-partitioning. *J. Parallel and Distributed Comp.*, 13:257–272, 1991.

[10] M. Schwarz, B. Hosticka, M. Kesper, P. Richert, and M. Scholles. A CMOS-array-computer with on-chip communication hardware developed for massively parallel applications. In *Proc. IEEE Int. Joint Conf. Neural Networks*, pages 89–94, Singapore, November 1991.

20 MILLION PATTERNS PER SECOND VLSI NEURAL NETWORK PATTERN CLASSIFIER

P. Masa, K. Hoen, H. Wallinga

MESA Research Institute, University of Twente, P.O. Box 217, 7500 AE Enschede, The Netherlands

L. Larsson, H. J. Behrend, W. Zimmermann

DESY, Notkestrasse 85, D 2000, Hamburg 52, Germany

Abstract- A special purpose neural IC is described which will be utilised in a data-acquisition system in DESY (Deutsches Elektronen Synchrotron). The analog CMOS VLSI chip implements a 70x4x1 fully interconnected feed-forward network and is capable of classifying 70 dimensional data-vectors within 50 ns. The high speed is essential for the real-time data processing and data-reduction. The classifier has to perform fixed function, therefore programming is not essential. The neural chip is under fabrication with 2.5 μm double metal CMOS process, occupies 6.5x4 mm^2 silicon area, dissipates 2W at 5V power supply, performs 6 billion multiplications per second and has 1.5 GBytes/s equivalent input bandwidth.

Introduction

In order to investigate and understand the behaviour of fundamental particles and forces, at the HEP institute "Deutsches Elektronen Synchrotron" (DESY) in Hamburg, two large detectors namely H1 and ZEUS are installed within the particle accelerator HERA. The detectors consist of different components each specialised for detecting track, momentum or energy of particles coming from the interaction region, where electrons and protons collide. A great challenge for the data-acquisition is the resulting data-flow which is so large, that real-time processing is necessary to preselect physics events from the background. The background rate exceeds the physics rate by a factor of 10000. Approximately only 5 events per second can be stored on tapes for off-line analysis, which results still in a data rate of 500 KBytes per second for one detector.

As charged particles -generated by electron proton collision- penetrate through a drift cell in the detector, ionise gas molecules and leave a cloud of charge (track segment) along their path. Those charge clouds are detected and red out electronically. Linking the track segments of drift chambers allows reconstruction of the path of the charged particles. The "z-vertex histogram" is created by counting the tracks crossing the beam axis at certain intervals. All tracks of a physics event should come from one vertex, from the location of the electron proton collision. "Vertex cell histograms" and "non-vertex cell histograms" are generated as well, representing track segments coming from the electron-proton interaction region and from a region outside of it respectively. These histograms are characteristic for the two classes of detected events, the physics and background events. Analytical rules can be developed in order to separate the event classes and to make a preselection. However the high data-rate severely limits the complexity of computation that can be performed on each data-vector.

Classification with a feed-forward neural net is a very attractive alternative to the analytical method. It offers highly parallel computing, the computation time is independent of data-vector size in case of special-purpose hardware, and the classification boundaries for event classes can be obtained from training procedure with available examples. Another advantage of the neural network approach is that while the analytical method relies on the predictable behaviour of detector components, and read-out electronics, the neural net develops the classification rules for the detector with its imperfections.

Hardware

The network architecture and the required resolution of synaptic weights were determined empirically based on available training data and using the Back-Propagation algorithm. Since the classifier performs fixed function, programmability is not essential.

Chip layout is shown on Fig.1. The chip contains 4 hidden-layer neurons with non-linear activation function, the linear output neuron, 289 synapse circuits, 70 voltage inverters. Every synapse block is built up from 16 unity size transconductors, and the desired synaptic weight is programmed by parallel connecting appropriate number of unit cells. The synapse-matrix of the neural net classifier is "mask-programmable", it can be reconfigured with minor modification on the first metal layer.

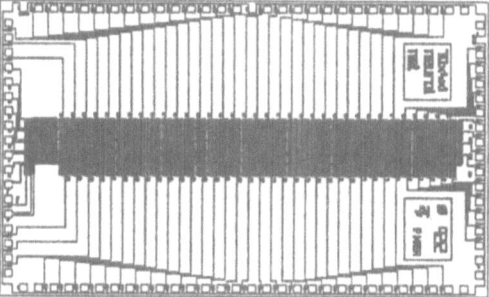

Fig. 1. Chip layout of the 70x4x1 feed-forward net

Principle of operation

The basic circuit blocks through the data-processing path are shown on Fig.2. The complete processing time of the classifying net is merely the time delay introduced by these circuit blocks, due to the parallelism. The synapse circuit is a VHF transconductor [1], which multiplies its input voltage by its transconductance resulting a current output. Different weight values are obtained by the number of parallel connected transconductors. The output currents of synapses are summed on the summing nodes according to the Kirchhoff law. The cell body has low input impedance, dominating the impedance of the summing node, and its current to voltage transfer has strong saturating characteristic. The output voltage of the cell body is multiplied by the second layer synapse also providing a current output. The output currents of second layer synapses are summed, generating the network output. Negative synapses are implemented by connecting their input to inverted input signal which is produced by the voltage inverter.

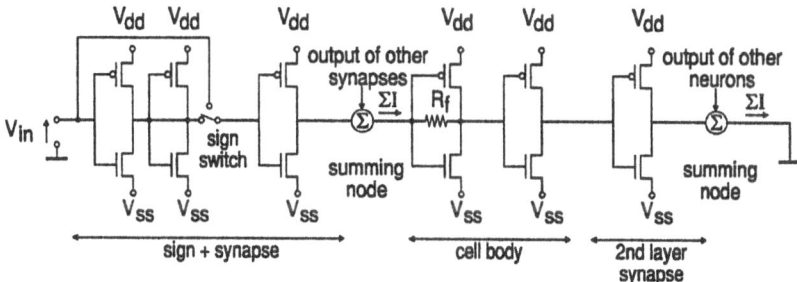

Fig. 2. Consecutive circuit blocks through the signal path from the input to the output

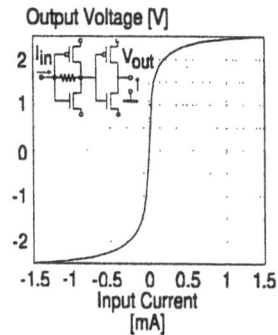

Fig.3. DC characteristic of cell body

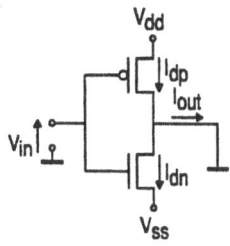

Fig.4. Synapse: VHF transconductor

Quasi-linear synapse

The input-output characteristic of the transconductor can be calculated in the following way:
Assuming that both transistors operate in saturation,

$$I_{out} = I_{dp} - I_{dn} = K_p(V_{dd} - V_{in} + V_{tp})^2 - K_n(-V_{ss} + V_{in} - V_{tn})^2 \tag{1}$$

$$K = \mu C_{ox}\frac{W}{2L}$$

V_{gsp}, V_{gsn}:   gate-source voltages of the p and n type transistors respectively

V_{tn}, V_{tp}:   threshold voltages of the p and n type transistors respectively

μ: carrier mobility

C_{ox}:   gate oxide capacitance per unit area

W/L: channel width / channel length

$$I_{out} = aV_{in}^2 + bV_{in} + c \tag{2}$$

with:

$$a = K_p - K_n \tag{3}$$

$$b = 2(K_n(V_{ss} + V_{tn}) - K_p(V_{dd} - V_{tp})) \tag{4}$$

$$c = K_p(V_{dd} - V_{tp})^2 - K_n(V_{ss} + V_{tn})^2 \tag{5}$$

The constant term, c can be zeroed by appropriate Vdd and Vss, or cancelled by the bias input of the neuron. From (2), (3) we can see, that the synapses are linear with respect to Vin if $Kp=Kn$. However perfect matching of K-factors is not feasible, because of process parameter variations and the discretized, "lambda-based layout" generation. Therefore the multiplication by the synapse will not be perfectly linear with respect to Vin. Rewriting (2) and assuming that c is cancelled by the neuron bias:

$$I_{out} = aV_{in}^2 + bV_{in} = b(V_{in} + \varepsilon V_{in}^2) \tag{6}$$

with: $\varepsilon = \dfrac{a}{b}$

the quadratic error term ε and the desired value of synapse, b is separated. Typical value for ε of the transconductor of Fig.4. for a 2.5 μm CMOS process, with $Wn/Ln = 5\mu m/2.5\mu m$, $Wp/Lp = 10\mu m/2.5\mu m$, is ε=0.01-0.1, and this error term is equal for all synapses.

The 2D separation boundary (where activation equals to zero) of a neuron with such non-linear synapses:

$$b_1(V_{in1} + \varepsilon V_{in1}^2) + b_2(V_{in2} + \varepsilon V_{in2}^2) = 0 \tag{7}$$

or:

$$V_{in2} = -\frac{1}{2\varepsilon} \pm \sqrt{\frac{1}{4\varepsilon^2} - \frac{b_1}{b_2}\left(\frac{V_{in1}}{\varepsilon} + V_{in1}^2\right)} \tag{8}$$

which is an ellipse or a hyperbola in contrast to the linear separation boundary of linear-synapse neuron. The shape of the separation boundary for different values of ε is shown on Fig.6. If the non-linearity error can be neglected or not, is depending on the magnitude of ε and the distribution of patterns of each class in the input space. As we can see on Fig.6., in case of non-overlapping pattern classes the non-linear classification boundary may separate just as well as the linear one. If the classes overlap, the effect of nonlinearity has to be taken into account when training the network.

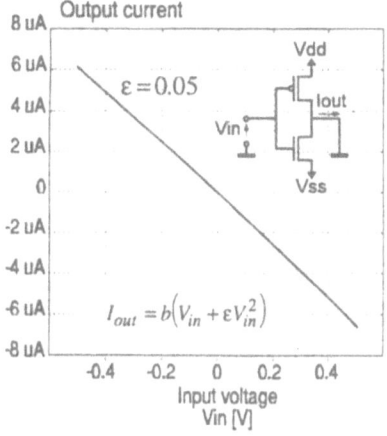

Fig.5. DC characteristic of the quasi linear synapse

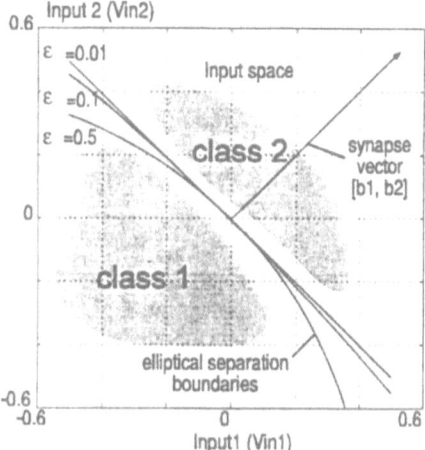

Fig.6. Elliptical separation boundaries of non-linear synapse

Power supply	5V (Vdd=2.5V, Vss=-2.5V)
Chip size	6.5x4 mm^2
Power dissipation	2W
Network architecture	70x4x1 feed-forward
Crosstalk between input lines	< -40 dB
Synapse resolution	integer values from -16 to 16
Synapse size	120x50µm^2
Synapse non-linearity error	<6%
PGA package	122 pins
Processing delay	50 ns
Computation speed	6 billion multiplications per second
Equivalent input bandwidth:	1.5 GBytes/second
Number of transistors:	17 000

Table 1. Chip specifications

Conclusions

The discussed integrated circuit clearly demonstrates the strong attributes of analog VLSI neural networks. Due to analog input the pins are used more effectively, and about an order of magnitude more information can be processed than with digital input. Thousands of compact analog synapses can be placed on a single chip, performing billions of multiplications per second without pipe-lining, which enables very short processing delay. Experimental results show, that discretisation down to approximately 5 bit resolution and 5% non-linearity of synapses had no significant effect on the classification performance [2].

Acknowledgement:
This project is financed by the Foundation for Fundamental Research on Matter (FOM)

[1] B. Nauta, *Analog CMOS Filters For Very-High Frequencies*, Ph.D. thesis, University of Twente, The Netherlands, 1991
[2] P. Masa, *On the Implementation of a Feed-forward Neural Network for Pattern Recognition*, Internal report

High-Density Analog-EEPROM Based Neural Network

Jerzy B. Lont

Electronics Laboratory
Swiss Federal Institute of Technology
ETH Zentrum, CH-8092 Zürich, Switzerland

Abstract

We have developed an experimental neural network utilizing analog-EEPROMs (floating-gate devices) for weight storage. Our integrated circuit contains three different-sized floating-gate devices, one neuron with three synapses and a threshold, and a dummy three-layer perceptron. Furthermore, we have investigated the influence of the size of the EEPROM cell on its programming characteristics and charge retention performance. Long-term charge retention measurements at room temperature are reported. Medium-sized EEPROM cells have been built in nonlinear synapses yielding a high density neural network. The size of the synapse is reduced by 54 % compared with our previous design incorporating capacitor weight storage. The experimental chip is realized in a 3 μm CMOS technology.

1 Introduction

One of the problems of analog neural networks is the weight storage. Most of implementations use dynamically refreshed capacitors for this purpose [1, 2]. This approach has serious disadvantages such as the complexity of the refreshing circuits, the need for external digital weight storage, weight decay, and the non-continuous operation caused by the refreshing periods. A solution to these problems is the use of analog-EEPROMs for weight storage [3]. Floating-gate devices are non-volatile, modifiable, highly precise, and compact, and can be programmed to any analog value of threshold voltage by controlling the voltage and width of the programming pulse [4].

2 Circuit Description

An analog-EEPROM cell, also called floating-gate device, is shown in Fig. 1. It consists of an n-channel transistor in a p-well and a metal-nitride-poly capacitor. The upper plate of the storage capacitor, the control gate (CG), is electrically accessible from the outside while the lower one, the

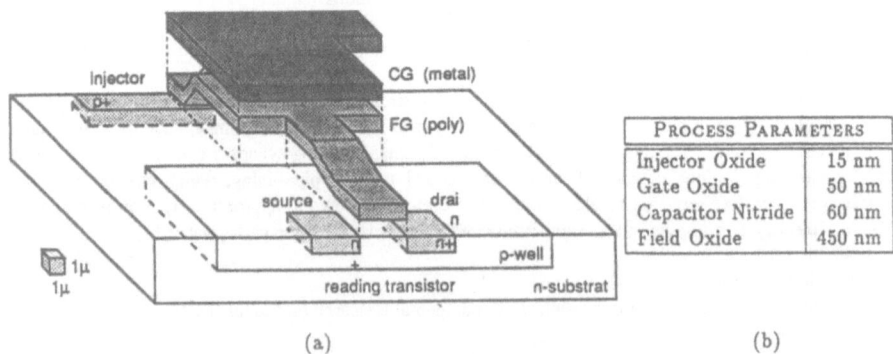

PROCESS PARAMETERS	
Injector Oxide	15 nm
Gate Oxide	50 nm
Capacitor Nitride	60 nm
Field Oxide	450 nm

(a)   (b)

Figure 1: (a) Simplified view of the floating-gate device. (b) Process parameters.

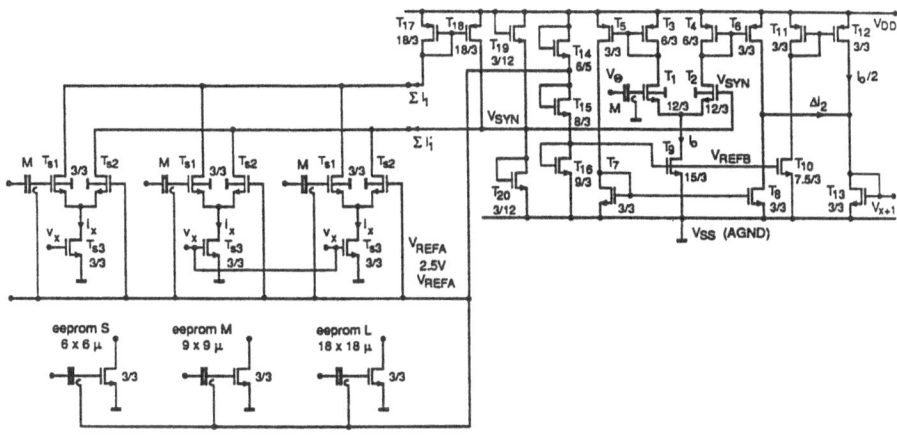

Figure 2: Schematic of the EEPROM cells and the neuron.

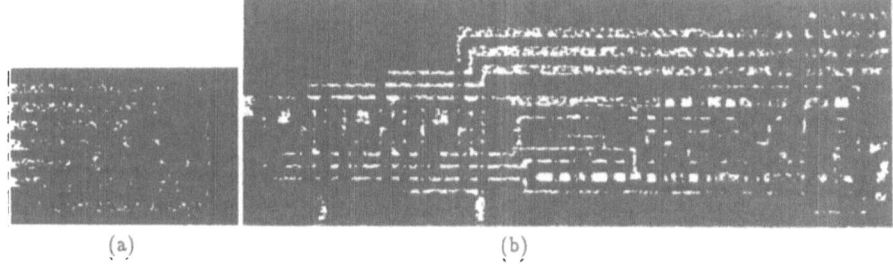

(a)　　　　　　　　　　　　　　　(b)

Figure 3: Microphotograph of (a) the EEPROMs and (b) the neuron.

floating gate (FG), is connected to the gate of the reading transistor only. A thin-oxide area (15 nm) between the floating gate and a p^+ implantation below it (INJ = injector) allows the floating gate to be charged or discharged by means of Fowler-Nordheim tunnel currents.

In order to investigate the influence of the size of the EEPROM cell on its programming characteristics and charge retention performance, we have integrated three different-sized floating-gate devices on the chip. All cells differ only in the size of the coupling capacitor. The size of this capacitor in the small cell (S) is 6 by 6 μm, in the medium one (M) 9 by 9 μm, and in the large one (L) 18 by 18 μm. A transistor with minimal dimensions ($W/L = 3\mu$m$/3\mu$m) is used for reading out.

Medium-sized EEPROM cells are built in nonlinear synapses composed of a differential pair (see Fig. 2. The circuit structure is basically the same as in our previous design [1]. The network has been implemented in a 3 μm single-poly single-metal SACMOS technology [5]. The 1.2 by 1.6 mm^2 chip (see Fig. 4) contains three different-sized floating-gate devices (Fig. 3(a)), one neuron with three synapses and a threshold (Fig. 3(b)), and a dummy three-layer perceptron. The last one is included for the comparison of effective area occupation only and is not connected to the bonding pads. A high density of 226 synapses/mm^2 has been reached (only the active area is counted, without pads and input protections). This figure is obtained for the dummy perceptron including somas. This is an improvement by 59 % compared with our previous design incorporating capacitor weight storage. A synapse has an area of 1638 μm^2 (39 by 42 μm) and a soma has an area of 7344 μm^2 (153 by 48 μm). Thus the occupied area has been reduced by 54 % for the synapse and by 23 % for the soma.

(a)

Chip Characteristics	
Neurons	1
Synapses	3
EEPROM cells	7
Dummy neurons	12
Dummy synapses	54
Dummy EEPROM cells	66
Synaptic density	226 syn./mm^2
Silicon area	1.2×1.6 mm^2
Technology	3-μm CMOS
Supply voltage	5 V
Package	16 DIL

(b)

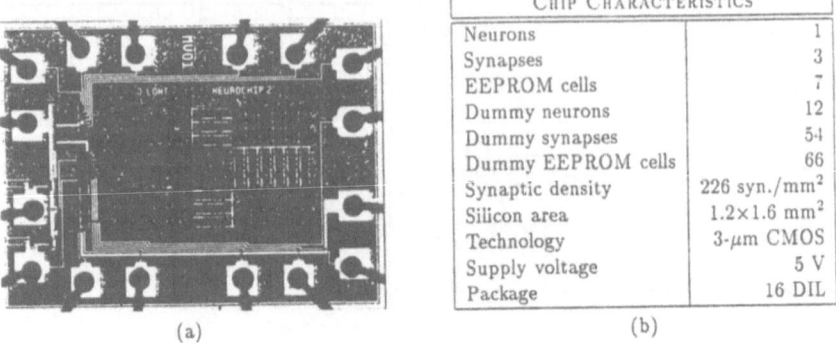

Figure 4: (a) Microphotograph of the chip. (b) Chip parameters.

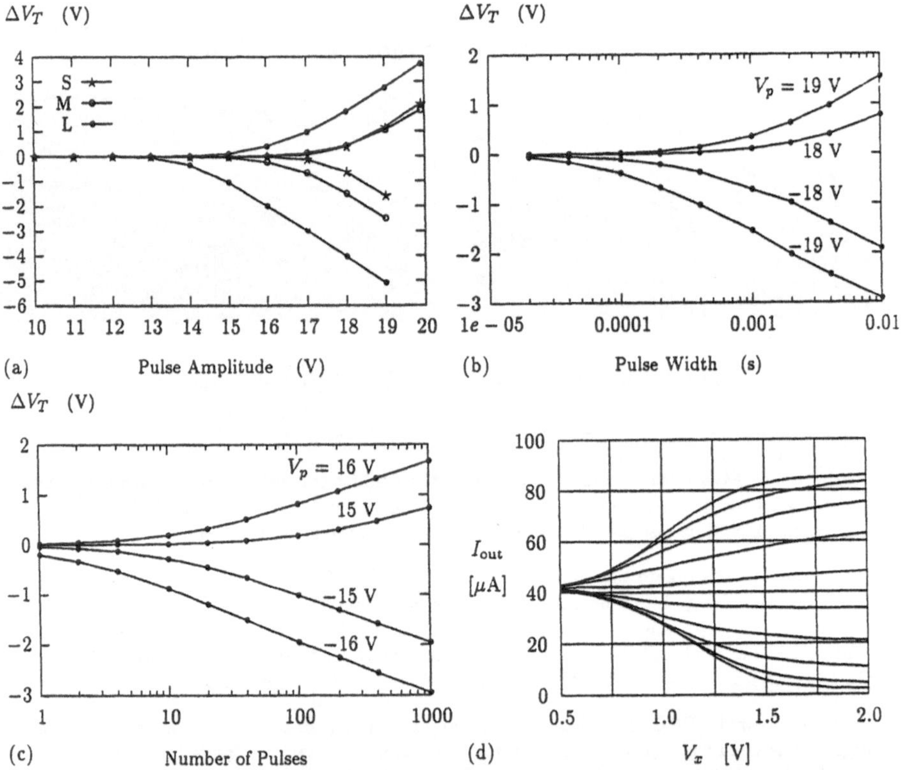

Figure 5: Threshold voltage change ΔV_{TH} as a function of (a) programming voltage V_p (note: in the case of the small (S) EEPROM cell, 10,000 pulses have been applied), (b) width of the programming pulse, and (c) number of pulses. (d) Measured transfer function of the synapse for different weights.

3 Experimental Results and Conclusions

Figure 5 shows measured characteristics for several programming methods and a transfer function of a synapse for different weights. The results of long-term charge-retention measurements at room temperature are plotted in Fig. 6. The measuring points are spaced uniformly in the logarithmic

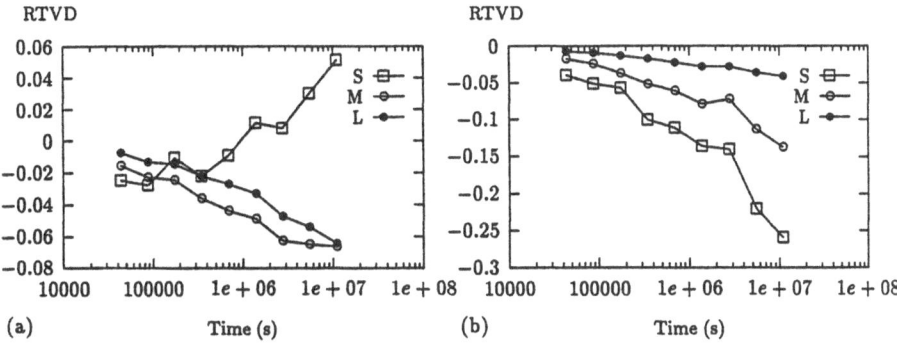

Figure 6: Normalized threshold voltage drift RTVD of different-sized floating-gate devices measured at room temperature for (a) positive and (b) negative programming voltages.

time scale. The quantity on the y axes is the relative threshold-voltage drift (RTVD) [4]:

$$\text{RTVD} = \frac{V_{TH}(t) - V_{TH\text{target}}}{V_{TH\text{target}} - V_{TH\text{init}}}. \tag{1}$$

It can be easily seen that the large cell (L) performs best. The somewhat strange behavior of the small cell (S) for positive programming voltages is still subject to research. It has been observed for all three S cells programmed with positive pulses. The charge retention performance seems to be worse than results published elsewhere [4, 6]. However, those measurements have been conducted at elevated temperatures. This could not only accelerate the voltage drift but also cause some annealing effects. The conclusion is that one should apply a retention bake [3] to the EEPROMs before starting long-term measurements.

Acknowledgments

We would like to thank CSEM SA, Neuchatel for mask preparation and Faselec AG, Zurich (subsidiary of Philips) for chip fabrication.

References

[1] J. B. Lont and W. Guggenbühl. Analog CMOS implementation of a multi-layer perceptron with nonlinear synapses. *IEEE Trans. on Neural Networks*, 3(3):457–465, May 1992.

[2] J. B. Lont. *Analog CMOS Implementation of a Multi-Layer Perceptron with Nonlinear Synapses*. PhD thesis, Swiss Federal Institute of Technology (ETH), Zürich, Switzerland, 1993. To appear as a book in the *Series on Microelectronics*, Hartung-Gorre Verlag, Konstanz, Germany.

[3] M. Holler, S. Tam, H. Castro, and R. Benson. An electrically trainable artificial neural network (ETANN) with 10240 floating gate synapses. In *Proceedings of the 1989 IEEE INNS International Joint Conference on Neural Networks*, pages II–191–II–196, June 1989.

[4] E. Säckinger and W. Guggenbühl. An analog trimming circuit based on a floating-gate device. *IEEE J. Solid-State Circuits*, 23(6):1437–1440, December 1988.

[5] R. E. Lüscher and J. Solo de Zaldivar. A high density CMOS process. In *IEEE International Solid-State Circuits Conference Technical Digest*, pages 260–261, 1985.

[6] C. Bleiker. *Analysis and Modeling of the Programming-, Retention-, and Endurance Characteristics of Floating-Gate EEPROM Cells*. PhD thesis, Swiss Federal Institute of Technology (ETHZ), Institut für Quantenelektronik, Hönggerberg, Zürich, 1987. (In German).

A Simple Training Law Suitable for On-Chip Learning.

V.Petridis, K.Paraschidis.

Dept of Electrical Engineering.Faculty of Engineering. Aristotle University of Thessaloniki.

BOX 438 54006 Thessaloniki - Greece. E-mail: petridis@vergina.eng.auth.gr

Abstract.

A very simple and efficient method for artificial neural networks training is proposed. Extensive simulation has established that it works very well as an ANN training law. It is faster than BackPropagation, it is suitable for on-chip learning and it can be implemented in parallel computers very easily.

1. Indroduction.

The backpropagation (BP) training method of artificial neural networks (ANN) has received much attention and many variations of the basic method have been proposed [1-4]. One of its main drawbacks is that it cannot be implemented in hardware easily.

In this paper we present a very simple method of training ANNs, what we call the FeedForward (FF) method. It involves only forward passes and it is based on one of the simplest search procedures. Its hardware implementation is very easy. It is faster than BP, its signals propagate only in the forward direction and it can be used even in the case of unmodeled activation function. Moreover, it is ammenable for parallel implementation.

We may say that the fact that simple training laws, like the one proposed here, work, might indicate the munificence of high dimensionality [6].

2. The Feedforward method of training.

The FF method is basically a very simple direct search method , a univariant search parallel to the axes [5] in the weight space. It is iterative and it is terminated when the total error becomes less that a predefined value Ecr. The total error is defined as

$$E = \frac{1}{M\,m} \sum_{p=1}^{M} \sum_{j=1}^{m} (d_{pj} - O_{pj})^2 \qquad (1)$$

where M and m are the number of input/output patterns and the number of output units respectively. d and O are the desired and actual output respectively for pattern p at the output j.

One of its basic features is that the weights are updated one at a time, within each iteration (iteration is defined as the cycle within which all the weights are updated once). The weights are numbered in arbitrary order from 1 to N. Their values define what we call the network weight point in the weight space. They are initially set to random values and are updated one by one from w_1 to w_N by an updating procedure presented below. That amounts to the network weight point moving along the w_i axis to a new location in the weights space, which is the output of the updating procedure. When all the weights have been updated the current iteration is completed.

The updating procedure of a particular weight w_i consists of three phases, namely, the perturbation, the forward steps and the backward steps phase .

a) **The Perturbation phase.** The direction of search is determined by perturbation. The network weight point moves along the w_i axis by an amount f or -f (f > 0 is a fraction of the stepsize s, e.g. f=0.1 s). If the total error reduces when the network weight point moves by f, the direction of search is positive, otherwise (change by -f) it is negative.

b) **The Forward steps phase.** During this phase, the network weight point moves in steps of size s along the direction of search, until, either the maximum allowable number of forward steps, d, is reached (in which case the updating procedure of this weight is completed and its output is the network weight point reached after the last step taken), or a total error larger than the one produced at the previous step is observed (in which case the network weight point reached after the last step taken is the starting point for the next phase).

c) **The Backward steps phase.** During this phase, the network weight point reached during the forward steps phase, moves in a direction opposite to the direction of search by a number of steps. The size of the first step is s/2 and it is decreased progressively by a factor of 2 for every following step. The backward steps continue until, either the maximum allowable number of backward steps, c, is reached (in which case the output of the updating procedure is the network weight point reached after the last step taken), or an increase in the total error is observed (in which case the output of the updating procedure is the network weight point reached before the last step taken).

Fig. 1 shows the updating procedure for the case that the maximum number of forward and backward steps (d=c=3) is taken.

3. Simulation results - Conclusions.

Extensive simulation has been carried out for various ANNs. The computer employed was a SparcStation IPC.

Table 1, compares the FF with the BP method. Three different ANNs have been trained over a set of 10 (or 20) random patterns. 10 runs have been used in each case, the training set being different for each run. Both the FF and BP started from the same initial weight settings. The learning rate in the case of BP was equal to 2 (the best found during a search in the range of 0.05 to 5.0). It can be seen that the FF method is much faster than BP.

Table 2, shows the performance of the FF method, for various ANNs. The ANNs have been trained over a set of 10 or 30 random associations. 50 runs have been used for all ANNs except the last one for which 10 runs have been used. It is seen that as the size of the network increases, the performance of the FF method for large stepsizes improves. Finally it can be seen that the FF method performs very well even in case of large networks (e.g the last one).

Table 3, shows the performance of the FF method when tested on various problems. In the case of the N-Parity (N=8) problem sequential training was used. In the case of the N-M-N (N=8,M=3 & N=16,M=4) encoder-decoder problem batch training was used. Finally the last examples show that FF performs well even in case low resolution arithmetic is employed (10 bits for sigmoid outputs and 8 bits for weigths).

The most important advantages of FF are summarized below :

1). Simulation results show that the FF method is faster than BackPropagation. 2).It does not involve the calculation of the derivative of the activation function. Therefore it can be applied to ANNs with various types of activation functions even in the case of no continous derivatives and unmodeled activation functions. 3).No signals propagate backwards during the training process. The only extra calculations required, apart from those involved in the forward passes, are the increase/decrease of the weights and the comparison of the total error for different weight vectors. 4).The FF method is directly applicable to other types of ANN architectures (generalized ANNs, recurrent ANNs e.t.c). and error functions. 5) Any kind of adaptation is easily incorporated in the training process (e.g. limitimg the sigmoids output to prevent it from reaching saturation). 6).As shown by experimental results, it can still operate under low resolution arithmetic. 7).It has inherent parallelism (different processors can run different groups of patterns or calculate in parallel the total error for different values (steps) of the weight under consideration). 8).It can be implemented in hardware very easily (see advantages 2,3,6).

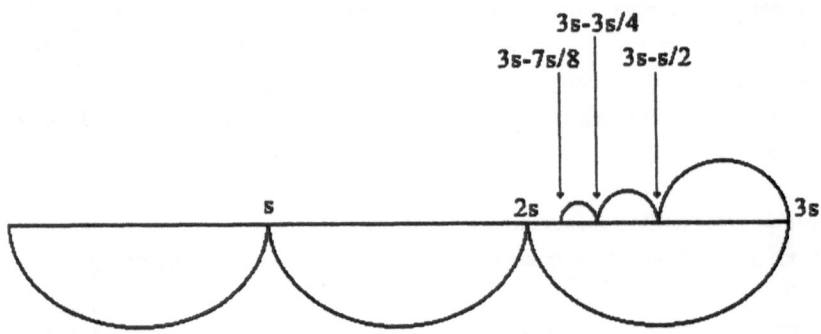

Figure 1. Schematical explanation of the FF method.

[TABLE 1] Comparison between the FF and the BP method.   10 runs have been used.

Type of ANN	2 7 7 2				2 10 10 2				2 15 15 2			
No. of patterns to be learned	10		20		10		20		10		20	
Type of training	FF	BP	FF	BP	FF	BP	FF	BP	FF	BP	FF	BP
No. of runs that converged	10/10	10/10	9/10	9/10	10/10	10/10	9/10	10/10	10/10	10/10	10/10	9/10
Mean convergence time in secs	66	193	964	1311	269	729	647	2198	84	401	746	2087

[TABLE 2] Simulation results of the FF method.   (Random Associations)

StepSize, s ANN & No. of patterns to be learned.	0.1		0.5		1.0		2.0	
2 24 24 2 (10)	50/50	116	50/50	45	49/50	32	49/50	20
2 15 15 2 (10)	50/50	73	50/50	36	50/50	30	49/50	20
2 8 8 2 (10)	49/50	59	45/50	31	42/50	27	30/50	19
4 15 15 4 (10)	47/50	91	47/50	43	47/50	31	47/50	26
8 15 15 8 (10)	50/50	138	50/50	79	50/50	74	50/50	77
2 52 52 2 (30)	10/10	2441	10/10	1415	10/10	1546	10/10	2312

First column: Number of runs that converged.

Second column: Mean convergence time in secs.

[TABLE 3] Simulation results of the FF method.   (All runs have converged).

Problem	ANN	Time (sec)	Comments
N-Parity, 8 bits, (10 runs)	8 8 1	338	1051 iterations
Sequential training			11810 iterations
Encoder-Decoder 8 bits	8 3 8	19	25 iterations
Encoder-Decoder 16 bits	16 4 16	385	59 iterations
Low Resolution Sigmoid 10 bits, Weights 8 bits			
10 random associations (20 runs)	2 7 7 2	81	154 iterations
15 random associations (20 runs)	2 7 7 2	910	314 iterations

REFERENCES.

1).Jacobs, R.A. (1988). Increased rates of Convergence through Learning Rate Adaptation. Neural Networks,vol.1.pp 295-307

2).Wasserman P.D.(1988).   Experiments in Translating Chinese Characters Using Back-Propagation.Proc.of 33 IEEE Comp. Soc.Int.Conf., IEEE Comp.Society Press, pp 399-402.

3).Tollenaere T. (1990). Super SAB: Fast Adaptive Backpropagation with Good Scaling Properties. Neural Networks, vol.3, pp.561-573.

4).Samad T. (1991). Back Propagation With Expected Source Values. Neural Networks, vol.4, pp.615-618.

5).Beveridge G.S.G. and Schechter R.S. (1970). Optimization: Theory and Practice. McGraw-Hill, ISE.

6).Robert Hecht-Nielsen The munificence of high dimensionality. Artificial Neural Networks, 2 (I. Aleksander & J.Taylor Eds) pp.1017-1030 North Holland, 1992.

SIMULATION OF NEURAL NETWORKS AND GENETIC ALGORITHMS IN A DISTRIBUTED COMPUTING ENVIRONMENT USING NeuroGraph

PETER WILKE

LEHRSTUHL FUER PROGRAMMIERSPRACHEN DER UNIVERSITAET ERLANGEN-NUERNBERG
MARTENSSTR. 3 • D-8520 ERLANGEN • GERMANY
☎ ++49+9131-857624 • 📠 ++49+9131-39388
💻 WILKE@INFORMATIK.UNI-ERLANGEN.DE

1 ABSTRACT

NeuroGraph is a simulation environment for design, construction and execution of neural networks and genetic algorithms in a distributed computing environment. The simulator parts either run on single computers or as distributed applications on UNIX/X-based networks consisting of personal computers, workstations or multi-processors. The parallelization component offers the possibility to divide computational tasks into concurrently executable modules, according to restrictions due to the neural net topology and computer net capabilities, i.e. NeuroGraph tries to select the best configuration out of the available distributed hardware environment to fit performance requirements.

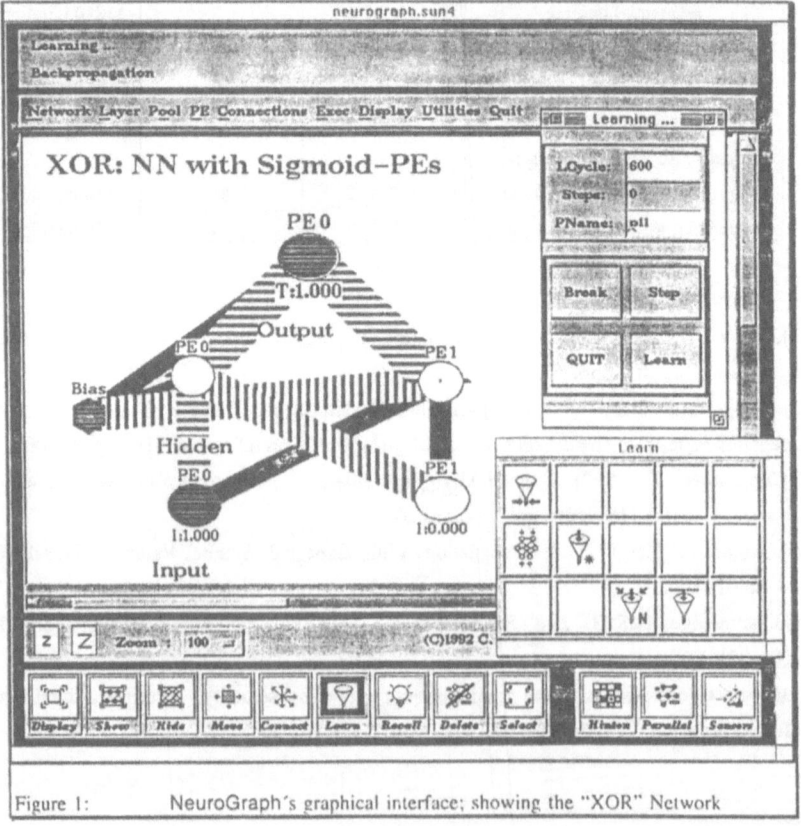

Figure 1: NeuroGraph's graphical interface; showing the "XOR" Network

2 THE NeuroGraph NEURAL NETWORK SIMULATOR FOR DISTRIBUTED COMPUTING ENVIRONMENTS

NeuroGraph is a neural network simulation environment for UNIX workstations. It is a software tool based on the X-Windows programming libraries and allows interactive design, training, testing and visualization of artificial neural networks [see fig. 1]. The simulator consists of four major components: a simulator kernel that operates on the internal representation of the neural network models, a graphical user interface to interactively design net topologies and functionality and control network dynamics, an analyzation component to get and evaluate performance information during net execution, and a real world interface component for easy integration of the simulated networks into practical application environments.

Furthermore, there are two special modules for advanced experimentation with neural nets: an easy to use toolbox for developers of neural nets offering interactive definition of specialized neuron functionality as well as net control strategies, and a parallelization component which is able to divide neural nets into parallely executable modules, and then map these modules to parallel hardware. With the parallelization module, time-consuming net control algorithms can be efficiently performed on multiprocessor platforms or workstation clusters, whereas online visualization of the network dynamics is handled by graphic workstations.

3 THE SIMULATOR KERNEL

The simulator kernel [see fig. 2] provides a library of functions which are responsible for the management of the internal representation of the neural network topology and functionality. The kernel routines are subdivided into two major parts: routines which offer model-independent manipulation of data structures (installing or deleting layers, adding neurons, defining pools of neurons, installing connections between groups of neurons etc.), and routines which essentially depend on different neural network models, their connectivity and control strategy, i.e. modules containing model-specific learn and recall algorithms and interface routines (e.g., feed-forward backpropagation nets, constraint satisfaction, interactive activation and competition, pattern-associator, competitive learning, for a description of these models see [3]). The whole kernel is written in C for efficiency and portability reasons.

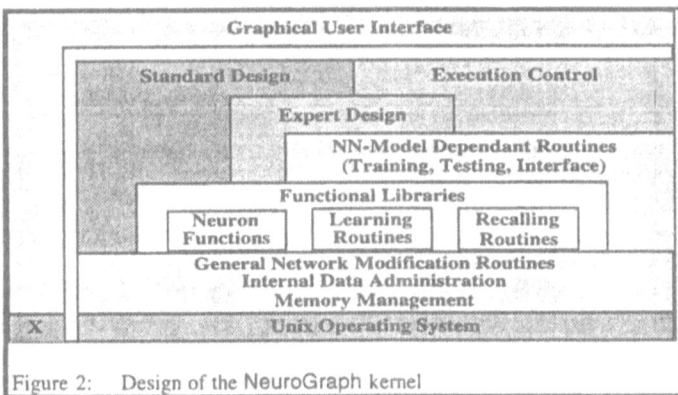

Figure 2:   Design of the NeuroGraph kernel

4 THE GRAPHICAL USER INTERFACE

The graphical user interface (GUI) is based on the X-Windows system together with the OSF/Motif Widget Libraries. The GUI is [see fig. 1] an easy-to-use tool to interactively design different standard net topologies and control strategies; however, it is also possible to define new net architectures as well as neuronal and net-global functional behaviour. There are a number of standard functions (scaled summation/product, linear threshold, sigmoid, stochastic sigmoid, step, linear) available to characterize the learn and recall functionality of each processing element. But there is also a module to define user-specific signal processing and learning functions for the neurons; description of the functions is done via entering mathematical expressions for these functions.

In order to get different views on a simulated network there are several modes to visualize its topology and/or connection structures, to show only parts of a network, display neuron parameters in graphical and/or textual form etc. One special module presents connection structures in the form of a diagram [see fig. 3], as proposed by Hinton and Sejnowsky [2].

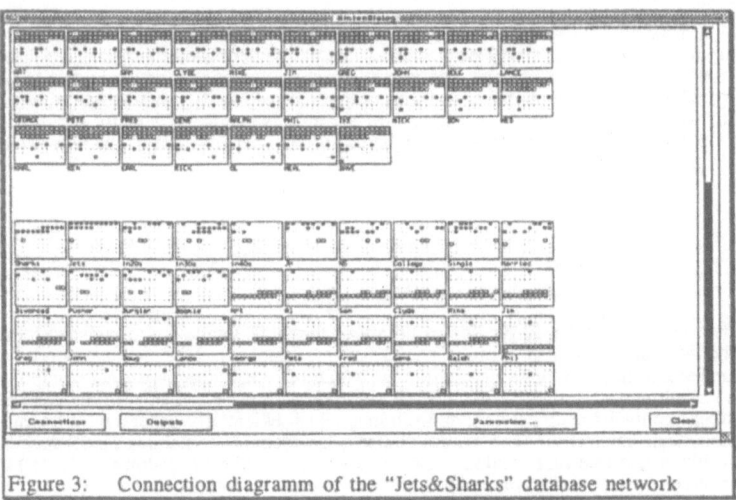

Figure 3:   Connection diagramm of the "Jets&Sharks" database network

5 SYSTEM COMPONENTS COMMUNICATION BETWEEN NEURAL NETWORK SIMULATOR AND GENETIC ALGORITHMS GENERATOR

As mentioned before, NeuroGraph is devided into separate modules. Common to all modules is the internal network representation which is a flexible data structure.

Using NeuroGraph on a single workstation this data structure is accessed via shared memory management. In a distributed computing environment NeuroGraph uses the X-Protocol [7] for communication between the software modules running on different computer network nodes [1,4,5,6]. Therefore NeuroGraph is able to visualize a neural net on a personal computer running an X-Server, using network parameters stored on a fileserver while the net is being trained on a multi-processor-system using back-propagation or any other learning rule.

This communication is also used for communication between the neural network simulator and the genetic algorithm generator. Fig. 4 shows the genetic algorithms generator at work. The simulated individuals enter the genetic process from the "start"-box in the right window. In the example given only one genetic operator is specified, i.e. the box in the middle. Individuals are manipulated according to the actuals settings of the sliders. The fitness is shown in the upper left window. The straight line shows that the fitness was improved only a little bit. The example shown is a genetic algorithm for finding a solution for the travelling salesman problem. The lower left windows shows the routes which are coded by the individuals. Via exchange of data these genetic algorithm generator can be used to optimize the neural network topology by evolution of network parameters.

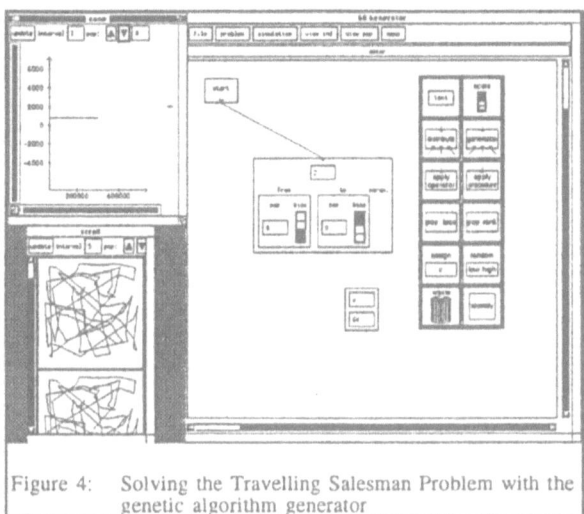

Figure 4: Solving the Travelling Salesman Problem with the genetic algorithm generator

6 COMMUNICATION WITH PARALLEL HARDWARE

In order to achieve a reasonable speedup in the performance of large and complex networks (with more than 10000 processing elements and more than 100000 connections), the functional network data (i.e., excluding the parameters needed for visualizing the net topology on the screen) is sent to a multi-processor system via X-Protocol mechanisms. Time-consuming learning and recalling is then performed in parallel (as far as possible, depending on the selected control strategy and connectivity) with specially designed algorithms that efficiently use the underlying hardware. Any network data changes of which the user wants to be informed are signaled to the supervising kernel or GUI (Graphical User Interface) module, which then decides whether to reflect the changes on the screen immediately (e.g., to see a refined step by step evaluation of the network) or to wait for a general update signal (e.g., at the end of a training epoch). Alternatively the described data interchange can take place via files if the learning process is to be performed independently of the visualization GUI module (e.g., when learning may last some hours).

7 LITERATURE

[1] C. Jacob, P. Wilke: A Distributed Network Simulation Environment For Multi-Processing Systems, Proc. Int. Joint Conf. on Neural Networks, IJCNN 1991, Singapore, p.1178pp

[2] D.E. Rumelhart, C.L. McClelland: Parallel Distributed Processing, Vol. 1&2, MIT Press, Cambridge, 1986

[3] J.L. McClelland, D.E. Rumelhart: Explorations in Parallel Distributed Processing, MIT Press, Cambridge, 1988

[4] P. Wilke, C. Jacob,: The NeuroGraph Neural Network Simulator, Proc. Int. Workshop on Modelling, Analysis and Simulation of Computer and Telecommunication Systems MASCOTS '93, Simulation Series Vol. 25, 1, San Diego, 1993, p. 341pp

[5] P. Wilke, C. Jacob,: Simulating Neural Networks in a Distributed Computing Environment Using Neuro Graph, Proc. Int. Workshop on Modelling, Analysis and Simulation of Computer and Telecommunication Systems MASCOTS'93, Simulation Series Vol.25,1, San Diego, 1993, p. 382pp

[6] P. Wilke: Simulation of Neural Networks in a Distributed Computing Environment using NeuroGraph, Proc. Int. Workshop on Artificial Neural Networks IWANN '93, Sitges / Barcelona, 1993

[7] Douglas A.Young: The X Window System - Programming and Applications with Xt OSF/Motif Edition, Prentice Hall, Englewood Cliffs, 1990

A Parallel Implementation of the Back-Propagation of Errors Learning Algorithm on a SIMD Parallel Computer

Antonio d' Acierno and Roberto Vaccaro
Istituto per la Ricerca sui Sistemi Informatici Paralleli, I.R.S.I.P. - C.N.R.
Via P. Castellino, 111 - 80131 Napoli -ITALY
Tel. + 81 5904217
Fax. + 81 5454330
email: antonio@irsip.na.cnr.it

Abstract
Training algorithms for artificial neural networks tend to be very time-consuming. It is therefore obvious to capitalise on the intrinsic parallelism of these systems in order to speed up the computations. This contribution describes the implementation of the back-propagation of errors learning algorithm on a SIMD parallel computer.

1. Introduction

The simulation of large artificial neural networks (ANNs) has become an important tool for solving real worlds problems and for studying the dynamic behaviour of large populations of interconnected processing elements. The behaviour of an ANN is determined by parameters denoted as weights and a learning procedure is used to compute these parameters. Such a learning procedure tends to be very time-consuming and it is therefore obvious to try to develop faster learning algorithms or to capitalise on the intrinsic parallelism of these systems in order to speed up the computations. Massively parallel computers have the potential to exploit the parallel nature of neural algorithms and to provide thus the computational power for high-speed simulations.

This paper presents a mapping model for the parallel implementation of the back-propagation of errors learning algorithm (BPA) [1][2]. This algorithm is typically used for the training of feed-forward neural networks. With reference to a network with one hidden layer, the algorithm can be described as follows. ($I[1..N_i]$, $H[1..N_h]$ and $O[1..N_o]$ represent the state vectors of, respectively, input, hidden and output neurons, while $W1[N_h][N_i]$ represent the connection matrix between input and hidden layers and $W2[N_o][N_h]$ represents the connection matrix between hidden and output layers).

In the first phase of the BPA (forward phase) an input vector to the network is provided and values propagate forward through the network to compute the output vector:

$$H_i = f(\sum_{j=1}^{N_i} W1_{ij} * I_j) \tag{1}$$

and

$$O_i = f(\sum_{j=1}^{N_h} W2_{ij} * H_j) \tag{2}$$

By the way, this phase is also the retrieving phase. The output vector O is then compared with a target vector T (which is provided by a teacher) resulting in an error vector:

$$E=T-O \tag{3}$$

In the second phase (backward phase) the error vector values are propagated back through the network by defining, for each node, the error signal (δ). The error signal depends on the derivative of the activation function. Specifically, the error signals are determined from a weighted sum of error signals of the next layer again using the connection weights (now backward); the weighted sum is then multiplied by the derivative of the activation function.

Formally, if we suppose:

$$f(x) = \frac{1}{1+e^{-x}} \tag{4}$$

being

$$\frac{df}{dx} = f(x)*[1-f(x)] \tag{5}$$

we have:

$$\delta_i^O = O_i*(1-O_i)*E_i \tag{6}$$

for output nodes, and

$$\delta_i^H = H_i*(1-H_i)*(\sum_{j=1}^{N_o} W2_{ji}*\delta_j^O) \tag{7}$$

for hidden nodes.

Finally, the weight changes are evaluated according to:

$$\Delta W1_{ij} = \alpha*\delta_i^H*I_j \tag{8.a}$$

$$\Delta W2_{ij} = \alpha*\delta_i^O*H_j \tag{8.b}$$

where α is the learning rate (empirically chosen between 0 and 1).

In the "on-line" version of the algorithm the weight changes are applied as they are evaluated while, in the "batch" version of the BPA, the weight changes are accumulated to compute a total weight change after all training patterns have been presented.

2. The Parallel Implementation

There are at least two degrees of parallelism in the batch BPA. First, there is the parallel processing performed by the many nodes of each layer; second, there is the parallel processing of the many training examples. A third parallel aspect of the batch BPA (maybe less obvious) stems from the fact that the forward and backward phases for different training patterns can be pipelined. Despite of this, the parallel implementation of the BPA is not a very simple task since such a parallel implementation must perform efficiently the product of a matrix (W2) by a vector as well as the product of the transpose of W2 by a vector (see formulas 2 and 7). This introduces a hard mapping problem, whose trivial solution requires or the doubling of the connection matrices (so introducing an overload for the updating of doubled weights) or the growth of the number of communications. In this paper, we propose a mapping scheme that overcomes such a trivial approach.

As a first step let us analyse the most obvious form of parallelism, i.e. the one due to the parallel processing of neurons belonging to the same layer. As regards to the hidden layer (for example) N_h processes can simulate the neurons belonging to this level by evaluating in parallel the activation values for the hidden neurons; this form of parallelism corresponds to viewing calculations as matrix-vector products and letting each row of the matrix map onto a processor (*neuron parallelism*, [3]). At each input to a neuron, however, the arriving activation value is multiplied by the weight of the specific input. This can be done simultaneously at all inputs to the neuron, while the subsequent summation of all the products can be also parallelized using a suitable communication structure; this form of parallelism (that uses N_i processes) corresponds to viewing calculations again as matrix-vector products but letting each column of the matrix mapped onto a processor (*synapse parallelism*, [3]).

In the implementation we propose, we use a mixture of synapse parallelism and neuron parallelism so that neither the doubling of connection matrices nor the growth of the number of communications is required.

More precisely, we use:

> neuron parallelism to evaluate the activation values of hidden neurons;
> synapse parallelism to evaluate the activation value of output neurons;
> neuron parallelism to evaluate the error terms of hidden neurons.

As regards to the on-line version of the BPA, it should be clear that N_h processes can be used; each process must know the incoming weights of the neuron of the hidden layer that it handles and the outgoing weights of such a neuron.

Supposing that all processes know all the examples, each process (except for one process that we call *master*) performs the following steps:

1. evaluates the activation value of the handled hidden neuron;
2. evaluates the partial sums for the calculation of the *net input* to output neurons;
3. sends the evaluated vector of partial sums to the master;
4. waits for the arriving of error terms of output neurons;
5. evaluates the error term of the handled hidden neuron;
6. updates weights.

The *master* performs the same steps and, besides, it evaluates the error terms of output neurons and sends such error terms to the other processes.

As it should be clear, the proposed implementation fully parallelizes the forward phase, the second step of the backward phase and the updating phase, while the first step of the backward phase (formula 6) is performed sequentially by the *master* process.

The proposed collection of processes communicating via message-passing can be mapped out on each parallel machine (SIMD, re-configurable MIMD, non re-configurable MIMD, shared memory machine, distributed memory machine). Such a collection of processes can be also replicated to implement the batch version of the BPA.

3. The Mp-1 Architecture

The MP-1 from MasPas Computer Corporation is a fine grained, massively parallel SIMD computer. It consists of a UNIX subsystem, an array control unit, the processor element array and some fast I/O facilities. The UNIX subsystem transmits programs and data to the array control unit, where scalar code is run and parallel instructions are decoded and broadcasted to the processor array. The processor array consists of 1024 up to 16384 (4096 in our machine) 4-bit processing elements (PEs) with up to 64-Kbytes of local memory (16-Kbytes in our configuration). The PEs are arranged in 2-dimensional cyclic mesh; local communications are provided via X-net, which connects each PE to its eight nearest neighbours. The mesh is divided in clusters of 4x4 PEs and each cluster has a bi-directional link to the global router, which can realise any communication pattern.

4. The Implementation

We implemented the batch version of the algorithm described in section 2 on the MP-1 and we used the MPL programming language (e.g. an extension of the standard C language). On each row of the 64x64 array of PEs we mapped a copy of the network and the training set is thus divided in 64 subsets; processors belonging to the *i-th* row knows all the training examples of the *i-th* subset, so that communications are required only to evaluate the net input of neurons belonging to output layer and

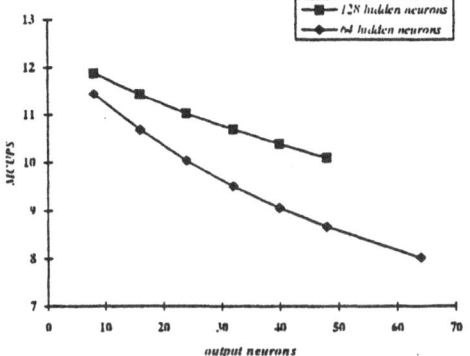

Figure 1. Experimental findings on the MP-1. The network has 256 input neurons and the training set contains 512 examples.

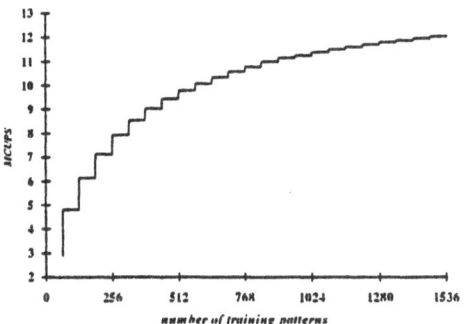

Figure 2. Experimental findings on the MP-1. The network has 128 input neurons, 64 hidden neurons and 16 output neurons.

to communicate the error terms of output neurons. To perform these operations, we used the *scanAddf* (that performs communications and computations with high bandwidth) and the *Xnetc* (that performs communications one-to-many with high bandwidth). Once the weight changes for all the patterns have been evaluated, the total weight changes are evaluated by using a *rotate-and-sum* algorithm; last, each process updates its weights and the learning process continues.

We tested our algorithm with reference to a network with 256 input neurons and 512 training examples. Figure 1 shows the obtained MCUPS (Millions of Connections Updated per Seconds) as the number of output neurons varies. Figure 2 shows the MCUPS for a small network (128-64-16) as the number T of training patterns changes. The stepwise shape of figure 2 is due to the fact that, for $64 < T < 129$ (for example), there is at least a row that has to elaborate two patterns. It is worth noting in figure 2, that the MCUPS grow up as T increases. While this effect does not seem surprising, it is a matter of fact that it is here amplified, since the evaluation of the total weight changes (e.g. the *rotate-and-sum* algorithm) is quite inefficient, being not possible to use the *scanAddf* function.

5. Conclusions

We proposed a new mapping scheme for the parallel implementation of the back-propagation of errors learning algorithm that overcomes the vertical slicing, that seems too communication-intensive. Such a scheme is based on the use of a mixture of neuron parallelism and synapse parallelism and it can be combined with the classical *data partitioning* [3]. The proposed scheme allows the definition of a parallel algorithm that fully exploits the parallelism inherent in the forward, backward and updating phases; besides, such algorithm can be mapped on each parallel machine (i.e. both SIMD and MIMD). In this paper we have tested the MP-1, a SIMD computer that gets 300 MFLOPS of peak performances in our configuration, and we have obtained encouraging results. Currently, the algorithm is being improved and generalised as well as mapped and tested on Transputer-based MIMD machines.

References

[1] D.E. Rumelhart and J. L. McClelland, Parallel Distributed Processing: Explorations in the Microstructure of Cognition, MIT Press, Cambridge, MA, 1986, Vols. I and II.
[2] D. E. Rumelhart, G.E. Hinton, R.J. Williams, *Learning Representation by Back-Propagation of Errors*, Nature, **323**, 1986, 533-536.
[3] T. Nordstrom and B. Svensson, *Using and Designing Massively Parallel Computers for Artificial Neural Networks*, Journal of Parallel and Distributed Computing, **14**, 1992, 260-285.

CONVIS, a Distributed Environment for Control and Visualization of Neural Network Simulation Programs

Louis Vuurpijl and Theo Schouten

University of Nijmegen, Faculty of Mathematics and Informatics,
Department of Real-Time Systems, Toernooiveld 1, 6525 ED Nijmegen,
e-mail: louis@cs.kun.nl, ths@cs.kun.nl

Abstract.

This paper introduces CONVIS, a software environment and set of interface definitions for managing control and visualization of neural network simulation programs. It is designed with the aim to allow fast integration of (new) simulation programs and corresponding tools within the environment. An action-oriented conceptual model of simulation programs is presented. If a program description is given following this model, CONVIS is able to access and control the corresponding simulation. Via a standardized communication and data interface, tools and simulation programs can be connected to CONVIS. Both tools and neural network simulations can be started up on any machine connectable to the machine on which CONVIS is running and therefore, the environment can be executed in parallel.

1. Introduction.

A neurosimulator is a set of software and/or hardware components that can operate together to support the construction, manipulation, visualization or (fast) execution of neural network simulations. Currently, many neurosimulators exist such as Pygmalion [5], The Rochester Connectionist Simulator [1], Genesis [8], PlaNet [3], Aspirine Migraines[2], etc. All have a number of features in common. A graphical or command line user-interface is supplied for controlling neural network simulations and starting up tools for visualizing or processing data contained in the neural network. Furthermore, an algorithm library is available containing a number of the most popular neural network models. A significant feature that often makes part of current neurosimulators is some kind of general neural network datastructure that is based on a hierarchical, object-oriented description of neural networks. Based on such a datastructure, a number of tools can be distinguished. A neural network description language can be defined that allows the specification of neural networks with arbitrary architecture. Network construction tools can be provided that support a user when making a network description. Via dedicated compilers, the latter can be generated into neural network simulation code, possibly decomposed and mapped on specific (parallel) hardware architectures. Furthermore, visualization tools can be available that display the status of data contained in the neural network, and dynamic manipulation of the neural network objects can be supported on several levels of abstraction. In [4], a more detailed taxonomy of neurosimulators can be found.

Unlike existing neurosimulators, CONVIS does not use general neural network description datastructures as a basis for neural network simulation programs. The programs do not have to be specified following the syntax of a specific neural network description language or using algorithm libraries supplied by the neurosimulator. In our approach, each neural network can be implemented specifically targeted on a certain machine architecture or using the most appropriate datastructures and flow of control. Instead of using a general *network* description, CONVIS uses a set of interface definitions for data exchange with and control of simulation *programs* and tools. These definitions are based on the assumption that neural network simulation programs implement a limited number of actions, such as loading of a network and patterns, and training, recall and operational actions. In the interface definitions, simple data structures are used that can be defined in terms of (multi-dimensional) arrays. By keeping the datastructures simple and general applicable, it is aimed that new or existing tools and simulation programs can be implemented (or adjusted) and interfaced with CONVIS relatively easy.

Unlike most other neurosimulators, CONVIS does not explicitly support the construction of networks. However, parameterizable simulation programs can be initiated and all other tools that other neurosimulators support can be interfaced with CONVIS. As CONVIS allows for control of efficient neural network simulations, it is suited for appliers (people who are no neural network experts but rather want to use them as an efficient means to solve their specific problem). Furthermore, it is a flexible environment for programmers who want to program their own neural network code (e.g. to test a new kind of model or achieve high performances on a specific hardware topology), but do not want be concerned with implementing the user-interface.

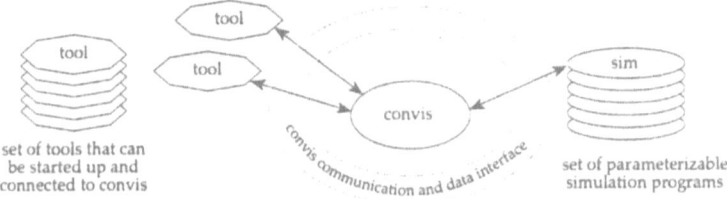

Figure 1. Convis and how it is interfaced with a simulation program and tools. Both could be running on any workstation, the simulation program also on a transputer system.

Another important difference with existing neurosimulators is that CONVIS fully exploits the availability of workstations in a network environment. A small set of communication and data interface routines is defined that can be used to equip any tools and simulation programs with a mechanism that is able to communicate with CONVIS. In this way these programs can be running on any machine that allows inter-process communications via sockets. Currently, CONVIS can handle tools and simulation programs running on workstations and transputer systems running under Helios. It has been succesfully tested on heterogeneous networks consisting of Dec and Sun workstations and a Parsytec transputer system.

2. Action-oriented control.

CONVIS uses the description of neural network simulation *programs*, rather than the description of a neural network. This can be justified by the fact that the implementation of the simulation code for a given type of neural network usually is a relatively simple task. This also holds for parallel implementations of certain classes of neural networks [7]. Furthermore, the amount of code actually needed to implement a neural network's dynamics in general is far less than the amount needed for user-interface and (file) I/O. So the main design issue for CONVIS was to accomodate a general purpose user-interface together with a means via which new or existing neural network simulation programs and associated visualization and processing tools can be integrated within the CONVIS environment.

Each neural network simulation program only implements a limited number of actions. These are divided in four categories, being initialization (loading, constructing and storing of a network and patterns), and training, recall and operation. The latter three represent distinct stages during the tuning of a neural network for a specific application. During the first stage, a network is trained with a set of patterns. In general, this stage is followed by a test phase in which the network performance (e.g. generalization capabilities) is examined. In the final stage the network is operational and can be used to do the job. As stated before, the implementation of such actions is relatively easy to do. However, managing the actions by initiating them, manipulating or monitoring objects associated with them, and interrupting and resuming execution, in most cases requires far more programming efforts. CONVIS provides an environment which takes care of the control of actions while making minimal assumptions about the way in which they are implemented. It provides a user-interface and manages any tools by supplying them with the associated neural network data.

CONVIS uses what we call an action-oriented model of simulation programs. A program can be described by specifying the actions that it implements. Each action can have a number of objects associated with it, e.g. parameters, variables, data, options and settings. The following picture depicts the conceptual model of CONVIS program descriptions:

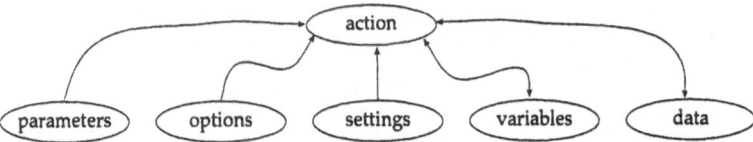

Figure 2. A program description consists of a number of actions. Each action has a number of objects associated with it. It has parameters, options and settings for installing the initial configuration of an action. It has variables and data whose values can be changing during execution of the action.

For example consider an action "learn". Typical parameters for this action would be the number of epochs that have to be iterated, the learning rate and the number of epochs after which a netsave has to be made. Options can be considered as switch buttons switching between a number of states of which only one may be true. An option could be the learning mode, which may switch between by pattern or by epoch, indicating whether weights have to be updated after each pattern or after a complete epoch. Settings can be considered as radio buttons or boolean values, they can be either true or false. For example, settings could indicate whether the network has to be saved after a number of iterations or whether training has to be performed with or without noise. These three objects form the inputs of an action, i.e. they are set before an action is executed. The other two objects, variables and data, can also be used as input, but may be changed during execution of the network. Variables are scalar values such as an error or current epoch number. Data can be any non scalar neural network object, such as the weights, input and output patterns or activation values.

3. Interfacing CONVIS with a simulation program or tools.

A set of interface definitions is provided that can be used to equip simulation programs and tools with a mechanism that handles communication with CONVIS. Interfacing a simulation program with CONVIS involves specifying its program description and installing a main loop that listens to CONVIS and "knows" about the simulation by examining its program description. The specification of a program description involves careful examination of the initialization and learn, recall and operation actions and programming of the corresponding action objects. A number of convenience routines are defined that can be used to make the description. Note that the specification of the program description can be a strong guide for program development. Once the actions are specified, the corresponding functions have to be implemented. The implementation of these is further guided by a template which has to be filled in. Finally, if an action has to be controllable, i.e. one must be able to interrupt it and view or edit the corresponding objects, the action_control routine has to be incorporated within the code. A typical action such as a training session will then look like:

```
while (error>err_crit&&epoch<nepochs&&action_control(action))
    train_one_epoch;
```

The action_control routine can also be used in nested loops, which is used if the action must be controlled at a finer level of detail (e.g. not per epoch but per pattern). It is up to the neural network programmer to decide at what points in his code this routine is placed. Note that no assumptions are made on the code that implements a training epoch. So for new or existing neural network simulation programs, indeed any datastructures or flow of control can be used, tailored for a specific neural network model or application.

The actions that are supported by CONVIS form the basic equipment that is needed to control most neural network simulation programs. However, some could have their own network specific actions. If these cannot be captured in general categories such as learn, recall, operation and initialization routines, CONVIS allows the specification of miscellaneous actions. The same procedure has to be followed as described above to interface these with CONVIS. Furthermore, the user-interface is implemented such that it can easily be changed and customized according to the specific needs of a user, so CONVIS can be changed such that it is oriented at any actions the user (or simulation program) requires.

Via the CONVIS user-interface, tools can be started on any machine to which CONVIS can connect to. For interfacing tools with CONVIS, a simple communication and data interface is defined. After the communication with CONVIS has been established, typically, tools perform a loop that iterates while waiting for CONVIS to send a new variable or data object. For example, consider the following codes of a tool that handles scalar variables (e.g. an error plot tool) and a tool that handles vector/matrix data (e.g. a weight monitor):

```
Variable v;                          Data d;
int fd = setup_communication();      int fd = setup_communication();
while (ReceiveVariable (&v,fd))       while (ReceiveData (&d,fd))
   PlotVariable (v);                    MonitorWeights (d);
```

Similar to neural network simulation programs, no further assumptions are made on the code of the tool programs. However, for tools handling data, the data has to be extracted from the datastructure as described by CONVIS. As this description is simple and general purpose (the data is structured in multi-dimensional arrays), in most cases the extraction is straight-forward. Amongst others, we have connected CONVIS to matlab [9], a powerful data analysis, data processing and data visualization tool. A more detailed description about how CONVIS works and how it can be interfaced with simulation programs and tools can be found in [6].

4. Conclusion.

Using the concept of program descriptions, CONVIS can relatively easy be interfaced with neural network simulation programs. Using the communication and data interface, tools can be added to the CONVIS tool set. As CONVIS makes few assumptions about how a neural network is implemented, the network simulation code can be tailored to the needs of a specific application or neural network model, and can thus be highly efficient and compact. Because of the exploitation of the computing resources available in a network, CONVIS operates in parallel with the tools and simulation program it is connected with, so it can operate faster and handle larger amounts of data than current neurosimulators. We believe that the design of CONVIS makes it suitable for appliers of neural networks to use it as an efficient platform to solve their specific problems. It can also very well be used as an extendable general purpose neurosimulator for controlling simulations written by neural network programmers.

References.

[1] N.H. Goddard, K.J. Lynne, T. Mintz, L. Bukys, *Rochester Connectionist Simulator*, Technical Report 233 (revised), University of Rochester, 1989.
[2] R. Leighton, A. Wieland, *The Aspirine/MIGRAINES Software Tools User's Manual Release V4.0*, The MITRE Corporation, Washington C3I Division, 7525 Colshire Drive, McLean Virginia 22102, 1991.
[3] Y. Miyata, *A User's Guide to PlaNet Version 5.6 A Tool for Constructing, Running, and Looking into a PDP Network*, Computer Science Department University of Colorado, Boulder, 1991.
[4] M.L. Recce, P.V. Rocha, P.C. Treleaven, *Neural Programming Environments*, in Artificial Neural Networks 2, I. Aleksander and J. Taylor (eds.), pp. 1237-1244, 1992.
[5] M.M.B.R Vellasco, *Pygmalion Neural Programming Environment: nC Tutorial*, Pygmalion v1.02 Documentation, Department of Computer Science, UCL, 1990.
[6] L.G. Vuurpijl, Th.E. Schouten, *CONVIS, an Action-Oriented Neurosimulator for Control and Visualization of Neural Networks*, Technical Description, draft version. University of Nijmegen, Faculty of Mathematics and Informatics, Toernooiveld 1, 6525 ED Nijmegen.
[7] L.G. Vuurpijl, Th.E. Schouten, *Suitability of Transputers for Neural Network Simulations*, in W. Joosen, E. Milgrom (eds.), Parallel Computing: From Theory to Sound Practice, pp. 528-537, IOS Press,1992.
[8] M. Wilson, J. Uhley, U. Bhalla,, D. Billitch, M. Nelson, J. Brower, *GENESIS & XODUS General Purpose Neural Network Simulation Tool*, Division of Computation and Neural Systems, California Institute of Technology, Pasadena California 91125.
[9] *matlab, High Performance Numeric Computation and Visualization Software*, Version 4.0 User's Guide. Mathworks, Cochituate Place 24, Prime Pathway, Natic, Mas 01760, 1992.

Mapping of Some Neural Network Algorithms to a General Purpose Parallel Neurocomputer

Petri Kotilainen, Jukka Saarinen and Kimmo Kaski
Tampere University of Technology, Microelectronics Laboratory
P.O.Box 692, SF-33101 Tampere, Finland
e-mail petri@ee.tut.fi (Internet), tel. +358 31 161 396, fax +358 31 162 620

The objective in our neurocomputer development project was to design a parallel general-purpose neuro-computer, which is capable to perform the network training as well as the forward mode operations. The modular structure and expandability would allow the system to be easily configured to meet various requirements of computation speed and system cost. The system is to be operated as a neural co-processor connected to a general-purpose computer.

The overall structure of our proposed architecture [1] consists of a number of identical processing units (PU), a tree-shaped network of communication units (CU) and a single host interface unit. The communication network is used to transfer the data between the processing units and the host interface using *broadcast*, *read* and *write* operations. Additionally the network can perform some global computation tasks. These tasks are the computation of the *sum* of all numbers sent from the processing units and finding the *minimum* or *maximum* values of the numbers sent by the PU's.

The most popular training method for *multilayer perceptrons (MLP's)* is the *error back-propagation method* [2], which consists of three phases, namely the forward phase, backward phase and weight updating phase. The computation of the forward phase can be performed by broadcasting one layer's input vector to all PU's, calculating the activation functions of the neurons on one layer in parallel in the PU's, reading the neuron outputs to the host interface unit and broadcasting them to all PU's as the input vector of the next layer. The computation of the backward phase can be performed by writing the error vector to the PU's calculating the delta factors within each PU and multiplying the delta vector with the transpose of the weight matrix. The communication network configured as an adder tree can be used for this purpose. These operations are repeated for each layer of the network. Finally, the weight updating can be performed in parallel within the PU's without additional communication between them.

Kohonen's *Self-Organising Feature Map (SOFM)* algorithm [3] is also suitable to be computed using the multiprocessor architecture. The most convenient way is to use neuron-parallel mapping by assigning one or more neurons to each PU. The recall phase of the trained SOFM involves broadcasting the input vector to all PU's and computation of the matching scores in parallel. The search of the best matching unit can be done by using the communication network configured for minimum search operation. In the training phase the index of the best matching unit is broadcast to all PU's, which can calculate the neighbourhood functions of the neurons of the network on the basis of this information. Finally the weights in the neighbourhood of the best matching unit are updated, which can also be performed in parallel.

Kanerva's *Sparse Distributed Memory (SDM)* [4] can be computed in the multiprocessor architecture by mapping one or more rows of the address and memory matrices to each PU. The calculation of the select vector in SDM involves broadcasting the address input vector to the PU's and computing the Hamming distances between the applied input vector and all rows in the address matrix. In the *memory write* operation the data input vector is broadcast to all PU's and added in parallel to the contents of each selected row of the memory matrix. In the *memory read* operation each PU calculates its own partial sum vector from the rows assigned to it. The final sum vector can then be calculated from the partial sums using the adder tree structure. The thresholding of the sum vector is then performed at the host.

Computation of the synaptic weighting factors and modifying the weights require most of the operations in the cases of multilayer perceptron networks and Self-Organising Feature Map. These operations can be effectively parallelised. In the case of Sparse Distributed Memory, the computation of the Hamming distances will become the most critical step in practice. In all three cases the communication delays between the host interface and the PU's depend mostly on the sizes of the vectors to be transferred and not so much on the number of PU's.

References

[1] P. Kotilainen, J. Saarinen and K. Kaski, "A Multiprocessor Architecture for General Purpose Neuro-computing Applications", *Proceedings of 3rd International Conference on Microelectronics for Neural Networks (MICRO-NEURO 1993)*, Edinburgh, Scotland (April 1993).

[2] R. Beale and T. Jackson, *Neural Computing: An Introduction*, Adam Hilger, Bristol (1990).

[3] T. Kohonen, *Self-Organization and Associative Memory*, Springer-Verlag, Berlin (1984).

[4] P. Kanerva, *Sparse Distributed Memory*, MIT Press, Massachusetts (1988).

Architecture of Low Cost, Large Scale Neural Networks

Stef Joosten

Design Methodology Group, Centre for Telematics and Information Technology
University of Twente, P.O. Box 217, 7500 AE Enschede, the Netherlands
e-mail: joosten@cs.utwente.nl

Abstract

This poster describes ideas about implementing a neurocomputer in hardware. The architecture can cope with large amounts of neurons (in the range of 10^3 - 10^6), with a large number of inputs per neuron (about 10^3 - 10^4). The architecture exploits wafer scale integration and integrated optics.

Keywords: neural networks, neural computer, wafer scale integration, optical computing, integrated sensors and actuators.

Problem

Large scale neural networks are hardly being implemented because the communication between neurons poses immense wiring problems. If n neurons have m inputs each, then the number of wires needed is $n \times m$. For a completely connected network of n nodes, this requirse n^2 connections. The solution is to get away from the two-dimensional space on a chip, and use a three dimensional space by letting neurons communicate optically.

Architecture

The architecture is that of a layered network, and is remarkably simple. Each layer corresponds to one wafer, on which a multitude of neurons reside in a grid format. The communication between two layers is done optically as shown in the figure on the right. Within one layer, the interaction between neurons (such

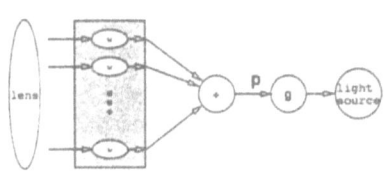

A neuron   Communication between layers

as neighbour inhibition) can be done electronically without wiring problems. The optical communication involves lenses and light sources. The input channels of one neuron are represented by photosensitive cells, arranged in a two-dimensional grid. The outputs of the neurons in one layer are arranged in a two-dimensional grid. on one side of a wafer. Every output is a light emitting device (led). This grid of leds is projected by means of a lens on the inputs of a neuron in the following layer. The optical communication has the big advantage that the 'wiring' is done in three dimensions instead of two. This solves the wiring problem.

In the poster session details can be found. Please contact the author for a copy of the paper.

A Generalized Recurrent Neural Network for Matrix Inversion

Jun Wang
Department of Industrial Technology
University of North Dakota
Grand Forks, ND 58202-7118, USA

Matrix inversion deals with computing the inverse matrix A^{-1} of a given $n \times n$ nonsingular matrix A, which has been widely used in a variety of applications such as control, robotics, and signal processing. This paper presents a generalized recurrent neural network for matrix inversion. The generalized recurrent neural network can use any monotone nondecreasing activation functions and its steady states are guaranteed to present the inverse matrix A^{-1}.

The dynamics of the recurrent neural network for matrix inversion can be described by the following matrix-valued differential and algebraic equations,

$$\frac{dV(t)}{dt} = -\mu A^T U(t),$$
$$U(t) = G[AV(t) - I],$$

where A^T denotes the transpose of A, $V(t)$ and $U(t)$ are $n \times n$ activation state matrices of the neurons in the output and input layers respectively, $\mu > 0$ is a scalar gain parameter, $G[X] = [g_{ij}(x_{ij})]$ is a matrix of activation functions, and I is the identity matrix.

The architecture of the recurrent neural network consists of two layers and each layer consists of an $n \times n$ array of neurons. Each output neuron represents an element in the inverse matrix A^{-1}. The connection weight matrix from the input layer to the output layer is defined as $-\mu A^T$ and the connection weight matrix from the output layer to the input layer is defined as A. The biasing threshold (constant input) matrix in the input layer is defined as I and there are no biases for the neurons in the output layer. There is a functional transformation for each neuron in the input layer and an integral transformation for each neuron in the output layer.

The dynamical equations show that $v_{ij}(t)$ is connected with $u_{1j}(t), u_{2j}(t), \ldots, u_{nj}(t)$ only and $u_{ij}(t)$ is connected with $v_{1j}(t), v_{2j}(t), \ldots, v_{nj}(t)$ only. This pattern of connectivity shows the proposed recurrent neural network can actually be decomposed into n independent subnetworks. Each subnetwork represents one column vector of A^{-1}. The dynamical equations also indicate that the connection weight matrices are identical for each subnetwork.

Main Result: If $g_{ij}(x_{ij})$ is monotone nondecreasing with respect to x_{ij} (i.e., $dg_{ij}/dx_{ij} \geq 0$) and $g_{ij}(x_{ij}) = 0$ is equivalent to $x_{ij} = 0$ (i.e., $g_{ij}(x_{ij}) = 0$ iff $x_{ij} = 0$) for $i, j = 1, 2 \ldots, n$, then the recurrent neural network for matrix inversion is asymptotically stable in the large and its equilibrium states represent the inverse matrix of a given nonsingular matrix (i.e., $\forall V(0) \in R^{n \times n}, \lim_{t \to \infty} V(t) = A^{-1}$).

The analytical result has been substantiated by simulation results.

On the Realization of Back-Propagation on a Transputer Based System

Petridis V.[1], Adamidis P.[1], Margaritis K.G.[2]
[1] Dept of Electrical Eng, Aristotle University of Thessaloniki, 54006 Greece.
Email: petridis@vergina.eng.auth.gr
[2] Informatics Centre, University of Macedonia, Greece

This paper presents some experimental results on the realization of a parallel simulation of an Artificial Neural Network (ANN) on a Transputer based system. The Back-Propagation algorithm has been used, and the parallelization technique utilized is the Processor Farm based on the partition of the training patterns between the available transputers so that each transputer (trainee) is trained on a different part of the training set, while another transputer (the controller) keeps track of the global network state.

Initially the controller sends to each trainee information on: network architecture, initial weights and a part of the training pattern set. Each trainee processes the patterns presented to it (forward flow, back propagation and weight update), and then it returns to the controller, the updated weights and the output error that it has computed. The controller computes the average error and the average of each weight received from each trainee. If the average error has not reached an acceptable level, then the controller sends the new weights and another part of the training pattern file to each trainee. This procedure continues, until the error reaches a specified level. Trainees update their connection weights each time they finish processing a training pattern, since it was found that this accelerates the convergence of the ANN.

The number of patterns sent to each trainee is kept constant and it is termed Update Rate, since the weight average is found after each trainee has finished processing its patterns.

The Telmat system used, is consisted of 24 20Mhz T800 transputers with 2MB RAM. The software was developed in C under Helios, and Component Distribution Language (CDL).

The main issues addressed are: (a) the speed-up of the parallel simulation against the simulation on one processor and (b) the existence of an optimal value of the Update Rate

A number of simulations has been performed. Fig.1 shows the results for cartesian to polar transformations. The simulation on one Transputer requires 324 iterations and 14194 seconds.

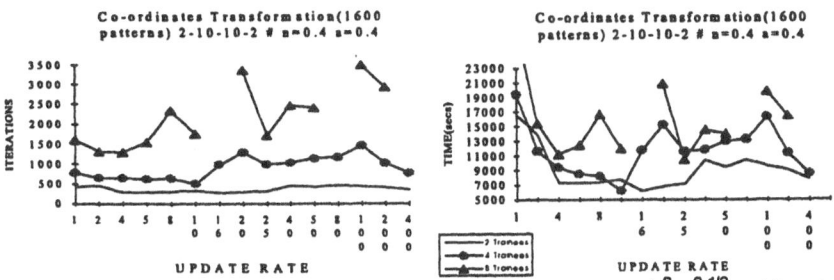

Fig. 1 Performance of the Parallel ANN simulation trained on the functions $z=(x^2+y^2)^{1/2}$ and $\theta=\arctan(y/x)$.

REFERENCES

1. Smith L.S., "Neural net simulation using Transputer nets", SERC/DTI Transputer Loan Report, 1988.
2. Tollenaere T, Orban G A, " Simulating modular neural networks on message-passing multiprocessors", Parallel Computing 17, 1991.
3. Wang C, Wu C, "Parallel simulation of Neural Networks", Simulation, 56, 1991.

Self-Organisation of Large Feature Maps using Local Computations: Analysis and VLSI Integration

Christian Lehmann

MANTRA Centre for Neuro-Mimetic Systems

MANTRA-DI-EPFL

1015 Lausanne (Switzerland)

lehmann@eldi.epfl.ch

Abstract

Self-organising feature maps are very powerful data analysis tools which are inspired by observations on the vertebrate cortex. If complex data are to be processed in such a way, it is most likely that specialised hardware will be required. In this case, it is very important to keep computations and communications local. A VLSI systolic implementation using only local communications has been built to process lateral interactions in the map. This text provides analytical results for the choice of the parameters involved.

In the field of ANN, the cortex is modelled as a grid of labelled units. Using such a model, the self-organisation of the visual cortical map was explained by von der Marlsburg [1]. Later, Kohonen presented an abstract implementation of this process. His algorithm is well suited to a sequential computer implementation and allows for very interesting analytical studies. Recently, Miikkulainen [2] presented simulations showing that the more "biologically plausible" implementation, using neurone competition, allows for self-organisation of feature maps (SOFM).

We are in the process of building a neuro-mimetic system based on synaptic systolic arrays which will provide the necessary power for large scale simulations. In our design each PE of the systolic array handles the computational tasks of one synapse. It is capable of similarity measure evaluation as well as weight adaptation. In the following, we describe a computation process for the selection phase of the self-organising algorithm which does not require additional hardware [3].

In this process, the initial map response $y(0)$ enters a recurrent lateral interaction layer according to the following rule :

$$y_i' = \Phi\left[\rho(y_i + (w^E + w^I - 1)\bar{y}_i - w^I \bar{y})\right] \quad (1)$$

where w^E and w^I are the excitation and inhibition factor respectively, $\bar{y}_i$ is the local sum L units around i, $\bar{y}$ is the sum of all N activities and $\Phi()$ clamps the values between 0 and 1.

The parameters found in equation (1) influence considerably the behaviour of the system. We have investigated the problem starting with $L = 0$ (fig. 1) for which we have shown correct winner-take-all behaviour in [3] with $\rho w^E > 1$,

$$\rho(w^E - w^I) \le 1 \text{ and } {w^E}/{w^I} > N - 1. \quad (2)$$

In the case where $L \ne 0$ the analysis is made more difficult by the fact that lateral excitation result in topological dependencies between the neurones; this is why self-organisation appears! We were however able to show correct organization of up to 20x20 SOFM with the following values :

$$w^E - 1 = w^I = 1 / (N - 2(2L + 1)).$$

The machine currently in development will allow experiments on larger feature maps.

References

[1] C. von der Marlsburg, "Self-Organization of Orientation Sensitive Cells in the Striata Cortex", in Kybernetik, Vol. 14, pp. 85-100, 1973.

[2] R. Miikkulainen, "Self-Organizing Process Based On Lateral Inhibition And Synaptic Resource Redistribution", in proceedings of ICANN, Vol. 1, pp. 415-420, Espoo, Finland, 1991.

[3] C. Lehmann, "Réseaux de Neurones Compétitifs de Grandes Dimensions pour l'Auto-Organisation : Analyse, Synthèse et Implantation VLSI", PhD Thesis, EPFL, 1993.

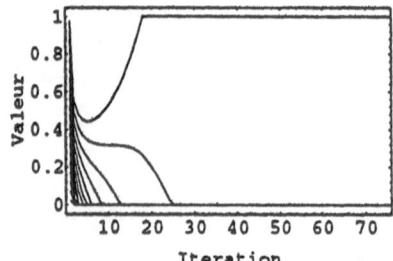

Figure 1 Winner-Take-All with local competitive process. The 20 initial values are uniformly distributed over the interval]0,1[. Lateral coupling parameters are such as in (2) with $\rho = w^E$.

NEUROCOBOL : A COBOL-like Neural Network Simulation
Language Based on the Layer Macro Definition

Gyu Wan Hong , Su Dong Lee
Department of Computer Engineering , Ulsan University,
Mugeo Dong NamGu, Ulsan 680-749 South Korea

ABSTRACT

Most of neural network simulation programs have been implemented with the conventional computer languages. It is difficult to implement simulation programs and is various to build neural network, because the laguages have not data structures and statements suitable for expressing neural network structures and controlling neural behaviors. This paper describes a neural network simulation language of NEUROCOBOL, specifically designed for the development and modeling of neural network applications. It has a layer macro definition concept to express the arbitray network structures and dynamic neural behaviors easily. It consists of the two divisions. First division describes the network structures. Second division controls the network behaviors.

1) Layer Macro Definition
A layer macro definition is the layer primitive unit of a neural network simulation language which define a dynamic behavior of neural network as a layer structure level. A simple syntax of a layer macro definition is following.

```
<Layer macro defintion> ::=
                    LMACRO <macro name>
                    SUMMATION FUNCTION IS <function name>.
                    ACTIVATION FUNCTION IS <function name>.
                    LAYER TOPOLOGY IS <topology type>.
                    LAYER LEARNING IS <learning rule name>.
                    INITIAL WEIGHT VALUE IS <value type>.
                    INITIAL THRESHOLD VALUE IS <value type>.
                    ENDLMACRO.
```

2)Program Structure
A NEUROCOBOL program structure consists of two successive divisions. A first division of NETWORK STRUCTURE DIVISION defines network topologies, the second division of PROCEDURE DIVISION controls a dynamic behabior of neural networks. The simple syntax of program is following.

```
<program> ::= NETWORK STRUCTURE DIVISION.
              <network structure division body>
              PROCEDURE DIVISION.
              <procedure division body>
```

Encapsulated Objects for Neural Network Simulation

Mark A. Rubin[1]

Physics Division, Naval Air Warfare Center[2]
China Lake, California 93555

Physics Department, Rockefeller University
New York, New York 10021

Abstract

We consider aspects of program design relevant to the creation of object-oriented software tools for research in neural network algorithms, and discuss these in the context of EONNS, a set of C++ classes for neural networks which is currently under development.

---

[1] email: mark@peewee.chinalake.navy.mil
[2] mailing address

A Harmony Theory Network Solution To The N-Queens Problem

T.Tambouratzis, Inst. of Informatics and Telecommunications, NCRPS "Demokritos", Aghia Paraskevi 153 10, Attiki, Greece.

A parallel implementation of the Artificial Intelligence N-Queens problem is presented. The problem consists of placing the greatest possible number of queens in an NxN chessboard in such a way that no queen can threaten or be threatened by any other queen. It has been proved that the maximum number of mutually non-attacking queens equals N. The N-Queens problem has traditionally been solved by depth-first search with chronological back-tracking [1]. This however is not very efficient since: a) a state can be proved invalid very late (e.g. while placing the Nth queen) whereas the error is due to the location of the first queen, and b) there exists a very limited number of solutions to the problem (e.g. for N=4, out of the 1820 combinatorially possible states with 4 queens placed on the 4x4 chessboard, only 2 constitute valid solutions). Recently, more economical search algorithms have been proposed [2].

Here, the N-Queens problem is expressed as a constraint-propagation task. The constraints are dictated by the locations of the NxN chessboard which are invalidated once a queen is placed on the chessboard. Hence, after placing a queen, no other queen is allowed to be located on the same: 1- row , 2- column, 3- right diagonal, and 4- left diagonal of the chessboard (such two queens would mutually threaten each other). These constraints can then be propagated in order to find a solution to the N-Queens problem.

A Harmony Theory network [3] is employed to solve the N-Queens problem, since Harmony Theory supports parallel implementation of tasks solved by constraint-propagation. Requiring no training phase, Harmony Theory is especially suited to problems whose regularities (constraints, requirements) are clearly stated. The lower layer of the network consists of NxN nodes, each node representing the state of one square of the NxN chessboard (whether it is occupied by a queen or not). The upper layer consists of nodes encoding each of the constraints 1-, 2-, 3- and 4- for every square of the chessboard. Connections are only enforced between the two layers and specify which squares of the chessboard may become simultaneously occupied. Weights are mathematically defined [3].

The presented Harmony Theory network implementation has been found to be able to optimally solve the N-Queens problem by achieving to place N queens in compatible positions. The parallel constraint-propagation formulation of search has furthermore been found to obliterate the need of employing heuristics or back-tracking.

1.   Kale, L.V., (1990). "An Almost Perfect Heuristic For The N Nonattacking Queens Problem". Information Processing Letters, 34, pp. 173-178.
2.   Sosic, R., Gu, J., (1991). "Fast Search Algorithms For The N-Queens Problem". IEEE-SMC, 21, pp.1572-1576.
3.   Smolensky, P., (1986). "Foundations Of Harmony Theory", pp. 194-281, in "Parallel Distributed Processing : Foundations".

Author Index